Wills' **Biochemical Basis of Medicine**

Wills' Biochemical Basis of Medicine

Third Edition

Brian Gillham, PhD
Reader in Endocrine Biochemistry

Despo K. Papachristodoulou, MSc, PhD
Lecturer in Biochemistry

J. Hywel Thomas, PhD, FIBiol, CBiol
Dean, Basic Medical Sciences
Senior Lecturer in Biochemistry

Division of Biochemistry and Molecular Biology
United Medical and Dental Schools of Guy's and
St Thomas's Hospitals, London

Butterworth-Heinemann
Linacre House, Jordan Hill, Oxford OX2 8DP
A division of Reed Educational and Professional Publishing Ltd

A member of the Reed Elsevier plc group

OXFORD BOSTON JOHANNESBURG
MELBOURNE NEW DELHI SINGAPORE

First published 1985
Second edition 1989
Third edition 1997

British Library Cataloguing in Publication Data
Gillham, Brian
Wills' biochemical basis of medicine – 3rd ed.
1 Biochemistry
I Title II Papachristodoulou, Despo K.
III Thomas, J. Hywel
574.1'9

ISBN 0 7506 2013 7

Library of Congress Cataloguing in Publication Data
Gillham, Brian.
Wills' biochemical basis of medicine. – 3rd ed./Brian Gillham, Despo K. Papachristodoulou, J. Hywel Thomas.
p. cm.
Includes bibliographical references and index.
ISBN 0 7506 2013 7
1. Biochemistry. 2. Clinical biochemistry. I. Wills, Eric D. II. Papachristodoulou, Despo K. III. Thomas, J. Hywel. IV. Title.
[DNLM: 1. Biochemistry. 2. Metabolism. QU 120 G479w]
QP514.2.W55 96–24866
612'.015–dc20 CIP

Composition by Scribe Design, Gillingham, Kent
Printed and bound in Great Britain by The Bath Press

Contents

Preface to the third edition *xiii*
Preface to the second edition *xv*
Preface to the first edition *xvi*
Acknowledgements *xvii*

Part 1 **Cellular and molecular biology**

Chapter 1 **What happens where in the living mammalian cell?** *3*
1.1 Introduction
1.2 Size of cells
1.3 Methods of studying cell structure and function
1.4 The cellular organelles

Chapter 2 **Roles of extracellular and intracellular membranes: membrane structure and membrane transport** *12*
2.1 Introduction
2.2 Membrane composition
2.3 Membrane structure
2.4 Types of membrane proteins: their functional roles

Chapter 3 **Biochemical communication** *20*
3.1 Introduction
3.2 The nature of cellular responses to external signals
3.3 Properties of molecules used for extracellular signalling
3.4 Receptors for molecules used for signalling
3.5 Diseases resulting from dysfunctions in biochemical communication

Chapter 4 **Cellular organelles: the nucleus** *30*
4.1 Introduction
4.2 Structure of the nucleus
4.3 DNA synthesis (replication)
4.4 RNA synthesis (transcription)
4.5 Reverse transcription
4.6 A dynamic model for the flow of genetic information
4.7 NAD^+ synthesis
4.8 The cell cycle

Chapter 5 **Cellular organelles: mitochondria and energy conservation** *41*
5.1 Introduction
5.2 Mitochondria in representative cells
5.3 Structure, composition and general functions of mitochondria
5.4 The oxidative pathways of mitochondria
5.5 Oxidative phosphorylation
5.6 Relation of mitochondrial oxidations to cellular metabolism
5.7 Replication of mitochondria
5.8 Mitochondrial diseases

Chapter 6 **The cytosol** *56*
6.1 Composition of the cytosol

6.2 Functions of the cytosol: intermediary metabolism
6.3 Protein synthesis

Chapter 7 **Cellular organelles: the endoplasmic reticulum and Golgi apparatus** *68*
7.1 Introduction
7.2 Protein targeting

Chapter 8 **Cellular organelles: lysosomes** *73*
8.1 Introduction
8.2 Origin of lysosomal enzymes
8.3 Nature and properties of lysosomal enzymes
8.4 The life cycle of the lysosome
8.5 Functions of lysosomes in the tissues
8.6 Lysosomal storage diseases

Part 2 **Human metabolism and nutrition**

Chapter 9 **Plasma glucose and its regulation** *79*
9.1 Introduction
9.2 Maintenance of glucose concentration in the fasting state
9.3 Insulin release in the fed state
9.4 Tissue response to increased plasma insulin and glucose concentrations
9.5 The role of glucagon
9.6 Carbohydrate metabolism in the fetus and newborn
9.7 Metabolism of fructose and galactose
9.8 Circulating glucose under stressful conditions

Chapter 10 **Plasma lipids and their regulation** *85*
10.1 Introduction
10.2 Classification of plasma lipids
10.3 Lipid transport in the fed state
10.4 Lipid transport in the fasting state
10.5 Fatty livers
10.6 Cholesterol homeostasis
10.7 Importance of elevated concentrations of plasma lipids in atheromatous vascular disease
10.8 Factors leading to raised plasma lipid levels
10.9 The use of cholesterol-lowering drugs
10.10 Lipoprotein (a): an association with atheromatous vascular disease

Chapter 11 **Plasma amino acids and utilization of amino acids by the tissues** *99*
11.1 Introduction
11.2 Utilization of amino acids by humans
11.3 The effect of dietary proteins on plasma amino acid concentrations
11.4 The utilization of branched-chain amino acids in muscle and formation of alanine and glutamine
11.5 The induction of amino acid catabolizing enzymes
11.6 Hormonal regulation of plasma amino acids
11.7 Intracellular turnovers of proteins: the role of ubiquitin

Chapter 12 **Plasma calcium and phosphate homeostasis** *106*
12.1 Introduction: importance of calcium and phosphate in the body
12.2 Plasma calcium concentration
12.3 Calcium and phosphate requirements and body balance
12.4 Calcium intake and absorption
12.5 Regulation of plasma calcium levels
12.6 Parathyroid hormone (PTH)
12.7 Parathyroid hormone related protein (PTH-rP)

12.8 Calcitonin
12.9 Vitamin D (cholecalciferol)
12.10 Inter relationships of vitamin D, parathyroid hormone and calcitonin in the regulation of plasma calcium concentration
12.11 Functions of vitamin D not related to calcium homeostasis
12.12 Disorders of calcium homeostasis

Chapter **13** **Digestion and absorption of foodstuffs** *116*
13.1 Introduction
13.2 The role of digestive organs
13.3 Digestive processes
13.4 Digestive secretions
13.5 Control of digestive secretions: the gastrointestinal hormones
13.6 Carbohydrate digestion and absorption
13.7 Protein digestion and absorption
13.8 Fat digestion and absorption
13.9 Bacterial flora in the gastrointestinal tract
13.10 Malabsorption syndromes

Chapter **14** **Nutrition: general aspects** *126*
14.1 Diet and health
14.2 The need for energy and nutrients
14.3 Dietary requirements
14.4 Nutritional problems in modern society and guidelines for a healthy diet
14.5 The toxicity of food: food additives and contaminants
14.6 Causes of malnutrition
14.7 Assessment of nutritional status

Chapter **15** **Nutrition: energy** *135*
15.1 Forms and units of energy
15.2 Energy supply and utilization
15.3 Energy expenditure
15.4 Energy requirements
15.5 Energy balance and control of body weight

Chapter **16** **Nutrition: proteins** *142*
16.1 The need for protein in the diet
16.2 Protein turnover
16.3 Nitrogen balance
16.4 Protein content of food
16.5 Protein utilization and protein quality
16.6 Protein requirements

Chapter **17** **Nutrition: lipids and carbohydrates** *146*
17.1 Dietary lipid
17.2 Essential fatty acids
17.3 Deficiency of essential fatty acids
17.4 Functions of essential fatty acids
17.5 Polyunsaturated fatty acids and lipid peroxidation
17.6 '*Trans*' unsaturated fatty acids
17.7 Dietary lipid and cardiovascular disease
17.8 Dietary lipid and cancer
17.9 Dietary carbohydrate
17.10 Dietary carbohydrate and disease

Chapter **18** **Nutrition: obesity and starvation** *158*
18.1 The incidence of obesity
18.2 Classification and measurement of obesity

18.3 Obesity as a health risk
18.4 The causes of obesity
18.5 The treatment of obesity
18.6 Starvation: metabolic adaptation
18.7 Clinical aspects of starvation: starvation as a result of a medical condition
18.8 Starvation in childhood: protein-energy malnutrition (growth failure, marasmus and kwashiorkor)

Chapter **19** **Nutrition: vitamins** *171*
19.1 Introduction
19.2 Vitamin classification and nomenclature
19.3 Vitamin A
19.4 The B group of vitamins

Chapter **20** **Nutrition: inorganic constituents of the diet** *182*
20.1 Introduction
20.2 Metals found in the human body
20.3 Factors affecting metal requirements
20.4 Dietary requirements for metal ions
20.5 Roles of metal ions
20.6 Anions in the diet

Chapter **21** **Nutrition: iron and iron metabolism** *186*
21.1 Introduction
21.2 Iron balance
21.3 The absorption of dietary iron and its control
21.4 Iron transport: transferrin
21.5 Iron storage: ferritin and haemosiderin
21.6 The kinetics of iron
21.7 Valency of iron during metabolism
21.8 Intracellular iron homeostasis
21.9 Iron pathology

Chapter **22** **Nutrition: folate and vitamin B_{12}** *196*
22.1 Introduction
22.2 Absorption and distribution of folate and vitamin B_{12}
22.3 Interrelationships between folate and vitamin B_{12}
22.4 Modes of action of folate and vitamin B_{12}
22.5 Causes and effects of deficiency of folate and vitamin B_{12}

Part 3 **Specialized metabolism of tissues**

Chapter **23** **Blood: metabolism in the red blood cell** *205*
23.1 Introduction
23.2 Role of glycolysis and the pentose phosphate pathway
23.3 The role of 2,3-*bis*phosphoglycerate
23.4 The role of glutathione and NADPH
23.5 Genetic disorders: enzyme deficiencies

Chapter **24** **Blood: blood clotting** *209*
24.1 Introduction
24.2 Physiological events in blood clotting
24.3 The opposing roles of thromboxane A_2 (TxA_2) and prostacyclin (PGI_2) in platelet aggregation: the anti-clotting effects of aspirin
24.4 An overview of the biochemical events in the clotting process
24.5 The intrinsic pathway
24.6 The extrinsic pathway

24.7 The common pathway of blood clotting
24.8 Factors that limit the growth of clots
24.9 Fibrinolysis
24.10 Diseases affecting blood clotting

Chapter **25** **The liver** *219*
25.1 Introduction
25.2 Structure of the liver
25.3 Metabolic roles of the liver: general considerations
25.4 The role of the liver in carbohydrate metabolism
25.5 The role of the liver in fat metabolism
25.6 The role of the liver in amino acid metabolism
25.7 The role of the liver in protein synthesis
25.8 The role of the liver in storage
25.9 The role of the liver in providing digestive secretions
25.10 The excretory role of the liver
25.11 Ethanol and the liver

Chapter **26** **The kidney** *230*
26.1 Introduction
26.2 Structure of the kidney
26.3 Functional activity
26.4 The glomerular capillary basement membrane in health and disease
26.5 Composition of urine
26.6 Energy provision in the kidney
26.7 Water absorption
26.8 Absorption of electrolytes
26.9 Absorption of glucose
26.10 Absorption of amino acids
26.11 Regulation of pH
26.12 The kidney and erythropoiesis

Chapter **27** **Muscle** *242*
27.1 Introduction
27.2 Structure of skeletal muscle
27.3 Composition of muscle fibres
27.4 Mechanism of skeletal muscle contraction
27.5 Sources of energy for muscle contraction
27.6 Cardiac muscle
27.7 Smooth muscle
27.8 Muscle disorders: muscular dystrophy

Chapter **28** **The endocrine tissues** *255*
28.1 Introduction
28.2 Structural relationship between the hypothalamus, the anterior pituitary gland and target organs
28.3 Biosynthesis of peptide hormones
28.4 Hormones of the hypothalamus
28.5 Hormones of the anterior pituitary gland
28.6 Hormones of the posterior pituitary gland
28.7 Hormones of the pancreas
28.8 Hormones of the thyroid and parathyroid glands
28.9 Hormones of the adrenal medulla
28.10 Steroidogenic organs
28.11 Disorders of steroid hormone production and action

Chapter **29** **Bone** *279*
29.1 Introduction
29.2 Bone structure

29.3 Bone mineral
29.4 Bone matrix: collagens
29.5 Bone matrix: non-collagen, calcium-binding proteins
29.6 Collagen biosynthesis
29.7 Collagen diseases
29.8 Bone formation and growth
29.9 Mineralization of bone
29.10 Bone remodelling and repair
29.11 Metabolic bone disorders

Chapter **30** **The brain** *289*
30.1 Introduction
30.2 The cell types of the brain
30.3 The synapse
30.4 Excitation and conduction
30.5 Chemical transmission and transmitters
30.6 Myelin
30.7 Metabolism in the brain

Chapter **31** **The eye** *308*
31.1 Introduction
31.2 The cornea
31.3 The lens
31.4 The retina

Chapter **32** **The immune system** *317*
32.1 Introduction
32.2 The organization of the immune system
32.3 Antigens and antibodies
32.4 Antibody structure: the immunoglobulins
32.5 Complement
32.6 Cell-mediated immunity
32.7 The inflammatory response

Part 4 Health, disease and the environment

Chapter **33** **Free radicals in health and disease** *343*
33.1 The nature of free radicals and the generation of reactive oxygen species
33.2 Targets for attack by reactive oxygen species: polyunsaturated fatty acids, proteins and DNA
33.3 Antioxidant defence mechanisms
33.4 Free radical involvement in disease
33.5 Antioxidants as nutrients: vitamin E
33.6 Antioxidants as nutrients: vitamin C

Chapter **34** **Toxicology: general principles** *355*
34.1 Introduction
34.2 Environmental chemicals
34.3 Biochemical damage caused by toxic substances
34.4 The ways in which the body handles foreign compounds

Chapter **35** **Toxic metals** *358*
35.1 Introduction
35.2 Metal ion complexes and chelates
35.3 Protein–metal ion complexes
35.4 Copper

35.5 Toxicity of mercury
35.6 Toxicity of lead
35.7 Toxicity of aluminium
35.8 Radionuclides
35.9 Removal of toxic metals: chelation therapy

Chapter **36** **Metabolism of foreign compounds** *367*
36.1 The nature of foreign compounds and their routes of entry into the body
36.2 General properties of the metabolites of foreign compounds
36.3 Central role of the liver in the metabolism of foreign compounds
36.4 Phase I and phase II reactions
36.5 Role of reductive processes
36.6 Hydrolysis of foreign compounds
36.7 Oxidative metabolism of foreign compounds
36.8 Conjugation reactions of foreign compounds
36.9 Induction of the metabolism of foreign compounds
36.10 Inter-ethnic variation in the metabolism of foreign compounds

Chapter **37** **Multiple environmental challenges: cancer** *378*
37.1 Introduction: what is cancer?
37.2 Cancer as a multi-step disease
37.3 The transformation of cells to a cancerous phenotype
37.4 The nature of oncogenes
37.5 The nature of tumour-suppressor genes
37.6 Viral oncogenesis
37.7 Chemical carcinogenesis
37.8 The progression to cancer
37.9 Testing for mutagens: the Ames test
37.10 The treatment of cancer
37.11 Human cancer risk assessment

Part 5 **A biochemical perspective on disease and its treatment**

Chapter **38** **Recombinant DNA and genetic engineering** *399*
38.1 Introduction
38.2 Techniques and enzymes used to manipulate DNA
38.3 Molecular analysis of inherited disease
38.4 Other medical applications of the polymerase chain reaction

Chapter **39** **Biochemical principles underlying chemotherapy and drug resistance** *411*
39.1 Introduction
39.2 The discovery of drugs active against infectious agents
39.3 Sources of new drugs
39.4 The modes of action of drugs effective against pathogens
39.5 The occurrence of drug resistance in bacteria
39.6 Prevention of the symptoms of infectious diseases
39.7 Approaches to cancer chemotherapy
39.8 Resistance to the drugs used in cancer chemotherapy

Chapter **40** **Metabolism in injury and trauma** *431*
40.1 Introduction: the causes and nature of injury
40.2 The phases of the metabolic response to injury
40.3 The ebb, or shock, phase
40.4 The flow, or catabolic, phase
40.5 The convalescent, or anabolic, phase

Chapter **41** **Diabetes mellitus** *439*
41.1 Introduction
41.2 Two types of diabetes mellitus
41.3 Natural history of the disease
41.4 Deranged metabolism in diabetes mellitus
41.5 The biochemical basis for the complications of diabetes mellitus
41.6 Treatment

Chapter **42** **Asthma** *444*
42.1 Introduction
42.2 The disease
42.3 The causes of asthma
43.4 Events triggered during the course of an asthmatic attack
43.5 Environmental agents in asthma

Chapter **43** **Amyotrophic lateral sclerosis** *449*
43.1 Introduction
43.2 The disease
43.3 Possible causes of ALS
43.4 Inherited forms of ALS
43.5 Prospects for treatment

Index *453*

Preface to the third edition

The aims of this third edition of *Wills' Biochemical Basis of Medicine* remain essentially the same as for the previous two editions: to present those biochemical principles that are fundamental to the study of medicine; to relate biochemistry to the other basic medical sciences such as histology, physiology and pharmacology; to discuss, in biochemical terms, studies on the structure and function of organelles, cells, tissues and the whole body; and to apply this knowledge to an understanding of health maintenance and of disease processes.

While this edition was being written, the General Medical Council of the United Kingdom published its recommendations on Undergraduate Medical Education. In common with similar initiatives in several other countries, the council is promoting an approach to medical education which differs significantly from those previously promulgated. Some of the curriculum themes advocated by the council were anticipated by the late Professor Eric Wills in the first edition, and we have taken care to ensure that these themes run throughout this new edition. We recognize the requirement for students to take an integrated approach when considering the organization and function of the body. This requires a synthesis of information and concepts about cellular, organ and whole-body processes. The development of such a synthetic approach should foster an understanding of the abnormalities of structure and function that underlie disease processes. Students will then be able to understand the environmental and genetic factors that determine disease, and thereby acquire a knowledge and understanding of disease in terms of processes such as the immune response, neoplasia and the metabolic disturbances that arise in diabetes mellitus and trauma, for example.

The book is intended primarily for undergraduate students in the first two years of their medical and dental courses, and the order of chapters reflects this, building on earlier teaching. In this edition, we have kept to the arrangement of organizing the chapters into a number of parts. A departure from former practice is that material covered in Appendices in the previous editions has now been fully integrated into the body of the text.

In the first part, basic cellular structures and functions, for example membrane transport and signalling mechanisms, are dealt with, together with the functions of the individual organelles. It is here that the subject matter often referred to as 'intermediary metabolism' is introduced. We recognize that 'metabolic pathways' are not popular with medical and dental students but, in deciding what to include in this section, we have always tried to apply the test, 'Is this material necessary to the development of an integrated view?', i.e. 'Is the subject essential for an understanding of what follows?' If the answer has seemed to us to be 'no', then we have usually omitted it. There are, of course, some exceptions to this and these arise when the logical scientific coherence of a section depends on information, itself not strictly required for an understanding of health and disease.

Part 2, entitled Human Metabolism and Nutrition, covers in a general manner the metabolic processes of the body, with particular attention being paid to the regulation of the main metabolic fuels. It was the far-sighted inclusion in the first edition of a major section on nutrition that was one of Professor Wills' major innovations. We have therefore sought to maintain, and to bring up to date, the strong nutritional content of the book.

As in the previous editions, Part 3 focuses on the specialized functions of the major organs, systems and tissues of the body. A chapter on the immune system has been restored; this seemed to be particularly important for, as the discipline of immunology comes of age, the insights that it provides tell us much about the operation of integrated systems. These are comparable with the insights derived from a study of the operation of the other two great signalling systems of the body: the nervous and the endocrine systems.

The fourth part, entitled Health, Disease and the Environment, acknowledges the important part played by a study of the environment at all levels of education. This whole section is devoted to a treatment of the effects of environmental, largely chemical, challenges on the functions of the body. An important new focus for this section is a chapter on free radicals in health and disease.

The text concludes with a set of chapters presenting a biochemical perspective on disease and its treatment. The first chapter in this part is designed to provide a basic overview of the many contributions the study of molecular biology is making to medicine, e.g. in the understanding of dysfunction in the nervous system or in the unravelling of the complex processes that underlie neoplasia. Finally, we have selected five 'studies of diseases'. These have been chosen to illustrate how a fully-developed view of biochemical processes can assist in the understanding of medicine. These were challenging chapters to write, because in no case is it possible to develop a full understanding of a disease process simply from a study of biochemistry alone. Nevertheless we believe that, in choosing the subject matter of these closing chapters, we have identified areas in which an understanding of biochemical principles does make a significant contribution to the appreciation of the pathology of disease.

Again we should like to express our deepest appreciation to the many colleagues who have given freely of their time to help in the preparation of this book. We are grateful to many members of staff at the United Medical and Dental Schools of Guy's and St Thomas's Hospitals, particularly to Professor C.A. Rice-Evans, Drs P. Evans, R.W. Evans, G.H. Mitchell, B.H. Moreland, M.A.N. Rattray and K.N. White, for reading and criticizing sections of the book. Dr J. Hinson, of Queen Mary and Westfield College, also kindly read two of the chapters and Dr M. Barac Nietto of University of Kuwait who read the chapter on the kidney. We would also like to thank Drs L. Bannister, J.V. Priestley and K.N. White for supplying figures used in the text. We would like to thank Tim Brown, Helen Gyde and all at Butterworth-Heinemann who helped in the production of this text, especially Bob (in the absence of Alf) Pearson, who was always constructive.

B.G.
D.K.P.
J.H.T.

Preface to the second edition

Professor Eric D. Wills, the author of the first edition, was particularly concerned that the subject of biochemistry should not be seen by medical students as simply another academic hurdle, remote from clinical practice, to be cleared on the way to obtain a medical qualification. There has been a tendency among medical students to regard biochemistry as being primarily concerned with metabolic pathways rather than providing a molecular explanation of physiological and pathological phenomena. It was his desire to emphasize the importance of biochemistry in medicine that was so characteristic of Professor Wills' approach to the subject and which formed the foundation on which he based the first edition. Unfortunately, Professor Wills died a few hours before the first copies of his book were received from the printer. The first edition will remain as a fitting tribute to his scholarship.

When the publishers decided to produce a second edition of Professor Wills' book, we readily accepted their invitation to carry out the revision as we are in sympathy with Professor Wills' approach to the presentation of biochemistry to medical students and feel that the first edition had met a distinct need.

In preparing this second edition, we have generally kept to the arrangement adopted for its predecessor of dividing the contents into five parts. All the chapters have been reviewed, updated, and in many cases almost completely rewritten. In Part 1, which deals with the biochemistry of the subcellular organelles, the chapters on the functions of the nucleus, mitochondria and the endoplasmic reticulum have been extensively reorganized to account for our increased understanding of the roles of these organelles. The chapters on the biochemistry of the hormones and the regulation of the metabolic fuels in Part 2 have been largely rewritten, whilst a new chapter, entitled 'The eye' has been included in Part 3. The chapter on 'The brain' in this section has been completely rewritten. Part 4 is devoted to the biochemistry of environmental hazards and the major change in this section concerns the chapter on 'Chemical carcinogenesis' now renamed 'Carcinogenesis' which includes a treatment of viral carcinogenesis. We have omitted the chapter on 'Immunology' from Part 5 as we believe that it is difficult to do justice to this rapidly evolving subject in a single chapter in a general textbook of biochemistry. The chapter on 'Biochemical genetics' has been revised to include a discussion of the results of application of recombinant DNA technology to the diagnosis of genetic disorders.

Much of the material originally covered in the Appendices in the first edition has now been incorporated into the main body of the text but a number of Appendices have been retained as essential memory aids to much of the material found in basic textbooks of biochemistry.

We should like to thank all those who contributed to this book. We specifically thank our colleagues in the Division of Biochemistry at the United Medical and Dental Schools of Guy's and St Thomas's Hospitals: Drs P. Callaghan, R.W. Evans, R.P. Hopkins and B.M. Moreland for reading and criticizing sections of the book. A very special note of appreciation is extended to Dr Michael Thorn who reviewed various chapters, making many excellent suggestions and corrections. We are grateful to Roy Baker and Sue Deeley who encouraged us to undertake the task of revising this book; to Ray French for the preparation of many of the illustrations; to Jane Sugarman for her care as a sub-editor; and to the many who have helped with the typing.

J. Hywel Thomas
Brian Gillham

Preface to the first edition

During the past 50 years, biochemical principles, concepts and technology have become increasingly important for the understanding of disease, in clinical treatment and in medical research. It is therefore essential that the medical course should include an extensive groundwork of academic biochemistry and demonstrations of the application of biochemistry to medicine.

Although many excellent textbooks of basic academic biochemistry have been published during the past two decades, the author, after many years of teaching biochemistry in the basic medical sciences course, has come to support the view that a radical new approach to the subject is required in the medical course. This is because, as taught in conventional medical courses, biochemistry has tended to become an essentially academic discipline, divorced from clinical medicine. It is thus difficult for students to appreciate the great importance and significance of the applications of the subject in medicine.

Furthermore, during their basic medical and science courses, it is common experience that students often find difficulty in relating biochemistry to other disciplines such as anatomy, physiology and pharmacology—for example, in applying the biochemical concepts to the micro- and subcellular structures of the cell and, more important, to the functioning of the body as a whole, and thus the application of these concepts to disease processes.

This book is written with the object of overcoming these problems and presents the subject, in five Parts, from a new viewpoint. In the first Part, the biochemistry of the subcellular organelles is described in detail and the second Part deals with the biochemistry of the body as a whole. The third Part includes descriptions of specialized metabolism which occurs in many of the tissues of the body, whilst the fourth and fifth parts are devoted to applications of biochemistry—the fourth to environmental hazards and the fifth to some biochemical aspects of diagnosis and treatment.

I have adopted an approach to the subject that involves the discussion of structural detail and metabolic processes in the context in which they occur in the cell or in the tissues. This has necessitated omitting detailed systematic descriptions of much basic material such as carbohydrate, lipid and protein structure, enzymology, basic metabolic processes such as glycolysis, citrate cycle, fatty acid oxidation, amino acid metabolism and protein synthesis, which subjects form the major proportion of current biochemical teaching and several biochemical texts. This decision is, I believe, quite justified because all these topics have been described repeatedly and so elegantly in many other text books that their inclusion is considered unnecessary in this book. A good knowledge of this basic material is, however, really essential to a full understanding of this book and students and their teachers are advised to consider how this may best be achieved before beginning the study of this book. However, in order to avoid the necessity of a second text being always available, basic biochemical information is summarized in a series of appendices for refreshment of memory of, for example, chemical structures or a metabolic pathway relevant to any particular chapter.

In addition to providing guidance for students, it is hoped that the approach adopted in the book may also be useful for teachers who wish to give serious thought to the type of basic biochemical syllabus that is desirable for a full understanding of the book and thus, also, for an understanding of the role of biochemical principles which are valuable in medicine. Although the book is written primarily with medical students in mind, it is hoped that many in the medical profession who wish to refresh their knowledge could find it useful and that it will be valuable for science students who wish to become acquainted with the applications of biochemistry to medical science.

Several have helped in the production of this book and the author would like to acknowledge the great help he has received from many associates and colleagues: to John Gillman who encouraged me to write the book and for much helpful advice; to the colleagues in my Department and especially Dr J.D. Hawkins, Dr D.M.G. Armstrong and Dr K. Brocklehurst who helped me by providing valuable information and for critical comments; to the many who have helped with the typing including my wife, Caroline Reddick, Alison Dowler and Sara Taylor; to Ray French for the preparation of many of the illustrations and especially to Jane Sugarman for her invaluable assistance as a subeditor.

Eric D. Wills

Acknowledgements

We would like to acknowledge the source of the following illustrations:

Figure 1.2 from C. Clarke and E.D. Wills (1980) *International Journal of Radiation Biology*, **38**, 21–30, by kind permission of the publishers, Taylor and Francis, Basingstoke.
Figure 1.6 from F.P. Altman (1979) *Progress in Histochemistry and Cytochemistry*, **9**, 40, by kind permission of the author and publishers, Gustav Fischer Verlag, Stuttgart.
Figure 2.4 from G. Weissman and R. Claiborne (1979) *Cell Membrane*, by kind permission of the artist and the publishers, H.P. Publishing Inc., New York.
Figure 2.9 from S.R. Goodman (1994) *Medical Cell Biology* by kind permission of the publishers, J.B. Lippincott Co., Philadelphia.
Figures 4.1, 4.2 from F. Beck and J.B. Lloyd (eds) (1974) *Cell in Medical Science*, Vol. 1., by kind permission of the publishers, Academic Press, London.
Figure 4.12 redrawn from T.M. Devlin (ed.) (1986) *Textbook of Biochemistry with Clinical Correlations*, by kind permission of the author and publishers, John Wiley and Sons Inc., New York.
Figures 5.1, 5.2, 27.1, 27.2 from K.E. Carr and P.G. Toner (1982) *Cell Structure*, by kind permission of the authors and publishers, Churchill Livingstone, Edinburgh.
Figures 5.14, 5.15 from M.D. Brand (1995) *Biochemist*, **16**, 20–21, by kind permission of the Biochemical Society, London.
Figure 10.4 from the *Annual Review of Biochemistry*, **49**, 1980, by kind permission of the publishers, Annual Reviews Inc., Palo Alto.
Figure 11.3 from R. Park, R. Paul, W. Radosevich *et al.* (1983) *American Journal of Physiology*, **245**, E94–E101, by kind permission of the authors and publishers, The American Physiological Society, Bethesda.
Figure 11.7 from G.L. Dohm, G.J. Kasperck, E.B. Tapscott and H.A. Barakat (1985) *Federation Proceedings*, **44**, 348–352, by kind permission of the *FASEB Journal*, Bethesda.
Figure 11.10 from D.M. Rubin and D. Findley (1995) *Current Biology*, **5**, by kind permission of the publishers, Current Biology Ltd, London.
Figure 12.1 from G.W. Dolphin and I.S. Eue (1963) *Physics in Medicine and Biology*, **8**, 197–207, by kind permission of the publishers, I.O.P. Publishing, Bristol.
Figure 12.7 from F. Bicknell and F. Prescott (1946) *Vitamins and Medicine*, 2nd edn, by kind permission of the publishers, Heinemann Medical Books, London
Figure 17.6 from A. Keys (1980) *Seven Countries: A Multivariate Analysis of Death and Coronary Heart Disease*, by kind permission of the publishers, Harvard University Press, Cambridge, MA.
Figure 17.7 from D.J. Barker and C. Osmond (1986) *Journal of Epidemiology and Community Health*, **40**, 37–44, by kind permission of the *British Medical Journal*, London.
Figure 21.3 redrawn from E.N. Baker, S.V. Rumball and B.F. Anderson (1987) *Trends in Biochemical Sciences*, **12**, 350, by kind permission of the publishers Elsevier Publications, Cambridge.
Figure 21.4 from J.R. Frausto da Silva and R.J.P. Williamson (1991) *The Biological Chemistry of the Elements*, by kind permission of the publishers, Oxford University Press, Oxford.
Figure 24.1 from N. Crawford and D.G. Taylor (1977) *British Medical Bulletin*, **33**, 199–206, by kind permission of the *British Medical Journal*, London.
Figure 25.2 from H. Elias and J.E. Pauley (1966) *Human Microanatomy*, by kind permission of the publishers, F.A. Davis Co., Philadelphia.
Figure 26.1, 26.4 adapted from R.F. Pitts (1974) *Physiology of the Kidney and Body Fluids*, 3rd edn, Mosby Yearbook Inc., Chicago.
Figures 27.3, 27.5, 27.7(a,b), 27.12 from J.M. Murray and A. Weber (1974) The cooperative action of muscle proteins. Copyright circa 1974 by *Scientific American, Inc.*, **230**, 58–71. All rights reserved.
Figures 29.1, 29.8 from F.G.E. Pautard (1978) *New Trends in Bioinorganic Chemistry*, by kind permission of the author and publishers, Academic Press, London.
Figure 29.2 from K. Simkiss (1975) *Bones and Biomineralisation*, by kind permission of the authors and publishers, Edward Arnold, London.
Figure 30.1 from H.S. Bachelard (1981) *Brain Biochemistry*, 2nd edn, by kind permission of the author and publishers, Chapman and Hall, London.
Figure 30.14 from J.A. Hardy and G.A. Higgins (1992) *Science*, **256**, 184, by kind permission of the publishers, American Society for the Advancement of Science,
Figure 31.1 from H. Shichi (1985) *Handbook of Neurochemistry*, Vol. 8, *Neurochemical Systems*, 2nd edn, edited by A. Lajtha, by kind permission of the author and publishers, Plenum Press, New York.
Figure 32.13 from L.E.H. Whitby and C.J.C. Britton (1957) *Disorders of the Blood*, by kind permission of the authors and publishers, Churchill Livingstone, London.
Figure 34.1 from J.V. Rodericks (1992) *Calculated Risk*, by kind permission of the Press Syndicate of the University of Cambridge.

Figure 37.14 from L.M. Franks and N.M. Teich (eds) (1990) *Introduction to the Cellular and Molecular Biology of Cancer*, 2nd edn, by kind permission of the publishers, Oxford University Press, Oxford.

Figure 38.1 from A.E.H. Emery (1981) *Lancet*, **ii**, 406, by kind permission of the publishers, The Lancet, London.

Figure 39.4 from H.C. Neu (1992) *Science*, **257**, 1064, by kind permission of the publishers, the American Society for the Advancement of Science, Washington, DC.

Figure 39.5 from N. Nikaido (1994) *Science*, **264**, 383, by kind permission of the publishers, the American Society for the Advancement of Science, Washington, DC.

Figure 39.14 from R. Stone (1994) *Science*, **264**, 367, by kind permission of the publishers, the American Society for the Advancement of Science, Washington, DC.

Figure 41.1 redrawn from D.A. White, B. Middleton and M. Baxter (1984) *Hormones and Metabolic Control* by kind permission of the publishers, Edward Arnold (Publishers) Ltd, London.

Figure 41.2 from A.L. Peter and M.B. Davidson (1995) *Current Opinion in Endocrinology and Diabetes*, **2**, 330, by kind permission of the publishers, Current Sciences, Philadelphia.

Figure 42.1, 42.2, 42.4, 42.5 from R. Davies and S. Ollier (1989) *Allergy: The Facts*, by kind permission of the publishers, Oxford University Press, Oxford.

Figure 42.3 from S.E. Webber and D.R. Corfield (1993) *Pathophysiology of the Gut and Airways–An Introduction*, edited by P. Andrew and J. Widdicombe, by kind permission of the publishers, Portland Press, London.

Figure 42.6 from P.J. Barnes (1991) *Trends in Biochemical Sciences*, **16**, 367, by kind permission of the publishers, Elsevier Science Publishers, Amsterdam.

Part 1

Cellular and molecular biology

Chapter 1

What happens where in the living mammalian cell?

1.1 Introduction

The objective of this chapter is to introduce the idea that it is possible to discuss the functions of cells in the context of their structures. This is a major theme that will be developed in the first eight chapters. Clearly the subjects of cell structure and function are closely related, but for present purposes it will be assumed that the reader already has a good working knowledge of cellular structure.

The major types of cell are classified as either prokaryotes (bacteria) or eukaryotes. The crucial difference between these two types of cell is the presence in the latter, but not the former, of internal membranous structures, most especially the presence of a nucleus. All cells, whether prokaryotic or eukaryotic, are bounded by a lipid bilayer structure referred to as a cell or plasma membrane (see Chapter 2). As a result of their sequestration from the environment, mammalian eukaryotic cells generally have no need of additional protection. Bacteria, on the other hand, have rigid cell walls (see Chapter 39) that serve to protect them from changes in the external environment, notably from osmotic changes.

1.2 Size of cells

Although there are interesting exceptions to this rule, most eukaryotic cells are 10–20 times longer in their linear dimensions than prokaryotic cells. For example, the microorganism *Escherichia coli,* found in the human gut, is a sausage-like prokaryotic cell about 1 μm in length, whereas a liver cell (hepatocyte) is about 30 μm in diameter. This difference means that the relative surface-to-volume ratios for these two cell types are about 6.0 and 0.2, respectively. For all cell types it is advantageous to have available membrane surfaces on which groups of enzymes that catalyse related reactions may be organized. Bacteria make extensive use of their cell membranes for this purpose, whereas eukaryotic cells have acquired internal membranes. The surface area of these internal membranes accounts for 96–98% of the total membrane area associated with hepatocytes or the cells of the exocrine pancreas, for example. There are some important exceptions to this general rule, and these always relate to specialist functions. Thus, the intestinal epithelial cells, which are specialized for the absorption of nutrients, have a very much extended plasma membrane that is invaginated in the form of microvilli

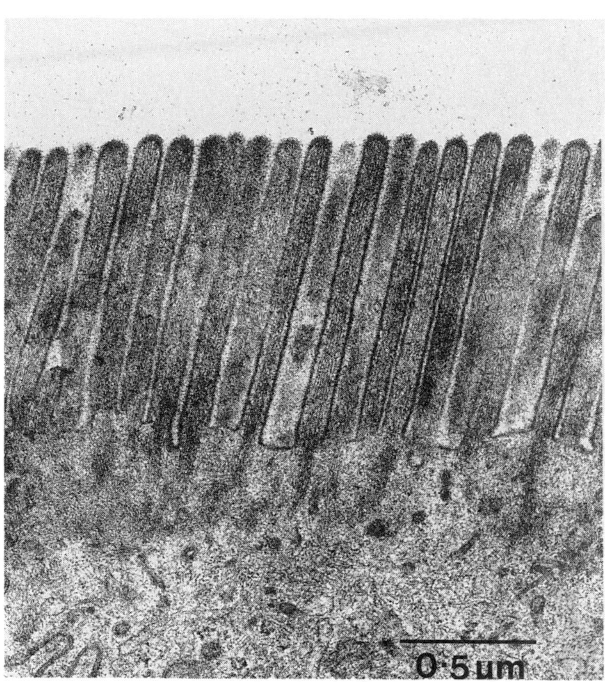

Figure 1.1 **An electron micrograph of a vertical section through the surface of a columnar epithelial cell (enterocyte) of the small intestine, showing a series of microvilli.** Note the presence of central cores of actin filaments within the microvilli, and the fuzzy glycocalyx on the external surface of each microvillus

(Figure 1.1). Likewise, the plasma membrane of the Schwann cells is greatly extended to wrap around peripheral nerve cells to form a myelin sheath (see Chapter 30).

1.3 Methods of studying cell structure and function

Microscopic studies

Light microscopy

Since the discovery, by Schleiden and Schwann, in 1838, that all animal and plant tissues contain cells, the application of light microscopic techniques has made great contributions to the understanding of cellular structure

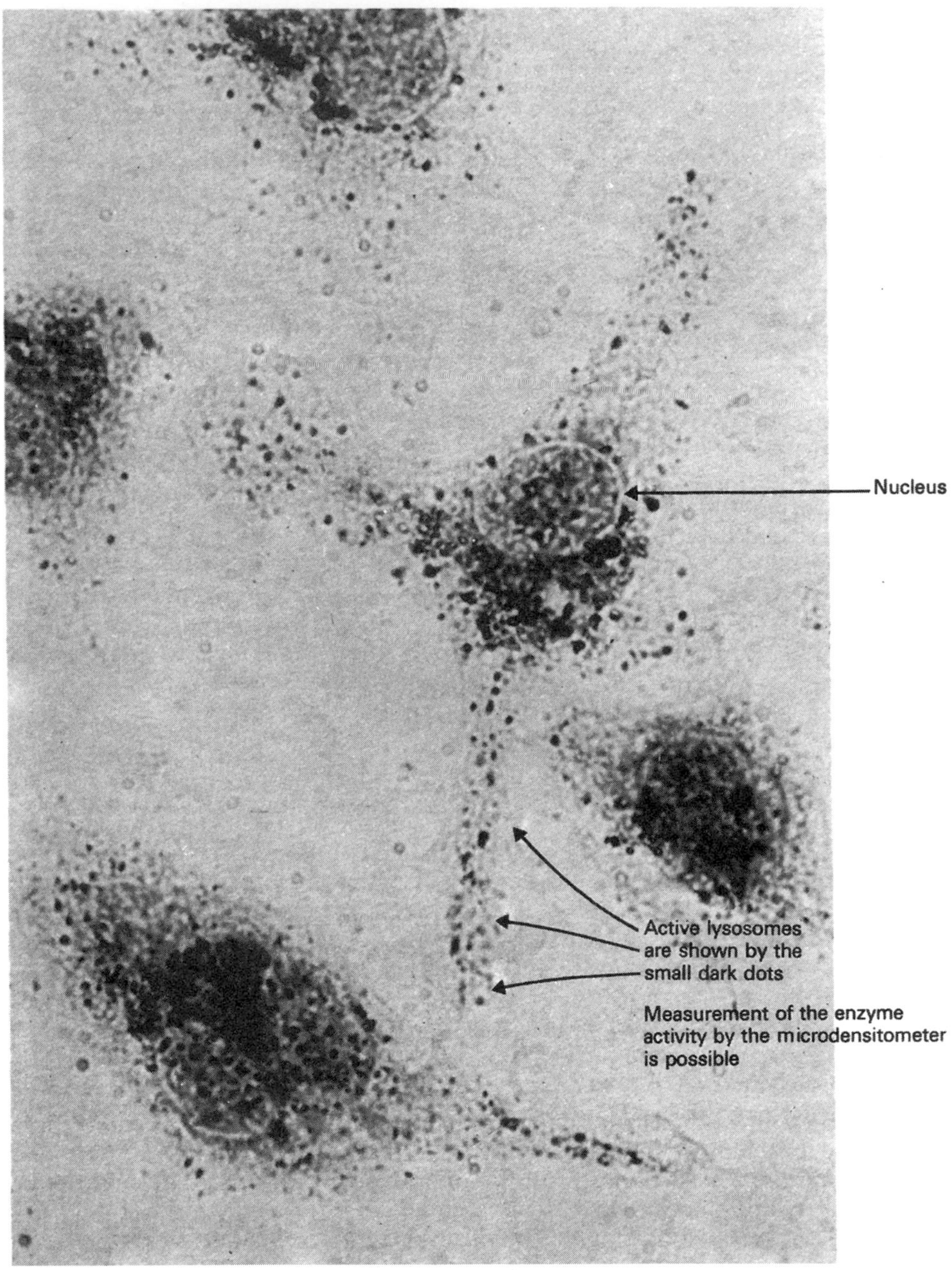

Figure 1.2 **The action of lysosomal acid phosphatase demonstrated by cytochemical methods in a cultured macrophage cell**

and its relation to function. Originally, doubtlessly as a result of the fixing techniques and stains used, cells were viewed as being essentially rigid structures, but now the combination of cinematic and computer techniques has shown many cells to be in a constantly dynamic state, flexible and often changing their shape.

Electron microscopy

After the advances in the understanding of cellular structure that followed the development of light microscopy, the next major step forward followed the introduction of the electron microscope for the study of biological tissues. Specimens for electron microscopy must be fixed, dehydrated and 'shadowed', using, for example, osmium tetroxide, prior to examination. To this extent the images obtained by this means must therefore be artefactual. Nevertheless, much detailed information about the ultrastructure of the cell has been gained. In recent developments of this technique, the disposition of gold particles attached to antibodies within specific subcellular structures has helped in relating structure to function (see next section).

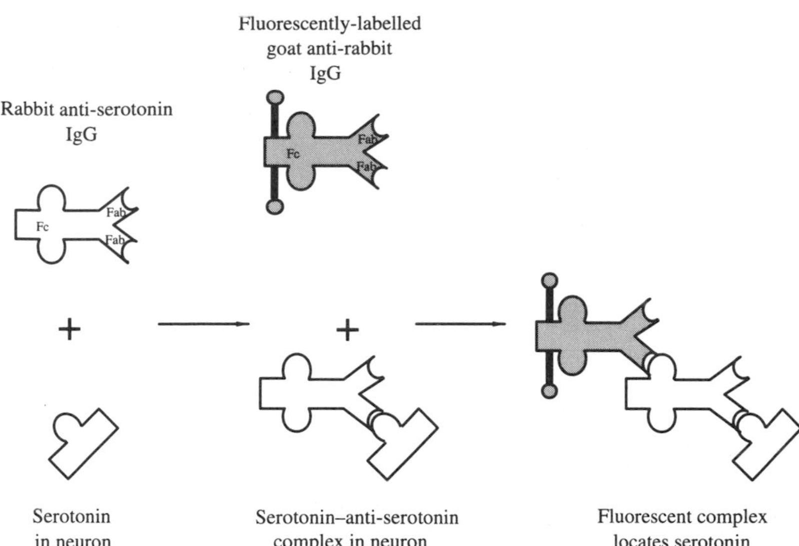

Figure 1.3 **The detection of serotonin in neurons using a first antibody specific for the neurotransmitter and a fluorescently labelled second antibody specific for the first**

Determining cellular specialization

An important goal in the biochemical study of the tissues is to determine which processes occur in which cell type. This has been a major problem in attempts to study the biochemical function of the brain, for example. A fundamental requirement that must be fulfilled if such studies are to be successful is one of specificity. Means have to be available that allow particular processes to be assigned to particular cells.

Histochemistry

Early attempts at achieving this goal depended on the specificity of enzymes in cells. Substrates were devised which, when taken up by intact cells in tissue sections, mounted on microscope slides, underwent enzyme-catalysed reactions in which the products were insoluble and coloured. By detecting the deposition of insoluble products in cells, the types of reaction they are engaged upon may be inferred (Figure 1.2). Unfortunately only a limited number of reaction-types, largely hydrolysis or oxidation-reductions could be detected in this way.

Immunological techniques

The introduction of immunological techniques has revolutionized the detection of enzymes, and other proteins in cells. The basis of immunohistological and immunocytological techniques, which respectively allow the cellular and subcellular location of protein and other molecules, is the ability of antibodies to bind to specific sequences or structures in proteins. Antibodies are important mediators of so-called humoral immunity in the body, and the cells that produce them, the B-lymphocytes, are capable of generating antibodies with almost infinite variety. This means that antibody molecules capable of binding with high affinity and specificity to any protein (or other molecule) may be generated as part of an immune response (see Chapter 32). Indeed antibodies may be raised in one species of animal, e.g. goat, that bind to antibodies produced by, say, rabbit (by binding to the Fc region; Chapter 32). The method of detecting in tissues a protein of interest using such antibodies is illustrated in Figure 1.3. The first antibody, produced in this case by the rabbit, is allowed to bind to the molecule that is being sought (in this case the neurotransmitter serotonin) and subsequently the second antibody is allowed to bind to the first. The second antibody is produced by the goat and binds to any antibody produced by the rabbit. Detection is achieved by the second antibody having a fluorescent 'label' attached covalently. When the tissue section is illuminated with light of the correct wavelength, regions in which the 'label' is located fluoresce, and may be observed by suitable microscopic techniques. The detection of serotonin-producing neurons is shown in Figure 1.4.

In situ *hybridization with nucleic acids*

As is explained in Chapter 4, the unique sequences of proteins are encoded in nucleic acid molecules designated messenger ribonucleic acids (mRNAs). An important property of nucleic acids, seen most clearly in the structure of deoxyribonucleic (DNA), is the ability of complementary strands of these polymeric molecules to bind via hydrogen bonds to form stable double-stranded structures. The rules that underlie this binding were discovered by Watson and Crick and require specific 'pairing' of the bases that form part of the nucleic acid structure (see Chapter 4). This provides a useful means of detecting a particular mRNA and hence of inferring which particular protein a tissue might produce. This is because, if an oligonucleotide is synthesized to be complementary (in the Watson–Crick sense) with the mRNA for a particular protein, the two may be caused

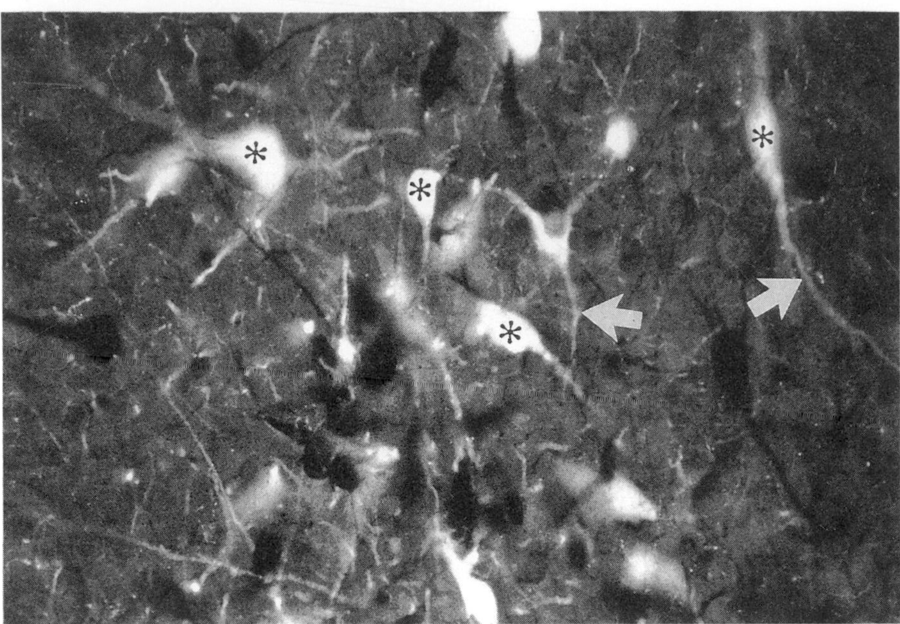

Figure 1.4 **The detection of neurons in the brain stem that contain the neurotransmitter serotonin** (*: cell body; ⇒: dendrites). Magnification ×400

to bind specifically to each other in the tissue. The process is known as *in situ* hybridization. If the oligonucleotide has a fluorescent 'label' attached to it, the cells containing the mRNA of interest can be detected by fluorescent microscopy. Figure 1.5 shows the detection, not of the mRNA for a protein, but rather the RNA associated with ribosomes (rRNA).

Although the methods described in this section are most useful in simply detecting cells (or organelles) with associated proteins or mRNA molecules, they may be adapted for quantitative analysis by using modern methods of image analysis.

Determining subcellular specialization

Microscopic methods

Under favourable circumstances the techniques of histochemistry described in the previous section can be adapted to the cytochemical localization of enzyme activity to a particular organelle. For example, the enzyme succinate dehydrogenase (see Chapter 5) catalyses an oxidation–reduction reaction in mitochondria and its activity may be demonstrated in tissue sections by application of the synthetic substrate tetrazolium which is converted to an insoluble formazan (Figure 1.6).

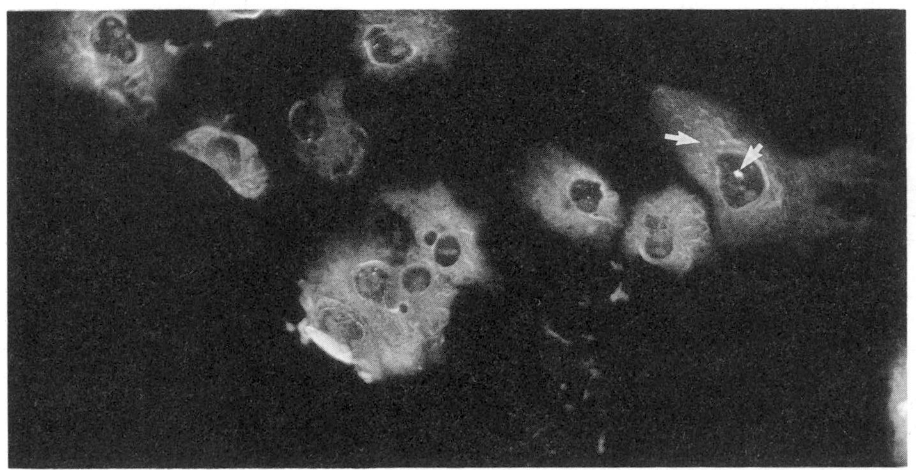

Figure 1.5 **The detection of ribosomal RNA (rRNA) in cells in culture by means of *in situ* hybridization** (note the highly fluorescent nucleolus (→) and the more diffuse fluorescence in the cytoplasm)

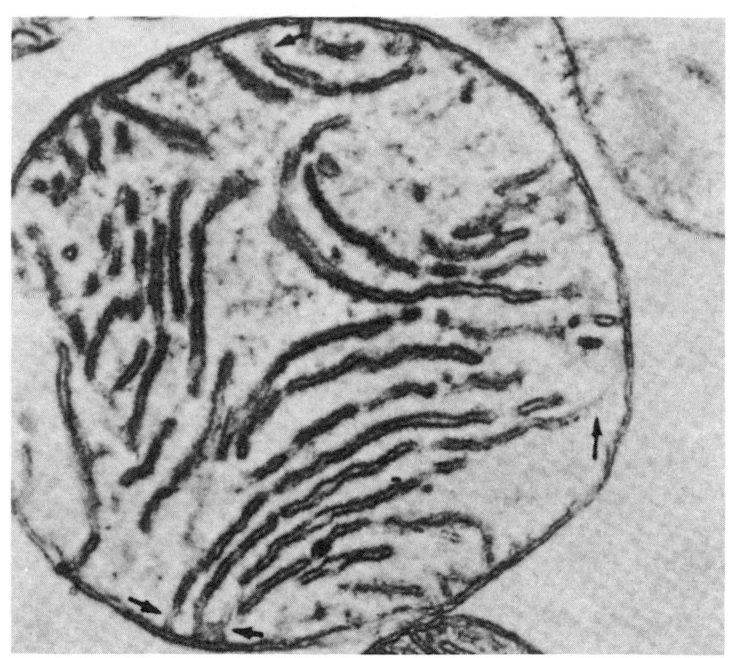

Figure 1.6 **Succinate dehydrogenase activity demonstrated cytochemically in heart mitochondria.** The enzyme reduces a tetrazolium salt to an insoluble coloured formazan which can be seen and measured by light microscopy and using the microdensitometer. For electron microscopy, the tissue section must be treated with osmium tetroxide. Magnification ×50 000

Similarly, the techniques that rely on the application of antibodies or synthetic, complementary oligonucleotides may be adapted so that the complex formed may be detected in the electron microscope. For example, electron-dense gold particles may be attached to the antibodies used for the immunological detection of proteins in subcellular organelles.

Homogenization followed by ultracentrifugation

The separation and isolation of organelles for biochemical studies was first achieved by Schneider, Hogeboom and Pallade in the USA in the late 1940s. They showed that the organelles of a homogenate of a tissue such as rat liver could be separated by a procedure known as 'differential centrifugation'. The tissue is suspended in 0.25 M sucrose, the cells disrupted, by means of the shearing forces generated in a Potter homogenizer, and then is subjected to high-speed centrifugation. Separation is achieved as a result of differences in size and density and, to some extent, shape of the subcellular organelles. Centrifugal forces of sufficient magnitude and duration are used to produce separation of the organelles.

Density gradient centrifugation may also be employed. In this procedure the homogenate is layered on top of a continuous or discontinuous density gradient of sucrose solution and centrifugation is continued until the subcellular particles are in density equilibrium with the surrounding medium. The procedure for differential centrifugation, the more common method, is shown in Figure 1.7.

Nuclei, being the heaviest particles, sediment readily, but they are not generally pure, being contaminated with whole cells and cell debris. Consequently special methods must be used to purify the nuclei.

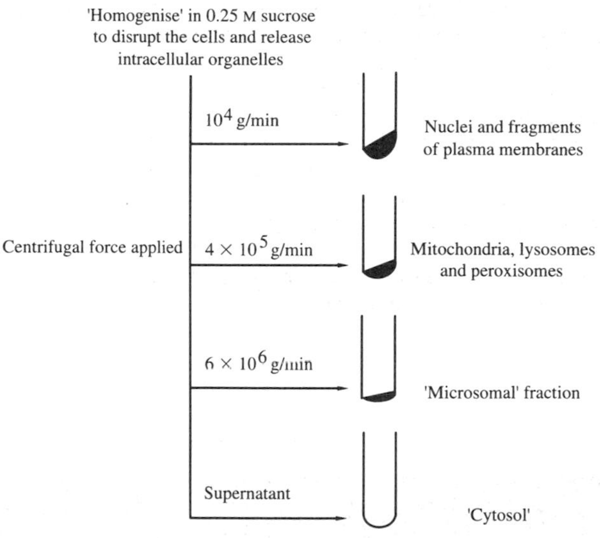

Figure 1.7 **Separation of subcellular components by differential centrifugation**

The second fraction spun down after 10–20 min at approximately 10 000 × g contains most of the mitochondria, but this fraction is usually contaminated with the lysosomes. Special methods, using density gradient centrifugation, must be used to separate mitochondria and lysosomes. If the supernatant from the mitochondrial centrifugation is spun for 1 h at 100 000 × g, a fraction is deposited which is described as 'the microsomes' and contains the fragmented membranes of the endoplasmic reticulum, Golgi apparatus and also fragments of the plasma membrane and the ribosomes.

The supernatant remaining after sedimentation of the microsomes is the cytosol or 'cell sap' and this contains most of the soluble protein and other molecules of the cell.

1.4 The cellular organelles

In view of the clear specialization of cells in complex organisms such as animals, one may doubt the validity of the question that forms the title of this chapter. However it is observed that most cells contain examples of all the major organelles, what varies is the relative proportions of the different structures present. Therefore it is useful to envisage a general cell and to consider how the structures present relate to function. In some cases, however, cells will possess a particular organelle, but will not carry out a particular function because an important enzyme is lacking. For example, smooth endoplasmic reticulum (SER) is present in both liver and muscle cells, but the former and not the latter can release glucose because only the SER of the hepatocytes contains the enzyme glucose-6-phosphatase.

Figure 1.8 shows the major organelles present in our 'typical' cell. It also shows the major processes that occur in each organelle.

The nucleus

The nucleus is the most prominent organelle in most cells. It is bounded by a double membrane (each membrane being a separate lipid bilayer) which come together in places to form nuclear pores. It is across these pores that movement of molecules into and out of the nucleus occurs. The inside of the nucleus consists of a prominent **nucleolus** and apparently less organized **nucleoplasm**. But just as with the cytoplasm (see later), the nucleoplasm has a complex of protein fibres termed **lamina** (the proteins they contain are laminins) which are bound both to the inner nuclear membrane and also to DNA.

Since the nucleus contains almost all of the DNA of the cell, the major functional role of the nucleus is that of **replication** (synthesis of new DNA). In addition, the nucleus is responsible for the synthesis of the three major forms of RNA: ribosomal RNA (rRNA), messenger RNA (mRNA) and transfer RNA (tRNA). All of the molecules operate functionally outside the nucleus and seem to leave via the nuclear pores. The process of synthesis of RNA requires a DNA template and is referred to as **transcription**. The most prominent area of transcription in the nucleus is the nucleolus where the synthesis of rRNA takes place. **Ribosomes** are ribonucleoprotein particles and the rRNA formed in the nucleolus interacts with proteins to form ribosomes and it is these that make the nucleolus so prominent (see Figure 1.5). The advantage of packaging DNA within the nucleus derives in part from the fact that the enzymes that catalyse reactions involving DNA, not only replication but also repair and transcription, are concentrated into an area in which they are required, not dispersed throughout the cell. This same consideration of concentration of

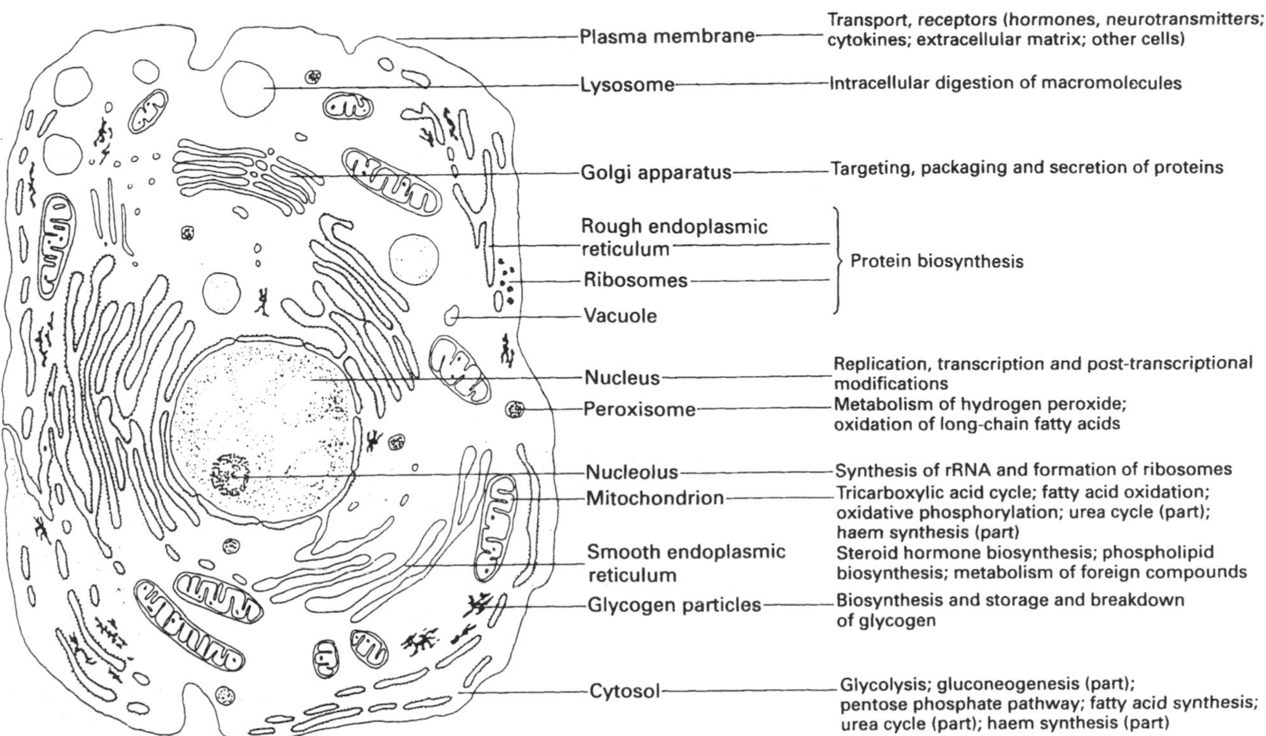

Figure 1.8 **A diagrammatic representation of a typical eukaryotic cell showing the biochemical functions of the subcellular components**

enzymes of related function applies equally to other organelles.

The endoplasmic reticulum

All mammalian eukaryotic cells contain a dense interlacing network of membranes called the 'endoplasmic reticulum'. These very extensive cytoplasmic membranes seem in parts of the cell to be continuous with the outer nuclear membrane. On purely morphological grounds they have been classified as smooth or rough, and the visible structural differences relate to function. Simply the rough endoplasmic reticulum (RER) appears 'rough' or granular because it has ribosomes attached to it and these ribosomes are engaged in the synthesis of specialized groups of protein. It should be noted at this stage that the synthesis of *all* proteins is initiated on free ribosomes in the cytosol. Once initiation has occurred, those ribosomes that are making particular proteins become attached to the endoplasmic reticulum. These proteins include almost all secretory proteins and proteins destined to take up residence in other organelles (except the nucleus, peroxisomes and mitochondria) and in the plasma membrane. Thus, cells involved in much secretion of proteins are rich in rough endoplasmic reticulum, e.g. cells in endocrine and exocrine organs. Many of these 'targeted' proteins have carbohydrate molecules covalently attached to them (indeed this glycosylation is one of the means of targeting) and this process starts in the RER. The process in which mRNA is used by ribosomes as a template for protein synthesis is described as **translation** and the subsequent additions of non-protein molecules are referred to as **post-translational modifications**.

The SER has several roles. As is shown in Figure 1.8 these include steroid biosynthesis, phospholipid synthesis and drug metabolism.

The Golgi apparatus

The Golgi apparatus is structurally very similar to the endoplasmic reticulum and functionally closely related to it. The Golgi consists of a stack of interconnected plate-like membranes, with associated membranous vesicles. The functional relationship between these two types of cytoplasmic membranes is that proteins sequestered in the lumen (cisternae) of the RER are packaged into membranous vesicles that 'bud off' from the RER and subsequently attach to and fuse with the Golgi stack. Thus, any protein in the vesicle membranes or in the vesicle lumen is transferred to the Golgi apparatus. Here glycosylated proteins undergo further glycosylation and other types of post-translational modifications In addition, proteins are 'sorted'. As a result of this sorting, proteins may be returned to the ER, retained in the Golgi, included in the lumen of vesicles that will fuse with the plasma membrane to release their (secretory) contents, or they may be included in the membranes of such vesicles so that the proteins become incorporated in the plasma membrane. Finally, vesicles budding from the Golgi may contain the membrane and luminal proteins that will eventually appear in the lysosomes.

Thus the specialized function of the ER and Golgi jointly is the packaging and targeting of proteins.

Lysosomes

These are membranous structures characterized by the presence in their lumen of a range of enzymes of hydrolytic function. All of these enzymes are most efficient as catalysts at acid pH values, and for this reason they are termed acid hydrolases. Things are well arranged because the lysosome has in its membrane a proton-pumping ATPase that allows the lumen to be acidified to about pH 5.5, at which pH the enzymes are highly active. The roles of the lysosomes include the hydrolysis of external molecules or particles taken up either by endocytosis or phagocytosis. Lysosomes are also involved in the breakdown of some endogenous molecules (or structures). The necessity to keep those destructive enzymes away from the rest of the cell contents clearly shows why this compartment is so necessary.

Peroxisomes

These organelles resemble the lysosomes in their appearance, but they differ both in function and biogenesis. They do not arise from Golgi membranes but, rather, they resemble mitochondria in that they are formed by the division of pre-existing peroxisomes. Peroxisomes are found in all eukaryotic cells, are spherical or spheroidal in shape with a diameter of 0.5–1.5 μm and are bounded by a single membrane. They are characterized by the presence of the enzymes peroxidase and catalase, both of which use hydrogen peroxide as substrates. Peroxisomes are also capable of carrying out the so-called β-oxidation of fatty acids, a process that is more commonly associated with mitochondria. The drug clofibrate acts to cause the proliferation of peroxisomes in cells.

Mitochondria

In common with the nucleus, mitochondria have a double membrane. They are tubular structures of about 0.1–0.5 × 1–2 μm. The inner membrane is extensively invaginated and these in-foldings are referred to as cristae. Mitochondria replicate independently of the nucleus. They also contain their own DNA and ribosomes. The DNA encodes some, but by no means all, of the proteins found in mitochondria, but also rRNA and tRNAs that differ from those from the nucleus.

From a functional point of view mitochondria act as the energy-producing, oxidative machines of the cell. The enzymes of an important metabolic pathway of the cell, the tricarboxylic acid cycle, are all found in the mitochondrial matrix. This pathway catalyses the oxidation of the 2-carbon acetate fragment (supplied in the form of acetylCoA) to CO_2 and H_2O. The oxidizing agent is molecular oxygen which is reduced to water by enzymes found in the inner membrane and referred to as the electron transfer chain. As a consequence of the movement of electrons in the plane of the inner

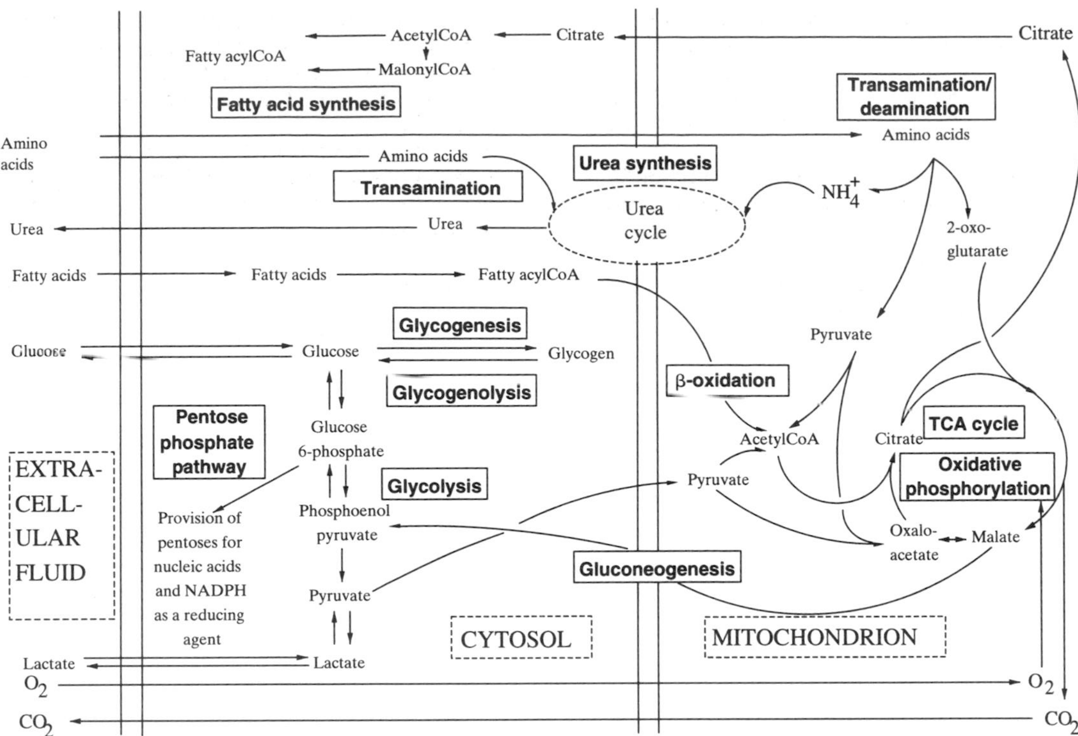

Figure 1.9 **The principal metabolic pathways and their intracellular locations**

membrane a proton gradient is set up across this membrane with the concentration of H^+ being higher in the inter-membrane space than in the matrix. This electrochemical gradient is made to do useful work when the protons are 'channelled' back into the matrix via specialized proteins. As part of their functional activity these proteins, designated F_0F_1, catalyse the synthesis of adenosine triphosphate (ATP) from adenosine diphosphate (ADP) and inorganic phosphate (Pi). ATP is used by the cell in performing a wide range of work processes. Other important oxidative reactions that occur in mitochondria include the β-oxidation of fatty acids.

The plasma membrane

The plasma membrane, which serves to sequester the cell contents from their environment, plays a key role in the interaction of the cell with its surroundings. It acts as a very selective barrier to the outside. Small, lipid-soluble molecules such as oxygen, carbon dioxide, ethanol and urea are able to diffuse into or out of cells across this membrane. Larger, more water-soluble molecules or ions cannot simply diffuse across the plasma membrane. Consequently transport proteins for molecules such as glucose and ion-channel proteins for both anions and cations, e.g. Ca^{2+}, allow these substances to enter or leave the cell. The plasma membrane also has a series of receptor proteins that permit extracellular molecules, e.g. peptide hormones, to signal their presence to the cell and thereby alter intracellular chemistry without having to enter the cell. Other plasma membrane receptors mediate the interaction of the cell with its neighbouring cells or with the extracellular matrix.

Cytosol

It is now appropriate to refer to the approximate 50% of the intracellular volume *not* enclosed by any of the internal membranes. This is referred to as the cytosol ('cell sap') and the implication from Figure 1.8 is that it is structureless, containing only soluble molecules. It is clear, however, that this is not the case in the living cell. There is a very complex network of protein fibres, which is referred to as the cytoskeleton. This includes the actin-containing microfilaments, referred to as 'stress fibres', and the tubulin-containing microtubules. These fibres play an important role in the maintenance of cell shape and mobility. They interact with specialized proteins located in the plasma membrane, for example, and thereby truly serve as 'skeleton' elements. Actin filaments are clearly visible in the microvilli of the intestinal epithelial cells (see Figure 1.1) and these contractile proteins are involved in the movement of the villi.

The benefit to eukaryotic cells in possessing organelles

In the discussion above, reference has been made to some of the advantages of there being a number of compartments in the mammalian cell. The most important of these are:

1. Enzymes that generate, as their products, substrates for other enzymes work much more efficiently if they are held in close proximity in or on a membrane or within a compartment. Such interacting systems are referred to as **metabolic pathways**.
2. When two metabolic pathways carry out apparently opposing groups of reactions, their occurrence can be conveniently controlled if the enzymes that catalyse each reaction sequence are located in different compartments. Thus the reaction of fatty acid oxidation occurs in the mitochondrion whilst fatty acid biosynthesis takes place in the cytosol.
3. Organelles can be maintained in disequilibrium with the rest of the cell, for example, with respect to their pH. This is the case in the lysosomes and other membranous vesicles derived from the Golgi apparatus.
4. Substances can be stored in high concentrations and these may be subsequently released as required, e.g. Ca^{2+} stored in the sarcoplasmic reticulum and released to promote muscular contraction.

The locations of the principal metabolic pathways in the 'general cell' are shown in Figure 1.9. These processes will be dealt with in the following chapters.

Chapter 2

Roles of extracellular and intracellular membranes: membrane structure and membrane transport

2.1 Introduction

The fact that all living cells are surrounded by a membrane, often called the 'plasma membrane', was apparent from the time cells were first studied by microscopic means. Its function appeared then to be to keep the contents of the cells from dispersing, and this has subsequently been amply confirmed. Thus the plasma membrane plays an important role both in cellular and in organism structure.

As microscopy, and in particular electron microscopy, developed to give improved resolution, it became apparent that membranes also surrounded all of the subcellular structures. In addition, almost all eukaryotic cells were shown to contain a complex internal network of membranes, the endoplasmic reticulum and the Golgi apparatus.

These internal membranous structures permit a range of specialized, discrete metabolic pathways to continue independently. The effectiveness of this metabolic compartmentalization is reinforced by an important property of the membranes : they are highly selective in their permeability. Biological membranes are able to facilitate the passage of certain molecules while preventing the movement of others. The protein molecules that facilitate the movement of molecules across membranes are called **transporters**. These transport processes are very important in the overall regulation of metabolism, in that they permit certain molecules to undergo metabolism in specific organelles. In addition to the transporters, many enzymes are also integral to specific membranes; for example, many of the enzymes involved in the metabolism of foreign compounds are found associated with the smooth endoplasmic reticulum, especially in the liver (see Chapter 36).

Permanent integrity of their membranes is of vital importance to all cells. Lytic agents, e.g. detergents and a range of venoms, will solubilize membranes and thereby cause the release of intracellular components. Thus, the disruption of the plasma membrane of the red cell caused by some snake bites leads to the release of their haemoglobin, with the clear-cut serious consequence of a haemolytic crisis. Damage to intracellular membranes can also lead to serious consequences. For example, damage to lysosomal membranes results in the leakage of hydrolytic enzymes, such as proteases and lipases. These will digest vital cellular proteins and lipids, despite being some way from their acid pH optima (see Chapter 8).

2.2 Membrane composition

All biological membranes are composed of lipids, proteins and a smaller proportion of carbohydrates.

Lipids

Lipids form from 40% to 80% of total membrane constituents and, of the lipids, **phospholipids** are essential and major components of all membranes. Thus 75% of the lipid of the endoplasmic reticulum of hepatocytes is phospholipid. The corresponding figure for the human red cell plasma membrane is 62%. The structures of the phospholipids are shown in Figure 2.1. Phosphatidylcholine is usually the most abundant phospholipid and, together with phosphatidylethanolamine, these may also be found as lysophospholipids in which the fatty acyl residue that should occupy position 2 of the glycerol has been lost by hydrolysis. Cholesterol, cholesteryl esters, triacylglycerols and non-esterified fatty acids all occur in membranes and are referred to collectively as **neutral lipids**.

When the composition of the lipids of the subcellular membranes are compared, several important differences are noted. The Golgi apparatus contains a high proportion of neutral lipids, particularly cholesterol, whereas, in contrast, mitochondrial membranes contain very little of the sterol. In addition, sphingosine is virtually absent from mitochondria but the endoplasmic reticulum contains significant quantities of this lipid component.

It is important to note that many different phospholipids (possibly 150–200) may be found in membranes. This diversity arises because each glycerophospholipid contains two fatty acids and the sphingolipids one fatty acid. These may be any one of a range of naturally-occurring fatty acids, the most important of which are shown in Figure 2.2. These molecules all contain an even number of carbon atoms (a consequence of the mode of biosynthesis; see Chapter 6) and these range from 12 to 22. In addition to the fully

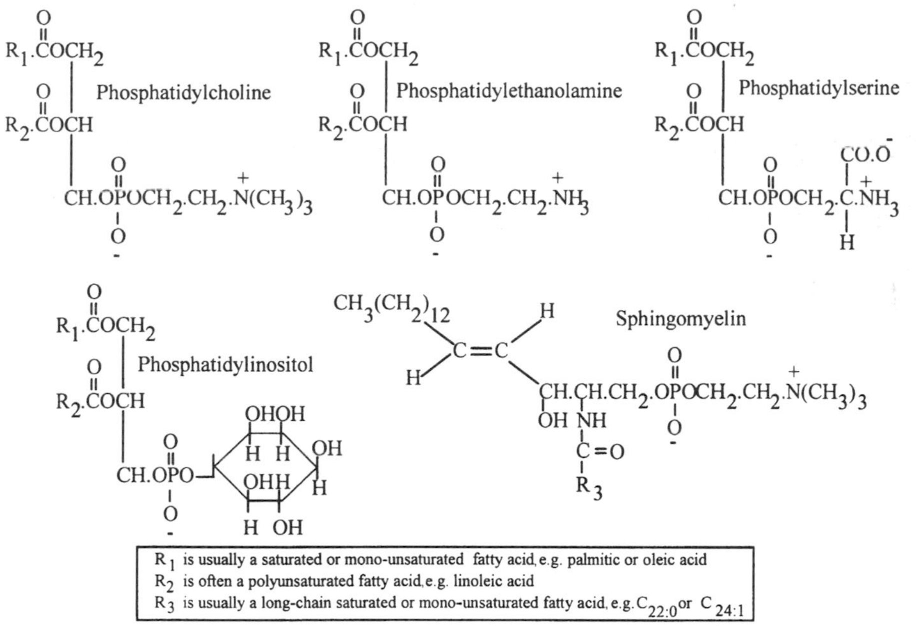

Phosphatidylcholine

Phosphatidylethanolamine

Phosphatidylserine

Phosphatidylinositol

Sphingomyelin

R_1 is usually a saturated or mono-unsaturated fatty acid, e.g. palmitic or oleic acid
R_2 is often a polyunsaturated fatty acid, e.g. linoleic acid
R_3 is usually a long-chain saturated or mono-unsaturated fatty acid, e.g. $C_{22:0}$ or $C_{24:1}$

Figure 2.1 **Examples of phospholipids: glycerophospholipids and a sphingophospholipid**

saturated fatty acids, a number are unsaturated, containing between 1 and 6 carbon–carbon double bonds. The location of the double bonds may be specified in one of two ways: either in relation to the carboxylate group (Δ) or back from the final methyl group ('n-'). Thus, for example, in order to designate the location of the double bonds in linoleic acid, which has 18 carbon atoms and two double bonds, we may write either $18{:}2^{\Delta 9,12}$ or 18:2(n-6). In the former case the location of the first carbon atom in each double bond is specified. In the 'n-' it is the location of the first atom in the first double bond only that is specified. The nomenclature of the essential fatty acids is further explored in Chapter 17. The geometrical isomerism found about the double bonds in all the major naturally-occurring unsaturated fatty acids encountered in mammals is of the *cis* type.

Membrane fluidity

As a consequence of the above considerations a very large number of variants of each type of phospholipid exist and many of these may be incorporated into membranes. The precise significance of the different fatty

Saturated fatty acids	Lauric acid (12:0)	$CH_3(CH_2)_{10}COOH$
	Myristic acid (14:0)	$CH_3(CH_2)_{12}COOH$
	Palmitic acid (16:0)	$CH_3(CH_2)_{14}COOH$
	Stearic acid (18:0)	$CH_3(CH_2)_{16}COOH$
Mono-unsaturated fatty acids	Palmitoleic acid (16:1)(n-7)	$CH_3(CH_2)_5CH{=}CH(CH_2)_7COOH$
	Oleic acid (18:1)(n-9)	$CH_3(CH_2)_7CH{=}CH(CH_2)_7COOH$
Poly-unsaturated fatty acids	Linoleic acid (18:2)(n-6,9)	$CH_3(CH_2)_4CH{=}CH.CH_2.CH{=}CH(CH_2)_7COOH$
	Linoleneic acid (18:3)(n-3,6,9)	$CH_3CH_2CH{=}CH.CH_2.CH{=}CH.CH_2.CH{=}CH(CH_2)_7COOH$
	Arachidonic acid (20:4)(n-6,9,12,15)	$CH_3(CH_2)_4CH{=}CH.CH_2.CH{=}CH.CH_2.CH{=}CH.CH_2CH{=}CH(CH_2)_3COOH$

Figure 2.2 **The major fatty acids found in phospholipids**

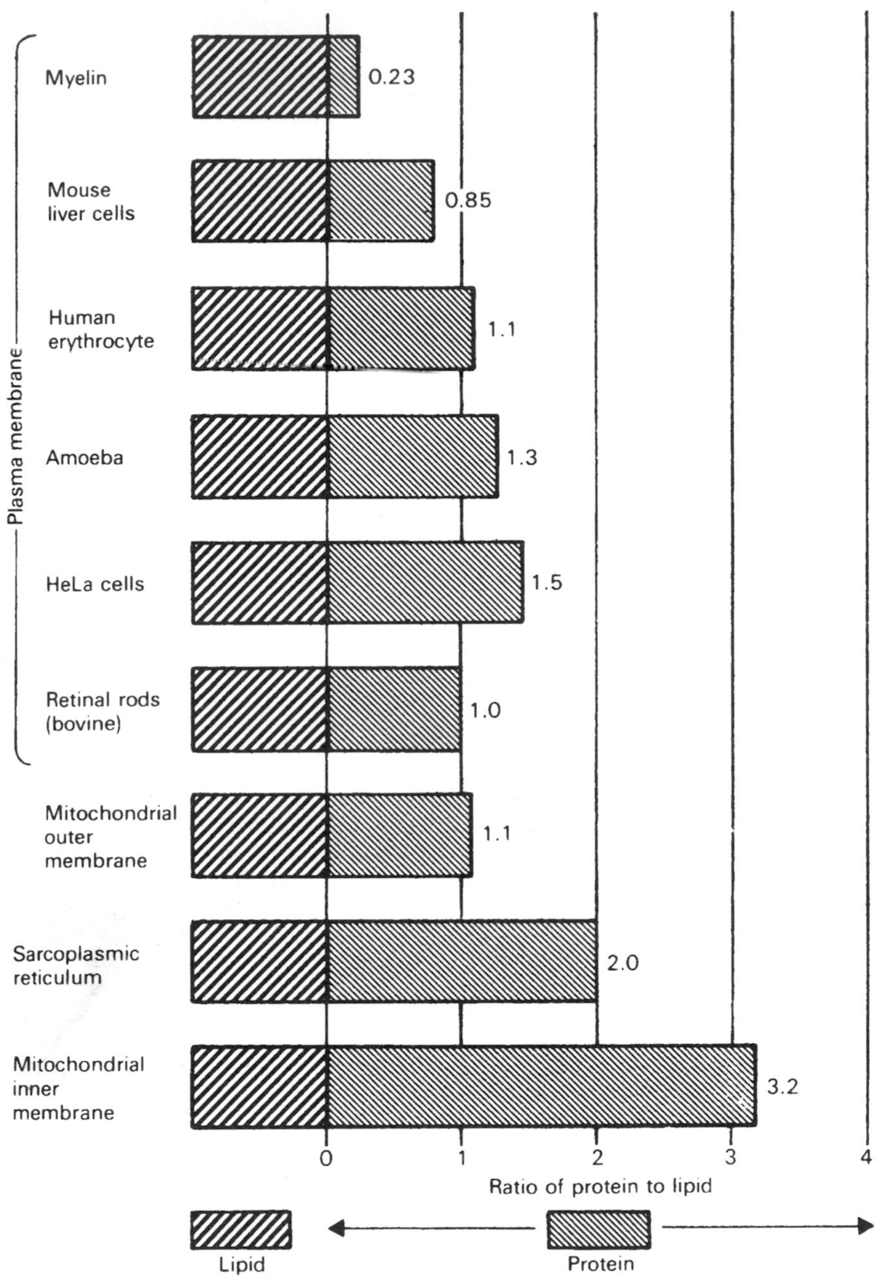

Figure 2.3 **Protein–lipid ratios in typical membranes**

acids which, as phospholipid components, are found in membranes is not entirely clear, but the exact composition does appear to be functionally important. Most importantly the fluidity, or flexibility, of membranes is dependent on the degree of unsaturation of the fatty acids found in the phospholipids. As the degree of unsaturation increases, the membranes become more flexible and fluid. Secondly, because some polyunsaturated fatty acids, e.g. linoleic acid, cannot be synthezised in the body, conditions of essential fatty acid deficiency are displayed when this fatty acid is omitted from the diet (see Chapter 17). An important consequence of this is altered membrane function. Thirdly, if polyunsaturated fatty acids undergo peroxidation following oxidative attack on the double bonds, they can be destroyed with consequent loss of membrane structure and function (see Chapter 33).

Studies on artificial membranes composed of phospholipids show that cholesterol exerts a 'condensing effect' on the phospholipids. As a consequence of the incorporation of cholesterol into these artificial membranes, the phospholipids occupy a smaller area at the lipid–water interface. These experimental findings make it quite likely that the presence of cholesterol will exert an important controlling influence on the structure of biological membranes. Generally, membranes with higher cholesterol contents tend to be less fluid than those with less of the sterol.

```
                 OH
                 |
  ┌─────────────C─COOH
  |              |
  |            H─C─H
  |              |
  O            H─C─OH
  |              |         Ac = CH3CO–
  |        Ac─NH─C─H
  |              |
  └──────────────C─H
                 |
               H─C─OH
                 |
               H─C─OH
                 |
                CH2OH
```

Figure 2.4 **The structure of sialic acid (***N*-acetylneuraminic acid)

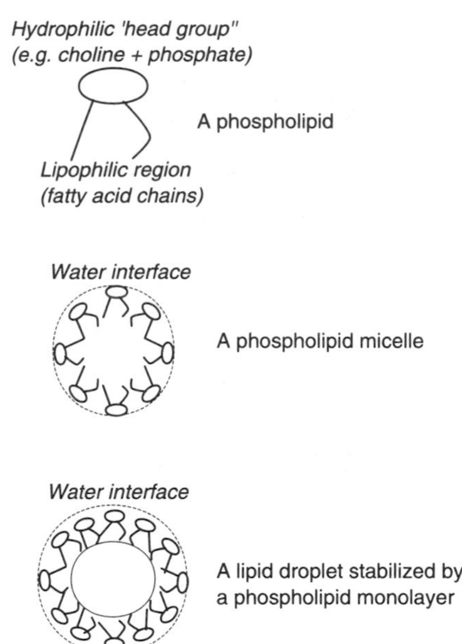

Figure 2.5 **Phospholipid micelles**

Proteins

Proteins normally form a major proportion of most membranes, usually 50–70% by weight. An exception is the myelin which ensheaths some of the neurons of the nervous system. This extension of the plasma membrane of certain glial cells (see Chapter 30) contains 90% lipid. The relation of lipid to the protein composition of a range of membranes is shown in Figure 2.3.

In membranes, proteins play a range of key structural and functional roles. In the form of the 'membrane skeleton' they help to maintain a degree of stability to the membrane. As receptors they act as recognition molecules and they constitute transporters.

A typical plasma membrane, such as that of the red blood cell, will contain many different proteins. Of these, a high proportion are glycoproteins, i.e. proteins to which oligosaccharides numbering from 4 to 15 carbohydrate residues are covalently attached. A range of monosaccharides are found in glycoproteins; these include the hexoses glucose, galactose (and its 6-deoxy derivative, fucose) and mannose and the pentose sugars arabinose and xylose. In addition, several amino sugars are found as their *N*-acetyl derivatives, these including *N*-acetylglucosamine, *N*-acetylgalactosamine and also *N*-acetylneuraminic acid (sialic acid; Figure 2.4). Although the component units of the oligosaccharides are fewer in number than those of proteins, further variations arise because the sugars can link at several different sites and branching can, and does, arise. For example, five different sugars in a 13-residue oligosaccharide can form 10^{24} different structures.

2.3 Membrane structure

Lipids

All biological membranes are composed of a lipid bilayer. Typically a phospholipid contains both a strongly lipophilic component and a strongly hydrophilic component. The former consists of the two fatty acid residues in the glycerophospholipids or the single fatty acid and the hydrocarbon portion of the sphingolipids, and the latter of the phosphate group with (a) strongly hydrophilic residue(s) attached (Figures 2.1 and 2.5). These are often referred to, respectively, as lipophilic or hydrophobic 'tails' and polar or hydrophilic 'head groups'.

In solution, phospholipids tend to form micelles; the lipophilic parts of several phospholipids associate by van der Waals attraction forces and the hydrophilic parts form a surface layer at the water interface (Figure 2.5). By associating not only with each other but also with a lipid droplet, phospholipids can stabilize and thus help emulsify the droplet.

Cells, however, are neither micelles nor stabilized lipid droplets, but rather consist essentially of an aqueous environment bounded by a lipid bilayer. This limiting bilayer may be regarded as being generated by the association of pairs of phospholipids via their lipophilic tails to form the centre of the membrane, whereas their hydrophilic head groups are presented to either the internal or the external aqueous environment. Predominantly lipophilic molecules, such as cholesterol, intercalate between the lipophilic residues of the phospholipids and consequently interact only minimally with the aqueous medium. In the case of cholesterol, this contact is restricted to the hydroxyl group attached at position 3 of the sterol (Figure 2.6).

At body temperature, membrane lipids are in a fluid state, and this fluidity of the membrane is essential for the normal functioning to occur, e.g. exocytosis, endocytosis (see Chapter 7) and lysosomal activity (see Chapter 8). Lipid molecules in the bilayer are free to move laterally through the phase of the membrane, but the movement of lipids between the leaflets of the bilayer ('flip-flop') is much more restricted.

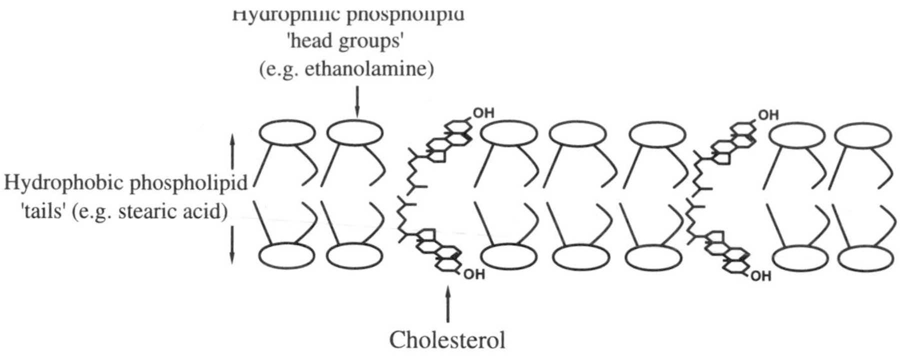

Figure 2.6 **The general organization of phospholipid bilayers to form cell membranes, showing the location of cholesterol molecules**

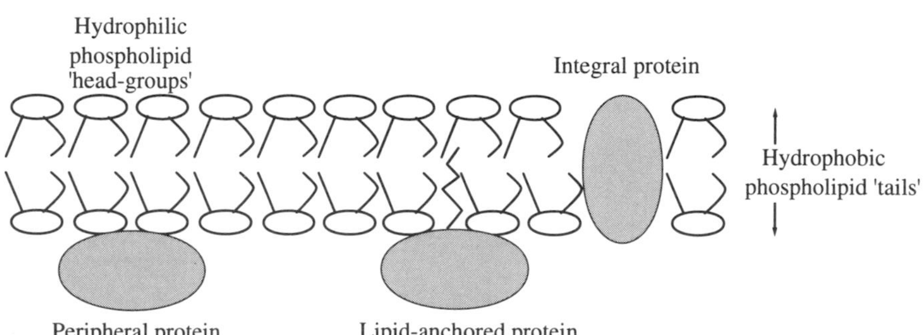

Figure 2.7 **Three ways in which proteins may associate with a biological membrane**

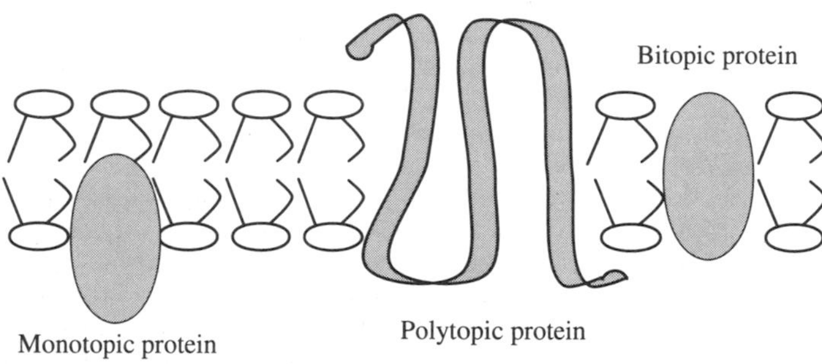

Figure 2.8 **Different ways in which integral proteins may be oriented in a biological membrane**

Proteins

Proteins may be associated with membranes in one of three ways: as peripheral proteins, as lipid-anchored proteins or as integral proteins (Figure 2.7). As the name implies, peripheral proteins interact with the surface of the membrane. Their attachment is rather weak, and they may easily be separated from the membrane (during incautious membrane preparation, for example). Superficially, the second class of proteins resemble peripheral proteins in their location; however, their binding to the membrane is much stronger. This is because they have a lipid structure covalently attached. The lipid may either be a simple fatty acid, e.g. palmitic or myristic acid, or a more complex glycolipid. These additions to the protein are able to insert into the lipid bilayer and hence the protein becomes firmly attached to the membrane. Integral proteins have exposed, relatively hydrophobic surfaces and these allow the protein itself to insert in the bilayer. Proteins that insert into just one leaflet of the bilayer are referred to as **monotopic**, those that span the whole membrane once are called **bitopic** whereas those spanning the membrane several times are **polytopic** (Figure 2.8).

It has proved very difficult to use physical techniques, such as X-ray crystallography, to ascertain precisely how bitopic or polytopic proteins are arranged so that they span the membrane. However, by the use of high-resolution electron microscopy, it has been established that the polytopic protein bacteriorhodopsin spans the cytoplasmic membrane of certain photosynthetic bacteria seven times. Each crossing of the membrane is in the form of a relatively hydrophobic α-helix, of some 20 amino acid residues. It seems likely that a number of bitopic and polytopic membrane proteins will prove to span the membrane by forming hydrophobic α-helices,

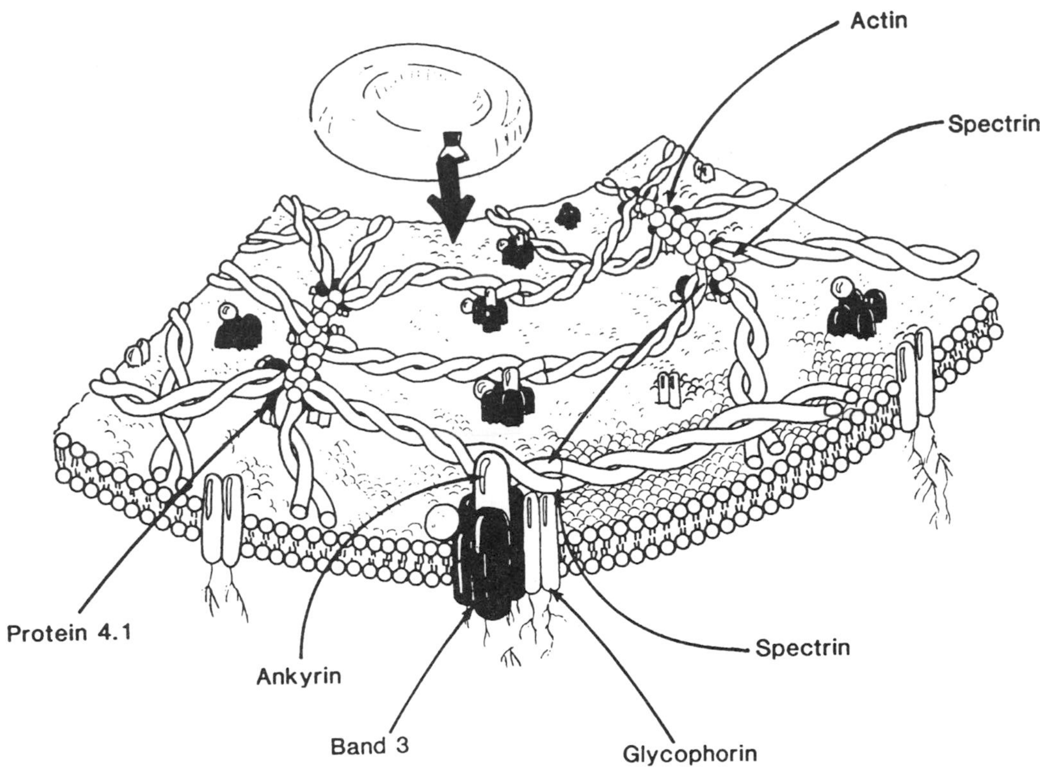

Figure 2.9 **The organization of the proteins that form the membrane skeleton of the red blood cell**

e.g. catalytic and G-protein coupled receptors (see Chapter 3). It should not be assumed, however, that this will always be the case. For example, an outer membrane protein (OmpC) of the bacterium *Escherichia coli is* known to form membrane-spanning β-pleated sheets. In other cases, transitions between α-helices and β-sheets may underlie the functional activity of a protein in the membrane. It should be borne in mind, therefore, that when texts (including the present one) indicate that a particular bitopic or polytopic protein spans a biological membrane as an α-helix, this is usually a prediction based on amino acid sequence, i.e. homology to other proteins in a family, not on experimental observation.

2.4 Types of membrane proteins: their functional roles

Membrane proteins are most usefully classified according to the functional role they play: they may be **structural molecules**, **transporters**, **ion channels**, **receptors** or **enzymes**. However, it should be remembered that these proteins may be bifunctional. Thus, the insulin receptor is a catalytic receptor, i.e. the binding of insulin to its receptor causes the latter to become active as an enzyme (see Chapter 3). Likewise, the so-called band III protein of the red cell membrane serves a structural role as an 'anchor' for the cytoskeleton and also as an exchange transporter for HCO_3^- and Cl^-.

Structural proteins

Three general types of structural membrane proteins may be recognized : those that mediate binding of one cell to another, those that allow the cell to interact with the matrix in which they are embedded, and those that act as 'anchors' for the fibrous filaments of the cytoskeleton. Thus the plasma membrane protein neural cell adhesion molecule (NCAM) of one neuron will interact by means of non-covalent forces with an homologous protein in a second nerve cell. Likewise, the plasma membrane protein fibronectin receptor serves to bind cells to the major adhesive protein of connective tissue, fibronectin. As mentioned previously, the membrane protein of the red cell designated band III protein acts as an adaptor to anchor the cytoskeleton protein spectrin, the interaction being facilitated by a further protein, ankyrin. Ultimately the spectrin fibres interact with actin filaments (Figure 2.9). The cytoskeletal network so formed serves to maintain the characteristic biconcave shape of the red cell. In addition, it confers on the cells the degree of flexibility to squeeze through the vascular capillaries.

Receptor molecules and ion channels

Receptor molecules are found in the plasma membranes of all eukaryotic cells, and they may be regarded as constituting 'windows' on the 'world' outside the cell. The binding of extracellular molecules to these receptors causes changes to occur in the functional activity of cells

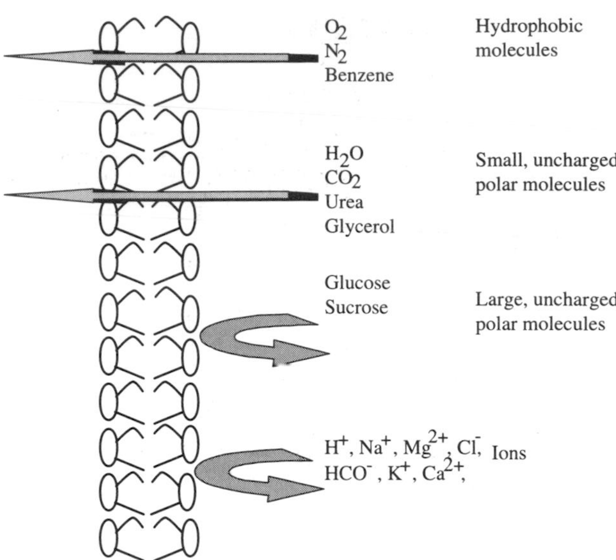

Figure 2.10 **The semi-permeable nature of biological membranes**

(these matters are dealt with in Chapter 3). Some receptors are also ion channels, i.e. they are bifunctional. This means that the polypeptide chain is so arranged in the membrane that when an extracellular molecule binds to its receptor the protein undergoes a conformational change. This has the result of generating a 'channel' or pore across the membrane, through which a specific ion may pass. The protein is said to 'gate' the ion in question. Some ion channels respond not to signalling molecules but to changes in the electrical potential that exists across all mammalian cell plasma membranes. These are referred to as 'voltage-gated' ion channels (see Chapter 3). Yet other ion channels are constitutive, i.e. they are 'open', allowing ions to pass across membranes at all times.

Transport proteins

Biological membranes are semi-permeable. They allow the passage of small uncharged or hydrophobic molecules, but not of large polar molecules (whether charged or not; Figure 2.10). Substances in the latter category do, however, cross biological membranes and they do so in a very specific fashion. The specificity of the transport processes involved points to the participation of some of the proteins of the plasma membrane. The protein may simply act to facilitate the transport or it may be actively involved.

Facilitated diffusion

Further analysis of the characteristic features of membrane transport processes reveals that, in some cases, the movement occurs only if the substance moves 'down' its concentration gradient. Such protein-mediated transport is referred to as facilitated diffusion. Facilitated diffusion may readily be distinguished from **passive diffusion** by experiment. If the movement of a molecule across a biological membrane occurs by passive diffusion the rate will increase in a linear fashion as the concentration gradient increases (Figure 2.11). This is observed when molecules such as oxygen diffuse into, or urea out of cells. If a transporter, or a carrier, facilitates the movement, then as the concentration of the solute increases, a point will be reached when all of the transporter molecules are occupied with solute, and no further increase in rate is possible: the process is said to be **saturable** (Figure 2.11).

In some cases two substances may be transferred by means of the same transporter. If both substances move in the same direction, the process is referred to as **symport**,

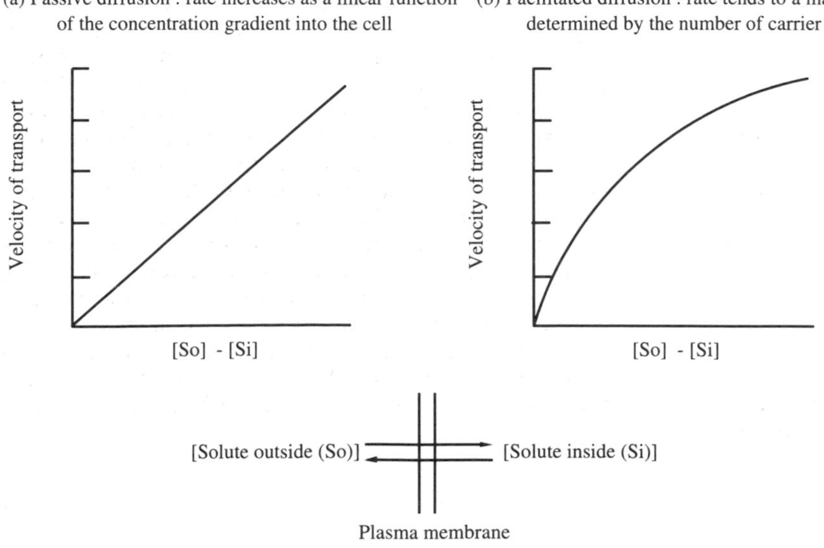

Figure 2.11 **The kinetics of the passage of molecules into cells by (a) passive diffusion and (b) carrier-facilitated diffusion**

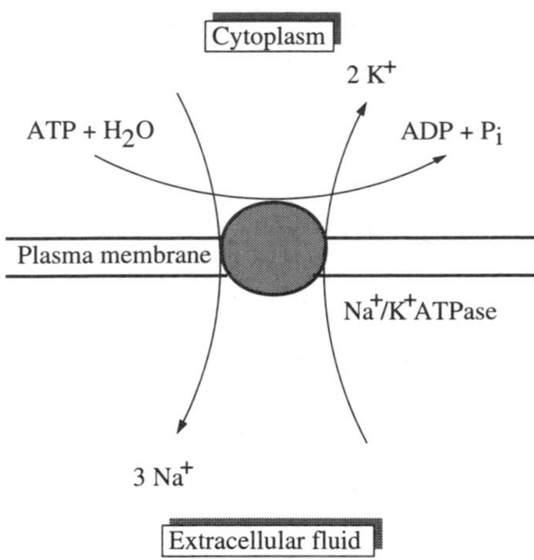

Figure 2.12 **The hydrolysis of ATP supplies the energy required to transport K$^+$ into cells and Na$^+$ out against the prevailing concentration gradients**

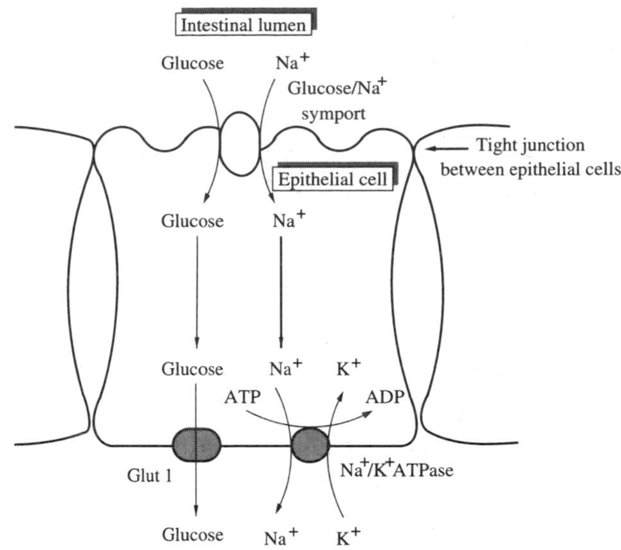

Figure 2.13 **Transport of glucose in the intestine is a 'secondary active process' that depends on the maintenance of a concentration gradient of Na$^+$ out of the epithelial cell**

whereas if an exchange occurs the process is referred to as **antiport**. Examples of both symport and antiport systems are encountered in mitochondria (see Chapter 5).

Important examples of simple (one substance handled) facilitated diffusion are provided by the family of polytopic glucose transporters (they are predicted to span the membrane 12 times). The properties of these transporters are well adapted to their functional roles. Thus, glucose transporter 1 (glut 1) is found in almost all cells where it provides for the basal uptake of glucose. Glut 2 is found only in the liver and in the β-cells of the islets of Langerhans in the pancreas. This transporter is saturated only at very high concentrations of glucose. As a consequence, the amount of glucose transported into cells in these tissues increases as the blood glucose concentration increases across a wide range. This is an important property of these glucose-sensing tissues. Another key glucose transporter is designated glut 4. This protein, found predominately in muscle and adipose tissue, differs from the other transporters in that it is insulin-responsive. When the circulating concentration of insulin (and hence of glucose – see Chapter 28) is low, this transporter is found, not in the plasma membrane of adipocytes and myocytes, but rather is associated with intracellular membranes. As the concentration of insulin in the circulation increases it causes the translocation of glut 4 to the plasma membrane, where it can begin to transport glucose. This is an important regulatory mechanism that ensures that adipose tissue and muscle use glucose when its concentration is high (as signalled by insulin), but fail to do so as the concentration falls.

Active transport

Certain substances may be accumulated inside cells, or inside intracellular organelles, against a concentration gradient. Since these are energetically unfavourable processes, work must be done to achieve this type of movement. The process is referred to as active transport: work must be done to accomplish the movement of the solute in question. The energy to carry out the work may come from the free energy liberated during the hydrolysis of adenosine triphosphate (ATP) or from the potential energy of the concentration gradient of a second solute. When the transport depends directly on the hydrolysis of ATP it is referred to as a **primary active transport process**. Thus the hydrolysis of ATP catalysed by a special plasma membrane ATPase leads directly to the movement of 3 Na$^+$ out of cells in exchange for 2 K$^+$ leaving the cell. Both types of cation move against their concentration gradients and the protein involved is a Na$^+$/K$^+$-ATPase.

Figure 2.12 shows diagrammatically how the system appears to operate. As a consequence of the operation of this process in all mammalian cells, the extracellular concentration of Na$^+$ always greatly exceeds that inside cells. This gradient can be made to do work. Thus, in the luminal plasma membrane of intestinal epithelial cells a symport transporter is found that co-transports both glucose and Na$^+$. This allows glucose to be accumulated against its concentration gradient from the lumen of the intestine, provided that the Na$^+$/K$^+$-ATPase remains active. For this reason the simultaneous uptake of glucose and Na$^+$ is referred to as **secondary active transport**. The glucose taken into the epithelial cells by this process is then transferred into the blood by a conventional glucose transporter located in the contraluminal plasma membrane of the cells (Figure 2.13).

Chapter 3

Biochemical communication

3.1 Introduction

Biologists have long appreciated that a characteristic attribute of the cells in a multicellular organism is their ability to communicate with one another. Until recently, however, biologists who studied the different aspects of such communication rarely communicated with one another. Those interested in the control of flow through blood vessels (for example) believed themselves to have little in common with others studying the way cells of the immune system set about killing invading microorganisms. Now that we are beginning to understand the means by which cells 'talk' to each other these restricted views are no longer sustainable. It is important, therefore, to give an overview of the operation of biological signalling mechanisms. This will then form the basis for understanding the particular ways different groups of cells make use of these systems in their functional activities.

The types of information exchanged by communicating cells are either spatial or temporal. For example, it is by the exchange of spatial information that cells organize themselves into tissues (developmental signals). Temporal information indicates to target cells when it is appropriate for them to alter their chemistry and for how long; for example, insulin signals to muscle cells that they should take up glucose. This point about the duration of the signal helps to emphasize that the means of terminating cellular responses to external stimuli is as important as the means of initiating them.

3.2 The nature of cellular responses to external signals

Cells may respond to external cues electrically or chemically. This is a useful generalization for the purposes of describing the nature of cellular responses to external influences. But it will become apparent that these two types of change are simply the extremes of a continuum of responses. The basis of the **electrical response** is a change in the movement of small ions across the plasma membrane. This is invariably a rapid response (milliseconds) and forms the basis of much of the signalling in the nervous system (see Chapter 30). The **chemical response**, on the other hand, may occur with timecourses that vary from moderately rapid (seconds) to rather slow (hours or even days). The basis of the chemical response is that the catalytic activity of one or several intracellular enzymes is changed in response to the presence of extracellular signalling molecules. Moderately fast responses arise when the enzyme that is activated influences the movement of ions across the plasma membrane. Processes that depend on the activation of enzymes that control the metabolic activity of cells usually occupy an intermediate time domain of minutes to hours. Finally, responses that frequently require hours or days arise because the activation of proteins results in changes in the rate of expression of specific genes or groups of genes (**transcriptional control**, the production of mRNA molecules – see Chapter 4). It must be borne in mind that events set in motion in any of the time domains described may ultimately lead to the subsequent prevention or facilitation of later processes. Thus, events of slow onset may render it easier or more difficult for fast events subsequently to occur (**modulation**).

It might appear from this that the mechanisms that allow cells to sense and respond to cues in their environment must be numerous and complicated. It turns out, however, that although the extracellular primary messengers or signalling molecules are numerous, the types of cellular **signal transducing mechanisms** (intracellular signalling) are rather few in number.

3.3 Properties of molecules used for extracellular signalling

Some of the signalling molecules are released from cells to influence the activities of other cells, whereas others appear on cell surfaces to mediate interactions with other cells or with the cell matrix. Molecules that are released by cells may be further classified as either lipid soluble or water soluble and the two types play roughly complementary roles by producing their effects in separate time-domains.

Lipid soluble molecules

The characteristic of this type of signalling molecule is the ability to diffuse freely into and out of cells. For the cells that produce them this means that there is no requirement for a secretory apparatus. For the target cell the molecule is free to enter and influence the cell chemistry directly. Lipid solubility does present a problem if the target cells are in other tissues, because these molecules must be transported in the blood to produce their effects. That is, they act as **hormones (endocrine effects)**. This problem of lipid solubility is overcome for transport in the blood by the binding of the

Table 3.1 **The lipid-soluble signalling molecules (first messengers)**

Examples	*Produced by*	*Biosynthesis*	*Transport in circulation bound to:*	*Target cells*	*Action*
GROUP 1 THE STEROID HORMONES					
Cortisol	Adrenal cortex	Multi-step biosynthesis from cholesterol involving cytochrome P-450	Cortisol-binding globulin	Liver, muscle	Acting via specific cytosolic receptors (homodimers) they effect changes in the rate of expression of certain tissue-specific genes
Aldosterone	Adrenal cortex		Albumin, but no specific binding protein	Distal nephron epithelia	
Progesterone	Corpus luteum, placenta		Albumin, but no specific binding protein	Uterus, placenta, mammary gland	
Testosterone	Testes		Sex-hormone binding globulin	Prostate, secondary reproductive organs	
Oestradiol	Ovaries		Sex-hormone binding globulin	Uterus, secondary reproductive organs	
GROUP 2 HORMONES PRODUCED FROM ESSENTIAL NUTRIENTS					
1,25,dihydroxychole calciferol [1,25-di(OH)CC]	The liver and then the kidneys	Hydroxylation of vitamin D first in the liver and then in the kidneys	Vitamin D binding protein	Bone, kidneys, intestine	Acting via specific cytosolic receptors (homo- or hetero-dimers) they effect changes in the rate of expression of certain tissue-specific genes
Retinoic acid*	Many cells, especially in developing cells	Oxidation of vitamin A	May only be produced from retinol in target cells	Developing cells, inc. some of those with receptors for 1,25-di(OH)CC and T_3	
Tri-iodothyronine (T_3)	Thyroid gland, or target tissues from T_4	From thyroglobulin	T_3-binding protein	Many cells, including developing brain cells	
Prostaglandins* Thromboxanes* Leukotrienes* Lipoxins*	Many cells, inc. endothelial cells, respiratory epithelium, leucocytes, platelets	From poly-unsaturated fatty acids, esp. arachidonic acid (addition of glutathione required for leukotrienes)	Largely autocrine or paracrine effects	Many cells	Binding to specific membrane receptors
GROUP 3 LOCALLY-ACTING PARACRINE AGENTS					
Nitric oxide	Many neurons, intestine, penis	From arginine by nitric oxide synthase	None, too reactive, very local action	Neurons, smooth muscle, platelets	Activates the enzyme guanylyl cyclase
Carbon monoxide	Some neurons in the central nervous system	From haem by haem mono-oxygenase	None known, probably only local action	Neurons	

*Not strictly hormones, not usually effective via the circulation.

hormone to a specific binding protein. The lipid-soluble signalling molecules are also capable of acting locally either on adjacent cells (**paracrine effects**) or back on the cells that produce them (**autocrine effects**). Indeed a molecule such as nitric oxide is so chemically unstable that it can only act locally.

The actions of molecules in this category are usually exercised at the genomic level (that is they influence gene expression) and therefore fall in to the time domain of slow effects. There are some notable exceptions to this rule: nitric oxide diffuses into target cells to activate the enzyme guanylyl cyclase which catalyses the formation of guanosine 3',5'-monophosphate (cGMP); certain neurosteroids react with membrane receptors, as do the group of molecules, such as the prostaglandins, that are derived from the polyunsaturated fatty acid arachidonic acid. Table 3.1 gives some information about the lipid-soluble signalling molecules: examples of the tissues that produce them, their biosynthesis, and some of their target cells.

Table 3.2 **The water-soluble signalling molecules (first messengers): classified according to receptor**

GROUP 1 RECEPTOR ITSELF AN ION-CHANNEL

Example of ligand	*Subunit molecular weight (Da)*	*Subunit organization*	*Subunits*	*Ion(s) gated*
Acetylcholine (nicotinic receptor)	c. 50 000	Four transmembrane α-helices	Five in number, usually a mixture of different subunits	Na^+
$GABA_A$				Cl^-
Glycine				Cl^-
Glutamate (NMDA)	c. 100 000			Ca^{2+}, K^+, Na^+
Glutamate (AMPA)				K^+, Na^+

GROUP 2 RECEPTOR ITSELF AN ENZYME (CATALYTIC RECEPTOR)

Family	*Examples of ligands*	*Subunit organization*	*No. of subunits*	*Catalytic activity*
Not classified	Atrial natriuretic factor	Single transmembrane α-helix	Probably two	Guanylyl cyclase
Growth factor	1. Insulin		Four, two α and two β; joined by S-S bridges	Protein tyrosine kinase
	2. Nerve-growth factor		Two different, both required for high affinity binding	

GROUP 3 RECEPTOR COUPLED DIRECTLY TO AN INTRACELLULAR ENZYME

Family	*Examples of ligands*	*Subunit organization*	*No. of subunits*	*Coupled enzyme*
Erythropoietin (JAK protein coupled)	1. Growth hormone	Single transmembrane α-helix	Two identical	JAK kinase (a cytosolic protein tyrosine kinase)
	2. Interleukin 2 (IL-2) 3. Interferon γ		At least two different	
Sphingomyelinase-coupled	1. Interleukin 1 (IL-1) 2. Tumour necrosis factor-α (TNFα)	Single transmembrane α-helix	Two	Sphingomyelinase (a plasma membrane enzyme)

GROUP 4 RECEPTOR COUPLED INDIRECTLY TO AN INTRACELLULAR ENZYME VIA A G-PROTEIN

Example of ligand	*Receptor*	*Structure*	*G-protein*	*Enzyme*	*Coupling*
Photon*	Rhodopsin	Seven transmembrane α-helices, single polypeptide	G_t	cGMP phosphodiesterase	Stimulatory (+ve)
Adrenaline	β-Adrenoceptor		G_s	Adenylyl cyclase	Stimulatory (+ve)
Somatostatin	Somatostatin receptor		G_i		Inhibitory (–ve)
Vasopressin	V_1-Receptor		G_p	Phospholipase C	Stimulatory (+ve)

GROUP 5 RECEPTOR COUPLED TO AN ION-CHANNEL VIA A G-PROTEIN

Family	*Examples of ligands*	*Subunit organization*	*G-protein*	*Ion channel*
Acetylcholine	Muscarinic (M_2)	Seven transmembrane α-helices, single polypeptide	G_o	K^+
Adenosine	Adenosine (A_1)			Ca^{2+}

*Water-solubility has no meaning for photons, which are included here for purposes of comparison

Water soluble molecules

By far the most numerous and diverse group of signalling molecules produced by cells are the water-soluble ones. Generally speaking, transport in the circulation does not require a binding protein, but there is a need for cells producing molecules in this group to have specialized secretory mechanisms. Equally, target cells must be able to recognize the extracellular presence of these molecules, and this is achieved by means of specific receptor proteins associated with their plasma membranes. Molecules in this group may act locally or at a distance and they may convey both spatial and temporal information. Table 3.2 gives some examples of the types of water-soluble signalling molecule. These range from small, charged organic molecules such as acetylcholine, adrenaline and glutamic acid to large proteins such as growth hormone. A wide range of group names have been assigned to these molecules. These depend on the effects they produce: they may be described as **hormones, cytokines, growth factors, neurotransmitters** or **neuromodulators**. Hormones generally act via the circulation, cytokines (originating from cells of the immune system) and growth factors may act either systemically or locally, while molecules in the last two categories act locally in nervous tissue.

Molecules immobilised on cell surfaces

This area is one of the most difficult to classify. It has been known for some time that cells express in their plasma membrane protein molecules that allow them to bind to proteins in the membranes of adjacent cells or with molecules in the extracellular matrix. Although these matters are not dealt with further, it is worth noting here that interactions between these groups of molecules can lead to the generation of intracellular **second messengers** and related signals just as occurs when the water-soluble signalling molecules bind to their receptors. Thus the interaction of epithelial cell integrin molecules with collagen, fibronectin or laminin results in the phosphorylation of intracellular proteins on tyrosine residues. This is also a typical response of cells to the presence of water-soluble growth factors.

It is by subverting this means of communication between the extracellular matrix and epithelial cells that pathogenic bacteria such as *Yersinia* spp. are able to enter and grow in cells of the reticuloendothelial system. For example, *Y. pestis* causes bubonic plague by entering the dermal lymphatic system via the bite of an infected flea.

Naming of extracellular signalling molecules

It is now apparent that the responses of cells to external stimuli depend not only on the presence of the appropriate receptor but more importantly on the cell phenotype. This means that the same signalling molecule can elicit different responses in different tissues. Thus insulin promotes the uptake of glucose into muscle cells but not into liver cells despite the presence of the insulin receptor in both tissues. This is because the phenotype of the former, but not the latter tissue includes the synthesis of a glucose transport protein (Glut-4, see Chapter 2). Acetylcholine causes skeletal muscles to contract by activating a nicotinic acetylcholine receptor (an ionotrophic receptor that is permeable to, i.e. 'gates', Na^+), whereas its action on heart muscle is via a muscarinic receptor that indirectly controls the movement of K^+.

This multiplicity of actions of signalling molecules has led to some confusion in naming. Thus gonadotrophin-releasing hormone (Gn-RH) was isolated due to its ability to cause the release of luteinizing hormone from the anterior pituitary gland, but it also acts as a neuromodulator in certain lumbar sympathetic ganglia and has paracrine effects in the ovary. The same name is used to describe the molecule wherever it acts. There are several examples of mediators having multiple separate roles, but bearing a single descriptive name.

3.4 Receptors for molecules used for signalling

As might be inferred from the foregoing description of the types of signalling molecules found in mammals, the lipid-soluble molecules have receptor proteins that are quite different from those used for the other two classes of molecules.

Receptors for lipid-soluble signalling molecules

Receptors for the lipid-soluble signalling molecules may be divided into two groups.

The steroid hormone family of receptors

The common features of this class of receptor are their intracellular location (cytosolic or nuclear) and their ability to bind to DNA to alter the rate of transcription of genes. Figure 3.1 uses the example of the glucose-providing action of cortisol (a glucocorticoid) to illustrate the way in which molecules binding to this family of receptors produce their effects. (The actions of the adrenal cortical hormone cortisol are further addressed in Chapter 28.) The receptor has a domain that allows the hormone to bind to it, the hormone–receptor complex, if not already there, is translocated to the nucleus where it binds to specific regions of the DNA known as **hormone responsive elements**. These are regions of DNA that control the rate of transcription of specific genes. Which genes are affected depends on the hormone in question and the target tissue. In the case shown, the cells in question will be those of the liver, and the gene that for phospho*enol*pyruvate carboxykinase. This enzyme catalyses a rate-determining step in the process known as gluconeogenesis (see Chapter 6). An increase in its catalytic activity will result in an increase in the rate of formation of glucose in the liver from non-carbohydrate precursors.

Hormones that act in this way include tri-iodothyronine (T_3, the thyroid hormone, Chapter 28), 1,25-dihydroxycholecalciferol (the hormonally-active metabolite of vitamin D3, Chapter 12) and 9-*cis*-retinoic acid (the autocrine derivative of vitamin A, Chapter 19) and the steroid hormones cortisol, aldosterone, progesterone, oestradiol and testosterone (Chapter 28).

Catalytic, cytosolic receptors

As indicated previously the lipid-soluble signalling molecule nitric oxide diffuses into target cells to bind to and thereby activate the enzyme guanylyl cyclase (a catalytic receptor). Some nerve cells in the central nervous system and also smooth muscle cells contain this soluble form of guanylyl cyclase. Recently it has been claimed that carbon monoxide, produced by action of the enzyme haem mono-oxygenase on haem in neurons in the brain, can also activate guanylyl cyclase.

Table 3.1 gives examples of the two types of receptor for the lipid-soluble signalling molecules.

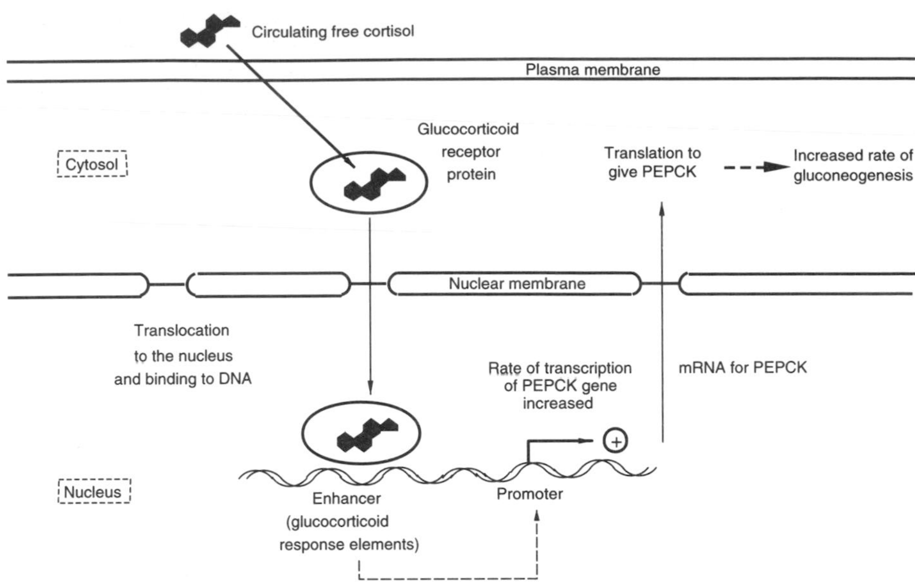

Figure 3.1 **Example of a steroid hormone acting to regulate gene transcription: cortisol acts as a glucocorticoid to induce the production of the enzyme phosphoenolpyruvate carboxykinase (PEPCK), a rate-limiting enzyme in gluconeogenesis**

Receptors for water-soluble signalling molecules

There are five groups of receptor types for the water-soluble signalling molecules, examples of which are given in Table 3.2.

Ion-channel receptors

The first of the types of plasma membrane receptor that respond to the water-soluble signalling molecules are themselves ion channels (sometimes referred to as **ionotrophic receptors**). That is substances binding to them from the outside of the cell cause the channel to 'open' and thereby allow the passage of a specific cation or anion. Such receptors are generally found in neurons or post-synaptically at neuromuscular junctions. Molecules that carry signals between the cells of the nervous system are given the special name **neurotransmitter** (although not all neurotransmitters act on ionotrophic receptors). The actions of these are discussed further in Chapter 30. Examples of ion-channel receptors include the nicotinic acetylcholine receptor and receptors for the amino acid neurotransmitters glutamate, γ-aminobutyrate (GABA) and glycine.

Catalytic, membrane receptors

These receptors consist of membrane-spanning polypeptide chains that bind hormones or neurotransmitters on an extracellular domain and this leads to the activation of an enzymic part of the molecule which is located on an intracellular domain of the protein. Thus the binding of the hormone atrial natriuretic factor (ANF) to its receptor results in an increase of the formation of cGMP catalysed by an intracellular domain of the protein. (The actions of ANF to promote natriuresis are described in Chapter 26).

A major family of receptors in this category are referred to as **protein tyrosine kinase receptors**. These operate in a fashion analogous to that described for the integrin molecules that respond to the presence of collagen and other molecules by undergoing activation of an intracellular protein kinase domain. A common feature of the activation of these receptors by the binding of their specific ligands is that the intracellular part of the receptor becomes capable of catalysing the transfer of a phosphate group from ATP to a series of acceptor protein molecules, which thereby become phosphorylated on certain tyrosine residues. If these acceptor proteins are themselves enzymes, phosphorylation causes a change in their catalytic activity. A characteristic of the protein tyrosine kinase class of receptor catalyst is that it is able to undergo self- (or auto-) phosphorylation and this frequently leads to self-activation. Important members of this group of catalytic receptors are the **growth-factor receptors**. There are separate growth-factor receptors for a wide variety of growth-promoting molecules and they all appear to be protein tyrosine kinases. Growth-factor receptors are dimeric proteins which may either be homodimers (two interacting copies of the same polypeptide chain) or heterodimers. Figure 3.2 shows the consequences of the activation of a typical growth-factor receptor by its ligand.

Dysfunctions of growth factors or their receptors are frequently encountered in tumour cells (see Chapter 37).

It is worth emphasizing that not all protein tyrosine kinase receptors are solely growth-promoting receptors, thus the **metabolic** effects of insulin on cells are mediated by the insulin receptor, itself a tyrosine kinase (Chapter 28).

Receptors coupled directly to a cytosolic enzyme

Erythropoietin-type receptors, named for the renal hormone that helps to control erythropoiesis (sometimes called **cytokine receptors** or **JAK protein-coupled receptors**) fall into this group. These receptors resemble the growth factor receptors in that they are hetero- or

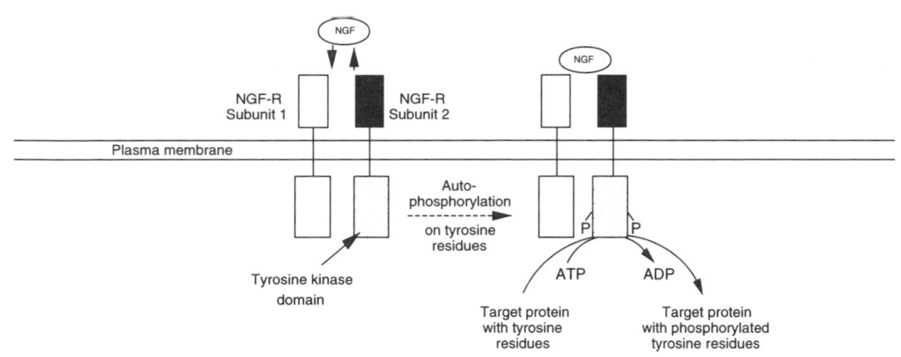

Figure 3.2 **Binding of nerve-growth factor to its receptor (NGF-R) causes the phosphorylation of a range of proteins, including the receptor itself**

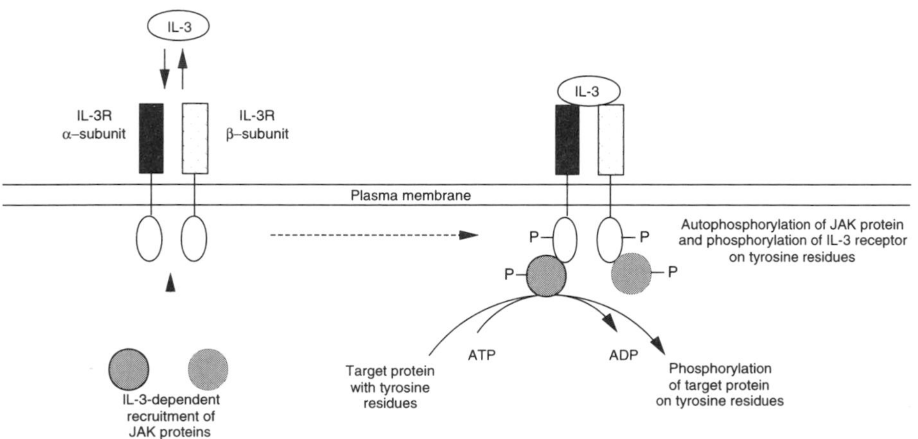

Figure 3.3 **The interleukin 3 (IL-3R) couples directly to a cytosolic JAK protein which, when activated, has protein tyrosine kinase activity**

homo-dimeric proteins with polypeptide chains that span the plasma membrane once. They differ from the former class of receptor, however, in that they have no intracellular catalytic domains. Instead, the binding of the ligand to both parts of the dimer results in them binding a cytosolic **JAK** (standing for **J**ust **A**nother **K**inase!) **protein** which itself is a protein, tyrosine kinase. This results in the activation of this enzyme, with the outcome being the same as for the growth factor receptors, i.e. phosphorylation of target proteins (including self-phosphorylation and phosphorylation of the receptor itself) on tyrosyl residues (Figure 3.3). Included in the long list of ligands for the erythropoietin family of receptors are growth hormone, prolactin, and the group of signalling molecules produced by lymphocytes termed **lymphokines** (a subset of cytokines). Included in the latter group are the **interleukins** (IL-2, IL-3, IL-4, IL-5 and IL-6) and **interferons** (interferons α, β and γ; see Chapter 32).

G-protein coupled membrane receptors

The archetypes of this type of receptor are the bacterial photon-harvesting protein bacteriorhodopsin and its mammalian counterpart rhodopsin (Chapter 31). The organization of the former molecule in bacterial membranes has been determined by high-resolution electron microscopy, and it has been shown that the protein spans the membrane by means of seven largely hydrophobic α-helices. The same is almost certainly true for the visual pigment rhodopsin. Rhodopsin is orientated in the membrane in such a way that its N-terminus is outside the membrane and the C-terminus is on the inside (see Chapter 31). Intracellular domains interact with a special type of GTP-binding protein called a **G-protein**. This particular arrangement of a seven membrane-spanning domain protein receptor interacting with a G-protein is found time and again in receptors for a wide range of extracellular signalling molecules. The G-protein serves to 'couple' the receptor to a second protein. For this reason G-proteins are sometimes referred to as **'transducing'** molecules. Indeed, the G-protein that interacts with rhodopsin in the visual process is called **transducin**. The proteins influenced by G-proteins are one of two enzymes or an ion channel. The enzymes are **adenylyl cyclase** that catalyses the formation of **adenosine 3',5'-monophosphate** (**cAMP**; Figure 3.4) from ATP, or a special form of **phospholipase C** that catalyses the hydrolysis of the membrane phospholipid **phosphatidylinositol *bis*phosphate (PIP$_2$)** to form **inositol *tris*phosphate (IP$_3$)** and **diacylglycerol** (**DAG**; Figure 3.5). It is the production of these three molecules, cAMP or IP$_3$ and DAG that serves to signal the changes in intracellular chemistry that occurs when many G-protein coupled receptors are activated. Because of their special position as the intracellular signalling counterparts of the extracellular molecules that bind to receptors, cAMP, IP$_3$ and DAG have been termed **'second messengers'**. Figures 3.4 and 3.5 also show the reactions that remove these second messengers, processes which are just as

Table 3.3 **Signal-transducing diseases**

1. TOXIN-ASSOCIATED DISEASES

Agent	*Source*	*Target molecule(s)*	*Cellular effects*	*Symptoms*	*Disease*
Bacterial endotoxin (lipopolysaccharide)	Gram-negative bacteria	CD-14 (an integrin):surface protein of macrophages	Secretion of cytokines (inc. IL-2), reactive oxygen species, arachidonic acid metabolites (inc. prostaglandins) and nitric oxide	High fever, hypotension, disseminated blood clotting	Lethal shock
Cholera toxin	*Vibrio cholerae*	GTP-binding proteins (G_S) of intestinal epithelial cells	Inappropriately elevated cAMP concentrations due to failure of G_s to hydrolyse GTP	Loss of electrolytes to the lumen of the intestine	Dysentery
Pertussis toxin	*Bordetella pertussis*	GTP-binding proteins (G_i) of airways epithelial cells	Inappropriately elevated cAMP concentrations due to failure of G_i to inhibit adenylyl cyclase	Catarrhal paroxysms	Whooping cough
Domoic acid	Toxin produced by the plankton *Nitzschia pungens* found in mussels in Prince Edward Island, Canada	Potent agonist on the glutamate (NMDA) receptor	Excitotoxicity due to sustained high levels of Ca^{2+} esp. in the CA3 area of the hippocampus	Vomiting, seizures, severe anterograde amnesia, 3 deaths recorded	Toxic encephalopathy

2. GENETIC DEFECTS OF INTRACELLULAR RECEPTOR FUNCTION

Receptor for:	*Inherited defect*	*Nature of inheritance*	*Syndrome*
Androgen	No receptor Receptor present in reduced number, or labile	X-linked recessive	Complete testicular feminization Incomplete testicular feminization
Triiodothyronine (T_3)	T_3-receptor present in reduced number and defective	Autosomal recessive	Severe T_3-resistance
Cortisol	Cortisol receptor with reduced affinity	Autosomal dominant	Cortisol resistance
Progesterone	Reduced receptor number	Autosomal recessive	Pseudo-corpus luteum insufficiency
1,25-dihydroxycholecalciferol	No receptor Reduced receptor number Reduced affinity for ligand Defective translocation to nucleus	X-linked recessive	Vitamin D-resistant rickets

3. GENETIC DEFECTS OF PLASMA MEMBRANE RECEPTOR FUNCTION

Hormone/ effector	*Secondary errors of receptor function (pre-receptor defects)*		*Primary errors in receptor function*		
	Variant hormone	*Receptor antibodies*	*Defects in bindings*	*Post-binding defects*	*Post-receptor defects*
Insulin	5 different mutations	Extreme insulin-resistance type B	1. Extreme insulin resistance: • type D (no receptor) • type A (reduced number) • type E (reduced affinity) 2. Seip syndrome (reduced number and affinity) 3. Rabson–Mendenhall syndrome (reduced number and affinity) 4. Leprechaunism (reduced number)	Extreme insulin resistance type C	1. Extreme insulin resistance type F 2. Seip syndrome 3. Leprechaunism

Hormone/ effector	*Secondary errors of receptor function (pre-receptor defects)*		*Primary errors in receptor function*		
	Variant hormone	*Receptor antibodies*	*Defects in bindings*	*Post-binding defects*	*Post-receptor defects*
Parathyroid hormone	Pseudo-parathyroid hormone resistance			Pseudo-hypoparathyroidism type Ia (inactivating G-protein mutation) sometimes accompanied by precocious puberty	Pseudo-hypoparathyroidism type II
Growth hormone			Laron dwarfism	Acromegaly due to somatotroph tumours (activating G-protein mutation)	
Vasopressin			Nephrogenic diabetes insipidus		
Thyrotrophin (TSH)			1. TSH-resistant inborn hypothyroidism 2. Thyroid adenoma associated hyperthyroidism (activating mutation of TSH receptor)		
Adrenocorticotrophin (ACTH)			Familial glucocorticoid deficiency (mutations in the ACTH receptor)		
Luteinizing hormone (LH)			1. Hypogonadotrophic hypogonadism (reduced binding) 2. Male precocious puberty (activating mutation of LH receptor)		
Photons			1. Retinitis pigmentosa (rod opsin mutations) 2. Spectral sensitivity variants of colour blindness (cone opsin mutations)		

4. AUTOIMMUNE DISEASES

Disease	*Immunological nature*	*Nature of defect*	*Sign and symptom*
Juvenile-onset or insulin-dependent diabetes mellitus (type I)	T-cell-mediated destruction of the β-cells of the pancreas. Believed to be an autoimmune response to glutamic acid decarboxylase (GAD)	Failure to secrete adequate amounts of insulin	Hyperglycaemia, weight loss, polyuria, polydipsia, ketosis, dehydration and sometimes coma
Myasthenia gravis	Circulating antibodies to the nicotinic acetylcholine receptor	Inability to sustain repeated muscular responses to acetylcholine released by a train of depolarizing signals in cholinergic neurons innervating neuromuscular junctions due to reduced receptor number	Progressive disease of severe muscular weakness and fatigability
Lambert–Eaton myasthenic syndrome	Antibodies produced to an abnormal voltage-gated calcium channel (VGCC) associated with a small-cell carcinoma of the lung also cause the internalization and destruction of normal VGCCs/immunoglobulin complexes in neurons presynaptic to muscles throughout the body	VGCCs fail to open in response to neuronal depolarization and hence there is widespread failure of Ca^{2+}-dependent release of neurotransmitters	Proximal muscle weakness, reduced tendon reflexes, dry mouth, sexual impotence and constipation
Graves' disease	A circulating antibody to the TSH-receptor (long-acting thyroid stimulator, LATS) causes profound, inappropriate stimulation of the thyroid gland	Hypersecretion of T_3/T_4 not subject to feedback control	Sequelae of a high metabolic rate: increased cardiac output, tachycardia, restlessness, irritability, sleeplessness, anxiety, diarrhoea, sensitivity to heat

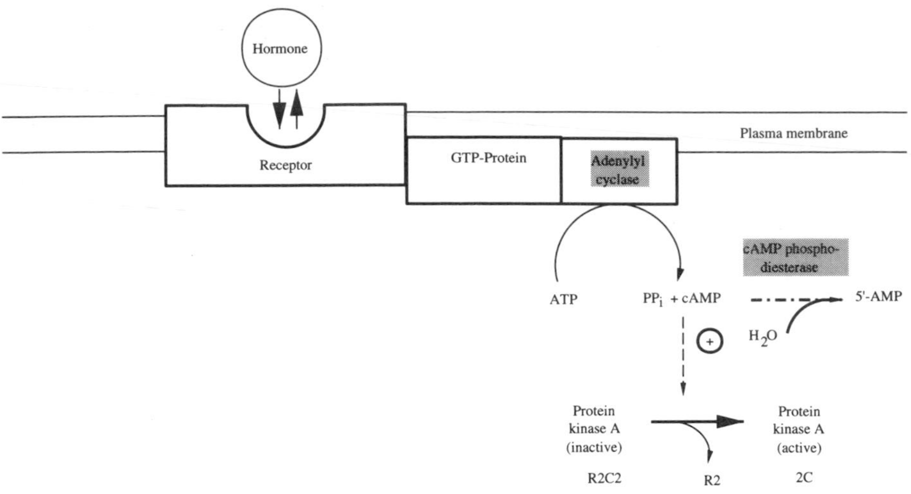

Figure 3.4 **The binding of some hormones to their receptors causes the activation of the enzyme adenylyl cyclase, and the cAMP produced, in turn, cause the activation of protein kinase A**

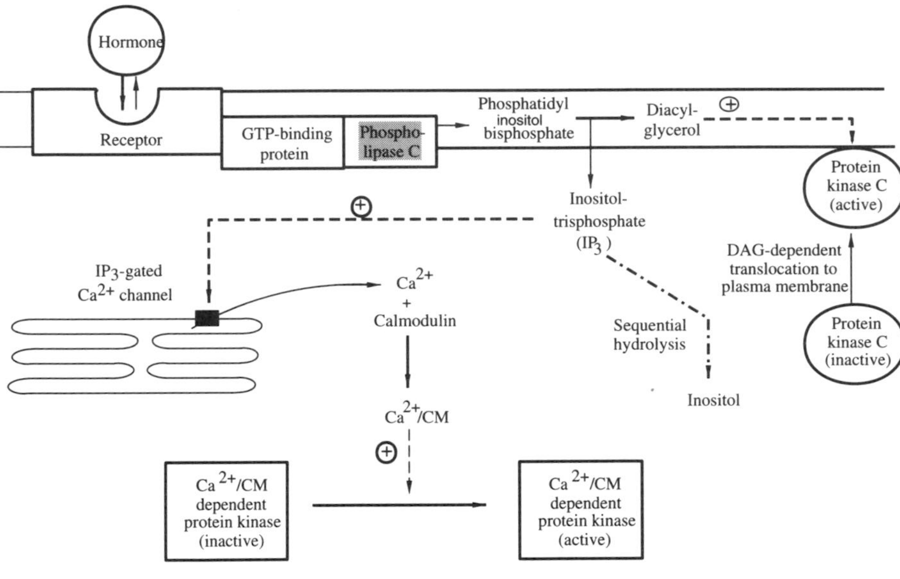

Figure 3.5 **The activation of receptor coupled via a G-protein to phospholipase C results in the production of two second messengers: inositol trisphosphate and diacylglycerol**

vital to intracellular homeostasis as those that generate the molecules in the first place. cAMP and IP_3 undergo hydrolysis and DAG is used for the synthesis of further membrane phospholipids.

The binding of signalling molecules to G-protein-coupled receptors may result in an increase or a decrease in the catalytic activity of adenylyl cyclase or an increase in that of phospholipase C. Since the negatively-coupled receptors simply cause a reduction in the amount of cAMP being formed, it is more convenient to describe the consequences of activation of the two G-protein-coupled enzymes (although as emphasized previously the means of terminating signals are as important as the means of generating them). For the purposes of the present discussion the consequences of the activation of adenylyl cyclase or phospholipase C will be described in general terms. Consideration of how G-protein-mediated signalling can give rise to responses in different time-domains will be dealt with in later chapters.

The general principle appears to be that for each of the second messengers (cAMP, IP_3 or DAG) a protein kinase becomes activated. For cAMP the appropriate enzyme is protein kinase A, which is capable of catalysing the phosphorylation of target proteins on serine residues (Figure 3.4). DAG activates protein kinase C, in part, by causing its binding to the plasma membrane of cells (Figure 3.5). The actions of IP_3 are rather more indirect and require the participation of calcium ions (Ca^{2+}). When IP_3 is formed, it binds to a membrane receptor in the endoplasmic reticulum to cause the release of Ca^{2+} stored there. These ions bind to a specific, cytosolic Ca^{2+}-binding protein, calmodulin, and the Ca^{2+}/calmodulin complex activates a third protein kinase Ca^{2+}/calmodulin-dependent protein kinase. Because of their special mode of action, calcium ions are sometimes referred to as '**third messengers**'.

In summary, therefore, second and third messengers are kept at very low intracellular (cytosolic) concentrations which can increase many-fold and fairly rapidly on the activation of G-protein-coupled receptors. Thus the concentration of Ca^{2+} in the cytosol of resting cells is about 10 nM and this can increase on activation to 100 nM or more in seconds or less, as a result of the release of the cation from stores in the endoplasmic reticulum. Alternatively Ca^{2+} may enter cells from outside via ionotrophic receptors that gate the cation.

In addition, some G-proteins can interact, not with an enzyme, but rather directly with an ion channel to cause it to open or to close. The muscarinic receptor (M_2) couples to a K^+ -selective ion channel.

3.5 Diseases resulting from dysfunctions in biochemical communication

With signalling systems being at the centre of so many of the functional activities of cells, it is not surprising that a large number of diseases have been recognized as arising owing to dysfunction or subversion of the processes involved. Table 3.3 lists some of these diseases, several of which will be dealt with in other chapters.

Chapter 4

Cellular organelles: the nucleus

4.1 Introduction

The observation that the cells of most multicellular organisms possess a large prominent organelle described as the nucleus was made in the early days of light microscopy. In the early 1900s the importance of the nucleus as a store of genetic information was understood, but it was not known whether the information was stored in the 'simple' structure of polynucleotides or in the more 'complex' structure of the associated proteins. The studies which proved DNA to be the genetic material were carried out in prokaryotes. In 1928, Griffiths, working with virulent and non-virulent strains of *Diplococcus pneumoniae* showed that injection of the non-virulent strain into mice had no effect on the animals, but injection of the non-virulent together with heat-killed virulent bacteria was fatal. Since the heat-killed virulent strain is not infectious on its own, it seemed that genetic information was passed from the dead, virulent to the live, non-virulent strain and 'transformed' the latter into virulent cells. Avery, in 1944, showed that the 'transforming' factor was DNA, not protein. Hershey and Chase in 1952 used bacteriophages (viruses which infect bacteria) radioactively labelled with either ^{32}P, which labels DNA, or ^{35}S, which labels protein. They found that when bacteria were infected they were labelled with ^{32}P, not ^{35}S. As these bacteria were capable of producing more viruses, it was clear that genetic information was carried in DNA and not protein. Evidence that DNA carries the genetic information in eukaryotes was initially indirect. Agents which caused alterations in DNA structure produced mutations which altered the structure and function of proteins. It is now possible to synthesize DNA corresponding to eukaryotic proteins, insert this DNA into bacteria and cause production of the eukaryotic proteins. The production of hormones such as insulin and growth hormone by these means provides indisputable proof that DNA is the eukaryotic genetic material.

4.2 Structure of the nucleus

The nucleus is surrounded by a **nuclear envelope** which is seen in the electron microscope to be composed of a double nuclear membrane (Figure 4.1). Each membrane is 2 nm thick and the perinuclear space between them 10–50 nm. About 20% of the area of the nuclear envelope consists of circular pores 60–90 nm in diameter (Figure 4.2). The size varies and is larger in some cells such as oocytes. The pores are surrounded by eight protein granules and act as molecular sieves. They allow material such as RNA to be transported through the nuclear envelope between the nucleus and the cytoplasm.

Chromatin

A prominent feature of the nucleus is the network of **chromatin fibres**. They are composed of approximately one-third DNA and two-thirds protein. The major protein components are the **histones**, which are basic proteins rich in arginine and lysine (Table 4.1). There are also non-histone proteins and some RNA. The ratio of histones to DNA is about 1:1. The amount of DNA found in the nucleus is frequently expressed as a number of nucleotide (base) pairs. In the case of the human haploid genome, this amounts to 2.8×10^9 base pairs (bp). In the human sperm there is 3.2 pg of DNA, corresponding to a total linear length of about 1 m. This DNA is distributed between the 23 chromosomes, with each chromosome containing a single DNA molecule of average length about 4 cm. Isolated DNA was shown by James Watson and Francis Crick to be present as a double-helical structure, held together by the hydrogen bonds which form specifically between the bases and by hydrophobic interactions between the stacks of bases (Figure 4.3).

Adenine and **thymine** are held together by two hydrogen bonds, whereas **guanine** and **cytosine** are held together by three of these (Figure 4.4). The two strands of DNA are said to be **complementary** to each other, for whenever an adenine (guanine) is present in one strand there must be a thymine (cytosine) in the other. The presence of the double-helical structure in cells was demonstrated by Maurice Wilkins. The diameter of the rather rigid rod predicted by Watson and Crick is 2 nm, but even in the interphase nucleus the chromosomes appear to have a minimum diameter of 11 nm with a characteristic beaded structure. Chromatin is highly

Table 4.1 **Groups of histones**

Type	*Name*	*Arg (%)*	*Lys (%)*
Lysine rich	H_1	2.5	26.3
Moderately lysine rich	H_{2a}	6.7	16–17
Moderately lysine rich	H_{2b}	11.5	10.5
Arginine rich	H_3	14.1	9.7
Arginine rich	H_4	13.4	8.8

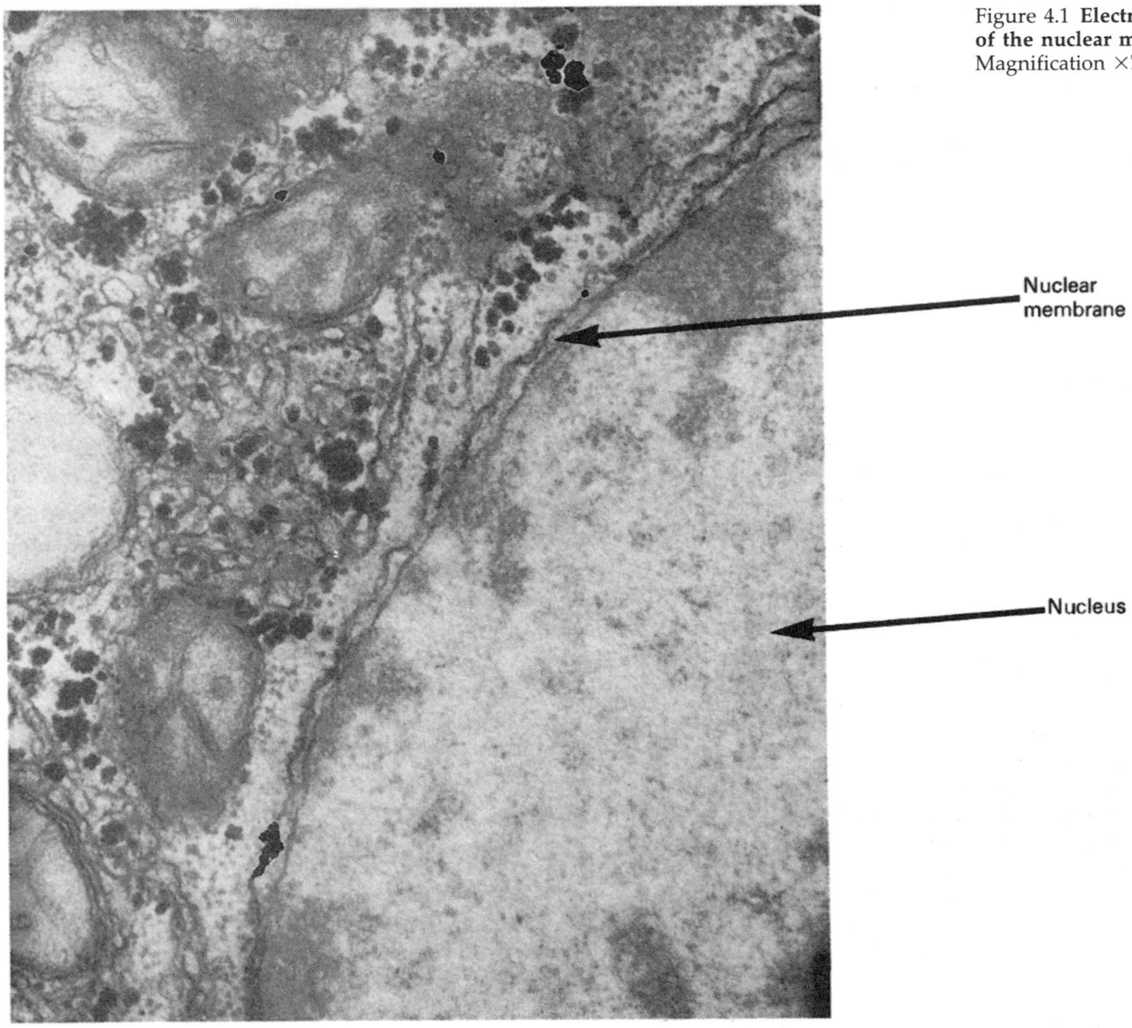

Nuclear membrane

Nucleus

Figure 4.1 **Electron micrograph of the nuclear membrane.** Magnification ×75 000

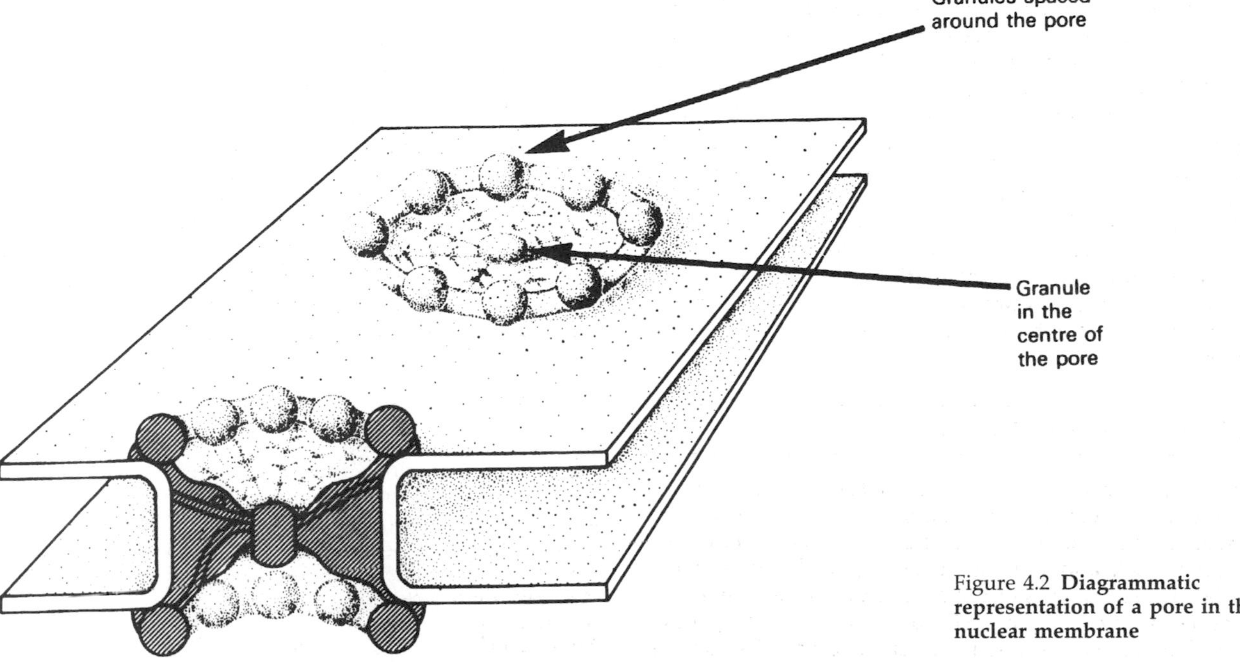

Figure 4.2 **Diagrammatic representation of a pore in the nuclear membrane**

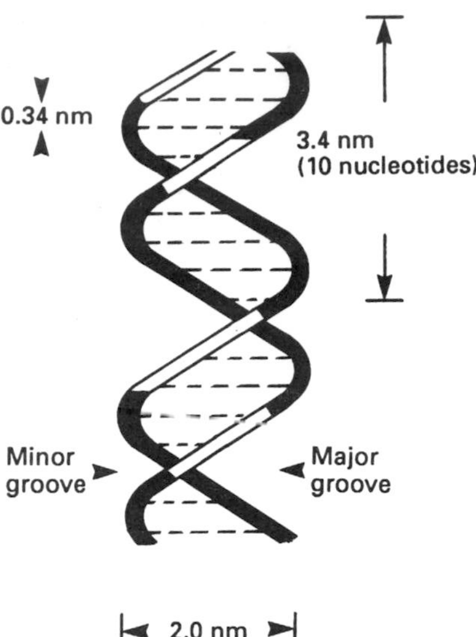

Figure 4.3 **The Watson–Crick model for the B-form of DNA**

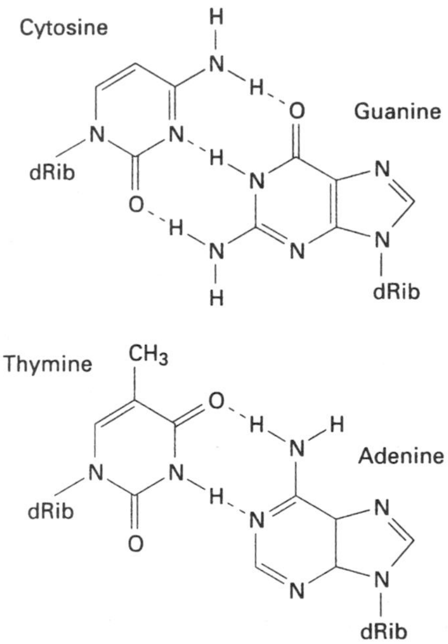

Figure 4.4 **The conversion of single-stranded retroviral RNA into double-stranded DNA**

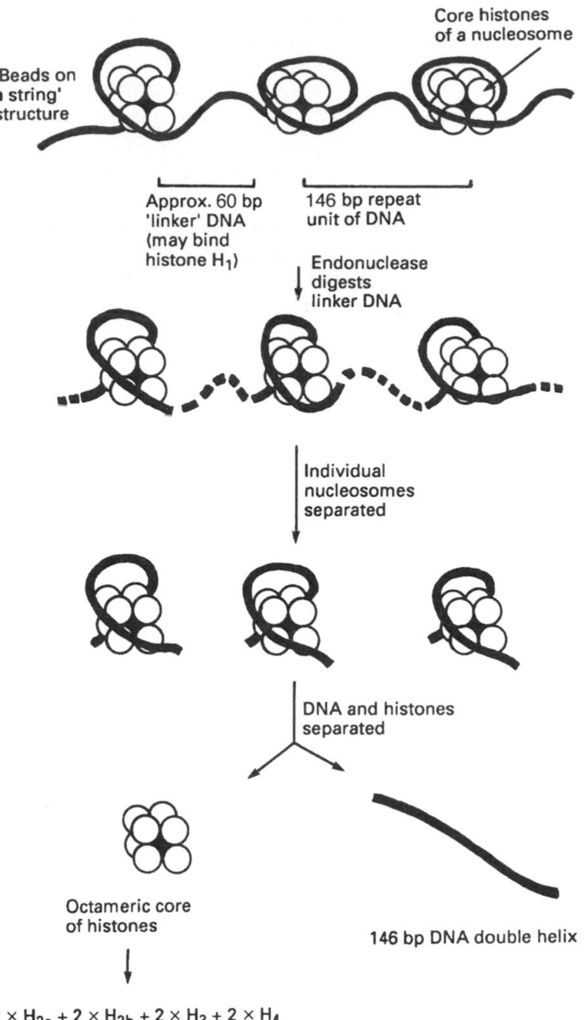

Figure 4.5 **The organization of eukaryotic chromatins into 'nucleosomes' and the composition of a 'nuclesome'**

folded. This is essential to allow the packaging of the long linear DNA structure into the nucleus. Chromatin is made up of repeating units known as **nucleosomes** (Figure 4.5). Each nucleosome consists of a core of eight histone molecules $2H_{2a}$, $2H_{2b}$, $2H_3$ and $2H_4$, round which is wound a length of DNA of about 146 base pairs. A molecule of histone H_1 is placed at the location where DNA enters and exits the core. The segments of DNA between nucleosomes are about 60 bases long and are known as 'linker' DNA.

The further folding required to produce the densely packed chromosomes seen in mitosis appears to involve the arrangement of the nucleosomes themselves into helical structures termed **solenoids**. The solenoids are further organised into loops 250–400 nm long which are attached to the nuclear lamina.

The obvious structural role of histones in chromatin led to the view that they were unlikely to play a functional role as well. This view was supported by the observation that regions of chromatin classified as **euchromatin** and **heterochromatin** have very similar histone composition but are functionally distinct. Euchromatin regions are functionally active, being involved in RNA production on the DNA template (transcription), whereas heterochromatin regions are inactive. These conclusions ignored two facts. First, the methods of analysis lacked the resolving power required to recognize transient dissociation/re-association of histones from specific nucleosomes. Secondly, histones associated with specific areas in euchromatin undergo covalent modification. Histones located prior to genes are prone to *N*-acetylation which results in the masking of the positive charge on lysine residues. This may well permit the dissociation of histones from the nucleosomes in that region, allowing the binding of proteins required for transcription. Another type of covalent modification of histones, the addition of ubiquitin, seems to occur specifically in areas corresponding to genes currently being expressed. The current view is that covalent modification of specific histones does play a role in the transcription of active genes, perhaps by facilitating the reversible dissociation of groups of these proteins from the nucleosomes.

Non-histone proteins are thought to be involved in DNA transcription, as transcription factors, and also in DNA replication, as scaffold proteins. One of these is now known to be a topoisomerase, involved in relieving the strain on the DNA helix during replication (see Section 4.3).

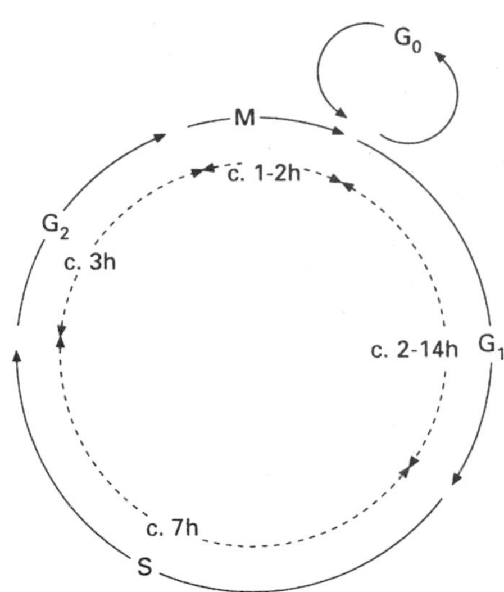

Figure 4.6 **The operation of the cell-cycle in the somatic cells of vertebrate animals**

The nuclear lamina

The nuclear lamina is located between the inner nuclear membrane and chromatin, with loops of chromatin bound to it. It is composed of three distinct polypeptides known as lamins A, B and C. They are arranged in a network of fibres bound to the inner nuclear membrane. The lamins are known to undergo phosphorylation and dephosphorylation which may be related to the disassembly and restructuring of the nucleus which occurs during the cell cycle.

The nucleolus

Nucleoli are granules in the nucleus which stain strongly for RNA and weakly for DNA. Nucleolar DNA codes for ribosomal RNA. The nucleolus is the site of synthesis of rRNA and also of the assembly of ribosomal subunits. Ribosomal RNA is transcribed in the nucleus and is bound, in the nucleolus, to ribosomal protein synthesized in the cytoplasm.

4.3 DNA synthesis (replication)

DNA synthesis, or replication, is an important function of the nucleus and must occur as a preliminary to every cell division. DNA replication does not occur continuously but during the 'S' or synthetic phase of the cell cycle (Figure 4.6). Histone synthesis also occurs during this phase. The rate at which eukaryotic cells divide varies enormously. Liver cells for example rarely divide unless the tissue is damaged, but cells of the intestinal mucosa and the bone marrow divide steadily.

The model proposed by Watson and Crick for DNA structure provides a basis for understanding DNA replication. Their DNA model was based on X-ray analysis of eukaryotic material, but it was found to apply to prokaryotes as well. Most of our understanding of DNA replication came originally from experiments on prokaryotic organisms.

Prokaryotic DNA replication

Meselson and Stahl showed, in 1958, that the process is **semi-conservative**. Bacterial DNA is circular and double-stranded. Each strand of the parental DNA acts as a template and the new DNA double helix consists of one parental strand and one new, or daughter, strand. (Since there is no simple verb meaning 'to make a complementary copy', the verb 'to copy' will be used to imply this process.) Replication is always initiated at a single point in the DNA molecule, the **origin** or **ori C** site and proceeds in both directions in both chains simultaneously. The areas at which replication is occurring are known as **replication forks**.

The requirements for DNA synthesis are:

(a) a template (a region of DNA to be copied)
(b) a primer (a short piece of RNA formed using DNA as a template)
(c) magnesium ions
(d) the four deoxyribonucleoside triphosphates
(e) DNA polymerases.

The addition of the new nucleotide can be represented by the reaction:

$$(\text{dNMP})_n + \text{dNTP} \rightarrow (\text{dNMP})_{n+1} + \text{PPi}$$

where dNMP represents a deoxyribonucleoside monophosphate and dNTP a deoxyribonucleoside triphosphate. The reaction is driven by the hydrolysis of pyrophosphate (PPi) to inorganic phosphate.

Escherichia coli DNA can replicate in 30–40 min. This means that about 1000 nucleotides are added per second. DNA synthesis occurs in the 5' → 3' direction, as a free hydroxyl group is needed at the 3' end of the growing chain in order to carry out a nucleophilic attack on the inner phosphate of the incoming nucleoside triphosphate (Figure 4.7).

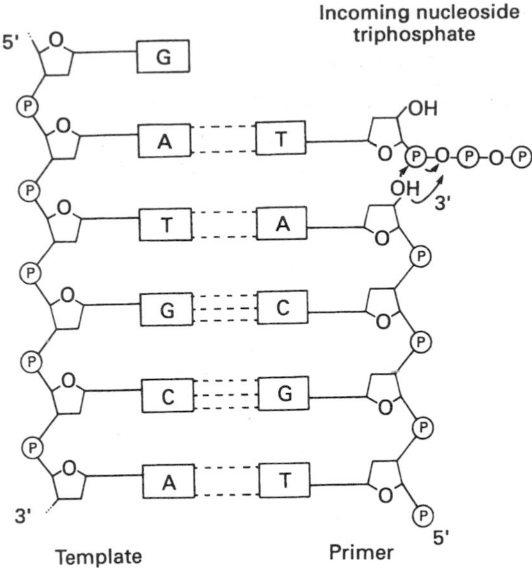

Figure 4.7 **The reaction catalysed by DNA polymerases**

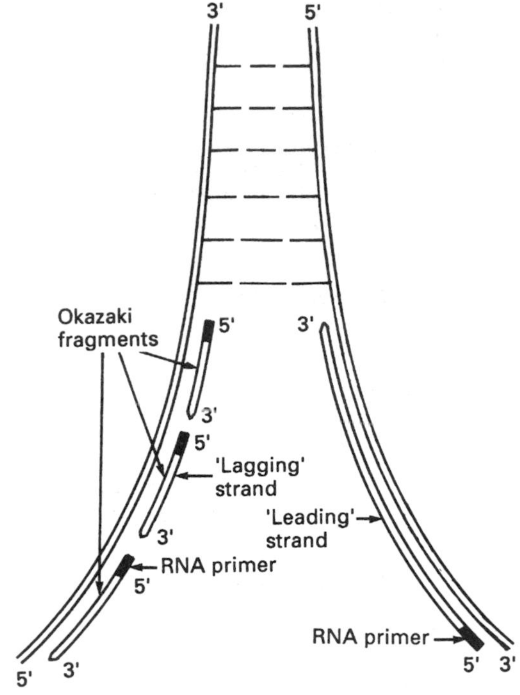

Figure 4.8 **The formation of 1000–2000 base sequences (Okazaki fragments) on the 'lagging' strand during replication**

There are three prokaryotic DNA polymerases. **Polymerase I** is the enzyme originally described by Kornberg. **Polymerase III** is the principal replicative enzyme in prokaryotic organisms. The physiological function of polymerase II is unknown. All three enzymes show 5' → 3' polymerase activity and 3' → 5' exonuclease activity; that is, they are able to hydrolyse single stranded DNA in the 3' → 5' direction. Polymerases I and III also have 5' → 3' exonuclease activity using double-stranded DNA as a substrate.

The primer which is required for DNA synthesis is a short piece of RNA, some 10 nucleotides in length, formed by a DNA-dependent RNA polymerase known as a **primase**, which uses DNA as a template and does not itself require a primer. The primase forms a complex with N-proteins which can recognize a specific site on DNA where primer synthesis can occur. The primase/N-protein complex is known as a **primosome**.

A model for bacterial DNA replication is shown in Figure 4.8. The DNA chain which runs in the 3' → 5' direction is copied by polymerase III in the 5' → 3' direction as a continuous strand, requiring one primer. This new strand is known as the **leading strand**. The other strand, known as the **lagging strand**, must be synthesized in a discontinuous manner as synthesis can only proceed in the 5' → 3' direction. This also implies the need for numerous RNA primers at specified intervals followed by synthesis of segments of DNA. These segments, referred to as **Okazaki fragments** after their discoverer, are 1000–2000 nucleotides long. The DNA strand which acts as a template for the lagging strand is looped at the replication fork so that it physically points in the same direction as the template for the leading strand.

The RNA primers are excised mainly by polymerase I acting as a 5' → 3' **exonuclease**. As it removes each ribonucleotide, polymerase I replaces it with a deoxyribonucleotide. The resulting segments are then joined by the action of **DNA ligase** (Figure 4.9) and a continuous strand is formed.

For DNA replication to occur, the two chains have to unwind and separate. Single-stranded DNA would tend to reform double strands by H-bonding unless the single-stranded structure is stabilized. Several proteins are known to be involved in the preparation of the replication fork (Figure 4.10). **Helicases** unwind the DNA inducing supercoiling. The strain produced by supercoiling is released by **topoisomerases**. They cut either one (type I topoisomerases) or both (type II topoisomerases II or **gyrases**) DNA strands, allowing a segment of DNA to pass through the break and then they reseal the break. The strands, whilst unwound, are held apart by **single-strand binding proteins** (SSB) thus allowing new strand synthesis to occur.

Proof-reading

DNA is copied by DNA polymerase with high fidelity. Incorrect nucleotides are incorporated with a frequency of one in 10^8–10^{12} bases. Mismatches, that is incorrect insertions, occur more frequently but do not lead to stable incorporations because all three prokaryotic polymerases have **3'–5' exonuclease** activity. Polymerases I and III are known to excise erroneous nucleotides before the introduction of the next nucleotide. This process is known as **proof-reading** (Figure 4.11).

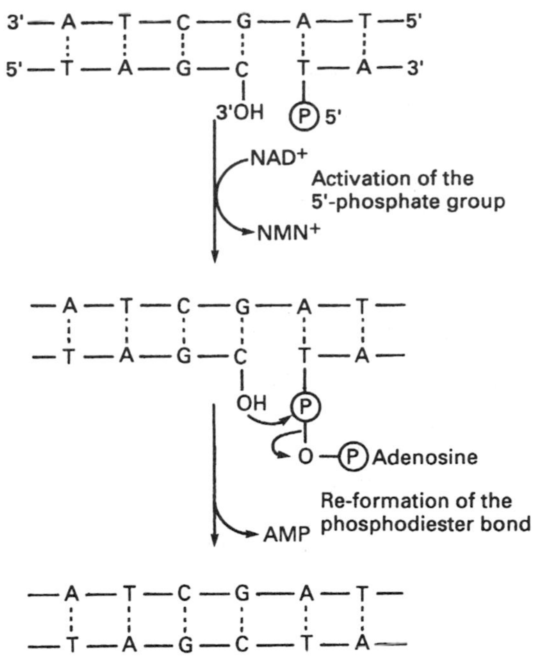

3'—A—T—C—G—A—T—5'

5'—T—A—G—C T—A—3'

3'OH (P)5'

NAD+

Activation of the 5'-phosphate group

NMN+

—A—T—C—G—A—T—

—T—A—G—C T—A—

OH (P)

O—(P)Adenosine

Re-formation of the phosphodiester bond

AMP

—A—T—C—G—A—T—

—T—A—G—C—T—A—

Figure 4.9 **Reaction catalysed by DNA ligase**

(P) = ⁻O–P(=O)(O⁻)–; NMN^+ = Nicotinamide mononucleotide

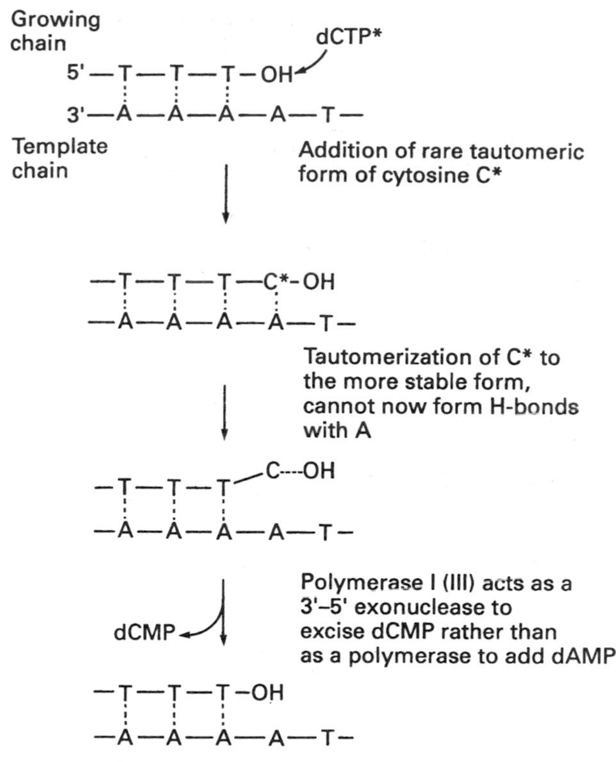

Figure 4.11 **Proof-reading by DNA polymerases I and III**

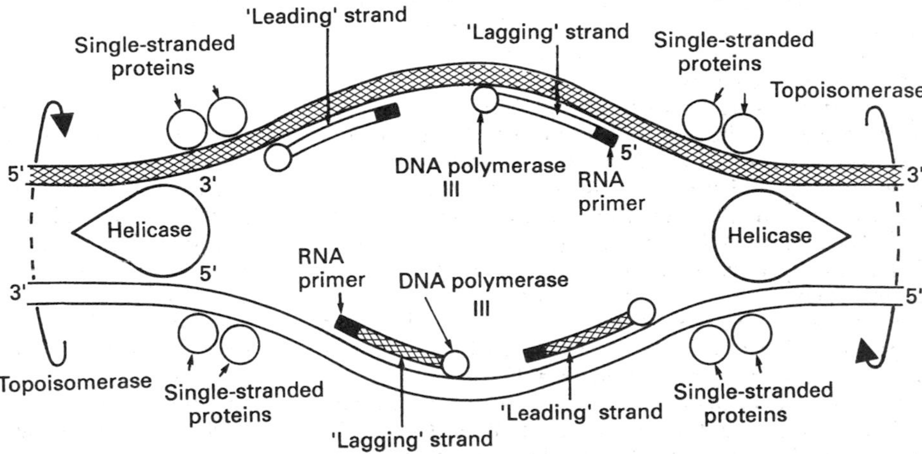

Figure 4.10 **Summary of the roles of the various enzymes and other proteins involved in replication in prokaryotes**

Eukaryotic DNA organization

The organization of eukaryotic DNA is different to prokaryotic DNA. Eukaryotic DNA is not circular but arranged in linear **chromosomes**. It contains many repeated base sequences. More than 30% of human DNA contains sequences repeated 20 times. Only a small part of the human genome consists of genes, which are defined as sequences of DNA necessary to synthesize a functional polypeptide or RNA molecule. The function of the major part is still unknown. The repeated sequences show such variation that their presence and position in the genome forms the basis of identification of individual human beings by a process known as **genetic finger-printing**. Some genes, such as that encoding lysozyme, exist as a single copy, whereas others are duplicated. Frequently the multiple copies exist in tandem with one copy next to another in the genome. Examples of tandemly repeating genes are those for the α- and β-globin families, the immunoglobulins, myosin heavy chains, histones, and ribosomal and transfer RNAs. Repeated DNA sequences are of various types. Some are short sequences of 5–10 nucleotides known as **simple sequence** or **satellite DNA**. Other sequences of varying length are interspersed in many places on the genome and are known as **SINES** or **LINES** for Short or Long Interspersed elements. Large amounts of satellite DNA are found at centromeres which are regions of the metaphase chromosome essential for segregation during mitosis and meiosis. They are likely to form protein binding sites and are necessary for centromere activity.

Eukaryotic DNA replication

DNA replication in eukaryotic organisms resembles that in bacteria but is not identical. The rate of nucleotide incorporation by eukaryotic DNA polymerases is 20 times slower than by the prokaryotic enzymes. The Okazaki fragments formed are about 40–300 bases long and are, therefore, much smaller than the corresponding fragments in prokaryotes.

Five eukaryotic DNA polymerases have been recognized: α β, γ, δ and ϵ. **Polymerase** α copies the lagging strand and **polymerase δ** copies the leading strand. The latter possesses 3'–5' exonuclease activity ensuring adequate proof-reading. **Polymerases δ** and **ϵ** are involved in repairing damaged DNA. **Polymerase γ** copies mitochondrial DNA. It has no known exonuclease activity and this may account for the observation that mitochondrial DNA shows a much higher rate of mutation than nuclear DNA (see Chapter 5).

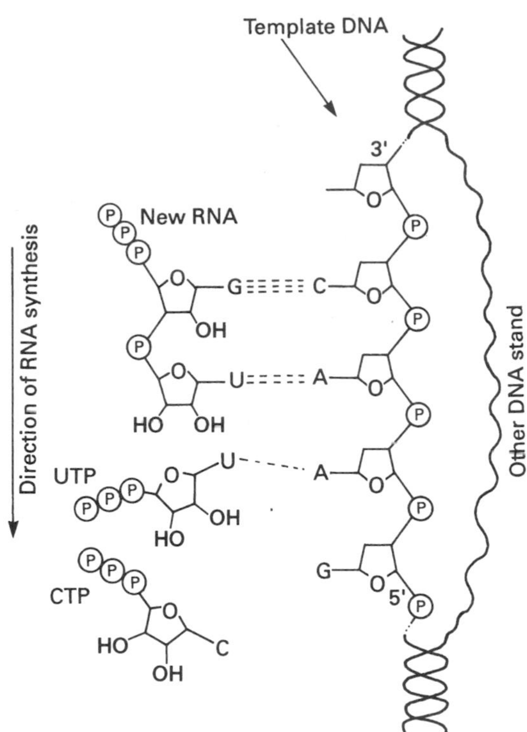

Figure 4.12 **The mechanism of RNA synthesis.** The chain elongation reaction is catalysed by RNA polymerase. Only one strand of the DNA template is transcribed in a particular region of the genome

Mutation, damage and DNA repair

DNA can mutate spontaneously or can be damaged by various agents. Cytosine, for example, can deaminate spontaneously to uracil. Ionizing radiation can cause strand breaks or base alterations. Ultraviolet light irradiation may produce pyrimidine dimers which distort DNA structure. Various chemical agents may induce strand breaks, base alterations, or cross-linking of strands. Normally a system of endonucleases, polymerases and DNA ligase operates to excise the damaged section and replace it by the correct sequence. The importance of DNA excision-repair can be seen in cases where the repair systems are defective. In the case of **xeroderma pigmentosum**, a rare autosomal recessive condition, an endonuclease (**excinuclease**) which removes pyrimidine dimers formed by ultraviolet light exposure, is missing. This results in multiple skin cancers and early death. It is now known that the risk of some colon, ovarian, uterine and kidney cancers is dramatically increased in individuals who have mutations involving gene p53. The product of this gene helps to arrest cells in the G1 phase of the cell cycle while mismatch repair occurs (Chapter 37).

4.4 RNA synthesis (transcription)

Cellular RNAs include messenger or mRNA, ribosomal or rRNA, transfer or tRNA and several small nuclear or snRNAs. All are transcribed from DNA by means of DNA-dependent RNA polymerases. RNA synthesis, known as transcription, like DNA replication, occurs in the 5'→3' direction and involves the condensation of nucleoside triphosphates (ATP, GTP, CTP and UTP), the sequence of the bases in RNA being dictated by the base sequence of the DNA template (Figure 4.12). The addition of the new nucleotide is rendered thermodynamically favourable by the hydrolysis of the released pyrophosphate, as in DNA synthesis.

Prokaryotic RNA synthesis

Prokaryotic RNA synthesis is initiated at specific regions of the DNA template called **promoter sites** to which RNA polymerase can bind. They are located closely upstream, that is, in the 5' direction, from the point where transcription starts. Unlike DNA synthesis, no primer is required for RNA synthesis and only one of the strands is transcribed. A short segment of DNA–RNA hybrid, about 12 bp long, is formed during transcription. The 5' end of the RNA chain then detaches itself from the DNA allowing re-formation of the DNA duplex. RNA synthesis continues until a **terminator** sequence is reached. RNA polymerase has no 3'→5' exonuclease activity, therefore proof-reading is not possible. Errors, however, are not critical. mRNA molecules in prokaryotes are degraded within a few minutes of synthesis and, in this way, errors are not perpetuated. There is only one type of prokaryotic RNA polymerase. It consists of a tetrameric core and a dissociable σ subunit which allows the enzyme to recognize promoter sequences in DNA. Variant forms of the σ subunit are sometimes produced by *E.coli*. In this way, expression of genes not normally transcribed by the polymerase becomes possible, as the variant subunit recognizes promoter sequences not recognized by the normal subunit. For example, *E.coli* subjected to heat stress produces a subunit which recognizes promoter regions on genes coding for '**heat-shock**' proteins. These are proteins which are thought to protect the cell from damage caused by increased temperatures.

Shortly after initiation the σ subunit dissociates from the core enzyme and binds another RNA polymerase tetramer which then begins to transcribe DNA. This enables many RNA polymerases to carry out transcription on a strand of DNA at the same time.

Eukaryotic RNA synthesis

Transcription in eukaryotes is a more complex process than in prokaryotes. It takes place mainly in the nucleus and to a limited extent in mitochondria. In the nucleus, only one of the two DNA strands is transcribed. In mammalian mitochondria, both strands are transcribed and one is subsequently degraded. Most genes have sequences, referred to as **exons**, which will be transcribed and ultimately be present in the functional RNA molecule. These are interrupted by sequences, known as **introns**, which are first transcribed but subsequently removed during the production of functional RNA molecules. The length of the introns generally exceeds that of the exons. Figure 4.13 shows the relationship between the organization of a hypothetical eukaryotic gene and the formation of its corresponding functional mRNA molecule.

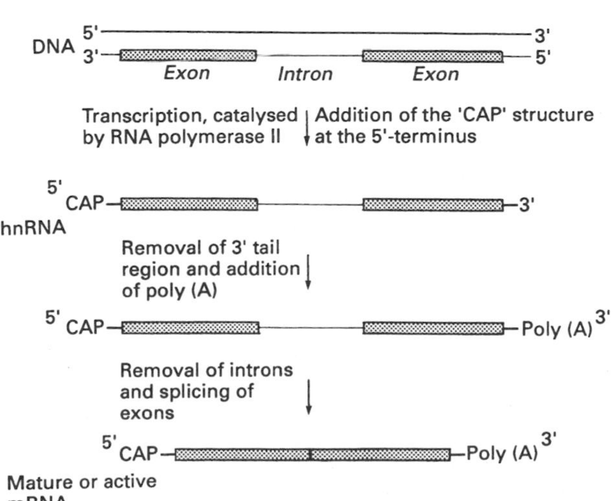

Figure 4.13 **Formation of a functional mRNA molecule in eukaryotes**

Table 4.2 **Some properties of eukaryotic RNA polymerases**

Type	*Location*	*RNA product synthesized*	*Sensitivity to α-amanitin*
I	Nucleolus	5.8-S, 18-S, 28-SrRNA (as 45-S precursor)	Insensitive
II	Nucleo-plasm	mRNA precursors (hnRNA)	Extremely sensitive
III	Nucleo-plasm	tRNAs, 5-S rRNA	Sensitive
Mito-chondrial	Mito-chondria	Mitochondrial RNAs	Insensitive

There are four types of eukaryotic RNA polymerase, three nuclear and one mitochondrial (Table 4.2) Proteins known as **transcription factors** are also required. Unlike prokaryotic factors they do not bind RNA polymerase, but associate directly with chromatin and thereby facilitate the initiation of transcription. **Polymerase I** is found in the nucleolus and synthesizes a large (45S) rRNA precursor which is subsequently cleaved to three smaller molecules (18S, 28S and 5.8S rRNA) **Polymerase II** transcribes protein coding genes. The primary transcript formed is known as **heteronuclear RNA** (hnRNA) and corresponds to the entire gene, including introns. It is cleaved at the intron–exon boundaries and the sequences corresponding to the exons spliced together. This process requires **small nuclear ribonucleoproteins** (snRPs) which are complexes of small nuclear RNA (snRNA) and protein. About 35% of the total hnRNA synthesized is processed to mRNA, the rest remains in the nucleus and is gradually degraded. A **cap** of 7-methyl guanosine is added to the 5' end, and a **polyA tail** is added to the 3' end of the molecule. These modifications are essential for the stabilization and ultimate translation of the mature mRNA by ribosomes. The promoter regions for polymerase II commonly include the following sequences: a TATA sequence (box) about 25 bp from the starting point of transcription, a CAAT sequence 80 bp from the start and a variably located CCGGG sequence. They function in binding transcription factors and so allowing RNA synthesis to begin. Separate sequences known as enhancers, which may be located at a considerable distance from the transcribed gene, are involved in regulating gene expression and allow certain genes to be expressed only in some cells. Receptors for steroid hormones, thyroid hormone, vitamin D and retinoic acid bind **enhancer sequences** (also referred to as **responsive elements**) on DNA and, in this way, regulate the initiation of transcription (see Figure 3.1).

Polymerase III is responsible for the formation of small species of RNA such as 5S rRNA, tRNA and some small nuclear and cytoplasmic RNAs such as **signal recognition particle RNA**. This forms a complex with proteins needed for the recognition of ribosomes on the rough endoplasmic reticulum and the translocation of the growing polypeptide chain through the membrane.

4.5 Reverse transcription

The flow of information from DNA to RNA to protein was considered to be unidirectional until it was found that some viruses, referred to as 'retroviruses', carry RNA as their genetic material and can synthesize double-stranded DNA from their genomic RNA by a process known as **reverse transcription** (Figure 4.14). An example of a retrovirus is the human immunodeficiency virus (HIV) which causes AIDS. The synthesis of DNA using an RNA template is catalysed by the viral enzyme **reverse transcriptase**. A primer is required, in the form of a host tRNA, which binds the 3' end of the viral RNA. A DNA–RNA hybrid is initially formed. Reverse transcriptase has a second catalytic activity, referred to as **RNAse H**, and this permits the removal of the RNA. (RNAse H is used in genetic engineering to cause limited hydrolysis of RNA in a DNA/RNA heteroduplex - Chapter 38.) A second primer then interacts with the 3′-end of the single-stranded DNA present, and DNA polymerase I then catalyses the formation of a second strand of DNA using the first strand as a template. The resulting double-stranded DNA is incorporated into the host genome and it is subsequently transcribed by host RNA polymerase to produce viral RNA.

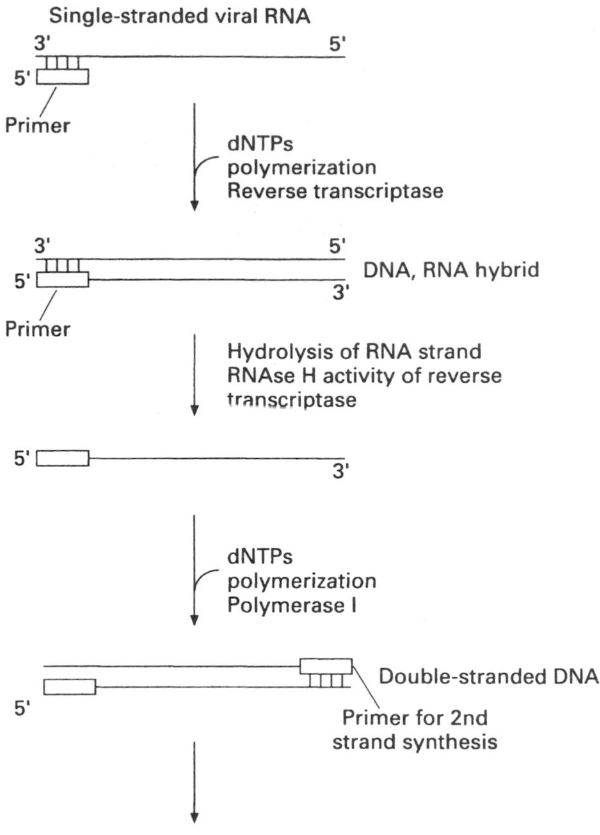

Figure 4.14 **The conversion of a single-stranded retrovirus RNA into double-stranded DNA**

4.6 A dynamic model for the flow of genetic information

'Genes in DNA are transcribed as mRNA and directly translated into protein.' This is the traditional or 'static' model for the flow of genetic information. In the light of present knowledge this model is inadequate and a new 'dynamic' model has emerged (Figure 4.15).

DNA is not simply transcribed in a linear fashion. Extensive reorganization of DNA takes place before transcription. Genes can be rearranged, deleted or amplified. In the development of the lymphoid B cell, for example, genes coding for the variable part of an immunoglobulin light chain are brought to a position closer to the genes coding for the joining segment and the constant region of the light chain (Chapter 32). Intervening genes are deleted from the genome. Gene amplification can be seen in cancer cells grown in culture, in the presence of methotrexate, an inhibitor of dihydrofolate reductase (DHFR). This enzyme is needed by the cells in order to produce tetrahydrofolate which is necessary for nucleic acid synthesis. In the presence of methotrexate, the cancer cells show an increase in the amount of DNA encoding DHFR and a concomitant increase in the amount of enzyme produced. This is also the basis for the development of resistance to methotrexate that can arise in cancer patients treated with the drug (see Chapter 39).

RNA is normally formed using DNA as a template, but reverse transcription occasionally occurs. Many eukaryotic repeat sequences, known as **retroposons,** appear to be DNA copies of RNA molecules which have become incorporated into the genome by reverse transcription. The mechanism has probably been similar to that by which the DNA corresponding to the RNA in retroviruses becomes integrated into the host genome.

The sequence in mRNA often does not correspond to the translated polypeptide. The linear expression of a gene is interrupted at different levels by introns. The best studied introns are **RNA introns**, which are transcribed and subsequently excised from the primary transcript, as in the case of the β-globin gene. Here modification takes place at the transcriptional level. There are introns which interrupt expression at the post-translational level by causing the excision of a segment from the polypeptide product. These are known as **protein introns** and are also excised through splicing mechanisms. Some introns are not excised; they persist in a coding segment of mRNA but are bypassed by the translational machinery of the cell. Examples of the latter **translational introns** include those found in the *E.coli* tryptophan repressor gene and the bacteriophage T4 gene 60 which encodes a subunit of DNA topoisomerase.

4.7 NAD⁺ synthesis

Another important function of the nucleolus is the supply of NAD^+ to the cell. The final reaction involved in the synthesis of NAD^+ from nicotinamide mononucleotide (NMN^+) and ATP, i.e. :

$$NMN^+ + ATP \rightleftharpoons NAD^+ + PP_i$$

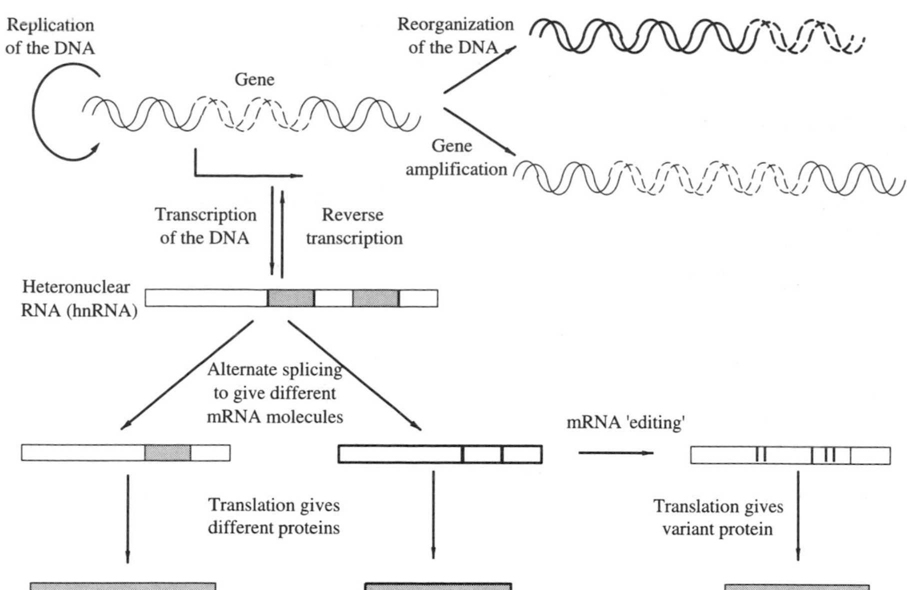

Figure 4.15 **Dynamic model for the storage and expression of genetic information**

is catalysed by an ATP: NMN^+ adenylyl-transferase which is exclusively located in the nucleolus to which it is tightly bound.

NAD^+ is an important cofactor for oxidation–reduction reactions in cellular metabolism. It also acts as the donor of ADP-ribose units in **ADP-ribosylation** reactions. The modification of proteins by covalent association of ADP-ribose units was first noted on histones in chromatin. ADP-ribosylation is now recognized as a means by which living organisms can modify the structure and functions of their own or others' proteins. Substrates for ADP-ribosylation include RNA and DNA polymerases, topoisomerases and many other proteins involved in the control of gene expression in events such as the cell cycle, cell differentiation and DNA repair.

Cholera toxin produces its effect by directing the ADP-ribosylation of the G_s protein associated with adenylate cyclase in cells of the intestinal mucosa. This results in continuous activation of the enzyme and, therefore, the production of high levels of cAMP which in turn give rise to the secretion of large amounts of fluid into the small intestine causing the severe dehydration which is characteristic of cholera.

4.8 The cell cycle

A fundamental attribute of the cells of a multicellular organism is their capacity for growth and division. At appropriate times, and in response to specific cues, certain cells are capable of many rounds of duplication, during which their genetic potential is conserved.

The process which leads to the formation of a pair of daughter cells from a parental eukaryotic cell, is referred to as the **cell cycle**. One of the triumphs of molecular biology has been the description, in molecular terms, of the events which constitute this cycle. A long time prior to this, however, careful observations of cells in tissues had led to the understanding that the processes of growth and division are very rigorously controlled. At the same time, it was established that the DNA of the genome is duplicated and consigned to daughter cells. The times during which these two processes occurred were referred to as **S-** (for **DNA Synthesis**), and **M-** (for **Mitotic**) phase. Intervening between these two phases, were two **G-** (for **Gap**) phases, termed G_1 and G_2 (see Figure 4.6)

The total duration of the cell cycle, as observed in a wide range of somatic cells in invertebrate animals, ranges from 12 to 24 h. Within this, the S-phase usually requires about 7 h, the M-phase occupies 1–2 h, and the G_2-phase about 3 h. The duration of the G_1-phase is more variable. Some of the biochemical events that dominate the various phases have been elucidated. Thus, G_1 is a preparative phase, in which the enzymes necessary to replicate the genome are produced, genomic DNA is checked for integrity, and the building materials required in mitosis begin to accumulate. In the S-phase, the biosynthesis of new DNA on pre-existing templates is carried out. In the G_2-phase the successful completion of replication of the genome is confirmed, as evidenced by the presence of daughter chromosomes. In the M-phase, the nuclear membrane is disrupted, and the duplicate chromosomes are aligned and distributed equally to either side of the parental cell. This segregation of the chromosomes is achieved by means of elements of the cytoskeleton, referred to as the **spindle**.

Crucial to the correct operation of these complex processes is the activity of a range of enzymes which act as 'clocks', or regulators. These enzymes, which signal the initiation or the termination of the ordered events involved in the cell cycle, are protein kinases. In this way, they closely resemble other signalling molecules, described in Chapter 3, as being involved in the transduction of extracellular signals. The cell cycle kinases undergo cyclic variations in their catalytic activities as a result of the periodic appearance of proteins called **cyclins**. As the cyclins accumulate during the cell cycle, they promote the phosphorylation activity of the protein kinases, and this signals the onset of mitosis. Specific ubitiquin-dependent proteases (Chapter 11) then arise to remove the cyclins in the M- to G_1-transition.

It should be noted that cells leaving the M-phase are not absolutely committed to entering the G_1- phase; they may instead be diverted into a quiescent phase designated G_0. The duration of the G_0 phase is subject to large-scale variation. It may range from hours to years. The entry to, or exit from, the G_0 phase is frequently determined by the presence or absence of growth factors.

It can be seen that there are a number of critical decision points in the operation of the cell cycle. At these points, frequently at the transitions between the phases, rigorous conditions, such as, for example, lack of damage to the DNA, must be fulfilled before the cycle may proceed. The restriction point which determines the passage from the G_1- to the S-phase is referred to as 'start'.

With such a sensitive arrangement it is not surprising that defects in the system of checks and balances controlling the cell cycle have serious consequences in terms of human health. The occurrence of cancer provides an important example in which proper control of the cell cycle is lost or impaired.

Chapter 5

Cellular organelles: mitochondria and energy conservation

5.1 Introduction

By the end of the last century it was known from microscopic studies that most cells contained small elongated bodies to which the name 'mitochondria' was given. Although several researchers had indicated the possibility that these organelles may play a role in cellular oxidation, their true function remained uncertain for nearly 50 years after their discovery. An important development came in 1948 when it was demonstrated that the subcellular organelles of homogenized tissues could be obtained by differential centrifugation (see Chapter 1).

It had been known for over 200 years that most tissues and cells consumed oxygen, but the general belief was that the consumption was uniform throughout the cell. Studies of isolated mitochondria showed, however, that this was not the case and that a large percentage of the oxygen consumed by cells occurred in these organelles. It is now estimated that, in a hepatocyte (liver cell), some 75% of the total cellular consumption of oxygen occurs in the mitochondria. The remainder is used by oxidative processes that take place largely in the endoplasmic reticulum (see Chapters 29 and 36).

Further studies showed that the oxygen-dependent oxidation of metabolites leads to the conservation of the energy produced in the form of adenosine triphosphate (ATP), with mitochondria being the main site for the bringing together of adenosine diphosphate (ADP) and inorganic phosphate (HPO_4^{2-}; Pi) to yield ATP.

Mitochondria are complex structures, possessing outer and inner membranes, and even their own DNA that encodes some, but by no means all, of their own proteins. These proteins are formed on mitochondria-specific ribosomes (see Chapter 6). So great is the apparent autonomy of mitochondria that it has been proposed that they were originally primitive organisms (perhaps the forerunners of the bacteria) which, in the course of evolution, became incorporated into eukaryotic cells bringing with them their capacity for oxidative metabolism and thereby setting up a kind of symbiosis.

A typical mammalian cell, as might be found in the liver of a rat, will contain some 800 mitochondria, but the range found in specialized cells runs from just 24 in sperm cells to around 10 000 in the oocytes of the sea-urchins.

5.2 Mitochondria in representative cells

When viewed under the electron microscope mitochondria usually appear as sausage-shaped bodies (Figure 5.1), but these become spherical under adverse conditions, for example those necessary for their

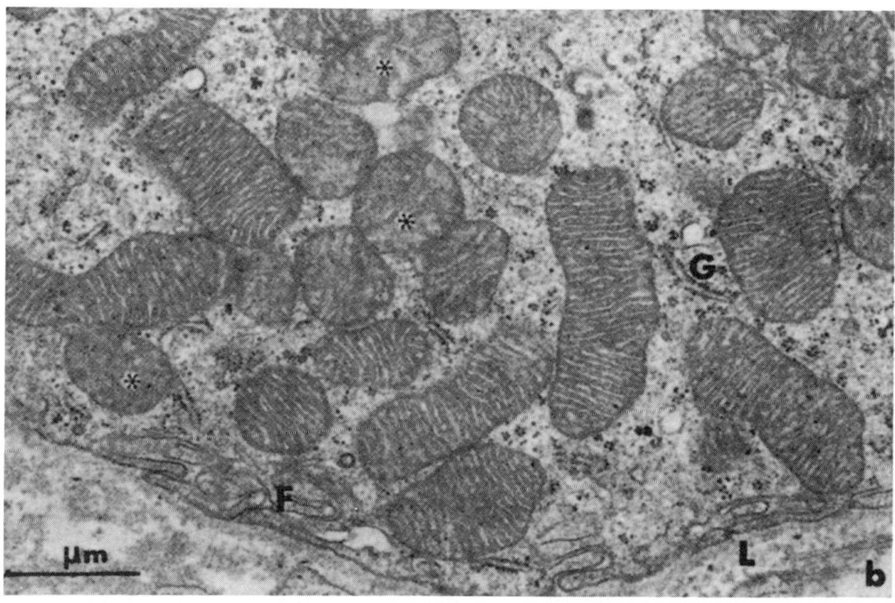

Figure 5.1 **An electron micrograph showing mitochondria in kidney cells**

isolation by centrifugation. In the cells of some tissues, such as the hepatocytes or the gastric parietal cells, the mitochondria are distributed rather haphazardly throughout the cytoplasm, but in others, such as those in muscle, they are packed systematically around the contractile apparatus (the myofibrils; see Chapter 29). In the cells of adipose tissue, a sort of connective tissue found throughout the body, the mitochondria are restricted to the periphery of the adipocytes, most of the rest of the intracellular space being occupied by a large lipid droplet that contains triacylglycerols.

5.3 Structure, composition and general functions of mitochondria

Appearance

Typical mitochondria appear to be about 3 μm long and 0.5–1 μm in diameter. The functionally more important 'inner' membrane is surrounded by an 'outer' membrane. The inner membrane has many folds (invaginations) and these are referred to as **cristae**. The part of the mitochondrion located between the two membranes is referred to as the peripheral or intermembrane space whereas that region of the mitochondrion enclosed by the inner membrane is referred to as the **matrix** (Figure 5.2). If the inner membrane is examined carefully it is found to be studded with a number of spherical particles which consist of proteins (called F_1-proteins) and these are attached, on the matrix-facing surface of the membrane, to other membrane-spanning (integral) proteins (called F_0). These proteins, which jointly constitute the F_0–F_1 complex, are responsible for the formation of ATP (see Section 5.5). In this process the oxidation of substrate is 'coupled' to the phosphorylation of ADP to form ATP. This process is called **oxidative phosphorylation**.

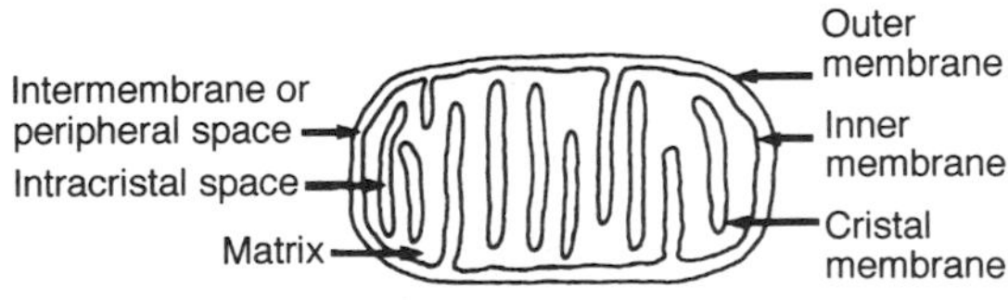

Figure 5.2 **The various mitochondrial compartments**

Composition

When analysed, the membranes of whole mitochondria differ in significant ways from other cellular membranes, and this is due almost entirely to the unusual composition of the functionally highly specialized inner membrane. (Analytically the outer membrane resembles the endoplasmic reticulum membrane.)

Three observations about the inner membrane are particularly noteworthy:

1. It has a lower ratio of lipid to protein than any other cell membrane, with 75% of its dry weight being protein and most of the rest being accounted for by lipids.
2. Analysis of the lipids shows them to contain a much higher proportion of the unusual lipid cardiolipin (diphosphatidylglycerol) than is found in other cell membranes.
3. Cholesterol is virtually absent.

Of the proteins associated with the inner membrane, about half have a defined enzymic role whilst the rest play structural or 'transport' roles. A variety of proteins are located in the matrix compartment, including all the enzymes of the tricarboxylic acid cycle except for succinate dehydrogenase, which is found in the inner membrane (see section 5.4). An important outer membrane enzyme is monoamine oxidase (see Chapter 30).

Functional roles

The general functional roles of mitochondria are summarized in Table 5.1. These are largely concerned with the conservation of the energy made available by the oxidation of many molecules produced from ingested or stored fuels. Generally, this conservation involves the formation of ATP that can be utilized for processes of cellular synthesis. It should not be forgotten, however, that in a particular tissue mitochondria will perform roles specific for that tissue, e.g. in gluconeogenesis in the liver and kidney (see Chapter 6); in lipogenesis in the liver (see Chapter 10); in the formation of steroid hormones in endocrine organs (see Chapter 28); in urea synthesis in the liver; and in haem synthesis principally in the liver and erythroid tissues.

Table 5.1 **Summary of major functions of the mitochondria**

1. Operation of the tricarboxylic acid cycle and production of NADH, i.e. the generation of reducing equivalents
2. β-Oxidation of fatty acids producing acetylSCoA, reduced flavoproteins and NADH
3. Oxidation of NADH by means of electron transport coupled with synthesis of ATP
4. Accumulation of divalent cations such as Ca^{2+}
5. Synthesis of metabolic intermediates for the provision of haem, urea, glucose, steroids, etc.

5.4 The oxidative pathways of mitochondria

The complete oxidation of the major fuels in the body gives rise to carbon dioxide and water. Whether the fuels are carbohydrates or fatty acids their conversion to these two ultimate products occurs in the mitochondria. Specifically, the reactions occur in the matrix or are associated with the inner membrane. The conversion of glucose to pyruvate was dealt with in Chapter 6, so to complete the description of the full oxidation of glucose it is simply necessary to describe the oxidation of the oxo-acid. It turns out that pyruvate must undergo oxidative decarboxylation to produce acetylCoA in a complex reaction catalysed, in the mitochondrial matrix, by pyruvate dehydrogenase. This reaction requires thiamin pyrophosphate, NAD^+, coenzyme A and lipoic

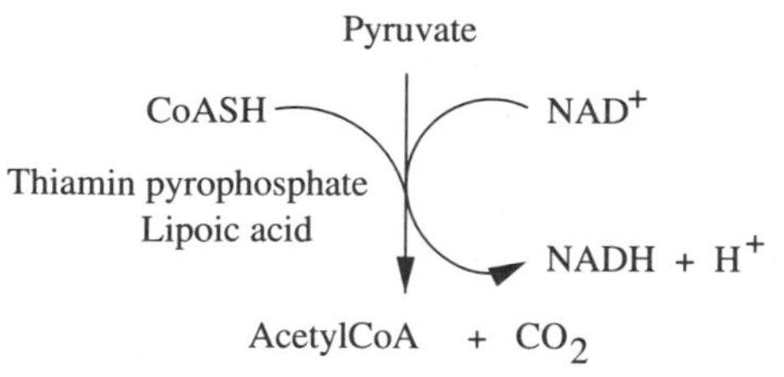

Pyruvate

CoASH → NAD$^+$

Thiamin pyrophosphate
Lipoic acid

NADH + H$^+$

AcetylCoA + CO_2

Figure 5.3 **The reaction catalysed by pyruvate dehydrogenase**

acid. (The last named molecule has two thiol groups capable of undergoing reversible oxidation–reduction reactions). The reaction catalysed by pyruvate dehydrogenase is shown in Figure 5.3.

The formation and further oxidation of acetylCoA plays a central role in the energy–conserving reactions of mitochondria, because the oxidation of fatty acids in the mitochondrial matrix also gives rise to this compound. The catabolic process giving rise to acetylCoA from fatty acids is referred to as β-oxidation, because the oxidative reactions affecting fatty acid derivatives are focused on carbon atom number 3 (the β-carbon). It is appropriate therefore first to describe the process of β-oxidation and then to describe the full oxidation of acetylCoA as occurs in the tricarboxylic acid cycle.

β-oxidation of fatty acids

Fatty acids taken up into cells that will ultimately carry out their oxidation are not direct substrates for this process, but they must first be converted into coenzyme A derivatives (acylCoA). This reaction occurs in the cytosol, is catalysed by acylCoA synthase and requires ATP, which is converted into adenosine monophosphate (AMP) and inorganic pyrophosphate (PPi). The reaction is shown in Figure 5.4. This 'activation' of the fatty acid facilitates its transfer into the mitochondrial matrix by the carnitine–acylcarnitine exchange carrier (see section 5.6), where it undergoes β-oxidation. Figure 5.5 shows the essentials of this process, which involves two oxidative steps. The first of these sees the formation of a carbon–carbon double bond and is catalysed by acylCoA dehydrogenase, an enzyme of the inner membrane that has flavin adenine dinucleotide (FAD) as its prosthetic group. Hydration of the newly-formed double bond catalysed by enoylCoA hydratase gives rise to a 3-hydroxyacylCoA derivative which acts as the substrate for the second dehydrogenase. This enzyme, 3-hydroxyacylCoA dehydrogenase, catalyses the oxidation of the hydroxy- to an oxo-group using NAD$^+$ as the acceptor for the hydrogens removed. The 3-oxoacylCoA derivative undergoes cleavage in the presence of coenzyme A to give rise to acetylCoA and acylCoA with two fewer carbon atoms than the original fatty acid (usually now 14 or 16 carbon atoms). This molecule is now able to undergo the same series of reactions again to give rise to another molecule of acetylCoA and so on until the penultimate product of the whole pathway 3-oxobutyrylCoA (also known as acetoacetylCoA) is produced. This molecule has four of the original carbon

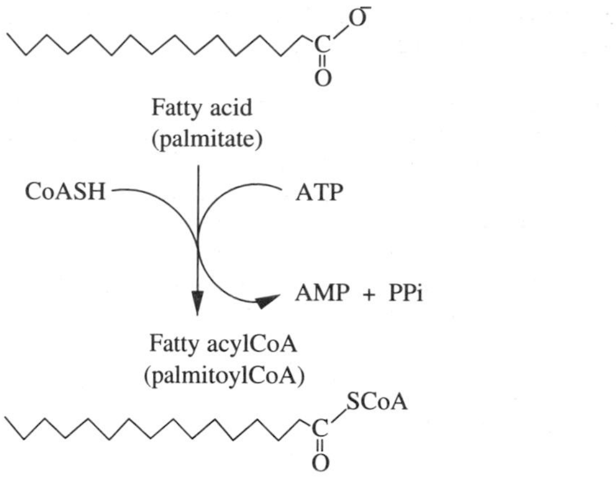

Fatty acid
(palmitate)

CoASH → ATP

AMP + PPi

Fatty acylCoA
(palmitoylCoA)

Figure 5.4 **The reaction catalysed by acylCoA synthase**

Fatty acyl-SCoA

(1) FAD → $FADH_2$

Enoyl-SCoA

(2) H_2O

L-Hydroxyacyl-SCoA

(3) NAD$^+$ → NADH + H$^+$

Oxo-acyl-SCoA

(4) CoASH

Fatty acylSCoA shortened by two carbon atoms + CH_3–CO–SCoA AcetylSCoA

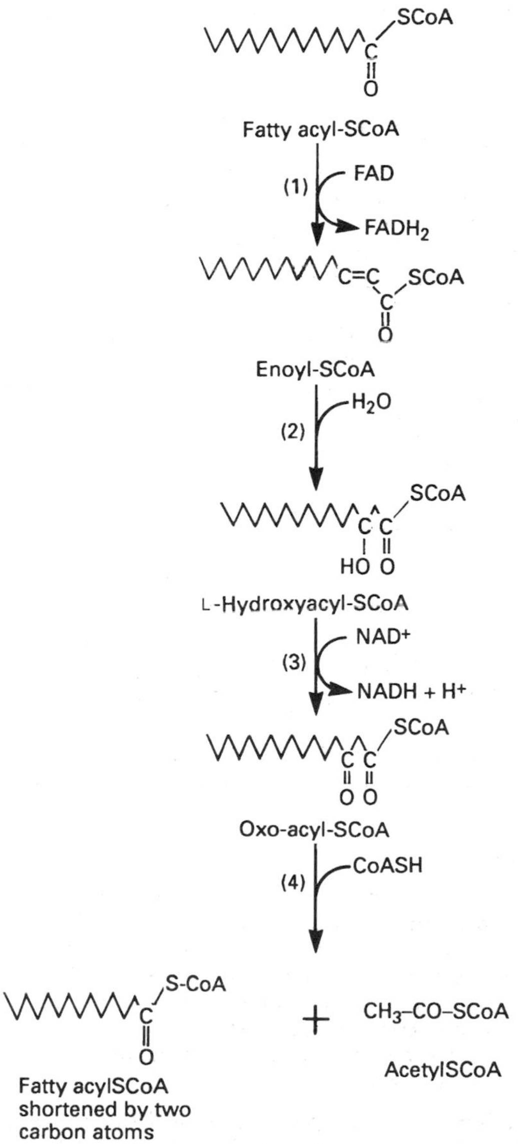

Figure 5.5 **β-Oxidation of fatty acids in mitochondria**

atoms of the parent fatty acid and when it is cleaved in the reaction catalysed by 3-oxothiolase, two molecules of acetylCoA are produced.

Thus during the complete oxidation of a fatty acid such as palmitate (C16), 8 molecules of acetylCoA will be produced in the mitochondrial matrix together with 8 molecules of NADH and 8 of FAD.2H (associated with acylCoA dehydrogenase). The reduced cofactors must be reoxidized and the consequences of this are dealt with in Section 5.5. The further metabolism of acetylCoA occurs in the tricarboxylic acid cycle.

Tricarboxylic acid cycle

In the complete oxidation of acetylCoA the two carbon atoms of the acetyl group are both effectively converted into carbon dioxide and the intermediates of the tricarboxylic acid cycle may be regarded as acting as catalyst in the process (Figure 5.6). The reason for this is that the cycle is initiated when acetylCoA (either derived from pyruvate or via β-oxidation) reacts with oxaloacetate to form citrate (a tricarboxylic acid). After a cyclic series of reactions, oxaloacetate is regenerated and is able once more to react with acetylCoA. The initial reaction is catalysed by citrate synthase. This is followed by a series of reactions in which the citrate is converted into its isomer isocitrate via cisaconitate in a reaction catalysed by aconitase. Isocitrate then undergoes oxidative decarboxylation catalysed by isocitrate dehydrogenase to form 2-oxoglutarate, a reaction that requires NAD^+. In a reaction that closely resembles that catalysed by pyruvate dehydrogenase, in cofactor requirements, 2-oxoglutarate is converted into succinylCoA with the release of the second molecule of carbon dioxide. The enzyme for this reaction is 2-oxoglutarate dehydrogenase, an NAD^+-dependent enzyme. In a complex reaction, succinylCoA is converted into the dicarboxylic acid succinate with the concomitant formation of guanosine triphosphate (GTP). The enzyme required for this reaction is succinylCoA thiokinase. The remaining three reactions in which succinate is converted into oxaloacetate, thereby completing the cycle, closely resemble those occurring in β-oxidation. The first of these involves the formation of a carbon–carbon double bond that converts succinate into fumarate. As in β-oxidation the enzyme involved has FAD as its prosthetic group and is called succinate dehydrogenase. Fumarase then catalyses the hydration of fumarate to yield malate which, in turn acts as a substrate for the NAD^+-dependent malate dehydrogenase to regenerate oxaloacetate.

Consequently, the fate of the acetyl moiety of acetylCoA in the TCA cycle is to become oxidized to two molecules of carbon dioxide. At three stages in this process NADH is generated, while at a fourth stage the primary dehydrogenase uses FAD as the cofactor. If the cycle is to continue, then evidently the NADH and FAD.2H must be reoxidized and the way in which this occurs and how this relates to the energy economy of the cell is the subject of the next section.

5.5 Oxidative phosphorylation

ATP as a key intermediate between energy-yielding and energy-utilizing reactions

The discovery that ATP played a role as an energy source for muscular contraction came from the pioneering work of Meyerhof and Lohman in Germany in the 1920s, but it was Lipmann who proposed that ATP played a central role in energy transactions in all cells. This view derived from the observations (Figures 5.7 and 5.8) that energy is conserved as ATP when food is oxidized in the body and this conserved energy can be channelled into work performance. Essentially, Lipmann's proposal can be represented by Figure 5.9 In

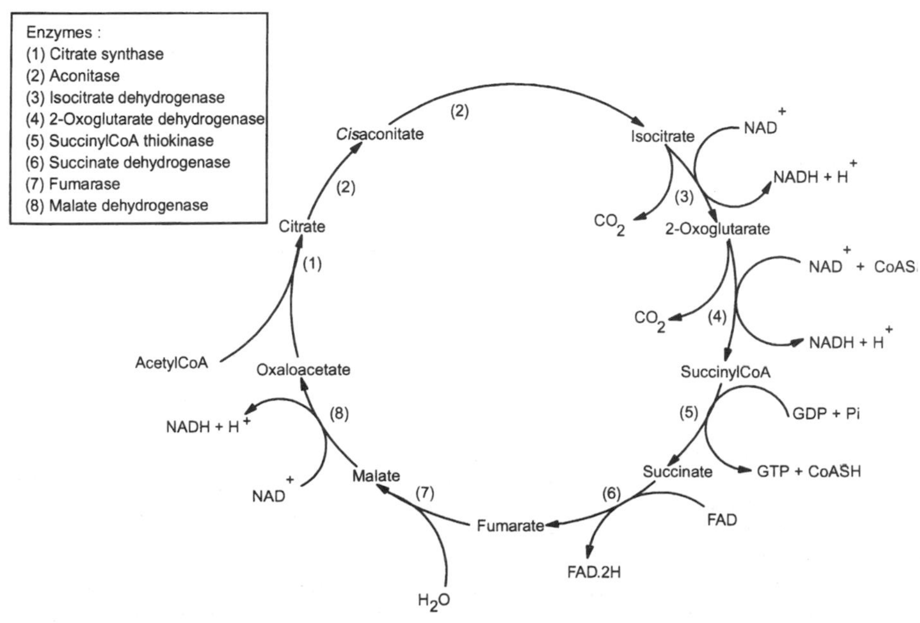

Figure 5.6 **The tricarboxylic acid cycle**

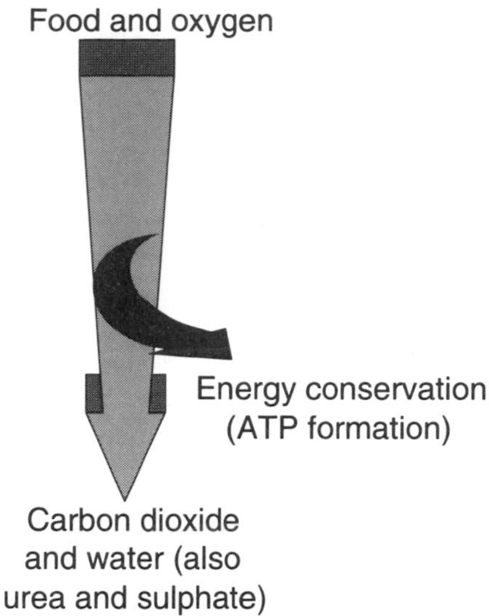

Figure 5.7 **The provision of energy for work performance in biological systems**

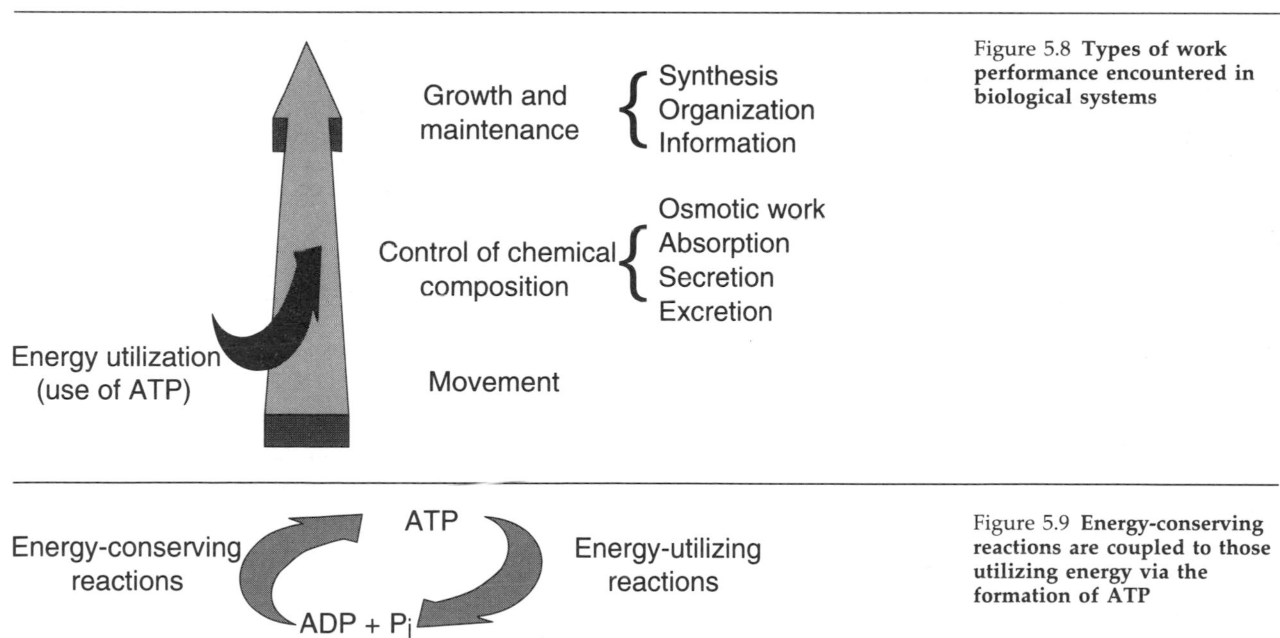

Figure 5.8 **Types of work performance encountered in biological systems**

Figure 5.9 **Energy-conserving reactions are coupled to those utilizing energy via the formation of ATP**

this, the cyclic formation of ATP and its breakdown to yield ADP is shown as 'coupling' energy-yielding (**catabolic**) reactions to energy-consuming (**anabolic**) reactions. The types of anabolic reaction supported in various cells either directly or indirectly by ATP are summarized in Figure 5.10. Many of these reactions and processes are described elsewhere in the text, but it can be seen that the synthesis of all the classes of macromolecule encountered in cells, the movement of ions and molecules across membranes against their concentration gradients as well as in the various contractile processes all depend upon a supply of ATP.

The importance of ATP depends on the fact that it is a compound with a high phosphate-transfer potential (sometimes referred as a '**high-energy**' phosphate compound). That is, the free energy of hydrolysis of ATP to ADP and inorganic phosphate (Pi) is large and negative. When phosphate esters, for example glucose 6-phosphate, undergo hydrolysis, the free energy change occurring is still negative but much less so than is the case for the hydrolysis of acid anhydrides such as ATP. Phosphate esters, such as glucose 6-phosphate, are therefore sometimes referred to as being 'low-energy' phosphate compounds. It should be noted, however, that it is not correct to refer to the acid anhydride bond that undergoes hydrolysis in ATP as a 'high-energy bond' as has been done in the past. This is because the free energy of any reaction is the thermodynamic property of that reaction as a whole, and, in this case, it is determined by all the molecular changes that occur when ATP and water react to give ADP and Pi, at the prevailing pH value.

The utilization of ATP occurs in all subcellular locations, including the mitochondria themselves, but the production of ATP from ADP and Pi takes place primarily within the mitochondria. It should not be overlooked, however, that ATP is also synthesized in the cytosol, when the process is referred to as '**substrate-level phosphorylation**'. This occurs in two separate reactions of glycolysis catalysed by the enzymes glyceraldehyde 3-phosphate dehydrogenase and pyruvate kinase (see Chapter 6). For mammalian cells that lack mitochondria, such as the erythrocytes and the cells of the lens in the eye, the operation of glycolysis is the sole means of forming ATP.

In cells that possess them, the mitochondria perform the vital function of coupling the process of oxygen-dependent oxidation of molecules derived from food to

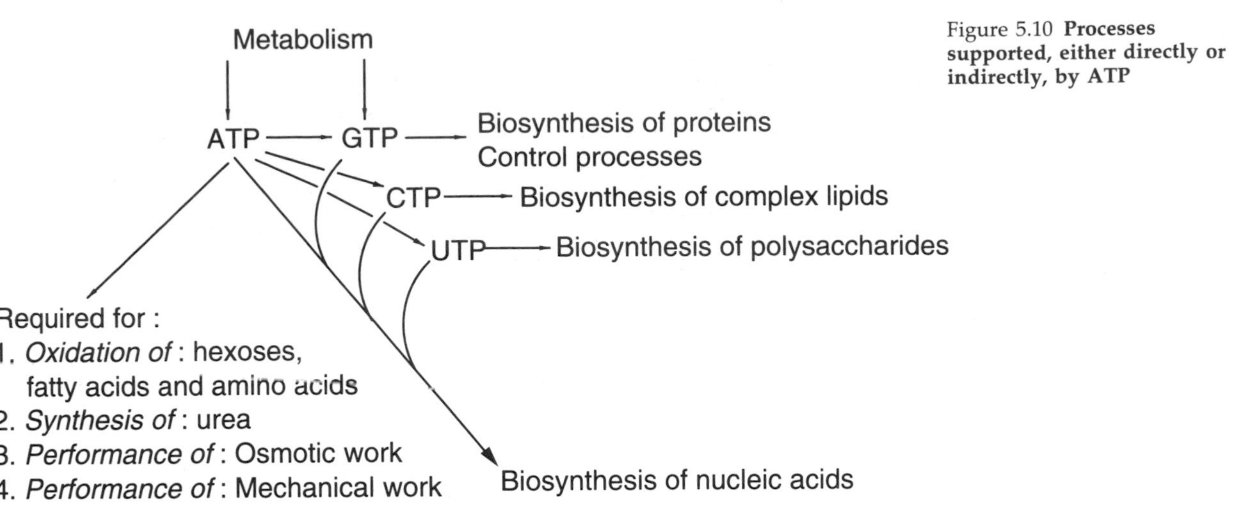

Figure 5.10 **Processes supported, either directly or indirectly, by ATP**

that of forming ATP and any interruption of this process leads rapidly to cellular damage.

Molecular oxygen and biological redox reactions

The importance of the utilization of oxygen by all animals was established in the eighteenth century by Joseph Priestley. Early biochemists assumed that any molecule that became oxidized in cells would do so by the addition of oxygen to it. Such reactions do occur (in the biosynthesis of steroid hormones, for example), but more typically, cellular oxidations involve the removal of pairs of hydrogen atoms from selected substrates in reactions catalysed by dehydrogenases. The action of dehydrogenases was first observed by Thunberg and Weiland, working in Germany in the 1930s. Dehydrogenation constitutes an oxidation because the hydrogen atom can be regarded as being composed of a proton (H^+) and an electron (e^-) and, in current usage, the loss of an electron constitutes oxidation. Conversely the gain of an electron is referred to as reduction. Because the processes of oxidation and reduction must go hand in hand (electrons must be transferred from one species to another), the reactions are referred to as **redox reactions**. A simple, but biologically important, redox reaction occurs during the interconversion of two forms of iron: Fe^{2+} (ferrous, FeII) and Fe^{3+} (ferric, FeIII). The iron part of the cytochromes found in the inner mitochondrial membrane undergoes such reversible redox reactions, the process being referred to as '**electron transfer**'. On occasion, electrons may be transferred associated with the negatively-charged hydride ion (H^-). Such a transfer occurs when the co-enzyme NAD^+ is reduced to NADH. It should be noted that this reaction requires the transfer of the equivalent of one proton and two electrons, and since two 'H' are removed, it follows that the second proton must be discharged into the medium. Reactions that result in the reduction of NAD^+ are of major importance in energy metabolism. One such reaction is catalysed by glyceraldehyde 3-phosphate dehydrogenase in the glycolytic pathway (see Chapter 6); a second reaction is catalysed by pyruvate dehydrogenase; in the tricarboxylic acid cycle the three dehydrogenases using isocitrate, 2-oxoglutarate or malate as substrates, all generate NADH (see Section 5.4). A final example of an NAD^+-requiring enzyme is encountered in β-oxidation, the reaction being catalysed by 3-hydroxyacylCoA dehydrogenase (see Section 5.4). It has been noted that, with the exception of the first of these NADH-generating reactions (which constitutes a special problem to be dealt with in Section 5.6), all of the enzymes are located in the mitochondrial matrix. As a consequence, the NADH generated in the course of each of these reactions is free to diffuse and react with the redox enzyme complex called NADH dehydrogenase located in the inner mitochondrial membrane. Thus NADH produced in the matrix is funnelled to the inner membrane for reoxidation.

Two other dehydrogenases of particular importance in energy metabolism, which have already been noted in Section 5.4, are succinate dehydrogenase and acylCoA dehydrogenase. They both use FAD as an hydrogen acceptor. FAD differs from NAD^+ in two important respects: it accepts two hydrogen atoms (i.e. two protons and two electrons) and it is a prosthetic group (i.e. it is covalently bound to the enzyme protein). Of the two enzymes mentioned above, succinate dehydrogenase catalyses a reaction in the tricarboxylic acid cycle whilst acylCoA dehydrogenase is found in the β-oxidation pathway. In view of the fact that FAD.2H is reoxidized by enzymes of the inner membrane, and yet is not free to diffuse from the surface of the enzymes, it is not surprising that these flavoproteins themselves are located on the inner membrane.

Organization of the electron-transport chain

It would be possible to use NADH or FAD.2H to reduce molecular oxygen to water directly with a large free energy change, but no practical way of harnessing the reaction to the formation of ATP has evolved. Instead, the ultimate reduction of molecular oxygen is carried out in a series of discrete redox steps, each catalysed by an enzyme complex, and which collectively make up the

electron transport chain. The chemical potential energy required to form ATP from ADP and Pi is generated as a result of the operation of each step. The reoxidation of NADH involves more such steps than that of FAD.2H and hence more ATP is produced per mol of NADH oxidized. Technically, NADH is said to have a greater redox potential (more negative) than FAD.2H.

As has been noted, the enzyme complexes catalysing the sequential electron transfer reactions are all located in the inner membrane, from which independent electron transfer complexes may be isolated. The four major complexes are given names that describe the overall reactions they catalyse and these are: NADH dehydrogenase (complex I), succinate dehydrogenase (complex II), ubiquinol dehydrogenase (complex III) and cytochrome oxidase (complex IV) :

Complex I (NADH dehydrogenase)

NADH + H^+ + CoQ ⇒ NAD^+ + CoQ.2H

Complex II (succinate dehydrogenase)

Succinate + CoQ ⇒ Fumarate + CoQ.2H

Complex III (ubiquinol dehydrogenase)

CoQ.2H +2(cytochrome c)⇒
$2H^+$ + CoQ + 2(reduced cytochrome c)

Complex IV (cytochrome oxidase)

2(reduced cytochrome c) + $1/2O_2$ + $2H^+$ ⇒ 2(cytochrome c) + H_2O

Ubiquinone (coenzyme Q; CoQ) is a lipid-soluble compound (Figure 5.11) and therefore freely diffusible within the inner membrane. It is capable of undergoing reversible redox reactions involving one or both oxygen atoms. Reduction of one oxygen atom yields a semiquinone, whereas reduction of both leads to the formation of ubiquinol. The cytochromes (of which cytochrome c is representative) are haem proteins, the single iron atom of which is capable of undergoing reversible, one electron, oxidation–reduction reactions.

The operation of these processes results in the formation of ATP from ADP and Pi.

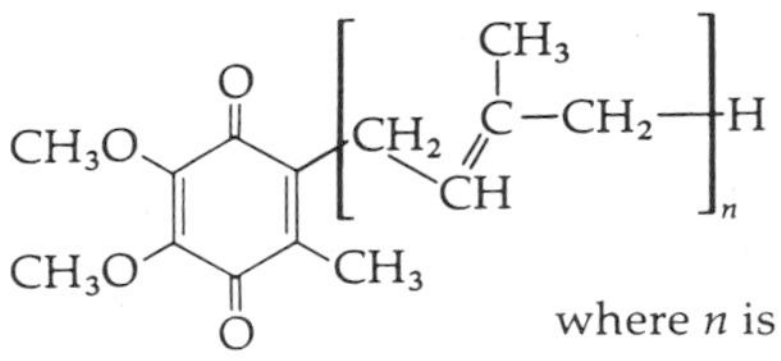

where n is frequently 10

Figure 5.11 **The structure of ubiquinone**

Inspection of the reactions catalysed by complexes I and II reveals that, in each case, their operation leads to the formation of fully-reduced ubiquinol. Thereafter, ubiquinol is a substrate for complex III, and the product of this reaction (reduced cytochrome c) is, in turn, a substrate for complex IV. It is possible, therefore, to write the reactions catalysed by the four complexes as a series of coupled reactions, with CoQ and cytochrome c acting as carriers between the complexes (Figures 5.12 and 5.13).

Closer analysis of the reactions catalysed by the complexes, shows that each may be further resolved into a series of coupled redox reactions. The NADH dehydrogenase stage involves an FMN-containing flavoprotein and a protein referred to as containing 'non-haem iron'. This iron is capable of redox reaction and is so named to indicate that the binding of the metal ion to the protein does not require a haem group. Deficits in the operation of complex I have been described in certain brain regions of patients suffering from **parkinsonism** (see Chapter 30). Complex II consists of a different non-haem iron protein, as well as the FAD-containing enzyme. There are two types of cytochrome in the ubiquinol dehydrogenase complex: b and c_1. Cytochrome oxidase has two related cytocromes, a and a_3 , and two further redox proteins containing copper (Cu_A and Cu_B).

Chemiosmosis and oxidative phosphorylation

Considering the types of carriers involved in the ordered series of reactions leading to the reduction of oxygen, it

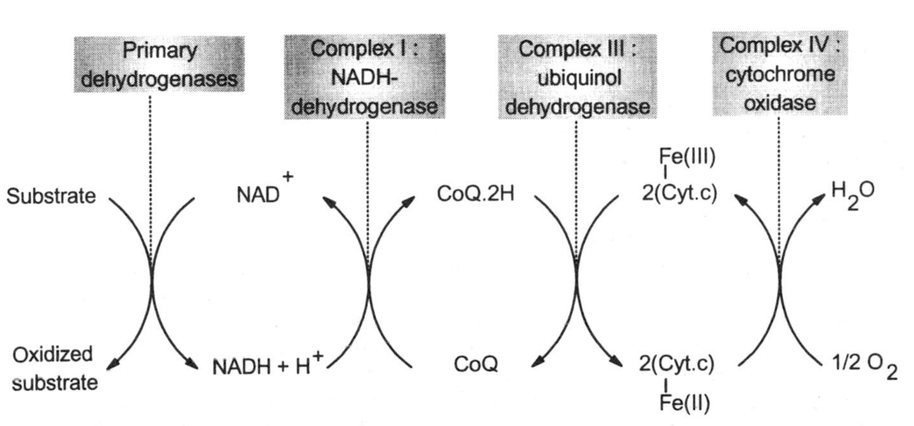

Figure 5.12 **Three electron-transfer complexes of the inner mitochondrial membrane act sequentially to reduce molecular oxygen to water at the expense of NADH**

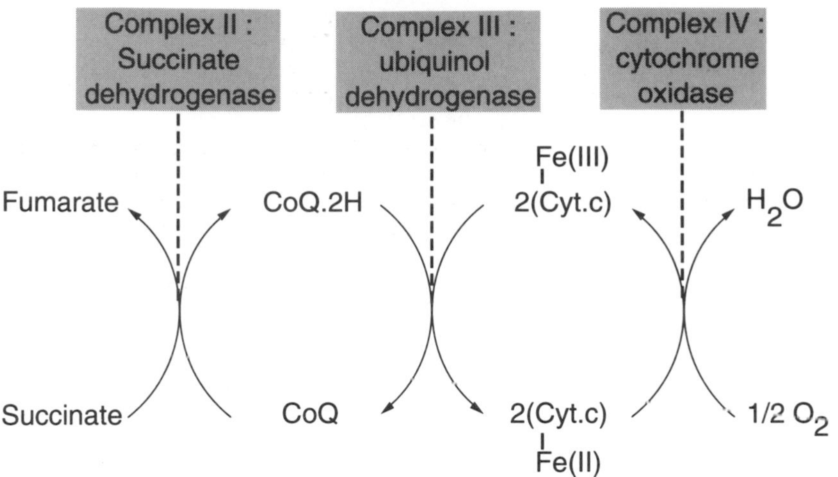

Figure 5.13 **The succinate dehydrogenase complex of the inner mitochondrial membrane replaces the NADH complex when molecular oxygen is reduced to water at the expense of succinate**

Table 5.2 **Carriers of the inner mitochondrial membrane**

Carry 2[H] $\equiv 2H^+ + 2e^-$	
1. NADH + H^+	Mobile coenzymes
2. Reduced coenzyme Q	
3. $FMNH_2$	Enzyme prosthetic groups
4. $FADH_2$	
Carry e^-	
1. Haem proteins: cytochromes b, c, c_1, a and a_3	
2. Non-haem iron proteins	

may be recognised that they fall into one of two classes: those that carry 2'H', i.e. 2 protons and 2 electrons, and those that carry one electron (Table 5.2). The scheme shown in Figure 5.14 envisages the reduction of the pure electron carriers by carriers of 'H'; it follows, therefore, that when this happens protons must be discharged. This, insight together with speculations about the reason for the location of these carriers in the inner membrane, provided the starting point for the English biochemist Peter Mitchell to formulate his '**chemiosmotic**' hypothesis of how the reduction of oxygen is coupled to

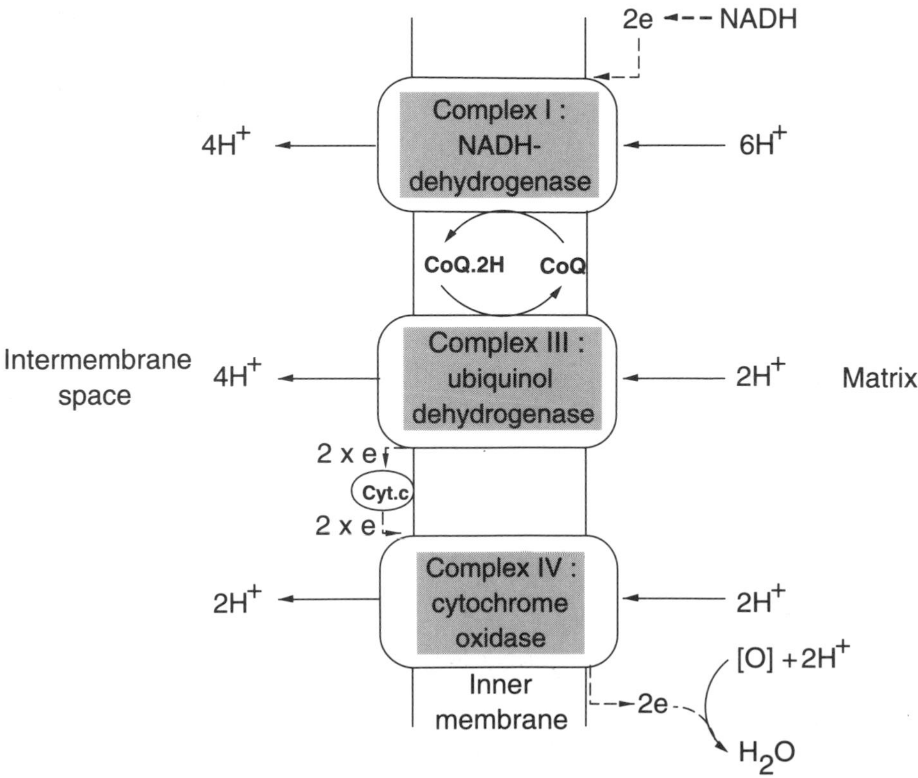

Figure 5.14 **For every oxygen equivalent reduced using matrix NADH, 10 protons are 'pumped' across the inner mitochondrial membrane**

the conversion of ADP and Pi into ATP in oxidative phosphorylation.

It is well known that chemical energy can be stored in a concentration gradient of any solute (sucrose, Na^+ or H^+, for example). With this in mind, Mitchell postulated that as electrons are transferred to oxygen in the inner membrane, the protons generated when 'H' carriers reduce electron carriers are discharged vectorially into the intermembrane space. This would have the effect of setting up both a potential difference and a concentration gradient across the inner membrane, i.e. this is a chemiosmotic process. It should be noted that such an energy gradient can, in principle, be used to perform types of work other than forming ATP. Thus, for example, mitochondria are able to use the energy to accumulate Ca^{2+} against their concentration gradient. This constitutes an important means for controlling the activity of intramitochondrial Ca^{2+}-dependent enzymes.

Current evidence strongly supports Mitchell's ideas, and it is now generally accepted that the transport of $4H^+$ occurs as a result of the operation of each of complexes I and III, while $2H^+$ are discharged at complex IV. It should be pointed out, however, that the switch from 'H' carriers to e carriers with the consequent discharge of H^+ cannot account for all the protons transported. This is seen particularly clearly at complex IV, where all of the carriers are electron transporters. Possible mechanisms for the movement of these protons are beyond the scope of the present text. The outcome of this process, however, is that for the oxidation of one molecule of NADH, produced in the matrix, a total of $10H^+$ are 'pumped' into the intermembrane space. The corresponding value for the FAD-dependent oxidations is $6H^+$ since the 2'H' generated at complex II (and also those generated by acylCoA dehydrogenase) enter the chain at complex III (Figure 5.15).

Of course, the model predicts a particular orientation of the proteins of the inner membrane, one that would favour the vectorial movements, and experiments have confirmed these predictions. The arrangement is now known to be as shown in Figure 5.14. In this, CoQ is shown as spanning the membrane (more accurately, it is free to diffuse across the membrane carrying 2 'H' each time), whereas cytochrome c is depicted as being loosely attached to the outer face of the inner membrane (consistent with its ready loss during incautious isolation of mitochondria).

The formation of ATP from ADP and P_i

The explanation of the mechanism whereby the energy conserved in the electrochemical gradient is used to form ATP depends on the fact that the inner membrane is generally impermeable to protons tending to move back into the matrix space as their concentration dictates. Instead, the protons are specifically conducted back by the F_o–F_1 protein complex of the inner membrane (Figure 5.16). In isolation, F_1 proteins act as ATPases, i.e. they catalyse the hydrolysis of ATP. However when F_1 is functionally coupled to F_0, the latter channels protons to the F_1 proteins in such a way that conformational changes reverse their ATPase activity and the synthesis

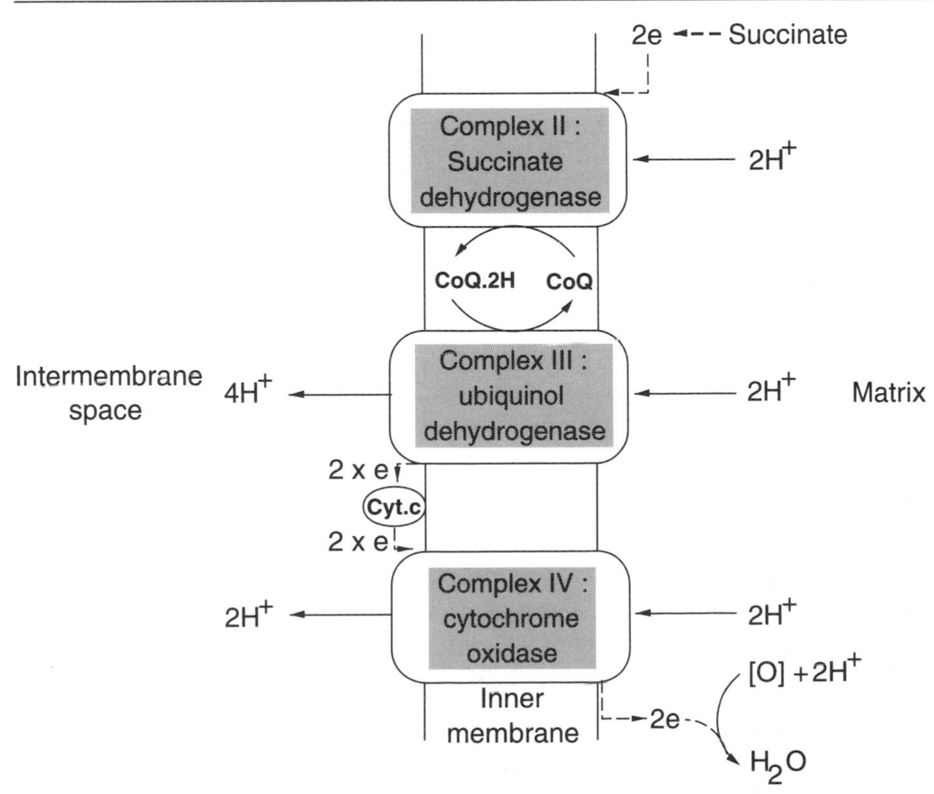

Figure 5.15 **For every oxygen equivalent reduced using succinate, 6 protons are 'pumped' across the inner mitochondrial membrane**

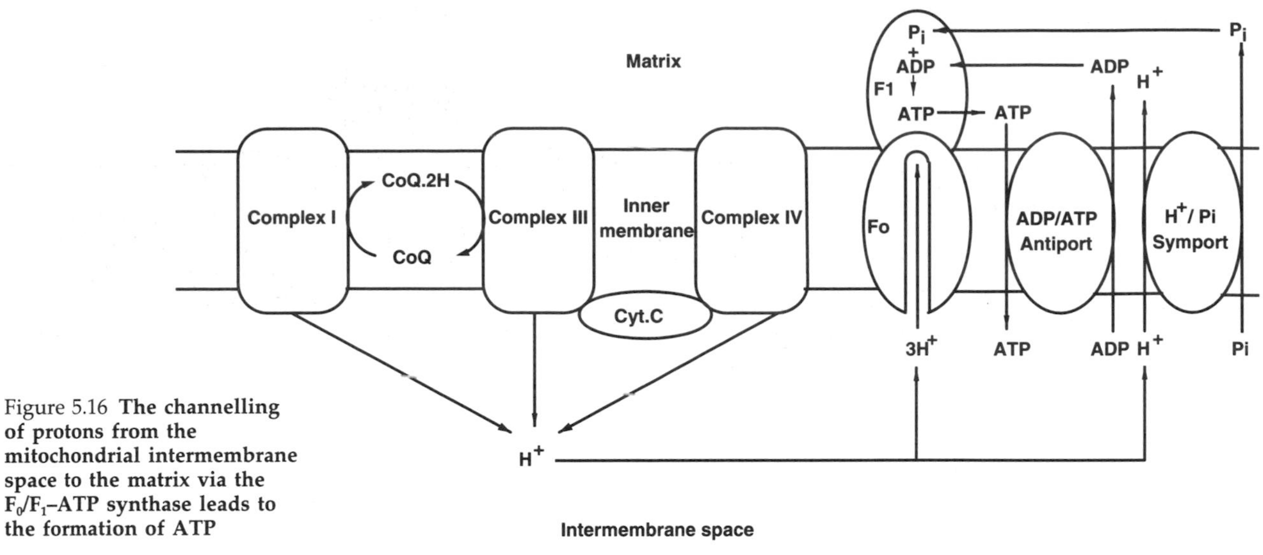

Figure 5.16 **The channelling of protons from the mitochondrial intermembrane space to the matrix via the F_0/F_1–ATP synthase leads to the formation of ATP**

of ATP is favoured. In fact, the F_1 proteins have three catalytic domains, each capable of forming ATP with, at any one time, one being in the process of binding ADP and P_i, one forming the acid anhydride link between these two substrates, and one discharging the product. The energy is actually expended in the discharge of the very tightly bound ATP from the catalytic domain.

It should be noted that both of the substrates for ATP synthesis must be transported into the matrix. An antiport transporter exchanges ATP for ADP while P_i is co-transported with one proton. Since ATP synthesis requires the transport of 3 protons via F_0 to F_1, it follows that a total of 4 mol of H^+ must be transported for 1 mol of ATP to be produced. In turn this means that the oxidation of 1 mol of matrix NADH at the expense of oxygen leads to the formation of 2.5 mol of ATP (driven by $10H^+$), while the oxidation of 1 mol succinate leads to the formation of 1.5 mol of ATP (driven by $6H^+$). In the older literature the values quoted are 3 and 2 for the oxidation of NADH and succinate respectively, but these values were derived by assuming that electron transfer generated 'high-energy' chemical intermediates that were then employed to form ATP (as happens in substrate-level phosphorylation). Such chemical reactions, in turn, required whole number stoichiometry. It may provide some consolation to previous generations of medical and dental students, who have used oxygen electrodes to determine the so-called P/O ratio (mol of ADP phosphorylated divided by mol of oxygen atoms reduced to water), that the non-integral values they obtained for these were closer to the real value than they were given credit for at the time! At the same time it should be appreciated that the exact numbers of protons transported out of mitochondria during electron transfer is very difficult to determine experimentally and the values given here simply represent the best estimates currently available (although further data are not likely to change these accepted values greatly).

Uncoupling and inhibition of electron transport

The proton-dependent formation of ATP is tightly 'coupled' to the transfer of electrons and vice versa. This means that the rate of respiration of mitochondria is determined largely by the availability of ADP. Thus, if the intracellular concentration of ADP is low, then the rate of respiration is correspondingly low. The process whereby the rate of respiration is regulated by the supply of ADP is referred to as '**respiratory control**'. The tightness of the coupling is not absolute and in resting hepatocytes, for example, 33% of the oxygen consumption is due to the 'leaking' of protons, but this value declines as the cells become more metabolically active. In some tissues the tightness of coupling is further relaxed, for example in **brown adipose tissue** the tissue-specific protein **thermogenin** allows protons to 'short-circuit' the F_0–F_1 protein complex. When this happens the energy is dissipated as heat. This is an important mechanism for maintaining body temperature in hibernating animals. Several compounds are able to promote the collapse of the proton gradient, and these are referred to as '**uncoupling**' agents because, in their presence, respiration is no longer coupled to, and therefore is not controlled by, phosphorylation. One such compound is **2,4-dinitrophenol (DNP)** which is effective because the pK_a value of its phenolic hydroxyl group is such that the 2,4-dinitrophenate anion can accept a proton in the intermembrane space. The resulting weak acid is sufficiently lipid soluble to be able to diffuse across the inner membrane into the matrix and there discharge the proton at the higher pH value (Figure 5.17).

Several compounds are able to inhibit the electron transfer process itself and these compounds prove to be extremely toxic. One of the best known inhibitors is **cyanide**, which binds avidly to the free ligand position of the iron in all the cytochromes, with cytochrome

Figure 5.17 **Sites of action of some compounds that interfere with oxidative phosphorylation**

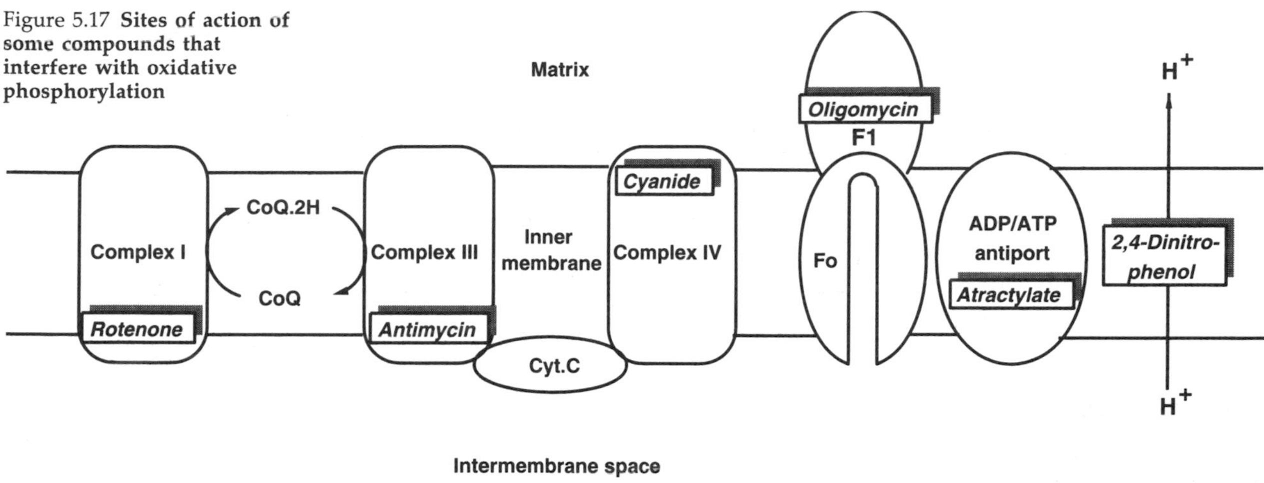

oxidase being the most vulnerable. Molecules such as **hydrogen sulphide** and **carbon monoxide** will also bind to the cytochromes to inhibit electron transport. Some naturally-occurring toxins such as **rotenone**, a toxic extract of plants used on arrows by indigenous South Americans, block electron transport at complex I, whilst the antibiotic **antimycin A** blocks at complex II. The net effect in each case is to block both oxygen consumption and the formation of ATP. This effect is also achieved by the compounds **oligomycin**, which blocks the phosphorylation of ADP by binding to the active sites of F_1, and **atractylate**, which prevents the exchange of ATP for ADP across the inner membrane. The sites of action of these compounds are shown in Figure 5.17.

5.6 Relation of mitochondrial oxidations to cellular metabolism

A simplified representation of the operation of the mitochondria in conserving energy is shown in Figure 5.18. This diagram addresses a problem only briefly touched on in the previous section, for it can be seen that many substrates that are destined to undergo oxidation in the matrix derive from the cytosol. This is so for fatty acids (strictly acylCoA derivatives) and also the NADH generated in glycolysis by the reaction catalysed by glyceraldehyde-3-phosphate dehydrogenase. The inner mitochondrial membrane is impermeable to these molecules, and to many others. As a result, in each case, a carrier mechanism is required to transport the molecules concerned from the intermembrane space to the matrix. For example, a carrier system for citrate plays a central role in the effective transport out of mitochondrial matrix of acetylCoA required for the biosynthesis of fatty acids and also of acetylcholine.

The carnitine–acylcarnitine exchange carrier

Fatty acids are important substrates for mitochondrial respiration in many tissues, for example the heart. When they enter cells, fatty acids are converted into acylCoA derivatives in the cytosol. However, the enzymes of β-oxidation are situated in the mitochondria and coenzyme A derivatives are not able to cross the inner membrane. The solution to this problem is provided by the operation of the carnitine–acylcarnitine exchange carrier. On the intermembrane and the matrix sides of the inner membrane are located two enzymes designated carnitine acyltransferases I and II (Figure 5.19). These interact directly with an integral membrane carnitine carrier protein. In the transport process, the first enzyme catalyses the formation of acylcarnitine from the acylCoA and the base carnitine. The acylcarnitine is transported by the transport protein and the original reaction is reversed by the second carnitine acyltransferase to reform acylCoA in the matrix, where it may be oxidized. The transport protein then carries the carnitine formed back out of the mitochondrion in exchange for acylcarnitine.

Effective transport of NADH into the mitochondria

A problem already highlighted is the reoxidation of NADH generated in the cytosol. The most efficient way of achieving this in terms of energy conservation would be to transfer NADH into the matrix, but no carrier exists for this purpose. Instead, two indirect mechanisms operate.

Aspartate–malate shuttle

In the first of the two mechanisms, which predominates in the liver, cytosolic NADH is used to reduce oxaloacetate to malate in a reaction catalysed by malate dehydrogenase. The malate is transported into the mitochondrion in exchange for 2-oxoglutarate. In the matrix, the malate dehydrogenase-catalysed reaction occurs in reverse, with the production of NADH and oxaloacetate. The process would be simple if oxaloacetate had a transporter, but none exists in mitochondria. Instead, the molecule undergoes an aminotransferase reaction with glutamate, catalysed by aspartate aminotransferase, to yield 2-oxoglutarate (that

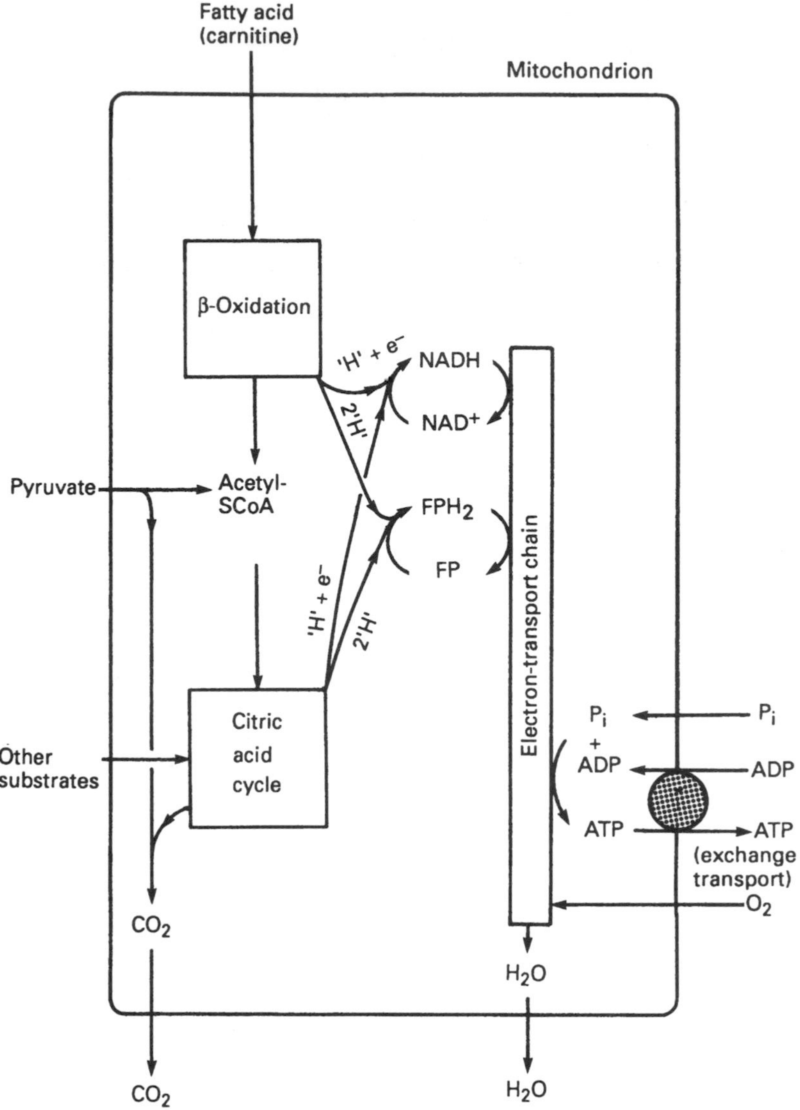

Figure 5.18 **Schematic representation of the role of the mitochondria in energy conservation**

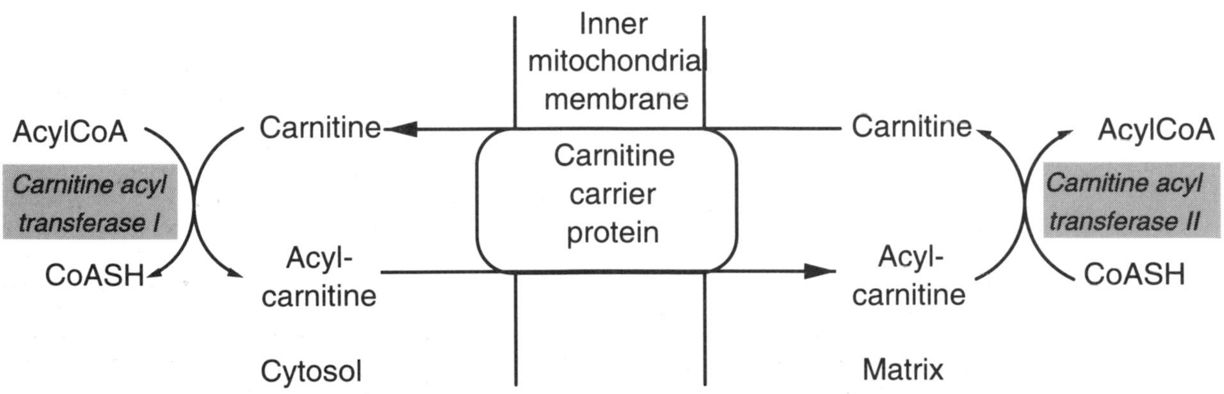

Figure 5.19 **The operation of the 'carnitine shuttle' for the transport of fatty acylSCoA into mitochondria**

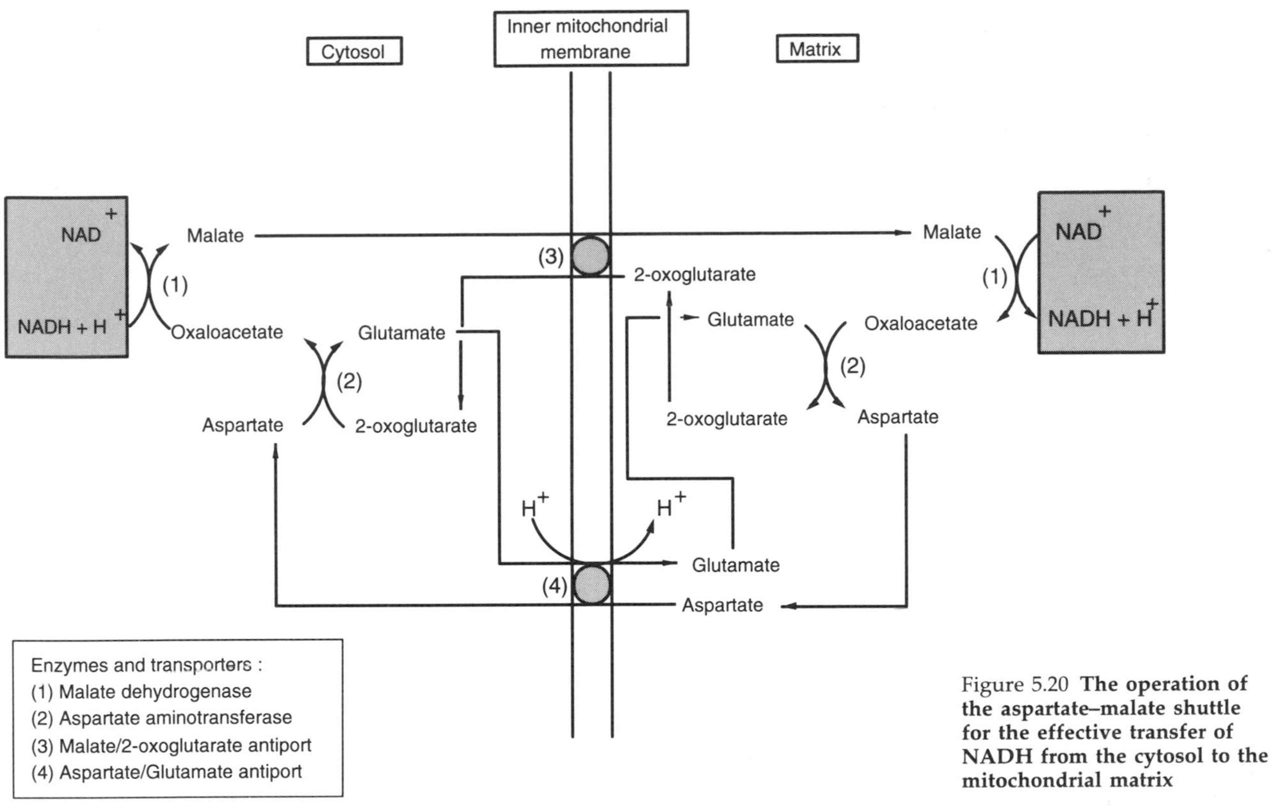

Figure 5.20 **The operation of the aspartate–malate shuttle for the effective transfer of NADH from the cytosol to the mitochondrial matrix**

can exchange for malate) and the amino acid aspartate. A transporter then exchanges the two amino acids aspartate and glutamate, and a reversal of the aspartate aminotransferase reaction in the cytosol gives rise to oxaloacetate and glutamate, thereby completing the 'shuttle-like' process (Figure 5.20). It should be noted that, since the NADH/NAD^+ ratio is higher in the matrix than in the cytosol, this process is thermodynamically unfavourable. That the process operates as indicated is ensured by the movement of protons down their concentration gradient when glutamate is exchanged for aspartate

The glycerophosphate shuttle

The second mechanism is more important in muscle, especially the flight muscles of insects. In this, cytosolic NADH reduces dihydroxyacetone phosphate in a reaction catalysed by glycerol-3-phosphate dehydrogenase. The product of this reaction, glycerol 3-phosphate diffuses into the intermembrane space of the mitochondrion to be reoxidized by an FAD-containing version of glycerol 3-phosphate dehydrogenase located in the inner membrane, but orientated towards the outside (Figure 5.21). The reducing equivalents of the FAD are then transferred to ubiquinone, as happens with succinate dehydrogenase. Evidently, this system differs from the others dealt with in this section, in that transport into the matrix does not occur (no carrier for glycerol 3-phosphate has been identified).

Transport of citrate from the mitochondria: relation to fatty acid biosynthesis and the formation of acetylcholine

In conditions of excess availability of substrates for oxidation and limited energy requirement, there is a need to store fuels to meet later energy demands. Conversion of other fuels to fats is one of the most effective ways of providing for the long-term storage of energy sources. The formation of fatty acids in humans occurs largely in the liver, but the fatty acids are ultimately stored as triacylglycerols in adipose tissue (see Chapter 11).

The operation of this system depends critically on the transport out of the mitochondrial matrix of citrate. Figure 5.22 depicts the circumstances in which excess glucose (in the fed state) is being converted into fatty acids. Metabolism of glucose via the glycolytic pathway gives rise to pyruvate which is transported into mitochondria in exchange for hydroxyl ions. Once in the mitochondrial matrix the pyruvate may converted into either acetylCoA (reaction catalysed by pyruvate dehydrogenase) or into oxaloacetate (reaction catalysed by pyruvate carboxylase). These two products interact, in the initial step in the tricarboxylic acid cycle, to give citrate. The citrate is then transported out of the matrix in exchange for malate (or other dicarboxylic acids) and, in the cytosol citrate is cleaved by citrate lyase to re-form oxaloacetate and acetylCoA. The latter may be reduced to malate in a reaction catalysed by malate dehydrogenase and requiring NADH as the coenzyme.

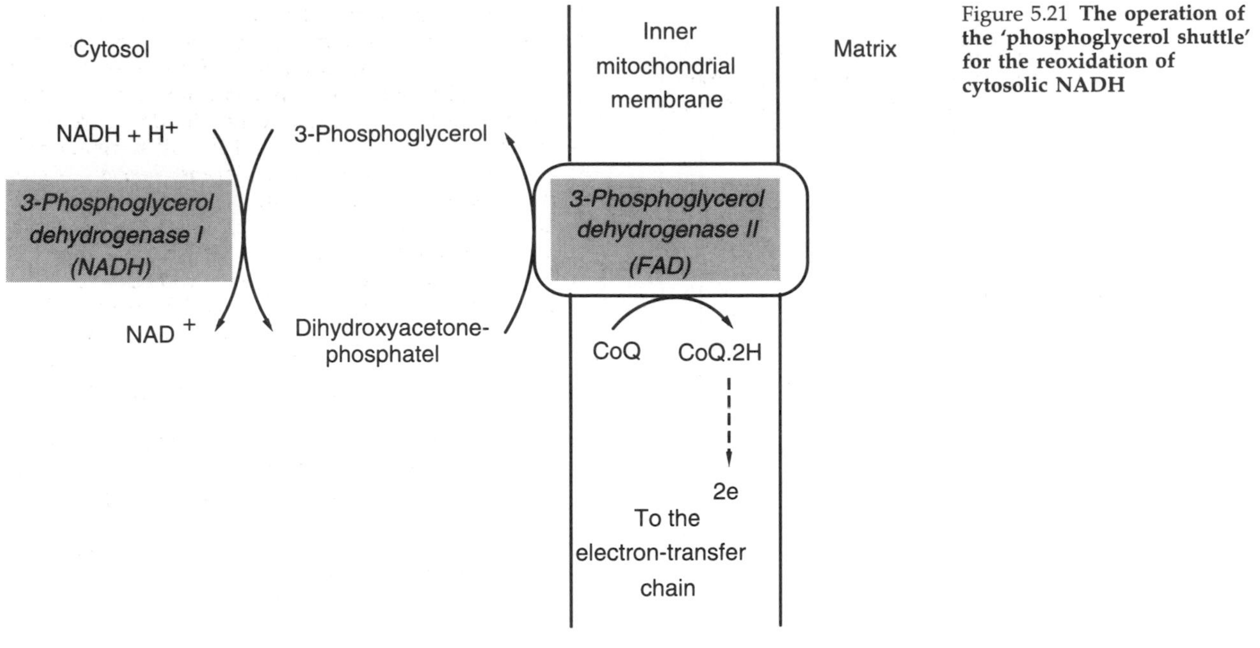

Figure 5.21 **The operation of the 'phosphoglycerol shuttle' for the reoxidation of cytosolic NADH**

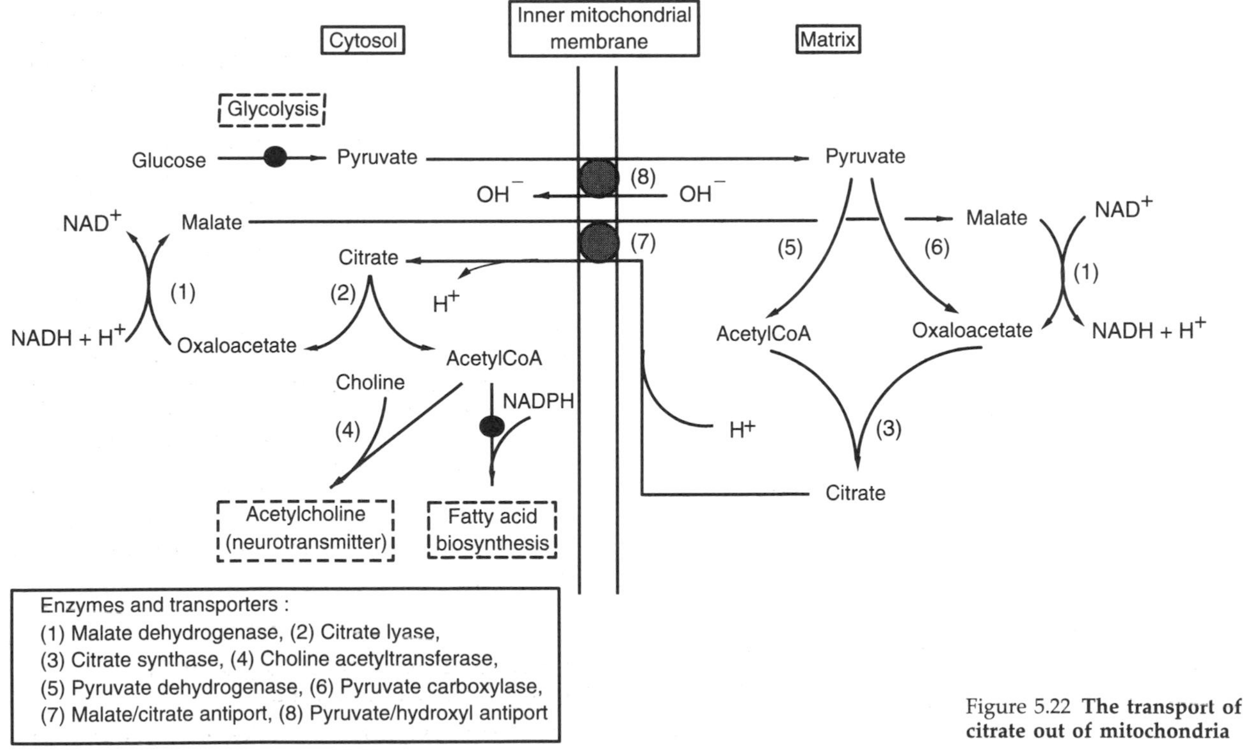

Figure 5.22 **The transport of citrate out of mitochondria**

The malate so formed may exchange with citrate (as is shown in Figure 5.22) or it may undergo oxidative decarboxylation in an $NADP^+$-dependent reaction catalysed by malate dehydrogenase (decarboxylating). For clarity this reaction is omitted from Figure 5.22, but it should be noted that its occurrence can account for the supply of up to 60% of the NADPH required for fatty acid biosynthesis. The other requirement for the biosynthesis of fatty acids is acetylCoA, and of course this is generated by the cleavage of citrate, catalysed by citrate lyase.

In neuronal tissue, the acetylCoA transported from the mitochondrion as citrate can be used for the synthesis of acetylcholine (see Chapter 30).

Summary

It will be appreciated that mitochondrial transport systems play an important part in metabolism, moving some metabolites into the matrix for oxidation and other metabolites into the cytosol where they may be used for biosynthetic purposes.

5.7 Replication of mitochondria

It is very doubtful if mitochondria ever appear *de novo* in a cell. They are believed to arise from the maternal zygote, and during cell division become partitioned between the daughter cells. However, if this process was all that happened, the number of mitochondria per cell would decrease from generation to generation. What appears to happen is that mitochondria in dividing cells (and probably those in other cells as well) accumulate more of the components of which they are made by accretion, thereby increasing in size, and eventually they undergo binary fission.

As has been noted, mitochondria contain a special form of DNA (mtDNA) which resembles that found in bacteria in that it consists of a closed circular, double-stranded molecule. A full set of enzymes exists in mitochondria for protein synthesis. These include an RNA polymerase and amino acid activating enzymes, plus the necessary transfer RNA molecules and ribosomes. This means that mitochondria are able to synthesize proteins encoded in their own DNA. The information available in the mtDNA is, however, very limited, only being sufficient to code for about a dozen proteins. Consequently, most of the mitochondrial proteins are encoded in nuclear DNA. Such proteins are synthesized on ribosomes that are free in the cytosol, their subsequent uptake into mitochondria being dependent on signals contained within their sequence.

5.8 Mitochondrial diseases

Mutations to mitochondrial DNA

In the past two decades, the occurrence of a range of rare, late-onset, inherited syndromes with unusual genetics has been recognized. These are tissue-specific, degenerative diseases characterized by mutations to the mitochondrial genome. In fact, the first example of a mitochondrial disease was described in 1953 by Boris Ephrussi who noted the occurrence of a '*petite*' mutation in yeast in which the cells relied exclusively on glycolysis for ATP production. This was due to a total lack of mitochondrial DNA. Syndromes in humans that involve mutations to mtDNA include **Laber's hereditary optic neuropathy (LHON), myoclonic epilepsy, ragged-red fibre disease** and the **Kearns–Sayre syndrome**.

It is a characteristic feature of mtDNA that it has a high rate of mutation and this is probably associated with inadequate repair mechanisms. As is the case with all genetic diseases, mutations in the germ line can give rise to familial diseases, whereas those that arise during development or in somatic tissue result in sporadic diseases. In practice, many inherited mitochondrial diseases emerge only as the result of the occurrence of further somatic mutations.

New mutations to mtDNA will usually affect only some of the mitochondria in a cell; consequently, affected cells will contain a mixture of mutation-bearing and normal mitochondria. This condition is referred to as **heteroplasmy**. As these cells divide there is a tendency for the daughter cells to have all normal or all mutation-bearing mitochondria (**homoplasmy**). Once inherited, mutant mtDNA continues to undergo somatic mutations which may affect the genes that encode the proteins of oxidative phosphorylation, for example. As these mutations accumulate, the capacity of the mitochondria bearing them to form ATP declines. This is especially the case for cells in tissues that place heavy reliance on oxidative metabolism: brain, heart, kidney and endocrine glands. As a consequence degeneration is seen in these tissues.

LHON has been recognized for some time as a maternally-inherited, adult-onset blindness associated with death of the metabolically very active optic nerve. The disease shows genetic heterogeneity in that several different point-mutations to mtDNA are observed. One such mutation affects a gene that encodes one of the proteins of the cytochrome oxidase complex.

A recent and unexpected finding is of a maternally-transmitted form of **diabetes mellitus**. In patients with this form of the disease, who often also suffer from deafness, there is a mutation to the mitochondrial gene that encodes the transfer RNA (tRNA) for the amino acid leucine. About 1–3% of patients with diabetes mellitus have this single mutation. Some patients with the same mutation present with so-called **MELAS** syndrome (**Mitochondrial Encephalomyelopathy, Lactic Acidosis and Stroke**), a neuromuscular disease, which may be accompanied by diabetes and deafness. It is suggested that the different phenotypic expressions, maternally-transmitted diabetes and deafness or MELAS, depend on the degree of heteroplasmy occurring in the tissues.

It should be borne in mind that it is perfectly possible for individuals to accumulate mutations to genes encoding different mitochondrial proteins. In addition, some of the mutations may affect nuclear genes that encode mitochondrial proteins. This can result in complex genetic inheritance patterns.

The reason for the late onset of the mitochondrial diseases is the same as that put forward for cancer: it takes time for sufficient mutations to arise to cause overt dysfunction (see Chapter 37).

Drug-induced diseases involving inhibition of mitochondrial DNA replication

Recently there have been disturbing reports of patients treated for long periods with antiviral nucleoside analogues (see Chapter 39) developing symptoms that are very similar to those associated with mutations to the mitochondrial genome. The clearest example of this has been reported in AIDS patients treated for long periods with zidovudine (AZT). In this group of patients there is a 20% prevalence of mitochondrial myopathy. It seems likely that the reported symptoms arise as a result of the inhibition of mitochondrial DNA polymerase (polγ) by the drug.

Chapter 6

The cytosol

6.1 Composition of the cytosol

The cytosol is the fluid compartment which permeates the whole internal environment of the cell and generally represents about 50–60% of the total cell volume. As the cytosol is in contact with all cell organelles, it is clearly an important vehicle for the transport of metabolites from one organelle to another. Conventional histology shows the cytosol as a uniformly stained background. In electron microscopy it appears clear.

The cytosol is not a simple aqueous solution but a viscous gel with a protein concentration of 200 g/l. The proteins are mainly enzymes involved in intermediary metabolism and include the enzymes of the glycolytic and pentose phosphate pathways, glycogen synthesis and degradation and fatty acid synthesis. There are also transport proteins such as carriers of metals and steroid hormones. Metabolic fuel is stored in the cytosol in the form of glycogen granules in liver (up to 10% of wet weight) and muscle cells and in the form of triacylglycerol droplets in adipocytes.

The cytoskeleton

The cytosol of eukaryotic cells contains a network of fibres known as the cytoskeleton. It is important in maintaining cell shape and mobility and provides points of anchorage for the cell organelles. It comprises three types of filaments: **microfilaments** containing **actin**, **microtubules** containing **tubulin** and **intermediate filaments**. The latter form rigid fibres and their function is at present under investigation. The components of the microfilaments and microtubules exist in two interchangeable forms, either as insoluble fibres or as soluble globular structures. Actin microfilaments maintain cell movement by interacting with myosin, in an ATP-dependent process, resembling skeletal muscle contraction. Microtubules allow movement of organelles and metabolites through the cytosol.

6.2 Functions of the cytosol: intermediary metabolism

Carbohydrate metabolism

Glycolysis

The enzymes of the glycolytic pathway are located in the cytosol (Figure 6.1). This pathway converts glucose into pyruvate and is initiated by the ATP-dependent phosphorylation of glucose in the 6-position. In most tissues the enzyme involved is hexokinase which has a low specificity and a high affinity for glucose, allowing tissues, such as brain and erythrocyte, which depend on glucose for their energy supply, to utilize glucose even when it is present at low concentrations. In the liver, this reaction is catalysed by the more specific glucokinase (hexokinase D) which catalyses phosphorylation of glucose at a half maximal rate only when the concentration of the monosaccharide is about 5.0 mM. As a consequence, the rate of phosphorylation of glucose in

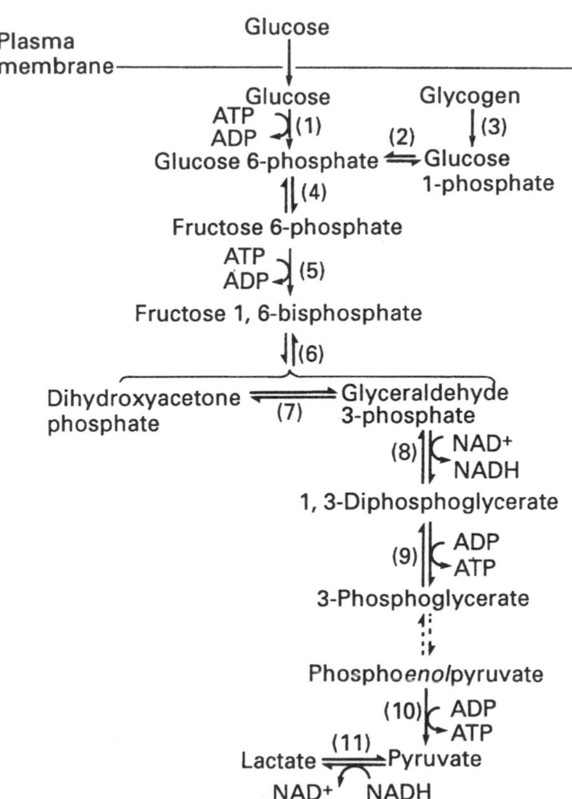

Figure 6.1 **The major steps in glycolysis**. The enzymes involved are (1) hexokinase or glucokinase; (2) phosphoglucomutase; (3) glycogen phosphorylase; (4) phosphohexose isomerase; (5) phosphofructokinase; (6) aldolase; (7) triose phosphate isomerase; (8) glyceraldehyde 3-phosphate dehydrogenase; (9) phosphoglycerate kinase; (10) pyruvate kinase; (11) lactate dehydrogenase

the liver can vary considerably in the physiological range of circulating concentrations of the sugar. One result of this is that the liver, which can act as a supplier of glucose for other tissues, will not itself utilize glucose when its availability is limited.

Glucose 6-phosphate is converted to fructose 6-phosphate and phosphorylated for a second time, by phosphofructokinase, using ATP. The resulting fructose 1,6-bisphosphate is split into two inter-convertible triose phosphates, glyceraldehyde 3-phosphate and dihydroxyacetone phosphate. A complex reaction then follows, catalysed by glyceraldehyde 3-phosphate dehydrogenase, in which glyceraldehyde 3-phosphate is oxidized, using NAD^+ as a coenzyme and inorganic phosphate as a second substrate, to form 1,3-bisphosphoglycerate and NADH. The former molecule has a high phosphate transfer potential by virtue of the fact that the phosphate at the 1-position is in a mixed-acid anhydride linkage with the carboxylate group. It can therefore be transferred to ADP to yield ATP (two molecules of ATP per molecule of glucose), an example of substrate-level phosphorylation. Subsequent reactions lead to the formation of phospho*enol*pyruvate, the phosphate of which is also readily transferable to ADP to yield ATP (another two molecules per glucose molecule). The enzyme involved is pyruvate kinase. The net energy yield in glycolysis starting with one molecule of glucose is, therefore, two molecules of ATP. The NADH generated has to be converted back to NAD^+ or the process will come to a halt. NADH can be reoxidized in the mitochondria via the electron transport chain using molecular oxygen. The reducing power of NADH can be transferred to the mitochondria by 'shuttle' mechanisms, such as the malate–aspartate and the glycerol phosphate shuttles (see Chapter 5). The reoxidation of NADH then depends on the rate of transfer of reduced intermediates into the mitochondria and the supply of oxygen. If the rate of glycolysis is substantially increased, or if the supply of oxygen is compromised, NADH can be re-oxidized in the cytosol. This happens in tissues such as rapidly contracting skeletal muscle and ischaemic tissues, which receive inadequate amounts of oxygen, and also in the erythrocyte, which lacks mitochondria. NADH and pyruvate act as substrates for the enzyme lactate dehydrogenase in a reaction which produces NAD^+ and lactate.

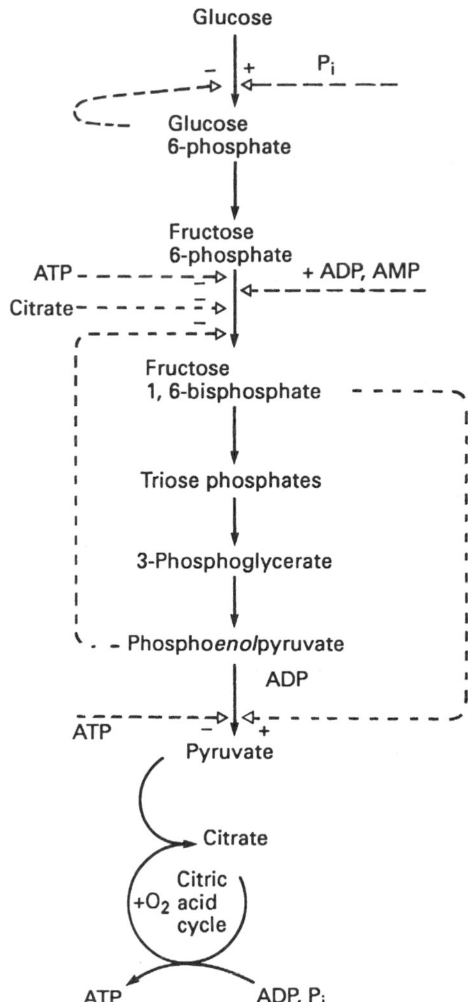

Figure 6.2 **Regulation of glycolysis.** Sites of control indicated by open arrow heads

Regulation of glycolysis

Glycolysis is controlled at several sites (Figure 6.2), and the mechanism depends on the tissue involved. In brain, skeletal muscle and erythrocytes, glycolysis is a catabolic, energy-producing pathway, whereas in the liver it is part of an overall anabolic process, the synthesis of fatty acids from glucose. The conditions under which glycolysis assumes importance in the liver (the fed state) are clearly different from those in which glycolysis is prevalent in muscle (exercise), and it would be surprising if the regulating factors were the same in both tissues. An important site of control is the enzyme phosphofructokinase which catalyses the conversion of fructose 6-phosphate to fructose 1,6-bisphosphate. In the brain, muscle and erythrocytes which use glucose as a source of energy, phosphofructokinase is controlled by molecules which reflect the energy state of the cell. A high ATP/AMP ratio inhibits phosphofructokinase. Phosphofructokinase inhibition results in the accumulation of glucose 6-phosphate which in turn inhibits hexokinase so decreasing the rate of glycolysis. The citric acid cycle is also inhibited when ATP/AMP ratios are high. This results in increased citrate levels which also inhibit phosphofructokinase. High ADP or AMP levels, indicating a need for energy, activate phosphofructokinase. In the liver, phosphofructokinase activity is ultimately controlled by the availability of glucose. Glucagon, released when blood glucose is low, and acting via cAMP-dependent protein kinases, inhibits phosphofructokinase by decreasing the level of the main activator of the enzyme, fructose 2,6-*b*isphosphate. Pyruvate kinase is also inhibited by glucagon via phosphorylation. This means that when glucose availability is low, as in starvation or when consuming a diet high in protein and low in carbohydrate, the liver does not utilize glucose itself and can act as a glucose supplier for other tissues.

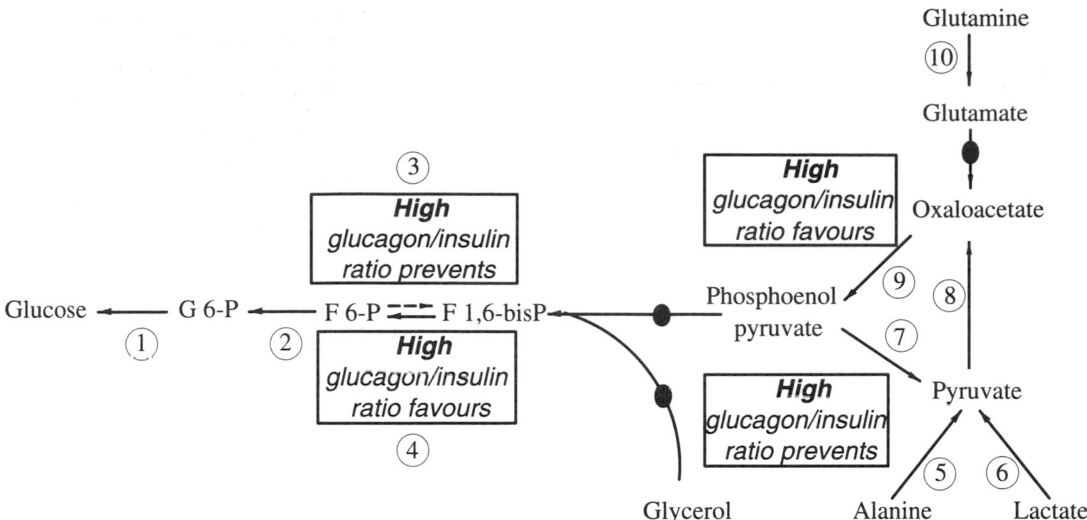

Figure 6.3 **The major precursors for gluconeogenesis and the main points at which glucagon and insulin regulate the process.** (1) Glucose 6-phosphatase; (2) phosphoglucomutase; (3) phosphofructokinase; (4) fructose 1, 6-bisphosphate phosphatase; (5) aminotransferase; (6) lactate dehydrogenase; (7) pyruvate kinase; (8) pyruvate carboxylase; (9) phospho*enol*pyruvate carboxykinase; (10) glutaminase; —●→, more than one enzyme reaction involved

Gluconeogenesis

Glycolysis is effectively reversed and glucose is produced in a process known as gluconeogenesis (Figure 6.3). Starting materials are lactate, released by muscle and erythrocytes, glycerol, released from adipose tissue following lipolysis, and glucogenic amino acids, particularly alanine and glutamine. Lactate and alanine are converted into pyruvate. Pyruvate is diverted to glucose formation instead of acetyl coenzyme A (acetylCoA), as the high glucagon/low insulin levels inhibit pyruvate dehydrogenase. Pyruvate is carboxylated in the mitochondria by the action of pyruvate carboxylase to form oxaloacetate which is reduced to malate in a reaction catalysed by malate dehydrogenase. Following its transport to the cytosol, malate is reoxidized to oxaloacetate. The action of pyruvate kinase cannot be reversed and phospho*enol*pyruvate is formed from the oxaloacetate in a reaction catalysed by phosphoenolpyruvate carboxykinase. The metabolic steps which follow are those of glycolysis in reverse except for two reactions: fructose 1,6-bisphosphate is dephosphorylated to fructose 6-phosphate by fructose 1,6-bisphosphatase, and glucose 6-phosphate is dephosphorylated by glucose 6-phosphatase which is not found free in the cytosol but bound to endoplasmic reticulum membranes. The decrease in the level of fructose 2,6-bisphosphate, as a result of high glucagon/ insulin ratio which prevails in starvation, leads to inhibition of phosphofructokinase and activation of fructose 1,6-bisphosphatase and, therefore, to an increase in glucose production.

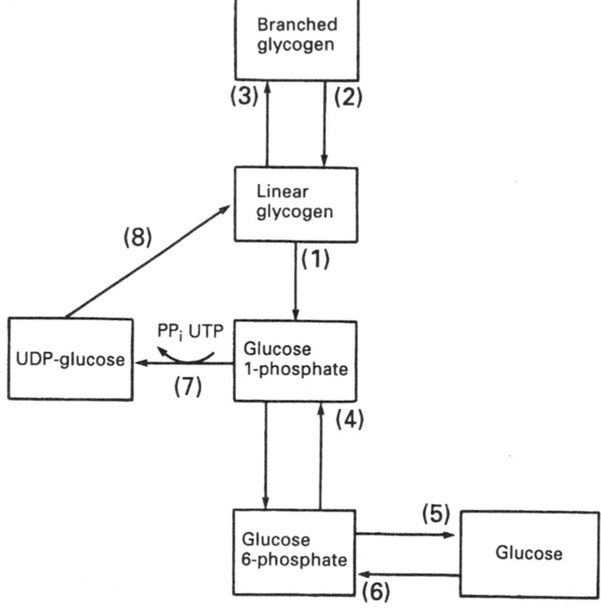

Figure 6.4 **Glycogen: synthesis and degradation.** Enzymes involved are (1) phosphorylases (hepatic and muscle); (2) debranching enzymes (glucan transferase and amylo-1,6-glucosidase); (3) glycosyl(4,6)transferase; (4) phosphoglucomutase; (5) glucose 6-phosphatase; (6) glucokinase (liver) or hexokinase (all tissues); (7) UDP-glucose pyrophosphorylase; (8) glycogen synthase

Glycogen synthesis **and** ***degradation***

The reactions that lead to the formation and degradation of glycogen are both cytosolic processes (Figure 6.4). Glycogen synthesis requires a supply of energy in the form of uridine triphosphate (UTP). It occurs in muscle and liver in the fed state when the insulin/glucagon ratio is high. The synthesis and degradation of glycogen are controlled by a series of elaborate, interrelated

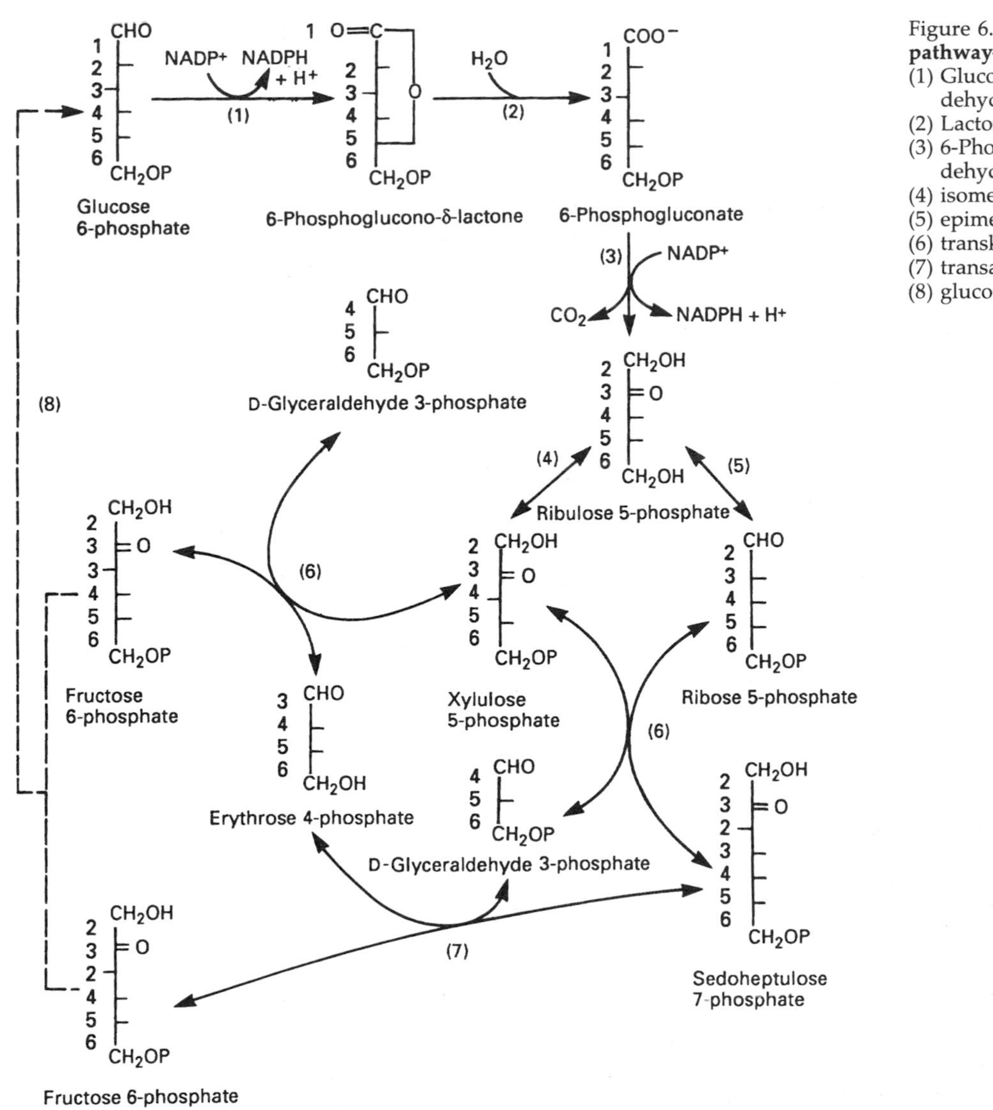

Figure 6.5 **Pentose phosphate pathway–outline of stages.**
(1) Glucose 6-phosphate dehydrogenase } Oxidative
(2) Lactonase } portion
(3) 6-Phosphogluconate dehydrogenase
(4) isomerase;
(5) epimerase;
(6) transketolase;
(7) transaldolase;
(8) gluconeogenic pathway

mechanisms whereby synthesis is inhibited when degradation is activated and in this way newly synthesized glycogen is stored and not broken down. Degradation of glycogen is stimulated by adrenaline in muscle and glucagon in the liver via activation of adenylyl cyclase. The consequent increase in levels of cAMP in turn activates protein kinase A (see Chapter 3). The latter enzyme eventually activates glycogen phosphorylase which releases glucose 1-phosphate units from glycogen. In the liver, glucose can be produced from glucose 1-phosphate via glucose 6-phosphate, but in muscle, glucose 6-phosphatase is absent and the glucose 6-phosphate produced from glycogen is channelled into the glycolytic pathway for provision of energy.

Pentose phosphate pathway

Glycolysis is not the only means of glucose oxidation. Glucose 6-phosphate can enter the cytosolic pentose phosphate pathway (also known as the **hexose monophosphate shunt**; Figure 6.5) in which it undergoes oxidation catalysed by a dehydrogenase to form 6-phosphogluconate and then a second oxidation to form a pentose sugar phosphate, ribulose 5-phosphate. In both of these reactions $NADP^+$ is reduced to NADPH which is an important biological reducing agent needed for biosynthetic reactions and other cellular processes. The rest of the pathway is non-oxidative and results in the production of ribose 5- phosphate which is needed for nucleic acid synthesis. The pentose phosphate pathway can operate in different modes depending on the tissue involved and on whether NADPH or ribose 5-phosphate is required. In the liver and the lactating mammary gland, NADPH is needed for fatty acid synthesis. Additionally, in the liver it is used in the reduction of molecular oxygen as a preliminary to the hydroxylation of endogenous and exogenous compounds by the cytochrome P-450 system (see Chapters 28 and 36). In the erythrocytes and the eye,

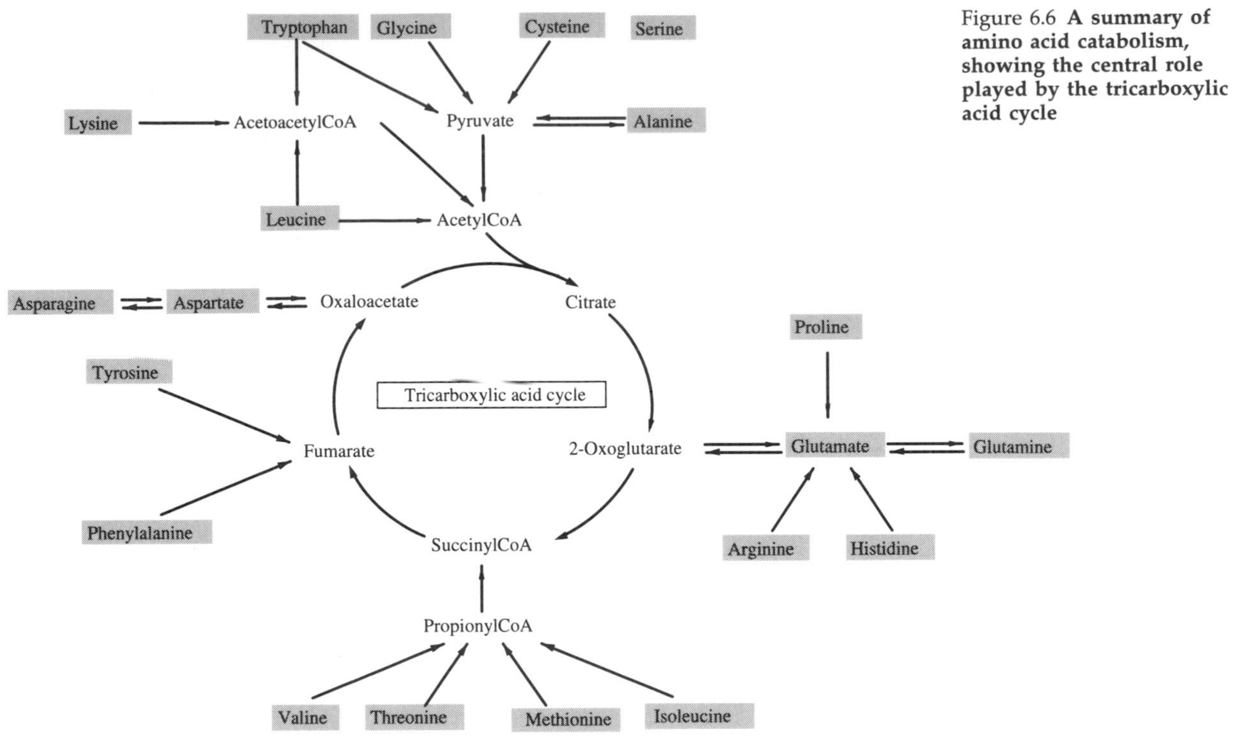

Figure 6.6 **A summary of amino acid catabolism, showing the central role played by the tricarboxylic acid cycle**

NADPH is needed for the reduction of glutathione which is essential for the protection of these tissues from oxidative damage to which they are vulnerable. In these cases the pentose phosphate pathway provides NADPH and is then linked to the glycolytic pathway by converting ribose 5-phosphate into glyceraldehyde 3-phosphate and fructose 6-phosphate. In tissues which are active in nucleic acid synthesis but have no particular requirement for NADPH, the oxidative part of the pathway is inactive and ribose 5-phosphate can be formed from the glycolytic intermediates glyceraldehyde 3-phosphate and fructose 6-phosphate.

Amino acid metabolism and haem biosynthesis: examples of interaction between cellular compartments

Amino acid metabolism

Extensive metabolism of amino acids occurs in the cytosol (Figure 6.6). Deamination, decarboxylation, transamination and oxidative decarboxylation of 2-oxoacids formed from the branched-chain amino acids take place in the cytosol, but are not confined to it. Some of these reactions, such as transamination, also occur within mitochondria.

For some metabolic pathways there is a complex interrelationship between the reactions occurring in the cytosol and in the mitochondria. A good example is that of the **ornithine cycle** which is the pathway for the synthesis of urea (Figure 6.7). Part of this pathway occurs in the cytosol, part in the mitochondria and part in the endoplasmic reticulum (Figure 6.8). In this system, the ornithine carrier plays an important role by transporting ornithine into the mitochondria and this is coupled with the transport of citrulline out into the cytosol.

Haem biosynthesis

A similar arrangement is seen in the biosynthesis of **haem**, where the initial step, the formation of 5-aminolaevulinic acid (ALA) from glycine and succinylCoA, and the final three steps, occur in the mitochondria. The intermediate steps are catalysed by enzymes located in the cytosol (Figure 6.9). Haem synthesis is most pronounced in the bone marrow and liver where the requirements for incorporation of haem into haemoglobin and cytochrome P-450 are high.

Fatty acid metabolism

Fatty acids are transported in the plasma bound to albumin and they are oxidized within the mitochondria. Transport of fatty acids from the cytosol into the mitochondria involves their conversion into carnityl derivatives. This occurs in two stages: the fatty acids are initially converted into coenzyme A derivatives which then take up carnitine and are transported across the mitochondrial membrane as carnitine complexes (see Chapter 5). In contrast to β-oxidation, the *de novo* synthesis of saturated fatty acids occurs in the cytosol. The requirements for fatty acid synthesis are: acetylCoA, usually derived from the metabolism of glucose, ATP as an energy source, and reducing equivalents in the form of NADPH. It is evident that a plentiful supply of carbohydrate is a prerequisite for the synthesis of fatty acids. Synthesis starts with the

Figure 6.7 **Ornithine cycle – synthesis of urea**

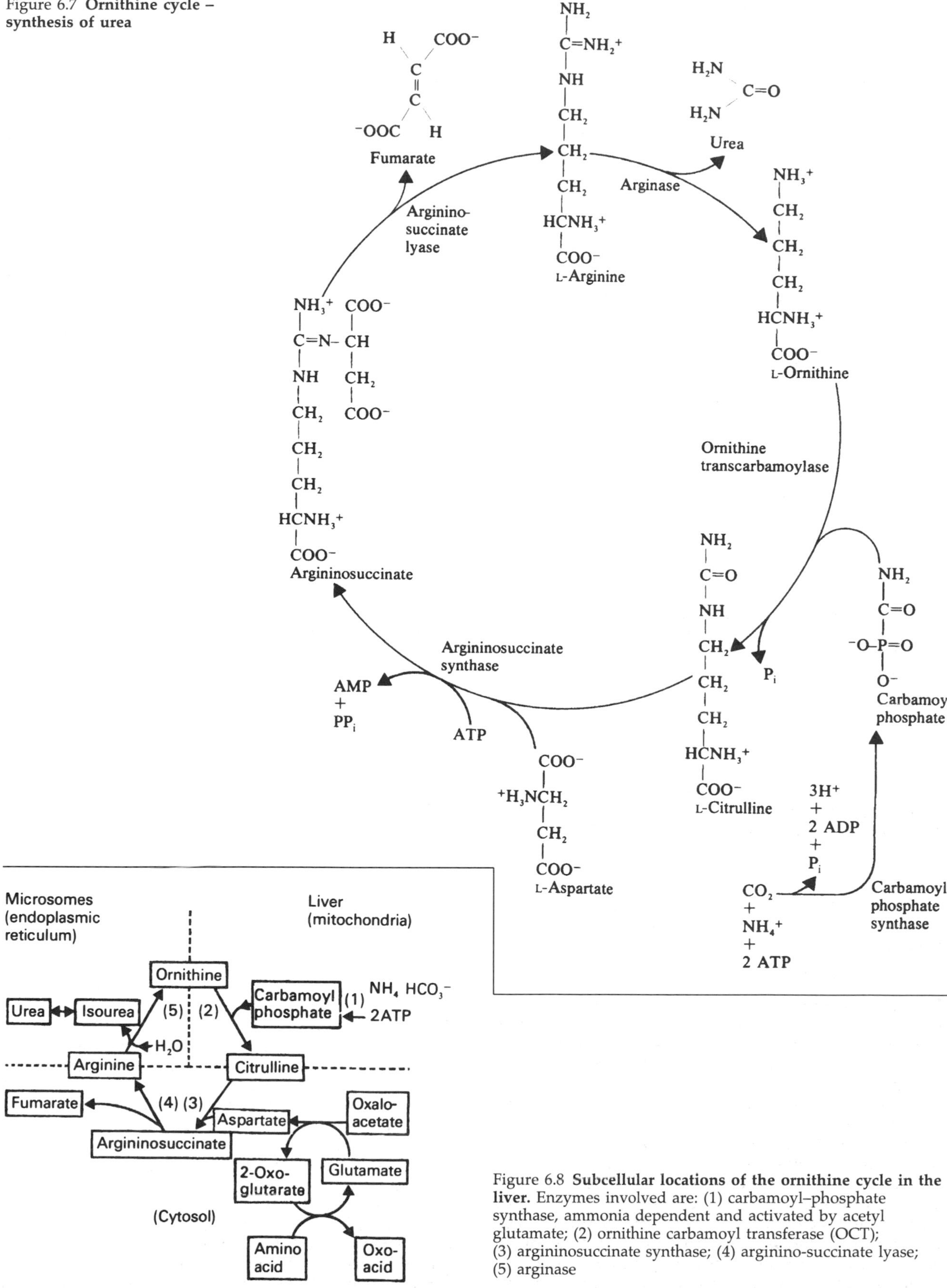

Figure 6.8 **Subcellular locations of the ornithine cycle in the liver.** Enzymes involved are: (1) carbamoyl–phosphate synthase, ammonia dependent and activated by acetyl glutamate; (2) ornithine carbamoyl transferase (OCT); (3) argininosuccinate synthase; (4) arginino-succinate lyase; (5) arginase

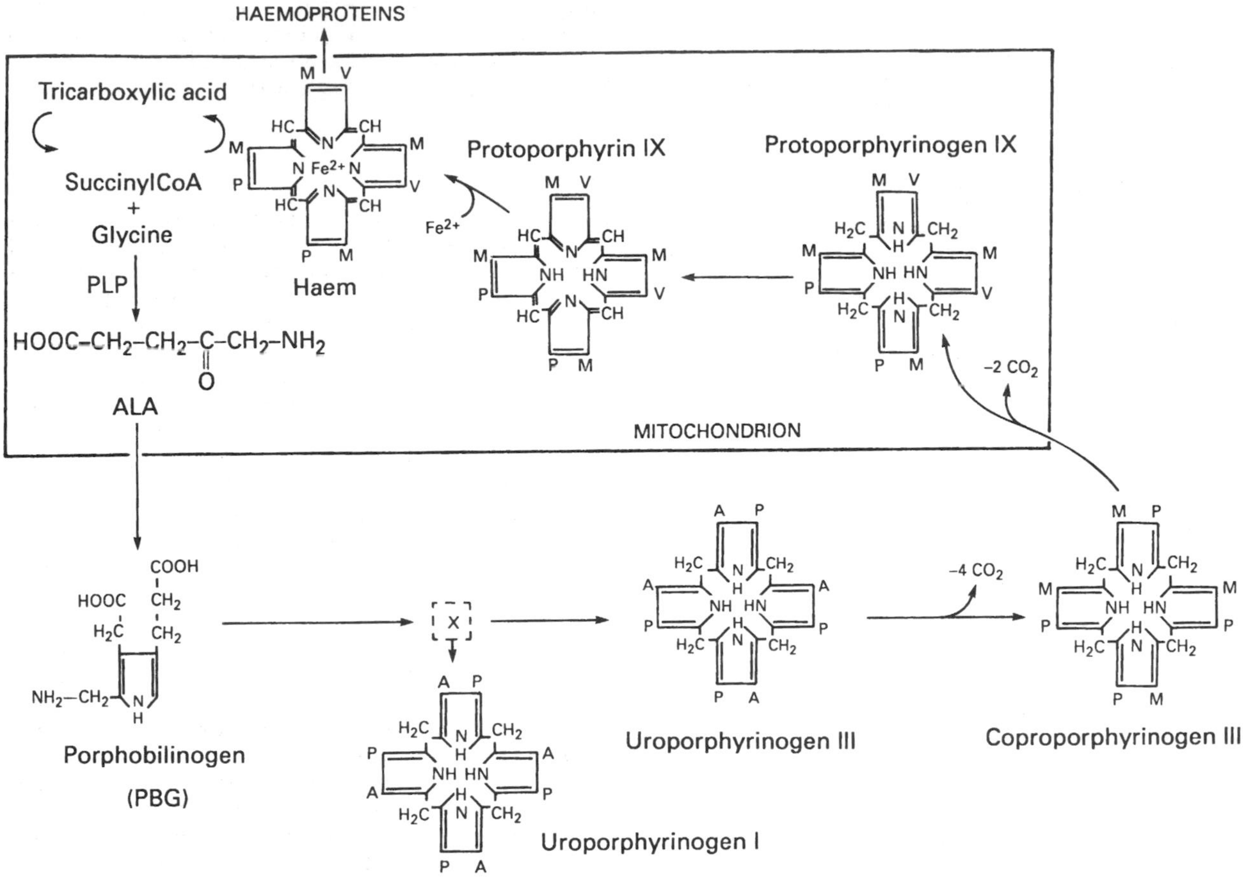

Figure 6.9 **Outline of haem biosynthesis.** ALA, 5-aminolaevulinic acid; PLP, pyridoxal 5'-phosphate; A, acetyl; P, propionate; M, methyl; V, vinyl; X, unstable intermediate

carboxylation of acetylCoA to malonyl CoA in a reaction catalysed by the biotin-containing enzyme acetylCoA carboxylase (Figure 6.10). All the subsequent reactions are catalysed by the enzyme fatty acid synthase (FAS). In eukaryotic systems this enzyme consists of two identical subunits, each of which has several active sites each catalysing a different step in fatty acid synthesis. The advantages of having an enzyme with multiple different active sites are that a high concentration of the intermediate metabolites can be maintained within the complex and, at the same time, the possibility of side-reactions is reduced. The major product of fatty acid synthase is the 16-carbon saturated fatty acid palmitate. Longer chain fatty acids are formed by elongation reactions, mainly on the endoplasmic reticulum, but also, to a limited extent, in mitochondria.

Purine, pyrimidine and nucleic acid metabolism

Many of the stages of purine, pyrimidine and nucleotide synthesis take place in the cytosol, but some stages of nucleotide synthesis, for example that of NAD^+, take place in the nucleus (see Chapter 4). Apart from the small amount of DNA and RNA found in the mitochondria, these molecules are synthesized in the nucleus. Nevertheless, the attachment of amino acids to tRNAs, as a preliminary to protein synthesis (see Section 6.3), catalysed by aminoacyl-tRNA synthases occurs in the cytosol, except, again, for a small amount of autonomous activity which takes place in the mitochondria. In cells in which protein synthesis is occurring at a rapid rate, the concentration of the synthases in the cytosol is very high and they can form a substantial proportion of the total cytosolic protein.

Ribosomes synthesizing proteins may be active free in the cytosol or attached to the endoplasmic reticulum. Proteins which are to be secreted from the cell, and certain membrane and organelle proteins are synthesized on the ribosomes bound to the endoplasmic reticulum (see Chapter 7). Other classes of proteins, including cytosolic proteins and many of those found in mitochondria and the nucleus, are synthesized by ribosomes which appear to be free in the cytosol, but which may be bound to cytoskeletal fibres.

6.3 Protein synthesis

Genetic information is stored in genes in chromosomes and is expressed through **transcription** to messenger RNA (mRNA) and subsequent **translation** into

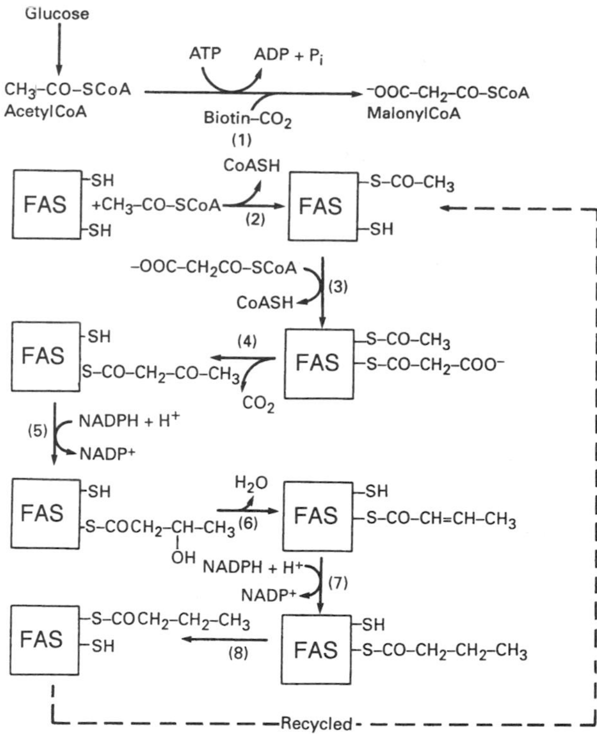

Figure 6.10 **Biosynthesis of fatty acids.** Fatty acid synthesis proceeds by addition of 2-carbon units to acetylSCoA. MalonylSCoA is the intermediate by which the 2-carbon units are added. (1) Reaction catalysed by acetylSCoA carboxylase; reactions (2)–(8) are catalysed by fatty acid synthase (FAS)

polypeptide chains. Translation depends on the **genetic code** which is the precise protocol according to which a specific sequence of nucleotide base pairs is translated to a specific sequence of amino acids.

The genetic code

DNA is made up of four mononucleotide units, the deoxyribonucleotides of adenine, cytosine, guanine and thymine, and RNA is made of the ribonucleotides of adenine, cytosine, guanine and uracil. The four nucleotides present in nucleic acids cannot specify the linear arrangement of the 20 common amino acids in a polypeptide chain in a one-to-one manner. A sequence of three nucleotides is the minimum size of a coding group that would specify 20 amino acids, as 64 different combinations are possible. A pair of nucleotides would not be able to code for all the amino acids as only 16 combinations can be made. Figure 6.11 shows the 64 possible combinations of nucleotide bases and the amino acids corresponding to each combination. Each base triplet is known as a **codon**. Sixty-one out of the 64 codons specify amino acids and the other three are **termination codons** signalling the cessation of polypeptide chain formation. These are UAG, UGA and UAA. As there are 61 codons for 20 amino acids, many amino acids have more than one codon, and the code is referred to as **degenerate** indicating that there are redundancies. The genetic code is **universal**, that is, the meaning of each codon is the same in almost all known organisms, a fact which supports the theory that life on earth evolved only once. The triplet code is '**comma-less**'

First letter	Second letter: U	C	A	G	Third letter
U	UUU Phe	UCU	UAU Tyr	UGU Cys	U
	UUC	UCG	UAC	UGC	C
	UUA Leu	UCA Ser	UAA*	UGA*	A
	UUG	UCG	UAG*	UGG Trp	G
C	CUU	CCU	CAU His	CGU	U
	CUC Leu	CCC Pro	CAC	CGC Arg	C
	CUA	CCA	CAA Gln	CGA	A
	CUG	CCG	CAG	CGG	G
A	AUU	ACU	AAU Asn	AGU Ser	U
	AUC Ile	ACC Thr	AAC	AGC	C
	AUA	ACA	AAA Lys	AGA Arg	A
	AUG Met	ACG	AAG	AGG	G
G	GUU	GCU	GAU Asp	GGU	U
	GUC Val	GCC Ala	GAC	GGC Gly	C
	GUA	GCA	GAA Glu	GGA	A
	GUG	GCG	GAG	GGG	G

* Termination code

Figure 6.11 **The genetic code**

which means that the 'reading frame' of the message can be shifted by moving the starting point of translation by one or two bases. Such 'frame shifts' would result in the production of an entirely different protein after transcription and translation or, indeed, to the production of no protein at all, if a stop codon is encountered. Mutations due to **deletions** or **insertions** of nucleotides in a gene can lead to a shift in the reading frame with catastrophic results. One such example is the myopathy known as Duchenne's muscular dystrophy, where deletions in the gene coding for the cytoskeletal protein dystrophin cause a shift in the reading frame of the message, with the result that this protein is not produced and muscle tissue degenerates (see Chapter 27).

If one nucleotide is altered, i.e. there is a **substitution** rather than a deletion or insertion, then there are three possible outcomes for translation. The first possibility is that the triplet may still code for the same amino acid. For example, if a mutation alters the codon AAA for AAG the amino acid incorporated into the polypeptide chain will still be lysine. This is known as a **silent** mutation. The second possibility is that a different amino acid will be encoded and, therefore become incorporated into the protein. If, for example, the codon GAG which codes for glutamic acid, is altered to GUG, valine will replace glutamic acid in the protein. A substitution of this kind in the haemoglobin β-chain is known to be the cause of sickle cell anaemia and the mutation is known as a **missense** mutation. The third possibility is that a triplet coding for an amino acid may be altered to one coding for a termination codon, for example, UCG which codes for serine may be altered to UAG which would stop translation at that point. This type of mutation is referred to as **nonsense** mutation.

Two further terms used to describe types of mutations may be mentioned here: **transitions** and **transversions**. In the former a pyrimidine is substituted for a pyrimidine, or a purine for a purine. Transversions are said to occur when a pyrimidine is substituted for a purine, or vice versa.

Components and processes of translation

Translation of a mRNA message into a polypeptide chain requires the following materials: the mRNA to be translated, all the amino acids found in the polypeptide, transfer RNA (tRNA), ribosomes, energy in the form of ATP and GTP, enzymes, and specific factors needed for a four-stage process. The stages are **activation, initiation, elongation** and **termination**.

Activation

The triplet codons on mRNA do not select amino acids directly but require the presence of an 'adapter' molecule which can both recognize and bind the codon and also carry the appropriate amino acid to the growing polypeptide chain. This role is performed by **transfer RNA (tRNA)** which is a fairly small species of RNA, consisting of 70–80 nucleotides forming distinct domains or loops and folded into an L-shaped structure (Figure 6.12). There is at least one type of tRNA for each

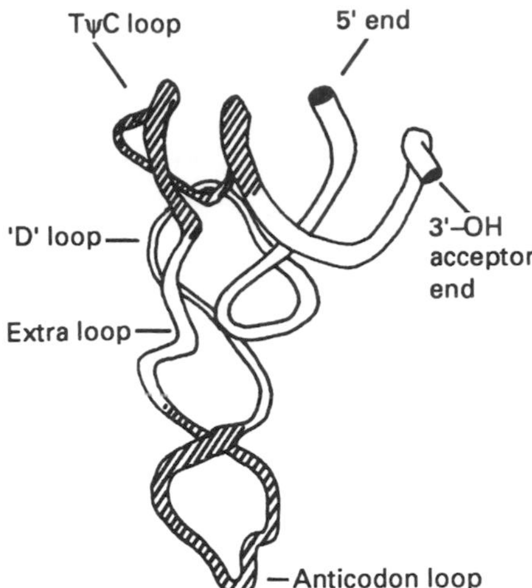

Figure 6.12 **A representation of the three-dimensional (folded) structure of a typical tRNA molecule.** The loops 'TψC' and 'D' are named for the unusual nucleosides they contain in their structures. The first invariably contains a sequence in which the nucleosides are thymidine, pseudouridine, cytidine, while the second always contains a dihydrouridine. The 'extra loop' is found in only some tRNA molecules.

amino acid. Amino acids are added to their respective tRNAs by 20 different enzymes known as **aminoacyl tRNA synthases (AAS)**, each enzyme specific for a particular amino acid. The amino acid is attached via its carboxyl group to the 3' hydroxyl group of the tRNA in a two-step process, requiring ATP which is hydrolysed into AMP and inorganic pyrophosphate (PP_i) :

$$\begin{array}{c}\text{amino acid + ATP + AAS}\\ \downarrow \\ \text{amino acid–AMP–AAS + tRNA}\\ \downarrow \\ \text{aminoacyl–tRNA} + PP_i + \text{AAS}\end{array}$$

When the amino acid is attached to the tRNA, the latter can recognize a codon on mRNA by virtue of an **anticodon** triplet of bases on one of the loops in the tRNA structure. When the tRNA is carrying an amino acid it is referred to as being 'charged' and the amino acid as being 'activated'.

It is not known how each AAS recognizes its corresponding tRNA, but the process is highly specific and largely accounts for the fact that translation is a very accurate process.

One tRNA can recognize more than one codon because of non-standard base pairing, often referred to as 'wobble', between the third nucleotide on the codon and its corresponding (first) nucleotide on the anticodon. Examination of the genetic code (see Figure 6.11) points to the relative unimportance of the third nucleotide in the codons for several of the amino acids, e.g. UC**U**, UC**G**, UC**A** and UC**C** all encode serine.

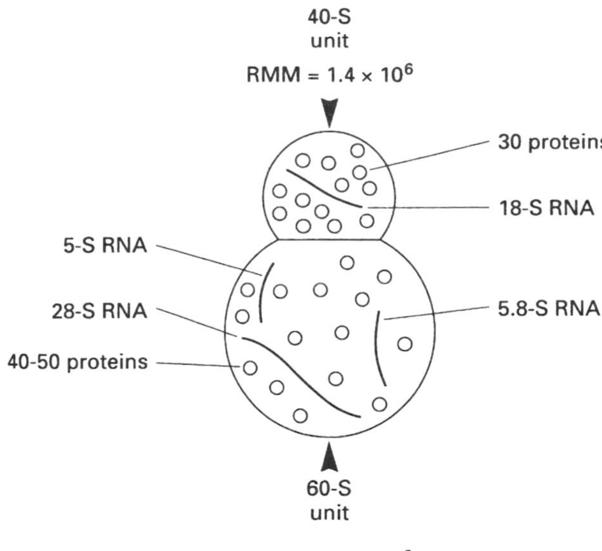

Figure 6.13 **Components of eukaryotic ribosomes**

Ribosomes

The polypeptide chain is assembled on ribosomes. In eukaryotic cells, ribosomes are either free in the cytosol or bound to the membranes of the endoplasmic reticulum to form the so-called 'rough endoplasmic reticulum'. On the whole, proteins which remain soluble in the cell are synthesized in the cytosol, whereas proteins destined for export from the cell, or incorporation into cellular membranes, are assembled on the rough endoplasmic reticulum. The major exceptions to this general rule are those proteins destined for the nucleus and mitochondria. These proteins are initially formed on 'free ribosomes'.

All ribosomes are composed of two subunits (Figure 6.13). Their properties are often described by sedimentation coefficients, or 'S' values (Svedberg units) because extensive studies on ribosomes were originally carried out by ultracentrifugation. Prokaryotic ribosomes consist of a unit of 70 S composed of two subunits, of 30 S and 50 S, respectively. Sedimentation coefficients depend on the size and shape of a particle as well as its molecular mass, so that the values given for the two subunits cannot simply be added up to give the value for the whole ribosome.

The ribosomes found in mammalian cells are larger than their bacterial counterparts. They are 80 S particles composed of two subunits of 40 S and 60 S, respectively. The smaller subunit of a liver ribosome, for example, contains one molecule of 18 S RNA and about 30 proteins, so that the ratio of RNA to protein is one to one. The larger subunit contains three molecules of RNA, 5 S, 5.8 S and 28 S, respectively and 40–50 proteins. Not all eukaryotic ribosomes have exactly the same size, although the 40 S subunit has remained relatively constant during the course of evolution. Despite the differences in size and composition, both prokaryotic and eukaryotic chromosomes carry out essentially identical processes in protein synthesis.

Messenger RNA (mRNA)

The mRNA molecules are translated from the 5' end to the 3' end, and the resulting polypeptide chain is synthesized from the amino-terminus to the carboxyl-terminus.

Eukaryotic mRNA is referred to as **monocistronic**, meaning that there is only one coding sequence on each mRNA molecule producing only one polypeptide chain, whereas prokaryotic mRNA is **polycistronic**, often encoding more than one polypeptide on the same mRNA.

Initiation

The steps involved in the initiation of translation are outlined in Figure 6.14 (stage 1). The process of translation begins with the assembly of the various components of the system. The first step is the formation of **methionyl-tRNA** by the attachment of methionine to its corresponding tRNA, by the enzyme methionyl-tRNA synthase. There are two types of methionyl-tRNA that can be formed as a result of the action of this synthase. One of them is known as **Met-tRNA$_i^{Met}$** and it is the form that can bind to the small ribosomal subunit to initiate protein synthesis, whereas the other is known as **Met-tRNA^{Met}** and is the form used to incorporate methionine in the interior of a polypeptide chain. They are formed by the attachment of methionine to two different tRNA molecules, tRNA$_i^{Met}$ and tRNA^{Met}, respectively.

In bacteria, the equivalent of tRNA$_i^{Met}$ is known as tRNA$_i^{fMet}$ because the amino group of the methionine is modified by the addition of a formyl group, derived from N^{10}-formyltetrahydrofolate, via the action of the enzyme transformylase. The actual steps in translation are, in fact, very similar in prokaryotes and eukaryotes.

The next step in translation is the binding together of the Met-tRNA$_i^{Met}$, the mRNA and the small ribosomal subunit. The energy needed in order to position the initiation codon AUG (which codes for methionine) correctly on the ribosomal subunit is provided by the hydrolysis of GTP.

In bacteria, the correct positioning of the small ribosomal subunit relative to the mRNA is achieved by the presence on the mRNA of a sequence of nucleotide bases known as the **Shine–Dalgarno sequence**, found 5–10 nucleotides 'upstream' (i.e. in the 5' direction) of the initiation codon AUG:

5' UAAGGAGG————AUG—

This sequence is complementary to part of the 3' terminus of the small rRNA molecule, thus allowing the formation of base pairs and the correct positioning of the mRNA relative to the ribosome.

In eukaryotes, recognition of mRNA by ribosomes is achieved in two ways. The first recognition signal is the following sequence of nucleotides on the mRNA:

5' ACCAUGG ———AUG—

The second recognition signal is the presence of a methyl group on the 5' end of the mRNA, sometimes referred to as the **methyl cap**. This group seems to be very important for the recognition of the initiation code which is usually the AUG codon nearest to the 5'-end, as it has

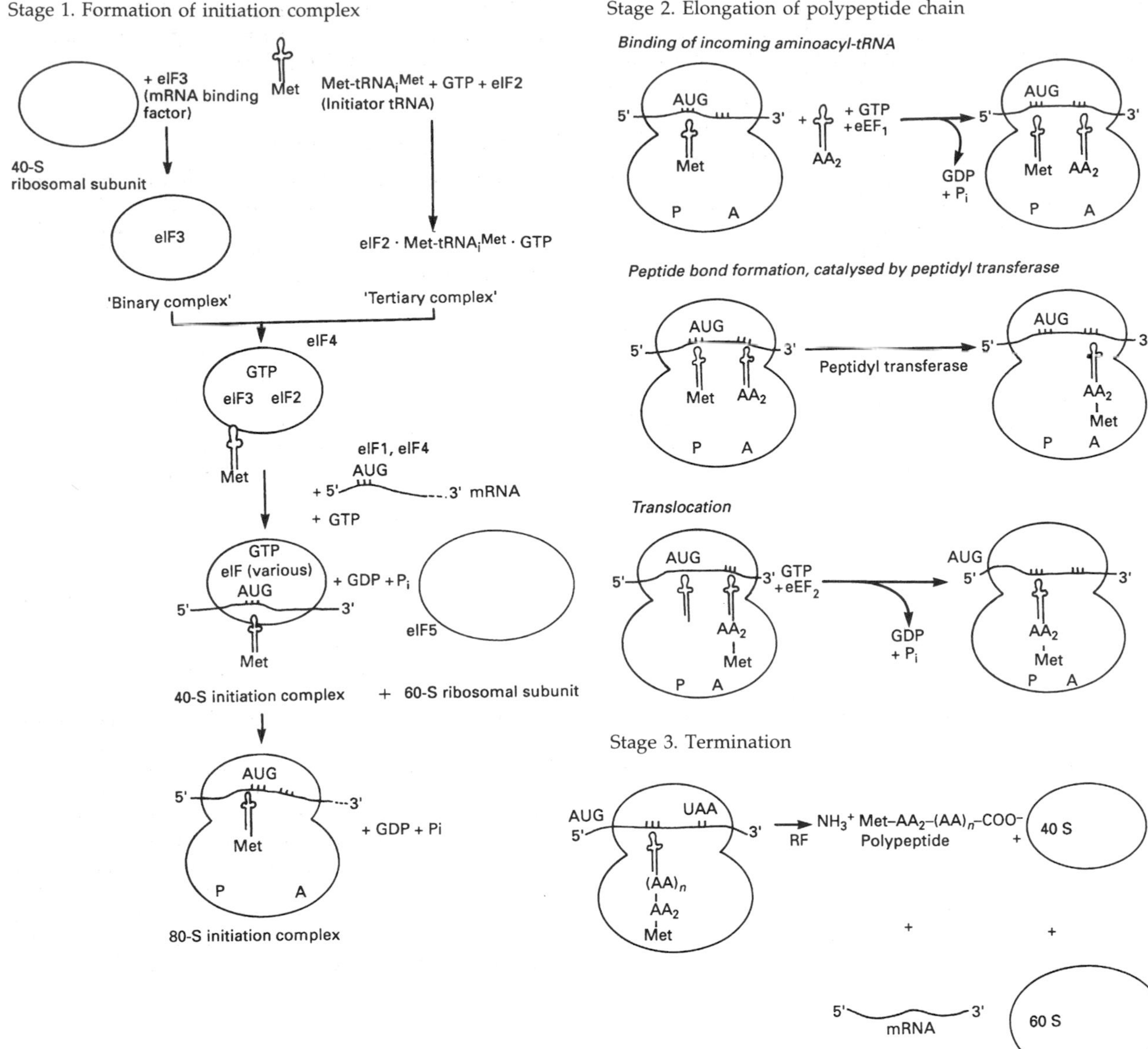

Figure 6.14 **Steps in eukaryotic protein synthesis**

been shown that unmethylated mRNAs are very poorly translated.

The formation of the complex consisting of mRNA, Met-tRNA$_i^{Met}$, and the small ribosomal subunit requires the presence of certain proteins known as **initiation factors (IF)**. Three of these are known to be involved in prokaryotic protein synthesis: IF1, IF2 and IF3. At least nine are known to be present in eukaryotic systems and they are designated eIF1, eIF2, etc. eIF2 is involved in the recognition of the initiation codon AUG, whereas eIF4 seems to facilitate the binding of the mRNA to the small ribosomal subunit. The large ribosomal subunit binds the complex of Met-tRNA$_i^{Met}$, GTP and mRNA after the AUG codon has been located. This binding, which results in the formation of the 80 S ribosomal complex is facilitated by various initiation factors.

Elongation

The process of elongation is outlined in Figure 6.14 (stage 2). Elongation involves the sequential addition of amino acids to the carboxyl end of the growing polypeptide chain, by the formation of peptide bonds. Various proteins, referred to as **elongation factors**, are required for this process and they are designated EF for prokaryotic and eEF for eukaryotic systems.

There are two sites on mRNA where amino acyl-tRNA (aa-tRNA) can bind, one known as the **A-site** which receives the new incoming aa-tRNA and another, known as the **P-site**, which is occupied by the tRNA still linked to the already formed peptide chain. Met-tRNA$_i$Met enters the P-site. The second amino acid, shown in figure 6.14 as AA_2, is introduced in the form of AA_2-tRNA which becomes attached to the A-site. A peptide bond is formed in a reaction catalysed by the enzyme **peptidyl transferase**, an activity intrinsic to the ribosome, and the dipeptide moves from the A-site to the P-site, leaving the A-site free to accept the third amino acyl-tRNA. Energy for this translocation is provided by the hydrolysis of GTP. In each translocation the ribosome, with the attached peptidyl-tRNA, moves three nucleotides along the length of the mRNA from its 5' to its 3' end. The third amino acyl-tRNA then attaches itself to the A-site and the process continues until the stop codons UAA UAG, or UGA are reached, whereupon termination of translocation occurs.

mRNA molecules are long enough to allow more than one ribosome to carry out translation at the same time. The complex of mRNA and a number of ribosomes simultaneously translating the message is termed a **polysome**. In the case of haemoglobin, for example, as many as 5 ribosomes may simultaneously be engaged in the translation of an mRNA message for the globin chain.

Termination

The events taking place during termination of translation are outlined in Figure 6.14 (stage 3). The termination codons UAA, UAG and UGA, mentioned above, signal the release of the polypeptidyl-tRNA from the ribosome. These codons are recognized by proteins known as **termination (TF)** or **release factors (RF)**. Three such factors are known to exist in bacteria, whereas only one is known to be involved in eukaryotic termination. After the release of the polypeptidyl-tRNA from the complex, the mRNA dissociates from the ribosome which divides into its two subunits. The process of termination also requires energy in the form of GTP. mRNA, the ribosomal subunits and the initiation, elongation and termination factors can be recycled and used to synthesize another polypeptide chain. The length of life of mRNA molecules is determined partly by nucleotide sequences not encoding proteins, e.g. non-translated sequences, such as the 'poly A tail'.

Inhibitors of protein synthesis

A number of commonly used pharmaceutical agents, such as a variety of antibiotics, act by inhibiting selectively the process of prokaryotic translation. The actions of these drugs are outlined below and they are further considered in Chapter 39.

Streptomycin inhibits the process of initiation by altering the structure of the 30 S ribosomal subunit and preventing it from forming the initiation complex.

Tetracyclines inhibit the process of elongation by blocking access of the aa-tRNA to the mRNA–ribosome complex through their interaction with the 30 S subunit.

Chloramphenicol also interferes with prokaryotic elongation by acting as an inhibitor of peptidyl transferase.

Puromycin inhibits both prokaryotic and eukaryotic translation by becoming incorporated into the growing polypeptide chain, by virtue of its similarity in structure with aa-tRNA. When puromycin is incorporated into the chain, translation stops.

Erythromycin inhibits prokaryotic elongation by binding to the 50 S ribosomal subunit and inhibiting translocation.

Eukaryotic protein synthesis can be the target for the action of some bacterial toxins. The devastating effect of **diphtheria toxin**, for example, arises through inhibition of the eukaryotic process of translation. The toxin is produced by the bacterium *Corynebacterium diphtheriae*. The toxin from this organism enters host cells and prevents eEF2 from functioning. It is noteworthy that eEF2 is a GTP-binding protein and, to this extent, diphtheria toxin resembles both cholera and pertussis toxins, because both of these exert their effect by acting on other GTP binding proteins (see Chapter 3).

Chapter 7

Cellular organelles: the endoplasmic reticulum and Golgi apparatus

7.1 Introduction

The study of electron micrographs has shown most eukaryotic cells to contain a complex network of cytoplasmic membranes. Within this set of membranes several subtypes may be distinguished, based upon morphology and function. Thus, the endoplasmic reticulum (ER) consists of membranes that are folded extensively to form vesicles, tubules and flattened sacs. These membranes appear as parallel pairs (Figure 7.1). The structures that constitute the ER may be further subdivided into 'rough' or 'smooth', depending on whether the membranes have ribosomes attached. Rough endoplasmic reticulum (RER) is granular in appearance due to the adherence of the ribosomes.

Another type of cytoplasmic membrane that may be distinguished on morphological and functional grounds is referred to as the Golgi complex or apparatus. These structures are usually found quite close to the nucleus and consist of agranular cisternae that look rather like a 'stack of plates' (Figure 7.1). The number of such stacks varies between cell types, and may consist of one large stack or hundreds of small ones. A characteristic feature of the Golgi apparatus is its associated vesicles. These Golgi vesicles play a vital role in protein 'trafficking' within the cell. Some of these vesicles have special names relating to their function. Thus endosomes are involved in the process of endocytosis, whereby external molecules and larger fragments may be taken up into the cell. Other vesicles will ultimately give rise to lysosomes. Yet another group of vesicles is concerned with the transfer of proteins to the plasma membrane and also with the secretion of proteins. These are referred to as **secretory vesicles**, a subpopulation of which are called **dense-core vesicles**. The latter are involved in the **regulated secretion of hormones**, such as adrenaline.

Proteins synthesized in association with the RER are transferred to the Golgi complex in another set of vesicles which 'bud' from the RER and fuse with the so-called *cis*-face of the Golgi. Many of these proteins exit via the *trans*-face in the Golgi vesicles just described. It is now appreciated that the ER and the Golgi apparatus constitute a functional unit that acts to 'target' the proteins formed by ribosomes attached to the RER into secretory pathways, into lysosomes, into plasma membranes or, indeed, into residence in the ER or Golgi apparatus themselves. These functional relationships are shown diagrammatically in Figure 7.2.

Much of the rest of the present chapter will be concerned with the processes leading to 'targeting' of proteins. It should be borne in mind however, that the ER has roles not directly related to protein targeting. In this respect the **smooth endoplasmic reticulum (SER)** is of particular interest. These other roles are best understood in the liver, and are summarized in Table 7.1. Many of the functions will be described in detail in other chapters, but it is useful at this stage to review the overall picture of the processes that are located in this component of the cell. Of special interest is the enzyme glucose 6-phosphatase that catalyses the hydrolysis of its substrate to yield glucose and inorganic phosphate. This enzyme (also found in the kidney) is of vital importance for the regulation of the blood glucose concentration since almost all of the glucose leaving the liver, where it is stored as glycogen or is synthesized from non-carbohydrate precursors (gluconeogenesis), must first be converted into glucose 6-phosphate prior to hydrolysis to give glucose. In addition, the endoplasmic reticulum plays an important role in plasma lipoprotein formation (see Chapter 11), the formation of bile salts (see Chapter 25) and the metabolism of foreign compounds (see Chapter 36).

Another important role of the ER is in the sequestration of Ca^{2+}. The total surface area of the ER

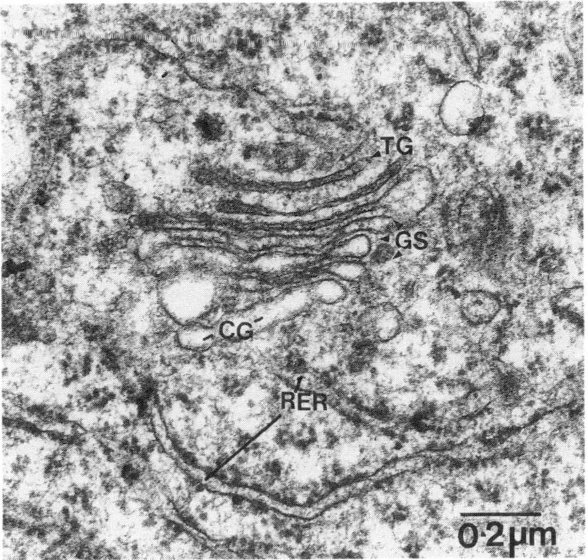

Figure 7.1 **An electron micrograph of a section through the Golgi complex and related structures (from a serous gland cell).** Visible are: rough endoplasmic reticulum (RER), *cis*-Golgi cisternae (CG), the main Golgi stack of cisternae (GS) and the *trans*-Golgi face (TG) of the complex

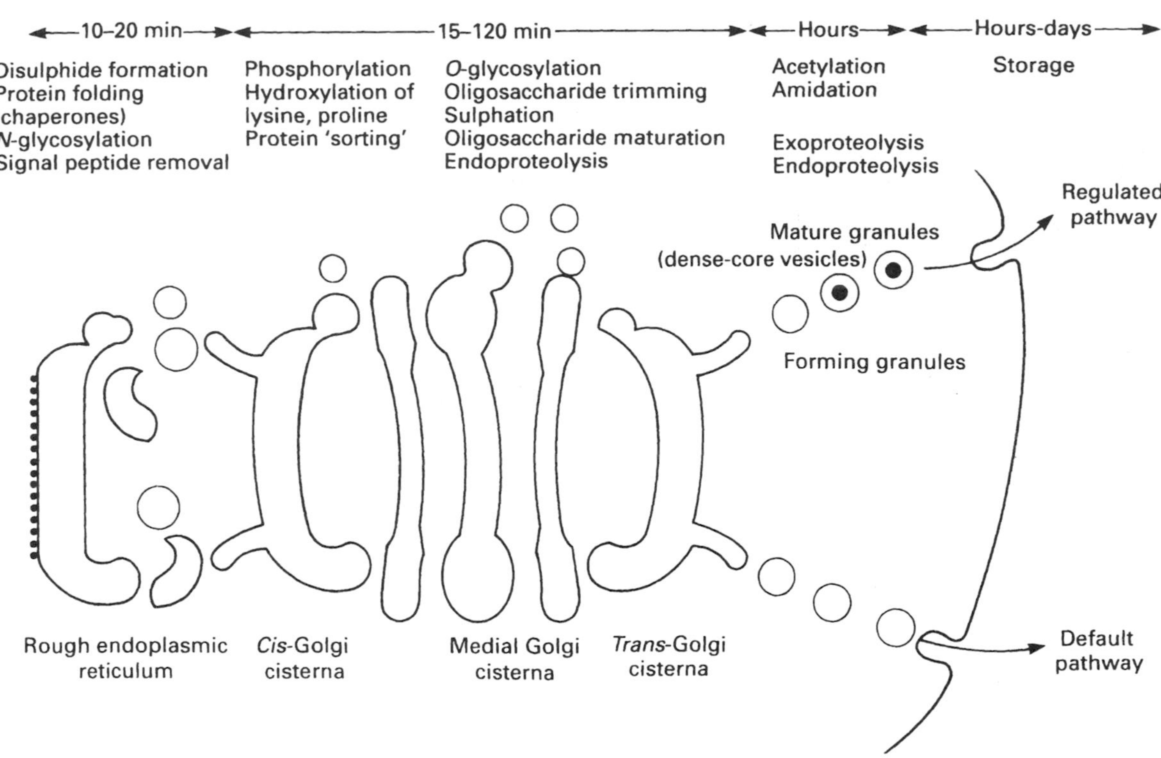

Figure 7.2 **Multiple post-translational modifications of proteins are possible as they pass through the endoplasmic reticulum/Golgi apparatus**

Table 7.1 **Functions of smooth endoplasmic reticulum in liver**

Conversion of glucose-6-phosphate into glucose by glucose 6-phosphatase
Plasma lipoprotein synthesis
Synthesis of triacylglycerols, phospholipids and part of the synthetic pathway for cholesterol
Bile salt synthesis and secretion – 7α- and 12α-hydroxylation of cholesterol and conjugation of cholic acid
Inactivation of steroid hormones, which sometimes involves hydroxylation
Detoxication of foreign compounds
Conjugation reactions, e.g. UDP-glucuronyl transferase
Oxidative metabolism by cytochrome P450 system
Oxidative metabolism of carcinogens, e.g. polycyclic hydrocarbons to active epoxides and diols

exceeds that of the plasma membrane by between one and two orders of magnitude. Consequently the uptake of substances, such as Ca^{2+}, can be more rapidly achieved by their transfer into the ER than by their extrusion across the plasma membrane. Many intracellular events are signalled by a transient rise in the concentration of Ca^{2+} in the cytosol (see Chapter 3), and energy-dependent uptake systems for the cation in the ER serve rapidly to return the cytosolic concentration to resting levels. The '**sarcoplasmic reticulum**' of muscle is a specialized form of ER membrane in which the role of controlling Ca^{2+} is particularly well-developed (see Chapter 27).

7.2 Protein targeting

The passage of proteins through the ER, Golgi apparatus and related vesicles may be regarded as being controlled by a series of 'filters' (F1, etc., in Figure 7.3). The passage through the first of these filters is an absolute requirement for entry into the complex; thereafter the filters are conditional and whether a protein passes 'through' them is determined by its structure. Any protein that passes through filter F1, thus having gained access to the lumen of the RER, but which is not recognized by any of the succeeding 'filters', will inevitably be secreted from the cell in a passive fashion. This 'unfiltered' pathway, that leads to the secretion of molecules such as serum albumin, is referred to as '**default**' or '**constitutive**'. The operation of this pathway will not be discussed further here.

'Filter' F_1 : the attachment of ribosomes to the endoplasmic reticulum

There is no evidence that ribosomes attached to the RER differ from those engaged in protein synthesis in the cytosol. This leads to two conclusions:

1. The sequence of the protein being synthesized on ribosomes free in the cytosol must 'signal' that the ribosome is to be attached to the RER.
2. Since this binding will require the diffusion of the ribosome/nascent protein to the RER, there must be some means of arresting protein synthesis while 'docking' with the RER occurs.

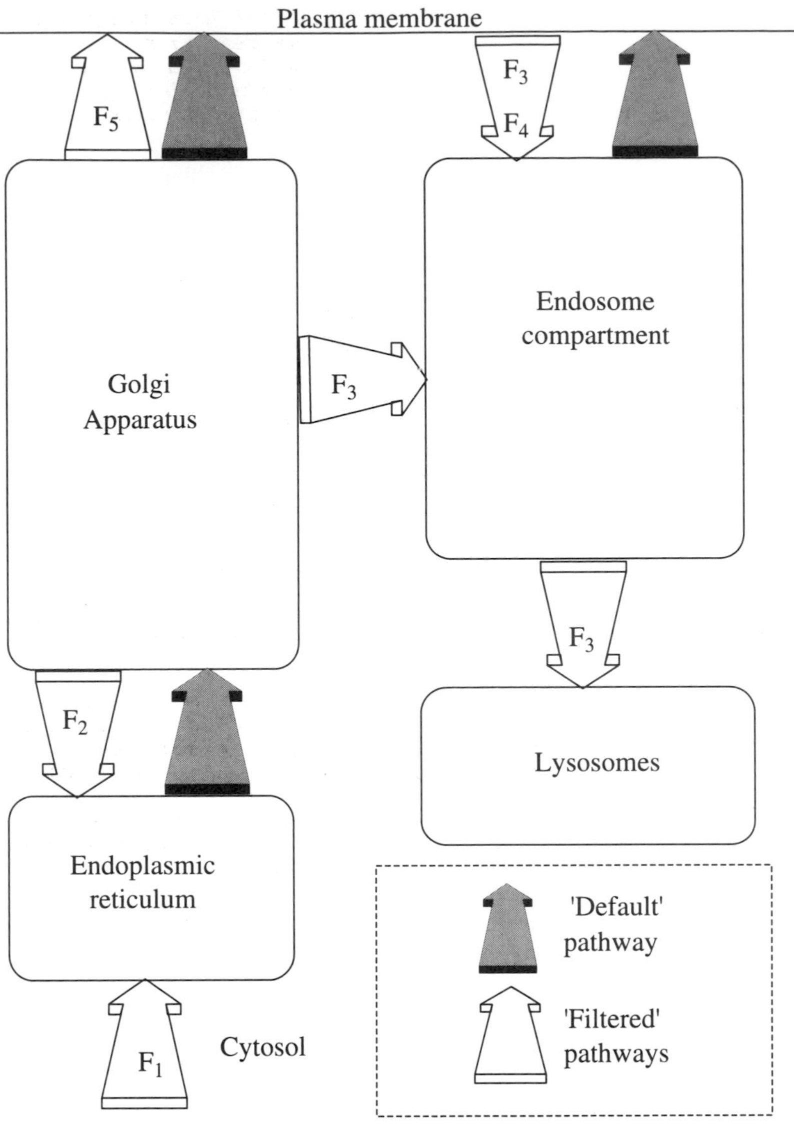

Figure 7.3 **Protein targeting requires a series of 'filters' (usually receptors)**

Both of these conclusions are borne out by observation. The growing polypeptide chain produces a sequence called a '**signal sequence**' that consists of about 20 amino acids. The signal sequence is not the same for all proteins whose synthesis is completed on the RER, but in each case it contains a core of amino acids with hydrophobic residues. When such a sequence appears in the growing polypeptide, the binding of a '**signal-recognition particle' (SRP)** is facilitated. This is a ribonucleoprotein (containing RNA and proteins) that has the function of arresting protein synthesis until the ribosome–SRP complex 'docks' on the RER. The 'docking' is facilitated by the presence in the RER membrane of an '**SRP receptor**'. Once the process of docking is complete synthesis of the protein recommences (Figure 7.4).

There are two types of location for the signal sequence within the growing polypeptide chain (Figure 7.5). The first, referred to as a cleavable signal sequence, is located at the extreme N-terminal end of the nascent polypeptide chain. As synthesis resumes after docking, this sequence 'loops' back through the membrane and consequently the growing chain is introduced into the lumen of the ER. While this is happening a '**signal peptidase**' catalyses the hydrolysis of the peptide bond joining the signal peptide to the rest of the protein. If this happens, then the protein will be released into the lumen of the ER. On the other hand if a nascent protein has both an N-terminal signal sequence and a '**stop-transfer**' sequence elsewhere in its structure, the passage of the latter through the membrane will serve to anchor the protein in the membrane, with the N-terminus in the lumen. Such membrane proteins with luminal N-termini are referred to as type I. Sometimes the signal sequence is neither N-terminal nor cleavable. These sequences also target proteins to the ER membrane, but they serve, additionally, to anchor the proteins in the membrane (**signal-anchor sequences**). Such proteins may then be orientated as type I proteins or type II (C-terminus in the lumen). Such signal-anchor sequence proteins may span

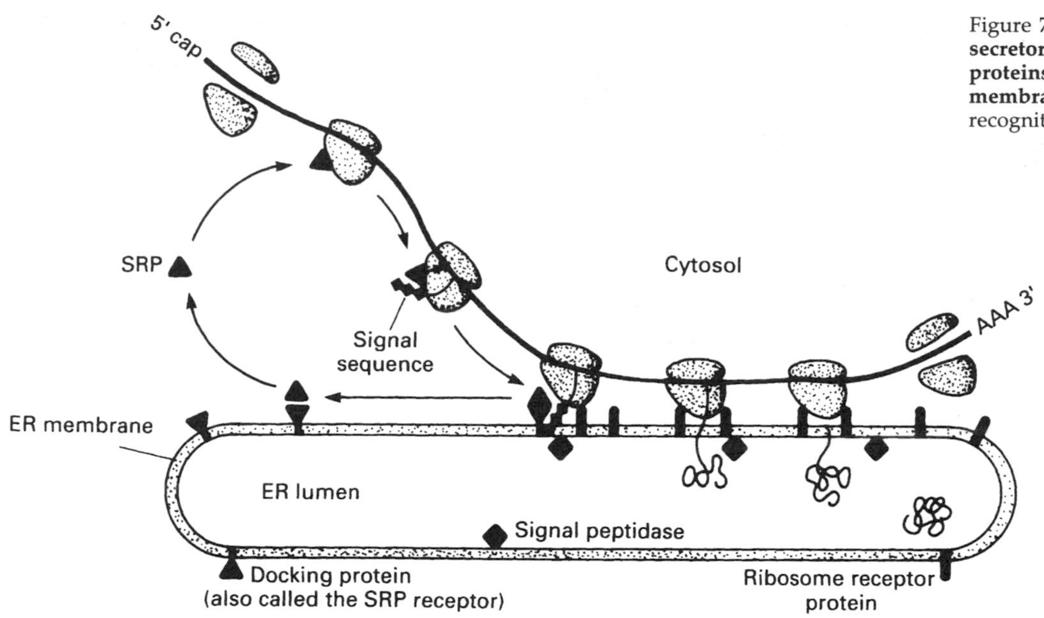

Figure 7.4 **The transport of secretory and membrane proteins across the ER membrane** (SRP, signal recognition particle)

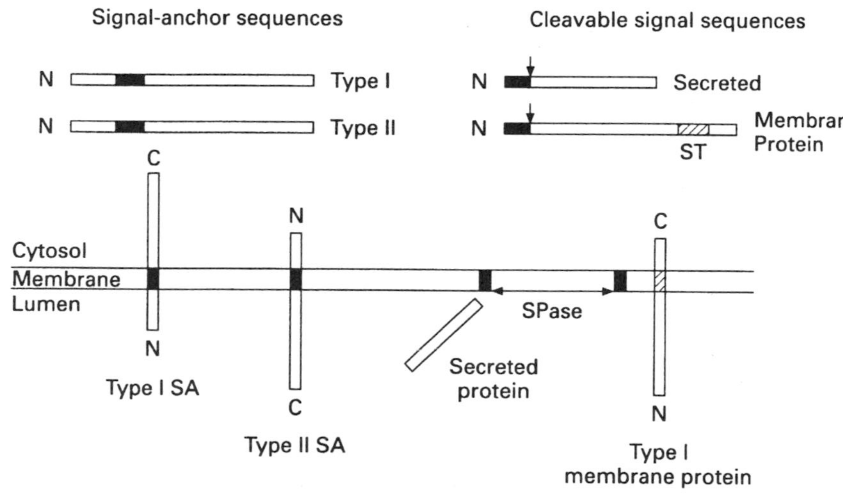

Figure 7.5 **Sequences of growing polypeptide chains that allow ribosomes to bind to the membranes of the endoplasmic reticulum (■,** Signal sequence; SA, signal anchor sequence; ST, stop-transfer sequence; SPase, signal peptidase)

the membrane several times (i.e. a polytopic membrane protein; see Chapter 2) and are then referred to type III proteins.

'Filter' F_2: retention in the endoplasmic reticulum

All proteins delivered to the ER, whether to the membrane or the lumen via filter F1 are transferred to the *cis*-Golgi network via their incorporation into vesicles that 'bud' from the ER and fuse with the Golgi. Any protein having a residential function in the ER, e.g. the SRP receptor, must be retrieved from the Golgi apparatus. This is achieved by the means of a special receptor that, in the *trans* -Golgi network, binds proteins with the sequence lysine–aspartic acid–glutamic acid–leucine (KDEL) towards their C-terminal ends. Both the KDEL receptor and any protein bound to it are returned to the ER by means of vesicles. This constitutes filter F2. The same mechanism handles both luminal and membrane proteins of the Golgi.

Post-translational modification of proteins in the ER/Golgi apparatus pathway

Covalent post-translational modification of proteins occurs extensively in the ER and Golgi apparatus. These modifications include disulphide bond formation (both inter- and intra-chain disulphide bridges) and proteolysis as well as covalent modification of specific amino acyl residues.

Some of the changes that occur relate to 'targeting' of the proteins involved, whereas others have to do with the functional activation of the protein. Thus the hormone cholecystokinin (CCK) is produced by proteolysis as it passes through the Golgi and also undergoes sulphation on a tyrosine residue. The sulphation of CCK is required for it to bind to its receptors in the brain. Figure 7.2 summarizes the alterations that arise in the ER/Golgi complex. Of major importance, both in terms of targeting and functioning of proteins, is the process of **glycosylation**. This can occur by means of the attachment of oligosaccharides to asparagine residues in proteins (*N*-linked) or to serine (or threonine) residues (O-linked). The former type of modification commences in the ER, even before the formation of the polypeptide chain is complete. The oligosaccharide is assembled attached to a membrane associated, lipid-soluble carrier molecule called **dolichol**. The whole of the preformed oligosaccharide is then transferred as a unit so that it becomes attached by covalent linkage to certain asparagine residues in the growing polypeptide chain. After the transfer of the glycosylated protein to the Golgi apparatus, the oligosaccharide chains undergo processes referred to respectively as '**trimming**', in which some monosaccharides are removed by hydrolysis and '**maturation**', in which additional monosaccharides are added. Prominent among the latter is the hexose mannose. In addition, *O*-glycosylation occurs in the Golgi apparatus.

Filter F_3: targeting to the lysosomes

One particular post-translational modification is essential for the delivery of the precursors of hydrolytic enzymes to the lysosomes. Such proteins have had carbohydrate added in *N*-linkage as they pass through the ER and the Golgi apparatus. Their 'default' passage to the cell surface is diverted by the phosphorylation of the sugar mannose in the 6-position. As was indicated previously, mannose is one of the types of monosaccharide added to carbohydrates, *N*-linked to proteins, during maturation in the Golgi apparatus. This organelle has in its *trans* membrane a mannose 6-phosphate receptor. Binding of proteins with the mannose 6-phosphate 'signature' to the receptor diverts them into vesicles which will ultimately give rise to lysosomes (see Chapter 8). This constitutes filter F_3.

Because it is so important that these potentially highly destructive precursors of lysosomal hydrolytic enzymes do not accumulate outside the cell, a 'fail-safe' mechanism exists. In this, a second, distinct mannose 6-phosphate receptor, located in the plasma membrane, retrieves any lysosomal proenzyme that may have inadvertently followed the 'default' pathway to be released from the cell.

Filter F_4 : regulated secretion

Not all proteins that are secreted from cells follow the 'default' pathway: the release of peptide hormones, for example, occurs on demand. There is therefore a filter (F_4) that ensures that such proteins are incorporated into vesicles, which after budding from the *trans* Golgi stack, fuse with the plasma only in response to external stimulation of the cell. These storage vesicles have recognizable 'dense cores'. The nature of filter F_4 is not fully resolved, but the presence of the dense cores in the vesicles provides a clue, because these contain precipitates consisting of the hormone to be secreted together with proteins known as **chromogranins**. It is believed that the *trans*-Golgi compartment undergoes acidification as the result of the action of a proton-pumping ATPase. (This enzyme is also responsible for the acidification of the various vesicles that derive from the Golgi apparatus, e.g. the lysosomes.) At acid pH values, the hormone is believed to co-precipitate with the chromogranin in nascent, 'budding' vesicles.

Filter F_5 : the cell surface retrieval filter

Some plasma membrane proteins, e.g. receptors, are believed to cycle between the plasma membrane and a so-called **'endosomal compartment' (endosomes)**. Such molecules reach the plasma membrane via the 'default' pathway. When necessary, the protein is retrieved to the endosomes. The nature of filter F_5 is complex, because it recognizes a particular sequence of part of the protein that is exposed to the cytoplasm and also special protein adapters known as **clathrins**. Very often the binding of a ligand to its receptor marks the complex for endocytosis. In this process, so-called '**clathrin-coated pits**' form in the region of the membrane proteins marked for endocytosis. These 'pits' bud from the membrane to form endosomes. This process is well-illustrated by the case of the transferrin receptor (see Chapter 21). Iron is transferred into cells when its circulating binding-protein, transferrin, binds to its receptor and a clathrin-coated pit forms. This leads to the endocytosis of the complex and ultimately to the acquisition of iron by the cell and the recycling of the receptor–transferrin complex to the cell surface. The internalization of immunoglobulin A (IgA) that precedes its **transcytosis** across mucosal cells, and secretion, occurs by a similar process, but in this case part of the IgA receptor also constitutes the 'secretory piece' that is added to the immunoglobulin to signal its release at the mucosal surface and protect the molecule from proteolysis (see Chapter 32).

Chapter 8

Cellular organelles: lysosomes

8.1 Introduction

The lysosomes are small sac-like organelles found in most eukaryotic cells, but are much more abundant in some cell types and tissues than in others. For example, they are abundant in macrophages, but lymphocytes have relatively few lysosomes. Lysosomes characteristically contain a battery of hydrolytic enzymes with a range of specificities.

Lysosomes are not static organelles, but move through a 'life-cycle' which is described in Section 8.4. For this reason the term 'lysosome' may refer to the organelle at any stage in the cycle.

Although the lysosomes are not found in bacteria, they emerged very early in evolution and they play an important role in protozoa such as amoeba, where they form the digestive system in the cell and thus play a vital role in the cellular nutrition. During the course of evolution this role of lysosomes has been partially retained, but often modified to suit other requirements. For example, lysosomes of macrophages play an important part in the body's defence against disease (see Chapter 32).

8.2 Origin of lysosomal enzymes

Lysosomal enzymes are proteins and, like many other proteins, are synthesized on ribosomes attached to the rough endoplasmic reticulum (see Chapter 7). The proteins destined to become lysosomal enzymes then receive an *N*-linked oligosaccharide, which is modified in the Golgi apparatus, most importantly to receive residues of the hexose sugar mannose. Some of these mannose residues undergo phosphorylation in the 6-position and this serves to 'target' the enzymes to the lysosomal compartment. A mannose 6-phosphate receptor located in the Golgi apparatus membrane binds lysosomal proteins that are then entrapped in a vesicle that eventually leaves the Golgi as a nascent lysosome (Figure 8.1). This 'targeting' of lysosomal enzymes is dealt with in Chapter 7.

8.3 Nature and properties of lysosomal enzymes

The battery of enzymes found in the lysosomes is largely hydrolytic in nature. (As is the case for their counterparts, found in the gastrointestinal tract, these enzymes are produced as inactive proenzymes.) Their major substrates are polymeric macromolecules; thus included in the set of lysosomal hydrolases are nucleases, proteases (usually referred to as cathepsins), glycosidases and also several lipases. In general, the lysosomal enzymes show relatively low specificity, although the cathepsins do show a degree of selectivity for peptide bonds involving particular amino acids. There is also an amidase that is specific for the amide link by which the *N*-acetylglucosamine of the carbohydrate chain is attached to the amino acid asparagine in glycoproteins.

The hydrolysis of some of the substrates for the lysosomal enzymes is, in some cases, facilitated by the substrates binding to accessory proteins. Thus the hydrolysis of sphingolipids is greatly accelerated by their binding to sphingolipid activating proteins (SAPs).

The enzymes of lysosomes generally have their pH optima in the range 4.0–5.5. This coincides with the occurrence in the lysosomal membrane of a proton-

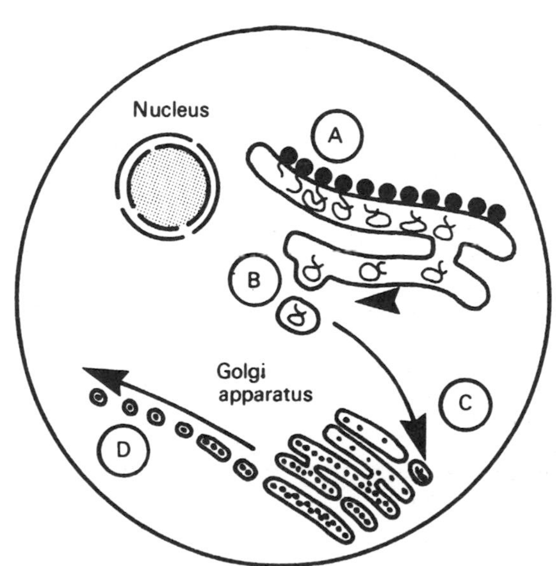

Figure 8.1 **Schematic representation of formation of lysosomes.** (A) Lysosomal enzymes are synthesized on the rough endoplasmic reticulum as glycoproteins. (B) The complete lysosomal enzyme complement is then transported to the smooth endoplasmic reticulum where enclosure in a portion of membrane occurs. (C) The enzymes, now contained in a membrane sac, are transported to the Golgi apparatus where mannose residues are phosphorylated. (D) A specific receptor for mannose 6-phosphate in the Golgi membrane binds the lysosomal protein and directs it to lysosomes

pumping ATPase that acts to acidify the matrix of the organelle to pH 5.5.

8.4 The life-cycle of the lysosome

Three mechanisms are recognized for the presentation of their substrates to the lysosomal enzymes, and these are illustrated in Figure 8.2. Two of these sources are exogenous and depend upon the uptake by cells of materials from their surroundings. They differ only in the size of the particles handled. Macromolecules and small particles undergo a process of endocytosis (sometimes known as pinocytosis) with the formation of endosomal vesicles. These fuse with nascent lysosomes (derived from the Golgi apparatus) to form endolysosomes. Large particles, e.g. a bacterium, undergo a process of phagocytosis and the phagosome so produced fuses with nascent lysosomes to yield phagolysosomes. Endogenous substrates arise as intracellular organelles, e.g. mitochondria, become damaged and are engulfed, probably by the endoplasmic reticulum, to yield autophagosomes. Fusion of these structures with nascent lysosomes affords autophagolysosomes. It is in these fusion structures that the macromolecules, particles and bacteria undergo digestion. Having completed the hydrolysis of their contents, fusion structures are often referred to simply as lysosomes. Any material that fails to be digested in this way, accumulates in the lysosomes, and this can lead to serious diseases (see section 8.6).

The final products of lysosomal digestion, but not the intermediates, are released into the cytosol. This is not just a question of passive diffusion, but rather requires the participation of membrane transport proteins, e.g. for glucose and cystine.

Of course cells cannot continue to accumulate lysosomes in this way and therefore an equilibrium exists whereby lysosomal membrane 'recycle' to the Golgi apparatus, taking with them the mannose 6-phosphate receptor. The hydrolases do not accumulate because, being proteins, they are ultimately digested by the cathepsins.

8.5 Functions of lysosomes in the tissues

Lysosomes perform important roles in various tissues. These roles show certain common features, but they differ from tissue to tissue and are related to tissue function.

General

In all tissues lysosomes play an important 'autolytic' role. Dying or dead cells rapidly acidified as the metabolism of the pyruvate formed by glycolysis is diverted, from its oxidation via the tricarboxylic acid cycle, into the formation of lactic acid (as the oxygen supply fails). The fall in pH inside cells means that lysosomal enzymes that begin to 'leak' from the organelle become maximally active. This view of lysosomes as 'suicide bags' was first put forward by their discoverer, the Belgian scientist Christian de Duve. This is, indeed, is one of the roles of lysosomes, but now many more rather more specialized roles have been recognized.

Kidney

Relatively large numbers of lysosomes are found in the endothelial cells of the glomerulus and in the lining cells of all the other structures of the kidney (see Chapter 26). Of these, lysosomes in the proximal tubules are likely to

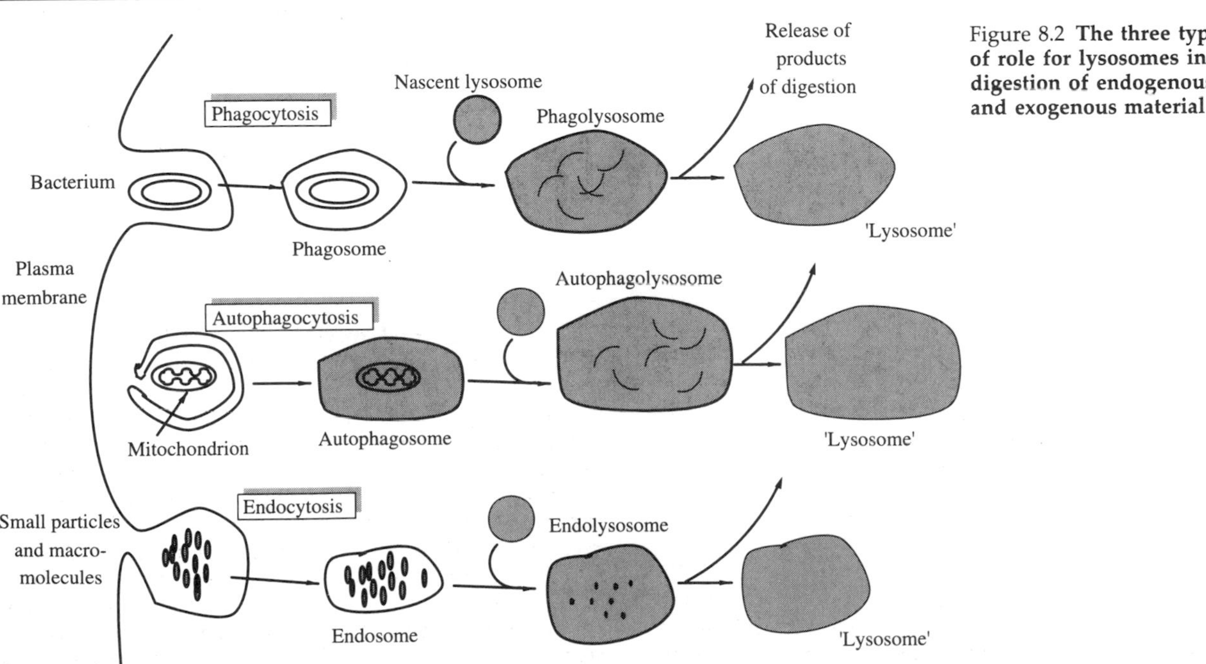

Figure 8.2 **The three types of role for lysosomes in the digestion of endogenous and exogenous material**

play a special role. Here protein molecules such as albumin that have adventitiously passed through the glomerulus undergo pinocytosis followed by degradation in endolysosomes catalysed by cathepsins. It has been estimated that up to 10–15% of total serum albumin degradation occurs in the kidney.

The 'professional phagocytes'

The so-called professional phagocytes, macrophages and neutrophils, are important cells of the immune system (see Chapter 32). The activity of lysosomes in macrophages have been most thoroughly investigated and they are known to subserve two related functions:

1. They take up foreign cells and proteins by phagocytosis and pinocytosis, respectively. In addition, in the spleen, they engulf aged or damaged red cells. The material taken up then undergoes lysosomal hydrolysis.
2. The macrophages are examples of antigen-presenting cells (see Chapter 32). In the mounting of an immune response, peptides generated in the endolysosomal compartment from foreign proteins are brought together with specific binding proteins which are part of the major histocompatibility complex (MHC II). This occurs when vesicles derived from the Golgi apparatus, and possessing MHC II proteins in their membranes, fuse with the endolysosomes. The foreign peptide–MHC II complex moves to the plasma membrane where the peptide is 'presented' to other cells of the immune system, thereby provoking an immune response to the invading organism.

The roles of neutrophils in the immune response are dealt with in Chapter 32.

Nervous system

It has only recently been appreciated that lysosomes are ubiquitous in the brain, both in neurons and in glial cells. Whether they have any special functional role in nervous tissue is unclear, but the accumulation of undigested substrates for the lysosomal enzymes seems to underlie a wide range of neurodegenerative disorders (see Section 8.6).

Bone

Lysosomal enzymes play key roles in the resorption and remodelling of bone (see Chapter 29). These processes involve the lysosomal-rich osteoclasts (and probably the osteocytes). Enzymes such as hyaluronidase and collagenase are active during bone resorption, but this only follows their release into the extracellular matrix.

8.6 Lysosomal storage diseases

With lysosomes playing such a central role in the turnover of the constituents of cells, thereby facilitating the removal of damaged organelles, it is not surprising that disorders of lysosomal function give rise to serious disorders, known as lysosomal storage diseases. Prominent among these are the approximately 30 inherited diseases that involve lysosomal dysfunction. All aspects of the life-cycle of the lysosome are known to be affected by such mutations, which may:

1. Render the enzymes inactive.
2. Cause a failure to target the enzymes correctly to the lysosomes.
3. Result in a failure to transport the products of hydrolysis out of the lysosome.
4. Result in the production of substrates which fail to undergo the expected lysosomal hydrolysis.

Some examples of inherited diseases associated with lysosome dysfunction are given below and others that specifically affect brain function are given in Chapter 30.

Type II glycogenosis: a glycogen storage disease

Patients with this condition have extensive deposits of glycogen in their tissues and they suffer from muscle weakness due to the deposition of the polysaccharide in the muscle fibres. The deficient enzyme is an α-glucosidase. This enzyme catalyses the hydrolysis of glycogen taken up into lysosomes. The normal pathway for glycogenolysis is intact in these patients and they are, therefore, not subject to periods of hypoglycaemia.

I-cell disease

The study of I-cell disease (also known as mucolipidosis II) has provided much information on the targeting of proteins to lysosomes. The name reflects the occurrence of large inclusion bodies in the lysosomes of tissues. These contain undegraded glycosaminoglycans and glycolipids. It has been observed that the lysosomes of patients with this condition, who present with psychomotor retardation and skeletal deformities, are deficient in about eight hydrolases. The defect is a mutation to the enzyme that catalyses the phosphorylation of mannose in glycoproteins in the 6-position. Without this targeting signal, the proenzymes that should be delivered to the lysosomes are released from the cell instead.

Cystinosis

This is an inherited condition in which the transport properties of the lysosomal membrane are affected rather than one of the hydrolases. The disease was first described by Fanconi in Switzerland who noted that children affected showed signs of stunted growth, rickets and glycosuria. It became apparent that these observed effects are associated with progressive renal glomerular damage. Close examination showed that large intracellular stores of cystine accumulate in affected individuals and the stores are largely lysosomal. It is now known that the efflux of cystine out of the lysosomes is coupled to their ATP-dependent uptake of protons. In patients with cystinosis, this process is deficient. Consequently, in the tissues of affected individuals cystine is formed in lysosomes as a result of cathepsin-mediated protein catabolism, but the amino acid cannot be released into the cytosol. Cystine is sparingly soluble, so that crystals readily form and ultimately these will disrupt the lysosomes and in turn the lysosomal enzymes will destroy the cell.

Neurodegenerative diseases

Recently, a range of neurodegenerative disorders have formed the focus of much attention. These include Alzheimer's disease, Huntingdon's chorea, Kennedy's disease and the prion diseases (e.g. Creutzfeld–Jakob disease). At first it was believed that these diseases could in no way be related because different brain regions or systems were affected in each. Further, when the molecular nature of the defects involved were resolved there appeared to be no common factor. It is now appreciated that the variant proteins produced in each disease are resistant to hydrolysis in the lysosomes, in which they accumulate causing disruption of the membrane with consequent release of the battery of hydrolases. These, it is postulated, contribute to the degeneration of the groups of neurons that produce the aberrant protein (see Chapters 30 and 43).

Part 2

Human metabolism and nutrition

Chapter 9

Plasma glucose and its regulation

9.1 Introduction

The concentration of glucose in blood in all mammals is maintained at a very constant level (Figure 9.1). In the normal human subject the blood glucose concentration, even after meals, rarely rises more than 3–4 mmol/l from the overnight-fasting value of about 4.5 mmol/l, and prolonged exercise causes it to fall by less than 30%. This relatively narrow range can be maintained even with a glucose intake as high as 4 mol/day, a remarkable feat when one considers that total body free glucose is approximately 0.1 mol (18 g).

The precise maintenance of the glucose concentration must possess strong survival value because, during the course of evolution, the control mechanisms have become progressively more effective and accurate. In reptiles, for example, the fluctuations of blood glucose are much greater. The reason for this very effective homeostasis of glucose is probably associated with the needs of the brain which uses between 60% and 80% of the glucose released in the fasting state. A very serious clinical condition, hypoglycaemia, develops if the blood glucose concentration falls dramatically, for example to below half its normal value. At these low concentrations, transport across the blood–brain barrier and into brain cells becomes rate limiting for glucose use and, as a result, cerebral function is impaired leading to coma, convulsions and even death.

Conversely, while a raised glucose concentration in blood, for example to double its normal value, is not in itself dangerous over a short period, over a longer period, as in poorly controlled diabetes mellitus (see Chapter 41), glucose in high concentrations in the blood leads to glycosylation of proteins, for example in the basement membranes of blood vessel walls, and this can lead to significant pathological disturbances. Glycosylation of the crystallins of the lens of the eye is associated with cataract formation (see Chapter 31), whilst that of haemoglobin gives rise to a new form of haemoglobin HbA_{1c}.

The measurement of HbA_{1c} can indicate the severity of the diabetes. A raised fasting glucose concentration in the plasma (>8.0 mM) indicates an existence of clinical disease, usually diabetes. The high circulating concentration of glucose causes an osmotic diuresis accompanied by loss of water and electrolytes as well as glucose in the urine. Thus, it is possible to carry out a simple test for diabetes by testing urine for glucose. In normal subjects a negligible quantity of glucose should appear in the urine. Some other pathological conditions, for example kidney disease, can cause an increase in glucose concentration in the urine; confirmation of the diagnosis of diabetes can, however, be made by the 'oral glucose tolerance test'. The fasted subject is given a test dose of 75 g glucose, the blood glucose being measured just before the test and at half-hourly intervals afterwards. In normal individuals, the raised glucose concentration in blood, resulting from the intestinal intake is rapidly lowered to a normal value but, in diabetic individuals who secrete little or no insulin, the glucose concentration remains at a much higher level for a prolonged period (Figure 9.2). The category of 'impaired glucose tolerance' has been created to accommodate equivocal responses and spares the patient from being categorized unjustifiably as 'diabetic'.

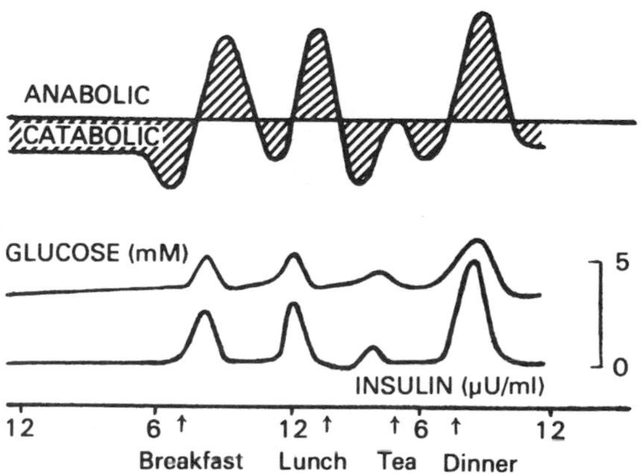

Figure 9.1 **Changes in glucose and insulin levels during a typical day showing responses to meals**

9.2 Maintenance of glucose concentration in the fasting state

Between meals, the blood glucose is maintained by the catabolism of liver glycogen by the process shown in Chapter 6. The total body content of free glucose has been estimated to be only 0.1 mol (18 g) or sufficient for 1 hour's worth of fuel for basal energy requirements. In contrast, about 100 g glycogen is stored in the liver and as the human liver can produce about 125–150 mg glucose/minute or 180–220 g/24 h, this store can, therefore, last about 12–16 h. It is unlikely, however, that stores of glycogen are normally run down to zero during the overnight fasting period, and gluconeogenesis (see Chapter 6) is switched on to provide blood glucose.

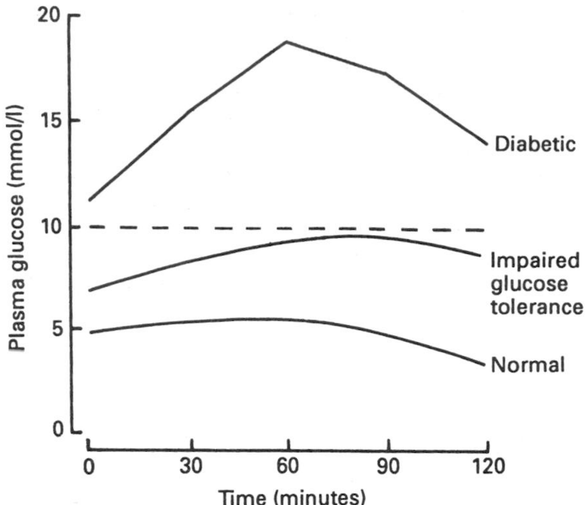

Figure 9.2 **Oral glucose tolerance test.** Plasma glucose concentrations following an oral 75 g glucose load in subjects suffering from diabetes mellitus, subjects with impaired glucose tolerance and subjects with normal glucose tolerance. ----Indicates the maximal tubular reabsorption rate for glucose (T_{mg}) which occurs in most normal subjects at a blood glucose concentration of about 10 mmol/l

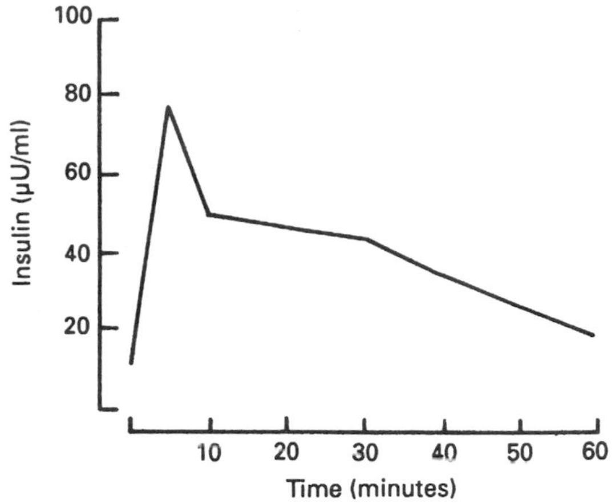

Figure 9.3 **Insulin release in man following the intravenous infusion of a dose of glucose**

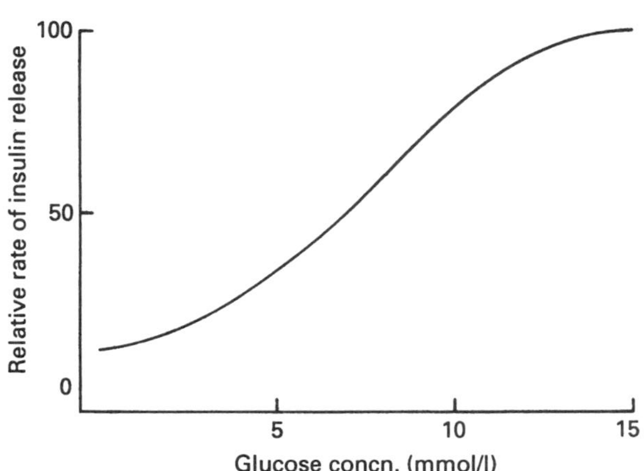

Figure 9.4 **Effect of glucose concentration on the rate of insulin release from isolated islets of Langerhans**

The major gluconeogenic precursors during this period are recycled lactate and pyruvate, which are derived predominantly from red blood cells and skeletal muscle. Glucose can also be produced from amino acids released from the muscle and liver proteins, and from glycerol, released following the hydrolysis of adipose tissue triacylglycerols. During fasting the plasma concentration of insulin is lowered and this permits the breakdown of muscle proteins and the hydrolysis of triacylglycerols in adipose tissue. If, however, the fasting condition is prolonged for several days, the release of amino acids from muscle is depressed. This adaptation to starvation is discussed in Chapter 18.

The major proportion of the glucose, about 70%, is utilized by the central nervous system and the remainder by the red blood cells, bone marrow and renal medulla. However, unlike the brain where most of the glucose is completely oxidized to CO_2, the other tissues convert it to lactate which circulates to the liver where it is resynthesized into glycogen ('the Cori cycle').

9.3 Insulin release in the fed state

One of the most characteristic responses to the ingestion of a carbohydrate meal is the rise in blood glucose concentration which is accompanied by a rise in the concentration of plasma insulin (see Figure 9.1). The response of the pancreatic B-cells in secreting insulin is extremely rapid, a fact which can be demonstrated by infusing glucose intravenously. When this is carried out, an increase in insulin concentration can be demonstrated to occur within a minute (Figure 9.3). The first rapid 'spike' of insulin released is followed by a slow second phase of release. It has been suggested that the first phase results from the release of insulin from granules which are near the plasma membrane of the B-cells. It has been known for a number of years that more insulin appears to be released into the circulation when glucose is administered orally than when the same quantity of glucose is infused intravenously. It has been assumed that this is caused by the release of a gastrointestinal hormone following the interaction of glucose with cells in the gastrointestinal tract. The hormone responsible for potentiating the release of insulin has not been conclusively identified but the most likely candidate is the gastric inhibitory peptide.

Studies with isolated islets of Langerhans have shown that the secretory response to glucose follows a sigmoidal pattern (Figure 9.4). The response is most sensitive over a comparatively narrow range of glucose concentration just above the fasting level, and is clearly designed to ensure that blood glucose concentrations return to fasting values as soon as possible after ingestion of carbohydrate. In addition to these direct responses of the pancreas to glucose, certain regions of the brain are 'glucose sensitive' and neural mechanisms can also act to control insulin secretion.

The rapid release of insulin is much more important in the control of blood glucose than the total quantity of insulin release over a period of time. This fact can be demonstrated in patients suffering from severe diabetes of the juvenile onset form who produce very little insulin. The effects observed after administration of insulin to these individuals clearly show that the efficient utilization of glucose correlates most closely to the amount of insulin available during the first 10 min of dosing with glucose rather than the total insulin release. These observations have led to the replacement of the former practice of injecting a large dose of insulin for the treatment of diabetic ketoacidosis by a low-dose-infusion or injection procedure (see Chapter 41).

9.4 Tissue response to increased plasma insulin and glucose concentrations

Glucose is rapidly taken up into liver cells by a passive carrier system (Glut 2) which has a high V_{max} for glucose. Once inside the cell, glucose is converted to glucose 6-phosphate by the enzyme glucokinase (hexokinase type D). This isoenzyme of hexokinase, which is found only in hepatocytes and the B-cells of the pancreas, is unique in that it has a low affinity for glucose, and is not inhibited by the product of the reaction, glucose 6-phosphate. Thus in the liver, hyperglycaemia results in increased glucose phosphorylation, and hence glucose uptake and utilization. The glucosyl residue of glucose 6-phosphate can be incorporated into glycogen, or undergo glycolysis to pyruvate, with the pyruvate being subsequently oxidized to acetylCoA which is used both for energy requirements in the citric acid cycle and also for fatty acid synthesis. Fatty acid synthesis is dependent on a source of NADPH which is obtained by the metabolism of glucose 6-phosphate by means of the pentose phosphate pathway.

Recently, however, evidence has been emerging that, unlike fructose (see Section 9.7), glucose is in fact a relatively poor substrate for liver metabolism and that during the postprandial phase, most of the glycogen and fatty acids synthesized in the liver are derived from three-carbon intermediates such as lactate, and not directly from the plasma glucose itself. Studies of the effects of glucose infusion in man show that only a small quantity of glucose, about 10%, is recovered in additional liver glycogen, whereas some 30–40% is recovered in muscle glycogen. Of the remaining 50%, about half is probably oxidized and half converted into

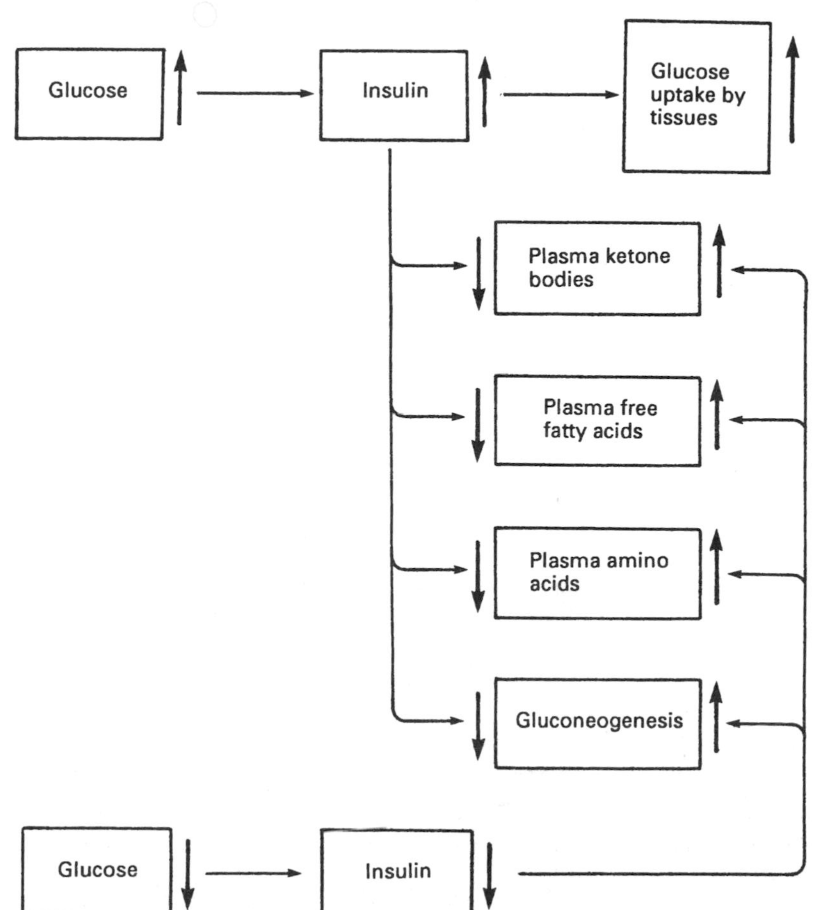

Figure 9.5 **The role of insulin in the regulation of plasma glucose, amino acids, fatty acids and ketone bodies.**

triacylglycerol stores. The peripheral metabolism of glucose produces lactate which, together with other gluconeogenic precursors, is then taken up by the liver. There is also evidence that some of the absorbed glucose is converted to lactate in the epithelial cells of the small intestine, and this lactate is also transported to the liver, via the hepatic portal vein. The lactate can then be efficiently converted to glucose 6-phosphate and hence glycogen by means of the gluconeogenic pathway.

The uptake of glucose into muscle and adipose tissue is controlled by insulin. The hormone stimulates glucose transport by increasing the number of glucose transporters (Glut 4) available at the plasma membrane by promoting their transfer from intracellular membranes. In the muscle, glucose is incorporated into glycogen as well as being used as fuel for contraction, while in adipose tissue, glucose is converted into glycerol 3-phosphate which is subsequently used to make triacylglycerols.

In addition to its effect on the uptake of glucose by muscle, insulin has an important regulatory action in the supply of amino acids from muscle protein for gluconeogenesis. The release of amino acids is dependent on a low level of circulating insulin. When insulin concentration is raised as the result of glucose intake, the uptake of amino acids into muscle is promoted, as is their conversion into protein. The release of amino acids therefore ceases immediately.

Insulin also switches off the release of fatty acids from adipose tissue by decreasing the activity of triacylglycerol lipase. The control of the lipase by insulin is very complicated. The essential sequence of events is shown in Figure 9.5.

9.5 The role of glucagon

Glucagon which is released from the A-cells of the pancreatic islets, also plays a very important role in the regulation of the plasma glucose concentration. Its effects are virtually opposite to those of insulin, stimulating liver glycogenolysis, inhibiting glycogen synthesis and stimulating gluconeogenesis in the liver. Glucagon is, therefore, more active during the fasting condition when the concentration of insulin is low and its release is inhibited by an influx of glucose.

Glucagon exerts its effects on metabolic processes through activation of adenylyl cyclase and a rise in cAMP levels. Glycogenolysis is triggered by the activation of phosphorylase which catalyses the phosphorolysis of glycogen producing glucose 1-phosphate (see Chapter 6).

It is also likely that glucagon, in addition to and in conjunction with insulin, plays an important role in regulating glucose production from amino acids following a large protein meal. An influx of amino acids under such conditions, like glucose causes a stimulation of insulin release which immediately switches off the liberation of glucose from liver glycogen. This could clearly result in a disastrous hypoglycaemia. Amino acids however, also stimulate the secretion of glucagon which, by promoting glycogenolysis and gluconeogenesis, can compensate for the hypoglycaemic action of insulin.

9.6 Carbohydrate metabolism in the fetus and newborn

It is generally accepted that a continuous supply of glucose is of particular importance to the developing embryo, the fetus and the young animal after birth. The fetus *in utero* receives a continuous supply of glucose from the maternal circulation via the placenta but during the third trimester of pregnancy, several fetal tissues accumulate large amounts of glycogen. Thus the concentration of liver glycogen at term is usually two to three times that of the adult. Fetal animals also tend to accumulate large amounts of glycogen in their skeletal and cardiac muscles during the latter part of gestation.

The liver glycogen concentration falls very rapidly after birth; the level usually reaches 10% or less of the initial value within 12–24 h, and remains low for several days before rising gradually towards the adult level which is reached within 2–3 weeks. The glycogen concentration in the skeletal muscles also falls after birth, though more slowly than that in the liver.

There is an interval between birth and the establishment of feeding, digestion and absorption from the gut, during which the newborn animal is entirely dependent on its own reserves of energy and this helps to explain the approximate 10% loss of body weight which occurs *post partum.* During this period, there will be an increase in the metabolic rate after birth owing to increased physical activity, the work of breathing etc. The blood glucose concentration falls to about 3 mmol/l within a few hours of birth in normal full-term babies and usually stays low for some days (Figure 9.6).

The rapid postnatal fall in liver glycogen suggests that it may play an important part in the maintenance of blood glucose concentration during the first few days of independent life. This implies that the glycogenolytic

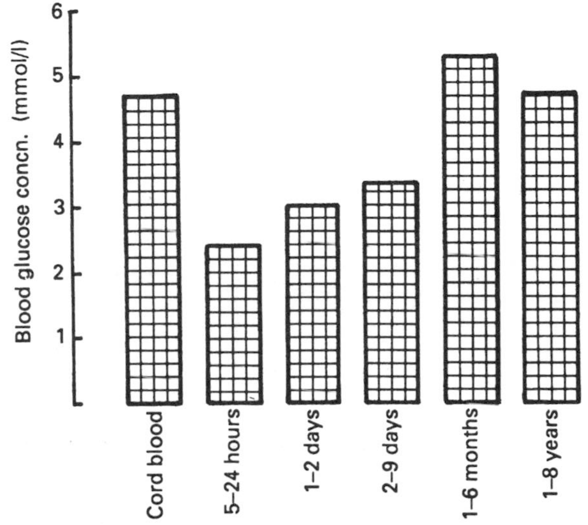

Figure 9.6 **Glucose concentrations in the blood of neonates, infants and children**

enzymes must be functional at birth and that glucose 6-phosphatase is also active. The subsequent rise in blood glucose is probably due to the development of gluconeogenesis and the ability to convert galactose and fructose into glucose (see below).

9.7 Metabolism of fructose and galactose

In many societies, a considerable proportion of dietary carbohydrate is in the form of sucrose, which is a disaccharide composed of fructose and glucose. It has been estimated that a Western diet may contain on average about 50–70 g of fructose, mainly in the form of sucrose. Fructose is absorbed in the gut by facilitated diffusion and is taken up from the hepatic circulation by the liver where it is phosphorylated by a specific fructokinase. The fructose 1-phosphate formed is split by fructose bisphosphate aldolase into dihydroxyacetone phosphate and glyceraldehyde. The latter is phosphorylated to glyceraldehyde 3-phosphate by the enzyme triokinase, so that fructose metabolism results in the formation of the two triose phosphate intermediates of glycolysis. The fate of the triose phosphates is determined by the hormonal state of the liver. In the fed state fructose is very rapidly converted to glucose (gluconeogenesis). Some fructose is directly converted to glycogen and acetylCoA which can be used for fatty acid synthesis, and some is oxidized to CO_2. The pathway is summarized in Figure 9.7.

The metabolism of fructose, unlike that of glucose does not involve the enzyme phosphofructokinase. Since this enzyme is thought to play an important role in

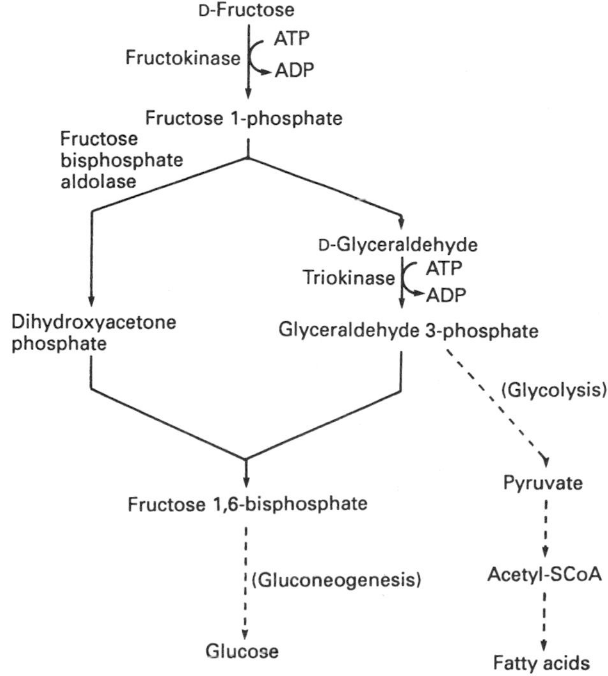

Figure 9.7 **Metabolism of fructose**

D-Glucose		L-Sorbitol		D-Fructose
CHO		CH_2OH		CH_2OH
HCOH	Aldose reductase →	HCOH	Sorbitol dehydrogenase →	CO
HOCH	NADPH + H^+ → $NADP^+$	HOCH	NAD^+ → NADH + H^+	HOCH
HCOH		HCOH		HCOH
HCOH		HCOH		HCOH
CH_2OH		CH_2OH		CH_2OH

Figure 9.8 **The sorbitol (polyol) pathway**

controlling the rate of glycolysis, the ability of fructose to bypass this reaction may explain the observation that fructose will be converted to acetylCoA more readily than glucose. This may account for the link between sucrose ingestion and increased levels of triacylglycerol in the bloodstream.

Fructose is not only metabolized by man, it is also produced from glucose by some tissues. Spermatozoa live on fructose which is produced by the prostate and seminal vesicles. The interconversion of the hexoses involves sorbitol, one of a number of polyols (polyhydric alcohols) of metabolic and clinical importance. Glucose undergoes reduction to sorbitol catalysed by the NADPH-dependent aldose reductase, followed by the oxidation of sorbitol to fructose in the presence of NAD^+ and sorbitol dehydrogenase. The sorbitol or polyol pathway (Figure 9.8) is also found in the liver, lens, nerve tissue, blood vessels and muscle. During hyperglycaemia, sorbitol has been shown to accumulate in some of these tissues and may be involved in the pathogenesis of diabetic complications (see Chapter 41).

Galactose is derived from the hydrolysis of lactose. It is synthesized from glucose in actively secreting mammary glands, and the blood and urine of lactating women may contain both galactose and lactose. Galactose has its own specific enzyme, galactokinase, which converts the hexose to galactose 1-phosphate. Then, in a somewhat unusual reaction, galactose 1-phosphate reacts with uridine diphosphate glucose (UDP-glucose) to form UDP-galactose and glucose 1-phosphate; the reaction is catalysed by galactose-1-phosphate uridyltransferase. The glucose 1-phosphate can be converted to glucose by means of the reactions catalysed by phosphoglucomutase and glucose 6-phosphate, while UDP-galactose is re-converted to UDP-glucose by the enzyme UDP-glucose 4-epimerase. These metabolic reactions are summarized in Figure 9.9.

9.8 Circulating glucose under stressful conditions

So far we have discussed the responses of glucose in the blood to varied conditions of nutrition; marked fluctuations can also, however, occur for short periods under conditions of stress which are unrelated to the nutritional state.

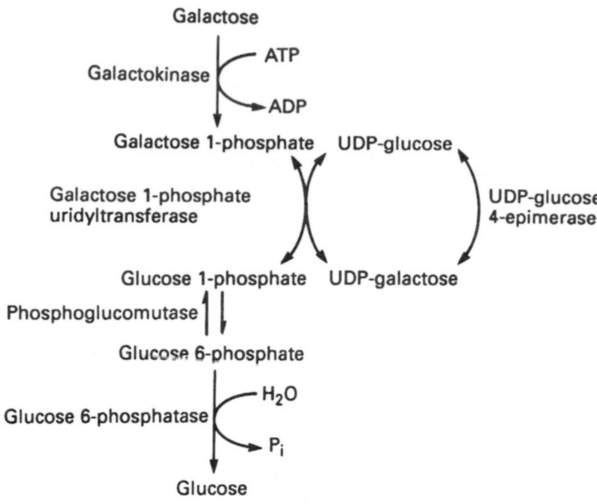

Figure 9.9 **Metabolism of galactose**

One of the best studied hormones to cause this effect is adrenaline. This hormone, together with some other catecholamines, is released in response to a very wide variety of stressful situations or pathological conditions, such as hypoxia, asphyxia, acidaemia, hypoglycaemia, hypothermia, hypotension, haemorrhage, exercise, fear, anger or excitement. The release is triggered by the sympathetic nervous system and was described in 1929 by Cannon as the 'fright, fight or flight' syndrome.

Adrenaline causes a wide variety of important physiological effects particularly on the circulatory system, but it is also a powerful stimulator of glycogenolysis and causes a rapid increase in the concentration of blood glucose. The action of adrenaline is mediated by the formation of cAMP in a similar manner to the action of glucagon. Adrenaline also causes release of fatty acids from adipose tissue and the glucose and fatty acids clearly provide the energy for the 'flight' or 'fight' (see Chapter 40).

In addition to adrenaline, the glucocorticoids released from the adrenal cortex also have important effects on blood glucose (see Chapter 28). Their effects are, however, generally responses to a more prolonged stress condition, such as a debilitating illness, rather than the immediate response which occurs following adrenaline release. The glucocorticoids have important effects on gluconeogenesis and these are exerted in two ways. First, they stimulate catabolism of muscle protein releasing amino acids for uptake by the liver, and secondly they stimulate the production of enzymes, such as aminotransferases, in the liver, these enzymes aiding the conversion of amino acids to glucose in the liver.

In summary, it can be stated that the concentration of glucose in the plasma is dependent on a wide range of nutritional and hormonal factors, but these are carefully interregulated to effect a remarkably accurate control, and reliable homeostasis.

Chapter 10

Plasma lipids and their regulation

10.1 Introduction

Whereas glucose and amino acids are transported in aqueous solution in the blood, the transport of the water-insoluble lipids involves the participation of a range of complex molecules. All lipids are transported as part of lipoprotein complexes and the majority of these contain triacylglycerols, phospholipids, cholesterol and its esters. The proportions of these vary considerably between the different lipoprotein classes. A partial exception to this rule are the non-esterified fatty acids (NEFA), which are transported bound to albumin.

The biosynthesis and utilization of the components of these complexes are now being elucidated and it is appreciated what a central role they play in both health and disease. In particular, the circulating concentrations of certain of the lipoproteins is strongly positively correlated with the incidence of atheromatous vascular disease.

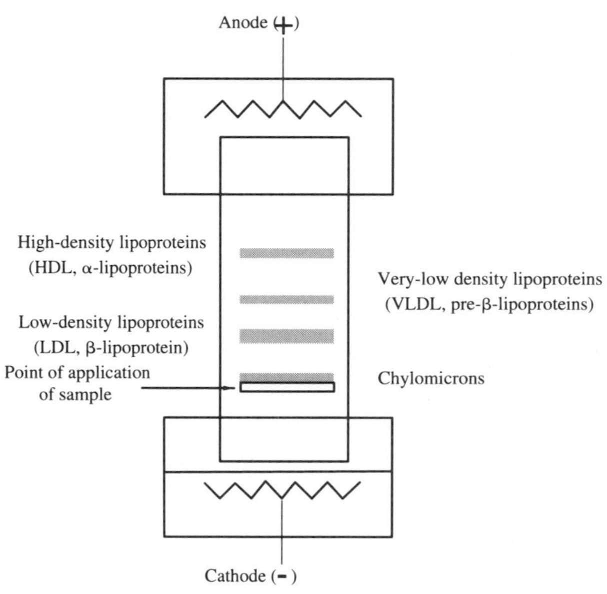

Figure 10.1 **Electrophoretic mobility of the plasma lipoproteins on cellulose acetate at pH 8.3**

10.2 Classification of plasma lipids

It is usual to classify plasma lipids into one of two groups; those associated with:

1. Albumin, i.e. NEFA
2. Lipoproteins

Non-esterified fatty acids

Non-esterified fatty acids are often referred to as 'free fatty acids', but it should be noted that they are neither free nor acids. As indicated, they are bound to albumin (at least 99%) and at pH 7.4 they are present as anions. Consequently the abbreviation NEFA is much to be preferred. The total circulating concentration of NEFA ranges from about 0.2 to 0.6 mM (depending on nutritional status), but may rise as high as 2.0 mM in severe stress, for example.

Lipoproteins

Following their discovery, great efforts went into the development of ways of separating the plasma lipoproteins. Two methods proved to be successful: electrophoresis on agarose gels and ultracentrifugation. The first method, which separates the complexes according to their charge and mass, yields four major bands. The second allows separation according to density and also yields four major fractions. The electrophoretic bands found are indicated in Figure 10.1 which also shows the alternative names used to specify them. Currently the names used to designate the density of the lipoproteins are commonly used in naming them. These are, in order of increasing density, very-low-density lipoproteins (VLDL), low-density lipoproteins (LDL) and high-density lipoproteins (HDL). These forms of lipoprotein are all derived from endogenous lipid components, whereas the fourth type of lipoprotein, termed chylomicrons, derive their lipids from dietary sources. The density of chylomicrons is comparable to that of VLDL. Another lipoprotein class is now recognized as being intermediate in density between VLDL and LDL and is termed IDL. IDL lies on the pathway by which VLDL is converted into LDL.

Although much effort went into the separation and chemical characterization of plasma lipoproteins, it is now appreciated that a keynote of their existence is their ability to exchange materials, both lipids and proteins. In this respect the HDL complexes play a central role.

Table 10.1 **Composition of the lipoprotein families**

Lipoprotein	*Composition (% dry wt) of*					
	Protein	*Triacylglycerol*	*Unesterified cholesterol*	*Esterified cholesterol*	*Phospholipids*	*Total lipid (% dry wt)*
Chylomicrons	2	86	2	3	7	98
VLDLs	5–12	50–60	6–8	10–13	13–20	88–95
IDLs	19	23	9	29	19	81
LDLs	22	6–10	10	47–55	28–30	78
HDLs	50	6	3	14	25	50

VLDLs = very-low-density lipoproteins
IDLs = intermediate density lipoproteins
LDLs = low-density lipoproteins
HDLs = high-density lipoproteins

Constitution of the lipoproteins

Lipids

Typical analyses of the lipid composition of plasma lipoproteins are shown in Table 10.1. The following observations are especially noteworthy:

1. All lipoproteins carry all types of lipid, although the quantities and relative proportions vary widely.
2. The density is directly proportional to the protein content and inversely proportional to the lipid content.
3. The bulk of plasma triacylglycerol is transported associated with chylomicrons or VLDL.
4. The bulk of the plasma cholesterol is transported associated with LDL.

Proteins

The proteins of lipoproteins confer specificity. Some apoproteins serve to define the type of receptor with which the lipoprotein particle may bind. Other apoproteins serve to activate lipoprotein-specific enzymes or, in some cases, they are themselves enzymes. Five broad groups of apolipoprotein (A–E) have been identified as being present in lipoproteins. In some cases there is more than one gene for a particular class of apolipoprotein and these are specified by post-scripts (e.g. apo A-I, C-II). In addition, different alleles at a particular locus are recognized in the population, i.e. there are polymorphisms. The possession of particular alleles is sometimes found to be associated with increased susceptibility to certain diseases. For example, for individuals expressing the ϵ4 allele for apoE there is a greatly increased risk of developing Alzheimer's disease (see Chapter 30; the role of lipoproteins in brain biochemistry is not understood). According to current nomenclature, group A apolipoproteins are those found in highest amounts in HDL while group B are associated with VLDL, LDL and chylomicrons. ApoA activates the HDL-associated enzyme lecithin: cholesterol acyltransferase (LCAT), while apoB is required for the recognition of LDL by cells. Members of groups C and E exchange between several classes of lipoproteins, with HDL acting as the distributor. ApoD acts to facilitate the transfer of esters of cholesterol between lipoproteins. Two forms of apoB are recognized: apoB-100 and apoB-48. The former is found in VLDL and the latter in chylomicrons. It has been found that apoB-100 is large protein (c. 4500 amino acid residues), while apoB-48 consists of the N-terminal 48% of apoB-100 (which, according to this nomenclature, is 100%). Those apolipoproteins for which a definite role has been assigned are listed in Table 10.2.

Table 10.2 **The functional roles of some of the apolipoproteins**

Apolipoprotein	*Function*
A-I	Activates lecithin:cholesterol acyl transferase (LCAT)
B-100	Directs binding of LDL particles to the LDL receptor
B-48	Together with apoE directs binding of chylomicron remnants to hepatic lipoprotein receptors
C-II	Activates lipoprotein lipase
D	Acts as a cholesteryl ester transfer protein to facilitate the movement of cholesteryl esters from HDL_3 to other lipoproteins
E	Together with apoC-II directs binding of chylomicron remnants to hepatic lipoprotein receptors. In higher concentrations, directs binding of LDL particles to the LDL receptor

Structure of lipoproteins

By the use of X-ray diffraction techniques progress, has been made in elucidating the very complex structures of the lipoproteins. In general the structures of the different plasma lipoproteins are very similar, being more or less spherical. The hydrophobic components, triacylglycerols and cholesteryl esters are located in the core of the complexes surrounded by a monolayer of amphipathic surface constituents which helps maintain water stability.

The larger particles (VLDLs and chylomicrons) are composed of predominantly triacylglycerol cores surrounded by a monolayer of phospholipids and cholesterol into which some of the apolipoproteins are embedded and to which others bind. In the more compact LDL, the hydrophobic core consists largely of cholesteryl esters, while the surface consists of phospholipids and

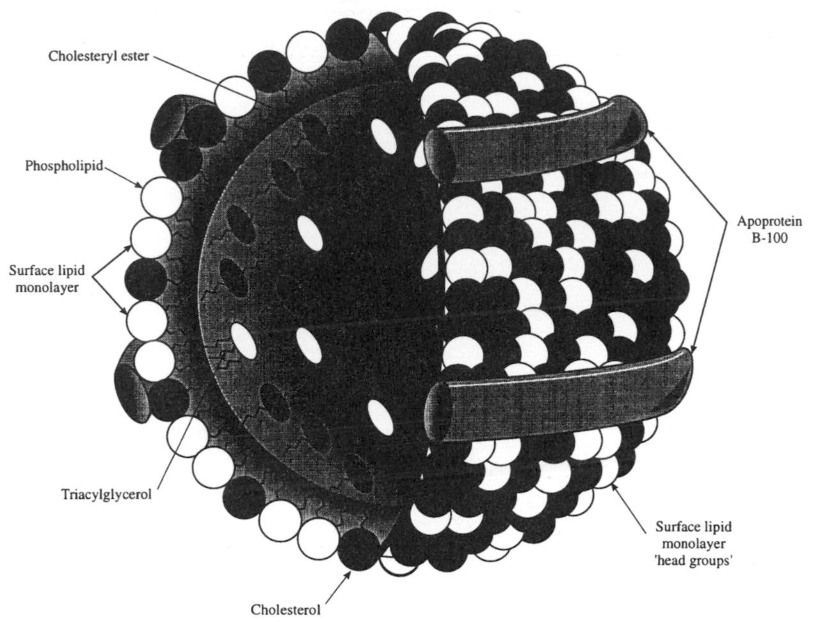

Figure 10.2 **The postulated structure of the LDL particle, showing apoprotein B-100 encircling the structure and a 'cut-away' to reveal part of the predominant cholesteryl ester core of the LDL and the lipid monolayer surface**

cholesterol and some of the apolipoproteins inherited from the VLDL. Two structural variants of HDL are recognized. Nascent HDL particles, released following synthesis in the liver, are discoid phospholipid/cholesterol bilayers (somewhat resembling membrane bilayers). These undergo refinement in the circulation when cholesterol is esterified and the resulting esters force the two layers of the lipid bilayer apart as they accumulate between them. Ultimately this converts the discoid nascent particles into spherical structures with a typical amphipathic lipid/apolipoprotein monolayer surface and a viscous core consisting largely of cholesteryl ester. The exposed outer surface area at this stage consists in roughly equal proportions of polar heads of phospholipids and polar side-chains of proteins, with a little cholesterol. A representation of the structure of an LDL particle, as determined by X-ray analysis is shown in Figure 10.2.

Sources and utilization of plasma lipids

As a preliminary to the discussion of the metabolism of lipoproteins it is useful to consider the sources of plasma lipids and also their sites of catabolism. These are summarised in Table 10.3. Generally, NEFA are found to have a very short plasma half-life (of the order of minutes), this stretches to 5–10 h for triacylglycerols and to days for cholesterol and its esters.

In considering lipid homeostasis it is convenient to follow the fate of plasma lipids generated in the 'fed state', i.e. immediately after a meal, and in the 'fasting state', e.g. after an overnight fast. Secondly, although the metabolic fates of cholesterol and triacylglycerols are interconnected by virtue of the presence of both types of lipid in the lipoproteins, it is convenient to deal with each lipid type separately.

Table 10.3 **Sources and metabolic fates of lipoproteins**

Plasma component	*Source*	*Metabolic fate*
Non-esterified fatty acids – albumin	Adipose tissue	Uptake of non-esterified fatty acids by many tissues, especially liver and muscle
Chylomicrons	Small intestine	Peripheral catabolism to remnants, fatty acids enter peripheral tissues
Chylomicron remnants	In the circulation, from chylomicrons	Uptake into the liver followed by catabolism
Very-low density lipoprotein (VLDL)	Liver (some from the small intestine)	Peripheral catabolism to IDL – fatty acids enter peripheral tissues, especially the heart
Intermediate-density lipoprotein (IDL)	In the circulation, from VLDL	Uptake into the liver followed by catabolism, or conversion into LDL in the circulation
Low-density lipoprotein (LDL)	In the circulation, from IDL	Uptake into peripheral tissues and the liver followed by catabolism in both
High-density lipoprotein (HDL)	Liver (some from the small intestine)	Cholesteryl ester transfer to other lipoproteins (mainly LDL), ultimately uptake into liver or steroidogenic tissues

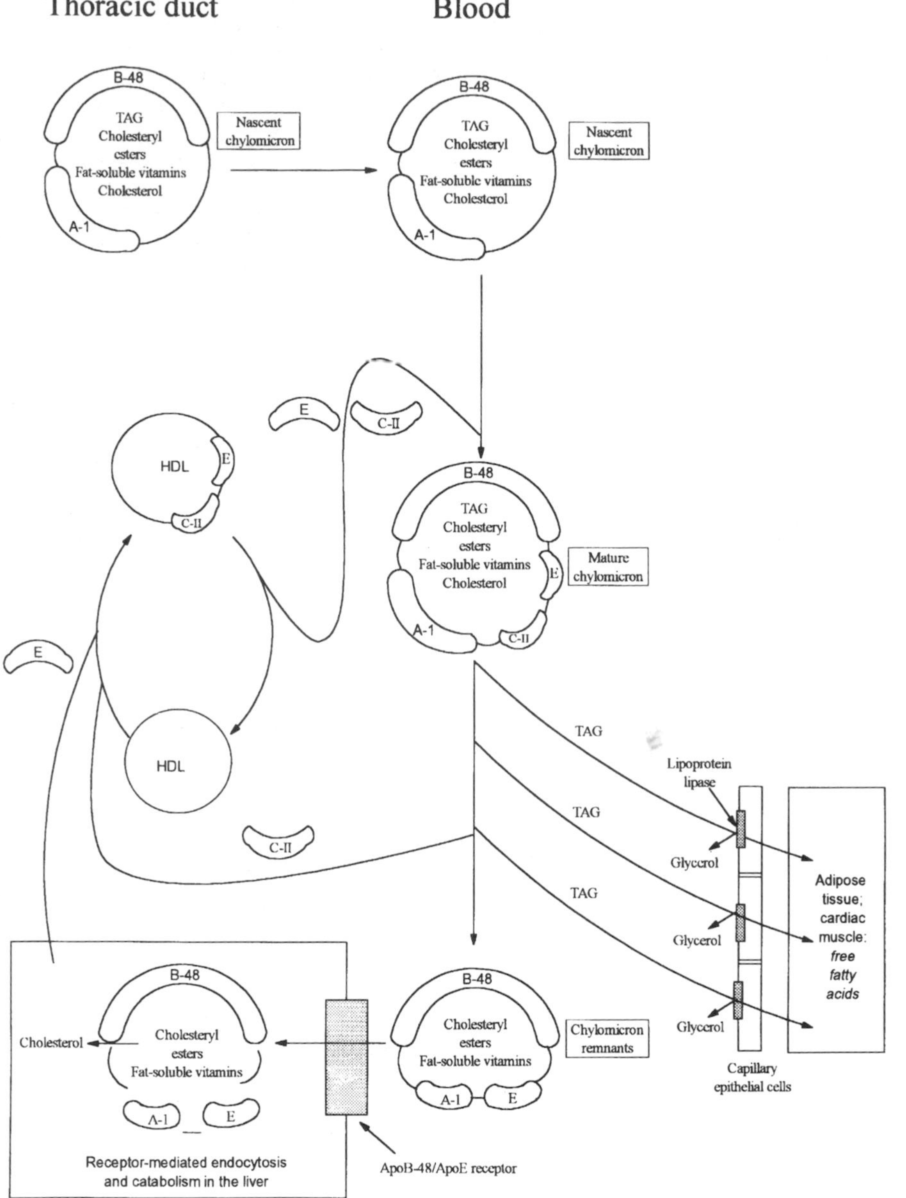

Figure 10.3 **The disposal of chylomicrons**

10.3 Lipid transport in the fed state

The formation of chylomicrons

Following a fatty meal, the bulk of ingested triacylglycerols are reconstituted within intestinal mucosal cells (see Chapter 13). In the Golgi apparatus of these cells the triacylglycerols are joined with phospholipids, dietary cholesterol and cholesteryl esters together with apoA-1 and apoB-48 to form nascent chylomicrons. This process closely resembles the one that occurs in the liver during the formation of VLDL (see Figure 10.7), except that VLDL contains apoB-100 rather than apoB-48. The nascent chylomicrons are delivered to the blood via the lymphatic system, and thus these particles (in a mature form) reach peripheral tissues before being presented to the liver. This contrasts strongly with the other main dietary fuels, carbohydrates and amino acids, that are delivered directly to the liver via the hepatic portal system.

On entering the circulation the nascent particles acquire apoC-II and apoE from plasma HDL. The acquisition of these apolipoproteins renders the mature chylomicrons functionally competent. In particular, the arrival of apoC-II allows the mature particles to activate the enzyme **lipoprotein lipase** (Figure 10.3).

In the fed state, large amounts of dietary monosaccharides, including fructose, are presented to the liver. These are converted, in part, into fatty acids which are incorporated into VLDL. Nevertheless, the release of VLDL from the liver is inhibited in the fed state and consequently the particles accumulate in a 'pool' in the liver, to be released in the fasting state.

It should be remembered in the description that follows that many of the facts relating to the disposition of chylomicrons apply equally to VLDL.

Mode of action of lipoprotein lipase

The adipose tissue is largely responsible for the removal of triacylglycerols derived from chylomicrons from the circulation (a significant proportion is taken up by muscle, predominantly cardiac muscle, but very little by the liver). The clearance of chylomicron-derived triacylglycerol from the circulation is achieved by **lipoprotein lipase**, an enzyme that catalyses the hydrolysis of triacylglycerols to release NEFA and ultimately glycerol (although significant quantities of mono-acylglycerol also enter the adipocytes, where they undergo hydrolysis). Lipoprotein lipase is a dimeric, *N*-linked glycoprotein, and both glycosylation on asparagine and dimerization are necessary for catalytic activity. The specificity of the enzyme is such that only triacylglycerols found in lipoproteins undergo hydrolysis. Specifically the enzyme is activated by apoC-II. This fact indicates that lipoprotein lipase must be located in the extracellular compartment. In fact, it is located bound to the outer aspect of the plasma membrane of the endothelial cells that line the capillaries supplying blood to adipose tissue and other peripheral tissues. The binding is to heparan sulphate proteoglycans that are associated with the surface of these cells. The appearance of lipoprotein lipase in association with the endothelium is subject to hormonal regulation, with insulin acting to promote transfer to the cells and noradrenaline (and adrenaline) inhibiting it. It is now known that although the enzyme resides in the endothelium, the hormones exert their control via the adipocytes (or, in general, in parenchymal cells). It is in these cells that the proenzyme is synthesized, subsequently to be transferred to the endothelial cells. The final association is not firm and the enzyme may be released by the systemic administration of heparin, which presumably interferes with the normal binding to heparan sulphates. Since the enzyme is released into the circulation, the administration of small amounts of heparin is a convenient method for assessing lipoprotein lipase activity in patients (synthesis in adipocytes will soon replace any enzyme that has been displaced).

The nature of the hormonal control of lipoprotein lipase expression means that the enzyme activity in adipose tissue will increase in the fed state and will decline in the fasted state. This arrangement is ideally suited for dealing with diet-derived triacylglycerols. On the other hand, the reverse is true for lipoprotein lipase in other tissues bearing the enzyme, so that, for example, the activity of the enzyme in cardiac tissue increases in fasting. This means that triacylglycerols from VLDL tend to be targeted to non-adipose tissues in the fasting state. The actions of lipoprotein lipase in adipose tissue is shown in Figure 10.4.

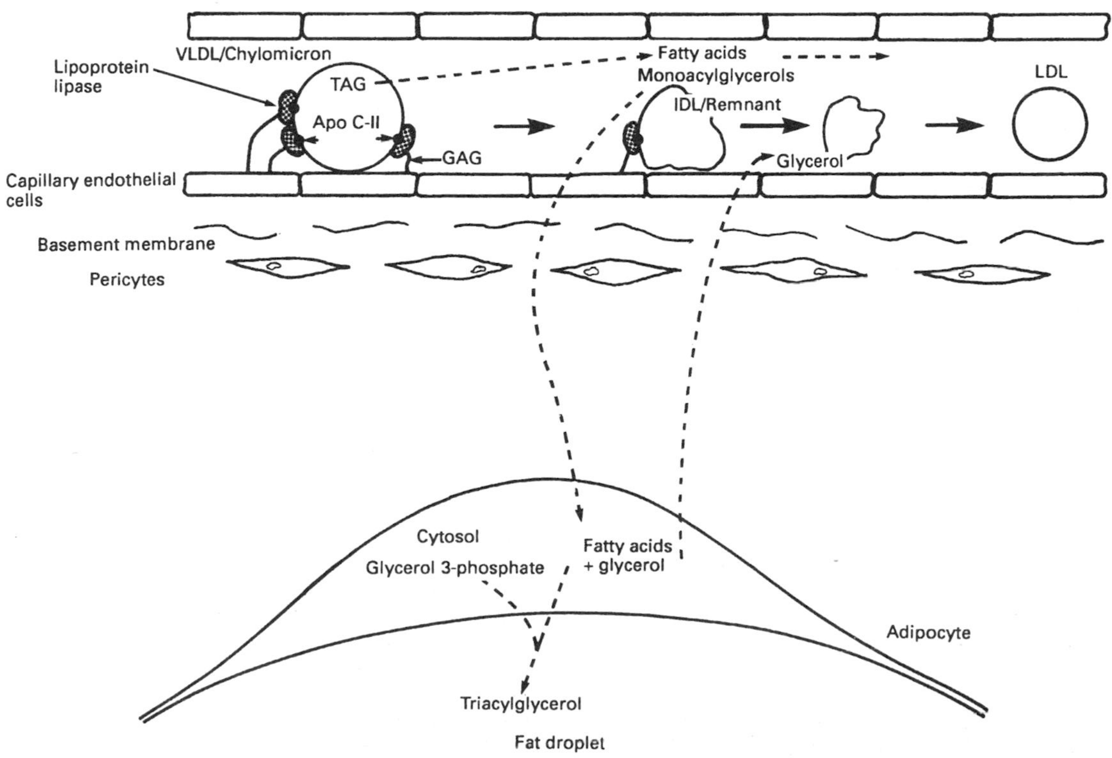

Figure 10.4 **Mode of action of lipoprotein lipase at the endothelial cells of adipose tissue.** Lipoprotein lipase, located on glycosaminoglycan residues (GAG), catalyses the hydrolysis of lipoprotein triacyglycerols producing monoacyl-glycerols and fatty acids, which then enter the adipocytes. Apoprotein C-II is required for the interaction between the lipase and the lipoprotein particle

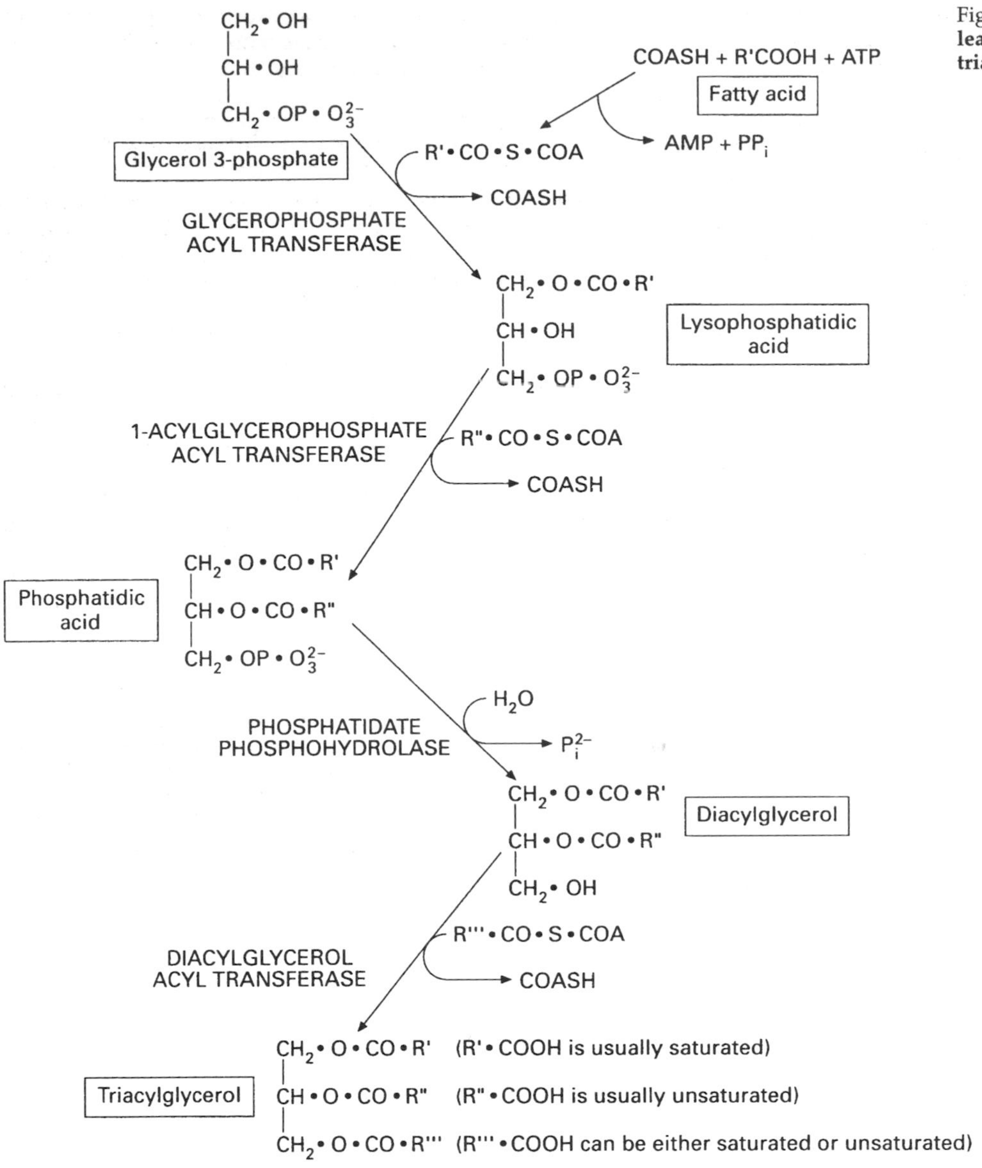

$CH_2 \cdot OH$
|
$CH \cdot OH$
|
$CH_2 \cdot OP \cdot O_3^{2-}$
Glycerol 3-phosphate

COASH + R'COOH + ATP
Fatty acid
AMP + PP_i
$R' \cdot CO \cdot S \cdot COA$
COASH
GLYCEROPHOSPHATE ACYL TRANSFERASE

$CH_2 \cdot O \cdot CO \cdot R'$
|
$CH \cdot OH$
|
$CH_2 \cdot OP \cdot O_3^{2-}$
Lysophosphatidic acid

1-ACYLGLYCEROPHOSPHATE ACYL TRANSFERASE
$R'' \cdot CO \cdot S \cdot COA$
COASH

$CH_2 \cdot O \cdot CO \cdot R'$
|
$CH \cdot O \cdot CO \cdot R''$
|
$CH_2 \cdot OP \cdot O_3^{2-}$
Phosphatidic acid

PHOSPHATIDATE PHOSPHOHYDROLASE
H_2O
P_i^{2-}

$CH_2 \cdot O \cdot CO \cdot R'$
|
$CH \cdot O \cdot CO \cdot R''$
|
$CH_2 \cdot OH$
Diacylglycerol

DIACYLGLYCEROL ACYL TRANSFERASE
$R''' \cdot CO \cdot S \cdot COA$
COASH

$CH_2 \cdot O \cdot CO \cdot R'$ ($R' \cdot COOH$ is usually saturated)
|
$CH \cdot O \cdot CO \cdot R''$ ($R'' \cdot COOH$ is usually unsaturated)
|
$CH_2 \cdot O \cdot CO \cdot R'''$ ($R''' \cdot COOH$ can be either saturated or unsaturated)
Triacylglycerol

Figure 10.5 **The reactions leading to the formation of triacylglycerols**

For a long time it was believed that the uptake of NEFA into adipocytes occurred solely by passive diffusion, but recently a fatty acid transport protein has been found in the plasma membrane of these cells and it is believed that this facilitates the uptake. Much of the NEFA entering the adipocyte is used for the biosynthesis of triacylglycerols. The glycerol released with the NEFA cannot be used for this purpose, as there is no glycerokinase present in adipose tissue to convert glycerol into its 3-phosphate ester. Glycerol, therefore, passes into the circulation, where it is transported to the liver to undergo gluconeogenesis (see Chapter 6).

The fate of chylomicron remnants

The stripping away of triacylglycerols from chylomicrons results in the formation of **chylomicron remnants**. By this stage these particles are relatively enriched in cholesterol and its esters and depleted of triacylglycerols. The presence of apoE on the surface of the remnants is recognized by an apoE receptor found in the plasma membranes of hepatocytes. Consequently these particles undergo receptor-mediated endocytosis leading to their catabolism in hepatic lysosomes (see Chapter 8). Hence much of the cholesterol taken in the diet is delivered to the liver (together with the fat-soluble vitamins).

The roles of glucose and insulin

Invariably, in the fed state, the appearance of chylomicrons in the circulation is accompanied by that of glucose (with appropriately elevated concentrations of insulin). It is important, therefore, to consider the interplay of these two dietary fuels. As has already been described, insulin directly favours the appearance of lipoprotein lipase on endothelial cells in adipose tissue with the consequent release of NEFA and uptake into adipocytes. These NEFA are then used for the synthesis of triacylglycerols. The precursors required for the process of lipogenesis are fatty acylCoA molecules and glycerol 3-phosphate. Fatty acylCoA derivatives are formed in a reaction catalysed by fatty acylCoA synthetase which requires ATP (Figure 10.5). The sole source of glycerol 3-phosphate in adipose tissue is glucose (hence lipogenesis in adipose tissue is both glucose- and insulin-dependent). Glycolysis leads to the formation of glyceraldehyde 3-phosphate which undergoes reduction catalysed by glycerol 3-phosphate dehydrogenase. Two fatty acyl groups are then transferred sequentially to the 1- and 2-positions of glycerol 3-phosphate to yield phosphatidic acid. Subsequently the phosphate group of the molecule is removed by hydrolysis catalysed by phosphatidate phosphohydrolase to give a diacylglycerol to which a third acyl group is transferred to form a triacylglycerol (Figure 10.5). It should be appreciated that in adipose tissue triacylglycerols are in a dynamic state, undergoing both synthesis as described and also hydrolysis. In the fed state lipogenesis is favoured over lipolysis by the prevailing hormonal background (high insulin, low glucagon and catecholamines).

10.4 Lipid transport in the fasting state

As the circulating concentration of chylomicrons declines in the plasma, adipose tissue may begin to release its stores of triacylglycerol (TAG). This is accomplished by a pair of lipases that are distinct from lipoprotein lipase and these are found within adipocytes. The major enzyme is subject to endocrine control and is termed '**hormone sensitive lipase**'. This catalyses the hydrolysis of a terminal fatty acyl residue from TAG. The resultant diacylglycerols are subject to hydrolysis by a less specific lipase that catalyses hydrolysis to release the remaining two fatty acids and glycerol.

In the fed state the hormone-sensitive triacyglycerol lipase is of low activity but in the transition to the fasting state its activity increases. This depends on the cAMP-dependent phosphorylation of the enzyme as shown in Figure 10.6. In humans, noradrenaline (and adrenaline)

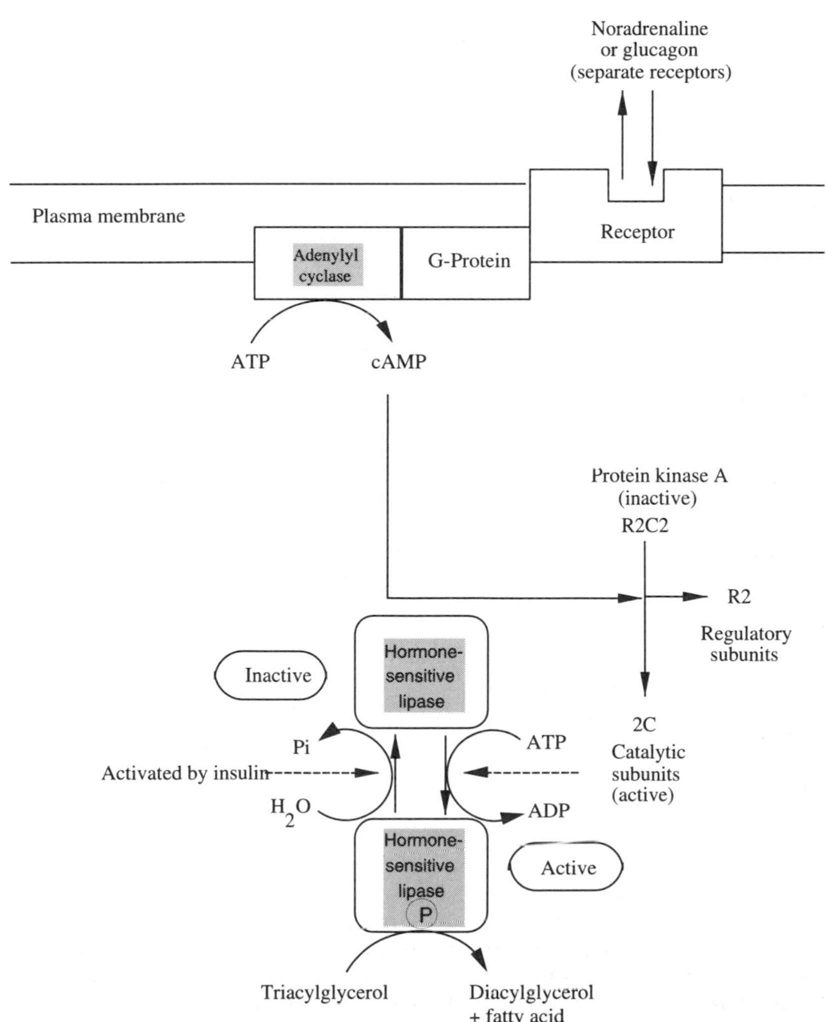

Figure 10.6 **Activation of the hormone-sensitive lipase of adipose tissue by phosphorylation** (TAG, triacylglycerol; DAG, diacylglycerol; ADP, adenosine diphosphate; ATP adenosine triphosphate)

are the major effectors that cause an increase in the activity of adenylyl cyclase via a G-protein-coupled β-adrenoceptor. This activation of hormone-sensitive lipase is opposed by insulin which favours dephosphorylation of the enzyme. Consequently, in the fasting state as the circulating concentration of insulin declines so does its inhibitory influence on hormone-sensitive lipase, thus favouring lipolysis.

Fate of the non-esterified fatty acids and glycerol and the formation of VLDL

The NEFA released from adipose tissue may be taken up and oxidized by many tissues. There is no doubt, however, that a proportion of NEFA (and most of the glycerol) enter the cells of the liver. Although the β-oxidation of the NEFA will help to meet the energy requirements of the liver, a substantial fraction undergoes esterification. Indeed the enzyme phosphatidate phosphohydrolase that converts phosphatidic acids to diacylglycerols (see Figure 10.5) is activated in the presence of NEFA. As was noted earlier, fatty acids may also be synthesized *de novo* in the liver from monosaccharides and incorporated into TAG during feeding. The assembly of newly synthesized TAG, phospholipids, cholesterol (and its esters), together with apoA-I and apoB-100 to form VLDL in the Golgi apparatus of the hepatocyte is shown in Figure 10.7 and, as has been noted (see Section 10.3), this process closely resembles the assembly of chylomicrons in enterocytes.

In the fasting state, therefore, the liver will release newly-formed VLDL and also those particles already stored in the 'pool' previously formed during feeding.

The fate of VLDL and the genesis of LDL

The peripheral handling of VLDL broadly resembles that of chylomicrons in that the action of lipoprotein lipase results in the progressive stripping of TAG from the particle (Figure 10.8). There are, however, some important differences in detail: hydrolysis of VLDL-derived TAG occurs largely in the muscles rather than in adipose tissue. In particular, the Michaelis constant of cardiac lipoprotein lipase for TAG is lower than that in adipose tissue by an order of magnitude. In addition, the appearance of the enzyme in the heart is favoured in the fasting state, whereas in the adipose tissue the enzyme appears in an insulin-dependent fashion (i.e. in the fed state). Thus the heart has both a high affinity and a high capacity for VLDL in the fasting state.

As with chylomicrons, nascent VLDL particles become functional physiologically following the transfer of apoC-II and apoE from HDL, apoC-II being recycled following TAG depletion. The counterparts of chylomicron remnants are IDL (retains more TAG and also apoE) and LDL, the latter being formed from IDL. One consequence of the removal of TAG from VLDL is that IDL and especially LDL are enriched in cholesterol and its esters when compared with their parent VLDL.

Relation between non-esterified fatty acids and VLDL

In view of the ready oxidation of NEFA by peripheral tissues the complex scheme whereby the liver forms VLDL may appear to be redundant, the more so because TAG must undergo hydrolysis to yield NEFA prior to their oxidation. No complete explanation can be given for this apparent paradox, but it may be important that

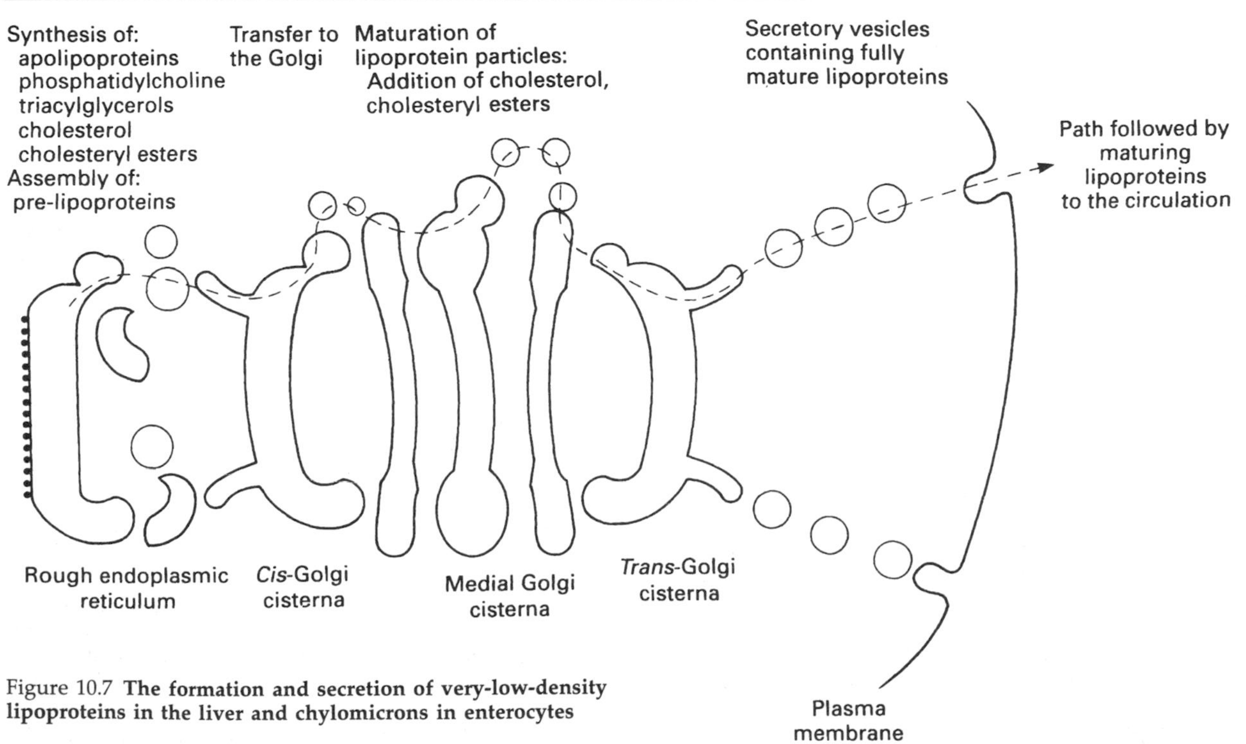

Figure 10.7 **The formation and secretion of very-low-density lipoproteins in the liver and chylomicrons in enterocytes**

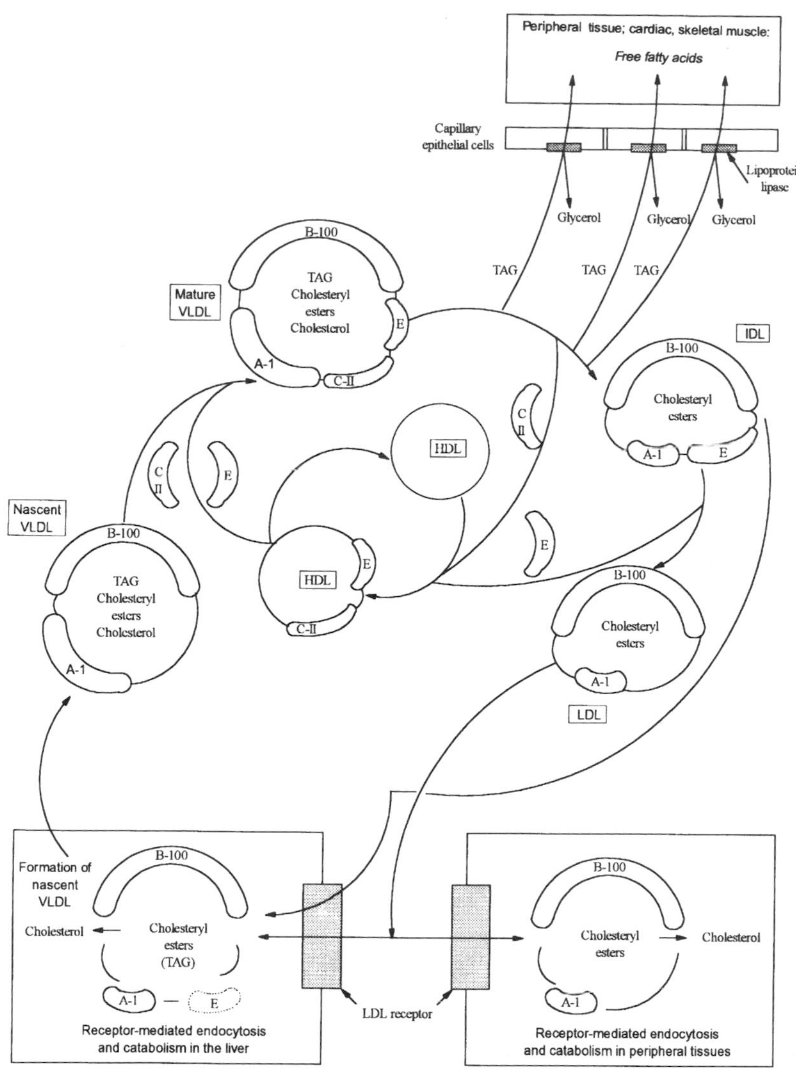

Figure 10.8 **The metabolism of VLDL and LDL**

in reassembling TAG the liver is able to control their fatty acid composition more closely to meet all the cellular needs of extrahepatic tissues, both for membrane maintenance and for energy supply.

The fates of IDL and LDL

IDL and about half of the LDL formed from it share a similar fate: uptake into and degradation in the liver. The remaining 50% of LDL is taken up by cells in peripheral tissues, especially fibroblasts, vascular smooth muscle and lymphocytes. These uptake processes are all examples of receptor-mediated endocytosis occurring via clathrin-coated pits (see Chapter 7). The receptor in question is the LDL receptor, sometimes referred to as the apoB-100/apoE receptor because its discrimination is achieved by the recognition of apoB-100 (or high concentrations of apoE). The subsequent degradation of LDL (and IDL) in the lysosomal compartment of cells results in the release of cholesterol.

Receptor-mediated endocytosis of LDL is evidently an important facet of whole-body cholesterol homoeostasis. This conclusion is amply supported by the finding of a group of patients suffering from **familial hypercholesterolaemia**, who have mutations to their LDL receptor gene, which render the receptors non-functional. Such patients frequently present with coronary atherosclerosis (see Section 10.7).

Although the metabolism of TAG and cholesterol is clearly intertwined as a result of their mutual presence in the lipoproteins, for convenience, a discussion of the cholesterol homoeostasis will be deferred until Section 10.6.

It should be noted that chemical modification of LDL apoB-100 that may occur in the extracellular spaces, for example the formation of Schiff's bases on lysine residues (see Chapter 33) precludes uptake by the normal LDL receptor, but favours uptake by a so-called **'scavenger'** receptor found in the plasma membrane of cells such as macrophages. The possible role of this receptor in the pathogenesis of atherosclerosis is currently under intense investigation (see Chapter 33).

Summary of events occurring in the fasting state

A summary of tissue interrelationships in the handling of lipids in the fasting state is shown in Figure 10.8.

Formation and roles of HDL

HDL particles are formed mainly in the liver, although a limited quantity appears to derive from the intestine. The pathway for the assembly of HDL resembles that of VLDL.

One of the principal roles of HDL is to carry and distribute apoC-II and apoE, that in turn regulate the disposal of lipoproteins. In addition, they play a central part in cholesterol homoeostasis and this will be explored in Section 10.6.

10.5 Fatty livers

Fat, mainly triacylglycerol, can accumulate in the liver to such an extent that it may be observed with the naked eye. The condition is known as '**fatty liver**' or sometimes as '**fatty infiltration of the liver**'. This usually occurs as a result of defective lipoprotein metabolism and hence it is appropriate to discuss this condition at this stage.

Disruption of lipoprotein synthesis can be caused by a wide variety of pathological or disturbed physiological conditions. Starvation, diabetes mellitus, other hormonal dysfunctions, diets deficient in certain essential amino acids or poisoning with ethanol, carbon tetrachloride or ethionine can all cause fatty liver.

Although the exact mechanisms underlying the occurrence of fatty liver are not established for each type of insult, the effects can be classified into two categories, described as 'physiological' and 'pathological'. The pattern of effects is shown in Figure 10.9. It will be noted that 'physiological' fatty liver can result from an excess supply of fatty acids to the liver, as may occur in diabetes mellitus or in starvation. In these circumstances the fatty acids may be oxidized, converted into ketone bodies or incorporated into TAG. Synthesis of TAG occurs rapidly and should there be a mismatch between their rate of synthesis and that of other components of VLDL, the former molecules will accumulate in the liver. This mismatch may arise as the result of diets deficient in one or more essential amino acids required for apolipoprotein synthesis or in choline required for phosphatidylcholine synthesis. Choline can also be formed endogenously, but a supply of methionine is required as a donor of methyl groups, and diets deficient both in choline and methionine constitute a particular risk for the development of fatty liver.

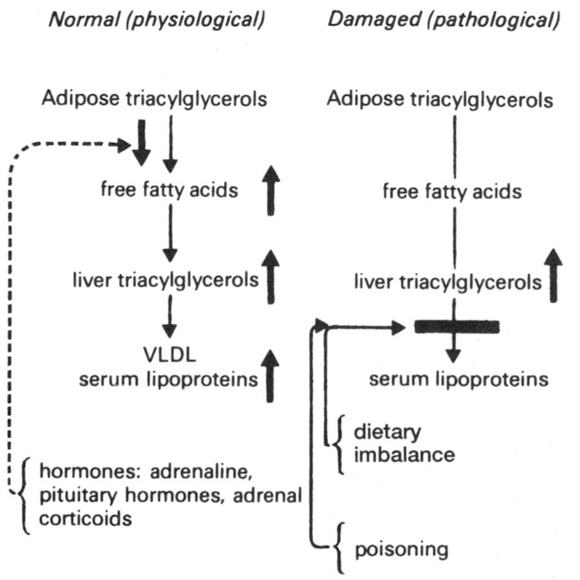

Figure 10.9 **Elevated fat contents of normal and damaged livers**

Ethionine inhibits hepatic protein synthesis because it causes a drain on the ATP required for many of the synthetic reactions underlying lipoprotein formation. This seems to arise by the formation of large quantities of *S*-adenosylethionine which thereby sequesters the adenine required in ATP.

The mode of action of carbon tetrachloride is more complex and may result from its metabolism to yield free radical species that cause damage to the plasma membrane and also to intracellular membranes. This view is supported by the observation that administration of vitamin E helps protect from the damaging effects of carbon tetrachloride intoxication (see Chapter 33).

10.6 Cholesterol homeostasis

A central player in the cholesterol economy of the body is HDL and it is appropriate, at this stage, further to consider its structure and functions.

Nascent HDL derives largely from the liver and on entering the circulation is found to be discoid in shape. The disc consists of a phospholipid/cholesterol bilayer associated with which are a number of apolipoproteins (A-I, C-II, D and E) and the enzyme lecithin (phosphatidylcholine):cholesterol acyltransferase (LCAT). An important role of apoA-I is to function as an activator of LCAT. In the circulation LCAT catalyses the transfer of fatty acyl residues, from position 2 of lecithin, to cholesterol to form lysolecithin and cholesteryl esters (Figure 10.10). The lysolecithin is taken up by albumin, while the hydrophobic esters are concentrated between the two leaflets of the bilayer of the discoid, thereby forcing them apart. This favours the formation of a spherical, mature HDL particle with an amphipathic lipid monolayer on its surface (HDL_3; Figure 10.11). The cholesteryl esters may then be transferred to other lipoproteins, e.g. VLDL or chylomicrons, in a process facilitated by apoD (also known as **cholesteryl ester transfer protein**). While in the circulation, HDL_3 acquires cholesterol from tissue sources and phospholipids from lower density lipoproteins. This has the effect of reducing the density of HDL (HDL_2). The cholesterol acquired in this manner is esterified in the reaction catalysed by LCAT and the resulting esters may be transferred to lower density lipoproteins. These acceptors of cholesteryl esters will ultimately be converted into remnants or LDL which will undergo receptor-mediated endocytosis and hydrolysis in the

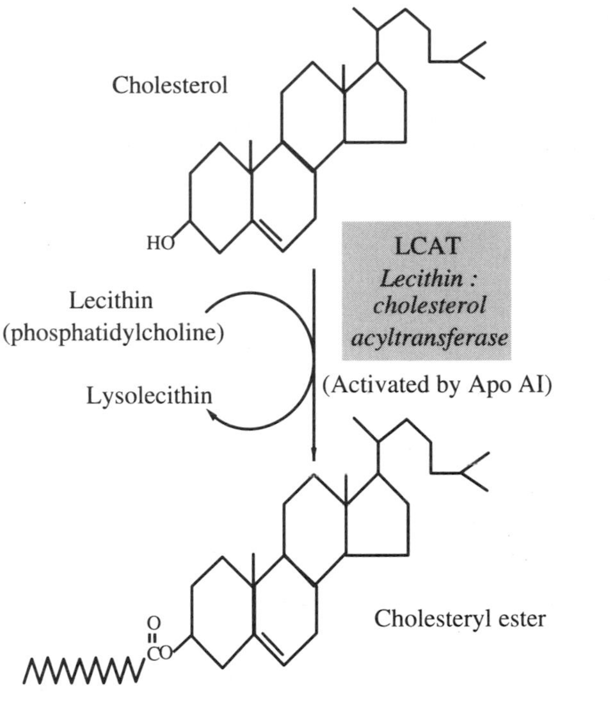

Cholesterol

HO

LCAT
Lecithin : cholesterol acyltransferase

Lecithin (phosphatidylcholine)

Lysolecithin

(Activated by Apo AI)

O
CO

Cholesteryl ester

Figure 10.10 **The reaction catalysed by lecithin : cholesterol acyltransferase**

tissues, with the liver being prominent (dealing with chylomicron remnants, IDL and half of the LDL). It can be seen, therefore, that HDL, and especially HDL2, play a key role in the movement of cholesterol from the tissues (or obtained in the diet) to the liver. Evidently this is a crucial mechanism that prevents the inappropriate accumulation of cholesterol in peripheral tissues. The accumulation of cholesterol in the tissues is strongly associated with the development of atherosclerosis. Indeed, studies have shown there to be an inverse relationship between the circulating concentration of HDL_2 and the incidence of coronary atherosclerosis. This is a life-threatening disease, in which the supply of blood to the heart itself may be compromised as a result of the occlusion of the coronary vessels. Figure 10.11 shows the roles of the various lipoproteins in cholesterol homeostasis.

The sources of cholesterol

Except in the description of the handling of the chylomicrons no attention has been paid, in the present discussion, to the sources of cholesterol. Dietary sources are undoubtedly of importance, but to complement this, many tissues are capable of carrying out *de novo* synthesis of the sterol. It is now understood that there is tight control of biosynthesis so that it is responsive to changes in dietary supplies. In Western societies *de novo* synthesis and dietary sources make roughly equal contributions of about 500 mg each to daily requirements. In humans the types of cell making major contributions to cholesterol biosynthesis are those in the liver (50%), the intestinal epithelial cells and the fibroblasts (especially those in the skin).

Although the biosynthesis of cholesterol is a complex process its initiation will be familiar to those who have studied ketogenesis. The first two reactions involved are identical with those leading to ketone body formation. However, whereas the enzymes that catalyse the reactions of cholesterol synthesis are located in the cytosol and endoplasmic reticulum, those involved in ketogenesis

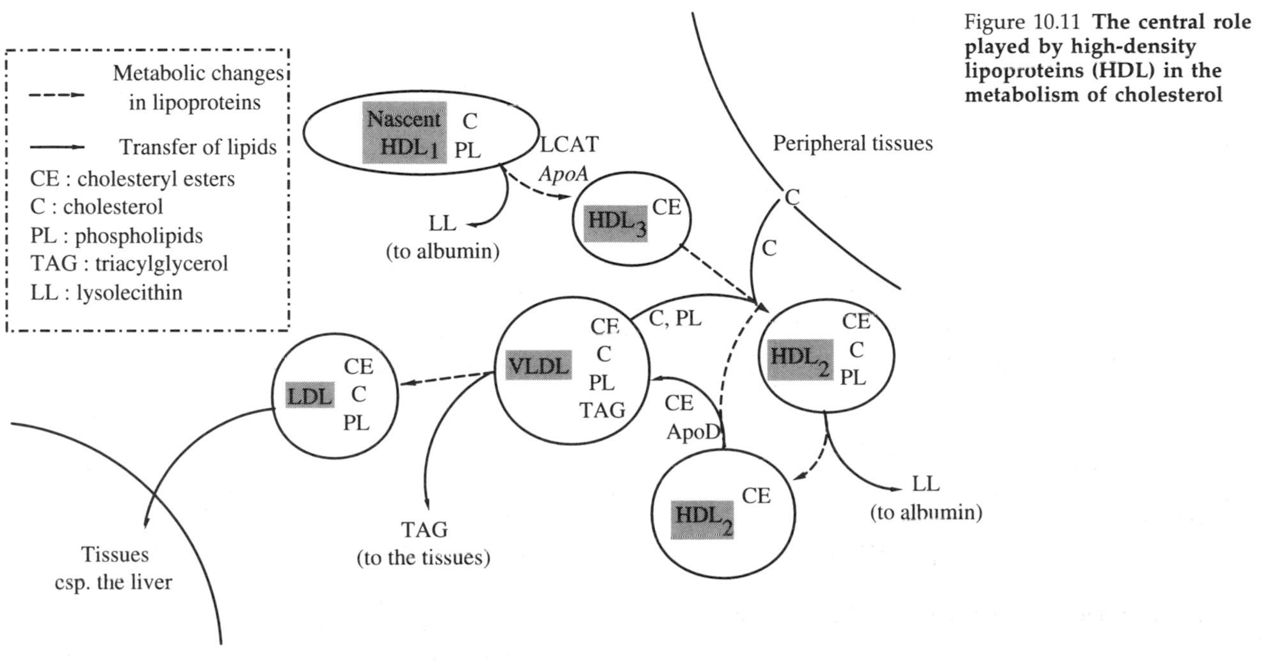

Figure 10.11 **The central role played by high-density lipoproteins (HDL) in the metabolism of cholesterol**

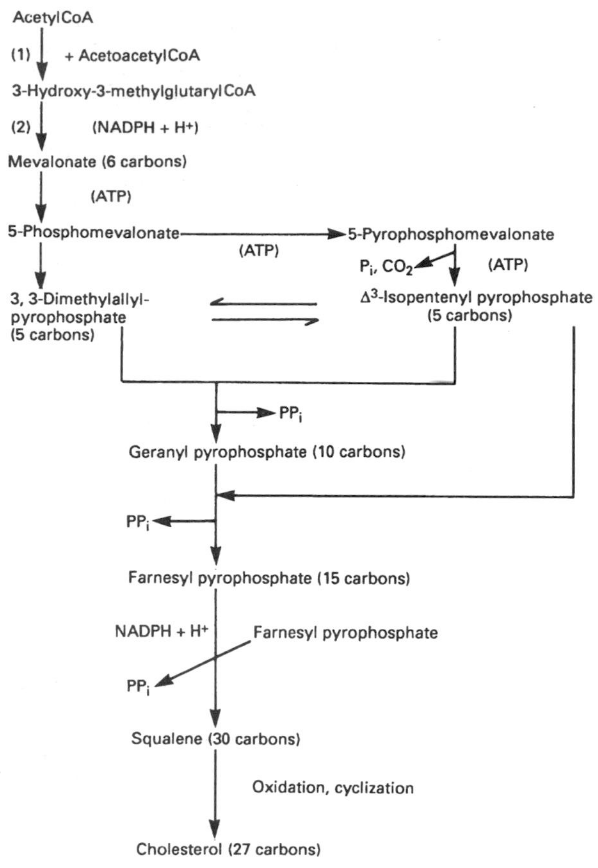

Figure 10.12 **Biosynthesis of cholesterol**

occur in mitochondria. The key intermediate is 3-hydroxy-3-methylglutarylCoA (HMGCoA). This is formed by the cytosolic enzyme HMGCoA synthase. HMGCoA is then reduced to mevalonate with the concomitant release of co-enzyme A. Thereafter a complex series of reactions lead to the formation of the polyunsaturated hydrocarbon squalene. This 27-carbon atom-containing molecule undergoes cyclization to yield cholesterol (Figure 10.12).

Regulation of cholesterol biosynthesis

A key enzyme in the pathway is HMGCoA reductase and this activity is subject to multi-faceted regulation. An important aspect of this is that the enzyme is subject to phosphorylation/dephosphorylation cycles, with the dephospho-form being the more active. This activation is favoured by insulin and opposed by glucagon (although the role of the latter is only minor). Another important measure of regulation is exercised by cholesterol itself (or possibly an oxidation product), which inhibits the biosynthesis of LDL receptor protein (Figure 10.13).

Newly-synthesized cholesterol in cells equilibrates with lipoprotein-derived cholesterol. The cholesterol in cells may then undergo one of several possible reactions. It may be esterified with a fatty acid to yield cholesteryl esters in a reaction catalysed by **acylCoA : cholesterol acyltransferase (ACAT)**, an enzyme of the outer aspect of the smooth endoplasmic reticulum, when the fatty acyl donor is fatty acylCoA. The ester may be stored until mobilized by hydrolysis catalysed by cholesteryl esterase. In certain tissues, e.g. the zona fasciculata of the adrenal cortex, cholesteryl ester stores may be so extensive that the cells resemble adipocytes in possessing

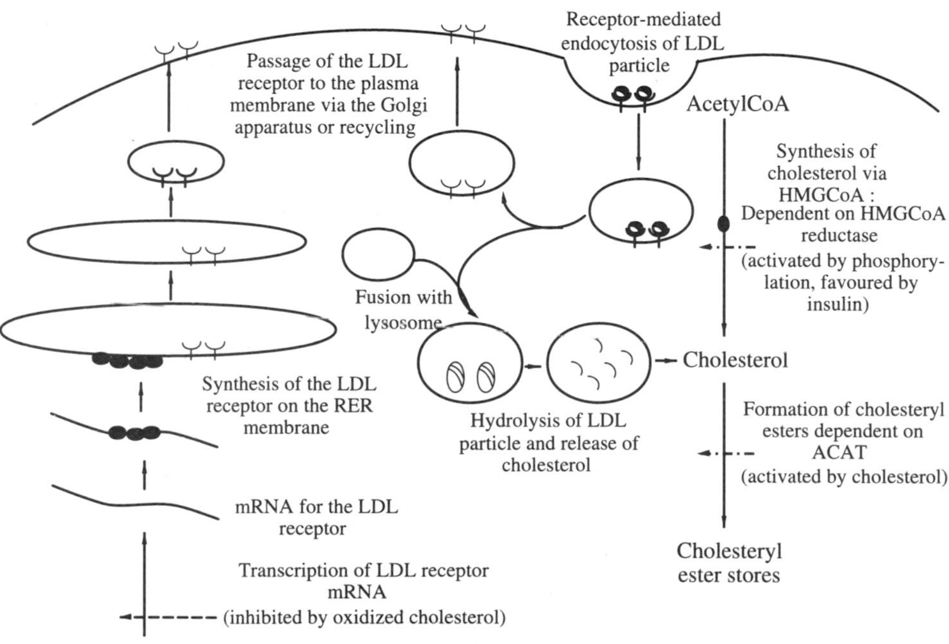

Figure 10.13 **Control of cholesterol homeostasis in cells** (LDL, low-density lipoprotein; HMGCoA, 3-hydroxy-3-methylglutarylCoA; ACAT, acylCoA : cholesterol acyltransferase)

visible lipid droplets. In the disease **adrenoleucodystrophy**, abnormally large quantities of cholesteryl ester accumulate with the fatty acyl residues being unusually long (22-carbon atoms). This disease, which was the subject of the film 'Lorenzo's oil', arises due to a mutation to the fatty acid transport protein that moves long chain-fatty acids into the peroxisomes, where they would normally undergo β-oxidation (see Chapter 1).

Disposal of cholesterol via the biliary route

Two other destinations for intracellular cholesterol should be noted. Cholesterol in the liver is used extensively in the biosynthesis of bile salts (see Chapter 25). In a healthy individual the bile salts and cholesterol are exported from the liver in the bile. A majority of the bile salts excreted in this way are recovered via the enterohepatic circulation, but daily faecal losses equivalent to 500 mg of cholesterol occur. This loss of bile acid corresponds to about 50% of the total daily cholesterol load from diet and biosynthesis. The remaining 500 mg of cholesterol passes out in the bile (although much is reabsorbed) and from epithelial cells desquamated into the intestinal lumen. Cholesterol derived from both of these sources is converted by the microflora of the intestine to coprostanol, which is the main sterol found in the faeces.

Summary of cholesterol homeostasis

The body has to deal with about 1 g of cholesterol a day, derived from both endogenous and exogenous sources. This quantity remains fairly constant because, if dietary intake changes, compensatory changes occur in the biosynthesis of the sterol in the tissues. Homeostasis is maintained by the loss of cholesterol metabolites in the faeces. In the whole body cholesterol economy it is the role of LDL to distribute the sterol to the tissues and the role of HDL to reverse this by favouring its transfer to the liver although, in apparent paradox, much of the cholesterol reaching the liver from endogenous sources is carried by LDL.

10.7 Importance of elevated concentrations of plasma lipids in atheromatous vascular disease

During recent years much attention has been focused on the relationship of plasma lipoproteins to disease, particularly arterial disease, and some aspects of these matters have already been touched upon. Atheromatous vascular disease, ultimately resulting in coronary thrombosis or brain embolism (stroke), is a subject of great concern in developed countries. For example, vascular disease is the leading cause of death in the UK.

Many studies throughout the world have provided substantial evidence that there is a strong correlation between raised plasma lipid levels (especially cholesterol) and death from atheromatous vascular diseases.

Raised concentrations of plasma lipids are found in a series of inherited diseases affecting lipoprotein formation and distribution and a study of these has provided an insight into the deleterious affects of deranged lipid metabolism. The classification, the observed abnormalities and the effects of these conditions collectively termed **familial hyperlipoproteinaemias**, are shown in Table 10.4. Patients presenting with type II diabetes mellitus (see Chapter 41), with hypothyroidism or with atherosclerosis are often found to have circulating lipid patterns resembling those found in patients with familial hyperlipoproteinaemias.

10.8 Factors leading to raised plasma lipid levels

As raised plasma concentrations of cholesterol or triacylglycerols are linked to atheromatous vascular disease it is important to try to assess what factors might lead to an increase in plasma lipid levels.

Age and gender

The concentrations of plasma cholesterol and TAG increase gradually with age, with the level being lower in females than males between menarche and the menopause. This suggests that the female hormonal background in these years affords a measure of protection. The reason for the slow increase that occurs with ageing in both sexes is uncertain, but it may relate to some of the factors listed below, e.g. an age-related decline in physical activity.

Exercise

It is generally agreed that the more sedentary and inactive the lifestyle, the more likely is the occurrence of

Table 10.4 **Hyperlipoproteinaemias**

Type	*Lipoprotein abnormalities*	*Likely disorder*	*Estimated frequency*
I	Chylomicrons markedly increased	1. Lipoprotein lipase deficiency	Rare
		2. Apo C-II deficiency	Rare
IIa	LDL increased	Defect in receptor LDL	Common
IIb	VLDL, LDL increased	Overproduction of VLDL	Common
III	Presence of VLDL (IDL and chylomicron remnants)	Abnormal apo-E protein, reduced clearance of remnant particles from circulation	1%
IV	VLDL increased	Unknown	Adulthood: 70%
V	Chylomicrons present, VLDL increased	Unknown	0.2–0.3%

arterial disease. One must realize, however, that it may be that the beneficial effects of exercise are not causally related to any lipid lowering. They may relate to other complex physiological processes favoured by exercise, i.e. dilatation of the arteries of the heart or other muscular tissues and the increased flow of blood that occurs in exercise. Nevertheless, inactivity, particularly when accompanied by a high energy intake, tends to cause a rise in the levels of circulating TAG. Raised TAG levels must occur when large quantities of fat are consumed and chylomicrons appear in the circulation. The resulting obesity that can occur is nearly always accompanied by an increase in plasma NEFA. Much of these will be incorporated into VLDL in the liver and these particles will give rise to LDL which tend to deposit cholesterol in the tissues.

Stress

There is much evidence that exposure to chronic stress plays an important role in the development of arterial disease. Although there are many complex interacting factors involved in the adverse effects of stressful lives, it is likely that the level of circulating lipids is one of the more important factors. A possible sequence of events is shown in Figure 10.14. As in obesity, the hypothesis is that chronically-elevated NEFA, this time caused by the catecholamines, result in an increase in VLDL synthesis. This is proposed to lead to the deposition of LDL-cholesterol in the tissues.

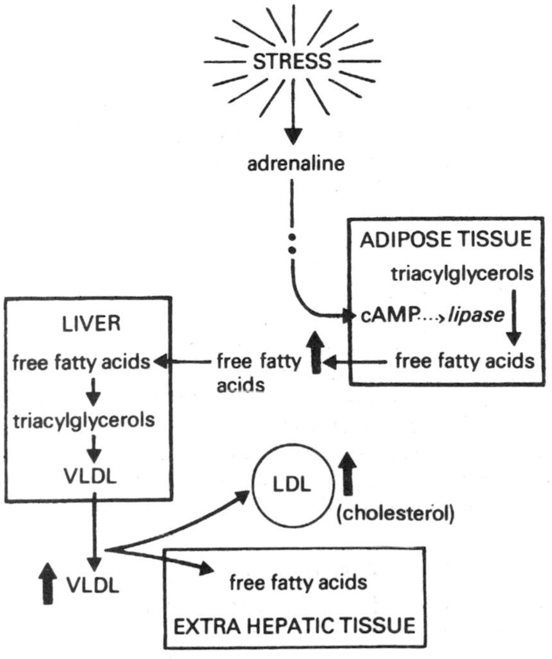

Figure 10.14 **Effect of stress on plasma lipoprotein concentrations.** Stress causes raised plasma free fatty acids, very-low-density lipoproteins (VLDL), low-density lipoproteins (LDL) and, if continued, cholesterol because LDL is the main vehicle for cholesterol transport

Diet

Many studies on the impact of diet on cardiovascular disease have been undertaken and the results of these and the conclusions that may be drawn from them are discussed in Chapter 17.

10.9 The use of cholesterol-lowering drugs

If attention to the above physiological factors fails to cause a fall in the circulating concentration of cholesterol (and by implication its deposition in extrahepatic tissues), certain pharmacological means may be tried.

A wide range of possible targets have been identified as possible sites for cholesterol-lowering strategies. Some of these, however, are associated with toxic side-effects. Drugs that have given encouraging results include **levastatin**, an inhibitor of HMGCoA reductase (the enzyme is rate-limiting for cholesterol biosynthesis). The oxidation of fatty acids in the liver can be increased by the drug **clofibrate**, with the effect that they are diverted from incorporation into TAG and then into VLDL. Clofibrate is a so-called **peroxisome-proliferator drug** and it acts on the liver to enhance fatty acid oxidation in these organelles (but also of oxidation in the mitochondria). Cationic resins such as **cholestyramine** bind strongly to bile salts in the gut and thereby interrupt their enterohepatic circulation. In turn this leads to the faecal excretion of these molecules. The effect of this intervention is to cause more bile salts to be formed from cholesterol in the liver.

10.10 Lipoprotein (a): an association with atheromatous vascular disease

Recently a sub population of LDL has been identified in which an additional protein is covalently attached to apoB-100 by a disulphide bridge. This accessory protein has been called apolipoprotein (a) and the particles carrying it, lipoprotein (a). Structural analysis of apolipoprotein (a) has shown it to be highly homologous with another blood protein, plasminogen. This protein is involved in the control of blood-clot dissolution (see Chapter 24). The physiological significance of the modification of LDL in this way is not certain, but if apolipoprotein (a) shares with plasminogen the ability (when activated) to bind to fibrin in blood clots, this could serve to target LDL cholesterol to regions where tissue repair mechanisms are at work. The cholesterol could then be used to form the plasma membrane of regenerating cells in these areas.

Whatever is the physiological role of lipoprotein (a), its circulating concentration has been found to be strongly positively correlated with the occurrence of atheromatous vascular disease. It may be that elevated concentrations of lipoprotein (a) compete with activated plasminogen for binding sites on fibrin and thus prevent clot dissolution.

In view of the dual nature of the proposed physiological and pathological roles of lipoprotein (a), it may prove difficult to devise interventions aimed at 'normalizing' plasma lipoprotein (a) levels. Nevertheless, the finding of increased circulating levels of the particle should prompt more vigorous attempts to reduce the cardiovascular risk factors identified in Section 10.8.

Chapter 11

Plasma amino acids and utilization of amino acids by the tissues

11.1 Introduction

The plasma normally contains all the 20 amino acids commonly found in proteins, but other amino acids, such as citrulline, ornithine, taurine and 3-methylhistidine, are also present.

It is clear that a discussion of the various factors which regulate the concentration of each of these amino acids separately will pose complex problems, and it is not surprising that these factors have only been partially resolved. Fortunately, it is not essential to discuss the metabolism and regulation of the concentration of each individual amino acid, since certain amino acids, e.g. the dicarboxylic amino acids or the branched-chain amino acids, can be considered in groups.

Study of the blood amino acids is very important because it provides valuable information about the state of nutrition of the body and about pathological changes which may be occurring in the tissues. However, this study often presents difficulties.

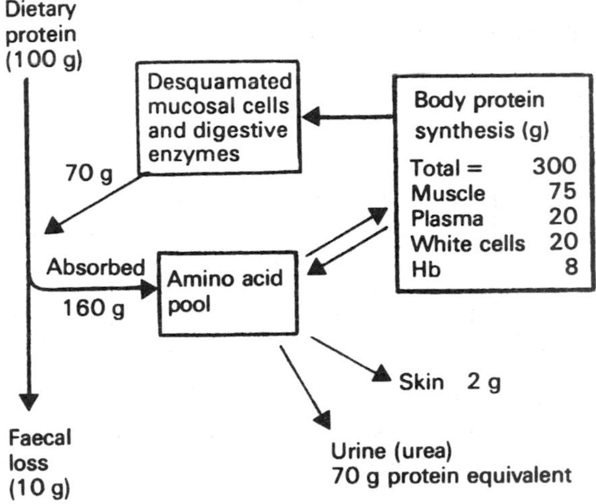

Figure 11.1 **Daily amino acid flux in humans**

11.2 Utilization of amino acids by humans

A summary of the utilization of amino acids in the adult human body is shown in Figure 11.1. The average adult in Western societies ingests about 90–100 g protein a day. The protein in the digestive tract is, however, substantially increased by the secretion of digestive juices in the form of enzymes, which, together with protein released when mucosal cells 'slough off' in the intestine, break down and release their contents, can add up to 50–70 g protein per day to the intestinal tract. Approximately 160 g amino acids are absorbed each day and 10 g lost in the faeces, either as protein or amino acids.

The amino acids are absorbed from the intestine and become incorporated into the 'amino acid pool' which is a useful concept and is represented by the summation of all free amino acids in all cells of the body. If necessary, we can restrict the confines of the 'amino acid pool' to a particular organ, such as the liver, but in this section the 'total body amino acid pool' is discussed. The 'pool' is being constantly replenished by ingested amino acids or by degradation of body proteins, and steadily loses amino acids by incorporation into newly synthesized protein or gains amino acids as a result of the action of proteolytic enzymes. Approximately 300 g body proteins are synthesized and degraded each day, as shown in Figure 11.1. An average adult produces some 22 g of urea per day, the two nitrogen atoms of which derive from amino acids (directly or indirectly), so it is possible to calculate that some 70 g of protein are catabolized per day. The overall turnover, 300 g, is 210 g greater than the net intake, 90 g, and this difference emphasizes the large contribution made by the dynamic state of the body proteins to the free amino acid pool.

The concepts of an amino acid pool and a dynamic state of body proteins emphasize the important roles that protein synthesis and degradation play, in different tissues, in the regulation of the nature and concentration of the various amino acids in the blood. The ingested amino acids contribute to this pool and dynamic state, but add only a proportion of the total amino acids involved in the daily turnover in the body.

Four non-essential amino acids – alanine, glutamic acid, glutamine and glycine – form the major proportion of amino acids in the pool.

11.3 The effect of dietary proteins on plasma amino acid concentrations

Typical concentrations of amino acids in the blood of normal adults are shown in Table 11.1. The majority of amino acids are found in approximately equal concentrations in the plasma and red blood cells, the main exceptions being aspartic and glutamic acids which are transported almost entirely in the red cells, and

$$^{-}O_2C\text{–}CH_2\text{–}CH(OH)CO_2^{-} + NADP^{+} \xrightarrow[\text{dehydrogenase (decarboxylating)}]{\text{Malate}} CH_3\text{–}CO\text{–}CO_2^{-} + NADPH + H^{+} + CO_2$$

Malate → Pyruvate

Figure 11.2 **The reaction catalyzed by malate dehydrogenase (decarboxylating)**

Table 11.1 **Typical amino acid concentrations in human blood in the fasting state**

	Plasma (μmol/l)	*Erythrocytes* (μmol/kg)
Alanine	345	350
Arginine	82	–
Asparagine	43	–
Aspartic acid	16	370
Cystine	74	–
Glutamic acid	59	320
Glutamine	568	–
Glycine	232	370
Histidine	80	140
Isoleucine	54	40
Leucine	101	400
Lysine	174	130
Methionine	21	–
Phenylalanine	58	40
Proline	236	170
Serine	112	150
Threonine	163	160
Tryptophan*	48	–
Tyrosine	50	50
Valine	170	330

*Largely bound to albumin in plasma.

Table 11.2 **The activities of the branched-chain amino acid aminotransferase (BrCAAT) and the branched-chain oxoacid dehydrogenase (BrCOAD) in various rat tissues**

	Enzyme activity in extracts	
Tissue	BrCAAT (nmol/min per g tissue)	BrCOAD (nmol/min per g tissue)
Liver	95–116	500–650
Skeletal muscle	450–1070	25–50
Heart	1210–3750	210–330
Kidney	1050–2300	300–470

From Harper *et al.* (1984) *Annu. Rev. Nutr.* **4**, 414. © Annual Reviews Inc.

glycine whose cell concentration is greater than the plasma concentration.

It might be anticipated that the ingestion of a protein meal would cause a large and significant increase in the concentration of all amino acids in systemic blood but, for several reasons, this does not occur.

Mucosal cells of the intestine actively transaminate the dicarboxylic amino acids, glutamate and aspartate, using pyruvate as the acceptor for the amino groups, thus forming alanine. The oxoacids resulting, 2-oxoglutarate and oxaloacetate, are both convertible via the tricarboxylic acid cycle into malate which can act as a substrate for malate dehydrogenase (decarboxylating) to produce pyruvate (Figure 11.2). Thus the intestine effectively converts glutamate and aspartate into alanine which is found in high concentration in the hepatic portal blood. In addition, the mucosal cells, being rapidly dividing, require glutamine as a nitrogen donor for the synthesis of the bases of the purine nucleotides which are to be incorporated into nucleic acids. The resulting glutamate is dealt with in the way outlined above. Even in the fasting state mucosal cells require glutamine for nucleotide synthesis, and under these circumstances part of the glutamate may be oxidized for energy provision, but release of alanine still occurs (Figure 11.3).

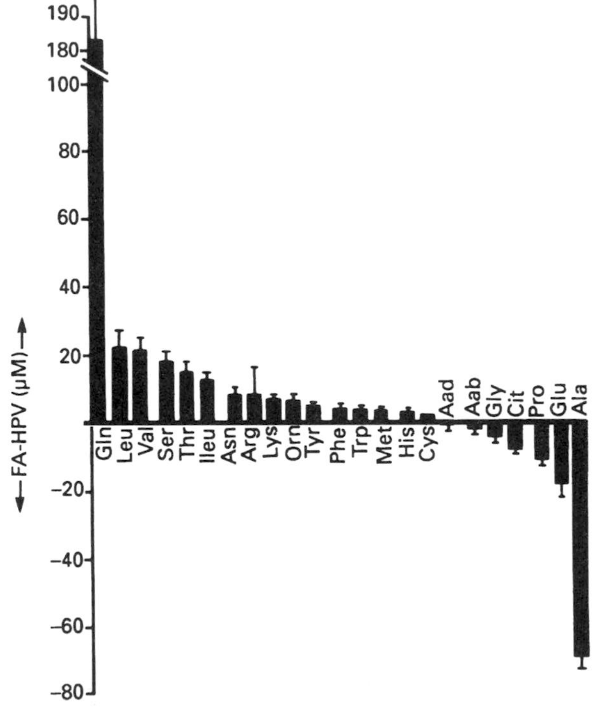

Figure 11.3 **Profile of plasma amino acid uptake and release by non-hepatic splanchnic bed or gut (FA-HPV) of fasted diabetic dogs (*n*=9).** Gut removes about two-thirds of circulating amino acids, but glutamine (Gln) alone accounts for more than half of the total amino acids removed. Gut releases seven amino acids, but alanine (Ala) accounts for more than half the total amino acids released

A second tissue which plays an important role in the control of the plasma amino acids is the liver. The fate of the amino acids of a protein meal, as shown in Figure 11.4, has been demonstrated definitively in dogs and is believed to be the same in man. It will be observed that only a relatively small proportion of the absorbed amino acids reaches the systemic circulation. The major fraction of amino acids entering the liver from the portal blood after

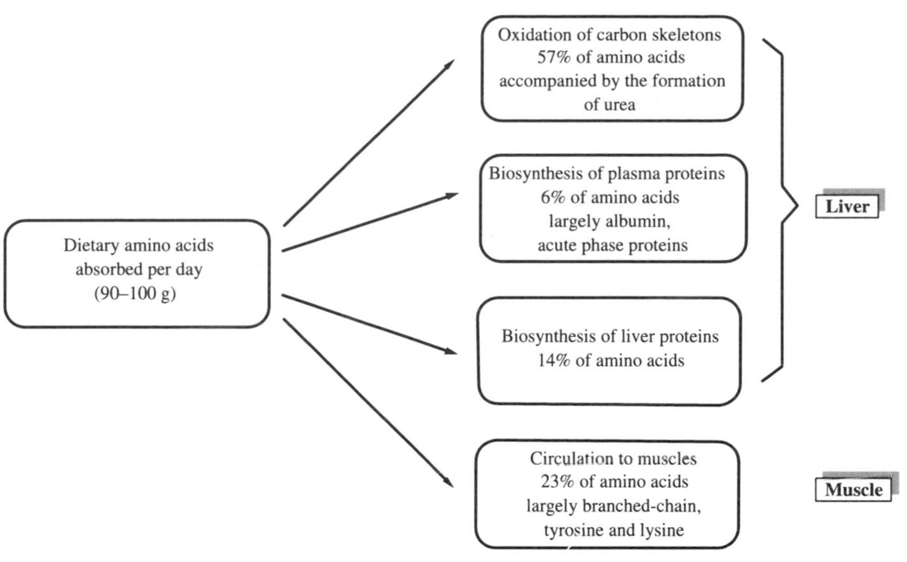

Figure 11.4 **The fate of dietary amino acids in the liver**

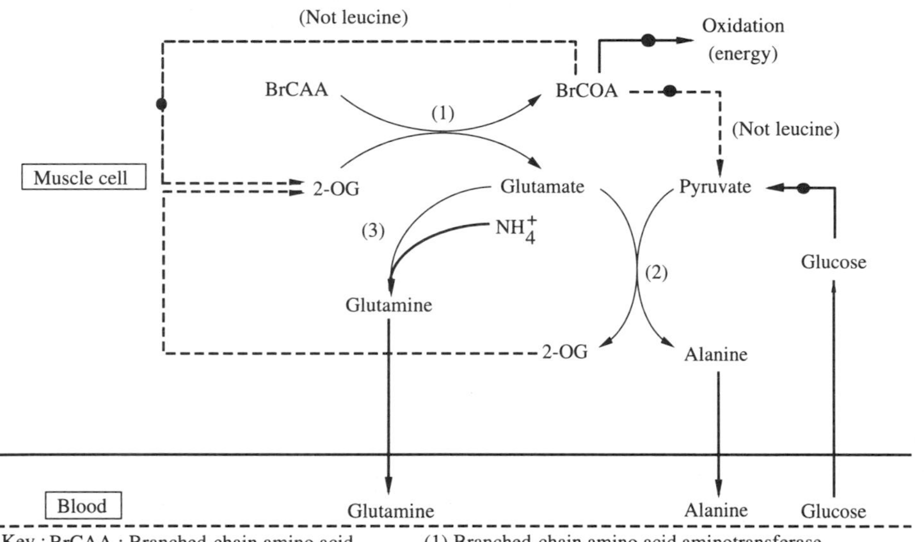

Figure 11.5 **Metabolism of branched-chain amino acids in muscle**

a meal is either catabolized or incorporated into protein.

The liver is relatively inefficient at oxidizing tyrosine, lysine and the branched-chain amino acids, leucine, isoleucine and valine (BrCAAs). In the case of the BrCAAs this is due to the low catalytic activity of the aminotransferase which transfers the α-amino group to 2-oxoglutarate, thus initiating the catabolism of these molecules (Figure 11.5 and Table 11.2). Note, however, that the activity of the second catabolic enzyme which catalyses the oxidative decarboxylation of the resulting 2-oxoacids (Figure 11.5 and Table 11.2) is high in the liver. The BrCAAs therefore pass through the liver to be taken up by other tissues in which both the aminotransferase and the decarboxylating dehydrogenase are active (Table 11.2). Quantitatively most important because of their large mass are the muscles, in which the aminotransferase is found both in the cytosol and the mitochondria. Any 2-oxoacids formed by the cytosolic enzyme may either be transferred to the mitochondria to be oxidized (but the activity of the dehydrogenase is low in resting muscle) or they may be released into the circulation where they are bound to albumin. The plasma oxoacids may then be taken up by the liver to be oxidized in the mitochondria.

11.4 The utilization of branched-chain amino acids in muscle and formation of alanine and glutamine

When the oxidation of BrCAAs does occur in muscle, e.g. during exercise in the post-absorptive state, the amino nitrogen which is released is incorporated into alanine or glutamine by the reactions shown in Figure 11.5.

Transamination of the branched-chain amino acids

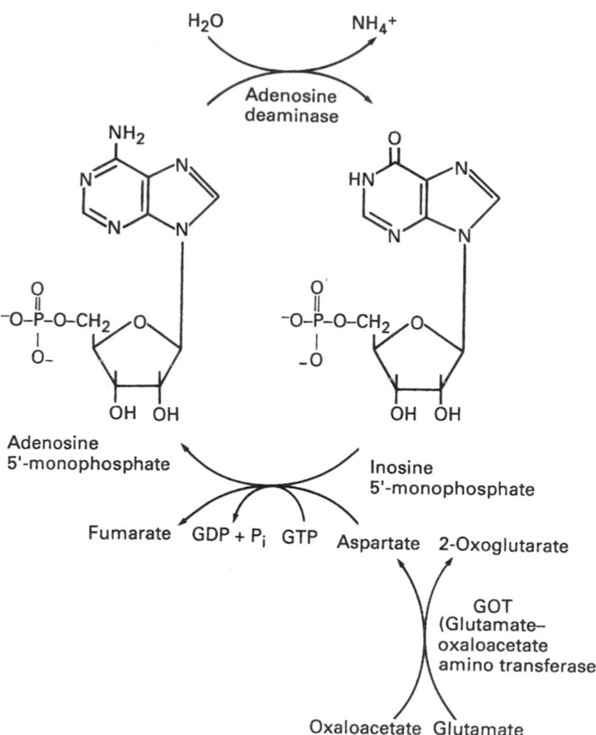

Figure 11.6 **Possible pathway for the effective release of NH^+_4 from glutamate in skeletal muscle**

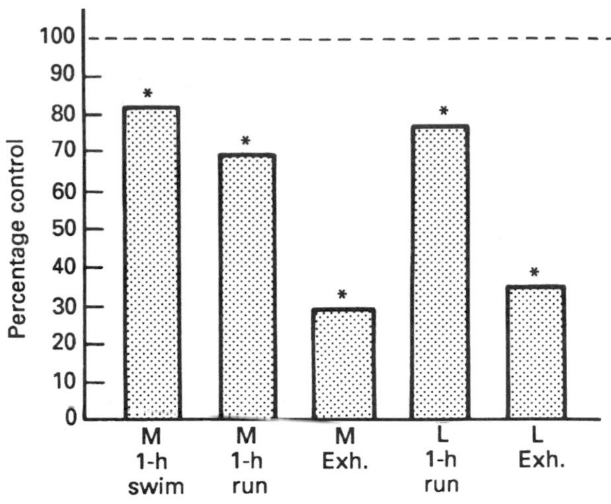

Figure 11.7 **The effect of exercise on protein synthesis in muscle (M) and liver (L).** Protein synthesis was measured in a perfused hindquarter preparation for the 1-hour swimming experiment. Muscle protein synthesis was measured *in vivo* for the running exercise (Exh., exhaustion). *Significant difference ($P < 0.05$) between rested and exercised groups

occurs almost exclusively with 2-oxoglutarate which is converted to glutamate. The glutamate can then either donate its amino group to pyruvate, with the formation of alanine, or be converted to glutamine by the incorporation of an additional amino group. The ammonia required for the formation of glutamine can arise from extracellular sources or by deamination of amino acids or purines. Of these possibilities it may be that deamination of adenosine (or its 5'-monophosphate, AMP) to form inosine (or its 5'-monophosphate, IMP) may be most important, because the activity of the major enzyme for the removal of ammonia from amino acids, glutamate dehydrogenase, is very low in muscle. It is possible to envisage the IMP being reaminated at the expense of aspartate and thus the AMP would act catalytically to provide ammonia for glutamine synthesis (Figure 11.6).

Provision of BrCAAs to muscle causes the level of intracellular glutamate to rise and production of alanine and glutamine is stimulated. The relative quantities of alanine and glutamine formed will depend on the concentration of ammonia within the tissue. Increase of ammonia concentration will cause increase of glutamine synthesis and decrease of alanine synthesis. The oxoacids formed from the BrCAAs can be completely degraded to CO_2 with the liberation of energy as ATP.

The source of the pyruvate used to form the alanine is not fully resolved, but in the fed state it is likely to be circulating glucose.

The key role played by BrCAAs, and in particular of leucine, in muscle energy metabolism and protein synthesis is nicely illustrated by studies of exercise in the post-absorptive state. It has been known for some time that exercise results in the inhibition of protein synthesis in muscle and also in the liver, the degree of inhibition being proportional to the degree of exercise and its duration. This is illustrated by Figure 11.7 for experiments on rats caused to swim or to run for 1 h or to exercise to exhaustion. The inhibition of protein synthesis is most pronounced in fast-twitch glycolytic white muscle fibres (type II; see Chapter 27), and this results in an increase in intracellular amino acid available for oxidation. The cause of this change has not been resolved, but the decrease in synthesis of proteins correlates well with a fall in the concentrations of phosphocreatine and ATP in the tissue. So it may be that a reduction in energy provision results in less protein synthesis. As the amino acids begin to accumulate there is an increase in the rate of oxidation of BrCAAs, especially leucine, and also in the efflux of alanine from

Table 11.3 **Plasma concentration of metabolites and insulin at rest and following 90 min exercise in obese subjects. A measure of leucine oxidation is also given**

		Rest	*Exercise*
Lactate	(mM)	0.39	0.84*
Glucose	(mM)	3.78	3.54
3-Hydroxybutyrate	(mM)	0.20	0.23
Leucine	(mM)	129	122
Alanine	(mM)	277	454*
Insulin	(μU/ml)	8.1	4.2
Leucine oxidation (units/h)		474	967*

*Indicate significant changes from resting values.
From S.A. Hagg, G.L. Morse and S.A. Adibi (1982) *Am. J. Physiol.* 242, E409.

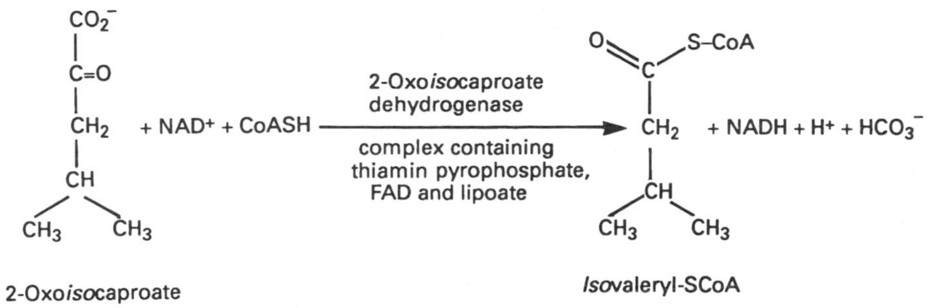

2-Oxo*iso*caproate ($(CH_3)_2CH{-}CH_2{-}C(=O){-}CO_2^-$) + NAD^+ + CoASH → *Iso*valeryl-SCoA ($(CH_3)_2CH{-}CH_2{-}C(=O){-}S{-}CoA$) + NADH + H^+ + HCO_3^-

(2-Oxo*iso*caproate dehydrogenase complex containing thiamin pyrophosphate, FAD and lipoate)

Figure 11.8 **The reaction catalysed by the 2-oxoisocaproate dehydrogenase**

muscle tissue (Table 11.3). The rate-limiting enzyme for BrCAA catabolism is the one which catalyses the oxidative decarboxylation of the branched-chain oxoacids. This enzyme complex closely resembles pyruvate dehydrogenase in the type of reaction it catalyses and its cofactor requirement (Figure 11.8), and, like that enzyme, is inactivated by a protein kinase-mediated phosphorylation and the phosphoprotein is activated by a phosphatase. The protein kinase is inhibited by 2-oxo*iso*caproate formed from leucine (and to a much lesser degree by the oxoacids formed from isoleucine and valine); consequently as leucine begins to accumulate following reductions in protein synthesis rates, more of its oxoacid is produced and this stimulates its own oxidation. Once more the amino groups from the BrCAAs are ultimately transferred to pyruvate to form alanine but now the major source of the pyruvate is other amino acids in the tissue (including isoleucine and valine). The increased production of alanine, as evidence by the changes in circulating concentration shown in Table 11.3 under-represents the *flux* of the amino acid released into the circulation, because gluconeogenesis from alanine in the liver increases in exercise; the true increase is probably threefold.

The changes outlined above have been ascribed to changes in amino acid and protein metabolism in the muscle, but in addition, hormonal effects and the metabolism of other fuels may also be important. Thus, adrenaline infusions in the post-absorptive state cause a fall in protein synthesis, an increase in BrCAA oxidation and an increase in alanine transfer to the liver (plasma alanine concentrations did not change in this study because of the stimulation of gluconeogenesis of the liver).

Finally it should be borne in mind that, although oxidation of amino acids in muscle is biochemically important for energy provision in medium term exercise, it never accounts for more than 15% of all the ATP required. In prolonged exercise, fatty acid oxidation will play an increasing role, and oxidation of fatty acids results in an increase in the NADH/NAD^+ ratio in mitochondria, which in turn will tend to limit the oxidation of branched-chain oxoacids as their dehydrogenase requires NAD^+ as a coenzyme.

The ability of fatty acids to spare BrCAA catabolism is also probably an important adaptation to starvation, in which a falling concentration of insulin results in fatty acid release from adipose tissue.

This regulatory mechanism only takes into consideration the BrCAAs, but it appears likely that total protein conservation in the body is regulated by the rate of catabolism of these amino acids.

11.5 The induction of amino acid catabolizing enzymes

When the concentrations of amino acids reaching the liver are relatively low, the major proportion of these amino acids is incorporated into protein. However, as the concentrations increase, a proportion of the amino acids is catabolized. The relative K_m values of the two-enzyme system are of major importance in regulating the fate of the amino acids (Figure 11.9); this is clearly a mechanism to prevent the waste of valuable essential amino acids by catabolism.

When animals are maintained on a high protein diet, many of the amino acid catabolizing enzymes, e.g. tryptophan pyrrolase, phenylalanine hydroxylase, 2-oxoacid dehydrogenase and serine dehydratase, are rapidly induced. This effect occurs mainly in the liver and is much less marked in other tissues, e.g. the kidney or heart.

The induction mechanism for these enzymes has yet to be clarified. It has been demonstrated with some of the enzymes that increased quantities of the mRNA for the enzyme are synthesized, indicating that the control is exercised at the level of transcription of the genes for the relevant enzyme, being controlled by several hormones. Glucose strongly inhibits the induction process, maybe resulting from insulin release, whereas glucagon is a potent stimulator of the induction of several of the amino acid catabolizing enzymes.

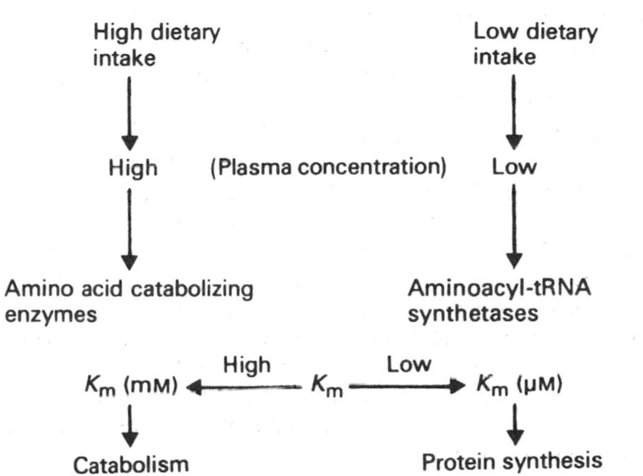

Figure 11.9 **Fates of dietary amino acids**

Table 11.4 **A summary of the principal hormonal effects on amino acid metabolism**

	Tissue affected		
Hormone	*Liver*	*Muscle*	*Other*
Insulin	Stimulates amino acid uptake and protein synthesis	Stimulates protein synthesis, inhibits proteolysis and promotes amino acid uptake in the led state	1. Facilitates the transfer of glutamate from erythrocytes to muscle 2. Activates branched-chain oxoacid dehydrogenase in adipose tissue
Glucagon	Stimulates proteolysis and inhibits protein synthesis	–	Stimulates BrCAA oxidation in the heart in the post-absorptive state
Adrenaline	–	Depresses protein synthesis and the plasma concentration of BrCAA. Increases BrCAA oxidation and the release of alanine and glutamine	Stimulates BrCAA oxidation in the heart
Cortisol	Promotes glutamine synthesis in the fasting state. Induces amino acid catabolic enzymes, e.g. tyrosine aminotransferase	Promotes proteolysis and inhibits protein synthesis. Inhibits amino acid uptake. Promotes efflux of glutamine	

Most of the experiments on induction of amino acid catabolism have been carried out on animals and, as yet, there is nothing to contradict the belief that similar regulatory mechanisms operate in the human. On this assumption, the activities of several of the enzymes which catabolize essential amino acids, such as tryptophan, will be very low before meals, but will be induced by a meal with a high protein concentration.

11.6 Hormonal regulation of plasma amino acids

Many of the hormones produced in the animal body regulate the concentrations of plasma amino acids. Some of the effects produced by them are summarized in Table 11.4. Discussion of their actions can be simplified by division into two main categories: the anabolic hormones and the catabolic hormones. The anabolic hormones, e.g. growth hormone, tend to promote the incorporation of plasma amino acids into muscle protein. Insulin may also be included in this group, although its action appears to be mediated mainly through inhibition of muscle proteolysis rather than by promotion of increased uptake. In fact one of the main effects of dietary glucose is the release of insulin which, in turn, inhibits the muscle protein degradation. Amino acids are clearly not required for energy supply when adequate glucose is available.

The catabolic hormones, particularly catabolic steroid hormones such as cortisol act antagonistically to the anabolic hormones. Release of these hormones causes degradation of muscle protein, increase of amino acid oxidation in muscle and amino acid release into the plasma. Glucagon plays an important role in stimulating the production of amino acid catabolizing enzymes in the liver.

The plasma amino acid concentration is, therefore, dependent to an important extent on the delicate balance between these two groups of hormones and the control of their release.

11.7 Intracellular turnover of proteins: the role of ubiquitin

It has been emphasized in the preceding discussion that intracellular proteins are in a dynamic state, being formed and undergoing proteolysis. Much of this turnover takes place in the lysosomes (see Chapter 8), but these organelles are not the only sites of intracellular proteolysis. Eukaryotic cells have evolved a second system for carrying out hydrolysis of proteins in cells: one that is ATP dependent. At the heart of this process is the 76 amino acid residue protein, **ubiquitin**. This is evolutionarily one of the best conserved proteins known, in that there is only one amino acid difference between the protein found in humans and that of the nematode – two taxa which are believed to have diverged 1.1 billion years ago. Such rigorous conservation of structure clearly indicates a vital function of this protein.

The intracellular concentration of ubiquitin is very high and its role seems to be the 'tagging' of proteins for intracellular proteolysis, by special supramolecular structures termed **proteosomes**. Proteins subject to ubiquitin 'tagging' include several key proteins in the control of the cell cycle and also the major histocompatibility class I antigens (MHC I antigens; see Chapter 32). Ubiquitin is activated by its sequential attachment to three accessory proteins E_1, E_2 and E_3 (Figure 11.10). The final protein in this system is able to catalyse the transfer of multiple molecules of ubiquitin to proteins to be marked out for catabolism, e.g. partially denatured proteins. The ubiquitin molecules are added 'head-to-tail' to each other following an initial addition to the protein to be 'tagged'. Proteins with such a ubiquitin 'tail' are able to bind to the proteosome complex, and the ATP-dependent hydrolysis of the protein follows. This may involve partial hydrolysis, as is the case for MHC I antigens, or complete hydrolysis as occurs with the cell cycle protein cyclin. Finally, the polyubiquitin

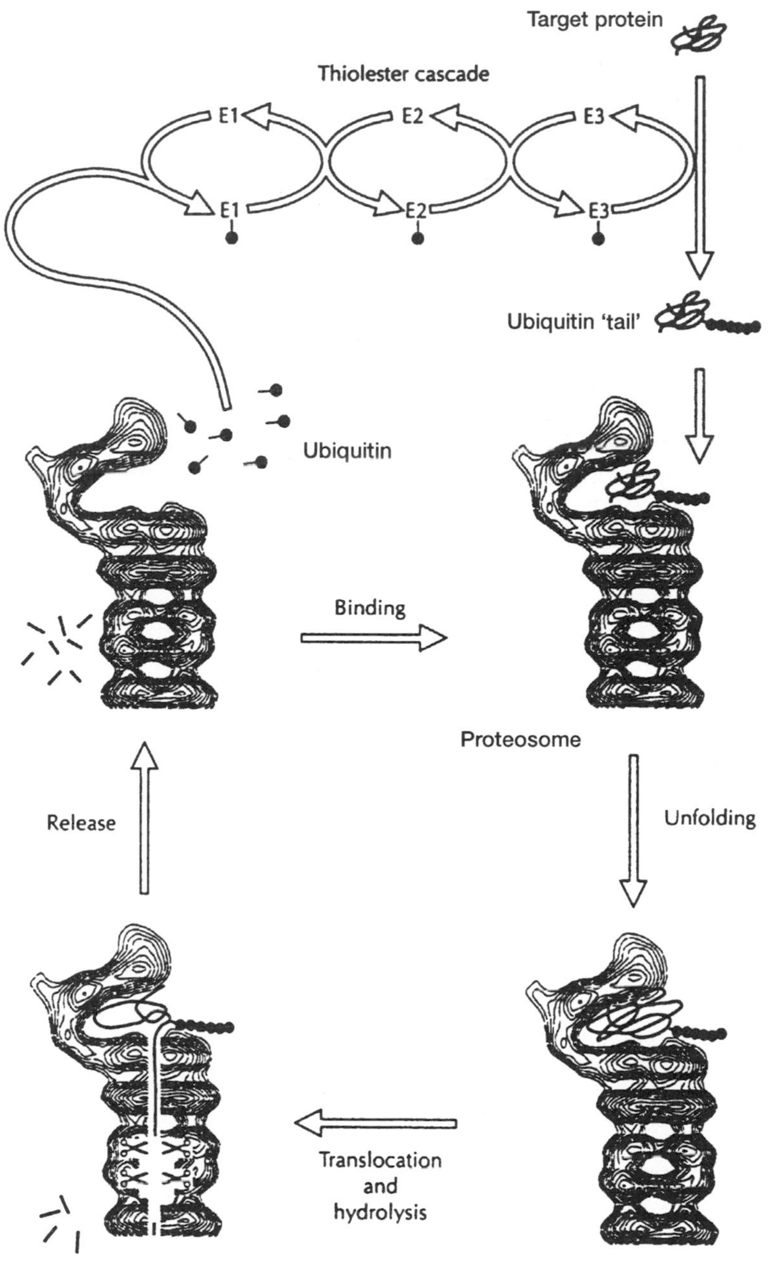

Figure 11.10 **The proposed proteosome cycle**

undergoes hydrolysis to yield monomeric units and these may be recycled.

In addition to its role in normal cell homeostasis, the amount of ubiquitin in cells increases rapidly following cellular insults, e.g. those laying great metabolic demands on tissues, such as temperature changes ('heat shock'), suggesting a role for ubiquitin in the response to such challenges.

Recently, several diseases have been associated with dysfunction of the ubiquitin pathway, although whether as a cause or an effect of the disease is not clear. For example, 'spheroid bodies' containing precipitates of proteins 'tagged' with ubiquitin have been shown to accumulate in the cell bodies and the proximal axons of the motor neurons of patients with amyotrophic lateral sclerosis (ALS; see Chapter 43).

Chapter 12

Plasma calcium and phosphate homeostasis

12.1 Introduction: importance of calcium and phosphate in the body

Calcium is known to play an important role in many processes and systems in the body. Some of these functions are closely related to those of phosphate and it is often useful to consider calcium and phosphate together.

Calcium is essential for the following processes to occur:

1. Formation of bone mineral as hydroxyapatite (see Chapter 29). About 1 kg of calcium is found in the skeleton of the average adult man.
2. Muscle contraction, initiated by the binding of Ca^{2+} to troponin (Chapter 27).
3. Activation of a number of enzymes or binding proteins which require Ca^{2+} as a specific cofactor.
4. Release of all hormones (with the possible exception of some steroid hormones; Chapter 28), neurotransmitters and neuromodulators (Chapter 30).
5. Responses to most hormones and to all neurotransmitters and neuromodulators.
6. The function of certain plasma membrane K^+ channels.

12.2 Plasma calcium concentration

In view of its great importance in the regulation of ionic movement, in transmission of nerve impulses and in muscle contraction, the control of the plasma calcium concentration within a very limited range is of paramount physiological significance.

The plasma calcium concentration in normal individuals is approximately 2.5 mmol/l, and it seldom falls outside the range 2.2–2.6 mmol/l The major proportion of the ions, 50%, is bound to plasma proteins: 45% is in the ionic, freely diffusible form, and the remaining 5% is complexed with molecules such as citrate.

In **hypocalcaemia** the calcium concentration falls below the normal range and the nervous system becomes progressively more excitable, as the permeability of membranes increases. The increased excitability causes spontaneous discharges which initiate nerve impulses, and these are transferred to the skeletal muscles, thus causing tetany. Plasma concentrations of calcium below 1.5 mmol/l are lethal.

Hypercalcaemia occurs when plasma concentrations of calcium rise above 3.75 mmol/l. The nervous system is depressed, reflexes are sluggish, and the muscles become weak as a result of the effects of calcium on muscle cell membranes. Hypercalcaemia leads to **ectopic calcification** of many tissues especially the synovial fluid membranes, the kidneys, the myocardium, the pancreas and the uterus. This can be a serious matter and may prove lethal.

12.3 Calcium and phosphate requirements and body balance

The daily requirement of calcium in the human adult depends on three main factors:

1. The percentage of the intake absorbed from the gastrointestinal tract.
2. The quantity of calcium excreted into the tract via digestive secretions.
3. The quantity of calcium excreted daily in the urine.

If the intake of calcium falls below about 200 mg/day the net absorption of calcium is negative, because of the large amount which is secreted in the gastrointestinal tract, very little of which is reabsorbed. In addition there is an obligatory loss of about 120 mg calcium per day in the urine, making the total obligatory loss of calcium about 320 mg/day. These values are only approximate and depend on other dietary factors and the individual's physiological state. It is known, for example, that urinary calcium excretion is higher on high protein and sodium diets and that it is decreased in old age with decreased glomerular filtration. In any case, it seems that a minimum daily intake of about 600 mg calcium per day is essential to provide sufficient calcium to make up for these losses.Phosphate reabsorption by the kidney, unlike that of calcium, is extremely efficient. **Phosphate** is widely distributed in foods and problems of lack of phosphate in the diet are virtually unknown. A minimum intake of 600 mg/day is, however, recommended.

In the UK, the reference nutrient intake (RNI; see Chapter 14) for calcium for infants and children up to 10 years of age is about 500 mg/day, for males 11–18 years, 1000 mg/day, and for females 11–18 years, 800 mg/day. Above the age of 18 the RNI for calcium is 700 mg/day for both men and women. The RNI for lactating women is 1250 mg/day.

Figure 12.1 shows the movements of calcium through the main systems of the body in a normal adult on an intake of 1200 mg calcium per day over a 24-hour period. The major routes of calcium loss are in the faeces, from non-absorbed calcium and via digestive juices, and in the urine, owing to inefficient reabsorption by kidney tubules. Smaller quantities are lost in sweat.

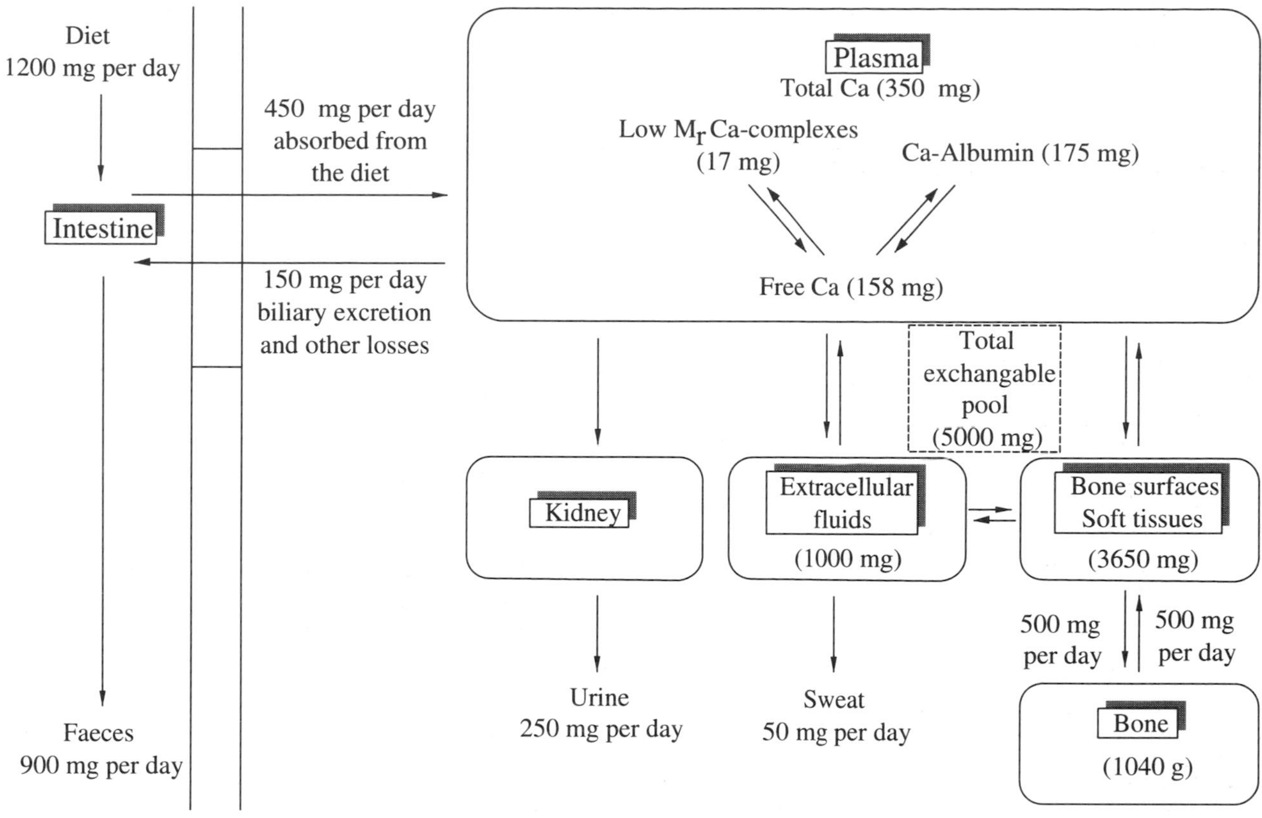

Figure 12.1 **A summary of the kinetics of calcium**

The large reserves of bone calcium are in dynamic equilibrium with plasma calcium. Bone calcium is mobilized under hormonal control when plasma calcium levels fall.

12.4 Calcium intake and absorption

The main dietary sources of calcium are milk and dairy products. Half a litre (roughly equivalent to a pint) of milk contains approximately 1000 mg of calcium, which would cover most individuals' requirements.The absorption of calcium from the intestine depends on several factors. The percentage of a test dose absorbed depends on the subject's customary calcium intake; subjects whose normal intake is high will absorb a relatively small percentage of the dose and vice versa. Factors which stimulate calcium absorption include lactose, basic amino acids and vitamin D and parathyroid hormone.

Lactose is known to increase the absorption of calcium, probably by forming soluble complexes with the ion. This is clearly important, especially in the young animal, as both lactose and calcium are supplied in milk. Amino acids with **basic side-chains**, such as lysine and arginine, and some other amino acids including histidine, tryptophan, methionine and isoleucine, increase calcium absorption. This means that calcium is more efficiently absorbed from high protein diets. Its excretion, however, is also increased on such diets.

Vitamin D is the most important factor regulating calcium absorption. Its mode of action is discussed in Section 12.9.

Parathyroid hormone (PTH) stimulates calcium absorption indirectly via activated vitamin D metabolites.

Factors which inhibit calcium absorption include dietary fibre and especially phytic acid, obstruction of the bile duct, defective bile salt supply, coeliac disease and renal failure.

Phytic acid or **inositol hexaphosphate** is a common constituent of many cereals. It binds dietary calcium forming insoluble complexes which are not absorbed by the intestine. This is particularly important in areas of the world where people exist on a mainly vegetarian diet, and calcium malnutrition frequently develops, particularly in infancy and pregnancy.

Obstruction of the bile duct or a **defective bile salt supply** leads to inadequate vitamin D uptake from the intestine and unless sufficient vitamin D is synthesized in the skin, calcium absorption will be impaired. Pathological conditions which affect the integrity of the intestinal mucosa, such as coeliac disease, will also interfere with vitamin D absorption and consequently the absorption of calcium.

Chronic renal failure is associated with inadequately calcified and weak bones. The reason is that the conversion of vitamin D to its active metabolite occurs in the kidney, and if this process is impaired, the absorption of calcium will also be compromised.

Table 12.1 **Biological actions of hormones regulating the metabolism of calcium and phosphate ions**

Action on:	*Parathyroid hormone*	*Calcitonin*	*1,25-Dihydroxycholecalciferol*
Serum calcium	Increases	Decreases	Increases
Serum phosphate	Decreases	Decreases	Increases
Kidney	Increased 1,α-hydroxylase activity Increased reabsorption of Ca^{2+} Decreased reabsorption of P_i	Decreased reabsorption of Ca^{2+} Decreased reabsorption of P_i	Increased reabsorption of Ca^{2+} Increased reabsorption of P_i
Bone	Increased resorption of bone	Decreased resorption of bone	Ca^{2+} mobilization from bone
Gastrointestinal tract	Indirectly causes increased absorption of Ca^{2+} and P_i by increased production of 1,25-dihydroxycholecalciferol	No effect	Increased absorption of Ca^{2+} Increased absorption of P_i

12.5 Regulation of plasma calcium levels

Homeostatic control is achieved mainly through the actions of parathyroid hormone and vitamin D. Other hormones such as parathyroid hormone-related protein, calcitonin, oestrogens and prolactin are also involved, to a smaller extent, under special physiological conditions.

There are three major processes by which plasma calcium levels are maintained:

1. Absorption of calcium from the intestine, mainly through the action of vitamin D.
2. Reabsorption of calcium from the kidney, mainly through the action of parathyroid hormone and vitamin D.
3. Demineralization of bone, mainly through the action of parathyroid hormone, but facilitated by vitamin D.

Table 12.1 summarizes the biological actions of the main hormones regulating the metabolism of calcium and phosphate ions.

12.6 Parathyroid hormone (PTH)

Parathyroid hormone is a peptide hormone consisting of 84 amino acids. Its biosynthesis is described in detail in Section 28.3. It is derived from enzymic cleavage, first, of pre-proparathyroid hormone (115 amino acids) and subsequently of proparathyroid hormone (90 amino acids). It is released from the chief cells of the four parathyroid glands, usually found on the posterior side of each lobe of the thyroid gland, in response to low plasma calcium. Cells of the parathyroid gland possess a 'Ca^{2+}-sensing' receptor which appears to regulate the secretion of PTH. The release is very rapid, occurring within 20 s following a decrease in the calcium concentration.

Mode of action of PTH

PTH acts in a complex manner on two main target organs, bone and kidney and indirectly, via the activation of vitamin D, on the intestine. The end result is an increase in plasma calcium concentrations.

Parathyroid hormone acts through binding to cell surface receptors and activating adenylyl cyclase to produce cAMP.

Kidney

In the kidney, PTH acts on two separate sites. It acts in the distal tubule to increase reabsorption of Ca^{2+}, and it acts in the proximal tubule to cause inhibition of the reabsorption and, therefore, an increase in the excretion of inorganic phosphate.

It has another effect on calcium homeostasis, namely to activate the enzyme 1,α-hydroxylase, one of the enzymes involved in producing the active form of vitamin D (Figure 12.2).

Bone

In the bone, PTH has an important catabolic effect, with the release of both calcium and phosphate. The effect is

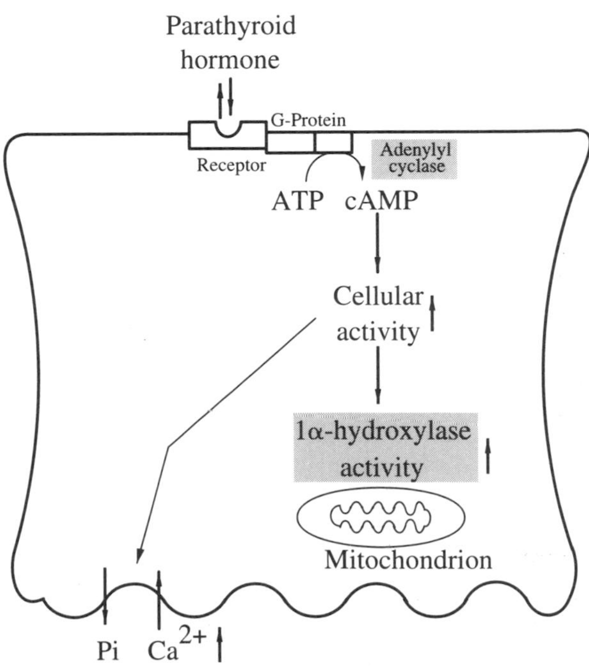

Figure 12.2 **The action of parathyroid hormone in the kidney.** Parathyroid hormone acts on the proximal nephron to increase cyclic AMP, causing the activation of 1-hydroxylase and also the transport of Ca^{2+} into cells and phosphate (P_i) out

biphasic and involves all three types of bone cells, namely osteoblasts, osteocytes and osteoclasts.

Osteoblasts are the cells involved in the formation of the matrix of new bone. **Osteocytes** are derived from osteoblasts and become embedded in the mineralized bone matrix. **Osteoclasts** are largely concentrated on the surface of bone and are involved with bone resorption (bone loss).

The first stage in the action of PTH on bone involves the rapid release of calcium and phosphate from osteoclasts and osteocytes. The process can be initiated within a few minutes of the exposure of bone to PTH and does not involve protein synthesis nor resorption of the actual bone matrix.

The second stage is initiated between 15 min and 2 h of exposure and it involves extensive RNA and protein synthesis within osteoclasts. The proteins synthesized include lysosomal enzymes, such as collagenase, and this is released onto the bone matrix together with hydrogen ions. The action of collagenase in the presence of a low pH results in erosion of calcified bone and the release of both calcium and phosphate into the bloodstream.

The effect of parathyroid hormone on osteoclastic function may, at first sight, seem a little surprising, as parathyroid hormone receptors are only present on osteoblasts and not on osteoclasts. It is now known that the binding of parathyroid hormone to osteoblast receptors results in the release by osteoblasts of soluble factors, such as prostaglandins and interleukins, which act in a paracrine manner to influence osteoclastic metabolism.

12.7 Parathyroid hormone related protein (PTH-rP)

PTH-rP is a 141 amino acid peptide, with actions similar to parathyroid hormone with which it shares a receptor. It is not found in the circulation in appreciable amounts, except in patients with **malignancies**, particularly squamous cell tumours of the head and neck, and also those derived from the kidney, breast and lymphoid tissue. It is produced in the placenta and the amnion and also in the lactating breast.

Its physiological role is still not clear. It probably acts to regulate placental calcium transfer, ensuring the supply of calcium for fetal bone formation and maintaining the fetus hypercalcaemic with relation to the mother. Experiments on mice which do not express the PTR-rP gene *in utero* show abnormalities in endochondral bone formation, suggesting that PTR-rP may be important in the control of fetal calcium homeostasis and bone formation.

12.8 Calcitonin

Calcitonin is a 32 amino acid peptide, secreted by the C- or parafollicular cells of the thyroid when the plasma concentration of Ca^{2+} rises significantly above normal levels. The synthesis and release of the hormone is directly controlled by plasma calcium levels and by the release of gut hormones such as gastrin.

Calcitonin acts directly on osteoclasts to inhibit bone resorption and, in the kidney, stimulates the excretion of many electrolytes, particularly phosphate and calcium.

Calcitonin acts through the same second messenger as parathyroid hormone, yet its action is antagonistic. The apparent paradox can be explained by the fact that they act primarily on different bone cells. Parathyroid hormone receptors are found on osteoblasts, whereas calcitonin receptors are found exclusively on osteoclasts.

Despite knowledge of the actions of calcitonin, its physiological significance is difficult to establish except perhaps in pregnancy and neonatal life.

Patients who have undergone thyroidectomy require thyroxine replacement therapy, but not calcitonin, and do not exhibit any changes in their calcium balance. Furthermore, malignancies of the parafollicular cells of the thyroid, producing calcitonin up to 20 000 times the normal amount, do not seem to upset calcium balance.

In pregnancy, however, its inhibitory effect on osteoclastic bone resorption seems to be important, as it provides a mechanism whereby the maternal skeleton is protected against the potentially excessive demands of the fetus. Calcitonin may also be important in the new-born, suckling animal, in buffering plasma calcium levels which may rise excessively during the intake of milk.

12.9 Vitamin D (cholecalciferol)

The disease **rickets**, which is characterized by poor bone calcification and which affects young children, has been known in Europe for many hundreds of years. It was first described in 1650 by Glisson, Professor of Medicine at Cambridge. The disease occurred in countries with temperate climates, especially in poor, urban areas, and was virtually unknown in the tropics and in Scandinavia. Rickets was widespread in the nineteenth century in Britain, particularly in communities living in the cramped conditions of the developing industrial society. The clinical features of the disease will be discussed in Section 12.12.

In the early 1900s, it was observed that the disease could be treated successfully by either incorporating fish, and particularly fish oils, into the diet, or by exposing the children to ultraviolet light. The relationship between the two treatments was completely baffling until the 1920s, when it was discovered that an essential factor for the body was missing in rickets, and that this factor was present in fish oils, but could also be formed in the skin by the action of ultraviolet light It was called vitamin D, because it was the fourth vitamin to be discovered, after A, B and C.

The findings provided the explanation for the absence of rickets in Scandinavia, where large quantities of fish are consumed, supplying a dietary source of vitamin D, and in tropical countries, where there is adequate exposure to sunlight and, therefore, no need for the supply of preformed vitamin D in the diet.

There are two forms of vitamin D, the naturally produced form D_3, or **cholecalciferol**, and the artificially produced form D_2, or **ergocalciferol**, differing slightly in the structure of their side-chains. Cholecalciferol is the form obtained from animal sources in the diet, or made

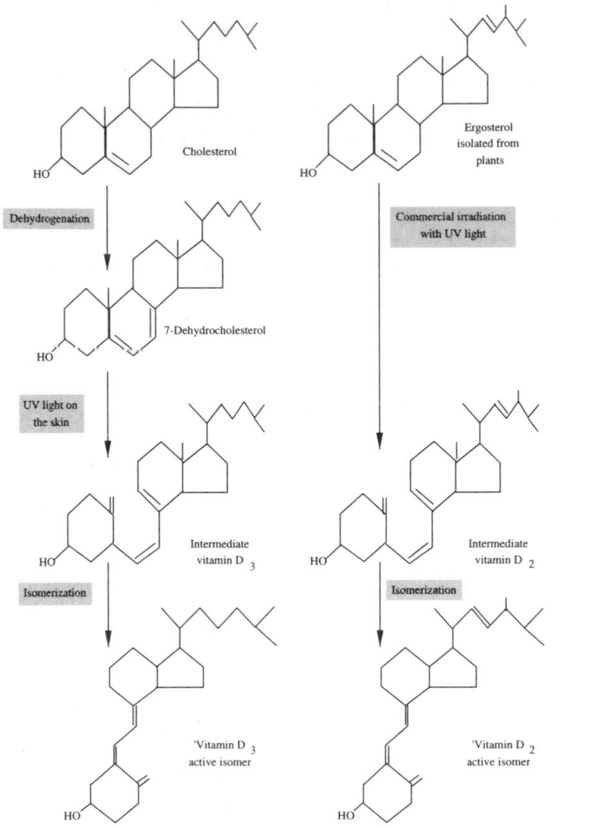

Cholesterol

Ergosterol isolated from plants

Dehydrogenation

Commercial irradiation with UV light

7-Dehydrocholesterol

UV light on the skin

Intermediate vitamin D_3

Intermediate vitamin D_2

Isomerization

Isomerization

'Vitamin D_3 active isomer

'Vitamin D_2 active isomer

Figure 12.3 **The formation of vitamin D_3 in the body and vitamin D_2 commercially**

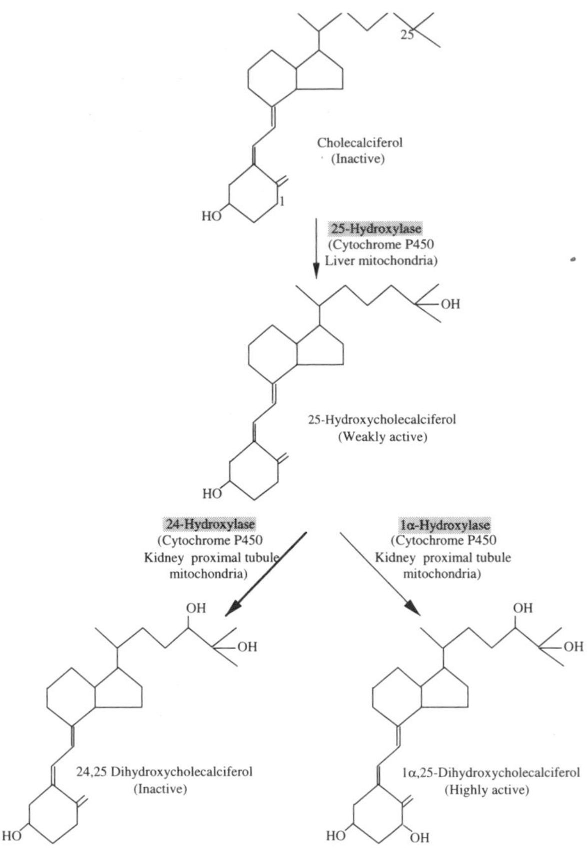

Figure 12.4 **The metabolism of vitamin D_3 – (cholecalciferol) to 1,25-dihydroxy-vitamin-D_3**

in the skin by the action of ultraviolet light on **7-dehydrocholesterol**. Ergocalciferol is the form made in the laboratory by irradiating the plant sterol **ergosterol** and it is the form most readily available for pharmaceutical use. The biological actions of the two forms are comparable.

Vitamin D synthesis and activation

7-Dehydrocholesterol is synthesized in the body from cholesterol. In the skin, ultraviolet light causes scission of the B-ring of 7-dehydrocholesterol to form pre-vitamin D_3 (Figure 12.3). There follows a spontaneous, but slow, rearrangement to form cholecalciferol (CC), or vitamin D_3. This process takes about 36 h at 37°C. It is not clear how cholecalciferol is transported from the skin to the liver. Its circulating form is bound to 'vitamin D binding protein', an α-globulin produced in the liver. Exogenous vitamin D is absorbed in the duodenum and jejunum from bile salt micelles and it appears in the circulation as a constituent of chylomicrons. It is transported to the liver in chylomicron remnants (see Chapter 10).

Vitamin D should be thought of as a hormone rather than a vitamin, as it can be synthesized in the body, is released in the circulation and has distinct target organs, but it is still included in the list of vitamins, as it becomes an essential dietary factor when endogenous synthesis is inadequate to meet the physiological requirements. Cholecalciferol is, in fact, a prohormone, as it needs further metabolism to produce the active form of the hormone.

The first step is the conversion to 25-hydroxy-cholecalciferol, 25-(OH)D, or 25-(OH)CC. (These terms are interchangeable, but 25-(OH)D will be used here for simplicity and consistency.) Figure 12.4 shows an outline of the metabolism of vitamin D. The 25-hydroxylation occurs in liver mitochondria, and is catalysed by a cytochrome P450-dependent hydroxylase which requires oxygen and NADPH. The 25-(OH)D formed is transported in the plasma, bound to a specific α-globulin, to the kidney where it is further hydroxylated in the 1-position to 1,25-$(OH)_2$D. This hydroxylation, which is also dependent on NADPH, oxygen and cytochrome P450, is an important site of regulation (Figure 12.5).

The 1,25-$(OH)_2$D is the most active form of the hormone, and it is 500 times more potent than 25-(OH)D. Another form of the hormone produced in the kidney, is 24,25-$(OH)_2$D which is not active, and is formed by the action of 24-hydroxylase on 25-(OH)D. The plasma concentration of the 24,25 metabolite is normally 100 times that of the 1,25 form. It is believed

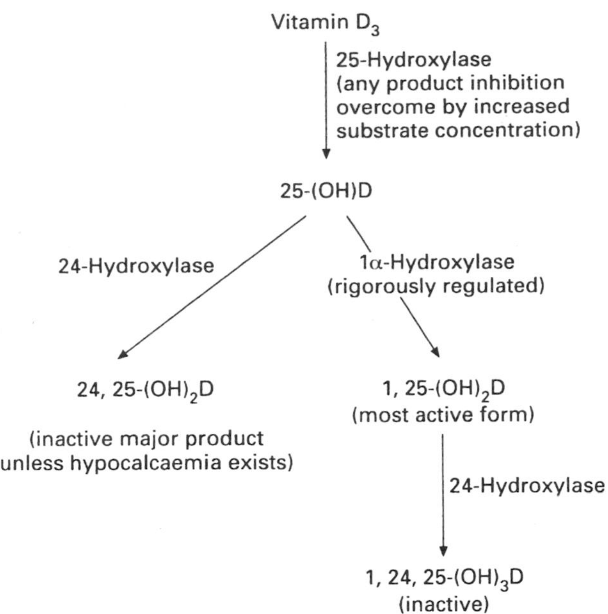

Figure 12.5 **Regulation of vitamin D metabolism to its active form** (the ratio of 24,25-$(OH)_2$D : 1,25-$(OH)_2$D in the plasma is normally 100 : 1)

that no effective control exists for the activity of the 25-hydroxylase in the liver, and all the vitamin D available in the body is converted to the 25-hydroxy derivative. The 1α-hydroxylase in the kidney, however is very rigorously controlled, both acutely (regulation of activity), and in the long term (regulation of synthesis). 25-(OH)D entering the kidney is normally metabolized to 24,25-$(OH)_2$D unless the 1α-hydroxylase is activated.

The activity of the 1α-hydroxylase depends on parathyroid hormone, plasma calcium and phosphate concentrations and the product of the hydroxylation, 1,25-$(OH)_2$D. Insulin, growth hormone and prolactin are also known to affect (increase) the production of 1,25-$(OH)_2$D.

A low concentration of plasma calcium stimulates 1α-hydroxylation, through the secretion of parathyroid hormone. The latter stimulates the production of 1,25-$(OH)_2$D, both by activating the 1α-hydroxylase, and also by increasing the rate of synthesis of the enzyme. The acute effect is thought to be exercised through increased cAMP levels and, probably also, via changes in intracellular calcium concentration.

Reduced plasma phosphate concentration also activates the 1α-hydroxylase, but independently of parathyroid hormone. In pregnancy and lactation, there is an increased production of 1,25-$(OH)_2$D, possibly mediated by oestrogens and prolactin respectively.

The most potent inhibitor of the 1-hydroxylation is the product of the reaction, 1,25-$(OH)_2$D, via inhibition of 1α-hydroxylase synthesis. As the concentration of 1,25-$(OH)_2$D increases, its production is halted, allowing more of the inactive form of the hormone 24,25-$(OH)_2$D to be produced. This allows an accurate control to be exerted over plasma calcium concentration.

High concentrations of calcium and phosphate in the plasma, inhibit the activity of the 1α-hydroxylase directly and, in the case of calcium, also through suppression of parathyroid hormone secretion.

Mechanism of action

The 1,25-dihydroxy derivative of vitamin D causes the increase of plasma calcium and phosphate concentrations by stimulating the following processes:

1. Absorption of calcium and phosphate from the intestine
2. Reabsorption of calcium and phosphate from the kidney
3. Mobilization of calcium and phosphate from bone.

Inorganic phosphate always accompanies calcium through the membranes of intestinal or renal tubule cells, but it is not clear whether this is a passive process, with the negatively charged phosphate simply balancing the charges on the positively charged calcium ions, or an active process controlled by vitamin D.

1,25-$(OH)_2$D is a powerful regulator of calcium homeostasis, and also a hormone with important roles in cell growth and differentiation. Its biological effects are produced in both a classical genomic manner (transduction of a signal to the genome), as well as in a non-genomic fashion (a direct action on cell membranes). The vitamin D receptor (VDR) found on 1,25-$(OH)_2$D target cells plays an important role in the genomic effects. It is a 50 kD protein belonging to the family of receptors which includes the steroid, thyroid and retinoic acid receptors, all of which are regulators of transcription. The VDR is thought to bind 1,25-$(OH)_2$D and form a heterodimer with a retinoid X or another, as yet uncharacterized, nuclear receptor. The complex then binds regions of DNA which correspond to promoter elements of genes which respond to 1,25-$(OH)_2$D. These regions are known as vitamin D responsive elements (VDRE). The binding of the complex to the DNA results in either an increase or a decrease in gene expression, depending on the gene in question. It is known, for example, that the expression of genes coding for calcium binding proteins, such as the calbindins and osteocalcin, is stimulated, whereas the expression of the gene coding for parathyroid hormone is inhibited, in response to vitamin D.

Role of vitamin D in calcium absorption from the intestine

This is the most important effect of vitamin D on calcium homeostasis, as it allows the entry of calcium into the body, so increasing the total amount of available calcium, whereas the actions of both vitamin D and parathyroid hormone on bone and kidney involve rearrangements of already existing calcium pools.

In intestinal epithelial cells, the non-genomic effect of 1,25-$(OH)_2$ D, is a rapid stimulation of calcium transport, known as **transcaltachia**. It is thought that, at the brush border membrane, 1,25-$(OH)_2$D causes the rapid opening

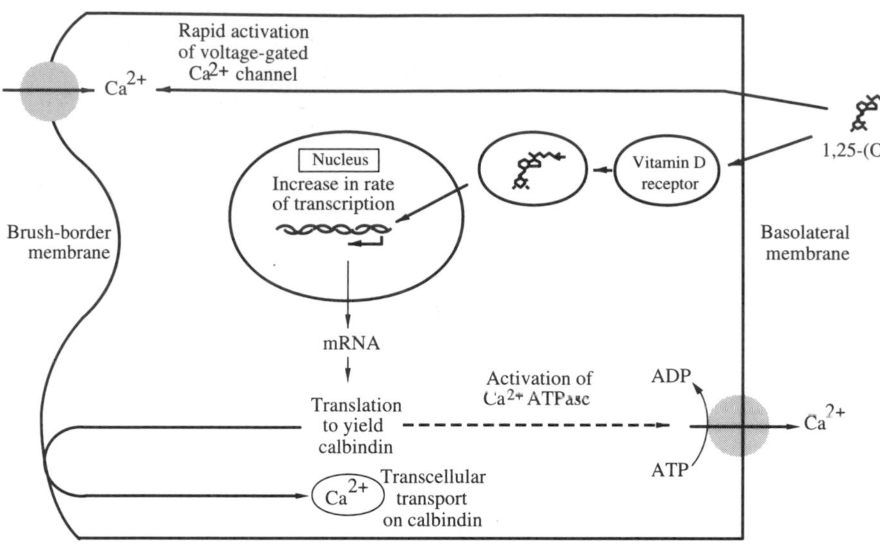

Figure 12.6 **1,25-$(OH)_2$D acts on intestinal epithelial cells to promote the absorption of Ca^{2+} by changing the level of expression of certain genes**

of 1,25-$(OH)_2$D-activated voltage-gated calcium channels, thus transporting calcium into the cells. There is also evidence that 1,25-$(OH)_2$D activates a vesicular mechanism of calcium transport, involving microtubules and microtubule-associated calbindin, but the subcellular components of this pathway have not yet been clearly identified.

The genomic effects of 1,25-$(OH)_2$D on calcium absorption in the intestine are thought to be brought about by the stimulation of the synthesis of calcium binding proteins known as **calbindins**. The effect can be observed a few hours after 1,25-$(OH)_2$D administration. The calbindins are members of the troponin-C superfamily of proteins, and they are found, not only in tissues involved with calcium transport, such as the intestine, the kidney tubule and the placenta, but also in other tissues, such as the brain and the peripheral nervous system, bone, the pancreas and the parathyroid glands. Calbindin 28K and calbindin 9K are thought to be involved in calcium absorption from the intestine, in the following manner: they have a great affinity for calcium ions at the brush border membrane, and facilitate cytosolic transport of calcium from the brush border membrane to the basolateral membrane, where they activate a Ca^{2+} ATPase which pumps calcium out of the cell. These 1,25-$(OH)_2$D dependent events result in active transport of calcium across the intestinal epithelia (Figure 12.6)

Reabsorption of calcium from the kidney tubule

In the distal convoluted tubule, events similar to those described above for the intestinal absorption of calcium are thought to operate, resulting in the reabsorption of calcium from the glomerular filtrate. Phosphate reabsorption from both the proximal and the distal tubule is also stimulated.

As mentioned above, in the proximal tubule, 1,25-$(OH)_2$D also regulates its own synthesis, by regulating the synthesis of the 1α-hydroxylase.

Role of vitamin D on bone

The role of vitamin D on bone seems paradoxical as it can be both anabolic and catabolic. Vitamin D is needed for the proper calcification of bone, and in its absence, poorly mineralized bone is seen, as in rickets and osteomalacia. Whether calcification is a direct effect of vitamin D on bone, or simply the consequence of the increased levels of calcium and phosphate in the plasma, achieved as a result of vitamin D action on the intestine, is not clear. It had been generally believed that the latter was the case, but some recent studies on calcium binding proteins in bone, especially calbindin 9K, indicate that vitamin D may have a more direct effect on bone mineralization. Calbindin 9K, which is synthesized in response to 1,25-$(OH)_2$D, is found, in bone, in the matrix vesicles near the mineralizing front of both trabecular and cortical bone, and it remains in position in mineralized bone. Its synthesis and distribution in normal and rachitic bone suggest an involvement of vitamin D in mineralization. Various other calcium binding proteins are also known to be synthesized in bone in response to 1,25-$(OH)_2$D. These include bone GLA protein (γ-carboxyglutamic acid containing protein), osteopontin and osteocalcin, whose functions are obscure, and the third component of complement (C_3; see Chapter 32).

The best established action of vitamin D on bone is in osteoclastic resorption. This has been studied in organ cultures, and can also be observed in studies on animals on calcium-free diets. This effect was thought to be due to unopposed parathyroid hormone action in the absence of exogenous sources of calcium, but it seems that vitamin D is more directly involved. Administration of 1,25-$(OH)_2$D causes an increase in osteoclast number and activity. Collagen synthesis is inhibited, and lysosomal enzymes including collagenase are synthesized and released, leading to bone erosion. Osteoclasts do not have vitamin D receptors and isolated osteoclasts do not respond to vitamin D. Osteoblasts do have receptors, and it seems that they are the primary target cells which, in turn, affect osteoclastic activity through the release of cytokines, in a manner comparable to that of parathyroid hormone action.

Apart from regulating plasma calcium levels, 1,25-$(OH)_2D$ is a hormone important for cell growth and differentiation. It acts on osteoblast and osteoclast progenitors which contain vitamin D receptors, and promotes the differentiation of these immature cells into osteoblasts and osteoclasts.

Vitamin D requirements

There are not many natural dietary sources of vitamin D except fish oils and egg yolk, and it must be noted that vitamin D, being a cholesterol derivative, is not present in plants. In the UK, margarine is required by law to be fortified with vitamin D. Individuals on strict vegetarian diets (vegans) rely on endogenous vitamin D synthesis.

Exogenous supply of vitamin D becomes necessary when synthesis is less than 10 μg /day. In the UK, there is no RNI set for vitamin D for individuals living a normal life style with adequate exposure to sunlight. Various groups of the population have however been identified as being at risk of deficiency and an intake of 10 μg/day is recommended, to ensure that this does not happen.

There is no solar ultraviolet radiation of the appropriate wavelength (290–310 nm), in the UK, from the beginning of November to the beginning of April. For the rest of the year, the effective radiation depends on the latitude (it is less powerful in the north), the time of day (it is highest between 11.00 and 15.00 h), and the weather (it is much reduced by cloud). Skin pigmentation also affects ultraviolet light absorption, with darker skins absorbing less light. Therefore, an individual's vitamin D status in the UK depends on many factors, including seasonal variation and time spent outdoors, mobility, skin pigmentation and cultural considerations on skin exposure. Babies born in autumn and breast fed in the winter will have low plasma 25-$(OH)_2D$, an indicator of vitamin D status, unless the mothers receive supplements, as the milk will be very low in vitamin D. It has been shown that women in Scotland who did not take supplements in pregnancy had infants with a higher incidence of hypocalcaemia and hyperparathyroidism, and a dental enamel defect related to low vitamin D status.

The elderly are another group at risk, especially if they are housebound, or institutionalized, or living alone on a poor diet. Histologically proven osteomalacia has been shown in up to 5% of elderly people attending hospital in the UK. Up to 50% of institutionalized elderly people have low vitamin D status.

A low vitamin D status is more common among the Asian community, especially women and children. This is thought to be due to a combination of risk factors, including higher skin pigmentation, limited exposure to sunlight for cultural reasons, and vegetarian diets which are usually low in calcium, and from which calcium is poorly absorbed. Asian women and children, especially in the north of England and Scotland, are, therefore, advised to take vitamin D supplements to achieve an intake of 10 μg/day.

12.10 Inter relationships of vitamin D, parathyroid hormone and calcitonin in the regulation of plasma calcium concentration

The precise mechanism of the control of plasma calcium levels is still partially unresolved, but the sequence of events thought to occur when plasma calcium levels fall is summarized below:

1. Release of calcitonin from the thyroid is depressed
2. Release of parathyroid hormone from the parathyroid glands is increased
3. Parathyroid hormone:
   (a) causes a release of calcium and phosphate from bone
   (b) activates the renal 1α-hydroxylase which converts 25-$(OH)_2D$ into 1,25-$(OH)_2D$
   (c) causes the reabsorption of calcium from the kidney tubule
   (d) causes the excretion of phosphate by the kidney.
4. 1,25-$(OH)_2D$ causes an increase in plasma calcium and phosphate by:
   (a) stimulating the intestinal uptake of calcium and possibly phosphate
   (b) acting in conjunction with parathyroid hormone to release calcium and phosphate from bone
   (c) increasing calcium and phosphate retention by the kidney.

The regulation of plasma calcium and phosphate is therefore dependent on both parathyroid hormone and vitamin D, and their relative proportions are of major importance.

12.11 Functions of vitamin D not related to calcium homeostasis

Many aspects of vitamin D function do not seem to be related to calcium homeostasis. Vitamin D receptors are found in many tissues which are not classic target tissues for regulating calcium metabolism, and 1,25-$(OH)_2D$ can exert its effects on most cell types. The actions of vitamin D extend to the regulation of differentiation and proliferation of various cells, including immunoregulatory cells, epidermal cells and malignant tumour cells. These discoveries have opened a new field of vitamin D study and have led to the development of vitamin D analogues for therapeutic use, directed to the non-classic sites of action, with little effect on calcium metabolism.

Vitamin D analogues have been used, for example, in the treatment of psoriasis, a disease of abnormal epidermal growth and differentiation. The rationale for the treatment is based on observations that vitamin D induces differentiation and inhibits proliferation in many cells, including keratinocytes. A rise in intracellular free calcium seems to be required to produce this effect, but the exact mechanism by which 1,25-$(OH)_2D$ brings this about, is not clear.

Vitamin D and calcium deficiencies have been shown to be associated with an increased incidence of colon and breast cancers. Studies on breast cells *in vitro* have shown that vitamin D analogues inhibit the proliferation of these cells.

Vitamin D may also be involved in the regulation of hormone secretion in various tissues. Vitamin D receptors are found, for example, in the pancreatic B-cells, and vitamin D deficiency was found to be associated with decreased insulin secretion. The parathyroid glands also contain vitamin D receptors, and it is known that the synthesis and secretion of parathyroid hormone is inhibited by 1,25-$(OH)_2D$.

There is increasing evidence that vitamin D is involved in the modulation of the immune system. 1,25-$(OH)_2D$ depresses immunoglobulin production, most probably by inhibiting the function of helper T-cells, which in turn affect B-cell function.

Vitamin D deficiency has also been associated with an increased risk of infections. Studies on patients with defective vitamin D receptors have shown a decreased ability of phagocytes to cope with certain micro-organisms, such as *Candida albicans*.

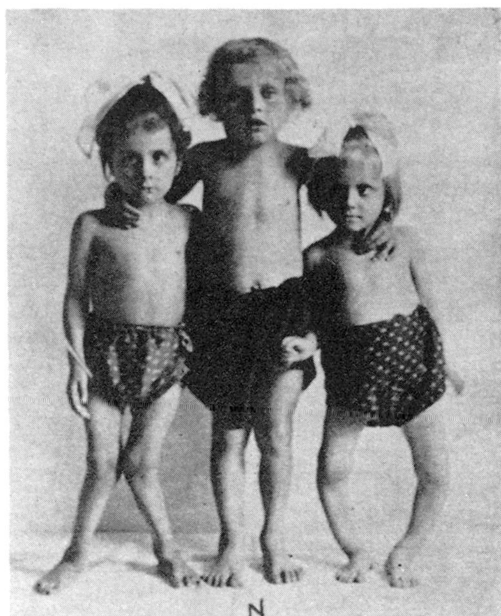

Figure 12.7 **Symptoms of severe rickets.** This photograph shows rickets in Vienna in the postwar famine of 1920. The affected children, 6 years of age, show severe rachitic deformities compared with the normally grown child of the same age in the centre. Note enlarged wrists and ankles, and bow legs

12.12 Disorders of calcium homeostasis

Parathyroid hormone disorders

Hypoparathyroidism

This results in hypocalcaemia. The commonest cause is neck surgery, for thyroidectomy, or the treatment of hyperparathyroidism. Non-iatrogenic causes include atrophy of the glands or an autoimmune disorder.

Pseudohypoparathyroidism

This refers to a group of X-linked dominant genetic disorders characterized by hypocalcaemia in the presence of high levels of parathyroid hormone. The defect is at the receptor or post-receptor level (defective G protein), causing tissue resistance to parathyroid hormone.

In the condition known as **pseudoidiopathic hypoparathyroidism**, hypocalcaemia develops as a result of an inactive parathyroid hormone due to a mutation on the gene.

Parathyroid hormone disorders: hyperparathyroidism

Hyperparathyroidism (which affects 0.1% of the UK population) results in hypercalcaemia. Eighty-five per cent of the cases are caused by benign tumours of the glands, 1% by carcinoma, and the remainder by hyperplasia of the glands. A small number of patients who present with life-threatening, inherited neonatal hyperparathyroidism are homozygous for a mutation at the 'calcium sensing' receptor of the parathyroid gland, which controls PTH secretion. Patients with familial hypocalciuric hypercalcaemia are heterozygous for the same mutation. Treatment of all severe cases of hyperparathyroidism consists of removal of the glands and PTH replacement therapy.

Vitamin D deficiency: rickets and osteomalacia

Prolonged vitamin D deficiency in children leads to the appearance of **rickets**, characterized by poor mineralization of bone. The main features of the disease include a large head with protruding forehead, pigeon chest and depressed ribs, a protruding abdomen, kyphosis (abnormal curvature of the spine) and curved long bones. Although the humerus, radius and ulna are affected, the deformity is more apparent on the weight-bearing bones – the femur, tibia and fibula – especially when the child begins to stand, resulting in the characteristic bowing of legs (Figure 12.7). Plasma calcium may be low or normal. Rachitic children are usually anaemic and prone to infections. Rickets can be fatal when severe.

A condition analogous to rickets, and known as **osteomalacia**, develops in adults as a result of vitamin D deficiency. Bones are poorly calcified with a decreased mineral : matrix ratio, and patients commonly complain of muscle weakness and pain in the joints and the spine.

Osteoporosis

Osteoporosis is a metabolic bone disease frequently seen in postmenopausal women is characterized by loss of total bone mass, without alteration in the mineral : matrix ratio. It is not due to vitamin D deficiency, although vitamin D supplements are sometimes used therapeutically, and it will be discussed in Chapter 29.

Vitamin D resistant rickets

As the name implies, this is a disease which does not respond to treatment with vitamin D. There are various possible causes of this condition, and all involve a defect in the metabolism or mechanism of action of 1,25-$(OH)_2D$.

At least two inherited disorders have been characterized. One is due to a defective VDR, and the second due to a defective 1α-hydroxylase in the kidney.

Liver disease and kidney failure may also give rise to vitamin D resistant rickets, as the production of 25-$(OH)_2D$ and 1,25-$(OH)_2D$, respectively, will be inefficient in the damaged tissue.

Vitamin D toxicity

High doses of vitamin D over a long period are toxic. This situation arises because the rising plasma concentration of vitamin D easily overrides any product inhibition of the 25-hydroxylase in the liver, and the vitamin is converted into 25-(OH)D. Although the latter is 500 times less potent than the 1,25-$(OH)_2D$ form, in high concentrations there is sufficient activity to cause hypercalcaemia and calcification of many tissues apart from bone. This has serious consequences on health and can prove fatal.

Chapter 13

Digestion and absorption of foodstuffs

13.1 Introduction

Natural or prepared foods consumed by the average human adult are very complex mixtures, although the majority of the components of a typical human diet can be subdivided as shown in Table 13.1. Ingested foods can also be classified depending on whether they are absorbed unchanged, e.g. glucose, or whether they require extensive enzymic degradation as for muscle proteins.

The handling of foods in the first three groups of Table 13.1 forms the subject of the present chapter, while the last four groups of substances are dealt with in separate chapters.

Enzymic digestive processes are by now well understood, but the detailed mechanisms of the absorption of the digested fragments from the intestinal lumen into the blood have only been partially elucidated.

Table 13.1 **Dietary components and absorption**

Dietary intake	*Absorbed as*
Protein	Amino acids Dipeptides
Carbohydrates	Monosaccharides (?Disaccharides)
Fat (triacylglycerols)	Monoacylglycerols Fatty acids
Vitamins	Vitamins (esters of vitamins)
Electrolytes	Electrolytes
Trace metals	Trace metals (complexes)

13.2 The role of the digestive organs

The digestive processes can be considered from two different viewpoints, biochemical or physiological. In the biochemical approach the processes of digestion and absorption of an ingested food, e.g. starch, are discussed; the physiological approach focuses attention on the role of each organ, e.g. the stomach or the pancreas, in the process.

It is desirable that the student should acquire a coordinated view of the whole digestive process and both approaches will be used in this chapter.

13.3 Digestive processes

The majority of enzymes involved in the digestive process are hydrolases, i.e. they split bonds of esters, glycosides or peptides by addition of the elements of water.

The digestive enzymes catalyse reactions very similar to those occurring in the lysosomes (see Chapter 8). The two groups of enzymes are not, however, identical since most lysosomal enzymes have an optimum pH on the acid side of neutrality (about pH 5.0), whereas most digestive enzymes are most active within the pH range 6.5–7.5.

Many of the digestive enzymes have trivial names, such as pepsin, trypsin or amylase, because they were the first enzymes to be discovered, and their discovery occurred long before systematic nomenclature had come into existence. Most of these names have been retained in common use, since they are of such long standing.

13.4 Digestive secretions

The enzymes involved in digestion are present in a number of secretions released at specific locations into the digestive tract.

Saliva

Saliva is secreted into the mouth from three main pairs of glands; the submandibular pair, the parotid pair and the sublingual pair. Approximately 1500 ml saliva are secreted daily in man. It is a dilute secretion containing only 0.3–1.4% solid material which is composed mainly of protein, proteoglycans, glycoproteins and electrolytes.

The composition of the secretions from different glands is not identical, a greater proportion of glycoproteins and proteoglycans being secreted by the submandibular glands. One important enzyme contained in the secretion is salivary amylase and this initiates the hydrolysis of food starch into maltose. The major function of the saliva is, however, to lubricate the food by means of its proteoglycans and glycoproteins and thus render swallowing easier. It also protects the epidermal cells.

Saliva, by digesting starch particles that may be lodged in the teeth, may play an important role in cleansing the teeth. In support of this concept, the secretion of saliva by patients who have received X-ray treatment for tumours in the mouth region is often severely impaired and this can rapidly lead to severe tooth decay. The buffering action of saliva is also important in neutralizing acids produced by bacteria which also helps to prevent caries.

Gastric juice

Gastric juice is a composite secretion from three different cell types: mucus-producing cells, the oxyntic or parietal cells which secrete hydrochloric acid and the peptic or chief cells which produce pepsinogen. The juice normally contains about 1.1% non-aqueous material; of this hydrochloric acid of concentration 170 mM makes up about one-half by weight. Consequently, gastric juice is the most acid secretion in the body. The remaining solids are pepsinogen, lipase and glycoproteins (mucus).

Some of the glycoproteins present are large molecules of molecular weight 1.5×10^6–2.0×10^6 and contain 80–85% carbohydrate. Among the smaller glycoproteins (mol. wt 45 000–60 000) is the intrinsic factor which plays a very important role in the absorption of vitamin B_{12} (see Chapter 22).

The main function of the gastric juice is to provide both the powerful proteolytic enzyme, pepsin, and an acid environment for its action. Pepsin is extremely unusual in having such an acid optimum pH (≈1.5–2.0) and it will attack nearly all proteins.

The mechanism of hydrochloric acid secretion has been controversial for many years and is still not fully understood.

Currently, it is believed that a membrane ATPase is involved, as illustrated in Figure 13.1. An ATP-driven H^+/K^+ exchange pump has been shown to be present in the luminal region of the oxyntic cell. It couples the hydrolysis of ATP to the exchange of H^+ for K^+, secreting H^+ into the lumen and taking K^+ back into the cell. The hydrochloric acid can only be secreted if the luminal membrane is permeable to both K^+ and Cl^-. Chloride ions may be provided at the serosal surface by exchange with HCO_3^- (or by means of coupled NaCl entry). The generation of HCO_3^- from CO_2 and OH^- by carbonic anhydrase could also be a role for this enzyme.

Bile

Bile secreted by the human liver is a clear brownish-yellow or green fluid and about 500 ml are produced daily by the human adult. It is stored in the gallbladder where water is removed and the total solid content proportionally increased (Table 13.2).

The bile serves two functions: an excretory function and a digestive function.

Excretory function

The major excretory products are the bile pigments formed by the degradation of haemoglobin and are described in Chapter 25. In addition, insoluble solid material, such as sand, soot or metal particles which accidentally enter the bloodstream, are disposed of through the bile. The particles are transported to the

Table 13.2 **Constitution of bile**

Constituent	Percentage composition: Liver	Gallbladder
Water	96.5–97.5	83–90
Total solids	2.5–3.5	10–17.5
Bile salts	1.0–1.8	6–11
Glycosaminoglycans	0.4–0.5	1.5–3.0
Bile pigments, Cholesterol, Phospholipids	0.2–0.4	0.5–5.0
Electrolytes	0.7–0.8	0.6–1.0

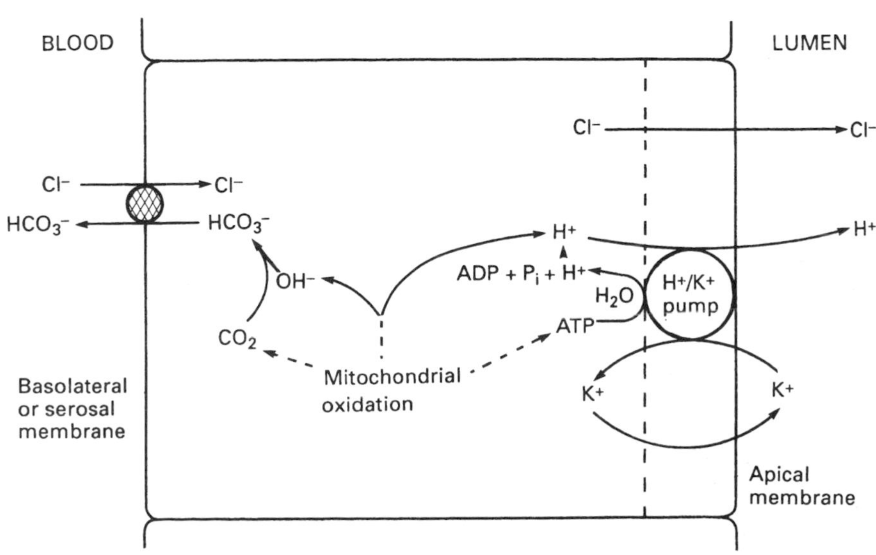

Figure 13.1 **Model for secretion of hydrochloric acid by gastric epithelial cells**

liver where they are taken up by the Kupffer cells; they then enter the lysosomes, any digestible material is hydrolysed by the hydrolytic enzymes and undigested material is then excreted into the bile.

Cholesterol, in excess, may also be regarded as an excretory product and several drugs and their metabolites are excreted in the bile (see Chapters 25 and 36).

Digestive function

Bile plays three important roles in digestion. First, due to its bicarbonate content it is an alkaline secretion with a pH as high as 8.0, and it is therefore important in neutralizing the acidity of the gastric juice and in increasing the pH of the intestine contents to a value close to pH 6.5–7.0. The enzymes of the pancreatic juice and the intestine can then function under optimal conditions. Secondly, bile is essential for the digestion and absorption of fats. Bile salts, together with the phospholipids, emulsify the fats into small droplets so that they are readily attacked by pancreatic lipase. They then aid absorption by taking part in micelle formation together with the monoacylglycerols formed by digestion. Finally, bile salts facilitate the activation of pancreatic hydrolase precursors.

Pancreatic juice

The pancreatic juice is secreted from the pancreas by the duct joining the common bile duct at the ampulla of Vater, so that both the pancreatic juice and the bile mix before entering the intestinal tract.

The pancreatic juice contains about 1.8% solid material; it is, in fact, really composed of two separate secretions: (1) the electrolyte secretion (≈1.2% solid) and (2) the organic and mainly enzyme protein secretion (≈0.1% solid). The separate identity of these secretions is indicated by the fact that they are under entirely different hormonal controls.

The cations of the juice are mainly Na^+, K^+ and Ca^{2+} at concentrations close to that of the plasma. The concentration of HCO_3^- is normally about 80 mM, but it can increase to 135 mM; it has an important regulatory effect on the pH of the intestinal contents, pancreatic juice being capable of neutralizing an equal volume of gastric juice.

Table 13.3 **Pancreatic juice enzymes**

Enzyme	*Substrate*
Chymotrypsin(ogen) Trypsin(ogen) Elastase Carboxypeptidase A and B	Proteins
Amylase	Starch
Lipase Esterase Phospholipase Cholesterol esterase	Acylglycerols/esters
Ribonuclease Deoxyribonuclease	Nucleic acids

The pancreatic juice supplies a large range of precursors of powerful hydrolytic enzymes. Once activated these proenzymes are capable of digesting the majority of ingested food-stuffs and these are listed in Table 13.3.

13.5 Control of digestive secretions: the gastrointestinal hormones

The gastrointestinal hormones are polypeptides which play an important part in the regulation of the secretions of gastric and pancreatic juices, and of bile. During the passage of a meal they are released from the gastrointestinal tract, absorbed and passed through the bloodstream to act on their respective target organs, although some of them may exert local (paracrine) effects. For reference, the various structures of the gastrointestinal tract are shown in Figure 13.2.

Secreted by the stomach into the circulation

Gastrin

Gastrin release is stimulated by the presence of food in the stomach. It is secreted by the antral region of the gastric mucosa and by the duodenal and jejunal mucosae. Several different forms of gastrin exist, but the secretagogue activity resides in the terminal five amino acid sequence and synthetic pentagastrin has strong gastrin activity.

The main function of gastrin is to stimulate secretion of acid into the stomach, but it also stimulates pepsin secretion and increases the motility of the gastric antrum.

Secreted by the duodenum and jejunum into the circulation

Secretin

Secretion of secretin is stimulated by the presence of acid in the duodenum and jejunum and it is released from the mucosae of these tissues. It is a polypeptide composed of 27 aminoacyl residues and its major functions are to stimulate pancreatic secretion of water and electrolytes, to potentiate the effect of cholecystokinin on the pancreas, to increase the secretion of bicarbonate by the liver, and to delay gastric emptying.

Its structure is very similar to that of glucagon and vasoactive intestinal polypeptide. No fragment containing parts of the sequence of secretin has been shown to possess biological activity, which must therefore be a property of the whole molecule.

Cholecystokinin

This hormone is produced by the duodenal and jejunal mucosae; it is released by distension of these organs and by the digested products of fat and protein in the duodenum and jejunum. It is composed of 33 aminoacyl residues, and the entire activity is found in the C-terminal eight amino acids. Five of these amino acids are identical with the five terminal amino acids of gastrin.

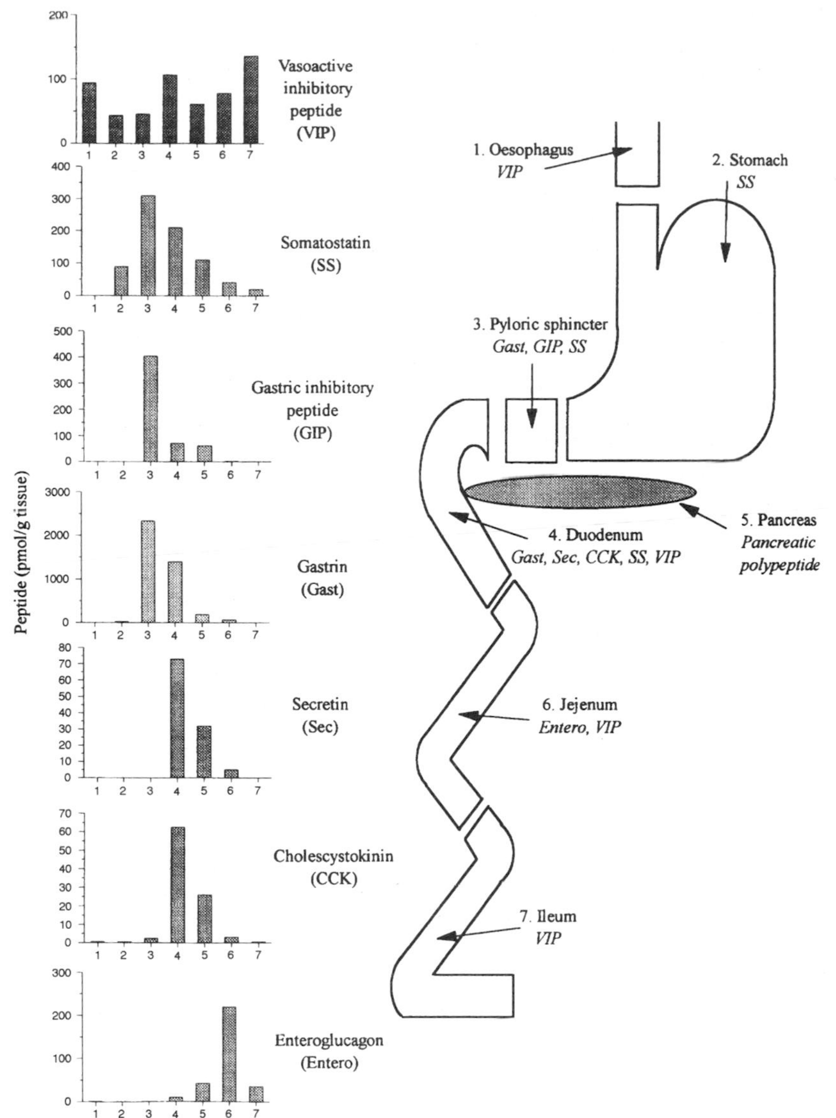

Figure 13.2 **The distribution of hormones in the gastrointestinal tract.** The hormones identified after the name of each anatomical region are those present in the highest concentration)

An interesting structural feature of cholecystokinin (and also of gastrin) is the presence of sulphate esterified to the tyrosine residue closest to the C-terminus of the molecule. In the intestine the major molecular form of the hormone (with 33 aminoacyl residues) may be sulphated, but the sulphate is not essential, whereas in the brain where a peptide consisting of the C-terminal eight residues of the larger molecule is produced, the smaller peptide is invariably sulphated.

The main function of this hormone in the intestine is to stimulate enzyme secretions from the pancreas and to cause contraction of the gallbladder.

Enteroglucagon

Enteroglucagon is secreted to a limited extent by the stomach mucosa, but it is also secreted from the mucosa of the small intestine and colon. As its name implies, it is very similar in structure to glucagon itself (see Chapter 28). Its main functions are to delay the passage of food from the stomach, through the intestine and to release pancreatic insulin. It also serves to stimulate the growth of mucosal cells.

Gastric inhibitory peptide

Gastric inhibitory peptide, a polypeptide composed of 43 aminoacyl residues, is secreted by the duodenal and jejunal mucosae as a result of the action of digested products in the duodenum and jejunum. It inhibits gastric acid secretion, gastric emptying and causes the release of insulin.

Pancreatic polypeptide

Pancreatic polypeptide is produced by the D or PP cells of the pancreas and inhibits pancreatic enzyme secretion and gallbladder contraction.

Secreted locally as paracrine agents

The hormones described to this point are all distributed by the bloodstream but some hormones are produced locally in the gastrointestinal tract, possibly at nerve terminals, and they act locally. In this group are the following:

Somatostatin (growth hormone release inhibiting hormone)

This hormone is found in the hypothalamus, the adrenal medulla, the pancreas and in the gastrointestinal tract and is discussed in Chapter 28. It has many complex actions on the gastrointestinal tract, including inhibition of release of gastric acid, pepsin and gastrin from the stomach, and it thus acts as a powerful antacid. It also inhibits the release of glucagon, insulin and exocrine secretions from the pancreas and the release of motilin, enteroglucagon and gastric inhibitory peptide from the duodenum and jejunum.

Vasoactive intestinal polypeptide

Vasoactive intestinal polypeptide is a polypeptide composed of 28 aminoacyl residues and is secreted along the whole length of the gastrointestinal tract. Like gastric inhibitory peptide and somatostatin, it inhibits gastric secretion, but stimulates intestinal blood flow and secretion from the small intestine.

In summary, it will be seen that hormones play an important role in the regulation of digestion and absorption over the whole gastrointestinal tract. This regulation is accomplished by two main methods:

1. By regulating the quantities of the various secretions of the gastrointestinal tract and, therefore, the concentrations of enzymes to which the various foodstuffs are exposed.
2. By regulating the rate of movement of foodstuffs along the gastrointestinal tract, this in turn controlling the time for which any particular foodstuff is exposed to the enzymes and the time available during which digestive products are exposed to the intestinal mucosa for absorption.

13.6 Carbohydrate digestion and absorption

A typical daily intake of carbohydrate for a human adult in Western society is shown in Table 13.4.

For practical purposes, it will be noted that the human digestive system is concerned mainly with the digestion and absorption of starch, sucrose, lactose and, usually, some fructose.

Table 13.4 **Typical daily carbohydrate intake for a human adult consuming 11 000 kJ/day**

	Intake	
Carbohydrate	(g/day)	(%)
Polysaccharides		
Starch	200	64
Glycogen	1	0.5
Disaccharides		
Sucrose	80	26
Lactose	20	6.5
Maltose	≈0	≈0
Monosaccharides		
Fructose	10	3
Glucose	≈0	≈0

Digestion

The ingested starch is first attacked by an amylase in the saliva and later by a very similar amylase secreted in the pancreatic juice. The amylase attacks and almost completely digests the unbranched component of starch (amylose) which contains only 1 : 4 linkages to maltose; the branched component (amylopectin) is resistant to digestion around the 1 : 6 branching links (Figure 13.3).

The value of salivary amylase in digesting starch has often been considered to be negligible, its role being primarily involved with teeth cleansing as described earlier. This view is based on the fact that, after swallowing, food rapidly comes into contact with the acid environment of the stomach, so that the salivary amylase, which is most active at pH 7.0, is inactivated. In practice, however, the role of the enzyme may be more important in the digestion of starch than is generally assumed. Its activity will depend on the time and extent to which the food is chewed and on the type of bolus swallowed. If this is relatively compact, the acid of the stomach may take some time to penetrate it so that digestion of the starch can continue after the food has entered the stomach. Experiments on volunteers have demonstrated that the percentage of starch digested may reach 40%, but is usually within the range 15–20%.

After the food leaves the stomach, the remaining starch is attacked by pancreatic amylase. In addition, the highly branched starch fragments formed by amylase, sometimes known as 'limit dextrans', are hydrolysed by *iso*maltase which splits the 1 : 6 glucose linkages (Figure 13.3). Digestion of short fragments containing 1 : 4 linkages may then be continued by amylase so that nearly all the original starch is converted to maltose. A small quantity of glucose is also formed.

After a mixed meal, the intestine will, therefore, contain a mixture of disaccharides, maltose, lactose and sucrose, although some non-enzymic hydrolysis of sucrose can occur in the acid conditions of the stomach. The disaccharidases located in the brush border of microvilli (sucrase, maltase and β-galactosidase or lactase) hydrolyse the disaccharides into glucose + fructose, glucose, and glucose + galactose, respectively. The disorder congenital hypolactasia is characterized by an inability to digest lactose, due to a deficiency of lactase. Patients present with a history of milk intolerance, i.e. nausea, diarrhoea and abdominal cramps, following the intake of milk. The symptoms disappear completely if milk is excluded from the diet, or if lactase is added to the milk prior to ingestion. It is

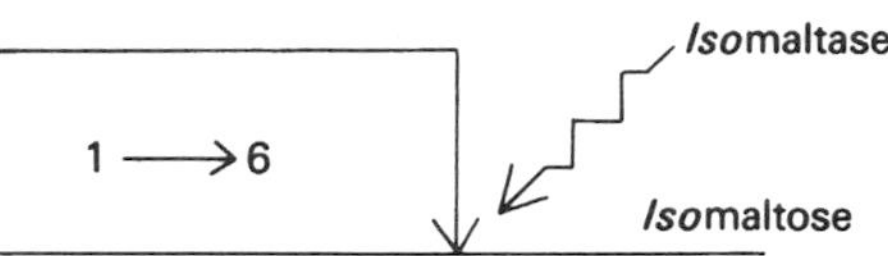

Figure 13.3 **Hydrolysis of branched-chain polysaccharides by amylase and *iso*maltase.** Note that the bonds in the vicinity of the 1,6-branching points are resistant to amylase, but hydrolysis of the 1,6-link by isomaltase releases unbranched chains. ↓ Split by amylase; X, resistant to amylase; G, glucose unit

believed that hydrolysis of the disaccharides is closely associated with their absorption.

Absorption

The rate of absorption of glucose, fructose or galactose depends on their infused concentrations in the intestine, and the kinetics of their absorption are very similar to those described for enzyme actions described by Michaelis–Menten kinetics. A translocation rate (equivalent to v_1) is related to the concentration infused and a constant equivalent to the Michaelis contant (K_m) can be calculated. This will effectively give a measure of the affinity of the sugar for the transport system.

The system for the transport of glucose has been intensively studied and two important concepts have been developed which are applicable to the absorption of some other nutrients and indeed to some transport systems elsewhere in the body.

1. The transport or carrier system has a high degree of specificity, as is indicated in Figure 13.4. Thus D-glucose, D-galactose and α-methyl-D-glucose are good substrates for the luminal transport system, whereas D-fructose is not. (There is a separate facilitated diffusion system for fructose, but this is clearly distinguished from that handling glucose by its independence of Na^+.)
2. The transport of glucose is an example of a secondary active process (see Chapter 2). In this, the movement of glucose is coupled to that of Na^+.

Figure 13.4 **Specificity of glucose transport systems.** Note effect of size of X (transport depends on size of group). 3-Methylglucose – active; 3-ethylglucose and 3-propylglucose – inactive

Table 13.5 **Composition of a solution given to patients with cholera**

Constituent	*Concentration* (mM)
Glucose	110
Na^+	99
Cl^-	74
HCO_3^-	39
K^+	4

From S.E. Nixon and G.E. Mawer (1970) *British Journal of Nutrition*, **24**, 241.

As a result of the discovery of the Na^+ dependence of the intestinal uptake of glucose, a simple means of combating the water and electrolyte losses which make cholera such a serious condition has been introduced. An oral solution containing both glucose and Na^+ together with other anions and cations is given (see Table 13.5 for the composition of such a solution).

13.7 Protein digestion and absorption

Proteins undergoing digestion

As discussed in Chapter 16, adult man requires only 40–45 g protein/day, but typically ingests about 90 g. This forms the exogenous intake, but it should be remembered that the endogenous protein in the gastrointestinal tract and the digestive enzymes themselves form a substantial proportion of the total protein undergoing digestion. It is especially large in some animals, such as the dog or rat, but in man is normally about 10–12 g for an intake of 30 g protein in a typical meal.

Characteristics of proteolytic enzymes

All proteolytic enzymes show the following characteristics:

1. They are secreted as inactive proenzymes (zymogens).
2. They catalyse the hydrolysis of peptide bonds exposing NH_3^+ and COO^- groups.
3. They are classified as 'endo' or 'exo' peptidases depending on whether they attack peptide linkages within the protein molecule (endo) or split off terminal amino acids (exo).
4. They are specific for peptide linkages in which the amino- or carboxy-group is supplied by certain amino acids.

Proenzymes

Pepsin is secreted as inactive pepsinogen in the stomach, which is converted to active pepsin by the cleavage of 44 amino acids from the amino terminus. Cleavage of the peptide bond between residues 44 and 45 of pepsinogen is triggered first by the acidity of the stomach and then by an autocatalytic effect of pepsin itself. The liberated peptide undergoes further degradation by pepsin to produce large peptide fragments and a few free amino acids. The proenzymes secreted in the pancreatic juice are activated by cleavage of a single specific peptide bond. Trypsinogen is converted to trypsin with the splitting off of a hexapeptide in a reaction catalysed by enteropeptidase (or enterokinase), a specific protease produced by duodenal epithelial cells. The activated trypsin so formed can autocatalytically activate more trypsinogen as well as the other proenzymes, chymotrypsinogen, proelastase and the procarboxypeptidases. The activation of trypsin is thus a key step in the activation of the pancreatic proteases and, in order to regulate the concentration of any excess trypsin, pancreatic juice normally contains a low-molecular-weight peptide, trypsin inhibitor, which inhibits trypsin by binding very tightly to its active site.

Endo- and exopeptidases

There are two types of exopeptidase. Aminopeptidases split off amino acids at the N-terminal part of the peptide chain, whereas carboxypeptidases split the terminal amino acid with a free carboxyl group. Endopeptidases attack peptide links within the main peptide chain.

Specificity

All endopeptidases, and also some exopeptidases, exhibit a high degree of specificity. This is directed by the side-chain of the amino acids adjacent to the peptide linkage being hydrolysed and is illustrated in Figure 13.5.

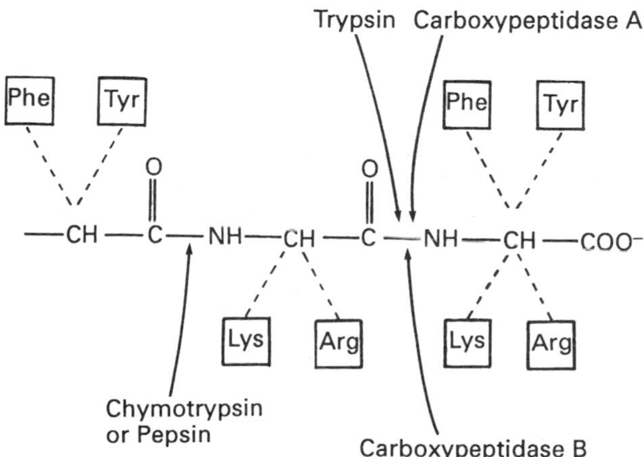

Figure 13.5 **Specificity of peptidases.** Pepsin and chymotrypsin will attack peptide links adjacent to aromatic residues, whereas trypsin attacks peptide bonds close to positively charged lysine and arginine residues. Analogous specificity is shown by the carboxypeptidases which are described as belonging to the A or B group.

Mechanisms of protein digestion

The action of pepsin in the stomach followed by that of trypsin, chymotrypsin and exopeptidase in the intestine rapidly digest the ingested proteins into a mixture of amino acids and many short-chain molecules, e.g. dipeptides. The intestinal mucosa possesses many dipeptidases, although it is not clear how many enzymes exist and how specific each dipeptidase is. Dipeptidases so far studied usually appear to be specific for one amino acid and will, for example, hydrolyse leucine aminopeptides. Theoretically, considering just two amino acids, 380 different dipeptides exist and it is not impossible, although unlikely, that a separate peptidase exists for each dipeptide. These dipeptidases can thence finally convert all the ingested protein into free amino acids.

Absorption of amino acids

The problem of amino acid absorption involving 20 different amino acids is clearly much more complex than that of monosaccharides, where only three are of major importance.

The following facts have, however, been established concerning the absorption of amino acids:

1. They are absorbed at different rates.
2. The natural or L series of amino acids are absorbed much more quickly than the unnatural D series, but the stereospecificity of absorption is not absolute.
3. For those amino acids which have been studied, the mechanism of absorption appears to be similar to that described for D-glucose, in that Na^+-dependent transport systems have been described in the luminal membrane and Na^+-independent ones in the basolateral membrane of the intestinal epithelial cells. The energy for amino acid transport appears to be only indirectly dependent on the hydrolysis of ATP.
4. Some amino acids compete with others for absorption. This has led to the concept of a limited number of transport systems each of which is specific for a number of amino acids. Although this concept is so far not precisely defined, there appears to be at least five different systems for amino acid absorption:
   (a) for neutral amino acids (e.g. alanine, valine, phenylalanine);
   (b) for basic amino acids (lysine, arginine) and cystine;
   (c) for glycine and imino acids (proline, hydroxyproline);
   (d) for acidic amino acids (aspartate, glutamate);
   (e) for other amino acids (e.g. taurine).

Absorption of intact proteins

Under certain conditions and in certain individuals, intact proteins can be absorbed. These proteins often cause undesirable immunological responses and are responsible for the symptoms of food allergies. In contrast to the situation encountered in the adult, there is evidence that the small intestine cells of fetal and neonatal animals can absorb intact proteins through the process of pinocytosis (see Chapter 8). This mechanism is thought to be largely employed for the transfer of maternal antibodies to the offspring but it has also been suggested that this form of absorption may play a role in the nutrition of the immature animal before the establishment of efficient luminal digestion and absorption.

13.8 Fat digestion and absorption

In most Western societies, man obtains a very large proportion, between 40% and 50%, of his daily energy requirements from fat, of which the bulk is triacylglycerol. It may, therefore, be calculated that the total daily intake of triacylglycerol is about 150 g. Other lipid-soluble components form a much smaller proportion of the total lipid intake. As discussed in Chapter 17, cholesterol is not desirable in the food, but cannot be completely avoided. The fat-soluble vitamins A, D, E and K are, however, very important dietary constituents and if fat absorption is impaired, symptoms of vitamin deficiences may follow as a consequence.

Digestion

The digestion of triacylglycerol presents a problem not posed by other constituents of the diet, since these molecules are completely immiscible with water. Churning of the food in the stomach helps to disperse the fatty globules.

Before fat digestion can occur, it must be dispersed in fine droplets as an emulsion and this is accomplished by the secretions of the bile. After concentration in the gallbladder, human bile contains between 5% and 10% bile acids. These are composed of a mixture of trihydroxy acids, the cholic acid group and the dihydroxy acids, the chenocholic or deoxycholic acid group. Conjugates of glycine or taurine are formed in the liver from these acids (Figure 13.6). The formation of bile is described in detail in Chapter 25.

Molecules of the bile acids are composed of a lipophilic part, the sterol ring, and a hydrophilic part, the three hydroxyl groups and the side-chain. As a consequence, bile acids are amphipathic molecules and powerful emulsifying agents. Phospholipids composed of two lipophilic fatty acid chains, hydrophilic phosphate and a hydrophilic molecule, e.g. choline, are also secreted in the bile and, in conjunction with the bile acids, emulsify the triacylglycerols into small droplets of 200–5000 nm in diameter (Figure 13.7).

Emulsified triacylglycerols are readily attacked by pancreatic lipase secreted in the pancreatic juice. Lipase is a remarkable enzyme in that it is only effective at the lipid aqueous phase interface.

Lipase hydrolyses fatty acid in the 1- and 3-positions of the triacylglycerols, rapidly producing, from each molecule of triacylglycerol, 2-monoacylglycerols and two molecules of fatty acid. Subsequent slow isomerization of the 2-monoacygylycerols to 1- or 3-monoacylglycerols occurs and these are then hydrolysed to glycerol and a third molecule of fatty acid (Figure 13.8). However, under most conditions in the digestive tract, minimal formation of glycerol occurs, most of the hydrolysed triacylglycerols forming monoacylglycerols.

Absorption

The formation of monoacylglycerols in the presence of bile acids, fatty acids and phospholipids causes the production of micelles that are of very much smaller

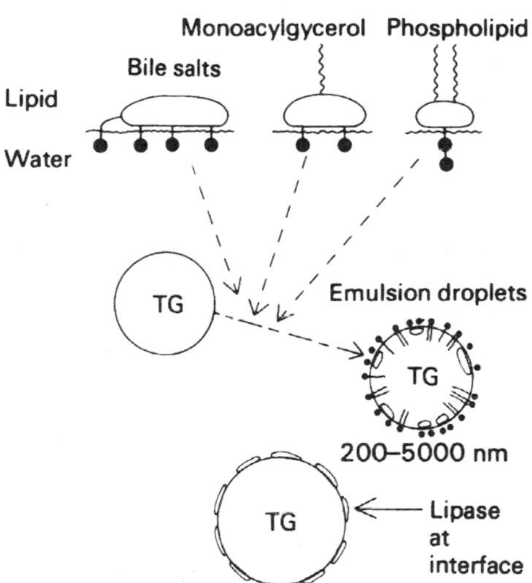

Figure 13.7 **Role of the bile salts, monoacylglycerols and phospholipids in the emulsification of triacylglycerols**

Figure 13.6 **Structures of the bile salts.** Percentage of bile salts in bile is: cholic, 3α, 7α, 12α30–40%; chenocholic, 3α, 7α30–40%; deoxycholic, 3α10–30%. Bacterial 7α dehydroxylation forms deoxycholic and lithocholic acids. A diagrammatic representation of the hydrophilic groups (●) of the bile salts is shown on the left

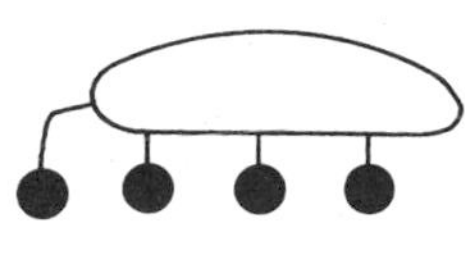

HO

COOH

HO OH

Cholic acid

$CO–NH–CH_2CH_2–SO_3^-$

Taurine

$CO–NH–CH_2–COO^-$

Glycine

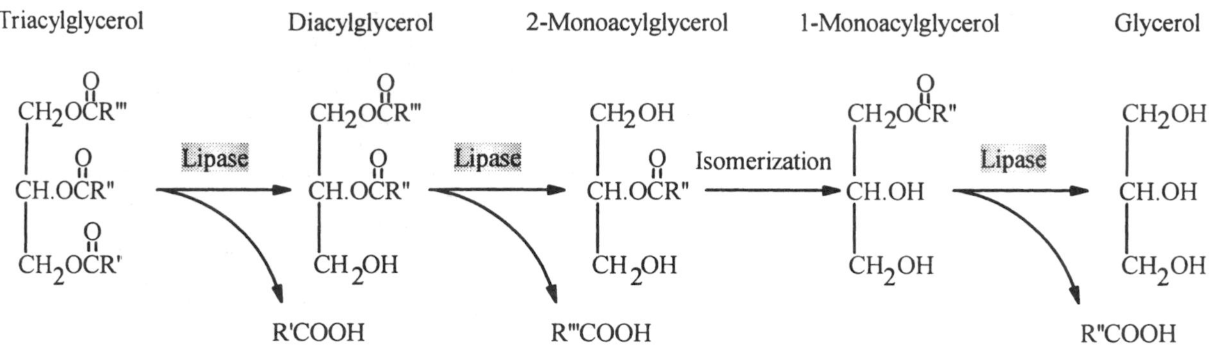

Figure 13.8 **Formation of monoacylglycerols as a result of the hydrolysis of triacylglycerols by lipase.** (Monoacylglycerols may undergo further hydrolysis following isomerisation)

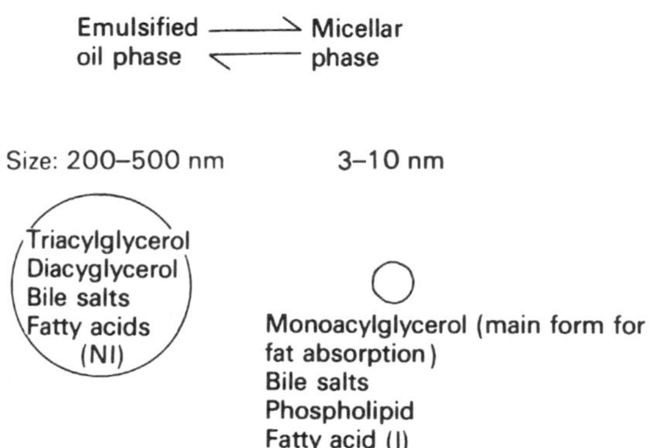

Figure 13.9 **Relation between emulsified fat droplets and micelles. Fatty acids are ionized (I) in the micelles but are not ionized (NI) in the emulsion**

dimensions than emulsion globules, having a diameter within the range 3–10 nm, i.e. comparable to the dimensions of a large molecule (Figure 13.9). Micelles are stable in a relatively clear solution and are transported into the mucosal cells. The net result is the transfer of monoacylglycerol and fatty acid molecules into the cell. In the endoplasmic reticulum, the monoacylglycerols are reconverted to triacylglycerols. The fatty acids required for this synthesis can arise from three sources:

1. Absorbed from the lumen.
2. Produced by hydrolysis of absorbed monoacylglycerols.
3. Synthesized in the mucosal cells.

Fatty acids are converted to acylCoA derivatives by the action of thiokinase in the presence of coenzyme A and ATP; these are then utilized to form triacylglycerols (see Chapter 6). Instead of utilizing the monoacylglycerols for fat synthesis, the mucosal cells can metabolize absorbed glucose to glycerol 3-phosphate and synthesize phosphatidic acid by utilization of two fatty acylSCoA (FACoA) derivatives. Phosphatidic acid can then be converted either to phospholipids or to triacylglycerols according to the scheme shown in Figure 10.5. This pathway is believed to be concerned mainly with the biosynthesis of phospholipids and not of triacylglycerols.

It is, therefore, apparent that triacylglycerols may be synthesized in the mucosal cells by different pathways. The precise pathway used at any particular time will depend on several factors, such as the rate of absorption of free fatty acids by the monoacylglycerol micelles and the rate of absorption of glucose if the meal is a mixed one, as it normally would be. It will be noted that the overall process has transferred triacylglycerol from the intestinal lumen to the mucosal cells, but the molecules within the cells are not identical to those originally attacked by lipase, due to degradation and resynthesis.

The triacylglycerols are transported from the mucosal cells into the lymph in the form of lipoprotein globules about 75 nm in diameter known as 'chylomicrons' (see Chapter 10). These are composed of a very large (85–90%) proportion of triacylglycerols with small amounts of cholesterol (5%), protein (2%) and phospholipid (7%). The phospholipid and protein surround the globule and are essential for its stability in the lymph and plasma. Although the amounts of protein and phospholipid in the chylomicrons are small they are, nevertheless, essential for the liberation of fat from the mucosal cells and, if their synthesis is impaired, serious reduction in the transport of fat will occur.

The chylomicrons pass from the lymph into the blood through the thoracic duct and, after a fatty meal, the plasma is distinctly milky in appearance due to the presence of these particles. The fate of the chylomicrons is discussed in Chapter 10.

13.9 Bacterial flora in the gastrointestinal tract

The human gastrointestinal tract contains at least 150 different species of bacteria. Most are anaerobic or facultative anaerobes, the genera being listed in Table 13.6.

There are very few bacteria in the stomach or duodenum, more in the lower regions of the small intestine, while the large intestine contains a very large population. These bacteria have a profound effect on the foodstuff in the intestinal lumen, on the mucosal cells and

Table 13.6 **Genera of bacteria* in human gastrointestinal tract**

Group	*Bacteria*
1	Facultative or aerobic (Gram-negative)
	Enterobacter
	Pseudomonas
2	Facultative streptococci
3	Lactobacilli
4	Anaerobic cocci
5	*Bacteroides*
6	*Eubacterium*
7	*Bifidobacterium*
8	Clostridia

*150 species, mainly anaerobic.

Table 13.7 **Consequences of removal of microflora (gnotobiotic animals)**

Wall of intestine thinner – reduced in weight
Villi longer – slender
Surface area of intestine reduced
Rate of epithelial cell renewal – reduced
Animals able to withstand ionizing radiation more effectively
Enzyme complement of mucosa altered
  Alkaline phosphatase increased
  α-Glucosidase increased
Absorption of glucose increased
Absorption of vitamins increased
Greatest response is in small intestine when bacterial count is low

particularly on the brush border. Their effects can be demonstrated by the study of gnotobiotic animals, which are born and kept under completely sterile conditions so there are no bacteria in their intestines. These animals show many differences from control (normally bred) animals, and these are listed in Table 13.7.

Bacteria can have desirable or undesirable effects on human metabolism. They synthesize some vitamins, e.g. biotin and vitamin K, that can be absorbed and utilized by the hosts. They can be harmful by utilizing valuable nutrients so preventing absorption and by producing toxic compounds that damage the mucosa and inhibit absorption; bacteria also possibly metabolize bile acids to carcinogenic compounds.

The study of the effects of bacteria in human intestine is extremely complex, since there are many variations between population groups, types of diet consumed, times of day and year and even between individuals.

13.10 Malabsorption syndromes

Defects of absorption can occur as a result of:

1. Deficiencies of enzyme secretions.
2. Bile deficiency on account of blockage of the bile duct.
3. Damage to the brush border and mucosal cells.

Deficiency of gastric or pancreatic juices can result in a reduced efficiency of the digestion of proteins, lipids and carbohydrates, but the effects of lack of bile are most clearly shown in the reduced efficiency of fat digestion and absorption. **Lactase deficiency** (see Section 13.6) is an example of enzyme deficiency, deficiencies of this enzyme in the adult being widespread in many parts of the world. It appears to be genetically linked and is very common in Black Americans, Africans and in Asia. Damage to the intestinal mucosa occurs as a result of coeliac disease or gluten enteropathy. This condition is relatively common and between 1 : 1000 and 1 : 2000 of the population suffer from it in the UK. Serious damage is caused to the mucosal brush border by the gliadin of wheat protein, but the exact mechanism of the damage is still obscure. Immunological reactions with the mucosal cell may be involved. As a consequence of this condition, the absorption of most digested foods, particularly fat and vitamins, e.g. folate or vitamin B_{12}, is severely impaired and deficiency symptoms will develop (see Chapter 22).

Chapter 14

Nutrition: general aspects

14.1 Diet and health

The importance of nutrition in relation to health has been recognized since antiquity. Hippocrates was advising patients who suffered from night blindness to eat liver, and doctors throughout history have tried to cure various illnesses with changes in the diet. Nutrition as a science, however, did not exist until Lavoisier carried out his first experiments at the end of the eighteenth century and concluded that 'life is a combustion'. **Nutrition** is the science of food, and the nutrients and other substances contained in food. It is the study of their actions, interactions and balance in relation to health and disease. **Health** can be thought of as the state of the organism when it functions optimally and **disease** the state of the organism when function is impaired. **Nutritional status** refers to the state of the organism in relation to the consumption, utilization and storage of nutrients. It is only one of many factors which determine the achievement, or maintenance, of a state of health in an individual. **Diet** is the regular course of eating and drinking adopted by an individual.

A reliable, abundant and often excessive supply of nutrients is a new condition for the human race, and only applies to the last half of the century and to the populations of Europe, North America and Australasia. Much of Africa, Asia and South America still suffer from inadequate food supply.

Malnutrition is the term commonly used to denote **under-nutrition**, that is, insufficient intake of nutrients and/or energy. This term can, however, be used to describe dietary practices inconsistent with health, which are characterized by an **excessive** rather than an inadequate intake of a dietary component. Malnutrition, in this more general sense, may be a significant factor contributing to the 'diseases of affluence' seen in the last century in the Western world. These include disorders such as cardiovascular and cerebrovascular disease, hypertension, cancer, obesity and diabetes mellitus and will be dealt with in other chapters of this book.

14.2 The need for energy and nutrients

Food supplies the body with energy and specific nutrients. The term 'nutrient' has acquired a specific meaning in recent years. **Nutrients** are essential dietary factors, such as vitamins, minerals, essential amino acids and essential fatty acids which cannot be synthesized by the human body. Sources of **energy** (metabolic fuels) are not classed as nutrients. Metabolic fuels include carbohydrates, lipids, proteins and alcohol (ethanol). Additional components of the diet include **inert** substances, such as dietary fibre (non-starch polysaccharides) which cannot be digested or absorbed but are thought to be of considerable value in the diet.

The oxidation of metabolic fuels to carbon dioxide, water and other products provides energy needed to sustain the organism. Energy is constantly needed for muscular contraction, ion transport and synthesis of macromolecules. To a large extent, sources of dietary energy are interchangeable and there is no requirement for a dietary source of carbohydrate, fat (except for a small amount of essential fatty acids) or ethanol. Carbohydrate can be synthesized from protein, but most diets will not provide enough protein to cover the need for carbohydrate. In practice, most diets provide at least 30% of their energy content in the form of carbohydrate and in the developing countries this can be as high as 90%. Similarly, apart from providing some polyunsaturated fatty acids, known as the essential fatty acids, fat is not a dietary essential, as it can be made from carbohydrate. Fat-free diets, however, would be unpalatable, might not provide an adequate amount of energy, as carbohydrate and protein are much less concentrated forms of energy, and might lead to fat-soluble vitamin deficiencies.

In contrast to carbohydrate, most of the fat that we eat, and ethanol, there is a definite dietary requirement for protein. Apart from the obvious need to synthesize new body protein, the nitrogen can be used in the synthesis of a large variety of nitrogen-containing substances such as nucleic acids, certain neurotransmitters and complex lipids. Very little nitrogen is taken into the body in any other form than protein. The importance of protein in the diet is to act as a source of **essential amino acids**, which cannot be synthesized by the body, and to provide enough nitrogen to allow the synthesis of the non-essential amino acids. Apart from the obvious requirement for protein by the growing child and the pregnant woman for the accumulation of new tissue, there is a need for dietary protein in adults in order to replace body protein which is continuously degraded.

Three polyunsaturated fatty acids – linoleic, linolenic and arachidonic – are known as the **essential fatty acids.** They must be provided in the diet as the body cannot synthesize them and they are needed as structural components of all cell membranes and also as precursors for a series of compounds known as eicosanoids, which include prostaglandins and leukotrienes. If the essential amino acids and the essential fatty acids are taken in amounts which exceed the minimum requirement by the body, they will be metabolized in the same way as the rest of the metabolic fuels and provide energy. In this way, they differ from nutrients such as **vitamins** and **minerals** which do not become sources of energy, even when consumed in amounts excessive to requirement.

Water and oxygen, although essential to life, are not classified as nutrients.

14.3 Dietary requirements

Consumption of metabolic fuels and nutrients in the correct daily quantities would provide an adequate diet consistent with good health. The criteria for adequacy are not accurately defined. An adequate diet would provide enough energy and nutrients to prevent the appearance of deficiency symptoms. Clearly, this is the minimum criterion of adequacy, and allowances should be made for tissue saturation or storage of a certain nutrient, so that an individual can withstand periods of inadequate intake or increased demand without impairment of function and risk to health. In the case of some nutrients, the physiological effects differ at different levels of intake. A choice of criteria of adequacy is therefore available, and the goal will determine the requirement. The question is further complicated by the fact that the actual determination of requirements is not always based on precise experimental data, but often on epidemiology, where variation in the dietary intake must occur within a population, and little control is possible. Finally, the requirement for a nutrient varies from individual to individual, and for each individual it varies with their physiological state and with changes in other components of the diet, as one dietary component may affect the absorption, utilization, storage or excretion of another component.

Estimates of requirements are generally based on the following information:

1. The intakes of a nutrient needed to cure the signs of clinical deficiency. In this category, the information is usually obtained from animal experiments. Laboratory animals, showing signs of deficiency having been fed on a diet lacking a particular nutrient, are fed graded quantities of that nutrient to ascertain the amounts needed to cure the deficiency.
2. Intakes of individuals and groups associated with absence of symptoms of clinical deficiency. This involves the assessment of the nutritional status of individuals, groups or communities by analysis of their dietary habits.
3. Intakes required to maintain acceptable circulating levels or tissue concentrations of a nutrient. In this category, biochemical measurements are often used, and blood or urine samples from individuals can give an indication of their nutritional status and therefore an estimate of their requirement for a particular nutrient.

It is generally assumed that the requirement for a nutrient in a population is normally distributed (Figure 14.1).

The **estimated average requirement (EAR)** of the population for energy or protein or a vitamin or mineral represents the intake which will satisfy the needs of 50% of the population, but will be insufficient for the other 50%.

The **reference nutrient intake (RNI)**, set at two standard deviations above the EAR, represents a level of intake which will be sufficient or more than sufficient for about 97% of the population. If the average intake is set at the RNI, then the risk of deficiency in the population is very small, although, in theory, 2–3% of the population will be inadequately catered for.

The **lower reference nutrient intake (LRNI)**, set at two standard deviations below the EAR, will be sufficient for only the few people in a group who have low needs and, therefore, represents a level of intake almost certainly inadequate for most individuals.

The term **safe intake** is used to indicate the intake or a range of intakes of a nutrient for which there is not enough information to estimate RNI, EAR or LRNI. It represents an amount which is more than adequate for almost every member of the population, but not so large as to cause toxicity or other undesirable effects.

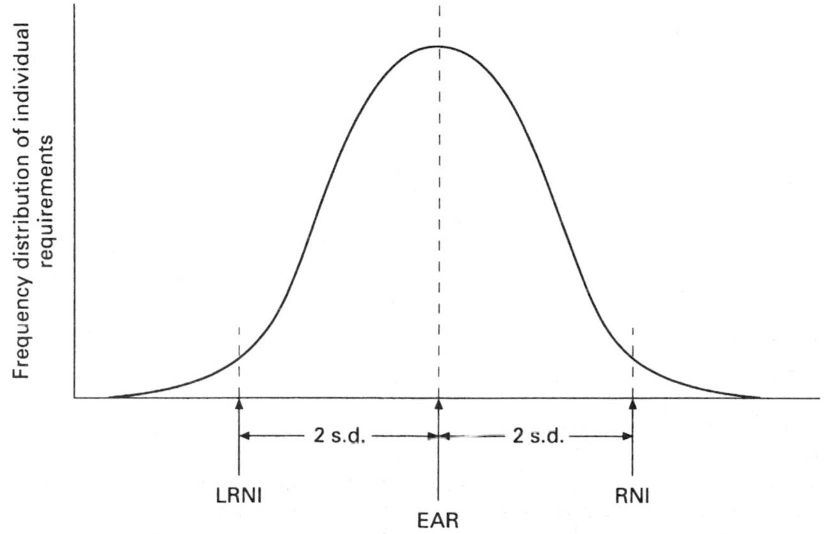

Figure 14.1 **Definitions of dietary reference values** (EAR, estimated average requirement; RNI, reference nutrient intake; LRNI, lower reference nutrient intake)

The terms described above have been set by the Committee on Medical Aspects of Food Policy of the UK Department of Health in 1991 and are collectively known as **dietary reference values.** They are intended to replace terms such as **recommended daily intake** used in the UK until recently, although they are not recommendations but reference standards, against which the intakes of a population can be compared and dietary guidelines can be formulated. The term **recommended daily allowance** is still used in the USA.

It must be appreciated that many errors are inherent in these kinds of estimates, as they are often based on individuals' recall of dietary intake, which is often inaccurate. Food composition tables also lack precision, as various assumptions have been made in their construction.

14.4 Nutritional problems in modern society and guidelines for a healthy diet

Animal studies and human metabolic studies have traditionally formed the basis of dietary recommendations. The realization in the last few decades that certain diets are associated with certain diseases which may develop over years, whereas some other diets seem to have a protective effect, led to a different view being taken on diet and health. The original guidelines on healthy eating aimed at prevention of deficiencies, whereas recently, the focus of attention has shifted towards the effects of diet on health in the absence of deficiency. Diets and disease rates certainly vary among different populations, but as these groups differ in many other respects, such as living conditions and environmental factors, not to mention heredity, it is difficult to identify the effects on disease of diet alone.

Many recent studies, however, mainly epidemiological, have led to the accumulation of a strong body of evidence implicating dietary factors in the cause and prevention of disease associated with life in the developed countries, or with a 'Western' pattern of lifestyle in affluent groups of populations in less developed countries. These studies recognize that diet may be only one of many factors responsible, but one which it is possible to change. Diseases, or disorders in which diet seems to have a causative or contributory role are diverse. In addition to cardiovascular and cerebrovascular disease, cancer, hypertension, diabetes mellitus and obesity, previously mentioned, the list includes dental caries, diseases of the colon, and birth defects.

Cardiovascular disease

Mortality from cardiovascular disease in the UK is among the highest in the world. Atherosclerosis is an early event in coronary heart disease (CHD) and is characterized by the accumulation of cholesterol-rich deposits in the arterial wall, eventually leading to endothelial damage, plaque formation and narrowing of the arterial lumen. Cholesterol taken up by macrophages in the arterial wall in the form of low-density lipoprotein (LDL) is implicated in the pathogenesis of atherosclerosis. Thrombosis, which is the final event in coronary occlusion, involves platelet aggregation and fibrin formation. Smoking, hypertension and increased serum cholesterol in the form of LDL are major risk factors. In contrast, plasma concentrations of high-density lipoprotein (HDL) are inversely associated with CHD mortality.

Diet is the major non-genetic factor responsible for high levels of serum cholesterol and, in different countries, average national saturated fat consumption correlates closely with average national serum cholesterol levels. Death rates from CHD are similarly associated with consumption of saturated fat. Polyunsaturated fatty acid intake seems to decrease the level of total and LDL cholesterol independently of changes in saturated fat but the role of mono-unsaturated fatty acids is unclear. Countries, such as Greece and southern Italy, have low rates of CHD and high consumption of mono-unsaturated fatty acids, but they also have diets characterized by low levels of saturated fat and high levels of fruit and vegetables. The picture is further complicated by evidence that antioxidant nutrients such as vitamins E and C may have a protective effect against CHD by preventing the oxidation of LDL which seems to be an important step in atherogenesis (see Chapter 33). LDL particles formed on a diet high in mono-unsaturated fatty acids are relatively resistant to oxidation.

Naturally occurring unsaturated fatty acids have a '***cis***'-configuration about the double bond. The introduction of '***trans***'-fatty acids in the diet, mainly as partially hydrogenated vegetable and fish oils, has raised concerns about their safety, as they are thought to behave metabolically in a manner similar to saturated fatty acids, and might, therefore, present the same risk to health as the latter (see Section 17.6). Their biological effects, the subject of extensive study in recent years, remain to be elucidated. Although it is felt that their consumption should not go unchecked, and should not increase from the present level, there is insufficient evidence, at present, to warrant advice for decreasing their consumption from the current average intake of 5 g/day, or 2% of the total dietary energy intake in the UK.

Cancer

Countries with low fat intake (the less affluent countries) have low rates of colon, breast, large bowel, pancreas and prostate cancers. The correlations are seen mainly with animal fat and meat consumption rather than with vegetable fat. Animal studies support epidemiology in many cases. Breast tumour incidence is double in animals fed 40% energy as fat compared to those fed 20%. Total fat intake and breast cancer mortality show a close association in cross-country studies, but not within one country. The risk of breast and endometrial cancer in postmenopausal women increases with obesity. The risk of colon cancer is increased with increased fat intakes, while dietary fibre seems to have a protective effect. Prostate cancer was shown to be related to high fat intake within USA counties. The mechanism by which dietary fat may exert its effect is obscure.

Hormones, especially oestrogens, have been implicated as they are known to be promoters of breast cancer, but no clear relationship has ever been established between dietary fat and female sex hormone levels. Breast cancer studies involved middle-aged women and the effect of high fat diets in early life has not been investigated. The large differences in breast cancer rates among various countries are probably due to a number of factors, including reproductive patterns, obesity, alcohol consumption and use of exogenous oestrogen. Adult height, which is in part a measure of childhood weight gain, is positively associated with breast cancer rates internationally. Greater childhood weight gain leads to a lower age for menarche, which in turn increases the risk of breast cancer. In rural areas of China, where the age of menarche is about 18 years, the incidence of breast cancer is low.

Obesity

Obesity, the excess of body fatness due to both genetic and environmental factors, can be assessed in various ways. A commonly used, even if not totally accurate, means of assessment is the **body mass index (BMI)**, which is expressed as the body weight in kilograms divided by the square of the height in metres:

$$BMI = wt/ht^2$$

A ratio of 20–25 is considered normal or desirable, 25–30 is regarded as overweight (obesity grade I) and over 30 indicates obesity (obesity grade II). From 1980 to 1987, in the UK, the proportion of adults with BMI over 25 has increased from 35% to 40% and those with BMI over 30 has increased from 7% to 10%.

Obesity is recognized as an important cause of both morbidity and mortality. It is a risk factor for coronary heart disease and a contributory factor in the development of non-insulin-dependent diabetes mellitus, respiratory problems, gallbladder stones, arthritis and possibly gout. It has also been implicated in the development of some cancers. No single dietary component has been alleged to be responsible for the development of obesity, although the inclusion of a large amount of dietary fat would imply a high energy intake.

Populations on high fat diets are, on the whole, more obese than those on low fat diets, but this may simply reflect a higher total amount of dietary energy intake. Indeed, southern European countries have a lower fat intake than northern European countries, yet show a greater incidence of obesity.

Diabetes mellitus

This condition is characterized by increased blood glucose due to a deficiency or lack of action of insulin (see Chapter 9). Insulin-dependent diabetes mellitus (IDDM), which affects approximately 0.5% of the UK population, usually starts in childhood. Non-insulin-dependent diabetes mellitus (NIDDM), is commoner and is often associated with obesity. Diet seems to play no role in the development of IDDM, but may contribute to the development of NIDDM in so far as it contributes to obesity.

Guidelines for a healthy diet

Recommendations of the UK Department of Health

The Committee on Medical Aspects of Food Policy (COMA) of the UK Department of Health, having considered current evidence for the involvement of diet in health and disease, has made a number of recommendations for changes to the British diet. These were published in 1991, and are summarized in Table 14.1.

1. Dietary energy intake should be adjusted so that a constant body weight is maintained, within the range indicated by a BMI of 20–25, which is consistent with the longest life expectancy.
2. The intake of total fat should be reduced from the present level of 40% of total dietary energy intake to 33%, and especially the saturated fatty acid component from 16% to 10%. There is not enough evidence to warrant a change in the intake of mono-unsaturated fatty acids from the current level of 12% nor that of *cis*-polyunsaturated fatty acids from 6%. Similarly, there is no basis for recommending a decrease in the intake of *trans*-fatty acids from 2% of the dietary energy, but this level should not be

Table 14.1 **Recommendations made by the UK Department of Health in 1991 for changes in the nation's diet**

Dietary components	*Current diet (1991)*	*Recommendations*
Total fat	40% of total energy (88 g/day)	Decrease to 33%
Saturated fat	16% of total energy	Decrease to 10%
Mono-unsaturated fatty acids	12% of total energy	No change
Cis-polyunsaturated fatty acids	6% of total energy	No change
Trans-polyunsaturated fatty acids	2% of total energy	Should not increase
Protein	15% of total energy (men 84 g/day, women 64 g/day)	Should not increase
Starch	20% of total energy	Increase to 37%
Non-starch polysaccharides	13 g/day	Increase to 18 g/day
Sucrose	104 g/day	Should not exceed 60 g/day
Alcohol	–	Should not exceed 24 g/day for men and 16 g/day for women
Sodium (as salt)	2–10 g/day	Decrease to 1.6 g/day

increased until more light is thrown on the possible involvement of these fatty acids in cardiovascular and other diseases.

3. Alcohol should not be consumed in excess of 24 g of pure alcohol per day (3 units) for men, and 16 g/day (2 units) for women.
4. Non-starch polysaccharides or NSP (the term to replace fibre) should be taken in at approximately 18–24 g/day. NSP may be beneficial for preventing disorders such as diverticular disease, appendicitis, colon cancer, haemorrhoids, constipation and increased serum cholesterol.
5. Sugar (sucrose) intake should not exceed 60 g/day, for the prevention of dental caries.
6. Starches should provide the bulk of the dietary energy (37%), a point vigorously pursued by the UK potato marketing board.
7. Protein, which at present provides 14–15% of dietary energy, should not be increased. Adults should not exceed an intake of 1.5 g/kg body weight/day as this may lead to demineralization of bone.
8. Sodium intake (as salt) should be reduced from the current range of 2–10 g/day to 1.6 g/day if possible, and should not exceed 3.2 g/day, as it is a contributory if not causative factor in the development of hypertension.

Recommendations of the USA National Research Council

The USA National Research Council recommended similar changes to the American diet, in 1989, and these are summarized as follows:

A reduction in the total fat intake to 30% of the dietary energy intake, a reduction of saturated fat to less than 10% of the dietary energy, and a reduction in dietary cholesterol to less than 300 mg/day.

The consumption of five helpings of fruit and vegetables per day.

An increase in the consumption of starch and complex carbohydrates.

A moderate intake of dietary protein.

A balanced dietary energy intake to maintain the appropriate body weight.

A reduction in salt intake to less than 6 g/day.

An adequate calcium intake.

An optimal intake of fluoride, especially during the years of tooth formation.

A limitation in the consumption of alcohol to 25 g of pure ethanol per day.

14.5 The toxicity of food: food additives and contaminants

Nearly all foods contain small amounts of potentially toxic substances.

Intentional additives

These agents are intentionally introduced during food processing and include preservatives, colourings, antioxidants, added to prevent peroxidation of polyunsaturated fatty acids, and substances such as emulsifiers which are used to improve the texture of food. There are now, in most countries, strict laws and regulations governing the use of all food additives. Lists of legally permitted additives are available, and the use of compounds, not included in these lists, and therefore not considered safe, is prohibited.

Incidental additives

These are compounds which gain entry into food incidentally. They include pesticides and fertilizers which are introduced into food during crop spraying, and plant and animal growth promoters such as antibiotics. It should perhaps be noted here, that levels of pesticide residues found in bran are nine times higher than those found in white flour.

Naturally occurring toxic substances

The fear of harmful effects of food additives sometimes leads to the erroneous assumption that all 'natural' foods are fine for health. Many plants, for example, contain toxins which are extremely dangerous and have caused many human deaths and illnesses. Cyanogens are present in many plants such as yams, peas and maize and these are capable of producing cyanide as a result of enzymic hydrolysis, depending on the method of food preparation. Goitrogens, which are thyroid antimetabolites, occur in the brassica family. Flavonoids, found in the skins of oranges and tangerines, have cytotoxic effects on embryonic cells. Lectins, which are plant glycoproteins found mainly in leguminous plants, can, if not denatured first by heat, agglutinate red blood cells.

Contaminating microorganisms

Many foods are contaminated with microorganisms which produce dangerous toxins. *Clostridium botulinum*, for example, produces two of the most powerful exotoxins known to man, with neurotoxic and blood coagulating effects, respectively. *Staphylococcus aureus* and *Streptococcus pyogenes* toxins have haemolytic effects, cause connective tissue breakdown and are fibrinolytic. *Vibrio cholerae* toxin causes the severe diarrhoea and haemolysis seen in cholera. *Bordetella pertussis* toxin causes the paroxysmal coughing characteristic of whooping cough (see Table 3.3). Aflatoxins, produced by the fungus *Aspergillus flavus*, frequently on mouldy peanuts, are potent inducers of hepatic carcinogenesis (see Chapter 37).

The Department of Health, the Ministry of Agriculture, Fisheries and Food and other public health bodies closely monitor standards of food safety and take measures to prevent or contain cases of outbreak of disease through bacterial or other contamination of food.

Salmonella typhimurium was responsible for most outbreaks of food poisoning in the UK until recently, when it has been overtaken by *Campylobacter jejuni*, and also by *Salmonella enteriditis*. The presence of the latter in raw poultry and eggs has been a cause for concern in the UK in the past few years. Thorough cooking of these foods, which destroys these bacteria, should eradicate the problem and such advice has been given to the public by the Department of Health. Warning has also been given recently about the consumption of dairy products, especially soft cheeses, made from unpasteurized milk, as they are often contaminated with *Listeria monocytogenes*. This organism is particularly harmful to the developing fetus and, therefore, pregnant women should be especially cautious about the consumption of these dairy products.

14.6 Causes of malnutrition

Deficiency of nutrients in the diet will give rise to symptoms of malnutrition, but it is not the only cause. Dietary insufficiency is described as a **primary deficiency**.

Secondary deficiencies are also common and may be due to poor absorption, an increased demand due to a change in the physiological state, or the development of a pathological state, decreased utilization due to antagonists, or increased destruction and excretion. Pancreatic insufficiency, bile duct obstruction and chronic liver disease will affect fat digestion and absorption.

Malabsorption syndromes, characterized by changes in the morphology of the small intestine, as seen in coeliac disease, will also lead to decreased fat absorption and consequently a high risk of deficiencies in energy, essential fatty acids and fat-soluble vitamins. Failure of production of a glycoprotein known as 'intrinsic factor' by the parietal gastric cells will lead to inability to absorb vitamin B_{12} from the intestine and is the commonest cause of pernicious anaemia. An autoimmune response to the 'intrinsic factor' may also prevent its action on vitamin B_{12} absorption. Ethanol inhibits the absorption of thiamin by inhibiting its active transport into the cells of the intestinal mucosa. Infection and trauma increase the requirement for protein and lead to negative nitrogen balance. Long-term antibiotic therapy can lead to vitamin K deficiency by destroying the intestinal flora which is a source of a significant proportion of this vitamin in some individuals. Drugs, such as isoniazid, prescribed for the treatment of tuberculosis, may lead to pyridoxine deficiency by binding to and therefore effectively decreasing the availability of this vitamin. Phytic acid (inositol hexaphosphate), present in unrefined cereals, will bind to and decrease the absorption of dietary calcium. Smoking increases the catabolism of vitamin C, and can increase the requirement for this vitamin to twice the normal level.

Both primary and secondary causes of malnutrition will lead to the gradual depletion of nutrients from the tissues followed by biochemical lesions and finally anatomical or clinical signs of deficiency (Figure 14.2).

14.7 Assessment of nutritional status

The nutritional status of patients is an important determinant of morbidity, mortality and length of hospital stay. Methods available for the assessment of nutritional status vary, and although they are on the whole adequate to determine gross deficiencies, they are often inadequate to detect borderline deficiencies in large populations. As shown in Figure 14.3, clinical deficiency signs and anatomical lesions are often the end result of long-term deficiency, whereas subclinical deficiency may remain undetected.

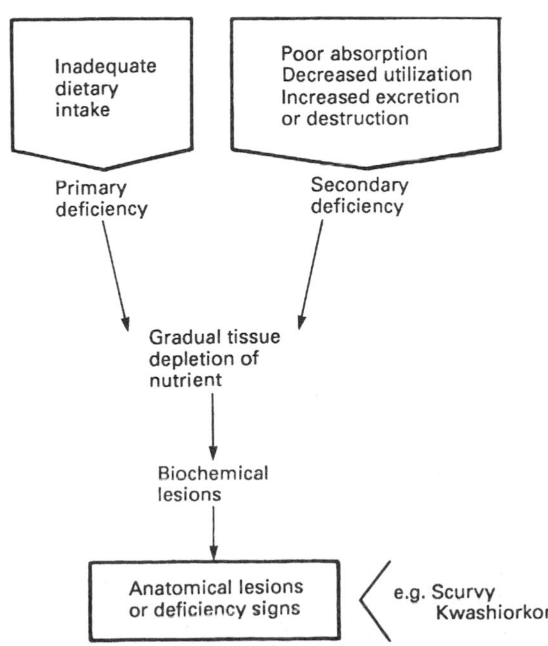

Figure 14.2 **Causes of malnutrition**

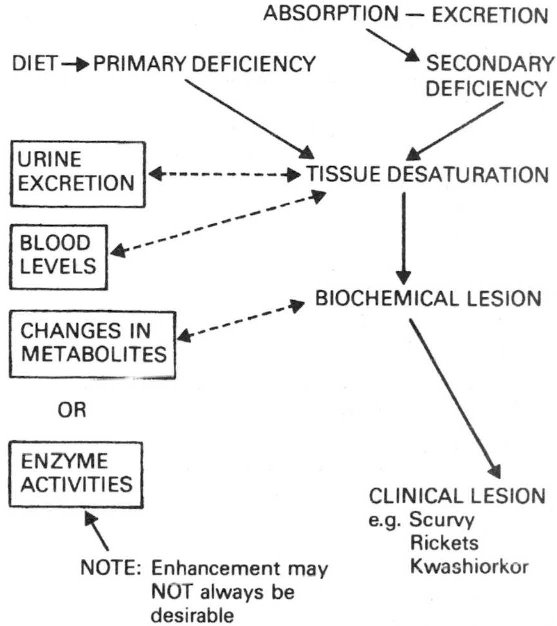

Figure 14.3 **The development and progression of malnutrition**

Methods of assessment should include:

(a) medical, social and dietary history
(b) anthropometric measurements
(c) clinical examination
(d) biochemical tests.

It has been shown that clinical examination can only detect about 25% of moderate to serious cases of malnutrition. Apart from hospitalized patients who may be malnourished, there are other groups of the population at risk of dietary deficiencies even in developed countries. A Department of Health survey of the diets of British schoolchildren in 1989 indicated low intakes of iron, calcium and riboflavin in teenage girls and low intakes of pyridoxine in both boys and girls. A study of 172 successive admissions to a psychiatric hospital in the UK in 1982 showed that 53% of the patients were deficient in thiamin, riboflavin and pyridoxine. Histologically proven osteomalacia is seen in 2–5% of the elderly in hospitals, but as many as 40% of the elderly in homes and hospitals have low vitamin D status.

Of the methods listed above, only the biochemical investigations can detect malnutrition before it is overtly obvious, as changes in tissue concentrations of nutrients, or their metabolites and alterations in enzyme activity, precede any clinical signs of malnutrition. They are also precise, sensitive and specific, in the way that physical examination and anthropometric measurements are not.

Medical and dietary history

Information should be obtained on recent weight changes, anorexia or dysphagia (difficulty in swallowing), fever and infection (both of which increase metabolic needs), wounds, burns or recent major surgery, as these conditions increase the likelihood of malnutrition. The social and economic status is important here. The elderly, especially if living alone, lacking ability or facilities to prepare food, and suffering from poor dentition as well as an apathy towards food, are particularly at risk of malnutrition.

Anthropometric measurements

Anthropometric measurements are quick and easy and do not involve complicated apparatus. They are valuable in assessing growth in children, and reserves of body fat and protein in children and adults. A reasonable estimate of these parameters can be obtained from four simple measurements:

height
weight
triceps skinfold thickness
mid-arm circumference

These measurements can provide the following information:

1. Body weight can be expressed as a percentage of a reference standard, taking height into consideration, and also age in the case of children.
2. The triceps skinfold thickness (TSF), measured with callipers at the midpoint between the acromial process of the scapula and the olecranon process of the ulna, provides a simple estimate of fat mass.
3. The arm muscle circumference (AMC), can be calculated from the mid-arm circumference (MAC) and the triceps skinfold thickness as follows:

$$AMC = MAC - 4.18\ TSF$$

It provides a simple estimate of muscle mass and, therefore, somatic protein stores. The midpoint of the upper arm is chosen for both measurements because of its accessibility and the fact that oedema is not usually seen at this site.

Clinical examination

Most clinical signs and symptoms of malnutrition are non-specific, that is, they could be due to more than one cause. Also, early signals of malnutrition cannot, on the whole, be detected by clinical examination and, therefore, clinical signs and symptoms are only useful in recognizing the severely malnourished patient. Table 14.2 gives a summary of signs and symptoms of malnutrition in adults.

A wasted appearance would indicate a deficient energy, or energy and protein intake.

Table 14.2 **Clinical signs and symptoms of malnutrition in adults**

Sign or symptom	*Deficiency*
General	
Wasted appearance	Energy, protein
Skin	
Follicular hyperkeratosis	Vitamin A
Petechiae	Vitamin C
Flaky dermatitis	Protein, energy, riboflavin, niacin, zinc
Bruising	Vitamin C, vitamin K, essential fatty acids
Hair	
Sparse and thin	Protein
Eyes	
Night blindness	Vitamin A
Mouth	
Glossitis	Riboflavin, niacin, folate, vitamin B_{12}
Angular stomatitis, cheilosis	Riboflavin
Tongue atrophy	Riboflavin, niacin, iron
Bleeding gums	Vitamin C
Neck	
Goitre	Iodine
Abdomen	
Distension	Protein-energy
Extremities	
Oedema	Protein, thiamin
Bone and joint pain	Vitamin D, vitamin C
Muscle wasting	Protein, energy, vitamin D, thiamin
Nails	
Spooning (koilonychia)	Iron
Nervous system	
Paraesthesias	Thiamin
Loss of position sense	Vitamin B_{12}
Dementia	Thiamin, niacin

Skin changes, including follicular keratosis, drying and thickening, are non-specific, and most probably, in this country, unlikely to be due to malnutrition.

Angular stomatitis may be due to riboflavin or pyridoxine deficiency, and bleeding gums may be indicative of scurvy, but in the UK the most likely cause of both is badly fitting dentures. Acute glossitis (inflammation of the tongue) may be due to the lack of nicotinic acid, riboflavin, folate, vitamin B_{12} or iron.

Xerophthalmia (dryness and inflammation of the cornea), caused by vitamin A deficiency, a common cause of blindnesss in the tropics, is unknown in the UK.

Hair changes, such as depigmentation, are seen in protein-energy malnutrition, but examination of hair is of little diagnostic value in the UK.

Koilonychia (spoon-shaped nails) may be due to iron deficiency, or may be genetic.

Loss of vertebral height, is more likely to be due to osteoporosis, whose causes are probably non-nutritional rather than osteomalacia due to vitamin D deficiency.

Psychosis or dementia may be due to thiamin or niacin deficiency. Peripheral neuropathy may be associated with thiamin or other vitamin B deficiencies, but would not be common in this country.

Biochemical tests

Biochemical tests have advantages over other methods of assessment of nutritional status, as they are precise, specific and sensitive and can detect signs of malnutrition long before physical signs appear.

The specificity and sensitivity of these tests have increased in recent years by the use of analytical techniques such as high-performance liquid chromatography (HPLC), mass spectrometry (MS) and competitive protein binding assays. The use of modern analytical instruments and microprocessors in diagnostic laboratories has greatly increased the speed and efficiency with which biochemical tests can be performed.

Biochemical tests can be divided into two broad categories:

(a) measurement of a nutrient or its metabolites in body fluids or tissues
(b) measurement of a biochemical function depending on the nutrient in question.

The choice of tissue is important and, often, accuracy has to be sacrificed for the sake of convenience and accessibility. Small laboratories in rural areas, especially in developing countries, will clearly not be able to employ the same techniques as well-equipped laboratories in big medical centres.

Urine is the most accessible body fluid, but its analysis is the least informative. The concentration of urine varies, and collection over 24 h is recommended, but may not always be practicable. When using random samples, correction for urine osmolarity can be made by expressing the results in terms of amount of metabolite per mg of creatinine excreted. The assumption is that creatinine excretion reflects lean body mass and should be fairly constant from day to day, but creatinine excretion is known to vary by 12–15% daily. Twenty-four hour urine samples are therefore preferable, if less convenient. Serum analysis is more informative than urine, but can also be misleading as for some nutrients the serum level reflects recent dietary intake, rather than stores. Analysis of tissues, such as the liver, might answer the question of storage and therefore nutritional status, but liver biopsy is clearly inappropriate in routine testing. For example, the most common test for vitamin A status, is serum retinol concentration. Circulating retinol levels will not drop until hepatic stores are depleted. Hepatic retinol ester determination would, therefore, provide the most accurate estimation of vitamin A status, but as mentioned above, sampling difficulties prohibit this type of assay.

Micronutrients whose concentrations in serum are used as an index of nutritional status include:

vitamin A (serum retinol)
vitamin E (serum tocopherol)
vitamin D (serum 25-hydroxycholecalciferol)
vitamin B_{12}

For other nutrients, more indirect (but not necessarily more inaccurate) measurements are employed:

Vitamin C

Serum ascorbate reflects recent intake of this vitamin. Leukocyte ascorbate is a better index of vitamin C status, but is technically more difficult to determine. **Saturation tests** are sometimes employed, but although informative, they are tedious and not suitable for large groups of people. A large dose of ascorbic acid is given orally, and the excretion of ascorbic acid measured over a set period of time. The amount excreted reflects tissue saturation, as a deficient subject will retain a larger proportion of the administered dose than an individual whose body stores are high.

Thiamin (vitamin B_1)

Urinary thiamin excretion reflects changes in dietary intake of thiamin, but is not a good indicator of the extent of tissue depletion. The best test for the assessment of thiamin status is a functional test, the measurement of erythrocyte transketolase activity, an enzyme of the pentose phosphate pathway (see Chapter 6), dependent on thiamin pyrophosphate (TPP) for activity. The procedure involves measuring the activity of this enzyme in two identical samples of erythrocyte haemolysate, in the presence or absence of added TPP. Deficiency is indicated if the activity of the enzyme is higher, by 20% or more, in the sample with the added cofactor. This test has the added advantage of not requiring comparison to standard ranges of values, as the split sample provides its own control.

Riboflavin (vitamin B_2)

The same principles which apply to the assessment of thiamin status also apply to riboflavin. The best functional test for the latter is the assay of erythrocyte glutathione reductase activity in the presence or absence of added riboflavin-containing cofactor FAD. Deficiency is indicated by an increase in activity by 40% or more in the supplemented, or 'stimulated', sample.

Pyridoxine (vitamin B_6)

Two types of functional test may be used here. The first is along the same lines as for thiamin and riboflavin, and involves measurement of pyridoxal phosphate-dependent aminotransferases in erythrocytes. The second, is a **loading test.** It is based on the fact that the metabolism of tryptophan involves many pyridoxine-dependent steps, and is performed by giving an oral load of tryptophan, and measuring abnormal tryptophan metabolites in the urine.

Folic acid and vitamin B_{12}

The haematological profile seen in folate deficiency is identical to that seen in B_{12} deficiency and it is, therefore, important to distinguish between the two possible causes of the disorder. This is done by measuring erythrocyte folate and serum vitamin B_{12} levels.

Vitamin K

Deficiency of vitamin K results in prolonged blood clotting time (prothrombin time) and measurements of clotting time have been used to estimate vitamin K status. A more specific test has recently been introduced which involves the measurement by radioimmunoassay of abnormal prothrombin resulting from defective carboxylation of prothrombin precursors owing to vitamin K deficiency.

Iron

The assessment of iron status is very important, as iron deficiency anaemia is common throughout the world, even in affluent countries, especially among women of reproductive age. Decreases in haemoglobin and haematocrit indicate severe iron deficiency. Serum ferritin measurements are more useful in detecting lowered iron stores.

Protein-energy malnutrition

Some biochemical tests are useful in distinguishing different forms of protein-energy malnutrition. These include measuring serum albumin levels, the ratio of essential to non-essential amino acids, and the ratio of urinary creatinine to height. All three of these indices are lowered in primary protein deficiency.

Macrominerals

The status of minerals such as sodium, potassium, calcium, magnesium and phosphate is routinely assessed by measuring serum levels.

Chapter 15

Nutrition: energy

15.1 Forms and units of energy

The first law of thermodynamics states that 'energy can neither be created nor destroyed'. All systems must be in energetic equilibrium with their environment, and this statement applies to all living organisms, including man. As a consequence, the energy for most living processes which require energy must be supplied from the environment in the form of chemical energy, that is as food, or as sunlight.

The international unit of energy is the **joule** (J), which is defined as the work done when one kilogram (kg) is moved one metre (m) by a force of one newton (N). Large quantities of energy are usually expressed in kilojoules (1 kJ = 1000 J). In nutritional studies, the unit of energy used traditionally, was the **calorie**, equivalent to 4.185 J and there is still a certain resistance to change to the international units, especially in everyday usage by non-specialists. In this chapter, energy values will be expressed as kilojoules or megajoules (1 MJ = 1000 kJ), but reference will be made to conversion to kilocalories.

15.2 Energy supply and utilization

The chemical energy supplied in the form of food is converted by the human body into mechanical work, such as muscle contraction, electrical work, such as the maintenance of ionic equilibria across cell membranes, or is used to perform chemical work, such as synthesis of macromolecules. These interconversions result in the loss of a substantial proportion of the energy intake, as heat. Figure 15.1 outlines the fate of dietary energy. The energy intake, E_1, is utilized for several purposes which may be subdivided into: E_2 (body maintenance), E_3 (stores, such as glycogen or fat), E_4 (work, which produces heat as a by-product), and E_5 (energy lost in urine, sweat and faeces):

$$E_1 = E_2 + E_3 + E_4 + E_5$$

If the energy intake is excessive, then the energy stores will increase, and obesity will develop. If the intake is insufficient, or ceases, as in fasting or starvation, energy must be supplied from the stores, as the requirement for maintenance is continuous.

Energy content of food

Carbohydrate, fat and protein may be oxidized to provide human energy requirements. The maximum amount of energy these foods could theoretically supply can be measured in the laboratory, by means of a bomb calorimeter. A measured weight of foodstuff is placed in a sealed container and burnt in oxygen. The heat produced is measured by the calorimeter, and is termed the **gross energy** of food. It represents the total chemical energy in the food. Not all of this is, however, available to the body. Some of the energy is lost, because some components of the diet are not digested. The energy available to the body, after allowing for these losses in digestion, is known as **digestible energy**. After absorption of the digested product, some of the energy is lost in the urine, sweat and faeces. Protein is completely oxidized in the bomb calorimeter, and the nitrogen is converted to nitrogen oxides such as NO_2. In the human body, these are formed to a very limited extent, and nitrogen is excreted largely as urea, which is not fully oxidized. Also, the urine always contains small quantities of amino acids and lactic, and other acids, which could provide energy after oxidation.

The energy actually available to the body, after allowing for the losses described above, is termed **metabolizable energy**. The exact values for metabolizable energy for the major foodstuffs vary with the type of food. In theory, starch and glycogen have a higher energy content than their constituent monosaccharide, glucose, but this is a small discrepancy, compared to the different energy content of different

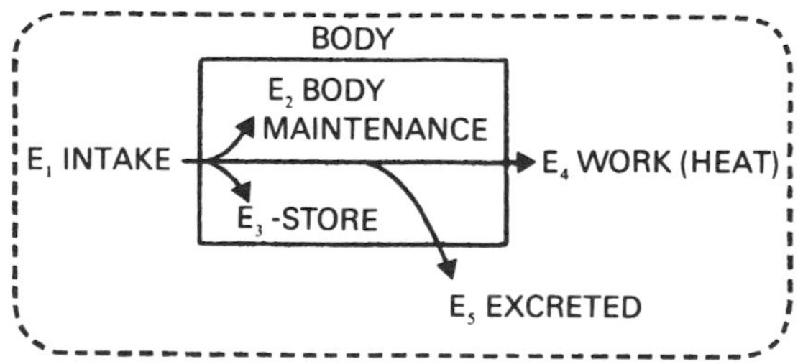

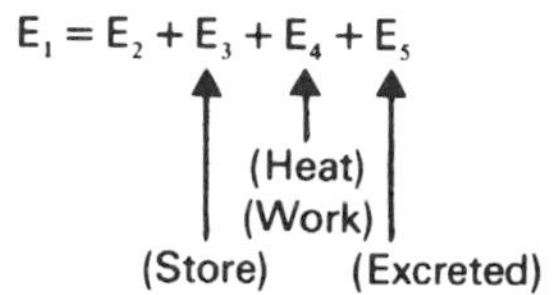

Figure 15.1 **Energy balance in animal body.** E_2 = 'basal metabolism'

Table 15.1 **Energy content of the major dietary fuels**

Fuel	*Gross energy (kJ/g)*	*Digestible energy (kJ/g)*	*Metabolizable energy (kJ/g)*
Starch	17.5	17.3	17.3
Glucose	15.6	15.4	15.4
Fat	39.1	37.1	37.1
Protein	22.9	21.1	15.9
Ethanol	29.8	29.8	29.8

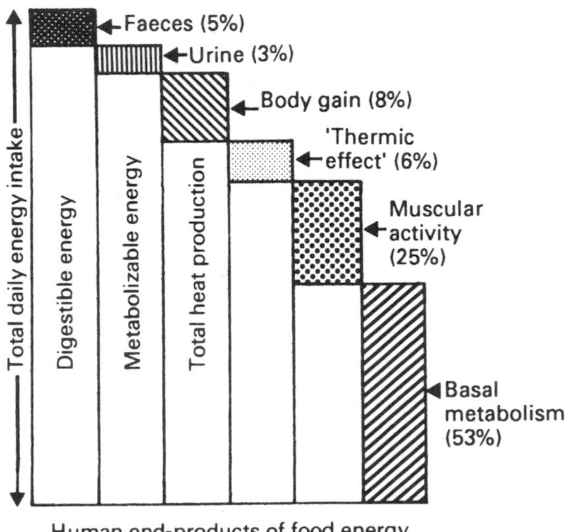

Figure 15.2 **Typical disposal of food energy**

samples of the same foodstuff. In addition, although starch is theoretically higher in energy than glucose, it is often incompletely digested. When great accuracy is required, for example for research purposes, the metabolizable energy of a diet can be determined by bomb calorimetry of the diet, urine and faeces. For practical purposes, the values shown in Table 15.1 can be used. It should be noted that food composition tables show metabolizable energy values.

The metabolizable energy is used in various ways, as shown in Figure 15.2. The greatest proportion, about 50%, is normally used for basal metabolism, about 25% is used for mechanical work, a small amount is stored, and some is dissipated as heat, generated as a result of ingestion, absorption and metabolic processing of food, and is known as **the thermic effect** of food, **diet-induced thermogenesis** or **postprandial thermogenesis**.

15.3 Energy expenditure

Energy intake has to be adequate to meet energy expenditure, otherwise the body's reserves will be utilized without being properly replenished, resulting in loss of body weight, impairment of function and finally death. If, on the other hand, the energy intake is excessive compared to the energy expenditure, the body's fuel reserves will increase, resulting in obesity which is also a health risk. Determinations of energy requirements depend, therefore, on measurements of energy expenditure. The principle is very simple, as energy expenditure is equivalent to a person's heat output. If the total heat production of an individual can be measured, then it is simple to calculate the amount of energy required to match this energy expenditure. In practice, this is not so simple. Various techniques are available, usually offering a choice of either accuracy or feasibility.

Direct calorimetry

The total amount of energy expended by a human can be measured by placing him or her in a large calorimeter and measuring the total heat output, having excluded anything which produces heat, other than the subject. Such measurements have been achieved, but the operation of these calorimetric chambers is clearly technically difficult, tedious and costly, and is certainly not the method of choice for large numbers of determinations.

Indirect calorimetry

Indirect calorimetry, which does not involve measurement of heat, is more practical, and most measurements of energy expenditure are conducted in this way. It is based on the observation, derived from direct calorimetry, that when dietary fuel is oxidized in the body, the amount of oxygen consumed, and carbon dioxide produced, are proportional to the heat generated. As a result of painstaking experiments, measuring both heat output by direct calorimetry, as well as oxygen consumption, it has been shown that the consumption of 1 litre of oxygen is equivalent to the production of 20.7 kJ of heat, when glucose is oxidized, and 19.3 kJ of heat, when fat or protein are used as fuels. This similarity among carbohydrate, fat and protein, allows the approximation to a value of 20.3 kJ per litre of oxygen consumed, when considering mixed diets.

Using the above assumptions, measurements of energy expenditure become measurements of oxygen consumption. Various instruments known as **respirometers** are available for such measurements. The most useful are portable, and allow the measurement of energy expenditure while the subject pursues various activities. Their use is, however, limited, as some activities are difficult or impossible to perform while the subject is carrying these instruments and wearing a face-mask, or a mouthpiece and nose-clips, although some versions have a ventilated hood which is worn over the head and so allow more freedom of movement. Apart from considerations of discomfort, the period over which these instruments can be used is limited to a few hours at a time.

The doubly-labelled water method

A refinement in indirect calorimetry has recently been added to the techniques employed for measuring energy expenditure, and allows measurements over a period of weeks, rather than hours. It involves the administration of a sample of water, labelled with the stable isotopes, deuterium (^{2} H), and ^{18}oxygen (^{18}O). The rate of loss of the two labels from the body is then measured. The rate of loss will be different for ^{2}H and ^{18}O, being faster for the latter. This is because hydrogen is lost from the body only as water, but oxygen is lost both as water, and as carbon dioxide. The amount of carbon dioxide produced can then be calculated, and from that, and a knowledge of the **respiratory quotient (RQ)**, the amount of oxygen consumed can be estimated. This is, as explained above, equivalent to the energy expenditure.

The respiratory quotient is the ratio of the amount of carbon dioxide produced to the amount of oxygen consumed, and is different for carbohydrate, fat and protein, being 1.0, 0.7 and 0.8 respectively. The RQ can be estimated from a knowledge of the proportions of carbohydrate, fat and protein consumed during the study. The main advantage of the doubly-labelled water method is that it allows measurement of total energy expenditure in free-living subjects without constraining their activities. The disadvantages are that it is expensive, as it requires mass spectrometry, and also that it gives an average of energy expenditure over many activities over many days, and it is not possible to be used for measuring the energy cost of a single activity for a short time.

Components of energy expenditure and factors affecting them

The main components of energy expenditure in the adult are :

1. Energy used for body maintenance (basal metabolism)
2. Energy used for physical activity
3. Diet- and cold-induced thermogenesis.

For children and for pregnant and lactating women, energy is also needed for the deposition of tissues and the production of milk. The largest component of the daily energy expenditure is, for most people, the **basal metabolism (BM)**, or the **basal metabolic rate (BMR)**. It is the energy cost of body maintenance, including processes such as the activity of the heart, conduction of nerve impulses, ion transport across membranes, reabsorption in the kidney, and metabolic activity, such as synthesis of macromolecules. It is defined as the energy expenditure at rest, but not during sleep, in a thermoneutral environment, 8–12 h after the last meal and 8–12 h after any significant physical activity. The **resting metabolic rate** is sometimes measured. This is also measured at rest, but not under standardized conditions, and tends to be somewhat higher than the basal metabolic rate.

Basal metabolic rate differs among different individuals. It depends on many factors, some invariable, such as gender, age and genetic factors, and some variable, such as diet, body size and composition, environmental temperature, the individual's hormonal state, stress and disease. The major determinants of BMR are **body size** and **body composition**. Table 15.2 shows the effect of body weight on BMR. Adipose tissue is not as metabolically active as lean body mass, and BMR is often expressed per kilogram of **lean body mass** or **fat-free mass**. **Age**, affects BMR, mainly through a decrease in lean body mass after adulthood (Table 15.3). **Gender** differences in BMR usually reflect the lower proportion of lean body mass in women, when compared to men of the same body weight.

Genetic differences have been studied extensively. BMR varies by up to 10% between subjects of the same sex, age, body weight and fat content. Studies on differences between ethnic groups, such as Asian, African and Caucasian, have so far proved conflicting and inconclusive. **Diet**, in terms of low or high energy, affects the BMR to a degree not explained by changes in body weight. It is known that starvation leads to an adaptive decrease in the BMR, over and above that which results from the decrease in lean body mass. The converse is true when energy intake is increased. The **endocrinological** state is important. In hyperthyroidism, for example, the BMR is increased, and in hypothyroidism, it may be decreased by up to 40%, leading to weight gain. **Stress, anxiety** and **disease states**, especially infections, fever, burns and cancer, also increase the BMR (Chapter 40). **Drugs** affect the BMR in different ways. Smoking (nicotine) and coffee (caffeine and theophylline) increase the BMR, whereas β-blockers tend to decrease energy expenditure.

Table 15.2 **Effect of body weight on BMR**

Weight (kg)		*BMR* (kJ/day)	
Men	*Women*	*Men*	*Women*
–	35	–	6920
–	40	–	7630
45	45	10240	8315
50	50	11060	8980
55	55	11850	9625
60	60	12635	10260
65	65	13400	10875
70	70	14140	11480
75	–	14870	–
80	–	15590	–

Note: Effects on the total energy requirements for 'reference' man and woman are shown.
Adapted from R.W. Swift and K.H. Fisher (1964) in *Nutrition*, edited by G.H. Beaton, Vol. 1, Academic Press, New York

Table 15.3 **Effect of age on BMR**

Age (years)	*Percentage of reference*	*BMR* (kJ/day)	
		Men	*Women*
20–30	100.0	13390	9625
30–40	97.0	12990	9335
40–50	94.0	12590	9050
50–60	86.5	11585	8330
60–70	79.0	10580	7605
70	69.0	9240	6640

Note: Effects on the total energy requirements for 'reference' man and woman are shown.
Adapted from Swift and Fisher (1964) – see Table 15.2.

Table 15.4 **Physical activity ratios (PAR) for different kinds of activities**

PAR range 1.0–1.4
Sitting or lying at rest, e.g. watching television, reading, writing, listening or eating
Standing

PAR range 1.5–1.8
Driving a car, sewing, peeling vegetables, washing up, doing general office work

PAR range 1.9–2.4
Cleaning the house, cooking, bowling, washing clothes by hand

PAR range 2.5–3.3
Vacuum cleaning, making beds, walking 3–4 km/h, painting walls, playing cricket

PAR range 3.4–4.4
Gardening, sailing, playing golf, bricklaying, making furniture, cleaning windows

PAR range 4.5–5.9
Chopping wood, cycling, jogging, swimming, dancing, playing volleyball, digging and shovelling

PAR range 6.0–7.9
Walking uphill, climbing stairs, playing football, playing tennis, cycling or swimming fast, skiing

Physical activity

For most people, physical activity represents 20–40% of their total energy expenditure. The energy cost of various activities can be measured by direct, or more conveniently, indirect calorimetry, as described above. The energy cost of various activities for any individual can be expressed as a multiple of the BMR and is referred to as the **physical activity ratio (PAR)**. For example, sitting at rest has a PAR value of 1.2. This means that there is an increase in the energy expenditure to 1.2 × BMR, when sitting at rest. Similarly, digging and shovelling, increase the metabolic rate to 3.7 × BMR, and are assigned a PAR value of 3.7. Examples of the physical activity ratios of different types of activity are shown in Table 15.4.

Physical activity level (PAL) is the ratio of the overall daily energy expenditure to BMR. It is calculated as follows: the time spent for each activity is multiplied by PAR for that activity e.g.

3 hours walking:
$3 \times 2.8 = 8.4$
2 hours watching television:
$2 \times 1.2 = 2.4$
etc.

The sum of the products over 24 h expressed as a ratio of the BMR gives the physical activity level for each person.

PAL = sum (PAR × time in each activity) over 24 h, as a ratio to the BMR.

A man, in 'light occupational activity', and indulging in 'moderately active' non-occupational activity, will have a PAL of about 1.5. Examples of PAL and the way in which they are derived are shown in Table 15.5.

The BMR can be measured or predicted from tables, taking into account body weight, age and gender. The energy cost of physical activity can be measured or predicted from tables, and if the duration of each activity is known, the **total energy expenditure (TEE)**, for a 24 h period can be calculated:

TEE = BMR [time in bed + sum (PAR × time in each activity)]

Diet induced thermogenesis

Another component of energy expenditure in man is diet-induced thermogenesis (DIT), also known as post-prandial thermogenesis (PPT). This is the energy expended in the digestion, absorption and subsequent

Table 15.5 **Examples of physical activity levels for men and women in light activity occupations with inactive, moderately active or very active non-occupational activities**

		Non-occupational activity					
		Not active		*Moderately active*		*Very active*	
	time (h)	*PAR*	MJ/day	*PAR*	MJ/day	*PAR*	MJ/day
MEN							
Bed	8	1.0	2.4	1.0	2.4	2.4	2.4
Occupation	6	1.7	2.9	1.7	2.9	1.7	2.9
Non-occupational	10	1.5	4.8	1.7	5.4	1.9	6.1
Total energy expenditure (TEE)			10.1		10.7		11.4
PAL (TEE/BMR)			1.4		1.5		1.6
WOMEN							
Bed	8	1.0	1.9	1.0	1.9	2.4	1.9
Occupation	6	1.7	2.2	1.7	2.2	1.7	2.2
Non-occupational	10	1.5	3.7	1.7	3.7	1.9	4.7
Total energy expenditure (TEE)			7.8		8.3		8.8
PAL (TEE/BMR)			1.4		1.5		1.6

PAR, physical activity ratio; PAL, physical activity level; TEE, total energy expenditure; BMR, basal metabolic rate.

processing of food. This effect was originally attributed solely to the metabolic handling of protein, and was termed 'specific dynamic action', but it is now recognized as an effect produced by the consumption of all dietary fuels. The consumption of protein does, however, produce the greatest increase in energy loss (20–30% of intake), compared to fat (2.5–4%), or carbohydrate (5–6%). The percentage of the total energy lost in this way is not constant, but varies with the time after the ingestion of a meal, the nutritional state of an individual, and the composition of the diet. It also varies considerably from individual to individual and has been implicated in the mechanism of energy and weight balance. It has been suggested that, in some people, a high degree of postprandial thermogenesis may be a factor which allows them to maintain their normal body weight after overeating.

Table 15.6 **Estimated average requirements (EAR) for energy for children and adults**

Age	*EAR* (MJ/day)	
	Male	*Female*
0–3 months	2.3	2.2
4–12 months	3.4	2.7
1–3 years	5.2	4.9
4–10 years	7.7	6.9
11–18 years	9.9	7.4
19–59 years	10.6	8.1
60–74 years	9.8	8.0
75 plus	8.8	7.6
Pregnancy		+0.8
Lactation		+2.0

Cold-induced thermogenesis

Environmental temperature affects the metabolic rate. Low temperature increases energy expenditure by inducing shivering- and non-shivering thermogenesis. The site of non-shivering thermogenesis is the brown adipose tissue, so called because of its dark colour, a result of a high blood supply and a high concentration of cytochromes. A GDP-binding protein, known as thermogenin, is present in the inner mitochondrial membrane. Non-esterified fatty acids, liberated during lipolysis as a result of adrenergic stimulation, displace the GDP bound to **thermogenin**, and this event leads to the dissipation of the proton gradient normally established across the mitochondrial membrane (see Chapter 5). The end result is uncoupling of oxidative phosphorylation, with fatty acids being oxidized without ATP formation, but with generation of heat. Brown adipose tissue is important as a thermogenic tissue in hibernating animals, and also in newborn animals including human babies, but its contribution to thermogenesis in the adult human is debatable.

High temperature also has an effect on energy expenditure, by increasing heat loss through sweating.

15.4 Energy requirements

The energy requirement of an individual is defined by the Food and Agriculture Organisation of the United Nations and the World Health Organisation (FAO/WHO) as 'the energy intake which will balance energy expenditure in an individual whose body size and composition and level of physical activity are consistent with long-term good health'. For children and for pregnant and lactating women, allowances are additionally made for growth of tissue and production of milk.

A variety of methods are available for estimating energy requirements. Most of the data have come from dietary surveys, so that estimates of requirements have been based on estimates of intakes, assuming that the population studied is in energy balance at the time of the study. This is perhaps not a valid assumption in countries such as the UK and other developed countries, where obesity is common. Another problem is the inaccuracy of dietary recall by most subjects. As more information has become available from studies of energy expenditure, more accurate estimates of energy requirements can be made. The UK Department of Health (1991) has set dietary reference values (DRV) for energy, based on current estimates of energy expenditure. Estimated energy requirements for children and adults are shown in Table 15.6. It must be noted that reference nutrient intakes (RNI) for all nutrients are set at the estimated average requirement (EAR) plus two standard deviations. This covers the needs of almost all members of the population and provides about half the population with a moderate excess. This consideration was not applied to energy, in view of the fact that overweight and obesity are increasing in the British population, and recommendations for energy have, therefore, been set at the estimated average requirement.

For infants, measurements of energy expenditure are not appropriate, and estimates of energy requirements have been based on energy intakes of breast-fed or formula-fed infants showing an appropriate rate of growth and general good health.

Recommendations for pregnancy are an additional intake of 0.8 MJ/day, in the third trimester only, and for lactation, an additional 2.0 MJ/day. There is no evidence for a need to increase the energy intake in the first two trimesters of pregnancy, as the extra energy needed for the development of the fetus and the supporting tissues is usually counterbalanced by a small reduction in the BMR and a decrease in energy expenditure through decreased physical activity.

15.5 Energy balance and control of body weight

Energy balance is achieved when the energy intake matches the energy expenditure. It seems to be regulated with a fair degree of precision, as most adults maintain their body weight over many years, having consumed tonnes of food without calculating precisely how much to eat every day. There are numerous theories on the mechanisms which control energy intake and energy expenditure, but it is unclear how these mechanisms are coordinated to control energy balance and to monitor body weight. The operation of these mechanisms is further complicated by the fact that energy expenditure affects energy intake, and energy intake affects energy expenditure, through changes in metabolic rate, and especially through diet-induced thermogenesis.

The feeling of a need to eat is generally described as **hunger**. The desire to eat a specific type of food, rather than another, is described as **appetite. Satiety** is the feeling of having consumed a satisfactory amount of food, and is felt at some point between the relief of hunger and the feeling of discomfort which follows having eaten in excess.

Much information on the control of food intake comes from experiments using laboratory animals. The physiological control of food intake, in man, is further complicated, however, by psychological and sociological factors which do not operate in laboratory animals, which are given the same type of food every day and consume the same amount every day. If we were to eat, repeatedly, exactly the same type of food, we would also, most probably, consume the same amount of food every day, an amount adequate to fulfil our physiological needs. It is well known, however, that a four-course meal can be eaten more pleasantly and easily, if the four courses are different, than if they are identical. Satiety seems to be produced at a lower level of energy intake in the second case and is, therefore, as will be discussed later, not a purely physiological phenomenon, but to some extent a conditioned event.

Control of energy intake

The hypothalamus was considered to be the main centre controlling hunger and satiety. The lateral hypothalamic area was thought to be specific for controlling initiation of feeding, and the ventromedial nucleus specific for satiety and for controlling cessation of feeding. Recent work has suggested that this is a greatly simplified view, and the precise role of the hypothalamus in hunger and satiety remains to be established although undoubtedly there is one.

Neuropeptides, neurotransmitters, and hormones certainly affect feeding behaviour. Cholecystokinin (CCK) is known to produce satiety in experimental animals, but it is not clear whether the physiological effect is through CCK released from the gut, subsequently acting on the brain, or CCK released within the central nervous system. Opioid peptides may also be involved, and it has been shown that opioid antagonists induce satiety in experimental animals and in man. Serotonin (5-hydroxytryptamine, 5-HT) levels also seem to be involved, and fenfluramine, which is a 5-HT uptake blocker, is used as an appetite suppressant in the treatment of obesity. Signals from the periphery are also thought to be involved in the control of energy intake, but the mechanisms whereby these signals are integrated to produce a response are far from clear. Hypoglycaemia certainly produces a feeling of hunger, but hyperglycaemia does not necessarily inhibit feeding. Pancreatic and gastrointestinal hormones, such as insulin, glucagon and cholecystokinin, released during digestion and absorption of food, are important in affecting feeding behaviour. Mechanical signals, such as gastric distension, are detected by baroreceptors in the stomach, and the signals are transmitted to the central nervous system. This is consistent with the observation that high fibre diets which are bulky, produce an earlier feeling of satiety than diets of low bulk.

Changes in the circulating concentrations of metabolites, such as glycerol, free fatty acids, ketone bodies, amino acids and glucose, are also detected, as is the rise in body temperature which follows feeding. Mathematical models have been produced, to predict food intake, from a knowledge of the circulating concentrations of glucose, insulin and adrenaline, the amount of liver glycogen, the rate of glucose uptake by the lateral hypothalamus, the body weight and the environmental temperature. All these factors are known to be relevant to the control of food intake, but their relative contributions in causing initiation and cessation of feeding are not known. None of the changes in these physiological parameters occurs rapidly enough, while eating, to decide when to stop. Satiety, as mentioned above, is to a certain extent learnt through experience. The perception of the energy content of food is fairly poor in man, and certainly not immediate. The quantity of food which relieves hunger and fulfils the physiological requirements has to be learnt, to a certain degree, by repeated meals of the same content. Eating habits and customs can override physiological control mechanisms.

It has been shown, for example, that in groups of young cadets, the regulation of energy intake in relation to energy requirements was very imprecise over 24-hour periods. In another study, only 28 out of 69 subjects ingested an amount of food which corresponded to their energy expenditure of the same day or, indeed, the preceding two days. Over a week, a further 23 subjects achieved a balance between their energy intake and expenditure, but even after this period, 18 subjects still showed either positive or negative balance. The energy content of a new food, cannot be judged immediately by physiological mechanisms. The composition of a meal may also be important. It has been suggested that dietary carbohydrate is better at signalling satiety than is dietary fat, and that this may account for the fact that individuals on high fat diets tend to be more obese than individuals on high carbohydrate diets. Not only is satiety a conditioned event, but it is also specific to different foods. As mentioned above, an individual may refuse another helping of the main course of a meal on the grounds of feeling 'full', but may happily eat a dessert of comparable energy content. Another person may refuse the second helping and the dessert, not because it would be unpleasant, but because of being conscious of the effect that the extra quantity of food might have on his or her body weight. Apart from conscious choice, other psychological factors are also important in determining food intake. Stress and depression can affect eating patterns, leading to either anorexia or excessive and compulsive eating, depending on the individual case.

Control of energy expenditure and energy balance

The factors affecting energy expenditure have been discussed in Section 15.3, but it is worth restating that the change in body energy stores is the difference between energy intake and energy expenditure:

$$\Delta \text{ E stores} = \text{E intake} - \text{E expenditure}$$

It is clear from the above that body weight (and composition) depend on the balance between energy intake and energy expenditure, but this simple equation is complicated by the fact that changes in energy intake affect energy expenditure and vice versa.

Energy expenditure through physical exercise causes a reduction in food intake in rats, but there is no comparable evidence from human studies. A decrease in energy intake is usually observed with age, as physical activity decreases. One component of energy expenditure can also affect another component. It has been shown, for example, that severe exercise leads to an increase in basal metabolic rate.

The temporary effect of energy intake on energy expenditure has been studied extensively. A decrease in energy intake, as seen in starvation, leads to a decrease in basal metabolic rate, to a degree not justified by the loss of lean body tissue. Conversely, excessive eating leads to an increase in basal metabolic rate. Energy intake affects energy expenditure via diet-induced thermogenesis. Overfeeding studies have shown that a large proportion of the excess energy intake is lost as heat. There is a great variation among individuals in the degree to which this wastage occurs. Some people seem to be better than others in dissipating the excess energy, and maintaining, or only slightly increasing, their body weight. Although these individuals are considered fortunate in Western society, they can also be viewed as inefficient at storing excess energy for later use, and they would be at a disadvantage, if scarcity of food were to occur for prolonged periods.

Energy balance over the years seems to be achieved by the body buffering large sporadic changes in energy intake, and also by small consistent changes in energy intake. Large consistent changes are not buffered efficiently and obesity develops. The overweight and obese do not increase their weight continuously, but usually maintain it, albeit at a higher level. It seems that the mechanisms which operate to monitor energy balance do operate again, after a new weight standard has been set. The subject of obesity will be discussed in Chapter 18.

Chapter 16

Nutrition: proteins

16.1 The need for protein in the diet

A regular and adequate supply of protein in the diet is essential for cell integrity and function. In most unrestricted diets, 10–15% of the energy is provided in the form of protein. The proportion of dietary energy derived from fat and carbohydrate varies throughout the world, but the proportion of energy derived from protein is remarkably constant. This is because there is a physiological requirement for a minimum quantity of certain amino acids, whereas there is no specific requirement for dietary carbohydrate or fat, except for small quantities of the essential fatty acids, and as long as an adequate amount of energy is provided in the diet, the relative amounts of fat and carbohydrate acting as supplies of metabolic fuel are not important.

The importance of an adequate supply of dietary protein must have been apparent from very early times, and the cult of hunting animals for food began many thousands of years ago. This practice was scientifically endorsed by Liebig in Germany in the early nineteenth century. He believed that muscle action consumed muscle protein which, therefore, has to be replaced by a dietary intake of protein. He advocated intakes of 100–200 g a day. The American nutritionist Chittenden, experimenting on himself, showed in 1905 that such high intakes were unnecessary and that an intake of 40–50 g/day would maintain perfect health. Subsequent research has vindicated Chittenden's findings, but the cult of high protein intake is still ingrained into our way of life.

16.2 Protein turnover

The importance of dietary protein becomes obvious when considering the concepts of 'protein turnover' and 'nitrogen (or protein) balance'.

The concept of **protein turnover** arose primarily from the work of Schoenheimer in the USA in the 1940s. Studies on the metabolism of proteins and amino acids labelled with the newly available stable isotope of nitrogen, ^{15}N, showed that all proteins in the body were in a constant state of flux, being constantly synthesized and degraded. He suggested that an '**amino acid pool**' existed in cells, in dynamic equilibrium with the amino acids formed from the degradation of body protein and those taken in the diet, and that losses from this pool occurred when no amino acids were ingested. Figure 16.1 gives a simple representation of this concept of protein turnover. The amino acid pool can be regarded as a tank with a drain, X, which cannot be plugged, and an overflow, Y. The loss of amino acids from the pool cannot be prevented, but if the intake of amino acids is excessive, the excess is metabolized and the nitrogen excreted as urea, represented by Y in the diagram.

The constant loss of amino acids from the pool makes the regular intake of protein essential. This loss occurs even if the body is able to recycle and reuse most of the amino acids produced by protein degradation. Amino acids are needed as gluconeogenic precursors and also as precursors for compounds such as catecholamines, thyroid hormones, neurotransmitters, haem and glutathione.

The amino acid pool is not large. It is made up of about 100 g of amino acids, which is small compared to the 10 kg of protein present in the average 70 kg man.

About 300 g of protein are synthesized and degraded daily in the human adult. The turnover is high in infancy and decreases with age. It also differs between tissues. Skeletal muscle accounts for approximately 50% of the body's protein content but is responsible for about 25% of the turnover, whereas the liver and the gut, which account for less than 10% of the protein, contribute 50% to the turnover. Protein turnover also differs within tissues and even within cells.

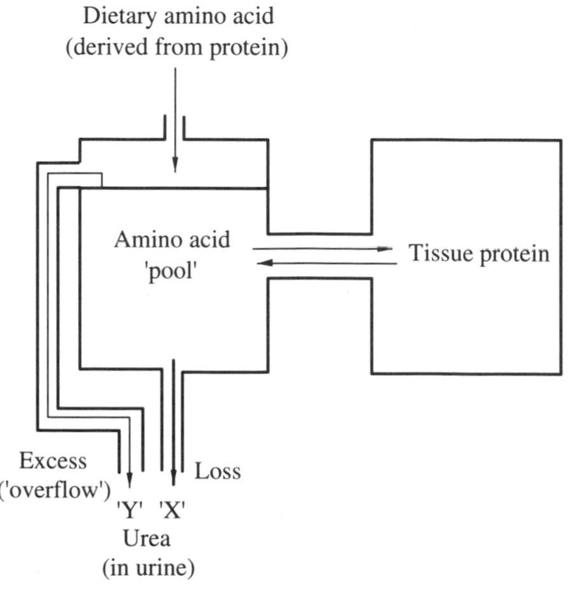

Figure 16.1 **Protein turnover**

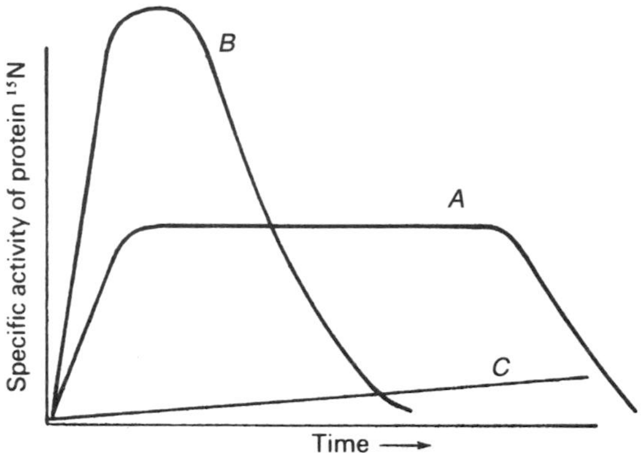

Figure 16.2 **Protein turnover: measurement of the rate of incorporation of ^{15}N of an amino acid, such as labelled leucine, into typical proteins and the loss of label during protein degradation**

Generally, the pattern of turnover falls into one of three groups, as shown in Figure 16.2. Group A proteins are rapidly synthesized, have a finite life and are rapidly degraded. Typical of these is haemoglobin, which is normally stable during the 120 days of life of the human erythrocyte. Group B proteins are rapidly synthesized and rapidly degraded. Typical of these proteins are enzymes which regulate metabolic pathways. Such enzymes are usually found at low concentrations in cells, and their concentrations fluctuate rapidly in response to increases or decreases in the rate of protein synthesis and degradation. Proteins of group C, on the other hand, turn over very slowly and have a very long half-life. Collagen is an example of this group of proteins.

The heterogeneity in the turnover of different proteins, even in the same cell, suggests that the process is selective. Apart from lysosomal degradation of cell proteins, various cytoplasmic enzymes are responsible for the turnover of cell proteins and removal of damaged proteins. The degradation of cytoplasmic proteins does not take place indiscriminately. Proteins whose NH_2-terminal amino acid consists of Met, Ser, Thr, Ala, Val, Cys, Pro, or Gly, are resistant to proteolysis. Proteins with any of the other amino acids at the NH_2-terminus are destabilized and are tagged with **ubiquitin**, a small heat stable protein which binds to a lysine residue at the NH_2 - terminus. A number of ubiquitins are subsequently added to the first one and the complex serves as a signal for degradation of the tagged protein (see Chapter 11). Another signal for degradation is the presence in proteins of '**PEST**' sequences. These are sequences containing proline (P), glutamic acid (E), serine (S) and threonine (T), and proteins rich in these sequences have short half-lives.

The addition of ubiquitin to membrane proteins (e.g. those of the plasma membrane) also 'tags' these proteins for proteolysis. In this case, however, the ubiquitin serves to direct the protein to the endolysosomal pathway (see Chapter 8) and degradation occurs in the lysosomes.

Typical values for the turnover of proteins of the body as a whole, are shown in Table 16.1. Human body protein is turned over every five to six months. The rate of turnover can be markedly increased by increasing the dietary protein intake.

Table 16.1 **Rates of turnover of body protein**

	Half-life (days) *of*	
	Total body	*Liver/serum proteins*
Rat	17	6–7
Man	158	10–20
	Effect of diet (rat)	
Diet	*Half-life of plasma protein* (days)	
No protein	17	
25% protein	5	
65% protein	2.9	

16.3 Nitrogen balance

Nitrogen balance studies evaluate the relationship between the nitrogen intake (strictly nitrogen containing compounds, which are mainly proteins) and the nitrogen excretion. In normal adults:

nitrogen intake = nitrogen excretion

The subject is then said to be in **nitrogen balance**. In this situation, the rate of body protein synthesis is equal to the rate of degradation.

Two other situations are possible:

1. Nitrogen intake > nitrogen excretion. This is known as **positive nitrogen balance**. It shows that nitrogen is retained in the body, which usually means that protein is laid down. This occurs during growth, in pregnancy and in convalescence after a serious illness.
2. Nitrogen intake < nitrogen excretion. This situation is known as **negative nitrogen balance** and occurs in starvation, during serious illness and in injury and trauma (see Chapter 40). If prolonged, it will ultimately lead to death.

16.4 Protein content of food

Most proteins contain approximately 16% nitrogen, and protein supplies nearly all the dietary nitrogen. If the nitrogen is determined in any foodstuff, and the figure obtained multiplied by 6.25, a very close approximation of the protein content of any food can be obtained.

The protein content of various foodstuffs (expressed as the proportion of energy derived from protein) is shown in Table 16.2. Although meat, fish and eggs are known to be good sources of protein, some vegetable foods, such as peanuts or soybeans, have a similar percentage of protein to meat. Also, the value of bread should not be underestimated, as it can, in theory, supply the total protein requirement of an adult if eaten in quantities high enough to satisfy the daily energy requirements.

Table 16.2 **Proportion of energy derived from protein in various common foods**

Foodstuff	*Per cent of energy derived from protein*
Fish (fatty)	45
Beef (lean)	38
Beans and peas	25
Cow's milk	22
Cow's milk (skimmed)	40
Peanuts	20
Wheat flour	13
Rice	8
Potatoes	7
Bananas	4
Cassava	3

16.5 Protein utilization and protein quality

In the early 1900s, as a result of work carried out by Osborne and Mendel in the USA and Hopkins in the UK, it became apparent that not all proteins were of the same nutritional value, as some could support the growth of laboratory animals better than others. In general, proteins of animal origin were more effective than those derived from plants. Animal proteins were described as being high-quality proteins, whereas those derived from plants were low-quality proteins. The reason for this was subsequently shown to be the different amino acid composition of animal and plant proteins which results in their different degree of utilization in the body. In 1921, it was shown by Rose in the USA that the need for dietary protein was in fact the need for certain amino acids, and if these were provided in the diet in adequate quantities, protein could be omitted altogether. These are described as '**essential amino acids**' and must be provided in the diet because they cannot be synthesized in the body. All other amino acids were referred to as '**non-essential**' because they can be synthesized in the body, provided enough amino acid nitrogen is available. Non-essential amino acids are now more rigidly defined as those that can be synthesized by simple transamination, using a carbon chain synthesized by a pathway of intermediary metabolism. This means that only glutamate, aspartate and alanine are truly non-essential, and another class of amino acids has recently been defined, the '**conditionally essential**'. These amino acids cannot be produced by simple transamination reactions, but have specific amino acid precursors. Table 16.3 show the essential and conditionally essential amino acids and the precursors of the latter group.

The discovery of the essential amino acids clarified the basis for the division of proteins into high- and low-quality groups. High-quality proteins contain all the essential amino acids in proportions similar to those needed for daily protein turnover, whereas low-quality proteins have a relative deficiency in one or more of these essential amino acids.

The nutritional value of any protein can be assessed quantitatively by measuring the protein nitrogen intake and relating it to the nitrogen excretion during the time that the protein under test is consumed. The nitrogen retained in the body expressed as a percentage of the nitrogen intake is known as **net protein utilization (NPU)** and gives a measure of the nutritional value of that protein:

Table 16.3 **The essential amino acids, and the conditionally essential amino acids and their precursors**

Essential amino acids	
Valine	
Methionine	
Lysine	
Leucine	
Isoleucine	
Phenylalanine	
Threonine	
Conditionally essential amino acids	*Precursors*
Glycine	Serine, choline
Histidine	Adenine, glutamine
Cysteine	Methionine, serine
Proline	Glutamate
Arginine	Glutamate, glutamine, aspartate
Tyrosine	Phenylalanine

$$NPU = \frac{N_I - N_E}{N_I} \%$$

where N_I = nitrogen intake and N_E= nitrogen excreted.

The NPU value will be 100% if the protein is completely utilized and none is excreted, and 0% if none is utilized and all is excreted. In practice, most proteins fall between 40% and 100%, with animal proteins having high values (albumin 91%, beef 67%,) and plant proteins much lower (wheat gluten 42%).

A protein which supplies all the essential amino acids in the correct proportions can be chosen as a standard against which all other proteins can be compared. A few animal proteins, such as hen's egg albumin and casein, approach this perfect pattern of amino acid composition and are known as **reference proteins**. Albumin has, in fact, a surplus of sulphur-containing amino acids, and casein a surplus of lysine and a relative deficiency of sulphur-containing amino acids and of tryptophan.

In any protein, the essential amino acid present in the lowest concentration compared to the reference protein is known as the **limiting amino acid**. The nutritional value, or quality, of a protein can be expressed in terms of the concentration of the limiting amino acid in that protein, compared to the concentration of that amino acid in the reference protein. The ratio is known as the **chemical score**, and gives a measure of the nutritional value of a protein.

Chemical score = % of limiting amino acid in test protein / % of that amino acid in reference protein.

Table 16.4 shows some examples of the use of chemical scores in assessing the nutritional value of some proteins. Wheat gluten, for example, is low in sulphur-containing amino acids, lysine and tryptophan. Lysine is lowest, compared to the reference protein, and so the chemical score is calculated on the basis of lysine content

Table 16.4 **Use of 'chemical scores' to measure the nutritional values of proteins**

Protein	Concn (mg/g N) Sulphur	Tryptophan	Lysine	Chemical score (%)
Ideal	270	90	270	100
Egg protein	342	106	396	>100
Casein	(215)	85	497	$\frac{215}{270} = 80$
Beef muscle	237	(75)	540	$\frac{75}{90} = 80$
Wheat gluten	223	60	(107)	$\frac{107}{270} = 40$

alone, and has a value of 40. Similar, but not identical, ratings are obtained whether the nutritional value of a protein is assessed using NPU or chemical score as the criterion.

The nutritional value of a protein can, of course, be improved by mixing it with proteins which have different limiting amino acids, so that the relative deficit in one amino acid can be made good by the relative excess of that amino acid in another protein. The combination of bread (low lysine and tryptophan) and pulses, such as beans or peas (low in sulphur containing amino acids), for example, is common to many traditional diets based on 'low-quality' proteins'. This improvement of the nutritional value by mixing different proteins is known as **protein complementation**.

16.6 Protein requirements

The high levels of daily protein intake recommended by various workers towards the end of last century are now known not to be necessary. It is possible to calculate the requirements accurately by measuring nitrogen excretion. If healthy adults are fed a diet adequate in all other respects but devoid of protein, the nitrogen excretion declines steadily over 5–6 days, but does not reach zero. It continues at approximately 2.7 g of nitrogen per day. For the normal 70 kg man, the daily loss of nitrogen corresponds to 16 g of protein. This figure, however, cannot be used directly to infer a minimum required intake of protein because the utilization of protein is not 100% efficient. For typical mixed Western diets, for example, the utilization of protein is about 70% efficient. The figure of 16 g must therefore be multiplied by 10/7 to give an estimate of a minimum intake of approximately 24 g per day. The expert committees of the United Nations and the World Health Organisation recommend a level which is higher than the bare minimum. The recommendations of the UK Department of Health, published in 1991, are based on the FAO/WHO/UNU recommendations and are shown in Table 16.5.

Table 16.5 **Dietary reference values for protein**

Age	Estimated average requirement (g/day)	Reference nutrient intake (g/day)
0–12 months	11.0	13.0
1–3 years	11.7	14.5
4–6 years	14.8	19.7
7–10 years	22.8	28.3
Males		
11–14 years	33.8	42.1
15–50+ years	44.4	55.5
Females		
11–14 years	33.4	41.2
15–50+ years	37.0	45.3
*Pregnancy**		0.6
*Lactation**		
0–6 months		0.11
6+ months		0.8

*To be added to the adult requirement.

The dietary reference values (DRV) were derived by nitrogen balance studies, measuring the amounts of high-quality egg or milk protein needed to achieve equilibrium. For infants, the recommendations were based on the milk intakes of breast-fed infants who showed a satisfactory rate of growth, positive nitrogen balance, and who were in good health. Allowances were made for growth in children. As the rate of growth is not constant, and different amounts of protein are laid down from day to day, enough protein must be provided for periods when growth is rapid.

For pregnancy, allowances were made to accommodate the protein deposition in the fetus and maternal tissues compatible with the production of a 3.3 kg infant. Allowances for lactation are based on the protein content of human milk.

The average consumption of protein in the UK is consistently higher than the DRV. No group in the UK was identified as likely to be consuming food which would not supply adequate levels of protein and energy. For adults consuming mixed diets, including vegetarian diets, there is no reason for concern about amino acid composition and adequate intake of essential amino acids.

The question of high protein intake has been addressed in recent years. Are high protein intakes simply unnecessary, or actually harmful? The UK Department of Health report cited above recommends a maximum intake of protein by adults of 1.5 g/kg body weight/day. This corresponds to twice the DRV. Although evidence on the harmful effects of high protein intakes is insufficient, there are indications that excessive protein intakes may contribute to bone demineralization and may also lead to further deterioration of renal function in patients with renal disease.

Chapter 17

Nutrition: lipids and carbohydrates

17.1 Dietary lipid

The average British diet provides 88 g of fat (lipids) per day, which corresponds to 40% of the total energy in the diet. Similar figures apply to the rest of Europe and the USA. The lipid intake is mainly in the form of triacylglycerols. A small amount of lipid is necessary in the diet, in order to provide the polyunsaturated fatty acids known as **essential fatty acids**, which are important both structurally and metabolically, and cannot be synthesized in the body.

Apart from providing essential fatty acids, fat is not an essential component of the diet, but its consumption is, nevertheless, important for a number of reasons.

It represents the most concentrated form of energy available, supplying 38 kJ/g, compared to 16 kJ/g supplied by carbohydrate and protein. It imparts palatability to food, by acting as a lubricant and also by providing flavour. A fat-free diet is dry, difficult to swallow, tasteless and unattractive. It is not surprising to see that 'junk' food, which many children and adolescents find attractive, is high in fat, and even laboratory animals soon learn to consume large quantities of such 'cafeteria-type' food.

Dietary fat is necessary for the absorption of fat-soluble vitamins. These normally occur in association with fat, but if they are supplied in adequate quantities in a fat-free diet, they will be very poorly absorbed, as their absorption across the intestinal mucosa takes place during lipid absorption. Fat-soluble vitamin deficiency is known to occur in cases of fat malabsorption.

Fatty acid deficiency is very rare, as most diets, even if they are very poor, usually contain an adequate amount of essential fatty acids. Essential fatty acids are present in all natural lipid structures, but particularly good sources are vegetable and fish oils. Of greater concern, in Western society, is the excessive consumption of fat, and especially saturated fat, in view of its association with the 'diseases of affluence' such as cardiovascular disease, cancer and obesity (Sections 17.7 and 17.8; Chapters 18 and 37).

17.2 Essential fatty acids

In the late 1920s, two American nutritionists, Burr and Burr, observed that rats fed on a fat-free diet failed to grow and lost large amounts of water through their skin. These abnormalities could be cured only by the addition of specific polyunsaturated fatty acids, namely linoleic, linolenic and arachidonic acids. The structures of these compounds, which became known as 'the essential fatty acids', are shown in Figure 2.2.

Other signs of fatty acid deficiency subsequently became apparent. These included swelling of liver mitochondria, decrease in prostaglandin synthesis, alterations in the fatty acid pattern in many tissues, and a defective immune response.

According to current usage, the three essential fatty acids are said to contain either n-6 (linoleic and arachidonic) or n-3 (α-linolenic) double bonds. In this system, 'n-' refers to the carbon of the terminal methyl group in a fatty acid. Thus in n-6 unsaturated fatty acids, the first double bond occurs between the sixth and

$C_{18:2}$ $CH_3\text{–}(CH_2)_4\text{–}CH{=}CH\text{–}CH_2\text{–}CH{=}CH\text{–}(CH_2)_7\text{–}COOH$

Linoleic acid

↓

$CH_3\text{–}(CH_2)_4\text{–}CH{=}CH\text{–}CH_2\text{–}CH{=}CH\text{–}CH_2\text{–}CH{=}CH\text{–}(CH_2)_4\text{–}COOH$

γ-Linolenic acid

↓

$CH_3\text{–}(CH_2)_4\text{–}CH{=}CH\text{–}CH_2\text{–}CH{=}CH\text{–}CH_2\text{–}CH{=}CH\text{–}(CH_2)_6\text{–}COOH$

Dihomo-γ-linolenic acid

↓

$C_{20:4}$ $CH_3\text{–}(CH_2)_4\text{–}CH{=}CH\text{–}CH_2\text{–}CH{=}CH\text{–}CH_2\text{–}CH{=}CH\text{–}CH_2\text{–}CH{=}CH\text{–}(CH_2)_3\text{–}COOH$

Arachidonic acid

Figure 17.1 **Metabolism of linoleic acid to arachidonic acid**

seventh carbon atoms, the first carbon atom being that of the terminal methyl group.

Desaturation of fatty acids in mammalian tissues is possible only between carbon atoms present from position n-9 to the terminal carboxylic group, hence explaining the essential nature of linoleic and linolenic acids. Linoleic acid, referred to as C18:2 n-6, to indicate that it has 18 carbon atoms and two double bonds, the first of which is found between carbon atoms six and seven from the methyl group, can, in fact, be converted to arachidonic acid (C20:4 n-6) by a series of reactions outlined in Figure 17.1, and the inclusion of arachidonic acid in the list of essential fatty acids has often been questioned. As arachidonic acid is a very important cellular component, and as the efficiency of conversion of linoleic to arachidonic acid is not documented in humans, it is probably desirable to include small quantities in the diet.

The status of α-linolenic acid (C18:3 n-3) as an essential fatty acid had also been questioned. Its importance, however, is now recognized, and it has recently been added to some infant formula feeds in addition to linoleic acid. This is because linoleic, α-linolenic and their respective metabolites, arachidonic and docosahexanoic (C22:6 n-3) acids, are found in brain, and they are important in the development of brain and retina in infants. As mammalian tissues cannot form double bonds in the n-3 position, it seems certain that either α-linolenic, or some other n-3 acid, must be present in the diet. The quantities required are likely to be very small, as the retention of n-3 acids by the mammalian body is very efficient.

17.3 Deficiency of essential fatty acids.

The UK Department of Health in 1991 recommended that linoleic acid should provide at least 1%, and α-linolenic 0.2% of the total energy of the diet, which corresponds to an intake of 2–5 g of essential fatty acids per day. The average UK diet provides 11.7 g of n-6, and 1.6 g of n-3 fatty acids, and most adult Western diets provide 8–15 g of essential fatty acids per day. Healthy people have a reserve of 500–1000 g of essential fatty acids in the triacylglycerol stores of adipose tissue.

Essential fatty acid deficiency is observed when the intake of essential fatty acids is less than 2–5 g per day. Only a few cases of essential fatty acid deficiency have been reported world-wide. A link between essential fatty acid deficiency and eczema in children was shown in 1947, but the first unequivocal cases of essential fatty acid deficiency were not described until the early 1970s. A patient who had most of his small intestine resected and was fed intravenously (parenterally) on a standard preparation containing amino acids, glucose, electrolytes and vitamins, but no fatty acids, developed dermatitis after about 3 months. Similar observations were reported on patients who were not fed intravenously, but were fed a very low fat diet because of small bowel disease. The symptoms were alleviated by the addition of linoleic acid to the intravenous solution in the first case, and by local application of essential fatty acids to the skin in the second case.

After only a few days on a fat-free diet, the blood fatty acid pattern changes considerably. The proportions of linoleic and arachidonic acids decline sharply, and the triene, eicosatrienoic acid (C20:3 n-9), formed from oleic acid, rises. The ratio of the concentration of this triene to that of the tetraene, arachidonic acid, in the plasma can be used as an index of essential fatty acid deficiency. It should normally be very small but rises sharply when the diet is deficient in linoleic acid. The synthesis of eicosatrienoic acid, in the absence of an intake of essential fatty acids, leads to its incorporation into cellular membranes, causing malfunction of organelles, for example mitochondria.

17.4 Functions of essential fatty acids

Essential fatty acids are needed for two reasons:

1. They are important constituents of membrane phospholipids
2. They are precursors of an important family of regulators, collectively known as **eicosanoids**, which include prostaglandins, thromboxanes, prostacyclins and leukotrienes.

Membrane composition

The majority of these long-chain polyunsaturated fatty acids are incorporated into the phospholipids of membranes of nearly all tissues in the body. If these fatty acids are deficient in the diet, the pattern of fatty acids in the membrane is altered. They are required in the membranes to achieve the correct fluidity and conformation. Defects in the structure of cell membranes can lead to defective function, as enzymes, receptors and other proteins embedded in the membranes depend on the correct composition of the membrane for optimal function (see Chapter 2).

Eicosanoids

The eicosanoids, as the name implies, are derived from C20 fatty acids, the most important precursor being arachidonic acid.

Most of the arachidonic acid found in cells is in the form of phospholipids in the cytosolic face of the plasma membrane. The action of phospholipase A_2 liberates arachidonic acid from carbon C-2 on the glycerol backbone on which it is esterified, and releases it into the cytosol.

Eicosanoid structure and nomenclature

Figure 17.2 shows the essential structural features of prostaglandins, prostacyclins and thromboxanes. They can all be considered to be derivatives of the hypothetical parent compound, prostanoic acid.

The **prostaglandins** are oxygenated C20 unsaturated fatty acids containing a cyclopentane ring. They are divided into groups A–J depending on the substituents on the cyclopentane ring, and further subdivided, by being assigned a number corresponding to the number of double bonds in the side-chains (Figure 17.3). For example, PGE_2 has two double bonds.

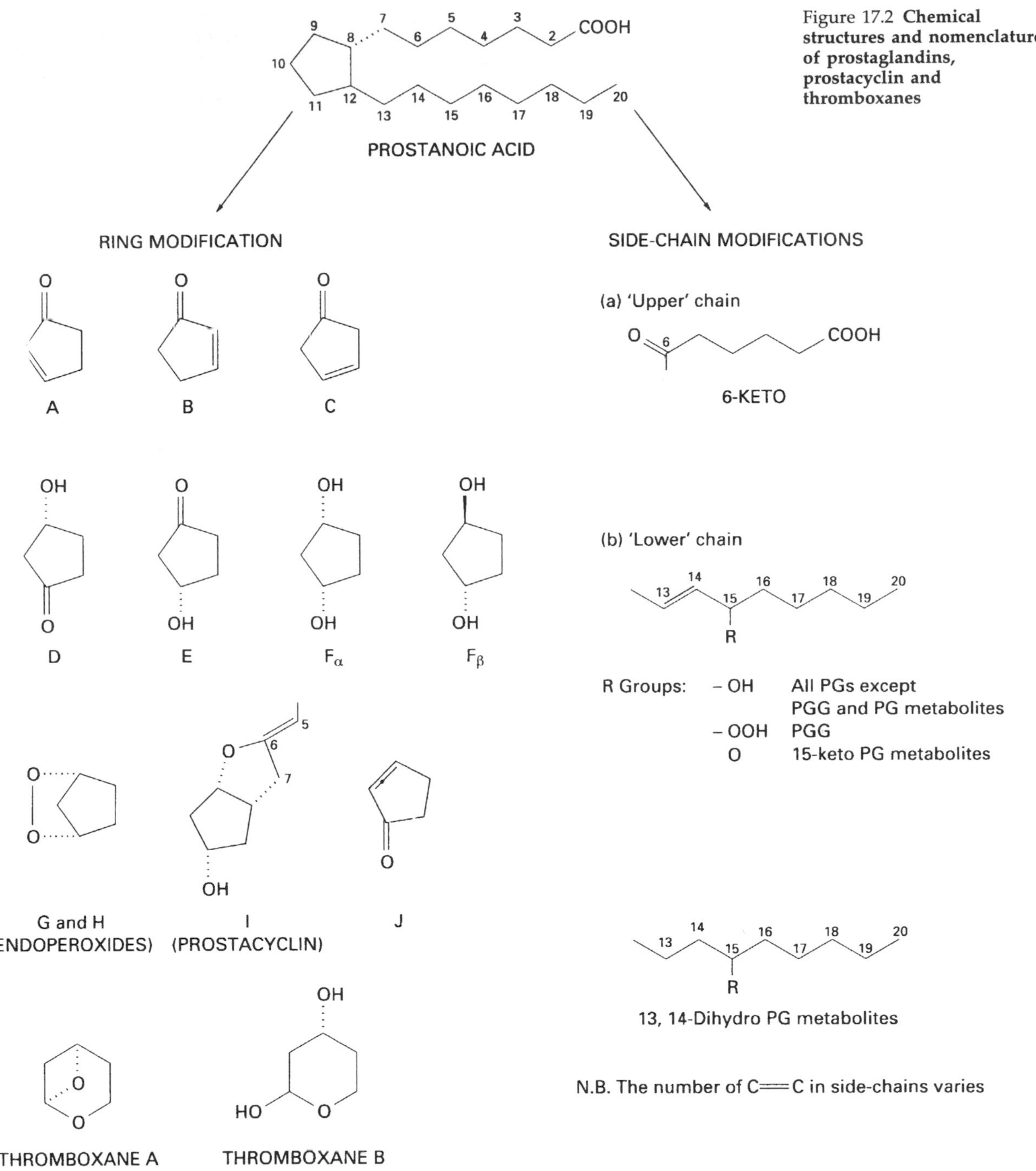

Figure 17.2 **Chemical structures and nomenclature of prostaglandins, prostacyclin and thromboxanes**

Thromboxanes are so named because their synthesis was first demonstrated in platelets (thrombocytes) and they contain an oxane ring.

Leukotrienes were so named because they were initially described in leukocytes and are characterized by a conjugated triene system but no ring structure such as found in prostaglandins and thromboxanes. They are also divided into groups (A–F) depending on structural features, such as the attachment of peptide residues, and subdivided into groups bearing a number corresponding to the number of double bonds in the side-chains.

The two major routes for the conversion of arachidonic acid to its metabolites, the cyclooxygenase pathway and the lipoxygenase pathway, are outlined in Figure 17.4.

The cyclization of arachidonic acid catalysed by **cyclooxygenase** occurs in all cells except erythrocytes, and eventually leads to the production of a variety of prostaglandins, such as prostaglandin E_2 (PGE_2) and prostaglandin $F_{2\alpha}$ ($PGF_{2\alpha}$), prostacyclins such as PGI_2, and thromboxanes such as TxA_2. All these compounds act as local regulators. Unlike hormones, which are produced by glands and are released into the blood-stream in order

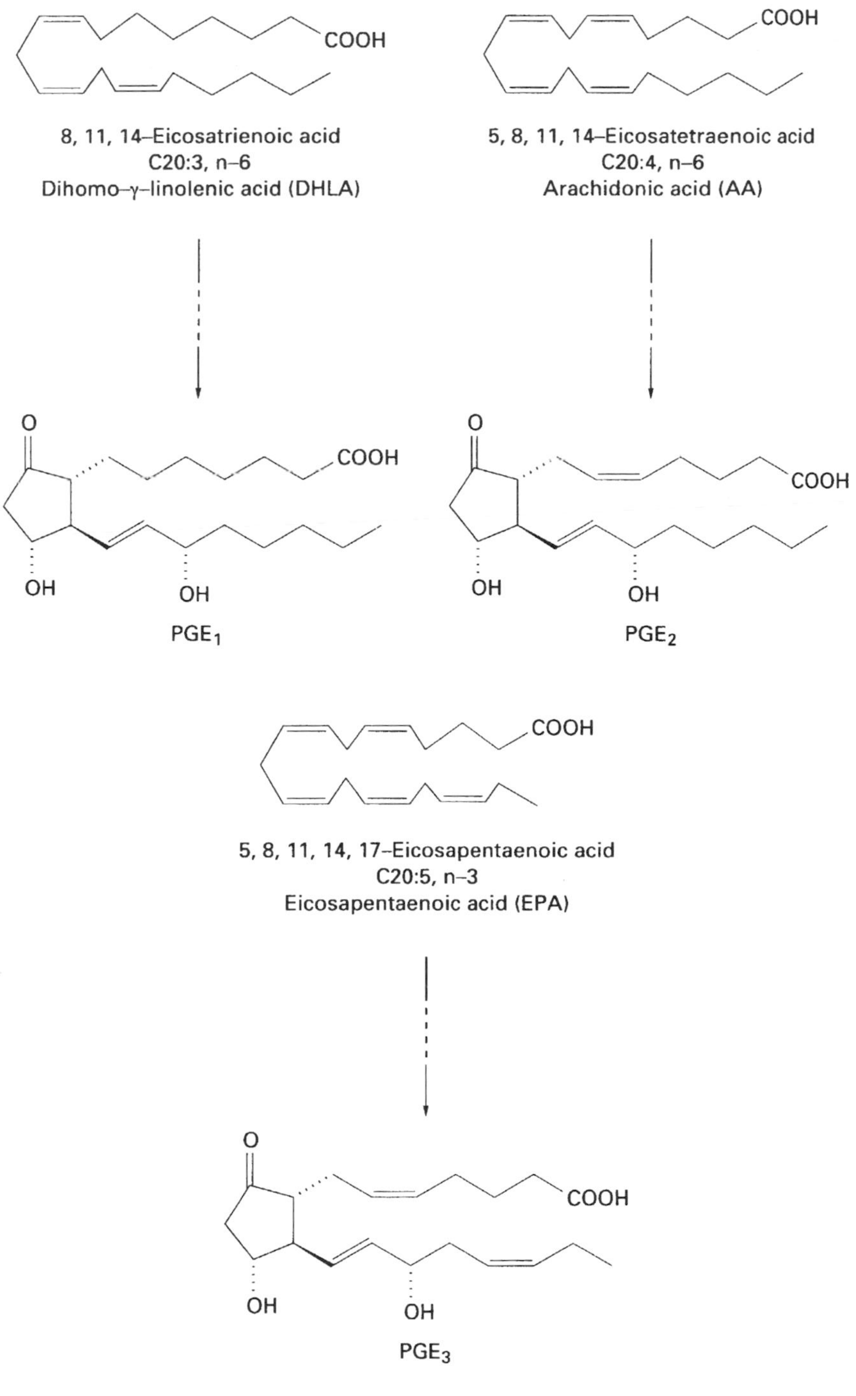

8, 11, 14–Eicosatrienoic acid
C20:3, n–6
Dihomo–γ–linolenic acid (DHLA)

5, 8, 11, 14–Eicosatetraenoic acid
C20:4, n–6
Arachidonic acid (AA)

PGE$_1$

PGE$_2$

5, 8, 11, 14, 17–Eicosapentaenoic acid
C20:5, n–3
Eicosapentaenoic acid (EPA)

PGE$_3$

Figure 17.3 **Subdivision of the prostaglandins according to the number of carbon–carbon double bonds in the side-chain**

to reach their target tissues, prostaglandins, prostacyclins and thromboxanes act in the immediate vicinity of the cells which produce them, i.e. they act as paracrine agents.

Prostaglandins were discovered in the 1930s when it was observed that fatty-acid-derived molecules in seminal plasma could cause contraction or relaxation of smooth muscle. It soon became apparent that prostaglandins were involved in a wide variety of processes and conditions summarized below:

1. In pregnancy, they are produced in response to oxytocin and act to promote uterine contraction. Because of this effect, they have been used to terminate unwanted pregnancies
2. They are released in response to neurotransmitters such as noradrenaline, and they act presynaptically as feedback inhibitors, limiting further release of the neurotransmitter
3. They are synthesized in response to various hormones. For example, in some tissues, such as the

Figure 17.4 **Pathways for the biosynthesis of eicosanoids from arachidonate**

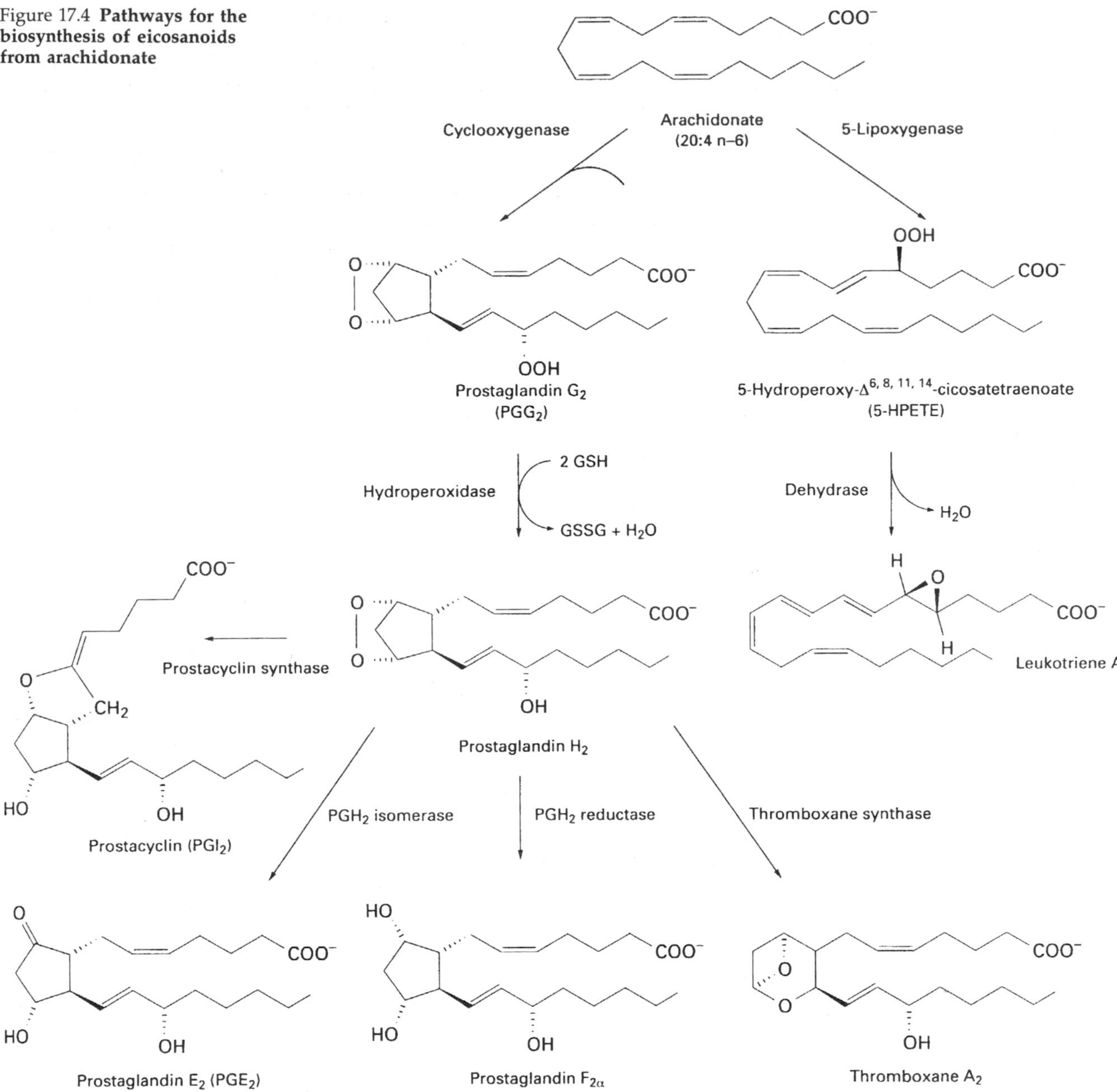

myometrium, corticotrophin releasing hormone (CRH; Chapter 28) stimulates prostaglandin synthesis, whereas in macrophages, for example, glucocorticoids inhibit prostaglandin synthesis

4. They can lower systemic arterial pressure through their vasodilatory effect
5. They are involved in inflammatory responses, causing oedema, swelling and prolonged erythema, by increasing capillary permeability
6. In the thermoregulatory centre of the hypothalamus, their biosynthesis and release is activated by pyrogens. Aspirin can block these effects, and can, therefore, act as an antipyretic drug, by virtue of its ability to inhibit prostaglandin synthesis. The active ingredient in aspirin, acetylsalicylic acid, inhibits cyclooxygenase irreversibly, by acetylating a serine residue in the active site of the enzyme. Indomethacin and ibuprofen also act at this level, as competitive inhibitors of cyclooxygenase
7. Prostaglandins additionally have an effect on platelet aggregation. PGE_2 promotes aggregation and thus clotting (see Chapter 24).

Blood platelets produce **thromboxane A_2**, which promotes platelet aggregation, causes contraction of the smooth muscles of the arterial wall and, therefore, changes in local blood flow and increase in blood pressure. Prostacyclins and thromboxanes have opposing effects in the maintenance of vascular function. **Prostacyclins**, such as PGI_2, produced by the vascular endothelium, inhibit platelet aggregation, relax the arterial wall and lower blood pressure. An increase in the dietary intake of n-3 polyunsaturated fatty acids, such as that achieved by increasing the consumption of fish and fish oils, leads to an alteration in the proportions of the different types of thromboxanes in platelets. Fish oils contain eicosapentanoic acid (C20:5 n-3) which can also act as a precursor for eicosanoid synthesis. TxA_2, which is derived from arachidonic acid (n-6), is decreased, and TxA_3, derived from n-3 precursors, is increased. TxA_3 is a less potent activator of platelet aggregation than TxA_2, and so the tendency for thrombus formation is decreased. This is one of the reasons for the recommendation for the consumption of fish oils in the prevention of cardiovascular disease. Another possible benefit is the reduction in the level of VLDL in blood.

The **lipoxygenase** pathway produces leukotriene A_4, the precursor of other leukotrienes which are mainly involved in inflammatory responses such as erythema, oedema and hyperthermia. In skin, they are responsible for the 'weal and flare' reaction seen in some allergic responses. A group of these leukotrienes are thought to be responsible for the occasional fatal side-effects of vaccinations.

Steroid drugs, such as prednisolone, have both antipyretic and anti-inflammatory action by inhibiting phospholipase A_2, the enzyme responsible for the release of arachidonic acid from membrane phospholipid.

17.5 Polyunsaturated fatty acids and lipid peroxidation

Membrane lipids, and especially, polyunsaturated fatty acids, are susceptible to oxidation and peroxidation. Reactive oxygen species (ROS), such as hydrogen peroxide, and hydroxyl and superoxide free radicals, can oxidize polyunsaturated fatty acids in membranes and disrupt cellular function (see Chapter 33).

Normal mammalian physiology depends on the controlled formation of free radicals. The control is exercised both at the level of production of these highly reactive species, and also at the level of their disposal, but is never absolutely successful, and oxidation of tissue components inevitably occurs. Indeed, in some pathological circumstances, the control systems may be overwhelmed and extensive tissue damage may ensue. A pathogenic role for free radicals and lipid peroxidation has been established for processes such as reperfusion injury and tissue transplant rejection, but the contribution of free radical damage to conditions like cardiovascular disease and cancer is very difficult to quantify.

There is no evidence that a high intake of polyunsaturated fatty acids, which are potential substrates for peroxidation, is associated with any disease. However, as the pattern of fatty acids in phospholipids and triacylglycerols can be modified in the body by modifying the dietary intake of polyunsaturated fatty acids, and as these acids offer themselves as substrates for peroxidation, it would seem reasonable to try to achieve a balance between free radical activity and antioxidant status. Some of the antioxidant defences of the body can be manipulated by diet, and it is recommended that diets high in polyunsaturated fatty acids should include adequate amounts of antioxidant nutrients, for example, vitamin E, vitamin C and carotenes.

17.6 '*Trans*' unsaturated fatty acids

Naturally occurring unsaturated fatty acids from all plant and most animal sources, are of the '*cis*' configuration. '*Trans*' fatty acids are not common and their main source in the diet is industrial hydrogenation of long-chain fatty acids, mainly from fish oils, in the manufacture of margarine. Fish oils contain substantial amounts of fatty acids with more than 20 carbon atoms and the possibility of producing unnatural isomers during their processing is higher than in the processing of shorter chain fatty acids from vegetable oils. In Europe and the USA, mainly vegetable oils are used in the production of margarine, whereas in the UK, the commonest raw materials are fish oils. '*Trans*' fatty acids are also produced naturally, by bacteria in the gut of ruminants, and in this way, they find their way into milk, dairy products and meat. Unnatural isomers can also be produced by the high temperatures achieved during frying and especially deep frying of foods.

The possible hazards of consuming '*trans*' fatty acids are currently the subject of research and there is much controversy. The average UK intake in 1991 was 5 g per day. A relationship between '*trans*' fatty acids and atherosclerosis has been suggested, but not confirmed. High levels of '*trans*' fatty acids in the diet can lead to an increase in serum cholesterol, but to a lesser extent than the equivalent amounts of C12 and C16 saturated fatty acids. There are also indications that HDL cholesterol levels are decreased with consumption of high levels of '*trans*' fatty acids. The UK Department of Health concluded that the information available in 1991 was insufficient to recommend a decrease in the current level of '*trans*' fatty acid consumption, but in view of the possibility of adverse effects on health, recommended that the intake of these fatty acids should not be increased (see Chapter 14).

17.7 Dietary lipid and cardiovascular disease

At least 1.5 million people are killed by cardiovascular disease each year. In the UK, and especially in Scotland, mortality from coronary heart disease (CHD) is one of the highest in the world. The disease was relatively uncommon at the beginning of the century, but increased steadily until the 1970s. Although the mortality rates have been decreasing in the past 15 years in the UK, CHD still accounts for more deaths in men than any other single cause. The mortality rates for women are lower than for men, but they are still equal to the rates from all types of cancer combined. Death usually occurs as a result of a myocardial infarction, which involves thrombus formation (thrombosis) in a coronary artery already narrowed by atheroma (atherosclerosis), thus interrupting blood supply to cardiac muscle.

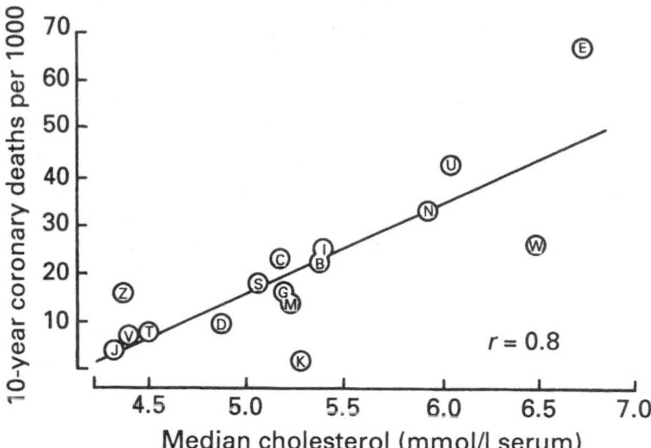

Figure 17.5 **The relation between serum cholesterol concentration and the incidence of deaths from coronary heart disease in the following places:** B, Belgrade; C, Crevalcore; D, Dalmatia; E, east Finland; G, Corfu; I, Italian railroad; J, Ushibuka; K, Crete; M, Montegiorgio; N, Zutphen; S, Slavonia; T, Tanushimara; U, US railroad; V, Velika Krsna; W, west Finland; Z, Zrenjanin

Atherosclerosis refers to the deposition of cholesterol and other lipids in the arterial wall which leads to the formation of plaque and results in endothelial damage and narrowing of the lumen. Thrombosis involves platelet aggregation on a fibrin matrix and is the final episode in myocardial infarction.

Many risk factors are involved in the development of cardiovascular disease apart from genetic susceptibility, and these include smoking, a sedentary life-style, high blood pressure and high serum cholesterol. Diet is an important factor in the prevention of CHD, as it can affect, to a certain degree, blood pressure, serum cholesterol levels and blood clotting.

It has been demonstrated consistently, in epidemiological studies within and between populations, that there is a continuous increase in risk of death from CHD with increasing serum cholesterol. Figure 17.5 shows the relation between serum cholesterol concentration and the incidence of deaths from CHD in various countries.

Cholesterol is transported in the plasma in the form of lipoproteins, the major classes being chylomicrons, very-low-density lipoprotein (VLDL) low-density lipoprotein (LDL), and high-density lipoprotein (HDL; see Chapter 10). A major risk factor for CHD seems to be a high concentration of LDL, which carries about 70% of the total circulating cholesterol. In contrast, high levels of HDL cholesterol are thought to have a protective effect, and lower the risk of CHD.

The first experiments that related diet to CHD were carried out by the Russian investigator Anitischkow who, in 1913, reported the production of atheromata in the arteries of rabbits fed high levels of cholesterol. Interest in the effect of diet on heart disease was resurrected as a result of studies on populations of countries such as Norway and Holland, which were occupied during the World War II. Most of the inhabitants lived on very poor diets, low in saturated fat in particular, and the incidence of cardiovascular disease declined sharply in the 1940s and 1950s. Later studies, in developing countries, also showed that populations consuming lower amounts of total fat, and especially saturated fat, had much lower incidence of cardiovascular disease than populations consuming typical Western-type diets.

The evidence for the relationship between LDL cholesterol levels and the incidence of cardiovascular disease is quite strong, but the exact mechanism whereby deterioration of the arteries and development of arterial disease occurs, is not entirely resolved. Diet is the most important non-genetic determinant of blood lipoprotein levels. Changes in dietary lipid affect to a certain extent the pattern of blood lipoproteins, but the mechanisms by which these changes are brought about are far from clear.

Increases in dietary cholesterol have only a small effect on blood cholesterol, and mainly in the LDL fraction. Increases in saturated fatty acids in the diet lead to increases in LDL and total cholesterol. The changes are mainly due to myristic acid (C14), while longer fatty acid chains are less effective. When saturated fatty acids in the diet are replaced by n-6 polyunsaturated fatty acids, LDL cholesterol decreases. Adding n-3 polyunsaturated fatty acids to the diet, does not affect LDL cholesterol, but decreases VLDL. In contrast, reduction in the amount of saturated fat and increase in the amount of carbohydrate in the diet, leads to a decrease in blood LDL concentrations, and sometimes an increase in VLDL levels. Reduction in the amount of n-6 polyunsaturated fatty acids and replacement by carbohydrate leads to an increase in circulating LDL concentrations.

The effect of mono-unsaturated fatty acids is less clear. Epidemiological studies have shown that the incidence of heart disease is lower in countries where the consumption of mono-unsaturated fatty acids, such as oleic acid (C18:1), is higher. Olive oil, which contains substantial amounts of oleic acid, is consumed widely in countries such as Greece and Southern Italy, where the incidence of heart disease is, indeed, lower than the rest of Europe, and this has led to an interest in 'Mediterranean diets' as a means of preventing heart disease. The difference in oleic acid consumption, however, is only one difference between traditional Mediterranean and typical northern European diets, and until more specific studies are completed, the value of mono-unsaturated fatty acids in the prevention of heart disease will remain controversial. The greatest difference in the two types of diet is the consumption of saturated fat, which is much lower in the Mediterranean diet. The consumption of fruit and vegetables is also different, being much higher in the Mediterranean countries.

Serum cholesterol can be lowered to some extent by diet, and to a considerable extent by drugs. Decrease in serum cholesterol has been shown to be associated with a decrease in CHD events, in many studies, most of which were carried out on men considered to be at high risk. The effect of lowering blood cholesterol in women, or in the general population, has not been demonstrated.

17.8 Dietary lipid and cancer

The distribution of different types of cancer is not uniform throughout the world. In some cases, and after a few generations, immigrants are found to suffer from the same types of cancer as the host population, rather than the types of cancer more common to their country of origin, suggesting that environmental factors are important. Diet is an obvious possibility, and dietary lipid has been implicated in a variety of cancers in the UK. Animal and epidemiological studies have shown that cancers of the breast, colon, pancreas and prostate are associated with high fat intakes.

Animal studies have shown that increasing the level of fat in the diet, from 20% of the total energy to 40%, results in a 100% increase in the appearance of breast tumours. A certain degree of suppression was observed when n-3 polyunsaturated fatty acids were included in the diet. Studies across various, but not all countries (Chapter 37) have suggested an association between total fat intake and breast cancer and it is known that obesity is a risk factor for both postmenopausal breast and endometrial cancer. Intake of saturated fat was shown to be linked with increase in the incidence of colon cancer, and total fat with cancer of the prostate.

The mechanism by which dietary fat may have an effect on the development of cancer is not understood. Oestrogens are known to be tumour promoters, in both animals and humans, but it is not clear how blood or tissue levels of oestrogens are affected by dietary lipid. Alternative mediators have been suggested, such as the effect of dietary lipid on the immune system, on the susceptibility of membrane lipids to peroxidation and on the synthesis of eicosanoids, but without convincing evidence as yet.

Difference in fat consumption is only one of the differences seen in breast cancer rates in different countries. Reproductive patterns, obesity, consumption of alcohol and use exogenous oestrogens vary as well.

The contribution of dietary lipid to the development of obesity and to the development of cancer is discussed in Chapters 18 and 37 respectively. The dietary reference values for fat, and the basis for these recommendations, are shown in Chapter 14.

17.9 Dietary carbohydrate

The proportion of dietary energy derived from carbohydrate varies considerably throughout the world. In affluent societies, the intake can be as low as 40%, whereas in poor populations, it can reach 80–90%. The main types of carbohydrate in the diet are **starch, non-starch polysaccharides** (fibre) and **sucrose**, and the proportions of these components also varies greatly among different diets.

Most dietary carbohydrate is of plant origin. Although animal tissues and animal products do contain small amounts of carbohydrate (mainly in the form of **glycogen**), the amount is too small to be of practical significance, except in the case of the disaccharide **lactose** found in milk.

In theory, there is no biological need for dietary carbohydrate, as it can be synthesized from protein. Protein, however, only provides a small amount of the total energy in any diet, and the supply of carbohydrate allows dietary protein to be used as a source of amino acids, for growth and maintenance of tissues, rather than as a source of metabolic fuel. Dietary carbohydrate is said to have a **protein-sparing** effect. Apart from this, the liver has a limited capacity for handling and disposing of amino acid nitrogen. Volunteers who ate lean meat as their only source of metabolic fuel suffered from nausea and severe abdominal pain after a few weeks, and lost a considerable amount of weight, as they could not metabolize enough protein to cover their needs for both amino acids and energy.

There is another reason why carbohydrate is a desirable dietary component. If a diet is low in carbohydrate, the liver will use fat as metabolic fuel, and the excessive conversion of fatty acids to ketone bodies will result in ketoacidosis.

The main types of dietary carbohydrate are shown in Figure 17.6, and their structures are shown in Figure 17.7.

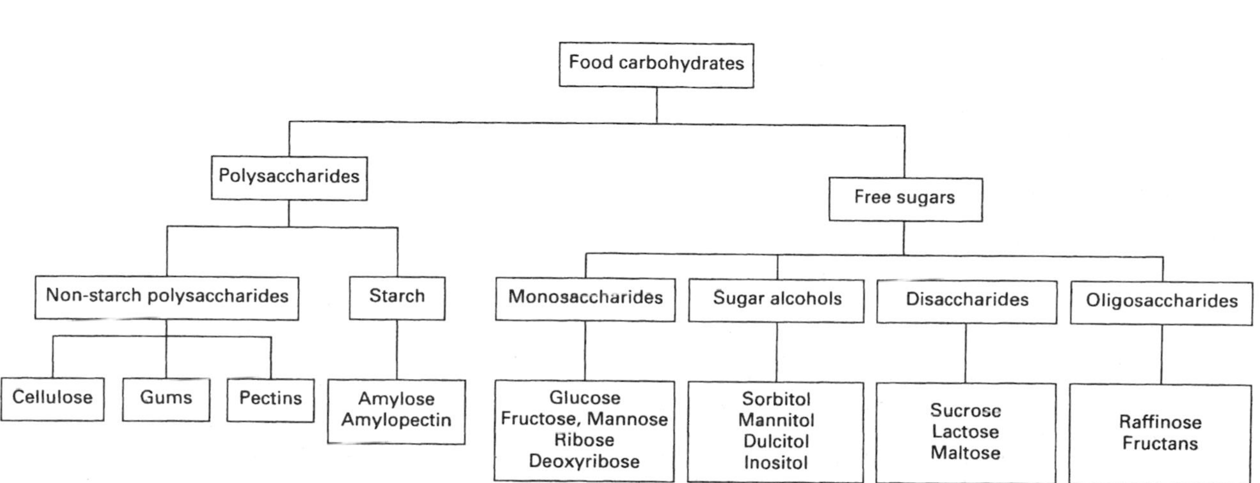

Figure 17.6 **The major types of dietary carbohydrate**

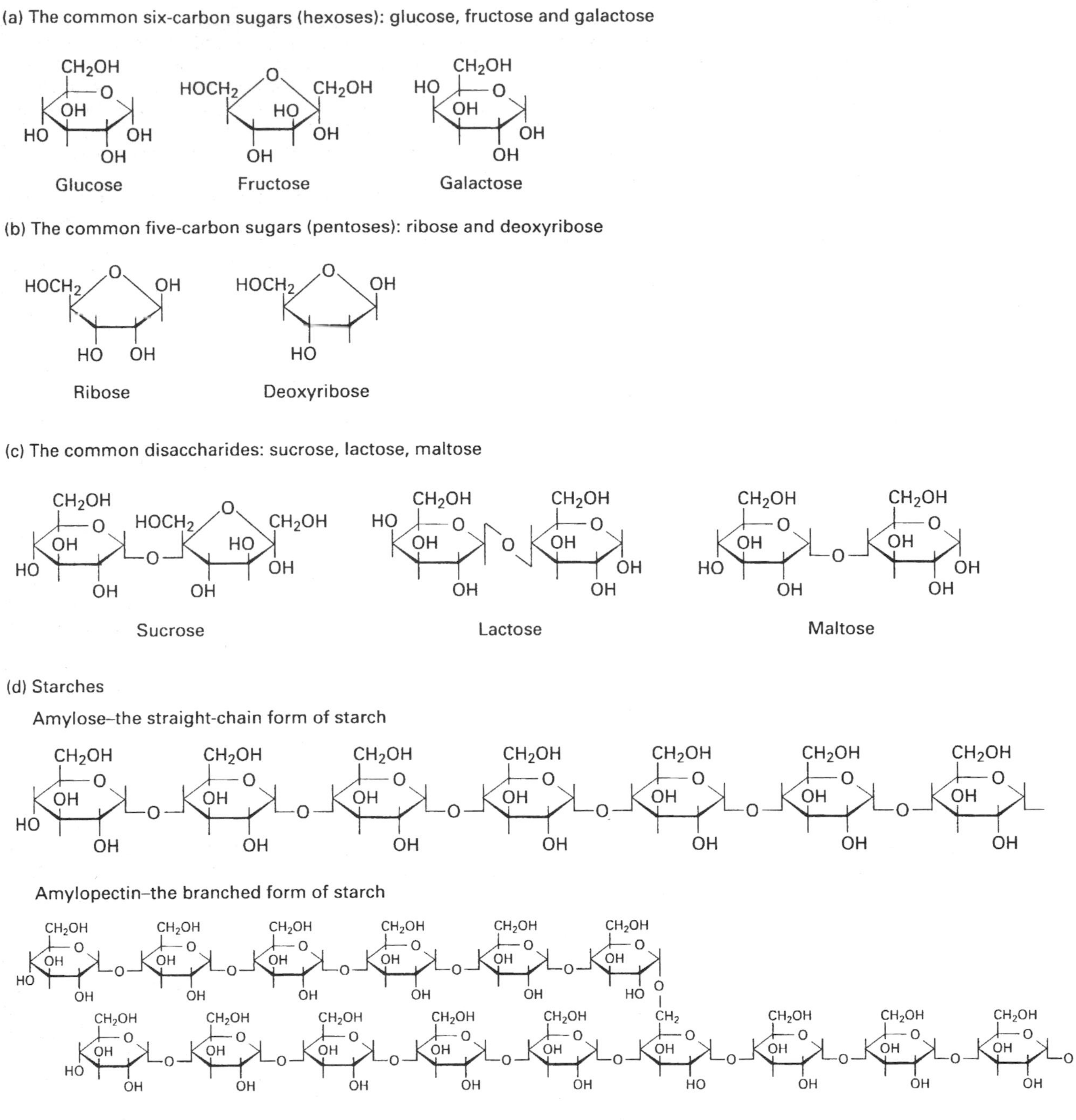

(a) The common six-carbon sugars (hexoses): glucose, fructose and galactose

Glucose Fructose Galactose

(b) The common five-carbon sugars (pentoses): ribose and deoxyribose

Ribose Deoxyribose

(c) The common disaccharides: sucrose, lactose, maltose

Sucrose Lactose Maltose

(d) Starches

Amylose–the straight-chain form of starch

Amylopectin–the branched form of starch

Figure 17.7 **Structures of major dietary carbohydrates**

Monosaccharides are taken in the diet in small amounts. The main **hexose sugars** are glucose and fructose. **Glucose**, which is the main monosaccharide in the bloodstream, is found in food in small amounts only, in fruits such as grapes and some vegetables. **Fructose** is also found in small amounts in fruit and vegetables, and both glucose and fructose together are the main constituents of honey.

Some sugar alcohols are also present naturally in the diet, but the main sources of them are commercial preparations. **Sorbitol,** for example, is found in some fruits, but the dietary intake of the commercially prepared form is much more significant. It is less sweet than sucrose, and is metabolized via fructose rather than glucose from which it is prepared by hydrogenation. Its consumption, therefore, does not lead to the same increase in blood glucose that would be seen with the equivalent amount of sucrose. For this reason it is used widely in preparations suitable for diabetics, such as jams, soft drinks and chocolates. It must be noted, however, that it has the same energy value as glucose, a fact sometimes ignored by slimmers who consume diabetic preparations in the hope that they are poorer in energy than ordinary foods.

The cyclic alcohol **inositol** is found naturally, especially as the hexaphosphate derivative, in the husks of cereals. Inositol hexaphosphate, known as **phytic acid**, binds metal ions such as iron and calcium and interferes with their absorption from the intestine. This is the reason for the supplementation of brown bread with calcium salts in the UK during the World War II, a practice which has persisted to this day.

Pentose sugars, such as **ribose** and **deoxyribose**, are present in the cells of all natural foods, mainly as components of nucleic acids, and their contribution to the dietary energy content is very small.

Disaccharides, such as sucrose and lactose, are much more abundant. **Sucrose**, which is a dimer of glucose and fructose, linked by an α 1–2 glycosidic bond (Figure 17.8), is the commonest disaccharide in the diet. Some is found in fruit and vegetables, but the main source, by far, is refined table sugar. **Lactose**, which is a dimer of glucose and galactose linked by a β1–4 glycosidic bond, is found in milk and some milk products. The first step in the metabolism of lactose depends on the presence of **lactase**, an enzyme found in the brush border of the small intestine, and which is necessary for the conversion of lactose into glucose and galactose. People from most ethnic groups, other than those of European origin, gradually stop producing lactase in adolescence, and may suffer from a condition known as **lactose intolerance**. Lactose cannot be absorbed by the intestine and it remains in the lumen where it is subjected to bacterial fermentation, resulting in abdominal pain and severe diarrhoea.

The Committee on Medical Aspects of Food Policy of the UK, reporting on dietary sugars and human disease in 1989, classified sugars as intrinsic or extrinsic, irrespective of chemical structure. **Intrinsic** sugars are those present in cells, and are consumed as such, an example being glucose in fruit. **Extrinsic** sugars are not incorporated in cellular structures, as for example glucose and fructose in honey, and lactose in milk. A further distinction was made between milk and non-milk extrinsic sugars. The significance of this classification will be discussed in Section 17.10.

Polysaccharides can be divided into starch and non-starch polysaccharides, the latter term to replace the term 'dietary fibre' or 'non-digestible carbohydrate' (carbohydrate resistant to hydrolysis by human digestive enzymes), on the grounds that the term 'dietary fibre' lacks precise scientific definition, and that a greater quantity of starch than non-starch polysaccharide escapes hydrolysis by human digestive enzymes in some foods.

Starch is the major dietary polysaccharide. The main sources are cereals, potatoes, yams and plantains. Starch is found in granules in plant cells and in the raw state these granules are crystalline and insoluble, and the starch unavailable to digestive enzymes. Heat and moisture (cooking) render the granules soluble and the starch accessible to digestive enzymes such as pancreatic amylase.

There are two types of polysaccharide chains in starch, amylose and amylopectin, both of which are composed entirely of glucose units. **Amylose** is an unbranched chain of glucose units linked together by α1–4 glycosidic bonds. **Amylopectin** is branched, the branches formed by α1–6 glycosidic linkages as shown in Figure 17.8. Most starches contain 65–85% amylopectin and 15–35% amylose.

Glycogen, the animal equivalent of starch, is also composed of glucose units and its structure resembles that of amylopectin, but it is more highly branched. Although glycogen is an important storage fuel in muscle and liver in the living animal, very little remains in animal tissues by the time they are consumed, and therefore makes no significant contribution to dietary carbohydrate. Oysters, 6% of the wet weight of which consists of glycogen, may be the only significant dietary source of glycogen.

Non-starch polysaccharides form the major component of the plant cell wall. **Lignin** is also a cell wall component normally included in the term 'dietary fibre', but it is chemically different from plant wall polysaccharides and its contribution to the diet very small. The major types of non-starch polysaccharides (NSP) and their main sources are shown in Table 17.1. NSP can be classified in many ways, but a useful classification, from the nutritional point of view, is that based on solubility, because of the different implications on health of soluble and insoluble NSP.

Cellulose is insoluble, whereas **non-cellulose NSP** are soluble to varying degrees. Cereals, such as wheat, rice and maize, especially where the whole grain is used, are a rich source of mostly insoluble NSP. Wheat bran contains 36% NSP, whereas white bread only 1.6%. Vegetables have a higher water content than cereals and their NSP content is therefore lower. They offer a mixture of soluble and insoluble NSP. Potatoes, for example, contain 1.2% NSP, and the values for carrots and green beans are 2.4% and 3.1%, respectively. Fruit has, on the whole, more soluble than insoluble NSP, with apples containing 1.7% and oranges 2.1% NSP.

Cellulose is the most abundant natural organic compound. It is composed of glucose units joined by β1–4 glycosidic bonds to form long unbranched polymers, with up to 10 000 glucose units. Rich sources of this insoluble NSP are vegetables and pulses such as peas and beans. The non-cellulose family of NSP are mostly soluble polymers.

Pectins are branched polymers containing a variety of monomers, such as galactose, galacturonic acid, rhamnose and arabinose. Pectin is extracted commercially and used extensively in the food industry, for example as a gelling agent in jams.

Glucans are branched polymers of glucose, containing β1–4 and β1–3 glycosidic links. They are soluble and they are taken in the diet mainly in oats and barley.

Gums are viscous polymers consisting of glucose, arabinose, mannose, rhamnose, and their glucuronic acids. They are also extracted commercially and used in the food industry as emulsifiers, stabilizers and thickening agents.

Partly soluble NSP, such as **arabinoglycans** and **arabinoxylans** containing arabinose, galactose, glucose and xylose, are found mainly in wheat and rye.

Table 17.1 **Major types of dietary non-starch polysaccharides sources**

Class	*Monomers*	*Sources*
1. Cellulose, insoluble	Glucose	Vegetables, peas, beans
2. Non-cellulose		
(a) Soluble		
Pectins	Galacturonic acid	Fruit and vegetables
Gums	Galactose, arabinose	Plant gums (additives)
Glucans	Glucose	Oats, barley, rye
(b) Partly soluble		
Arabinogalactans/ arabinoxylans	Arabinose, galactose Glucose, xylose	Wheat, rye

17.10 Dietary carbohydrate and disease

Role of sugars

The classification of sugars into intrinsic and extrinsic, and the further subdivision into milk- and non-milk extrinsic sugars, was considered useful, because the consumption of these different classes of carbohydrate has different implications on health. There is no evidence that intrinsic sugars, or extrinsic sugars found in milk and some milk products, have adverse effects on health. There is also no evidence suggesting that non-milk extrinsic sugars are involved in the pathogenesis of cardiovascular disease, diabetes mellitus or behavioural abnormalities, up to a level of intake providing 30% of the total dietary energy. The consumption of sucrose above this level, however, may lead to increases in blood cholesterol, glucose and insulin concentrations. It should be remembered that even at levels which are not associated with metabolic problems, sucrose provides no essential nutrients to the diet, hence it is often referred to as a source of 'empty calories'. If high levels of sucrose are consumed, the intake of some micronutrients may be compromised, as the rest of the diet may not provide sufficient amounts of these nutrients to ensure optimum health.

Epidemiological evidence does, however, point to non-milk extrinsic sugars, and especially sucrose, as the major dietary factor responsible for the development of dental caries. Individuals in societies consuming less than 20 kg per person per year, which corresponds to 60 g/person/day, or 10% of the total dietary energy, rarely develop dental caries.

In 1987, the consumption of sucrose in the UK was 104 g/person/day, providing 14% of total food energy, and in 1989 the corresponding figure fell slightly to 95 g/person/day. As dental caries is a fairly common problem in the UK, the Department of Health recommended that the consumption of non-milk extrinsic sugars should not exceed 60 g/person/day, or 10% of the total dietary energy intake. It must be noted, however, that the frequency of sucrose consumption, rather than the total intake, is the major factor in the development of dental caries. On the other hand, as large amounts of sucrose are unlikely to be eaten in one single dose during the day, the total amount of sucrose consumed will probably give a reasonable idea of the risk.

Role of non-starch polysaccharides.

The intake of NSP in the UK is estimated to be, on average, 11–13 g/person/day (range 5–26 g/person/day). Half of this intake is supplied by vegetables, and about 40% comes from cereals, and these sources provide roughly equal mixtures of soluble and insoluble NSP.

There has been considerable interest for the last twenty or thirty years in the importance of NSP in the diet. Foods rich in NSP are considered beneficial to health for a number of reasons. They are generally lower in energy than refined foods and produce a feeling of satiety at a lower energy intake than fat or available carbohydrate. Low intakes of NSP have been related to the prevalence of constipation, appendicitis, diverticular disease and cancer of the colon. Diets low in NSP tend to be high in fat or sugar or both, and it is difficult to ascribe precise and specific effects on health to the presence of NSP rather than the relative absence of fat or sugar, and it is not possible to identify NSP as a factor in the development of most of these diseases.

One effect which is well documented is that on large bowel function, especially the weight of stools (increased) and the transit time in the intestine (decreased). Insoluble forms of NSP, such as those found in cereals, are more effective than soluble forms, such as those found in fruit and vegetables. Epidemiological studies have, in fact, shown a relationship between low stool weight, long transit time and constipation, and increased risk of bowel cancer, diverticular disease and gallstone formation.

Dietary NSP can lead to a decrease in blood glucose, but the value of this finding for diabetic patients is doubtful, because the levels of NSP needed to achieve a glucose lowering effect are well outside the range for most UK diets.

There are some reports of an inverse relationship between dietary NSP and mortality from coronary heart disease. Sources of soluble NSP, such as oats and beans, and fruits containing pectin, are reported to have a cholesterol-lowering effect, especially in the LDL fraction. There seems to be no effect on blood VLDL. There is no evidence, however, directly pointing to NSP as a major determinant of blood lipid levels. It may be that diets high or low in NSP are indicative of other dietary habits and lifestyles which may be more directly related to levels of blood lipids and risk of coronary heart disease.

Some adverse effects of high levels of dietary NSP should also be considered. The absorption of some minerals, such as iron, copper and calcium, may be compromised by the presence of NSP in the intestine, as these polysaccharides bind cations and render them unavailable. Most members of the population would not seem to be at risk, but some groups, such as the elderly, whose intake of these minerals is known to be low, may actually become deficient in these nutrients.

Also, a diet high in NSP is bulky and relatively dilute in energy and nutrients, and children may find it physically difficult to consume the large quantities of such a diet which would be required to provide them with an adequate quantity of essential nutrients and energy necessary for their needs during growth.

It can be seen that recommendations on the level of intake of NSP are not easy to make, owing to the lack of reliable and specific information on the levels which may have beneficial or detrimental effects on health.

The UK Department of Health proposed in 1991 that diets should contain an average NSP of 17 g/person/day, and no more than 32 g/person/day, as there was no evidence for any beneficial effect beyond this level, and that children less than 2 years of age should take less than the adult quantities, so that they would not replace foods rich in energy and nutrients with carbohydrate which is not available for metabolism.

Chapter 18

Nutrition: obesity and starvation

18.1 The incidence of obesity

It is sad to reflect that, in the modern world, where millions of people are suffering from serious malnutrition, many in the affluent societies are suffering from obesity. Affluence is an important cause of the condition, since it is clearly established that it correlates closely with the incidence of obesity. An increased availability of palatable food and a more sedentary lifestyle lay the foundations for an obese society.

The degree of obesity is commonly assessed by means of the body mass index (BMI) as described in Chapter 14:

$$BMI = weight (kg) / height (m)^2$$

Values of BMI between 20–25 are considered normal or desirable, the range 25–30 denotes 'overweight' or 'obesity grade I', 30–35 indicates overt obesity (grade II), and values exceeding 35 indicate gross obesity (grade III).

The problem is widespread. In Germany, one-third of the adult population has been classified either as overweight (obesity grade I) or obese (grade II and above), and the incidence is sex dependent: 42% of women , but only 19% of men. In the USA the figures are closer, 42% of women and 36% of men are overweight or obese. The contribution of ethnic origin is difficult to assess. Eastern Europeans are heavier than western Europeans, and black women in the USA are heavier than white women, but white men are heavier than black men.

A study of 5000 men and 5000 women in the UK, in 1980, showed that 34% of men and 24% of women were overweight (obesity grade I), and that 6% of men and 8% of women were obese (grade II). A similar study in 1987 showed that the incidence of obesity had increased in the intervening seven years, with 45% of men and 36% of women in the 'obesity grade I' band and 8% of men and 12% of women in the 'obesity grade II' category. Interestingly, the average energy intakes recorded by the population were 10.3 MJ/day for men and 7.0 MJ/day for women, both of which are slightly lower than the estimated average requirements (EAR) for energy for the population. The EAR has been set at 10.6 MJ/day for men, and 8.1 MJ/day for women.

In the more affluent societies, the incidence of obesity varies inversely with socioeconomic class. In the UK, 21% of socioeconomic classes A and B could be classified as overweight or obese, 19% of class C, but the figure is as high as 52% in the lower socioeconomic classes D and E.

The detailed 'Framingham study', carried out in the USA in order to attempt to establish factors responsible for coronary heart disease, noted that a strong correlation existed between obese husbands and wives. Either the obese find obesity an attractive feature when choosing their partners or, more likely, common unwise eating patterns and lifestyle, established after marriage, lead to obesity in both partners concurrently. A Finnish study has shown that the risk factors for obesity were, a low level of education, a sedentary lifestyle, heavy alcohol consumption, marriage and giving up smoking. Obesity also increases with age, and this is well documented by life insurance companies in the USA and the UK.

18.2 Classification and measurement of obesity

Obesity is the result of a disturbed relationship between three components of energy: the input as food, the expenditure of energy, such as in body maintenance and in physical activity, and energy storage, primarily as fat. The problem of obesity, therefore, arises from an imbalance of energy intake in relation to energy expenditure. Obesity can be broadly classified as **primary** (or **simple**), or **secondary obesity**. Primary or simple obesity is that which is not associated with a demonstrable clinical condition. This, of course, does not mean that the causes of simple obesity are themselves simple. Secondary obesity is that which is associated with an identifiable medical disorder, such as a congenital syndrome, a hypothalamic or other endocrinological disorder, or drug therapy. The majority of cases of obesity belong to the primary, or simple, category. Obesity is not a straightforward increase in weight as occurs in growth in a child, or body building in an athlete, but can be described as an increase in weight, accompanied by a change in body composition, characterized by an increase in body energy stores, mainly fat. Although a certain amount of reserved energy in the form of fat is desirable in the body in order to survive times of food scarcity without impairment of function, and also in order to survive debilitating illness which may not allow access to food, or lead to an inability to eat, an excessive accumulation of body fat is undesirable, because it constitutes a health risk.

Obesity is said to exist when body fat content is greater than 28% of total body weight in men and 32%

Table 18.1 **Body composition of a normal weight man and a normal weight woman compared to an obese man**

	Normal man		*Normal woman*		*Obese man*	
	wt (kg)	(%)	wt (kg)	(%)	wt (kg)	(%)
Body weight	70		60		100	
Fat	10.5	14.0	15.0	25.0	40.0	40.0
Protein	11.0	16.0	9.0	15.0	12.0	12.0
Carbohydrate	1.0	1.5	0.9	1.5	1.0	1.0
Water	44.0	63.5	32.1	53.5	44.0	44.0
Minerals	3.5	5.0	3.0	5.0	3.0	3.0
Total stored energy	636 MJ		790 MJ		1880 MJ	

of total body weight in women. For most people this is somewhere in the range for obesity grade I as classified by means of the BMI. It must be remembered that BMI is a more arbitrary and inaccurate way of assessment of obesity, but of greater practical value and use than the measurement of body composition. A short and muscular man, for example, may be classified as overweight or obese by means of the BMI, but as normal or lean if the percentage of body fat were to be measured.

Body composition

The body is made up of water, fat, protein, carbohydrate and minerals. It is useful, especially when considering obesity and starvation, to divide the body into **fat** and **fat-free mass**, or **adipose tissue** and **lean body mass**. It must be noted that fat mass is not exactly equivalent to adipose tissue mass, as the latter contains connective tissue components in addition to triacylglycerols and, similarly, lean body mass is not exactly equivalent to fat-free mass, although the two terms are often used indiscriminately.

Table 18.1 shows examples of figures of body composition of normal weight men and women compared to an obese man.

Measurement of body composition is not a simple operation. For practical purposes, not all components are measured, and reasonable estimates can be obtained from measurements of either fat or fat-free mass, although some methods, such as densitometry and measurement of electrical conductivity, can give a measure of both of those parameters. The methods commonly used for the measurement of fat and fat-free mass are listed in Table 18.2

Table 18.2 **Methods for estimating body composition**

Body fat
Skin fold thickness at standard subcutaneous sites and arm muscle circumference
Body scanning (MRI or CT)
Photon absorptiometry
Dual X-ray absorptiometry

Lean body mass (or fat-free mass)
Total body water (deuterated or tritiated water dilution)
Total body potassium (^{40}K counting)
Body electrical conductivity
Urinary creatinine excretion
Urinary methyl histidine excretion

Both fat and fat-free mass
Body density (weighing underwater)
Anthropometry: body mass index, waist–hip ratio, waist–thigh ratio, etc. (information is not precise but gives a general idea)

18.3 Obesity as a health risk

As obesity constitutes a health risk, it is not surprising that a great deal of our information on desirable body weight ranges, and the classification and measurement of obesity, comes from life insurance company figures. Body weight is a fairly meaningless parameter, unless body height is taken into account, and various formulae have been devised which relate height to weight.

The desirable range of body weight, that is, the range associated with the longest life expectancy, corresponds to a BMI of 20–25.

The value above which a person is considered obese, is arbitrary, but it is related to figures for morbidity and mortality. Table 18.3 shows the classification of obesity based on values of BMI.

Table 18.4 shows the relative risk of mortality from various diseases in obese people, compared to people of normal weight. Obesity has a significant effect on both mortality and morbidity. The main cause of premature death in the obese is heart disease. This is because obesity is associated with hypertension, hyperlipidaemia, hyperinsulinaemia, non-insulin dependent diabetes mellitus (NIDDM), coronary thrombosis and congestive heart failure, all of which are independent risk factors for the development of ischaemic heart disease. NIDDM is a major cause of death in the obese, but not in people of normal weight. It is thought to be the result of a reduced sensitivity to insulin coupled with a failure of the pancreas to support compensatory increases in insulin secretion (see Chapter 41).

Digestive complaints are common in obesity. The concentration of cholesterol in the bile of obese subjects is high, and this predisposes to the formation of gall stones. Obesity is also associated with an increased

Table 18.3 **The classification of obesity**

	Body mass index	*Excess weight* (kg)	% *desirable weight*
Underweight	<20		<100
Normal	20–25	<5	100–110
Overweight (obesity grade I)	25–30	5–15	110–120
Obese (obesity grade II)	30–35	15–25	120–160
Grossly obese (obesity grade III)	>35	>25	>160

Table 18.4 **Mortality from various causes of premature death in obese compared to non-obese subjects**

Cause of death	%
Cardiovascular disease	160
Diabetes mellitus	400
Cirrhosis of the liver	220
Gallstones	220
Appendicitis	210
All cancers	100
Suicide	75

incidence of cancer of the colon, prostate and rectum in men, and breast, cervix, endometrium and ovary in women. This may relate to the fact that the high levels of the enzyme aromatase found in excessive adipose tissue mass may lead to increased conversion of androgens to oestrogens which are known to be tumour promoters (see Chapters 28 and 37).

Respiratory problems are also common in obesity and the risks from anaesthesia and surgery are greater than in normal weight patients.

The obese are also more prone to osteoarthritis, especially of the weight-bearing joints, such as the hips and knees.

In Western societies, the obese suffer social discrimination and often find themselves the objects of ridicule. Personal relationships may be problematic, and depression and neurosis are commonly encountered. As seen from Table 18.4, however, the incidence of suicide is lower in the obese than in persons of normal weight.

18.4 The causes of obesity

An obese man, of average height, with a body weight of 100 kg, carries an extra 30 kg of weight. Of this, 22.5 kg (75%), would be fat, and the rest fat-free tissue. The extra fat corresponds to an excessive energy intake of about 833 MJ. If this man has increased his body weight from 70 kg to 100 kg over 20 years, then the excess intake corresponds to 114 kJ per day (27 kcal). This is roughly equivalent to a teaspoonful of sugar, or half an apple, and it is unlikely that such a small increase in energy intake every day for 20 years would lead to the accumulation of 22.5 kg of excess body fat. Small increases in energy intake are usually buffered by the body. It is much more likely that obesity was the result of a greatly excessive energy intake over a shorter time, followed by the maintenance of a new equilibrium of body weight, set at a higher level. As has been mentioned in Chapter 17, the obese are able to maintain their body weight and do not increase it perpetually.

How and why does obesity develop? The explanation of the aetiology of secondary obesity is simpler than that of primary, or simple, obesity. The Prader–Willi syndrome, for example, which is caused by a gene defect on chromosome 15, is characterized by mental retardation and also by obesity. Obesity develops through compulsive, uncontrolled eating. No gene product has, however, been identified that would cause this eating disorder. The condition is extremely rare, and defective gene products cannot explain the high incidence of obesity seen in the general population.

Endocrinological abnormalities can lead to obesity. The polycystic ovary syndrome, is almost always accompanied by obesity, as a result of high circulating levels of insulin, oestrogen and testosterone. Use of the contraceptive pill may result in weight gain, due to the oestrogen content. Cushing's disease is associated with central obesity (obese trunk and thin limbs), caused by the excessive secretion of cortisol through adrenal hyperfunction (see Chapter 28). In a similar manner, the use of synthetic glucocorticoids as anti-inflammatory agents, often results in weight gain. NIDDM is also often characterized by obesity, but in this case obesity seems to be the cause rather than the result of diabetes.

The disorders described above are rare causes of obesity, and the majority of cases in the population are not associated with an identifiable medical condition and are therefore classified as simple or primary obesity. There does not seem to be one single cause of primary obesity. Clearly, energy intake must be, or must have been, in excess of energy expenditure, but it is not clear which component of the energy equation is more at fault in producing the energy imbalance, and why.

Genetic and environmental factors

The contribution of familial factors to the development of obesity is well established, but the importance of genetic factors is by no means resolved. Extensive twin, family and adoption studies have not been able to separate the effects of genes from those of the environment. If members of a family live together and lead a similar lifestyle, it is difficult to decide to what extent the common genes, or the common eating patterns, are responsible for the obesity observed in the members of that family.

Is energy intake genetically determined? Studies on twins have shown that there is a greater similarity between the dietary intakes of monozygotic rather than dizygotic twins. They also noted that monozygotic twins tend to eat together a lot more regularly than do dizygotic twins, and therefore the greater similarity could be explained by the shared environment. When total body fat is taken into account, the regional fat distribution shows a strong familial resemblance, with some individuals storing more fat on the trunk and others on the lower body.

It seems that the occurrence of obesity is phenotypically complex and cannot be reduced to simple Mendelian inheritance patterns. Several genes have the potential to cause obesity in humans, and recently attention has centred on the *ob* gene, mutations in which result in profound obesity in mice. The gene product has recently been characterized and named **leptin**. The leptin gene is stongly expressed in the adipose tissue and the protein is released into the circulation where it may well act as a hormone. It is possible that some grossly obese humans will prove to have a failure in the production of or response to leptin. Irrespective of this possible genetic component, it is likely that the development of the condition of obesity depends strongly on behavioural and lifestyle characteristics.

Energy intake

A number of questions need to be answered here: do the obese eat more than the lean during the development of obesity, and if so, do they have a disorder in the mechanisms which regulate food intake? Certain families and certain societies tend to eat more than others. Often, the abundance of available food, and its subsequent consumption, are seen as desirable signs of affluence. Hospitality and expression of friendliness or care through offering food are common to most societies. Social events and celebrations usually involve some degree of overeating by most people. Some studies on the energy intakes of the obese have suggested that the mechanisms which signal satiety may be defective. In one such study, obese and lean subjects were fed liquid meals without being able to see how much they consumed. The obese group consistently consumed more of the liquid than the control group, and so had a higher energy intake than the lean, without being aware of it.

Energy expenditure

Some individuals become obese despite an energy intake apparently similar to that of individuals who maintain their normal body weight. Is their energy expenditure much lower? Three components of energy expenditure have been studied: basal metabolic rate (BMR), physical activity and diet-induced thermogenesis. The BMR of the obese is not lower than that of people of normal weight, but is actually higher, because in the development of obesity lean body mass increases as well as fat, and BMR is known to depend on the amount of lean body mass.

Physical activity may be important, but although the obese often lead sedentary lives, it has not been shown, in affluent societies, that they are more inactive than their lean counterparts. Sometimes, the level of physical activity is lower as a result of obesity, because the effort involved in carrying the extra weight during exercise, is considerable.

A lot of attention has been focused, in the past 20 years, on diet-induced thermogenesis and its contribution to dissipating excessive energy intake as heat, without increasing the body's fat stores. Overfeeding studies have shown that individuals vary in the extent to which they store or waste the same excess of energy intake, but the contribution of this to the development of obesity has not been established. Studies on transgenic mice, with a primary deficiency of brown adipose tissue, have shown that obesity develops in the absence of hyperphagia, and support a critical role for brown adipose tissue in the energy balance of mice. Human studies have, however, failed to show any significant differences in diet-induced thermogenesis and brown adipose tissue activity, between obese and control groups.

18.5 The treatment of obesity

Dietary restriction

Reduction in energy intake, or increase in energy expenditure, or both, are the obvious forms of treatment of obesity. The success of the treatment depends on the individual case. The severity of obesity and any accompanying psychological and social problems must be taken into account when designing diets and advising changes in lifestyle. Dietary advice on healthy eating is important, so that long-term weight maintenance can be achieved once the desired weight has been reached. The rate of weight loss is not constant on a consistently reduced energy intake. The initial weight loss is mainly accounted for by liver and muscle glycogen, which are stored with three times their mass of water. The initial weight loss seems rapid, but it consists mainly of body water. At a later stage of energy restriction, more body fat is lost, and although the rate of weight loss is less impressive, the loss of body fat is considerable. Frequently, this is not explained to dieters, with the result that they lose faith before they lose weight and return in disappointment to their previous eating practices. This results in rapid regain of the lost weight. This can be explained on sound physiological grounds. The restricted energy intake leads to an adaptive decrease in BMR, to a degree higher than expected from the reduction in lean body mass. The sudden increase in food intake is not immediately matched by an increase in the BMR, and re-feeding causes a generalized increase in substrate utilization, with adaptive hyperlipogenesis.

For the same reason, even dieters who achieve their target weight find it difficult to maintain it on the energy intake they had before attempting to lose weight. Studies on fasting and re-feeding in rats have shown that animals deprived of food for 4 days could only maintain their new, reduced weight on 60% of their original energy intake. It is, therefore, not surprising, that about 90% of slimmers regain their lost weight. This is a sad fact for them, but it keeps the 'slimming industry' in business, as their clients return, and go through the same cycle of weight loss and weight gain. The majority of cases belong to the band with a BMI of 25–30, which is

Table 18.5 **Effect of total starvation and two isocaloric, restricted energy diets on loss of body weight**

Diet	*Duration* (days)	*Intake* (MJ/day)	*Weight loss* (g/day)	*Water loss* (g/day)	*Fat loss* (g/day)	*Protein loss* (g/day)	*Ketone production* (g/day)
Starvation	10	0	750	457	243	50	9
Low CHO diet (70% fat, 5% CHO)	10	3.36	483	300	165	18	3
Mixed diet (30% fat, 45% CHO)	10	3.36	275	100	165	10	0

CHO, carbohydrate.

not seriously associated with increased mortality, except at the higher end.

'Slimming diets'

Many types of diet are on offer to the obese. Most of these are not based on scientific principles, and are at best harmless, often ineffective, but sometimes dangerous. Diets with names such as the 'low carbohydrate diet' (which, by necessity, has to be a high fat diet, although not advertised as such), the 'drinking man's diet', the 'hip and thigh diet', or diets based on one food only, such as the 'grapefruit diet', seem to imply a magic formula for weight loss and can mislead and disappoint. Any diet, sensible or bizarre, will of course lead to weight loss, if it provides a restriction of energy intake, and all the diets mentioned above will work, but claims that one particular unbalanced diet is better than another are unsubstantiated.

An example of the effect of three different diets is shown in Table 18.5. In this, total starvation is compared to a low carbohydrate diet and a mixed diet of the same energy content (3.3 MJ/day). Starvation causes a large total weight loss, but also causes substantial loss of body protein and the increased production of ketone bodies. The low carbohydrate diet appears to cause a greater weight loss than the mixed diet, but the difference is almost entirely due to the loss of body water. The quantity of fat loss is identical to that achieved by the mixed diet. Almost twice as much body protein is lost through the low carbohydrate diet than the mixed diet. Although it would appear to be the least effective in terms of weight loss, the mixed diet is the best regimen, because it offers the same rate of fat loss as the low carbohydrate diet, with the advantage of both minimum loss of body protein and minimum ketone body formation.

The undesirability of body protein loss in starvation has been recognized by sections of the food industry dealing with foods for 'slimmers', and regimens known as 'protein supplemented modified fasts' (PSMF), have been designed. They usually provide 1.7–2.9 MJ/day (400–700 kcal) from protein only, in order to avoid the loss of lean body mass which accompanies the loss of fat. They usually consist of liquid mixtures of hydrolysed protein (usually collagen, which has a biological value of zero). Such diets are harmless if followed for a short time, but 60 deaths have been recorded so far, associated with their long-term use. Patients developed refractory ventricular arrhythmias, myocardial protein deficiency and myocarditis, possibly due to hypokalaemia and hypomagnesaemia, since electrolytes lost during protein hydrolysis are not usually replaced in these mixtures.

Surgery

Surgery has been used in extreme cases of obesity. Bowel resection and lipectomy are sometimes used and liposuction is increasing in popularity in cosmetic surgery, even in cases of moderate obesity. Wiring of jaws to prevent eating has been tried in severe cases, giving the impression that the surgeon's role is to punish the sinner, rather than to treat the patient. These practices do not provide a long-term solution to the problem of obesity.

Drug therapy

The use of drugs in treating obesity has met with varying success. Most drugs on offer have undesirable side-effects.

Fenfluramine, a serotonin (5-HT) reuptake inhibitor (see Chapter 30), has been used to produce a feeling of satiety, and therefore to act as an appetite suppressant. It may also cause nausea and diarrhoea while in use, and occasionally depression upon withdrawal. Amphetamines increase the BMR, but are addictive and cause psychological problems.

Thyroxine, used because it also increases the BMR, leads to body protein loss and bone demineralization.

Uncouplers of oxidative phosphorylation (see Chapter 5), such as dinitrophenol (DNP), have been tried in the past, but have been banned because of their toxic effects.

18.6 Starvation: metabolic adaptation

Starvation describes the situation of total food deprivation. The words **starvation** and **fasting** are often used indiscriminately, but the latter should be used to a describe voluntary abstinence from food. The word 'starvation' is also loosely used, to describe severe, rather than total, food deprivation, a situation better described as **undernutrition**. Starvation, in a strict biochemical sense, begins immediately after the absorption of a meal is complete, and is certainly a marked feature of metabolism during sleep.

Starvation and undernutrition are nowadays encountered mainly in south-east Asia, Africa, and parts of South America, but it must be remembered that until relatively recently, when sophisticated systems of food production and distribution were introduced, humans in all parts of the world were in danger of undernutrition.

The central problem in metabolism is that the demand for energy is continuous, but the supply intermittent, and it is not surprising that elaborate biochemical mechanisms have evolved, which can ensure survival during short or long times of food deprivation.

Utilization of body stores

In starvation, energy has to be derived from the body's own stores. Table 18.1 shows the available energy contents in a normal weight and in an obese adult man. It can be seen that the total energy immediately available from the plasma is very small, and would only supply the basal metabolic requirements of about 7.5 MJ/day for about 80 min. Liver glycogen is the main provider of energy during short-term starvation, whereas adipose tissue fat, and muscle protein, provide the main sources during longer periods of starvation. Humans, as long as they are provided with water, can survive from 1 to 3 months on these stores, and some obese individuals much longer. The obese person described in Table 18.1 seems to have three times as much stored energy as the person of normal weight, and in theory should be able to survive three times as long.

It must be borne in mind, however, that as only about 30–50% of body protein can be lost without serious impairment of function, the obese person described above will not survive for as long as the energy stores would allow in theory. Death from loss of lean body tissue (failure of respiratory muscles), exacerbated by vitamin and mineral deficiencies and infection, will occur before the energy stores are exhausted.

Starvation can be divided into three phases, characterized by distinct metabolic patterns: the **intraprandial** phase, which occurs between meals in a normal day, the **post-absorptive** phase, which covers the 12 h overnight fast and can extent to 24 h, and the phase of **prolonged starvation**, which lasts longer than 24 h and can extend into several days or weeks. The changes in metabolism from one phase into the next are not abrupt but gradual.

Short-term starvation (intraprandial and post-absorptive phases)

The first phase of starvation begins four or five hours after a meal. The brain, the erythrocytes, the bone marrow, the renal medulla and the peripheral nerves have to be supplied with glucose for their energy needs (Figure 18.1), and this glucose must be derived from body components. In this phase, the main source of blood glucose is liver glycogen. Not all tissues have an absolute requirement for glucose, however, and tissues such as muscle can readily use fatty acids released from adipose tissue. The utilization of fatty acids becomes more important as starvation proceeds, but fatty acids are metabolized by a number of tissues during the early phase.

Glycogenolysis in the liver (Figure 18.2) is mainly controlled by the circulating concentrations of glucose, insulin and glucagon, the ratio of the concentration of insulin to glucagon being particularly important. Glucagon stimulates cAMP formation, which activates the protein kinase essential for glycogen phosphorylase activation. The fall of blood glucose and insulin concentrations, and the rise of glucagon concentration that occur during starvation, are shown in Figure 18.3. At the same time there is a rise in the circulating

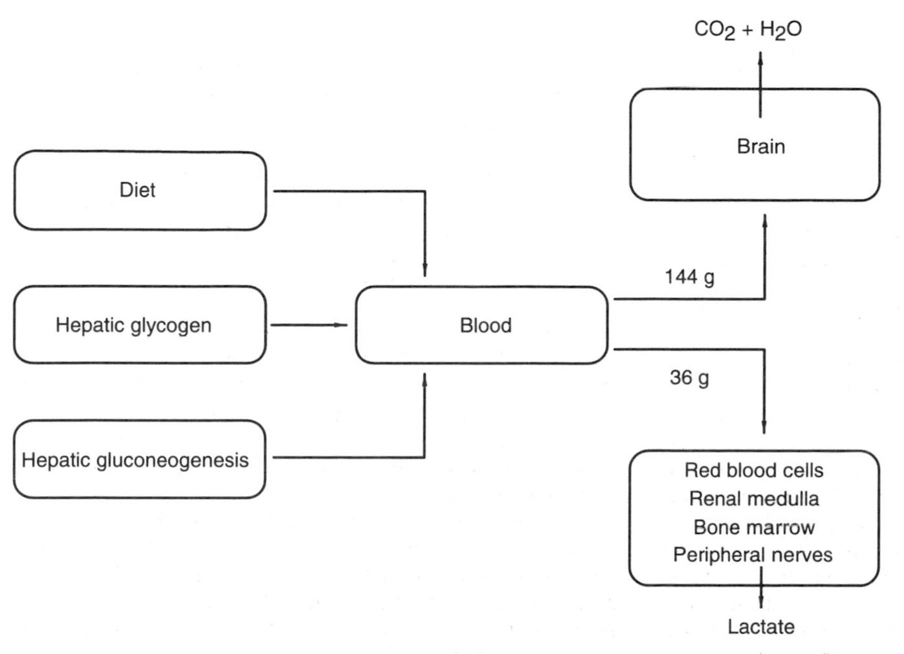

Figure 18.1 **The provision, during a 24 h period, of glucose for those tissues for which the monosaccharide is essential**

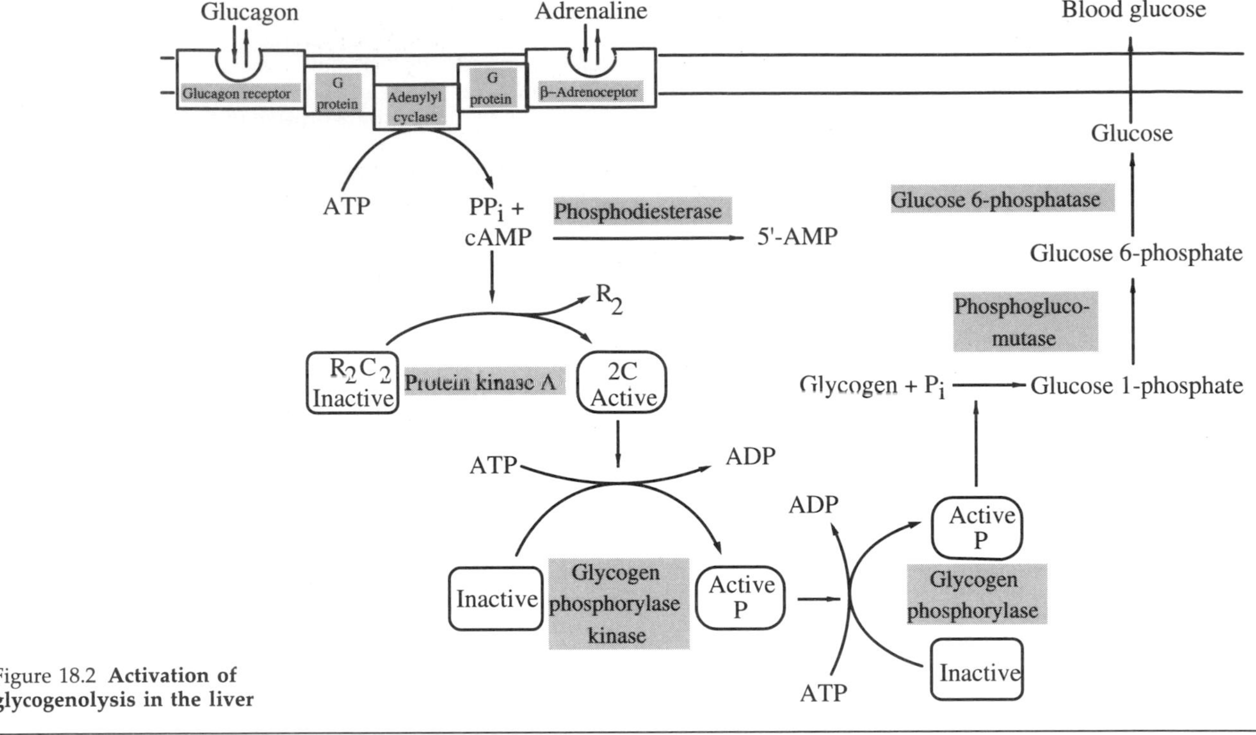

Figure 18.2 **Activation of glycogenolysis in the liver**

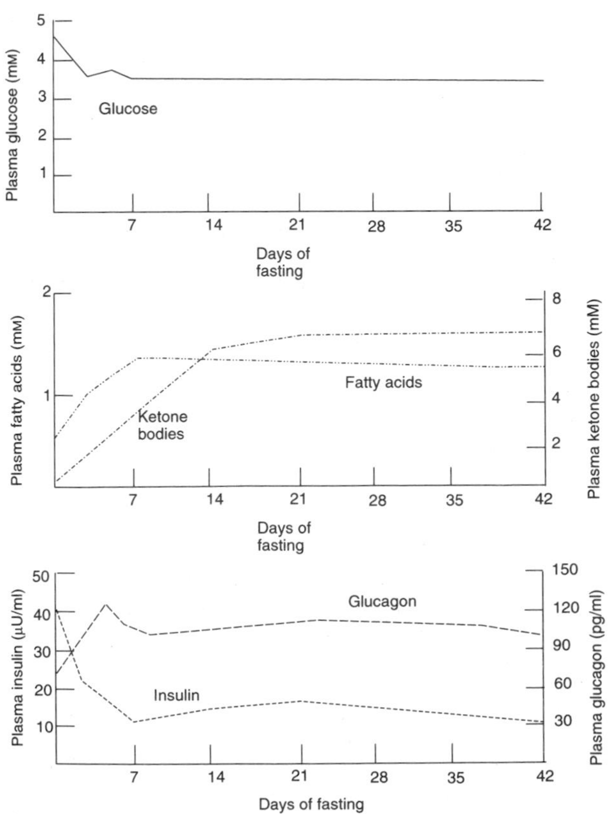

Figure 18.3 **Changes in circulating fuel and hormone concentrations in starvation**

concentrations of non-esterified fatty acids (NEFA) and ketone bodies.

Insulin also acts on the adipose tissue. The fall in the concentration of insulin and the increased release of noradrenaline lead to the release of NEFA into the bloodstream (see Chapter 28). As this phase of starvation continues, peripheral muscles and adipose tissue consume progressively less glucose, and switch over to fatty acids as their main energy source, so that after 8–10 h, more than half the muscles' need for energy is met by NEFA.

In the normal adult human, the liver glycogen is usually considered to form about 4–5% of the total liver weight, but occasionally it may reach 10% of this weight, shortly after a meal. This quantity of glycogen is capable of maintaining the blood glucose concentration at normal values for 12–16 h. As the glycogen store begins to be depleted, metabolic changes take place, which ensure a continuous supply of glucose to the brain and the other tissues which normally depend on glucose as a metabolic fuel. The body continues to synthesize glucose from lactic and pyruvic acids, but in addition, glycerol, released by the degradation of triacylglycerols, and glucogenic amino acids, released from muscle protein, become more important precursors of glucose. The formation of glucose from non-carbohydrate precursors is known as **gluconeogenesis** (see Chapter 6). The energy needed for glucose synthesis can be derived from the oxidation of fatty acids. Approximately 36 g of glucose per day can be synthesized from lactic and pyruvic acids produced by the tissues carrying out active glycolysis and about 16.5 g from glycerol.

The main source of substrates for gluconeogenesis is, however, tissue protein, in the form of amino acids

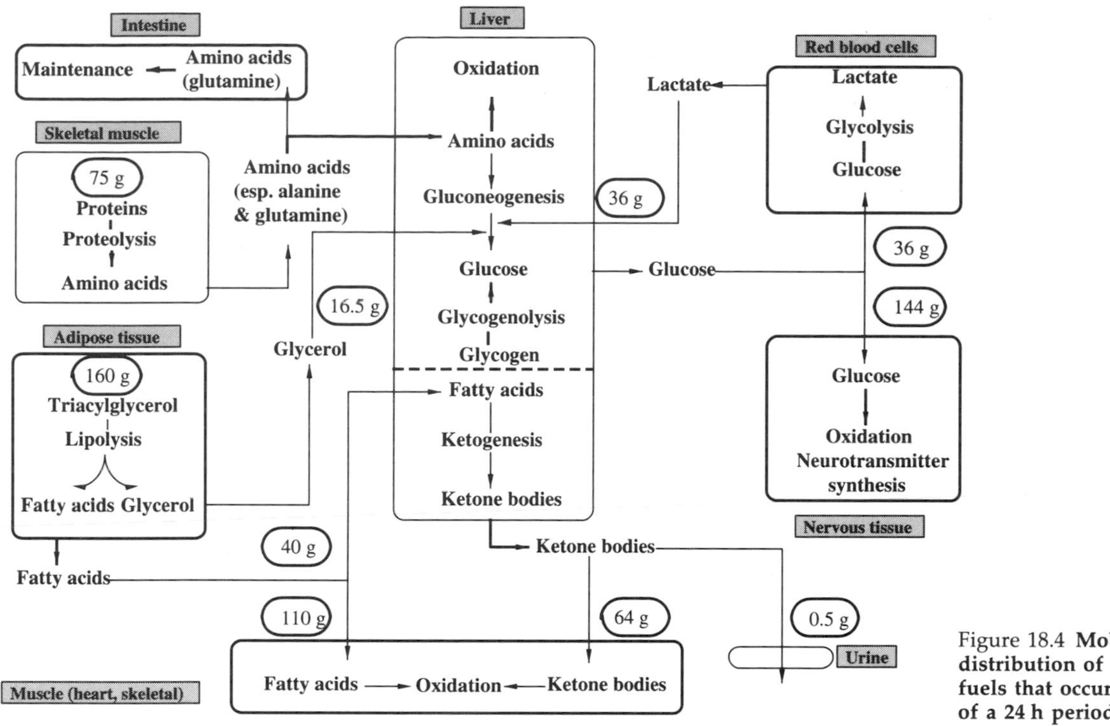

Figure 18.4 **Mobilization and distribution of metabolic fuels that occur in the course of a 24 h period of fasting**

released after proteolytic degradation. This is made possible by the low levels of insulin, which normally inhibits proteolysis (see Chapter 28). The major gluconeogenic amino acids coming from muscle are alanine and glutamine. They form 16% of the amino acids of muscle protein, but account for up to 46% of the amino acids released from the muscle during starvation. Alanine and glutamine must, therefore, be synthesized in the muscle. Branched chain amino acids are believed to be involved as donors of the required amino groups (see Chapter 11). These processes form branched-chain oxoacids, which are available for oxidative metabolism. Alanine is formed from pyruvate via transamination (see Figure 11.5). Another 2-oxoacid, 2-oxoglutarate, may receive an amino group to form glutamate, which in turn accepts an ammonium ion to form glutamine. This reaction is catalysed by glutamine synthase (see Chapter 11). Glucagon increases the uptake of alanine by the liver where it is converted to pyruvate. It is important that the pyruvate formed in this way is channelled to glucose and not to acetylCoA, because the latter process would result in the loss of glucose carbon, as the conversion of pyruvate to acetylCoA is an irreversible reaction. The conversion of pyruvate to oxaloacetate and hence to glucose, rather than to acetylCoA, is made possible by the oxidation of another metabolic fuel. The decrease in plasma insulin seen after an overnight fast, and the increased release of noradrenaline are sufficient to allow lipolysis to occur (see Chapter 28). NEFA are produced, and although they do not contribute to glucose production, they can be used by some tissues, such as skeletal muscle, as alternative sources of fuel, so sparing glucose. The oxidation of NEFA results in the production of acetylCoA. In the liver, acetylCoA has two regulatory effects, with respect to glucose production and utilization. It is an allosteric activator of pyruvate carboxylase which converts pyruvate to oxaloacetate in the gluconeogenic pathway, and it is an allosteric inhibitor of pyruvate dehydrogenase (PDH), which would convert pyruvate into acetylCoA and prevent it from being converted to glucose. The NADH produced during β-oxidation of fatty acids also inhibits PDH, and in this way drainage of carbohydrate carbon into an irretrievable form is prevented.

Figure 18.4 shows the pattern of fuel utilization during 24 h of fasting.

The metabolic fate of glutamine is more complex than that of alanine and is dealt with in Chapter 11. In essence, however, it may act as a substrate for gluconeogenesis in the liver or the kidneys and it may also be used in the intestine or by lymphocytes.

Prolonged starvation

The processes which take place in short-term starvation cannot go on indefinitely, because, although they provide efficiently for the body's energy requirements, they will soon deplete the body of a substantial proportion of its protein, and it is known that death ensues when 30–50% of body protein is lost. Adjustments to metabolism are made after 24–48 h, which conserve body protein. Blood glucose levels drop, from about 5 mM to about 3 mM after about 24 h, but are thereafter maintained. Conservation of body protein is accomplished by a reduction in glucose production, so that the liver glucose output falls from 150–250 g/ day after 1–3 days, to 40–50 g/day after 4–6 weeks of

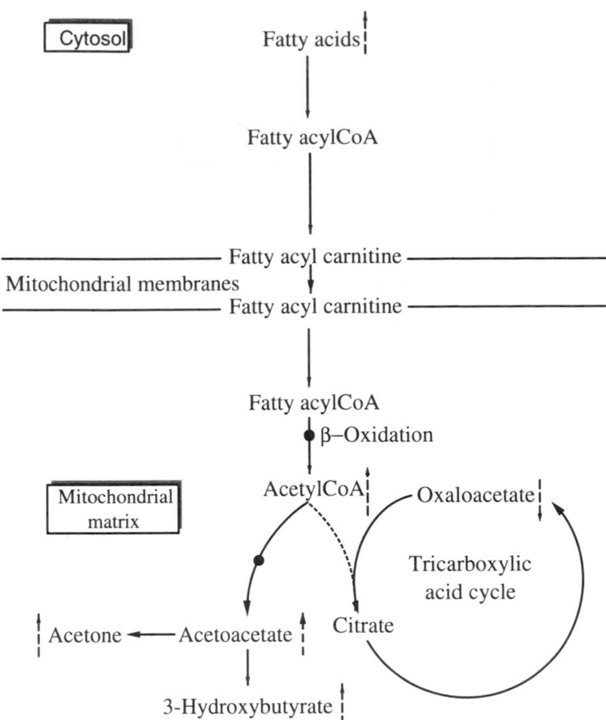

Figure 18.5 **Formation of 'ketone bodies' in the liver**

starvation. Also, the kidney becomes an important gluconeogenic tissue, supplying 50% of the total glucose production. The drop in glucose production is dependent on the fall in the output of muscle alanine and glutamine, which may be controlled by the rising concentration of **ketone bodies**, produced in the liver as a result of greatly increased fatty acid oxidation. The oxidation of fatty acids results in the production of large quantities of acetylCoA which cannot be oxidized by the citrate cycle. Entry into the citrate cycle is further prevented by the fact that oxaloacetate levels are low because oxaloacetate tends to be reduced to malate owing to the presence of large amounts of NADH produced in β-oxidation. In the liver, molecules of acetylCoA are condensed to produce ketone bodies. These are acetoacetate (3-oxobutyrate) and 3-hydroxybutyrate and acetone (Figure 18.5). The liver is not capable of utilizing the oxoacids, and they are released, in high concentrations, into the plasma. Their concentration can increase from 1–2 mM after 3–4 days' starvation, to 6–10 mM, after 10–14 days without food. They can be used as a source of energy by heart and skeletal muscle which oxidize them to carbon dioxide and water.

With prolonged starvation, the supply of glucose diminishes. As the supply of ketone bodies increases, the brain reduces its utilization of glucose and increases its oxidation of ketone bodies for its energy supply. Under these conditions, the brain can consume over 80% of the ketone bodies produced by the liver, although it should be noted that the brain still has a residual glucose requirement, about 40% of that in the early stages of starvation. This requirement is not only for provision of energy, but also for synthesis of neurotransmitters. The ketone bodies can therefore provide 50–60% of the brain's energy requirements, and the ability of the brain to switch over to ketone bodies as a source of energy must clearly have been of great importance for survival in serious conditions of food deprivation. The increased use of ketone bodies by the brain is a simple mass-action effect of the increased availability of the substrate.

Ketone bodies act on the B-cells of the pancreas to stimulate the release of insulin, seen between 7 and 21 days of starvation, which helps reduce the extent of mobilization of muscle protein. This is reflected in a decrease in nitrogen excretion, from around 12–15 g/day to 3–4 g/day, following 4–6 weeks of starvation. At the same time, the pattern of excretion of nitrogenous compounds is altered. Urea nitrogen diminishes markedly, suggesting that fewer amino aids undergo gluconeogenesis in the liver. In contrast, the need to increase the excretion of H^+, resulting from raised levels of ketone bodies in the blood, leads to an increased excretion of NH_4^+ ions, to preserve the acid-base balance. It should be noted that it is the switch from the acid generating process of urea formation in the liver to the NH_4^+ formation in the kidney that helps combat the acidosis. The formation of NH_4^+ also means that Na^+ is retained in the body, and extracellular volume can be maintained. Because of the decreased production of urea, there is very little need for the excretion of water, and the volume of urine may fall to 200 ml/day. This means that in a temperate climate a starving man need drink very little water, that produced by metabolism being approximately balanced by the loss of water in the urine, and by evaporation from the skin and lungs.

Apart from the increase in the ratio of glucagon to insulin concentrations, thyroxine (T_3) production is reduced, and this contributes to the decreased BMR and protein degradation observed in prolonged starvation. A decrease in sympathetic nervous activity is also observed, and this decreases the rate of lipolysis to a moderate level. If sympathetic activity was not reduced, the fall in insulin seen in prolonged starvation (beyond 21 days), would result in very high rates of lipolysis and a higher loss of energy stores. The metabolic changes in long-term starvation are shown in Figure 18.6

Starvation also leads to inadequate shivering- and non-shivering thermogenesis, which means that hypothermia and death from cold are more likely in starving than in fed individuals. Cardiovascular function is also affected. Blood pressure drops, especially when standing, and this adds to the impairment of body function, through inefficient blood flow. The wasting of cardiac muscle seen in starvation can lead to heart failure. Skeletal muscular function is also severely impaired. The wasting of respiratory muscle leads to breathing problems, and exacerbates the problem of hypoxia caused by inefficient blood flow. The body is more vulnerable to infections, and pneumonia is often the immediate cause of death in prolonged starvation. Given the inability of severely malnourished individuals to regulate their body temperature, survival times in periods of starvation are longer in warm climates.

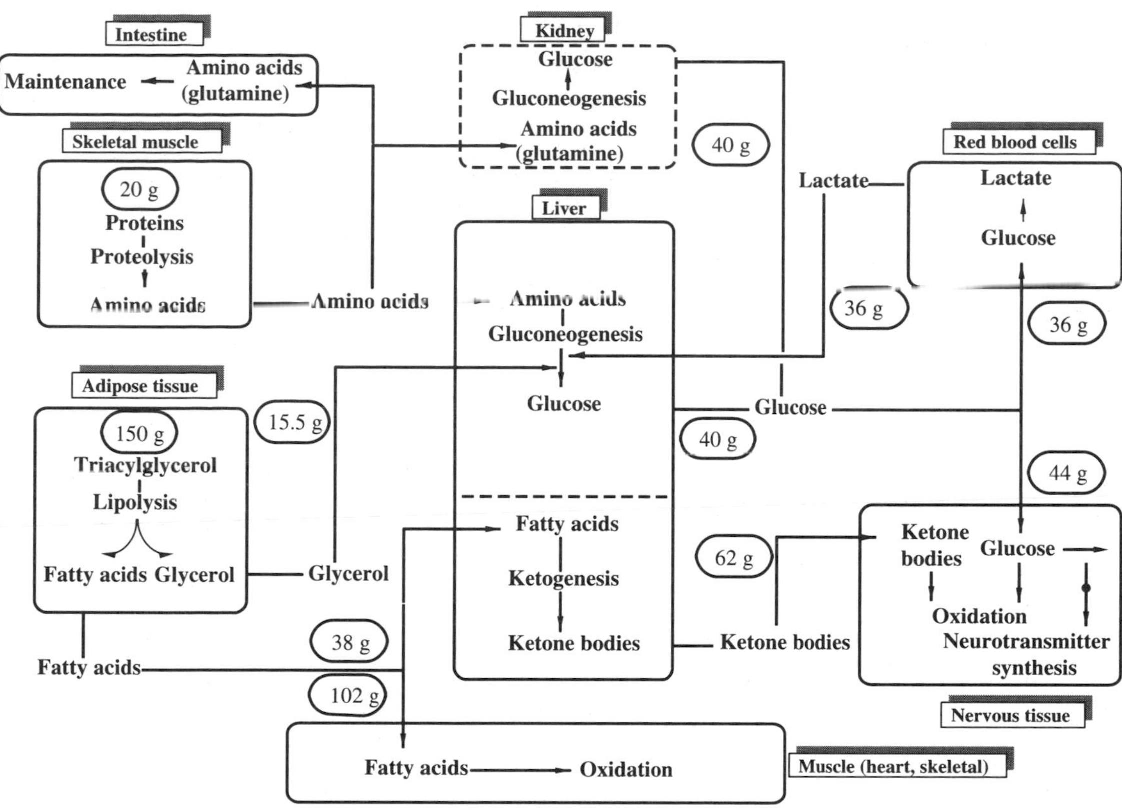

Figure 18.6 **Mobilization and distribution of metabolic fuels that occur after 4–6 weeks of starvation**

18.7 Clinical aspects of starvation: starvation as a result of a medical condition

Starvation is not always the result of unavailability or scarcity of food. Any medical condition which prevents consumption or utilization of available food will effectively lead to starvation, or severe undernutrition.

Malabsorption and food intolerance

Malabsorption syndromes (see Chapter 13), such as tropical sprue and coeliac disease, affect the absorption of many nutrients and metabolic fuels. Fat malabsorption, sometimes a result of pancreatitis or bile duct obstruction, will lead not only to a reduced energy intake, but also to essential fatty acid and fat-soluble vitamin deficiencies.

The intestinal mucosa can sometimes be damaged as a result of an intolerance, or allergy, to a particular food. Coeliac disease, for example, is an allergy to **gliadin**, a component of the wheat protein **gluten**. Ingestion of gluten leads to atrophy of the intestinal mucosa, with disappearance of the villi, and therefore a dramatic reduction in the absorptive surface. There is currently no treatment other than the avoidance of the offending protein. Many foods are now labelled to indicate whether they contain gluten.

Lactose intolerance is common in adults. In fact, lactase, the enzyme which catalyses the conversion of lactose to galactose and glucose before absorption, is found to persist into adulthood in people of European origin only. If lactose is not absorbed, it remains in the intestine where it forms the substrate for bacterial fermentation, resulting in diarrhoea accompanied by abdominal pain. The only treatment available is the avoidance of milk, which is almost the sole source of lactose in the diet.

Infections

Malabsorption and diarrhoea can be the result of infectious diseases. These are usually temporary conditions, but prolonged infection may damage the intestinal mucosa permanently, leading to undernutrition.

Some infections, especially if they cause fever, may not damage the intestinal mucosa, but lead to an increase in the metabolic rate, and so lead to undernutrition, on an apparently normal energy intake (hyperthyroidism has a similar effect). The BMR may increase up to twice its normal value in severe cases, and will lead to considerable weight loss if the energy intake is not increased to a comparable extent.

Cancer cachexia

Cancer patients often look seriously malnourished. The condition is known as **cachexia** and is the result of two main factors. First, cancer itself, and also some types of treatment, such as chemotherapy, can lead to a loss of appetite (anorexia) and to a feeling of nausea, when food is taken, so that the energy intake is reduced. Secondly, some tumours secrete the protein **cachexin** (also known as **tumour-necrosis factor-**α TNFα) which increases the metabolic rate and the loss of body protein leading to the emaciated appearance of cancer patients.

In severe cases of malabsorption, and in advanced cancer, parenteral or intravenous feeding is necessary in order to improve the nutritional status which may also be improved by the administration of ketone bodies (see Chapter 37).

Eating disorders: anorexia nervosa and bulimia

The word 'anorexia' simply describes a lack of appetite. **Anorexia nervosa** is a condition of refusal or reluctance to eat, characterized by psychological problems. It usually affects adolescent females and can cause very serious illness or even death. It is rare, but not unknown, in men and in older women. A variety of constitutional and growth factors interact with psychological factors concerning adolescence and obesity. The condition seems to reflect a psychologically necessary stifling of biological maturation through a reversal of the process of puberty, back to the biological state of childhood. This is achieved to a considerable degree, as many cases of anorexia are characterized by amenorrhoea. A reduction of body weight below 45 kg is usually associated with infertility. Anorexics usually have a distorted and exaggerated view of their body dimensions, and see themselves as obese, even when they know that their body weight is well below normal. They are often ravenously hungry and generally obsessed with food, although they successfully avoid its consumption, especially that of carbohydrate.

It is not entirely clear what causes this mental attitude. Over-reaction to mild obesity, fashion and social pressures, the cult of weight-watching and dieting, and early menarche, have all been suggested as possibilities. The hypothalamus and the extensive subcortical systems are important in the control of appetite, and although these systems are likely to play a major role in anorexia nervosa, exactly what this role is, remains to be established. It has been observed that during the weight loss phase of the disease the plasma cortisol concentration is elevated, despite normal levels of ACTH. This may suggest that an adrenal abnormality may be a component of the condition. As patients recover and regain weight, the hypercortisolaemia resolves.

Some individuals will indulge in 'binge eating' and relieve the feeling of guilt and physical discomfort as well as getting rid of the excess energy intake, by inducing vomiting soon after food consumption. This condition is known as **bulimia nervosa**. This also mainly affects females, but they tend to be older than anorexic patients, and often maintain their weight at the normal, or just below the normal, value for their height and their condition may remain unsuspected and undetected for years.

18.8 Starvation in childhood: protein-energy malnutrition (growth failure, marasmus and kwashiorkor)

Protein-energy malnutrition (PEM) is said to occur when the body's requirement for protein or energy, or both, is not met by the diet. PEM can affect all age groups, but is more frequently seen among infants and children, as their dietary requirements for protein and energy, per kilogram of body weight, are higher than those of adults, and also because they are unable to obtain food for themselves.

Most malnourished children are underweight, short for their age and often thin. Childhood malnutrition is characterized by growth failure, that is, low height for age. The weight for height may be normal, and the condition is then known as **stunting** of growth. More serious cases of malnutrition lead to **marasmus** (characterized by severe emaciation with loss of body fat and protein) or **kwashiorkor** (characterized by oedema and skin lesions), or a combination of the two, known as **marasmic kwashiorkor**. Kwashiorkor is also known as **oedematous malnutrition**. The term 'kwashiorkor' was introduced into the nutritional literature in 1932 by Cicely Williams, who used the word of the Ga tribe (in what is now Ghana) to describe 'the disease the older child gets, when the next baby is born'. The native word recognizes and gives a fair description of the conditions under which kwashiorkor generally develops, in that, as a result of the birth of the new baby, the older child is weaned onto an inadequate diet and becomes severely undernourished.

Severe malnutrition is said to occur when the weight for age is less than 60% of normal. The presence of oedema, however, must be taken into account because the weight of the excess fluid will mask the loss of body tissue and causes the severity of the condition to be underestimated.

The distinction between growth failure, marasmus and kwashiorkor is not clear-cut, in terms of clinical features, pathology or aetiology. Marasmus and kwashiorkor may be seen as the two extreme forms of a continuous spectrum of abnormalities produced by malnutrition. Although there are areas in the world where one form is more prevalent than the other, kwashiorkor being more common in rural areas of Africa, for example, and marasmus being more common in south-east Asia, the two forms are sometimes found to coexist in the same family, and a marasmic child may suddenly develop oedema to become a child with kwashiorkor.

Incidence and prevalence of PEM

The magnitude of PEM is difficult to establish. Figures from the World Health Organisation (WHO), the Food and Agriculture Organisation of the United Nations (FAO) and the World Bank estimate that 5 million children die every year from malnutrition, 300 million children have growth retardation because of malnutrition and, in developing countries, 20–75% of children under 5 years of age suffer or have suffered from PEM. To add to this, 800 million to 1 billion of people of all ages have some degree of PEM. The figures from FAO show that most malnourished people live in developing countries: 30% in Africa, 30% in south-east Asia, 15% in Latin America and 15% in the Near East.

In general, marasmus develops in infants prematurely weaned onto an inadequate diet of insufficient and often unhygienically prepared food, and is the prevalent form in children under one year of age. The oedematous form of the disease is more common in children over 18 months, living in rural areas and fed an inadequate diet of starchy gruels. Oedematous malnutrition has, however, been recorded in Jamaica, in infants no more than 4–5 months old, described as 'sugar babies'. The appearance of oedema is frequently preceded or accompanied by infection and acute diarrhoea.

Oedematous malnutrition tends to prevail in wet rather than dry areas of Africa, and during the wet rather than the dry season. It is more common in rural areas, where the staple foods are yams, cassava, plantains and rice, and it is very rare in wheat-eating areas.

Clinical, biochemical and pathological features of PEM

The main feature of PEM is weight loss. Chronic PEM results in growth retardation in terms of height (stunting) and in terms of weight (wasting).

In marasmus, the absence of subcutaneous fat and the severe muscle wasting give the patient an emaciated appearance. The skin is thin, non-elastic and wrinkled, and the hair is thin and dry. Some children are anorexic and some ravenously hungry. Heart rate, body temperature and blood pressure are low. Dehydration, respiratory and gastrointestinal (GI) tract infections and vitamin deficiencies are common complications.

In oedematous malnutrition, the characteristic feature is painless, pitting oedema. Skin lesions lead to exposure of the underlying tissues, which often become infected. Muscle wasting may be severe, but is masked by the presence of oedema and subcutaneous fat, and weight loss may not be seen, owing to retention of fluid. The liver is enlarged and often infiltrated with fat, and there is gaseous abdominal distension as a result of GI tract infections. The hair is dry, brittle and depigmented. This is particularly obvious in African children, where hair which is normally black and curly becomes brownish and straight. Alternate periods of severe and less severe malnutrition give rise to bands of pigmentation and depigmentation on the hair, known as 'flagging'.

Respiratory, GI tract and skin infections, and diarrhoea, are extremely common and severe. The most common cause of death in these children is pneumonia, with pulmonary oedema, and water and electrolyte imbalance.

Marasmic kwashiorkor is characterized by the oedema of kwashiorkor and the muscle wasting and loss of subcutaneous fat of marasmus. Some children have an emaciated upper half of the body and an oedematous lower half.

The concentration of serum proteins, especially albumin, is very low in kwashiorkor, but normal, or only slightly decreased, in marasmus. The ratio of non-essential to essential amino acids in the blood is high in kwashiorkor and normal in marasmus. There is an adaptive decrease in haemoglobin concentrations in both conditions.

In kwashiorkor, there is an increase in fatty acid and triacylglycerol synthesis, possibly from an adequate level of carbohydrate in the diet, and a decrease in the synthesis of circulating transport proteins, and this may account for the fatty infiltration of the liver (see Chapter 11). Body potassium levels are low in both conditions.

Adrenaline and corticosteroid secretion is high, being higher in marasmus than in kwashiorkor. Insulin, glucagon and thyroxine levels are low, and so is the responsiveness to T_3.

The immune system becomes defective, especially the T-lymphocyte response, but the B-cell response and immunoglobulin production are much less affected. The weakened immune system allows greater predisposition to infection, and infections which are normally mild can become very serious. The children are often apathetic, and show little response to pain or comfort. Severe PEM leads to decreases in brain growth, nerve myelination and neurotransmitter production and function: the consequent mental retardation may be permanent.

Causes and development of PEM in childhood

The traditional concept that marasmus is the end result of severe energy deficiency, and kwashiorkor the result of protein deficiency in the presence of an adequate energy supply, is too simplistic. Many other endogenous and exogenous factors play a role in the development of marasmic or oedematous malnutrition.

The conventional explanation of the development of kwashiorkor is that severe protein deficiency leads to a decrease in protein synthesis, including that of plasma albumin, and the decreased concentration of albumin leads to a drop in oncotic pressure and therefore leakage of fluid from the vascular bed into the interstitial space, producing oedema. This theory has been challenged in recent years by a number of findings. In a study in India, workers found no differences in protein intake between children who developed oedema and those who did not. It has also been shown that oedema often disappears with treatment, while the child is still on a low protein diet, and the plasma albumin still at a low level.

It must be remembered that simple protein deficiency, in the presence of an otherwise adequate and balanced diet, does not occur naturally, nor, of course, can it be tested experimentally in children. When experimental animals are fed diets adequate in energy and all nutrients except protein, they do not develop oedema, but instead fail to grow. Kwashiorkor occurs where protein intake is low and the protein of low quality, but the total food intake is also inadequate. Low protein diets are also low in vitamins, and minerals such as potassium, zinc, magnesium, sulphur and phosphorus. It is difficult to isolate the effect of protein deficiency from that of the associated micronutrients.

The causes of oedema in kwashiorkor have not yet been resolved, although various theories have been put forward. One of the most recent is that oedematous malnutrition is a result of a deficiency of antioxidant nutrients, such as selenium, zinc, magnesium and vitamins C and E, which normally provide defence against damage by free radicals. If antioxidant defence mechanisms are inadequate, free radicals generated as a result of infections and the presence of food toxins would damage cell membranes and would lead to deranged fluid balance and oedema. It is true that free radical generation increases in infection, and it has been observed that in kwashiorkor circulating glutathione levels are low, and ferritin levels are high, but it is difficult to say whether free radical damage is the single cause of oedematous malnutrition. The condition is most probably the end product of a number of challenges, each of which might have been dealt with adequately by the body, but the combination of which presents an insurmountable metabolic assault. Poverty, poor sanitation and infection always accompany malnutrition.

Oedematous malnutrition often develops after an infectious disease. Bacterial infections are associated with free radical generation, but they also have other effects. They may divert an already depleted amino acid pool to the synthesis of acute phase proteins and immunoglobulins instead of albumin and circulating transport proteins. Protein catabolism is increased in infection, and up to 2% of muscle protein per day can be lost. If this is superimposed onto a protein intake already close to the minimum requirement, then protein deficiency will develop. Deficiency of other nutrients usually accompanying protein may also be important. Low plasma potassium can lead to oedema, and zinc deficiency produces skin lesions similar to those seen in kwashiorkor.

Oedematous malnutrition seems to be a result of a combination of an inadequate intake of energy, protein and micronutrients which accompany protein, in the face of an increased demand for these dietary components, aggravated by a metabolic challenge to the system, through destructive free radical generation and inadequate antioxidant defences.

Treatment of PEM

The success of the treatment depends on the severity of the condition and its duration. Weight for height can be restored. Stunting is not easy to correct, and a growth-retarded child usually grows into a stunted adult, with low work capacity.

On the whole, fluid and electrolyte balance must be restored before nutritional rehabilitation. Oral rehydration treatment should start immediately, under supervision. The simplest oral rehydration solutions contain sugar and salt (8 teaspoons of sugar and 1 teaspoon of salt in 1 litre of water). More sophisticated mixtures are of course available, formulated and recommended by international agencies. Potassium, magnesium and zinc should be given orally, to correct any deficit.

Nutritional support should follow closely, aimed at providing sufficient energy and protein to prevent further tissue loss, and to allow catch-up growth. It is important not to overfeed, as concentrated diets cannot be tolerated and may cause death. Treatment can start with a dextrose solution followed by dilute milk, followed by full strength milk. Infant formula feeds can then be given. Lactose is often not tolerated, due to lactase deficiency, and glucose may have to be given as a substitute. After about a week of successful treatment, mood and appetite improve and the children can deal with normal food for their age. Recovery is normally rapid, and growth soon resumes.

Chapter 19

Nutrition: vitamins

19.1 Introduction

Diseases associated with vitamin deficiencies have been described as early as 2000 BC, but it was not until the beginning of the twentieth century that the concept of a deficiency disease in general, and the concept of vitamins and vitamin deficiencies in particular, was developed and accepted. The traditional concept of disease implied the presence of a toxic factor as the cause of malfunction, rather than the absence of an essential dietary component.

Beriberi (the result of thiamine deficiency) was first described in China about 2000 BC. It was noted as a serious condition characterized by numbness of legs, muscular pain, severe tiredness, difficulty in breathing and heart failure often leading to death.

Hippocrates described night-blindness (a result of vitamin A deficiency), in about 500 BC, as an inability to adapt the eyes to poor illumination after exposure to bright light, and advocated the consumption of liver as a cure. The application of liver on the eyes, as a cure for night-blindness, was used by the Chinese and Egyptians before 100 BC.

The voyages of exploration in the fifteenth and sixteenth centuries led to serious outbreaks of scurvy (vitamin C deficiency) among sailors embarking on these long journeys. It was not until the mid-eighteenth century that the Scots physician James Lind treated scurvy in the British Navy by issuing every sailor with two oranges and one lemon every day. Almost miraculous recovery from symptoms was reported after a week on this regimen.

In the seventeenth century, rickets (a bone disease caused by vitamin D deficiency) was described in England by Glisson, and it was noticed that it occurred largely in temperate zones in Europe but not in Scandinavia, that it was absent from the tropics, and that living in cramped conditions in cities, an inevitable consequence of the development of industrial societies, increased the prevalence of the disease.

About the same time, pellagra (a disease due to niacin deficiency and characterized by dermatitis, diarrhoea and dementia) was described in Spain, Italy, Romania and Egypt. The most serious outbreaks of the disease were, however, observed in the southern states of the USA at the beginning of this century, with 170 000 cases described in 1917, and 120 000 cases in 1927.

The first recognition of beriberi as a nutritional deficiency disease came from the Japanese Surgeon-Admiral Takaki, who had trained at Guy's Hospital, London. He compared the diet of the Japanese Navy, 60% of whom suffered from beriberi in 1880, to that of the British Navy, where beriberi was unknown. He concluded that the Japanese Navy suffered from protein deficiency, as their diet consisted mainly of rice, and, although wrong in his identification of the nutritional factor involved, he managed to eradicate the disease in 2 years, by adding more meat and fish to the sailors' diet. The Dutch physician Eijkman, working in Java, tested the hypothesis that beriberi was a dietary deficiency disease, in 1897. He showed that symptoms similar to those seen in human beriberi could be induced in domestic fowl by feeding a diet of polished rice, and that they could be cured if the rice polishings (husks) were added back to the diet.

The concept of disease resulting from inadequate diet was becoming clearer by the beginning of this century, but the idea that very small quantities of organic chemicals could have such a profound effect on health followed from the classic experiments of Gowland Hopkins in 1912 in Cambridge. He observed that rats fed on synthetic diets adequate in carbohydrate, fat, protein and minerals and providing more than sufficient energy, ceased to grow. Growth could be resumed by the addition of small quantities of milk, as little as 2–3 ml, to the daily diet. He concluded that small quantities of 'accessory food factors' were essential for health and growth. The term 'vitamine' was invented by Casimir Funk in 1911 when he isolated the anti-beriberi factor from rice polishings, and the word was coined to reflect the structural and functional properties of this compound, that is, a 'vital amine'. The anti-beriberi factor was subsequently named 'thiamine' and the term 'vitamin' has passed into general usage, although all vitamins are not amines, to describe 'an organic compound needed in the diet in small quantities and whose absence leads to a deficiency disease'.

19.2 Vitamin classification and nomenclature

The basis of vitamin nomenclature, using the letters A, B, C, D, E, was established by McCollum and Davis in 1915 in the USA. They described the presence of two vitamins, a fat-soluble component described as A, and a water-soluble component, B. It was soon discovered that B consisted of two fractions, a heat-labile, anti-beriberi factor, B, and another heat-stable, antiscorbutic factor, C. Factor B turned out to be a complex mixture and was later subdivided into B_1, B_2, etc., with various gaps in the numbering, as vitamins were thought to be discovered, but turned out to be mixtures of already established factors, or not vitamins at all. Some vitamins bearing a name, but no number, are now included in the B group (niacin, pantothenic acid and biotin), as are some bearing both a name and a number (B_1 is thiamin, B_2 riboflavin, B_6 pyridoxine and B_{12} cobalamin).

Factor A was also shown to be heterogeneous, consisting of an anti-night-blindness factor, A, and an antirachitic factor, D.

Another fat soluble vitamin, quite different from A and D, was discovered in the USA in 1922, by Evans and Bishop in California, and Matill in Iowa. As its deficiency was found to lead to reproductive failure in animals, it was named 'tocopherol' which derives from the Greek words *tokos* (birth) and *pherein* (bearing) and it was also assigned the letter E.

A further fat-soluble vitamin was discovered in the mid-1930s. Its deficiency was found to lead to defective blood clotting, and consequently the Danish scientist Dam, who discovered it named it 'Koagulation Faktor', hence vitamin K, and not F, as might have been expected in logical sequence.

The distinction between **fat-soluble** and **water-soluble vitamins** has persisted, although the compounds in each group have widely different structures and functions. They do, however, on the whole, share some general characteristics, which make this classification useful from a practical point of view. One reason for collecting the functionally diverse vitamins A, D, E and K together for consideration is that their common lipid solubility means that all of these molecules are handled in more or less the same way with respect to their absorption from the gut, their transport, distribution and storage. As bile is required for the absorption of all of them from the upper part of the small intestine, a failure of bile to reach the gastrointestinal tract may lead to a deficiency in all four vitamins. An important difference between fat- and water-soluble vitamins is that the fat-soluble vitamins A and D are toxic and even lethal when taken in excessive quantities, whereas water-soluble vitamins, on the whole, are not toxic, as the excess can be excreted. They are not stored extensively, however, and so their intake has to be more frequent than that of the fat-soluble vitamins, which are stored. A well-nourished adult, for example, may have three years' supply of vitamin A but only three months' supply of vitamin C.

Vitamin A and the majority of the B group of vitamins will be discussed in this chapter. Vitamin D is discussed in Chapter 12 which deals with calcium and phosphate homeostasis. Vitamin K is discussed in Chapter 24 which deals with blood clotting. Vitamin E and vitamin C are discussed in Chapter 33 which covers free radicals, in the section on antioxidant nutrients. Finally, vitamin B_{12} and folate are discussed together in Chapter 22.

19.3 Vitamin A

Structure, nomenclature and dietary sources

Vitamin A (all-*trans* retinol, vitamin A_1 alcohol) contains a single 6-membered ring to which is attached an 11-carbon side-chain, as shown in Figure 19.1. The main form is vitamin A_1, with the dehydro-form A_2 (found in some sea fish liver oils), making only a small contribution to the total amount of naturally occurring vitamin A.

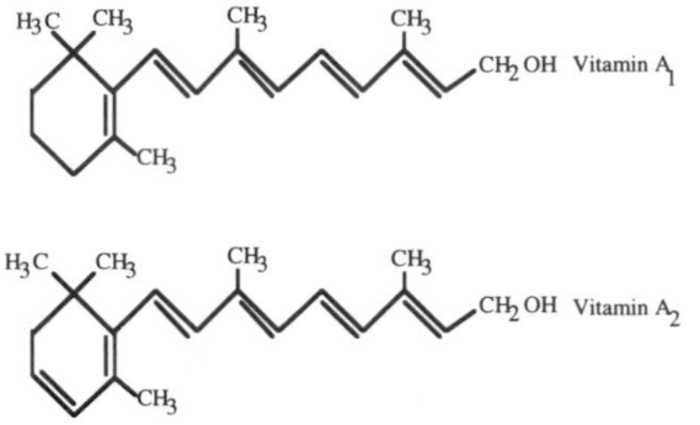

H_3C CH_3 CH_3 CH_3 CH_2OH Vitamin A_1 CH_3

H_3C CH_3 CH_3 CH_3 CH_2OH Vitamin A_2 CH_3

Figure 19.1 **Structures of the vitamins of the A group**

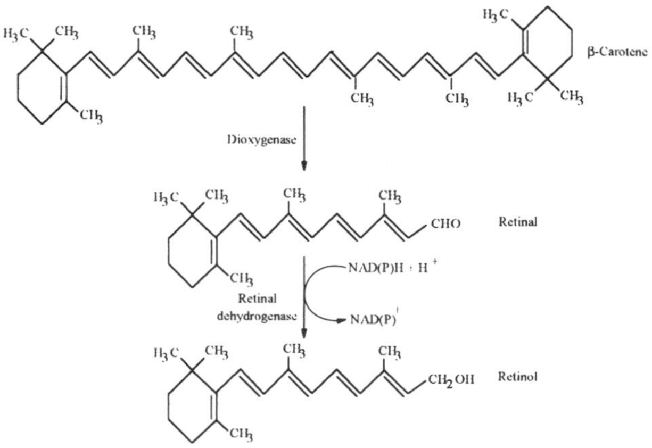

H_3C CH_3 CH_3 CH_3 H_3C β-Carotene CH_3 CH_3 CH_3 H_3C CH_3

Dioxygenase

H_3C CH_3 CH_3 CH_3 CHO Retinal CH_3

Retinal dehydrogenase $NAD(P)H + H^+$ $NAD(P)^+$

H_3C CH_3 CH_3 CH_3 CH_2OH Retinol CH_3

Figure 19.2 **Conversion of β-carotene to vitamin A**

Vitamin A is an alcohol (**retinol**), but can be converted into an aldehyde (**retinal**) or an acid (**retinoic acid**).

The provitamin A carotenoids are also included in the vitamin A family. The most important, quantitatively, is β-carotene, which occurs extensively in plants and is responsible for the orange colour of carrots. β-Carotene consists, effectively, of two molecules of vitamin A joined end to end, as shown in Figure 19.2. It can be converted into vitamin A in the intestinal mucosa. The efficiency of conversion is low, so that 6 μg of β-carotene are needed to produce 1 μg of retinol. This means that if an individual's total vitamin A intake were derived from β-carotene, instead of retinol, six times as much would be required. Most other carotenoids are not converted into retinol. Lycopene, for example, which forms most of the carotenoids in tomatoes, is absorbed intact from the intestine and is found in the circulation. The physiological function of carotenoids is different from that of retinol. Most carotenoids, unlike retinol, serve as antioxidants, and as such may be important in the prevention of some diseases (Chapter 33).

The vitamin A requirement is expressed as **retinol equivalents (RE)** in order to take into account the variability of the efficiency of conversion of carotenoids into retinol:

1 μg of RE = 1 μg retinol
= 6 μg β-carotene
= 12 μg other carotenoids
= 3.3 IU retinol.
(IU = International Unit)

In Europe and North America, about 1000 μg RE are consumed per person per day, approximately 50% of which come from retinol and 50% from carotenoids. In Africa and Asia the proportion of carotene-derived retinol equivalents is 4–5:1. The total intake in Africa is similar to that in Europe, but it is much lower in Asia (approximately 600 μg RE / person/ day), and vitamin A deficiency is fairly common in India, for example.

The richest dietary sources of preformed vitamin A (retinol) are fish liver oils, such as those derived from cod and halibut. Animal livers are also rich sources. Other good sources are milk and dairy products, but meat is rather low in vitamin A.

The richest dietary sources of carotenoids are palm oils. Good sources are dark-green leaves, such as spinach and watercress, and yellow and red fruit and vegetables, such as carrots, tomatoes and peaches. Table 19.1 shows the vitamin A content of some common foods.

Vitamin A is heat stable, and the losses occurring in normal cooking are not serious. It is, however, sensitive to ultraviolet light, and for this reason vitamin A preparations should be kept in dark or opaque containers away from direct sunlight.

Table 19.1 **The vitamin A content of some common dietary sources**

	μg retinol/100g
Animal sources	
Halibut liver oil	900 000
Cod liver oil	18 000
Sheep liver	15 000
Meat	0–4
Butter	800
Margarine (fortified)	900
Eggs	150
Milk	40
Plant sources	
Red palm oil	30 000
Carrots	2 000
Spinach	700
Peaches	250
Tomatoes	100

Absorption, transport and storage

Vitamin A is found in foods mainly as retinyl esters. These are hydrolysed to retinol in the upper part of the small intestine and are transferred across the intestinal mucosal cells. They are esterified and incorporated into chylomicrons together with dietary lipid. The presence of lipid in the intestine ensures efficient absorption of retinol, up to 80% of the intake. Carotenoid absorption is less efficient, about 40% of the intake. Any conversion of carotenoids into retinol occurs in the intestinal mucosal cells.

An outline of the transport of retinol to the liver and target cells is shown in Figure 19.3. Chylomicrons reach the periphery, and chylomicron remnants carrying almost all the absorbed retinol are cleared by liver parenchymal cells. Retinyl esters are hydrolysed to retinol, and are reversibly bound to **retinol binding protein (RBP)** in a one-to-one proportion. The complex is transferred to the stellate cells which can secrete the retinol–RBP complex directly into the circulation, and can also store excess retinol in the form of retinyl esters. Most retinol–RBP in the plasma is reversibly complexed

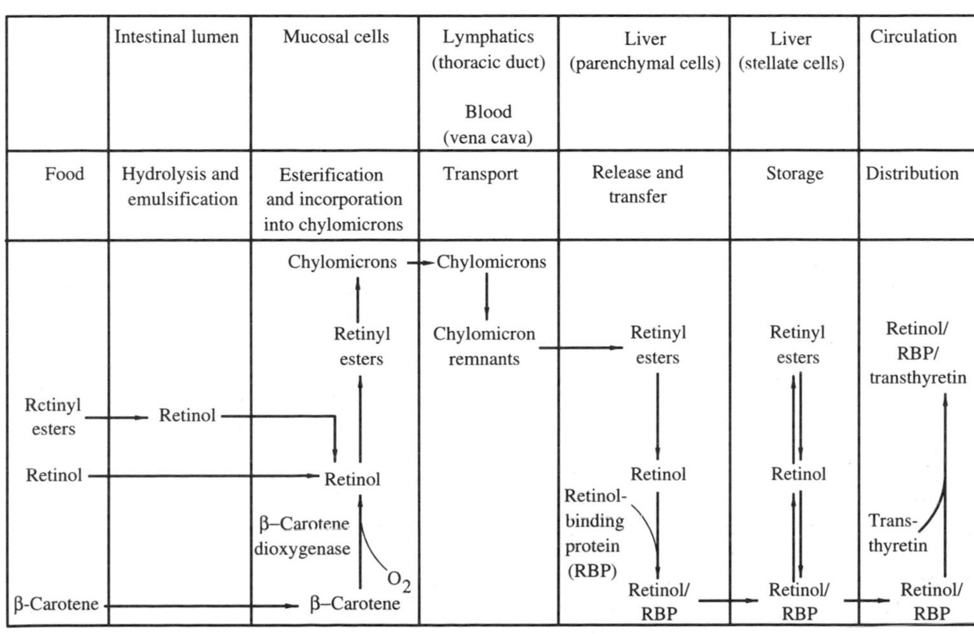

Figure 19.3 **The absorption, storage and distribution of vitamin A**

with transthyretin (T_4-binding protein). Target cells take up uncomplexed retinol–RBP by a RBP-specific receptor mediated process.

Carotenoids circulate in the plasma as components of various classes of lipoproteins.

Functions of vitamin A

The role of vitamin A in vision has been known since the 1940s mainly through the studies of G. Wald, who received the Nobel prize in 1943 for his work in this field. The importance of vitamin A in cellular function, apart from vision, has been recognized for many years, as animals deficient in vitamin A are known to die from metabolic disturbances unrelated to the visual process, but the mechanism of its action has only recently began to be unravelled. Different forms of the vitamin have different functions. The aldehyde form (retinal) is involved in vision, the acid form (retinoic acid) is involved in cellular differentiation, as a regulator of gene expression, and the related carotenoids have functions *per se* and not through their conversion into vitamin A.

Vision

The role of vitamin A in vision is well characterized and is described in detail in Chapter 31. Vitamin A, in the form of 11-*cis*-retinal, is the light absorbing chromophore found in the rhodopsin of the rods and cones of the retina.

Cellular differentiation

Retinoic acid is an important regulator of gene expression especially during growth and development, and in neoplasms. Retinoic acid, derived from maternal retinol, is essential for normal gene expression during embryonic development. Deficiency or excess can cause embryonic defects, such as in the development of the central nervous system. The abnormalities induced by deficiency or excess are different, even if they affect the same organ systems, and there seem to be intra-embryonic tissue differences in the requirement for retinoic acid.

Retinoic acid is synthesized from retinol in the endoplasmic reticulum of target cells and is transported to the nucleus on retinoic acid binding proteins CRABP I and CRABP II. In the nucleus, three different isomeric forms of retinoic acid (all *trans*, and 9- or 13-*cis*) bind to, and activate, two families of nuclear receptors, which cooperate as heterodimers in initiating the transcription of target genes. The retinoic acid receptors belong to a large family of intracellular proteins which include the thyroid and steroid hormone receptors, and also a group of 'orphan' receptors for which no specific ligands are known (Chapter 3).

Processes of differentiation which depend on, or are affected by, retinoic acid include:

Cell differentiation in spermatogenesis
The differentiation of epithelial cells (both mucous and keratinizing)

Metabolic effects

Retinoic acid exerts a number of metabolic effects on tissues. These include:

Control of the rate of gluconeogenesis in the liver (Chapter 6)
Control of the biosynthesis of membrane glycoproteins and glycosaminoglycans
Control of the biosynthesis of cholesterol.

The precise mechanisms by which most of these effects of retinoic acid on differentiation and metabolism are mediated are not clear. The same is true of the mechanism by which vitamin A therapy seems to boost immune responses of a range of groups of people, including the elderly, people who are exposed to ultraviolet light, patients after surgery, and patients with parasitic infections.

One exception to the general lack of understanding of mechanism relates to the role of retinoic acid in gluconeogenesis. This has only recently been elucidated, although it has been known for over 30 years that the liver glycogen depletion, which is observed in vitamin A deficiency, results from decreased gluconeogenesis. It has been established that retinoic acid, bound to its nuclear receptor, stimulates transcription of the gene for phosphoenolpyruvate carboxykinase (PEPCK), a rate limiting enzyme in gluconeogenesis, by binding to a short element of the promoter region of the PEPCK gene (a similar effect is exerted by cortisol - Chapter 3).

Vitamin A supplementation is also known to reduce mortality and complications in patients with measles, especially children, and this may be due to the repair of epithelial surfaces which are rapidly destroyed by measles.

The use of retinoic acid preparations as topical dermatological agents in the treatment of psoriasis, acne vulgaris and several other skin diseases is also most probably related to its involvement with epithelial cell differentiation and integrity.

Currently, acute promyelocytic leukaemia (APL) is treated by the administration of all-*trans* retinoic acid as a preliminary to, or adjunct with, chemotherapy. This disease is, in virtually all cases, associated with a translocation between chromosomes 15 and 17. The break involves the gene coding for the retinoic acid receptor α in chromosome 17.

Antioxidant function

Carotenoids are distributed in the body largely in lipoproteins in membranes, and in the lipid components of various intracellular structures, most often together with vitamin E with which they are known to interact. The antioxidant properties of the carotenoids are the subject of active study in many laboratories, but the significance of their role as antioxidant nutrients *in vivo* is still not established.

High levels of dietary carotenoids have been associated with a decreased risk of several chronic conditions, such as cardiovascular disease. Also, some precancerous lesions seem to respond to treatment with carotenoids and they are, therefore, currently studied as anti-cancer agents.

Table 19.2 **Reference nutrient intakes for vitamins**

Age	*Thiamin* (mg/d)	*Riboflavin* (mg/d)	*Niacin* (mg/d)	*Pyridoxine* (*vitamin B_6*) (mg/d)	*Vitamin B_{12}* (μg/day)	*Folate* (μg/day)	*Vitamin C* (mg/day)	*Vitamin A* (μg/day)	*Vitamin D* (μg/day)
0–3 months	0.2	0.4	4	0.3	0.4	50	25	350	8.5
1–10 years	0.7	0.8	10	0.9	0.8	100	30	500	7
Males, 11–50+ years	1.0	1.3	17	1.4	1.5	200	40	700	0 (10 after 65 years)
Females, 11–50+ years	0.8	1.1	13	1.2	1.5	200	40	600	0 (10 after 65 years)
Pregnancy	+0.1	+0.3	+0	+0	+0	+100	+10	+100	10
Lactation	+0.2	+0.5	+2	+0	+0.5	+60	+30	+350	10

Vitamin A requirements and intakes

The reference nutrient intake is 700 μg/day for men and 600 μg/day for women. (Table 19.2 shows the RNI values for vitamins for infants and adults for different ages and physiological states.)

In the UK, the median intake is well in excess of the RNI, with a value of 1100 μg/day for men and 900 μg/day for women. No group in the UK population has been found to be at risk of primary deficiency.

Vitamin A deficiency

Vitamin A in excess of immediate requirements is stored in the liver. The size of the liver stores is a good indicator of vitamin A status, but it is not easily determined in living subjects. The plasma retinol concentration does not fall until reserves are exhausted, and a value below 0.7 μM is indicative of gross deficiency, by which time other signs become obvious.

Vitamin A deficiency may be primary (dietary insufficiency) or secondary. The causes of secondary deficiency may include:

(a) Impaired absorption of lipids, as in coeliac disease, tropical sprue, gastrectomy, or obstructive jaundice
(b) Failure to synthesize apoB 48 and therefore inability to form chylomicrons into which vitamin A is normally incorporated after absorption
(c) Lack of lipases, as in pancreatitis
(d) Failure in converting β-carotene to retinol, because of an enzyme defect
(e) Impaired storage in hepatic cells in liver disease
(f) Failure to synthesize retinol binding proteins, thus affecting transport to target tissues.

The first sign of vitamin A deficiency is impaired dark adaptation. Xerophthalmia follows, the first stage of which is conjuctival xerosis which later gives rise to plaques, consisting of layers of keratinized epithelial cells, known as Bitot's spots. As deficiency progresses, a condition known as keratomalacia develops in which the cornea becomes soft and milky in appearance and finally disintegrates, resulting in blindness.

Vitamin A deficiency is the major cause of blindness in children, in south-east Asia, some parts of Africa, Latin America and the Middle East.

Other symptoms of vitamin A deficiency are:

(a) Failure of growth in children
(b) Faulty bone modelling, producing thick cancellous (spongy) bones instead of thinner, more compact ones
(c) Nerve lesions, often occurring with bone lesions
(d) Increased pressure of the cerebrospinal fluid, independent of, or associated with, deformity in skull bones
(e) Abnormalities of reproduction, including degeneration of the testes, abortion, or the production of malformed offspring
(f) Certain forms of skin disease.

Vitamin A toxicity

Arctic explorers were the first recorded victims of vitamin A toxicity. They soon learnt not to eat polar bear liver, as it caused nausea, sickness and dizziness. This is due to its high vitamin A content, which is in turn the result of consumption by the bears of seals and cod, both of which concentrate vitamin A in their livers.

Toxicity can be the result of a large single dose (300 mg or more for adults and 100 mg or more for children), but is more commonly the result of chronic ingestion of amounts grossly in excess of requirements. High levels of retinol cause headaches, double vision, nausea and vomiting, hair loss, liver and bone damage and finally death.

Retinol is teratogenic, and pregnant women in the UK are warned against taking vitamin A supplements, unless they are advised to do so by their doctor. The frequent consumption of animal liver is also to be discouraged, because animal livers in the UK have high levels of vitamin A, on average 13–40 mg per 100 g.

Carotenoids do not seem to be toxic, although high intakes lead to a yellow or orange discoloration of the skin, known as xanthosis cutis.

19.4 The B group of vitamins

The functions of the B group of vitamins are fairly well established, and all are known to be components or precursors of coenzymes which play vital roles in the metabolism of all cells. Table 19.3 shows the structures and a summary of the metabolic functions of this group of vitamins.

Thiamin (vitamin B_1)

Beriberi was first recognized as a nutritional deficiency disease by Eijkman, a Dutch physician working with beriberi patients in Java. Thiamin was isolated from rice polishings in his laboratory in 1926, it was tested in small birds, and was shown to be the anti-beriberi factor.

Table 19.3 **The vitamin B complex**

Designation	*Formula*	*Coenzyme form*	*Metabolic role(s) of coenzyme*
B1	Thiamin 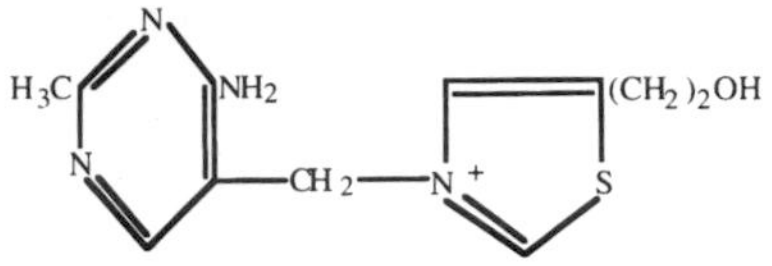	Thiamin pyrophosphate 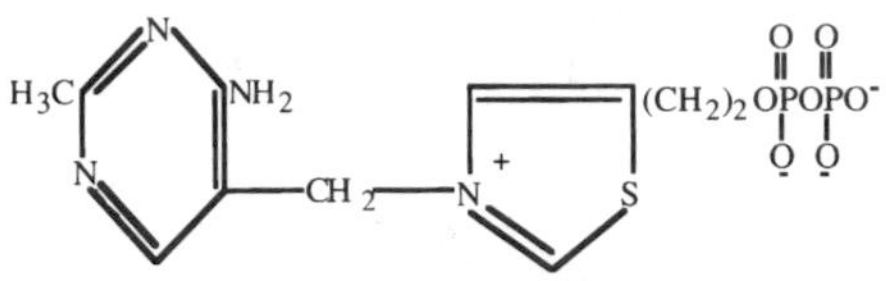	1. Decarboxylations of oxoacids, e.g. in the reaction catalysed by pyruvate dehydrogenase 2. Transketolase
B2	Riboflavin 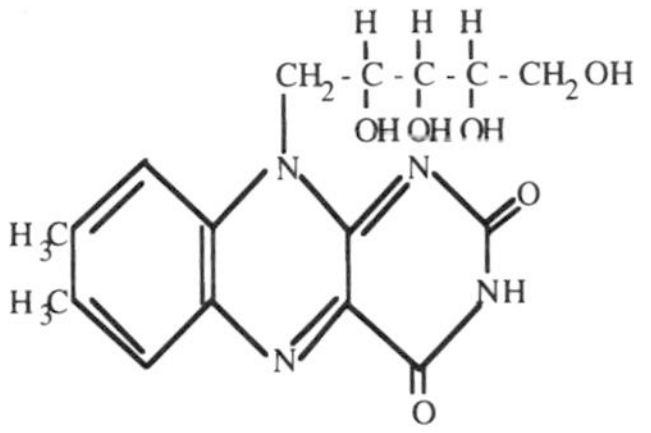	1. Flavin mononucleotide (FMN) 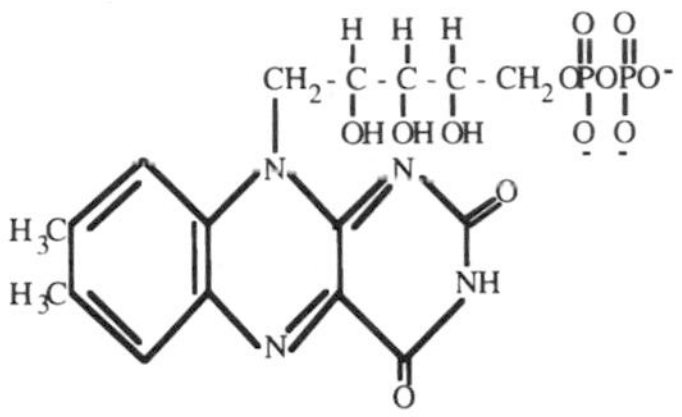 2. Flavin adenine dinucleotide (FAD) 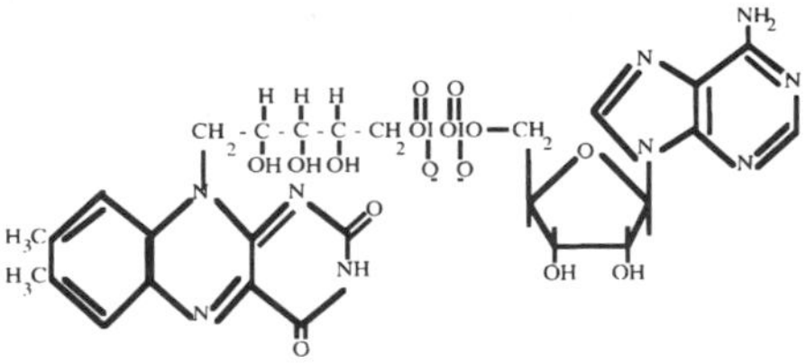	Prosthetic groups for flavoproteins that catalyse redox reactions e.g. succinate dehydrogenase, NADH dehydrogenase, xanthine oxidase.
Niacin	Nicotinamide 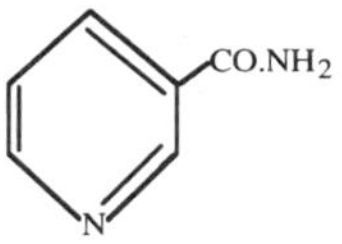	1. Nicotinamide adenine dinucleotide (NAD+) 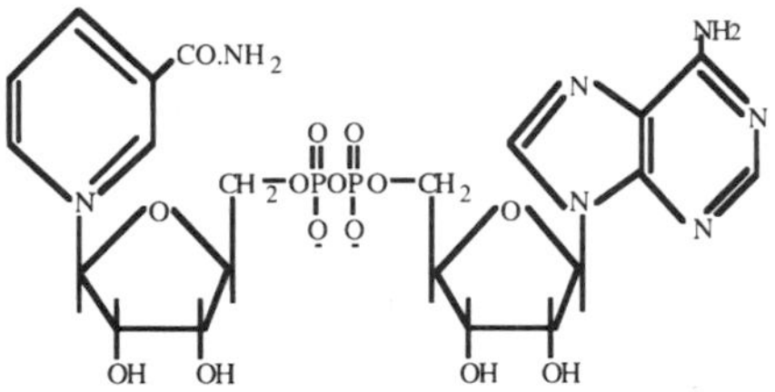 2. Nicotinamide adenine dinucleotide phosphate (NADP+) 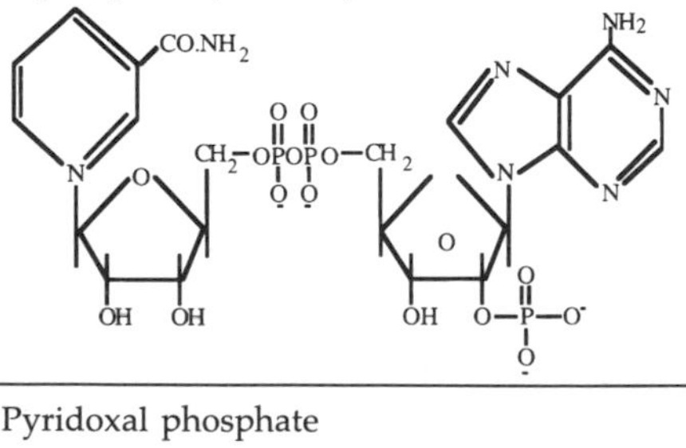	Hydrogen acceptors for dehydrogenations catalysed e.g. by malate, lactate and isocitrate dehydrogenases
B6	1. Pyridoxine 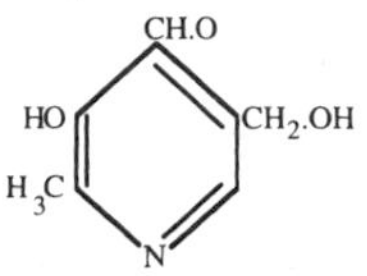 2. Pyridoxal 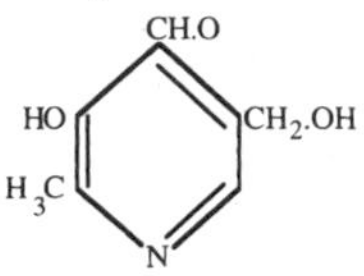 3. Pyridoxamine 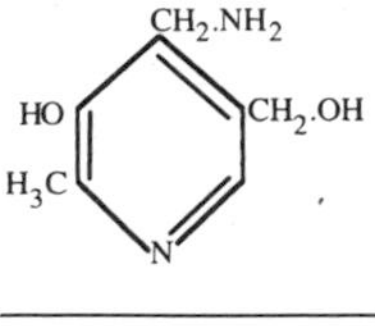	Pyridoxal phosphate 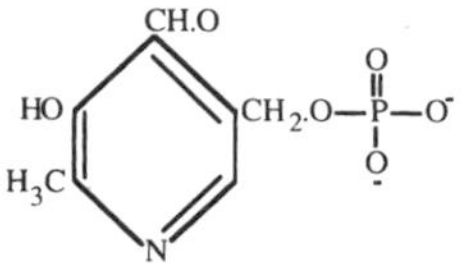	1. Transamination 2. Amino acid decarboxylation 3. Cysteine desulphydrase 4. Aminolaevulinate synthase 5. Glycogen phosphorylase

Table 19.3 **(continued)**

Designation	*Formula*	*Coenzyme form*	*Metabolic role(s) of coenzyme*
Pantothenic acid	Pantothenic acid 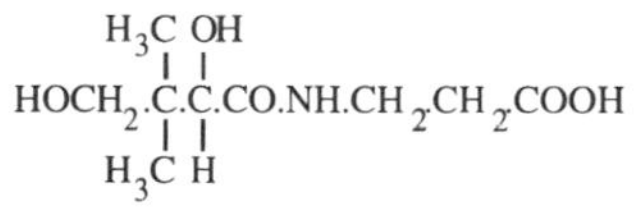 $HOCH_2.C(CH_3)(H).C(CH_3)(OH).CO.NH.CH_2CH_2COOH$	Coenzyme A 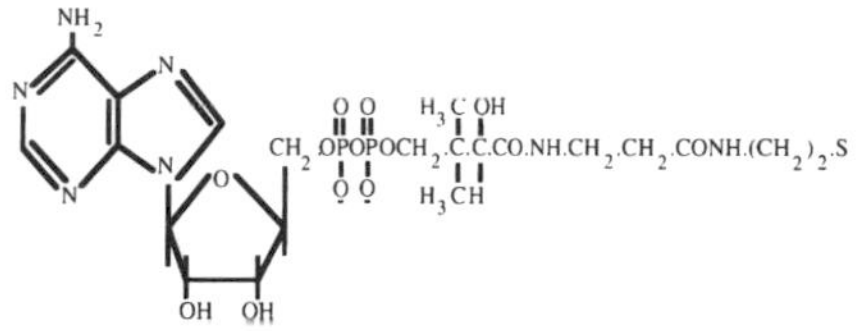	1 Acetyl group carrier 2. Formation of acetylCoA from pyruvate 3. Oxidation and synthesis of fatty acids
Biotin	Biotin 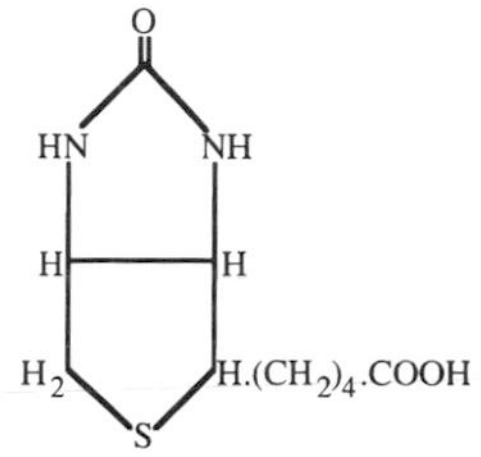	Acts as the prosthetic group for carboxylases	1. Pyruvate carboxylase 2. AcetylCoA carboxylase
Folic acid	Folic acid 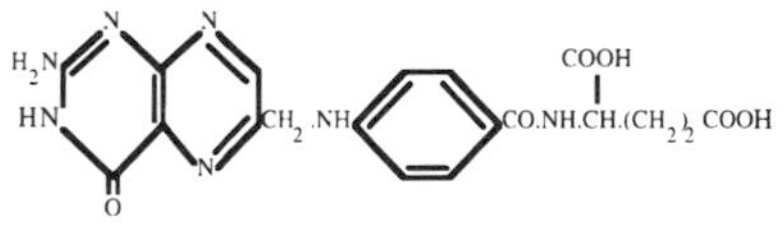	Tetrahydrofolic acid 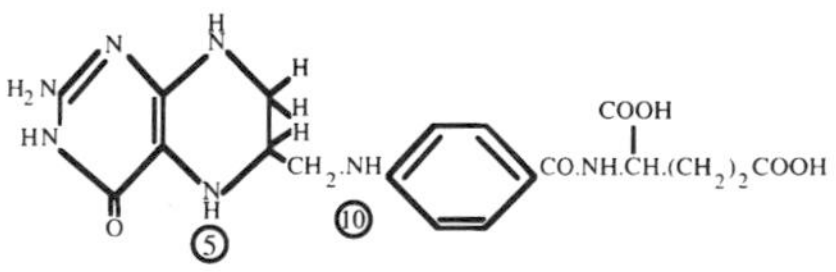	Carrier of '1-carbon units' - CHO, -CH2OH, -CH3 and -CH=NH on N5 , N10 or between both.
B12	Cyanocobalamin 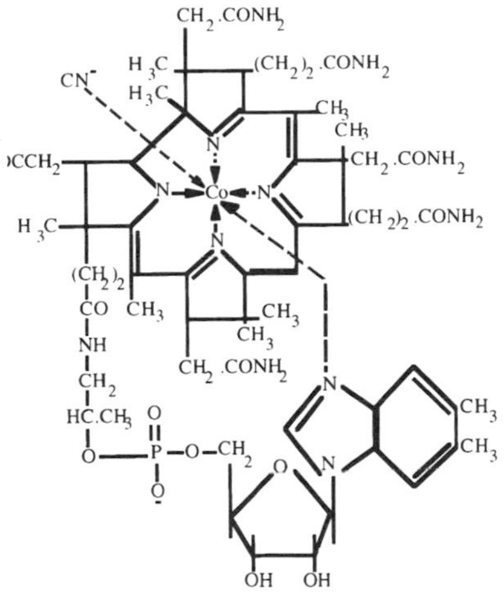	Deoxyadenosylcobalamin 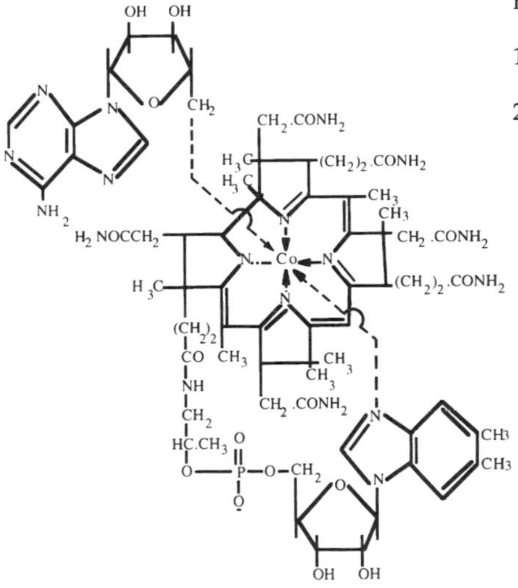	In mammals the established roles are as a cofactor for : 1. MethylmalonylCoA isomerase 2. Transfer of -CH3

The epidemic form of beriberi is found primarily in areas where white (polished) rice is the staple food, as in Japan, China, the Philippines, India and other countries of south-east Asia.

Structure and dietary sources

Thiamin consists of a pyrimidine ring attached to a thiazole ring. It is present in all natural foods, but particularly good dietary sources are unrefined cereals, pork meat, offal and nuts. White bread and polished rice are very poor sources of the vitamin.

Absorption, transport and causes of deficiency

Thiamin is absorbed from the proximal small intestine by two different mechanisms, depending on the level of intake. At levels below about 5 mg/day, thiamin is taken up by an active process requiring ATP, but at higher doses the uptake is passive. Ethanol inhibits the active transport, and this can be one of the causes of thiamin deficiency in alcoholics, whose thiamin intakes are usually also low.

Patients with a history of alcoholism, malnutrition, anorexia, malabsorption syndromes, inflammation and parenteral nutrition should be checked for thiamin deficiency.

After absorption, thiamin circulates in the blood in the free form and, following uptake by tissues, is converted into its active form, thiamin pyrophosphate (TPP).

Thiamin can be destroyed if the diet contains thiaminases. These are present in raw fish and seafood, and they are thought to contribute to the incidence of beriberi in areas in Japan where thiamin intakes are low and raw fish (sushi) is a common dietary constituent.

Metabolic functions

Thiamin is required mainly during carbohydrate metabolism, which explains why symptoms of the deficiency usually arise where much of the dietary energy is derived from carbohydrate. In fact, thiamin deficiency can be precipitated in a malnourished patient with low thiamin stores if excessive carbohydrate is administered therapeutically.

TPP is a coenzyme involved in several enzyme reactions. It is a component of the pyruvate dehydrogenase complex which catalyses the conversion of pyruvate into acetylCoA. AcetylCoA is needed for the function of the tricarboxylic acid cycle in mitochondria. In thiamin deficiency, pyruvate is converted into lactate leading to lactoacidosis. Apart from being a substrate for the TCA cycle, acetylCoA is a precursor for the synthesis of the neurotransmitter acetylcholine, and also for the synthesis of lipid, including myelin, which may explain the importance of thiamin in the correct function of the nervous system.

Thiamin deficiency affects tricarboxylic acid cycle function at another point: TPP is a coenzyme for 2-oxoglutarate dehydrogenase, which catalyses the conversion of 2-oxoglutarate into succinylCoA. Malfunction of the tricarboxylic acid cycle in the absence of thiamin, results in defective energy metabolism.

TPP is also a coenzyme for the enzyme transketolase, in the pentose phosphate pathway of glucose oxidation, which produces ribose sugars and supplies NADPH necessary for a wide variety of redox and biosynthetic reactions. It is also involved in decarboxylation reactions in the metabolism of branched-chain amino acids (BrCAAs). Patients presenting with 'Maple syrup disease' have a defect in the decarboxylase that catalyses the metabolism of BrCAAs, but in some cases they may be successfully treated with large doses of thiamine.

Another thiamin derivative, thiamin trisphosphate, is known to be involved in nerve conduction, but the exact role of the vitamin in this respect is not clear.

Dietary requirements

Thiamin requirements are related to energy metabolism and, therefore, energy intakes, and they are often expressed in terms of the latter. In the UK, the RNI for thiamin is 0.4 mg/1000 kcal (4.2 MJ) for adults, with the recommendation for a minimum absolute intake of 0.4 mg/day.

It is generally accepted that water-soluble vitamins are not toxic if taken in excess, but it has been recorded that chronic intakes of thiamin in excess of 3 g/day are toxic to adults, causing headaches, irritability, insomnia and dermatitis, and can lead to death.

Manifestations of deficiency

Two distinct clinical conditions are associated with thiamin deficiency, beriberi and Wernicke–Korsakoff syndrome.

Beriberi

Early descriptions of beriberi refer to 'wet' and 'dry' forms, but the two forms often coexist in a patient.

The wet form is characterized by oedema, cardiomyopathy, dyspnoea, pulmonary congestion, and a high-output cardiac failure.

The characteristic features of dry beriberi are a symmetrical ascending peripheral neuropathy, severe muscle weakness, ataxic gait, deep muscle pain and pain on contact with the skin.

Infantile beriberi still remains a significant cause of infant death in rice-eating populations in rural areas in the developing world. The disease is due to the low thiamin content of breast milk from deficient mothers. It is characterized by anorexia, vomiting, oedema, aphonia and encephalopathy.

Wernicke–Korsakoff syndrome

The main manifestation of thiamin deficiency in the Western world is the Wernicke–Korsakoff syndrome, usually associated with alcoholism. In the USA, alcohol-related thiamin deficiency is the third commonest cause of dementia. Wernicke's encephalopathy was considered to be the result of the acute effects of ethanol on the nervous system, until it was recognized as the cerebral form of beriberi. Thiamin deficiency, as mentioned above, is common in chronic alcoholics, and is exacerbated by the inadequate storage of thiamin in the cirrhotic liver. The main signs and symptoms of Wernicke's encephalopathy are a global confusion state, nystagmus, ophthalmoplegia, ataxia and polyneuropathy. If untreated, it usually progresses to Korsakoff's psychosis which is irreversible and is characterized by loss of memory of recent events and inability to retain new information.

There may be a genetic predisposition to the Wernicke-Korsakoff syndrome. It is more common, for example in white, rather than black, alcoholics. A possible involvement of the enzyme transketolase was suggested by findings that variant forms of the enzyme are found in patients with Wernicke–Korsakoff syndrome, but there is no evidence, as yet, for a specific role for transketolase in the production of the brain lesions seen in thiamin deficiency.

Riboflavin (vitamin B_2)

Dietary sources, structure and functions

The main dietary sources of riboflavin are milk and milk products, eggs, liver and meat. Cereals are poor sources.

Riboflavin is a yellow compound consisting of an isoalloxazine ring with a ribitol side-chain. It is a constituent of the coenzymes flavin mononucleotide (FMN) and flavin adenosine dinucleotide (FAD), essential for the activity of flavin-dependent (flavoprotein) enzymes which are involved in a wide variety of oxidation and reduction reactions in metabolism.

Absorption, transport and storage

Riboflavin is ingested in the form of flavoproteins. The FAD and FMN components are released from the protein complex in the stomach, and free riboflavin is released in the intestine from which it is absorbed by an active, ATP-dependent process. The main storage form of the vitamin, found mainly in the liver, is FAD.

Requirements and deficiency

Few epidemiological studies can be used to assess riboflavin requirements. This is mainly because the clinical signs of deficiency are non-specific. Also, riboflavin is intimately involved with dietary protein, so inadequate protein interferes with estimates of riboflavin requirements.

The symptoms of riboflavin deficiency include angular stomatitis (inflammation of the mouth), cheilosis (fissures at the angles of the mouth), glossitis (inflammation of the tongue) and a form of peripheral neuropathy. Although flavoproteins are essential for life, riboflavin deficiency symptoms are relatively mild and certainly not life threatening. There are two main reasons for this. One is that specific primary riboflavin deficiency is difficult to achieve, because riboflavin is associated with protein in the diet and any diet providing protein will also provide a fair amount of riboflavin. The other reason is that the recycling of riboflavin released from FAD and FMN is extremely efficient, and therefore only small amounts need be ingested.

Like thiamin deficiency, riboflavin deficiency can, however, be seen in conditions such as malabsorption, malnutrition, anorexia and chronic alcoholism. Drugs, such as barbiturates, may also cause riboflavin deficiency by inducing microsomal oxidation of the vitamin.

The best index of riboflavin status is the measurement of the activity of a riboflavin-dependent enzyme, usually erythrocyte glutathione reductase.

Niacin

The appearance of pellagra, the disease caused by niacin deficiency, followed the introduction of maize to Europe by the Spanish explorers returning from America. Pellagra became widespread in maize-eating areas in Europe, such as north Italy and Spain, and in parts of North Africa. In the USA, it was a major cause of death in the nineteenth and early twentieth century. The Spanish physician Casal described pellagra in the 1730s as a disease of malnutrition seen in maize-eating people. A characteristic feature of this disease was a rash, resembling sunburn, on areas of the skin exposed to the sun (Casal's collar).

In the 1900s it was also recognized that pellagra could be cured not only by eating mixed cereals instead of maize, but also by providing animal protein in the diet.

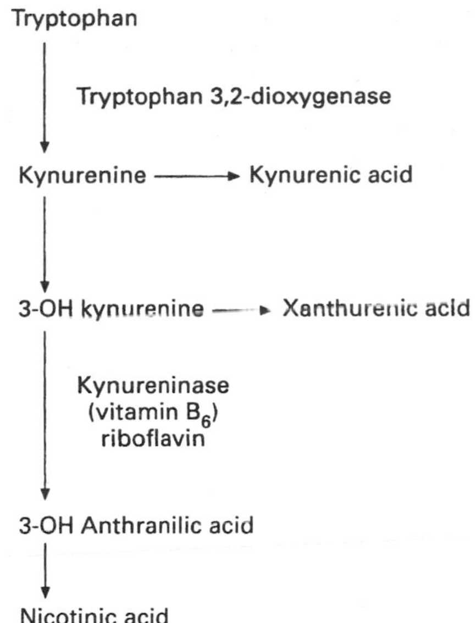

Figure 19.4 **The conversion of tryptophan into nicotinic acid**

Pellagra was not seen in maize-eating areas in America, such as Mexico, where maize is treated with lime before cooking. Apart from the fact that there is less niacin in maize compared to other cereals, its bioavailability is very low and treatment with alkaline agents results in the release of niacin from the bound unavailable form.

Structure, dietary sources and requirements

Niacin is the generic descriptor used to refer to the **vitamers** (different structural forms of a vitamin with the same biological activity) nicotinic acid and nicotinamide.

Niacin is found in cereals in small amounts and its bioavailability is low. Maize is much poorer in niacin content than wheat or rice, which explains why pellagra could be cured by replacing maize with other cereals. The cure of pellagra by increasing the protein content of the diet can be explained by the fact that nicotinic acid can be formed in the body from the amino acid tryptophan. The conversion is inefficient, and on average 60 mg of tryptophan give rise to 1 mg of nicotinic acid. It does, however, mean that a diet high in protein can provide all the niacin needed by the body, and only at low levels of protein intake does niacin constitute a dietary essential. The pathway of tryptophan metabolism to nicotinic acid is shown in outline in Figure 19.4 .

One of the steps involved, the conversion of 3-OH kynurenine to 3-OH anthranilic acid, is riboflavin and pyridoxine dependent, so that deficiencies in these vitamins may produce a secondary niacin deficiency in individuals whose niacin requirements have to be met through tryptophan metabolism.

Oestrogens are known to reduce tryptophan metabolism, which may explain why the incidence of pellagra is much higher in women rather than in men in maize-eating areas.

Niacin content in food as well as niacin requirements are expressed as 'niacin equivalents'. These are calculated as follows:

mg of niacin equivalents = mg of preformed niacin + 1/60 mg of tryptophan

The RNI for niacin in the UK is 6.6 mg of niacin equivalents. The average UK diet provides 84 g of protein for men and 62 g of protein for women. These levels can provide 17.6 mg of niacin equivalents for men and 13 mg for women, thus removing the requirement for preformed niacin in the diet. For individuals whose diets are lower in protein, preformed niacin becomes an essential dietary component.

Metabolic functions

Niacin, in the form of nicotinamide, is incorporated into the structure of the nicotinamide dinucleotides NAD^+ and $NADP^+$, and is, in these forms, involved in a great variety of oxidation and reduction reactions in intermediary metabolism.

NAD^+ also functions as an ADP-ribose donor for ADP-ribosylation reactions in the nucleus. These are eukaryotic post-translational modifications catalysed by the enzyme poly(ADP-ribose) polymerase. The primary function of this enzyme is unknown, but it seems to play a role in DNA repair and other cellular responses to DNA damage.

Clinical deficiency

Deficiency of niacin leads to pellagra, a disease characterized by dermatitis, diarrhoea and dementia. The dermatitis, which resembles sunburn, is seen in areas of the skin exposed to the sunlight. It is probably related to the role of NAD in DNA repair reactions following damage through exposure to UV light.

Pellagra is also seen in conditions where dietary amino acids, including tryptophan, are not properly absorbed, as in Hartnup's disease, a genetic disorder of amino acid transport. Severe pellagra can be fatal. It is now rare in Europe and America, but it is still seen in India and parts of Africa.

Therapeutic uses and toxicity

Nicotinic acid, used at high doses (1–2 g daily) has been shown to lower total plasma cholesterol, LDL-cholesterol and VLDL-triacylglycerols in patients with hyperlipoproteinaemias. It appears to lower the synthesis of VLDL by blocking adipose tissue lipolysis, thus reducing the free fatty acid flux to the liver. A reduction in hyperglycaemia has also been observed and the use of nicotinamide in the prevention of diabetes mellitus (type I, IDDM) is currently under investigation.

The large doses necessary to produce the lipid- and glucose-lowering effects, have undesirable side-effects, mainly vasodilatation and flushing. Cases of hepatic toxicity have also been reported after as little as a week's treatment in some patients.

Pyridoxine (vitamin B_6)

Structure and dietary sources

Vitamin B_6 consists of a mixture of three different vitamers, pyridoxine, pyridoxamine and pyridoxal, all of which can be phosphorylated and converted to the active form, pyridoxal phosphate.

Pyridoxine occurs widely in both animal and plant tissues and primary deficiency of the vitamin through poor intake is rare.

Absorption, transport, storage and dietary requirements

The absorption of pyridoxine from the intestine is quite efficient, about 80% of the intake being absorbed. If ingested in the phosphorylated form, it is dephosphorylated before absorption and it is rephosphorylated in various tissues to produce the active form. Pyridoxal phosphate (PLP) is transported in the blood bound to albumin.

Most of the body stores of pyridoxine are associated with the enzyme glycogen phosphorylase for which it is a coenzyme.

The dietary requirement, however, relates to protein intake rather than energy metabolism. Deficiency of pyridoxine can be produced in a shorter space of time in subjects on high, rather than low, protein diets. The RNI for pyridoxine is 15 μg/g dietary protein.

Metabolic functions

Pyridoxal phosphate is a coenzyme for a large number of enzyme reactions in intermediary metabolism and especially amino acid metabolism, and this is why pyridoxine requirements are expressed in terms of dietary protein intake.

PLP is a coenzyme for aminotransferases, where it acts as an amino group carrier. It is also a coenzyme for amino acid decarboxylases. Particularly important in this respect are glutamate- and DOPA-decarboxylases which are involved in the production of γ-aminobutyric acid (GABA) and the catecholamine neurotransmitters in the nervous system (see Chapter 30). PLP is also a cofactor for aminolaevulinic acid synthetase, involved in the production of haem, for glycogen phosphorylase in both liver and muscle, and for the enzymes responsible for the specific deamination of serine, threonine and cysteine.

Clinical deficiency

As pyridoxine occurs in most foods, the dietary deficiency of this vitamin is rare. Deficiency was, however, documented in the 1950s in infants fed on a milk preparation which had been overheated during manufacture, thereby resulting in the loss of almost all its pyridoxine content. These infants suffered from convulsions, and presented with a microcytic hypochromic anaemia that would not respond to iron but responded to administration of pyridoxine.

The commonest cause of pyridoxine deficiency is drug antagonism, and the main clinical symptoms of deficiency are anaemia and peripheral neuropathy.

Isoniazid, used in the treatment of tuberculosis, and penicillamide, used in the treatment of Wilson's disease and rheumatoid arthritis, can combine with pyridoxal phosphate thus rendering it unavailable. Patients on these medications are given vitamin B_6 supplements.

There is no evidence that oral contraceptives increase the requirement for vitamin B_6. The idea originally arose because of the method of assessing B_6 status through tryptophan metabolism. B_6 deficiency affects the metabolism of various amino acids, including tryptophan, and results in the production of abnormal tryptophan metabolites. Oestrogens and progesterone also interfere with tryptophan metabolism and lead to the appearance, in urine, of tryptophan metabolites similar to those observed in pyridoxine deficiency, but without interfering with pyridoxine function *per se*.

Therapeutic uses and toxicity

Many neurological conditions, e.g. carpal tunnel syndrome, may be responsive to pyridoxine treatment, but it is difficult to isolate the conditions which are most likely to respond. Pyridoxine is used for the treatment of seizures, Down's syndrome and autism, and premenstrual tension syndrome (PMS) with variable success. Unless there is a known or suspected pyridoxine deficiency, as in chronic alcoholism, for example, its value in the treatment of various neurological conditions is doubtful.

Pyridoxine seems to be safe at levels of 100–150 mg/day, as shown by studies on large population groups with carpal tunnel syndrome, but higher values are toxic. Women, self-medicating for PMS, taking 500–5000 mg /day have shown peripheral neuropathy within 1–3 years.

Biotin

Biotin was first discovered as a growth factor for yeast. The structure of the vitamin was determined in the 1940s.

Structure, dietary sources and causes of deficiency

Biotin consists of a tetrahydrothiophene ring bound to an imidazole ring and a valeric acid side-chain. It is widely distributed in foods and its dietary deficiency is unknown, except in some cases of patients on long-term total parenteral nutrition with inadequate biotin supply, and in a small number of people with the unusual dietary habit of consuming large amounts of uncooked eggs. Egg white contains the glycoprotein **avidin** which binds the imidazole group of biotin and renders it unavailable. Experimental biotin deficiency is difficult to achieve using natural foods, unless large amounts of raw egg white are fed.

Biotin is also synthesized by intestinal bacteria and is known to be absorbed in the colon, but the contribution from this source to the total available biotin is unknown.

Metabolic functions

Biotin is a carrier for carbon dioxide in a number of carboxylation reactions. The most important are the conversion of acetylCoA into malonyl CoA (catalysed by acetylCoA carboxylase) in fatty acid synthesis, and the conversion of pyruvate into oxaloacetate (catalysed by pyruvate carboxylase) in gluconeogenesis.

It is also involved in the catabolism of branched-chain amino acids.

Clinical deficiency

Patients on total parenteral nutrition devoid of biotin developed a scaly dermatitis and alopecia with loss of hair follicles. In experimental biotin deficiency, produced by feeding egg white at levels that provided 30% of the dietary energy, subjects suffered from nausea, depression, sleepiness as well as hair loss and dermatitis.

Pantothenic acid

Pantothenic acid was also discovered as a growth factor for yeast.

Structure and dietary sources

Pantothenic acid is a derivative of butyric acid bound to alanine and phosphate. As phosphopantetheine it forms part of the structure of coenzyme A and also of the acyl carrier protein in fatty acid synthesis.

The name pantothenic acid is derived from the Greek word *pantothen*, meaning 'from everywhere', and gives an indication of the wide distribution of the vitamin in foods.

Metabolic functions

As part of coenzyme A, it is involved in many reactions in the metabolism of carbohydrate, lipid and protein in the form of acetyl-, succinyl- and acyl- CoA derivatives.

Clinical deficiency

There is no documented primary dietary deficiency of pantothenic acid, except perhaps in observations of malnourished prisoners of war in the Far East in the 1940s, where a neurological condition known as the 'burning feet syndrome' was reported and attributed to pantothenic acid deficiency. As these people were severely malnourished and deficient in other vitamins as well, it is not possible to attribute this specific effect to pantothenic acid deficiency, and indeed they were treated with a mixture of all B vitamins.

Experimental pantothenic acid deficiency leads to paraesthesiae with the sensation of 'burning feet', headache, dizziness and gastrointestinal malfunction.

Chapter 20

Nutrition: inorganic constituents of the diet

20.1 Introduction

The inorganic constituents of the animal tissues, including human, are determined by the process of ashing, i.e. incinerating the tissue until all the organic matter is burned off. The material left behind contains large numbers of different metals and smaller numbers of anions such as sulphate, carbonate and chloride. It is understood that a range of metals play major roles in cellular function and these have been investigated very thoroughly, although several metals are found in the tissues for which no known function exists. These metals are believed to be absorbed, mainly in the diet, and are unavoidable contaminants of the environment. Some probably do little harm, whereas others such as lead are very toxic (see Chapter 35), and metals such as zinc and copper are important cellular constituents, but are toxic in excess.

20.2 Metals found in the human body

Metals in the human body are listed in Table 20.1 and it will be seen that they are divisible into two groups, the major metals forming the main proportion of the total metal content of the body and trace metals which are found in low, or even minute, quantities. Iron, although usually placed in the trace metal group, really falls into an intermediate group between the major and trace metals. Typical metal contents in the human adult are shown in Table 20.2.

Table 20.1 **Metals in the human body**

Major metals	*Trace metals*	
	Established function	*No established function*
Na	Fe	Ni
K	Cu	Al
Ca	Mn	Sn
Mg	Co	Ti
	Zn	Pb
	Mo	Li
	Cr	Ba
		Sr
		V
		Ag
		Au
		Ce

Table 20.2 **Metal content of the human body**

Metal		*Weight of metals/ fat-free tissue* (g/kg)	*Total body content* (g)
Major metals	Na	1.84	105
	K	3.12	245
	Ca	22.4	1050
	Mg	0.47	35
		(mg/kg)	(μg)
Minor metals	Fe	74	3000
	Cu	1.7	0.300
	Zn	28	0.100
	Mn	1	0.200
	Co	3 μg	120

Calcium is the most abundant metal in the whole body, mainly on account of the high concentration in the skeleton, but potassium is more concentrated than other metals in the soft tissues. Of the trace metals, only seven have been shown to possess clearly defined biological functions and these metals play vital roles, forming components of active enzymes. No function has yet been assigned to any of the long list of metals shown in the third column. For several of the metals in this category, it is extremely difficult to be certain whether or not a true function exists; the quantities in the tissues are very small and it is very difficult to carry out feeding experiments with all traces of the metal under study removed from the diet and removed from all materials which are likely to come into contact with the experimental animals. Unless a clearly defined deficiency disease can be shown to be associated with lack of a particular metal, it cannot be stated to be essential for life.

20.3 Factors affecting metal requirements

The daily dietary requirements of any metal are directly related to losses and the factors involved are illustrated in Figure 20.1. Metals taken in the diet are not absorbed with equal efficiency. For some, such as sodium and potassium, the efficiency is very high but for others it is very low (Table 20.3).

After absorption, the metals are transported in the blood, often carried on a special protein, and then transferred to the tissues for use or kept in a stored form, sometimes bound to protein (Figure 20.1). Circulating metals absorbed from the digestive tract or released from

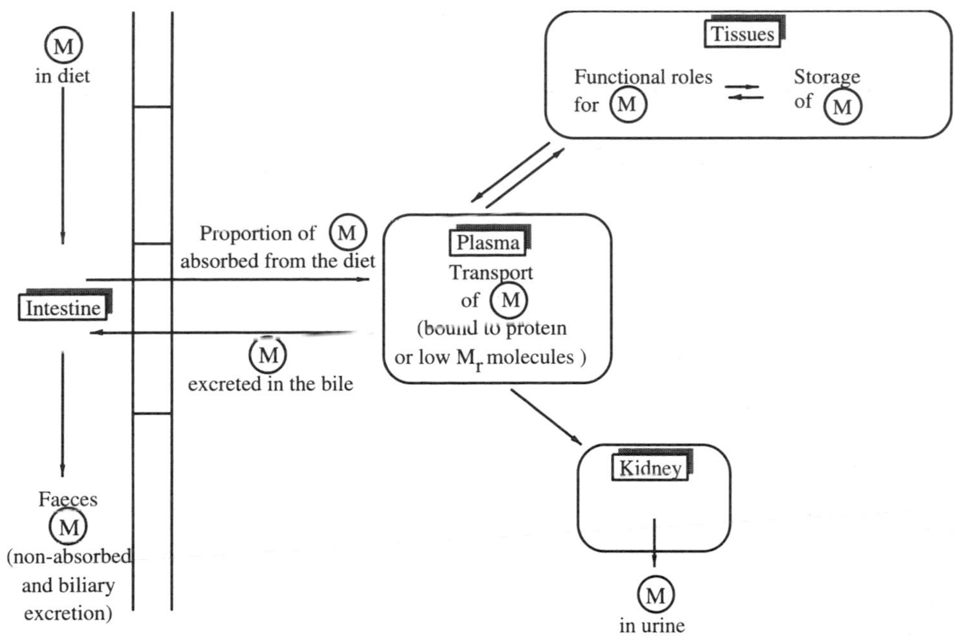

Figure 20.1 **Factors affecting metal requirements.** M=metal

Table 20.3 **Faecal loss of ingested metals**

	Metal	*Faecal loss* (% intake)
	Na, K	1–5
	Ca	35–80
	Mg	65
	Fe	30–95
	Zn	50
Extensive biliary excretion occurs	Co	80
	Mn	97
	Cu	99

the tissues can be lost in two ways, either by biliary excretion through the liver, or through the kidney.

The type of excretion varies with the metal. Most of the sodium and potassium is lost through the kidneys and a large proportion, normally about 50% of calcium and magnesium, is also lost by this route. Very small amounts of trace metals, such as copper or zinc, are excreted in the urine, but the major proportions of these metals are lost in the faeces.

20.4 Dietary requirements for metal ions

A steady intake of all essential metals is required to balance the losses from the body. Under some circumstances, for example those of hot humid conditions, losses of sodium can increase substantially and must be compensated by increased dietary intake. Potassium is very widely distributed in many foods and no special measures are normally necessary to ensure adequate intake. Sodium chloride is incorporated into many foods during preparation, cooking or during eating. It is now believed that many individuals consume excess salt which must be excreted and, in older individuals, retention of salt can lead to oedema and high blood pressure. Excessive consumption of sodium chloride is not, therefore, desirable. Magnesium is also widely distributed in the diet and is a constituent of chlorophyll. Deficiency of this metal in the human diet occurs very rarely, if at all.

Some cases of zinc deficiency in humans have been occasionally reported. Zinc plays several vital roles in most cells because it is an essential component of several important enzymes, such as carbonic anhydrase, alcohol dehydrogenase, carboxypeptidase and alkaline phosphatase. In addition, several proteins that bind to DNA and thereby influence the rate of its transcription have special zinc-binding domains known as *'zinc fingers'*; this is the case for the glucocorticoid receptor protein, for example. Rats deprived of zinc die within 2–3 weeks on a deficient diet. Taking all these aspects into consideration, only two metals are of major concern in the human diet both in children and adults: calcium and iron. These metals are discussed fully in Chapters 12 and 21. Table 20.4 shows the reference nutrient intakes of minerals currently recommended in the UK.

20.5 Roles of metal ions

Many of the biological roles of metal ions can be seen to arise as a consequence of their abilities to bind more or less strongly to proteins, especially enzymes (or their substrates). Table 20.5 lists some of the metal ions and the groups in proteins to which they bind. It is important to note, however, that the abilities of these metals to bind to proteins are not the same in all cases. There seems to be some correlation between the strength of the binding and the function that the metal subserves, as may be seen from Table 20.6. Some of the terms used in this table

Table 20.4 **Average reference nutrient intakes for the minerals at different ages**

Age (years)	*Calcium* (mmol/d)	*Phosphorus* (mmol/d)	*Magnesium* (mmol/d)	*Sodium* (mmol/d)	*Potassium* (mmol/d)	*Chloride* (mmol/d)	*Iron* (μmol/d)	*Zinc* (μmol/d)	*Copper* (μmol/d)	*Selenium* (μmol/d)	*Iodine* (μmol/d)
0–1	13.1	13.1	2.8	12.5	19.5	12.5	97.5*	67.5	5.0	0.4	0.5
1–10	11.3	11.3	5.4	34	32.7	34	130	95	8.7	0.3	0.8
Males											
11–18	25	25	11.9	70	85	70	200	145	15	0.8	1.0
19+	17.5	17.5	12.3	70	90	70	160	145	19	0.9	1.0
Females											
11–18	20	15	11.9	70	85	70	260	125	15	0.8	1.1
19+	17.5	17.5	10.9	70	90	70	260†	110	19	0.9	1.1

*RNI for iron increases from 30 to 140 μmol/day during the first year of life.
†RNI for iron reduces to 160 μmol/day for postmenopausal women.

Table 20.5 **Groups in proteins interacting with various metal ions**

Metal ion	*Protein group*
K^+	Singly charged oxygen or neutral oxygen
Mg^{2+}, Mn^{2+}, Ca^{2+}	Carboxylate, phosphate, nitrogen-containing groups
Cu^+, Mo^{2+}, Cd^{2+}	Thiol
Cu^{2+}	Amino groups ≫ carboxylates
Zn^{2+}	Imidazole groups, thiol groups
Fe^{2+}/Fe^{3+}	Thiol and amino groups, tyrosine hydroxyl groups, porphyrin rings

From M.N. Hughes (1974) *The Inorganic Chemistry of Biological Processes*, by kind permission of John Wiley & Sons Ltd.

Table 20.6 **The key roles of metal ions in biological processes**

Metal	*Role*	*Metal binding strength*
Na^+, K^+	Charge carrier	Weak
Mg^{2+}, Ca^{2+}	Trigger and control mechanisms; Stabilization of structure	Medium
Zn^{2+}	Strong Lewis acid in hydrolytic enzymes*	Strong
Fe^{2+}, Cu^+, Mo^{2+}	Redox catalysts	Very strong

(*Lewis acids are compounds or atoms able to accept a pair of electrons.)

require some explanation. An example of the operation of a 'trigger' mechanism is that which occurs when Ca^{2+} is released from the sarcoplasmic reticulum to initiate muscle contraction (see Chapter 27).

Another way of classifying the binding of metal ion to donor groups is according to the 'hard–soft' theory of acids and bases. In this scheme, 'soft' metals are large and easily polarized, whereas 'hard' metals are small and less easily polarized. In this context, Na^+, K^+, Mn^{2+} and Fe^{3+} are hard; Fe^{2+}, Zn^{2+} and Cu^{2+} intermediate; and Cu^+, Hg^{2+} and Pb^{2+} are soft.

20.6 Anions in the diet

The diet must contain certain essential anions. Of these, the most important quantitatively are chloride and phosphate, but smaller quantities of sulphate, carbonate, fluoride and iodide also occur. **Chloride** normally accompanies sodium as sodium chloride and **phosphate** is widely distributed in food. **Sulphate** and **carbonate** are metabolic end-products and not really essential in the diet although consumed incidentally. **Iodide** is essential for the biosynthesis of the hormones triiodothyronine and thyroxine and is oxidized to iodine before being incorporated into the thyroid hormones (see Chapter 28).

In earlier times, before the need for iodide was understood, many people living well away from the sea, such as in Derbyshire, UK and in Switzerland, suffered severe iodine deficiency. This was because the main source of iodine was from the sea, either concentrated in seaweed or from fish which had stored it as a result of a natural food chain. Iodides are normally added to table salt and very little deficiency occurs currently in modern societies. Table 20.4 shows the relevant reference nutrient intakes.

Fluoride is not an essential constituent of the diet but was accidentally found to be effective in the prevention of tooth decay. Studies, mainly in the USA and Canada, showed that children in areas where the fluoride content of the water was relatively high had far fewer decayed teeth than those in areas where fluoride was absent. If, however, the fluoride content of the water was high, a type of discoloration in the teeth occurred, called 'fluorosis', but this did not appear to have adverse effects on teeth or health (Figure 20.2). The concentration of fluoride in bones also increased as the fluoride content of the water increased (Figure 20.3). In both teeth and bones the F^- ion replaces some of the OH^- in the hydroxyapatite crystal lattice. Experiments on animals showed that fluoride could be toxic if added to the drinking water in concentrations of 50–100 ppm or more, but these were very much greater than the very small additions of 1 ppm required to prevent tooth decay (Table 20.7). These observations led to controlled

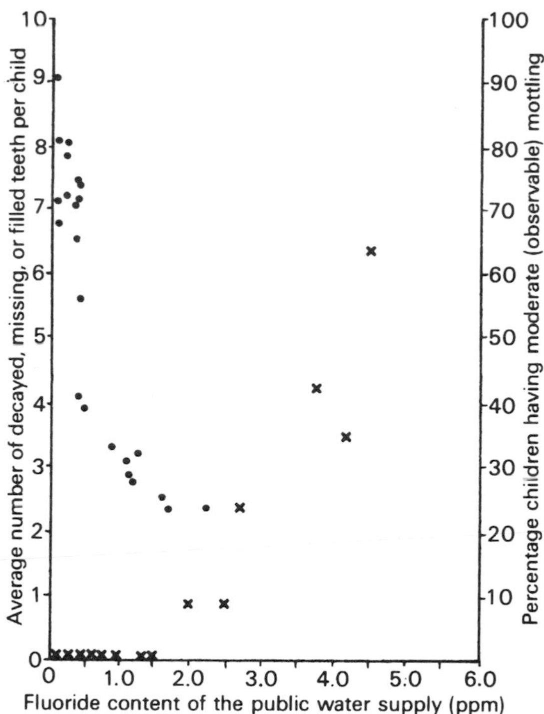

Figure 20.2 **Relation of fluoride content of public water supply to dental caries and fluorosis.** (×) Percentage fluorosis; (●) mean number of decayed, missing or filled teeth

experiments in which fluoride was added in a concentration of 1 ppm to the water supply of towns in the USA and Canada which were free of fluoride. When tested over a 10-year period, dramatic reductions in the tooth decay of young children within the age group 6–10 years were observed. This led to the advice by government health authorities in the USA, Canada and Europe that all water supplies should be treated with fluoride to bring the concentration up to 1 ppm if the existing natural concentration was less. Surprisingly, both in the USA and the UK, this proposal led to strong opposition led by the 'pure water' lobby who insisted that the drinking water should not be adulterated. In the

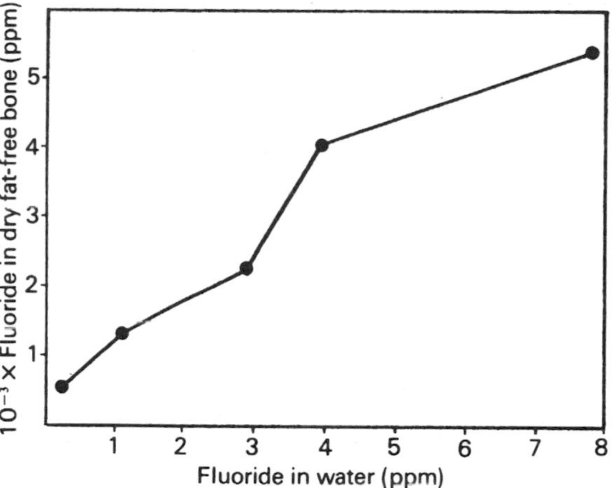

Figure 20.3 **Relation of fluoride in drinking water to fluoride content in human bones**

Table 20.7 **Effects of fluoride in different concentrations on man and experimental animals**

Concentration of fluoride (ppm)	*Vehicle*	*Effect*
1 and over	Water	Dental caries reduction (in man)
2 and over	Water	Fluorosed enamel (in man)
8 and over	Water	Osteofluorosis (in man)
50 and over	Food or water	Thyroid changes (experimental animals)
100 and over	Food or water	Growth retardation (experimental animals)
125 and over	Food or water	Kidney changes (experimental animals)

UK, the decision as to whether or not to treat water with fluoride has been left to local authority administrations so that, at present, approximately half the water supplies are treated and half are not. Addition of fluoride is strongly supported by the dental profession and most scientific authorities.

Chapter 21

Nutrition: iron and iron metabolism

21.1 Introduction

Metal ions, in association with proteins, play important catalytic roles in biochemistry, especially in acid-base and redox reactions. The metal ion used most frequently in the former type of reaction is zinc, whereas biological redox catalysts often include iron in their structures (sometimes copper, occasionally manganese).

Iron is found in the first row of the *d*-block of transition elements of the periodic table. It has the ability reversibly to take up or release an electron. The use of iron as the metal in biological systems for this purpose seems, in part, to have arisen due to its great abundance in the environment. Iron is the most abundant transition metal in the earth's crust. As noted, the chemistry of iron is utilized in biological systems when its ions are bound to proteins. Consequently its selection over other transition metals, such as cobalt or chromium (also capable of redox reactions), probably relates not only to its abundance but also to its ready association with, and dissociation from, its binding sites on proteins, i.e. its kinetic lability. The complexes formed, however, are thermodynamically stable.

It should be noted that the bioavailability of iron is much less with today's oxygen-rich atmosphere than it would have been in former times, when the earth's atmosphere was reducing. This is because, in the presence of oxygen and at pH 7, iron is precipitated out of aqueous solutions as the very insoluble ferric hydroxide. Consequently multicellular organisms have had to evolve complex systems for the uptake, transport and storage of iron. These systems simultaneously keep the iron from precipitating from solution and also help prevent it from catalysing damaging free radical-based reactions (see Chapter 33).

The major iron-bearing proteins in the body are not enzymes, however, but the dioxygen-binding and dioxygen-transport proteins myoglobin and haemoglobin. The relative amounts of iron associated with the various proteins of iron metabolism are given in Table 21.1. From this it may be seen that the total mass of iron in the adult human body amounts to some 4–4.5 g, i.e. that found in a 7 cm nail.

21.2 Iron balance

For an adult to stay healthy, the amount of iron lost each day must be replaced by an equivalent intake. Young, rapidly-growing children, on the other hand, will require a positive iron balance. It is useful, therefore, first to examine the factors that influence daily iron loss and, secondly, the daily iron intake and absorption from the gastrointestinal tract.

Table 21.1 **Distribution of iron in a typical adult (weight 70 kg)**

Tissue	*Form*	*Total weight* (g)	*Iron* (% total)
Red blood cell	Haemoglobin	2.72	70.5
Muscle	Myoglobin	0.12	3.2
Store (liver and spleen)	Ferritin (haemosiderin)	1.0	26.0
Blood plasma	Transferrin	0.003	0.1
Various	Cytochromes	0.0035	0.1
Various	Catalase	0.0045	0.1
Various	Other iron proteins		

From J.D. Cook, C.A. Finch and N.J. Smith (1976) *Blood*, **48**, 449–455

Iron losses

Iron is unique among the trace elements in that there is no excretory system for eliminating the metal from the body. It is, however, lost from the body when cells containing iron are lost. This most frequently involves the loss of erythrocytes in bleeding, but may involve the loss of other cells. The routes for loss of iron may be classified as either physiological or pathological.

Physiological iron losses

The adult male and the non-menstruating female lose about 1 mg iron per day. Of this total, 50–60% is accounted for by loss from the gastrointestinal tract, due either to biliary excretion or to the shedding of mucosal cells. The remainder is lost with hair, sloughing of skin or in the urine.

Losses during menstruation typically range, on average, from 1.4 to 3.2 mg iron per day. A survey conducted in Sweden in the early 1980s showed that the losses of blood do not, in any one individual, vary from one period to the next, but large interindividual differences do occur. Of the women studied, 50% lost iron at the bottom of the range stated, 25% lost 1.7 mg/day whilst 2.5% lost 3.2 mg/day.

Losses associated with underlying pathology

A range of conditions give rise to blood loss, but it is appropriate to defer the discussion of these until Section 21.9.

Table 21.2 **Relation between iron intake and energy intake**

Individual	*Recommended energy intake per day* (kJ)	*Iron* (mg/day)	
		Recommended	*Probable*
Reference man	13 400	5	10.5
Reference woman	10 000	14	7.7
Young female	10 500	12	8.8
Young male	10 500	5	8.8
Child (18 months old)	5 000	5	4.2

Iron intake

In the normal population, requirements for iron replacement range from 1.0 to 3.2 mg/day, and this must derive from dietary sources. The iron in food is either in the form of haem complexes, as found in myoglobin and haemoglobin, or in the form of non-haem iron. The latter still constitutes a bound form of the metal with binding being to proteins (usually) or polysaccharides. Although iron is widely distributed in foodstuffs, only a proportion, which may be 10% or less, is absorbed from the diet. For this reason, the minimal amount needed to be ingested daily is set at 10 times the amount actually lost daily. Table 21.2 shows the daily requirements of different groups in the UK.

The control of the absorption of iron is a central feature of whole-body iron homeostasis and it is appropriate to consider this process next.

21.3 The absorption of dietary iron and its control

It should be admitted at the outset that, despite intensive investigation of this process, which is so vital to human well-being, several aspects of the absorption of dietary iron and its control are poorly understood. The reasons for this undoubtedly relate, in part, to the problem alluded to previously: the great tendency of iron to precipitate from solution at pH 7 in the presence of oxygen. In order to try to combat this problem in the experimental setting, iron is frequently kept in solution by using (non-physiological) small organic chelating molecules (see Chapter 35). This renders the assessment of the physiological significance of any results difficult. An additional complicating factor in attempting to understand iron absorption is the great multiplicity of iron-containing compounds found in the diet.

Despite these areas of uncertainty concerning the absorption of iron from the duodenum and the jejunum, there is a consensus about many of the steps involved. It is universally accepted that the process of absorption is extremely tightly regulated. If an individual is iron replete then, almost irrespective of dietary intake, little absorption of iron occurs. This is a very important protective measure in view of the small losses of the metal that occur daily. Conversely, as iron deficiency begins to develop, absorption increases and may then be facilitated by optimal presentation in the diet.

When iron-containing foods enter the stomach, the proteins associated with the metal begin to be denatured by the acid conditions and to undergo pepsin-catalysed hydrolysis. This leads to the release of iron from its binding sites. The iron is kept in solution under these circumstances by its acid-dependent association with gastric mucins. (The requirement for this binding of iron in the acid of the stomach may help to explain why patients with **achlorhydrasia** tend to be iron-deficient.) Mucins are complex *O*-linked glycoproteins which appear in several secretions including saliva and those of the stomach. Iron bound to mucins remains in solution even as the pH rises to 7, as the partially-digested food enters the duodenum

As has been mentioned, the process of absorption of iron via the enterocytes in the intestine must fulfil two important criteria: it must favour iron uptake when the subject is depleted in iron, but must switch off as a state of iron repletion is achieved. This implies that the process of intestinal absorption of iron must have two cardinal features: it must be capable of the specific transport of iron in depletion and it must be responsive to some indicator of iron repletion. This degree of control involves the participation of a range of proteins each capable of the specific binding of the metal. It seems likely that the full range of these has not yet been determined, but significant advances have been made. The first of these proteins is integral to the luminal plasma membrane of the enterocytes and is a member of the family termed **integrins**. These molecules are most frequently encountered as mediating the interaction of cells with their supporting matrix. The enteric integrin strongly binds iron (probably transferred from the mucins), and by an unknown mechanism it helps effect the transfer of the iron to the cytoplasm of the enterocyte. It is proposed that the iron reaching the cytoplasm becomes bound to the iron-binding protein, termed **mobilferrin**. This protein, which has a single iron-binding site, appears to act as the 'iron sensor' of the body. In the iron-depleted state (Figure 21.1) it accepts iron from the integrin system and may transfer it (reversibly) to intracellular stores in association with a third iron binding protein, ferritin (see Section 21.5) or to the extracellular iron-transport protein transferrin (see Section 21.4). It is the amount of iron bound to transferrin that determines whether the metal is absorbed from the intestine. This comes about because the contraluminal enterocyte plasma membrane has transferrin receptors (see Section 21.6). These carry out the receptor-mediated endocytosis of the plasma iron-binding protein and, following uptake, the iron is released from its binding to transferrin. If large quantities of iron are released in this way, i.e. in iron repletion, the metal binds to mobilferrin, thereby preventing the uptake by this protein of integrin-bound iron (Figure 21.1).

Of course, whether the homeostatic system just described is able to maintain an iron-replete state must

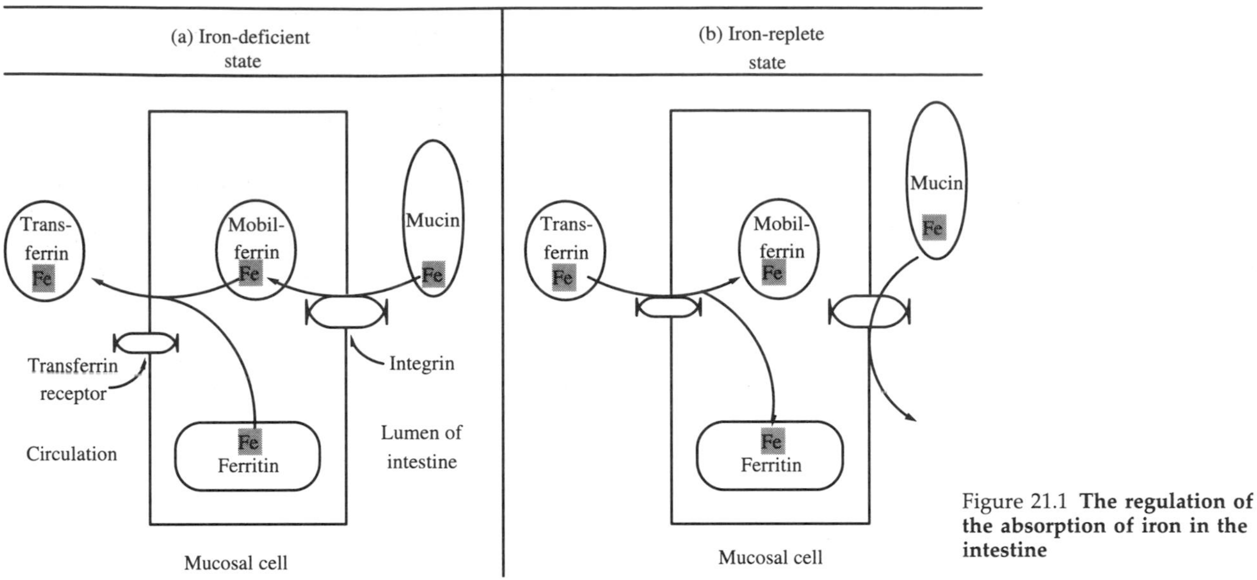

Figure 21.1 **The regulation of the absorption of iron in the intestine**

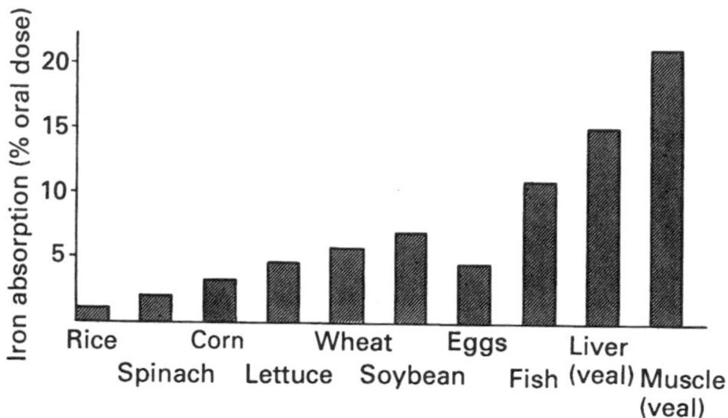

Figure 21.2 **Absorption of iron from vegetable and meat foods in human subjects:** 11–137 subjects used in each test

depend on the supply of dietary iron. If this is inadequate there is a real danger of iron-deficiency anaemia developing, i.e. the supplies of the metal are less than required for the formation of the correct quantities of haemoglobin. Humans, like most mammals, normally eat to satisfy their energy requirements and may easily select a diet that leads to an iron-deficiency anaemia. As is shown in Table 21.2, this is especially true of females. As previously noted, the recommended iron intake is 10 times the estimated loss, which is based on the assumption that only about 10% of dietary iron is absorbed. But, as is shown in Figure 21.2, the actual iron absorbed may be even less than this, depending on the nature of the food source. One consequence of the relatively low energy intake of females is that dietary iron intake is often inadequate, rendering them more prone to iron-deficiency anaemia.

21.4 Iron transport: transferrin

Iron is transported in the plasma bound to the glycoprotein **transferrin** which, on electrophoresis of plasma, migrates with the β-globulin fraction. The binding serves the dual function of keeping the iron in solution and preventing it from promoting reactions that give rise to deleterious free radicals (see Chapter 33). Humans also produce a closely-related protein, **lactoferrin**, which is found in milk and tears (it is also produced by neutrophils). These sources of lactoferrin suggests that it plays a protective role as part of the innate immune system (see Chapter 32), and indeed it has been shown to have antibiotic actions. For a while it was believed that this arose because lactoferrin, in binding iron, denied bacteria access to the metal. However, recently, non-iron-binding fragments of lactoferrin have been shown to be bacteriocidal.

Plasma transferrin is composed of a single polypeptide chain with a molecular mass of about 80 kDa. The carbohydrate moiety, whose molecular mass is about 4.5 kDa, consists of two branched heterosaccharide chains attached to asparagine residues, i.e. they are *N*-linked, in the C-terminal half of the molecule. There are two iron-binding sites in the transferrin molecule, one in each of two halves (lobes) of the protein (Figure 21.3).

Iron is always transported in the ferric form. In the normal, healthy human adult, approximately one-third of the total iron-binding sites is occupied. A convenient means of assessing the saturation of transferrin depends on the fact that the protein is stabilized by binding iron. As a consequence it is less easily denatured by exposure to urea as the number of iron ions bound increases. Denaturation changes the electrophoretic mobility of the protein on polyacrylamide gels as the exposure of charged groups and the overall shape changes. Consequently four bands are detected, corresponding to transferrin with no iron bound, with iron bound to the N- or the C- terminal lobe or bound to both lobes. From this the degree of its saturation may be deduced.

The degree of saturation of transferrin may be changed in pathological conditions such as iron-deficiency anaemia, pernicious anaemia or haemochromatosis, and hence saturation determinations give a sensitive indication of anaemia and iron overload.

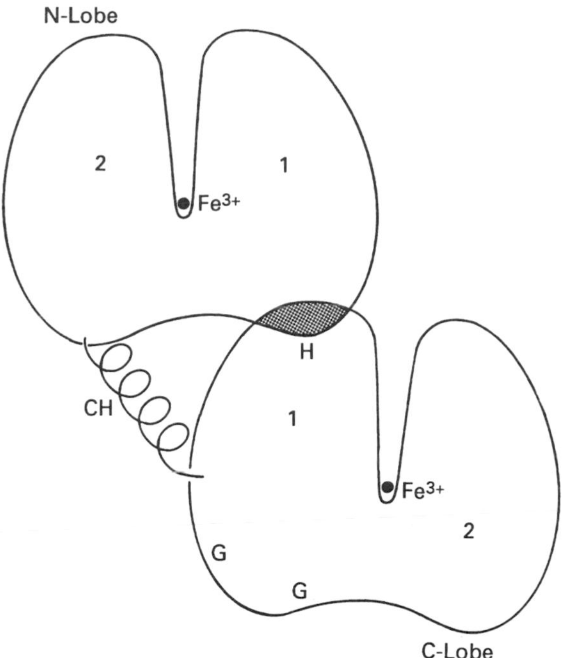

Figure 21.3 **Schematic diagram of the lactoferrin molecule.** Note: The transferrin molecule is thought to have a very similar structure. The two domains in each lobe are labelled (1) and (2). The site of the iron binding in each lobe is shown. A region where hydrophobic interactions between the two lobes are made is indicated (H). The helix connecting the two lobes is labelled CH. Carbohydrate attachment sites are labelled G

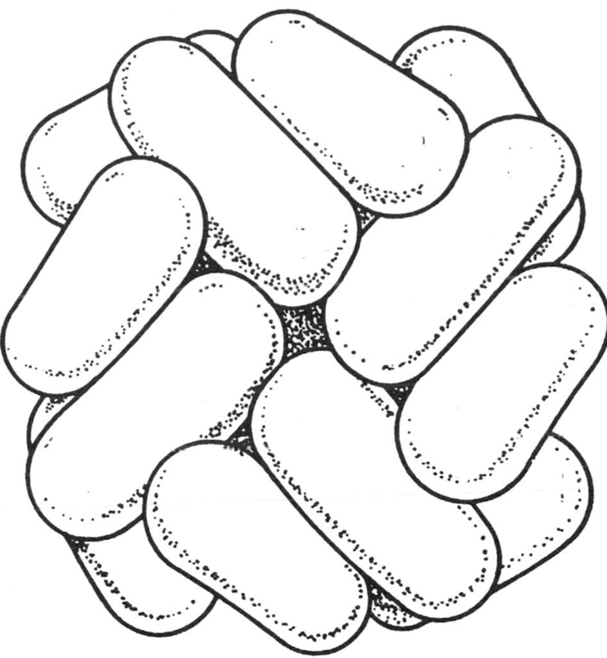

Figure 21.4 **The organization of the polypeptide chain of ferritin**

Transferrin and several related molecules have been crystallized and their structures determined by X-ray crystallography. This has revealed, that in both lobes, four of the coordination positions of the iron are occupied by side-chains of amino acids in the protein. These are two phenol side-chains of tyrosine residues, one carboxylate group from aspartate and one imidazole group from histidine. The fifth and sixth positions are occupied by a bidentate carbonate ion.

21.5 Iron storage: ferritin and haemosiderin

As mentioned previously, iron is incorporated into the iron-storage protein ferritin, in enterocytes, but the main stores of iron in the mammalian body are in the liver and spleen.

Ferritin

The protein component of **ferritin** (i.e. **apoferritin**) is composed of 24 subunits arranged spherically. Two types of subunit have been described, an H (heart) form of molecular mass 21 kDa and an L (liver) form with a molecular mass of 19 kDa. Although heteropolymers of ferritin occur, the predominant form in the heart is H_{24} while in the liver and spleen it is L_{24}. Consequently the L form of ferritin is the major one in the body. In iron overload, hepatic ferritin increases and some of the protein may escape into the circulation. Consequently the extent of iron overload in the liver may be assessed by measuring circulating concentrations of ferritin.

Each apoferritin molecule can take up a very large amount of iron. A saturated ferritin molecule will carry about 4500 iron atoms, i.e. one for each amino acid residue. More commonly, normal human hepatic ferritin has about 2000 iron atoms bound. Uptake of iron by ferritin is accompanied by its oxidation to the ferric form and the protein has ferroxidase activity. Once uptake of iron has commenced, however, the oxidation appears to happen spontaneously. The iron is not stored at specific sites, as is the case with transferrin, but it exists principally as a ferric oxyhydroxy-phosphate complex in the centre of a protein sphere (Figure 21.4).

Haemosiderin

In addition to storage as ferritin, iron can also be found in a form described as '**haemosiderin**'. Particles of haemosiderin are very large and can be detected under the light microscope. However, the precise nature of haemosiderin is obscure. It appears to be an ill-defined insoluble agglomerate of hydrated iron oxide with several organic constituents. It may be formed as a result of the partial degradation of the protein of ferritin by lysosomal proteases, followed by the release of iron oxide mixed micelles to form insoluble aggregates trapped in secondary lysosomes. Normally, very little haemosiderin is to be found in the liver, but the quantity increases steadily during iron overload. The iron in

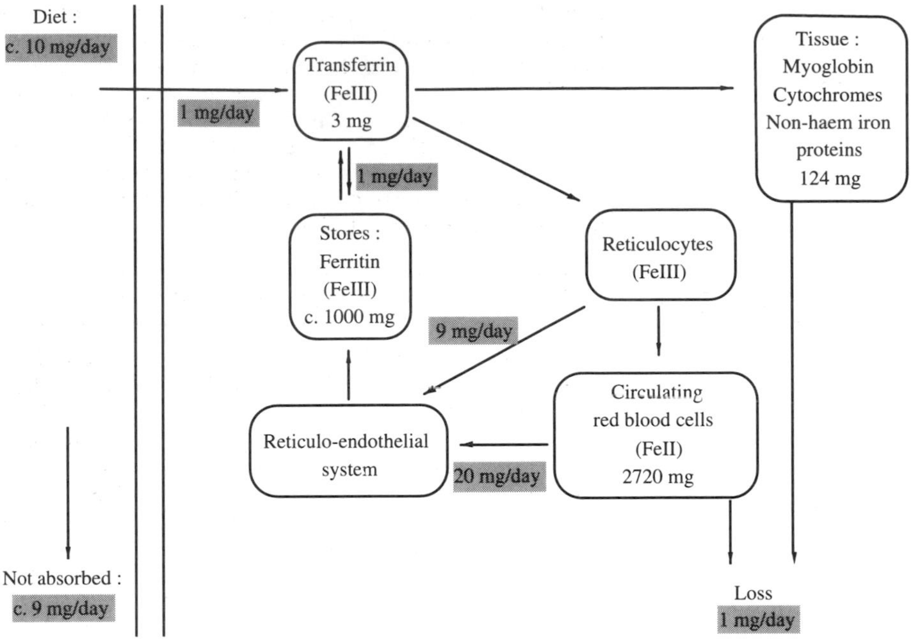

Diet :
c. 10 mg/day
Transferrin (FeIII) 3 mg
1 mg/day
1 mg/day
Tissue : Myoglobin Cytochromes Non-haem iron proteins 124 mg
Stores : Ferritin (FeIII) c. 1000 mg
Reticulocytes (FeIII)
9 mg/day
Circulating red blood cells (FeII) 2720 mg
Reticulo-endothelial system
20 mg/day
Not absorbed :
c. 9 mg/day
Loss
1 mg/day

Figure 21.5 **A summary of the kinetics of iron**

haemosiderin is less accessible than that of ferritin, and haemosiderin formation may represent a secondary protective mechanism against iron overload.

21.6 The kinetics of iron

It is now appropriate to consider the movement of iron from tissue to tissue and from complex to complex (Figure 21.5).

The catabolism of erythrocytes that occurs continually in the cells of the reticuloendothelial system results in the daily release of some 20–25 mg of iron. This may be taken up by transferrin but may also be diverted to ferritin if catabolism of the red cells is excessive. As explained, a small amount of iron is derived from the diet (1 mg) and this only constitutes about 10% of the total dietary intake. The iron is then transported to the bone marrow bound to transferrin where it is taken up by the reticulocytes (developing red cells). The uptake of iron bound to transferrin by reticulocytes and by many other cell types depends on a process of receptor-mediated endocytosis (Figure 21.6). This happens in a four-stage process:

1. Binding of transferrin to its specific receptors, which are integral proteins of the erythrocyte plasma membrane.
2. The transferrin–receptor complex undergoes clathrin-dependent endocytosis (see Chapter 7).
3. As part of the endosomal compartment, the vesicles formed during endocytosis are acidified down to about pH 5.5. This causes any bound iron to dissociate from its binding site on transferrin, but the iron-free (apo) transferrin remains bound to the receptor.
4. The apotransferrin–transferrin receptor complex then recycles back to the plasma membrane, where the former protein is released into the circulation.

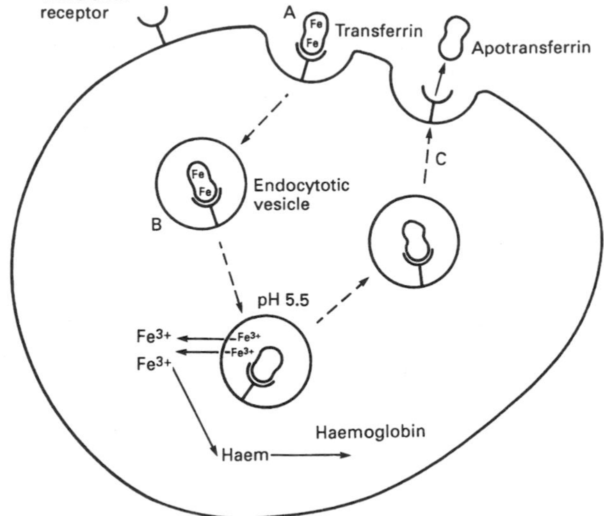

Figure 21.6 **Transfer of transferrin iron to reticulocytes.** The process involves three distinct stages: (A) the binding of transferrin onto receptors on the reticulocytes; (B) the entry of the transferrin–receptor complex into the cell by endocytosis and the release of the bound Fe^{3+} atoms; (C) recycling of the apotransferrin–receptor complex to the cell surface and the release of apotransferrin

The overall process is illustrated in Figure 21.6. This represents the general way in which iron is presented to cells. These include intestinal mucosal cells where uptake of transferrin helps to regulate the absorption of dietary iron (see Section 21.3).

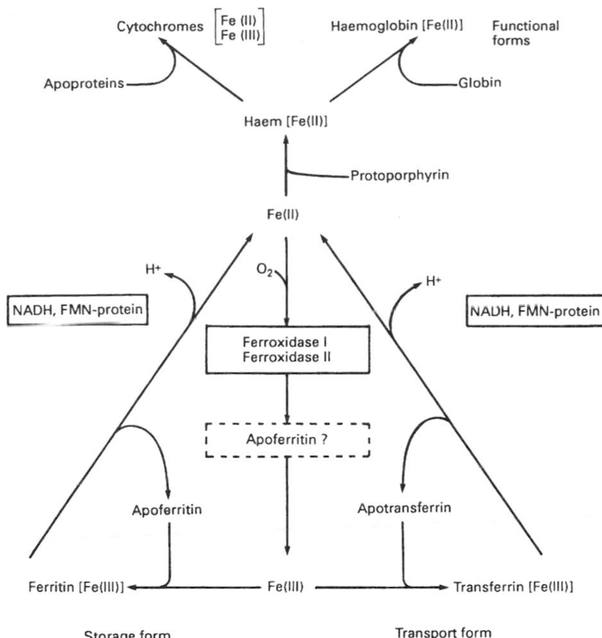

Figure 21.7 **Valency states of iron and ferrous–ferric cycles in iron metabolism.** Note that although iron is stored and transported in the ferric (III) form, it must be reduced to the ferrous (II) form before release from the store and then reoxidized to Fe(III) for the uptake by transferrin. A similar process occurs when iron is transferred from transferrin to ferritin (FMN, flavin mononucleotide)

21.7 Valency of iron during metabolism

It is well known that iron in its main physiological form, haemoglobin, is in the reduced or FeII state. Yet iron is transported on transferrin and bound to ferritin in the oxidized form FeIII.

The valency of iron is physiologically of great importance. Thus neither haemoglobin nor myoglobin will bind dioxygen if the haem iron is present in the FeIII form as in metmyoglobin or methaemoglobin. Conversely, although FeIII is strongly bound to ferritin it is released rapidly *in vitro* when reducing agents such as ascorbic acid are present.

From these considerations it is clear that the tissues must possess well-regulated mechanisms for setting the oxidation state of iron. The nature of these processes is still under investigation and current knowledge is summarized in Figure 21.7. As indicated, the release of iron from ferritin requires its reduction to the FeII form, but how this is achieved is still to be resolved.

The binding of FeII to transferrin or ferritin requires catalytic activity termed '**ferroxidase**'. Both proteins have intrinsic ferroxidase activity and consequently FeII can be oxidized after its binding. In the case of transferrin, however, a second protein, the copper-containing protein **ceruloplasmin**, appears to oxidize the iron prior to it binding to transferrin. This route to oxidation of FeII appears to be more important than that catalysed by transferrin itself. (It should be noted that caeruloplasmin may also act as a plasma copper-transport protein.) The role of caeruloplasmin in oxidizing iron prior to its binding to transferrin helps to explain the well-known observation that copper deficiency may be associated with the development of anaemia. In the absence of its oxidation in the intestine, iron could not be transported from the enterocytes. As mentioned above (see Section 21.5), apoferritin has intrinsic ferroxidase activity.

21.8 Intracellular iron homeostasis

The question posed about how whole-body iron homeostasis is achieved is equally pertinent when asked about events at the cellular level. In principle, and in practice, control of the amount of iron in a cell may be achieved by two means: regulation of its uptake or control of the extent to which it is stored. The former implies that cells will be able to regulate the number of transferrin receptors on their surfaces and the latter that the amount of apoferritin they produce is regulated. Recent studies have shown that the mRNAs for the two proteins in question have, as part of their structures, sequences referred to **iron regulatory elements** (**IREs**). Cells also contain an iron-binding protein designated an **iron-response factor** (**IRF**). The IRF is a cellular iron-sensor and evidence shows it to be a cytosolic protein very similar to the tricarboxylic acid cycle enzyme aconitase. Aconitase is itself an iron-sulphur-containing protein. When intracellular iron concentrations are low, little of the metal binds to IRF which is free to bind with high affinity to the IREs of transferrin and ferritin. The functional effect of this is to stabilize the transferrin receptor mRNA, thereby prolonging its half-life. This results in the synthesis of more receptor protein and consequently more iron uptake into the cell. Conversely the binding of IRF to the IRE of ferritin mRNA inhibits the translation of the message, which results in a decline in the intracellular concentration of ferritin. In the iron-replete, state IRF/FeIII complexes form and these have a low affinity for the IREs of the two mRNAs. This results in an unstable mRNA for the transferrin receptor, but enhanced translation of the ferritin mRNA. The effect of these two changes is to decrease cellular uptake of transferrin and to divert any iron that is taken up on to apoferritin and hence into storage (Figure 21.8).

21.9 Iron pathology

If the iron content of the tissues is too low, **iron-deficiency anaemia** will develop; if too high, the disease **haemochromatosis** is seen.

Iron-deficiency anaemia

Anaemia is a condition characterized by a reduced number of circulating red blood cells, or a reduced amount of haemoglobin in the cells, or both. Anaemias may be classified into three main types, iron deficiency not being the only cause.

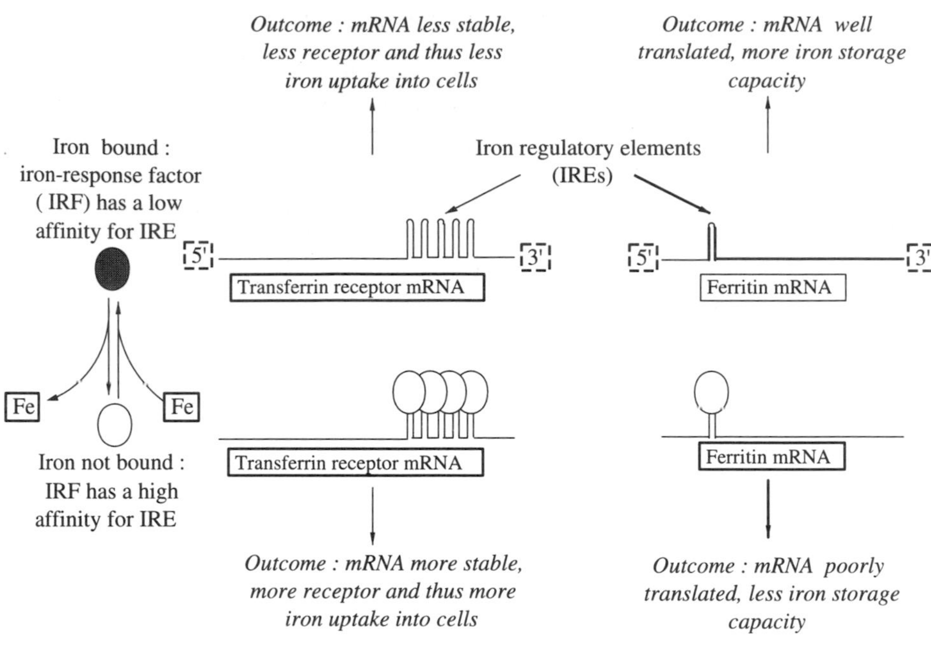

Figure 21.8 **Intracellular iron homeostasis is maintained by reciprocal changes in the rate and extent of translation of the mRNAs for the transferrin receptor and ferritin**

Normocytic

The red blood cells are unchanged in size and haemoglobin content, but their number in the circulation is reduced. This type of anaemia is often the result of external or internal haemorrhage, e.g. the 'quiet', i.e. undetected, bleeding that can occur with carcinoma of the colon.

Macrocytic

The size of the red blood cells is increased, and the numbers are reduced. This type of anaemia is usually caused by deficiency of folic acid or vitamin B_{12} (see Chapter 22). This, therefore, is a characteristic of **pernicious anaemia**.

Microcytic

The cells are much smaller than normal and have a much reduced haemoglobin content. The number of cells may be reduced, but to a lesser extent than in other forms of anaemia. Microcytic anaemia is caused by iron deficiency.

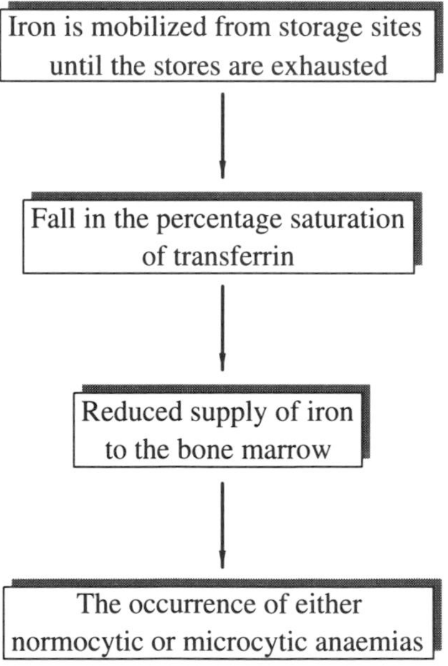

Figure 21.9 **A summary of the events that lead to the development of iron-deficiency anaemia**

Causes of anaemia

Iron-deficiency anaemias can result from:

1. Inadequate dietary intake or absorption of iron.
2. Abnormal losses of iron.

The development of anaemia can be traced through a sequence of events as is shown in Figure 21.9 and the observed changes associated with these events are recorded in Table 21.3. From this it is noteworthy that 'anaemia' is often the final end-product of iron deficiency and that many earlier indicators of iron deficiency short of anaemia may be recognized.

Clearly, it is desirable to diagnose, and thus prevent, iron deficiency before anaemia develops. This is usually accomplished by measuring the concentration of iron in the plasma and the total iron-binding capacity of transferrin. Normally the percentage saturation of transferrin is about 33%, but if it is less than 15% the subject is regarded as suffering from an incipient iron-deficiency condition.

Table 21.3 **Sequential stages in iron deficiency**

	Normal	*Iron depletion*	*Iron-deficient erythropoiesis*	*Iron-deficiency anaemia*
Iron stores	+ + + +	[+]	[0]	[0]
Haem iron	+ + + +	+ + + +	+ + +	+ +
Transferring (μg/dl)	330	360	390	410
Plasma ferritin (μg/l)	100	[20]	[10]	[<10]
Iron absorption	Normal	↑	↑	↑
Saturation of transferrin (%)	35	30	[<15]	[<10]
Erythrocytes	Normal	Normal	Normal	[Microcytic Hypochromic]

[ ] indicates significant deviation from normal values.

As indicated previously, the measurement of circulating concentrations of ferritin gives a sensitive measurement of total body iron stores. The normal circulating concentration of ferritin is about 100 μg/ml for males and 40 μg/ml for females.

Inadequate intake or absorption

This may arise either because the diet is low in iron or because the iron present in the diet is poorly absorbed. Poor absorption may occur, for example, as the result of components of the diet binding iron and thus rendering it unavailable. Following complete or partial gastrectomy an iron-deficiency anaemia may arise and this probably relates to the absence of acid-dependent binding of iron to mucins. Similarly, subjects with a failure to produce gastric acid (achlorhydrasia) are frequently iron deficient. In many tropical and sub-tropical countries, hookworm infections are a major cause of iron-deficiency anaemia. The hookworms appear to compete with the intestinal uptake system for the iron. As many as 500 million people may be infected and, for some of these, daily losses may be equal to 500 mg of iron. Clearly these losses are not accounted for simply in terms of a failure to absorb dietary iron: they must include iron lost from the body, including that from shed intestinal mucosal cells.

Abnormal losses

There are three principal causes of abnormal loss:

1. Excess menstrual blood loss.
2. Losses from the gastrointestinal tract caused either by blood loss or by an excess sloughing off of mucosal cells.
3. Pregnancy, which makes considerable demands on maternal iron stores, and in some cases this is only partially compensated by cessation of menstruation.

In summary, although the condition of iron-deficiency is described as 'iron-deficiency anaemia', we now know that full-blown anaemia is the end result of an 'iron-deficiency state'. Iron deficiency is commonly treated by the oral administration of an iron supplement, for example, 200 mg of iron daily as $FeSO_4$ is given to adults. It has been shown that, in deficiency, the daily absorption of iron can be doubled or tripled simply by taking the iron salt with orange juice or other foods rich in vitamin C.

Haemochromatosis

At the opposite end of the spectrum, there are clinical conditions in which excessive deposits of iron are present in the tissues, particularly in the normal iron-storage organs – the liver and the spleen. In severe cases the total iron in these organs can increase from the normal value of about 1 g to 40–50 g. Many complex clinical symptoms can result from this overload, the chief of which are :

(a) Hepatosplenomegaly and cirrhosis
(b) Skin pigmentation, presenting in Caucasians as a dusty brown coloration
(c) Diabetes mellitus in 60–80% of patients
(d) Hypogonadism and especially atrophy of the testes
(e) Joint disease
(f) Heart disease.

The causes of this multiple organ failure are not established, but probably relate to the free-radical-related effects of free iron (see Chapter 33). The most reliable diagnosis of iron overload may be achieved by the determination of serum ferritin. The concentration of this protein may increase from 40–100 μg/l to 3–6 mg/l in severe cases.

Development of haemochromatosis

Haemochromatosis can arise from several different causes, which can be subdivided into primary and secondary groups.

Primary causes

Genetic abnormality

Hereditary haemochromatosis has an estimated frequency of homozygosity of 0.2–0.45% in Caucasian populations. These individuals absorb slightly more iron than is normal. Consequently, there is a progressive overload of iron which, if undetected, leads to massive deposition of the metal in parenchyma, especially in the liver. When the disease reaches the stage of large-scale deposition in the organs, a number of manifestations may be apparent. These include many of the signs and symptoms listed above: cirrhosis, diabetes mellitus, arthritis and cardiac disease. Early diagnosis is, therefore, important and may be achieved by measuring transferrin saturation or the circulating levels of ferritin. If necessary, the diagnosis may be confirmed in adults by liver biopsy.

The nature of the mutation(s) involved is not resolved, but the affected gene seems to reside on chromosome 6 at a locus that is close to the genes for the class III HLA-antigens (see Chapter 32).

Excess dietary intake

The normal homeostatic mechanisms controlling iron absorption from the gut appear to be overridden in some individuals and as a consequence haemochromatosis develops. This is seen in striking form in some black South Africans. In this case the iron is present in traditional alcoholic beverages brewed in iron vessels, the iron being dissolved from the containers as a result of the low pH of the beverage. The presence of ethanol and also fructose appears additionally to facilitate the intestinal absorption of iron. As a result, it is not uncommon to find individuals whose daily iron intake may be 100 mg.

Secondary causes

Alcoholic cirrhosis

In alcoholics who have developed cirrhosis of the liver, iron overload is commonly encountered and is likely to relate to an ethanol-induced increase in iron absorption, especially when the beverage is high in iron, as is the case of some ciders and wines and the home-brew of the South Africans referred to above.

Portacaval shunt

Excess deposits of iron are frequently found in the livers of patients who have had a 'portacaval shunt'. This operation is performed to divert some of the flow of blood around the liver when the circulation is obstructed in cases of liver disease. The mechanistic basis for this observation, however, is not clear.

Chronic pancreatitis

Experiments in animals and observations in humans have both indicated that pancreatic juice may play an important role in the regulation of iron absorption, because a reduced secretion of pancreatic juice leads to a marked increase in iron absorption, and thence to iron deposition in the tissues. The mode of action of pancreatic juice in restricting iron absorption is not known, but the increase in the pH of the chyme caused by pancreatic juice may favour the oxidation and precipitation of iron in the duodenum.

Repeated blood transfusions

Excess deposits of iron are almost bound to occur in patients who have to receive regular blood transfusions for (usually inherited) blood disorders, e.g. haemoglobinopathies or thalassaemias. The reason for this is that a transfusion of 250 ml of blood delivers about 500 mg iron into the patient's bloodstream and, as explained, humans have no means of excreting iron.

Treatment of haemochromatosis

Treatments are based on one of two principles: venesection or chelation therapy.

Venesection therapy

Patients with inherited haemochromatosis are initially treated by the initial weekly withdrawal of 500 ml of blood. Once the iron stores are reduced to near normal, lifelong maintenance therapy is by the removal of further blood at 3–4 month intervals. It is usually considered to be necessary, in addition, to place these patients on a low-iron diet.

Chelation therapy

In this type of treatment (see Chapter 35) attempts are made to remove iron from its stores by means of powerful iron-chelating drugs. One such agent that has been employed is desferrioxamine (Figure 21.10). Removal of the stored iron is not easy to accomplish, because the binding in ferritin and haemosiderin

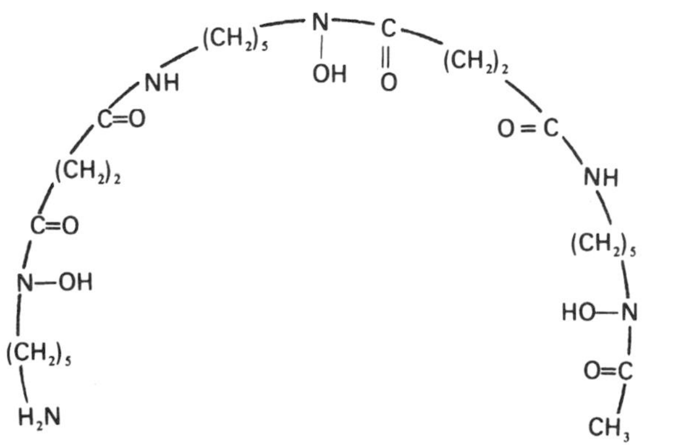

Figure 21.10 **Structure of desferrioxamine.** Desferrioxamine is a very powerful iron-chelating agent which has been used extensively to remove excess iron deposits from the body. It is a naturally occurring compound synthesized by the mould *Streptomyces* sp.

complexes is very tight. As has been noted, however, iron may be released from the former if the iron is reduced to FeII. *In vitro* this may be achieved by using ascorbic acid, and there is evidence that treatment with desferrioxamine results in more urinary excretion of iron if the therapy is combined with the administration of large doses (grams) of vitamin C.

One problem associated with the use of desferrioxamine in dealing with iron overload is that the compound is not effective orally, due to its poor absorption from the gastrointestinal tract. Thus the drug is normally infused into patients, leading to very low compliance. Even when patients are able to receive the drug by this route, there is now evidence that treatment with desferrioxamine can give rise to neurological complications. For these reasons a good deal of interest is focusing on the use of relatively simple, water-soluble iron-chelating drugs such as the hydroxypyridinones for treating patients presenting with iron overload.

Chapter 22

Nutrition: folate and vitamin B_{12}

22.1 Introduction

It is appropriate to delay a consideration of the two important water-soluble vitamins, folic acid and vitamin B_{12}, until the chapters on the biochemistry of blood, because their discovery is very much associated with investigations of the occurrence of the macrocytic anaemias.

Pernicious anaemia, although no doubt widespread for many centuries, was first clearly described by Thomas Addison in 1849. This is sometimes overlooked because the monograph also described the disease of adrenal hypofunction that Charles Brown-Sequard named 'Maladie d'Addison'. Pernicious anaemia was usually fatal, but with a characteristic insidious onset after the age of 40. It is characterized by a generalized weakness, loss of weight, diarrhoea and vomiting. Clear neurological symptoms are apparent in most cases, with tingling of the extremities and unsteadiness of the gait: the spinal cord and peripheral nerves are clearly involved.

It was not until 1926 that any form of treatment was available. In that year Minot and Murphy, working in the USA, showed that the feeding of raw liver or extracts from the tissue brought about a considerable improvement in the condition. A dietary factor present in the liver thus appeared to be essential for the prevention of the disease.

Further observations relating to the occurrence of macrocytic anaemias were made in the 1930s when Indian patients suffering from tropical sprue were investigated. In this condition there is serious malabsorption from the gastrointestinal tract but, in addition, patients present with an anaemia very similar to pernicious anaemia, in that the blood was found to contain many macrocytes. However, the neurological complications were absent.

By 1945 a 'green-leaf' factor named folic acid (Figure 22.1) had been isolated and characterized. Folate cofactors consist of a bicyclic pteridine ring with a hydoxyl group on position 4 and an amino group on position 2. This is connected via a methylene bridge at position 9 to a *para*aminobenzoic acid residue which, in turn, has a variable number of glutamyl residues attached by peptide bonds (Figure 22.1 shows just 1). Folic acid is the oxidized form of the vitamin and arises in nature only when the biologically active reduced forms di- and tetra-hydrofolate undergo spontaneous chemical oxidation.

When patients with tropical sprue were injected with folic acid, there followed a complete recovery from anaemia. Indeed, oxidized folic acid is the form of the vitamin in current therapeutic use. Similar trials on pernicious anaemia patients were, however, less satisfactory: in many, the anaemia improved, but the neurological symptoms remained. It was clear that folate was not the factor whose deficiency gave rise to this disease.

This caused the work to refocus on liver extracts and by 1956 the purification of a new vitamin, B_{12} or '**cobalamin**', had been achieved and in 1964 its structure was determined by Dorothy Hodgkin and her colleagues (Figure 22.2). The elucidation of the chemical structure of vitamin B_{12} by X-ray crystallographic means was a major achievement in biochemistry, because this cobalt-containing compound is by far the most complex of the vitamins. The heart of the structure consists of a porphyrin-like ring with cobalt at its centre. This is referred to as a corrin ring, and at right angles to it is a nucleotide group. The sixth coordination position of the cobalt is usually occupied either by a methyl group or by a deoxyadenosyl group, but there may be a hydroxyl or a cyanide group. In the first structure (methyl-cobalamin), CoIII is present whereas in deoxyadenosyl-cobalamin it is CoII.

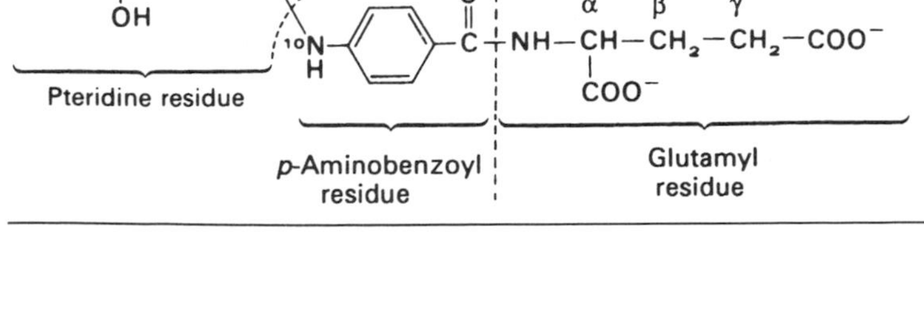

Figure 22.1 **Folic acid structures: note that this is the monoglutamate form of folic acid.** Additional glutamyl residues are added by the formation of peptide bonds with the terminal carboxylate group and the amino group of the next glutamate residue

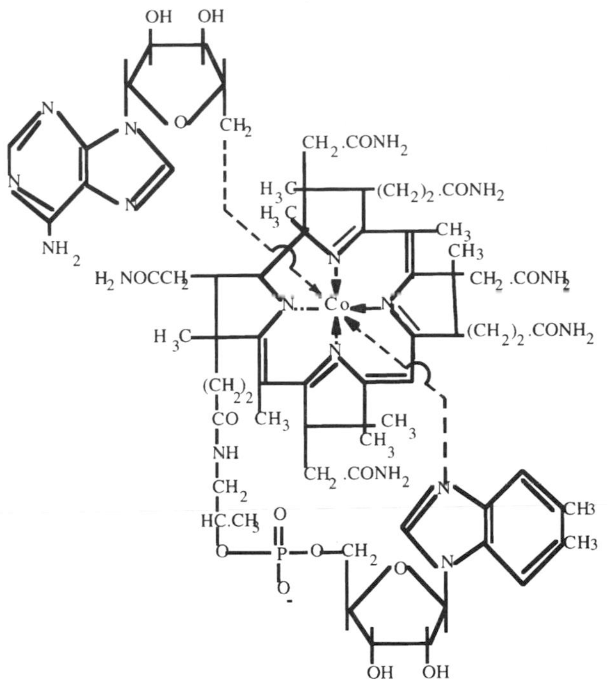

Figure 22.2 **Vitamin B_{12}–cobalamin.** The substituent is usually a methyl- or a deoxyadenosyl group

When injected into patients with pernicious anaemia the cobalamin derivatives cause a complete remission.

22.2 Absorption and distribution of folate and vitamin B_{12}

Folate

Folate is found mainly in green vegetables, liver and whole grains. The reference nutrient intake (RNI; see Chapter 14) is 50 μg/day.

In the diet, folic acid may possess 1, 3, 5 or 7 glutamyl residues attached to the *para*aminobenzoic acid in peptide linkages. Consequently the molecules have multiple negative charges and cannot traverse biological membranes by passive diffusion. This is a virtue in the tissues, where polyglutamylation serves to sequester folate in the cells in which it is required. In the lumen of the intestine, however, all but one of the glutamyl residues must be removed by hydrolysis prior to absorption. This is achieved by two γ-glutamyl hydrolases (sometimes referred to, misleadingly, as 'folate conjugases'). One is associated with the outer aspect of the lumen mucosal cell plasma membrane and the other is bound to the inner aspect of the membrane (Figure 22.3). The major circulating form of folate in the blood, whether derived from mucosal cells or from liver stores, is ***N^5*-methyltetrahydrofolate** and the normal concentration range is 5–15 ng/ml. Following its uptake into the mucosal

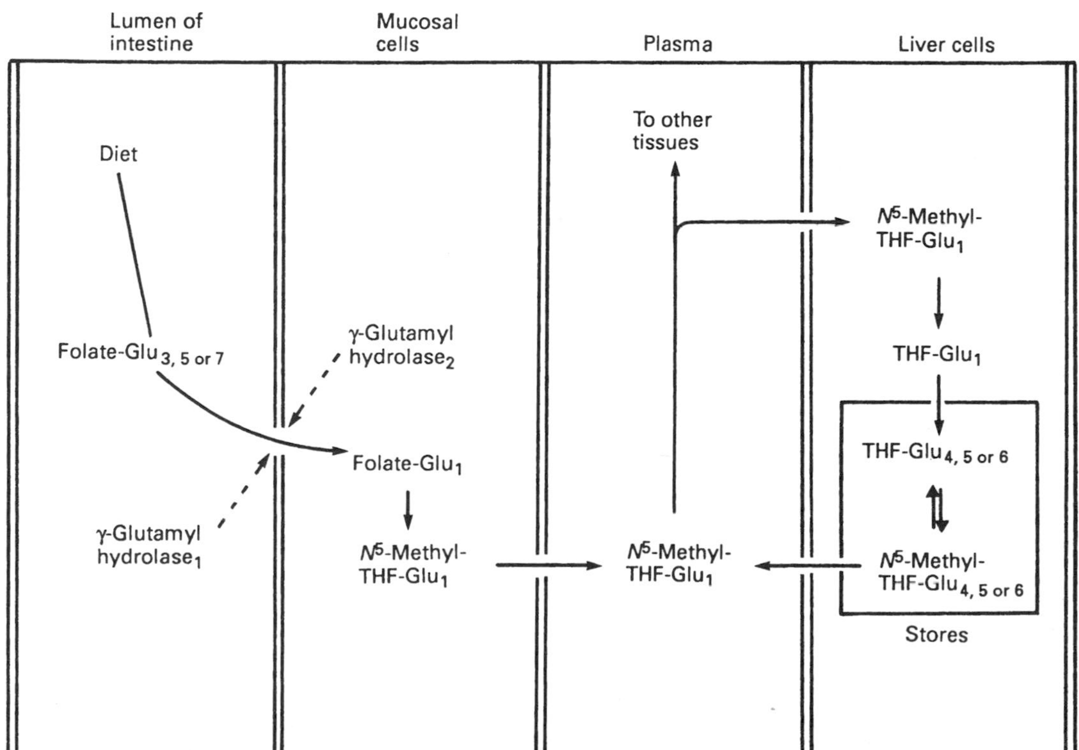

Figure 22.3 **The absorption, distribution and storage of dietary folate.** ($Glu_{1,3}$ etc., number of glutamyl residues; THF, tetrahydrofolate)

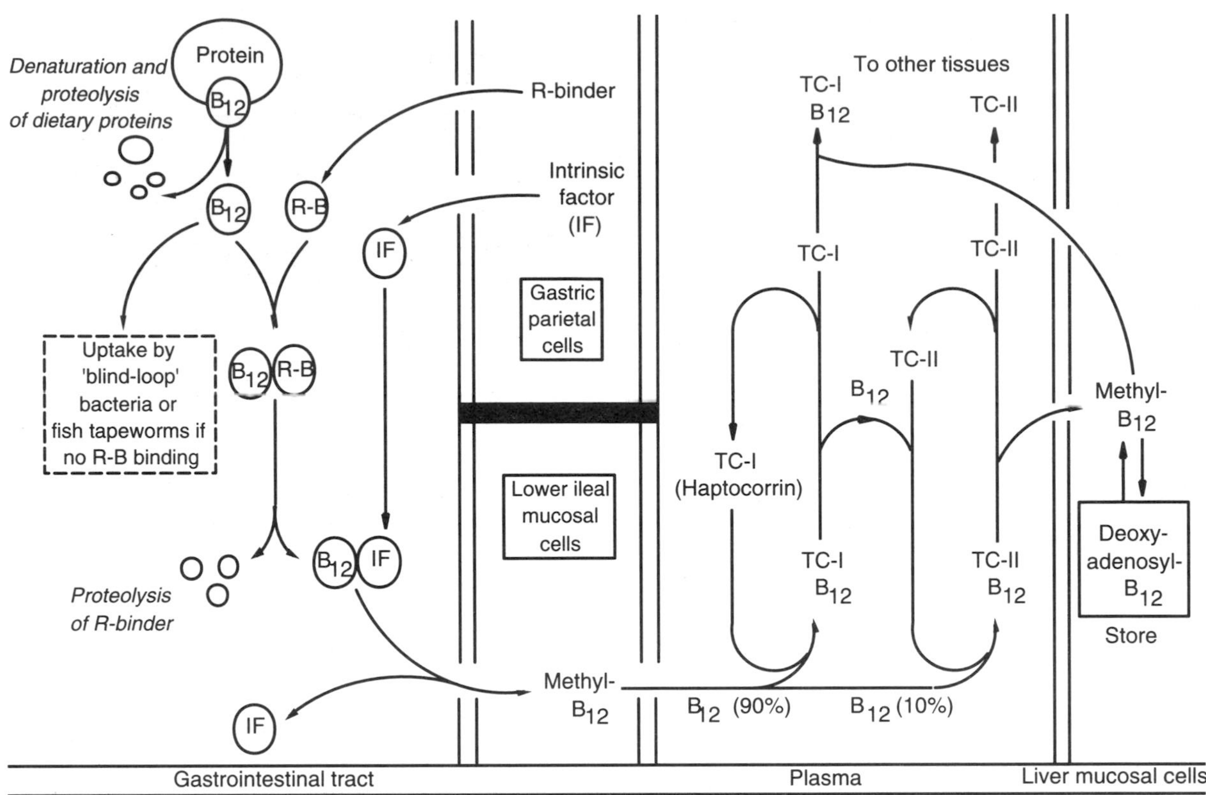

Figure 22.4 **Absorption, distribution and storage of vitamin B_{12}**

cells, folate must undergo reduction and methylation. Once it arrives in the liver, the methyl derivative is taken up, undergoes demethylation and the tetrahydrofolate is converted into polyglutamyl forms with 4, 5 or 6 residues. The polyglutamyl forms constitute a store of the vitamin. Prior to release from the store, and into the circulation, the compound is converted to the N^5-methyl-derivative, from which all but one of the glutamyl residues is removed.

Vitamin B_{12}

Vitamin B_{12} in most diets considerably exceeds the SRNI of 1 μg/day. The vitamin is synthesized by microorganisms, but it enters into the food chain only in food of animal origin.

The handling of dietary vitamin B_{12} is more complicated than that of most of the water-soluble vitamins, in many ways resembling that of iron. Thus, the acid conditions and the protein hydrolysis that commences in the stomach jointly cause the release of vitamin B_{12} (Figure 22.4). In saliva and also in gastric secretions are a group of mucins, termed **R-binders**. These glycoproteins bind vitamin B_{12} with high affinity in the acid environment of the stomach. In the more neutral surroundings of the duodenum the R-binders undergo hydrolysis, thereby releasing their vitamin B_{12}. The released vitamin is then bound by a second glycoprotein of gastric origin, known as **intrinsic factor**. It is stable in the duodenum and, having a high affinity for vitamin B_{12} at neutral pH values, binds it strongly and carries it to the terminal ileum. Intrinsic factor binds only biologically active forms of the vitamin. It has been proposed that it also serves to prevent the uptake of the vitamin by bacteria, as may be found in 'blind' or 'stagnant' loops of the intestine. In the ileum the intrinsic factor–vitamin B_{12} complex undergoes specific receptor-mediated endocytosis. This is a process that is dependent on Ca^{2+}. Once in the mucosal cells of the ileum, the vitamin is converted, if necessary, into its main plasma transport form of methylcobalamin.

The transport of vitamin B_{12} in the plasma occurs in association with two major binding proteins: **transcobalamins I** and **II** (TCI and TCII; the former is also known as **haptocorrin**). In the circulation, haptocorrin carries 90% of the vitamin, with the rest being associated with TCII; however, whereas vitamin B_{12} is rapidly cleared from the circulation when it is bound to TCII, the same is not true for the haptocorrin-bound form. For this reason it is believed that TCII mediates tissue uptake of vitamin B_{12}, while haptocorrin acts as a repository. Transfer of vitamin B_{12} from TCI to TCII, therefore, occurs in the circulation. Excess vitamin B_{12} is taken up by the liver, where it is stored as the deoxyadenosyl form of the vitamin.

22.3 Interrelationships between folate and vitamin B_{12}

The similarities of the pathologies associated with the deficiencies of folate and B_{12} pointed strongly to a

Table 22.1 **Relationships of folate and vitamin B_{12} deficiencies**

Features associated with folate or vitamin B_{12} deficiencies	*Folate deficiency*	*Vitamin B_{12} deficiency*
Urine formiminoglutamate	↑	↑
Serum folate	↓	↑
Serum vitamin B_{12}	↑ ↓	↓
Red blood cell folate	↓	↑ ↓
Response to large-dose folate	+ +	+ ?
Response to large-dose vitamin B_{12}	+ ?	+ +
Neurological damage	Absent	+ + + +
Methylmalonate excretion	Absent	+ + + +

relationship between the functions of the two vitamins. Before considering their individual roles it is useful to compare their effects on disease and on metabolism. These are summarized in Table 22.1 and explanations are presented in the following sections. One of the most significant observations is that serum folate actually rises during vitamin B_{12} deficiency and this has an important bearing on interpretations of the modes of action of the two vitamins.

22.4 Modes of action of folate and vitamin B_{12}

Folate

An outline of the metabolism of folate is presented in Figure 22.5. Folate is reduced in a two-stage process by the NAD⁺-dependent enzyme **dihydrofolate reductase**. This yields initially the dihydro- and then the tetra-hydro-derivative. The latter is the active coenzyme form. The tetrahydrofolate carries one-carbon units on the N^5- or N^{10}-positions in the form of formyl (-CHO), methyl (-CH_3) or formimino (-CH=NH) groups, or it can also carry a one-carbon unit that bridges the N^5- and N^{10}-positions via methylene- (>CH_2) or methenyl- (CH) groups (Figure 22.6). The tetrahydrofolate coenzymes serve as acceptors or donors of one-carbon units in a variety of reactions involved in amino acid and nucleotide metabolism (Figure 22.5).

Quantitatively the major source of the one-carbon units that become attached to tetrahydrofolate is the amino acid serine. The conversion of this amino acid to glycine is accompanied by the formation of N^5,N^{10}-methylenetetrahydrofolate. The other amino acid that acts as a source of a one-carbon unit is histidine which, in the course of its catabolism, is converted into formiminoglutamate (FIGLU). This molecule can donate the formimino group to the N^5-position of tetra-hydrofolate. These two one-carbon unit intermediates can be converted into the other derivatives shown in Figure 22.5. The requirement for folate for the catabolism of histidine to continue beyond formiminoglutamate means that suspected cases of folate deficiency may be confirmed by the detection of FIGLU in the urine.

Figure 22.5 also summarizes the major biosynthetic processes in which the tetrahydrofolate coenzymes participate. Both N^5- formyltetrahydrofolate and the N^5, N^{10}-methylene derivative are used in the biosynthesis of purines and therefore in the formation of both DNA and RNA. The enzyme thymidylate synthase that converts deoxyuridylate into thymidylate uses N^5, N^{10}-methylenetetrahydrofolate as the methyl donor for this reaction. Clearly, therefore, folate coenzymes play a central role in the biosynthesis of nucleic acids. The drugs **trimethoprim** and **methotrexate** inhibit the reduction of folate first to the dihydro- and then to the tetrahydro- form, and their abilities thereby to interfere with nucleic acid biosynthesis have made them useful as antibiotics or antitumour agents (see Chapter 39).

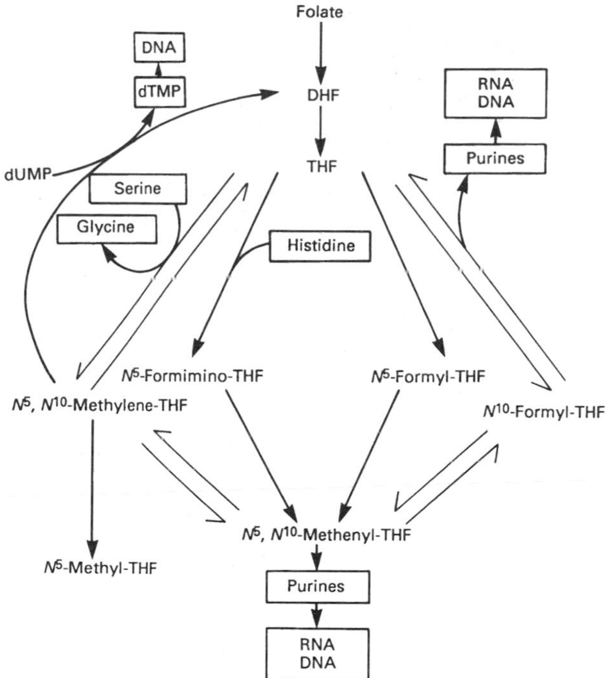

Figure 22.5 **Tetrahydrofolate as carrier of one-carbon units** (DHF, dihydrofolate; THF, tetrahydrofolate). Note that practically all the folates in mammalian tissues are present as the polyglutamate derivatives

Figure 22.6 **Tetrahydrofolate and one-carbon adducts:**

One-carbon adducts	*R group*
N^5-Formyltetrahydrofolate	—CHO
N^{10}-Formyltetrahydrofolate	—CHO
N^5-Formiminotetrahydrofolate	—CH=NH
N^5, N^{10}-Methenyltetrahydrofolate	⋗CH
N^5, N^{10}-Methylenetetrahydrofolate	>CH_2
N^5-Methyltetrahydrofolate	—CH_3

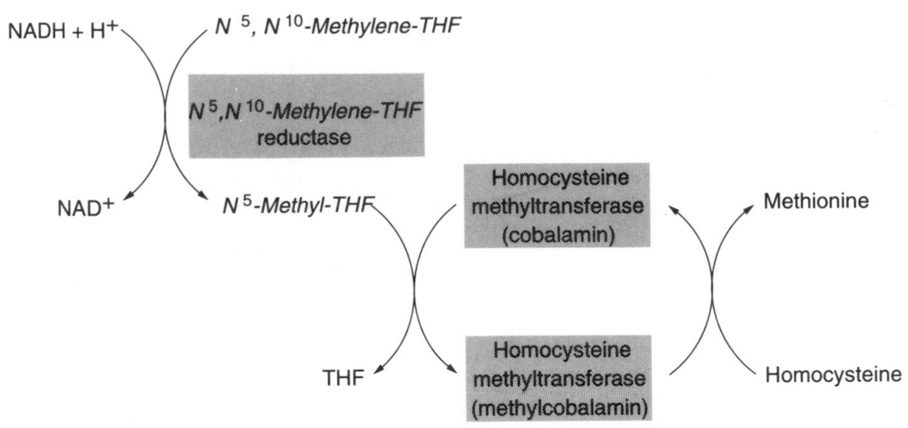

Figure 22.7 **The joint roles of vitamin B_{12} (cobalamin) and folate in the synthesis of methionine**

There is one other key reaction involving a tetrahydrofolate coenzyme that requires attention, especially as it provides a connection between the metabolic roles of folate and those of vitamin B_{12}. This is the reaction catalysed by homocysteine methyltransferase (methionine synthase; Figure 22.7). This enzyme catalyses the methylation of homocysteine to form methionine in a reaction for which methylcobalamin is the methyl donor and N^5-methyltetrahydrofolate is a coenzyme, recharging cobalamin with its methyl group. Since homocysteine arises when **S-adenosylmethionine** is used as a coenzyme for a range of methyl transferases of the type that catalyse the methylation of the bases in DNA, for example, it is clearly vital that methionine is constantly regenerated from homocysteine if these processes are to continue. It is known that hypomethylation of DNA is one of the changes observed in cellular transformations leading to cancer.

Vitamin B_{12}

Whereas vitamin B_{12} is known to be required by microorganisms for a variety of reactions, its role in mammals appears to be restricted to two reactions. The first of these has already been referred to and concerns the methylation of homocysteine to produce methionine, a reaction that depends upon methylcobalamin. The second reaction requires 5'-deoxyadenosylcobalamin and is concerned with the catabolism of the branched-chain amino acids valine and isoleucine (Figure 22.8). During the catabolism of these amino acids, methylmalonylCoA is formed and this is isomerized in a reaction catalysed by methylmalonylCoA mutase, which requires 5'-deoxyadenosylcobalamin as a cofactor. The product of this reaction is succinylCoA and this may be converted into succinate by one of two reactions. In the tricarboxylic acid cycle the enzyme succinylCoA synthetase catalyses the conversion of the former into the latter compound, and in the utilization of ketone bodies (see Chapter 10) the coenzyme A is transferred to 2-oxobutanoate prior to its oxidation. SuccinylCoA is also a precursor in the biosynthesis of porphyrin (see Chapter 6).

The requirement for vitamin B_{12} in the catabolism of valine and isoleucine beyond methylmalonate means that suspected cases of deficiency of the vitamin may be confirmed by detecting the latter compound in the urine. The diagnosis is facilitated by giving patients a 'load' of the branched-chain amino acids prior to urine collection.

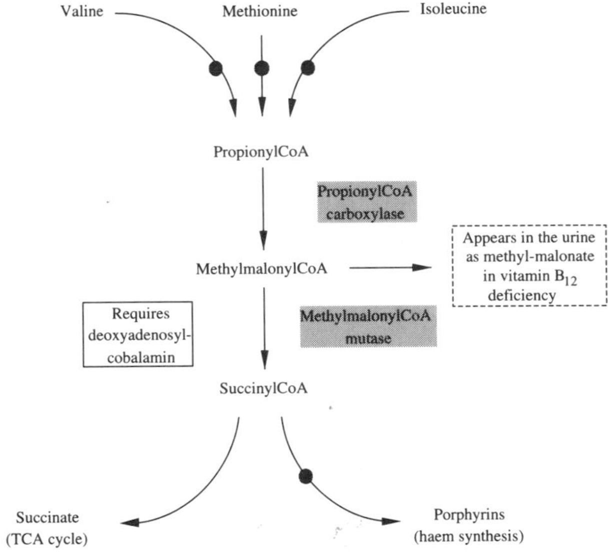

Figure 22.8 **The role of vitamin B_{12} in the metabolism of methylmalonyl CoA produced by the catabolism of three common amino acids**

22.5 Causes and effects of deficiency of folate and vitamin B_{12}

Folate deficiency

The members of society most liable to folate deficiency are pregnant women and the elderly. It is now recognized as being good practice to administer folate supplements as soon as a pregnancy is confirmed (or before) and to continue until term. In the first trimester the supplementation is prophylactic, benefiting the fetus, because there is clear evidence that the occurrence of neural tube defects such as spina bifida and anencephaly is reduced by as much as two-thirds. Later in the pregnancy the supplementation is to prevent folate deficiency in the mother, which can easily arise as fetal demands cause depletion of maternal folate stores.

Less commonly, folate deficiency may be seen in alcoholics, in patients being treated with anticonvulsants and in individuals with intestinal malabsorption conditions (tropical sprue has already been mentioned, but patients with coeliac disease may also be affected).

Biochemical explanation of the symptoms associated with folate deficiency

The central role played by folate in the biosynthesis of both DNA and RNA means that any deficiency state will influence the ability of cells to divide, the most vulnerable being the rapidly dividing cells in bone marrow and intestinal mucosa. Folate deficiency is associated with the development of macrocytic anaemia because the red cells in the bone marrow are megaloblastic. They have large, diffuse nuclei resulting from slower than normal rates of cell division. It should be borne in mind that the formation of other blood cells, e.g. platelets and granulocytes, is also compromised.

Vitamin B_{12}

In developed countries the amount of vitamin B_{12} in the diet greatly exceeds requirements. The only exception to this is found among strict vegetarians (vegans), since the only source of the vitamin in developed countries is from animals. (Vegans in developing countries may, nevertheless, obtain sufficient vitamin B_{12} as a result of bacterial contamination of their diets.) Alcoholics may also become deficient.

However, the most common cause of deficiency is malabsorption. In turn, the most common cause of malabsorption is the autoimmune disease **pernicious anaemia**. There are several points at which inappropriate activation of the immune system can prevent the absorption of vitamin B_{12}. The gastric parietal cells that secrete intrinsic factor may themselves be destroyed, a condition known as **atrophic gastritis**. Alternatively (or perhaps additionally), antibodies against intrinsic factor may prevent its binding the vitamin. Previously, gastrectomy for peptic ulcers was associated with failure to secrete intrinsic factor and thus to absorb vitamin B_{12}, but the introduction of drugs that block gastric acid secretion has eliminated this problem. Nevertheless, there are reports that the drugs themselves may interfere with the secretion of intrinsic factor.

Since the intrinsic factor–vitamin B_{12} complex is absorbed from the ileum, degenerative diseases of the mucosae of this organ, such as Crohn's disease, may give rise to vitamin B_{12} deficiency. This may also arise as a consequence of surgical resection of the ileum aimed at treating Crohn's disease.

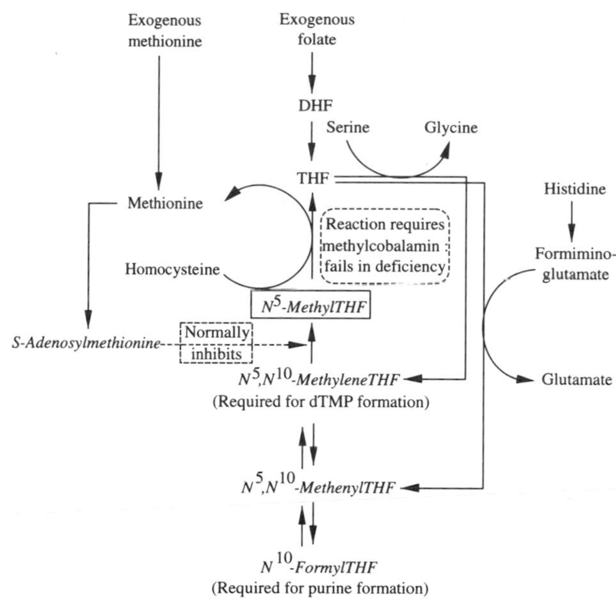

Figure 22.9 **A simplified version of the 'methyl-trap' hypothesis:** the only reaction that can lead to the formation of N^5-methylfolate is irreversible and this proceeds unrestrained if S-adenosylmethionine is not present to act as an inhibitor

Biochemical explanation of the symptoms associated with vitamin B_{12} deficiency

As mentioned previously, the clinical manifestations of vitamin deficiency states include a megaloblastic anaemia that closely resembles that seen in folate deficient patients, and the neurological syndrome called **subacute combined degeneration (SCD)**.

The occurrence of megaloblastic anaemia is explained by a hypothesis referred to as the **'methyl-trap hypothesis'** (sometimes referred as '**formate starvation**'). This hypothesis depends on the observation that the methylation of homocysteine depends on methylcobalamin and N^5-methyltetrahydrofolate, and this is the only mammalian reaction known to require both vitamins. It is postulated that when methylcobalamin is deficient, the catalytic activity of methionine synthase is greatly reduced. Consequently, N^5-methyltetrahydrofolate cannot be reconverted to tetrahydrofolate (Figure 22.9). No other reaction uses N^5-methyltetrahydrofolate, and its conversion to N^5,N^{10} methylenetetrahydrofolate is precluded because *in vivo* the reaction occurs strongly in the other direction. This is particularly the case in vitamin B_{12} deficiency. The reason for this is that the enzyme that catalyses the conversion of N^5,N^{10}-methylenetetrahydrofolate into N^5-methyltetrahydrofolate is inhibited by *S*-adenosylmethionine (SAM). But the formation of SAM requires prior formation of methionine and yet the enzyme that should do this is vitamin B_{12}-dependent. Thus, as the concentration of methionine, and therefore that of SAM, continues to fall, the enzyme that catalyses the conversion of N^5,N^{10}-methylenetetrahydrofolate into N^5-methyltetrahydrofolate increases in activity. This is because the inhibition

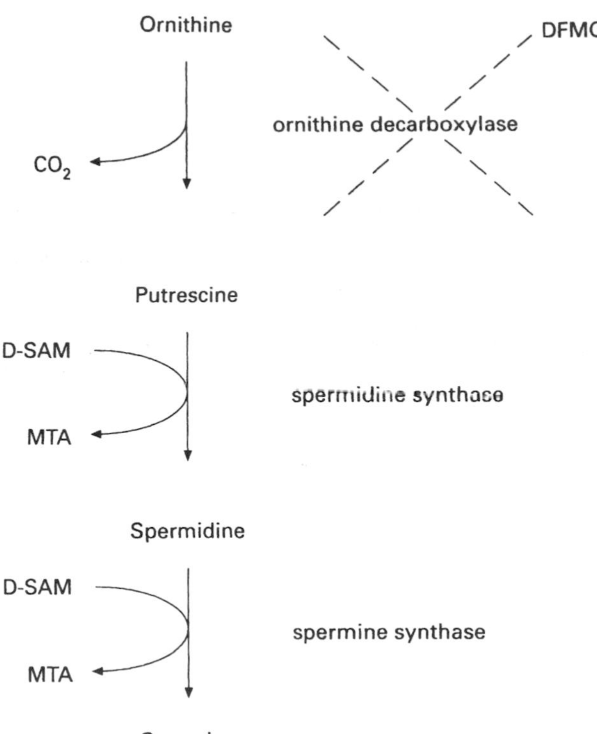

Figure 22.10 **The biosynthesis of polyamines** (DFMO, difluoromethylornithine; D-SAM, decarboxylated *S*-adenosylmethionine; MTA; methylthioadenosine)

caused by SAM becomes less pronounced. As a consequence, increasing amounts of N^5,N^{10}-methylenetetrahydrofolate are diverted into the 'dead end' formation of N^5-methyltetrahydrofolate, and thus the folate is 'trapped'. This effectively creates a deficiency state resembling true folate deficiency, because N^5,N^{10}-methylenetetrahydrofolate and N^{10}-formyltetrahydrofolate (formate starvation) are not available for DNA and RNA synthesis.

Apparently the administration of large doses of folate initially allows enough of the methylene form of the vitamin to be produced because such therapy, given when vitamin B_{12} deficiency is misdiagnosed as folate deficiency, can cause the anaemia to remit. But after a limited number of cycles this too would be trapped as N^5,N^{10}- methylenetetrahydrofolate and the anaemia recurs. Such inappropriate administration is serious because the folate will have no effect on the SCD that will continue to develop.

The occurrence of SCD is almost certainly due to a deficiency in *S*-adenosylmethionine required for a range of methylation reactions. Why a failure in methylation results in neuropathy is unknown, but may relate to a failure correctly to methylate neuronal DNA. It should be noted, however, that the biosynthesis of the important DNA-binding polyamines spermidine and spermine from putrescine requires *S*-adenosylmethionine; in this case SAM first undergoes decarboxylation and the decarboxylated form acts as an aminopropyl donor (Figure 22.10). The concentrations of both polyamines in the brain is about 2 mM.

Part 3

Specialized metabolism of tissues

Chapter 23

Blood: metabolism in the red blood cell

23.1 Introduction

The erythrocyte or red blood cell is unique in that, unlike other cells of the body, it is devoid of a nucleus, of DNA, RNA and intracellular organelles, and in particular the organized cytochrome system and the capacity for oxidative phosphorylation of mitochondria. The major task of the erythrocytes, the transport of haemoglobin, is performed with neither an expenditure nor a gain of energy.

The metabolic activity of the erythrocyte is directed at the provision of energy required for maintenance of the correct ion balance, brought about by the pumping out of sodium in exchange for potassium, for protection of haemoglobin against oxidative denaturation, for maintaining the correct conformation of the cell and for protection against the formation of methaemoglobin. In addition, other metabolic reactions ensure the production by the red blood cells of metabolites, such as 2,3-diphosphoglycerate required for haemoglobin function and NADPH which is required for the action of glutathione reductase. The energy requirements of the red cell are supplied by glycolysis, while the requirement for reducing power is provided by both glycolysis (NADH) and the pentose phosphate pathway (NADPH) (Figure 23.1).

23.2 Role of glycolysis and the pentose phosphate pathway

The energy requirements of the cell are met by the conversion of glucose to lactate which, as in other tissues, utilizes 2 ATP molecules and produces 4 ATP molecules with a net gain of 2. The production of

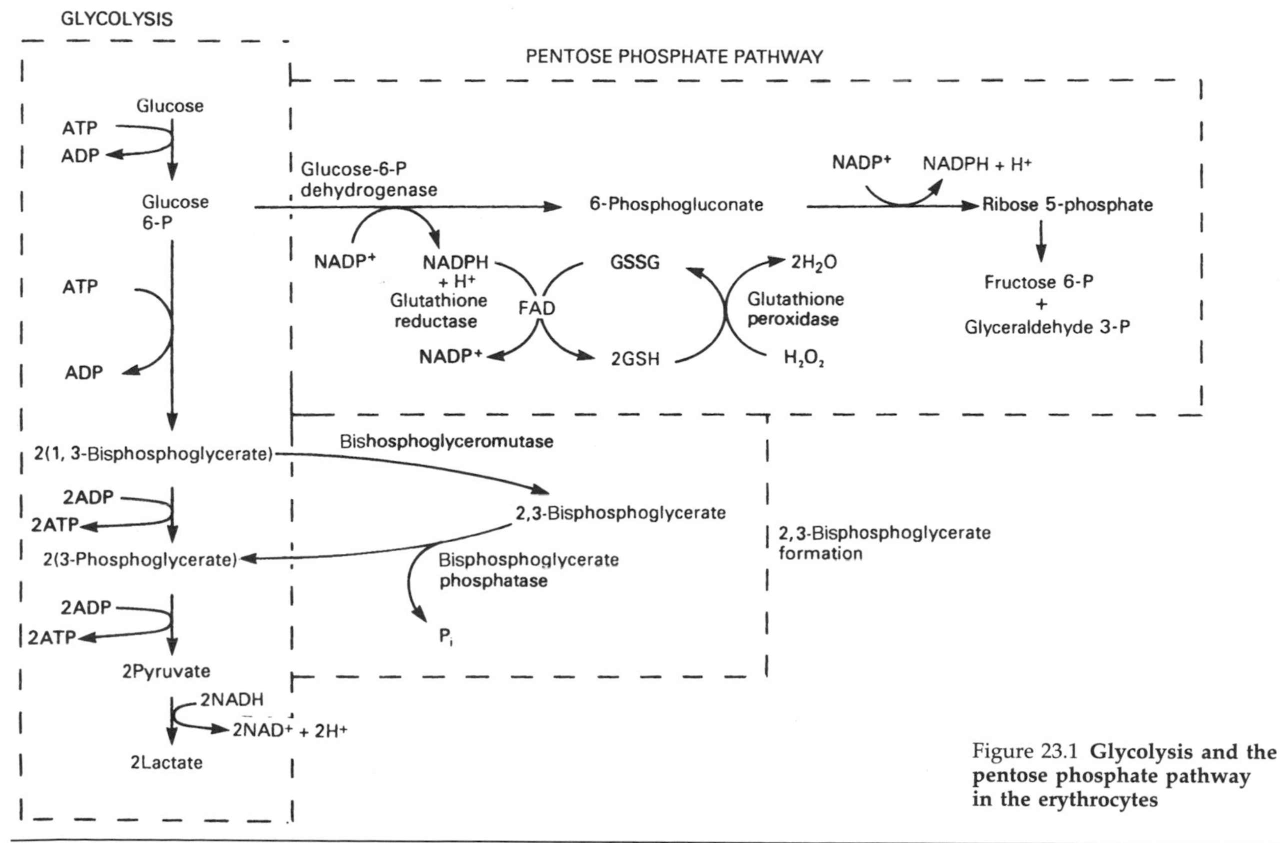

Figure 23.1 **Glycolysis and the pentose phosphate pathway in the erythrocytes**

2,3-*bis*phosphoglycerate, however, bypasses one of the stages producing ATP, so that the net energy produced by the system is zero for each molecule of 2,3-*bis*phosphoglycerate produced.

One of the chief functions of the pentose phosphate pathway (see Chapter 6) is to produce the NADPH which is essential for the regeneration of reduced glutathione from oxidized glutathione, a reaction catalysed by glutathione reductase:

```
Glu-Cys-Gly
     |
     S
     |          + NADPH + H+
     S
     |
Glu-Cys-Gly
  Oxidized
 glutathione
   (GSSG)
            |  Glutathione
            |  reductase
            ▼  (FAD)
2 Glu-Cys-Gly
      |
     SH          + NADP+
   Reduced
 glutathione
    (GSH)
```

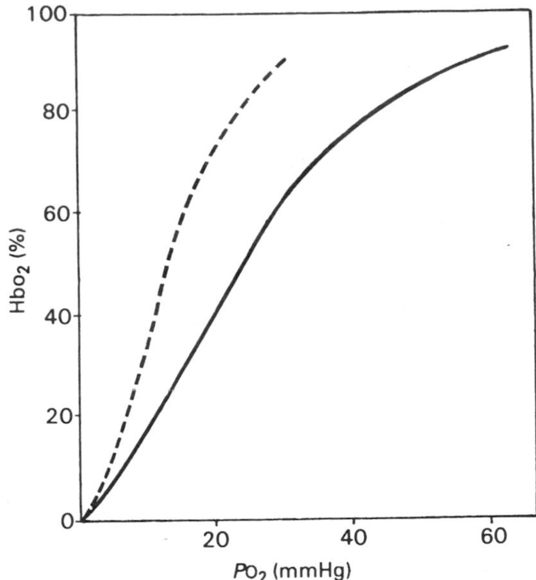

Figure 23.2 **The effect of 2,3-*bis*phosphoglycerate on the oxygen equilibrium curve of haemoglobin** (– – –Haemoglobin; —, haemoglobin +2,3-*bis*phosphoglycerate)

23.3 The role of 2,3-*bis*phosphoglycerate

2,3-*Bis*phosphoglycerate plays an important role in the binding of oxygen to haemoglobin in the red blood cell. The number of 2,3-*bis*phosphoglycerate molecules in each red blood cell is about 280 million which is approximately equal to the number of haemoglobin molecules (a concentration of about 2.2 mM).

Increase in the concentration of 2,3-*bis*phosphoglycerate causes the oxygen equilibrium curve for haemoglobin to shift to the right, meaning that haemoglobin will have a reduced oxygen affinity and so release more oxygen (Figure 23.2). The metabolism of the red blood cell thus switches over to produce more 2,3-*bis*phosphoglycerate in conditions of oxygen deficiency, for example, at high altitude.

The 2,3-*bis*phosphoglycerate molecule forms salt bridges between negatively charged phosphate groups and the positively charged residues in the haemoglobin chains; it is located in a cavity between two β-chains and, on oxygenation, the cavity contracts and the molecule is thus expelled (Figure 23.3). The histidyl residues 2 and 143, and the lysyl residue 82 of the two β-chains are believed to be involved in these salt linkages with the 2,3-*bis*phosphoglycerate molecules.

It has long been known that the affinity of oxygen for haemoglobin is affected by the pH, affinity decreasing as the pH becomes more acid. This is known as the 'Bohr effect' after its discoverer Christian Bohr, father of Neils the atomic physicist; it is clearly advantageous, because under conditions of oxygen lack, glycolysis will occur producing lactate and protons that in turn lower the tissue pH and causes release of more oxygen. 2,3-*Bis*phosphoglycerate increases the Bohr effect (Figure 23.4), since the salt links formed between the molecules increase the p*K* values of the positively charged groups

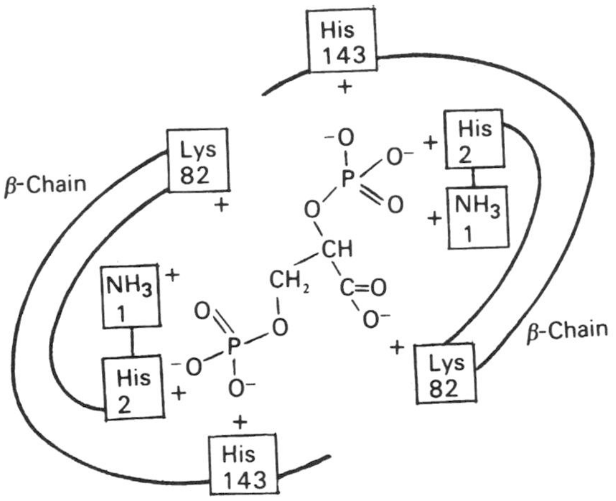

Figure 23.3 **Binding of 2,3-*bis*phosphoglycerate by human deoxyhaemoglobin β-chains**

on haemoglobin and correspondingly decrease the p*K* values of the phosphate and carboxyl groups in 2,3-*bis*phosphoglycerate. In fact, the combined effects of 2,3-*bis*phosphoglycerate and increased acidity lead to a marked reduction in oxygen-binding capacity of haemoglobin and, therefore, to oxygen release.

2,3-*Bis*phosphoglycerate plays an important role in blood that has been stored for transfusion. The concentration of 2,3-*bis*phosophoglycerate in the red blood cells stored in a citrate–glucose medium can fall from 2.2 mM to one-tenth of this concentration within a few days. This results in the haemoglobin having an inappropriately high affinity for oxygen, so that when the blood is transfused it is ineffective in releasing

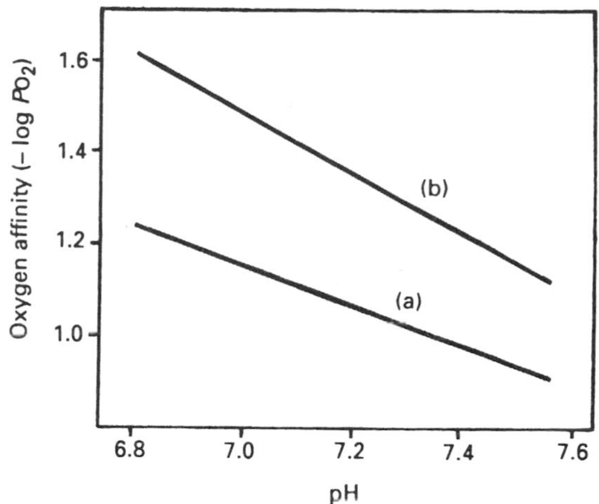

Figure 23.4 **Effects of pH and 2,3-*bis*phosphoglycerate on the binding of oxygen by haemoglobin:** (a) without 2,3-*bis*phosphoglycerate; (b) with 2,3-*bis*phosphoglycerate

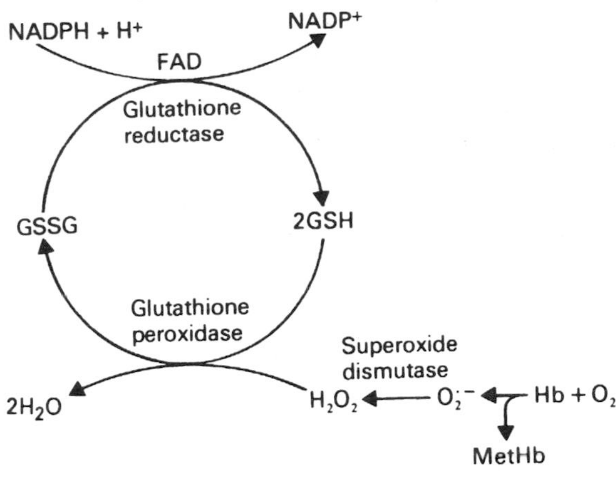

Figure 23.5 **Metabolism of glutathione and its role in the destruction of $O_2^{\cdot-}$ (the superoxide anion) and H_2O_2 in the red blood cell**

oxygen at the tissues. The problem cannot be solved by incorporating 2,3-*bis*phosphoglycerate into the medium because this polar molecule cannot be transferred across the cell membrane. The most effective method of restoring the 2,3-*bis*phosphoglycerate concentration within the red blood cell is by the addition to the medium of inosine which can be transported into the cell: here the ribose moiety is split off from the purine base and the former can then be metabolized via a section of the pentose phosphate pathway and via glycolysis to produce 2,3-*bis*phosphoglycerate.

23.4 The role of glutathione and NADPH

Earlier (see Section 23.2), the generation of NADPH and its role in the reduction of oxidized glutathione (GSSG) to reduced glutathione (GSH) was shown.

Haemoglobin in the red blood cell is always at risk since oxygen is an oxidizing agent tending to oxidize vulnerable groups in the protein and the haem group.

Oxygen, however, is not a powerful oxidizing agent until it acquires an electron, forming the superoxide anion, $O_2^{\cdot-}$; and this can happen during oxygenation of deoxyhaemoglobin in a process catalysed by Cl^- ions which are present in high concentration in red blood cells:

$$\underset{(Fe^{2+})}{Hb} + O_2 \rightarrow \underset{(Fe^{3+})}{\text{Methaemoglobin}} + O_2^{\cdot-}$$

About 10^7 superoxide anions are believed to be formed each day in the red cell, equivalent to about 3% of haemoglobin being converted to methaemoglobin daily, and they are extremely toxic to vulnerable groups such as the -SH of proteins and the unsaturated lipids of the cell membranes by initiating lipid peroxidation (see Chapter 33). Superoxide must, therefore, be destroyed rapidly and this is accomplished by the enzyme **superoxide dismutase**, that converts it to hydrogen peroxide:

$$2O_2^{\cdot-} + 2H^+ \xrightarrow{\text{Superoxide dismutase}} H_2O_2 + O_2$$

Hydrogen peroxide is itself a strong and toxic oxidizing agent and must therefore be destroyed rapidly. This is achieved by means of two enzymes, **catalase** and **glutathione peroxidase**; the latter converts reduced glutathione (GSH) to oxidized glutathione (GSSG) using a range of peroxides as substrates, including hydrogen peroxide and lipid peroxides:

$$ROOH + 2GSH \xrightarrow{\text{Glutathione peroxidase}} GSSG + ROH + H_2O$$

The interrelationships of the metabolic pathways involving glutathione are shown in Figure 23.5. In addition, the methaemoglobin generated must be reduced back to haemoglobin and this is achieved by an NADH-dependent **methaemoglobin reductase**:

$$MetHb + NADH + H^+ + \tfrac{1}{2}O_2 \xrightarrow{\text{Methaemoglobin reductase}} Hb + NAD^+ + H_2O$$

23.5 Genetic disorders: enzyme deficiencies

During World War II, many troops were stationed in malaria-infested areas and were treated regularly with drugs such as mepacrine or primaquine, as a prophylactic against malaria. A proportion of those treated suffered from a severe type of haemolytic anaemia which was often fatal and described as 'blackwater fever'. The term described the high concentration of haemoglobin in the urine resulting from haemolysis; the fever was often precipitated by an attack of the malarial parasite.

Not until about 10 years after the war period was the cause discovered. In certain individuals, it was found that the reduced glutathione in their red blood cells was very unstable and this was shown subsequently to be caused by an inherited defect of the first enzyme of the pentose phosphate pathway **glucose-6-phosphate dehydrogenase**. It will be clear from Figures 23.1 and 23.5 that a deficiency of this enzyme will lead to a failure of the red blood cell to restore GSSG to GSH, a step essential for the removal of superoxide. Cell damage is likely to result from oxidation of the membranes by the superoxide anion.

Deficiency of glucose 6-phosphate dehydrogenase is now known to be common, more than 100 million people in the world being affected. The incidence appears to be particularly high in black or Caucasian people living around the Mediterranean. Numerous mutant forms of the condition have now been described. In about 2% of affected subjects haemolysis is chronic, but in the remaining 98%, haemolytic episodes are precipitated by an infection, as with malaria referred to above, or by drugs capable of directly or indirectly producing oxidative denaturation of haemoglobin: these include aminoquinolines, sulphonamides and some vitamin K derivatives. In this group normally only the older, more susceptible, cells are destroyed.

The ingestion of the fava bean or the inhalation of pollen from the *Vicia faba* plant (fava) by people deficient in glucose 6-phosphate dehydrogenase brings about a haemolytic anaemia, described as **favism**, which can cause death, especially in infants and children. In Mediterranean countries, small children are forbidden to eat fava beans.

Deficiency of glutathione peroxidase can also occur, but is much rarer than glucose 6-phosphate dehydrogenase deficiency. The anaemic conditions that result are similar to those observed in glucose 6-phosphate dehydrogenase deficiency. Deficiency of glutathione reductase causes a wide range of haematological and other clinical symptoms. The activity of this enzyme may, however, be impaired because of the lack of availability of the coenzyme, flavin adenine dinucleotide (FAD). This requires riboflavin (vitamin B_2) for its synthesis and nutritional deficiencies of the vitamin are known to cause haemolytic diseases similar to those resulting from the enzyme deficiency.

The most common deficiency of the glycolytic pathway enzymes is that of pyruvate kinase. The anaemia resulting from such a deficiency varies from a very mild form to one necessitating lifelong transfusions. The deficiency is most commonly found in northern Europeans but is also found widely among black Americans, Syrians, Mexicans, Japanese and Italians. The disorder is transmitted as an autosomal recessive trait with homozygotes exhibiting a haemolytic syndrome, the heterozygotes being clinically and haematologically normal. In the latter group, enzyme assays on the red blood cells will indicate a partial enzyme deficiency, many mutant forms of the enzyme displaying reduced efficiency. The main metabolic consequence, which may be deduced from Figure 23.1, will be the inefficiency of ATP synthesis and an enhanced synthesis of 2,3-*bis*phosphoglycerate resulting from the metabolic blockage, and this leads to some of the symptoms described in Section 23.3.

Deficiencies of other enzymes of the glycolytic pathway have also been described but are much less common. These include deficiency of hexokinase, an enzyme that is very active in young reticulocytes but that normally shows decreased activity as the red cells age. In severe enzyme deficiency, cells will therefore become old before their normal life-span has elapsed.

Phosphohexose isomerase, phosphofructokinase, triose phosphate isomerase and lactate dehydrogenase deficiencies have also been demonstrated but are relatively rare. It should be noted that in some cases the haemolytic component is relatively severe, whereas in others it is mild. Furthermore, because these deficiencies are the result of genetic abnormalities, defective metabolism in other tissues may be of more importance than that in the red blood cell, for example, neurological defects appear early in triose phosphate isomerase deficiency.

These enzyme deficiencies in human red blood cells eloquently support the extensive genetic polymorphism of man. For example, more than 100 variants of glucose 6-phosphate dehydrogenase have been detected, 2% of this group showing chronic anaemia.

Polymorphism has also been demonstrated for most other enzymes for which genetic abnormalities have been found and the mutant enzymes are liable to differ from each other in terms of electrophoretic migration, kinetic effects of allosteric modifiers and stability to heat.

Chapter 24

Blood: blood clotting

24.1 Introduction

Clotting is the most striking and best known property of blood. Early in history, man must have appreciated the vital and often life-saving necessity of clotting to stem the loss of blood from wounds.

The processes involved in blood clotting have, therefore, been studied for many years, one of the major problems being to attempt to understand how blood remains fluid within the vessels but clots when they are damaged. Although some of the fundamental and correct ideas concerning the mechanisms of the process were first described over 100 years ago by Schmidt, the details have since been demonstrated to be extremely complex and are only now being unravelled by the most modern research.

During the past few years years a strong impetus to a much more intensive study of blood clotting has been provided by the realization that, although the occurrence of blood clotting on the body surface is a vital necessity, if it occurs within the vessels it can cause serious tissue damage and even death. Blood clot formation is extremely serious in the brain, where it my result in a 'stroke', or in the heart where it causes coronary thrombosis.

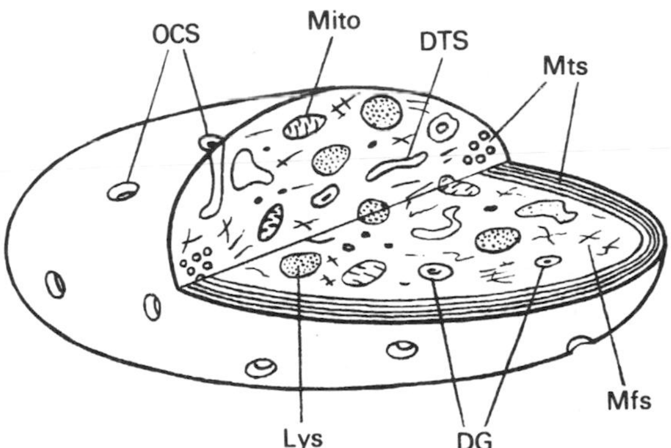

Figure 24.1 **The structure of the platelet.** This is a diagrammatic representation of a platelet sectioned through equatorial and transverse planes. (DG, dense granules – amine-storage bodies; DTS, dense tubular system; Lys, lysosome-like organelles; Mfs, microfilaments; mito, mitochondria; Mts, microtubules; OCS, open canalicular system)

24.2 Physiological events in blood clotting

The platelets

The blood platelets play a vital role in the clotting of blood. They are small cells, 2–4 μm in diameter and approximately 1 μm thick, and are produced by the explosive disintegration of the very large megakaryocytes that are often up to 40 μm in diameter. The cytoplasm of each megakaryocyte can liberate 2000–3000 platelets. The normal count of platelets in the blood is usually about 250 000 per mm^3, although the number can vary greatly even after normal physiological activity such as strenuous exercise.

Despite their small size, platelets have a very complex internal structure containing mitochondria, lysosomes, storage granules, and many microtubules and microfilaments, but no nucleus (Figure 24.1).

Sequence of physiological events

Injury to a blood vessel results in changes to the characteristics of the vessel's surface that cause platelets to activate the coagulation sequences. Contraction of the vessel also helps to reduce blood loss, but the formation of the haemostatic plug is vital. The sequence of events involving platelets is as follows:

1. The platelets adhere to the damaged blood vessel wall. This process is facilitated by the exposure of subendothelial structures: basement membrane, elastin and especially collagen. The binding of platelets to collagen is mediated by the plasma protein, von Willebrand's factor, for which there are specific receptors in the platelet plasma membrane.
2. Activation of receptors in the plasma membrane of platelets causes biosynthesis and release of thromboxane A_2 (TxA_2).
3. TxA_2 has several key roles in clot formation. It acts in an autocrine fashion to cause self-degranulation. In this process serotonin (5-HT) and ADP are expelled from the amine storage granules. It also promotes further aggregation of platelets, and it acts as a vasoconstrictor (two functional roles shared by the ADP and 5-HT released).
4. Thrombin, formed in the region of the forming platelet plug from prothrombin, activates its specific receptors in platelet membranes to cause further degranulation of these cells.
5. Thrombin-dependent deposition of fibrin provides an additional matrix to which further platelets adhere, and in which erythrocytes become entrapped. This ensures a rapid build-up of a blood clot.

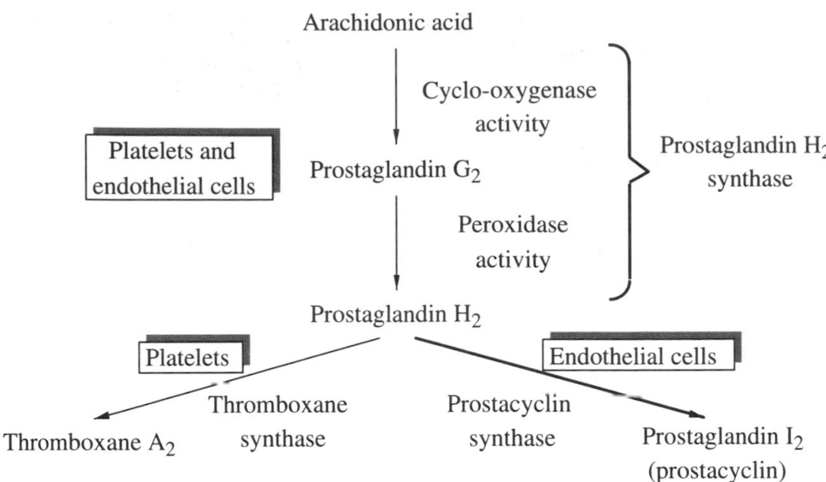

Figure 24.2 **The formation of prostaglandin H_2 and its conversion into thromboxane A_2 or prostaglandin I_2**

24.3 The opposing roles of thromboxane A_2 (TxA_2) and prostacyclin (PGI_2) in platelet aggregation: the anti-clotting effects of aspirin

In normal homeostasis, the fact that platelet plugs do not form in the absence of vascular damage is as important as their formation following trauma to the vessels. In this balance, two 'lipid-mediators' derived from arachidonic acid play complementary roles. Their production depends on the activation of the cyclic pathway of arachidonic acid metabolism (see Chapter 17) in which the enzyme prostaglandin H_2 synthase catalyses the formation of prostaglandin H_2 from arachidonic acid (Figure 24.2). The fate of this intermediate is then cell-specific. In the platelets a second enzyme, thromboxane synthase, converts PGH_2 into thromboxane A_2, which has profound procoagulating activities, as described previously. Thromboxane synthase is active only when the platelets are bound to subendothelial structures, i.e. in response to vascular damage. In the endothelial cells lining the blood vessel, PGH_2 is used as a substrate for a second enzyme, prostacyclin synthase, which catalyses the formation of PGI_2 (also known as prostacyclin) which acts to prevent platelet aggregation. Some of the PGH_2 used as a substrate for prostaglandin synthase may be released by platelets when their thromboxane synthase is inactive. As a consequence, inappropriate aggregation of platelets is prevented by the secretion of PGI_2, and this may be derived from platelet or endothelial PGH_2. Following tissue damage, however, platelet PGH_2 is diverted to the formation of TxA_2 which acts to promote rapid clot formation (Figure 24.3).

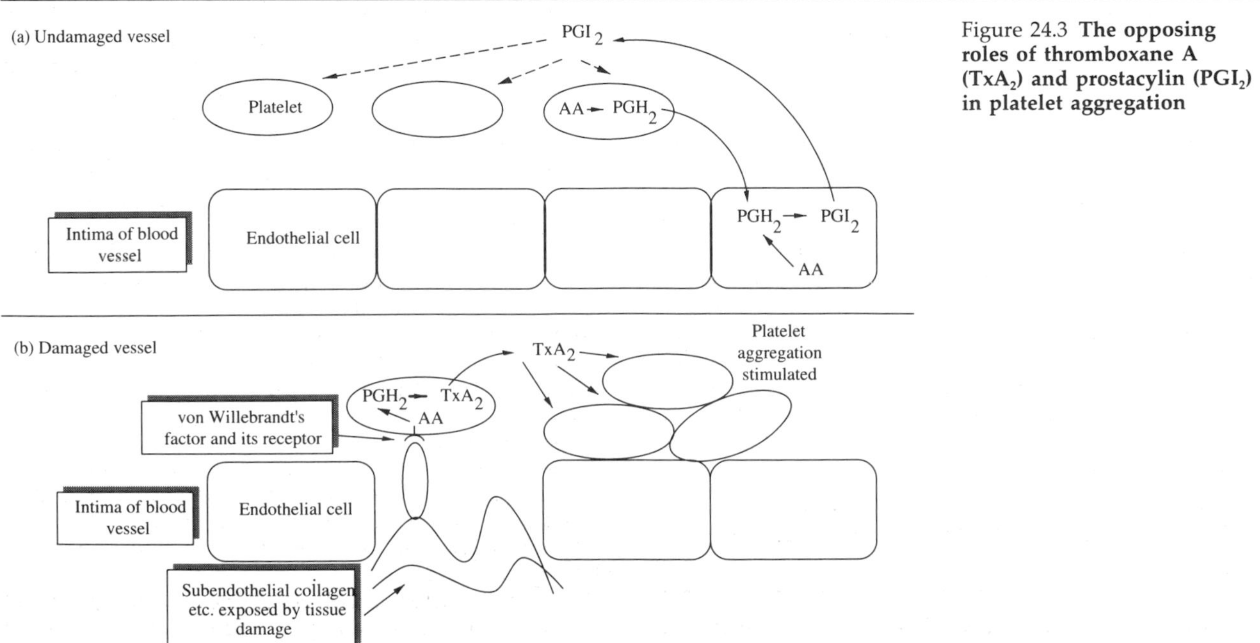

Figure 24.3 **The opposing roles of thromboxane A (TxA_2) and prostacylin (PGI_2) in platelet aggregation**

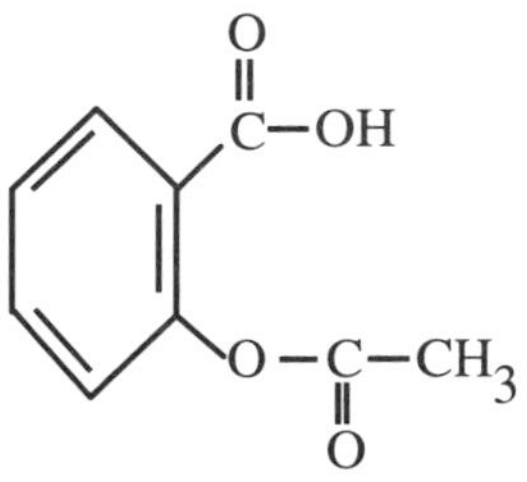

Acetylsalicylic acid (aspirin)

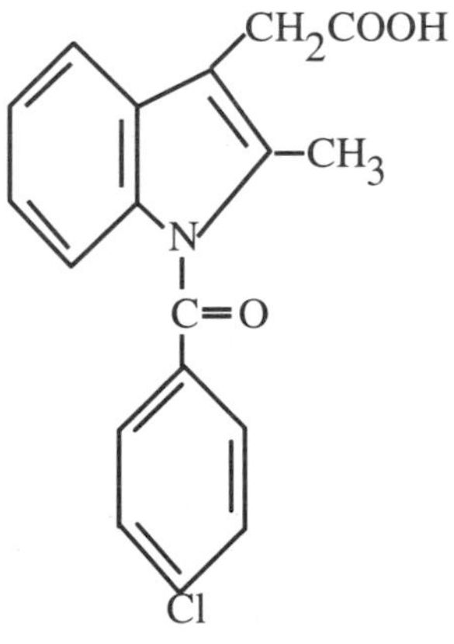

Indomethacin

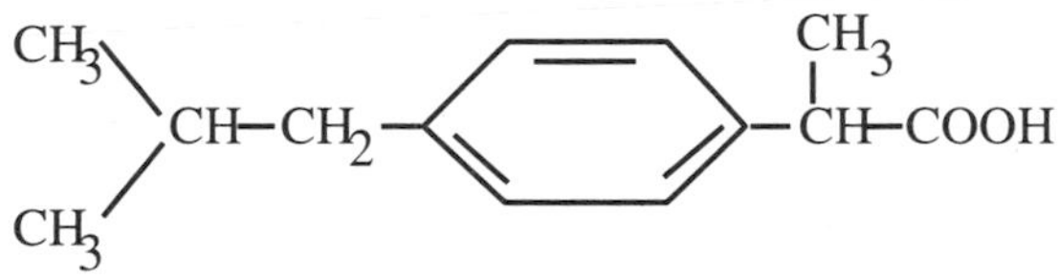

Ibuprofen

Figure 24.4 **Some examples of anti-inflammatory drugs (so-called non-steroidal anti-inflammatories)**

Drugs such as **aspirin**, **indomethacin** and **ibuprofen** (Figure 24.4) are all inhibitors of prostaglandin H_2 synthase. Aspirin, for example, donates an acetyl group which becomes covalently attached to the enzyme, i.e. an irreversible inhibition. Indomethacin and ibuprofen, on the other hand, are both competitive inhibitors of prostaglandin H_2 synthase.

The ability of aspirin to interfere with the procoagulant production of TxA_2 via PGH_2 in platelets means that it is a useful anticoagulant drug. Aspirin might be expected to inhibit the formation of PGH_2 equally in both platelets and vascular endothelial cells to the same extent. This suggests that the biosynthesis of TxA_2 and prostacyclin should be affected equally. However, because platelets lack a nucleus they are not able to replace covalently inhibited prostaglandin synthase by *de novo* synthesis, and therefore acetylation of the enzyme means its irreversible loss from those cells, which are consequently affected more than the nucleated endothelial cell. It is claimed that in low doses, i.e. 40 mg/day, aspirin significantly reduces the risk of heart attacks and strokes. It is suggested that, at this low dose, aspirin selectively inhibits prostaglandin H_2 synthase in platelets. The idea is that minor vascular damage and activation of the clotting process is, in part, responsible for the pathological changes that lead to such vascular accidents and the administration of aspirin helps to prevent this. In higher doses (100 mg/day), aspirin is given to patients following aortocoronary bypass surgery to help prevent graft occlusion.

24.4 An overview of the biochemical events in the clotting process

In addition to the changes relating to the clotting of blood that occur in platelets, a closely-related complex sequence of enzyme-catalysed reactions takes place involving plasma and vascular proteins. These lead to the formation of a fibrin clot and the overall process is summarized in Figure 24.5. This consists of two converging pathways, described as **intrinsic** and **extrinsic.** The proteins required for the functioning of the intrinsic pathway are all contained in the blood (platelets and plasma, i.e. *intra*vascular); key proteins of the extrinsic pathway derive from damaged tissues (cellular plasma membrane proteins, i.e. *extra*vascular). Nevertheless, the activation of the intrinsic pathway also involves contact with novel surfaces. These may be artificial, e.g. charged surfaces such as may be associated with glass or kaolin, or they may be exposed by tissue damage, e.g. collagen fibres. At several stages in both pathways the rate is greatly enhanced by the presence of Ca^{2+} or negatively-charged phospholipids. Phospholipids and Ca^{2+} jointly have a massively synergistic effect (more than merely additive). This synergism depends on the ability of the Ca^{2+} to trap certain of the blood-clotting proteins and help bind them to the plasma membrane of exposed and damaged subendothelial cells (see Figure 24.3). The proteins bound have the unusual amino acid **γ-carboxyglutamic** acid (see section 24.7), which possesses two negative charges, and the membrane phospholipids are also negatively charged, e.g. phosphatidylserine. Such phospholipids are normally found in the inner leaflet of cell plasma membranes and only become exposed following tissue (and cellular) damage. In addition, Ca^{2+} stabilizes clotting factor V and fibrinogen and it helps to activate one form of clotting factor XIII.

The major factors involved in blood clotting are all proteins and are generally described by the use of Roman numerals (as shown in Figure 24.5), although

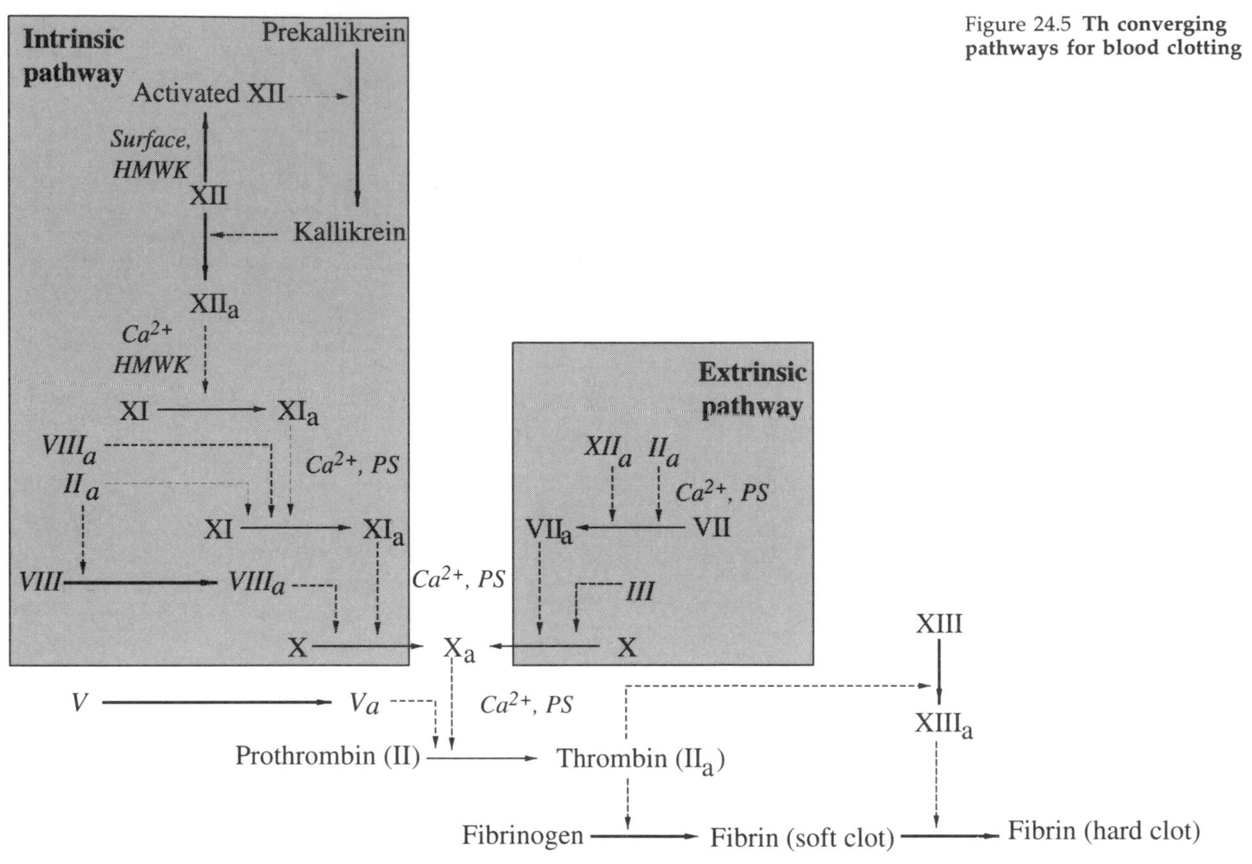

Figure 24.5 **Th converging pathways for blood clotting**

Table 24.1 **The names and functions of the principal clotting factors**

Factor number	*Common name*	*Function when activated*
I	Fibrinogen	Forms the clot network
II	Prothrombin	Converts fibrinogen into fibrin; a protease
III	Thromboplastin or tissue factor	Required by factor VII
IV	Calcium ions	Required for several clotting reactions
V	Proaccelerin	Accelerates the formation of thrombin
VI	Subsequently found to be an artefact	
VII	Proconvertin	Accelerates the formation of factor X_a
VIII	Antihaemophilia factor	Accelerates the formation of factor X_a
IX	Christmas factor	Accelerates the formation of factor X_a
X	Stuart factor	Accelerates the formation of thrombin
XI	Plasma thromboplastin antecedent	Accelerates the formation of factor X_a
XII	Hageman factor	Accelerates the formation of factor X_a
XIII	Fibrin-stabilizing factor (transglutaminase)	Catalyses the cross-linking of fibrin monomers
–	Prekallikrein	Accelerates the formation of factor XII_a
–	High-molecular-weight kininogen	Accelerates the formation of factor XII_a

several of the factors have common names. These names are given in Table 24.1. Unfortunately the numerals reflect the order of discovery of the proteins and consequently do not indicate the order in which they participate in clotting.

The processes leading to blood coagulation closely resemble those that occur in the activation of complement (see Chapter 32) in that a cascade is set up in which succeeding proteins are converted from an inactive to an active form. This activation is indicated by adding the subscript 'a' to the number of the factor, e.g. II_a. The activation usually arises due to proteolysis (occasionally a conformational change in the protein suffices). At certain stages there is a requirement for proteins that have no proteolytic activity and these are referred to as accessory factors.

24.5 The intrinsic pathway

Activation of factors XII and XI

A complex of three protease precursors (or zymogens) and an accessory protein factor jointly constitute the so-called **contact system** for the initiation of clotting via the intrinsic pathway, following surface contact. Attachment of factor XII to a surface, in the presence of **high-molecular-weight kininogen** (**HMWK**, which is also the precursor of the potent vasodilator **bradykinin**), causes the activation of factor XII, which is a protease zymogen. The substrate for the active protease is **prekallikrein** which is itself transformed into an active protease, **kallikrein**. The interaction between factor XII and kallikrein then becomes reciprocal because factor XII acts proteolytically on kallikrein to cause its further activation in a rapidly amplifying process. The hydrolysis of factor XII catalysed by kallikrein causes the release of a large fragment (factor E) and the emergence of XII_a as a functional protease. In turn, the substrate for factor XII_a is factor XI, with HMWK acting as a facilitating accessory factor in the cleavage. Although proteolysis of factor XI undoubtedly occurs, factor XI_a has the same size as its precursor: the individual chains formed are kept together by disulphide bridges.

Activation of factors IX and X

The powerful amplifying properties of immobilizing a protease precursor on surface via Ca^{2+} then come into play, because the next zymogens in the cascade, factors IX and X, both have multiple γ-carboxyglutamyl residues to mediate their Ca^{2+}-dependent binding to exposed cell plasma membrane phospholipids.

Hydrolysis of factor IX catalysed by factor XI_a involves two proteolytic steps, resulting in the formation of factor IX_a. It should be noted that this same activation can be brought about by factor VII_a (**proconvertin**), formed in the extrinsic pathway. Similarly, factor X can be activated by proteases generated in both pathways. In the intrinsic route, surface-bound factor X acts as a substrate for factor IX_a with factor VIII_a acting as an accessory protein. Factor VIII has excited a good deal of interest as an 'acute-phase protein' that increases in concentration in response to trauma and also in pregnancy. Factor VIII itself is activated by thrombin in one of the autocatalytic sequences that see the formation of the latter from its precursor, prothrombin (see Section 24.7).

As the formation of factor X_a coincides with merging of the two clotting pathways, it is convenient to consider its formation in the extrinsic pathway before describing its action.

24.6 The extrinsic pathway

As in the intrinsic pathway the attachment of factor X to a surface massively enhances its rate of activation via the extrinsic pathway. A key to the activation of this cascade is the activation of factor VII. This is achieved by proteolysis in a reaction catalysed either by thrombin or by factor XII_a, which require a Ca^{2+}-dependent association with phospholipid. As mentioned, factor VII_a can augment the action of factor XI_a in activating factor IX in the intrinsic pathway, but in addition it can act as a protease to activate factor X. This action is greatly facilitated by factor III (**tissue factor**), an integral protein of the plasma membranes of cells found in a wide range of tissues, but predominant in lungs, the walls of blood vessels and in the brain. Consequently any vascular damage rapidly results in the activation of the extrinsic pathway of clotting.

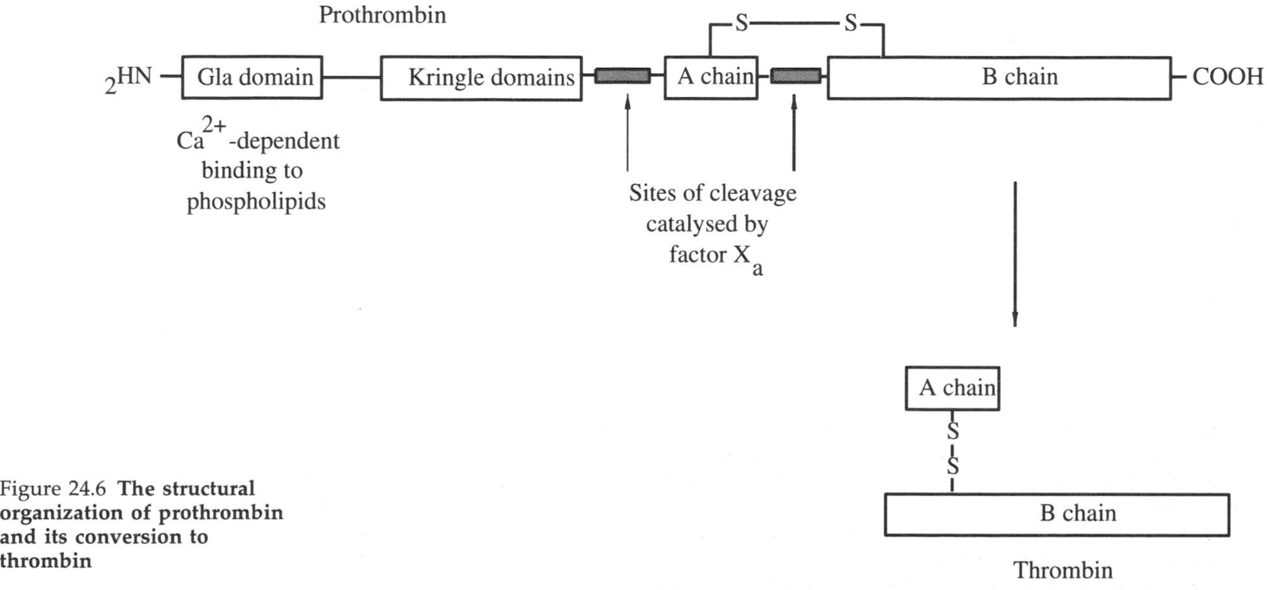

Figure 24.6 **The structural organization of prothrombin and its conversion to thrombin**

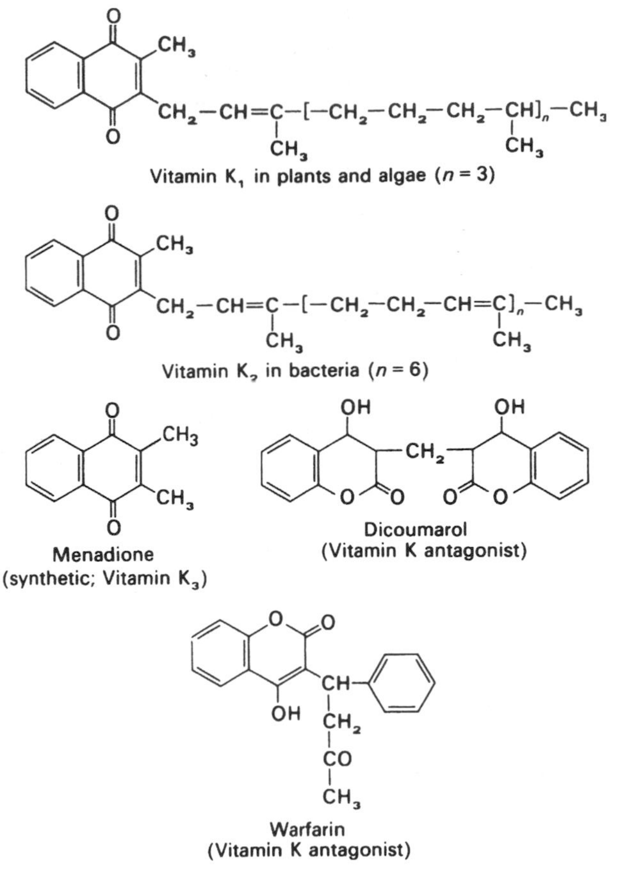

Vitamin K_1 in plants and algae ($n = 3$)

Vitamin K_2 in bacteria ($n = 6$)

Menadione
(synthetic; Vitamin K_3)

Dicoumarol
(Vitamin K antagonist)

Warfarin
(Vitamin K antagonist)

Figure 24.7 **The structures of the K vitamins and their antagonists**

24.7 The common pathway of blood clotting

In this part of the clotting process, two well-established reactions occur: the activation of the zymogen **prothrombin** (factor II) to unmask the protease thrombin, and the formation of the fibrin clot from fibrinogen.

Prothrombin and vitamin K

The structure of the single polypeptide chain that constitutes prothrombin is shown in Figure 24.6. This shows the molecule to be one of those cascade protease precursors that contain γ-carboxyglutamyl residues (Gla). These are located towards the N-terminal end of prothrombin. A second characteristic feature of the N-terminal region of prothrombin is the presence of two 'kringle' domains. These are compact globular domains containing three pairs of anti-parallel β-sheet structures maintained by three disulphide bridges.

Since the genetic code contains no codon for Gla, it is appropriate to consider the nature of the post-translational mechanism that leads to the introduction of the second carboxyl group into glutamyl residues. This is important, not only for an understanding of the activation of prothrombin and other cascade zymogens that contain Gla, but also because certain proteins found in bone, e.g. oestocalcin (see Chapter 29), undergo the same modification. In addition, this provides an opportunity to focus on the nature of the dietary requirement for vitamin K.

In 1929, H. Dam, working in Denmark, observed that chicks fed on fat-free diets were, within 3–4 weeks, severely affected by impaired blood clotting which led to internal haemorrhaging. Following replacement therapy with plant lipid extracts, Dam correctly surmised that he had found evidence for a new vitamin which he termed **vitamin K** (for koagulation).

Chemical analysis demonstrated the existence of two groups of vitamins K, one series deriving from plants (K_1) and the other from microorganisms (K_2). The simple synthetic compound, menadione, was also shown to have vitamin K activity, although it was subsequently shown that this was alkylated *in vivo* before expressing any biological activity (Figure 24.7).

It is noteworthy that there is extensive synthesis of the vitamin by gut microflora and sufficient of this can find its way into the blood to render dietary intake unnecessary for most animals, including adult humans. Very young children, prior to colonization of their guts by microorganisms, may require supplementation with the vitamin. Indeed, the supplements may be given to the mother before birth.

Vitamin K from the gut is transported, as a component of chylomicron remnants (see Chapter 10), to the liver where it plays a central role in the carboxylation of glutamyl residues in nascent clotting factors II, VII, IX and X in the Golgi apparatus. In this reaction, the ability of the vitamin to cycle between reduced hydroquinone and oxidized quinone forms is crucial. The stages of this reaction are (Figure 24.8):

1. The hydroquinone form of vitamin K is activated in an oxygen-consuming reaction which allows it to abstract one of the protons attached to the γ-carbon of a glutamyl residue in the target protein. This leads to the generation of a carbanion of the glutamyl residue and the formation of a 2,3-epoxide of vitamin K.
2. Carbon dioxide carries out an electrophilic attack on the carbanion to form the Gla structure.
3. The active hydroquinone form of vitamin K is regenerated as a result of two reductive reactions, both catalysed by the same enzyme. The nature of the reducing agent used in these reactions is not certain, but may consist of the reduced form of a low molecular weight *bis*thiol-containing compound, e.g. lipoic acid. In addition, NADPH may act as a reducing agent for the second reaction.

Anticoagulant drugs

The regeneration of the active, hydroquinone form of vitamin K is inhibited by **dicoumarol**, and also by **warfarin**, which are therefore termed anticoagulant drugs. The former occurs naturally in sweet clover and has caused serious haemorrhagic disease in cattle grazing on pastures containing such clover. Warfarin is

Figure 24.8 **The vitamin K cycle of the liver which results in the carboxylation of factors II, VII, IX and X**

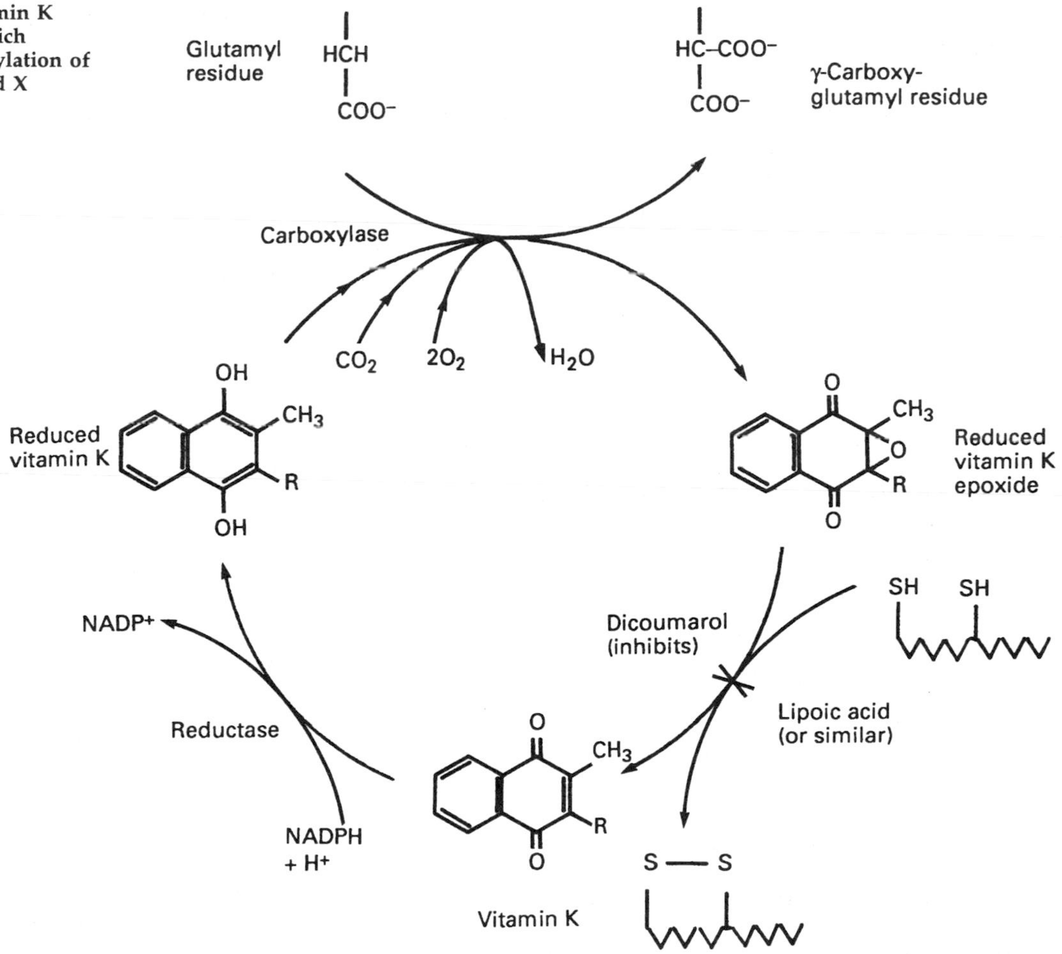

a synthetic anticoagulant and is used extensively as a rat poison; it is believed to act by causing extensive internal haemorrhaging. In controlled, moderate doses it is frequently used for the management of patients in danger of unwanted intravascular clotting, for example, those in danger of, or recovering from, coronary or pulmonary thrombosis.

Conversion of prothrombin to thrombin

The activation of prothrombin requires two proteolytic steps, each catalysed by factor X_a. These reactions are greatly accelerated by the Ca^{2+}-dependent binding of prothrombin to phospholipid, via its N-terminal cluster of Gla residues, and the presence of factor V_a (factor V is also known as **proaccelerin**). An autocatalytic loop is set up because thrombin is able to act as a protease to convert factor V into its active form (thrombin activates factor VII in a similar fashion; see Section 24.6). Although two peptide bonds are broken by factor X_a-mediated hydrolysis of prothrombin, only two separate peptides are released. The first consists of the N-terminal portion of the molecule (which includes the region containing the Gla residues and two kringle domains) and the second consists of two peptides (A and B) joined·by a disulphide bridge and derived from the C-terminal end. Peptide A undergoes further trimming and the resulting two-chain molecule is the catalytically-active protease, **thrombin**. Note that the action of factor X_a results in the release of thrombin into the plasma (the anchorage to phospholipids via the N-terminal end of prothrombin being lost). In addition, thrombin itself also can carry out a proteolytic attack on surface-bound prothrombin, hydrolysis occurring between the two kringle domains. The physiological significance of this last process is unclear, however, because the C-terminal fragment formed lacks the N-terminal anchoring sequence of prothrombin, but still requires further proteolysis to convert it into thrombin.

Whichever way it is produced, the final, catalytically-competent thrombin is a two-polypeptide chain protease, with serine at its active centre, i.e. a serine-protease. It resembles trypsin both in this regard and also in its specificity, that is, for peptide bonds in which arginine is found on the N-terminal side. The specificity is higher than that of trypsin, however, because lysine is only poorly recognized and the amino acid on the C-terminal side of the peptide bond must be glycine.

Conversion of fibrinogen to fibrin

The final stages of the blood coagulation process involve the formation of fibrin monomer from **fibrinogen**,

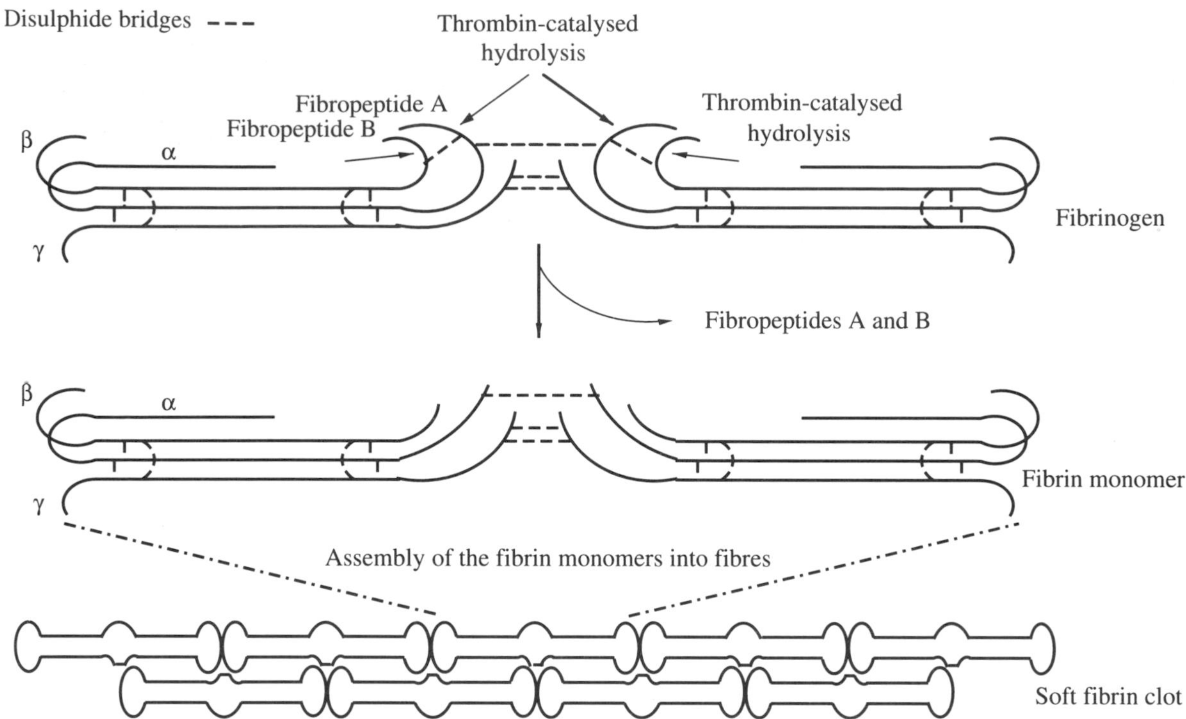

Figure 24.9 **The conversion of fibrinogen into fibrin catalysed by thrombin and the formation of a soft clot**

followed by the polymerization and stabilization of the fibrin clot. The process is initiated by the action of thrombin on fibrinogen. The latter is a rod-like molecule, some 45 nm in length, consisting of two arrays of three polypeptide chains designated Aα, Bβ and γ. In each array the three chains are interconnected by disulphide bridges (Figure 24.9) and the two arrays are also cross-linked back-to-back (N-termini of the individual chains) by disulphide bridges. Consequently fibrinogen may be formulated as $(A\alpha)_2(B\beta)_2(\gamma)_2$.

The peptides Aα and Bβ in each array form the foci for attack by thrombin, and hydrolysis at their N-terminal ends results in the release of the fibrinopeptides A and B (two of each). The removal of these short peptides causes no appreciable change in the length of fibrin compared with that of fibrinogen. The fibrin monomers so generated, spontaneously aggregate to form ordered fibres. These are seen by electron microscopy to be banded structures, with the bands repeated at regular intervals of 22.5 nm. This finding suggests that fibrin interacts in a staggered arrangement with a 50% overlap between the monomers.

These newly-formed fibres are referred to as forming a 'soft-clot' because the structure has very little strength or rigidity. However conversion to a stable, 'hard-clot' quickly follows initial deposition of the fibres. This involves the formation of unusual peptide bonds between the fibrin monomers catalysed by factor XIII_a. Two different forms of factor XIII (also known as **fibrin-stabilizing factor**) are known. One derives from plasma and the other from platelets. Both have two polypeptide chains designated *a*, but in addition the platelet-derived zymogen has a further two *b* chains, i.e. is an a_2b_2 tetramer. Cleavage by thrombin of an arg-gly bond near the N-termini of the *a* chains in both forms of factor XIII is required for their activation. For the plasma enzyme, proteolysis is sufficient for its full activation, but for the platelet-derived enzyme the Ca^{2+}-mediated dissociation of the two *b* chains is an additional requirement. Factor XIII_a is a **transamidase.** This enzyme catalyses the formation of an isopeptide bond between lysine residues in the γ chain of one fibrin molecule and glutamine residues in the γ chain of an adjacent fibrin in the fibre. These rapid reactions, which are illustrated in Figure 24.10, are followed by slower, but similar cross-linking of the α chains. These reactions greatly strengthen and stabilize the fibrin in clots.

24.8 Factors that limit the growth of clots

In Section 24.3, attention was drawn to the production of prostacyclin as a means of preventing the inappropriate aggregation of platelets. Similar mechanisms exist to ensure that the deposition of fibrin clot is not initiated when not required, or to limit the process to the region of vascular damage once it is activated. An important focus for such mechanisms is thrombin.

The most straightforward means of limiting the activation of thrombin is by the protease acting on itself to cause self-inactivation. Thrombin is also rapidly removed from the circulation, largely in the liver. These relatively simple means of limitation are augmented by the presence in the plasma of **antithrombin**. This protein binds to thrombin, and indeed to all of the proteases of the clotting cascade (except factor VII_a). The complex of

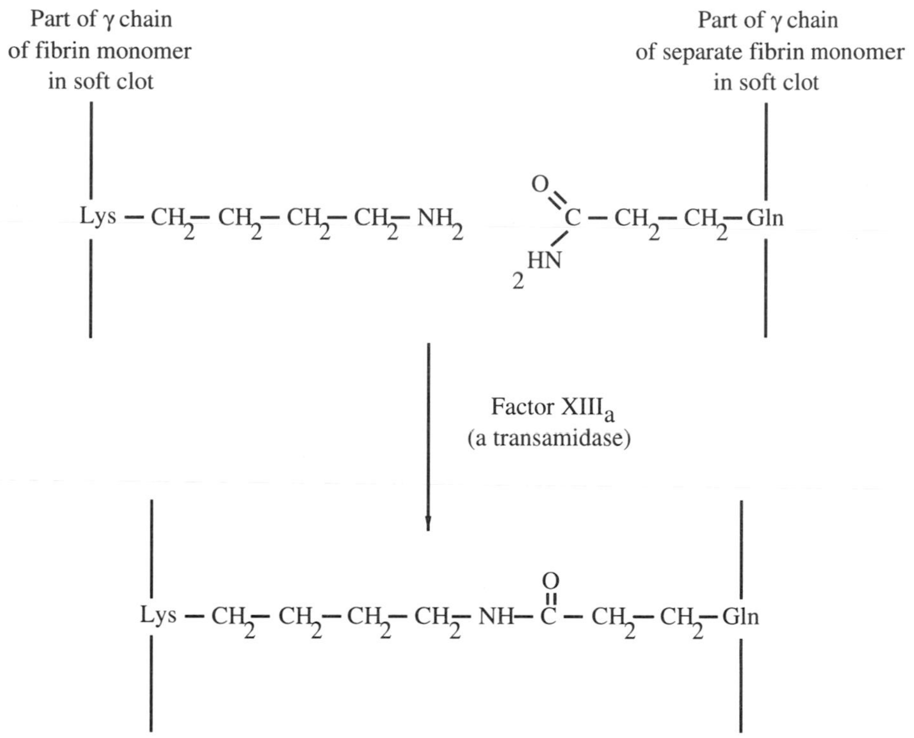

Part of γ chain of fibrin monomer in soft clot

Part of γ chain of separate fibrin monomer in soft clot

$$\text{Lys}-CH_2-CH_2-CH_2-CH_2-NH_2 \quad + \quad H_2N-C(=O)-CH_2-CH_2-\text{Gln}$$

Factor XIII_a (a transamidase)

$$\text{Lys}-CH_2-CH_2-CH_2-CH_2-NH-C(=O)-CH_2-CH_2-\text{Gln}$$

Figure 24.10 **The stabilization of soft fibrin clots is promoted by factor XIIIa and involves the formation of an isopeptide bond between lysine and glutamine residues**

thrombin and antithrombin has no protease activity. The avidity with which antithrombin binds to thrombin is greatly enhanced by the glycosaminoglycan **heparin**. It is for this reason that blood for analysis is collected in 'heparinized' syringes, when clotting must be avoided. Heparin is also given prophylactically to patients prior to surgery to avoid problems of intravascular clotting during the operation. Infusion of heparin-containing solutions is usually continued for up to 7–12 days postoperatively, or until the patient is ambulant.

Thrombin also activates plasma **protein C** by proteolysis which, in its activated form is a protease that has a high specificity for the accessory proteins factors V_a and VIII_a, which are rapidly inactivated. Activated protein C has no effect on factors V and VIII, that is, the precursor forms.

The methods described to date are intravascular; the limiting factors are present in the blood or are released into the blood following injury (the latter is the case for the mast-cell-derived heparin). These methods are complemented by at least one extravascular means of limitation of clotting. **Thrombomodulin**, which forms an inactivating complex with thrombin, is found as an integral protein of the plasma membranes of vascular endothelial cells. The thrombomodulin–thrombin complex, once formed, is very active as an activator of protein C (greatly exceeding the ability of thrombin acting alone to do so).

24.9 Fibrinolysis

Although the fibrin clot resists the action of most proteases it is attacked by a special protease called **plasmin**. This enzyme is generated from a precursor, **plasminogen**, which resembles the inactive forms of the clotting cascade protease zymogens. In fact, the name 'plasminogen' describes a family of closely-related proteins. The proteolytic activation of the plasminogens is catalysed by several enzymes. The best studied of these is of renal origin which, because it was first isolated from the urine, is called **urokinase** (which is unfortunate because it is no more of a kinase than enterokinase!). A very similar enzyme is associated with vascular tissues and is called **tissue-type plasminogen activator (t-PA)**. Some bacteria also produce a special plasminogen activator called **streptokinase** (another misnamed 'kinase').

t-PAs produced by recombinant DNA techniques (see Chapter 38), or complexes of streptokinase with plasminogen, have been given to patients as thrombinolytic agents. These plasminogen-activating enzymes appear to have intrinsic activity, i.e. there appears to be no requirement for their activation, but their physiological regulation has not been determined.

In addition to plasminogen activators, two groups of plasmin inhibitors have been described. Members of the first group act directly to prevent the activation of plasminogens and are referred to as 'anti-activators'. The second group constitute direct inhibitors of plasmin (**antiplasmins**). Antiplasmins form inactive complexes with plasmin and include a **platelet-derived antiplasmin** and also **vitamin E.**

24.10 Diseases affecting blood clotting

In view of the complex interactions that arise between the platelets and numerous clotting proteins, it is not surprising that a range of bleeding disorders arise due to dysfunction at a number of levels. Table 24.2 lists the

Table 24.2 **Some abnormalities of platelet function**

Disease	*Nature of defect*
Thrombocytopenia	Deficient platelet formation in bone marrow, due to: leukaemia aplastic anaemia ionizing radiation toxic chemicals, inc. certain drugs
Storage pool disease	Impaired aggregation due to inherited lack of dense-core granules
Congenital afibrinogenaemia	Inherited failure to produce fibrinogen
von Willebrand's disease	Inherited lack of protein that binds to collagen and platelets
Bernard–Souler (giant platelet) syndrome	Inherited lack of platelet receptor for von Willebrand's factor
Aspirin toxicity	Inappropriate inhibition of thromboxane A_2 synthase

major abnormalities of platelets and the principal functions rendered deficient as a consequence. Some of these abnormalities result from an acquired dysfunction, but several of them are familial in origin.

Similarly, several genetic diseases which affect the clotting mechanism have been recognized. The best documented and well known of these is a deficiency of factor VIII, giving rise to the most common of the clotting disorders **haemophilia A.** This haemorrhagic disorder is characterized, in severe cases, by bleeding into the joints and muscles. The disease is inherited in a sex-linked fashion and affects about 1 in 1000 male children. Fetuses affected by the condition can now be recognized following amniocentesis, by recognition of a restriction fragment length polymorphism (RFLP) present within the factor VIII gene (see Chapter 38). The second most common bleeding disorder, **haemophilia B**, derives from a deficiency in factor IX. Indeed, the alternative name for factor IX, Christmas factor, reflects that its absence was first recognized in Stuart Christmas, who suffered from haemophilia.

Until very recently the only way of treating patients with familial haemophilia was by means of blood transfusions, either of whole blood or individual factors isolated from human-donated blood. As a consequence this group of unfortunate patients has been very susceptible to viral infections, most especially AIDS and various forms of hepatitis. Thankfully the required factors are now produced by means of DNA-based technology (see Chapter 38).

Chapter 25

The liver

25.1 Introduction

Most of the important functions carried out by the liver are described in detail in other chapters, the objectives of this chapter being to summarize the known functions of the liver to help the reader comprehend the wide range of consequences that can arise as a result of liver damage or disease.

Some metabolic processes that have not been dealt with in detail, such as the synthesis of the bile salts, are discussed in this chapter.

25.2 Structure of the liver

Although the structure of the liver is described in detail in anatomy textbooks, there are certain aspects of the structure that bear an important relationship to its biochemical function and which are, therefore, emphasized in this chapter.

The liver is the largest organ in the human body and, in the adult, usually weighs approximately 1.5 kg. Despite the large size, the cells are remarkably uniform and parenchymal cells or hepatocytes form 60% of the total cell population. The other important cell type in the liver is the Kupffer or littoral cell and these cells form 15–33% of the total liver cell population in man.

The main blood supply to the liver, about 1.51 l/min, is received from the portal vein, the majority of the portal blood coming from the gut and, after a meal, containing high concentrations of digested food, such as glucose, amino acids and some short-chain fatty acids. A proportion, approximately 30%, of the portal blood comes from the spleen.

The hepatic artery delivers much less blood to the liver than the portal vein, arterial blood delivered at a rate of approximately 0.4l per min being the major source of oxygen for the liver. The blood supply to the liver is shown diagrammatically in Figure 25.1. The blood provided by these two sources mixes in the liver sinusoids. The system of sinusoids (Figure 25.2) differs from capillaries in having

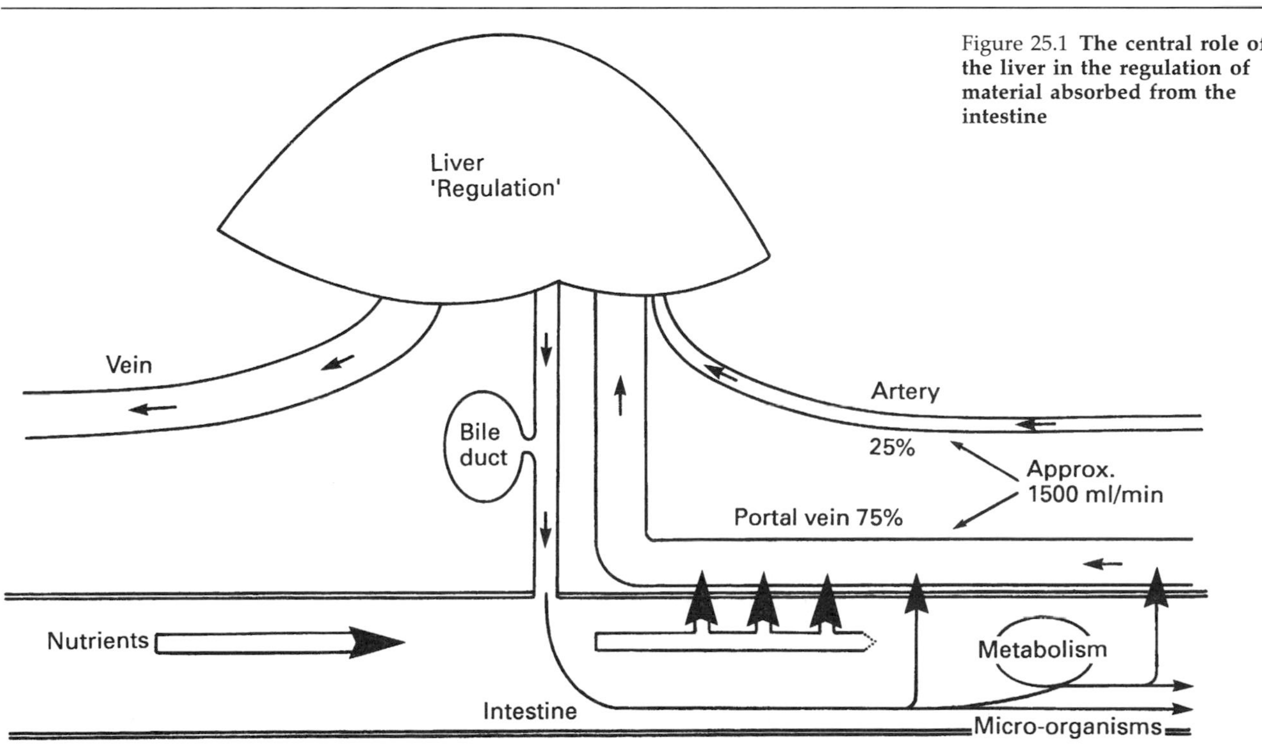

Figure 25.1 **The central role of the liver in the regulation of material absorbed from the intestine**

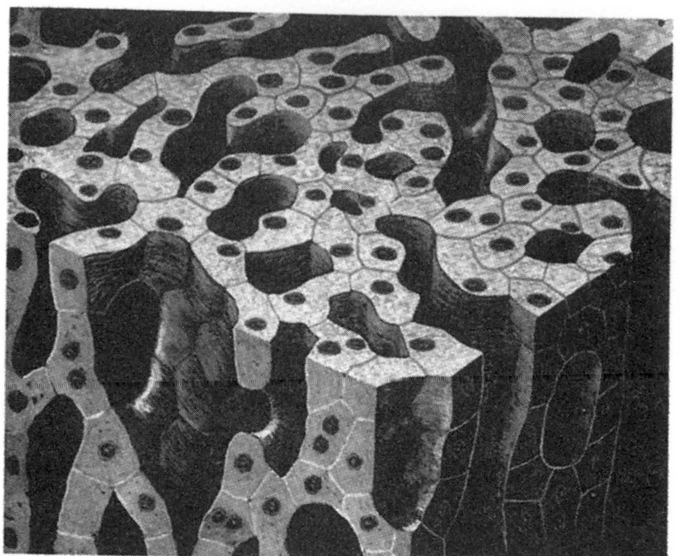

(a)

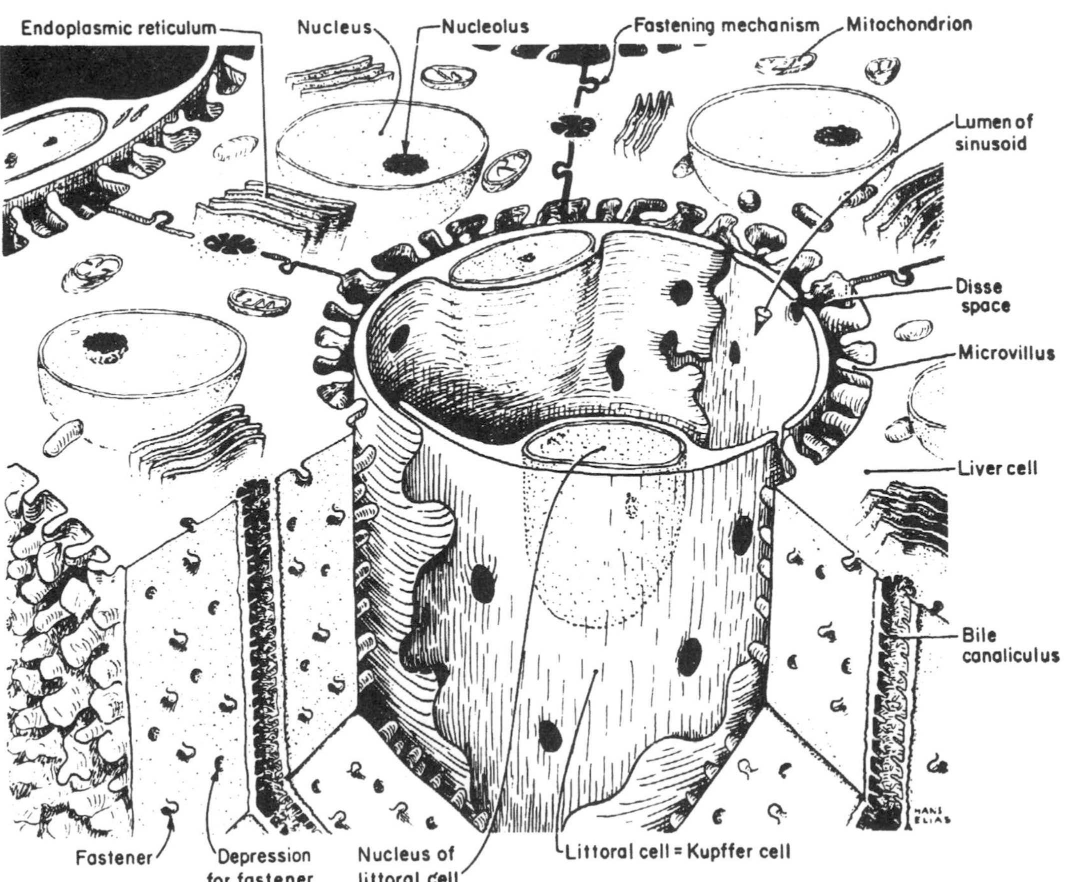

(b)

Figure 25.2 **Structure of human liver.** (a) Sinusoids and parenchymal cells; (b) ultrastructure

no collagen in their walls. Sinusoids are lined by the Kupffer cells, and are ideally suited for the efficient diffusion of metabolites into liver cells.

The Kupffer cells contain large numbers of lysosomes (see Chapter 8) that phagocytose very actively and can, therefore, swell to block the passage of blood through the sinusoids if ingested particles are not digested and dispersed.

25.3 Metabolic roles of the liver: general considerations

The liver plays a very important role in regulating the nutrition of the body tissues, which it does in both the fed and fasting states. In the fed state it will utilize the ingested glucose and amino acids, metabolizing them to

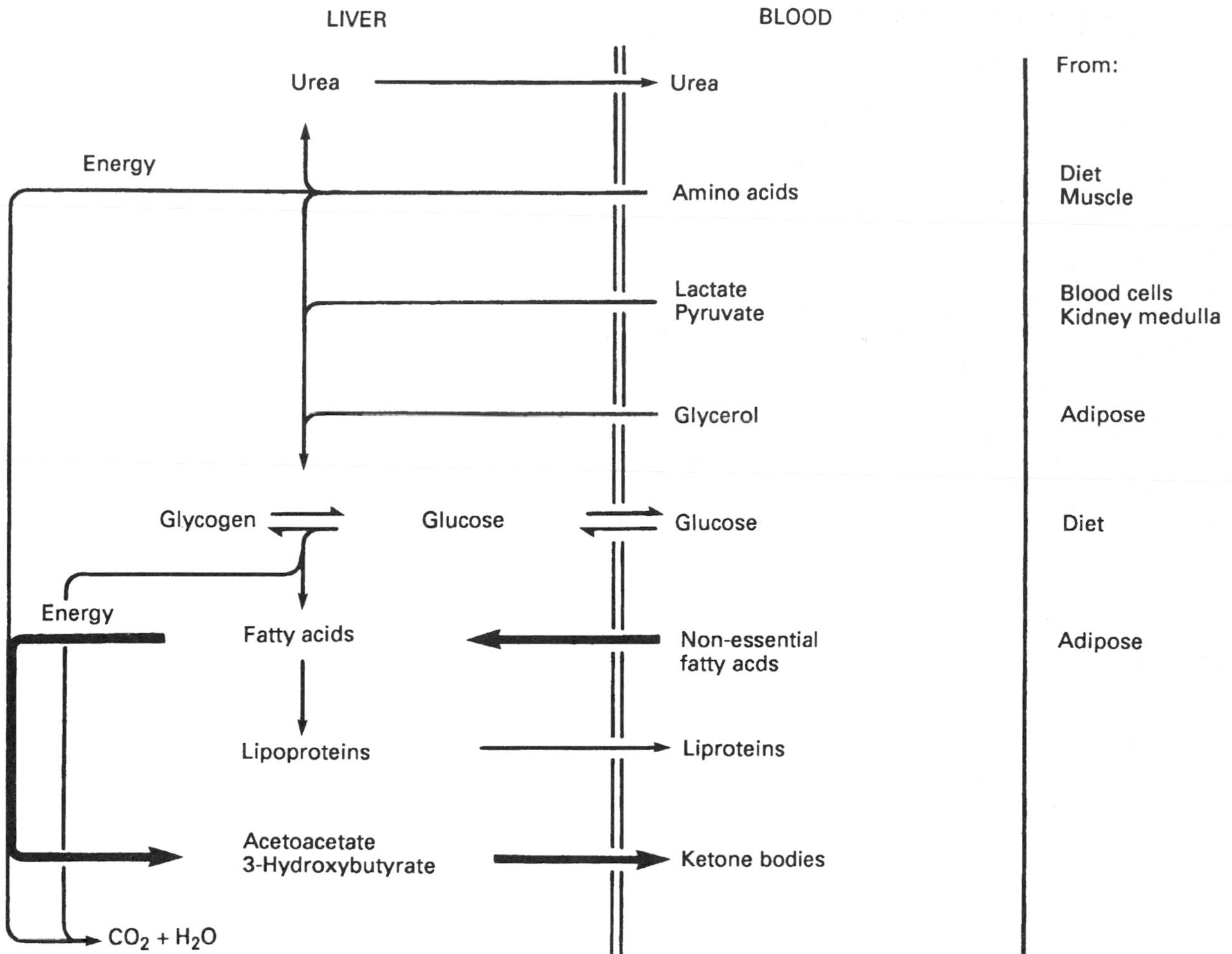

Figure 25.3 **The uptake and release of metabolic fuels by the liver**

form new amino acids. There is, therefore, extensive interconversion of amino acids by transamination and other pathways, so that the normal pattern of amino acids in the plasma is reached as soon as possible after a meal.

During the fasting state, the liver plays a similar role in processing nutrients. In this situation, however, nutrients are received from the storage tissues, for example amino acids from muscle and fatty acids from adipose tissues. The liver will convert the glucogenic amino acids, and particularly alanine (see Chapter 9), into blood glucose and glycogen, and the fatty acids into very-low-density lipoproteins that are transported to provide the tissues with fatty acids as fuel (see Chapter 10).

In summary, therefore, the liver can be described as an important nutrient regulating organ. These metabolic interrelationships are shown in Figure 25.3.

25.4 The role of the liver in carbohydrate metabolism

The functions of the liver in carbohydrate metabolism can be illustrated diagrammatically by the triangle shown in Figure 25.4. This represents the supply of monosaccharides, glucose, fructose and galactose received from the portal vein, and the supply of pyruvate and lactate from the other tissues, such as muscle, primarily via the hepatic artery. In summary, the liver:

1. Interconverts monosaccharides (pathway A of Figure 25.4; see Chapter 9):

   Glucose ⇌ Fructose

   Galactose ⇌ Glucose

2. Converts glucose to pentoses by the pentose phosphate pathway (pathway B in Figure 25.4).
3. Synthesizes glycogen from glucose and lactate, stores the glycogen and reconverts the glycogen to blood glucose (pathway C in Figure 25.4; see Chapter 5).
4. Converts glucose to pyruvate by the glycolytic pathway; synthesizes glucose from lactate or pyruvate by reversing the glycolytic pathway (pathway D in Figure 25.4; see Chapter 5).
5. Oxidizes pyruvate completely by the tricerboxylic cycle (pathway E in Figure 25.4; see Chapter 5).

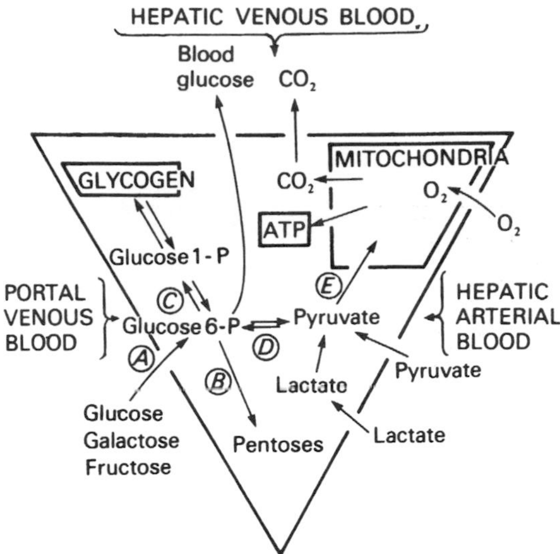

Figure 25.4 **Roles of the liver in carbohydrate metabolism**

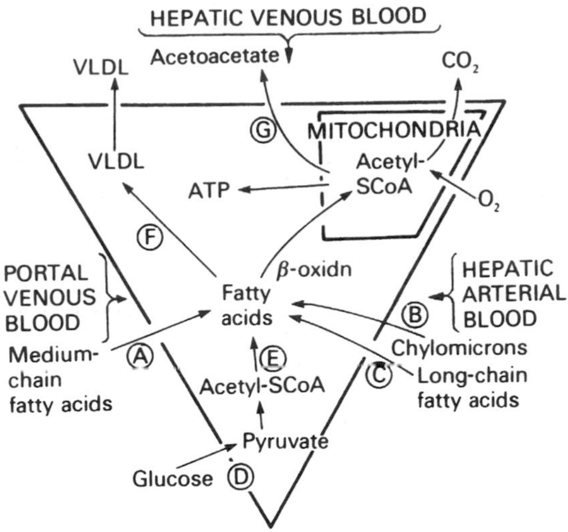

Figure 25.5 **Roles of the liver in fat metabolism.** (A) Medium-chain length fatty acids from the digestive tract; (B) chylomicrons not taken up by adipose or other tissues; (C) free fatty acids released from adipose tissue particularly during periods of fasting (see Chapter 10); (D) glycolysis; (E) fatty acid synthesis; (F) synthesis of lipoprotein, in particular very-low-density lipoproteins (VLDL); (G) production of acetoacetate

25.5 The role of the liver in fat metabolism

The fat, mainly triacylglycerol (triglyceride), content of the liver can vary widely depending on the state of nutrition. On an average mixed diet, high-carbohydrate or high-protein diet, fat will normally be relatively low and in the range 2.0–5.0% of the liver weight. After feeding a high-fat diet, however, the triacylglycerol content of the liver can rise to very high values, and form between 25% and 50% of the total weight. The fat content will also increase to 5–10% of the liver weight during fasting.

In the normal, well-fed adult in Western societies, a large proportion of the total energy requirements (45–50%) is provided in the diet as triacylglycerol, and thus the need for the synthesis of fatty acids is relatively small. Nevertheless, glucose, fructose and some amino acids can be converted to fatty acids in the liver, adipose tissue and some other organs (see Chapters 6 and 9). These conversions will be particularly important on a low-fat diet or with high intakes of carbohydrate or protein. About 50% of the fatty acids in human adipose tissue are unsaturated. Although desaturation can be carried out in the liver, certain polyunsaturated fatty acids cannot be made in this manner and must be obtained from the diet (see Chapter 17).

The liver plays a very important role in the synthesis of triacylglycerols and phospholipids from both endogenous and dietary fatty acids. As discussed in Chapter 10, the triacylglycerols and phospholipids are packaged into lipoproteins (very-low-density lipoproteins) for transport from the liver. If this process is overloaded, for example by the supply of very large quantities of fatty acids from the diet (high-fat diet) or of large quantities of fatty acids from the adipose tissues (starvation), the liver may not be able to synthesize the requisite phospholipids and protein at a rate sufficient to maintain lipoprotein synthesis. The supply of precursors, for example essential fatty acids for the phospholipids and amino acids for protein, will naturally be critical. Deficiencies of these precursors in relation to the supply of fatty acids will lead to the deposition of large amounts of triacylglycerols in the liver, resulting in the condition of 'fatty liver'.

The liver can also oxidize fatty acids by β-oxidation to conserve energy as ATP, but under conditions of excess production and under-utilization, the end-product of β-oxidation, acetylCoA, cannot be further utilized. AcetylCoA molecules are condensed to form 3-hydroxy-3-methylglutarylCoA. Loss of an acetylCoA as a result of the action of the 3-hydroxy-3-methylglutarylCoA lyase causes the formation of acetoacetate. Acetoacetate is readily reduced to 3-hydroxybutyrate by NADH-dependent 3-hydroxybutyrate dehydrogenase, so that both sides accumulate together:

$$\underset{\text{Acetoacetate}}{CH_3COCH_2COO^-} + NADH + H^+ \rightleftharpoons \underset{\text{3-Hydroxybutyrate}}{CH_3CHOHCOO^-} + NAD^+$$

The ratio of the two acids in plasma depends on the NADH : NAD^+ ratio in mitochondria and is usually three to six in favour of 3-hydroxybutyrate.

When produced in moderate quantities, these acids ('ketone bodies') can be utilized by extrahepatic tissues such as heart, muscle or kidney, but if extensive fatty acid oxidation occurs in the liver, as will happen during starvation or diabetes mellitus, large quantities of ketone bodies will be formed. In uncontrolled diabetes, the high concentration of 3-hydroxybutyric and acetoacetic acids produced can result in a severe metabolic acidosis. A summary of the role of the liver in fat metabolism is shown in Figure 25.5.

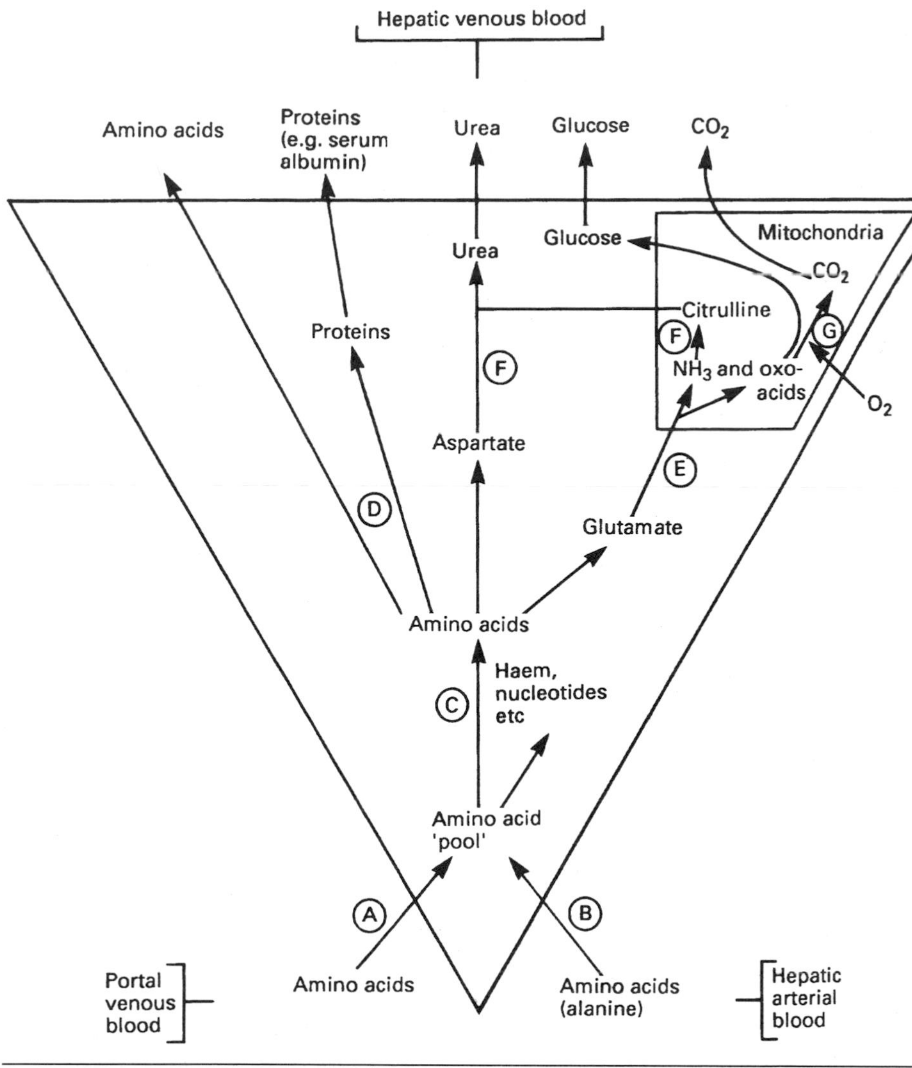

Figure 25.6 **Roles of the liver in amino acid metabolism.** (A) Amino acids received from the digestive tract after a meal; (B) amino acids (particularly alanine) produced by the muscle during a period of fasting; (C) interconversions of amino acids by transaminations and other reactions; (D) protein synthesis; (E) oxoacid formation; (F) urea formation; (G) oxidation of oxoacids by the citrate cycle

25.6 The role of the liver in amino acid metabolism

The essential functions of the liver in amino acid metabolism are summarized in Figure 25.6. Amino acids are received by the liver from two sources: during digestion and absorption of protein food via the portal blood (A) and during postabsorption and fasting conditions via the extrahepatic tissues, mainly the muscles (B). Alanine forms a major proportion of the amino acids released from muscle during fasting (see Chapter 11).

Amino acids from either source enter the amino acid pool where extensive interconversion can occur as a result of the operation of the tricarboxylic acid cycle and by transamination (C). Depending on the requirements of the animal at the time, the amino acids are received from the portal blood, they may be recirculated following interconversion, incorporated into other molecules such as purines, pyrimidines or porphyrins or into liver or plasma proteins (D). During fasting conditions, most of the amino acids received from the hepatic artery will be converted to blood glucose, but a proportion may be used for protein synthesis or be oxidized to produce energy after conversion to oxoacids (E).

It will be noted that an important function of the liver is to synthesize urea from the ammonia arising during deamination (F). The quantity of urea produced will, therefore, depend on the surplus amino acids from the diet not required for protein synthesis and that are being used for energy or, during fasting, to produce blood glucose.

25.7 The role of the liver in protein synthesis

As indicated in Figure 25.6, the liver plays a vital role in the synthesis of two categories of proteins, those of the liver tissue itself and those that will be exported into the plasma or the plasma proteins. The liver is a very active organ in the biosynthesis of proteins and uses the conventional pathways of tissue protein synthesis. It should be noted, however, that the total liver proteins are much more responsive to changes in the diet than those of other tissues in the body. No special storage protein appears to be laid down and the liver stores proteins by hypertrophy and hyperplasia.

Many of the major proteins of the plasma are synthesized in the liver, including albumin, α- and β-globulins, fibrinogen and, as discussed in Chapter 24, the blood clotting factors II (prothrombin), VII, IX and X, all of which are very unusual in requiring vitamin K for their synthesis. Many special transport proteins are also synthesized in the liver, including transferrin, ceruloplasmin and transport proteins for vitamin B_{12} (transcobalamins). In addition, the 'acute phase' proteins of the innate immune system are formed in the liver (see Chapters 32 and 40).

25.8 The role of the liver in storage

The liver's role in glycogen storage is well known and the glycogen can vary from about 0.3% of the total liver weight in fasting conditions to about 10% when a high carbohydrate diet is fed. As discussed in Chapter 9, the stored glycogen will normally supply the blood glucose for about 12 h after a meal. In addition, the liver plays an important role in the storage of many other valuable nutrients. As discussed in Section 25.7, it can act as a general store of protein and it can also store the protein blood-clotting factors, such as prothrombin.

Many vitamins are stored in the liver and the capacity is frequently very large. Thus, vitamin A, stored in the liver in rats, can, after feeding diets high in vitamin A, reach levels so that it would supply the animal for 70 years! The liver is also an important store of iron, which is largely present as ferritin in the parenchymal cells. In haemochromatosis (see Chapter 21), the excess iron is stored as haemosiderin in the Kupffer cells.

These stores are clearly of great value to the animal. They will provide supply in time of nutritional inadequacy, and also be very valuable for the female during pregnancy.

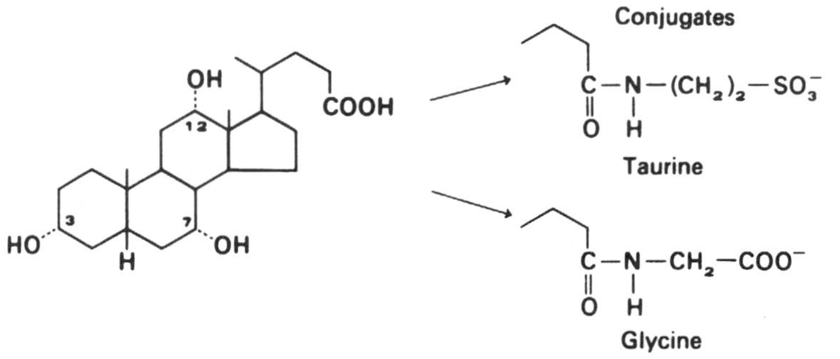

Conjugates

OH 12 COOH HO 3 H 7 OH

C(=O)–N(H)–$(CH_2)_2$–SO_3^-

Taurine

C(=O)–N(H)–CH_2–COO^-

Glycine

Figure 25.7 **Chemical structures of bile acids**

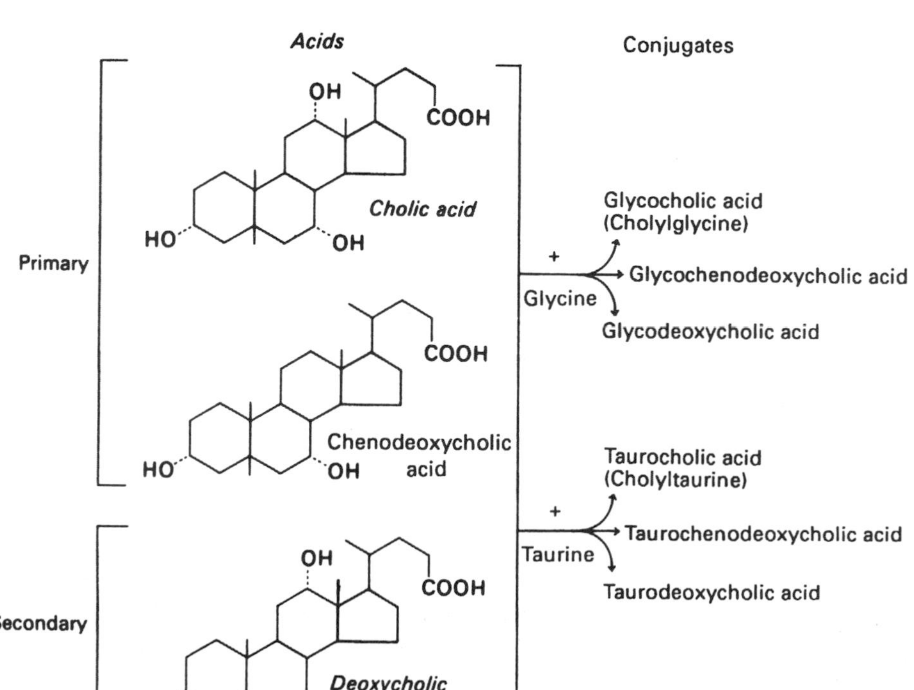

Figure 25.8 **Bile acids of man.** Normal composition of bile in man: 30–40% cholyl conjugates; 30–40% chenodeoxycholyl conjugates; 10–30% deoxycholyl conjugates

Cholesterol

(A) 7α-Hydroxylation

(B) 12-Hydroxylation

3α, 7α, 12α-Triol

(C) 26-Hydroxylation

(D)

Oxidative shortening of side chain

Cholic acid

Figure 25.9 **Biosynthesis of cholic acid from cholesterol.** (*A*) 7α-Hydroxylation is a regulatory step occurring in the endoplasmic reticulum. (*B*) Formation of 3α,7α,12α-trihydroxy-5-cholestane occurs in three stages: (1) oxidation at C-3 to form the ketone and isome
occurring in the endoplasmic reti
in the endoplasmic reticulum; (3)
bond and reduction of the keton
hydroxyl, occurring in the cytos
occurring in the mitochondria. (
side-chain occurring in three sta
carboxyl in the cytosol; (2) C-24
the endoplasmic reticulum; (3) β
mitochondria or cytosol

25.9 The role of the liver in providing digestive secretions

The digestive secretion produced by the liver is the bile described in Chapter 13. The major components that aid the digestion of fats are the bile salts, but the phospholipids also aid fat emulsification and the bicarbonate helps to neutralize the gastric acidity.

Bile salts in the liver contain 30–40% cholylconjugates (Figures 25.7 and 25.8). The deoxycholyl conjugates are secondary products formed by bacterial reduction in the gut.

Bile acids are synthesized by a complex series of reactions from cholesterol and the process is outlined in Figure 25.9. Several oxidative stages are involved and the side-chain of the cholesterol is removed, the double bond is saturated and two additional hydroxyl groups are introduced. The cholic acid which is formed is conjugated with glycine to form glycine conjugates (e.g. glycocholic acid) or with taurine to form taurine conjugates (e.g. taurocholic acid; Figure 25.8).

The bile acids are secreted into the lumen of the gut where they are acted on by bacterial enzymes which catalyse the removal by hydrolysis of glycine or taurine and then a 7α-dehydroxylase removes the 7-hydroxyl group. A complex mixture of different bile acids is then reabsorbed, the mixture containing unchanged conjugates and unconjugated bile acids, e.g. cholic and deoxycholic acid, which are returned to the liver where these unconjugated bile acids may be reconjugated with glycine or taurine to form new conjugates of, for example, deoxycholic acid. This process is part of the enterohepatic circulation and is illustrated in Figure 25.10.

The synthesis of bile acids is under control at two points. The feedback of the bile acids into the liver inhibits their synthesis from cholesterol, but in addition control of cholesterol synthesis itself is important. This is dealt with in Chapter 10.

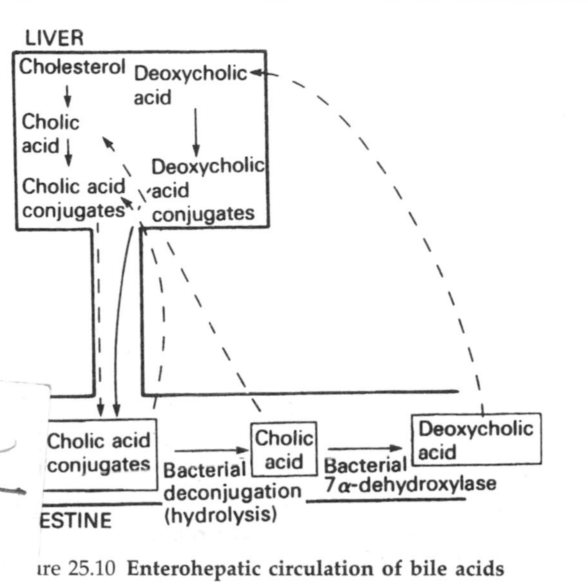

ıre 25.10 **Enterohepatic circulation of bile acids**

The metabolism of cholesterol to bile acid has many important medical implications. The plasma cholesterol plays a very important role in arterial disease (see Chapter 10), so that its excretion in the bile or conversion to bile acids is of major importance. It is estimated that approximately 350 mg of cholesterol per day is converted into bile acids. In addition, about 650 mg cholesterol is lost daily in the faeces. The relation between the cholesterol and bile acid concentrations in the bile is also important in maintaining the cholesterol solubilized in bile acid micelles. Precipitated cholesterol is the major cause of 'gallstones'.

25.10 The excretory role of the liver

The liver plays an important role in the processing of metabolites for excretion, using two routes. Water-soluble compounds are passed out into the blood for excretion by the kidney, and insoluble or lipid-soluble materials, following modification, are excreted through the bile and thence to the faeces. This is summarized in Figure 25.11.

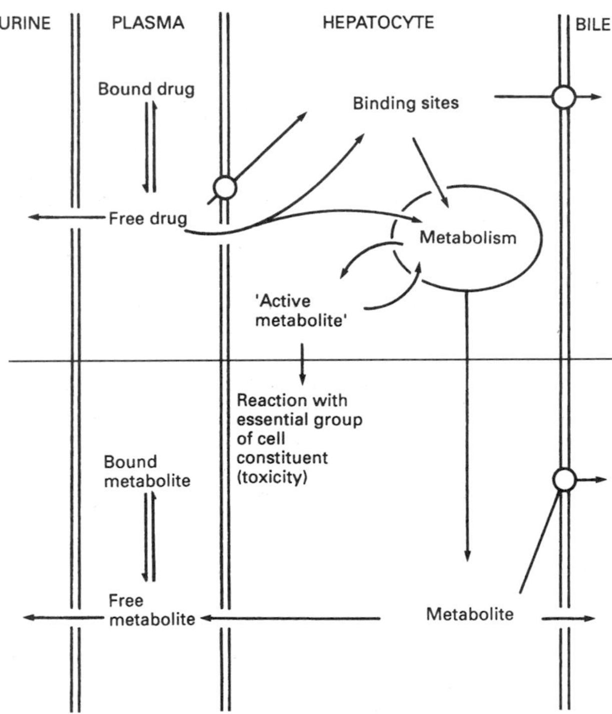

Figure 25.11 **The role of the liver in the processing of metabolites for excretion**

Formation of water-soluble metabolites for urinary excretion

In the former category are included the formation of urea described in Section 25.6 and the very important role of the liver in the metabolism of foreign compounds or the detoxication processes. As described in Chapter 36, the liver converts a vast range of toxic, often lipophilic, molecules into relatively non-toxic hydrophilic molecules that can easily be excreted by the kidney. The efficiency with which the liver metabolizes a drug will be a very important factor in determining its lifespan and its action in the body.

The liver is also the main site of metabolism of many steroid hormones, and this metabolism usually results in the inactivation of the hormone. Consequently the metabolism and subsequent excretion in the urine plays an important part in the control of hormonal activity. Steroid hormones with a Δ^4,3-oxo structure (all of them except dihydroepiandrosterone and the oestrogens) are metabolized chiefly by reduction of this structure. Thus tetrahydrocortisol is the major urinary metabolite of cortisol (Figure 25.12) and together with the hexahydro derivative accounts for 50–60% of the daily secretion of the hormone. For most of the steroid hormones, an acidic grouping is usually attached to the newly formed 3-hydroxy grouping yielding sulphate or glucuronic acid conjugates. Oestradiol is metabolized in part to oestrone and oestriol and all three are conjugated with sulphuric or glucuronic acids. These conjugation reactions resemble those undergone by foreign compounds and are described in detail in Chapter 36.

Biliary excretion

The liver is the main organ of the body involved with the excretion of the bile pigments. These lipophilic molecules are conjugated to give water-soluble glucuronides that are excreted in the bile. Many foreign compounds are also excreted by the liver through the bile and faeces. The extent of the biliary excretion of a compound is influenced by a number of factors. A common characteristic of many compounds excreted in bile is their amphipathic character, i.e. they contain both

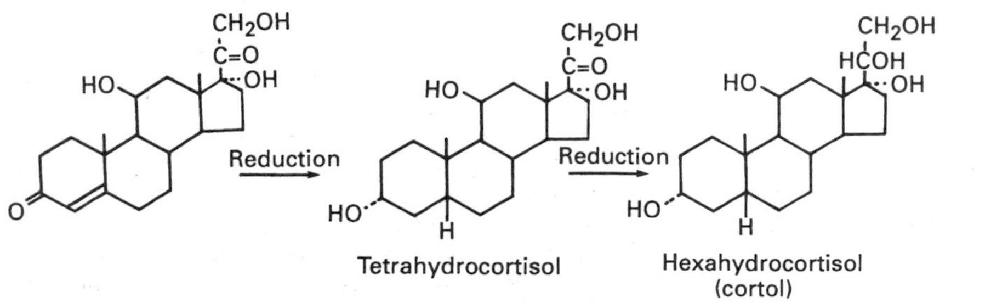

Reduction → Tetrahydrocortisol Reduction → Hexahydrocortisol (cortol)

Figure 25.12 **The major role for the metabolism of cortisol in the liver**

polar and non-polar groups in their molecular structure (compare bile salts). The polar group, as with the steroid metabolites, is frequently introduced by metabolic conjugation (see Chapter 36) and thus one commonly finds that compounds excreted in the bile are formed of conjugates with glucuronic acid, sulphate, glycine or glutathione.

The molecular size of a compound also influences the extent of biliary excretion. It appears that there is a minimum 'threshold' molecular weight below which little biliary excretion occurs. For many species this threshold is around 300–500. There is probably an upper limit to the molecular size of compounds which can be excreted in the bile but this remains to be determined.

If any particulate material, e.g. carbon particles, gains access to the blood as a result of a wound, the particles are actively phagocytosed by the Kupffer cells lining the sinusoids and, after digestion into lysosomes, excreted into the bile. The Kupffer cells can also recognize partially degraded proteins, such as serum albumin, and in fact their ability to recognize proteins of this type surpasses all physiochemical methods. These proteins, or damaged red blood cells, are phagocytosed but the proteins are degraded by the lysosomal cathepsins (see Chapter 8) and the amino acids made available to the circulation or to the liver parenchymal cells. These cells will also phagocytose bacteria provided that they have first been attacked by agglutinating antibodies.

Some particulate matter is, however, toxic after phagocytosis by the Kupffer cells. Thus some inert material, such as silica or methylcellulose or lipopolysaccharides produced by Gram-negative bacteria, are toxic. The reason why some particles or molecules are harmless and others toxic to these cells is at present unknown.

25.11 Ethanol and the liver

The liver is the major site of ethanol oxidation and accounts for over 90% of the ethanol metabolized at a blood concentration of approximately 200 mg/100 ml (43 mmol/l); the remainder of the ethanol absorbed is excreted directly through the lungs, kidneys and skin. In the liver, ethanol is first oxidized to ethanal (Figure 25.13). The principal enzyme responsible for this is alcohol dehydrogenase, a zinc-dependent enzyme found in the cytosol. Alcohol dehydrogenase exists in the form of six isoenzymes, made up of three types of subunits, with a combined effective K_m for ethanol of around 0.5–2.0 mM. The enzyme is very unspecific and will catalyse the oxidation of other primary alcohols (e.g. methanol) as well as polyhydric alcohols (e.g. ethylene glycol).

A second liver system for catalysing the oxidation of ethanol is the mixed-function oxidase known as the 'microsomal ethanol-oxidizing system' (MEOS) described by Lieber. This system (Figure 25.13) involves a specific cytochrome, P450 (Cyp 2El; see Chapter 36), and animal experiments have shown that the activity of this smooth endoplasmic reticulum system is increased by the chronic administration of ethanol. This is of interest when considering the development of tolerance to ethanol, drug–ethanol interactions (due to the induction of drug-metabolizing enzymes) and the

Alcohol dehydrogenase

$$C_2H_5OH + NAD^+ \rightleftharpoons CH_3CHO + NADH + H^+$$

Ethanol → Ethanal

'Microsomal ethanol-oxidizing system'

$$C_2H_5OH + NADPH + H^+ + 2O_2 \xrightarrow{\text{Cytochrome } P_{450}} CH_3CHO + NADP^+ + 2H_2O_2$$

Aldehyde dehydrogenase

$$CH_3CHO + NAD^+ + H_2O \longrightarrow CH_3COO^- + NADH + 2H^+$$

Ethanal → Acetate

Figure 25.13 **Enzymic pathways of alcohol metabolism**

formation of potential carcinogens (see Chapter 37) in alcoholics. The K_m of the microsomal ethanol-oxidizing system is about 10 mM and it has been estimated that around 5–20% of the ethanal production is accounted for by this reaction, most of the ethanol being oxidized by alcohol dehydrogenase.

The bulk of the ethanal produced is rapidly oxidized to acetate by liver aldehyde dehydrogenases (Figure 25.13). Several different isoenzymes of aldehyde dehydrogenase, which have different subcellular distributions and kinetic properties, are present in human liver. Kinetic studies indicate that the enzyme associated with the mitochondrial matrix (aldehyde dehydrogenase, $ALDH_2$) plays the dominant role in the metabolism of ethanal derived from ethanol. This enzyme can be inhibited by **disulfiram** and this property is used in the treatment of alcoholism. The administration of disulfiram gives rise to unpleasant reactions, including facial flushing, headaches and nausea, after the ingestion of even small amounts of alcohol because it leads to the accumulation of ethanal in the body. Ethanal is toxic because it interacts or binds with many compounds such as proteins, amino acids, amines, etc., to form 'aldehyde adducts'.

One of the obvious consequences of alcohol ingestion in many individuals is facial flushing and increased heart rate triggered off by the ethanal. In marked contrast to the near absence of ethanal in the peripheral blood of most Caucasians, alcohol ingestion results in a marked elevation of blood ethanal in a large part of the Oriental population. This in turn results in the facial flushing which is thought to be caused by an ethanal-induced release of catecholamines. The affected individuals possess a variant aldehyde dehydrogenase in the liver mitochondria which destroys the ethanal more slowly. The enzyme defect has been identified as being caused by a single amino acid substitution near the carboxyl terminal of the enzyme protein.

About 80% of the acetate produced by the aldehyde dehydrogenase leaves the liver (Figure 25.14) and undergoes further metabolism in tissues such as the

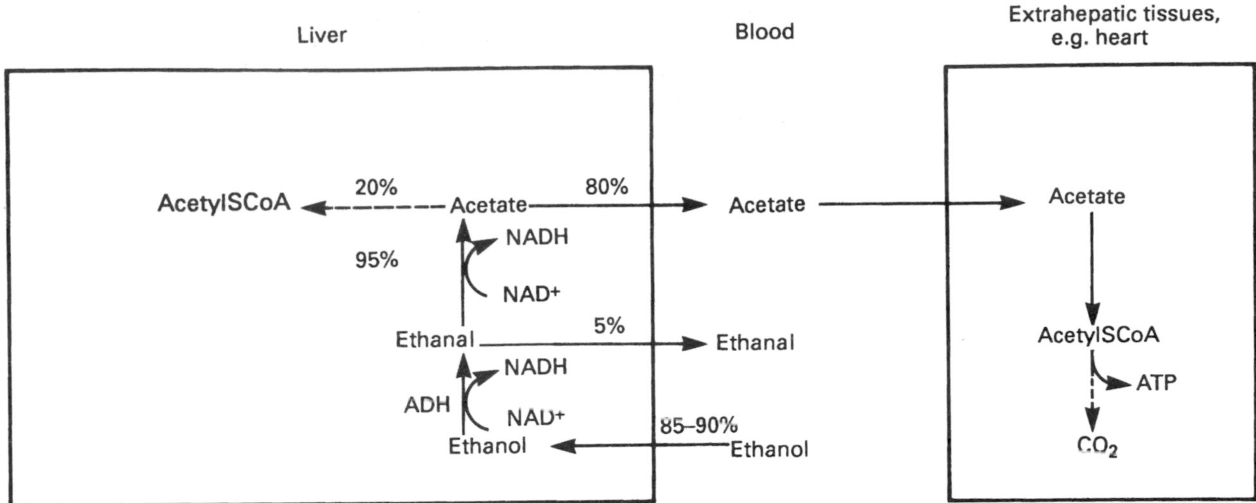

Figure 25.14 **Summary of alcohol metabolism** (ADH, alcohol dehydrogenase; ALDH aldehyde dehydrogenase)

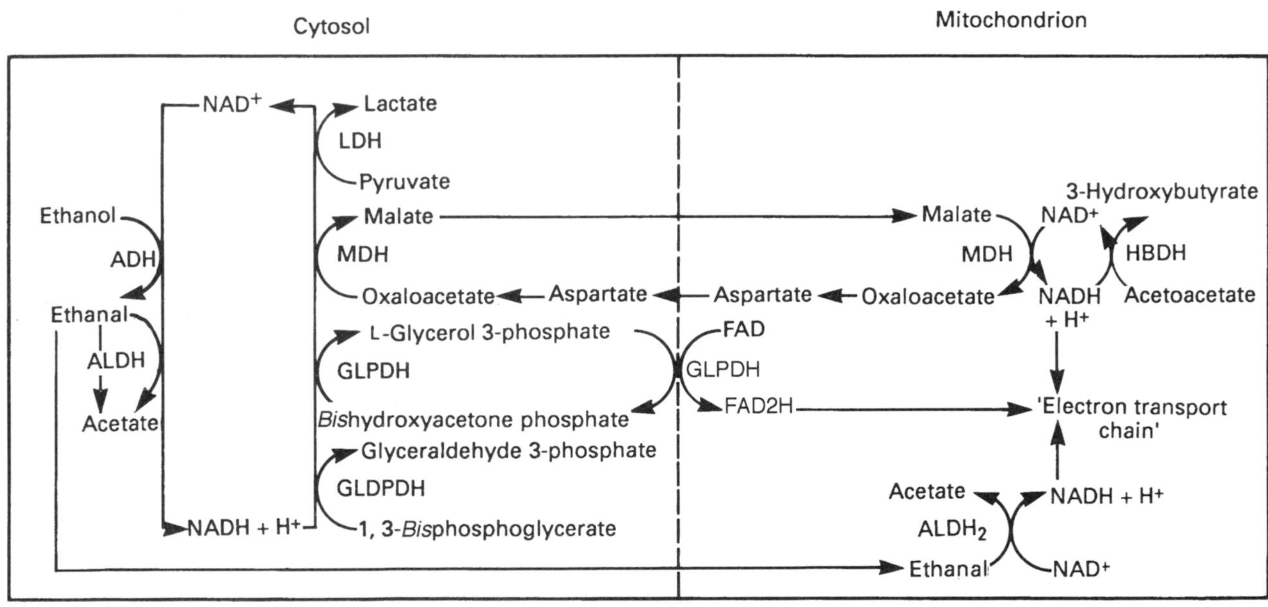

Figure 25.15 **The production and oxidation of NADH during the metabolism of ethanol** (ADH, alcohol dehydrogenase; ALDH, aldehyde dehydrogenase; GLPDH, glycerol 3-phosphate dehydrogenase; GLDPDH, glyceraldehyde 3-phosphate dehydrogenase; HBDH, 3-hydroxybutyrate dehydrogenase; LDH, lactate dehydrogenase; MDH, malate dehydrogenase)

heart and skeletal muscle. The first step in acetate metabolism is the formation of acetylCoA, which is to a large extent, oxidized to CO_2 through the tricarboxylic acid cycle yielding ATP, but experiments with [^{14}C]ethanol have shown that the acetate can also be converted to fatty acids, cholesterol and amino acids.

A significant aspect of the metabolism of ethanol is the action it exerts on the redox levels of the cytosol and mitochondria of liver cells. The NADH generated in the oxidation of ethanol in the cytosol can be reoxidized to NAD^+ by a variety of reactions, including the reduction of pyruvate to lactate, and the reduction step in gluconeogenesis, i.e. the conversion of 1,3-*bis*phosphoglycerate to glyceraldehyde 3-phosphate. The major route for the reoxidation of NADH is the electron transport system in the mitochondria. It was noted in Chapter 5 that the mitochondrial membrane is impermeable to NADH, but the transport of metabolites such as malate through the membrane enables NADH to be generated within the mitochondria with the concomitant regeneration of cytosolic NAD^+.

Even though all these potential pathways for the reoxidation of cytosolic NADH exist, a very characteristic effect of ethanol is to change the ratio of free NAD^+ to free NADH in the cytosol from about 1000 in the absence of ethanol to about 250–300 in the presence of ethanol. This change is due to a three- to fourfold increase in free cytosolic NADH, whilst the corresponding decrease in free NAD^+ is negligible. As a consequence of this, the ratios of lactate to pyruvate and 3-hydroxybutyrate to acetoacetate increase substantially in the venous blood leaving the liver. The metabolic reactions involved in the oxidation of ethanol in the hepatocyte are summarized in Figure 25.15.

Chapter 26

The kidney

26.1 Introduction

The kidney is a very important regulatory organ for mammals. Underlying its obvious role as a producer of urine are the following functions:

1. Regulation of the body water content.
2. Regulation of plasma electrolytes.
3. Regulation of plasma pH.
4. Excretion of unwanted blood-borne, water-soluble, low-molecular-mass compounds, including the non-volatile waste products of intermediary metabolism and a range of 'foreign' compounds (drugs, pollutants, food additives), frequently as their metabolites.

In addition the kidney is an endocrine organ. It secretes the hormone erythropoietin in response to a fall in oxygen tension. Cells of the juxtaglomerular apparatus release the enzyme renin, which is responsible for initiating the formation of angiotensin (see Chapter 28). The kidney also converts circulating 25-hydroxycholecalciferol into the hormonally-active form of vitamin D, 1,25-dihydroxycholecalciferol (see Chapter 12). In addition the kidney is also active in the production of eicosanoids (prostaglandins and thromboxanes - see Chapter 17)

26.2 Structure of the kidney

The functional unit of the kidney is the **nephron** which consists of a **glomerulus** (situated at the cortex of the organ) from which leads a tubular system with clearly-defined zones: the **proximal tubule**, the **loop of Henle** (which descends deep into the medulla and then returns), the **distal tubule** and the **collecting duct** from which passes the ureter (Figure 26.1). The glomerulus is a filtration unit comprising a hollow goblet-shaped Bowman's capsule. The internal space of this invaginated cavity, into which is sealed a network of blood capillaries, is continuous with the tubular system. Efferent blood from the glomerulus is conducted to other capillary networks in close contact with absorptive areas of the tubules. The functional role of the glomerulus is to filter the blood, retaining cells and proteins, while the tubules are involved in the recovery of water, small organic molecules and electrolytes, as well as in the secretion of NH_4^+, K^+, H^+ and also of organic anions and cations.

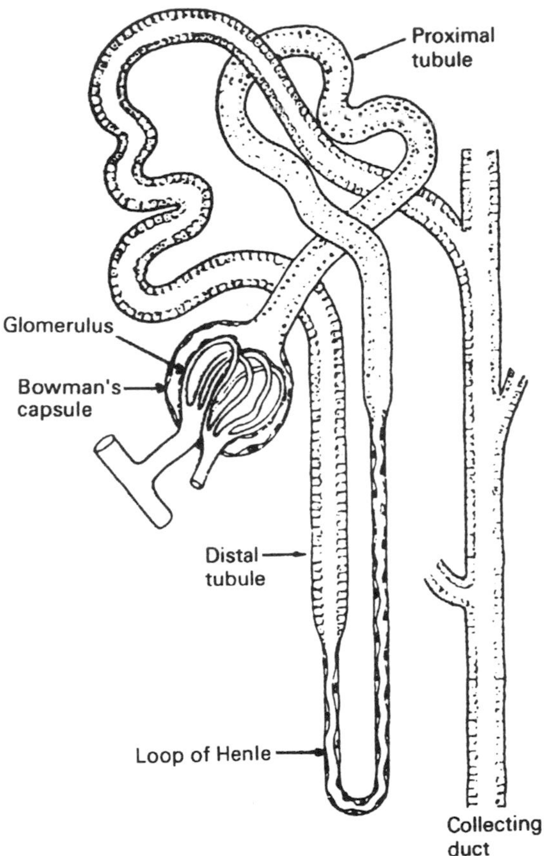

Figure 26.1 **The structure of the nephron**

26.3 Functional activity

A diagram of the functions of a single nephron (Figure 26.2) serves to represent the actions of the whole kidney. Each glomerulus filters a portion of the blood passing through it (about 20% of the plasma volume), retaining cells and protein. The net pressure for filtration in the kidney is 10 mmHg (1.33 kPa) through the semi-permeable basement membrane of the glomerulus. Large proteins (those of molecular mass of about 60 kDa or greater) are retained. Albumin (molecular mass 68 kDa) is normally almost totally retained, but haemoglobin (molecular mass 64.5 kDa), if it has escaped from the erythrocytes, passes through to a greater extent. It seems

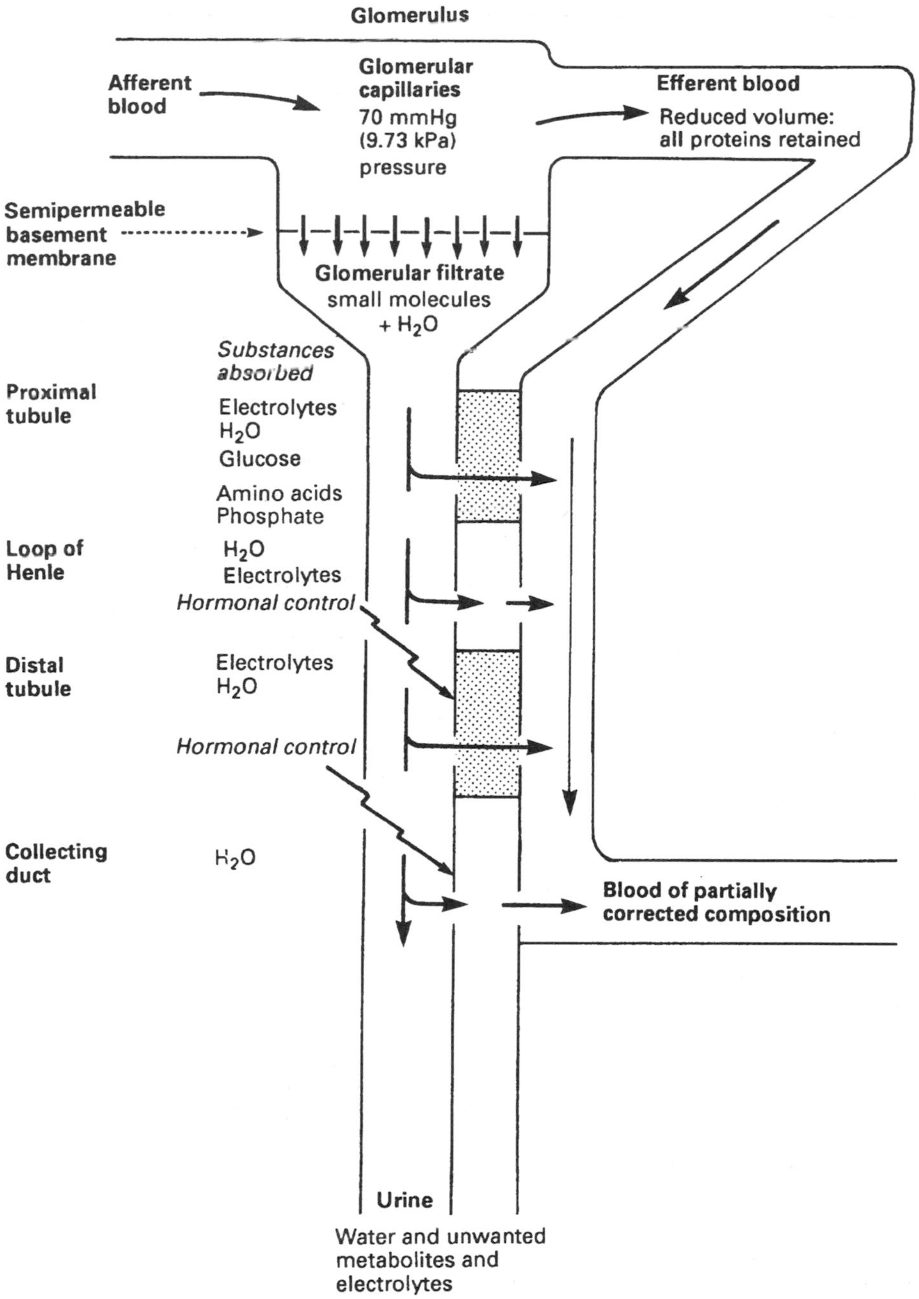

Figure 26.2 **Diagrammatic representation of the action of the nephron**

likely that that this degree of discrimination is not due solely to the small difference in mass, but also to overall charge and possibly shape.

Small ions and molecules pass through in the glomerular filtrate which then courses along the tubular system. They are of two main types: waste products (e.g. urea, NH_4^+, sulphate ions) or valuable metabolites (e.g. glucose, amino acids). The latter are reabsorbed, very little of them being found normally in the urine. Electrolytes are largely recovered, with the hormones aldosterone, atrial natriuretic factor, angiotensin II and dopamine being important determinants of the extent to which this happens. (The role of the recently-discovered adrenal cortical molecule ouabain is still to be evaluated – Section 26.8.) Water reabsorption also takes place, and in the later (distal) stages this is controlled by antidiuretic hormone. These recovered (reabsorbed) materials are 'added back' to the blood from which they had previously been depleted (cleared) by filtration. The enormous resorptive capacity of the kidney is demonstrated by the observation that, under normal circumstances in humans, about 180 litres of protein-free filtrate are produced per day, yet the daily output of urine is of the order of 1 litre.

The tubules also have specific secretory functions, located in various regions. The distal, regulated secretion of H^+, in correcting the body's acid-base balance, usually leads to the acidification of the urine, but the final pH of the urine may range from 5 to 8 and is also influenced by presence of buffers etc. e.g. HCO_3^-, $HPO_4^{2-}/H_2PO_4^-$, creatinine).

In summary, in one 'pass' of the blood through the kidneys about 20% of the plasma volume is 'examined' and its composition is corrected towards the optimum.

26.4 The glomerular capillary basement membrane in health and disease

There are three layers through which molecules must pass from the glomerular capillaries into the urinary (Bowman's) space (Figure 26.3) :

(a) the capillary endothelium
(b) the glomerular basement membrane
(c) the epithelial cells that form the inside wall of Bowman's capsule.

The capillary endothelium is pierced by pores called 'fenestrae', which allow the free passage of all molecules. The barrier to protein molecules seems to be the basement membrane, although there may also be a role for the slit pores between the epithelial foot processes. The glomerular basement membrane is a highly specialized type of basement membrane. In the mature human kidney it is between 310 and 380 nm thick, with a trilamellar structure consisting of a thick lamina densa sandwiched between two thinner laminae rare.

The glomerular basement membrane consists of a highly organized supra-molecular network, the mechanical stability of which derives from the presence a **type IV collagen** network. Other proteins that are found in basement membranes include the large glycoproteins **laminin**, **nidogen** and **entactin**. The functional role of the first two types of molecule is believed to be in mediating the interactions between the collagen network and glomerular cells. The role of enactin has yet to be established. As is often the case, when it is found in tissues, collagen in glomerular basement membranes interacts with proteoglycans. In the glomerulus this is largely heparan sulphate. These are polyanionic molecules (due to the presence of many sulphate and carboxylate groups), and this negative charge is of crucial functional importance, helping to account for the semi-permeable nature of the membrane. Negatively charged molecules are more readily retained than molecules having a similar mass and shape, but with an opposite charge. This probably explains the discrimination that allows positively-charged haemoglobin to pass into the urine while negatively-charged albumin is retained (see Section 26.3).

Alterations in the structure of the glomerular basement membrane can lead to kidney dysfunction. Thus impairment of kidney function may be seen in **diabetes mellitus**, a disease in which thickening of the basement membrane and proteinuria can occur. The membrane may also be weakened, with consequent loss of function following attack by proteases including those involved in complement activation (see Chapter 32). These may be released from tumour cells, cells involved in inflammatory responses (see Chapter 32), from bacteria or parasites. In addition a number of inherited glomerular diseases have been described. One of the most common of these is **Alport's syndrome** which has been shown to be caused by mutations to the gene encoding one of the chains of type IV collagen, the 'scaffolding' protein of basement membranes.

26.5 Composition of urine

Normal constituents

The composition and concentration of various end-products found in urine can vary throughout the day but, if samples collected for the whole of this period are pooled to give a '24-hour sample', a reasonable degree of constancy is achieved. Such sampling is useful when the production of the molecule to be measured undergoes episodic or circadian variations. Thus 24-hour sampling is used in the determination of 'urinary free cortisol' when patients are being investigated for Cushing's disease, for example (see Chapter 28).

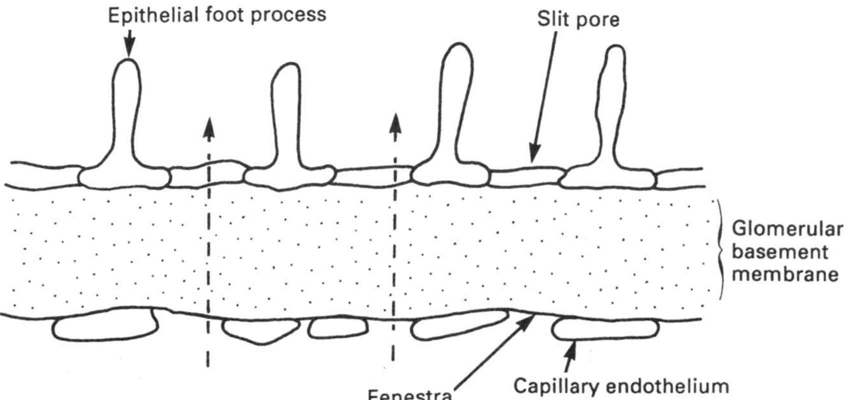

Figure 26.3 **Diagram of a cross-section of a glomerular capillary basement membrane.** The arrows indicate the direction of flow of fluid out of the capillaries

Table 26.1 **Summary of the composition of urine**

Components	*Constituents*
Water	
Inorganic substances	Electrolytes
	Trace constituents
Organic substances	Nitrogenous compounds
	Urea
	Uric acid (and other purines)
	Creatinine
	Indoles (metabolites of tryptophan
	Porphyrins
	Amino acids
	Organic acids
	Hormones
	Vitamins
	Enzymes

The volume, pH and composition of the urine can vary between wide limits depending mainly on dietary intake, including of water and electrolytes, but normally contains the constituents shown in Table 26.1. In the UK, a typical daily urine volume is 1–2 litres, but under conditions of severe water deprivation this may be reduced to as low as 200 ml. The daily rate of solute excretion ranges from 26 to 70 g. The major electrolytes and other minor inorganic constituents lost in the urine are shown in Table 26.2. The major source of sulphate ions in the urine is the dietary sulphur-containing amino acids which are oxidized in the course of metabolism. Nitrogenous compounds form another important group of urinary constituents and these are listed in Table 26.3. Total nitrogen excretion is closely related to dietary intake and the quantity of urea – the major end-product of nitrogen metabolism – varies directly with protein intake.

Small quantities of all the amino acids can be detected in the urine but the daily excretion of most of these amounts to no more than 10 mg (exceptions include glycine, histidine and glutamic acid, each of which appears in quantities in excess of 100 mg / day). Uric acid is the major purine excreted in the urine, resulting from the catabolism of adenine and guanine. Relatively large amounts of the porphyrin metabolite urobilinogen are present in the urine. This imparts the characteristic pale yellow colour.

Table 26.3 **Major nitrogenous constituents of urine after feeding on high- or low-protein diet – typical values**

Nitrogen	*Nitrogen excretion (g/24 h) after feeding*	
	High protein	*Low protein*
Total	16.8	3.60
Urea	14.7 (87.5)	2.20 (61.7)
Uric acid	0.18 (1.1)	0.09 (2.5)
Ammonia	0.50 (3.0)	0.42 (11.3)
Creatinine	0.60 (3.6)	0.60 (17.2)

Values in parentheses are percentages of the total nitrogen.

Abnormal constituents

A wide range of tests is available for the analysis of urine. These are designed to detect the occurrence of three types of circumstance as they may relate to underlying pathologies :

1. The presence of substances which, if they appear in the urine, are always indicative of pathology, e.g. haemoglobin, or the metabolites of substances of abuse.
2. The presence of substances which, if they appear, do not of necessity indicate pathology, but which will require further investigation, e.g. albumin found in the urine of pregnant women.
3. The presence of substances which are normal constituents of urine, but are present in abnormal amounts, e.g. water itself in **diabetes insipidus** and conjugates of pregnanetriol in patients with **congenital adrenal hyperplasia** (see Chapter 28).

The required analyses may be carried out on the wards and in consulting rooms where 'stix' tests can be used, otherwise the determinations are the province of hospital Departments of Chemical Pathology or Clinical Chemistry.

26.6 Energy provision in the kidney

As indicated, many of the biochemical events in the kidney relate to transport processes, particularly those

Table 26.2 **Inorganic constituents of urine**

Cation	*(mg/24 h)*	*(mmol/24 h)*	*Anion*	*(mg/24 h)*	*(mmol/24 h)*
Na^+	$(3–5) \times 10^3$	130–217	SO_4^{2-}	$(1–2) \times 10^3$	10.4–20.8
K^+	$(2–4) \times 10^3$	51–102	Organic sulphate about 10% of SO_4^{2-}		
NH_4^+	420–500	23–28			
Ca^{2+}	100–450	2.5–11.3	$H_2PO_4^-/HPO_4^{2-}$	70–835	0.7–8.7
Mg^{2+}	100–300	4.1–12.3	I^-	$(1.8–48.3) \times 10^{-2}$	$(1.4–38) \times 10^{-4}$
Zn^{2+}	0.3–0.5	$(5–8) \times 10^{-3}$			
Fe^{2+}/Fe^{3+}	0.02–1.1	$(3.5–200) \times 10^{-2}$	Br^-	2	2.5×10^{-2}
Ni^{2+}	0.15	2.5×10^{-3}	Cl^-	$(5–9) \times 10^3$	141–254
Cu^{2+}, Co^{2+}	All present in trace (μg)amounts		HCO_3^-	Insignificant below about pH 6.5,	
Pb^{2+}, Mn^{2+}	depending on exposure			but increases progressively rapidly	
Sn^{2+}				at higher pH values	

concerned with the recovery of water, electrolytes, glucose and amino acids. All of these process are energy-requiring. It is now established that the kidney is able to meet the energy demands placed on it, both by the oxidation of lactate and glutamine, and also of fatty acids (the kidney appears to have little dependence on the full oxidation of glucose for its energy requirements). The fatty acids may derive from circulating albumin or very-low-density lipoproteins. Neither of these types of protein passes through the glomerulus and consequently must reach the renal tubular cells from the peritubular capillary blood. In addition, fatty acids may derive from the stores of triacylglycerols found in lipid droplets in the medullary part of the organ.

26.7 Water absorption

Recovery of water by the kidney is closely dependent on that of electrolytes, particularly Na^+, Cl^- and HCO_3^-. In the following description of the mechanisms involved, repeated reference should be made to Figure 26.4.

Proximal tubule

As can be seen, about 80% of all of the water, Na^+ and Cl^- resorbed from the glomerular filtrate is removed in the course of passage through the proximal tubules. The fact that very little protein passes into the filtrate at the glomerulus means that, on entering the proximal tubules, the osmolarity of the filtrate has been reduced from about 300 to 295 mosmol/l. This provokes a movement of water out of the tubules to restore osmolarity. In the first few millimetres of the proximal tubules, Na^+ uptake is intimately dependent on the recovery of HCO_3^- and the slight acidification of urine. This mater is dealt with in detail in Section 26.11, but in essence, in the luminal face of the epithelial cells, the potential energy of the Na^+ gradient drives the expulsion of protons in exchange for the metal cation. The Na^+ is subsequently returned to the plasma via a basolateral Na^+/K^+ ATPase. The extruded H^+ permit the resorption of HCO_3^- (as CO_2), and hence at this stage $NaHCO_3$ is effectively removed together with an osmotic equivalent of water. This has the effect of increasing the concentration of Cl^- in the filtrate above their Donnan equilibrium value for plasma; consequently, as the

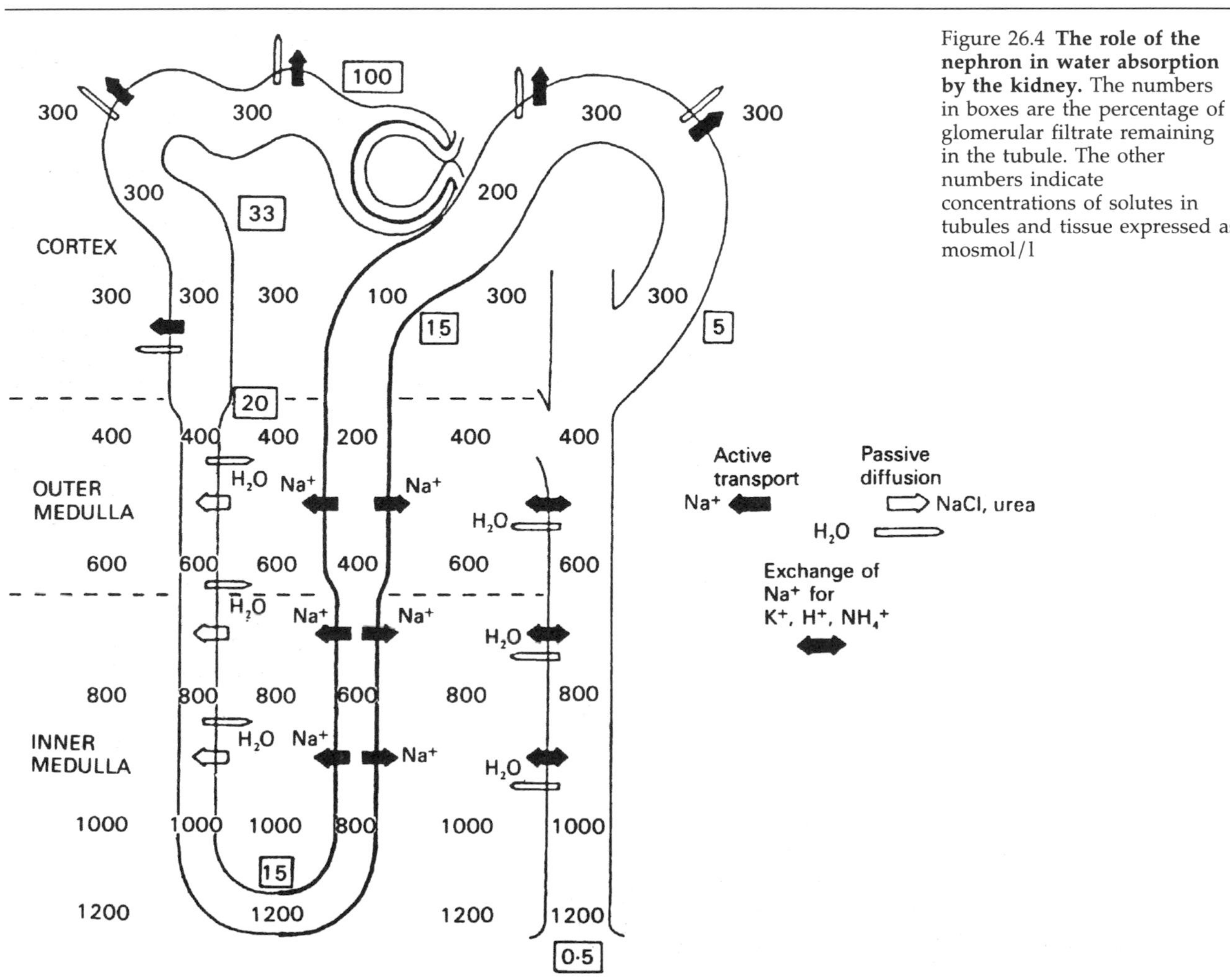

Figure 26.4 **The role of the nephron in water absorption by the kidney.** The numbers in boxes are the percentage of glomerular filtrate remaining in the tubule. The other numbers indicate concentrations of solutes in tubules and tissue expressed as mosmol/l

filtrate moves down the tubule, this anion moves into the interstitium of the cortex down its concentration gradient. The ion moves through Cl^--permeable regions at the cell junctions and must be accompanied by an equivalent amount of cations (mainly Na^+) and of water.

Since the combined effects of re-equilibration following protein removal, the absorption of $NaHCO_3$ and the Donnan movement of NaCl would not jointly explain the removal of 80% of the constituents of the filtrate known to occur in the proximal tubules – other mechanisms must come into play. As is shown in Sections 26.9 and 26.10, in their active absorption, glucose molecules and amino acids are co-transported with Na^+ and, once again, the Na^+ would be followed by an equivalent amount of Cl^-. But even when the Na^+ -dependent processes described above are taken into account, they seem unlikely to explain more than 50-60% of Na^+ absorption in this region of the kidney. It appears that the epithelium is rather 'leaky' to Na^+, and when the water flows across it, it draws Na^+ into the interstitium of the cortex.

Recently two water-selective channel proteins have been cloned from the kidney and termed **aquaporin-1** and **aquaporin-2**. The former is present in proximal tubules and in the descending thin limb of the loop of Henle, both of which structures have a constitutively high permeability for water. This protein, therefore, has the function of facilitating the osmotic movement of water associated with the solute movements just described. The second protein, aquaporin-2, is restricted to cells of the collecting duct and opens in response to antidiuretic hormone.

Loop of Henle

The balanced movement of solutes and water occurring in the proximal tubules results in the fluid entering the descending arm of the loop of Henle being iso-osmotic with plasma (300 mosmol/l). In the next phases, further recovery of water depends critically on the interstitial fluid of the medullary regions of the tissue being maintained hyperosmolar, with a gradient running from 400 mosmol/l in the outer medulla to 1200 mosmol/l in the inner medulla. This osmolar concentration gradient is paralleled in the fluid in the descending limb of the loop, whilst the one running in the opposite direction from 800–200 mosmol/l arises along the ascending limb. The mechanism underlying these observations is that while the descending limb is freely permeable to water (due to the presence of aquaporin-1) and allows passive movements of Na^+ and Cl^-, the thick ascending limb is essentially impermeable to water but actively moves both ions. The mechanism underlying the movement of Na^+ and Cl^- is shown in Figure 26.5. Thus NaCl is removed from the filtrate in the ascending limb of the loop, thereby rendering it hypotonic (in the absence of accompanying movement of water). Water is consequently attracted from the descending limb, rendering its fluid increasingly hypertonic. The simple removal of water does not fully explain the approximately fourfold rise in the osmolarity observed by the turn of the loop of Henle, and it must be borne in mind that the NaCl is free to equilibrate between the interstitium of the medulla and the fluid in the descending limb and, in addition, that urea also enters the tubule lumen from the interstitium in this region of the nephron (loop).

Distal tubule and collecting duct

The fluid emerging from the ascending limb into the distal tubule does so through a second cortical region permeable to water, so further water leaves until the tonicity is restored to normal (300 mosmol/l), but only in the presence of antidiuretic hormone. In addition, further active transport of Na^+ may occur and this is promoted by aldosterone, the production of which is regulated by angiotensin II, the peptide hormone responsive to changes in plasma volume, arterial pressure, etc. (see Chapter 28).

The interstitial fluid in the medulla surrounding the collecting duct is, like that surrounding the loop of Henle, hypertonic with respect to the filtrate entering the duct and consequently there is a tendency for water to move into the interstitium, but the extent to which this occurs depends on the presence of **antidiuretic hormone** (ADH). This hormone is released from the posterior pituitary gland in response to an increase in plasma osmolarity, small changes (1–2%) resulting in its secretion at a maximal rate. ADH binds to V_2-receptors coupled to adenylyl cyclase with the production of cAMP. This, in turn, causes the recruitment of

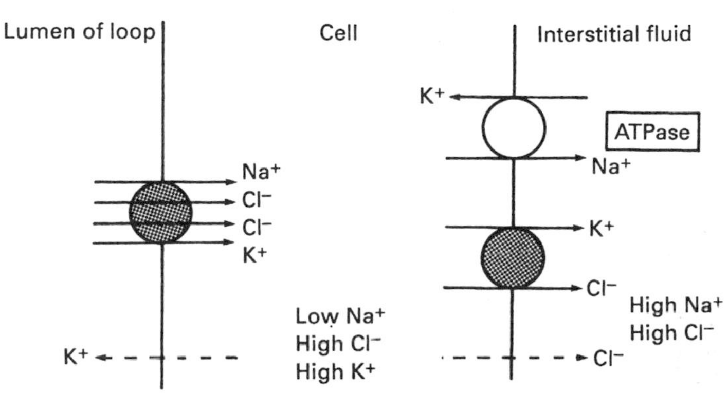

Figure 26.5 **The accumulation of NaCl in the interstitium of the loop of Henle.** ○, ATP-dependent exchange of Na^+ for K^+ (Na^+/K^+ ATPase); ●, electroneutral transporters; --->, passive diffusion (or specific channels)

aquaporin-2 from intracellular membranes to the luminal plasma membranes of the cell lining the distal tubule and collecting duct. This protein facilitates the recovery of water. ADH, therefore, plays a central role in controlling the amount of water that is recovered from the hypotonic fluid reaching the distal tubules and collecting duct. Failure to produce ADH, as occurs in **diabetes insipidus**, is associated with a reduced plasma volume, high plasma osmolarity and a copious dilute urine. Some patients, however, present with an inherited, nephrogenic version of the disease and they prove to have mutations to one of two genes: one encoding the V_2 receptor and the other the aquaporin-2 molecule.

The use of aldosterone (and also of natriuretic hormones – see Section 26.8) allows the appropriate retention of NaCl and water to balance the 'budget' of intake and other inevitable outputs elsewhere (e.g. through sweating) under a wide range of normal circumstances. Hence, the osmolarity of urine can vary greatly. It should be noted, however, that these mechanisms have difficulty in coping with extreme demands made on them, e.g. massive losses of water and salts in haemorrhaging, persistent diarrhoea and vomiting and diuresis of the type encountered in **diabetes mellitus**. In severe haemorrhage, the strong pressor action of ADH (also known as vasopressin) can itself be life-threatening and the drive to secrete the hormone can be overridden by cortisol (an example of the ability of the glucocorticoid to act as an 'anti-stress hormone'; see Chapter 28).

26.8 Absorption of electrolytes

Sodium and chloride ions

These two ions are absorbed with a very high degree of efficiency. Most of their absorption occurs in relation to the recovery of water and has been described in Section 26.7. As is apparent from the description of water absorption, the key enzyme in the uptake of Na^+ in both the proximal tubule and in the ascending limb of the loop of Henle is the Na^+/K^+ ATPase. This acts to establish Na^+ gradients that favour the movement of the cation in the desired direction.

Role of aldosterone

Sodium ion absorption in the distal tubule is under the direct control of the hormone **aldosterone**, the release of which is controlled indirectly by the enzyme **renin**. The latter is released from the kidney in response to reduced arterial pressure, reduced plasma volume, depletion of Na^+, or renal nervous stimulation. Renin catalyses the conversion of circulating α_2-globulin (angiotensinogen) into angiotensin I which then acts as a substrate for angiotensin converting enzyme (ACE) to form active **angiotensin II**. This hormone stimulates the adrenal cortex to produce the hormone aldosterone which, in turn, stimulates Na^+ resorption in the distal tubules (see Chapter 28).

The general mechanism of action for steroid hormones, described in Chapter 3, is known to apply to aldosterone acting in the kidney. In the cells of the aldosterone-responsive parts of the nephron there is a cytosolic mineralocorticoid receptor which, when occupied by the steroid, is translocated to the nucleus where the hormone–receptor complex binds to steroid-responsive elements in several genes to increase their rates of transcription. The full range of target genes have not been identified, but those for the Na^+/K^+-ATPase and a Na^+-channel, sensitive to the drug amilioride, are probably included.

The drug **spironolactone** competes with aldosterone in its binding to the cytoplasmic receptor, but once bound to the receptor the complex does not affect transcription, i.e. it acts as an antagonist. The drug is therefore used in treating the hyperkalaemia and hypertension suffered by patients who over-secrete the hormone, e.g. in primary hyperaldosteronism (see Chapter 28).

As is also noted in Chapter 28, the circulating concentration of cortisol exceeds that of aldosterone by about 100-fold, yet both hormones bind with roughly

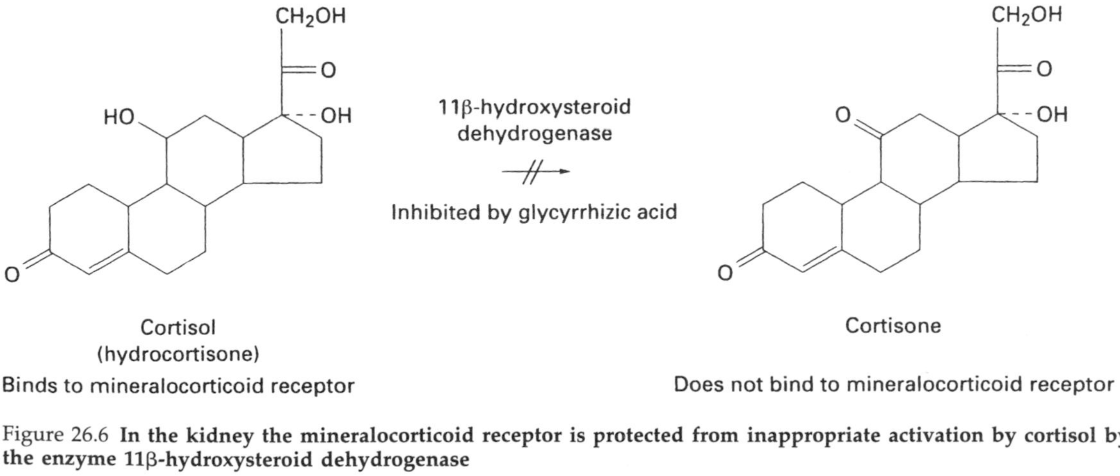

Figure 26.6 **In the kidney the mineralocorticoid receptor is protected from inappropriate activation by cortisol by the action of the enzyme 11β-hydroxysteroid dehydrogenase**

equal avidity to the mineralocorticoid receptor. These observations made it difficult to understand how specific responses of the kidney to aldosterone could be achieved in the face of overwhelming amounts of cortisol (which physiologically is a relatively weak mineralocorticoid). There appear to be two mechanisms whereby specificity is assured. In the kidney is found a very high concentration of the enzyme 11β-hydroxysteroid dehydrogenase. This enzyme catalyses the oxidation of cortisol to cortisone (Figure 26.6), which binds with very low affinity to the mineralocorticoid receptor. By this means the renal aldosterone receptors are 'protected' from cortisol. It should be noted that this protection may be impaired by ingestion of the compound glycyrrhizic acid, found in liquorice, and a potent inhibitor of 11β-hydroxysteroid dehydrogenase. Consequently those consuming large quantities of liquorice may be subject to hypertension, as cortisol is better able to activate the aldosterone receptor.

A second, unexpected, mechanism serves the dual purpose of ensuring both a rapid and also a specific response to aldosterone. In this, aldosterone acts via a membrane (not a cytosolic) receptor which binds neither cortisol nor spironolactone. The binding of aldosterone to this receptor leads to the rapid activation of a Na^+/H^+-exchanger channel. The actions of aldosterone in the kidney are summarized in Figure 26.7.

Other factors that may influence sodium retention

It has long been felt that the precision of the maintenance of plasma volumes and Na^+ concentration pointed to the intervention of factors in addition to aldosterone and 'physical' factors. These have sometimes been referred to as 'third factors'. These 'natriuretic factors' were proposed to promote the active renal excretion of Na^+, in appropriate circumstances. In the event, two such factors have been suggested. The first proposed naturietic molecule is a peptide hormone, produced by the atrium of the heart and termed '**atrial natriuretic peptide** (factor)' **(ANP)**. This peptide seems to be able to promote a natriuresis at several points:

1. It promotes a natriuresis by causing an increase in glomerular filtration rate, with a simultaneous fall in arterial blood pressure.
2. It inhibits renin secretion from the kidney, and as a consequence of this it indirectly:
   (a) inhibits aldosterone secretion by adrenal glomerulosa cells
   (b) interferes with the vasoconstrictive effects of angiotensin II.

The evidence that this atrial peptide plays a role in physiology is now strong. The evidence for a second natriuretic molecule must be regarded as less well established. The molecule in question is the Na^+/K^+ ATPase inhibitor **ouabain**. This steroid derivative, referred to as a **cardiac glycoside**, has a long history of medical use. The compound, which is isolated from the Ouabio tree of East Africa, is given to patients to increase the intensity of heart muscle contraction. Recently, however, the production of ouabain by the human adrenal cortex has been claimed and the molecule is proposed to play a role in normal Na^+ homeostasis, both at the cellular and the whole-body level. Its effect on whole-body Na^+ homeostasis would be expected to be that of promoting a natriuresis. This interesting suggestion, however, must await further validation.

Potassium ions

The kidney does not appear to be as efficient in its regulation of plasma K^+ as it is in the control of Na^+. Any tendency to hyperkalaemia (high circulating concentrations of K^+) is well compensated, but the organ is less efficient in dealing with hypokalaemia. Even in the complete dietary absence of K^+, some 10-20 mmol of the cation are lost per day in the urine.

In common with Na^+, some 80% of the K^+ in glomerular filtrate is resorbed in the proximal tubules by

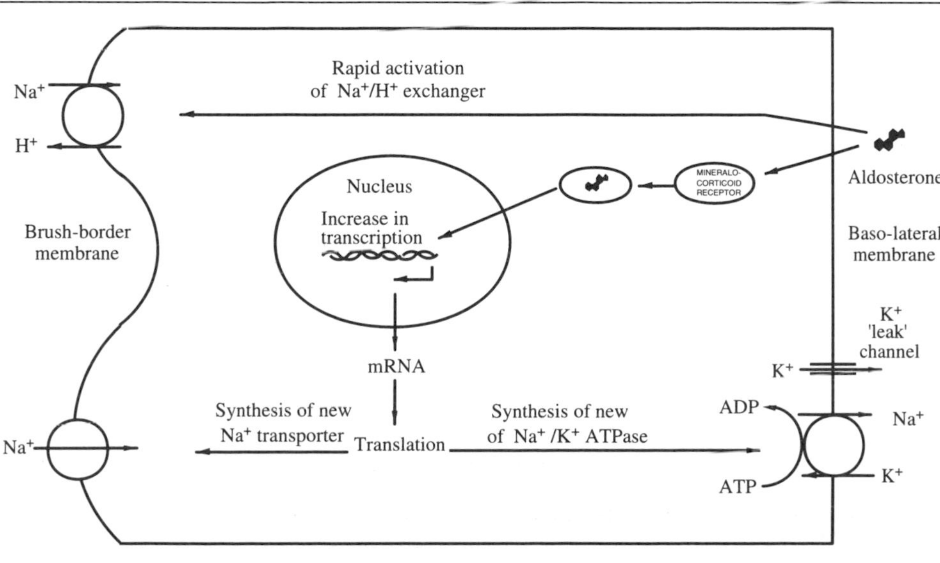

Figure 26.7 **Aldosterone acts in three ways to promote the transport of sodium ions across kidney epithelial cells**

an active-transport process. Resorption of the cation may also occur in the distal tubules, but usually the movement is in the other direction, i.e. into the luminal fluid. Secretion across the luminal membrane seems to be by a passive process, but the extent to which it occurs depends on several factors. These include the presence of aldosterone and the concentration of K^+ in the distal tubule cells.

26.9 Absorption of glucose

Under normal circumstances, virtually no glucose is excreted by the kidney, because all of the filtered monosaccharide is resorbed along the proximal tubule. However, as its plasma concentration is increased above a critical value (about 10 mm), the **renal transport maximum**, glucose does appear in the urine. After the concentration of glucose exceeds this maximum the amount appearing is approximately proportional to any further increase in the plasma concentration. It should be appreciated, however, that in addition to the plasma concentration of glucose, the rate of flow of blood through the kidney and the blood pressure can influence the renal excretion of glucose.

The recovery of glucose is an active process in which the potential energy of the Na^+ gradient into the proximal tubule epithelial cells is harnessed as glucose is co-transported with the cation. Two transporter proteins mediate this recovery of glucose. The first is found in highest concentration in the first few millimetres of the proximal tubules and transports glucose against a relatively shallow concentration gradient. The second can move the monosaccharide against a much steeper gradient and this operates further along the tubule.

The movement of glucose across the basolateral surface of the cells is facilitated by a glucose transporter that resembles, but has a much greater capacity than, that found in the liver (Glut 2). The overall process for glucose recovery, depending as it does on both a luminal (Na^+-dependent) and a contraluminal (Na^+-independent) transporter, closely parallels the one that operates in intestinal epithelial cells for the absorption of dietary glucose (see Chapters 2 and 13). **Phloridzin** is a powerful inhibitor of glucose resorption. This molecule binds to the Na^+/glucose co-transporter in the renal tubular epithelial cell luminal plasma membrane, so preventing the movement of the monosaccharide.

26.10 Absorption of amino acids

The glomerular filtrate contains amino acids in concentrations very close to their individual plasma concentrations. The rate of resorption of different amino acids is not equal and glycine, for example, is resorbed much more rapidly than arginine or lysine. It is now established that no fewer than seven Na^+-dependent transporter proteins are found in the kidney for the recovery of amino acids. These are specific for:

(a) dibasic amino acids, i.e. arginine, lysine and ornithine
(b) bulky neutral amino acids, e.g. leucine
(c) acidic amino acids, i.e. glutamic and aspartic acids
(d) glycine
(e) imino acids, e.g. proline
(f) cysteine and cystine
(g) less common β- and γ-amino acids, e.g. β-alanine (found in coenzyme A).

The existence, in the kidney tubules, of absorption systems specific for groups of amino acids was first inferred by Garrod at the beginning of the present century. He described the occurrence of the inherited disease **cystinuria**. This condition was characterized by the deposition of crystals of cystine in the urine of patients with this disease. To describe this and some other inherited diseases Garrod introduced the term '**inborn errors in metabolism**'. Subsequently, when analytical methods had improved, it was found that the urine of those subjects with cystinuria also contained relatively large quantities of lysine, arginine and ornithine, as well as cystine. There is now good evidence to suggest that the amino acids share a common transport system which is impaired in cystinuria, as well as being transported by separate systems as indicated above. Progress in the molecular cloning of the amino acid transporters will help to characterize and better define the nature and the exact roles of each of the tubular amino acid transport systems.

In addition to the conventional transporters for amino acids, there is also an H^+-dependent transporter, although the physiological significance of this is still to be established. Another possible amino acid transport system was proposed by Meister. In his proposal, a central role was ascribed to the tripeptide glutathione. In this the trapping of the amino acid to be transported is considered to involve a transpeptidation that utilizes the glutamyl residue of glutathione to convert the amino acid into a dipeptide. Glutamate is eventually released and combines once more with cysteine and glycine to reform glutathione, the latter two amino acids having been released from their dipeptide formed during the transpeptidation reaction (Figure 26.8). Although it is well established that a glutathione transpeptidase is active in the brush border of the kidney lumen, it is now not accepted that the enzyme plays a part in amino acid transport. This is because it has been demonstrated that the γ-glutamyl transpeptidase is extracellular and therefore not accessible to intracellular glutathione. The physiological role of the enzyme is likely to be as a stage in the process in which glutathione conjugates of foreign compounds are converted to mercapturic acids (see Chapter 36).

26.11 Regulation of pH

The kidney plays a very important role in regulating the pH of the plasma and of the whole body. The pH of urine can vary from as low as 4.8 to as high as 8.0, a much wider range than is tolerable in the plasma. From this it can be inferred that the kidney is capable of excreting H^+ or base (HCO_3^-) as circumstances may demand. Normally an acid urine is produced, with no HCO_3^-, and its measured acidity (i.e. H^+ buffered by HPO_4^{2-} and combined in the form of NH_4^+) is largely due to the acidotic effect of the metabolic oxidation (largely

Lumen of tubule
Plasma membrane
Amino acid AA
γ-Glutamyl transpeptidase
γ-Glu-Cys-Gly (glutathione)
ADP + Pi
ATP
Cys-Gly
Glycine
Peptidase
γ-Glu-AA
Glutathione synthetase
γ-Glutamyl cyclotransferase
Cysteine
γ-Glu-Cys
5-Oxoproline
ADP + Pi
ATP
AA
Oxoprolinase
γ-Glu-Cys synthetase
ATP
ADP + Pi
Glutamate

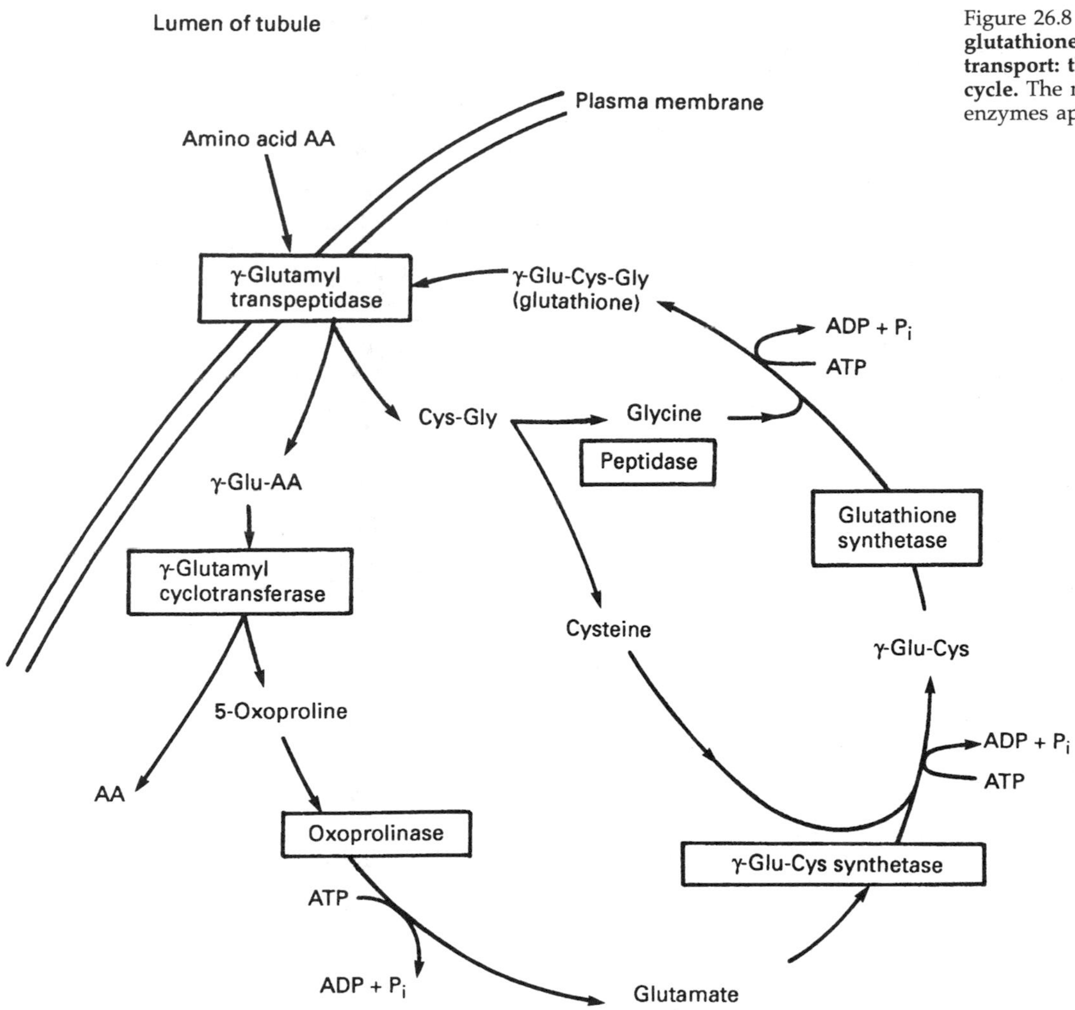

Figure 26.8 **Postulated role of glutathione in amino acid transport: the γ-glutamyl cycle.** The names of the enzymes appear in boxes

Glomerular filtrate
Cell
Blood
Metabolism: substrates + O2
PCO2 = 40 mmHg
H2O + CO2
CO2
CO2 + H2O
H2CO3
Carbonic anhydrase
Na+
Na+
H2CO3
HCO3−
H+
H+
HCO3−
HCO3−
K+
ATPase
Na+
K+

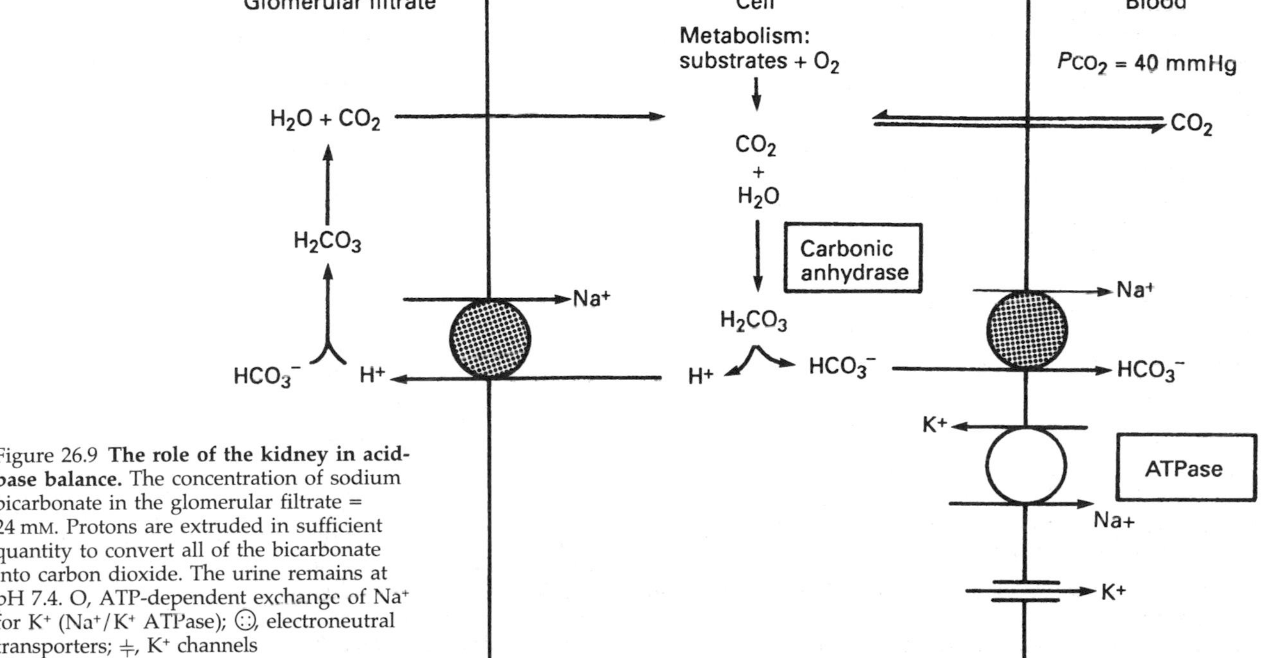

Figure 26.9 **The role of the kidney in acid-base balance.** The concentration of sodium bicarbonate in the glomerular filtrate = 24 mM. Protons are extruded in sufficient quantity to convert all of the bicarbonate into carbon dioxide. The urine remains at pH 7.4. O, ATP-dependent exchange of Na^+ for K^+ (Na^+/K^+ ATPase); ⦾, electroneutral transporters; ⏚, K^+ channels

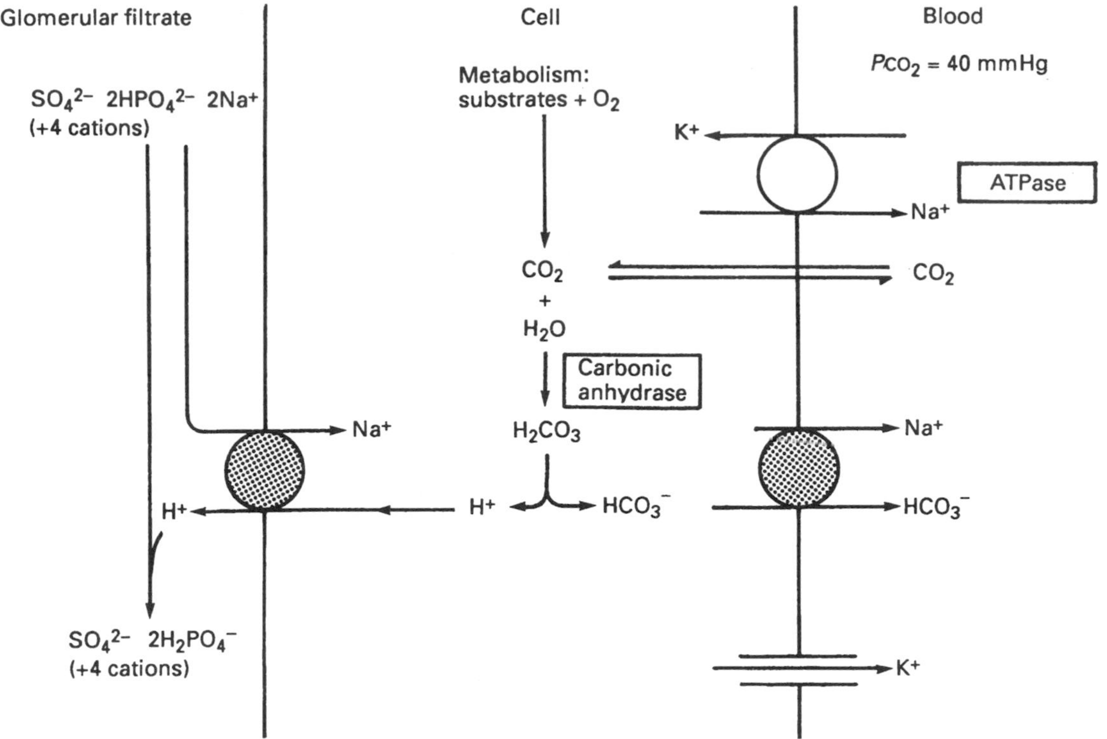

Figure 26.10 **The role of the kidney in acid-base balance.** The concentration of sodium bicarbonate in the glomerular filtrate is less than 24 mM as a result of the metabolic formation of acids other than carbonic, e.g. sulphuric. Protons sufficient to cause the conversion of 24 mM bicarbonate into carbon dioxide continue to be extruded, resulting in the acidification of the urine. The reduction of pH of the urine is minimized by the buffering action of HPO_4^{2-} in the filtrate. For the symbols, see Figure 26.9

in the liver) of dietary sulphur-containing amino acids. On the other hand, the excretion of an alkaline urine, with no NH_4^+, can occur if the plasma concentration of HCO_3^- rises: the tubules may resorb less than the total amount of the anion in the glomerular filtrate.

The mechanism of resorption of HCO_3^- may be explained initially by means of a simple model in which it is assumed that there has been no metabolic formation of permanent acids (e.g. protons and sulphate ions) and consequently the concentration of HCO_3^- in the filtrate is a normal one of 24 mmol/l (Figure 26.9). It seems that the kidney proximal tubule cells are 'tuned' to secrete just enough H^+ to decompose all of the HCO_3^- to CO_2 and H_2O when the pCO_2 is 40 mmHg (5.3 kPa), although how the tuning is achieved is uncertain. The secretion of these protons involves their energy-dependent exchange for Na^+, the required concentration gradient of the latter cation being set up by the action of a Na^+/K^+ ATPase in the contraluminal plasma membrane of the tubular cells. The CO_2 produced in this way diffuses into the cells and is converted into carbonic acid by the action of carbonic anhydrase and then into HCO_3^-. This HCO_3^- is returned to the plasma, being co-transported in a ratio of 3:1 with Na^+, the process being driven by the membrane potential of the cells. Thus, the net transfer of all of the HCO_3^- in the filtrate to the blood will have been achieved without any change in pH; the pH of the urine would be close to 7.4. However, the plasma and hence the filtrate are usually depleted of HCO_3^- to the same extent that SO_4^{2-} (and two protons) has been formed metabolically (2 mol HCO_3^- lost for each mol SO_4^{2-} formed). This is because the oxidation of acidic and basic amino acids from the diet make approximately equal and opposite contributions to the acidification of the urine. Despite the depletion of HCO_3^- from the filtrate, the tubular cells continue to secrete enough protons to decompose 24 mmol/l HCO_3^- to CO_2 and H_2O (Figure 26.10). By this means, any HCO_3^- is recovered, but the additional H^+ are buffered by HPO_4^{2-} to give $H_2PO_4^-$. Consequently an acidic urine is produced (although it does not fall in pH to the extent it would in the absence of HPO_4^{2-} and other buffer ions). Once again 24 mmol/l HCO_3^- is returned to the plasma; the additional HCO_3^- being generated via the action of carbonic anhydrase on 'metabolic' CO_2 formed in the cells.

The process of acidification that occurs in the proximal tubules is limited to the extent that the pH in the lumen is never more than 0.6 of a pH unit below that in the cells, i.e. about pH 6.4, a commonly encountered value for normal urine. There is a second proton pump in the collecting ducts which can achieve acidification down to pH 4.8 in circumstances of severe metabolic acidosis, e.g. untreated diabetes mellitus.

There is another mechanism for taking care of H^+, namely the secretion of NH_4^+ (Figure 26.11). The plasma amino acids, especially glutamine, are the major source of urinary NH_4^+. For example, in the tubular cells the enzyme glutaminase acts to catalyse the hydrolysis of glutamine to yield NH_4^+, which at intracellular pH

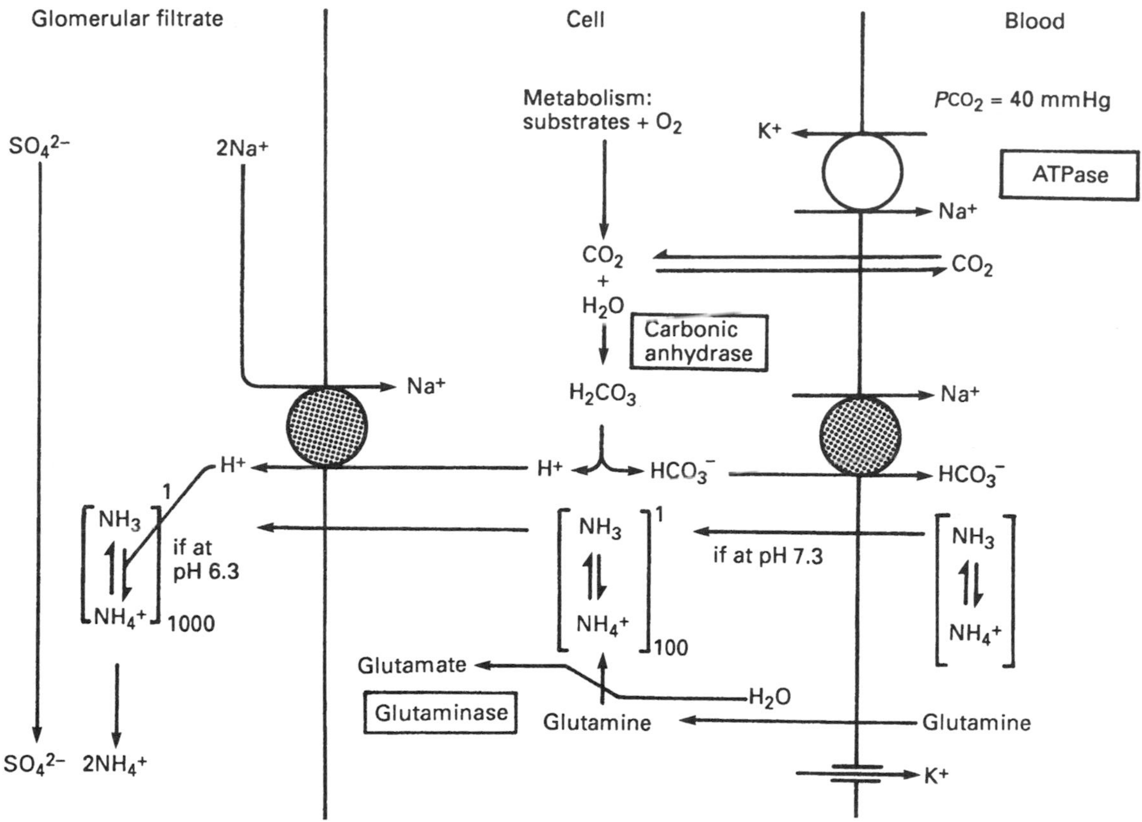

Figure 26.11 **The role of the kidney in acid-base balance.** Ammonium ions in the urine can spare the loss of metal ions and prevent the fall in pH that would otherwise occur. For the symbols, see Figure 26.9

values will be in equilibrium with about one-hundredth of their number of NH_3 molecules. Nevertheless, the latter diffuse into the filtrate, where they can once more accept a proton to yield NH_4^+, the ratio of NH_4^+ to NH_3 having risen to about 1000 : 1. Therefore NH_3 continues to move into the filtrate as long as protons continue to be exchanged for Na^+. The presence of NH_4^+ (cations) allows extra Na^+ ions to be recovered from the filtrate. In consequence, NH_4Cl is excreted and $NaHCO_3$ is recovered.

Sometimes NH_4^+ is referred to as a 'urinary buffer', but this clearly not the case, since the pH of the urine is more than two pH units away from its pK_a which is about 9.3. What it does is to neutralize the H^+ by combining with them, and thus there is no simultaneous drift of the pH. It can be appreciated, therefore, that ammonia secretion helps to reduce the extent to which urine needs to be acidified, thereby helping to minimize the pH drop that would otherwise result from the activity of the Na^+/H^+ exchanger.

The production of NH_4^+ is particularly advantageous, because it is adaptable to changing circumstances. It increases with development of metabolic acidosis and shuts down when the pH of the urine rises above 7.4.

26.12 The kidney and erythropoiesis

That the kidney has an important role in controlling the formation of erythrocytes in bone marrow was inferred from two observations. Polycythaemia (excessive production of red cells) is frequently associated with renal neoplasms, and conversely anaemia often arises in patients in renal failure. Subsequently a hormone, named **erythropoietin**, was isolated from the urine of patients with aplastic anaemia (those with defects in the erythropoietin receptor) and shown to act at an early stage in the differentiation of stem cells in the bone marrow to promote the emergence of erythrocytes (see Chapter 32), although the precise target cells have not been identified. The secretion of erythropoietin occurs in response to a fall in oxygen tension in the kidney. Erythropoietin is an extensively glycosylated protein hormone, and the gene encoding it has been cloned. Indeed, erythropoietin has provided one of the earliest examples of the power of the techniques of recombinant DNA technology. The therapeutic use of recombinant erythropoietin in treating anaemia arising in patients with renal failure is now well established.

Chapter 27

Muscle

27.1 Introduction

The vital importance of muscle, coupled with the relative ease of its study, has ensured that great attention has been paid to this tissue. As a consequence, muscle has been studied extensively by light and electron microscopy and by physiological and biochemical methods. The integration of these studies has led to an understanding of the relationships between structure and function.

There are three types of mammalian muscle. The most abundant is **skeletal muscle**, which is usually attached to the skeleton or bone via tendons, and is responsible for specific movements of the body. **Cardiac muscle** is that found in the walls of the heart and major blood vessels in the immediate vicinity. The muscle found in most blood vessels and visceral tissues such as the intestines is known as **smooth muscle**. The mechanism of contraction is similar in all three types in that it involves actin–myosin contractile elements, but the structural organization and the regulation of muscular contraction differs in the three types of muscle mentioned.

27.2 Structure of skeletal muscle

Skeletal muscle is composed of bundles of fibres, known as **myofibres** bound together by connective tissue rich in collagen and elastin, the **endomysium** which is found around individual fibres and the **perimysium** which covers groups of fibres together. Individual whole muscles are also covered with a layer of connective tissue known as the **epimysium** which is thicker and tougher than the connective tissue within muscles.

Each myofibre is a long cylindrical multinucleated cell and contains regular arrays of **myofibrils** which are composed of thick and thin **myofilaments**. The myofilaments are organized into the functional units of skeletal muscle known as **sarcomeres** which can be seen under the low-powered electron microscope as a number of longitudinal units making up skeletal muscle fibres (Figure 27.1). Each sarcomere is defined by two adjacent **'Z' lines** (or **'Z' disks**) and contains a dark **'A' band** and two halves of the adjacent lighter bands known as **'I' bands**.

Closer examination of the structure by electron microscopy (Figure 27.2) shows that the A band is formed of a large number of thick filaments and

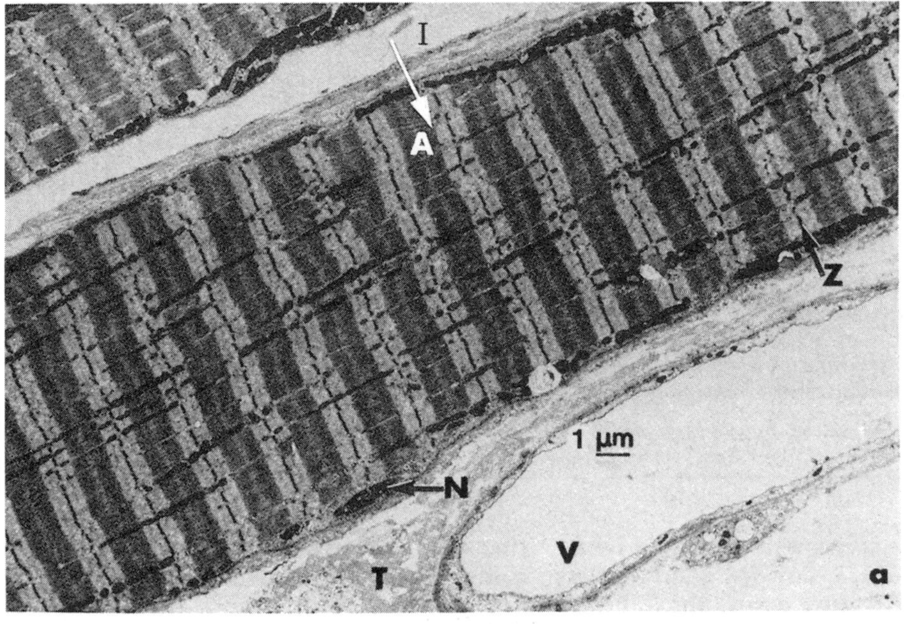

Figure 27.1 **Muscle fibres viewed under a low-powered electron microscope, showing thick and thin filaments.** Longitudinal section of a skeletal muscle (baboon). (A, A band; I, I band; Z, Z line; N, nucleus)

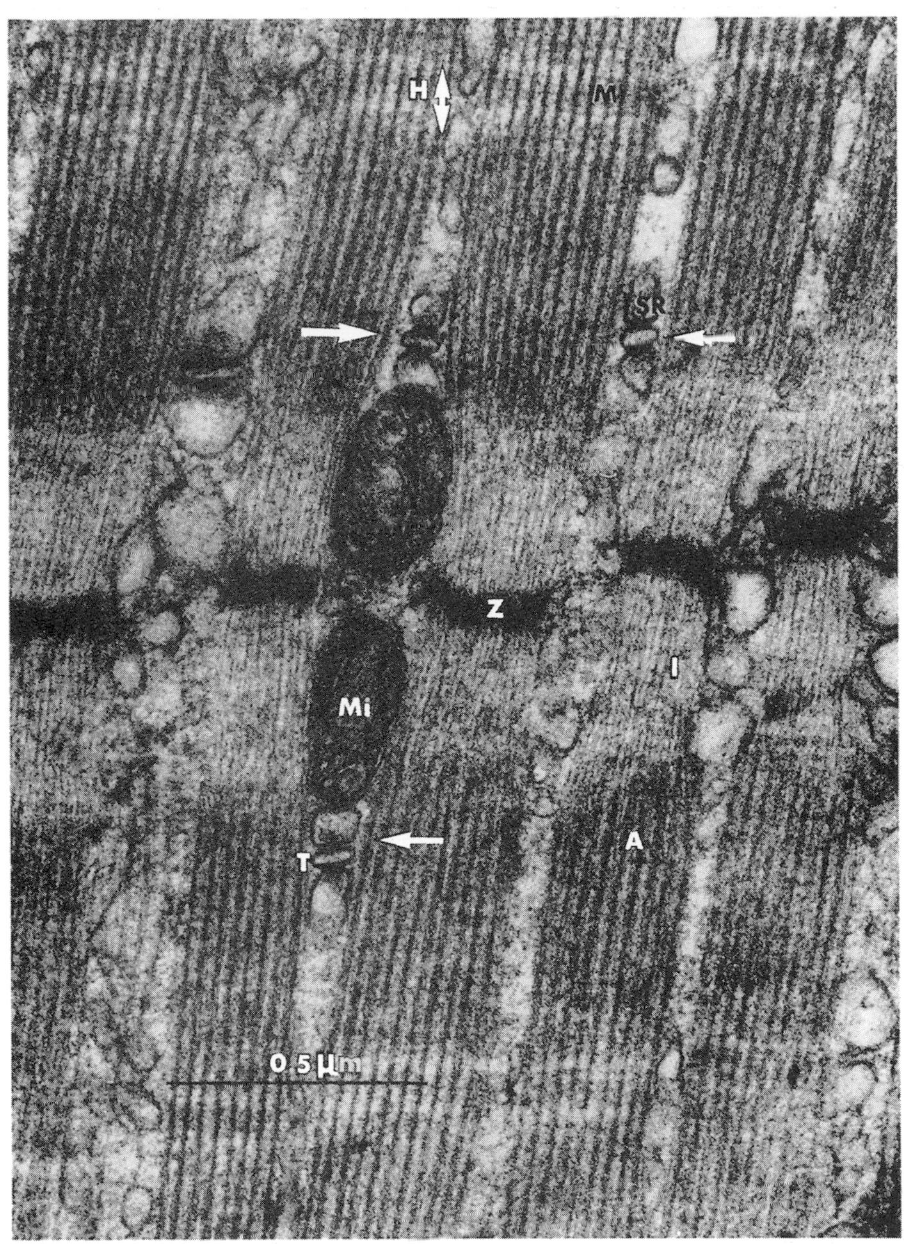

Figure 27.2 **Muscle fibres viewed under a high-power electron microscope.** Myofibrils of rat skeletal muscle cell (A, A band; Z, Z line, I, I band; Mi, mitochondria; SR, sarcoplasmic reticulum)

overlapping thin filaments, whereas the I band is composed of thin filaments only. The thin filaments which are mainly composed of the protein **actin** are interspersed between the thick filaments which are mainly composed of the protein **myosin II**. There is a region of lighter staining in the centre of the A band, which is known as the **'H' band**, and which corresponds to the location of assembly of myosin fibres. The Z disk anchors the thin filaments. During contraction the I band becomes much reduced in size owing to the extensive overlapping of the thick and thin filaments (Figure 27.3). Sections through the A band in regions where there is an overlap of thick and thin filaments shows that each thick filament is surrounded by six thin filaments.

27.3 Composition of muscle fibres

Skeletal muscle contains four main proteins: myosin II, actin, tropomyosin and troponin, and a number of minor or accessory proteins. These include actinin, capZ, titin and nebulin which are involved in stabilizing the architecture of the contractile complex of muscle, and also calsequestrin, which is involved in the regulation of contraction. Dystrophin is a minor muscle protein which has attracted a great deal of attention in the past few years because of its association with various types of muscular dystrophy (see Section 27.7)

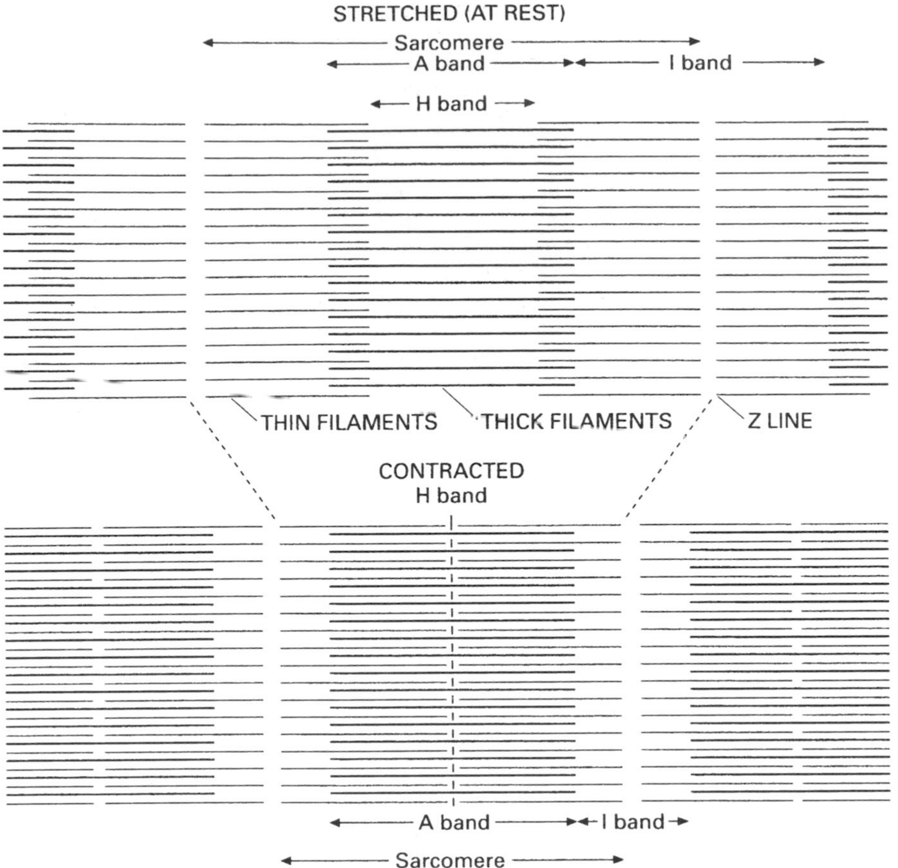

Figure 27.3 **Diagrammatic representation of a muscle fibre showing thick and thin filaments and their overlap resulting from contraction**

Composition of the thick filaments

The major protein in thick filaments is **myosin**. It is not unique to muscle but is known to exist in all cells in different forms. The commonest form of myosin in tissues, including skeletal muscle, is myosin II. It has a relative molecular mass of 460 kDa and is composed of two identical heavy chains (rMM 200 kDa) and two different pairs of light chains (rMM 20 kDa and 18 kDa, respectively). Each heavy chain consists of a long α-helical 'tail' segment and a globular 'head'. The helical segments of the two heavy chains are wound around each other to form a coiled-coil resulting in the type of structure shown diagrammatically in Figure 27.4. One of each type of light chain is found associated with each of the myosin heavy chain heads.

At physiological salt concentrations isolated myosin II aggregates spontaneously to form filaments which resemble the thick filaments seen in muscle. Thick filament assembly begins with the end-to-end association of myosin tails so that the heads protrude on either side of the centre of the filament. The H region seen in the centre of the A band described above, corresponds to this area of the thick filaments where there are no myosin heads. The myosin heads appear along the thick filament at 14 nm intervals and are orientated symmetrically around the filament (Figure 27.5).

Composition of the thin filaments

The thin filaments of skeletal muscle are composed of the proteins actin, tropomyosin and troponin. Actin, like myosin, is not a specialized muscle protein, but exists in all eukaryotic cells. The contraction of actin is responsible for a wide variety of motile processes, including cell locomotion, cytoplasmic streaming and transport, secretion, phagocytosis and cytokinesis.

F-actin (fibrous or filamentous) is the major protein in skeletal muscle thin filaments and is a polymer made up of hundreds of actin monomers. The monomeric form of actin is a globular protein of relative molecular mass of 43 kDa known as **G-actin**. At physiological concentrations of ATP and magnesium ions and also at concentrations of actin between 1 and 10 μM, G-actin polymerizes spontaneously to F-actin, with the concomitant cleavage of ATP to ADP and inorganic phosphate. F-actin filaments have the appearance of two intertwined strings of beads (Figure 27.6). There is a maximum number of actin monomers that can be accommodated in the F-actin polymer. When that critical size is reached, although polymerization is favoured by the high concentrations of actin in the cell (100 μM), it is thought that a kind of 'treadmilling' occurs, whereby for every monomer added to the polymer, another is removed so that F-actin maintains its size and is in equilibrium with the cellular pool of G-actin.

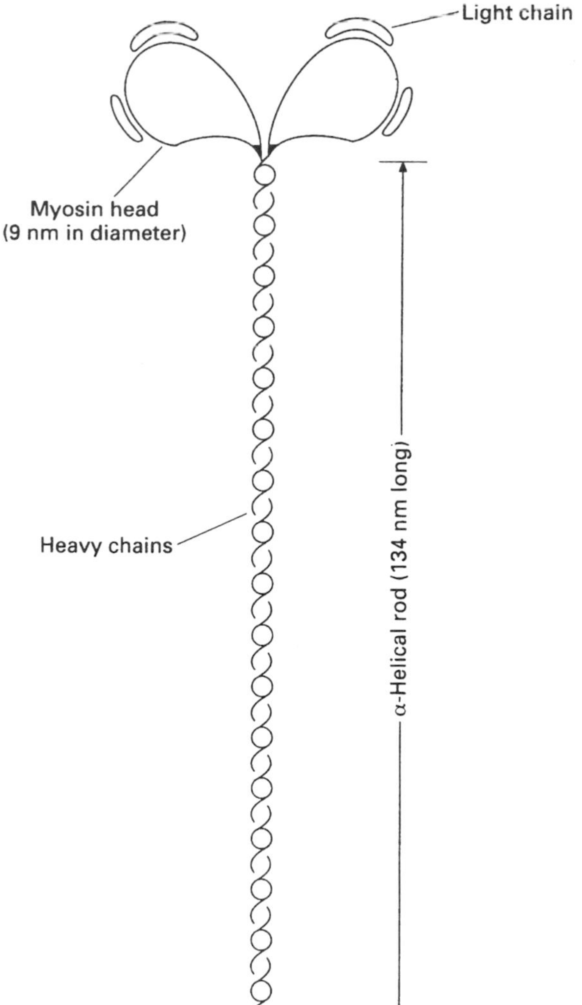

Figure 27.4 **Diagrammatic representation of a myosin molecule.** The two heavy chains consit of long α-helices wound round each other in the form of a superhelical rod

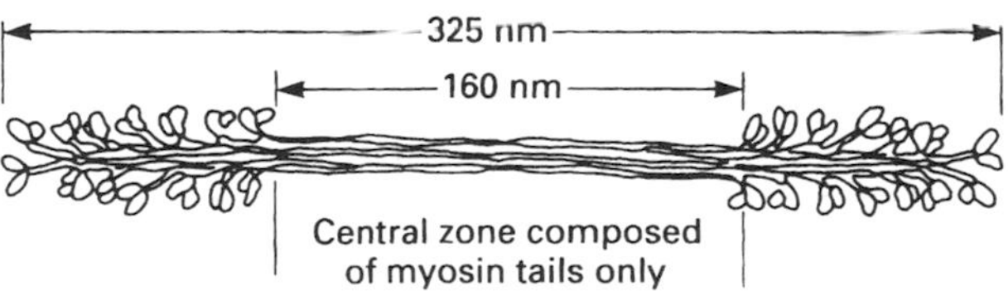

Figure 27.5 **The structure of the thick filaments of muscle showing the myosin heads protruding from the filament**

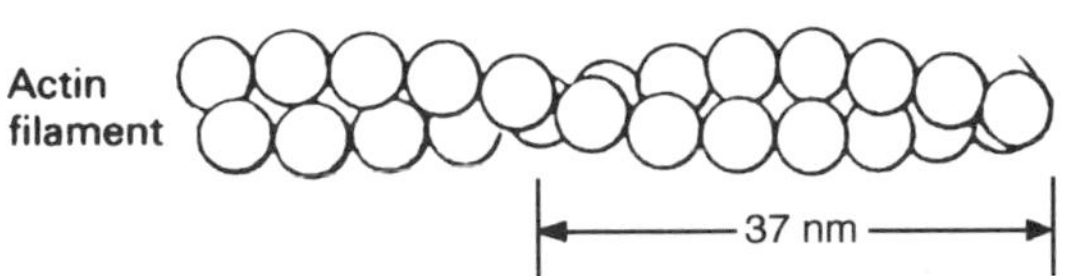

Figure 27.6 **Diagrammatic representation of assembly of G actin monomers**

Thin filaments contain two other proteins apart from actin, namely tropomyosin and troponin.

Tropomyosin is a long, rod-like molecule of rMM 70 kDa. It is composed of double-stranded α-helices which are attached end to end, forming a long thin filament. Tropomyosin lies in the groove between the two strands of actin. The thin filaments of skeletal muscle are approximately 1 μm in length and they contain 300–400 actin molecules and 40–60 tropomyosin molecules.

Troponin is a complex formed of three subunits, troponin C (TnC, rMM 18 kDa), troponin T (TnT, rMM 37 kDa) and troponin I (TnI, rMM 24 kDa). The letters T, C and I refer to the properties of these subunits. TpT binds tropomyosin, TnC binds calcium and TnI has an inhibitory role on the regulation of contraction by calcium ions. The complex of troponin subunits is an elongated structure with TnC and TnI forming a head and TnT forming a tail. The TnT domain binds tropomyosin and so positions troponin on the actin filament. There is one troponin complex for every seven actin monomers in the thin filaments (Figure 27.7).

The assembly of the thick and thin filaments in a muscle fibre is shown in diagrammatic form in Figure 27.8.

Accessory myofibrillar proteins

Muscle contraction depends on the correct structural organization of the thin and thick filaments. During contraction, the filaments slide relative to each other resulting in the shortening of the sarcomere. This is made possible by the fact that the thin filaments are anchored and, therefore, immobilized at the Z disk. The Z disk contains two sets of overlapping actin filaments of opposite polarity. They are anchored to the disk by the proteins actinin and capZ. **Actinin** is the major component of the Z disk. It is a fibrous protein made up of two identical subunits of rMM 190 kDa. **CapZ** protein is thought to prevent the growth and depolymerization of actin, thereby stabilizing the structure of the thin filaments as well as being involved with their immobilization at the Z disk.

Another two proteins are now known to be involved in maintaining the architecture of the contractile unit, namely titin and nebulin. **Titin** connects the thick filaments to the Z disk (Figure 27.9), thus maintaining the central location of the thick filament in the sarcomere, and seems to act as a rubber band during contraction. **Nebulin** is found extending from the Z disk along the thin filaments and is thought to be involved in the regulation of actin polymerization, and consequently the regulation of the length of the thin filaments as well as the maintenance of the regularity of filament structure during contraction.

Dystrophin is a cytoskeletal protein localized in surface membranes of muscle cells. Its role is not completely clear, but it seems to be involved in preserving the integrity and alignment of the plasma membrane relative to the myofibrils during muscle contraction and relaxation. Genetic defects in dystrophin synthesis or structure have severe consequences as they give rise to a group of related degenerative myopathies known as muscular dystrophy (see Section 27.8).

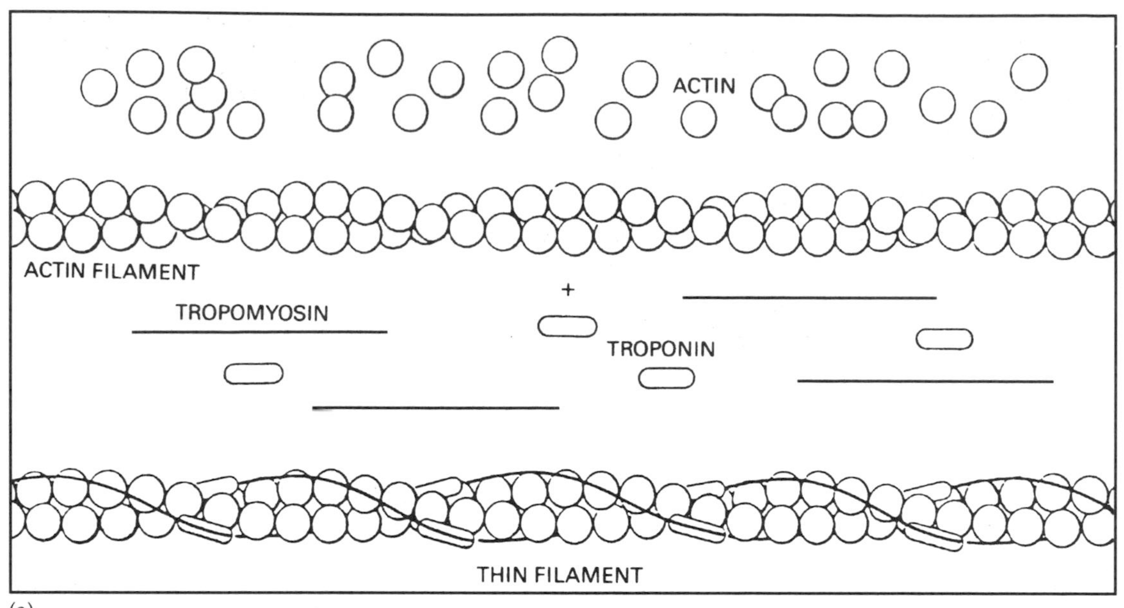

(a)

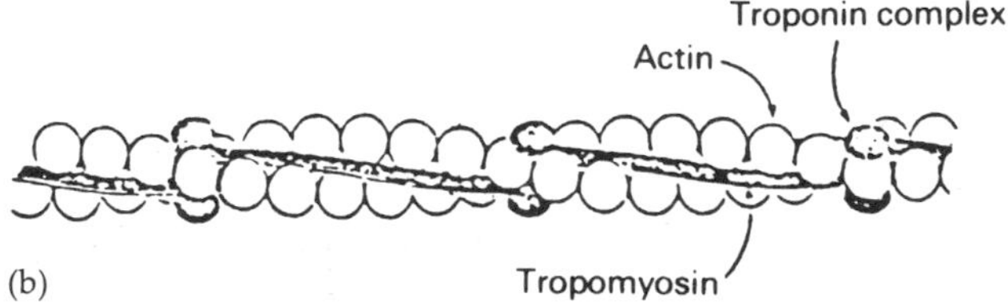

(b)

Figure 27.7 **Assembly of proteins in the thin filament.** One tropomysin molecule extends over seven actin molecules and 300–400 actin molecules are contained in the filaments that are about 1 μm in length. (a) Mode of assembly; (b) detail of structure of thin filament

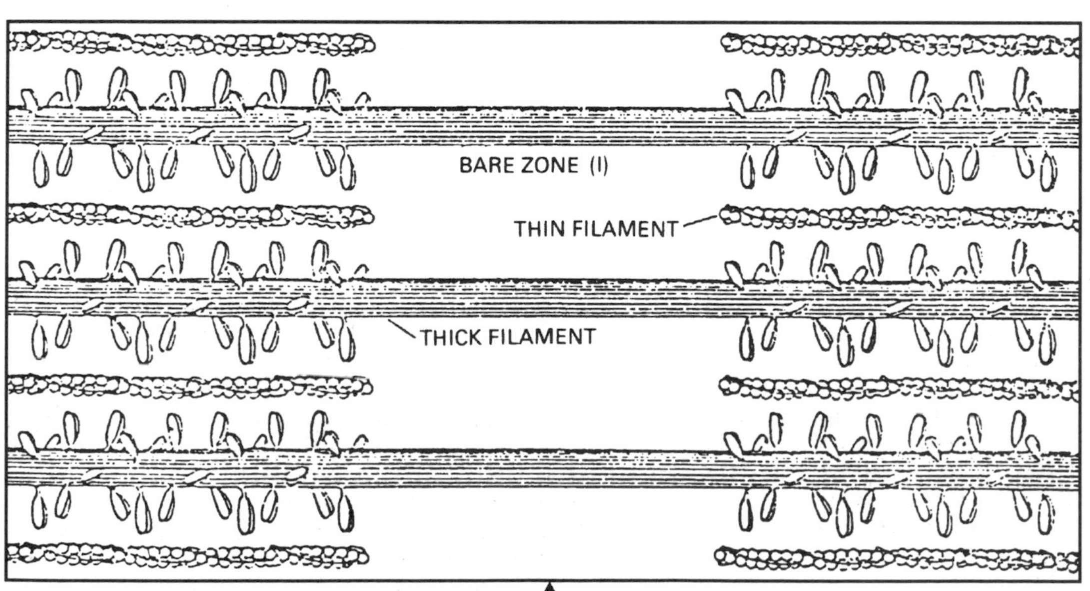

Figure 27.8 **Assembly of thick and thin filaments into fibres**

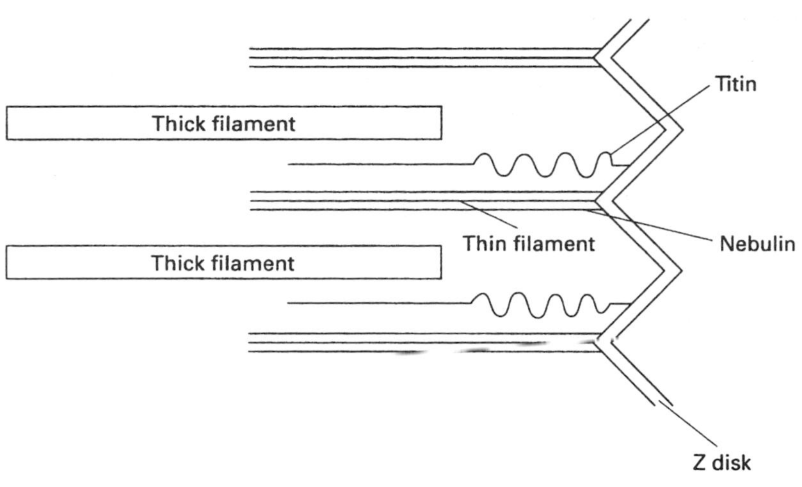

Figure 27.9 **The interaction of titin and nebulin with the Z disk and thin filaments in skeletal muscle**

27.4 Mechanism of skeletal muscle contraction

Muscle contraction is the process leading to the shortening of muscle tissue and the development of tension. The most generally accepted model of muscle contraction is 'the sliding filament model', first proposed in 1954. According to this, muscle contraction is achieved by the sliding of the thin filaments within the interspaces of the thick filaments so that the total length of the fibres is shortened. This model was based on observations, using electron microscopy, that although the length of the sarcomere decreased during muscular contraction, the lengths of the individual filaments did not change (Figure 27.10). The energy for muscle contraction is provided by ATP which is hydrolysed to ADP and inorganic phosphate. This takes place on the head region of the myosin molecules which possess ATPase activity and form the cross-bridges projecting from the thick filaments. The ATP-driven cycle of interactions between myosin and actin, which results in muscle contraction is shown in Figure 27.11.

When ATP binds a myosin headgroup, the conformation of the protein changes so that the myosin–actin association becomes weaker and the myosin dissociates from the thin filament (stage 1). The ATP is rapidly hydrolysed into ADP and P_i which remain bound on the myosin headgroup. This now associates with actin in the thin filament which is directly perpendicular to it (stage 2). The contact between the myosin and actin leads to the release of the P_i from myosin, which results in increasing the strength of the myosin–actin interaction (stage 3). This stronger binding results in a conformational change in the myosin head, which 'swivels' about the flexible hinge region of the molecule, thus generating the power stroke which pulls the thin filament along the myosin thick filament (stage 4). The myosin–actin complex remains intact until a new molecule of ATP is available. In normal muscle, this usually occurs within a millisecond. ATP replaces the ADP on the myosin head, leading to the dissociation of myosin from the thin filament and the initiation of a new cycle of events in the process of muscle contraction.

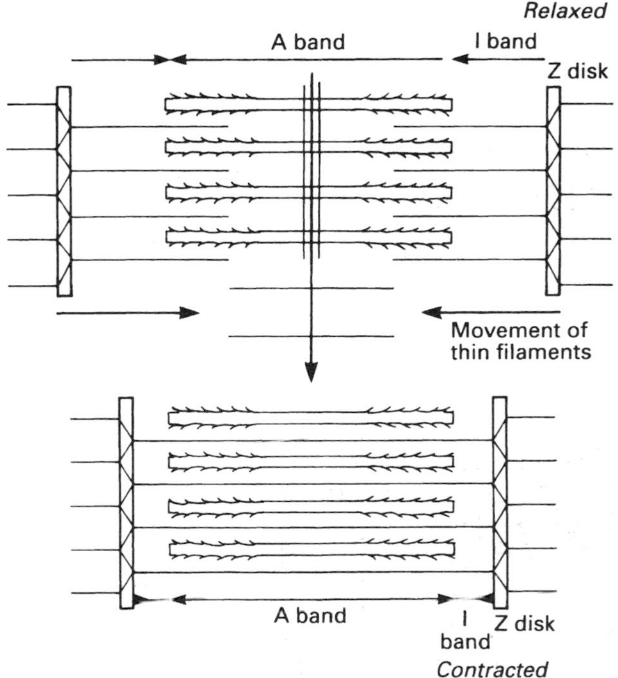

Figure 27.10 **A model showing the movement of the A band relative to the I band in skeletal muscle contraction**

If ATP is not available, myosin does not dissociate from the thin filament, as the complex of myosin–ADP with actin formed in stage 4 is very stable. It is known as 'rigor complex' and is in fact the form occurring in rigor mortis.

At the cellular level, the interactions of myosin and actin which produce muscle contraction are regulated by transient increases in calcium ion concentration.

The role of calcium in muscle contraction

The importance of calcium in muscle contraction was recognized more than 100 years ago by the physiologist

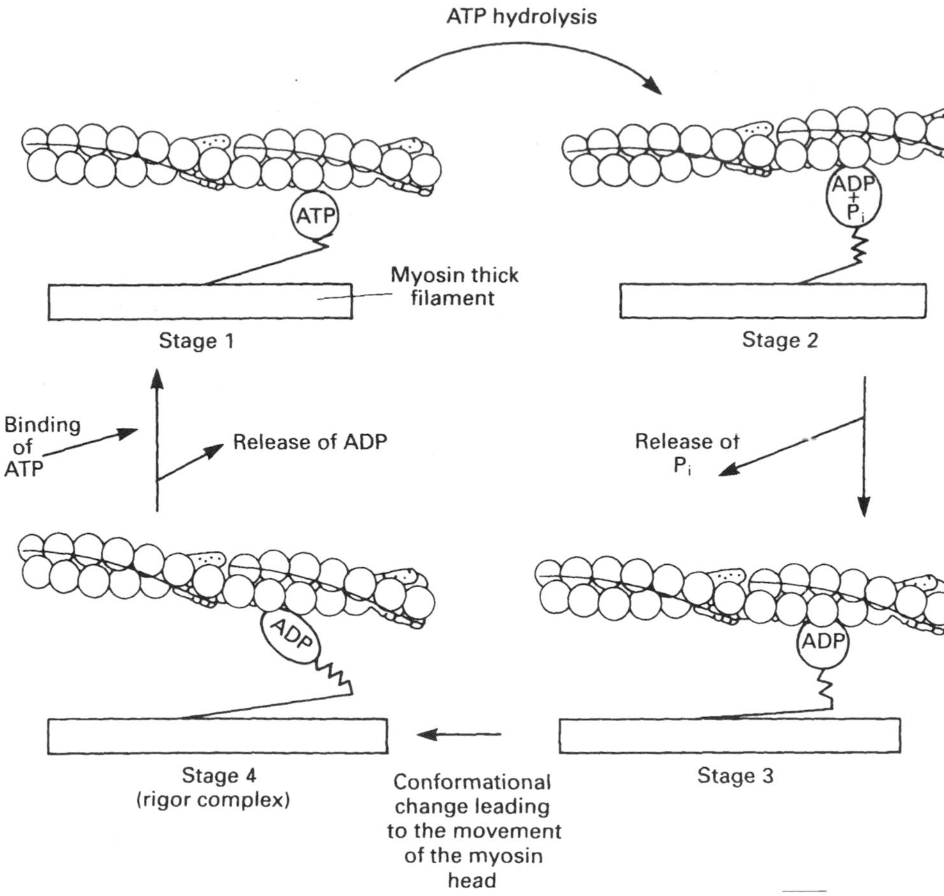

ATP hydrolysis
ATP
ADP + Pi
Myosin thick filament
Stage 1
Stage 2
Binding of ATP
Release of ADP
Release of Pi
ADP
ADP
Stage 4 (rigor complex)
Conformational change leading to the movement of the myosin head
Stage 3

Figure 27.11 **Myosin–actin interaction during contraction.** The effect of ATP

OFF (INACTIVE)
TROPONIN
+
CALCIUM IONS
Released from T-tubules and sarcoplasmic reticulum
ON (ACTIVE)
TROPONIN BOUND TO Ca2+
+
MYOSIN HEAD
ACTIN–MYOSIN COMPLEX

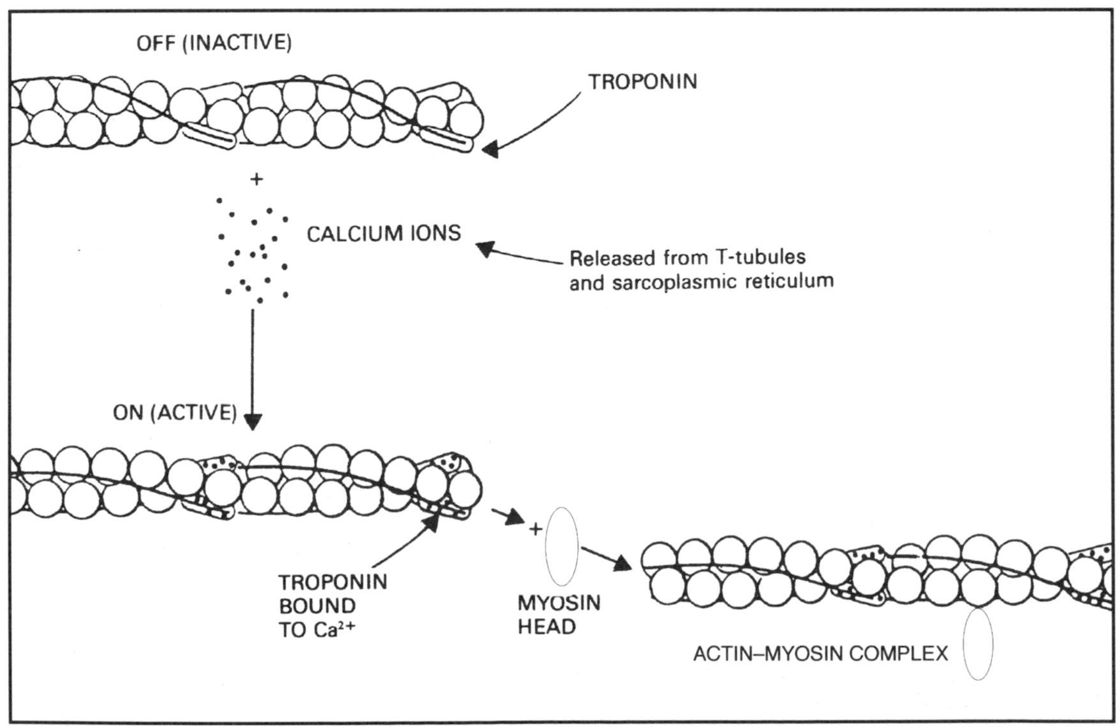

Figure 27.12 **Binding of Ca^{2+} to troponin, allowing the true association between actin and myosin**

Sydney Ringer who reported in the *Journal of Physiology* in 1882 that 'concerning the influence extended by the constituents of blood on the contraction of the ventricle, I discovered that the saline solution which I had been using had not been prepared with distilled water but with pipe water supplied by the New River Water Company. I tested the activity of the saline solution made with distilled water and did not get the effects described. It is obvious that the effects are due to some inorganic constituents of the pipe water'. Subsequent extensive research demonstrated that calcium was the constituent of pipe water essential for muscle contraction. The main action of the nerve impulse on muscle is to cause a release of calcium ions from the sarcoplasmic reticulum in the **sarcoplasm** (cytoplasm) of the muscle fibres. The **sarcoplasmic reticulum** is a specialized type of smooth endoplasmic reticulum found in muscle cells, surrounding the myofibrils, and it acts as a reservoir of calcium ions for contraction. In the resting state the concentration of calcium in the muscle cell is very low, but rapid changes occur during contraction. The sarcoplasmic reticulum is connected to the Z disks of the myofibrils and to the plasma membrane of the muscle cell (sarcolemma) via a network of transverse tubules (t-tubules) which allow the coupling of external signals such as motor neuron impulses to the release of calcium ions and muscle contraction. The role of calcium in contraction is to reverse the troponin/tropomyosin-dependent inhibition of myosin–actin association (Figure 27.12).

Tropomyosin physically blocks the attachment of myosin to actin. TnT, the elongated subunit of troponin, links TnC and TnI to tropomyosin. In the absence of calcium, TnI is bound to both TnT and actin in such a way that only weak interaction between actin and myosin head groups is possible. When calcium is released from the sarcoplasmic reticulum it binds TnC. This subunit of troponin has a structure similar to calmodulin and can bind four calcium ions. The binding of calcium results in a movement of the tropomyosin molecule towards the centre of the actin helix in such a way that the myosin headgroups can gain access to actin and contraction can occur.

Figure 27.13 is a schematic representation of a cross-section of a thin filament showing the relationship between the myosin heads, tropomyosin and actin in the absence and presence of calcium ions. Following contraction, the calcium ions are actively transported into the lumen of the sarcoplasmic reticulum by a membrane-bound Ca^{2+}-ATPase and the muscle returns to the relaxed state.

The sarcoplasmic reticulum contains a variety of calcium-binding proteins, the most important being **calsequestrin,** which binds the internalized calcium ions and so reduces the free calcium ion concentration against which the Ca^{2+}-ATPase must act. Each molecule of calsequestrin can bind up to 43 calcium ions.

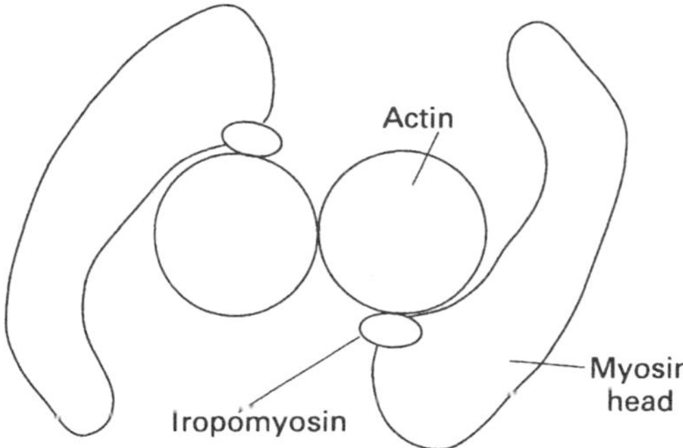

A. In the absence of Ca^{2+}, tropomyosin blocks the access of myosin to actin.

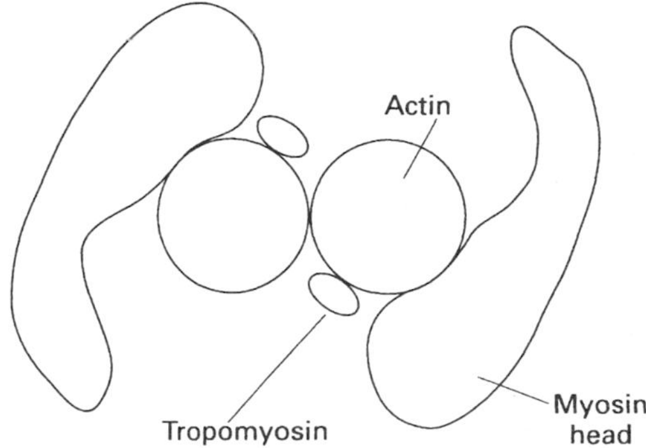

B. In the presence of Ca^{2+}, movement of tropomyosin exposes myosin-binding sites on actin.

Figure 27.13 **Schematic representation of the relationship of myosin heads, tropomyosin and actin.**

27.5 Sources of energy for muscle contraction

Resting muscle needs a continuous supply of ATP for maintenance of basic cellular activities. The requirement for ATP can increase up to 200 times in vigorous exercise. The amount of ATP present in a muscle cell would allow approximately 0.5 s of intense activity, corresponding to about 10 contractions. Clearly, there are systems in the cell which generate ATP as the need for energy is increased.

Muscle possesses the capacity to form ATP under anaerobic conditions, utilizing its store of glycogen, and metabolizing this by the glycolytic pathway. If no oxygen is present, or if the supply of oxygen is limited, then lactate is produced. A short period of strenuous exercise, for example running up a long flight of stairs, will cause a large increase in the concentration of lactate in the blood. During prolonged exercise, the oxygenation of muscle gradually increases to a maximum and then ATP may be produced by several different metabolic processes, glycolysis, β-oxidation of fatty acids, and the tricarboxylic acid cycle.

The importance of the oxidation of fatty acids (deriving from triacylglycerols or the NEFA bound to albumin) in energy provision in muscle is often overlooked, but it should be noted that muscle will oxidize fatty acids very efficiently. Measurements on human limbs have demonstrated that over 50% of the energy for contraction can be supplied by plasma-derived fatty acids.

Muscles obtain their energy for contraction by the general processes described above, but it should be noted that the fibres of which they are composed are by no means uniform in their metabolic activity.

Mammalian muscles generally contain a mixture of muscle fibres which differ in the manner in which they generate ATP. They can be classified into two types, I and II. **Type I fibres**, which are also referred to as 'red' muscle fibres, have a high capacity to produce ATP aerobically, owing to the presence of a large number of mitochondria and large amounts of the oxygen-binding protein myoglobin which is also responsible for the colour of the fibres. The high oxidative capacity of these cells enables them to generate ATP by the aerobic metabolism of both glucose and fatty acids. They are also known as 'slow twitch' fibres and they both contract and fatigue slowly. Their resistance to fatigue is thought to be due to their capacity to utilize fatty acids derived from triacylglycerol stores within the fibres.

Type II fibres, also referred to as 'white' fibres are 'fast twitch' fibres and they contract and fatigue quickly. They possess few mitochondria, and, therefore, have a low oxidative capacity. On the other hand, they have a highly developed glycolytic pathway. Most anatomical muscles contain a mixture of red and white fibres in different proportions. Examples of muscles consisting almost entirely of the type II fibres are the pectoral muscle of the domestic fowl and the psoas muscle of the rabbit. The properties of the two types of muscle fibre are summarized in Figure 27.14. It can be seen that they differ somewhat in the nature of their constituent contractile proteins. In particular, they contain slightly different myosin molecules known as **isoforms** which are encoded in separate, but related, genes.

In intense skeletal muscular activity, fast twitch muscles especially cannot generate ATP at the required rate. Two mechanisms, however, exist in the muscle cell whereby ADP can be rephosphorylated to form ATP, thus allowing contraction to continue. The first is a reaction catalysed by creatine kinase using phosphocreatine as the phosphate donor:

$$\text{ADP} + \text{phosphocreatine} \underset{\text{creatine kinase}}{\leftrightarrow} \text{creatine} + \text{ATP}$$

Phosphocreatine is the major phosphorylated compound in skeletal muscle. It is generally accepted that this reaction provides an emergency reservoir or 'buffer' of ATP when other sources of this compound are unavailable, as, for example during the first few seconds of strenuous exercise. Recently, however, studies on skeletal muscle using magnetic resonance spectroscopy have thrown doubt on this assumption, as they showed that there was no change in the rate of transfer of the phosphate group from phosphocreatine to ADP even when ATP utilization was increased several-fold, implying, therefore, no functional coupling of creatine kinase activity and ATP production. It has been suggested that the role of creatine kinase might be to 'buffer' the concentration of ADP which, in turn, indirectly controls oxidative phosphorylation.

The second means of ATP generation in muscle fibres is the reaction catalysed by the enzyme adenylate kinase (myokinase):

$$2\ \text{ADP} \underset{\text{adenylate kinase}}{\leftrightarrow} \text{AMP} + \text{ATP}$$

By means of this reaction, ADP formed during muscle contraction can give rise to ATP for further contractions.

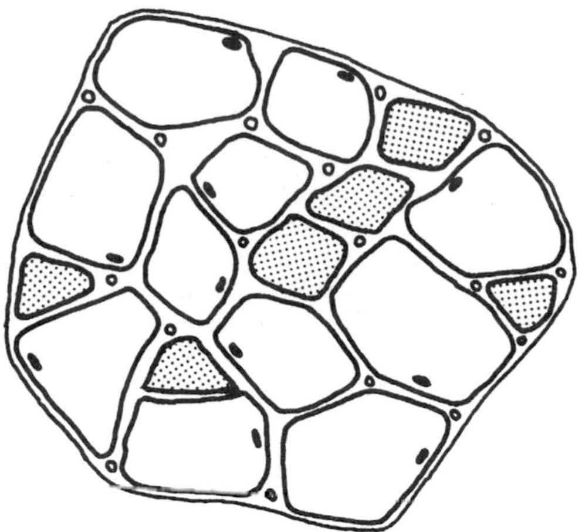

Figure 27.14 **The properties of human muscle fibre types.** Type I fibres are the dark (red) fibres, the type II fibres are the light (white) fibres.

Type I
Small
Red, oxidative
Myoglobin present
Many mitochondria
Acid-stable ATPase
Slow twitch (slow myosin isoforms)
Sparse sarcoplasmic reticular system
Broad Z bands

Type II
Large
White, glycolytic
Myoglobin absent
Few mitochondria
Alkali-stable ATPase
Fast twitch (fast myosin isoforms)
Elaborate sarcoplasmic reticular system
Narrow Z bands

Note: The relative proportions of type I and type II fibres in a muscle can change with development, exercise, training, electrical stimulation or thyroxine administration

27.6 Cardiac muscle

Cardiac muscle resembles skeletal muscle in that it is striated and has a similar contractile apparatus consisting of sarcomeres. There are, however, some important differences in the structural organization of the fibres. Skeletal muscle fibres are multinucleated cells which have arisen from fusion of mononucleated cells. Cardiac muscle fibres are arrangements of discrete mononucleated cells (Figure 27.15). The nuclei lie centrally in the cell, whereas in skeletal muscle they are found in the periphery of the cell. The sarcoplasmic reticulum does not cover each myofibril as in skeletal muscle, but the t-tubules from the sarcolemma run longitudinally along the myofibrils. The fibres are not aligned in parallel as in skeletal muscle, but branch to form an interlacing network. Inside the fibres, the myofibrils are less densely packed than in skeletal

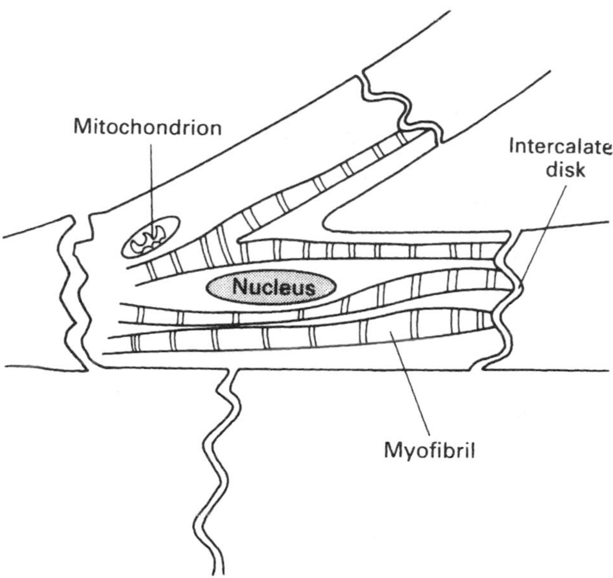

Figure 27.15 **Structure of cardiac muscle cells**

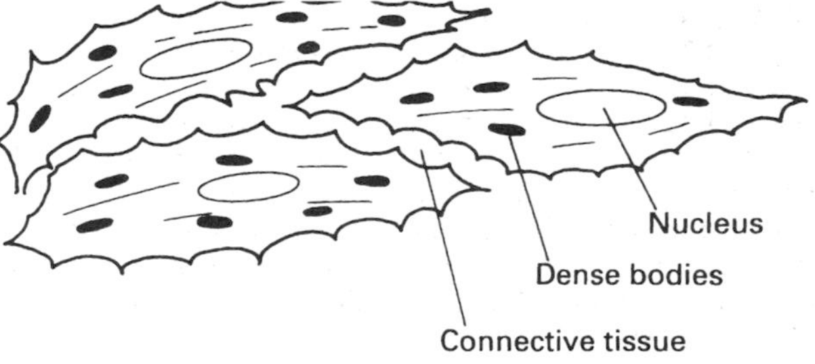

Figure 27.16 **Structure of smooth muscle cells**

muscle. The single cells comprising the cardiac muscle fibres are demarcated by a structure known as an **intercalated disk**. This has a step-like appearance and consists of a double sarcolemma containing both gap junctions and desmosomes. These form contacts between cells and allow communication between them. Apart from transport of various solutes, the junctions are important because they allow the electrical coupling of cardiac muscle cells so that contractions are synchronized and result in the regular rhythmic beating which is a characteristic feature of the heart. There are also some specialized fibres in cardiac muscle known as **Purkinje cells** which are grouped in bundles and carry and distribute electrical impulses to the myocardium.

The contractile elements in cardiac muscle fibres are very similar to skeletal muscle in that the thick filaments are made of myosin and the thin filaments of actin, tropomyosin and troponin, but all these proteins are isoforms of the corresponding skeletal muscle proteins. The mechanism of contraction is understood to be similar to skeletal muscle, i.e. via calcium-regulated association of thick and thin filaments.

27.7 Smooth muscle

Smooth muscle tissue is found in blood vessels and internal organs such as the intestines, the myometrium and the lungs. Smooth muscle generates waves of contraction rather than the abrupt shortening of fibres observed in skeletal muscle. The contractile elements of smooth muscle consist of elongated, usually spindle-shaped cells with a centrally located nucleus (Figure 27.16). The cells are bound together into sheets or bundles and lie parallel to each other, embedded in connective tissue. The cells themselves are very active in synthesizing and secreting connective tissue matrix which functions in preserving the shape of smooth muscle tissue. This is particularly important in tissues with tubular structures such as intestines and blood vessels, as it prevents distension of the tube, which would have adverse effects on function. The cells have projections on their surface (membrane spikes) which serve to connect them to other cells so allowing the waves of contraction to develop and propagate.

Unlike striated muscle fibres, smooth muscle cells do not have organized sarcomeres, although they contain the contractile proteins myosin, actin and tropomyosin. These proteins are isoforms of the corresponding proteins in skeletal muscle. Regular myofibrillar structures are absent, but filaments corresponding in size to the thick and thin filaments of skeletal muscle appear transiently. Contraction is effected by a mechanism similar to that operating in skeletal muscle in that it involves the sliding of filaments relative to each other, but the components of the filaments dissociate after contraction.

A characteristic feature of smooth muscle cells observed under the electron microscope is the presence of dark-staining structures referred to as **'dense bodies'**. They are probably equivalent in function to the Z disks of striated muscle. The major protein component of dense bodies is α-actinin, a protein which binds actin. The dense bodies are connected to the cytoskeleton by means of two proteins, **desmin** and **vimentin**.

This connection allows the plasma membrane to move towards the centre of the cell during contraction.

The thin filaments are similar to those seen in striated muscle, consisting of actin and tropomyosin but no troponin. The thin filaments are embedded in the dense bodies with opposing polarities, in the same way that they extend in two directions from the Z disks in striated muscle. In this way, contraction by formation of myosin–actin cross-bridges results in pulling the dense bodies towards one another and also in pulling the cell membrane inward so that the shape of the cell is altered.

Thick filaments are made from myosin whose structure is similar to that in striated muscle but whose properties are different. Smooth muscle myosin can only associate to form filaments when the 18 kDa light chains associated with the myosin headgroups are phosphorylated. Phosphorylation is catalysed by the enzyme **smooth muscle light chain kinase** (MLCK). When the light chain is not phosphorylated, the myosin 'tail' folds round the myosin heads so blocking both the actin binding site and the myosin ATPase activity. Smooth muscle contraction is regulated by the presence

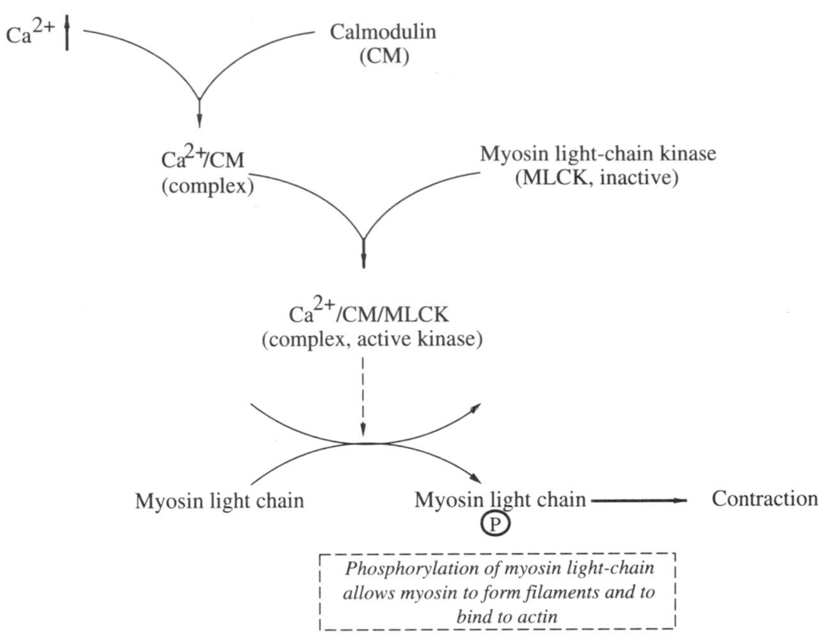

Figure 27.17 **The regulation of smooth muscle contraction by calcium ions**

of calcium ions, as is striated muscle contraction. Calcium binds not to troponin, which is absent, but to the ubiquitous and multifunctional protein **calmodulin** (CM). The principal function of CM in smooth muscle is to activate cross-bridge cycling and the development of force in response to calcium ion release. As calcium ion concentration increases in the cytosol, calcium ions bind to CM, and the calcium–calmodulin complex (Ca–CM) binds to and activates various proteins including MLCK. The activated kinase catalyses the phosphorylation of the myosin light chain, allowing the association of myosin molecules into filaments and the binding of actin (Figure 27.17). During muscle relaxation, calcium levels in the cell decrease and the myosin light chain becomes dephosphorylated. The phosphatase activity required to effect this dephosphorylation is not well defined at present.

Caldesmon is a calcium-binding protein which is found bound to actin at low calcium concentrations. It blocks the contact of myosin and actin so that cross-bridges cannot form. The inhibition is reversed when calcium levels increase and Ca–CM complexes are formed. Ca–CM binds caldesmon and causes its removal from the actin filament, thus allowing the formation of cross-bridges.

Calpontin also binds actin and inhibits the myosin ATPase activity, thus preventing ATP hydrolysis which provides the energy for muscular contraction. The inhibition can be reversed by the presence of Ca–CM, which causes the phosphorylation of calpontin by a Ca–CM-dependent kinase. Phosphorylation of calpontin is, possibly, also carried out by protein kinase C. Phosphorylation of calpontin blocks its interaction with actin and allows actin to bind myosin, so allowing contraction to take place. The ability of calcium ions to promote smooth muscle contraction is opposed by processes which tend to produce an increase in the concentration of the cyclic nucleotides cAMP and cGMP.

Stimulation of β-adrenergic receptors in smooth muscle causes an increase in the intracellular concentration of cAMP which activates protein kinase A. The target for phosphorylation by this enzyme is myosin light chain kinase (MLCK) which is rendered inactive as a result of this reaction. In turn, this prevents the formation of myosin filaments.

Two effectors act to cause an increase in cGMP in smooth muscle cells. **Atrial natriuretic peptide** (ANP) binds to and activates its receptor found in the plasma membrane of smooth muscle cells. The ANP receptor has, itself, guanylyl cyclase activity (see Chapter 3) and, when it binds its ligand, the intracellular concentration of cGMP increases. **Nitric oxide** also activates a guanylyl cyclase in smooth muscle cells, but this enzyme is found in the cytosol. Nitric oxide is produced by endothelial cells close to the smooth muscle cells (see Chapter 33) into which the gas diffuses. An increase in the intracellular concentration of cGMP causes the dephosphorylation of myosin light chain in smooth muscle cells and this would be expected to lead to relaxation.

27.8 Muscle disorders: muscular dystrophy

The term muscular dystrophy refers to a group of X-linked hereditary degenerative myopathies characterized by muscle weakness without nervous system involvement. The commonest and best characterized of these myopathies are **Duchenne muscular dystrophy (DMD)** and **Becker muscular dystrophy (BMD)**. Both disorders were recognized and described over 100 years ago, but their molecular basis was not identified until 1986. At the cellular level, they are characterized by the loss of individual muscle fibres leading to disruption in fibre arrangement and degeneration and fibrosis of the tissue. The clinical manifestations vary, from the mild calf hypertrophy with cramps seen in BMD, to the classical progressive degenerative myopathy of DMD.

DMD affects 1 out of 3500 live male births. It rarely affects females, an observation which gave the first hints that it is a X-linked disorder. There are no symptoms at birth, and few until the age of about 5 years, when it is usually reported as a weakness of the legs which leads to an inability to sustain physical exercise such as running or climbing. By the age of 12 years, muscle wasting is so severe that most sufferers are unable to walk. Muscle loss continues steadily until death occurs, usually from respiratory failure, in the late teens or early twenties. BMD is less common and occurs in 1 out of 35 000 live male births. As mentioned above, it is much less severe, the degree of disability is variable and the life expectancy much longer than in DMD.

The molecular basis of muscular dystrophy

The gene associated with DMD/BMD is located on the X-chromosome and it is the largest known human gene. It is composed of 2.5 million base pairs and includes at least 70 exons. The transcript consists of a 14 kB RNA molecule. The large size of the gene is thought to be responsible for its high mutation rate. The study of the molecular basis of DMD was helped by the fact that many individuals with the disease have deletions in this locus. By mixing separated strands of DNA obtained from a normal individual with strands from a DMD patient, it proved possible to identify the gene responsible for the disease and clone it from the normal subject. Cloning and expression of this gene led to the production of a protein which was given the name **dystrophin**.

Dystrophin is absent in patients with DMD but is present in an altered form in BMD. The structure of dystrophin is well characterized and its function in the muscle cell is gradually being unravelled. It should be noted that it is found not only in striated muscle, but also in smooth muscle, and in the brain where its function is at present unknown.

Skeletal muscle dystrophin is a 427 kDa protein localized in the plasma membrane of the cell (sarcolemma) and forms part of the cytoskeleton in association with a complex of glycoproteins. Dystrophin is a dimer of two identical subunits, each consisting of four structural domains (Figure 27.18). The first domain, at the N-terminus, is similar to the actin-binding domain of α-actinin. The second domain is triple helical and consists of 24 repeats of 109 amino acids in each chain. The helix is interrupted by two proline-rich regions which are thought to act as hinges providing flexibility to the structure. The third domain resembles the calcium-binding domain of α-actinin. The fourth domain is a glycoprotein-binding domain.

Dystrophin is associated with a large oligomeric complex of sarcolemmal glycoproteins collectively referred to as dystrophin-associated proteins. These include the recently characterized glycoprotein **dystroglycan**, which has a glycoprotein-binding domain homologous to that of dystrophin, and is thought to provide a linkage between the actin cytoskeleton and the protein laminin in the extracellular matrix. In DMD, the absence of dystrophin leads to the loss of other dystrophin-associated proteins, causing the disruption of the linkage between the sarcolemmal cytoskeleton and the extracellular matrix. This disruption is thought to initiate the process of muscle fibre necrosis. The importance of the sarcolemmal glycoproteins can also be demonstrated by the finding that specific deficiency of a component of the glycoprotein complex, a protein known as **utrophin**, also leads to a form of myopathy known as **severe childhood autosomal recessive muscular dystrophy**.

The molecular basis of DMD and BMD has been widely studied and various mutations in the gene have been identified. Sixty per cent of the total number of DMD and BMD patients have deletions, and 5% of the patients have duplications of one or more of the exons of the dystrophin gene. These alterations shift the reading frame of the message, and produce the severe Duchenne phenotype characterized by complete absence of the protein. In the remaining 35% of the patients, the mutations are less dramatic, being point mutations which do not shift the reading frame of the message and lead to the less severe BMD, where dystrophin is not totally absent, but is present in an altered form of either lower of higher molecular mass than normal, or is present in reduced quantities.

Diagnosis and treatment

The isolation of genomic and cDNA probes for the dystrophin gene has greatly facilitated the detection of DMD/BMD carriers and has enabled prenatal diagnosis of the disease. Recombinant DNA technology is usually used to diagnose DMD/BMD in male fetuses considered to be at risk. Fetal DNA can be extracted from cells obtained by amniocentesis or chorionic villus sampling and analysed using techniques such as the polymerase chains reaction (PCR), Southern blotting hybridization with cDNA markers and linkage studies with

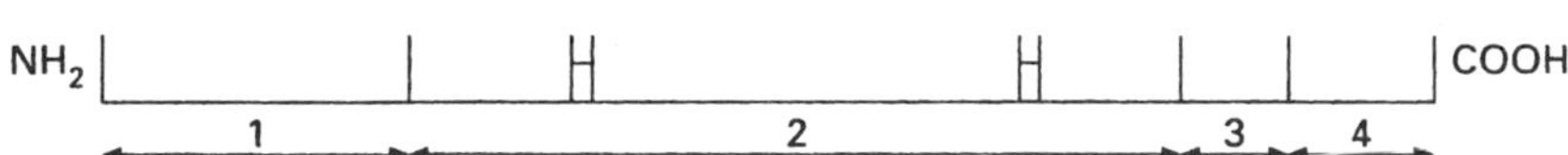

Figure 27.18 **Domain organization of dystrophin**
Domain 1: Actin binding
Domain 2: Triple helical repeats (H, proline-rich hinge regions)
Domain 3: Calcium binding
Domain 4: Glycoprotein binding

polymorphic DNA markers (restriction fragment length polymorphisms or RFLPs). The basis of the use of these techniques in prenatal diagnosis is described in Chapter 38. As mentioned above, the most common mutations are deletions and these can be detected in fetal DNA using PCR. When deletions cannot be detected, then RFLP analysis should be considered in order to perform carrier and prenatal diagnosis. Even this analysis cannot detect all the cases of DMD/BMD, because one-third of the patients develop the disease as a result of new mutations. In cases where the DNA analysis is uninformative, another, recently developed, method can be employed. This involves sonographically assisted fetal muscle biopsy for dystrophin protein analysis using immunological techniques such as immunofluorescence (see Chapter 1). It is possible to predict, with a high degree of accuracy, if the disease is going to develop, and if so which of the two forms of the disorder is involved. Absence of dystrophin predicts DMD, whereas low concentrations or the presence of an altered form of dystrophin (either larger or smaller than normal) predict BMD. Early recognition of the nature of the disorder will be of great use in the treatment of the disease. There is no successful treatment at present. The aim is to provide deficient cells with the capacity to synthesize normal dystrophin. This could, in theory, be achieved by either gene therapy or cell transplantation. Initial human experiments have proved unsuccessful. Another approach to treatment is to identify proteins similar to dystrophin and, by the use of pharmacological agents, increase the expression of the genes which code for them in the cell so that they may compensate for the lack of dystrophin.

Chapter 28

The endocrine tissues

28.1 Introduction

The discovery of the existence of 'hormones' or 'chemical messengers' as they were originally called, stems from the classic experiments of Bayliss and Starling at the beginning of the present century. They showed that extracts of duodenal mucosa, upon injection into the bloodstream, caused the flow of pancreatic juice. The duodenal factor(s) responsible must clearly have been (a) chemical(s) carried in the blood.

This discovery was of great and far-reaching importance because, up until that time, control between organs had always appeared to be mediated by the nervous system which, at the time, appeared to be a system solely of electrical control. Bayliss and Starling therefore laid the foundations of the discipline now known as endocrinology. The discovery of other hormone-secreting tissues and their glandular nature soon followed.

The wheel has come full circle, for it is now recognized that the endocrine system and the other two great signalling systems of the body – the nervous system and the immune system – have many features in common. The cells of each system release chemical messengers to influence cells within the same system or those in other systems. It is largely for historical reasons, therefore, that the secretions of the endocrine system are called hormones when they act via the blood and paracrines or autocrines (see Chapter 3) when they act locally. Similar molecules released by neurons are termed neurotransmitters or neuromodulators and the extracellular messengers of the immune system are called lymphokines. Major advances in the 1980s saw the emergence of disciplines such as neuroendocrinology and neuroimmunology that are specifically concerned with the study of intercommunication between the signalling systems.

As regards the hormones, the chemical structures of these molecules fall into three groups: those containing peptide bonds (peptides or proteins), those derived from cholesterol (steroid hormones, derivatives of vitamin D) and those arising from the amino acid tyrosine (catecholamines and thyroid hormones).

Several general aspects of hormones may be recalled. First they have 'target cells' on which they act and these may be located in a specific organ, such as the kidney, or be widely distributed in the body, forming special cell populations in various organs. Secondly, some hormones function largely to bring about the release of other hormones from different endocrine glands. Groups of tissues organized in this hierarchical way are referred to as endocrine axes. One such axis comprises the hypothalamus, the anterior pituitary gland and the adrenal cortex. Figure 28.1 shows the organization of several endocrine axes in which the hypothalamus and the anterior pituitary gland form the first two components. Characteristically the response in each succeeding component of an axis is greater than that in the previous one: there is amplification.

A third general point may be made: although the hormones may be divided into three structural groups, there are just two ways that they affect the biochemistry of their target cells. All of the peptide hormones and the catecholamines bind specifically to cell surface receptors to influence intracellular signalling systems indirectly. The steroid hormones, vitamin A and vitamin D derivatives and the thyroid hormones, being lipid soluble, are able to diffuse into cells and affect biochemical processes directly via intracellular receptors. Examples of both classes of receptor and their subclasses are given in Chapter 3.

Many of the peptide hormones, as well as the catecholamines, act by activating plasma membrane receptors that are coupled to one of two enzymes: adenylyl cyclase or phospholipase C via G-proteins (see Chapter 3). Table 28.1 lists the hormones that act on G-protein-coupled receptors and the enzyme that is activated as a consequence.

In this chapter it is not intended to describe the actions of all the hormones in detail, since many of these are dealt with in the chapters concerning the metabolic and other processes involved. Here the intention is to present a summary of the structures and major actions of the hormones for reference and to clarify aspects of other chapters.

28.2 Structural relationship between the hypothalamus, the anterior pituitary gland and target organs

Since many of the important peptide hormones are of pituitary gland origin and their secretion is regulated by peptides of hypothalamic origin, it is useful at this stage to have to hand a reminder of the anatomical relationship between the two tissues.

The hypothalamus is the most ventral part of the diencephalon or interbrain and is closely associated with the floor of the third ventricle. Several hypothalamic areas, called nuclei because of the presence of large numbers of perikarya (nerve-cell bodies), send axons in the direction of the pituitary gland. Some of these axons

Figure 28.1 **A summary of the major biological properties of the pituitary hormones**

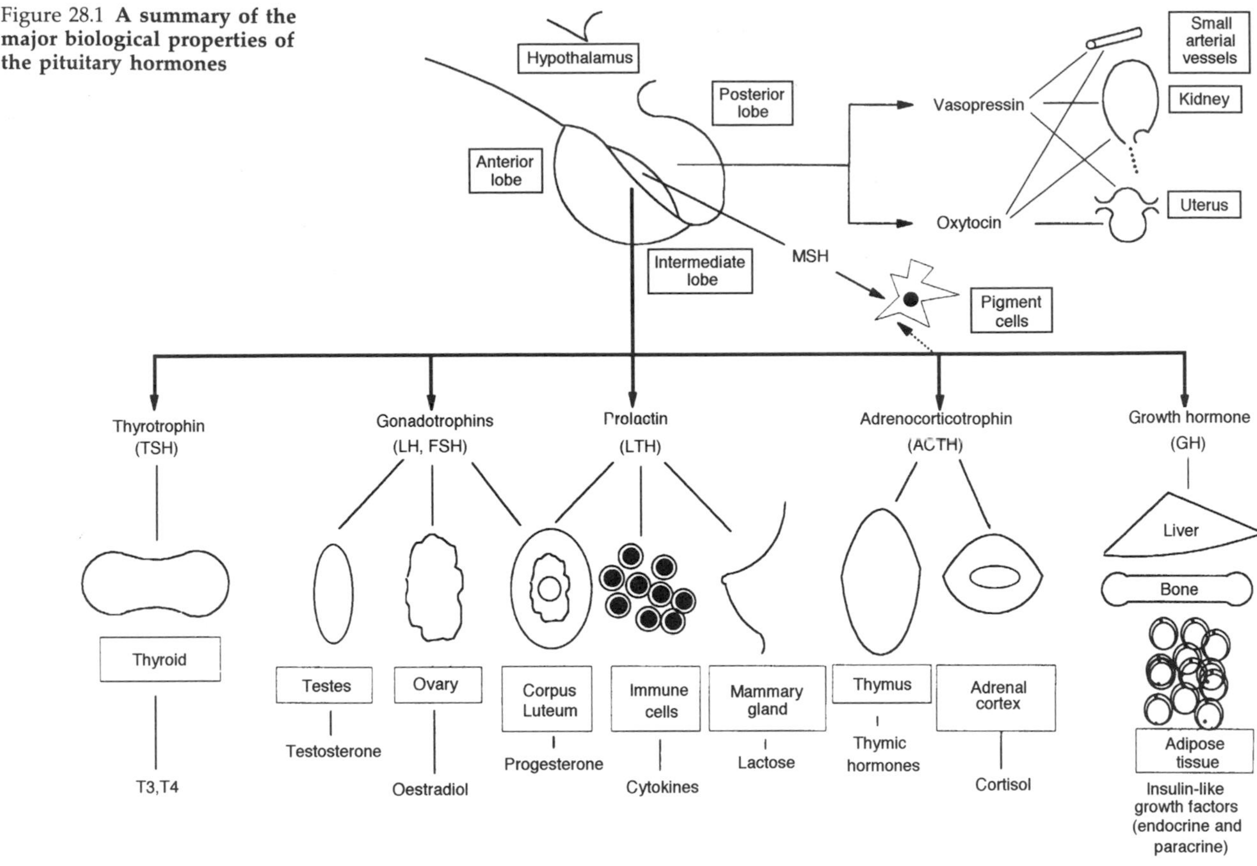

Table 28.1 **Hormones that cause the activation of adenylyl cyclase or phospholipase C**

	Adenylyl cyclase	*Phospholipase C*
Hypothalamic hormones	Growth hormone-releasing hormone (GHRH) Corticotrophin-releasing hormone (CRH) Vasopressin (V_2) (antidiuretic)	Thyrotrophin-releasing hormone (TRH) Gonadotrophin-releasing hormone (GnRH) Vasopressin (V_1) (pressor)
Anterior pituitary hormones	Adrenocorticotrophin (ACTH) Thyrotrophin (TSH) Gonadotrophins (LH, FSH)	
Adrenal medullary hormones	Adrenaline/noradrenaline (β-receptors)	Adrenaline/noradrenaline (α-receptors)
Other hormones	Glucagon Parathyroid hormone	Angiotensin II

have their terminals in the posterior portion of the gland and the secretions from these nerve endings are released directly into the circulation. The main hormones secreted via this route are the peptides vasopressin and oxytocin. The secretory terminals of other hypothalamic peptidergic neurons also release their hormones into the blood but do so in the median eminence of the hypothalamus. These hormones are trophic on the anterior pituitary gland whither they are carried in the hypophyseal portal blood. These hormones cause (or in two cases, inhibit) the release of anterior pituitary hormones. Figure 28.2 shows the structural relationship between these two tissues.

The secretions of the hypothalamus ultimately control the release of hormones from a variety of target tissues (see Figure 28.1). The hormonal secretions of these target tissues have the important ability to act back on the hypothalamus (and on other brain regions that control the functional activity of that tissue) and also the anterior pituitary gland (Figure 28.3). This process is referred to as negative-feedback inhibition. Under special circumstances positive-feedback facilitation may be observed. The best example of this underlies the action of oestradiol in helping to promote the 'surge' of luteinizing hormone (LH) that arises at the mid-point of

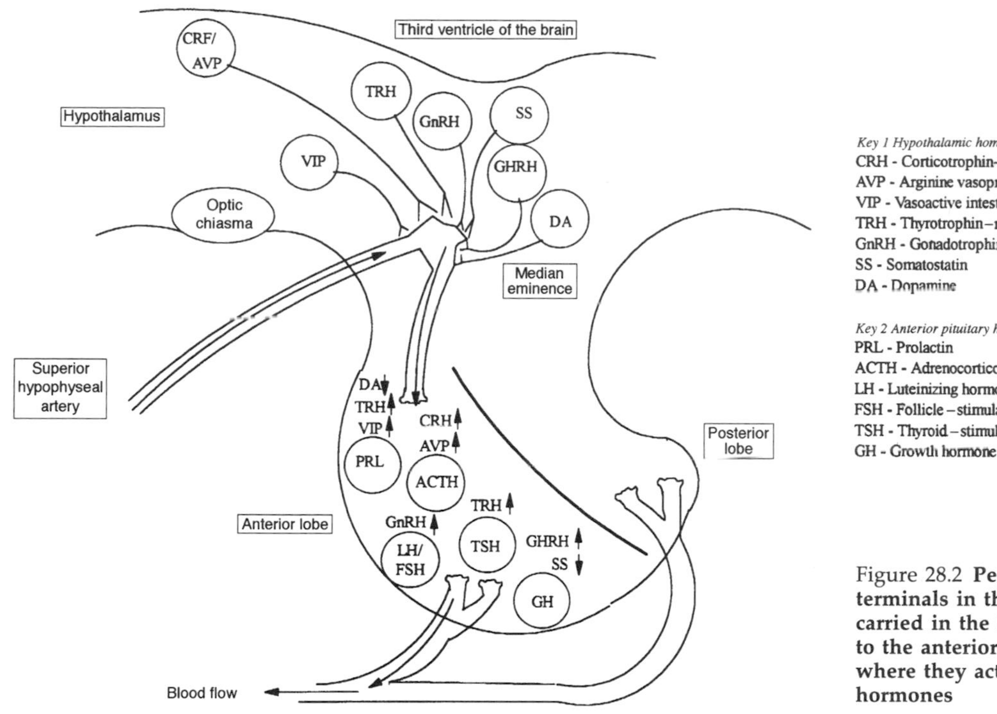

Figure 28.2 **Peptides released from nerve terminals in the median eminence are carried in the hypophyseal–portal blood to the anterior lobe of the pituitary gland where they act to control the release of hormones**

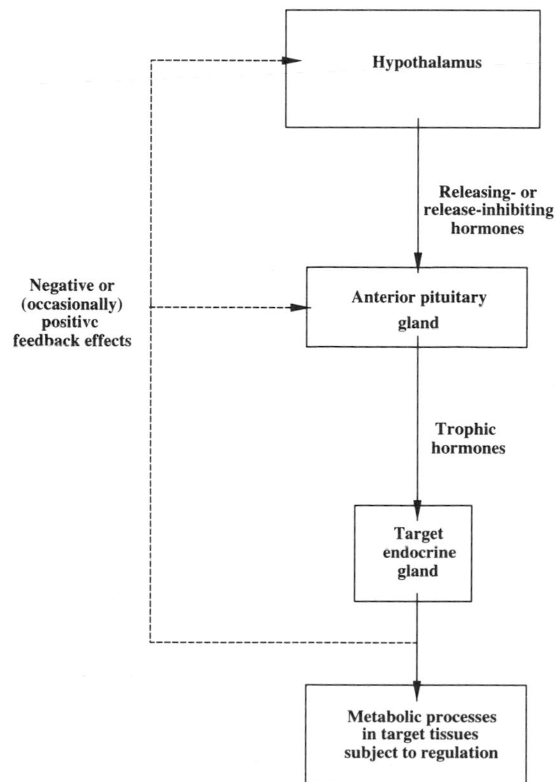

Figure 28.3 **Interactions within the hypothalamopituitary–target tissue axis**

the menstrual cycle. This facilitation is called positive feedback because LH helps to cause the synthesis and release of oestradiol.

28.3 Biosynthesis of peptide hormones

The biosynthesis of all the hormones classified as peptides and proteins shows a remarkable unity, whether the molecule is a tripeptide synthesized in hypothalamic neurons, an oligopeptide of pancreatic origin or a large dimeric glycoprotein formed by the placenta. For this reason, it is possible to choose parathyroid hormone to illustrate the steps involved. Figure 28.4 shows the processes; for the sake of clarity, the mRNA and ribosomes have been omitted. The processes are:

1. Initiation by the formation of a mRNA that specifies a polypeptide chain larger than the required hormone.
2. As the translation of the mRNA starts on ribosomes in the cytosol, a peptide of some 20 aminoacyl residues is formed. The residues are not identical for different hormones, but in every case they are predominantly hydrophobic in character. This N-terminal sequence is called a **signal peptide** (see Chapter 7).
3. In the presence of a '**signal recognition particle**', this facilitates the attachment of the growing polypeptide chain and the associated mRNA and ribosome to the endoplasmic reticulum (ER).
4. Growth of the polypeptide chain proceeds with the chain being introduced into the cisterna of the ER,

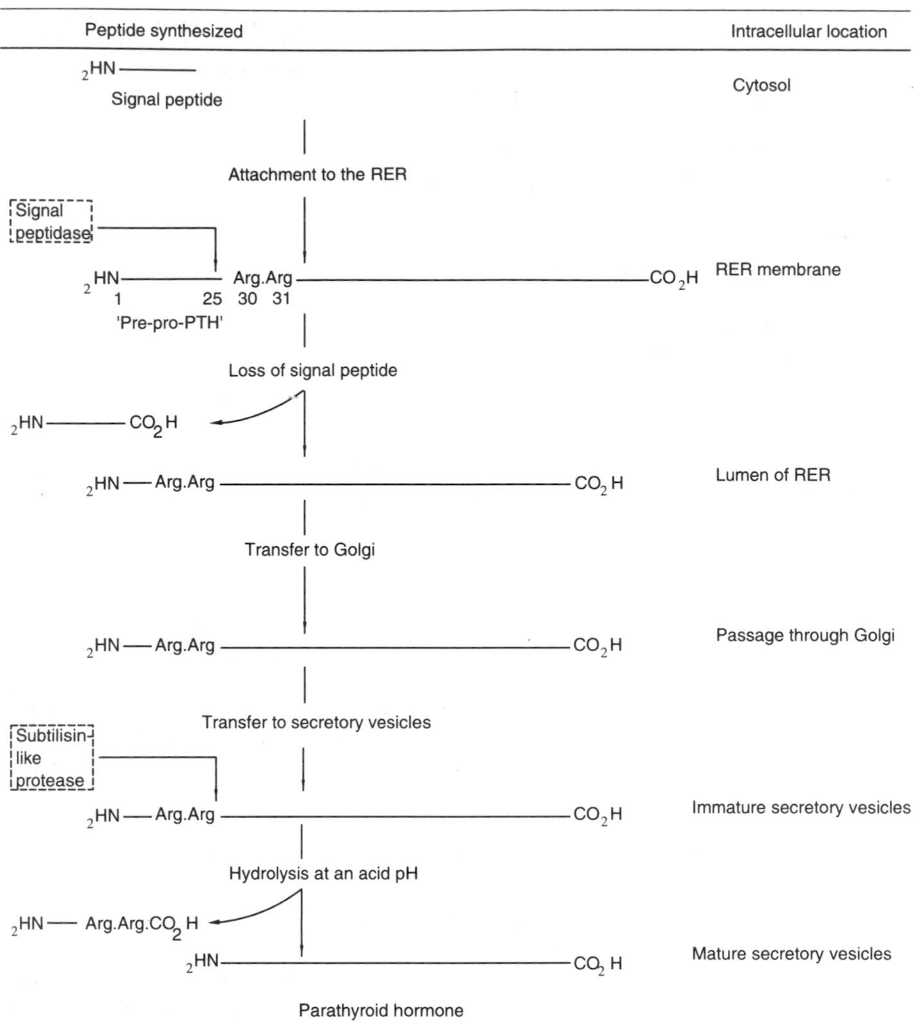

Figure 28.4 **Principal steps in the biosynthesis of parathyroid hormone**

and before the formation of the chain is complete a peptidase removes the signal peptide by hydrolysis.

5. Synthesis of the polypeptide continues to completion when the chain is sequestered in the lumen of the ER. If the hormone is glycosylated, this process is initiated as the growing peptide crosses the ER membrane (see Chapter 7). The sequestered molecule at this stage is a larger precursor of the active hormone and is termed a '**prohormone**'. For the dimeric glycoprotein hormones such as LH each chain is formed separately and the final protein is assembled in the secretory pathway.
6. By means of energy-requiring steps, involving guanosine triphosphate (GTP) and monomeric GTP-binding proteins, the prohormone is transferred in vesicles to the Golgi apparatus where glycosylation is completed (if necessary). It is in the Golgi apparatus that other post-translational modifications of the prohormone may occur. For example the gastrointestinal hormones cholecystokinin and gastrin undergo sulphation on tyrosyl residues. Subsequently the prohormone is packaged into further vesicles in which the conversion to the active form is effected. The final transformation is carried out by peptidases that generate a series of peptides. Important among these enzymes are the specific proteases designated PC1 and PC2. These enzymes, which resemble the bacterial protease subtilisin, catalyse the hydrolysis of the peptide bond on the C-terminal side of a pair of basic aminoacyl residues. In some cases only one of the resulting peptides has biological activity, but in others more than one active peptide is generated or two peptides may be produced with synergistic (cooperative) properties.
7. The functionally competent hormone remains in the vesicles in which they are formed until the cell is stimulated to secrete them. Stimulation of the cell then causes the Ca^{2+}-dependent fusion of the vesicle with the plasma membrane with the consequent release of the hormone(s).

28.4 Hormones of the hypothalamus

The hormones of the hypothalamus are either released from the posterior pituitary gland or from the median

Thyrotrophin releasing hormone pyroGlu.His.Pro.NH_2

Gonadotrophin-releasing hormone pyroGlu.His.Trp.Ser.Tyr.Gly.Leu.Arg.Pro.GlyNH_2

Somatostatin Ala.Gly.Cys.Lys.Asn.Phe.Phe.Trp.Lys.Thr.Phe.Thr.Ser.Cys (S—S bridge between Cys residues)

Vasopressin (Antidiuretic hormone) Cys.Tyr.Phe.Gln.Asn.Cys.Pro.Arg.GlyNH_2 (S—S bridge between Cys residues)

Oxytocin Cys.Tyr.Ile.Gln.Asn.Cys.Pro.Leu.GlyNH_2 (S—S bridge between Cys residues)

Figure 28.5 **The structures of some hypothalamic peptide hormones**

eminence to affect the anterior part of the gland. The latter hormones are called 'releasing' or 'release inhibiting' depending on how they affect their target cells. The aminoacyl sequences of the smaller members of this group of hormones are given in Figure 28.5.

Thyrotrophin-releasing hormone (TRH)

This is one of the simplest peptides, being a tripeptide with characteristic N-terminal pyroglutamate and C-terminal amide structures. TRH is responsible for stimulating the release of thyrotrophin (thyroid-stimulating hormone, TSH). The same peptide has also been shown to stimulate the release of a second anterior pituitary gland hormone, prolactin.

Corticotrophin-releasing hormone (CRH)

The control of the secretion of **adrenocorticotrophin** (adrenocorticotrophic hormone, ACTH) is complex. A 41 aminoacyl residue peptide designated CRH is independently able to stimulate the release of ACTH, but in the physiological setting this action is enhanced by vasopressin. The source of much of this vasopressin is anatomically distinct from that which supplies the posterior lobe, all the indications being that it is secreted from the same nerve terminals in the median eminence that release CRH.

CRH is also synthesized and secreted in increasing amounts by the human placenta during pregnancy and appears in the maternal circulation. The increase is greatest during the third trimester of pregnancy and reaches its zenith at parturition. That this increase does not lead, as might be expected, to maternal Cushing's disease (see Section 28.10) is due to the presence of a circulating binding protein specific for CRH. It used to be thought that binding proteins are only produced for lipid-soluble hormones, but recently several peptide-specific binding proteins have been identified. After parturition, the circulating concentration of CRH falls off steeply, except in ectopic pregnancies when the level is maintained until the placenta is reabsorbed. These facts clearly point to the placental origin of the hormone. In a fashion that appears to have its parallels in the interaction of CRH and vasopressin in the pituitary gland, CRH acts synergistically with oxytocin on human term myometrium to cause smooth muscle contraction, and thus seems to play an important role in parturition.

Gonadotrophin-releasing hormone (Gn-RH)

This decapeptide resembles TRH in that it has N-terminal pyroglutamate and C-terminal amide groups. It causes the synthesis and release of both the gonadotrophins (luteinizing hormone, LH and follicle-stimulating hormone, FSH). This explains the preferred designation of Gn-RH rather than LHRH or FSHRH.

Hormones influencing the secretion of prolactin

Prolactin differs from all the other anterior pituitary hormones in that the predominant control is by a tonic inhibitory mechanism. This is demonstrated by the sectioning of the hypophyseal portal vessel when prolactin is the only pituitary hormone found in the circulation in increased amounts. The molecule responsible for this inhibitory control is the catecholamine dopamine (see Section 28.9). Against this there are two peptides known to act as prolactin-releasing hormones: vasoactive intestinal polypeptide and TRH.

Hormones controlling the secretion of growth hormone

Growth hormone (GH) resembles prolactin in so far as it is subject to dual stimulatory and inhibitory control, but differs from the latter in that the predominant control is stimulatory. This is exercised by **growth hormone releasing hormone** (GHRH), a 44 aminoacyl residue peptide. The inhibitory counterpart to dopamine as far as GH is concerned is the peptide **somatostatin** (GH release inhibiting hormone) a 14 residue peptide (see Figure 28.5).

Shortly after its isolation it was realized that somatostatin is produced in a variety of extrahypothalamic tissues, both within the brain and in many peripheral tissues. Its presence in the D cells of the islets of Langerhans is of particular interest because it appears to be able to inhibit the secretion of both glucagon and (more potently) insulin.

28.5 Hormones of the anterior pituitary gland

The hormonal secretions of the anterior pituitary gland fall into three groups: oligopeptides related to ACTH, the monomeric proteins GH and prolactin and finally the dimeric glycoproteins LH, FSH and TSH. It should be borne in mind that close relations of these proteins are produced by the placenta. Hormones of placental origin are often given the prefix 'chorionic'.

Adrenocorticotrophin-related peptides

ACTH provides an excellent example of a hormone formed as one of several peptides produced when a precursor molecule (termed pro-opiomelanocortin, POMC) undergoes extensive proteolysis prior to secretion. Figure 28.6 shows that three biologically-active peptides are produced from POMC in the anterior pituitary gland. These are the N-terminal portion of the molecule (N-POMC), ACTH and β-lipotrophin (β-LPH). In the intermediate lobe of the pituitary of those animals that have such a structure (including fetal humans) and in the hypothalamus of many species, ACTH undergoes further hydrolysis to yield α-melanocyte-stimulating hormone (α-MSH) while β-LPH is converted into β-endorphin and γ-LPH. It is now understood that this differential processing depends on the types of subtilisin-like processing enzymes found in the different tissues. In the anterior pituitary gland, PC2 is found alone, while in the intermediate lobe and the hypothalamus this enzyme is found together with the closely-related PC1.

Adrenocorticotrophin

ACTH possesses 39 aminoacyl residues. The first 13 of these residues are the minimal requirements for expressing any steroidogenic activity, but for full activity a further 11 residues are required. ACTH acts on the adrenal cortex to stimulate the synthesis, and hence the release of a group of steroid hormones, the glucocorticoids, the chief of which in humans is **cortisol**. This stimulatory effect of ACTH is brought about by a receptor-mediated increase in the intracellular concentration of cAMP. This affects the chemistry of adrenal cortical cells in two main ways. Hydrolysis of cholesteryl esters stored in the organ is increased and the cholesterol released is used for the biosynthesis of cortisol. This process is further aided by an increase in the rate of transfer of the sterol into the mitochondria where steroid hormone biosynthesis is initiated. This transport is promoted by a sterol carrier protein.

Cortisol exerts a negative feedback inhibitory effect on ACTH secretion and this seems largely to depend on the inhibition of the secretions of CRH and vasopressin, i.e. by action on the hypothalamus rather than the pituitary gland.

Glycoprotein hormones

The hormones of this group comprise TSH, LH and FSH and the corresponding chorionic (placental) gonadotrophin and thyrotrophin.

These hormones are composed of two polypeptide chains designated α and β. All are glycoproteins. The biological activity of this group of hormones is determined by the nature of their β chains. Indeed, the aminoacyl sequences of the α chains from all these hormones appear to be identical (although the degree of glycosylation varies).

Thyrotrophin

The α and β chains of this molecule consist of 98 and 113 aminoacyl residues, respectively, and since the molecule overall is glycosylated to the extent of 18% the relative molecular mass is 20 kDa.

The principal function of TSH is to control the secretion of the thyroid hormones **triiodothyronine (T_3)** and **thyroxine (T_4)**. In addition, it influences several other aspects of thyroid function: these include the

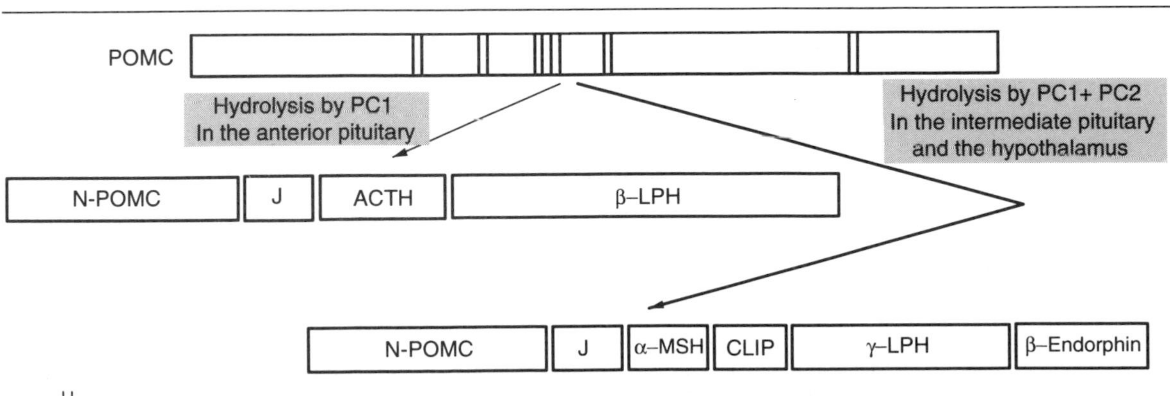

Figure 28.6 **Pro-opiomelanocortin gives rise to different peptides in different tissues**

uptake of iodide, the biosynthesis of T_3 and T_4 and the maintenance of thyroid mass. All these actions appear to result from TSH activating a G-protein-coupled receptor on the thyroid follicular cells. The G-protein is able, when activated, to increase the activity of adenylyl cyclase, so the effects of TSH are mediated by cAMP.

The feedback control of TSH is exerted largely by T_3 and T_4 acting directly on the anterior pituitary gland, although the thyroid hormones do also inhibit the secretion of TRH by the hypothalamus.

Luteinizing hormone

There is considerable species variation in the relative molecular mass of LH with the value ranging from 26 kDa to 34 kDa. The function of LH in the female is to stimulate both the formation, and secretion of progesterone by the corpus luteum in the ovary. It also acts jointly with FSH to promote follicular development, the secretion of oestrogens and subsequent ovulation. In the male the role of LH is to foster the morphological maturation of the Leydig cells of the testes and their secretion of testosterone.

Follicle-stimulating hormone

The relative molecular mass of FSH is about 30 kDa. In the female, the hormone stimulates follicular development beyond the stage of antrum formation and also jointly with LH. In the male it promotes spermatogenesis by the Sertoli cells.

Chorionic gonadotrophin

This hormone begins to be secreted by the syncytiotrophoblast cells soon after conception. It appears in the circulation and also in the urine, where its detection forms the basis for the early detection of pregnancy. The dimeric molecule most closely resembles LH in structure and its function in pregnancy is concerned with the enlargement of the corpus luteum.

The administration of hCG to athletes was once something of a problem. The rationale was that the hormone acts to promote the secretion of testosterone, thus avoiding the use of illegal anabolic steroids. However the detection of this hormone in healthy males is somewhat difficult to explain!

Somatomammotrophins

This closely-related group of proteins consists of the anterior pituitary gland hormones GH and prolactin (PRL) and the placental counterpart of these, chorionic somatomammotrophin (CS), also known as placental lactogen. They each have a single polypeptide chain of some 190 aminoacyl residues with two or three intra-chain disulphide bridges. Within humans, hGH and hCS show 85% homology, while hGH and hPRL are 35% homologous. It is not surprising therefore that all three hormones have both growth-promoting and lactogenic activities, although to different degrees.

Growth hormone

Structure

The human hormone has 191 aminoacyl residues and two disulphide bridges. The human molecule has growth-promoting properties in several other species, but the converse is not true. For a long time this meant that the only source of GH for the treatment of children of short stature due to GH deficiency was human cadavers. It is now appreciated (but not at the time) that this was a very risky procedure, because several children treated with GH obtained in this way have developed Creutzfeld–Jakob disease. This is a form of dementia believed to be caused by a prion, presumably extracted from pituitary glands together with the GH. Fortunately it is now possible to treat children of short stature with recombinant GH produced by bacteria.

Mode of action

GH has a range of actions, some of which are due to the hormone itself and others are ascribable to the paracrine or endocrine release of **insulin-like growth factors (IGFs)**. In all cases the primary response to GH is mediated by its binding simultaneously to two GH receptors in the plasma membrane of responsive tissues. This receptor has an intracellular domain that binds a JAK protein when the receptor is occupied (see Chapter 3). The actions of GH depend on the expression of the catalytic activity of the JAK protein with the phosphorylation of target proteins on tyrosyl residues.

IGF molecules released into the circulation act on both the hypothalamus and the anterior pituitary gland to cause negative-feedback inhibition of the release of GH.

Metabolic effects of GH – direct and indirect

Protein and amino acid metabolism

Growth hormone causes:

1. Stimulation of the uptake of amino acids into a wide variety of cells.
2. Stimulation of protein synthesis.
3. A decrease in the urinary excretion of urea (positive nitrogen balance).

Lipid metabolism

Growth hormone is ketogenic in fasted individuals, owing to its dual ability to facilitate lipolysis and inhibit liopogenesis.

Carbohydrate metabolism

Growth hormone causes:

1. A diabetogenic effect and an exacerbation of clinical diabetes mellitus. (Patients with this disease used to be treated by hypophysectomy in an attempt to save them from the monosaccharide-induced blindness that can occur.)

2. An increase in the circulating concentration of glucose, owing jointly to a decreased hepatic uptake and an increased hepatic output of the monosaccharide.

Mineral metabolism

Almost certainly by acting via IGF-1, GH has profound effects on the retention of several inorganic cations and anions. These include Ca^{2+}, Mg^{2+}, Na^+, K^+, $H_2PO_4^-$ and Cl^-. The effect on the first three ions is due the ability of IGF-1, released in response to GH, to promote long bone growth in children (prior to closing of the epiphyses) and appositional growth in adults.

Prolactin

The prolactin molecule found in humans has 199 aminoacyl residues and 3 disulphide bridges. Its role in female physiology has long been understood because of its well-established ability to act on the oestrogen/ progesterone-primed breast post partum to control the formation of milk. Nevertheless, the fact that the hormone is produced by males and non-parturient females has been puzzling. This has been particularly so because the circumstances that favour the secretion of prolactin closely parallel those that lead to the release of ACTH, a classic 'stress hormone'. The recent findings that prolactin receptors are to be found on about 85% of macrophages and both B- and T-lymphocytes suggests that prolactin may also mediate interactions between the endocrine and immune system. This suggestion is further supported by two types of observations – one in animals and one in humans. It is observed in mice that activation of immune responses results in an increase in the number of prolactin receptors found on B- and T-lymphocytes. In humans it has been noted that hyperprolactinaemia may be associated with some auto-immune diseases such as autoimmune thyroid disease (Hashimoto's disease). Indeed, the use of dopaminergic agonists (see Section 28.4) such as bromocriptine to reduce the rate of secretion of prolactin as a means of controlling the progression of autoimmune disease or preventing transplant rejection is currently undergoing clinical trials.

28.6 Hormones of the posterior pituitary gland

Relationship of the posterior pituitary gland to the hypothalamus

As is shown diagrammatically in Figure 28.2, the posterior pituitary gland is a downward growth of the floor of the brain, and nerve fibres arising from the cell bodies of neurons in the magnocellular nuclei of the hypothalamus project to the organ where the nerve endings abut capillaries. It is into these blood vessels that the neurons release the hormones **vasopressin** and **oxytocin**. Separate genes encode these two peptides, and the mRNAs when translated give rise to much larger prohormones, the sequences of which each include one binding protein, i.e. one for vasopressin and one for oxytocin. These binding proteins, called neurophysins I and II, are highly homologous. The former binds oxytocin and the latter vasopressin.

Vasopressin

Structure

This nonapeptide hormone (see Figure 28.5) shows species variation in its structure: in most species the basic aminoacyl residue present at position 8 is arginine, but in the pig and hippopotamus this is replaced by lysine. For this reason, the hormone is often referred to as arginine (or lysine) vasopressin (AVP or LVP). Sometimes the name argipressin is used for the former peptide.

Control of secretion

The rate of secretion of vasopressin is dependent on the osmolarity of the extracellular fluid, particularly that of the plasma. An increase in the concentration of solute in, or a decrease in the volume of, extracellular fluid promotes the secretion of the hormone. This causes water retention.

Mode of action

An alternative name for vasopressin is **antidiuretic hormone** (ADH), as this serves to emphasize that it acts primarily on the distal convoluted tubules and collecting ducts of the kidney to make them more water permeable. Consequently, it increases water retention by influencing the osmotic movement of water in the nephron. Conversely, whenever the synthesis or release of the hormone is depressed, a large volume of water is excreted. The failure either to produce vasopressin or of the kidney to respond results in the disease **diabetes insipidus**.

The mode of action of vasopressin on water retention is being uncovered. It appears to act by binding to a G-protein-coupled (V_2) receptor and thereby activating adenylyl cyclase with increased formation of cAMP. The cyclic nucleotide in turn promotes the recruitment of the 'water pore' protein aquaporin-2 from intracellular membranes to the luminal plasma membranes of the cells lining the responsive structures (see Chapter 26).

Vasopressin also has pressor actions: it causes the contraction of vascular smooth muscle. Once again, a G-protein-coupled receptor is involved (V_1), but in this case the second messenger is IP_3.

Oxytocin

Structure

This hormone is also composed of nine aminoacyl residues (see Figure 28.5). It is very similar to vasopressin, differing only in the presence of isoleucine for phenylalanine in position 3 and leucine for arginine (lysine) in position 8.

Mode of action

Oxytocin causes powerful contractions of the uterine smooth muscle once labour has begun, but it appears not to initiate the process. It is the principal hormone involved in the transit of the fetus during birth and is used pharmacologically to facilitate labour. Oxytocin appears to cause a reduction in the resting potential of the plasma membrane of responsive cells and, in conjunction with the oestrogens and CRH (see Section 28.4), it increases cellular contractility (this appears to be brought about by a reduction in the concentration of cAMP in uterine smooth muscle).

Oxytocin does not regulate the synthesis of the components of milk, but it is important in its release. Once again the effect seems to be on smooth muscle contraction, in this case the cells in question are those of the myoepithelium. These are arranged around the duct of the gland.

28.7 Hormones of the pancreas

Two hormones that play a central role in the control of fuel homeostasis are synthesized in the pancreas: **insulin** and **glucagon**. The endocrine pancreas also produces three other hormones: **somatostatin** (see Section 28.4), **pancreatic polypeptide** and the recently-discovered oligopeptide **amylin**. There is still much controversy surrounding the last of these, the 37 aminoacyl residue peptide amylin. There are suggestions that amylin opposes some of the metabolic effects of insulin, but this matter is by no means resolved.

These hormones are formed in a small group of cells, known as the islets of Langerhans, scattered in the acini of the organ. These islets comprise four histologically- and functionally-distinct types of cell. Some 60% of these cells (the B- or β-cells) synthesize and secrete both insulin and amylin (much more of the former than the latter, although the relative amounts vary). The A-cells are responsible for the production of glucagon, somatostatin comes from the D-cells, while the F- or PP-cells produce pancreatic polypeptide. These four cell types are arranged in specific topographic relation to each other, and this interrelationship may well be of importance in insuring coordinated responses to appropriate stimuli. It is now also appreciated the both A- and B-cells have receptors for neurotransmitters. A-cells are stimulated to secrete glucagon by adrenaline and noradrenaline. These catecholamines may reach the islet cells via the circulation or the sympathetic innervation of the pancreas (see Chapter 41). The B-cells have acetylcholine responsive muscarinic receptors, the secretion of insulin being stimulated in an IP_3-dependent fashion.

Insulin

Structure and biosynthesis

Insulin consists of two polypeptide chains – A (21 residues) and B (30 residues) – that are covalently joined by two disulphide bridges. Both chains derive from a single prohormone (**proinsulin**). The details of the formation of proinsulin and its packaging into secretory vesicles closely resemble those described for parathyroid hormone (see Section 28.3). Once in the secretory vesicles proinsulin undergoes proteolysis catalysed by the enzyme PC1. The activity results in the removal of a C-(connecting) peptide as the result of hydrolysis of two peptide bonds, each of which has a pair of basic aminoacyl residues on their N-terminal side. This yields insulin which retains one pair of the basic residues on the C-terminal end of the B-chain. These two residues are removed by hydrolysis catalysed by carboxypeptidase E to give the insulin molecule (Figure 28.7).

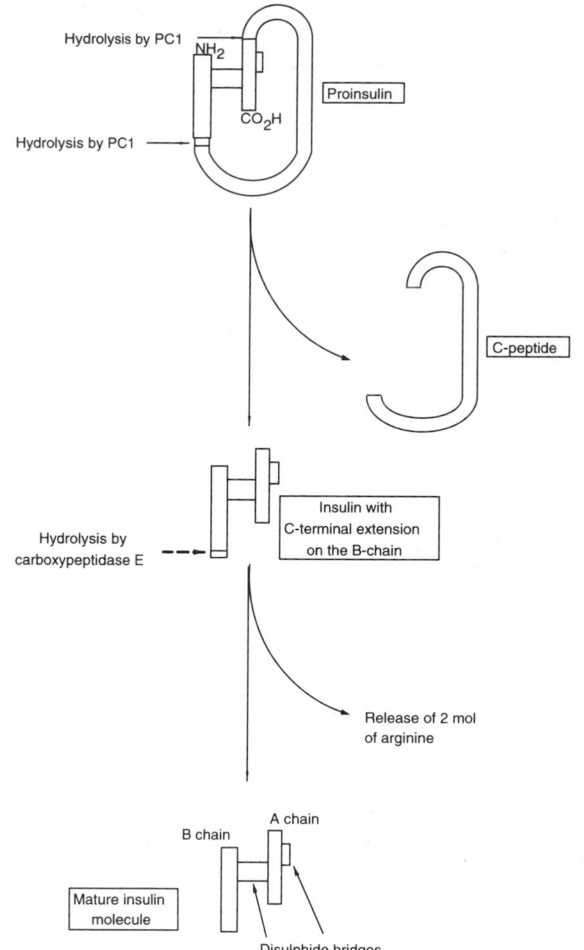

Figure 28.7 **The conversion of proinsulin into insulin**

Control of secretion

Insulin is stored in the form of a zinc–insulin complex in secretory granules prior to release. Increases in the circulating concentration of glucose are detected by B-cells in a mechanism that appears to depend in part on the phosphorylation of the monosaccharide catalysed by **hexokinase D** (this enzyme, which is also known as **glucokinase**, is also found in the liver, and has sigmoidal kinetics that make it responsive to changes of glucose concentration in the physiological range). The further metabolism of glucose 6-phosphate via glycolysis causes an increase in the ratio of ATP/ADP which then appears to act to promote the secretion of the hormone.

Mode of action

Insulin has a diversity of actions on a range of tissues, the chief of which are liver, muscle and adipose tissue. Two broad categories of types of action may be recognized: it acts as a growth factor to promote cellular growth and differentiation and it plays a central role in the regulation of intermediary metabolism. These two different types of effect are mediated by the binding of insulin to its receptor which is to be found in the plasma membranes of all responsive tissues. The effects of insulin, therefore, depend on the phenotype of the target cell or on the establishing of a new phenotype. In other words most of the effects of insulin arise either due to changes in the catalytic activity of enzymes already present in the tissue or to changes in gene expression.

The insulin receptor that can initiate such a diversity of effects is classified as a catalytic (growth factor) receptor (see Chapter 3). The receptor is a tetrameric protein with two membrane-spanning β-chains and two extracellular α-chains. The latter chains are linked to each other and also to the β-chains by disulphide bridges (Figure 28.8). The insulin receptor is a **protein tyrosine kinase**. This means that on binding to the α-chains the hormone causes the intracellular protein kinase domains of the β-chains to become active and thereby catalyse the ATP-dependent self-phosphorylation. This leads to the phosphorylation of a protein named insulin receptor substrate (IRS) on its tyrosine residues. In turn, phosphorylated IRS causes the activation of a cascade of phosphorylation/dephosphorylation reaction involving serine residues in a range of proteins. It is these secondary reactions that underlie the biological actions of insulin. One exception to this general rule relates to the way in which insulin acts to promote the transport into target cells. In this, IRS acts by a separate mechanism to cause a glucose transport protein to be translocated to the plasma membrane.

Carbohydrate metabolism

Insulin acts directly on muscle and adipose tissue to increase their ability to take up glucose, thereby tending to reduce the circulating concentration of blood glucose. This effect is fully established just 5 min after exposure of the tissue to insulin and depends upon the translocation of a specific glucose transporter (Glut 4) from intracellular membranes to the plasma membrane where it is able to facilitate the transport of glucose into the cell. The liver lacks such an insulin-responsive transporter.

Insulin also promotes the storage of glycogen in muscles and in the liver, and in the latter tissue it inhibits the conversion of glycogen into glucose and also the process of gluconeogenesis. The actions of insulin and those hormones that oppose it in respect of glycogen storage in the liver are summarized in Figure 28.9.

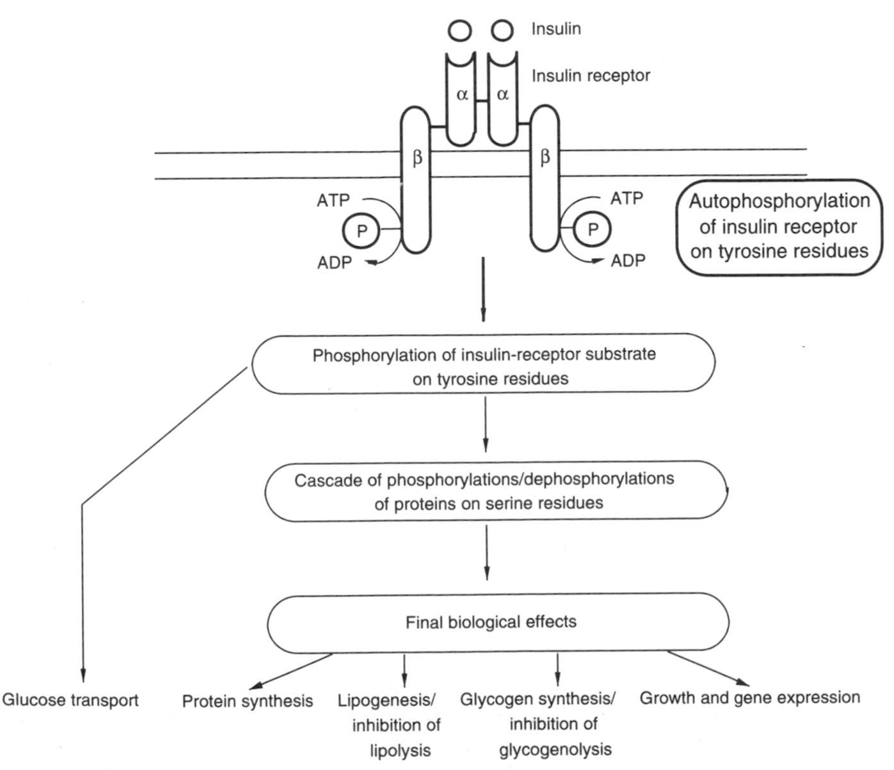

Figure 28.8 **Summary of the mechanism of action of insulin**

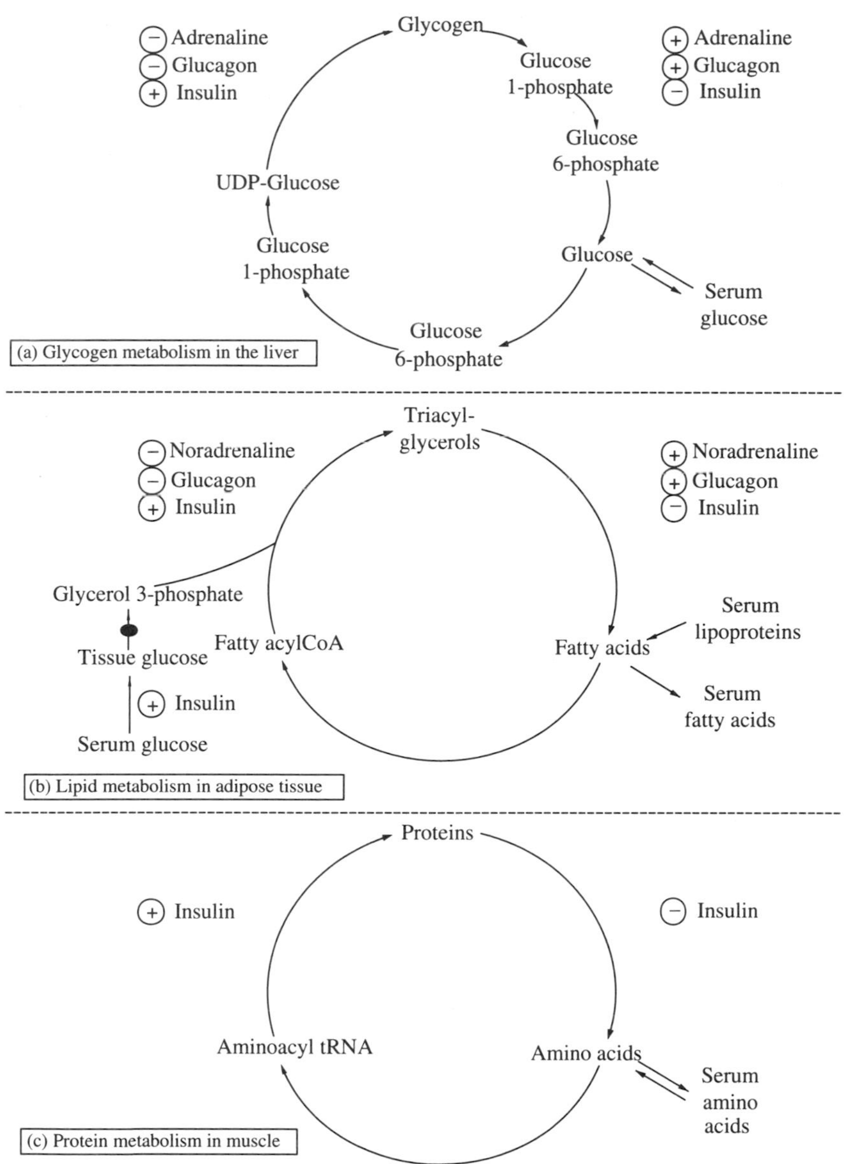

Figure 28.9 **The actions of the principal hormones that control fuel homeostasis**

Lipid metabolism

Insulin inhibits the release of fatty acids and glycerol from adipose tissue as the result of its actions both on that tissue and also on the liver. In the latter tissue the hormone promotes the synthesis of fatty acids and hence their export in the form of triacylglycerols associated with VLDLs (see Chapter 10). The clearance of these particles by the action of lipoprotein lipase in adipose tissue is then fostered by the insulin-mediated increase in the activity of that enzyme. This means that adipose tissue is presented with more fatty acids when the concentration of insulin is high. The other actions of insulin focus on reactions proceeding inside the adipocyte. These occur in three ways which jointly have the effect of favouring the synthesis of triacylglycerols over their hydrolysis: the provision of glucose that acts as a source of glycerol 3-phosphate and an acceleration of the conversion of this intermediate into triacylglycerols. Inhibition of lipogenesis arises because insulin acts to reduce the activity of hormone-sensitive lipase which catalyses the hydrolysis of triacylglycerols. The processes occurring in the adipose tissue and the counter-regulatory effects of glucagon and the catecholamines are shown diagrammatically in Figure 28.9.

The net effect of the inhibition of fatty acid release is that diminishing quantities are available from the circulation for the hepatic production of ketone bodies. Hence insulin is antiketotic.

Protein and amino acid metabolism

Just as insulin favours anabolic reactions over catabolic in the liver and in adipose tissue, so it does in muscle in which the focus of the effect is on protein biosynthesis, which is favoured over proteolysis in the presence of the hormone. Specifically, insulin promotes the uptake of amino acids into muscle (and also liver), it promotes polysome formation (and hence protein synthesis) and it acts to inhibit proteolysis. The actions of insulin on muscle protein are shown in Figure 28.9.

Inactivation

The circulating half-life of insulin is about 10 min, which indicates a fairly rapid clearance of the molecule. This occurs chiefly in the liver, but other insulin-sensitive tissues do contain catabolic enzymes for the hormone. The enzymes disposing of insulin are all intracellular so the hormone must be taken up into the cells by receptor-mediated endocytosis. This is followed by proteolysis and reduction of the disulphide bridges so that the two chains are separated. The reductive process is catalysed by the enzyme glutathione-insulin transhydrogenase.

Glucagon

Glucagon in its actions largely opposes those of insulin, so much so that glucose homeostasis is much better understood when the ratio of the concentration of the two hormones in the circulation is considered. A high ratio of insulin to glucagon favours anabolic processes while a low ratio favours catabolism.

Structure and biosynthesis

Glucagon is a simple oligopeptide which is formed biosynthetically as part of a larger pro-glucagon molecule (180 aminoacyl residues in humans) that is cleaved towards the N-terminal end to yield a 28 aminoacyl residue peptide.

Mode of action

Glucagon is a powerful hyperglycaemic agent and this effect is mediated by a receptor that is coupled positively to adenylyl cyclase via a G-protein. Its effects depend, therefore, on the production of cAMP and the activation of protein kinase A (see Chapter 3).

Carbohydrate metabolism

The principal actions of glucagon are on the liver where it acts to favour glycogenolysis over glycogen synthesis (see Figure 28.9) and also gluconeogenesis.

Lipid metabolism

There remains a good deal of uncertainty over the actions of glucagon on adipose tissues. In humans it seems to be of less importance as a lipolytic hormone than adrenaline and noradrenaline. However, the experimental basis for this statement has been questioned as it has been suggested that in the physiological setting the action of glucagon on adipose tissue to cause lipolysis is augmented by other factors, possibly including inhibition of the release of adenosine (see Figure 28.9).

28.8 Hormones of the thyroid and parathyroid glands

In the normal adult the thyroid gland weighs about 25 g and consists of two connected lobes, closely associated with the trachea. The blood flow through the gland is rapid, and exceeded only by that to the lungs and the carotid body. The gland is divided by connective tissue into clusters of follicles or acini. Each follicle consists of a single layer of cells surrounding a colloid containing the protein thyroglobulin; there are about a million follicles in the human gland. This colloidal thyroglobulin is the source of the two thyroid hormones **triiodothyronine** (T_3) and **thyroxine** (T_4).

A separate group of cells, the C-cells, which derive from the primitive nervous system, are responsible for the biosynthesis and release of the peptide hormone calcitonin.

Closely associated with the thyroid are the parathyroid glands, which number four in humans. The latter, which each weigh 30–40 mg, are situated on the dorsal surface of the thyroid gland and are responsible for the biosynthesis and secretion of parathyroid hormone, a polypeptide.

Structure and biosynthesis of T_3 and T_4

Iodide absorbed from the diet or drinking water is transported in the blood to the thyroid gland, where it is very efficiently transferred into the follicular cells of the gland by an 'iodide pump' system which is dependent on ATP. Uptake is inhibited by ions, such as perchlorate or thiocyanate, which act as competitive inhibitors of the transport process.

The enzyme that initiates the formation of T_3 and T_4 is thyroid peroxidase. The enzyme uses the large, dimeric glycoprotein **thyroglobulin**, iodide and hydrogen peroxide as its substrates. Thyroglobulin is a typical secretory protein (see Chapter 7) and as such its biosynthesis starts in the rough endoplasmic reticulum where its glycosylation is initiated. The latter process is completed in the Golgi apparatus before the molecule is transferred to secretory vesicles. The vesicles segregate to the apical plasma membrane of the follicular cells and it is this stage that thyroglobulin undergoes iodination prior to its release and storage in the colloid. This potential store of thyroid hormones is very large: it has been calculated that the quantities present in the follicles are sufficient to sustain normal circulating levels for up to 2 weeks.

The processes of iodination of thyroglobulin and formation of T_3 and T_4 are very complex, but undoubtedly involve the formation of free radicals, either from iodide (possibly I^+, the iodonium ion) and/or tyrosyl residues in thyroglobulin in reactions catalysed by the peroxidase. Iodination of tyrosyl residues in position 3 of the benzene ring gives rise to a mono-iodo derivative and this may be followed by a second iodination at position 5. The same enzyme then catalyses

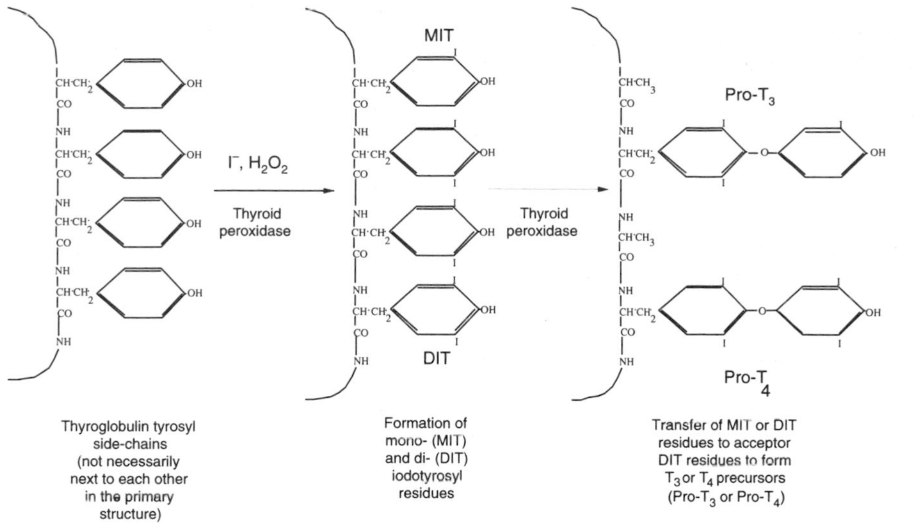

Figure 28.10 **Reactions occurring in the exocytotic vesicles of thyroid follicular cells**

Thyroglobulin tyrosyl side-chains (not necessarily next to each other in the primary structure)

I^-, H_2O_2 Thyroid peroxidase

Formation of mono- (MIT) and di- (DIT) iodotyrosyl residues

Thyroid peroxidase

Transfer of MIT or DIT residues to acceptor DIT residues to form T_3 or T_4 precursors (Pro-T_3 or Pro-T_4)

the transfer of either mono- or *bis*-iodophenyl residues to acceptor *bis*-iodotyrosyl residues. This leads to the formation of T_3 and T_4 still covalently part of the thyroglobulin structure (Figure 28.10). It is at this stage that the iodinated thyroglobulin is released from the cells to be stored in the colloid.

Control of secretion

Stimulation of follicle cells by TSH activates a cAMP-dependent response which leads to the reuptake of iodinated thyroglobulin from the colloid by a process of pinocytosis. The pinocytotic vesicles then fuse with lysosomes where proteolysis of the protein occurs to release T_3 and T_4. Any molecules of mono- or *bis*-iodotyrosine generated as a result of this process undergo de-iodination.

In most individuals the ratio of $T_4 : T_3$ secreted is in the range 20–30 : 1 but on a molar basis T_3 is 3–5 times more potent than T_4 and indeed the conversion of the former to the latter hormone occurs in target tissues in a reaction catalysed by a de-iodinase. These observations have led to the view that T_3 is the true hormone while T_4 serves as a prohormone.

Plasma transport of T_3 and T_4

In the plasma, both T_3 and T_4 are almost entirely (>99.5%) transported bound to one of three proteins. One is thyroxine-binding globulin (TBG) which has a high affinity for both molecules but rather a low capacity, another is thyroxine-binding pre-albumin (TBPA; lower affinity, higher capacity) and finally albumin itself (Alb; low affinity, high capacity). As a result of this, T_3 and T_4 are distributed between these three circulating proteins as indicated in Table 28.2.

The protein-bound forms of T_3 and T_4 are not biologically active and should, therefore, be regarded primarily as a reserve, because release of the free form is essential for expression of hormonal activity.

Table 28.2 **The binding of thyroid hormones to serum proteins in euthyroid subjects**

Hormone	*% hormone bound to:*			*[Total]*	*[Free]*
	TBG	TBPA	ALB	nM	pM
T_4	70	10	20	1.0	0.31
T_3	38	27	35	0.03	0.12

TBG, thyroxine-binding globulin; TBPA, thyroxine-binding prealbumin; ALB, albumin.

Mode of action of T_3 and T_4

As indicated, many tissues undertake the de-iodination of T_4 to yield T_3. The latter exerts its effects after binding to a specific cytosolic receptor protein. This is a member of the steroid-hormone receptor superfamily (see Chapter 3). This means that once the receptor is occupied by T_3, it moves to the nucleus and there the complex binds to control regions for certain target genes to modulate their rates of transcription. The full range of T_3 responsive genes has not been determined, but the protein product of several are involved in mediating the caloric effects of the hormone.

Metabolic effects T_3 and T_4

Caloric effects

Thyroid hormone increases heat production and oxygen consumption. It was originally considered that this effect was mediated by uncoupling of oxidative phosphorylation (see Chapter 5) in mitochondria, which would increase oxygen consumption and heat production. This hypothesis is now discounted since, in the experiments that appeared to support it, grossly supraphysiological concentrations of T_4 were employed, and these caused large-scale mitochondrial swelling (leading to uncoupling).

The only observation that has been made that might relate to mitochondria and heat production is that thyroid hormone causes an increase in the amount of the FAD-linked glycerol 3-phosphate dehydrogenase present in inner membranes. This enzyme is one of those that 'shuttles' reducing equivalents generated in glycolysis into the mitochondrion (see Chapter 5). It is not entirely clear that any consequent facilitation of the latter pathway would explain completely the thermogenic effect of thyroid hormone.

Effects on carbohydrate metabolism

Thyroid hormone:

1. Stimulates glycogenolysis in the liver, both directly and by potentiating the effect of adrenaline.
2. Stimulates intestinal absorption of glucose.

Effects on lipid metabolism

Thyroid hormone:

1. Stimulates the release of free fatty acids from adipose tissue by increasing the amount of hormone-sensitive lipase in adipocytes.
2. Stimulates the oxidation of fatty acids.
3. Reduces the circulating concentration of cholesterol by the stimulation of its hepatic conversion to bile salts (see Chapter 25).

Effects on protein metabolism

Thyroid hormone:

1. Stimulates the synthesis of proteins, mainly enzymes, involved in oxidative (energy-yielding) reactions that are required to support the normal rate of protein synthesis.
2. Stimulates protein catabolism. It is not clear whether this represents a physiological or pathological effect of the hormone.

Inactivation of T_3 and T_4

The de-iodination of T_4 to yield T_3 that occurs in target tissues yields equal amounts of two isomers of the latter. One biologically active (3,5,3′-T_3) and one inactive ('reverse T3', i.e. 3,3′5′-T_3). It is noteworthy, therefore, that the ratio of T_3 relative to rT_3 falls in febrile illness. Both T_3 (of thyroid or peripheral origin) and rT_3 are further de-iodinated to the biologically inactive *bis*-iodothyronines (3,5-T_2, 3,3′-T_2 and 3′,5′-T_2).

Disorders of thyroid function

A vital area of thyroid hormone function is in fetal and neonatal brain development, and the need for iodine to prevent the gross impairment of brain development, known as cretinism, has been known for a long time. It remains a distressing fact, therefore, that in 1990 the WHO estimated that in excess of 20 million people world-wide had suffered from brain damage, in varying degrees, due to maternal iodine deficiency during pregnancy. As a consequence this is the leading, and eminently preventable, cause of mental retardation in the world. For preventative interventions to be successful the mother must receive iodine supplements in both the first and the second trimesters of her pregnancy. It seems certain that thyroid hormones of maternal origin play key roles in early fetal brain development. Because of the increased demands for iodine in pregnancy, it is suggested that supplementation should be considered where the daily intake falls below 50 mg (see Chapter 20). Replacement with thyroxine after birth is not successful in reversing mental retardation. However, if the fetus has had the benefit of maternal thyroid hormones, but is born hypothyroid (perhaps due to a defect in one of the enzymes involved in the biosynthesis of thyroid hormones), then the administration of thyroxine is always beneficial in terms of preventing, or reducing the extent of, mental retardation. For this reason, in the UK and a number of other countries, there is a screening programme in which infant plasma TSH is measured. The aim of this test is to pick out concentrations that are too high, because this usually indicates that T_3/T_4 is too low and thus not exerting an appropriate negative feedback effect on TRH and TSH secretion. This gives a better perspective on the functional activity of the hypothalamo-pituitary–thyroid axis than would be obtained by measuring T_3/T_4. The occurrence of goitre, an enlargement of the thyroid gland caused by the excessive trophic effect of TSH, in areas of iodine deficiency is another instance of reduced negative feedback effects due to too low T_3/T_4. The 1990 WHO report suggested that, throughout the world, 1 billion people are iodine deficient due to low iodine in the soil, and as many as 20% have goitre. The problem is now being addressed by means of iodizing programmes for salt and oils. When seen in adults the hypothyroid state is often referred to as myxoedema.

The most common cause of thyroid disease, outside areas of iodine deficiency, is organ-specific auto-immunity. As is the case in insulin-dependent diabetes mellitus (see Chapter 41), this most frequently arises as a result of the T-lymphocyte-dependent destruction of endocrine (in this case thyroid follicular) cells. The condition, when it affects the thyroid gland, is known as **Hashimoto's disease** and it leads to hypothyroidism, with a need for lifelong replacement therapy with thyroxine. The nature of the antigenic determinants for this attack are not known, but in another immune-related disease of the thyroid, **Graves' disease**, there is a better understanding of the nature of the dysfunction of the immune system. In Graves' disease there is thyrotoxicosis, and this arises due to the binding of circulating antibodies to the TSH receptor, thus causing their unregulated stimulation. Finally, in a few cases, thyrotoxicosis has been associated with toxic nodules in the thyroid. In these there has been shown to be an activating somatic mutation to the TSH receptor.

Calcitonin and parathyroid hormone

Because of their key roles in Ca^{2+} and HPO_4^{2-} homeostasis and the relation of this to bone biochemistry, the details of the activities of these two peptide hormones are addressed in Chapters 12 and 29. The present section simply supplies a brief summary of some of the more important aspects of this pair of hormones.

Calcitonin

Structure and biosynthesis

In its circulating form calcitonin has a single oligopeptide chain of 32 aminoacyl residues. It is formed in the C-cells of the thyroid gland as part of a larger (136 residue) precursor, its sequence being located towards the C-terminus of the prohormone.

Mode of action

The actions of calcitonin remain the subject of some debate, with the view being put forward that as a hormone it only really comes to the fore in pregnancy. It is usually stated that calcitonin:

1. Reduces the concentration of Ca^{2+} in the blood by inhibiting bone resorption. This is an important function in pregnancy, helping to protect the maternal skeleton from what might otherwise prove to be excessive demands of the fetus for Ca^{2+}.
2. Increases the urinary excretion of both Ca^{2+} and $H_2PO_4^-$.

Parathyroid hormone

Structure and biosynthesis

The series of events that result in the formation of a 90 aminoacyl residue pro-parathyroid hormone and its conversion into the 84 residue hormone found in the circulation are described in Section 28.3.

A protein that has a degree of homology with parathyroid hormone, named parathyroid-hormone-related protein (PTH-rP), has recently been isolated and it appears to play an important role in fetal bone development.

Mode of action

The hormonal actions of parathyroid hormone are:

1. To promote bone resorption and thus to mobilize Ca^{2+} and HPO_4^{2-}.
2. To facilitate the reabsorption and thus the retention of Ca^{2+} by the kidney.
3. To decrease renal $H_2PO_4^-$ reabsorption – the so-called 'phosphaturic' effect. This has the effect of lowering the circulating concentration of HPO_4^{2-}.
4. To stimulate the conversion of vitamin D to its active metabolite (1,25-dihydroxycholecalciferol).

28.9 Hormones of the adrenal medulla

Structure of the adrenal gland

The two adrenal glands are situated close to the poles of the kidneys. Each gland is composed of two distinct parts, the **medulla** which forms and secretes catecholamines, and the **cortex**, which can be recognized as being organized into a number of histologically and functionally distinct zones (Figure 28.11) which secrete a range of steroid hormones.

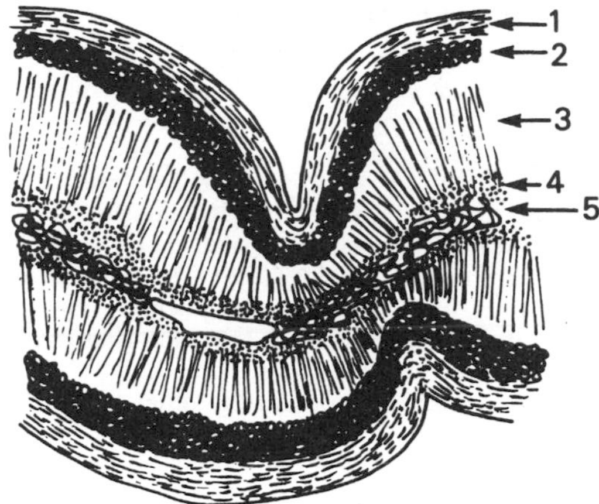

Figure 28.11 **The structure of the adrenal gland.** (1) Capsule; (2) zona glomerulosa; (3) zona fasciculata; (4) zona reticularis; (5) medulla

The medulla, that derives embryologically from the neural crest, remains a functional part of the nervous system and may be viewed as a specialized sympathetic ganglion innervated by preganglionic neurons. These neurons make cholinergic synapses with the characteristic cells of the tissue, the chromaffin cells. These are so named for the presence of cytoplasmic dense-core vesicles termed 'chromaffin granules' which contain the catecholamines adrenaline (80% of the total in adult humans) and noradrenaline (20%).

Biosynthesis of catecholamines

Three types of biologically active catecholamines are known: dopamine (dihydroxyphenylethylamine), noradrenaline and adrenaline. Dopamine acts largely as a brain neurotransmitter, and noradrenaline and adrenaline are released both from chromaffin cells and from peripheral and central neurons. The biosynthesis and inactivation of the catecholamines is described in Chapter 30.

Control of release

The release of the catecholamines is triggered by splanchnic nerve impulses. This cholinergic nerve releases acetylcholine that activates nicotinic acetylcholine receptors (see Chapter 3) in the plasma membranes of the chromaffin cells. The membrane depolarization that follows results in an influx of Ca^{2+} that causes the fusion of the granule membrane with the plasma membrane and the release of the vesicle contents.

Mode of action of adrenaline and noradrenaline

In order for them to exert their effects, the catecholamines must bind to plasma-membrane receptors, known as adrenoceptors. Their are two broad categories of adrenoceptor – α and β. The former respond preferentially to noradrenaline and are often involved in the contraction of smooth muscle. These receptors are coupled to phospholipase C via a G-protein and the production of IP_3 leads to the release of intracellular stores of Ca^{2+}. It is the increase in the cytosolic concentration of this cation that underlies many of the actions of the catecholamines mediated by α-adrenoceptors. β-Adrenoceptors, on the other hand, are coupled via a G-protein to adenylyl cyclase, and hence to cAMP production. Activation of these receptors is responsible for many of the metabolic effects of the catecholamines, e.g. the activation of hepatic glycogenolysis.

Metabolic and other effects of adrenaline and noradrenaline

The catecholamines have powerful effects on the cardiovascular system and also on several metabolic processes. In particular, they:

1. Increase the heart rate by their action on the sinoatrial node.
2. Increase the blood glucose concentration by actions on hepatic glycogenolysis (see Figure 28.9).
3. Increase circulating non-esterified fatty acids by activating adipose tissue hormone-sensitive lipase to increase the rate of hydrolysis of triacylglycerols (Figure 28.9).
4. Act on β-adrenoceptors to stimulate the release of glucagon.
5. Promote the rapid breakdown of muscle glycogen.

28.10 Steroidogenic organs

Sources and structures of steroid hormones

The steroidogenic organs include the adrenal cortex, the gonads and the placenta. In all of these tissues cholesterol is the precursor for the formation of the steroid hormones. It is useful to classify these hormones structurally, by the number of carbon atoms they contain, as well as functionally and by their tissue of origin (Table 28.3).

The **oestrogens** are unique among these hormones in the absence of the methyl group between the A and B rings, and in that the A ring is aromatic. Hence the hydroxyl group in position 3 is phenolic. The structures of the principal steroid hormones and the basis for their numbering are given in Figure 28.12. It can be seen from this that, with the exception of the oestrogens and dehydroepiandrosterone, all the biologically-active molecules have a Δ^4, 3-oxo- arrangement (double bond between carbon atoms 4 and 5 and an oxo-group on carbon 3).

Recently, evidence has been provided that humans produce the cardiotonic steroid **ouabain** which was previously believed to be solely of plant origin. The most likely source for this glycoside is the adrenal gland. Although it has been claimed that the circulating concentration of ouabain is elevated in patients with moderate to severe heart failure, its role in physiology and pathology has yet to be established: there are some indications that it may be involved in whole body Na^+ homeostasis, but this matter needs to be investigated further.

The structure of the steroid hormones is based on a (puckered) ring which has two faces: α and β. The orientation of a substituent attached to the ring is designated α if it is on the same face as the two methyl groups (carbon atoms 18 and 19) or β if on the opposite face.

Biosynthesis of the steroid hormones

Source of precursor

The cholesterol required to initiate the formation of the steroid hormones in all steroidogenic tissues derives predominantly from intracellular stores of cholesterol esters. The major exception to this rule is the placenta which, towards term, can secrete very large amounts of progesterone and for this the tissue draws directly on cholesterol obtained from the circulating lipoprotein LDL. On the other hand, the characteristic morphology of the zona fasciculata of the human adrenal cortex (see Figure 28.11) relates to the extensive presence of cholesteryl ester lipid droplets.

Formation of pregnenolone

The conversion of cholesterol into the steroid hormones involves some reactions that occur in the mitochondria

Table 28.3 **The steroid hormones and their tissues of origin**

Type	*Number of C atoms*	*Principal sources*	*Subsidiary sources*
Oestrogens	18	Ovary, placenta	Adrenal cortex, testis
Progestogen	21	Ovary, placenta	Adrenal cortex, testis
Androgens			
Testosterone	19	Testis	Adrenal cortex, ovary
Adrenal androgens	19	Adrenal cortex	Ovary, testis
Corticosteroids			
Glucocorticoids	21	Adrenal cortex	
Mineralocorticoids	21	Adrenal cortex	

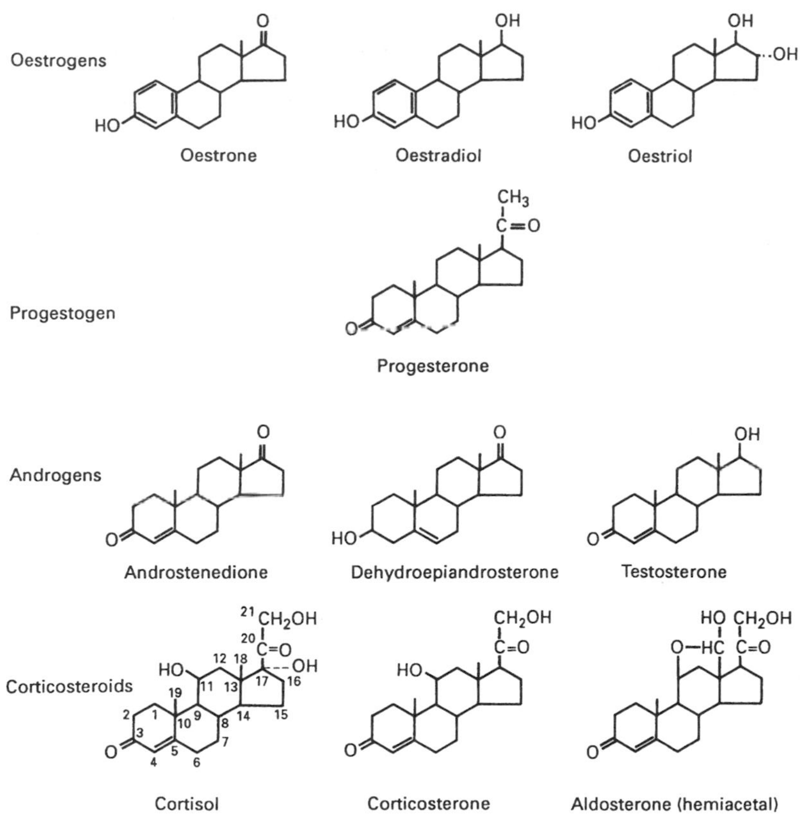

Oestrogens: Oestrone, Oestradiol, Oestriol

Progestogen: Progesterone

Androgens: Androstenedione, Dehydroepiandrosterone, Testosterone

Corticosteroids: Cortisol, Corticosterone, Aldosterone (hemiacetal)

Figure 28.12 **Structures of the steroid hormones**

and others that require the intermediates to be transferred to the endoplasmic reticulum. The process is initiated by the uptake of cholesterol into the mitochondria, a process that is facilitated by a protein called steroidogenic acute regulatory protein. It is in the mitochondria that the side-chain of cholesterol is removed in a complex oxidative process leading to the production of pregnenolone, which is released from the organelle (Figure 28.13). The enzyme that catalyses the formation of pregnenolone is a cytochrome P450-dependent mixed-function oxidase requiring oxygen and NADPH. In the reaction catalysed by this enzyme, often referred to as $P450_{scc}$ (for side-chain cleavage), sequential hydroxylations on the cholesterol side-chain lead to the production of pregnenolone and *iso*capraldehyde. To this stage the process described is common to the formation of all the steroid hormones. Thereafter, one of two reactions may occur as is shown in Figure 28.14, to which repeated reference will need to be made. It should be borne in mind that this figure is a composite: it does not represent the reactions that occur in any particular tissue. Rather, it shows all the possible reactions, some of which occur in particular tissues. Thus neither testosterone nor the oestrogens are more than very minor products of the adrenal cortex and cortisol is not secreted by the gonads.

The Δ^4- and Δ^5-pathways

Pregenolone may be converted into progesterone in a reaction catalysed by the endoplasmic reticulum enzyme NAD^+-dependent 3β-hydroxysteroid dehydrogenase. This reaction sees an oxo group formed at the 3-position and the carbon–carbon double bond moved to the Δ^4-position in conjugation with it. Alternatively the movement of the double bond may be delayed as long as possible in the reaction sequence. If this is the case, the route is referred to as the **Δ^5-pathway** (across the top of Figure 28.14, for example), otherwise it is the Δ^4-pathway (down the left-hand side of Figure 28.14). The Δ^5-pathway then commences and the **Δ^4-pathway** continues, with a cytochrome P-450-dependent hydroxylation in the 17α-position. The 17α-hydroxylase is an endoplasmic reticulum enzyme. The gene that encodes this enzyme is designated CYP17 and it has been established that the same gene-product also has 17,20-lyase activity, i.e. it catalyses the formation of dehydroepiandrosterone from 17α-hydroxypregnenolone in the pathways that lead to androgen and oestrogen biosynthesis.

Formation of oestrogens

In the formation of oestrogens, both the Δ^4- and the Δ^5-pathways are active. The former dominates in the human corpus luteum in pregnancy, whereas the Δ^5-pathway is followed in the Graafian follicle of the ovary, which is responsible for the mid-cycle peak of oestrogen production. As is shown in Figure 28.14, two androgens, androstenedione and testosterone, are produced by the oxidative removal of the side-chain at position 17 catalysed by the bifunctional enzyme, 17,20-lyase. The

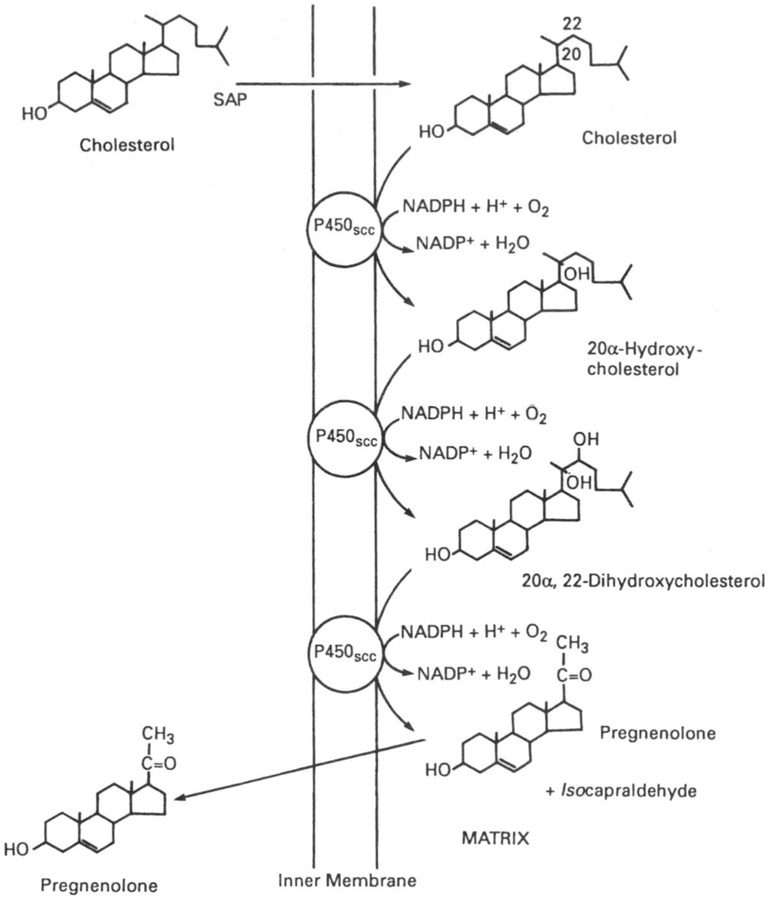

Figure 28.13 **The transport of cholesterol into a mitochondrion and its conversion into pregnenolone by the side-chain cleavage cytochrome P-450 (P-450$_{scc}$) (SAP - steroidogenic acute regulatory protein)**

two androgens then undergo aromatization by oxidative removal of the methyl group C-19, a further process catalysed by a cytochrome P-450-dependent aromatase complex. In fact, the androgens are produced by ovarian thecal cells, are released and taken up by granulosa cells to form the oestrogens. Androgens released in this way may escape into the circulation, thus explaining the appearance of these 'masculine' hormones in small quantities in the blood of females. The requirement for a supply of androgens for oestrogen synthesis is also seen in the placenta. This tissue must be provided with dehydroepiandrosterone, from either the maternal or fetal adrenal cortex, to produce oestriol.

Formation of aldosterone

The route to aldosterone, and also to corticosterone, follows the Δ^4-pathway. In this, successive hydroxylations occur at the 21-, the 11β- and then the 18-positions. The newly-formed alcohol group in position 18 undergoes further oxidation to an aldehyde. The aldehyde may then form a hemiacetal with hydroxyl group in position C-11. This last reaction serves to protect the 11β-hydroxyl group from further oxidation in aldosterone-target tissues. The remarkable observation has been made that the final three reactions leading to the formation of aldosterone, i.e. from deoxycorticosterone onwards, are catalysed by the product of a single gene designated CYP11B2. This mitochondrial enzyme therefore acts as an 11β-hydroxylase and also as corticosterone oxidases I and II shown in Figure 28.14. All of the aldosterone (about 100 μg/day in the adult human) and about half of the circulating corticosterone (nevertheless a minor glucocorticoid in humans) is produced by the zona glomerulosa of the adrenal cortex (see Figure 28.11). The rest of the corticosterone and all of the cortisol are products of the zona fasciculata.

Formation of cortisol

The Δ^5-pathway seems to dominate for cortisol, the order of cytochrome P-450-dependent hydroxylations being 17α, 21 and finally 11β. The enzyme responsible for the third hydroxylation, a product of the gene CYP11B1, closely resembles the one that encodes the multi-functional enzyme that produces aldosterone. It differs in that it has only 11β-hydroxylase activity.

Formation of the adrenal androgens

Adrenal androgens are formed largely in the zona reticularis of the tissue. The only possible route to

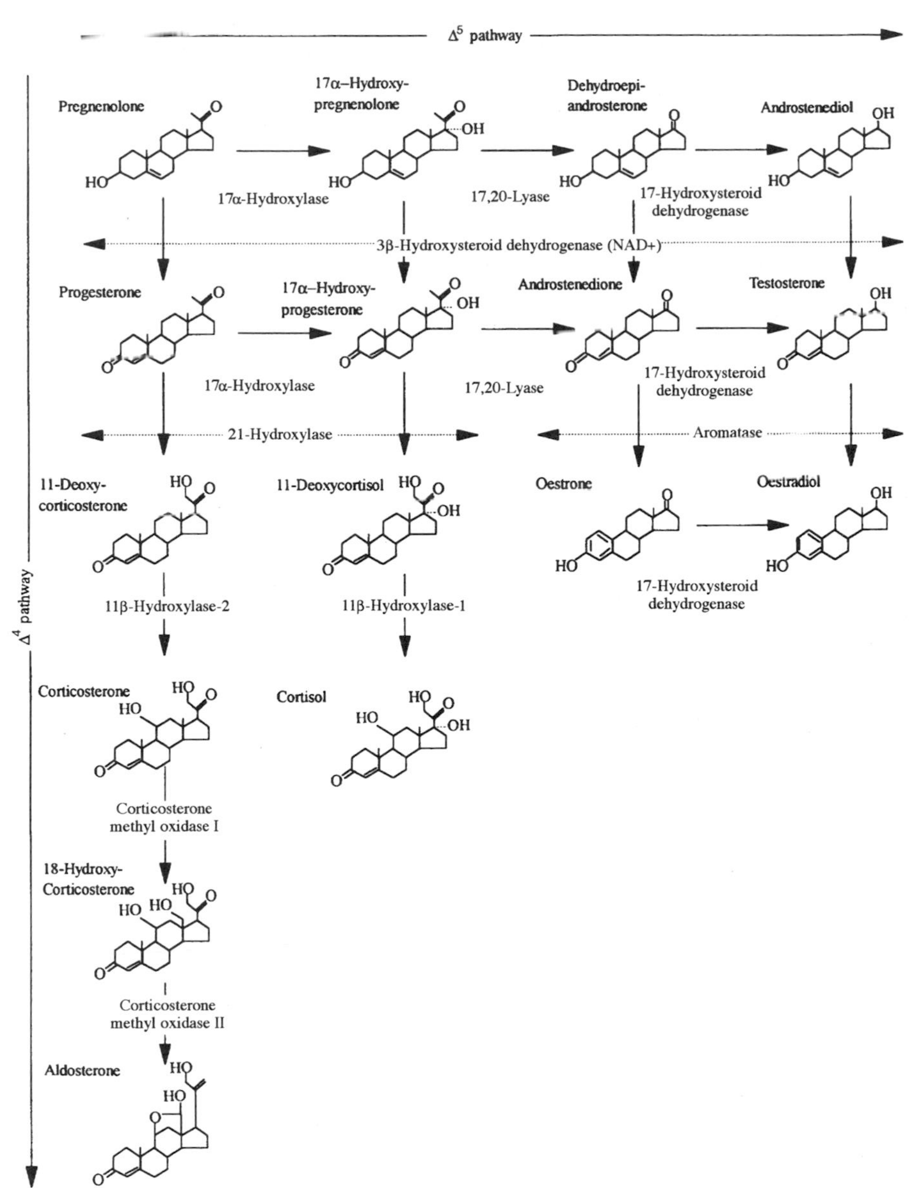

Figure 28.14 **The pathways of steroidogenesis.** Note that the diagram is a composite and does not represent any one steroidogenic tissue. The final three reactions leading to the formation of aldosterone are catalysed by a single, multifunctional enzyme. The reactions ascribed to 17α-hydroxylase and 17,20-lyase are catalysed by a single bifunctional enzyme

dehydroepiandrosterone is the Δ^5-pathway, while for androstenedione both routes are used. Much of the former steroid, which is a major product of the adrenal cortex, is sulphated in the 21-position prior to its release into the circulation. If it is used as a precursor for oestrogen synthesis, for example in postmenopausal women, it must undergo desulphation.

Formation of testosterone and 5α-dihydrotestosterone

The testicular androgens are formed in the Leydig cells of the testis where both pathways to testosterone are employed. There are grounds, however, for regarding testosterone as a prohormone in some tissues because in these it undergoes a reaction, catalysed by a NADPH-dependent 5α-reductase, to yield 5α-dihydrotestosterone. This metabolite binds more avidly to the androgen receptor than does its precursor. Not all target tissues, however, contain this enzyme.

Summary of steroid biosynthesis

The biosynthesis of the steroid hormones involves many hydroxylation reactions and all are catalysed by cytochrome P-450-dependent enzyme complexes. With the exception of the initial cleavage of the cholesterol side-chain, the hydroxylation at 11 and the oxidation at 18 by mitochondrial enzymes, all the other reactions take place on the endoplasmic reticulum. The only other type of reaction that occurs is that of NAD^+-dependent dehydrogenation, for which there are two distinct enzymes of the endoplasmic reticulum: 3β-hydroxysteroid dehydrogenase and 17α-hydroxysteroid dehydrogenase.

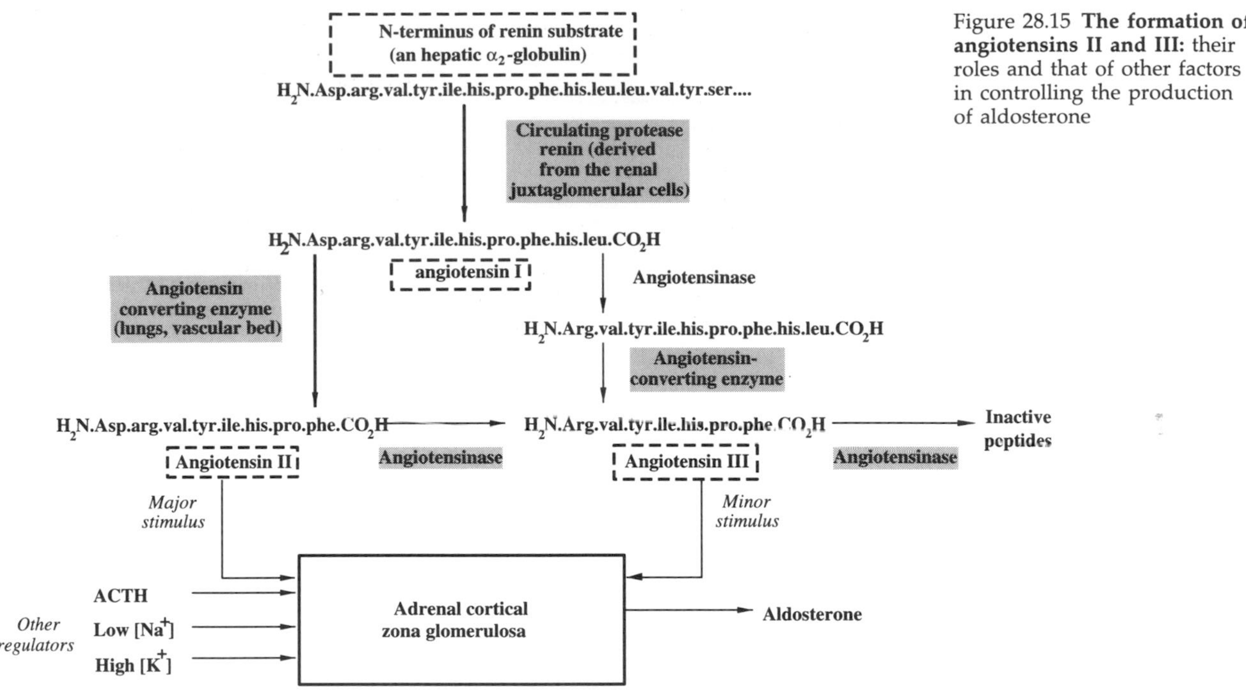

Figure 28.15 **The formation of angiotensins II and III:** their roles and that of other factors in controlling the production of aldosterone

Control of steroidogenesis

The actions of LH and FSH on the gonads resemble those of ACTH on the adrenal cortex described in Section 28.5. That is, they all bind to G-protein-coupled receptors to activate adenylyl cyclase. The effect of the cAMP produced is then to increase the provision of cholesterol for the mitochondrial cytochrome P-450_{scc}. In the ovaries, in which the task of forming oestrogens is divided between the cells of the theca interna and those of the granulosa, LH stimulates the production of androgens in the former cells, while FSH hastens the conversion of these precursors to oestrogens.

The biosynthesis of aldosterone, on the other hand, is controlled predominantly by the hormone **angiotensin II**. This is an octapeptide generated in the circulation in a two-step reaction. In the first, the proteolytic enzyme renin, which is released from the kidney, acts on an hepatic α_2-globulin to form the decapeptide angiotensin I. In turn, this serves as a substrate for an angiotensin-converting enzyme found in the lungs and in the vascular bed. This C-terminal peptidyl dipeptidase gives rise to the octapeptide angiotensin II. The biosynthesis of angiotensin II and its subsequent inactivation by angiotensinase are shown in Figure 28.15. In the adrenal zona glomerulosa the receptor for angiotensin II (which responds weakly to angiotensin III) is a G-protein-coupled to phospholipase C, i.e. IP_3 is the second messenger.

Figure 28.15 also indicates that the biosynthesis of aldosterone may be controlled by ACTH, circulating K^+ or Na^+.

Plasma transport of the steroid hormones

The transport of these lipid-soluble molecules closely resembles that of the thyroid hormones: there are specific binding proteins of high affinity and low capacity and non-specific binders (especially albumin) of high capacity and low affinity. The principal specific binding proteins are cortisol-binding globulin (CBG) and sex hormone-binding globulin (SHBG). The distribution of the steroid hormones between the various circulating binding proteins is shown in Table 28.4. It is noteworthy that neither aldosterone nor the adrenal androgen dehydroepiandrosterone (which circulates largely as a sulphate ester, DHEA-SO_4) has a specific binding protein. One consequence is that, although the total circulating concentration of cortisol exceeds that of aldosterone by a factor of 800, the difference in the free (effective) concentrations is reduced to 100.

Actions of the steroid hormones

Oestrogens

The name **oestrogen** was introduced to describe compounds able to induce oestrus in immature female rats. In humans, oestradiol is the most, and oestriol the least, potent oestrogen.

The biological effects of the oestrogens are:

1. Promotion of the development of female secondary sexual characteristics.
2. Initiation of endometrial and vaginal mitosis.
3. Normal negative-feedback inhibition of LH/FSH secretion changing to facilitation at mid-cycle.
4. Stimulation of ductal growth in the breast.
5. Helping epiphyseal closure.

Progesterone

Progesterone acts in the luteal phase to promote uterine mucosal proliferation prior to implantation of the fertilized ovum.

Table 28.4 **Circulating steroid hormone binding proteins**

Hormone	*Per cent hormone bound to:*			*Total concn* (nM)	*Free concn* (nM)
	CBG	SHBG	ALB		
Cortisol	80	0	14	400	24
Progesterone	86	0	12	1	0.02
Testosterone	0	60	38	15	0.3
Aldosterone	0	0	60	0.5	0.2
$DHEA\text{-}SO_4$	0	0	90	3900	390
Oestrogens	0	Occurs, but complex			

Cortisol-binding globulin (CBG) } Low capacity, high affinity
Sex-hormone binding globulin (SHBG) }
Albumin (ALB) – High capacity, low affinity

In pregnancy progesterone:

1. Promotes placental development.
2. Promotes mammary growth.
3. Has an anaesthetic effect on the myometrium.

Testosterone

Testosterone itself:

1. Acts with FSH to promote spermatogenesis.
2. Stimulates the development of male internal ducts (epididymis, vas deferens).
3. Promotes hypertrophy of the larynx.
4. Promotes a positive nitrogen balance (anabolic effect): increased size and mass of the heart, kidney, skeletal muscle and bones.
5. Helps to promote epiphyseal closure.
6. Exerts a negative-feedback inhibitory effect on LH secretion.
7. Exerts behavioural effect, e.g. promotes aggressiveness.

5α-Dihydrotestosterone

This product of testosterone reduction acts to:

1. Stimulate the development of male secondary reproductive organs: prostate (with prolactin), seminal vesicles, scrotum and penis.
2. Fosters the emergence of male secondary characteristics: increased facial hair, recession of the temporal hairline, diamond-shaped pubic escutcheon, sebaceous gland hypertrophy (acne).

Dehydroepiandrosterone

This adrenal androgen:

1. Is involved in the development of libido.
2. Promotes the growth of axillary hair.
3. In postmenopausal women acts as a precursor for oestrogen synthesis by adipose tissue and skin. This is the sole endogenous source of oestrogens in postmenopausal women.

Cortisol

For many years cortisol has been referred to as a 'stress hormone' which is unfortunate because there are many grounds for thinking of it as just the opposite; an **'anti-stress hormone**. This recent reappraisal of the functions of cortisol is due to the American physiologist Alvin Munck, and it has had the great virtue of harmonizing the well-established negative-feedback inhibitory effect of the hormone with its other actions. The essence of its action in the feedback context is that it acts to limit the secretion of its own controlling hormones so that hypersecretion does not occur. (When this feedback signal is not correctly transduced, the life-threatening Cushing's disease is seen.) In Munck's view, a whole series of powerful stimulatory agents are secreted in a wide range of stressful settings and if these are not subject to adequate control the effects may well be disastrous. It is precisely this type of effector hypersecretion that cortisol acts to prevent. Figure 28.16 shows the range of responses to different stresses subject to control by cortisol.

The glucocorticoids limit:

1. The immune response to infection by inhibiting the production of mediators such as interleukin-2 and also its receptor.
2. The inflammatory response by promoting the inhibition of the formation of pro-inflammatory lipid mediators derived from arachidonic acid.
3. The hormonal responses to fluid loss of the type that might arise following severe haemorrhage.
4. The hypersecretion of neurotransmitters in response to brain insults. These effects are likely closely to parallel the action of cortisol in inhibiting the secretion of CRH.
5. The hypoglycaemic effects of insulin.

Aldosterone

This very potent hormone acts on the renal tubules to promote the retention of Na^+ and the increased excretion of K^+.

As has been noted, the effective circulating concentration of cortisol exceeds that of aldosterone by a factor of about 100. Despite this, it is the latter rather than the former hormone that is the more important

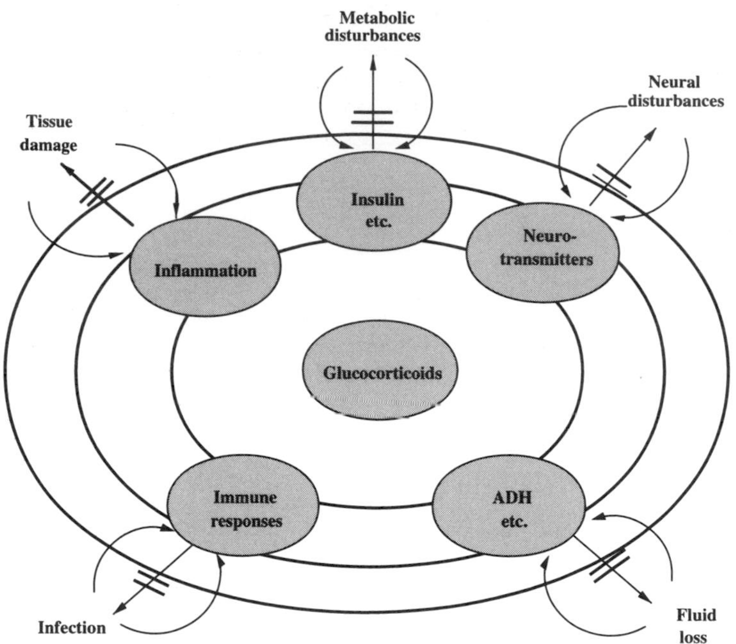

Figure 28.16 **Glucocorticoids act to prevent over-responses in the various protective systems of the body**

mineralocorticoid. This was surprising because both hormones bind to the mineralocorticoid receptor with about the same affinity. It is now appreciated that tissues, such as the kidney, that are selectively responsive to aldosterone have a very active NAD^+-dependent 11β-hydroxysteroid dehydrogenase that converts cortisol into cortisone and the latter has a very low affinity for the mineralocorticoid receptor. People who eat excessive amounts of liquorice may therefore be prone to hypertension because liquorice contains the 11β-hydroxysteroid dehydrogenase inhibitor glycyrrhizic acid (see Chapter 26).

Inactivation and elimination of steroid hormones

With such potently active molecules, an important aspect of their regulation lies in their inactivation and excretion. The chief route for all the steroid hormones having the Δ^4, 3-oxo-structure is via reduction at this point, in a reaction occurring in the liver. The carbon–carbon double bond can be reduced with the added hydrogen at position 5 in either the α- or the β-configuration. The latter reduction invariably results in total loss of biological activity, while some 5α metabolites retain residual activity. Other oxo-groups, e.g. that in position 20 of cortisol, may also be reduced. Frequently, further modification ensues, with the newly-formed 3β-hydroxyl group undergoing conjugation with glucuronic acid (the major route in humans) or sulphuric acid, much as drugs undergo 'phase II' metabolism (see Chapter 36). The conjugates are rapidly excreted in the urine.

Dehydroepiandrosterone sulphate may be excreted without further metabolism, while all three oestrogens are conjugated with either glucuronic or sulphuric acid.

28.11 Disorders of steroid hormone production and action

Three broad categories of steroid-related disorder are seen, associated with :

1. Inappropriate secretion.
2. Defects in biosynthetic pathways (which may well lead to inappropriate secretion).
3. Defects in end-organ sensitivity.

Excessive or deficient production of glucocorticoids or mineralocorticoids

Any condition leading to hyper- or hyposecretion of glucocorticoids is life-threatening, with the former being more difficult to manage clinically (replacement therapy usually suffices in the latter). In addition, hypersecretion of aldosterone, or other steroids with mineralocorticoid activity (e.g. deoxycorticosterone), can lead to serious hypertension.

Cushing's disease

Hypersecretion of cortisol may result from tumours of the adrenal gland in which steroid secretion is autonomous, or tumours of the pituitary gland (rarely of the hypothalamus), when the elevated plasma cortisol concentrations are secondary to elevated ACTH. In some cases, plasma ACTH (and hence cortisol) levels are high due to ectopic formation of the peptide hormone by tumours, e.g. oat-cell carcinomas of the lungs. If the cause of the disease is pituitary or ectopic ACTH production, the condition is known as Cushing's disease after Harvey Cushing the American surgeon who first described the condition early in this century. The condition arising due to autonomous secretion by the adrenal gland is usually referred to as Cushing's syndrome (although the names tend to be used interchangeably). The symptoms of both the disease and the syndrome (which may also be seen in alcoholic patients) are those to be expected in the face of excessive cortisol production and mainly focus on the metabolic effects of the hormone. Thus there is deranged carbohydrate and lipid metabolism resulting in 'steroid diabetes' and abnormal lipid storage. Oedema of the lower limbs and severe osteoporosis are also seen.

Addison's disease

The destruction of the adrenal gland leading to deficient production of cortisol was first described by Thomas Addison, then (in the 1820s) a physician of the United Hospitals of the Borough (Guy's and St Thomas's) and one of the 'famous triumvirate' of Bright, Addison and Hodgkin. A characteristic of the disease which bears Addison's name, when seen in Caucasian patients, is a bronzing of the skin due to excessive ACTH secretion. In high concentrations ACTH is able to activate MSH receptors in the skin. This may be detected in all patients by examination of the buccal cavity. In addition there is tiredness, hypotension, hypoglycaemia, gastrointestinal upset and depression. This latter observation also points to a role for cortisol in brain function (see Figure 28.16).

Hyperaldosteronism

The condition of **primary hyperaldosteronism** was first described by Conn. In most cases the disease arises due to a single autonomously-secreting adrenal cortical adenoma. Surgical treatment by adrenalectomy of patients with this condition almost always results in a cure both for their hypokalaemia and their hypertension.

Glucocorticoid-suppressible hyperaldosteronism is an apparently anomalous disease which presents as an autosomal-dominant form of inherited hypertension. It is characterized by the hypersecretion of aldosterone in the face of suppressed plasma levels of renin and angiotensin. A clue to the anomaly arises in that there is hypersecretion of 18-hydroxycortisol and 18-oxocortisol. The presence of these metabolites, and the fact that the condition is controlled by the administration of the synthetic glucocorticoid dexamethasone, strongly favours the idea that this is a disease of the zona fasciculata. It is now established that CYP11B2 activity (see Section 28.10) is conferred on this zone by a hybrid CYP11B1/CYP11B2 gene that is presumed to arise as a result of unequal crossing over at meiosis involving these closely-associated genes. This gene retains the CYP11B2-associated corticosterone methyloxidase activities necessary to form aldosterone while retaining the susceptibility of CYP11B1 to regulation by ACTH. This means that the formation of aldosterone from zona fasciculata corticosterone is prevented by dexamethasone as it acts in a negative-feedback fashion to inhibit the secretion of ACTH.

Congenital adrenal hyperplasia (CAH)

A number of diseases are ascribable to inborn errors of steroid biosynthesis. The nature and severity of the disease depends upon which biosynthetic enzyme is defective and whether the loss of activity is total or partial.

21-Hydroxylase defects

A deficiency in this enzyme constitutes the most common type of congenital adrenal hyperplasia and the defect can present in one of two ways: in 'classic' or 'late onset' forms. In the classic disease the symptoms are present *in utero*, whereas in the late onset form, symptoms arise only after puberty (although the enzyme defect is demonstrable before the symptoms). The incidence of the classic disease is 1 in 67 000, although in certain closed populations, such as the Yup'ik Eskimos of Alaska, it is about 1 in 500. About one-third of the cases are severe and characterized by salt loss. In these patients the enzymes of both the zona glomerulosa and zona fasciculata are affected, i.e. the synthesis of aldosterone as well as cortisol is defective. In the remaining cases it is only the enzyme required for cortisol biosynthesis that is deficient and hence salt loss is not seen. The consequent reduction in cortisol secretion, in all cases leading to a lack of negative feedback action, results in an increase in the output of ACTH and other peptides derived from POMC (see Section 28.5). This has two major effects. First, the POMC peptides act trophically on the adrenal cortex to cause hyperplasia. Secondly, the ACTH stimulates excessive production of steroids further back in the metabolic pathway; in particular large amounts of the adrenal androgens and 17α-hydroxyprogesterone are secreted (see Figure 28.14). The latter is reduced in the liver to pregnanetriol (Figure 28.17) the excretion of which in the urine as a glucuronide conjugate is a good diagnostic test for the disease (the levels are also elevated in previously undiagnosed late-onset subjects).

In most affected fetuses, the effects arise between weeks 12 and 20 of gestation, so that female development is arrested by adrenal androgens before the urogenital sinus is fully differentiated. Consequently the

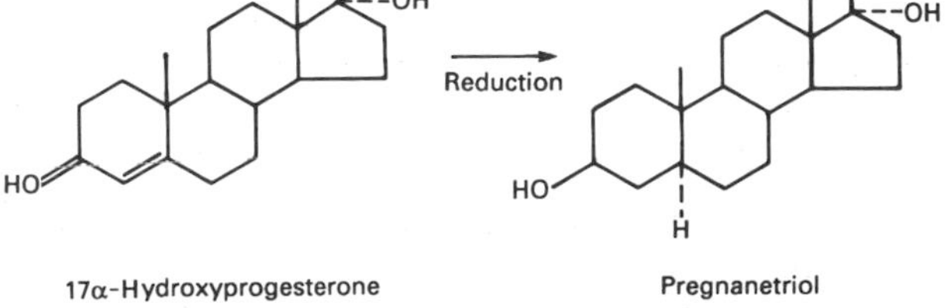

Figure 28.17 **The conversion of 17α-hydroxyprogesterone into pregnanetriol**

Table 28.5 **A summary of the defects in steroid hormone biosynthesis and their consequences**

	Enzyme defect or protein				
Consequence	*21-hydroxylase*	*11β-hydroxylase*	*3β-OH steroid dehydrogenase*	*17α-hydroxylase*	*Steroidogenic acute regulatory protein*
Salt loss	33% of cases	–	√	–	√
Hypertension	–	√	–	√	–
Virilism	√	√	√	–	–
External genitalia	♂	♂	Vagina and ♂ urinary tract	♀	♀
Oestrogen production	↑	↑	↓	↓	↓
Androgen production	↑	↑	↑↓	↓	↓
Glands affected	Adrenals	Adrenals	Adrenals/gonads	Adrenals/gonads	Adrenals/gonads
Late-onset form	√	√	–	–	–
Percentage of total cases seen	95	4	1		

external genitalia at birth are male. As indicated, patients who are salt losers synthesize neither cortisol nor aldosterone.

It is now established that in virtually all the cases of the 'classic' disease the mutations result from recombination between the normal gene CYP21 and the non-functional pseudogene CYP21P. Unequal crossing-over between these two genes in meiosis results in 20% of cases in total deletion of the CYP21 gene. In the rest of cases, non-functional elements of CYP21P are recombined with parts of CYP21.

Other enzyme defects

Only those defects that affect the biosynthetic pathway to cortisol later than the separation of androgen and glucocorticoid synthesis are virilizing, i.e. blocks at the levels of 21- and 11β-hydroxylases and 3β-hydroxysteroid dehydrogenase. Earlier blocks in the pathway, in addition, interfere with androgen and oestrogen biosynthesis. Thus in the case of cholesterol 20,22-lyase or 17α-hydroxylase/17,20-lyase deficiency, the external genitalia are female regardless of genotype (due to a failure to form androgens). If there is an 11β-hydroxylase defect due to a mutation to CYP11B2, hypertension is a result because the high circulating concentration of 11-deoxycorticosterone has a mineralocorticoid action (Na^+ retaining). Likewise a 17α-hydroxylase deficiency results in hypertension and hypokalaemia, because the corticosterone that accumulates is more potent than cortisol in promoting Na^+ retention. Recently, patients presenting with **congenital lipoid adrenal hyperplasia** have been shown to have mutations to steroidogenic acute regulatory protein which is responsible for promoting the transfer of cholesterol into mitochondria. A summary of some of these defects is given in Table 28.5.

Defects affecting steroid hormone receptors

Testicular feminization syndrome

The best-characterized disease associated with a steroid hormone receptor defect is called **testicular feminization syndrome**. Genetic males with this condition have testes and secrete testosterone, but have complete feminization of the external genitalia. In the cases that have been investigated to date, several point mutations to the androgen receptor gene have been identified; all of them have been found to involve changes in the exon that encodes the steroid-binding domain of the protein. Some of these result in complete failure to bind androgens, while others lead to low-affinity binding.

Oestrogens and breast cancer

Excessive exposure of the female breast to the mitogenic actions of the oestrogens is believed to underlie the development of female breast cancer, and the drug tamoxifen was undergoing clinical trials as a preventative agent. Tamoxifen is a competitive inhibitor of the oestrogens in their binding to their receptor, i.e. an antagonist (see Chapter 37).

Ectopic Cushing's disease

It has already been mentioned that patients with oat-cell carcinomas of the lungs may secrete large quantities of peptides related to ACTH. This leads to the hypersecretion of cortisol and the development of Cushing's disease. The hypersecretion of ACTH-related peptides from the tumour is not subject to negative-feedback inhibition by cortisol because the alleles for the glucocorticoid receptor in the tumour have been mutated in such a way that either the protein is not produced or, if formed, does not bind the steroid.

Chapter 29

Bone

29.1 Introduction

Soft tissues of vertebrates probably started to become hardened with minerals about 400 million years ago. Before that, animals had backbones consisting of cartilage and muscle, such as found today in sharks and lampreys.

The main mineral of the bones of higher animals is calcium phosphate. In many invertebrates it is calcium carbonate, and millions of generations of early invertebrates are responsible for the formation of white cliffs clearly visible along many coastlines.

Although bone consists largely of mineral material, it is a living tissue, and the calcium and phosphate found in bone are in dynamic equilibrium with the calcium and phosphate in plasma (see Chapter 12). Bone also contains 70% of the body's citrate, 60% of the total magnesium and 50% of the sodium.

Bone mineral is not a random precipitate, but part of complex subcellular structures. **Calcification**, the process by which organic tissue becomes hardened by deposits of calcium salts, is a very important physiological event during bone growth and also during repair after damage to the skeleton. Calcification can also occur, in certain pathological conditions, in soft tissues, such as the kidneys, the heart or the arteries, where it can cause serious damage; therefore an understanding of this process is of great value in medicine.

29.2 Bone structure

Bone consists of mineral salts deposited on an organic matrix. The mineral particles consist mainly of **calcium phosphate**. The matrix consists of proteins, of which **collagen** is the most important, proteoglycans and a smaller fraction of lipids and water. In general, bones are compact on the outside and cancellous (spongy) inside. Compact bone is not completely solid. It is permeated by Haversian canals containing blood vessels and bone cells.

The mineral particles are generally in the form of long cylinders, about 5 nm wide overlaid in the direction of the collagen fibres. Some cylinders aggregate to form thicker structures approximately 20 nm wide. Lamellar, trabecular and Haversian structures can be distinguished in different parts of bone (Figure 29.1).

Three types of cells have been described in bone: osteoblasts, osteoclasts and osteocytes. The **osteoblasts** are involved in the formation of the matrix of new bone. They are migratory cells which synthesize webs of collagen fibres and proteoglycans which are eventually calcified. The newly-formed matrix is referred to as **osteoid** tissue. It is gradually permeated by **osteocytes**, which are cells derived from osteoblasts and which possess long filamentous projections interlacing the osteoid tissue. The **osteoclasts** are large multinucleated cells, largely concentrated on the surface of bone. Their main function seems to be resorption (bone loss) and the release of calcium and phosphate into the plasma.

Figure 29.2 shows a diagrammatic representation of the cells which make up the structure of bone.

29.3 Bone mineral

Bone mineral is so closely associated with the organic matrix, especially with the collagen component, that it is difficult to prepare in a form free of contamination for study. In general, 19–26% consists of calcium, 9–12% of phosphates and 2–4% of carbonate. Some bones contain small amounts of magnesium (0.2–0.3%). The mineral is composed of narrow crystallites about 5 nm in diameter but of variable length, about 60–70 nm, arranged parallel to the collagen fibres. The crystal structure is similar to, but not identical with naturally occurring hydroxyapatite ($Ca_{10}(PO_4)_6(OH)_2$). Bone mineral has a higher content of carbonate and some phosphate may be replaced by sulphate or silicates. In dental enamel and dentine, up to 2% of the hydroxide ions may be replaced by fluoride ions.

As a result of this non-stoichiometry, bone mineral is poorly crystalline. The size of bone crystals varies in different tissues. In bone and dentine, the crystals are small. Large crystals of calcium phosphate can be found on external surfaces, such as dental enamel, and non-crystalline deposits also occur in tissues other than bone, often within subcellular organelles, such as mitochondria. Calcium phosphate is relatively insoluble in water and precipitation of this salt can occur spontaneously if the concentrations of calcium, phosphate, or both, cause the 'solubility product' of calcium phosphate to be exceeded. The relationship between the formation of calcium phosphate and its constituent ions can be represented by the following equation:

$$3Ca^{2+} + 2PO_4^{3-} \underset{K_{sp}}{\leftrightarrow} Ca_3(PO_4)^2$$

The solubility product for calcium phosphate is

$$K_{sp} = [Ca^{2+}]^3[PO_4^{3-}]^2$$

It is pH dependent, and at pH 7.0 its value is 25. The principle of the 'solubility product' is important in the regulation of bone mineralization, but it is not the only

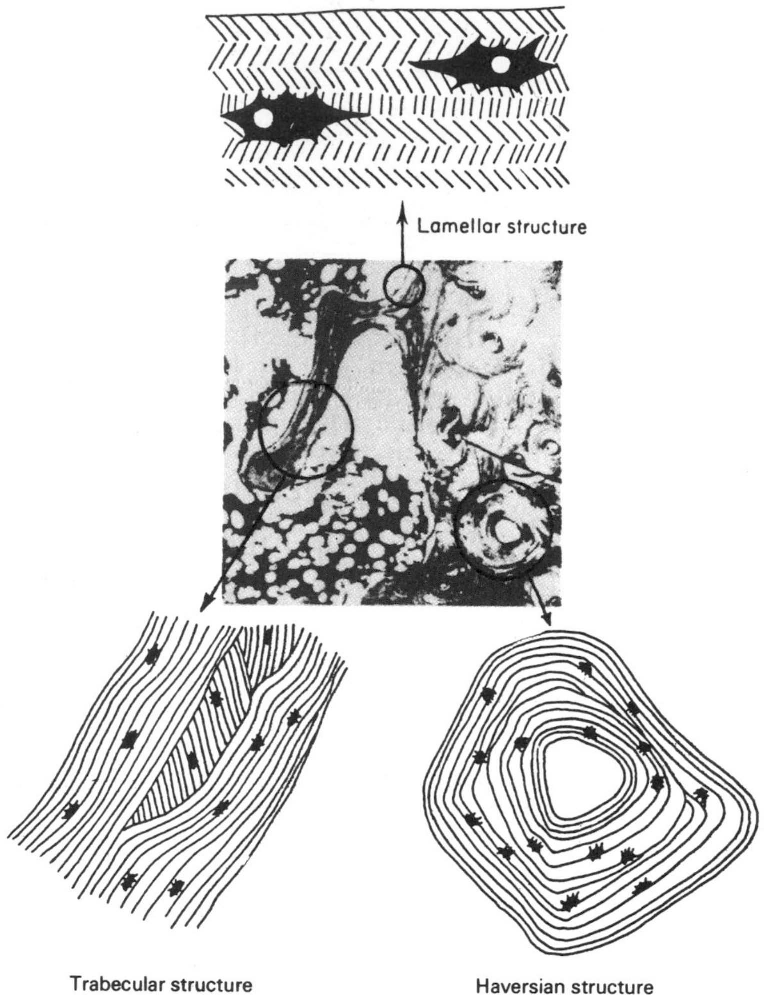

Figure 29.1 **Structures of bone**

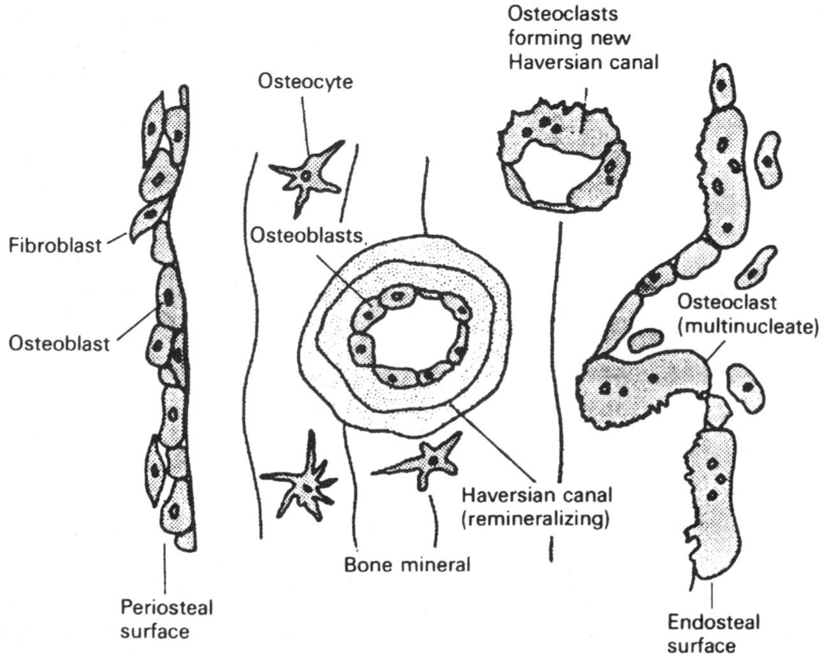

Figure 29.2 **Cells in bone structure**

factor on which this process depends. Calcium and phosphate ion concentrations in the aqueous medium may be altered by binding to proteins, and complex localized concentration changes may occur as a result of metabolic processes. Furthermore, the 'micro-environmental' concentrations of calcium and phosphate in biological systems may not be identical to those in the bulk of the solution.

29.4 Bone matrix: collagens

Collagens are the most abundant proteins in multicellular animals, including mammals. They constitute 25% of all vertebrate protein, 30% of the dry weight of bone and 50% of the dry weight of cartilage. Collagens have a structural function in most tissues and several types are usually found associated together in the same tissue, in different proportions. They are synthesized intracellularly but polymerize and mature extracellularly and associate with other components of the extracellular matrix and with minerals to form connective tissues.

Collagens form a family with at least 19 members. Strictly speaking, collagen is any protein with three polypeptide chains, each chain containing at least one stretch of repeating amino acid sequence (Gly–X–Y)$_n$, where X and Y can be any amino acid, but often are proline and hydroxyproline, respectively. The chains may, or may not, be identical and they combine to form a triple helix.

Collagens are classified on the basis of their size and mode of self-assembly as fibrillar or non-fibrillar. **Fibrillar** collagens include types I, II, III, V and XI and occur in almost all connective tissues.

Type I is the commonest collagen, forming about 90% of collagen in the human body. It is found in bone, including embryonic bone, tendon, dentine, lung, skin and muscle. Type II is found mainly in cartilage and in vitreous humour. Type III is often found associated with type I in extensible tissues such as the lung, fetal skin and the vascular system.

Non-fibrillar collagens form a diverse group and can be subdivided into **basement membrane** collagens (types IV and VII), found mainly in the extracellular matrix, **short-chain** collagens (VIII and X), found mainly in vascular and corneal endothelium, and **fibril-associated** collagens (IX, XII and XIV), found attached to existing fibrillar collagens. Collagen XIII has so far only been identified at the cDNA level. Not all proteins with the sequence Gly–X–Y are called collagens, however. The 'non-collagen' collagens include the complement protein C1$_q$, surfactant protein SP-A, acetylcholinesterase, mannose-binding lectins and conglutinin.

As mentioned above, the main type of collagen found in bone is type I. It is a heterotrimer of two α_1(I) and one α_2(I) chain. Each chain contains about 1050 amino acid residues and the molecule is a cylinder 1.4 nm in diameter and 300 nm in length (Figure 29.3).

The regular repeat of glycine is essential to the structure, because it is the only amino acid small enough to fit in the central core of the molecule. This allows the formation of a triple helix with three residues per turn which is less tight than the α-helix which has 3.6 residues per turn. The rise per residue along the collagen

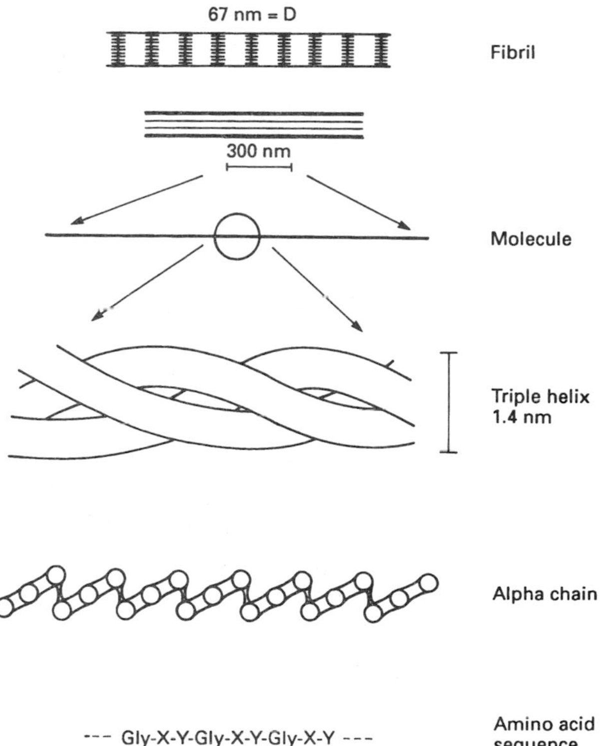

Figure 29.3 **Main features of the collagen molecule and its relation to the structure of the fibril**

helix is 0.32 nm compared to 0.15 nm for the α-helix. Each helix is left-handed and the triple helix is right-handed. This system of opposing twists gives the resulting molecule great strength. The ends of the molecule are not, however, in triple helical form, and they lack glycine. Proline and hydroxyproline together account for about 25% of the total amino acid residues. A high content of hydroxyproline is unusual in proteins, but it is needed for triple helix stability. About 90 hydroxyproline residues are needed per chain to preserve the triple helix at body temperature. If the formation of hydroxyproline is impaired, unstable collagen is formed in which the chains tend not to wind round each other. The presence of hydroxylysine residues is another feature seen in collagen but absent from most other proteins. It is essential for glycosylation of the chain, which in turn may affect water retention and cross-link formation. The carbohydrate content of different types of collagen is variable.

29.5 Bone matrix: non-collagen, calcium-binding proteins

A number of calcium-binding bone matrix glycoproteins have recently been characterized. They include osteopontin, osteonectin, osteocalcin, bone Gla protein and bone sialoprotein. They share a common characteristic feature, that of being anionic in nature. The negative charge arises in various ways, including the

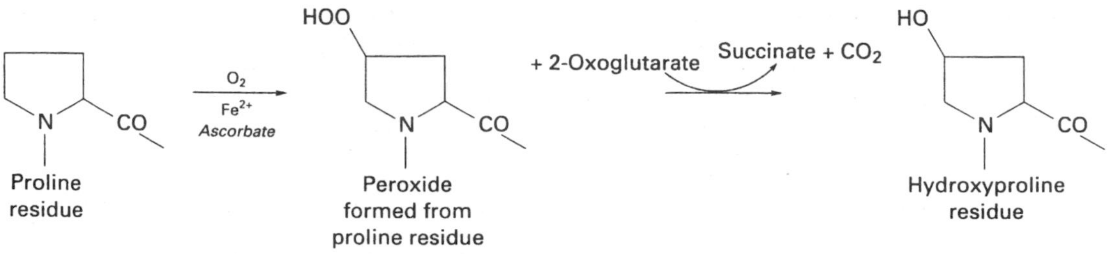

HOO

+ 2-Oxoglutarate Succinate + CO_2

HO

O_2 Fe^{2+} Ascorbate

N CO N CO N CO

Proline residue

Peroxide formed from proline residue

Hydroxyproline residue

Figure 29.4 **Mechanism of hydroxylation to form 4-hydroxyprotine**

presence of large numbers of aspartic acid residues (osteopontin), glutamic acid residues (bone sialoprotein), γ-carboxyglutamic acid residues (osteocalcin and matrix Gla protein), sialic acid side-chains (osteopontin and bone sialoprotein), or phosphorylated serine and sulphated tyrosine residues. This overall negative charge produces molecules with a high calcium-binding potential and they have, therefore, been implicated in bone calcification. No specific function has, as yet, been ascribed to any of them.

Osteopontin (RMM 32 600 kD) is known to bind hydroxyapatite and it appears to be associated with the attachment of osteoclasts to the matrix. Its synthesis is stimulated by vitamin D (see Chapter 12).

Osteonectin (RMM 35 000 kD) binds collagen and hydroxyapatite and its synthesis seems to be associated with bone formation and remodelling.

Osteocalcin (MW 11 000) contains two or three γ-carboxyglutamic acid residues at the centre of the molecule. Its synthesis is stimulated by vitamin D and the carboxylation is vitamin K dependent. It also binds to hydroxyapatite and may function in the regulation of crystal size. It is of diagnostic value as a marker of bone metabolism, as its production and levels in the blood reflect osteoblastic activity.

Matrix Gla protein (MW 12 000) is also synthesized in response to vitamin D and vitamin K and its function is obscure.

29.6 Collagen biosynthesis

Bone collagen is formed and secreted by chondrocytes and osteoblasts. All collagens, whether fibrillar or non-fibrillar, are synthesized on ribosomes on the rough endoplasmic reticulum (RER), with hydrophobic amino acid sequences at the N-terminus of each chain. These terminal sequences, known as 'signal peptides', are removed in the RER during translation of the polypeptide chain (see Chapter 7), and this is the first of a series of cotranslational or post-translational modifications necessary before functional collagen is formed.

The next modification step is the formation of hydroxyproline from proline residues after their incorporation into the peptide chain. Selective lysine residues are also hydroxylated. Three hydroxylases located in the RER catalyse these reactions. Two distinct proline hydroxylases convert proline into 4-

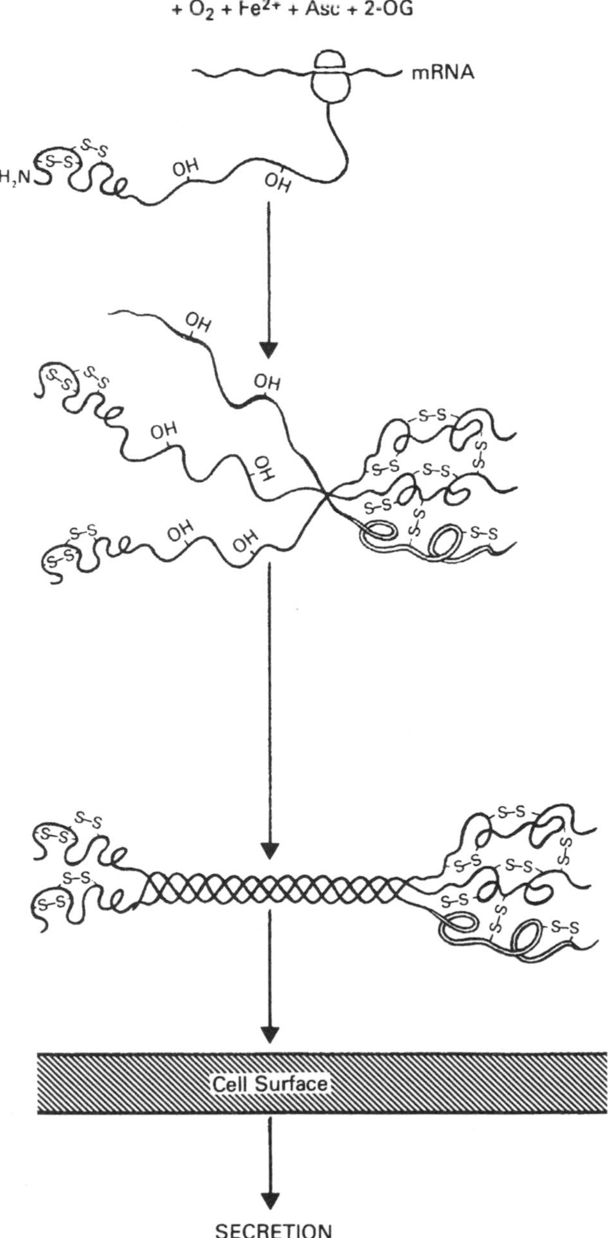

Figure 29.5 **Formation of the procollagen triple helix**

hydroxyproline (major form) or 3-hydroxyproline (minor form), and a lysine hydroxylase catalyses the formation of hydroxylysine. All three enzymes are mixed function oxidases requiring iron, ascorbic acid (vitamin C), molecular oxygen and 2-oxoglutarate which is oxidized to succinate. The formation of hydroxyproline is outlined in Figure 29.4. Ascorbic acid is needed to maintain the iron in the hydroxylases in the reduced (ferrous, Fe II) and active form. In vitamin C deficiency collagen is under-hydroxylated and the chains fail to form proper helices. They accumulate in the cells of scorbutic animals, they are secreted at a low rate and do not form functional collagens. (Vitamin C is discussed in Chapter 33.)

Glycosylation of selected hydroxylysine residues next takes place by the addition of galactosyl followed by glucosyl residues, catalysed by the appropriate transferases. Some asparagine residues are also glycosylated.

The three component chains of procollagen associate by the formation of inter-chain disulphide links at the non-helical carboxyl-terminal domain of the polypeptides (Figure 29.5). It is not clear whether this takes place while the chains are still attached to the ribosomes, or shortly after. The amino-terminal area contains only intra-chain disulphide links and does not contribute to the association of the three chains. The triple helix forms in a 'zipper' like manner from the C-end to the N-end.

Both proline and lysine hydroxylases and the gluco- and galactotransferases are only active on non-helical substrate. A delay in helix formation such as seen in some heritable collagen disorders results in over-modification of the polypeptide chains. As secretion depends on correct triple helix assembly, these mutant procollagen chains are not secreted but are degraded intracellularly. Normally, procollagen is assembled in the RER and transferred to the Golgi apparatus for packaging into vesicles for export.

Procollagen, in the form of cylindrical aggregates is secreted from the Golgi apparatus in vesicles The enzymes which subsequently act on fibrillar procollagen extracellularly to convert it to functional collagen are probably also packaged in the same vesicles for export. These are procollagen N-proteinases, procollagen C-proteinases and lysine oxidase. The N- and C-proteinases, as their name implies, catalyse the hydrolytic removal of the non-helical regions at the N- and C-terminals of the polypeptide chains (Figure 29.6) to produce tropocollagen. Mutations which affect the function of these proteinases lead to the production of abnormal collagens which do not align properly to form fibrillar collagens. Non-fibrillar collagens are not affected by these mutations, as they are normally not proteolytically processed and retain their non-helical domains.

Tropocollagen now becomes the substrate for reactions catalysed by lysine oxidase, a copper-containing enzyme. This results in the oxidative deamination of some lysine side-chains to produce the aldehyde form of lysine known as allysine. The acquisition of a reactive aldehyde group at various points along the length of the tropocollagen molecule allows the formation of cross-links with other tropocollagen molecules by Schiff's

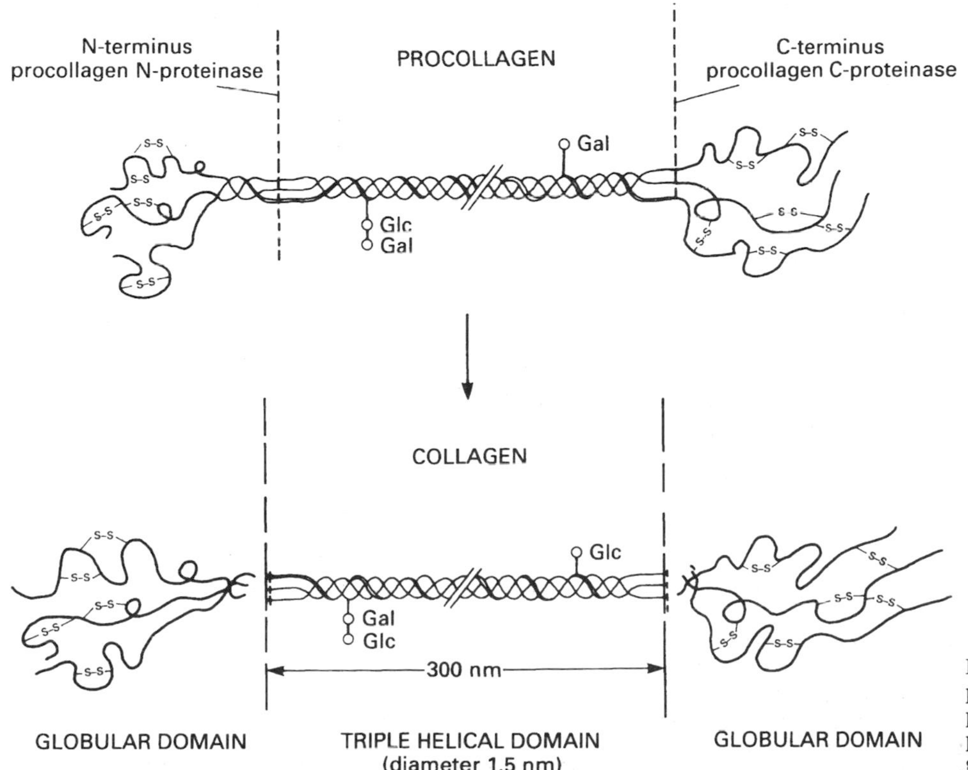

Figure 29.6 **The conversion of procollagen to collagen by procollagen N- and C-proteinases** (Glc, glucose; Gal, galactose

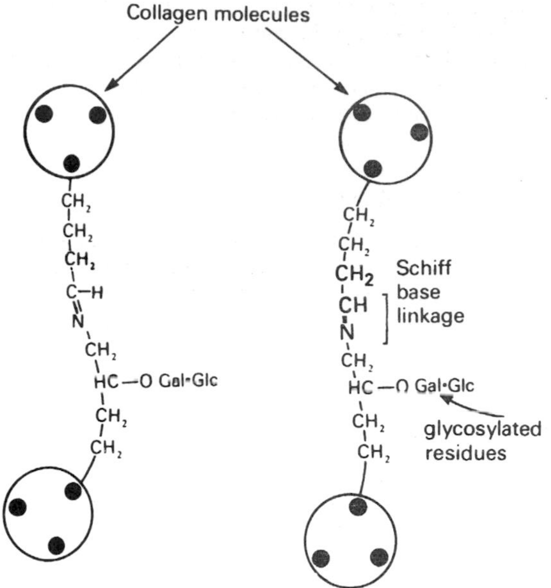

Figure 29.7 **Cross-linking of collagen chains.** Formation of Schiff base-type linkages between aldehyde groups of one chain and the lysyl amino groups of another chain

base-type linkages with other lysine side-chain amino groups. The number of links increases as mature collagen fibrils are formed with age (Figure 29.7).

The alignment of the collagen molecules within the fibrils gives rise to a characteristic pattern of striations (Figure 29.8), with a 64 nm repeat, which can be seen using electron microscopy. In bone, collagen fibres are arranged in layers which are orientated at right-angles to each other. This molecular arrangement gives bone great strength, as it allows it to withstand forces applied from different directions.

29.7 Collagen diseases

Abnormalities in collagen synthesis underlie many human diseases. The term 'collagen diseases' refers to a heterogeneous group of acute and chronic diseases affecting connective tissue. The original classification was based on the erroneous idea that 'collagen' was equivalent to 'connective tissue'. As knowledge about different types of collagen and other extracellular matrix proteins accumulated, the term 'collagen diseases' now refers to conditions where the primary defect is in the biosynthesis of collagen, its post-translational modifications, or its extracellular processing.

The original list of collagen disorders included Marfan's syndrome, the Ehlers–Danlos syndrome and osteogenesis imperfecta. The chondrodystrophies have now been added to the list as well as epidermolysis bullosa and some other less well-characterized disorders, whereas Marfan's syndrome has been removed, as it has been found to arise, not from a mutation to a collagen gene, but to a gene coding for **fibrillin**, an extracellular protein associated with elastin in elastic fibres. The word 'heritable' is often used to describe the collagen disorders, and is used to indicate that although they can be inherited, they can also arise by recent spontaneous mutations. Some collagen diseases are not caused by gene defects, however, but by nutrient deficiencies or the presence of enzyme inhibitors. These include vitamin C deficiency (scurvy), copper deficiency, and lathyrism.

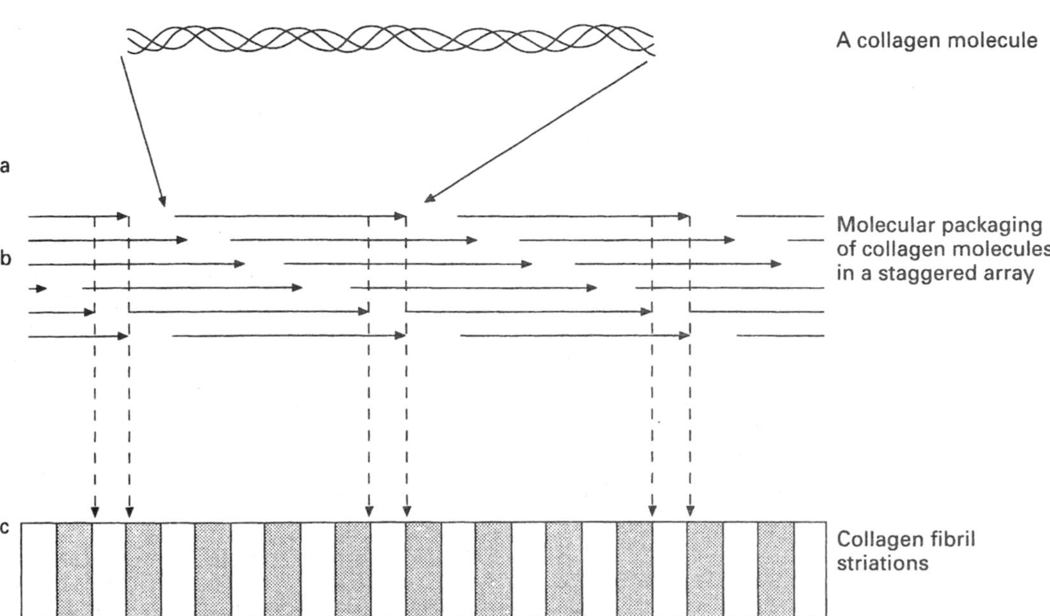

Figure 29.8 **The packaging of collagen molecules into fibrils**

Heritable disorders: osteogenesis imperfecta (brittle bone disease)

This is a collagen disorder resulting from defective biosynthesis of type I collagen, and is characterized by brittle, osteoporotic, deformed and easily fractured bone. It also presents with blue sclerae, loose joints, hearing loss and imperfect dentine formation. It affects about 1 in 10 000 individuals. The severity of the disease varies from mild to lethal. Over 80 different mutations have been identified to date in the genes coding for the two chains of procollagen I, namely COL1A1 and COL1A2. They are usually substitutions of glycine with other amino acids.

As osteogenesis imperfecta is usually inherited as an autosomal dominant condition, a mutation in only one allele can have devastating effects. Formation of the triple helix is impaired and the abnormal molecules are either degraded intracellularly, diminishing the type I collagen layer, or are secreted and interfere with the extracellular assembly of the fibres. There is a correlation between the position of the mutation in the type I molecule and the severity of the disease. Mutations near the C-terminus are lethal, and they have a progressively milder effect as the N-terminus is approached. This is thought to be because the triple helix begins to form at the C-terminus; therefore the effect of an abnormality on the correct formation of procollagen is greater at the C- rather than at the N-terminus.

Ehlers–Danlos syndrome (EDS)

This refers to a group of heterogeneous heritable disorders of connective tissue, varying in severity and transmitted as autosomal recessive, autosomal dominant, or X-linked recessive traits. They are characterized by hyperextensible skin and joints, a tendency to bruise, reduced wound-healing ability and visual defects mainly affecting the lens and cornea.

There are 10 types of EDS, classified on genetic, biochemical and clinical characteristics. The most serious and often lethal disorder is type IV, which can be caused by a variety of mutations in the genes coding for type III collagen. COL3A1 mutations affect the skin, which becomes hyperextensible but brittle, and result in weakness and rupture of arteries and intestines, through weakness of the supporting connective tissue.

EDS type III is associated with hypermobility of the joints.

EDS type VI is due to a defective lysine hydroxylase gene causing the formation of an unstable triple helix

In EDS type VII A and VII B the mutations are in collagen type I, whereas in type VII C, there is a loss of the procollagen N-proteinase cleavage site, resulting in improper processing of procollagen to collagen and, consequently, inability of the fibres to align properly.

In EDS type IX, there is decreased lysine oxidase activity.

Little is known about the genetic or biochemical defects associated with the rest of the EDS types.

Epidermolysis bullosa

This disease is characterized by severe blistering of the skin. There are three forms of this disorder: the simplex, junctional and dystrophic forms. The simplex form is the result of defects in keratin filaments, the junctional form is caused by defects in laminin 5, and the dystrophic form results from mutations of the gene coding for type VII collagen, an anchoring fibril collagen.

The chondrodystrophies

A number of these conditions are the result of mutations in the type II collagen gene (COL2A1) and others are due to defects in the genes for collagen type X or XI. They affect the correct formation and function of the growth plate cartilage.

Alport's syndrome

This disease mainly affects the structure of the kidney glomerular basement membrane (see Chapter 26) and is characterized by haematuria. The molecular basis of this disease is a number of mutations on the genes coding for type IV collagen, characteristically providing the mesh-like structure in basement membranes.

Environmental disorders: scurvy

Vitamin C is needed for maintaining the activity of proline and lysine hydroxylases, and in the absence of this vitamin, collagen is unstable, the skin bruises easily, wounds do not heal properly, gums bleed and gastrointestinal tract bleeding is common (see Chapter 33).

Copper deficiency

Copper deficiency leads to disruption and fragmentation of large arteries, owing to the production of collagen with few effective cross-links between molecules in a fibre. Primary dietary deficiency of copper is rare (see Chapter 35), and the disease is more commonly due to an inability to absorb copper from the gut. The enzyme lysine oxidase is copper dependent and its improper function compromises collagen strength and stability.

Lathyrism

Lathyrism was first described by Hippocrates who noted that eating the seeds of *Lathyrus odoratus* (sweet pea) gave rise to neurological symptoms. β-Cyanoalanine and β-aminoproprionitrile, present in these seeds, are competitive inhibitors of lysine oxidase, causing the production of collagen with defective cross-links.

29.8 Bone formation and growth

Bone formation, development and growth occurs by the coordinated activity of chondrocytes, osteoblasts and osteoclasts. Chondrocytes and osteoblasts are both derived from undifferentiated mesenchymal cells. They become committed to either the chondrogenic (cartilage-forming) or the osteogenic (bone-forming) line under the

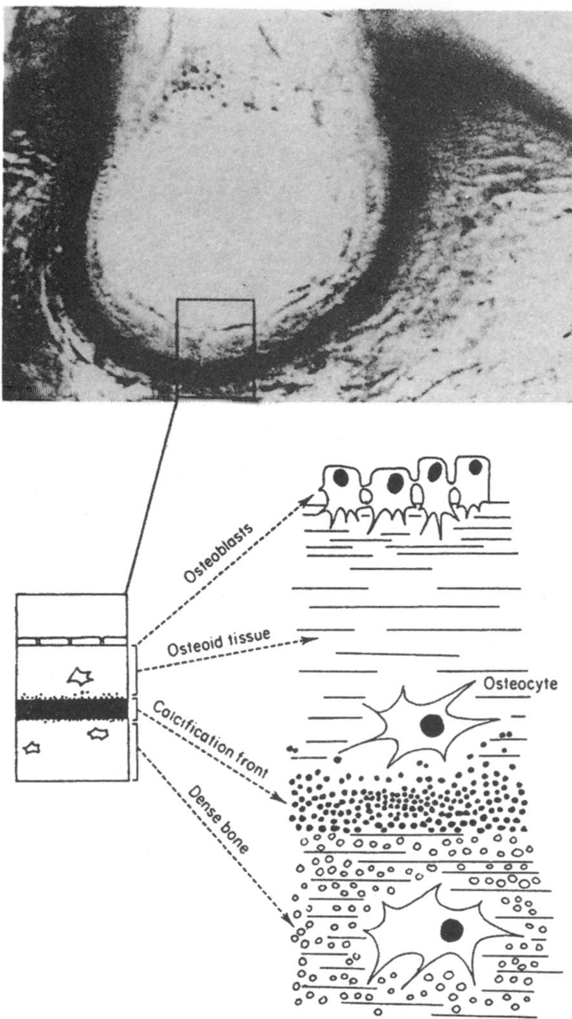

Figure 29.9 **Bone formation**

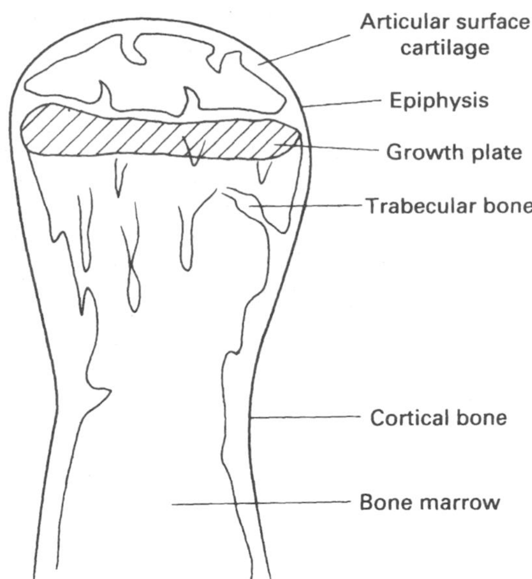

Figure 29.10 **Longitudinal bone growth by endochondral ossification**

influence of a variety of factors including transforming growth factors (TGFs), other growth factors, a protein known as CSA (chondrogenic stimulating activity), steroid and peptide hormones, and also collagens and other proteins of the extracellular matrix, by poorly understood processes. Osteoclasts are derived from the haematopoietic system and are involved in bone resorption.

Bone development and growth occurs in two ways. Most of the skeleton is formed by **endochondral ossification**, whereby cartilage is first formed and later replaced by bone. This type of growth involves both chondrocytes and osteoblasts. The flat bones of the skull are generated by a different mechanism known as **intramembranous ossification**, whereby mesenchymal cells differentiate directly into bone forming osteoblasts.

Embryonic long-bone formation is thought to occur by a process slightly different from that of the subsequent, growth-plate-associated longitudinal growth. Mesenchymal cells condense to form chondrocytes which produce a rod of cartilage which corresponds to the centre of a limb. Osteoid tissue, produced by osteoblasts derived from limb mesenchymal cells, is deposited outside the cartilage rod and serves as a focus of mineralization. Vascularization then occurs above the osteoid layer, followed by production of a further layer of osteoid, mineralization and vascularization in succession for about 10–15 layers. Osteoblasts become embedded in the osteoid and mature into osteocytes. The original cartilage rod is replaced by bone marrow as osteoblasts and osteoclasts (the latter derived from precursors in the blood stream) gradually excavate the bone cavity (Figure 29.9).

The mechanism of postnatal bone growth is complex. Longitudinal growth occurs by endochondral ossification, in a manner distinct from the formation of embryonic bone. Long bones consist of a tube of cortical bone surrounding the bone marrow space and capped at both ends (epiphyses) by the two epiphyseal growth plates. A growth plate consists of a cartilaginous template, located between the epiphysis and the diaphysis where bone growth occurs (Figure 29.10). At this site, cartilage is formed by proliferation and hypertrophy of chondrocytes and typical extracellular matrix is synthesized. The cartilage acts as a scaffold for the formation of new trabecular bone and marrow. It is calcified and then degraded and replaced by osseous tissue. In embryonic bone formation, the cartilage is not calcified but bone is formed outside it. In endochondral growth, the chondrocytes become hypertrophied, are calcified and finally die. Bone growth ceases when all the cartilage, except that present at the articular surface of the joint, is replaced by bone, and the epiphyses close. Oestrogen is required for the process of epiphyseal closure in both sexes and testosterone additionally in males.

The shape of bone depends, to a large extent, but not solely, on the shape of the original embryonic cartilage, which, in turn, depends on the correct condensation of mesenchymal cells. Paracrine factors are also important, however, and various abnormalities in skeletal patterning are associated with mutations of transforming growth factors. Inactivation of the gene coding for GDF-5 (growth and differentiation factor 5) in mice, for example, gives rise to a limb deformity known as 'brachypodism', characterized by short limbs. Skeletal growth is regulated mainly by growth hormone until the time of epiphyseal closure. Growth hormone probably acts by regulating the production of IGF-1 (insulin-like growth factor 1). Growth hormone seems to stimulate the slowly-dividing prechondrocytes in the germinative layer and promotes the clonal expansion of the proliferative cell layer. Hypopituitarism is characterized by a reduced rate of skeletal growth.

Thyroid hormone is also thought to affect bone growth by stimulating chondrocyte maturation. The process of endochondral ossification must be closely coordinated with growth of the epiphyses and longitudinal and radial growth. Incomplete coordination results in skeletal abnormalities such as seen in achondroplasia, the commonest cause of human short stature, characterized by short, wide bones. Many local mediators, such as fibroblast growth factors, are thought to be involved in the maintenance of skeletal growth and of the correct proportions of skeletal components. A mutation affecting FGFR-3 (fibroblast growth factor receptor 3), for example, is known to be associated with achondroplasia.

Transcription factors, controlling the formation of mRNAs, also seem to be involved. Abnormalities in craniofacial development, for example, are known to result from mutations affecting transcription factors msx1 and msx2. Dietary calcium, phosphate, vitamin D and its metabolites, and to a lesser extent, potassium, zinc, magnesium and vitamin K, are also important in the process of endochondral ossification. Experiments on growing dogs have shown that if these nutrients are inadequate, unbalanced or oversupplied, skeletal development is disturbed and results in severe pathological changes.

29.9 Mineralization of bone

Mineralization of bone is preceded by the formation of a web of collagen fibres and proteoglycans by the osteoblasts. Calcification of the collagen and proteoglycan matrix follows. The precise mechanism is not understood and two theories have been proposed to explain this process.

The first theory is that calcification involves some initial 'nucleation agent'. It has been proposed that the non-collagenous proteins of the matrix play a key role in initiating the precipitation of mineral, by providing 'nucleation' sites of the correct geometry for deposition of calcium phosphate.

The second theory is that cells themselves are responsible for mineralization by means of 'matrix vesicles'. These vesicles, which can be identified by electron microscopy, seem to be shed from the plasma membrane of osteoblasts or chondrocytes and they have been shown to contain alkaline phosphatase. It has been proposed that growth plate chondrocytes actively acquire calcium and phosphate ions, form concentrated calcium phosphate at the cell periphery and exfoliate it as calcium phosphate-rich matrix vesicles prior to its deposition on collagen.

There is at present no agreement as to which of these theories may be physiologically significant.

29.10 Bone remodelling and repair

Bone **remodelling** refers to the continuous turnover of bone matrix and mineral that involves an increase in bone resorption through activation of osteoclasts, followed by reactive bone formation through osteoblastic action. It takes place in the adult, at discrete foci, it involves both cortical and trabecular bone, and it ensures the mechanical integrity of bone throughout life. It also plays an important role in calcium homeostasis, as it allows bone to be used as a reservoir of calcium ions. An understanding of the process of bone remodelling is of great value in medicine, as the imbalance of the two events, resorption and formation results in many metabolic bone disorders, such as osteoporosis, osteopetrosis and osteitis deformans (Paget's disease of the bone).

A remodelling cycle consists of the following steps:

1. Activated osteoclasts resorb a discrete area of mineralized bone matrix.
2. Capillary endothelial cells provide the microvasculature.
3. Osteoblasts migrate into the resorption site.
4. Osteoblasts deposit new bone matrix (osteoid).
5. Osteoblasts become embedded in osteoid and mature into osteocytes.

The mechanisms involved in bone remodelling at the molecular level are, however, still poorly understood. The coordinated action of osteoclasts and osteoblasts seems to be regulated by an array of local factors produced in the environment of the remodelling cells. They include cytokines, prostaglandins, leukotrienes and growth factors, but the specific effects of these factors are not clear.

Bone **repair** refers to regeneration of bone following trauma. It is a complex process, controlled both by local factors derived from bone and by systemic factors delivered to the repair site as a result of the body's response to trauma. The systemic factors include mitogens such as PDGF (platelet-derived growth factor), heparin and CSA protein. These factors react with receptors on cells in the region of the break, the cells being both committed osteogenic cells and undifferentiated mesenchymal cells. The bone repair process is promoted both by an increase in biosynthetic events (biosynthesis of collagen, proteoglycans, etc.) in resident osteogenic cells, as well as differentiation of osteogenic precursor cells.

The main steps in bone repair are outlined below:

1. Bone break.
2. Acute inflammatory response including:
   (i) arrival of systemic factors such as PDGF
   (ii) arrival of phagocytic cells
   (iii) formation of clots
   (iv) demineralization of bone.
3. Mesenchymal cell attraction.

4. Mesenchymal cell multiplication.
5. Stem cell differentiation into chondrocytes.
6. Hypertrophy of chondrocytes.
7. Cartilage mineralization.
8. Osteogenesis.

These events are thought to occur following extensive damage of the bone, so that the surrounding tissue is also affected, which allows vascular-derived cells to invade the repair site. It is thought that in the case of hair-line fractures, when the surrounding tissue is not affected, repair is achieved by local cells and the intermediate formation of cartilage is bypassed. In this case, osteogenic cells are able to regenerate bone without the involvement of extrinsic systemic factors.

29.11 Metabolic bone disorders

Osteoporosis is a major health problem in the Western world, affecting primarily postmenopausal women. It is the result of abnormal bone remodelling, with a deficiency of formation relative to resorption, and is characterized by low bone mass and deterioration in the structure of bone tissue leading to increased bone fragility and, consequently, an increase in fracture risk.

Peak bone mass, which is the amount of bone tissue present at the end of skeletal maturation, usually between the ages of 20 and 35 years, is an important determinant of osteoporotic fracture risk. It is estimated that environmental factors, such as adequate calcium intake, can account for up to 20% variability in peak bone mass.

After the age of 35–45 years, bone mass begins to decline. Men lose bone mass at approximately the same rate over their lifetime. In women, the rate of loss increases dramatically after the menopause. Osteoporosis can be prevented by early introduction of **hormone replacement therapy** (oestrogen preparations combined with progesterone to avoid the risk of endometrial cancer), but treatment of the established condition is more difficult to achieve, as bone loss is accompanied by structural disintegration, and replacement of bone may not necessarily restore mechanical integrity.

Osteoporosis can also be drug-induced. Many conditions for which glucocorticoids are prescribed, such as rheumatoid arthritis, lupus erythematosus, asthma, inflammatory bowel disease, and Addison's disease, are associated with bone loss. Glucocorticoids are known to affect cellular responses within the bone microenvironment, and it is thought that they contribute to osteoporosis by inhibiting the secretion of a number of cytokines such as interleukin-1, tumour necrosis factor and insulin-like growth factor which normally act locally to regulate bone remodelling.

Treatment for osteoporosis is based on two types of therapeutic agents: those which inhibit bone remodelling, and those which stimulate bone formation. The former include calcium, vitamin D and its metabolites, gonadal steroids, calcitonin, and drugs, mainly bisphosphonates. Oestrogen replacement is quite effective in older women, and is the only therapy with proven anti-fracture effectiveness.

The mechanism by which oestrogen exerts its effect is far from clear. There is evidence to suggest that it may act by inhibiting the production of interleukin-6 (IL-6) by stromal and osteoblastic cells. IL-6 is thought to stimulate the formation of osteoclasts.

Calcitonin seems to stabilize bone mass and has the reported benefit of an analgesic effect. Of the therapeutic agents that stimulate bone formation, sodium fluoride is the most commonly used. It can dramatically increase bone density, but it has not been shown to have any effect in preventing bone fractures.

New therapies include intermittent injections of synthetic parathyroid hormone and bisphosphonates in order to activate and then repress resorption and formation. Treatment with growth hormone, anabolic steroids and growth factors, in order to increase osteoblastic activity, is currently being evaluated.

Osteitis deformans (Paget's disease of the bone)

This disease, which is the result of a primary increase in osteoclastic bone resorption, affects 3% of people over 40. It is marked by repeated episodes of bone resorption followed by excessive attempts at repair, resulting in weakened and deformed bones of increased mass. The resulting architecture of bone has a mosaic-like appearance in which the bone fibres lose their normal symmetry and assume a haphazard formation.

The role of viruses (paramyxoviruses) in the pathogenesis of the disease has been extensively investigated, but no causative role has been demonstrated between infective agents and Paget's disease.

Treatment is based on prevention of resorption using various agents, mainly bisphosphonates and calcitonin, both of which inhibit osteoclast action.

Paget's disease is believed to have been the cause of Beethoven's deafness.

Osteopetrosis

This condition is associated with decreased osteoclastic function. The external appearance of long bones is normal, but the bone marrow has dense trabeculations and the cortical bone is less compact than normal. A variety of mutations can cause osteopetrosis, including defects in genes coding for colony stimulating factor 1 (CSF-1), c-fos and c-src (see Chapter 37). All result in a block on osteoclastic differentiation and, consequently, a decreased osteoclast number and action.

Rickets and osteomalacia

These diseases are due to vitamin D deficiency, and are described in Chapter 12, which deals with calcium and phosphate homeostasis.

Chapter 30

The brain

30.1 Introduction

The complex function of the brain in regulating a vast range of physiological control mechanisms, intellectual processes and emotions makes it certain that the brain's biochemistry is the most complex of all the organs.

The most important function of the whole nervous system is the transmission of electrical impulses. The physiology of this process has been studied for many years and a detailed understanding has been developed, but the biochemical mechanisms underlying these processes have only begun to be elucidated more recently.

Several metabolic processes which occur in the brain have their exact counterpart in other tissues, such as the liver or kidney, but in this chapter emphasis is placed on the metabolism which is particularly relevant to brain function, including many specialized metabolic processes concerned with the transmission of information occurring in the brain.

30.2 The cell types of the brain

Under the microscope the brain is seen to contain many different types of cells as judged by their differing morphology. However, from a functional point of view, just two major categories of cell type may be recognized: **neurons** (the excitable nerve cells responsible for the transmission of electrical impulses) and the supportive **glia**. The latter cells are non-excitable, although in common with all cells they respond to extracellular signals.

As mentioned, the neurons are quite heterogeneous, with names such as pyramidal, stellate or bipolar which describe their morphological appearance, but some features are common to all neurons. A representative neuron is shown in Figure 30.1. It possesses a cell body (called a perikaryon) with a large nucleus, which in turn is characterized by a clearly-defined nucleolus. Other structures, such as ribosomes, rough and smooth endoplasmic reticulum and mitochondria, are present in large amounts in the perikaryon. The plasma membrane of the cell body is extended to form processes called the axon and dendrites. The **axon** is a long, often thin and occasionally branched process. The region of the plasma membrane from which it arises is termed the axon hillock. The axons of many neurons are wrapped in a myelin sheath that is itself an extension of the plasma membrane of specialized glia (see Section 30.6) and this sheath is interrupted periodically by breaks referred to as nodes of Ranvier. The **dendrites** are usually more numerous than the single axon; they also tend to be thicker and to be highly branched into a tree-like arrangement (hence their name). They constitute the 'receptive field' of the neuron, being able to transmit

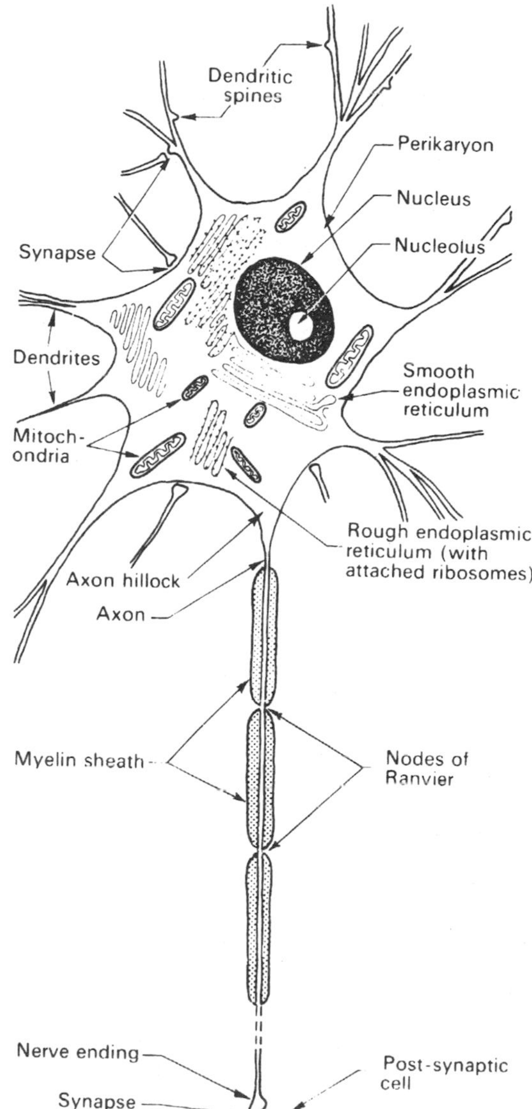

Figure 30.1 **Schematic drawing of a neuron**

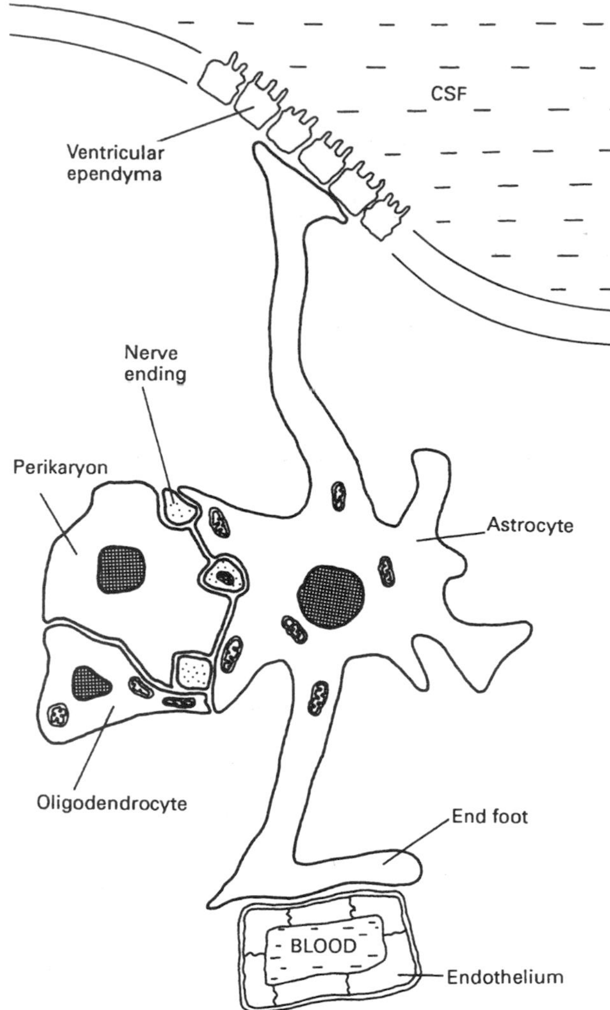

Figure 30.2 **A representation of some of the functional roles of glial cells** (CSF, cerebrospinal fluid)

passively-spreading (electrotonic) currents from synapses to cell body and the axon hillock (see Section 30.4).

Glial cells, which tend to be smaller than neurons, broadly seem to subserve one of two functions: investing axons with a myelin sheath or providing nutritional or metabolic support and protection for the neurons. In the brain, production of myelin is the function of oligodendroglia, although Schwann cells play this role for peripheral nerves. **Astrocytes** are glia which provide nutritional support and protection to neurons. When acting in their former role, they are frequently found close to blood vessels where specialized processes, termed 'end-feet' are believed to facilitate the transfer of blood-borne nutrients to the nerve cells and possibly they may also help to transfer waste products to the cerebrospinal fluid (CSF). Astrocytes are also found investing the region of a synapse (see Section 30.3) where their ability to act as 'buffers', taking up neurotransmitters and even K^+, helps protect neurons from overstimulation. Microglia closely resemble macrophages (see Chapter 32) and are believed to help protect the brain against infection. In addition, proliferation of glial cells occurs at sites of injury to the CNS with the formation of 'glial scars'. However, the function of this gliosis is unknown.

Figure 30.2 shows in diagrammatic form the relationship between the various cell types of the brain.

30.3 The synapse

The characteristic feature of the activity of the brain is that so many of the billions of functional units, the neurons, are in communication with each other. Electrical signals can pass along the axon and to a lesser extent the dendrites of nerve cells, but it is unusual for two neurons to lie close enough together so that these signals can be transmitted directly from cell to cell. Instead, neurons communicate with each other largely by chemical means via specialized structures termed 'synapses'. In its simplest form, a **synapse** consists of the extreme nerve ending of an axon, which may be somewhat swollen in appearance, in close proximity to part of another cell, termed the postsynaptic cell. Although the pre- and postsynaptic membranes are closely apposed, there is a distinct space of some 20 nm between the two, referred to as the synaptic cleft. The nerve ending is characterized by the presence of numerous vesicles which accumulate one or more chemical neurotransmitters that are released into the synaptic cleft on stimulation of that nerve. Because the release process and the subsequent reuptake of the transmitter both require energy, the nerve endings also contain mitochondria. The transmitter substances diffuse across the cleft, bind to specific membrane receptors and stimulate or inhibit the postsynaptic nerve. The number of synapses found in the nervous system is enormous; the 'dendritic tree' of a single neuron may receive inputs from tens of thousands of axons.

How electrical signals are transmitted from dendrites to cell body and thence along the axons, and how nerves communicate chemically at synapses, are the subjects of the next sections.

30.4 Excitation and conduction

For neurons, as for all eukaryotic cells, there exists across the plasma membrane a potential difference of about 75 mV, the inside being negative. Neurons (and muscle cells) differ from other eukaryotic cells in that they are additionally excitable. This means that a depolarization can be sustained and transmitted along their plasma membrane. These two properties of neuronal plasma membranes, polarization and the ability to sustain and propagate a depolarization, depend on the presence of membrane proteins that act as ion channels. It is appropriate, therefore, to consider the biochemical organization of neurons that leads to their special electrical properties.

The existence of a membrane potential

The potential difference that exists across the plasma membrane of all eukaryotic cells arises as a consequence

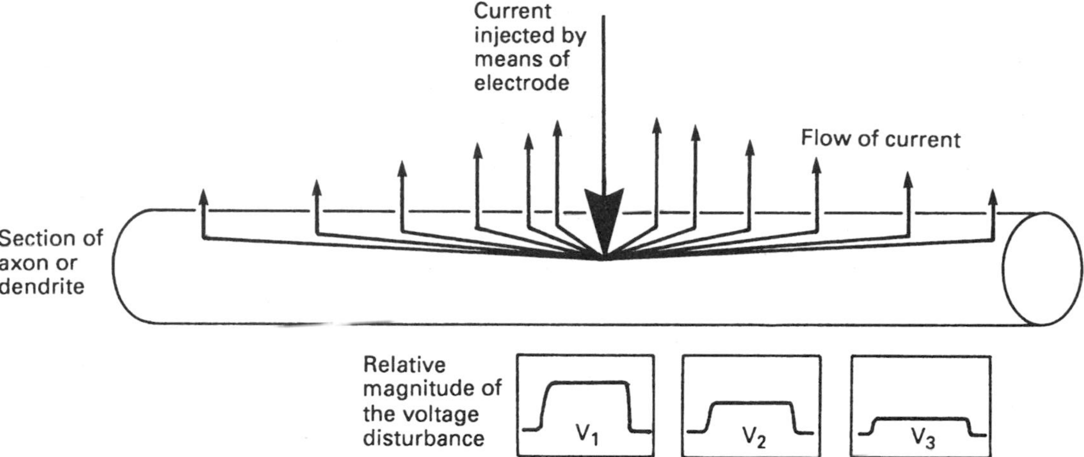

Figure 30.3 **Passive (electrotonic) spread of current injected into an axon or a dendrite.** The amplitude of the voltage disturbance falls exponentially as a function of the distance from the origin of the disturbance

of the asymmetrical distribution of anions and cations. This asymmetry is the result of several factors.

A Mg^{2+}-dependent Na^+/K^+ ATPase (the sodium 'pump') is found in all plasma membranes. This enzyme catalyses the accumulation of K^+ ions in cells against their concentration gradient and the extrusion of Na^+ ions against their concentration gradient (see Chapter 2). It may be thought that, since the 'sodium pump' is electrogenic (it catalyses the exchange of three Na^+ for two K^+), its action would itself explain the membrane potential, as more positive charge leaves the cell than enters. However, this is only a minor factor. The major reason for the charge difference is that plasma membranes are some 20 times more permeable to K^+ than to Na^+. In fact, there are specific K^+-'leak' (or resting) channels. These are membrane proteins that let this cation out of all cells and these greatly outnumber those which let Na^+ in. Thus K^+ ions tend to move down their concentration gradient and out of the cell. The consequent loss of positive charge cannot be offset by the movement of Na^+ in the opposite direction, nor can it be combated by the loss of anions from the cell, because the plasma membrane is very impermeable to the major intracellular anions: proteins and nucleic acids. Consequently, there is a small deficit of positive charge inside the cell. This amounts only to some 6 K^+ lost per 10 nm^2 patch of membrane. Nevertheless this suffices to the explain existence of the membrane potential.

Passive spread of depolarization

It has proved possible to 'inject' current into nerve cells, either into the dendrites or the axon. This may be controlled in such a way that a relatively small sustained increase (10 mV) in the resting potential towards zero is achieved (the membrane is said to be depolarized). At the site of depolarization there is now an excess of positive charge, principally carried by K^+, and these cations diffuse in both directions, within the dendrites, causing adjacent areas of membrane to depolarize. The range of spread of this passive sort of depolarization is very short, however, with the extent of the membrane perturbation falling off very rapidly as a consequence of K^+ 'leaking' across the plasma membrane via resting K^+-channels. This type of localized passive spread of current is shown in Figure 30.3. Thus nerve fibres are able to act rather like electric cables, although the distances over which such conduction occurs are relatively short. Nevertheless such passive spread of current, especially in dendrites, is an important aspect of information processing in the central nervous system. However, the present concern is with the all-or-none **action potential** which sweeps down the axon without appreciable attenuation. Such an action potential may also be induced in an axon if rather larger amounts of current are injected.

The action potential

It is the function of the cell body of the neuron and the adjacent axon hillock to summate into a 'grand postsynaptic potential' individual passively-spreading currents that invade them, from the dendrites, for example. These summed currents are sufficient to initiate an action potential at the axon hillock. Biochemical analysis has shown that in the plasma membrane of the axon hillock there is a high concentration of an integral protein termed a 'voltage-gated' Na^+-channel. This protein is sensitive to increases in membrane potential, and once these have reached 'threshold' values (about –40 mV) the channel opens and there is a very rapid influx of Na^+. This results in a positive inside membrane potential of about 30 mV. The voltage-gated Na^+-channel protein appears to be organized into four similar domains, each of which spans the membrane four times. It is not entirely clear how this protein senses the membrane potential to allow Na^+ to cross the membrane, but the presence of multiple glutamyl residues, each with a negative charge located on the cytosolic side of the membrane between the middle two domains is probably important.

A useful analytical tool is the shell fish toxin saxitoxin because it binds tightly to the voltage-gated Na^+-channel to prevent its opening. Because it binds so strongly to these channels in the axolemma, radioactive saxitoxin can be used to 'count' and to locate the channel proteins. In unmyelinated axons there are some 500 per μm^2 of membrane, evenly distributed along the whole axon, whereas for myelinated nerves there may be up to 12 000 per μm^2 at the nodes of Ranvier but very few under the myelin sheath.

Following their voltage-dependent opening, these channels close again. Thus the influx of Na^+ is only transient and, once closed, for a while the voltage-gated Na^+-channels are incapable of reopening in response to a subsequent perturbation in the membrane potential. This period when no opening is possible is referred to as the 'refractory period'.

Thus, the change in potential across the axon membrane at the hillock serves to cause the voltage-dependent opening of channels in adjacent areas of the membrane and the voltage perturbation caused by this in turn causes the opening of channels further along the axon. As a consequence the action potential sweeps down the axon, which then remains unable to transmit a second action potential during the refractory period of the channels. It can be seen, therefore, that the properties of the voltage-gated Na^+-channel form the basis of both the initiation and the perpetuation of the action potential.

The processes described above account for transmission of nervous signals in unmyelinated neurons, in which the voltage-gated Na^+-channels are evenly distributed along the length of the axon. In myelinated axons the perturbations of membrane potential jump in a so-called 'saltatory' fashion from one node of Ranvier to another, as might be predicted from the experimental observation that the voltage-gated Na^+ channels are located almost exclusively in those regions.

The rate at which the action potential is propagated depends upon the thickness of the axon and whether it is myelinated. For unmyelinated neurons the rates range from 1 to 10 m/s for thin axons up to 30–50 m/s for thick axons such as those found in the giant squid. A much more successful strategy for increasing the rate of conduction is not to increase the diameter of the axon, but rather to provide it with a myelin sheath with gaps at the nodes of Ranvier for voltage-gated Na^+ to propagate the action potential. Here the rate of saltatory transmission may be as high as 100 m/s. (For comparison, the velocity of sound in air at sea level is about 300 m/s.)

It is important to realize that the actual initiation and propagation of the action potential is not itself an energy-requiring process, and because the number of ions moving for each action potential is small, a neuron may fire repeatedly without the need for the energy-requiring sodium 'pump' to be called into play (although of course the 'pump' is required to set up the asymmetrical distribution of Na^+ and K^+ in the first place).

Figure 30.4 shows the timecourse of changes in the membrane potential and Na^+ conductance which occur as an action potential passes along a particular section of axon containing only voltage-gated Na^+-channels together with the sodium 'pump'. The form of this simple action potential, a rapid rise in the resting potential followed by a much slower return to the resting value, may be explained in terms of the operation of these two proteins: the Na^+-channels open, Na^+ ions move into the axon upon which the channels close and the movement ceases. Thereafter the Na^+ ions are extruded by the 'pump' and this process takes rather a long time as the ions are being moved against their concentration gradient. Therefore the restoration of a normal resting potential is slow. For most action potentials, however, the restoration of the resting potential is much faster, as is indicated by the sharpening of the 'spike' shown in Figure 30.4. This is

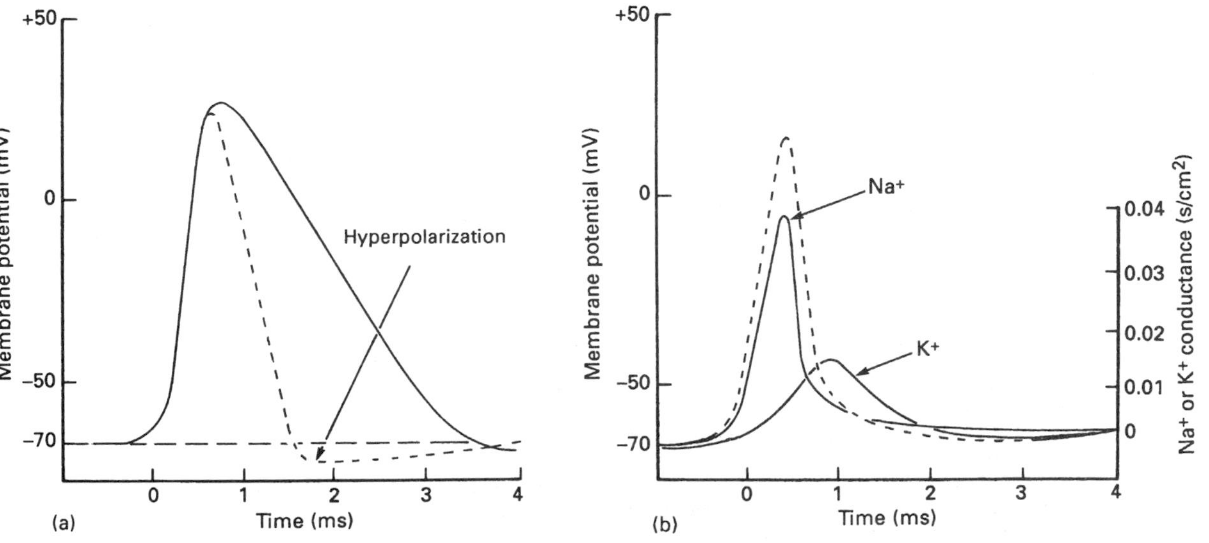

Figure 30.4 **(a) Action potential with (– – – –) or without (——) K^+ channels. (b) Changes in Na^+ and K^+ conductances**

achieved by allowing K^+ to move down their electrochemical gradient out of the cell and this is mediated by the slightly delayed opening of voltage-gated K^+-channels. The changes in membrane potential and the conductances of both Na^+ and K^+ for this more complex case are also shown in Figure 30.4.

When the action potential reaches the nerve terminal, the voltage perturbations cause the opening of 'voltage-gated' Ca^{2+}-channels and as the extracellular concentration of this cation so greatly exceeds that in the cytosol there is a rapid influx of Ca^{2+} into the terminal, resulting in the release of the neurotransmitter stored there. It is appropriate therefore to consider next the biochemistry of neurotransmitters.

30.5 Chemical transmission and transmitters

Those who study neurotransmitters (usually abbreviated to transmitters when the context makes it clear that the nervous system is being discussed) and neurotransmission have two aims in mind. The first is to map out the systems of neurons in the brain and the connections between them. The special contribution of the biochemist is in the characterization of synaptic events in terms of metabolic and other processes associated with the formation, transport, release, mechanism of action and removal of transmitter substances. The second objective has more direct medical implications because the concern is with the introduction of new methods for the prevention, diagnosis and treatment of diseases associated with transmitter dysfunction.

Criteria for the identification of transmitters

The classic work early in this century of Sir Henry Dale and of Otto Loewi on peripheral nerves led to the enumeration of five criteria for accepting that a substance does act as a transmitter:

1. The compound should be synthesized and/or stored in nerve terminals from which it is supposed to be released
2. Its presence in the extracellular fluid in the region of the synapse should be demonstrable following presynaptic stimulation. (The collection of samples suitable for analysis makes this criterion very difficult to establish for sites within the brain)
3. When applied postsynaptically it must mimic the action seen in response to presynaptic stimulation
4. The effects of both chemical and (presynaptic) electrical stimulation should be blocked by antagonists
5. Mechanisms that terminate the action of the transmitter should exist.

As the result of extensive investigations, a list of four main groups of chemical types of transmitter has been drawn up (Table 30.1).

The biosynthesis, release and termination of action of transmitters

Although there is a wide variety of transmitters, the basic details of the formation of many of them are very similar and straightforward. For each substance, four questions may be asked:

1. What is the mode of biosynthesis of the transmitter and if there are multiple steps, which one is rate limiting?
2. How is the requirement for precursor met?
3. What mechanisms underlie the storage and release of the transmitter?
4. What provisions are there for the termination of action of the transmitter?

In fact, this is merely a reformulation in more explicit biochemical terms of some of the criteria listed earlier. Details of the answers to these questions vary from substance to substance, but the general principles may be illustrated by reference to acetylcholine, glutamate, GABA, dopamine, noradrenaline and adrenaline.

Acetylcholine

For acetylcholine, Figure 30.5 illustrates the answers to all four questions posed above. Several points are particularly noteworthy. Choline, whether derived from acetylcholine or the circulation, is taken up by a high-affinity membrane transporter protein that utilizes the potential energy of the Na^+ concentration gradient that exists across the membrane for this purpose, the gradient being maintained by the Na^+/K^+-ATPase. The reuptake of all the 'simple', i.e. non-peptide, transmitters is effected in a similar way. In the particular case of acetylcholine, the uptake of the precursor choline is the rate-limiting step for the biosynthesis of the transmitter. The nerve terminal cytosolic enzyme **choline acetyltransferase** catalyses the reaction of acetylCoA with choline. The major source of the former substrate is

Table 30.1 **Examples of mammalian neurotransmitters and neuromodulators**

Monoamines	*Amino acids*	*Peptides*	*Others*
Dopamine Noradrenaline Adrenaline 5-Hydroxytryptamine (serotonin)	Glutamate Glycine γ-Aminobutyrate (GABA)	Somatostatin Gonadotrophin-releasing hormone (GnRH) Substance P Enkephalin	Acetylcholine Adenosine Adenosine triphosphate (ATP)

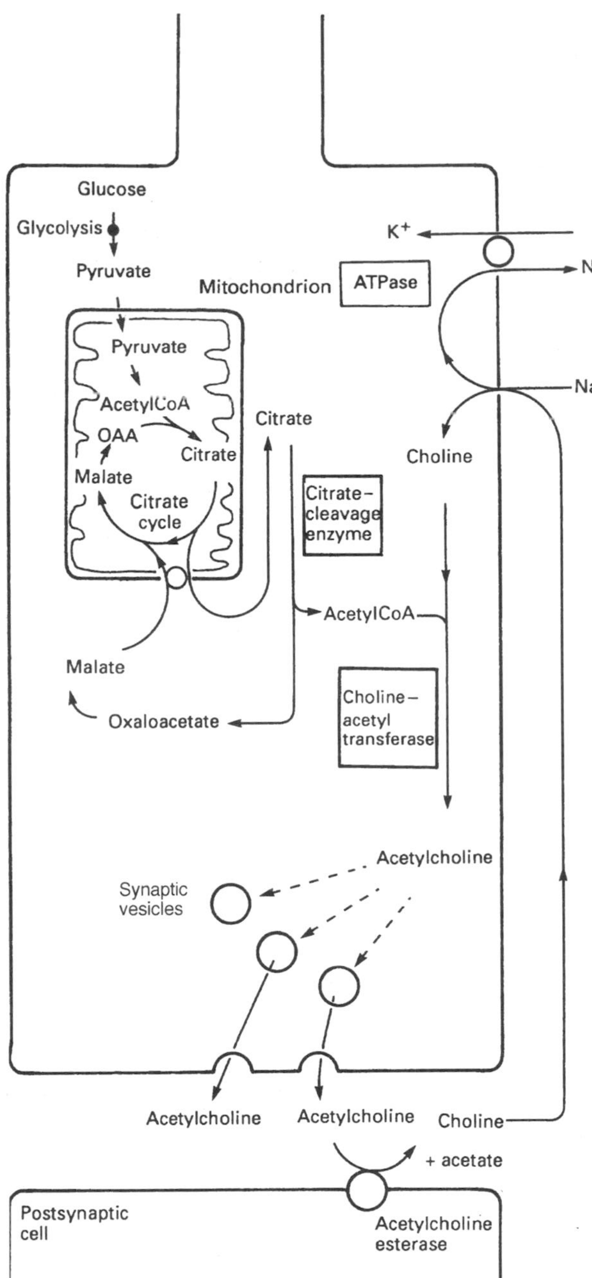

Figure 30.5 **The organization of a cholinergic nerve terminal** (OAA, oxaloacetate)

glucose which is converted via the (cytosolic) glycolytic pathway into pyruvate. However, the conversion of pyruvate into acetylCoA catalysed by pyruvate dehydrogenase occurs in the mitochondria, not in the cytosol. This problem is overcome by the formation of citrate, which is transported out of the mitochondria where the citrate cleavage-enzyme catalyses the reformation of acetylCoA. The enzymes that catalyse these reactions are described in Chapter 5.

Once formed, acetylcholine is taken up into synaptic vesicles, where it accumulates to a final concentration of about 900 mM. This process is energy-requiring, depending on the maintenance of a H^+ gradient across the vesicle membrane with the inside of the vesicle being at about pH 5.5. This is achieved by the action of a vesicular ATPase. This enzyme is to be found associated with vesicles in the nerve terminals of neurons.

Release of acetylcholine is caused by an action potential invading the nerve terminal. Voltage-gated Ca^{2+}-channels open, and the influx of the cation causes fusion of the vesicle with the plasma membrane with consequent release of the transmitter.

Finally the action of acetylcholine is rapidly terminated in the synaptic cleft. It is hydrolysed by **acetylcholinesterase** located on the outer aspects of both pre- and postsynaptic plasma membranes.

Glutamate and GABA

It is appropriate that these two amino acid transmitters are considered together, both because they bear a precursor–product relationship to each other and also in several brain regions they act in complementary excitatory–inhibitory fashions.

Considering glutamate first, its assessment as a transmitter is more difficult than that of acetylcholine. This is because glutamate is ubiquitously present in high concentration in the brain, whereas acetylcholine is found in discrete regions. The ubiquity of glutamate reflects the central role it plays in metabolic processes, ranging from protein synthesis to detoxification of ammonia. It has been established that at least 80% of the glutamate released from nerve terminals derives from the hydrolysis of glutamine catalysed by **glutaminase** (most of the remainder comes from glucose).

Figure 30.6 shows the series of events that occur at glutamatergic nerve terminals. The formation of glutamate is initiated when glutamine is taken up into nerve terminals and undergoes deamination catalysed by the cytosolic enzyme glutaminase (the presence of this enzyme in the nerve terminal is considered diagnostic for the transmitter being glutamate). The presence of a vesicular-ATPase in nerve terminals containing glutaminase strongly suggests that vesicular uptake and release of glutamate closely resembles that of acetylcholine.

Following its release, the process of uptake of glutamate is very similar to that for choline. However, the Na^+-dependent transporter is found largely in the plasma membrane of adjacent glial cells (astrocytes) into which the amino acid is transported. Following its sequestration in the astrocytes, the glutamate is inactivated by conversion into glutamine in an ATP-dependent reaction with ammonium ions catalysed by the cytosolic enzyme **glutamine synthase**. Glutamine may then be safely released from the glia without danger of inappropriate activation of glutamate receptors. In view of the serious consequences following from excessive activity of the glutamate receptors (as in stroke, see next section), this is an important safety mechanism. The CSF acts as a store of glutamine (concentration about 0.5 mM). Glutamine may diffuse into the CSF and thus provide a pool from which neurons may be replenished.

A similar, but not identical, picture obtains for GABA. Important differences relate to two transmitter-specific

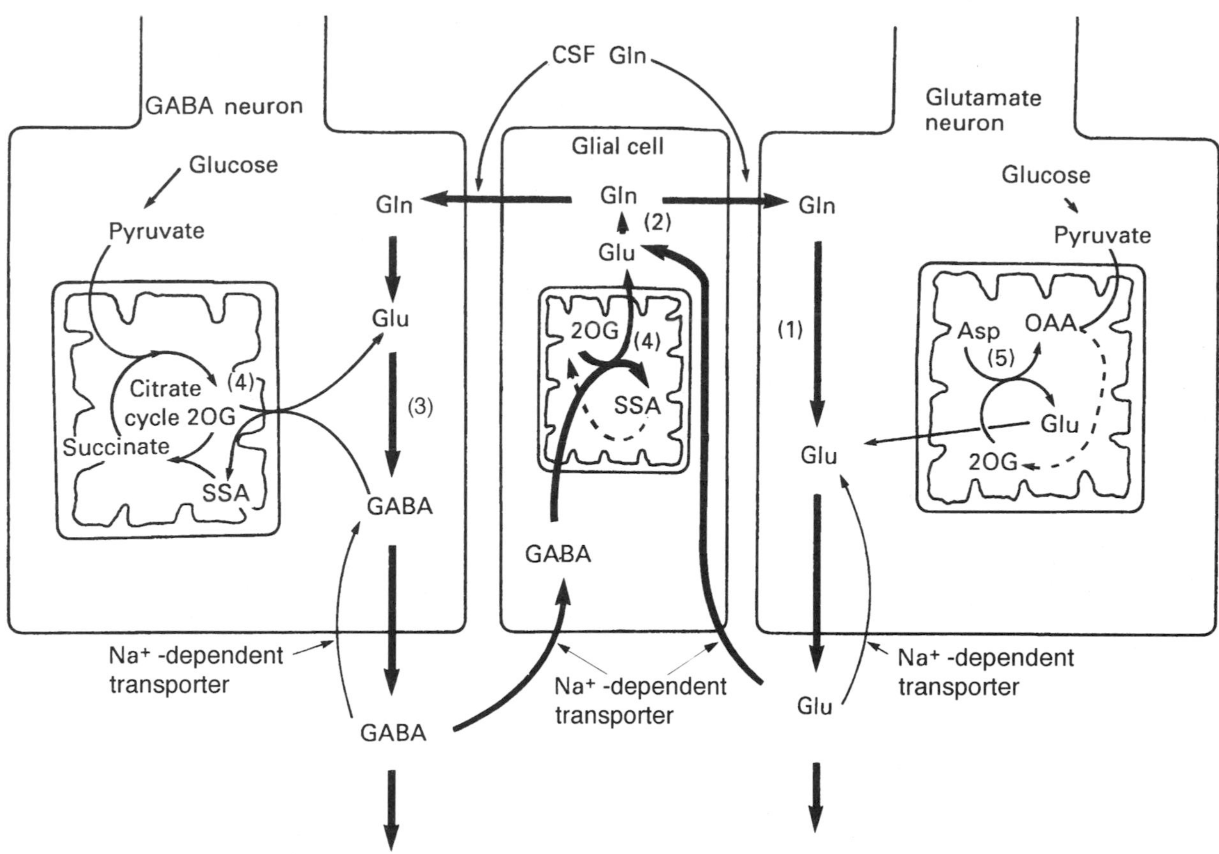

Figure 30.6 **The organization of nerve terminals synthesizing glutamate and GABA and their relation to glial cells.** Major pathways are shown by heavy arrows. Enzymes: (1) glutaminase; (2) glutamine synthase; (3) glutamate decarboxylase; (4) GABA : 2-oxoglutarate aminotransferase; (5) aspartate aminotransferase

enzymes found largely in GABAergic nerve terminals. The first, that catalyses the formation of the amino acid from glutamate, is **glutamate decarboxylase** (GAD, Figure 30.7). It should be noted that the insulin-secreting β-cells of the pancreas also form GAD, and this has a bearing on the occurrence of one type of diabetes mellitus (see Chapter 41). The second GABA-specific enzyme is **GABA : 2-oxoglutarate aminotransferase** (GABA-T). This enzyme catalyses a reaction that leads to the inactivation of GABA (Figure 30.7). These two enzymes require pyridoxal phosphate, but GAD is the more susceptible to drugs that interact with pyridoxal phosphate (probably because its K_m for the cofactor is the higher). An example of such a class of drugs are the hydrazides, e.g. hydralazine, that act as vasodilators on arteriolar smooth muscle. Consequently overdosage with such antipyridoxals can cause convulsions as the formation of inhibitory GABA is prevented preferentially over its breakdown.

The succinate semi-aldehyde formed by GABA-T undergoes an NAD$^+$-dependent oxidation to succinate. This, in turn, may be converted back into 2-oxoglutarate. The net effect of these reactions is that glia convert GABA into glutamate, and the latter amino acid is then converted into glutamine that is then transferred to the GABAergic neurons. Here glutamine is converted first into glutamate and then into GABA. Figure 30.6 shows the neuron–glial interactions that occur. For convenience, nerve terminals containing glutamate and GABA are shown in close proximity, but this will not always be the case.

GABA and Huntingdon's disease

Various movement disorders arise as a result of defects in the basal ganglia of the brain, two of which are parkinsonism and **Huntingdon's disease**. The former is dealt with in the next section, while a consideration of the latter is relevant here because it involves degeneration of GABAergic neurons. Specifically there appears to be a loss of neurons originating in the striatum that project to the globus pallidus. It is believed the loss of the inhibitory pathway results in the uncontrolled (choreic) movements that characterize the condition.

The disease was first described in clear, clinical terms in 1872 by the American physician Dr George Huntingdon (although his grandfather was reported to have studied the disease in the late eighteenth century). Huntingdon's disease is of late onset, is progressive and fatal. It is inherited as an autosomal dominant trait. The average age at onset is 36 and, if present, is always seen by age 65. It also shows the terrible characteristic of

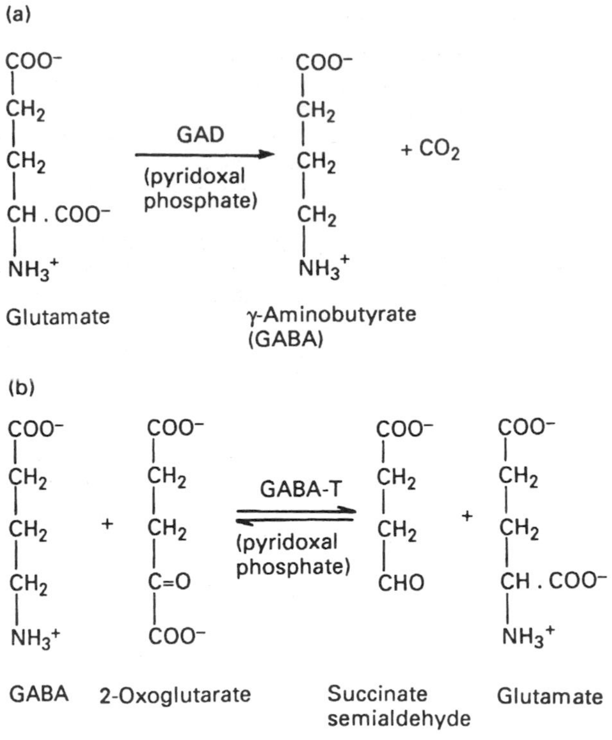

(a)

Glutamate $\xrightarrow[\text{(pyridoxal phosphate)}]{\text{GAD}}$ γ-Aminobutyrate (GABA) + CO_2

(b)

GABA + 2-Oxoglutarate $\underset{\text{(pyridoxal phosphate)}}{\overset{\text{GABA-T}}{\rightleftharpoons}}$ Succinate semialdehyde + Glutamate

Figure 30.7 **The formation (a) and disposal (b) of γ-aminobutyrate (GABA)** (GAD, glutamate decarboxylase; GABA-T, γ-aminobutyrate : 2-oxoglutarate aminotransferase)

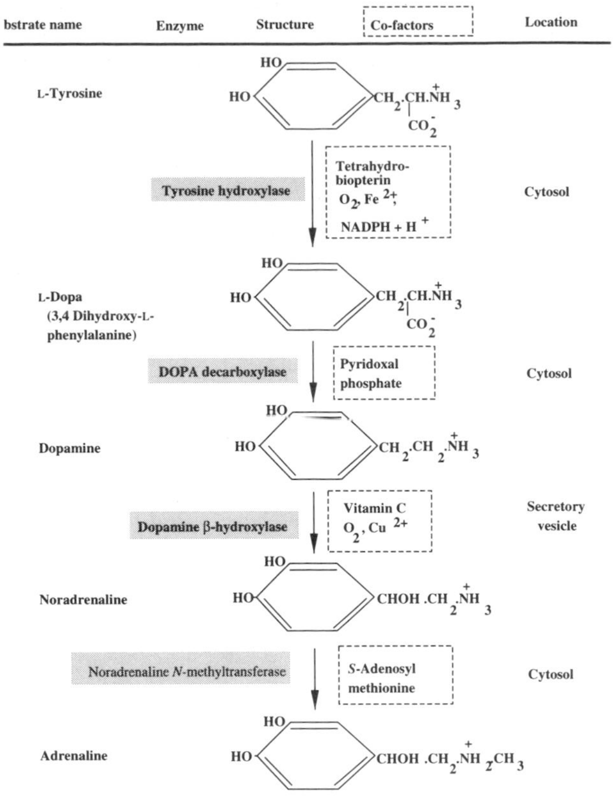

bstrate name	Enzyme	Co-factors	Location
L-Tyrosine			
	Tyrosine hydroxylase	Tetrahydro-biopterin O_2, Fe^{2+}, NADPH + H^+	Cytosol
L-Dopa (3,4 Dihydroxy-L-phenylalanine)			
	DOPA decarboxylase	Pyridoxal phosphate	Cytosol
Dopamine			
	Dopamine β-hydroxylase	Vitamin C O_2, Cu^{2+}	Secretory vesicle
Noradrenaline			
	Noradrenaline *N*-methyltransferase	*S*-Adenosyl methionine	Cytosol
Adrenaline			

Figure 30.8 **The biosynthesis of the catecholamines**

'anticipation', that is, its age of onset becomes earlier with successive generations. The gene associated with the disease has been cloned and the corresponding mRNA encodes a protein, called **'huntingtin'**, of some 3000 amino acyl residues. The function of this gene has not been determined nor is the gene expressed most markedly in the affected basal ganglia. The defect in the gene has been traced to a region in which the sequence 5'-CAG-3' is found in multiple tandem repeats, i.e. (CAG)n. In non-affected persons n=16–36, whereas people affected by the disease have 42-86 repeats. It is noted that 'anticipation' is associated with an increase in the number of repeats.

Since no role has been assigned to huntingtin it is not yet possible to associate the mutation with the deficit in GABA. It should be noted, however, that several disorders have been associated with trinucleotide expansions. These include the X-linked recessive Kennedy's disease (hereditary spinobulbar muscular atrophy). In this neurological disease the sequence 5'-CAG-3' is again expanded, this time within the androgen receptor gene. The increase is from 17–26 to 40–52 copies.

Recently it has been suggested that synthesis of proteins with large numbers of single amino acid repeats, e.g. glutamine in Huntingdon's disease, may lead to the overload of lysosomes with proteins that the organelles are not able to degrade. This is predicted to result in lysosomal degradation and destructive hydrolase release.

Dopamine, noradrenaline and adrenaline

It is convenient to describe the biosynthesis of the three catecholamine transmitters together, as they bear precursor product relationships. As is shown in Figure 30.8, the biosynthesis of adrenaline requires four enzymic steps, while three and two steps suffice for noradrenaline and dopamine, respectively. The common precursor is tyrosine and this undergoes hydroxylation to form L-dihydroxyphenylalanine (L-DOPA) in a reaction catalysed by tyrosine hydroxylase. This is a mono-oxygenase that requires molecular oxygen and tetrahydrobiopterin as a reducing agent. The latter is formed from dihydrobiopterin at the expense of NADPH. Tyrosine hydroxylase is a cytosolic enzyme found in catecholaminergic nerve terminals and is rate-limiting for the biosynthesis of the catecholamines, all of which exert negative-feedback control on the enzyme. In the second reaction, catalysed by the cytosolic pyridoxal phosphate-dependent enzyme L-DOPA decarboxylase, L-DOPA is converted to dopamine. Dopamine is then taken up into synaptic vesicles. This process is dependent on a H^+-gradient formed as the result of the action of a vesicular ATPase. Dopamine neurons lack the next two enzymes on the pathway to adrenaline and hence the formation of dopamine is the final biosynthetic step. In adrenergic neurons, vesicular dopamine β-hydroxylase, a copper-containing mono-oxygenase that requires ascorbic acid and molecular oxygen, catalyses the formation of noradrenaline. In the majority of adrenergic neurons in the CNS this completes the biosynthetic process. However, a few neurons in the brain and many of the chromaffin cells of the adrenal medulla also contain the cytosolic enzyme noradrenaline *N*-methyl transferase that catalyses the transfer of a methyl group from *S*-adenosylmethionine to form adrenaline. The cytosolic location of this enzyme means that noradrenaline must be released from the secretory vesicles and adrenaline must be taken back into them prior to its release.

The synaptic release and reuptake of the catecholamines resembles that of the other transmitters. The former process requires external Ca^{2+}, whereas the latter is dependent on the presence of specific transport proteins which promote Na^{+}-dependent reuptake of the transmitters.

Cocaine exerts its euphoric effects by interacting with, and thereby blocking, the **dopamine transporter**. Blockage occurs at the intramembrane phase of transport. This results in the potentiation of dopaminergic transmission, changes in the mesolimbic and mesocortical pathways being particularly important.

In addition two further enzymes catalyse the inactivation of catecholamines. These are **catecholamine *O*-methyl transferase (COMT)** and **monoamine oxidase (MAO)**. The former enzyme catalyses the transfer of a methyl group from *S*-adenosylmethionine to the 3-hydroxyl group of the catecholamines. The substrates for this enzyme include the aldehydes, alcohols and acids formed following the reaction catalysed by MAO. Conversely MAO, a flavoprotein found in the outer mitochondrial membrane, catalyses the oxidative deamination of catecholamines and their 3-*O*-methyl derivatives. The resulting aldehydes produced may be either oxidized to acids or reduced to alcohols.

Two isoenzymes of MAO are recognized: MAO-A and MAO-B. The former is the major form in the brain and shows preferential substrate specificity towards endogenous aromatic monoamines, such as serotonin, while MAO-B catalyses the metabolism of dietary amines, such as phenylethylamine (an abundant constituent in chocolate), more efficiently, although MAO-B is found in the brain.

The reactions catalysed by COMT and MAO are shown in Figure 30.9.

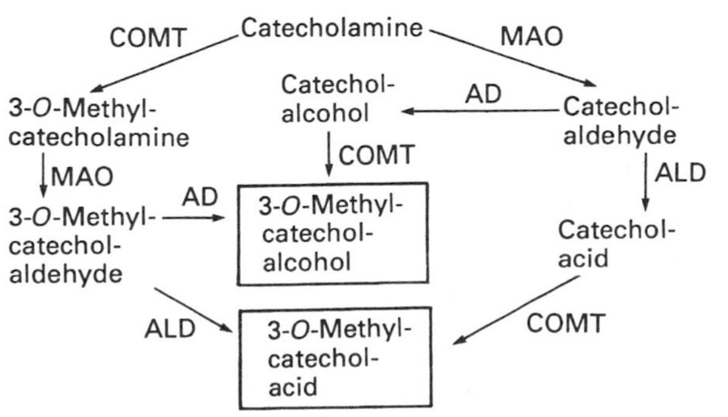

Figure 30.9 **The alternative pathways of catecholamine catabolism may lead to the formation of 3-*O*-methyl alcohols or acids** [COMT, catechol-*O*-methyl transferase (*S*-adenosyl methionine); MAO, monoamine oxidase; AD, alcohol dehydrogenase (NAD^{+}); ALD, aldehyde dehydrogenase (NAD^{+})]

Dopamine and parkinsonism

A second type of movement disorder that arises as a result of a defect in the basal ganglia is parkinsonism. The disease was first described by Dr James Parkinson in 1817 and is now known to involve the loss of dopamine neurons in the substantia nigra. The condition is relatively common in the second half of life with an incidence of 1 in 1000.

In the early 1980s a group of young patients presenting with what appeared to be parkinsonism was reported in California. It was established that the victims were all drug addicts who had taken a 'synthetic heroin' (a so-called 'designer drug'). Analysis of the material showed it to contain the compound 1-methyl-4-phenyl-1,2,3,6-tetrahydropyridine (MPTP) which proved to be selectively toxic to the zona compacta of the substantia nigra in experimental animals. Thus the tragic misadventure of the young Californians has led to the development of an animal model for the study of the course and possible prevention of the disease. It now established that both in patients dying with parkinsonism and the animal models there is a specific deficit in the activity of mitochondrial NADH dehydrogenase (complex I; see Chapter 5) in the substantia nigra.

In the case of MPTP-induced parkinsonism, the cause of this deficit is understood (Figure 30.10). MAO-B contained in glial cells in the substantia nigra catalyses the oxidation of MPTP to yield the toxin MPP^{+} which is avidly concentrated in dopaminergic neurons by the dopamine transport protein. This toxin acts on complex I to inactivate it and the resulting energy deficit leads to the selective death of dopamine neurons. It may well be that some, as yet unidentified, environmental toxin acts in a similar way to promote the occurrence of parkinsonism in humans. This view finds some support in the occurrence of a complex of diseases that includes parkinsonism, seen in the Western Pacific island of Guam, which appears to be associated with the ingestion of a natural neurotoxin. If the natural neurotoxin hypothesis proves correct, then MAO-B inhibitors such as **deprenyl** may be given to patients to prevent the progression of their disease. Currently the most effective treatment for parkinsonism is to give L-DOPA, which acts as a precursor for dopamine in surviving neurons.

Why are there so many different transmitters?

The analogy between the digital computer and the brain as processors of information led to the suggestion that only two types of transmitter should be necessary: one excitatory and one inhibitory. The model of the brain leading to this view was one of a series of discrete projection neurons with point-to-point neuronal circuitry. However, it is now clear that synaptic control can be exercised in more subtle ways. Some inputs may indeed cause rapid, postsynaptic events of the type first recognized as promoting muscular contraction, for example, but others may trigger long-lasting events. These may last for minutes rather than milliseconds and may thus act to regulate the excitability of the target cell rather than directly to control its rate of firing. Such

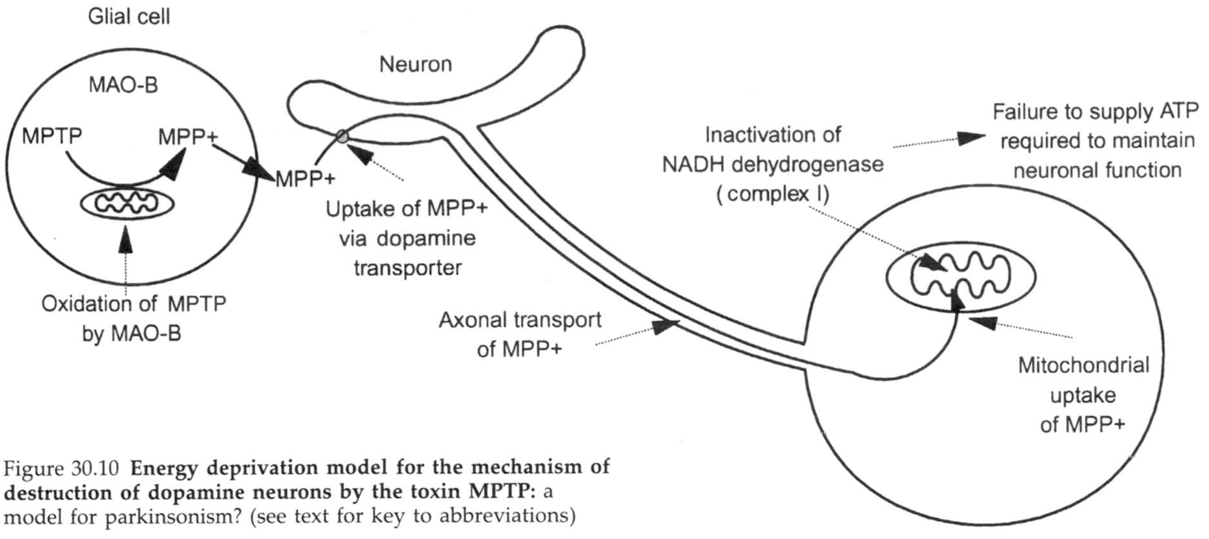

Figure 30.10 **Energy deprivation model for the mechanism of destruction of dopamine neurons by the toxin MPTP:** a model for parkinsonism? (see text for key to abbreviations)

substances are sometimes referred to as **neuromodulators**. The actions of neuromodulators tend to be more long-lasting and diffuse than those of 'classical' transmitters.

The division of signalling processes into rapid, point-to-point interactions and more diffuse events helps provide a useful means of classifying transmitters. Almost invariably it is the amino acid transmitters glutamate, glycine and GABA (together with acetylcholine in peripheral nerves) that mediate fast processes by binding to their specific receptors, while the activation of receptors for the monoamines and neuropeptides leads to diffuse, modulatory actions. The basis for this classification is clarified when the biochemical mechanisms producing postsynaptic effects are considered.

Transmitter receptors

The fact that certain neurons release particular transmitters and others respond to them implies the presence in the latter of specific receptors. These take the form of integral proteins of the plasma membrane (see Chapter 3). Sir John Eccles has proposed that these receptors be designated **ionotrophic** or **metabotrophic**. The basis for this division is that receptors of the former type themselves constitute ion-channels, whereas the effects of activating metabotrophic receptors result from changes in intracellular metabolism. It should be borne in mind, however, that the activation of ionotrophic receptors may lead to metabolic changes and the opening of ion-channels may be a secondary consequence of metabotrophic stimulation.

Ionotrophic receptors

Ionotrophic receptors are proteins which themselves constitute an ion-channel (they are usually referred to as ligand-gated ion-channels). Table 30.2. lists some of the receptors for ionotrophic neurotransmitters, the ions they gate and the functional effect their activation causes. The operation of three different types of ionotrophic receptor will be described: the nicotinic acetylcholine receptor and the $GABA_A$ receptor ('typical' ionotrophic receptors), and the atypical receptor for glutamate, designated NMDA.

Typical ionotrophic receptors: the nicotinic acetylcholine- and the $GABA_A$- receptors

Figure 30.11 shows the likely organization of the nicotinic receptor for acetylcholine in the plasma membrane. For nicotinic acetylcholine receptors in the

Table 30.2 **Some examples of ionotrophic receptors (receptors themselves constitute ion-channels)**

Receptor	*Ligand*	*Major ion(s) gated*	*Postsynaptic effect*
Nicotinic	Acetylcholine	Na^+	Fast EPSP
$GABA_A$	γ-Aminobutyrate	Cl^-	Fast IPSP
Glycine	Glycine	Cl^-	Fast IPSP
NMDA	Glutamate	$Na^+/K^+/Ca^{2+}$	Delayed EPSP
AMPA	Glutamate	Na^+/K^+	Fast EPSP

EPSP, excitatory postsynaptic potential; IPSP, inhibitory postsynaptic potential.

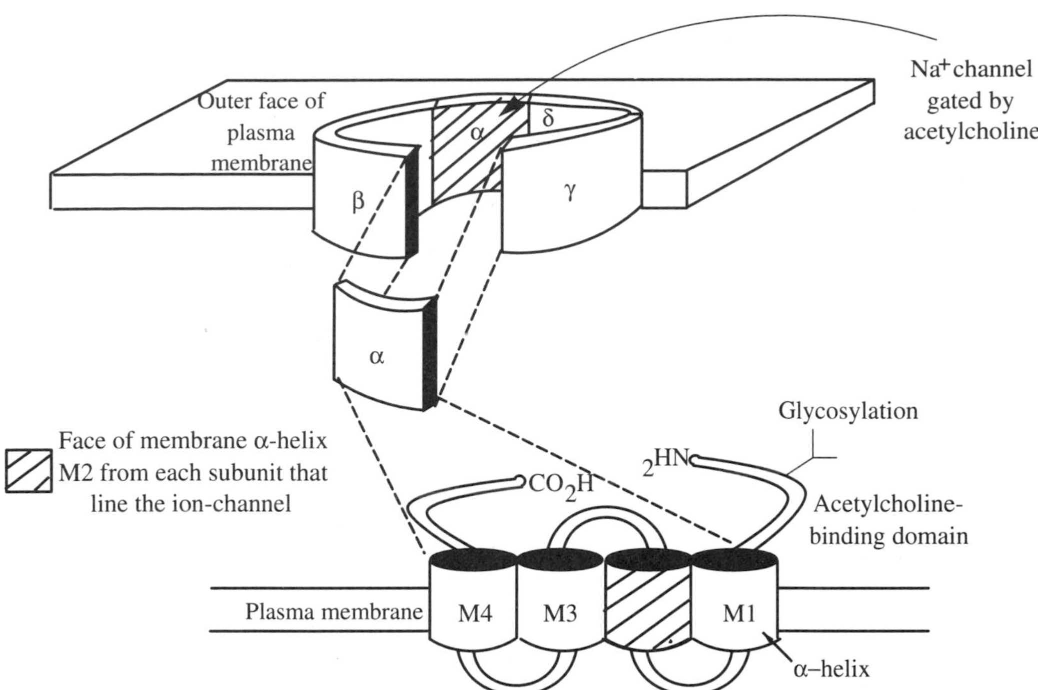

Figure 30.11 **The likely arrangement of the pentameric nicotinic acetylcholine receptor**

peripheral nervous system there are five separate polypeptides chains 2α, β, γ and δ. (The same number of subunits are found in the CNS but the isotypes differ.) Each subunit has the same general organization as the α-subunits in that four α-helical domains span the membrane with the N- and C-termini outside the cell. Binding of acetylcholine to the two α-subunits via the N-terminal domains causes a conformational change in the molecules such that a channel capable of allowing the passage of Na^+ into the cell is formed. The perturbation in the membrane potential caused is referred to as an **excitatory postsynaptic potential (EPSP)**. It should be noted that the activation of a single nicotinic receptor does not trigger an action potential: many EPSPs must be 'summed' at the axon hillock for this to happen.

The binding of GABA to its A-type receptor, which is also a protein with five subunits (usually 2α, 2β and γ) opens a channel that allows the inward movement of Cl^-. This influx of negative charge results in membrane hyperpolarization, an **inhibitory post-synaptic potential (IPSP)**.

Nicotinic acetylcholine receptor dysfunction: myasthenia gravis

Patients presenting with **myasthenia gravis** show characteristic profound muscle weakness. The disorder is an autoimmune disease in which circulating antibodies bind to the **nicotinic acetylcholine receptor** leading to its increased endocytosis and degradation. The consequent reduction in receptor number at neuromuscular junctions causes muscular weakness, as transmission is impaired.

An atypical ionotrophic receptor: the NMDA glutamate receptor

The **NMDA receptor** (so-called because it responds to N-methyl-D-aspartate) resembles the nicotinic receptor, since it is constituted of five subunits. It differs, however, in its cation specificity (it gates both Na^+ and Ca^{2+}) and its mode of operation. This second difference involves the regulation of the opening of the ion channel. Glutamate is able to cause the channel to open only if the membrane has already been partially depolarized. (This requirement is usually met by the activation of a second class of glutamate receptor, the typical ionotrophic AMPA receptor.) In a normally polarized plasma membrane (–70 mV) the channel is refractory to glutamate because it is blocked by Mg^{2+}. As the membrane depolarizes to –20 mV the Mg^{2+} is expelled and the amino acid becomes effective. The effect of glutamate is augmented by the allosteric action of glycine. The drug MK801 can prevent the action of glutamate (and glycine) at this stage. Because the membrane must already be depolarized (to –20 mV) for it to act, **MK801** is referred to as a **use-dependent inhibitor** of the receptor. The operation and inhibition of the NMDA receptor is shown in Figure 30.12.

Tissue damage associated with stroke is mediated by inappropriate activation of NMDA receptors: excitotoxicity

Much attention is currently focused on the NMDA receptor because it seems that to a large extent the damage to neuronal tissues that occurs in stroke victims is mediated by inappropriate activation of this receptor.

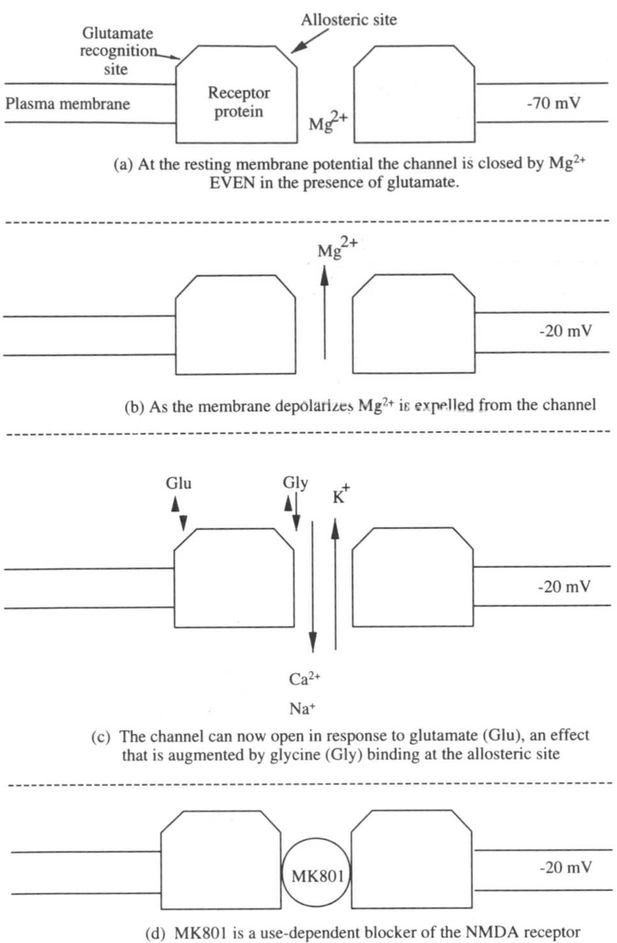

Figure 30.12 **The operation of the NMDA-type glutamate receptor**

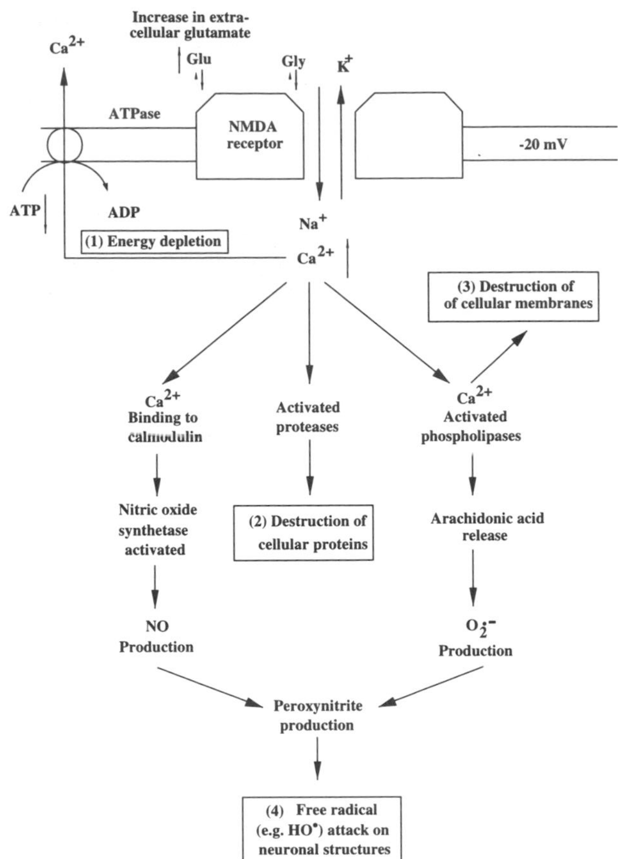

Figure 30.13 **Possible mechanisms underlying the brain-tissue damage in stroke**

In the period of ischaemia and during the subsequent re-establishment of blood flow associated with occurrence of stroke, a massive release of glutamate occurs in affected areas. This results from a reversal of the normal reuptake mechanism for glutamate. The glutamate binds to AMPA receptors and causes depolarization of the neurons. This is followed by NMDA receptor activation that leads to an influx of Ca^{2+} into the cells in an uncontrolled fashion. This proves to be toxic, cell death probably being caused by the Ca^{2+}-dependent activation of proteases and phospholipases. There is an associated increase in free-radical attack on the tissues which also become depleted in ATP, and are unable to carry out their normal metabolic reactions. Figure 30.13 suggests some of the processes believed to be associated with cell death during stroke. Since the influx of Ca^{2+} via the NMDA receptor ion channel lies at the heart of these processes it follows that use-dependent inhibitors of the channel, such as MK801, might prove protective in stroke. This appears to be the case in animal models of stroke, but unfortunately MK801 is not well tolerated in humans. There are hopes that manipulations of the NMDA receptor activity may be useful for treating stroke victims to limit tissue damage, but the rapidity of onset of damage following ischaemia will make this goal very difficult to realize. The prospects may be much better for using such drugs prophylactically to protect against damage from subsequent strokes.

Alzheimer's disease

Alzheimer's disease is a cause of dementia in many elderly people and also some of those patients with Down's syndrome who survive until their fifth decade. Examination by pathologists of the brain of those dying from Alzheimer's disease has revealed the presence of large numbers of **amyloid** ('starch-like' protein) deposits, termed **plaques**, around neurons that themselves contain so-called **neurofibrillary** tangles (abnormally organized neurofilaments). In addition, there is evidence of extensive neuronal cell death. The view is that that the deposition of amyloid β-protein (Aβ, sometimes referred to as 'protease-resistant forms of amyloid'), the main constituent of plaques, is causative, with the occurrence of neurofibrillary tangles, etc., following from this. The interaction of both genetic and environmental factors in the occurrence of the disease is complex, but recently a key observation has been made concerning the formation of Aβ in patients with hereditary, early-onset Alzheimer's disease.

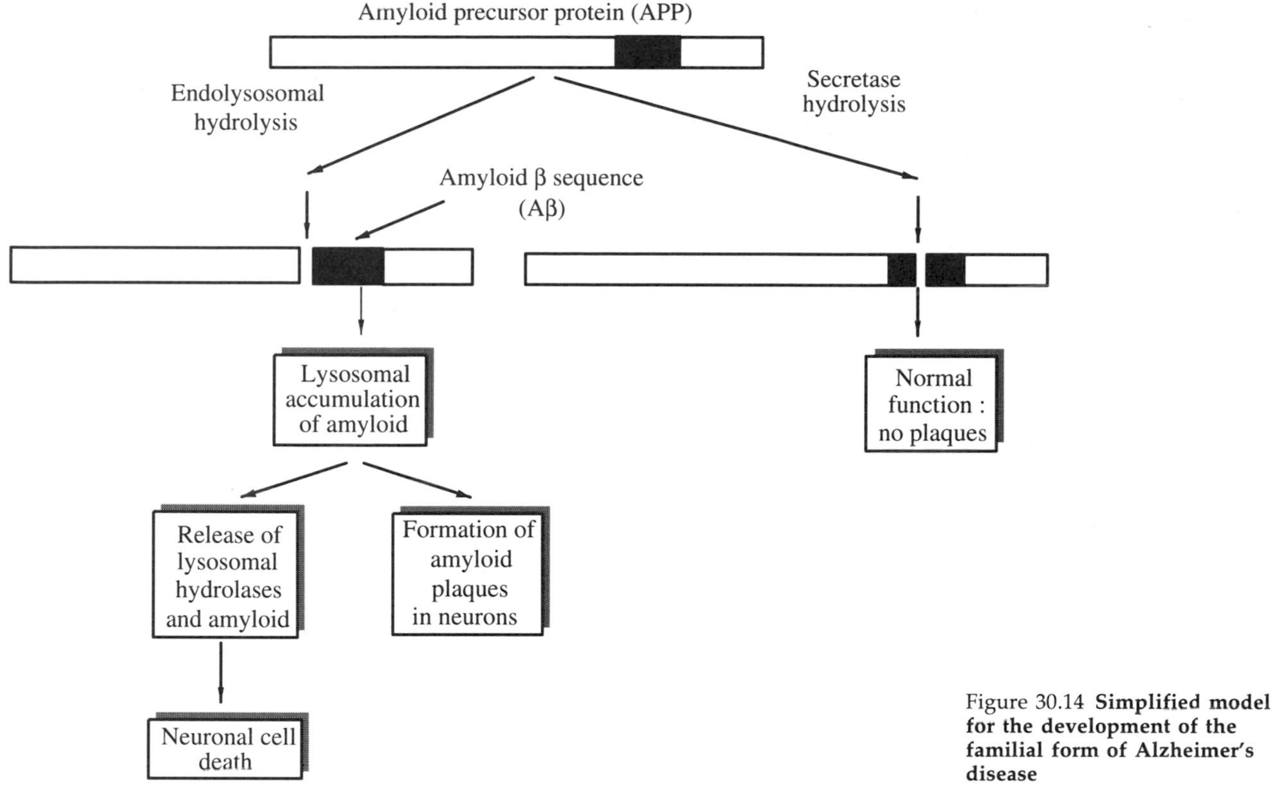

Figure 30.14 **Simplified model for the development of the familial form of Alzheimer's disease**

In these patients there have been identified mutations to a gene that encodes a protein designated **amyloid-precursor protein** (APP), part of the sequence of which corresponds to Aβ. In the course of its synthesis APP is inserted into the membranes of the endoplasmic reticulum. This is normally followed by proteolysis within the Aβ sequence, catalysed by the enzyme secretase. This proteolysis precludes amyloid formation. It seems that, in the patients with the inherited mutations, the APP is no longer a substrate for the secretase. Consequently, APP enters the endolysosomal pathway, where a different protease acts to cause large quantities of Aβ to accumulate in affected neurons. The Aβ produced is suggested to have several effects, all arising from its accumulation in lysosomes. The locally high concentration of protease-resistant amyloid in the acid environment of the lysosome may well result in its precipitation as plaques, thus disrupting the lysosomal membrane. The resulting escape of hydrolases and of amyloid itself would then lead to neuronal cell death due to attack by the hydrolases on cellular components. In addition, it has been suggested that amyloid renders neurons more susceptible to 'excitotoxic' damage (of the type that occurs in stroke), although the mechanism for this is unknown. A possible model for the changes seen in familial Alzheimer's disease is shown in Figure 30.14, but this should be seen as a simplified view of a very complex condition.

The possibility that APP plays a role in other forms of the disease, including its sporadic, late-onset form, is suggested by the finding that the gene for APP is located on chromosome 21: the chromosome that is present in three copies in Down's syndrome (trisomy 21).

Metabotrophic receptors

Metabotrophic receptors are G-protein-coupled receptors (see Chapter 3). Their activation results in changes in the concentration of three second messengers either cAMP or IP_3 and diacylglycerol (DAG). The receptors may be positively or negatively coupled to cAMP production or positively coupled to IP_3/DAG formation. Table 30.3 lists some metabotrophic receptors for transmitters and the nature of their coupling.

Almost invariably the effect of an increase in the concentration of cAMP in cells is to activate protein kinase A. Likewise IP_3/DAG set in train events that lead to the activation of other protein kinases. A calcium/calmodulin-dependent protein kinase (isoenzyme II in the brain) is activated as the result of Ca^{2+} release from intracellular stores. DAG, on the other hand, activates

Table 30.3 **Some examples of metabotrophic receptors (those that promote second messenger changes via G-proteins)**

1. Adenylyl-cyclase activating : cAMP↑
   (i) β_1-Adrenergic
   (ii) D_1-Dopaminergic
2. Adenylyl-cyclase inhibiting : cAMP↓
   (i) α_2-Adrenergic
   (ii) Somatostatinergic
3. Phospholipase C(β) activating : IP_3↑; DAG↑
   (i) α_1-Adrenergic
   (ii) M_1-Muscarinic

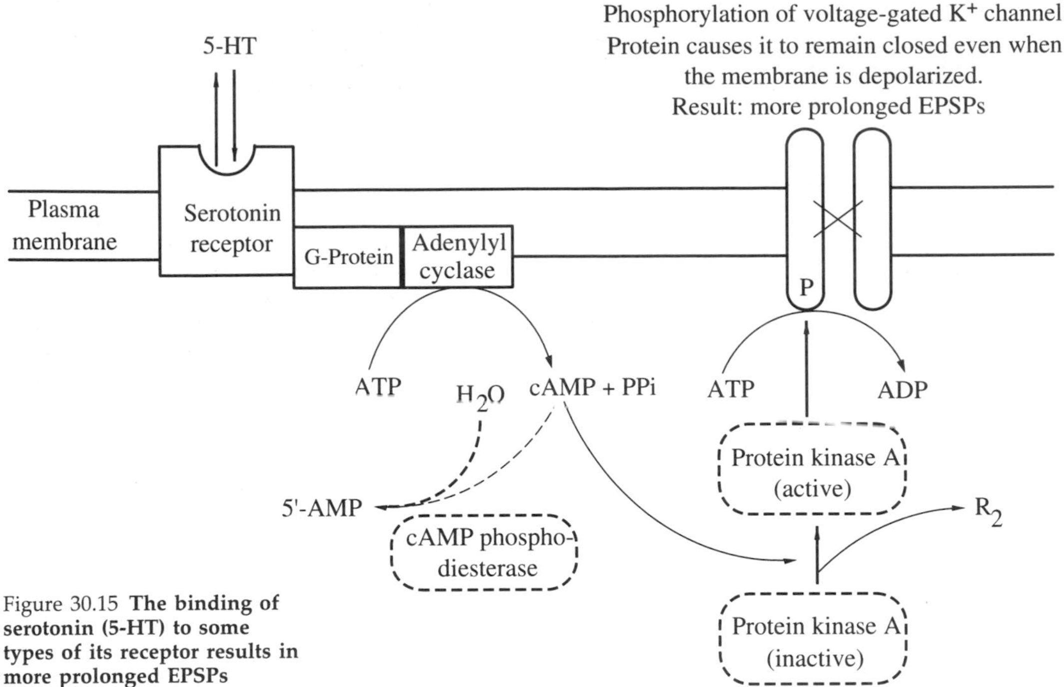

Figure 30.15 **The binding of serotonin (5-HT) to some types of its receptor results in more prolonged EPSPs**

protein kinase C. These three protein kinases catalyse the phosphorylation of specific target proteins and this covalent modification results in changes in the functional activity of the target. In some cases it is possible to trace the events occurring from the binding of the transmitter to its receptor to changes in cellular functional activity. An example of this is seen when serotonin binds to one of its receptors (one coupled positively to adenylyl cyclase, and thus to cAMP production). Activation of this receptor causes the depolarization associated with an EPSP to be more prolonged, a possible mechanism for which is shown in Figure 30.15. This proposes that the target for protein kinase A is a plasma-membrane K^+ channel, which on phosphorylation becomes less likely to open and thereby to terminate membrane depolarization. This effect of serotonin constitutes neuromodulation because, until the phosphate group is removed from the channel protein, the plasma membrane undergoes more prolonged depolarizations.

There are numerous target proteins for the range of protein kinases activated by second messengers. For example, in dopamine neurons protein kinase A causes the activation by phosphorylation of tyrosine hydroxylase, the rate-limiting enzyme for the biosynthesis of the catecholamines (see Section 30.4). The Ca^{2+}/calmodulin-dependent protein kinase II of some nerve cells promotes the phosphorylation of the protein synapsin I which is associated with vesicles containing transmitters to be released. The phosphorylation of this target protein renders it more able to facilitate the fusion of the vesicles with the plasma membrane with release of the transmitter. Another type of protein target for protein kinase A is a DNA-binding protein designated **cAMP response-element binding protein** (CREB). On phosphorylation this protein binds at specific positions 'upstream' (i.e. in the 5′ direction) of genes that are controlled by cAMP to enhance their transcription. One such gene is that for pro-opiomelanocortin, which encodes the neuropeptide β-endorphin (see Chapter 28). Proteins such as CREB are known as STAT (signal-transducing activators of transcription) proteins.

Table 30.4 summarizes the ways that occupation of metabotrophic receptors can influence neuronal activity.

Table 30.4 **Summary of the effects resulting from the activation of some metabotrophic receptors**

Receptor type	*Second messenger*	*Protein kinase activated*	*Target protein phosphorylated*	*Effect*
Adenylyl cyclase activating	cAMP	Protein kinase A	1. Some K^+ channel proteins	Prolonged depolarization of plasma membrane
			2. Tyrosine hydroxylase	Increased rate of synthesis of catecholamines
			3. cAMP response element binding (CREB) protein	Increased transcription of target genes, e.g. pro-opiomelanocortin (POMC) gene
Phospholipase C activating	IP_3	Ca^{2+}/calmodulin protein kinase II	1. Synapsin I	Increased rate of fusion of vesicles containing transmitter to plasma membrane

30.6 Myelin

Myelin is a very important constituent of the white matter of the brain where the myelin sheath forms 50% of the total dry weight. The presence of myelin largely explains the gross differences in chemical analysis found between white and grey matter, and accounts for the glistening appearance and high lipid content of white matter.

The myelin sheath consists of the extended plasma membrane of glial cells that wraps round axons in spiral fashion (Figure 30.16). The particular glia involved are the Schwann cells for peripheral nerves while oligodendrocytes invest nerves within the central nervous system. Oligodendrocytes and Schwann cells differ in that the former may send their plasma membrane around several axons, whereas the latter ensheath single axons. Myelin sheaths from single glia, whether in peripheral or central locations, do not cover the whole length of an axon. Instead the myelin sheath is segmented with portions of uncovered axon being found at intervals between individual sections of glial cell plasma membranes. These regions are known as nodes of Ranvier.

Function of myelin

Myelin acts as an electrical insulator surrounding the axon, thereby facilitating conduction. In particular, it reduces the capacitance per unit length of axon, thereby promoting the speed of spread of local current.

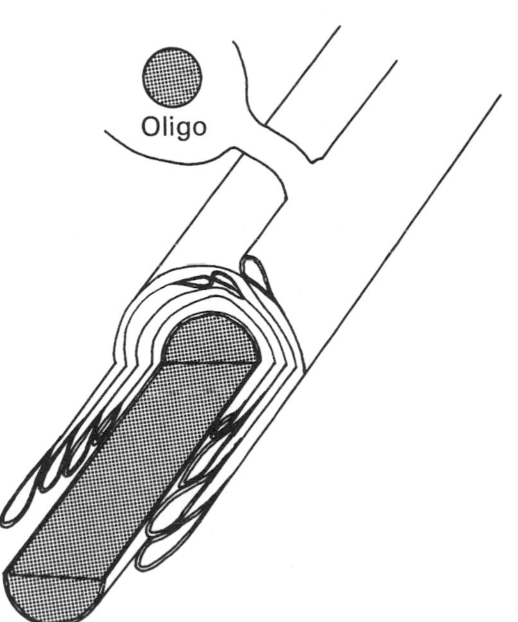

Figure 30.16 **Myelin layers surrounding an axon in the central nervous system.** Oligo: oligodendrocyte, the central nervous system equivalent of the Schwann cell. Oligodendrocytes may provide the myelin sheath for several axons

Ultimately, however, conduction in myelinated fibres depends on the sheath being interrupted periodically at the nodes of Ranvier, where voltage-gated Na^+ channels are concentrated (see Section 30.4).

Composition of myelin

Myelin is a lipid-rich structure with the predominant lipids being cholesterol, sphingolipids (sphingomyelin, gangliosides, cerebrosides, sulphatides) and phosphoglycerolipids (phosphatidylcholine, phosphatidyl ethanolamine and plasmalogens). The major fatty acid associated with these complex lipids is the mono-unsaturated oleic acid, with polyunsaturated fatty acids being poorly represented.

The predominant proteins found in myelin in the CNS are a proteolipid and myelin basic protein. The former is an integral and the latter a peripheral protein of the oligodendrocyte plasma membrane and their joint roles seem to be in the facilitation of the close apposition of the membrane stacks that form the myelin sheath. Some interest has focused on myelin basic protein because when it is injected into experimental animals it evokes an antibody response which is associated with a disease of the brain which involves focal areas of inflammation and demyelination of the type seen in multiple sclerosis.

The consequences of demyelination: multiple sclerosis

The great importance of myelin to the proper functioning of the CNS is vividly illustrated by those diseases in which hypomyelinatiom or demyelination occurs. **Multiple sclerosis** (MS) is by far the most common demyelinating disease seen in the UK. In the Orkney and Shetland Islands the disease is seen in 152 per 100 000 of the population and in Aberdeen the prevalence is 127 per 100 000. In the UK as a whole there are 50 000 registered sufferers (and many more not registered because they have not yet reached the stage of disablement). As a consequence, about 90% of all neurological diseases seen in the UK prove to be MS. The disease also shows a sexual preference with women (especially Caucasian women) developing the disease at an earlier age and in roughly double the numbers of Caucasian males. This bias towards females is especially marked in those parts of the world in which the prevalence rates are low, e.g. in Istria in former Yugoslavia

In MS there is an inflammatory response to some, as yet, uncharacterized factor in the CNS resulting in the breakdown of myelin in isolated unpredictable, regions. The demyelination is largely irreversible, and the outcome is that nerve conduction is impaired, or even lost entirely, giving rise to neurological symptoms such as tingling and numbness in the extremities, and blurred vision leading to blindness. Thus MS is a disease of conduction brought about by a specific membrane-directed pathogenesis (although peripheral nerves are not affected).

The cause of the disease is still to be fully resolved, but there are some clues. Rare 'epidemics' of MS suggest some sort of vector may be involved. In the Faroe Islands prior to 1939 the prevalence was zero. By 1950, during which time British troops had been stationed on the islands, it had risen to 42 per 100 000 population rising to 64 per 100 000 in 1961. Thereafter it followed a steady decline to 34 per 100 000 by 1977. These and other observations on the aetiology of the disease suggest that a latent viral infection may be a cause of the disease.

The course of the disease is well established. Macrophages are seen to invade areas of inflammation and there they promote the destruction of myelin. When the inflammation abates remyelination may occur to some extent, but the disease is characterized by relapses which progressively leave the sufferer with more disabilities.

The detection of lesions in the brain: magnetic resonance imaging (MRI)

The drive to develop non-invasive methods to study the metabolism of the brain in living subjects has led to the introduction of **magnetic resonance imaging**. The method has proved particularly useful in the detection of brain lesions associated with MS and tumour growth, for example. The basis of all nuclear magnetic resonance techniques is that the nuclei of certain atoms (those with an odd number of nucleons – protons and neutrons) have an intrinsic 'spin' and these include the proton (^{1}H) and ^{31}P. A consequence of the spin is that the nuclei generate a magnetic field. The field generated will 'line-up' parallel to an external, applied magnetic field (usually of several hundred millitesla). Such nuclei are referred to as being in a 'low-energy' state. However, by the application of a pulse of radiofrequency electromagnetic radiation (of frequency about 40 MHz) the field associated with the nuclei can be induced to lie antiparallel to the imposed field – a 'high-energy' state. The change from parallel to antiparallel alignment is referred to as 'resonance'; this may be detected by suitable means and the data analysed by computer-assisted methods (tomography). The particular resonances detected by MRI are those of the protons in water. The nature of the signals detected from such protons depends on the chemical environment of the water. This may be 'structured' in a particular region of a normal brain but less 'structured' in the corresponding region of a brain in which the presence of lesions is causing disruption, e.g. near the plaques in MS. The way the machines operate means that areas containing lesions show up as white patches on a grey background.

When the growth of a tumour has caused a breakdown in the blood–brain barrier, chelates of gadolinium may be given by systemic injection. This paramagnetic metal then serves to 'enhance' the contrast between normal tissue and the tumour.

30.7 Metabolism in the brain

Carbohydrate metabolism

In a 70 kg man, the mass of the brain is about 1.5 kg, yet this relatively small mass of tissue receives 15% of the cardiac output and accounts for 20% of the total oxygen consumption of the body at rest. Under normal nutritive conditions the respiratory quotient (*RQ*, mol oxygen consumed over mol carbon dioxide produced) is close to unity, indicating that carbohydrate is the major fuel for respiration.

In order to try to understand the nature of the particular requirement of the brain for blood glucose (stores of the hexose as the free molecule or in the form of the glycogen found in glia are sufficient only to meet energy requirement for a few minutes), it is useful to study the effects of hypoglycaemia. These are seen in most dramatic form in subjects accidentally given high doses of insulin, when convulsions can result. It is also known that recovery from these hypoglycaemic symptoms following the injection of glucose is remarkably rapid, intravenous glucose acting within 30 s. The reason for this critical need for glucose is not clear, because it has been shown in animals given insulin that there need not be a detectable change in respiration, ATP or phosphocreatine concentrations prior to the onset of seizures. It may be, therefore, that glucose is required critically for metabolic processes not related to energy provision. The rapid use of glucose as a precursor for the formation of transmitters such as acetylcholine may be the key to this.

Recently the introduction of **positron-emission tomography (PET)** has greatly facilitated the virtually non-invasive study of glucose and energy metabolism in the brain.

In PET studies the subject is given a small quantity of 2-deoxyglucose that has been labelled radioisotopically with ^{18}F in the 2-position (2-^{18}fluoro-2-deoxyglucose). The rationale of the method is that those areas of the brain that are metabolically most active will also be most active in their uptake of glucose, and hence of the fluorodeoxy analogue. As is indicated in Figure 30.17, both glucose and the analogue are substrates for hexokinase but the resulting 6-phosphates have quite different properties. Thus glucose 6-phosphate is both a

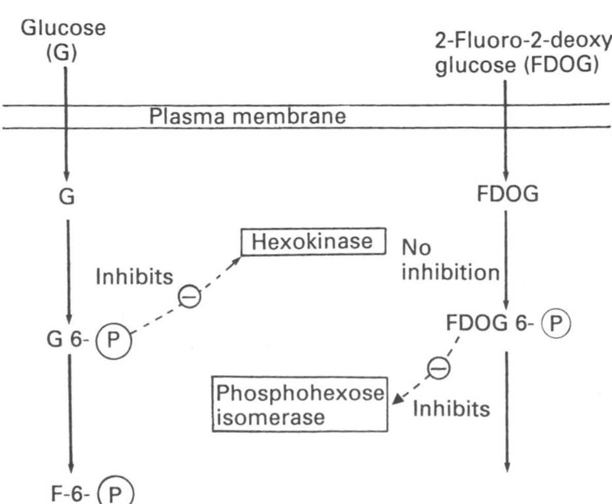

Figure 30.17 **The radioactive compound 2-^{18}F-2-deoxyglucose is taken into neurons, converted into its 6-phosphate which then accumulates.** The ability of the 6-phosphate of the analogue to inhibit the isomerase means that the cell continues to take up more of the fluoro compound

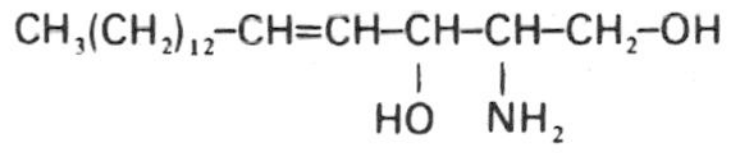

$CH_3(CH_2)_{12}$–CH=CH–CH–CH–CH_2–OH
(HO on the first CH, NH_2 on the second)

Sphingosine

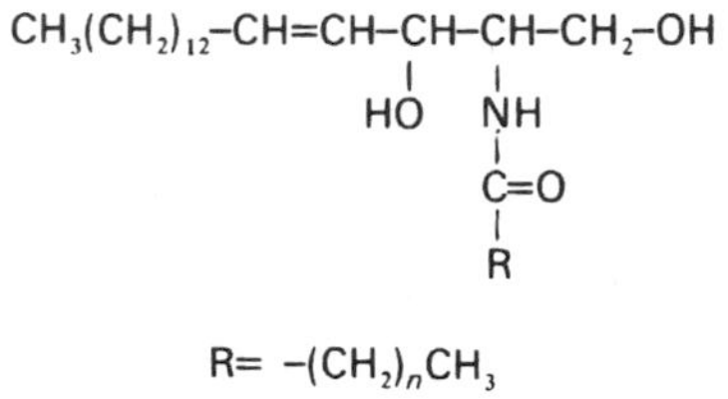

$CH_3(CH_2)_{12}$–CH=CH–CH–CH–CH_2–OH
(HO on the first CH; NH–C(=O)–R on the second)

R= $-(CH_2)_nCH_3$

1 Ceramide (*N*-acylsphingosine)

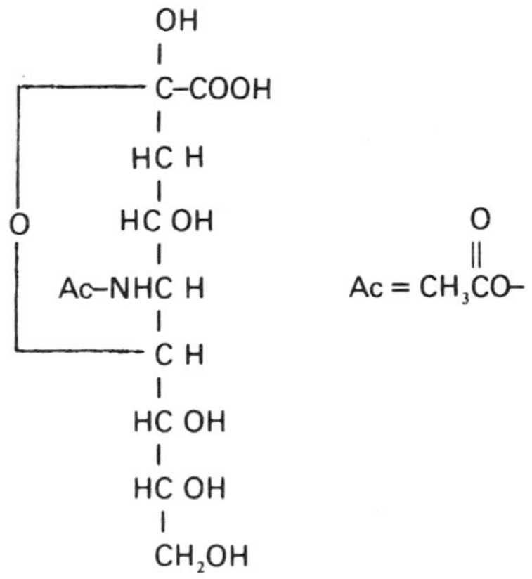

OH
|
C–COOH
|
HC H
|
HC OH
|
Ac–NHC H
|
C H
|
HC OH
|
HC OH
|
CH_2OH

(ring O linking C–COOH and C H)

Ac = CH_3CO– (C=O)

N-Acetylneuraminic acid (NAcNeu)

Figure 30.18 **Molecular units used in the formation of gangliosides**

Ceramide–Glc–Gal (NAcNeu on Gal) — *G_{M3}, haematoside*

Ceramide–Glc–Gal–NAcGal (NAcNeu on Gal) — *G_{M2}, Tay–Sachs ganglioside*

Ceramide–Glc–Gal–NAcGal–Gal (NAcNeu on first Gal) — *G_{M1}*

Ceramide–Glc–Gal–NAcGal–Gal (NAcNeu on first Gal; NAcNeu on terminal Gal) — *G_{D1a}*

Ceramide–Glc–Gal–NAcGal–Gal (NAcNeu–NAcNeu on first Gal) — *G_{D1b}*

Ceramide–Glc–Gal–NAcGal–Gal (NAcNeu–NAcNeu on first Gal; NAcNeu on terminal Gal) — *G_{T1}*

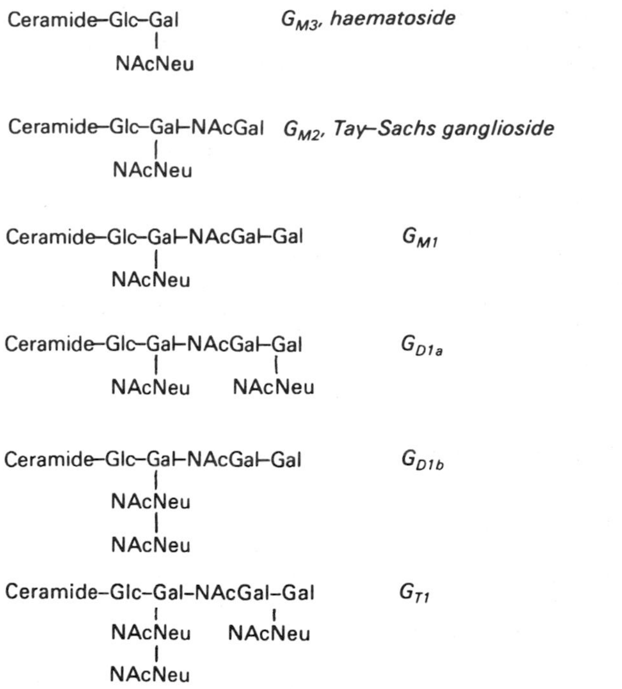

Figure 30.19 **Structures of the major gangliosides of nervous tissues** (Glc, glucose; Gal, galactose; NAcGal, *N*-acetylgalactosamine; NAcNeu, *N*-acetylneuraminic acid)

substrate for phosphohexose isomerase and a product inhibitor of hexokinase, while the fluoro-compound is neither. This means that the 2-deoxy derivative enters the glycolytic pathway to be 'trapped' as a phosphate ester. As a result, metabolically active regions will accumulate the isotope which decays with the emission of positrons. These antiparticles undergo local annihilation by encounter with electrons (their corresponding particles) to release a pair of gamma photons in directions 180° apart. These are readily detected by a conventional computerized tomographic (CT) scanner. With computer assistance the origin in the brain of the photons may be determined, as can the relative numbers coming from each brain area. This produces a 'metabolic map' of the brain.

A similar approach can be used to detect areas with particular concentrations of other metabolites in the brain. For example, fluoro-derivatives with a high affinity for dopamine receptors can be used to map out dopaminergic nerve terminals in the brain. This is potentially useful in recognizing early signs of degeneration in patients suspected of having parkinsonism, for example.

Lipid metabolism

The nature of the lipids found in the brain and the anabolic and catabolic pathways involving them are very similar to those found in other organs. However, the special role of lipids in brain function, e.g. in electrical processes, means that the brain is particularly vulnerable to disorders of lipid metabolism. The disorders seen almost invariably involve catabolic processes with abnormal accumulation of lipids in the brain (and other tissues, for example bone marrow). An important group of diseases of this type are the **sphingolipid storage diseases (sphingolipidoses)**

Sphingolipids are based on sphingosine (Figure 30.18) that is acylated to form ceramide which, in turn, is linked to simple carbohydrates to form **cerebrosides**. The further addition of sialic acid (*N*-acetylneuraminic acid) gives rise to the **gangliosides**. The presence of one silalyl residue is signified by a subscript 'M', two by 'D', etc., with the dispositions of the residue being specified by the number 1, 2 and letters a, b (Figure 30.19).

The first sphingolipidosis to be described clinically was recognized at the end of the last century by Tay and Sachs. As a consequence the condition has now become known as **Tay–Sachs disease**. The reason for the accumulation of the sphingolipid that characterized the disease was not understood until 1965 when it was shown that a related condition termed **Gaucher's disease** was associated with an inherited deficiency in β-glucosidase, a lysosomal hydrolytic enzyme required to catalyse the hydrolysis of glucocerebrosides. In the absence of this catabolic enzyme, its substrate accumulates in the brain. It is now apparent that a large

Table 30.5 **Metabolic diseases characterized by inabilities to degrade sphingolipids**

Disease	*Major sphingolipid accumulated*	*Enzyme defect*	*Clinical symptoms*
Niemann–Pick	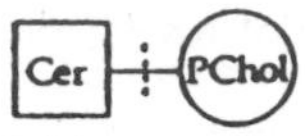 Sphingomyelin	Sphingomyelinase	Generally similar to Gaucher's disease; 30% with cherry-red spot in macula; marrow cells (foam cells) stain for both lipid and phosphorus
Gaucher	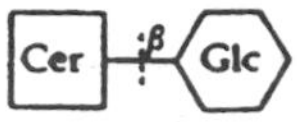 Ceramide glucoside (glucocerebroside)	β-Glucosidase	Mental retardation (infantile form only); hepatosplenomegaly; hip and long-bone involvement; oil red and periodic acid–Schiff positive lipid-laden (Gaucher) cells in bone marrow
Metachromatic leucodystrophy	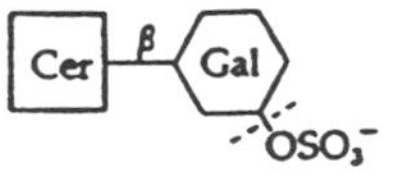 Ceramide galactose 3-sulphate (sulphatide)	Sulphatidase	Mental retardation; psychological disturbances (adult form); decreased nerve-conduction time; nerve biopsy shows yellow-brown droplets when stained with cresyl violet (metachromasia)
Ceramide lactoside lipidosis	 Ceramide lactoside	β-Galactosidase	Slowly progressing CNS impairment; organomegaly; macrocytic anaemia, leucopenia and thrombocytopenia due to involvement of bone marrow and spleen
Fabry	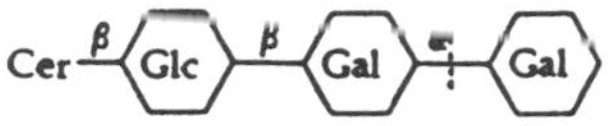 Ceramide trihexoside	α-Galactosidase	Reddish-purple maculopapular rash in umbilical, inguinal, and scrotal areas; renal impairment; corneal opacities; peripheral neuralgias and abnormalities of ECG
Tay–Sachs	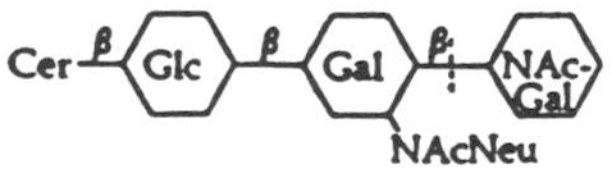 Ganglioside G_{M2}	Hexosaminidase A	Mental retardation; amaurosis; cherry-red spot in macula; neuronal cells distended with 'membranous cytoplasmic bodies'
Generalized gangliosidosis	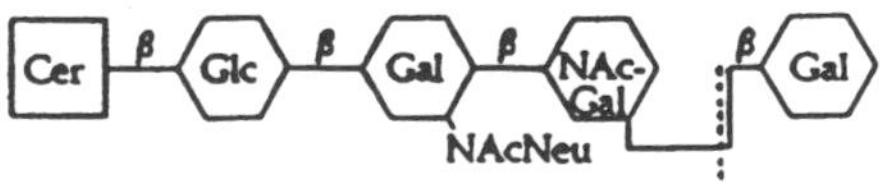 Ganglioside G_{M1}	β-Galactosidase	Mental retardation, cherry-red spot in macula; hepatomegaly; bone marrow involvement

Cer = *N*-acylsphingosine, or ceramide.
NAcNeu = *N*-acetylneuraminic acid.
Gal = galactose.
Glc = glucose.
PChol = phosphatidylcholine.
NAcGal = *N*-acetylgalactosamine.

group of sphingolipidoses are caused by mutations to the genes encoding sphingolipid-specific hydrolytic enzymes. More recently it has been recognized that the lysosomes contain a series of non-enzyme proteins, called sphingolipid-activating proteins (SAPs), that are required to render the sphingolipids suitable as substrates for the hydrolytic enzymes, and patients with mutations to these proteins have been described. In general, patients with these conditions present as having a partial defect in one or several of the hydrolytic enzymes. Conditions involving a failure of sphingolipid catabolism are listed in Table 30.5.

Amino acid and protein metabolism

Most of the processes involving the metabolism of amino acids and proteins have their counterparts in other tissues, so the present section will focus on brain-specific processes.

If the concentration of free amino acids in the brain and plasma are compared, it is found that certain amino acids, such as glutamate and glutamine, are present in much larger amounts in the brain. Indeed some two-thirds of the free amino nitrogen in the brain is accounted for in terms of glutamate and its immediate derivatives (glutamine and GABA).

N-Acetylaspartate occurs only in the brain where its concentration may exceed that of aspartate by 2–3-fold. Its appearance in the brain is developmentally controlled, being present in very low concentration at birth, but rising during development. Recent evidence suggests that its role in the developing brain is to act as a ready source of acetyl groups for lipid synthesis.

The homeostasis of amino acids in the brain is maintained by the activities of at least four transport proteins found in the luminal or contraluminal (brain side) plasma membranes of the endothelial cells that line the brain capillaries and constitute the blood–brain barrier. Two of these transporters are specific for neutral amino acids and one each handles the acidic and the basic amino acids, respectively. The acidic amino acid transport system operates to promote the efflux of glutamate (and to a lesser extent aspartate) and is located predominantly in the contraluminal membranes. In the case of glutamate this seems to be important: it is a means of controlling inappropriate exposure of the brain to this potentially excitotoxic amino acid. In view of its contraluminal location and specificity it is likely that the first of the two carriers of neutral amino acids facilitates the efflux of alanine and glycine. The remaining two carriers both seem to be able to promote bidirectional movement of their preferred molecules, but the net movement is inwards. The second of the carriers for neutral amino acids transports virtually all of the amino acids in this category, but it has a higher affinity for the more bulky members of the group: leucine, isoleucine, phenylalanine, tyrosine and tryptophan. This carrier is found in the luminal membranes, as is the fourth carrier for arginine, lysine and ornithine. The overall strategy seems to be that the systems operate to promote the efflux of potentially harmful neurotransmitters, amino acids such as glutamate, from the brain and the uptake of essential amino acids into the brain. In the latter category the transmitter precursors tyrosine and tryptophan are especially important and conditions that interfere with the uptake of these are very serious. This seems to be the case in phenylketonuria (PKU) in which the high circulating concentrations of phenylalanine act competitively to prevent the uptake of tyrosine and tryptophan into the brain in necessary amounts. Sufferers from untreated PKU have massive neurological deficits.

The biosynthesis of proteins in the brain closely resembles those in other tissues, although some unusual features are seen when considering the proteins found in axons, dendrites and nerve terminals. Transport of these proteins from the perikarya along axons does occurs as does that of some mRNA species. Studies with labelled proteins show that they move from the nerve cell body through the axon in two major 'waves', referred to as 'rapid flow' and 'slow flow'. Rapidly transported proteins move at a rate of several hundred millimetres per day, but most of the proteins move more slowly at a rate of several millimetres per day. The mechanism of transport involves microfilaments, microtubules and contractile proteins.

Chapter 31

The eye

31.1 Introduction

It is usual in discussion of the biochemistry of the eye to limit the scope to a consideration of the retina, but it is equally important to understand how the rest of the tissue is so constructed as to allow light to impinge on the photosensitive cells. As a consequence of this overriding requirement, those structures through which the light passes (Figure 31.1) must have minimal light-scattering and light-absorbing properties. This means that there can be no blood vessels, because haemoglobin absorbs light strongly, and the cells of the cornea and the lens can contain very few mitochondria with their complement of haemoproteins, the cytochromes. Thus nutrients are brought to the cornea and lens by the slowly circulating aqueous humour, and ATP is formed mainly by the glycolysis of glucose. Nevertheless, oxygen diffuses to these tissues (also the anterior surface of the cornea is in direct contact with the air); consequently there is a strong likelihood of photo-oxidation in both the lens and the cornea (and also in the retina).

Whereas the lens and cornea have little or no oxidative metabolism, the **retina** is an extension of the brain and shares with that tissue a vigorous oxidative metabolism and a great sensitivity to hypoxia. The respiratory rate of the retina is about 60 times that of the lens, but despite this the retina forms relatively large amounts of lactate. It has been estimated that in man 70–80% of the oxygen required by the retina comes from the choroidal circulation, and the remainder from the retinal vasculature. A blood–retinal barrier, analogous with the blood–brain barrier, is responsible for the maintenance of the microenvironment of the retinal cells. It is located in the endothelial cells of the retinal blood vessels and in the single layer of ependymal cells on the outer surface of the tissue, referred to as the pigment epithelium. In both these locations a mechanical barrier is formed by tight junctions between the cells. In addition, transport mechanisms are present which promote the movement of substrates into or from the retina.

31.2 The cornea

The **cornea** is a multilayered structure deriving the greater part of its thickness and its strength from a collagenous stroma. This stroma is a connective tissue of

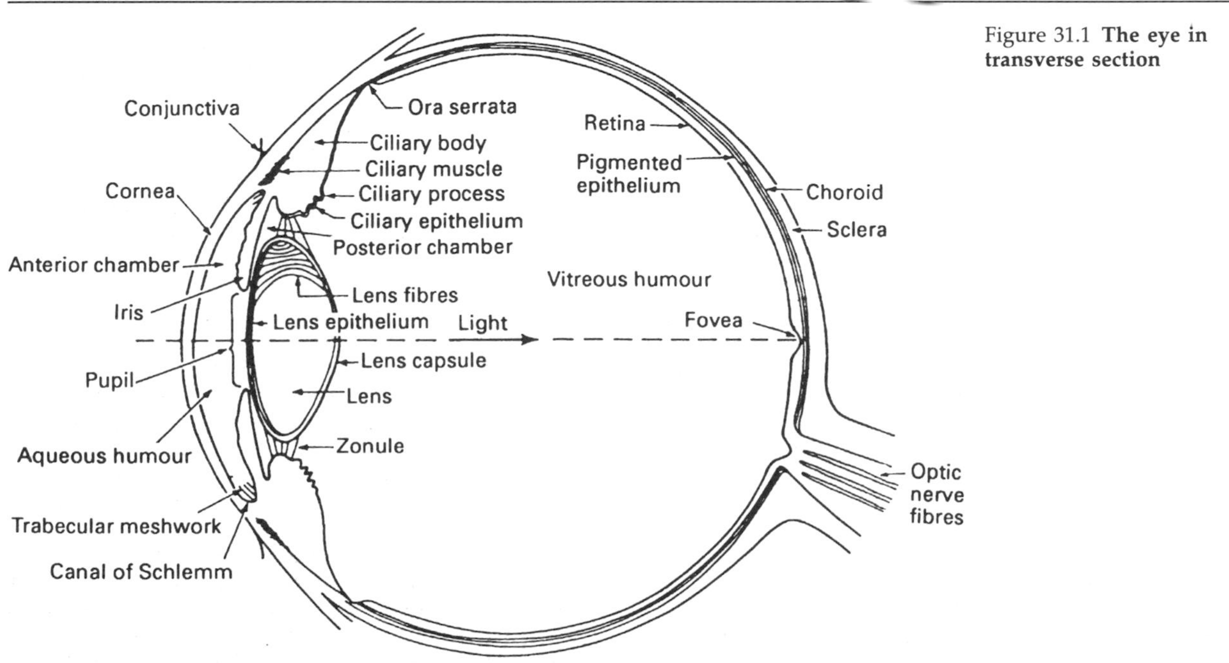

Figure 31.1 **The eye in transverse section**

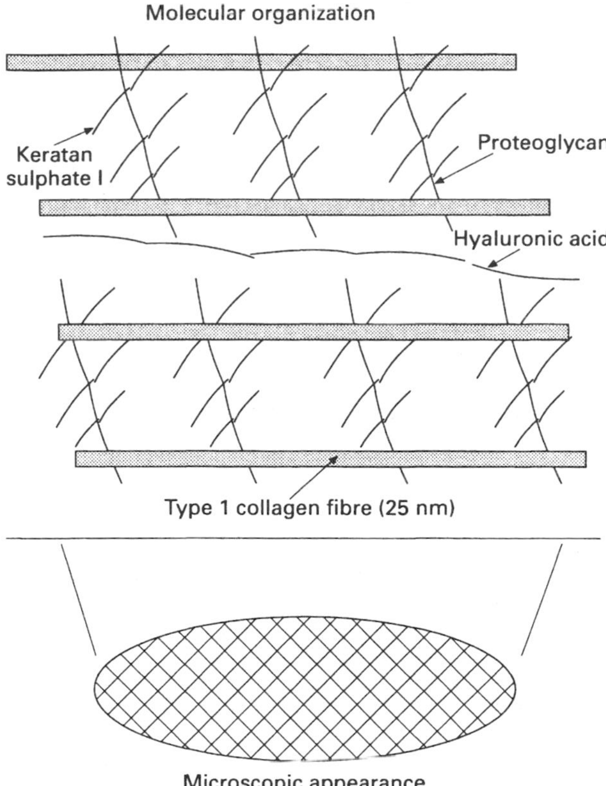

Figure 31.2 **Ordering of the collagen fibres of the cornea helps to ensure destructive interference of scattered light wavelets**

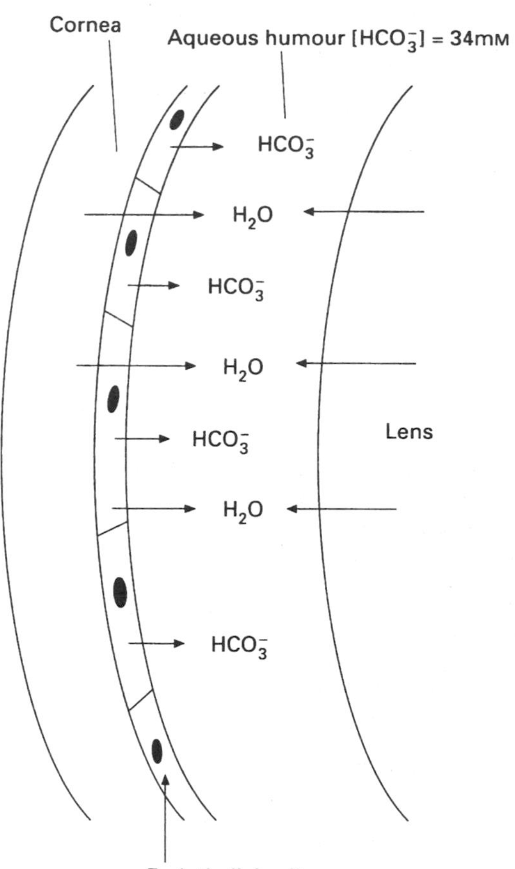

Figure 31.3 **An ATP-dependent bicarbonate 'pump', located in the endothelial cells lining the inner aspect of the cornea, helps to keep both the cornea and the lens relatively dehydrated**

mesenchymal origin. The type I collagen fibres are kept in orderly array by a special glycosaminoglycan, keratan sulphate I, which lies between them (Figure 31.2). The orderly arrangement of the fibres is an important feature of the cornea, helping to ensure transparency in that any scattered wavelets of light tend to remove each other by destructive interference. The presence of keratan sulphate I tends to cause the uptake of water into the cornea and this could lead to disruption of the fibre arrangement. This tendency is counteracted by the aqueous humour being kept hypertonic with respect to the cornea. The active 'pumping' of bicarbonate ions by the endothelial cells of the inner aspect of the cornea may help to maintain a hypertonic aqueous humour (Figure 31.3).

Glucose delivered to the cornea partitions almost equally between glycolysis to give lactate, and the pentose phosphate pathway. (see Chapter 6). The latter pathway is of importance in the generation of NADPH, the reducing power of which is used to prevent or reverse any photo-oxidation in the tissue. The role of NADPH will be discussed in more detail with respect to lenticular metabolism.

31.3 The lens

The lens is of ectodermal origin, the cells arising as a single layer of epithelium at the anterior surface. These cells differentiate into fibre cells, and as the lens ages they are displaced towards the interior to form what is known as the 'nucleus' of the lens, which consequently contains fibres formed by the fetus. The cells are never lost, so the nucleus enlarges with age. The periphery of the lens is the principal site of protein synthesis where the characteristic proteins, termed **'crystallins'**, are produced. There is evidence that the nature of the crystallins produced alters with age. This means that the lens is heterogeneous with respect to its protein composition, conferring on the tissue all of the optical benefits which heterogeneous (compound) lenses have over simple ones, e.g. elimination of aberration. The maintenance of high concentrations of relatively dehydrated crystallins (termed 'α', 'β' and 'γ crystallins') seems to be the strategy whereby the lens maintains transparency. When the concentration of these proteins in physiological saline solution is increased above a certain value, the light-scattering properties of the solution decline to low levels when the concentration approaches that found in the lens (Figure 31.4).

Experiments *in vitro* suggest that 90% of the glucose used by the whole lens undergoes glycolysis to lactate, while most of the remaining 10% enters the pentose phosphate pathway. Most of the NADPH generated as a

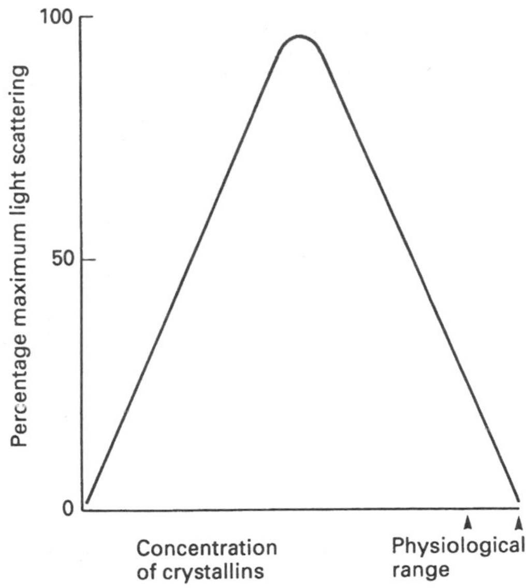

Figure 31.4 **At high concentrations (comparable to those found in the lens), the light-scattering properties of the crystallins are greatly reduced**

result of the latter process is used to maintain the high concentration of glutathione (20 mM) in the reduced form by the action of glutathione reductase (see Chapter 23). The lens also has a high concentration of a second reducing agent, ascorbic acid, which is present at 2 mM.

Cataract formation

The reason for the active pentose phosphate pathway and high concentrations of reducing agents in the lens can best be understood by considering a disease state, **cataract**, in which these systems are, very often, overwhelmed. The clouding and opacity of the lens which characterize the condition may arise in a variety of ways (Table 31.1). The relative importance of the various risk factors varies geographically, with combined malnutrition and chronic diarrhoea being major causes in India, for example. In the UK, more likely causes are high circulating concentrations of monosaccharides and radiation. The mechanisms which underlie the effects of these risk factors are usually oxidative and/or osmotic in nature.

Table 31.1 **Factors tending to lead to the development of cataract and their mechanisms of action**

Risk factors
Diabetes mellitus
Galactosaemia
Radiation (X-ray, infrared, microwaves (?))
Chronic diarrhoea
Malnutrition
Renal failure
Drugs (steroids, ecothiopate iodide (phospholine iodide), diuretics, major tranquillizers)
Genetic

Mechanisms
Oxidation
Osmotic
Phase separation
Chemical modifications of proteins (by steroids, glucose, glucose 6-phosphate, cyanate)

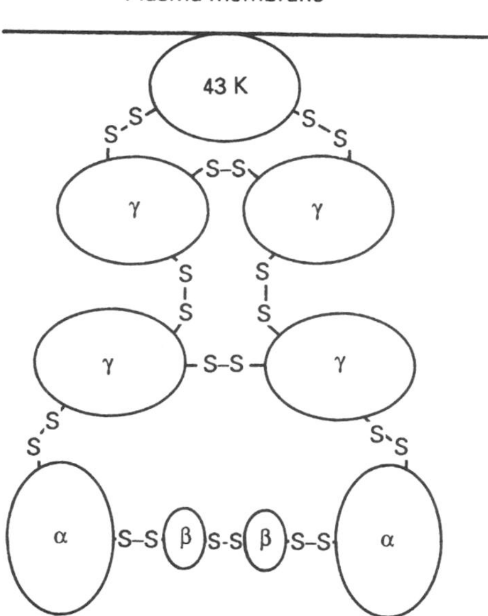

Figure 31.5 **A possible model for cataract formation in the presence of high concentration of Ca^{2+} and low concentrations of reduced glutathione.** (a) Step 1 – Ca^{2+} promotes the binding of protein of RMM 43 000 to the plasma membrane. (b) Step 2 – the crystallins become bound to this protein by disulphide bridges forming light-scattering aggregates

Exposure to X-rays or to other ionizing radiation can lead to cataract formation, and the ability of a given dose of radiation to induce cataract is related to the frequency of mitosis, i.e. young animals are most susceptible. In addition, there is a latent period before overt lesions are seen, but this period is shorter the higher the dose. Biochemically the first measurable event is a fall in the

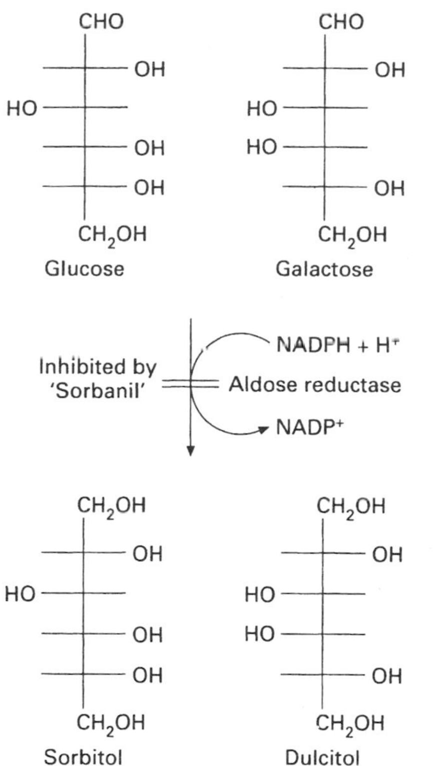

Figure 31.6 **Glucose and galactose in high concentration are converted into the sugar alcohols sorbitol and dulcitol in the lens of the eye.** Sorbitol and dulcitol are 'trapped' inside lenticular cells which thus retain an osmotic equivalent of water

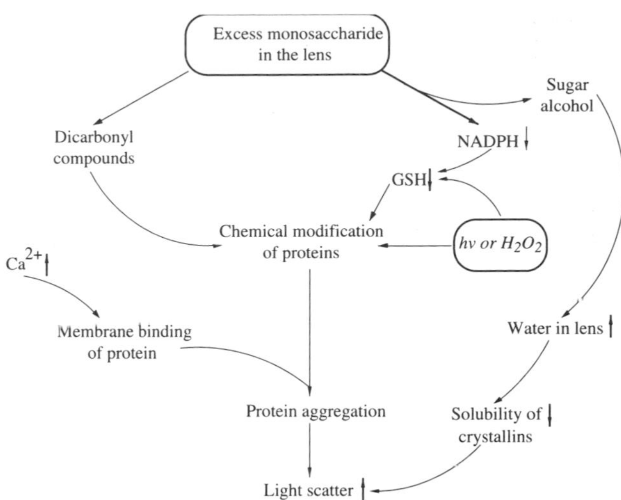

Figure 31.7 **A summary of some of the causes of cataract**

amount of reduced glutathione in the lens. This decrease precedes any overt signs of cataract or opacity, but continues as the lesion develops. With complete opacity the concentration of reduced glutathione will have dropped by 85%. As a secondary event, the concentration of free thiol groups in the crystallins decreases. It has also been noted that irradiation causes the activity of glutathione reductase to fall. Finally, measurements with microelectrodes have shown that the concentration of Ca^{2+} in cataractous areas is 10–100 times higher than that found in adjacent transparent areas. It has been proposed, therefore, that following exposure to radiation the concentration of reduced glutathione falls and that of Ca^{2+} rises. Calcium ions promote the binding of a protein of RMM 43 000 to the plasma membrane of lenticular cells. This forms a focus for the accumulation of crystallins (cross-linked by disulphide bridges) no longer maintained in the reduced form by glutathione. As they grow, the aggregates become increasingly light scattering (Figure 31.5).

Studies on the effects of monosaccharides serve both to emphasize the importance of reduced glutathione and also to point to the requirement of lenticular cells for a correct osmotic environment. It is well known that galactosaemic infants and diabetic patients are especially prone to developing cataracts. In addition, in experimental animals, cataracts may be induced by feeding diets rich in xylose or galactose, or by hyperglycaemia resulting from diabetes brought about by the administration of alloxan. In each case, the time for the cataract to form is inversely proportional to the circulating concentration of the monosaccharide. In the lens the sugars undergo both oxidation and reduction, and both these processes have serious consequences. Reduction is catalysed by **aldose reductase** which uses NADPH as the reducing agent (Figure 31.6), thus tending to cause a fall in the cellular concentration of the reduced coenzyme, and hence of reduced glutathione. In addition, the sugar alcohols produced (sorbitol from glucose or dulcitol from galactose) are trapped in the lens cells and cause the osmotic uptake of water. Oxidation of the sugars gives rise to several auto-oxidizing dicarbonyl compounds which cause the oxidation of glutathione and protein thiol groups, and they also act as inhibitors of pentose phosphate pathway enzymes. Thus many of the effects of monosaccharides are the same as those caused by ionizing radiation, but additionally the accumulation of intracellular polyols (and hence water, osmotically) causes the crystallins to become more hydrated and so more light scattering. One final process appears to contribute to the syndrome, and that is the glycosylation of the amino-terminal amino groups of the crystallins (analogous to the glycosylation of haemoglobin which occurs in diabetes). The drugs quercetin and sorbinil, which are aldose reductase inhibitors, have been given clinical trials as possible protective agents for those at risk of cataract formation due to high circulating concentrations of monosaccharides (Figure 31.6).

The various process which may give rise to cataract formation are summarized in Figure 31.7.

31.4 The retina

It has been estimated that about one-third of all the nerve fibres entering or leaving the central nervous system do so via the optic tract. In total about a million fibres, each receiving inputs from some 100 photosensitive cells, convey visual information to the brain.

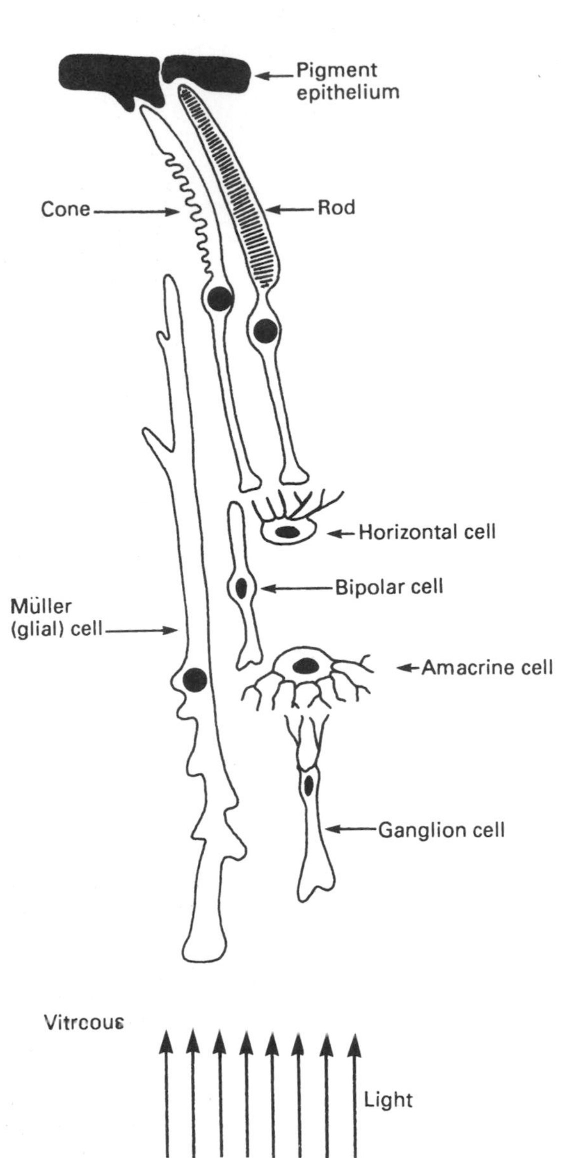

Figure 31.8 **The functionally inverted structure in the retina**

The retina is functionally 'inverted' (Figure 31.8) in the sense that, except at the fovea, light has to traverse all of the structures of the tissue before it impinges on the photoreceptors in the specialized nerve cells called rods and cones. A good deal more is known of the processes which occur in rods because they outnumber the cones by some 10 to 1 in the human retina. The **rods** are involved in vision at low light intensity (**scotopic vision**), although they adapt to vision at high light intensity. On the other hand, **cones** are concerned with high light-intensity vision and with colour vision. Indeed, cones have been shown to possess one of three distinct pigments that show light absorption maxima at 435 nm (blue sensitive), 535 nm (green sensitive) and 565 nm (often called red sensitive). In fact, of course, the last absorption maximum is located in the yellow part of the spectrum, but cones possessing this pigment are referred to as 'red sensitive' because they have appreciable absorption at 670 nm in the red part of the spectrum.

Scotopic vision

The question that any description of the visual process must address is: 'How does a single quantum of light impinging on the visual pigment result in the hyperpolarization of the plasma membrane of the cell containing the pigment?' A single photon will cause the plasma membrane of a rod cell to hyperpolarize by 1 mV for a few seconds, and about 50 photons will cause half-maximal hyperpolarization (a change such that the membrane potential becomes some 15–35 mV more negative inside). A related question is: 'How is this hyperpolarization translated into the depolarization which characterizes the action potential seen in the nerve fibres of the optic tract?' A coherent answer to the first question as it applies to rods is beginning to emerge, but the topic of the second question is much less well understood.

In order to establish the discussion of the biochemistry of the rod on a proper basis, it is useful to have in mind a general picture of its morphology (Figure 31.9). The cell has an outer segment packed with membranous disc structures. This is connected to an inner segment by a narrow cilium, and this second segment is rich in mitochondria. Next comes the cell body (perikaryon) with its characteristic axon leading to the nerve ending. The **discs** are closed membranous structures some 16 nm thick, and there are about 1000 packed into the outer segment. The discs arise in the

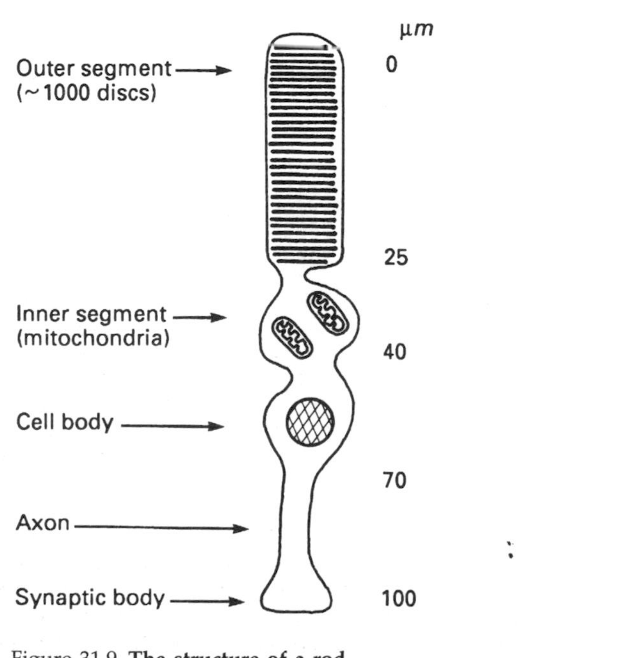

Figure 31.9 **The structure of a rod**

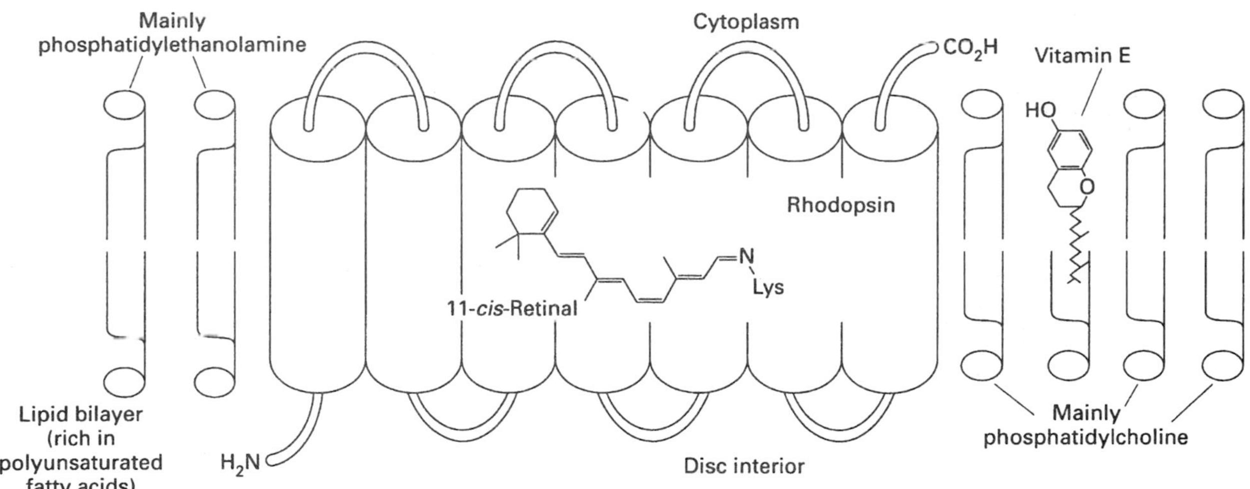

Figure 31.10 **The organization of rhodopsin in the disc membrane** (each cylinder represents an α-helix that spans the membrane)

region of the cilium from whence they migrate towards the pigment cells, where they are shed and undergo phagocytosis into those cells. The life of a disc is about 10 days in man. Analysis of the components of the outer segment shows it to contain lipid and protein in the ratio of 40% to 60%, respectively, most of which (95%) is present in the discs. Of the lipids, phospholipids (almost exclusively phosphatidylcholine and phosphatidylethanolamine, with a little phosphatidylserine) form 81% and cholesterol 9.8%. In addition, two fat-soluble vitamins, A and E, are present in high concentration. The latter is evidently there to protect the former from photo-oxidation, but it undoubtedly also prevents or reverses the photo-oxidation of the polyunsaturated fatty acids present in high concentration in the discs (30% of all the unsaturated fatty acyl residues are docosahexenoyl residues, $C_{22:6}$). The presence of these unsaturated fatty acyl residues, esterified predominantly to the central carbon atom of glycerol, together with the relatively low cholesterol content, means that the membrane is a very fluid one (see Chapter 2).

The chromophore, **rhodopsin**, which absorbs the light is a complex of opsin, a phosphoglycoprotein with a RMM of 38 000 (constituting 85% of all the protein in the outer segment) and a vitamin A derivative, 11-*cis*-retinal. In the discs, there are about 70 phospholipid molecules per rhodopsin and 1 molecule of vitamin E for 10 of rhodopsin. The organization of a section of a disc membrane is shown in Figure 31.10. The formation of the chromophore is dependent on the supply of vitamin A to the eye. In the retina, retinol (supplied from the liver—see Chapter 19) undergoes two enzyme-catalysed reactions: first retinol dehydrogenase converts the alcohol into an aldehyde using NAD^+ as the coenzyme and subsequently the double bond in the 11 position is isomerized from the *trans* to the *cis* form in a reaction catalysed by retinal isomerase. The 11-*cis*isomer reacts spontaneously with a lysyl residue towards the C-terminal end of an opsin molecule in the disc membrane to form a Schiff's base:

$$\underset{\text{11-}cis\text{-Retinal}}{R\text{—}C(=O)H} + \underset{\text{Lysyl residue of opsin}}{H_3{}^+N\text{-}(CH_2)_4\text{-opsin}} \longrightarrow$$

$$\underset{\text{Rhodopsin}}{R\text{-}CH{=}NH\text{—}(CH_2)_4\text{-opsin}} + H_2O + H^+$$

Rhodopsin is firmly bound in the disc membrane, the polypeptide chain folding back and forth across the membrane in 7 α-helical loops each of some 20–25 aminoacyl residues (Figure 31.10). The complex has a broad light absorption maximum centred on 498 nm. Because of the central role vitamin A plays in scotopic vision, a chronic deficiency of the vitamin leads to night blindness (hemeralopia) and ultimately to the deterioration of the outer segments (as does a deficiency of vitamin E).

The upper part of Figure 31.11 shows the events which are set in train when light of low intensity is absorbed by the rods. The quantum causes photo-isomerization of the 11-*cis* bond of the retinal and the progress of this process may be followed by optical methods because the retina 'bleaches', i.e. the pink colour changes to orange and then to yellow before fading completely. The loss of colour coincides with dissociation of all-*trans*-retinal from the opsin and this does not occur until about 1 min after exposure to light. Now this is too slow for the actual dissociation to be identified as the initiator of *the action potential* in the optic nerves, because it is known that the 'photocurrent latency', i.e. the time elapsed between a flash of light and the detection of a signal in optic tract axons, is about 10 ms. This finding strongly suggests that the conformational change that undoubtedly occurs in the 11-*cis*-retinal while it is still attached to the opsin is the event that leads to the hyperpolarization of the rod–cell plasma membrane. Nevertheless, exposure to light does inevitably lead to dissociation of all-*trans*-retinal from opsin and this process must be reversed if rhodopsin is to be regenerated. At low light intensity this is accomplished directly by the action of retinal isomerase.

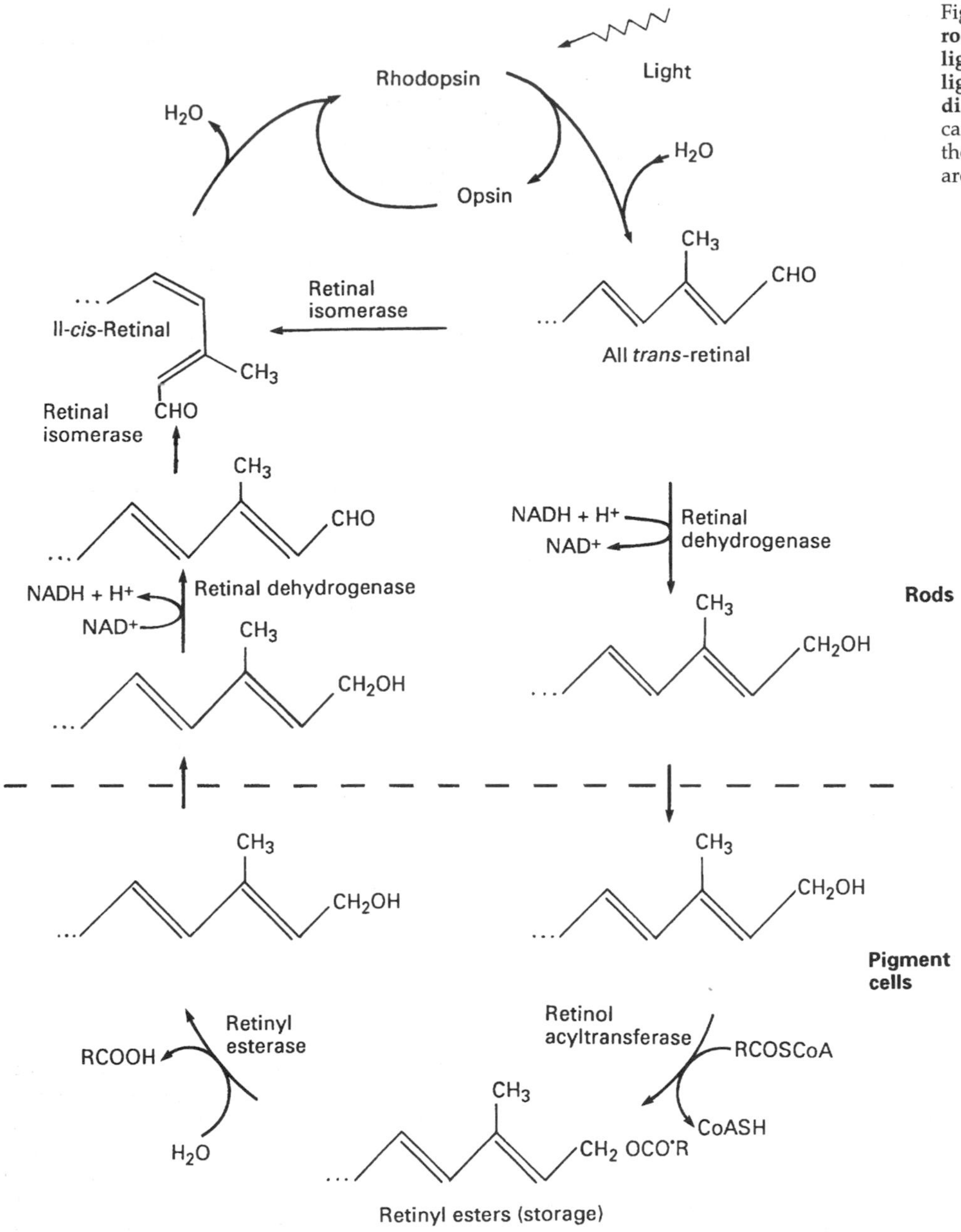

Figure 31.11 **Visual cycle of rods/pigment cells at low light intensity (top), high light intensity (whole diagram).** Note that only carbon atoms 10 onwards of the chain in retinol derivatives are shown

It is known, however, that this photodissociation system saturates functionally when about 1% of the rhodopsin has been converted to opsin, and on a reasonably bright day this is far exceeded. As a means of adaptation the following sequence which leads to a fall in the number of rhodopsin molecules per disc is set in train: the all-*trans*-retinal is reduced to retinol and the alcohol is transferred to the pigment cells where it is stored as an ester (Figure 31.11, lower section). This proceeds until the amount of rhodopsin has been diminished by about 50% of that present in the dark. On return to low light intensity the process is reversed. An additional benefit of this arrangement is that rod membranes are protected from the deleterious effects of high concentrations of free retinol.

A number of observations suggest the way in which events occurring in the disc membranes result in the hyperpolarization of the plasma membrane. This hyperpolarization seems to derive from a reduced permeability of the latter membrane to Na^+ in the light. The outer segment plasma membrane is unusually permeable to Na^+ due to the presence of some 1000 entry channels and consequently the resting membrane potential is rather low at −35 mV. Figure 31.12 shows that there is a large flow of Na^+ into the outer segment in the dark because there is still a steep concentration gradient due to the activity of a Na^+/K^+ ATPase (see Chapters 28 and 30) located in the inner segment plasma membrane. Thus, in the dark a current of Na^+ flows into the outer segment, diffuses to the inner segment to be

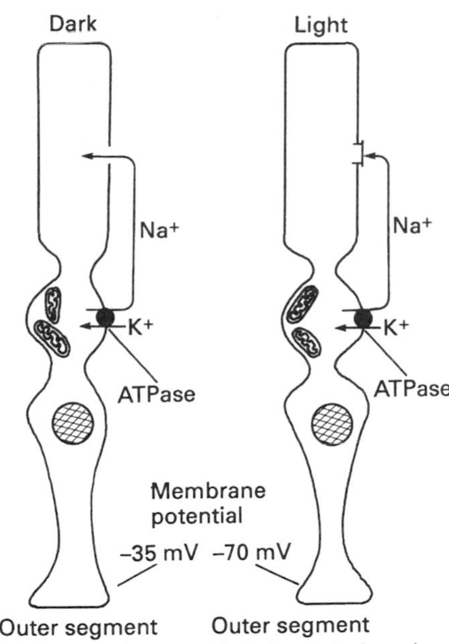

Figure 31.12 **Light absorbed by the rods causes the plasma membrane to hyperpolarize by causing the closing of the Na^+ channels**

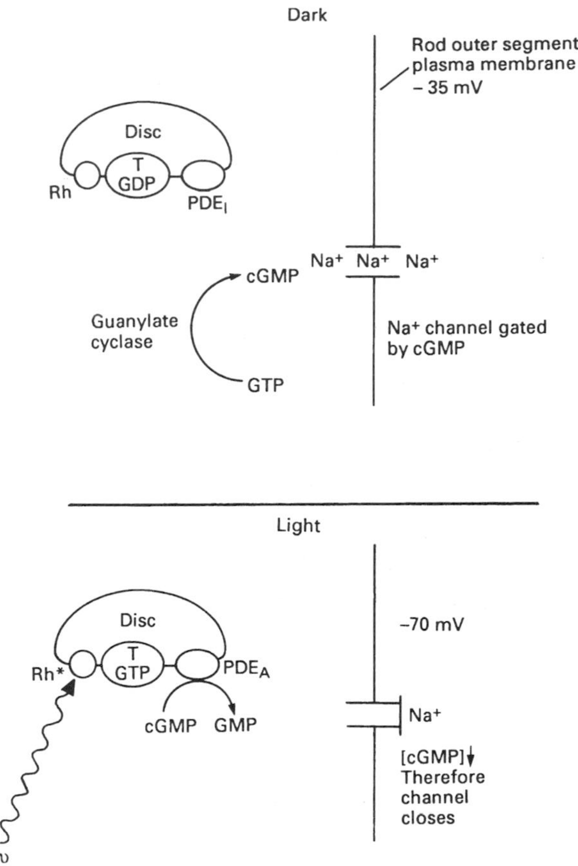

Figure 31.13 **Absorption of photons (hν) by rhodopsin results in the activation of a phosphodiesterase ($PDE_I \rightarrow PDE_A$)**

extruded by the Na^+ 'pump'. When exposed to light, the Na^+ channels close, leaving the 'pump' to operate unopposed and hence, the plasma membrane of the rod hyperpolarizes to a more normal value (compared to other neurons) of –70 mV.

The mechanism underlying the light-induced closing of the Na^+ channels seems to involve changes in the concentration of guanosine 3',5'-monophosphate (cGMP) in the outer segment (Figure 31.13). In the dark cGMP is generated by the catalytic action of guanylyl cyclase on GTP and the cyclic nucleotide acts directly to hold the Na^+ channels open. Light falling on rhodopsin in the discs causes conformational changes in the protein–chromophore complex which result in the activation of a cGMP phosphodiesterase also located in the disc membrane. When activated, this enzyme causes a rapid fall in the concentration of its substrate which in turn means that the Na^+ channels are no longer held open. The conformational change in rhodopsin is transmitted to the phosphodiesterase by a GTP-binding protein termed 'transducin', and this protein is highly homologous with other GTP-binding proteins which act in other systems to transduce signals, i.e. G proteins, which couple adrenoceptors to adenylyl cyclase and phospholipase C (see Chapter 3).

The final question which has yet to be resolved is how hyperpolarization of the rod plasma membrane results in the depolarization of the ganglion cells, the axons of which project in the optic tract. The neurotransmitter released by rods is glutamate. It is suggested, therefore, that this putative excitatory neurotransmitter (see Chapter 30) is released in the dark and that the light-induced hyperpolarization of the rods causes its release to cease. If this is the case it might be expected that the bipolar and horizontal cells (see Figure 31.8) in synaptic contact with the rods would hyperpolarize in the light, and this is indeed so. It may be that the bipolar cells (or the amacrine cells) release an inhibitory neurotransmitter in the dark and the release of this ceases in the light, thus causing the relative depolarization of ganglion cells. This question is not fully resolved, although it has been reported that the enzyme glutanate decarboxylase which catalyses the formation of the inhibitory neurotransmitter GABA (see Chapter 30) is present in amacrine cells.

Diseases of scotopic vision

Part of the remarkable sensitivity of rod vision depends on there being very low levels of 'noise' in the absence of light. This being the case, it is not surprising that even apparently minor defects arising from amino acid substitutions in the opsin gene can impair the visual process. The most common outcome of such defects is that the threshold for detecting light is higher than normal under dark-adapted conditions. This is referred to as **night-blindness**. Gene mutations give rise to two disorders of rod vision, both associated with night-blindness: **congenital stationary**

night blindness (CSNB) and **retinitis pigmentosa (RP)**. In the *latter* but not the former condition, vision is further impaired due to a progressive degeneration of the retina. Both diseases are genetically heterogeneous, i.e. several different mutations lead to the same signs and symptoms. In some cases of CSNB, the mutation is such that the rhodopsin molecule is able to activate the G-protein transducin even when no 11-*cis*-retinal is present. Consequently there is a large increase in 'noise' in the signals in the optic nerve from the retina of patients with this particular mutation.

Cone vision

The general arrangement seems to be the same in cones as in rods, but with some important difference in detail. Thus the 'disc' membranes are actually continuous with the plasma membrane in the cone outer segment and each contains one of three different rhodopsin molecules with differing sensitivities to light of different colours (see earlier). However, it is only the opsin part of the molecule which varies; each opsin combines with the same 11-*cis*-retinal to form the chromophore. The light absorption maxima of the pigments and their relative insensitivity to light, compared to that of the rhodopsin of rods, must derive from the particular structures of the opsin proteins and perhaps their degree of glycosylation and phosphorylation.

Inherited variations in colour vision

The colloquial term '**colour-blindness**' refers to variations in red-green colour vision that are due either to the loss of the red or the green colour pigment (**dichromy**) or to the production of a colour pigment with a shifted light absorption maximum (**anomalous trichromy**). Anomalous trichromy is encountered in the most common variation in human colour vision in which there is a polymorphism in the gene that encodes a protein found in the cone pigment. In Caucasian populations, an allele that gives rise to an opsin with a seryl residue at position 180 is found in 62% of cases, whereas in the remaining 38% the amino acid in that position is alanine. The position of the light absorption maximum for serine-variant is red-shifted by 5 nm relative to that found with the alanine variant.

In 1995, 141 years after his death, the wish of John Dalton that the cause of his self-diagnosed colour-blindness should be determined was met. This was because Dalton bequeathed his eyes for scientific investigation and consequently they had been preserved for this purpose. The application of the polymerase chain reaction technique (PCR; see Chapter 38) to a sample of the tissue recovered from the retinal region suggested the total absence of the gene that encodes the cone opsin, giving rise to a rhodopsin with an absorption maximum at 535 nm (green sensitive). Consequently, Dalton is now known to have suffered from the type of dichromy known as **deuteranopia**.

Chapter 32

The immune system

32.1 Introduction

In the introduction to their excellent textbook *Immunology: a Synthesis*, Golub and Green reviewed the development of our understanding of the operation of the immune system. They pointed out that an early reference to the concept of immunity is to be found in Thucydides' account of *The Peloponnesian War*. In this, Thucydides noted that those who had caught the plague and survived felt themselves to be safe, because 'no one caught the disease twice' or, if they did, the second attack was never fatal. Thucydides further shows that the concept of specificity in the immune response was not widely grasped at this time, for he notes that after recovery from the plague, sufferers 'fondly imagined that they could never die of any other disease in the future'. This second great step in the understanding of the operation of the immune system, the concept of the specific response, had clearly been taken by the eleventh century when we find the Chinese inoculating healthy individuals against smallpox using dried extracts prepared from the pustules of those infected with the disease. This hazardous procedure persisted until 1798 when Jenner intentionally caused a mild cowpox infection in the 8-year-old James Phipps in order to protect him against smallpox. The power of this technique, which was not immediately recognized by the medical profession, has been such that by the 1980s smallpox had been eradicated world-wide.

It needed another 100 years and the appearance of Louis Pasteur for it to be appreciated that vaccination, as the Frenchman named the procedure, thus acknowledging the work of Jenner, was a general method of protecting against infectious disease. This followed his chance observation that, on ageing, the infectious agent in cholera of chickens (now known as *Pasteurella antiseptica*) became avirulent, but protected the animals against subsequent exposure to fresh, virulent preparations. This led Pasteur to realize that it was possible to immunize individuals by using inactivated organisms. Such organisms retained their ability to provoke an immune response (they were **immunogenic**) while failing to induce disease (they were not **pathogenic**). The pioneering contributions of Jenner, Pasteur and many others have meant that by 1993, 80% of the 100 million children born annually in the developing countries of the world receive protection against six major infectious diseases: diphtheria, whooping cough, tetanus, poliomyelitis, measles and tuberculosis. With this comes the real hope that, despite some recent setbacks (the number of cases of tuberculosis in New York City more than doubled in the 1980s), two of these (measles and poliomyelitis) will follow smallpox in being eliminated by the year 2000.

There is no room for complacency, however, as the recent dramatic rise in the disease **acquired immune deficiency syndrome** (AIDS) shows only too clearly. In 1992, the Global AIDS Policy Coalition estimated that world-wide, some 19.5 million individuals (1 in 250 adults in the world), carry the causative agents of AIDS, human immunodeficiency viruses (HIVs). Some 2 million people are already suffering from the disease. The lives of those infected by HIV may be prolonged by currently available therapies (see Chapter 39), but almost all will die of AIDS or related diseases. The number believed infected in 1992 represents a threefold increase over the previous 5 years. The current view is that any vaccine developed in the near future is unlikely to afford more than 60–80% protection against the disease. This is a serious problem because it is inevitable that many receiving the vaccines will believe their protection to be absolute, with consequent changes in sexual behaviour. Attention to the sexual method of transmission of the disease is the most powerful preventative method currently available.

32.2 The organization of the immune system

The student coming to the study of the immune system for the first time will almost inevitably feel overwhelmed by the complexity of its organization. It is useful, therefore, to try to put the occurrence of this complexity into some sort of perspective. In his book entitled *Cosmos*, Carl Sagan, in attempting to explain the complexity of the brain, took an evolutionary view and compared the development of the brain with that of New York City. Sagan's analogy applies equally well to the immune system. He described the early evolution of the city as occurring from a small centre with slow growth and change. In this process, many of the old parts of the city continue to function. A similar course of events appears to have been followed by the immune system. There is no evolutionary way, in either case, for the old to be totally ripped out, because both the city and the immune system must continue to function during improvements and renovations. The old parts are in charge of too many vital functions for them to be replaced altogether. In the seventeenth century the traveller wishing to cross New York's East River had to do so by the ferry and 200 years had elapsed before advances in technology allowed first a suspension

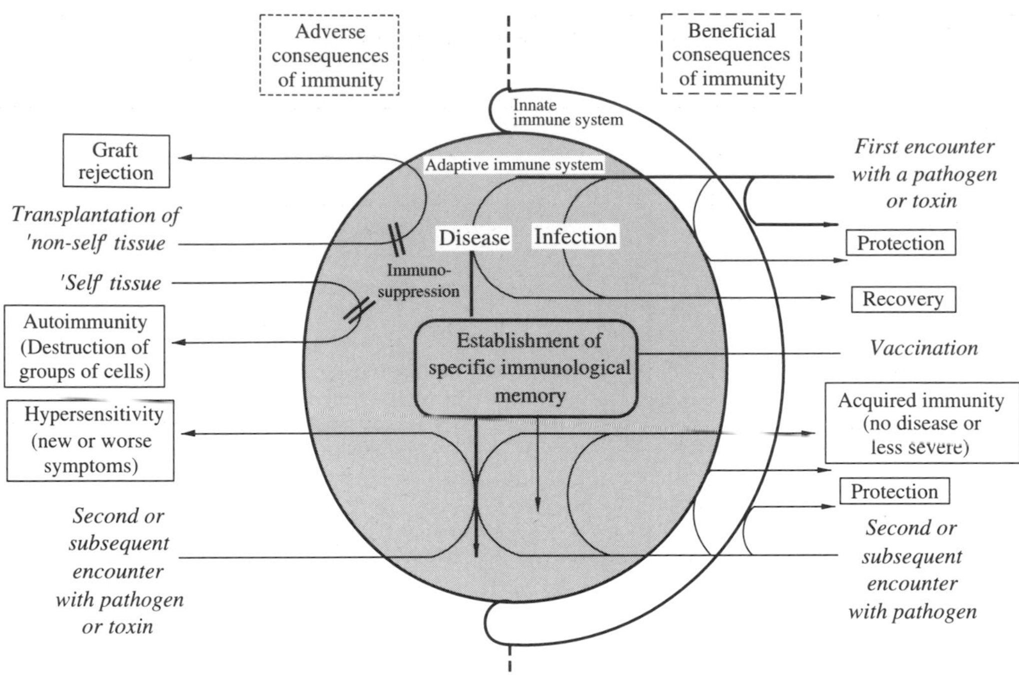

Figure 32.1 **The operation of the immune system depends on the interaction between its innate and adaptive arms, but is not always beneficial**

bridge and then a tunnel to be built. Both of these means of crossing were built at the ferry terminals, simply because the road transport system had been attracted by the placing of the ferry. Likewise the builders of the tunnel made use of the caissons emplaced during the building of the bridge as a basis of their construction.

In the same way the more ancient arm of the immune system, responsible for the so-called **innate immune response** (sometimes referred to as inborn, unchanging or natural), is still an important aspect of protection against infection in humans in whom it interacts with the more recent, more specific, **adaptive immune response** to provide excellent all-round protection against a variety of challenges from the environment. The interaction between the two types of response was first recognized in 1903 by two Englishmen, Wright and Douglas, who were working at St Mary's Hospital in London.

The protection afforded by the immune system

The scope of the joint operation of the innate and adaptive responses to foreign invaders (often referred to as **'non-self'**) is summarized in Figure 32.1. This shows that the innate system provides natural resistance by providing a barrier to invading organisms or by mounting a generalized, non-specific attack on those that gain access to the body. Generally such protection acting alone will not suffice to prevent infection whereupon the adaptive system comes into play. The immediate consequence of this is that the subject usually recovers but, crucially, a specific memory is laid down such that a second attack by the organism either fails to cause the disease or it does so in a very attenuated form: **acquired resistance (immunity)** is established. It is the laying down of the adaptive, specific memory in the absence of a primary infection that forms the basis of the process of vaccination.

The operation of the immune system is not always beneficial

It is inevitable with something as sophisticated as the immune system that errors will arise from time to time (Figure 32.1). This occurs when the complex mechanism that allows self to be distinguished from non-self fails and the ensuing condition is referred to as **autoimmunity** (Table 3.3 gives some examples of autoimmune diseases). The very ability of tissues to distinguish self from non-self is seen to be a major problem to the surgeon wishing to carry out transplant surgery, when **immunosuppression** is frequently the only method for preventing the rejection of grafts obtained from other donors. (Skin grafting from one part of the body to another in burns patients is invariably successful: self to self.) Occasionally the laying down of specific immune memory can cause problems, when a second invasion by the previously-encountered organism may lead to new or worse symptoms, with tissue damage. Such a defective response to reinfection is termed **hypersensitivity**.

Table 32.1 **The 'response–evasion' interaction between pathogens and the immune system**

Means of evasion		*Immune response*		
Viruses	*Bacteria*	*Innate*		*Adaptive*
	Saprozoic bacteria were able to:			
	1. Secrete hydrolases to promote tissue breakdown	Mechanical barriers Secretions: • Mucus • Tears (lysozyme) • Gut secretions		
	Facultative pathogens developed the ability to:	(Viruses)	(Bacteria)	
	1. Adhere to host cells 2. Invade host cells	Interferon Natural killer cells	Complement Phagocytes producing: • Hydrolases • Free radicals	
Were able to establish latency	3. Secrete toxin			
	Obligate pathogens developed the means of:			Antibody secretion
	1. Antigenic variation 2. Secretion of antibody-specific proteases 3. Survival in phagocytes by: • Inhibition of phagolysosome fusion • Coat resistant to hydrolases • Escape into cytoplasm 4. Disruption of lymphocyte–lymphocyte signalling			Cell surface scrutiny: • MHC-cytotoxic T-cells • Macrophage activation
Were able to integrate into the host genome				

A see-saw of 'response–evasion' has formed the basis for the evolution of the immune system

As has been already suggested, taking an evolutionary view is a useful way of trying to understand the multi-faceted nature of the operation of the immune system. As multicellular organisms evolved ever more sophisticated means of combating pathogens, so those organisms able to evade them were selected. Neither 'side' has been totally victorious in this struggle, although pathogens capable of taking up residence in and attacking the cells of the immune system provide a particularly severe threat.

Earlier in evolutionary times a major challenge to multicellular organisms came from the saprozoic organisms that largely lived on dead and rotting tissues (Table 32.1). These microorganisms secreted digestive enzymes to aid their work, but the innate defences of mechanical barriers (skin) and the production of secretions such as mucus, tears, gut secretions, etc., usually sufficed as protection by denying the invading organism access to potential host cells. These purely physical barriers to infection were supplemented by methods of killing: the production of acid by the stomach and the secretion of **lysozyme**. This enzyme, which was first described by Alexander Fleming, is capable of attacking and catalysing the hydrolytic destruction of the bacterial cell wall. Thus Fleming discovered both endogenous and exogenous antibiotics (lysozyme and penicillin).

At the same time, bacteria themselves were facing the challenge of bacteriophages. These bacterial viruses took over the apparatus of the host to express their own genetic material. The (innate) response of the bacteria was to produce restriction endonucleases (see Chapter 38) that destroyed bacteriophage, but not host, DNA.

Facultative pathogens, for example *Clostridium tetani*, the causative agent in tetanus, capable of evading these simple defences then developed. This was achieved by the presence on the bacterial surface of molecules permitting them to adhere to host cells (adhesion molecules). Thus the bacterium *Helicobacter pylori*, a common causative agent of stomach ulcers, binds preferentially to the surface of gastric epithelial cells via the so-called Lewis[b] antigens (part of the blood group antigens that determine blood group O). This may well help to explain the observation that people with blood group O are 1.5–2 times more likely to develop gastric ulcers (and also cancer of the stomach).

An additional response of pathogenic organisms was the secretion of **cytotoxins** that accelerated host cell death. Ultimately these pathogens were also able to invade, take up residence and replicate in host cells. This last property was also seen in viruses attacking eukaryotes, which in addition developed the ability to remain latent within host cells. The response of host organisms to these refinements by pathogens was several-fold. **Interferons**, produced by cells infected by viruses served to induce in other cells a general **non-specific virus-resistant state**. A subset of lymphocytes, called **natural killer** (NK) cells, also arose capable of destroying any cell infected by viruses. NK cells are themselves activated by interferons. Most of the defences against facultative pathogens were still largely non-specific. The other important, non-specific system that arose is a complex of some 15 serum components that constitute **complement**, first recognized in 1900 by the Belgian scientist Jules Bordet (although called 'alexine' by him). These serum proteins interact in a cascade mechanism (see Section 32.5), a process initiated by invading microorganisms. Products of the cascade can coat the foreign cells and hasten their destruction. At the same time, other products of the complement cascade set up an **acute inflammatory response**, by means of which the ensuing rise in temperature may kill the bacteria. In addition, a series of cells, themselves capable of killing the pathogens, are attracted to the site of infection. Important in this group of protective cells are the **macrophages**, which are examples of **phagocytes**. Phagocytotic cells can take up by means of endocytosis (see Chapter 8) bacteria adhering to their plasma membranes. The membranous endosomes then fuse with lysosomes to form endolysosomes containing collections of enzymes capable of destroying engulfed microorganisms. In addition, macrophages and other phagocytotic cells such as the **granulocytes** are able to kill bacteria by the use of cytotoxic free radical attack (see Chapter 33). Indeed, in the case of granulocytes, the oxidative attack on invading organisms is so vigorous that the cells themselves rarely survive, whereas macrophages can carry out many rounds of killing.

A further degree of sophistication was seen with the appearance in vertebrates of **adaptive immune responses** by using much more selective mechanisms. Thus complex molecules termed **antibodies** are secreted by B-lymphocytes in response to circulating foreign material and the former molecules are able to bind to very specific chemical structures in the foreign molecules. By virtue of this specific recognition, for example, bacterial cytotoxins are bound and prevented from gaining access to host cells (neutralized). The bacteria themselves are also bound by the antibodies and are thus marked out for accelerated phagocytosis (this is particularly important for the Gram-positive bacteria, for example, obligate pathogens such as *Pneumococci*, that had adapted to evade phagocytosis by changes in their capsules that stopped them from binding to phagocytes). Antibody complexes with bacteria cause the activation of the complement cascade resulting in the lysis of the pathogen. This is an excellent example of the interaction between the innate and adaptive arms of the immune response. Antibodies constitute very important weapons in combating infection, and it is not surprising to find that methods of evading them evolved in pathogens: viruses, bacteria and protozoans. Important in this type of evasion was the ability of pathogens to undergo what is known as **antigenic variation**. The protozoan *Trypanosoma cruzi* is able, during the course of an infection, to change its major surface glycoprotein (the *T. cruzi* genome has as many as 1000 genes for variants of this protein) and thus evade specific immune responses. A second mode of evasion arose in the pathogenic Gram-negative bacteria *Neisseria gonorrhoeae, N. meningitidis* and *Haemophilus influenzae*. All these organisms secrete specific proteases capable of catalysing the hydrolysis of antibodies. (In the case of these organisms the enzymes inactivate the secreted immunoglobulin IgA_1 – see Section 32.4)

A different strategy of evasion arose when the pathogenic organism became able to reside within phagocytes, either by inhibiting endolysosome fusion (the protozoan *Toxoplasma* for example), by producing inhibitors of lysosomal enzymes (the protozoan *Leishmania mexicana* for example), by resisting free-radical attack (another protozoan, *Trypanosoma dionisii* provides the example) or by escaping into the cytosol of the cells (*Trypanosoma cruzi* is able to do this). The host response to these challenges was the development of a system of cellular plasma membrane scrutiny. Cells infected with either bacteria or viruses acquired the ability to **present** fragments of proteins from the invading microorganisms on their surfaces by means of the plasma membrane proteins called the **major histocompatibility complex (MHC)**. Cells with foreign peptides bound to their MHC molecules were then recognized as being infected, and killed, by groups of T-lymphocytes expressing a **T-cell receptor** on their surfaces. In addition, the release of intercellular-signalling molecules (**cytokines**) by some T-cells promoted the activation of macrophages which were then able to deal with bacterial infections caused by *Listeria*, with the parasites causing malaria and even with *Leishmania* parasites. Not surprisingly, several groups of pathogens have developed the means of interfering with cytokine-mediated signalling; for example the Epstein-Barr virus (EBV) is able to disrupt the communication between T-lymphocytes.

The final evasive route taken by viruses was that of incorporation of their genetic material into the host genome (see Chapter 37). Thus the *Herpes simplex* virus that causes recurrent cold sores is able to insert into the DNA of neurons in the trigeminal nucleus.

Both humoral and cell-mediated responses underlie immune protection

It is clear from the foregoing discussion that the immune system (whether innate or adaptive) has two interacting means of attack on foreign material. One depends on the release into the circulation or the tissues of water-soluble molecules and is termed the **humoral response**. The other class of response is referred to as being **cell mediated**. The mounting of an adaptive, humoral response was regarded as being in the hands of the B-lymphocytes whilst the adaptive cell-mediated response was said to be undertaken by the T-lymphocytes. This is still a very useful concept on which to build, but it is important to realize that these cells interact with each

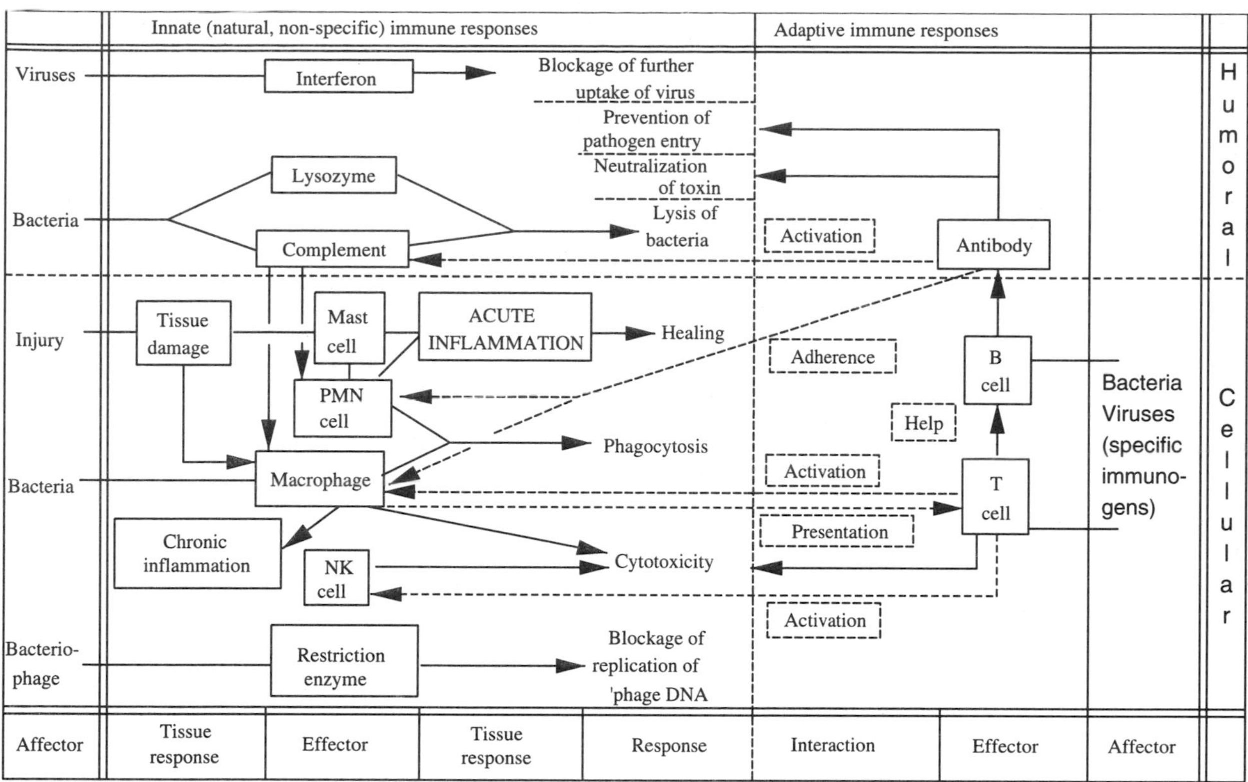

Figure 32.2 **Interaction between the innate and adaptive immune systems involves both cellular and humoral components**

other both by means of structures that resemble the synapses that arise between nerve cells and also by the secretion of cytokines.

Figure 32.2 shows in outline the interactions between the innate and adaptive arms of the immune system. Humoral responses are shown at the top of the diagram, while cell-mediated responses are at the bottom. The mechanistic bases of these interactions will be described in the rest of this chapter.

32.3 Antigens and antibodies

It is convenient to begin consideration of the operation of the immune system by a discussion of the molecules that participate in humoral responses. Within this area the antibody is the major component of adaptive immune responses.

The characteristics of the adaptive, humoral immune response

The term **antigen** was introduced to describe any substance capable of specific recognition by the immune system. Most frequently the substance is a protein, either pure or conjugated, but occasionally may be a polysaccharide (only rarely a lipid). In the normal course of events the protein will have been part of, or derived from, a pathogenic organism, but otherwise innocuous molecules such as egg albumin can also prove immunogenic. It should be noted that an antigen can also provoke a cellular immune response.

Once antigens gain access to an animal they cause a series of events that lead to the release of special proteins shown by Tiselius and Kabat to migrate on electrophoresis with the γ-globulin fraction of serum and termed **antibodies**. The key property of the antibody is its ability to bind non-covalently to an antigen. For many antigens, several antibodies arise that bind to different parts of the molecule. The particular part of an antigen molecule which is recognized by and interacts with an antibody is termed an **antigenic determinant** or **epitope**. It is now usual to refer to the ability to provoke a specific immune response as **immunogenicity**. On the other hand, the ability to bind to an antibody is referred to as **antigenicity**. At first sight it may appear that one of these two similar terms is redundant, and certainly all immunogens are antigens. The converse is not true: there are certain, usually small, molecules that themselves are not able to provoke an immune response but which can bind to antibodies. This is well-illustrated by the occurrence of **haptens**. It was observed by Landsteiner that when compounds, such as *p*-aminobenzene sulphonate are injected into animals there is no ensuing immune response unless the molecule is first attached covalently to a **carrier** (an immunogenic protein), when the animals produced a host of antibodies. These were largely specific for carrier epitopes, but some recognized the hapten, which they were able to bind even in the absence of the carrier. Landsteiner exploited the ability of haptens to induce a humoral immune response when

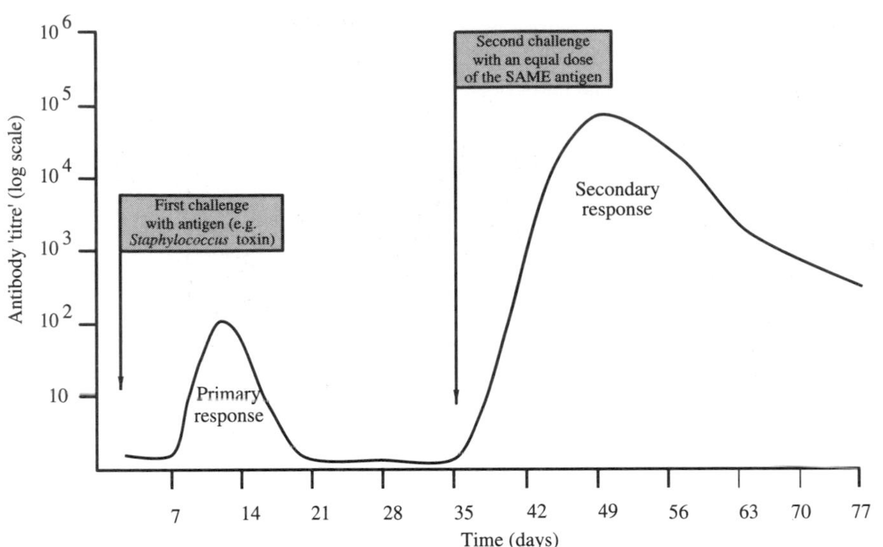

Figure 32.3 **Primary and secondary antibody responses**

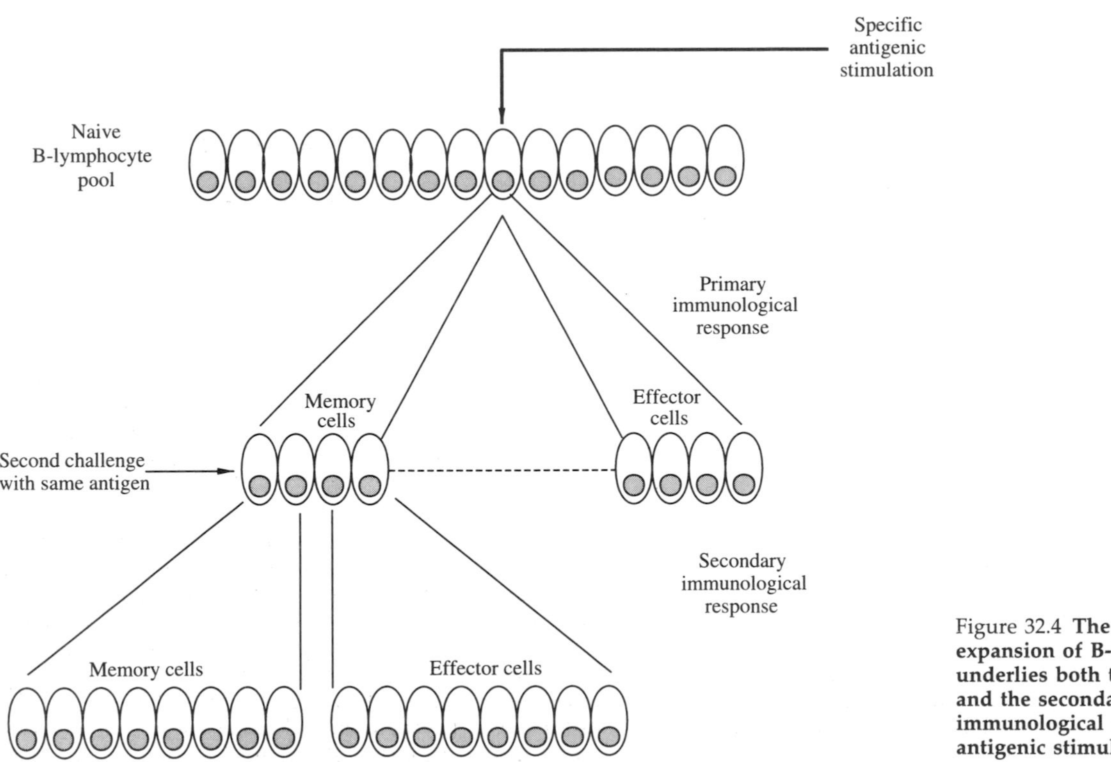

Figure 32.4 **The clonal expansion of B-lymphocytes underlies both the primary and the secondary immunological responses to antigenic stimulation**

bound to a carrier to demonstrate the exquisite specificity of such responses (and by inference of the responses to any immunogen). He showed, for example, that the antibody secreted in response to *p*-aminobenzene sulphonate failed to bind aminobenzene derivatives with a wide variety of other acid groups in the *para* position.

It is now understood that small peptides, that may form part of the sequence of a larger molecule, can also be conjugated to a carrier and act as haptens. The antibodies arising will include those capable of binding the whole molecule from which the peptide was derived. This is now leading to the generation of new vaccines by the use of haptenic (non-pathogenic) peptides synthesized in the laboratory with sequences deduced from the DNA code for the parent protein which may not itself have been isolated from the organism against which protection is desired.

Primary and secondary adaptive humoral immune responses

If an animal is injected with an immunogen, for example a toxin of *Staphylococcus*, the concentration of antibody found in the serum, and capable of binding (thereby neutralizing) this antigen, increases from zero to a maximum after about 10 days (Figure 32.3). This is followed by a decline in the antibody 'titre' to the toxin. If, however, the animal is again injected with the same dose of antigen, the (secondary) specific antibody response is much faster to develop, is of greater magnitude, is more effective and is longer lasting: the response is **adaptive**, some sort of **immunological 'memory'** is laid down during the primary response. It is now established that B-cells, which bind an antigen specifically, divide to give rise to a **clone** of similar cells, some of which help effect the removal of the antigen (**effector cells**) and some of which remain in the body to respond to the antigen during a subsequent encounter (**memory cells**). On a second exposure to the antigen, clonal expansion of these memory cells results in more effector and memory cells (Figure 32.4).

In addition, the **class** of antibody protein secreted changes between the primary and the secondary responses (**class switching**): the primary response is dominated (especially in its early phases) by antibodies of class M, while class G proteins come to the fore in the secondary response (see Section 32.4).

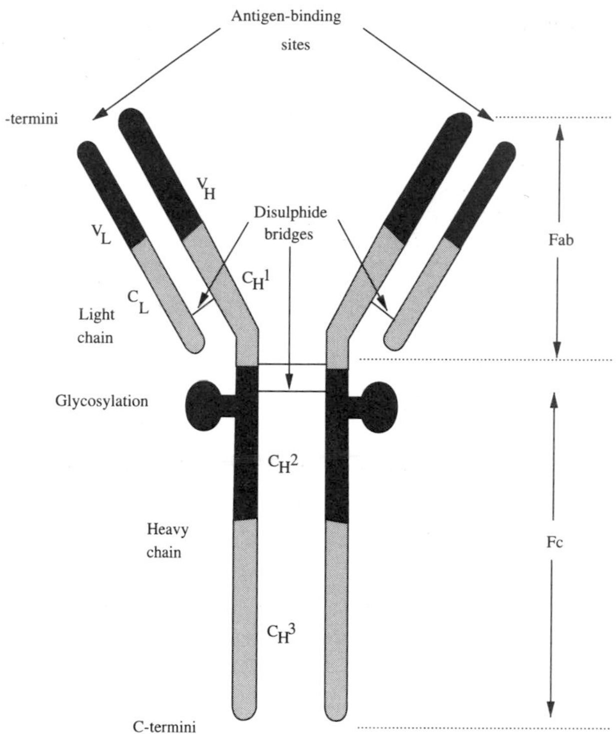

Figure 32.5 **The general structure of an immunoglobulin molecule**

32.4 Antibody structure: the immunoglobulins

The structures of the immunoglobulins

The work that led to the elucidation of the molecular structures of the antibodies, or **immunoglobulins (Ig)**, was carried out independently by Porter and Edelman. Their work, and that of others, now allows us to describe the chain structure of general immunoglobulin as shown in Figure 32.5. The protein molecule consists of two identical heavy chains of relative molecular mass c.50 kDA and two identical light chains of mass c.25 kDa. Thus, the simplest description of an immunoglobulin is H_2L_2. The chains are covalently linked by intra-chain disulphide bridges. The heavy chains are said to be 'hinged' in the vicinity of the two intra-chain disulphide bridges where the presence of several prolyl residues confers a degree of flexibility on the molecule in that region. Porter found that treatment of immunoglobulin with the protease papain produced three fragments. Two were identical, were able to bind antigen and were termed Fab (fragment with antigen binding). The third fragment did not bind antigen, was crystallizable and termed Fc.

This standard description of the structure of immunoglobulins appears at variance with the picture developed of molecules with a wide range of different specificities. The answer to this apparent paradox lies in the ability of the B-lymphocytes to synthesize a wide range of antibodies that differ in their amino acyl sequences. The variability of sequence is greatest in certain regions of the L and H chains of the molecule described as 'variable' (V), the remainder of the polypeptide chains retaining a relatively constant amino acyl sequence, the 'constant' (C) region. Both H- and L-chains possess V and C regions, and in different antibodies special parts of the V regions, referred to as 'hypervariable', show the most extensive differences in amino acyl sequence from molecule to molecule. These are associated with the antigen-binding sites, of which there are two in simple immunoglobulins.

There is one other structural 'motif' that is so characteristic of the immunoglobulin molecule that, when it is found in other molecules, it is referred to as the **immunoglobulin fold**, and constitutes a distinct **structural domain** within the molecule, with about 110 amino acyl residues. One defining feature of the immunoglobulin fold is the presence of an intra-chain disulphide bridge that contains a 'looped' structure. The secondary structure of the domain formed consists of two layers of β-pleated sheets linked by the disulphide bridge. Figure 32.6 shows the domain arrangement of a typical immunoglobulin molecule. Each L-chain has two such domains, one each in the variable (V_L) and constant (C_L) regions. The H-chains shown here have four domains V_H in the variable region and $C_H{}^1$, $C_H{}^2$ and $C_H{}^3$ in the constant region. Besides the immunoglobulins, many immunologically-important and other molecules possess one or more immunoglobulin folds and some of these are shown in Figure 32.7.

One final structural point needs to be made about the immunoglobulins: they are all glycoproteins. There is a variable degree of glycosylation in the region of domain $C_H{}^2$.

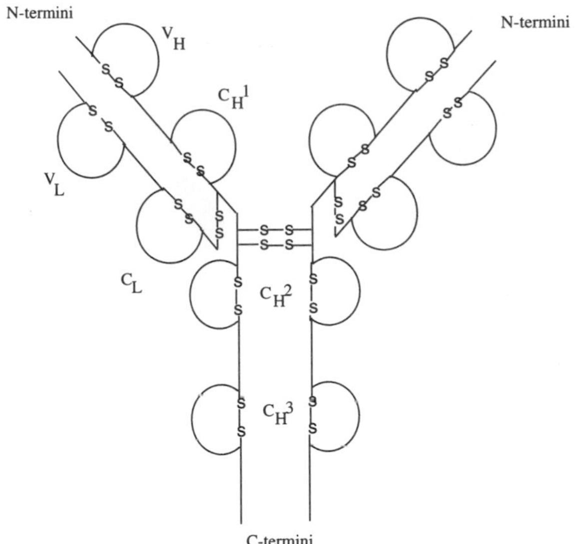

Figure 32.6 **General structure of an immunoglobulin molecule showing the disulphide-bridge limited domains**

Functions of immunoglobulins

The common function of all antibodies is that of binding to antigen via their antigen-binding regions (Fab). This can lead to the **precipitation** of soluble antigens by cross-linking, **agglutination** by the cross-linking of particulate antigens (e.g. viruses, bacteria or parasites) or **neutralization** (blocking of the attachment of viruses or bacterial toxins to membrane receptors). In addition, antibodies subserve a variety of **effector** functions which depend also on their Fc region. The activation of the **complement cascade** (see Section 32.5) is initiated by the binding of one of the complement molecules to the Fc portion of the immunoglobulin molecule. Antibodies with antigens bound to their Fab region are able to bind to phagocytotic cells having an Fc receptor on their surface, via the C_H2 domain of the immunoglobulin molecule. The coating of antigen with antibody leading to binding to phagocytes and phagocytosis is referred to as **opsonization**. The complement fragments C3b and, to a lesser extent, C4b (see Section 32.5) are **opsonins**, that is, they promote opsonization. Virus-infected cells (and probably tumour cells) may be killed by a process known as **antibody-dependent cell-mediated cytotoxicity (ADCC)**. In this process, ADCC effector cells (e.g. **natural killer cells – NK cells**) are able to bind antibody to their surface by Fc receptors, and if the Fab region of such molecules then interacts with an antigen they recognize on the surface of an infected cell, lysis of this cell is brought about by the secretion of lytic agents such as the perforins and the powerful proteases called granzymes (Figure 32.8). Alternatively, ADCC effector cells may bind, via their Fc receptors, to virally-infected cells already coated with antibodies.

The various roles of antibodies are summarized in Table 32.2.

Classes of immunoglobulin

Five classes of immunoglobulin molecule have been described which are recognized by the properties of their C_H region. The five classes are designated IgG, IgM, IgA, IgD and IgE. Each differs in the H-chain used, and the chains of each class are assigned a lower case Greek letter corresponding to the Roman letter of the class. Therefore, the heavy chain of IgM is μ, that of IgA α, IgG

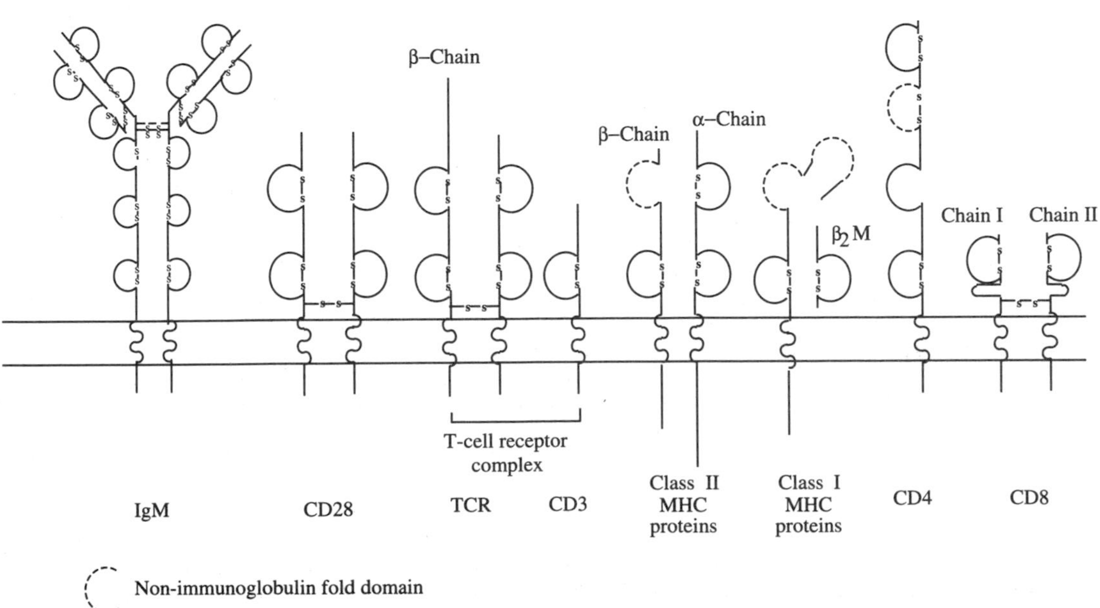

Figure 32.7 **Membrane proteins that have one or more immunoglobulin folds**

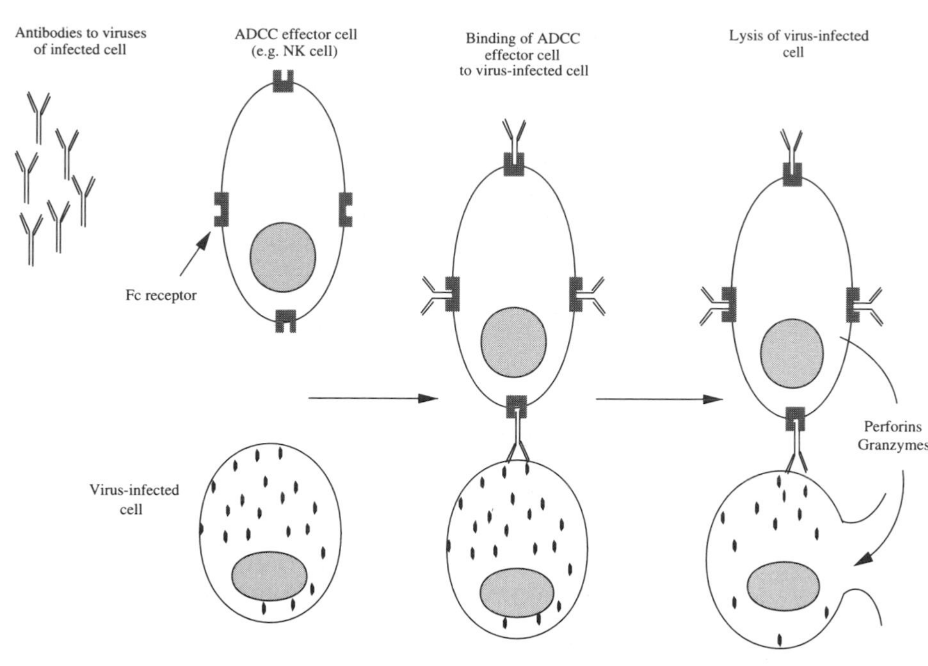

Figure 32.8 **An example of antibody-dependent cell-mediated cytotoxicity (ADCC)**

Table 32.2 **The various functional roles of the immunoglobulins**

Process supported	*Section of Ig molecule required*	*Mechanism of action*
Precipitation	Variable region of Fab	Cross-linking of soluble antigens
Agglutination	Fab	Cross-linking of bacteria, viruses and parasites
Neutralization	Fab	Prevention of the attachment of viruses or bacterial toxins to cell membrane receptors
Opsonization	Both Fab and Fc	The coating of antigen with antibody leading to binding to phagocytes via Fc receptors and consequent phagocytosis
Activation of complement	Both Fab and Fc	Complex formation between antigen, antibody and complement leading to antigen removal
Antibody-dependent cell cytotoxicity	Both Fab and Fc	Antibody binding to the surface of a target cell while Fc is engaging a killer cell
Passage across the placenta	Fc	Transfer of IgG from the mother to the fetus via the placenta

γ, IgD δ and IgE ϵ. Each of these is encoded by a separate C_H gene segment designated $C\mu$, etc. There are two varieties of light chain, kappa κ and lambda λ. All of the classes of immunoglobulin molecule use either of these two types of L chain. Thus each monomeric immunoglobulin molecule has two H chains of the same class and two L chains of the same type: IgG may be found as $\gamma_2\kappa_2$ or $\gamma_2\lambda_2$. In the case of IgG and of IgA there also exist four and two subclasses, respectively (γ_{1-4} and α_1 or α_2)and these differ due to slight variations in γ and α chain sequences. Table 32.3 summarizes the structures and some of the functions of Igs.

IgG

IgG is the major class of immunoglobulin found in the serum constituting some 70% of the total, with IgG_1 accounting for some 50% of this total. IgG, which dominates the secondary immune response, is the only class of immunoglobulin that crosses the placenta and therefore is the class of maternal antibody that protects the fetus. The transplacental movement of IgG is not without its problems, however, as is seen in **haemolytic disease of the newborn** (erythroblastosis foetalis). In this disease, a **Rhesus negative** (Rh^-) mother, lacking the D antigen on her red cells, produces antibodies to this antigen associated with the Rhesus positive (Rh^+) child she is carrying. This does not constitute a problem to the first pregnancy because the mother's immune system is most frequently exposed to fetal red cells during parturition, but a subsequent pregnancy with a second Rh^+ baby will be at risk. This is because the mother will now mount a secondary immune response to the fetal D antigen which will consist largely of IgG. This will cross the placenta and bind to D antigen on the surface of the fetal red cells. This leads to a haemolytic anaemia, resulting in the death of as many as 30% of infants affected in this way. The current therapeutic method of treating RhD negative mothers of an RhD positive first

Table 32.3 **Important chemical and biological properties of the immunoglobulins**

Property and activity of molecule	IgG_{1-4}	IgA_{1-2}	*IgM*	*IgE*	*IgD*
Nature of light chain	←		κ or λ		→
Nature of heavy chain	γ_{1-4}	α_{1-2}	μ	ϵ	δ
Molecular form of secreted protein	Monomer	Largely monomer (serum) and dimer (secreted)	Pentamer	Monomer	Monomer
Component of primary antibody response	√	–	√	–	–
Ability to activate classical complement pathway	√	–	√	–	–
Ability to cross the placenta	√ (IgG_2 – weak)	–	–	–	–
Is present in the plasma membranes of mature B-cells	–	–	√	–	√
Binds to the macrophage Fc receptor	√ (IgG_4 – weak)	–	√	–	–
Present in mucous secretions	–	√	√ (to a limited extent)	–	–
Induces mast-cell degranulation	–	–	-	√	–

child is via intra-muscular injection of antiRhD antibodies raised to D antigen in policemen who have volunteered to be deliberately sensitized to the antigen. The idea is that any fetal D antigen (red cells) transferred to the mother at birth will be rapidly neutralized.

IgM

IgM accounts for some 10% of normal serum immunoglobulin. The circulating molecule consists of a pentameric association of the basic monomeric immunoglobulin structures joined by disulphide bridges. The bridges are largely between immunoglobulin monomers, but the structure is closed in a ring by a J-chain that also forms disulphide links with two of the Igs. IgM is a major component of the primary immune response. In common with IgE (but in contrast to the other Igs), IgM has a fourth C_H domain. In a modified, monomeric, form IgM (together with IgD) is found in the plasma membrane of B-cells (Figure 32.7) and is sometimes referred to the B-cell receptor. Both IgM and IgD play important roles in the production of specific antibodies by these cells.

IgA

About 20% of serum immunoglobulin is in the form of IgA. Of this, 80% is a monomer, but the remaining 20% are dimers held together with the same J-protein that is found with IgM. IgA is the predominant immunoglobulin in **seromucous secretions**, being found in external secretions such as saliva, tracheobronchial secretions, colostrum, milk and genito-urinary secretions. (The IgA found in *internal* secretions such as synovial, amniotic, pleural and cerebrospinal fluids, is of the serum, not the secretory, type.)

The secretory molecule is a dimer, as is 20% of the serum IgA, but additionally it has a secretory piece covalently attached to it. Some epithelial cells express a surface IgA receptor (also known as the polyIg receptor) that recognizes the Fc region of this class of immunoglobulin and initiates receptor-mediated endocytosis of the IgA molecule as a first step in its transcytosis across the cell from the serosal to the mucosal side. The secretory protein forms part of the polyIg receptor and it becomes attached to the IgA molecule as it passes through the epithelial cells. It is noteworthy that the total amount of IgA in serum and secretions make it the most abundant immunoglobulin in humans.

IgD

The circulating concentration of IgD is very low and the molecule is particularly labile. These facts have made the study of IgD function difficult. The circulating molecule is a monomer. IgD, in a membrane-associated form, is found on the surface of B-cells appearing developmentally after IgM molecules.

IgE

Although present in very low concentrations in serum, much interest has focused on IgE for two reasons. The first is because, as the **homocytotropic** or **reaginic antibody**, it is involved in the processes of **immediate hypersensitivity** and **allergy**. Secondly, it seems to be the class of antibody most involved in providing protection in the disease schistosomiasis. In this disease infective larvae, shed by freshwater snails, penetrate the skin and after several stages of development egg-laying, adult worms are found in the venous system. Thus in an area of Gambia endemic for *Schistosoma haematobium* (which takes up residence in the genito-urinary venous system) a link has been established between the production of IgE against schistosomes and the acquisition of immunity during re-infection. In another study in Brazil, adolescents with high resistance to infection by *S. mansoni* (infecting the portal and mesenteric veins) have levels of IgE to the organism which are 6–8 times higher than those with low resistance; the concentration of other immunoglobulin classes was either similar in both groups or higher in the least resistant subjects.

Crucial to an understanding of the role of IgE in immediate hypersensitivity is the finding that it binds

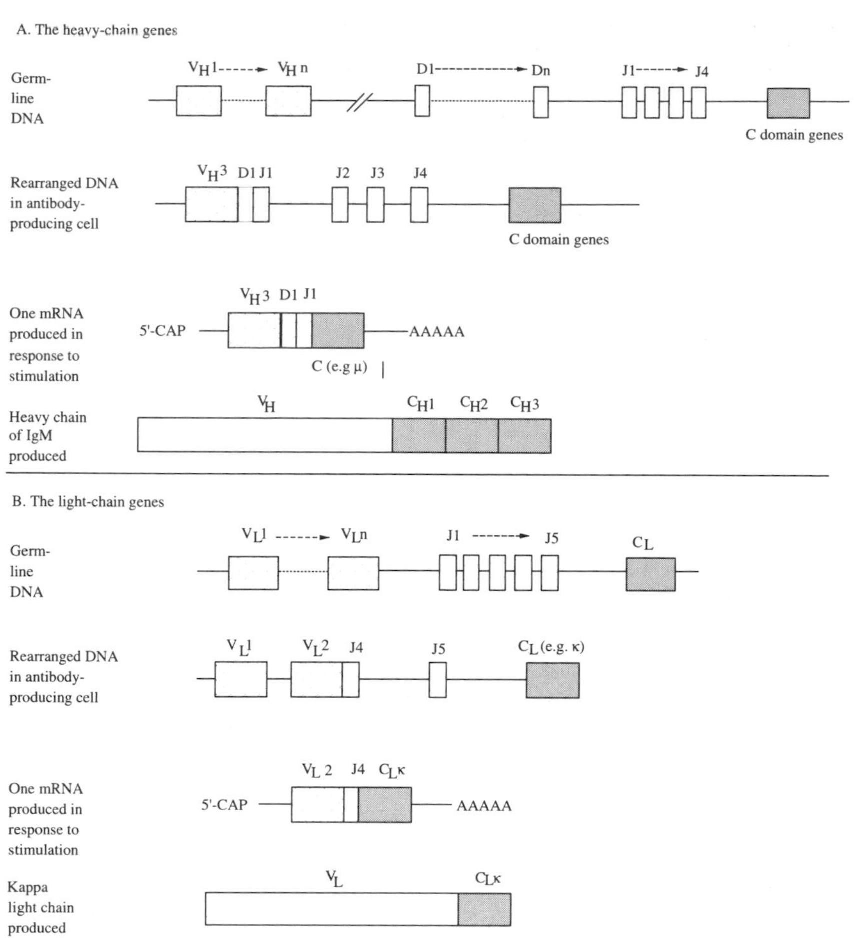

Figure 32.9 **The organization and expression of the genes for the immunoglobulin heavy and light chains**

strongly to receptors on the surface of mast cells via the Fc portion of the molecule. Occupation of these receptors by IgE molecules causes intracellular second messengers to be produced and these facilitate the degranulation of the mast cells with the release of pharmacologically active molecules that underlie allergic reactions (see Chapter 42), including food allergies. The role of IgE in affording protection in schistosomiasis remains to be established.

The source of antibody diversity

A central question of immunology, to which an answer has only recently been forthcoming, is how the immune system can generate antibodies of almost any specificity. This understanding has come about because it has been realized that in the germline there are antibody gene **segments** that are rearranged in lymphocytes. The presence of about 1000 such segments means that, by rearrangement, at least 10^7 different antibody coding genes may be generated. It was first shown by Hozumi and Tonegawa that two genes code for a single chain of an antibody molecule. In the DNA of the germ cells the genes coding for the V and C regions are separated, but are united in differentiated antibody-secreting B-cells. Subsequently, more detailed sequence analysis of germ-line cells has revealed that the V_L and V_H gene regions are segmented. The V_L region consists of some 100 V segments and 5 J (for joining) gene segments. This arrangement is shown in Figure 32.9. This figure also shows the DNA sequence that might arise in B-cells due to rearrangement and random selection of V and J segments during differentiation of a stem cell. The type of mRNA such a cell may then produce is also shown. A similar situation obtains with the germ-line DNA that gives rise to H-chain genes: there is a tandem array of V segments (some 100–200 in number) and 4 J regions, but in addition, between these are about 15 **diversity** (D) segments. As is shown in Figure 32.9, rearrangement and random selection of these gives rise to different H-chain genes in different cells. When activated, these genes can then give rise to functional mRNA molecules in which V, D and J segments are joined (**V-D-J joining**). In addition, there is also evidence that new segments are generated by **somatic mutation** during the processes of rearrangement of the stem cell DNA.

Thus the diversity of antibodies is generated at the genetic level by the presence in the germ-line of multiple pro H and pro L gene segments. In addition, diversity arises as a result of the way the segments rearrange, also by the combination of H and L chains to form the immunoglobulin molecule and by mutation. These sources of diversity are listed in Table 32.4.

Table 32.4 **The means of generation of antibody diversity**

Process involved	*Nature of diversity*
Selection from a large number of 'V' genes	Genetic
Random V-D-J joining	Recombinatorial
Independent diversification of H and L chains	Combinatorial
Somatic mutation	Mutational

32.5 Complement

The characteristic functions of antibodies are recognition and neutralization: they do not directly cause the death of invading pathogens. The other major component of the humoral immune response, **complement**, achieves this and additionally promotes inflammatory responses. There are two converging complement pathways. In the **classical pathway**, complement is activated by encounter with antigen–antibody complexes. The second pathway is designated **alternative**, but probably predates the 'classical' pathway because its activation is stimulated directly by pathogens. A variety of pathogenic bacteria can activate the complement pathways and this activation therefore constitutes an innate immune response. Approximately 15 liver-derived, plasma proteins constitute the complement system. These proteins are normally inert, but are activated by a cascade of reactions, many of which are proteolytic, and therefore resemble those of the blood-clotting cascade (see Chapter 24). Indeed one of the stages, in which the activated form of C1r is generated, is calcium ion-dependent.

The classical complement pathway

This pathway is initiated when the first complex of complement (C1) binds to an antibody–antigen (Ab/Ag) complex: for kinetic reasons antibody alone does not bind to C1. The explanation of this kinetic discrimination becomes apparent when it is realized that the component of C1, termed C1q, that binds to the Ab/Ag has the form of a 'bunch of tulips' with 5 'heads' each capable of binding to the CH_2 domain of IgG or CH_3 in IgM. C1q can bind to Fc on several immunoglobulin molecules bound to an antigen (Figure 32.10). For this reason the pentameric IgM is much more efficient than IgG in starting the cascade. The multimeric interaction between C1q and Ab/Ag means that the antigens are likely to be part of a bacterial cell wall. The binding therefore results in the activation of C1q attached to the target pathogen. This activation sequentially results in the activation of two further components of the C1 complex: C1r and C1s. A C1q-induced conformational change in C1r leads to its

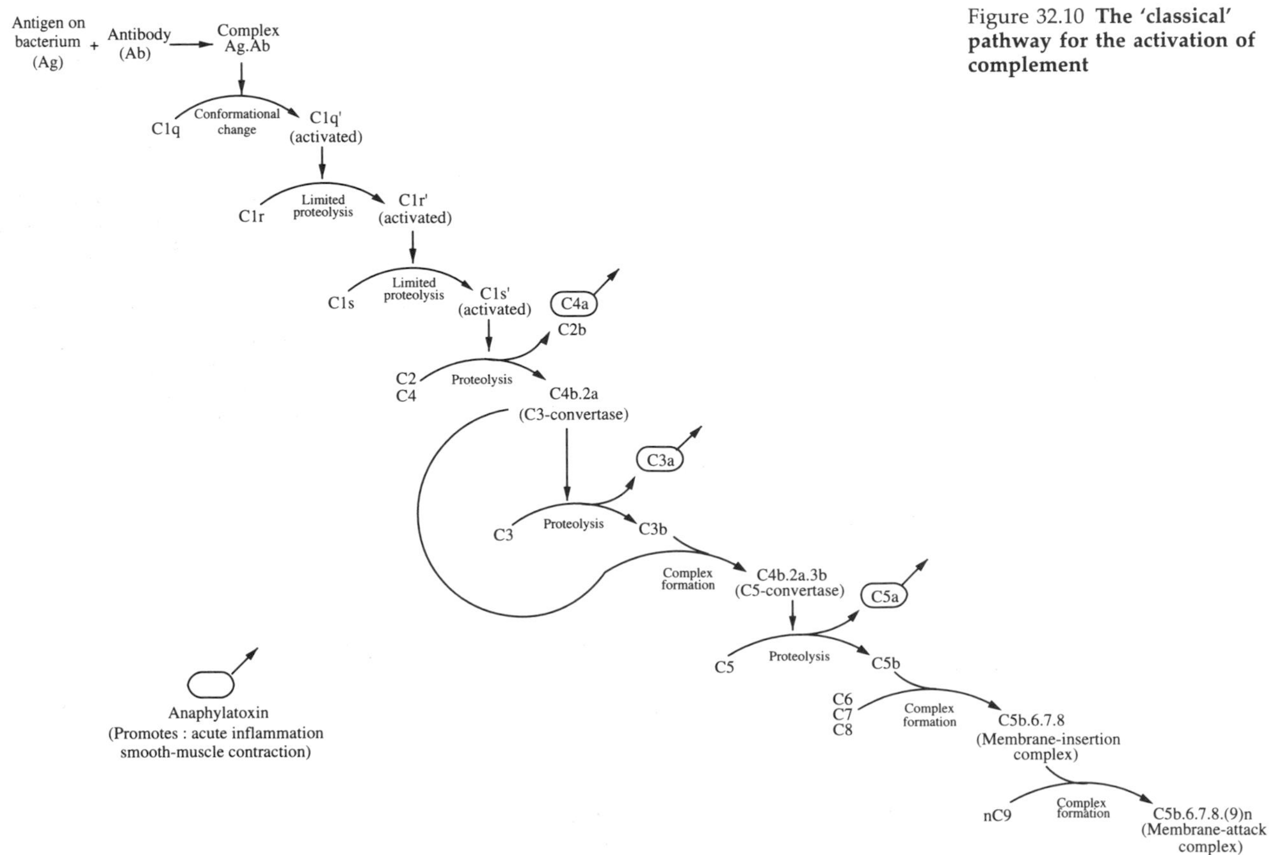

Figure 32.10 **The 'classical' pathway for the activation of complement**

activation as a protease which then acts autocatalytically to self-activate and also on C1s to cause its activation. In turn, C1s activates two further members of the cascade C2 and C4. This also involves proteolysis of C2 and C4 by activated C1s to generate two fragments – C2a and C4b – which bind covalently to the membrane to form a complex with proteolytic activity. The C2a.4b complex produced uses as its substrate the complement component C3 and for that reason is referred to as **C3 convertase**. A distinct C3 convertase, also using C3 as a substrate, is activated in the alternative pathway, so this reaction is the point at which the two pathways begin to converge. One fragment (C3b) from the attack of the convertase on its substrate then itself associates with the C2a.4b complex by binding covalently to the bacterial membrane to yield C2a.4b.3b and uncover yet another proteolytic activity. This is called **C5 convertase** (also referred to as the **'activation unit'**) because its substrate is component C5 of complement. It should be recalled that C4b and especially C3b also function as opsonins. It is also notewothy that the 'a' (soluble, unfixed) products of the cascade: C3a, C4a and C5a (produced with C5b during C5 convertase cleavage of C5) act as **anaphylatoxins**, in that they are able to promote acute inflammatory responses and smooth muscle contraction.

As a break in the pattern, C5b is not a protease, instead it binds jointly to C6 and C7 to form a **membrane insertion complex**. Once established in the membrane, C5b.6.7 is able to bind C8, an event that signals the rapid binding to this new complex of the final component of complement, C9, which itself polymerizes to form a membrane-spanning tubule. This collection of molecules is referred to as the **membrane attack complex**. The lumen of the tubule forms a channel that allows the passage of small molecules, metal cations, etc., across the bacterial membrane. The consequent osmotic disturbance results in water entering the bacteria which swell and burst. It has been suggested that as few as one such membrane attack complex suffices to destroy a bacterium.

The essence of this pathway is the binding of C1q to antibodies, themselves attached to bacteria. Two proteins have been described that are capable of replacing C1q in activating the complement cascade. Both are released from the liver during infections and are members of a group referred to as **acute-phase proteins**. These proteins differ from C1q in that they are able to bind directly to bacteria in the absence of an antibody. **C-reactive protein** binds to phosphorylcholine groups on the surface of bacteria, especially *Pneumococci*. The second protein is a **mannose-binding protein**. The production of this protein in the liver is prompted by the cytokine interleukin 6 which is secreted by macrophages on binding a bacterium. On encountering bacteria this protein binds to mannose residues on the surface of the organism.

The alternative complement pathway

The formation of C3b in the absence of Ab/Ag and in the fluid phase is the key feature of the alternative pathway (Figure 32.11). In this, C3 reacts with a serum component called **factor B** to form a C3.B complex. This complex now forms a substrate for the serum serine protease **factor D** which attacks the B subunit of the C3.B complex to yield C3.Bb (and inactive Ba). This latter is a second C3 convertase which serves to convert C3 into C3b. This last molecule binds to bacterial cell surfaces (e.g. via the lipopolysaccharide and teichoic acids associated with Gram-negative bacteria) where it binds further factor B. The surface-immobilized C3b.B is now acted on by factor D to form C3b.Bb which catalyses many rounds of C3 to C3b conversion (the so-called **'amplification loop'**). It is the presence of the foreign cell surface that allows the amplification to occur. So-called 'tickover' conversion of C3 to C3b occurs continuously, but at low level, in the absence of bacteria. 'Tickover' is also held positively in check by the presence in serum of factor I, an inactivator of the process, which interacts with a second factor H to catalyse the proteolytic degradation of soluble, but not membrane-bound C3b.

The complex of C3b with Bb is very unstable until it reacts with the serum protein **properdin**, a stabilizing

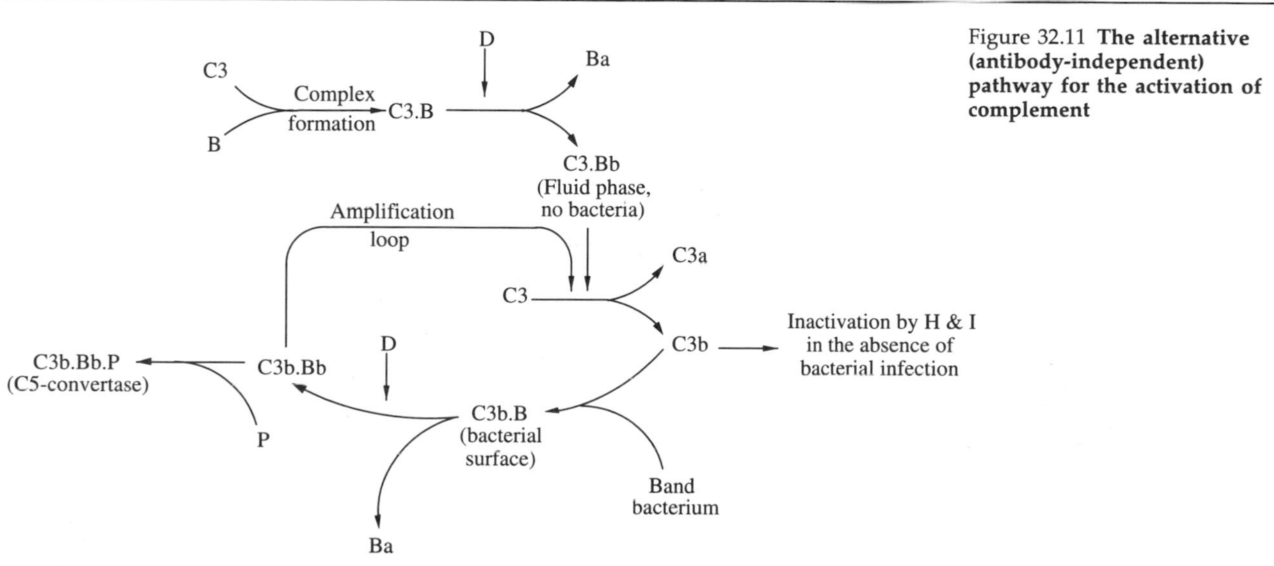

Figure 32.11 **The alternative (antibody-independent) pathway for the activation of complement**

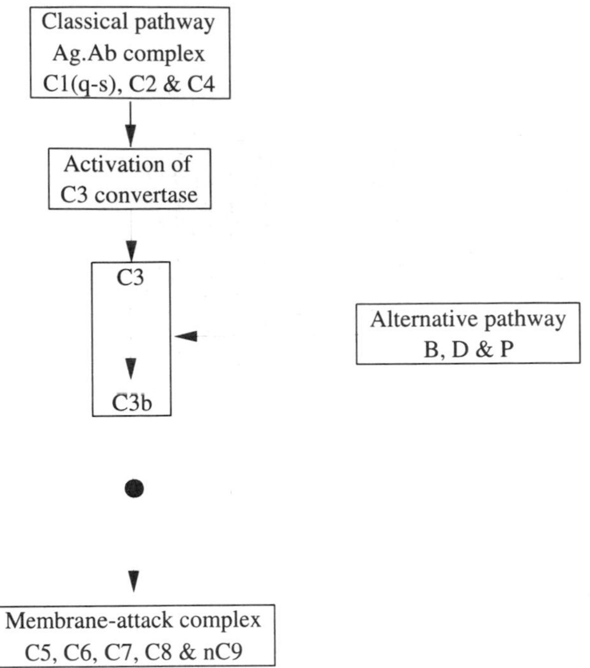

Figure 32.12 **The convergence for the 'classical' and alternative complement pathways**

molecule. The C3b.P.Bb complex now becomes active as a C5 convertase and thus the two complement pathways are united. Properdin was discovered in the 1950s by Pillemer, who believed it to initiate the alternative pathway.

The convergent complement pathways are shown in Figure 32.12. It is evident from this that the alternative pathway can also be activated by C3b generated in the classical pathway, thus the former pathway can also amplify a response initiated by the latter. The advantage of having the two pathways is clear: many bacteria are targeted by C3 or mannose-binding protein, but the discrimination is simply at the level of self–non-self. By combining the specificity of antibody binding with the innate response a range of parasites that evade the general defence mechanism are dealt with and rendered harmless.

32.6 Cell-mediated immunity

The term cell-mediated immunity refers to the adaptive response induced to combat intracellular infections and **allogenic** (i.e. foreign) 'antigenic insults'. Its defensive roles are listed in Table 32.5. The major components of this type of response are the lymphocytes, phagocytes and the cytokines.

In order to deal with the subject of cellular immunity it is first necessary to say something about the cells and structure of the lymphoid system, and also about the changes that occur as these cells develop and mature. This is necessary not only to explain further how diversity is achieved in the immune system, but also in order to show how self is distinguished from non-self, i.e. how **immunological tolerance** is achieved. With this will come an understanding of how this crucial recognition process may fail in autoimmune disease and how it may be suppressed in tissue transplantation.

Cells of the immune system

The cells of the immune system jointly constitute the leukocytes (white cells) and they are usually divided into five categories: **neutrophils** (sometimes called **polymorphonuclear** or **PMN** cells for their lobular nuclei), **eosinophils**, **basophils**, **lymphocytes** and **monocytes**. The first three cell types listed are sometimes referred to under the general name of **granulocytes**. The predominant granulocyte is the neutrophil which contains numerous lysosomes and secretory granules. Lymphocytes comprise B-cells, several subgroups of T-cells and the related **natural killer (NK)** cells. It is noteworthy that there are 2×10^{12} lymphocytes in the human body, the total mass of which is equivalent to that of the liver. Monocytes become **macrophages** when they move from the circulation into the tissues. Neutrophils and macrophages are sometimes described as being **'professional phagocytes'**.

The roles of T-cells are threefold. They kill host cells infected with pathogens which, because of their intracellular locations, are sequestered from antibodies. Where infections persist, they help maintain an inflammatory response. Finally, by secreting cytokines they act as regulators of the cells that mount both innate and adaptive immune responses.

Table 32.5 **Cell-mediated immunity**

Cell-mediated immunity		*Examples*
1. Provides defence against intracellular pathogens:	• viruses	HIV, herpes simplex, cytomegalovirus
	• bacteria	*Mycobacteria tuberculosis, Listeria monocytogenes*
	• parasites	*Leishmania mexicania, Trypanosoma cruzi*
2. Has as its major components:	• Lymphocytes	Helper T-cells, cytotoxic T-cells and natural killer (NK) cells
	• Phagocytes	Macrophages
	• Soluble secreted molecules	Interleukin 2 (cytokines)

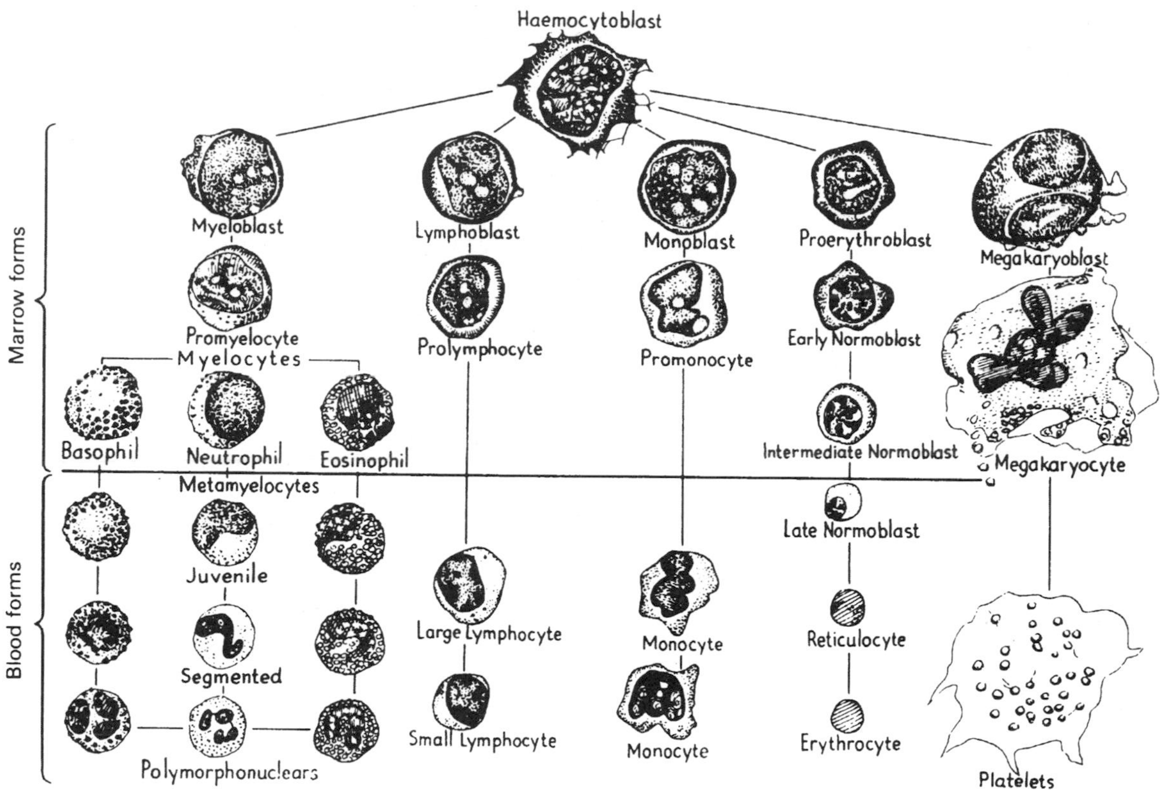

Figure 32.13 **Normal haematopoiesis and the formation of macrocytes in folate or vitamin B_{12} deficiencies**

Developmentally all the leukocytes, the red cells (erythrocytes) and the platelet precursor cells (megakaryocytes) derive from pluripotent stem cells, as is shown diagrammatically in Figure 32.13. These cells derive from the fetal liver and the adult bone marrow.

The functional organization of the lymphoid tissue

Anatomically, two types of lymphoid tissue are recognized (Figure 32.14). The **bone marrow** and the **thymus** constitute the major **primary lymphoid organs**. Stem cells in the bone marrow either differentiate to give rise to B-cells or they migrate to the thymus where their development into various types of T-cell is fostered. These lymphocytes then subserve many of their functional roles by moving to the **secondary lymphoid organs**. These are of two types: encapsulated, e.g. the lymph nodes and the spleen, and unencapsulated, the so-called mucosal-associated lymphoid tissues (MALT). Each of these secondary organs is concerned with a different type of immune response. In the lymph nodes the response to antigens in tissues is orchestrated, responses to antigens in blood are focused on the spleen, while the MALT system deals with antigens at mucosal surfaces.

Blood-borne lymphocytes enter the tissues and lymph nodes via high endothelial venules and leave via the draining lymphatics to return to the blood via the thoracic duct. While resident in the lymph nodes, B-cells congregate in the outer cortical area, whereas the T-cells are found more centrally in the paracortical area.

B-cell development

As indicated, B-cells undergo their early development in the bone marrow and the process is **antigen independent**, leading to the appearance in the circulation of cells most of which are self-tolerant, immature 'virgins' with diverse IgM and IgD molecules expressed on their surfaces. These molecules constitute the B-cell receptors or surface immunoglobulins (sIgs). The sequence of events leading to these changes is shown in Figure 32.15. In the first step, stem cells develop into pro-B-cells which then mature into pre-B-cells in response to a signal (interleukin 7) received from stromal cells in the bone marrow. These pre-B-cells undergo extensive proliferation and at the same time they undertake rearrangements of the gene segment encoding the immunoglobulin chains (see Section 32.4). This gives rise to cells that express one example of a vast range of monomeric IgM and IgD molecules on their surface. These sIgs are synthesized to incorporate a membrane-anchoring α-helix towards the C-termini of the two heavy chains. If these immature B-cells should happen to produce a sIg molecule that binds strongly to a self-antigen among the cells of the bone marrow, cell-death (**apoptosis**) occurs. This represents a process of **negative selection**. As a result, two types of B-cell

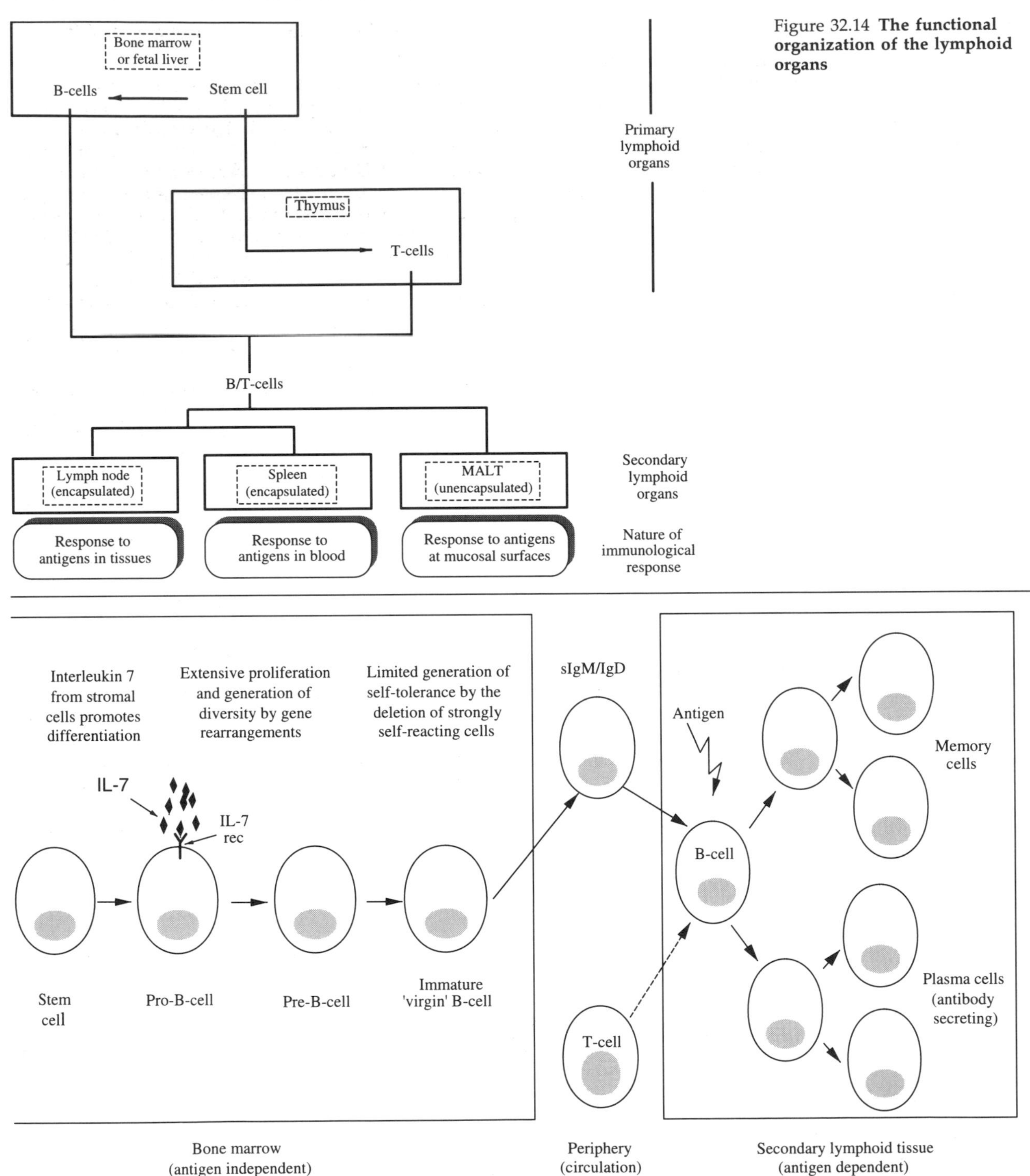

Figure 32.14 **The functional organization of the lymphoid organs**

Figure 32.15 **The course of B-cell development**

emerge from the tissue: **self-tolerant** or weakly anti-self acting. The presence of the latter cells means that the low-level production of auto-antibodies is quite normal, both in healthy individuals and in those having suffered recent tissue injury.

When virgin, self-tolerant B-cells reach the secondary lymphoid tissues and encounter foreign antigen they are stimulated to proliferate in a process that requires the help of certain T-cells. Proliferation gives rise to two types of cell: **plasma cells** that secrete the soluble antibody corresponding to the sIg found on the stimulated parent cell, and **memory cells** that persist until a second encounter with that antigen occurs, when they mediate the secondary immune response.

T-cell development

Because they are more diverse, the processes underlying T-cell development are somewhat more complex, although the general scheme of events is the same. Developing T-cells express diverse surface proteins as they mature and these molecules are the source of both functional diversity and self-tolerant selection. However, this last process has both positive and negative aspects.

The T-cell receptor

The T-cell receptor in some respects is the counterpart of the surface immunoglobulin of the B-cell. It consists of two polypeptide chains, usually α and β, but in some cells the alternative polypeptides γ and δ are found. Each α- and β-chain has one constant and one variable domain and a C-terminal transmembrane region. The two chains are joined by a disulphide bridge. The domains in each chain are organized into immunoglobulin folds. Completing the analogy with the immunoglobulins, there are gene-families encoding the two chains; these are segmented in the germ-line and undergo developmental diversification. Here the analogy ends because T-cell receptors are not able directly to bind foreign antigens. The antigen is recognized by the T-cell receptor only after it has been processed by other cells which present the antigenic determinant as a small peptide on their surface. These ancillary cells are called **antigen-presenting cells (APCs)**. The surface molecules responsible in APCs for binding and presenting antigen-derived peptides form part of the gene complex generally referred to as the **major histocompatibility complex (MHC)** and specifically in humans as the **human leucocyte associated antigen** or **HLA**. Evidently the products of these genes are very important proteins and will now be described.

The structure of the human MHC complex and its protein products

The HLA region is found on chromosome 6 and encodes three distinct classes of protein (I, II and III – Figure 32.16). The class I region is also designated HLA-ABC and consists of distinct loci: A, B and C. The class II region is designated HLA-D and also has three loci DP, DQ and DR. The class I and II proteins are concerned with antigen presentation. Somewhat confusingly, class I and II proteins are sometimes themselves referred to as MHC antigens. This reflects the fact that the presence on cell surfaces of these molecules was first recognized by immunological means: they were recognized as antigens by antibodies from other species. HLA antigens A, B and C are expressed on the surfaces of all nucleated cells, whereas HLA-D antigens are restricted to B-cells, activated T-cells (in humans), monocytes (and therefore macrophages) and on some epithelial cells (Table 32.6). A noteworthy feature of MHC proteins and the genes that encode them is their high **polymorphism**. For example, there are as many as 50 alleles for each of the MHC class I genes. Thus, while TCRs vary from cell to cell in an individual, MHC molecules are the same on all the cells of that person but differ from those of unrelated individuals. Consequently no two unrelated individuals are likely to express exactly the same HLA antigens. It is this fact that presents such a problem to the transplant surgeon because MHC molecules on donor cells presenting peptides generated from 'self-proteins' are recognized as foreign and destroyed by the T-lymphocytes of the recipient.

In the rare inherited disease **bare lymphocyte syndrome**, patients suffer from severe combined immunodeficiency as a result of the complete lack of MHC class II molecules on all cells. The defect appears to reside in the **regulation** of the expression of the MHC class II gene, not in the gene itself. Affected patients suffer from multiple infections and usually die very young.

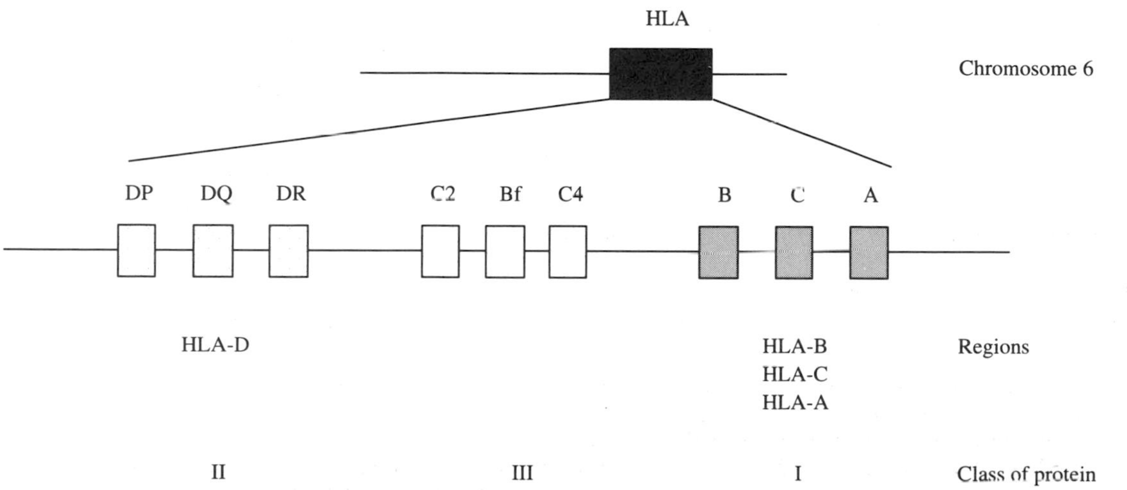

Figure 32.16 **The MHC regions on human chromosome 6**

Table 32.6 **The cellular distribution of MHC antigens: the nature of antigen-presenting cells**

MHC subregion expressed	*Cellular distribution of antigen*
HLA-A, B, C (class I antigens)	All nucleated cells and also platelets
HLA-D (class II antigens)	B-cells Macrophages Monocytes Epithelial cells Activated T-cells Dendritic cells

Class III genes encode the complement proteins and some of the cytokines.

Figure 32.7 shows the arrangement of both MHC class I and II proteins in the plasma membrane. Class I molecules have two polypeptide chains: a membrane-spanning subunit that has a single immunoglobulin fold, and an externally, non-covalently attached protein called β_2-microglobulin. This latter protein also has a single immunoglobulin fold. Class II proteins also have two polypeptide chains, α and β, each with a single immunoglobulin fold and single transmembrane regions.

Other members of the immunoglobulin super-family of molecules (having immunoglobulin folds) are central players in immune processes and should be mentioned here. These are also cell surface molecules specifically found on lymphocytes and named CD3, CD4, CD8 and CD28 (see Figure 32.7 for their structures). The CD nomenclature stands for **cluster designation** and simply refers to the fact that these proteins were first recognized as antigens by clusters of monoclonal antibodies.

CD3 always occurs as a complex with the T-cell receptor. CD4 and CD8 are important signals in T-cell differentiation and function. Cells expressing either one or both of these markers are designated $CD4^+$, $CD8^+$ and $CD4^+/CD8^+$.

The roles of MHC class I and II proteins in antigen presentation

There are two pathways for antigen presentation: one for peptides generated in the cytosol, either from endogenous proteins or from invading viruses; the other deals with peptides generated in endolysosomes and therefore deriving from organisms and molecules that have undergone phagocytosis or receptor-mediated endocytosis. Cytosolic peptides that are to be presented on the surface are processed (by proteolysis) and transported into the endoplasmic reticulum where MHC class I molecules fold round them. The peptide-loaded MHC molecule is then transported via the Golgi apparatus to the surface. Hence viral and cytosolic self-peptides are presented on the surface of all nucleated cells by MHC class I molecules. The MHC class II molecules are also assembled in the endoplasmic reticulum, but in this case around a specially-formed 'capping' protein which prevents the access of peptides to the binding site as the complex moves through the Golgi apparatus. It is only when vesicles carrying the capped MHC molecule fuse with endolysosomes that the acid environment inside the resulting vesicle causes the release of the capping protein. This frees the MHC class II molecule to bind peptides which have been generated by the proteolysis in the endolysosomes of bacterial (or other) proteins. If the APC is a macrophage the bacterium will have been taken up by phagocytosis. The uptake of proteins by B-cells, on the other hand, is via receptor-mediated endocytosis: both the protein and the surface immunoglobulin to which it is bound are taken up. Whether loaded in macrophages or in B-cells the peptide–MHC complex then moves to the surface to present the foreign peptide. The events involved in antigen processing to peptides and their presentation on the cell surface by class I and II molecules is shown in Figures 32.17 and 32.18. The peptide binding site of both class I and II molecules consists of a groove in the molecule that has a β-pleated sheet as a 'floor' and two α-helices as the 'walls'. This recognizes the presence of certain conserved residues in the peptide, but also the presence of peptide bonds.

The role of the thymus in T-cell development

With this knowledge of the several surface markers found on T-cells it becomes possible to show how the change from stem cell to naive T-cell occurs in the thymus (Figure 32.19). On entering this primary lymphoid organ the stem cells proliferate and rearrange their T-cell receptor gene segments (much as happens in B-cells for their immunoglobulin gene segments, with a similar outcome: cells with a diversity of TCRs arise). Most cells end up with receptors with α- and β-type chains. These cells then express both CD4 and CD8 antigens: they become $CD4^+/CD8^+$ cells (cells with $\gamma\delta$ TCRs become $CD4^-/CD8^-$). Cells with $\alpha\beta$ TCR only survive if they are able to recognize other cells with MHC molecules on their surface (positive selection). Cells that fail this test undergo apoptosis and are rapidly removed by phagocytes. This is followed by clonal expansion of self-MHC reacting cells. At the same time cells become segregated into those which express CD4 ($CD4^+$) and those expressing CD8 ($CD8^+$) surface markers. These events occur in the thymic cortex. A second round of selection now occurs in the medulla of the organ, but this time it is negative in nature. Any cell that recognizes the presence of MHC-bound peptides that are of self-origin are deleted by apoptosis. Surviving cells will emerge as 'naive' T-helper (T_H) cells if they are $CD4^+$ or as cytotoxic T (T_C) cells if $CD8^+$.

Following a process of negative (but not positive) selection similar to that described for the $CD4^+$and $CD8^+$ cells, $\gamma\delta$ TCR cells lacking either of these markers emerge as $\gamma\delta$ T-cells. These cells tend to be located in the skin and epithelial sites and subserve killer functions that have yet to be clearly defined.

Interactions between T-cells and other cells

The basis of T-cell function is their ability to recognize infected cells by 'scrutiny' of their surfaces and to influence the functional activity of these and other cells by the secretion of lymphokines. The rules for cell surface scrutiny are very simple: T-cells will bind to

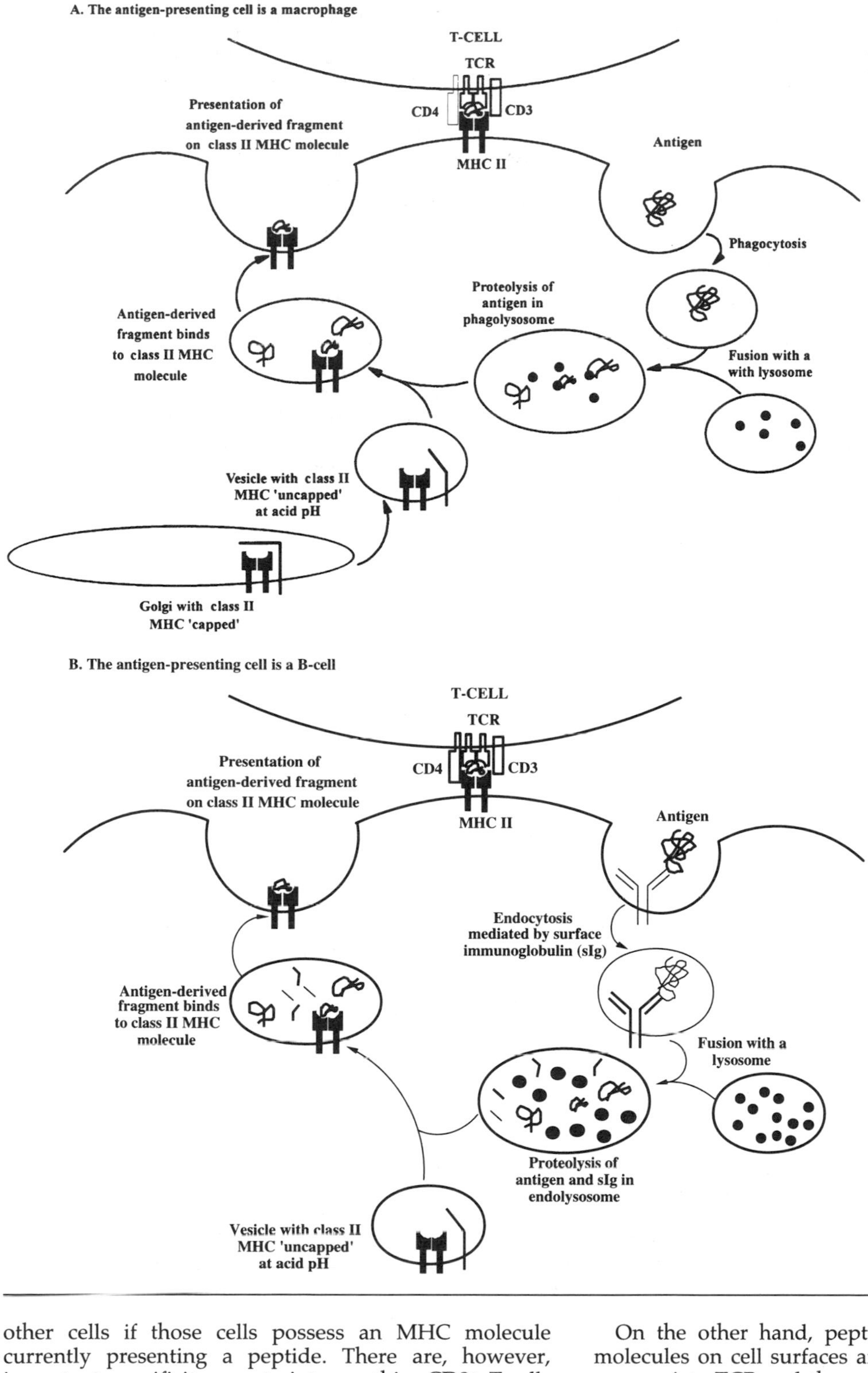

Figure 32.17 **The presentation of antigen-derived fragments with class II MHC molecules to CD4⁺ T-cells.**

other cells if those cells possess an MHC molecule currently presenting a peptide. There are, however, important specificity constraints on this. $CD8^+$ T-cells with T-cell receptors (TCRs) will bind to other cells only if the peptide the latter are presenting is associated with an MHC class I molecule. What this means is that, when dealing with infected cells, T_C cells will bind to them if their MHC class I proteins are presenting a peptide of viral origin. In addition, $CD8^+$ T-cells will also bind to tumour cells.

On the other hand, peptide-occupied MHC class II molecules on cell surfaces are jointly recognized by the appropriate TCR and the presence on the T-cell surface of $CD4^+$ molecules. Since helper T-cells fulfil this requirement (possession of $CD4^+$ and TCRs on their surfaces) these lymphocytes will bind to cells infected with bacteria that are presenting peptide fragments from them on MHC II molecules. As was shown in Table 32.6, such antigen-presenting cells can include other (activated) T-cells, B-cells, dendritic cells and

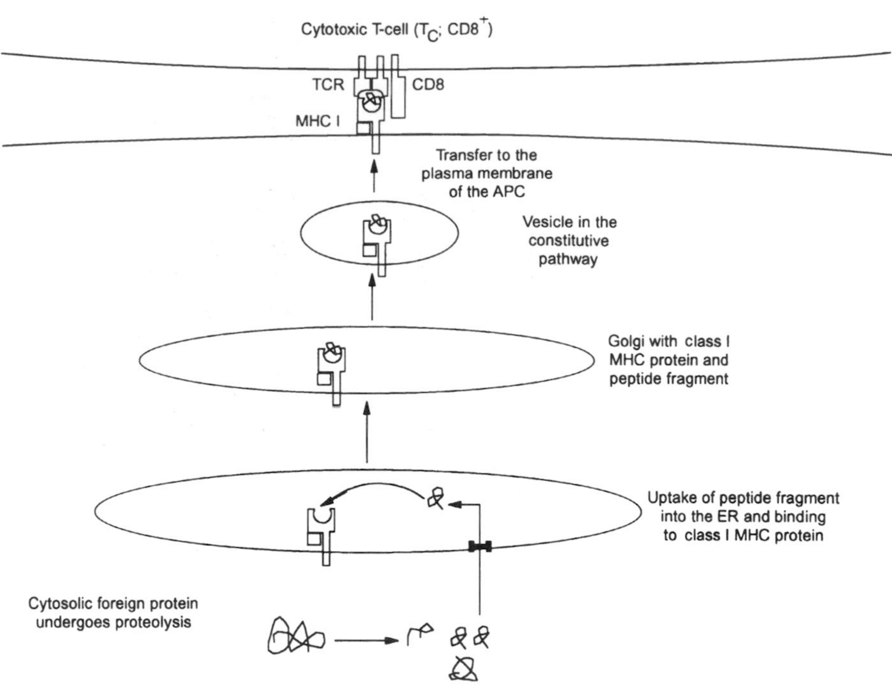

Figure 32.18 **The presentation of cytosolic foreign antigen-derived peptides with class 1 MHC molecules to CD8⁺ T-cells**

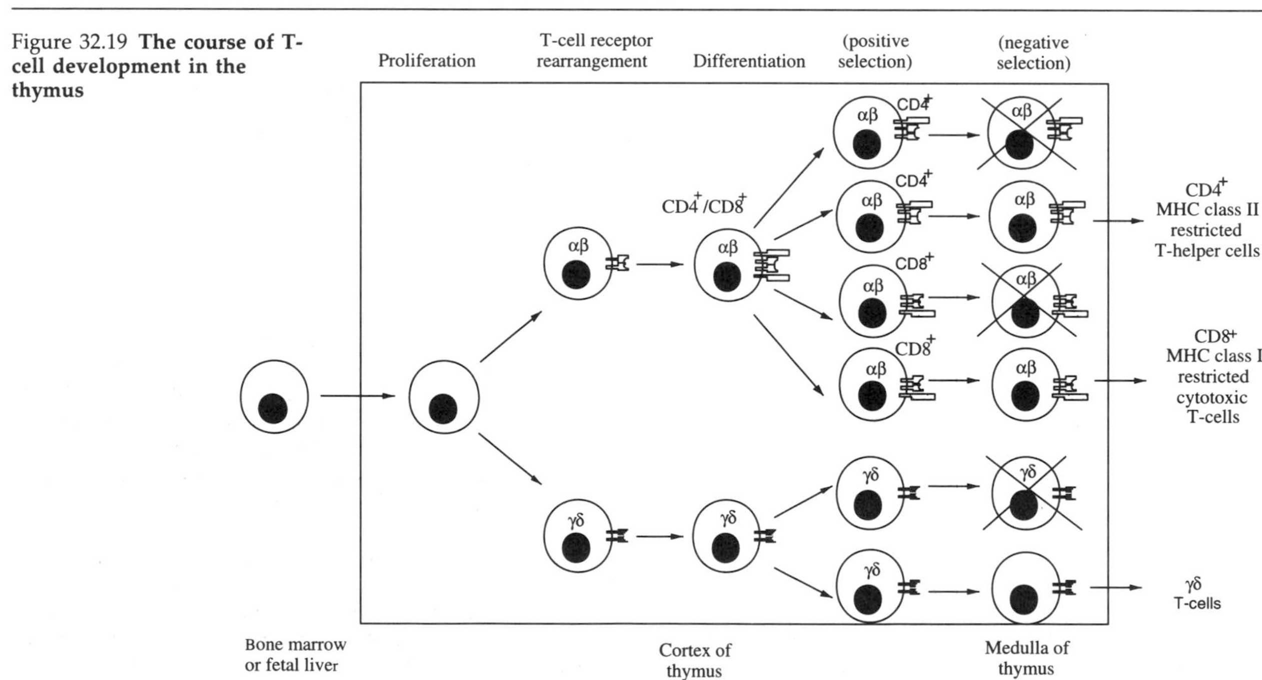

Figure 32.19 **The course of T-cell development in the thymus**

macrophages. These binding requirements are shown diagrammatically in Figure 32.20. Infected, but not infection-free, APCs additionally express on their surfaces the protein designated **B7**. There is a **B7 receptor**, CD28, on the surface of naive T_H cells and CD28 must be occupied by B7 at the same time that the TCR is engaged with the peptide-bearing MHC molecule for naive T_H cells to be activated ('armed'). If the TCR should happen to recognize a peptide presented on an MHC molecule by an APC, but that cell has no B7, then activation of the T-cell will not occur. In fact, because it is highly likely that the T-cell is recognizing a self-peptide it is rendered incapable of further response. This process is referred to as **anergy** and constitutes an extension of the self-tolerance initiated in the thymus.

A further mechanism helps to ensure that only appropriate immune responses are triggered and this involves the secretion of cytokines.

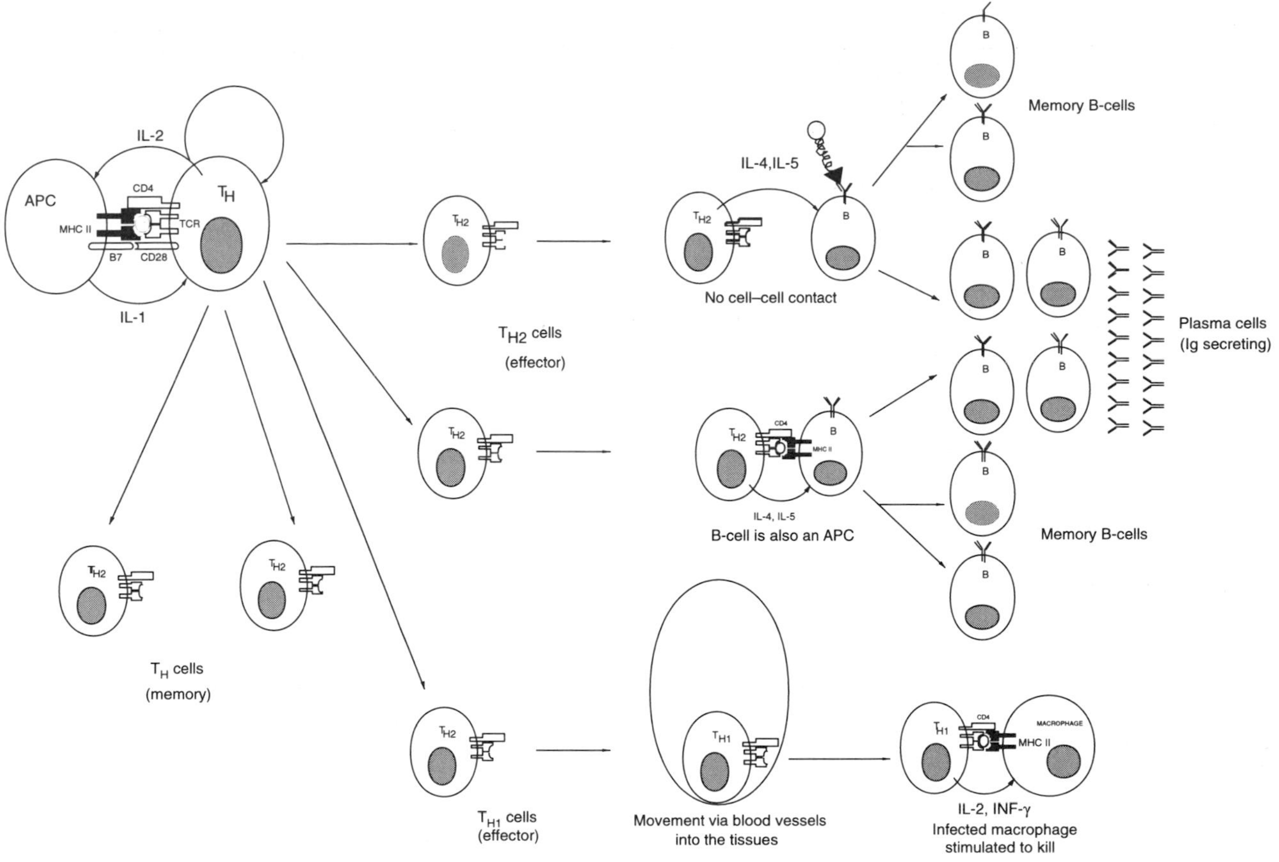

Figure 32.20 **The 'arming' and subsequent effector roles of T-helper cells**

Cytokines

The term **cytokine** is the collective name given to the proteins produced to transmit signals to and between cells engaged upon an immune response: in many ways they orchestrate the immune response. They are antigen-non-specific, transiently secreted, their half-lives are very short and their orbit of influence is usually of limited range.

Cytokines are sometimes given special names to indicate their source or actions. Thus cytokines produced by lymphocytes are termed **lymphokines**. **Interleukins** are secreted by one leukocyte to influence another. In the past the term **monokine** was used to describe macrophage-derived cytokines. **Interferons** α and β are produced altruistically by virally infected cells and help other cells to fight off the infection. **Interferon** γ is secreted by certain activated T_H cells and prompts killing by macrophages and natural killer cells. Table 32.7 lists some of the lymphokines secreted by activated T_H cells.

Many of the cytokines produce their effects via receptors that belong to the erythropoietin group (see Chapter 3). These receptors mediate the effects of their ligands by the direct activation of the cytosolic group of protein tyrosine kinases known as JAK proteins.

Table 32.7 **The major lymphokines secreted by activated T_H cells**

Lymphokine	*Target cells*	*Action*
IL-2	All T-cells and all B-cells	Promotes proliferation
IL-4, IL-5	All B-cells	Promote proliferation and differentiation into plasma cells (antibody-secreting)
IL-3 (also called colony-stimulating ractor, CSF)	Haemopoietic stem cells Mast cells	Growth and proliferation
Interferons (esp. interferon γ)	Natural killer (NK) cells	Promote the killing of virus-infected cells
	Macrophages	Action so that they become microbiocidal and tumoricidal
	B-cells	Promote proliferation and differentiation

Activation of T-helper cells

The antigen-dependent activation of T_H cells is at the heart of both cellular and humoral immune responses. As is shown in Figure 32.20, the binding of an **antigen-presenting cell (APC)** to a naive T_H cell causes the former to secrete interleukin 1 (IL-1). This lymphokine activates the T_H cell both to secrete IL-2 and to produce IL-2 receptor protein. The IL-2 then acts in an autocrine fashion to cause the further activation and also the proliferation of the T_H cells. This proliferative activity results in the production of more armed effector cells and also of memory cells.

The ability of IL-2 secreted by activated T_H cells to act in an autocrine fashion is an important feature of almost all immune responses. This observation is reinforced by two recent findings. First, **immunosuppressive drugs** such as **cyclosporin** and **FK506** both act to inhibit the enhanced production of IL-2 and its receptor by T_H cells normally signalled jointly via the activation of the TCR and the action of IL-1. These drugs play an important role in suppressing the recognition of transplanted tissues as non-self in transplant surgery. Secondly, individuals with the disease **X-linked severe combined immunodeficiency** (boy in the bubble disease) have inherited a gene that encodes a defective subunit of the IL-2 receptor. This same protein also forms part of the receptors for IL-4 and IL-7, thus explaining the severity of the disease (the maturation of B-cells requires the latter interleukin and the differentiation of virgin B-cells is prompted by IL-4).

T-helper cells assist other cells in making an immune response

Once they have been armed by their encounter with infected cells, activated T_H cells may undertake several tasks which involve them in interactions with other immune effector cells. To this end armed T_H cells develop one of two phenotypes characterized by the cytokines they produce: T_{H1} cells continue to secrete IL-2 but also begin to release interferon γ, while T_{H2} cells switch to the production of IL-4 and IL-5. These changes relate to different functions because T_{H1} cells promote cellular immunity, while T_{H2} cells effect humoral immunity.

B-cell–T_{H2} cell cooperation in making an immune response

The binding of antigen to most B-cells via their immunoglobulin surface receptor is insufficient to activate them: the lymphokines IL-4 and IL-5 secreted by T_{H2} cells are also required. This is shown in Figure 32.20 In a primary immune response, B-cells are stimulated to proliferate and differentiate into plasma and memory cells only when their binding of an antigen via an sIg is accompanied by their stimulation with IL-4 and IL-5 released by T_{H2} cells in their vicinity – the T_{H2} cell having been triggered by encounter with APCs presenting peptides derived from the protein bound to the sIg molecules of the B-cells. The requirement for B-cell proliferation helps explain why it takes 5 days for a primary immune response really to become established (see Section 32.3). In secondary immune responses the B-cells may also be the APC and then the interactions described may well be initiated by memory T_{H2} cells bound to B-cells.

T_{H1}–T_C cell interactions in cytotoxic cell killing

In common with all the cells of the immune system, T_C cells must receive at least two signals before they are activated (for T_C cells activation results in the killing of infected cells). They must be bound to their target cells via the peptide epitope/MHC class I-TCR complex (signal 1) and they must be activated by IL-2 (signal 2) secreted by T_{H1} cells (Figure 32.21). The targeted cell is then killed by the secretion of several destructive proteins, the chief of which are the perforins and granzymes. The former undergo calcium-dependent polymerization in the extracellular fluid and this allows them to form channels in the plasma membrane of target cells. This, in turn, provides a route by which the proteolytic granzymes gain access to cells and destroy them.

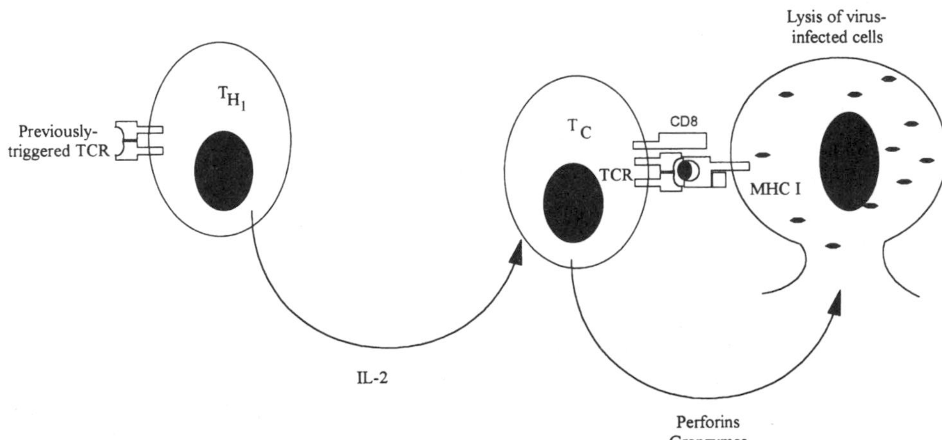

Figure 32.21 **Cooperation between T-helper cells and cytotoxic T-cells in the killing of virus-infected cells**

When the pair of signals are received, T_C cells also proliferate with production of memory T_C cells.

T_{H1} cells also stimulate the killing activity of macrophages

T_{H1} cells can also migrate through blood vessel walls into tissues where they activate macrophages to digest pathogens they have ingested. They bind to the macrophages via their TCRs and the interferon γ they produce acts as the stimulation (see Figure 32.20).

Delayed-type hypersensitivity

If T_H cells are armed as T_{H1} cells by interaction with infected dendritic APCs in the skin, for example, they secrete **chemotactic** cytokines which attract macrophages and neutrophils to the site of infection (or to the proximity of environmental toxins such as poison ivy) and then activate these professional phagocytes. Because these inflammatory responses take 24–48 h to evolve they are referred to as **delayed-type hypersensitivity**.

32.7 The inflammatory response

The establishment of an inflammatory reaction forms an important part of immune responses. The factors tending to promote an inflammatory reaction have been referred to previously, but it is convenient to gather all these together to show how they act in a coordinated fashion to produce this characteristic response.

The appearance of an inflammatory response has been observed by all who have suffered from a thorn under the skin. The tissue surrounding the splinter first becomes reddened, swelling occurs, there is pain and the affected tissue may feel warmer. These are the classical signs and symptoms ascribed to inflammation by Celsus, and are also observed with infections.

The vascular changes that underlie the observed response include an increased supply of blood to the affected site, with an accompanying constriction of the blood vessels that lead away from that area. Endothelial cells lining capillaries retract, causing these vessels to become more permeable to humoral immune mediators, in particular the complement proteins (Figure 32.22). The peptide fragments then released during the activation of complement (C3a, C4a and C5a) cause the degranulation of the resident 'sentinels' of submucosal and dermal tissues, the **mast cells**. These cells release a series of effector molecules including histamine, serotonin and hyaluronidase, that further increase vascular permeability. Consequently the professional phagocytes (most especially the neutrophils) are able to escape from the capillaries into the affected tissues. Once in the tissues the phagocytes move towards the centre of infection, moving up a concentration gradient of several **chemotactic** molecules. These include soluble complement fragment C5a (see Section 32.5) and **formylmethioninyl** peptides released by bacteria. Finally the phagocytes engulf and destroy the invading organisms. However, if the bacteria are able to resist the phagocytes, then the symptoms of the disease may be caused by the inflammatory response itself. For example, in tuberculosis the arrival of *Mycobacterium tuberculosis* prompts an inflammatory reaction. Although the phagocytes engulf the organism in the response, the bacterial cell wall resists the attack by the lysosomal enzymes, and the pathogen multiplies in the host cell ultimately killing it. Thus the inflammatory response

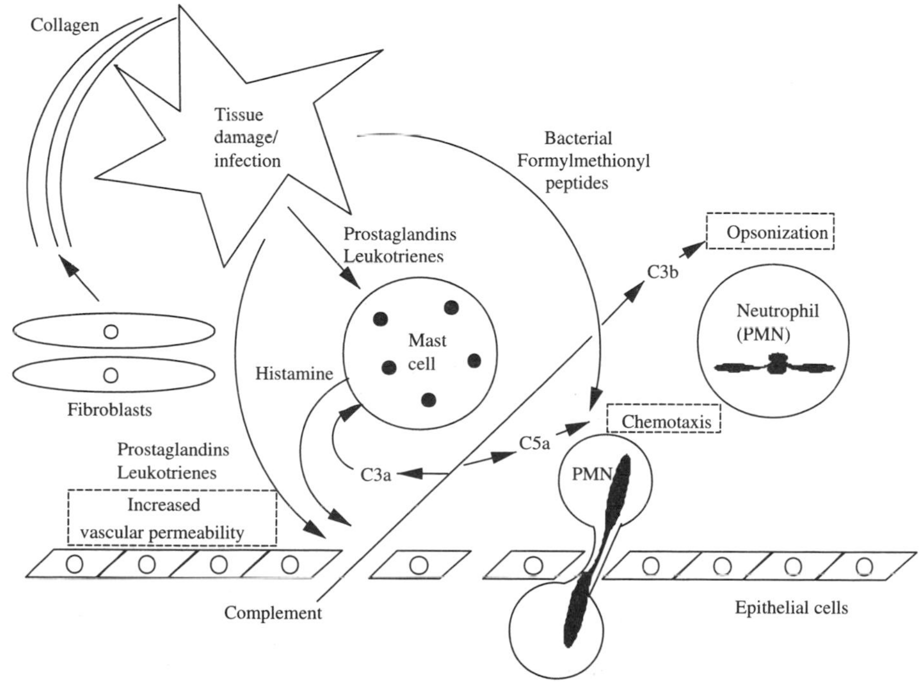

Figure 32.22 **The course of the inflammatory response**

becomes chronic with more and more phagocytes being attracted to the site. The bacteria-induced death of these cells causes the release of their lysosomal enzymes that begin to destroy the tissues of the lungs or bone, for example. Damage arising in this way is limited by the activation of fibroblasts which secrete collagen that 'walls off' or encapsulates the site of chronic inflammation.

Fortunately such outcomes are very much the exception and the process of inflammation is usually an excellent first-line defence against invasions that breach the skin.

Part 4

Health, disease and the environment

Chapter 33

Free radicals in health and disease

33.1 The nature of free radicals and the generation of reactive oxygen species

Free radicals are atomic or molecular species containing one or more unpaired electrons. They are generally highly reactive species and tend either to lose an electron, thereby acting as reducing agents, or to gain an electron, acting as oxidizing agents. It should be borne in mind that, according to the above definition, molecular oxygen (dioxygen) is a free radical, having two unpaired electrons, but it is not highly reactive as a result of its particular electronic arrangement.

The most important radicals which may be involved in disease processes are species which may be derived from molecular oxygen, and certain oxides of nitrogen, especially nitric oxide. There exist a number of enzymes which catalyse the formation of free radicals which are then released and may act as signalling molecules, or may provide defence against invading microorganisms. These species are potentially very active and special provisions need to be made to ensure that inappropriate and deleterious reactions do not occur. Failure of these protective measures may have severe consequences in terms of pathological conditions. It is also known that a range of enzymes generate free radical species as part of their catalytic mechanisms. These species usually have only a transient existence in the catalytic process, but very occasionally they may escape from the active site of the enzyme in question, with potentially destructive effects. Consequently, the possible involvement of free radicals in the aetiology and pathogenesis of disease is currently a very active area of research.

Some of the reactions which involve oxygen-derived free radicals give rise to compounds which are not themselves free radicals, e.g. hydrogen peroxide. They are, nevertheless, reactive and for this reason, oxygen-derived free radicals and related non-radical compounds are referred to as **reactive oxygen species (ROS)**.

The chemistry of ROS generation

When the dioxygen molecule is introduced into a reducing environment it may undergo a series of reactions leading to the formation of ROS.

The addition of a single electron gives rise to the **superoxide anion**, a free radical:

$$O_2 + e^- \rightarrow O_2{\cdot}^-$$

A further electron added to superoxide leads to the formation of the **peroxide anion**:

$$O_2{\cdot}^- + e^- \rightarrow O_2^{2-}$$

This may accept two hydrogen ions to yield **hydrogen peroxide**:

$$O_2^{2-} + 2H^+ \rightarrow H_2O_2$$

Production of hydrogen peroxide is potentially damaging to the cell, but the addition of two further electrons and two protons converts it into water:

$$H_2O_2 + 2e^- + 2H^+ \rightarrow 2H_2O$$

If a single electron is, however, transferred to hydrogen peroxide it forms the **hydroxyl radical**, one of the most reactive oxygen-derived radicals known. This electron may be provided by transition metal cations such as iron and copper. The reduction of hydrogen peroxide in the presence of these cations, which are themselves oxidized in the process, may result in the formation of a hydroxyl radical:

$$H_2O_2 + Fe^{2+} \rightarrow OH^- + {\cdot}OH + Fe^{3+}$$

This species is highly reactive and has a high rate of reaction with a range of biological molecules.

The concentration of these free metal cations, as distinct from protein-bound forms, is very low. Perhaps in some disease processes they become released from the association with proteins, i.e. become 'delocalized' as, for example, in the rheumatoid joint.

Enzyme reactions in which free radicals are by-products

A wide range of reactions occurring in cells depend on a supply of oxygen (dioxygen). The enzymes catalysing oxygen-requiring reactions may be divided into a number of groups:

1. **Oxidases** which catalyse the reduction of dioxygen to yield water or hydrogen peroxide. For example, the cytochrome oxidase enzyme complex found in mitochondria (see Chapter 5) has been shown to have superoxide bound to it during the reduction of dioxygen.
2. **Mono-oxygenases** which catalyse reactions in which one oxygen atom is reduced to water while the second is incorporated into an organic molecule. For example, the cytochrome P-450 complex found in the endoplasmic reticulum and responsible for the hydroxylation of a wide range of endogenous and exogenous molecules (see Chapters 28 and 36) generates superoxide as an intermediate.

3. **Dioxygenases** which catalyse the covalent attachment of oxygen to organic molecules to form derivatives which are hydroxyperoxy radicals. The reactions catalysed by cyclo-oxygenase and lipoxygenase (see Chapter 17) both lead to the formation of lipid hydroperoxides. Indeed, as part of the catalytic process, the former enzyme itself becomes a free radical in a reaction involving a tyrosine residue. The splitting of oxygen is coupled to the formation of a tyrosyl free radical enzyme intermediate in the reaction catalysed by ribonucleotide reductase.

The superoxide which is transiently produced in the reactions catalysed by enzymes in groups 1 and 2 above may occasionally escape from the enzyme surface with potentially deleterious consequences.

Enzyme reactions in which free radicals are functional products

Intercellular signalling

The free radical **nitric oxide** is generated in a wide range of cells in mammalian systems by a reaction catalysed by nitric oxide synthase (NOS):

$$\text{L-arginine} + \text{NADPH} + O_2 \downarrow \text{citrulline} + \text{NADP}^+ + \cdot\text{NO}$$

The reaction also requires four cofactors: haem, FMN, FAD and tetrahydropterin. There are three forms of NOS which differ in their cellular distribution. One isoenzyme is found in endothelial cells, a second in the brain and a third in macrophages. Nitric oxide generated by the first two types of isoenzymes diffuses out of the cells which form it and causes functional changes in neighbouring cells, i.e. it acts in a paracrine fashion (see Chapter 3), but with a rather special 'receptor'. The targets for endothelial-derived nitric oxide are vascular smooth muscle cells, which relax in the presence of the gas. A similar signalling role has been proposed for nitric oxide in the brain, but in this case different types of neurons produce and respond to the molecule. The 'receptor' target for nitric oxide in both locations is guanylyl cyclase. This enzyme is a haem protein found in the cytosol of nitric oxide responsive cells.

The generation of nitric oxide by the third group of cells which produce it, namely the macrophages, focuses attention on the possible adverse effects of the uncontrolled generation of the gas. This is because the cells generate the molecule as part of their range of mechanisms for killing both bacterial and tumour cells. It is here that the generation of nitric oxide coincides with production of ROS and it is, therefore, appropriate to consider these two processes together.

The toxic effects associated with the generation of nitric oxide are much more severe when the formation of the free radical is associated with formation of ROS. This appears to occur during macrophage-based killing and also in stroke (see Chapter 30). The keynote to these events appears to be the reaction of nitric oxide with superoxide to yield peroxynitrite :

$$\cdot NO + O_2\cdot^- \rightarrow ONOO^-$$

The latter may decompose spontaneously to yield two free radicals, the highly reactive hydroxyl radical and nitrogen dioxide:

$$ONOO^- + H^+ \rightarrow HONOO$$

$$HONOO \rightarrow \cdot OH + NO_2^{\cdot}$$

These species are potentially harmful as they can react with and alter the structure of cellular or extracellular macromolecules such as unsaturated fatty acids, proteins and DNA.

Cellular defence mechanisms

A range of phagocytic cells, such as monocytes, macrophages, neutrophils and eosinophils, form large quantities of superoxide as part of their panoply of mechanisms for the destruction and removal of foreign cells. During phagocytosis of foreign particles such as bacteria, the phagocytotic cells are activated and generate a **'respiratory burst'** (Figure 33.1, see also Chapter 32). In this, the NADPH oxidase complex reduces dioxygen to superoxide, which may itself be damaging to the cell, but more importantly, at the low pH which is maintained inside endocytotic vesicles, the superoxide may be protonated to the more reactive perhydroxyl radical ($HO_2\cdot$):

$$O_2\cdot^- + H+ \rightarrow HO_2\cdot$$

In some phagocytotic cells, superoxide undergoes a reaction catalysed by the enzyme superoxide dismutase (see Section 33.3) to yield hydrogen peroxide and oxygen:

$$2H^+ + 2O_2\cdot^- \rightarrow H_2O_2 + O_2$$

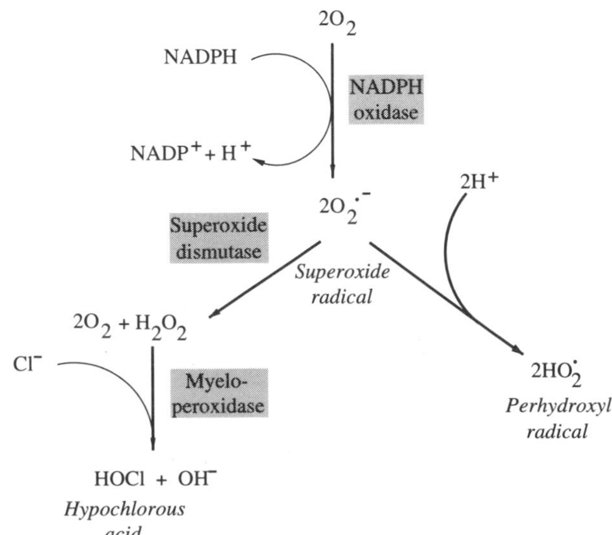

Figure 33.1 **The occurrence of the 'respiratory burst' during phagocytosis by a neutrophil**

In some instances the hydrogen peroxide may act as a substrate for the enzyme myeloperoxidase, a product of which is hypochlorous acid (HOCl):

$H_2O_2 + Cl^- \rightarrow HOCl + OH^-$

The importance of NADPH oxidase in the body's defences is emphasized by the observation that patients with **chronic granulomatous disease** have a deficiency in this enzyme complex. Such patients show persistent and multiple infections of the skin, lungs, bone, liver and lymphocytes. Deficiencies of myeloperoxidase, however, are not associated with any overt disease, indicating that the sequence of events leading to superoxide and hydrogen peroxide generation are more important for killing infectious agents than are those that lead to the formation of hypochlorous acid.

Exogenous causes of free radical generation

The potentially harmful generation of free radicals may also arise as the result of encountering a range of environmental challenges. Such challenges may arise, for example, as the result of exposure to ionizing radiation, to toxic chemicals (including pollutants) and to smoke.

Environmental challenges can be divided into three categories :

1. Exposure to ionizing radiation (e.g. X-rays or ultraviolet). It seems that the ability of electromagnetic radiation to damage tissues results largely from the splitting of water to generate hydroxyl radicals and a hydrogen atom, itself a radical species. It has been known for some time, however, that many of the damaging effects caused by exposure of tissues to ionizing radiation are dependent on the presence of oxygen. It seems likely that the effects arise as a consequence of the formation of superoxide. Ionizing radiation also has the effect of promoting molecular oxygen to excited states referred to as 'singlet oxygen' ($^1O_2$):

   $^3O_2$ --- $h\nu$ ---> $^1O_2$

   In these excited states dioxygen can react rapidly with cellular molecules, producing, for example, lipid hydroperoxides.
2. Exposure to a range of environmental anutrients including toxic chemicals, certain classes of drugs and also to alcohol. Exposure to all the above can result in tissue damage to which free radicals contribute.
3. Exposure to hyperbaric oxygen. This may also have deleterious effects, as observed, for example, in neonatal lungs.

Under most normal circumstances, free radicals arising in the various ways described above can be dealt with by an interrelated series of bodily defence systems. Pathological processes will only progress if the systems which normally handle unwanted free radicals are deficient or overwhelmed.

The main biological targets for attack by free radicals are polyunsaturated fatty acids, proteins and nucleic acids, especially DNA.

33.2 Targets for attack by reactive oxygen species: polyunsaturated fatty acids, proteins and DNA

Polyunsaturated fatty acids

The oxidation of polyunsaturated fatty acids by free radicals has been studied for many years. The main focus of research had been the modification of fatty acids in edible oils and, subsequently, in biological systems such as cellular membranes. Attack on the latter may result in disruption of membrane function and energy metabolism. A relatively new focus of research is the oxidation of polyunsaturated fatty acids (and proteins) in circulating low-density lipoproteins (LDLs), and the contribution of this process in the development of degenerative diseases such as cardiovascular disease.

The process of attack on a polyunsaturated fatty acid by a free radical, such as the hydroxyl radical, is described in Figure 33.2 and can be outlined as follows. The hydroxyl radical can abstract a hydrogen atom from the fatty acid to produce a lipid free radical (L$^\cdot$):

$LH + \cdot OH \rightarrow L\cdot + H_2O$

The interaction of molecular oxygen with these radicals gives rise to a peroxyl radical :

$L\cdot + O_2 \rightarrow LOO\cdot$

The peroxyl radical can react with another polyunsaturated fatty acid to produce a lipid hydroperoxide and another lipid free radical :

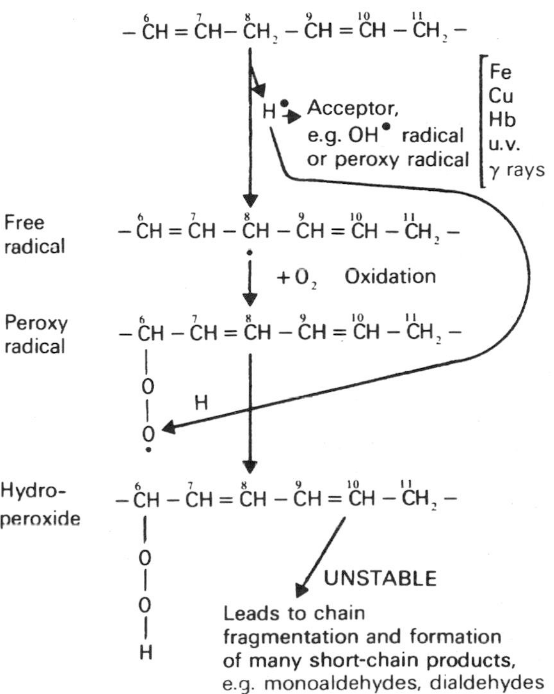

Figure 33.2 **Sequences of events in polyunsaturated fatty acid forming hydroperoxides.** A section of the carbon chain between carbon atoms 6 and 11 is shown which contains two double bonds

$$LOO\cdot + LH \rightarrow LOOH + L\cdot$$

This new lipid radical can, in turn, be converted into another lipid peroxyl radical, and a chain reaction is set up producing lipid hydroperoxides via lipid radicals. If the chain is not disrupted by a free radical scavenger, such as vitamin E, there will be degradation of the peroxidized lipid, with the generation of reactive and harmful products.

Vitamin E (EH) can intercept this process by reacting with the lipid peroxide radical, to form a tocopheroxyl radical (E·). This is possible *in vivo*, because the tocopherol molecules are incorporated into the membrane, in the vicinity of the polyunsaturated fatty acid (Figure 33.3):

$$LOO\cdot + EH \rightarrow LOOH + E\cdot$$

The resulting vitamin E radical is stable, as it is able to delocalize the unpaired electron within its structure, and does not therefore contribute to the propagation of the chain reaction. Although vitamin E is the major antioxidant in cell membranes, its concentration (one vitamin E: 2000–3000 phospholipid molecules) is not high enough to allow it to continue to protect enough phospholipid molecules from attack by free radicals, unless its reduced form is regenerated in some way. The tocopheroxyl radical can be reduced *in vitro* by vitamin C. Although vitamin E is located in membranes, and vitamin C is located in aqueous phase, vitamin C is able to regenerate vitamin E from the tocopheroxyl radical, permitting the former once more to act as an antioxidant. The extent to which this reaction is important *in vivo* has not been quantified. In addition, various flavo-proteins may have the same effect of regenerating native vitamin E *in situ*, and carotenoids have also been suggested as possible reducing agents. (Figure 33.4).

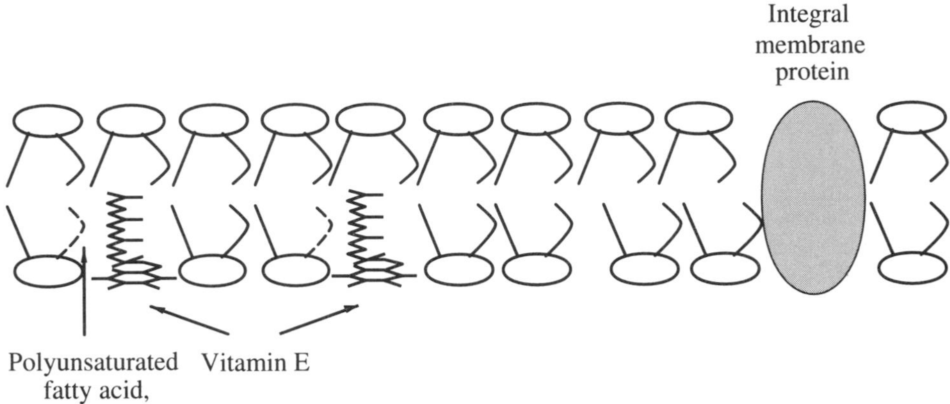

Figure 33.3 **Vitamin E helps protect membrane structures from severe oxidative damage, which tends to be initiated at polyunsaturated fatty acids in the phospholipids**

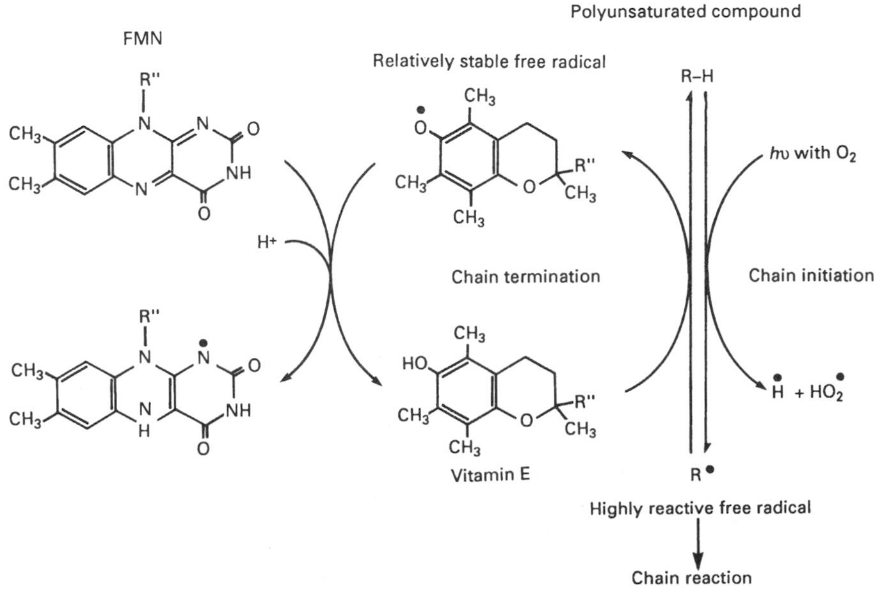

FMN
Relatively stable free radical
Polyunsaturated compound
R–H
hυ with O_2
H^+
Chain termination
Chain initiation
$\dot{H} + H\dot{O}_2$
Vitamin E
R•
Highly reactive free radical
Chain reaction

Figure 33.4 **The action of vitamin E as a terminator of chain reactions and its regeneration**

Lipid hydroperoxides generated in the reaction of lipid peroxyl radicals with vitamin E may themselves be converted back into peroxy radicals or into alkoxyl radicals in the presence of haem proteins, or delocalized iron or copper ions :

$$LOOH + Fe^{2+} \rightarrow LO\cdot + Fe^{3+} + OH^-$$

$$LOOH + Fe^{3+} \rightarrow LOO\cdot + Fe^{2+} + H^+$$

Even the absence of these metal ions, lipid hydroperoxides themselves are not stable. Cleavage of carbon–carbon bonds may take place during peroxidation, with the formation of degradation products such as malondialdehyde (an alkandial) and 4-hydroxynonenal (an alkenal; Figure 33.5). Alkandials are very reactive species. They react with protein thiol groups, and also cross-link amino groups on proteins, producing functionally defective aggregates. Alkenals are also harmful, because they react with amino acid side-chains in proteins and, in this way, compromise the function of a great array of enzymes within the cell. These include enzymes which control the cell cycle, for example.

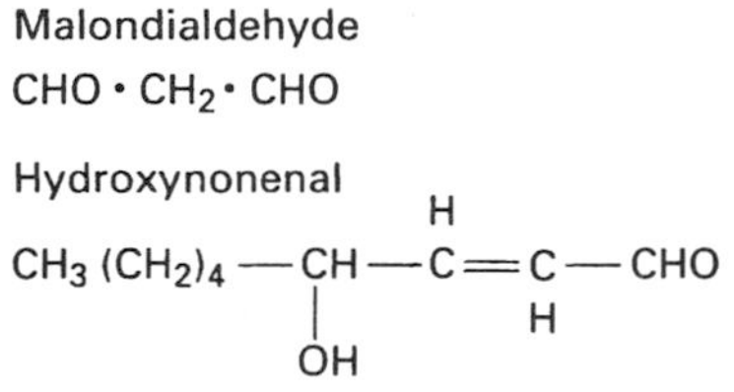

Malondialdehyde

$CHO \cdot CH_2 \cdot CHO$

Hydroxynonenal

$CH_3(CH_2)_4—CH(OH)—CH=CH—CHO$

Figure 33.5 **Aldehydes formed from lipid hydroperoxides**

The actions of alkandials and alkenals are not confined to cellular components, but also to lipoproteins such as LDL in the tissues. These molecules can modify the structure of the apoprotein B component of the lipoprotein (see Chapter 10), mainly by reacting with the ϵ-amino group of lysinyl residues. This modified apo B is not recognized by the normal LDL receptor whose function is the controlled uptake of LDL into cells. Instead LDL, modified as a result of oxidation and peroxidation in the arterial intima, is taken up by target macrophage receptors (scavenger receptors), thus contributing to the formation of cholesterol-laden foam cells and the development of atherosclerosis (see Section 33.4).

Proteins

Cellular proteins are also susceptible to attack by free radicals. The specific targets of attack are various amino acyl side-chains. Some side-chains are particularly vulnerable. Those belonging to the sulphur-containing amino acids methionine and cysteine, and amino acids such as tryptophan, histidine, arginine, lysine and proline can be oxidized by free radicals. The damaged proteins which result may form cross-linked aggregates leading to cellular malfunction. If the protein in question is a structural component of a membrane, for example, then membrane function will be impaired and various transport processes compromised. If the protein is an enzyme, the result will be loss of activity and impaired cellular metabolism. Amino acids may be attacked directly by newly generated hydroxyl radicals, or by free radicals produced as a result of lipid peroxidation, for example, alkoxyl (LO·) and peroxyl (LOO·) radicals produced in this way may, in turn, react with amino acyl side-chains. As mentioned above, degradation products of lipid peroxidation, such as malondialdehyde, are also reactive and attack side-chain amino groups on lysine residues.

Proteins which are structurally modified as a result of free radical attack are also more susceptible to proteolytic degradation, and therefore further disruption of cellular function ensues.

DNA

The major processes by which DNA can be damaged are oxidation, methylation, deamination and depurination. It has been estimated that oxygen-derived free radicals cause modification to 10 000 DNA bases per cell per day. The potential for DNA damage and, therefore, mutagenesis and carcinogenesis is clearly not insignificant if DNA repair mechanisms are not effective.

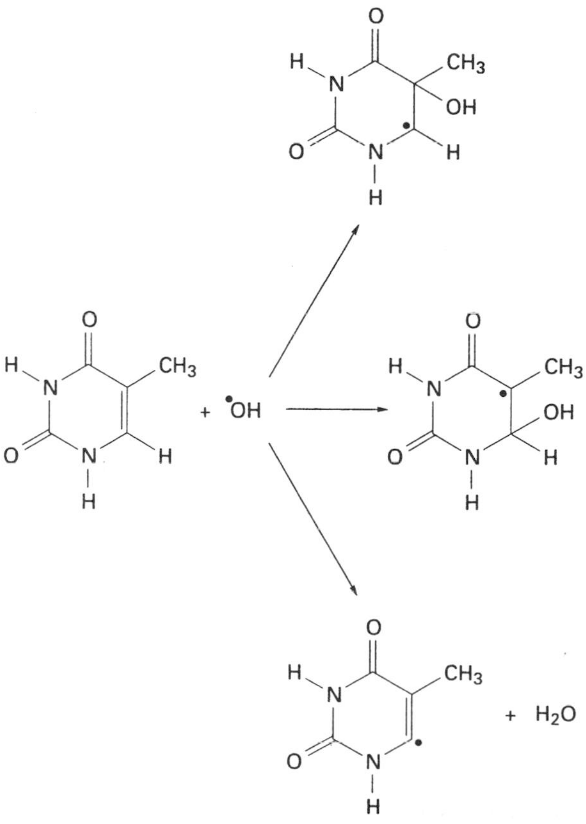

Figure 33.6 **Reaction of the hydroxyl radical (·OH) with thymine**

There are various mechanisms by which DNA can be damaged by free radicals. Purines, pyrimidines and sugars can be modified, single strand breaks can be introduced, bases may be removed, or DNA and nucleoproteins may be cross-linked. The hydroxyl radical, for example, can react with both purines and pyrimidines to form addition or abstraction compounds of these bases, at the same time converting them into free radicals. Figure 33.6 shows the modification of thymine by the hydroxyl radical. Similar reactions occur on other bases and also the deoxyribose component of DNA. The detection of specific degradation products of DNA is an indication that hydroxyl radical induced damage does occur *in vivo* (see Chapter 37), although the origin of free radicals such as the superoxide or the hydroxyl radical in the nucleus is not clear. These products of free radical damage are sometimes referred to as the 'footprints' of free radical attack on tissues.

33.3 Antioxidant defence mechanisms

Vitamin E has already been mentioned as an antioxidant capable of interrupting the chain-reaction sequence of lipid peroxidation, by acting as a scavenger of lipid free radicals. This is only one component of a complex antioxidant defence mechanism in tissues. Different antioxidants act at different levels:

(a) They may prevent the initiation of chain reactions, by removing free radicals
(b) They may scavenge free radicals generated in chain reactions, thereby interrupting the chain sequence
(c) They may remove peroxides, thereby preventing further generation of ROS.

There are two main lines of defence against ROS:

1. Formation of antioxidant enzyme systems. Such systems include superoxide dismutase, catalase and glutathione peroxidase. The activity of these enzymes, in turn depends a supply of manganese, copper, zinc or selenium.
2. Defence systems that depend directly on antioxidant nutrients: vitamins E and C, the carotenoids and the flavonoids.

It seems, therefore, that both types of defence mechanism depend ultimately on the supply of nutrients: manganese, copper, zinc, selenium, vitamin C, vitamin E and carotenoids. It is known, for example, that selenium deficiency is the cause of two human diseases. A cardiopathy, known as Keshan disease, and an osteoarthropathy, known as Kachin–Beck disease, have both been described in areas of China where the soil and, consequently, the native diet is deficient in selenium. In the case of Keshan disease, the cardiopathy may be caused by the Coxsackie virus which mutates to a more virulent form when dietary selenium is inadequate.

Antioxidant enzyme systems

Manganese, copper and zinc are involved with a group of closely related enzymes known as **superoxide dismutases (SOD)**. Copper and zinc are components of the cytosolic enzyme SOD1, and manganese is a component of a similar mitochondrial enzyme, SOD2. These enzymes catalyse the following reaction:

$$2H^+ + 2O_2^{\cdot -} \rightarrow H_2O_2 + O_2$$

The importance of SOD in the prevention of disease may be inferred from the finding that some (but not all) members of a group of patients with a familial form of motor neuron disease (amyotrophic lateral sclerosis, ALS) have a mutation to SOD1 (see Chapter 43).

The hydrogen peroxide which is generated by SOD can be removed by **catalase**, found in peroxisomes:

$$2H_2O_2 \rightarrow H_2O + O_2$$

Most of the hydrogen peroxide generated by superoxide dismutase is, however, removed by the selenium-containing **glutathione peroxidase** (Gpx), using glutathione (GSH) as the reducing agent:

$$H_2O_2 + 2GSH \rightarrow 2H_2O + GSSG$$

Glutathione itself becomes oxidized (GSSG), and it is reduced to GSH by the enzyme **glutathione reductase** using NADPH as the hydrogen donor:

$$GSSG + NADPH + H^+ \rightarrow 2GSH + NADP^+$$

Glutathione peroxidase can also act on lipid hydroperoxides as well as hydrogen peroxide:

$$LOOH + 2GSH \rightarrow LOH + H_2O + GSSG$$

The removal of lipid hydroperoxides is important in preventing the regeneration of lipid peroxyl radicals through the action of transition metal ions, and also in preventing the generation of toxic alkoxyl radicals (LO•). The relationship between glutathione, selenium and vitamin E is shown in Figure 33.7.

Antioxidant nutrients

Vitamin E is involved in intercepting the production of secondary radicals, as outlined above, itself forming a free radical. Carotenoids are known to have a function *per se*, and not through their conversion to retinol (see Chapter 31). For example, it has been shown recently that they may have a role in preventing some forms of

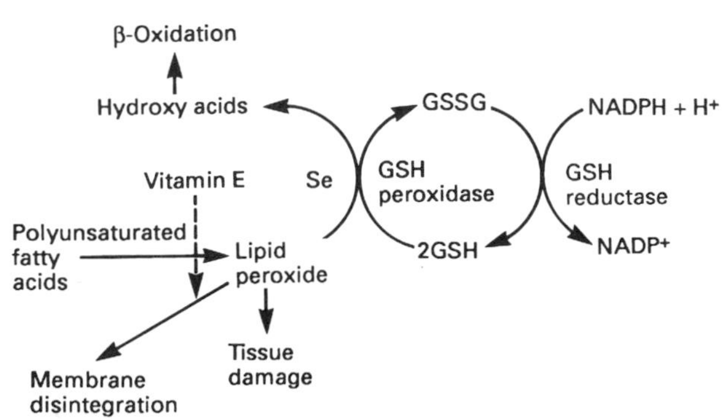

Figure 33.7 **The relationship between glutathione peroxidase, selenium and vitamin E**

cancer, and that they may have some protective effect in cardiovascular disease (see section 8, 33.4).

Another class of antioxidant molecules whose inclusion in the diet may be beneficial, are the flavonoids, which are present in apples, onions, red (but not white) wine and tea.

33.4 Free radical involvement in disease

In physiological circumstances, the balance between pro-oxidants and antioxidants is maintained. When the balance is tipped in favour of pro-oxidants, the cell or organism is said to be under **'oxidative stress'**. The stress may be caused by free radicals generated entirely endogenously or in response to exogenous challenges.

Examples of exogenous sources of free radicals were given in Section 33.1. Endogenous causes of excessive free radical formation include:

(a) Excessive activation of phagocytes, e.g. at sites of inflammation
(b) Disruption of the normally well-ordered mitochondrial electron transport chain, e.g. during reperfusion of ischaemic tissue
(c) Increased concentration of free transition metal ions, probably by delocalization as a result of escape from haem proteins, at sites of damage, e.g. in sickle cell disease and in rheumatoid arthritis
(d) Decreased levels of antioxidant defence mechanisms.

These seem to be the best documented examples of free radicals playing a major, possibly a causative, role. However, very often, it is difficult to determine whether, in human disease, free radicals do cause the disease rather than that they simply exacerbate the pathological process. Thus, tissue injury, by any means, including direct mechanical trauma, can elicit an inflammatory response (see Chapter 32), with secondary generation of free radicals by neutrophils and macrophages. Consequently, in each case, it is important to examine critically any evidence for the association of free radicals and disease, especially when considering the therapeutic potential of interventions aimed at the elimination of free radicals in the treatment of patients with inflammatory diseases. Interventions which might specifically block the initial injury would be different from those which would involve a subsequent, non-specific anti-inflammatory action.

Even with this in mind, as has been indicated, the involvement of free radicals in several disease states is becoming well established. These include atherosclerosis, some forms of cancer, cataract formation and other disorders of the eye, and diseases which involve the inflammatory response, such as rheumatoid arthritis, ulcerative colitis and Crohn's disease. Furthermore, there is highly suggestive evidence that increasing the antioxidant nutrient supply may be significant in the prevention of at least three of the diseases mentioned above, namely atherosclerosis, some forms of cancer and cataract.

Atherosclerosis

Mortality from coronary heart disease is very high in Europe and the USA, with UK figures among the highest in the world. Atherosclerosis is the primary cause of coronary heart disease, stroke and other vascular diseases. It is characterized by the formation, within the intima, of fatty streaks containing foam cells. These are lipid-filled cells derived from smooth muscle cells of the arterial wall, and from macrophages formed from

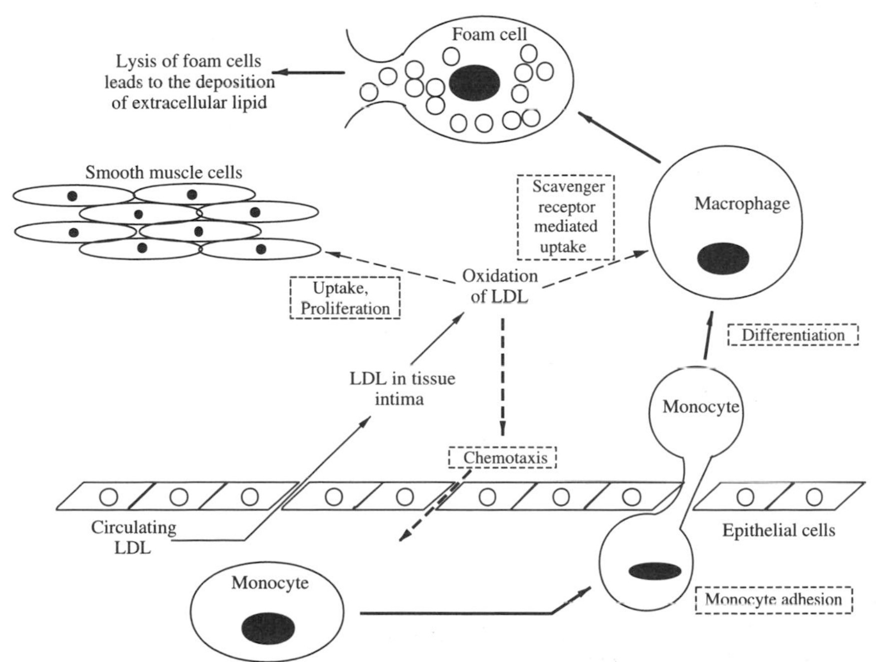

Figure 33.8 **Some of the events believed to occur in the early stages of atherosclerosis**

circulating monocytes. The appearance of foam cells is followed by the formation of fibrous plaques containing blood elements and calcium deposits which eventually occlude the arterial lumen and stop the blood supply. The contribution of free radicals to the process of atherosclerosis is thought to involve oxidation of LDL, and to occur during the early events that lead to the formation of the foam cells and fatty streaks.

The initial event in atherosclerosis seems to be damage to the endothelium, perhaps through mechanical trauma, and the deposition of immune complexes. Oxidation of LDL in the arterial wall is proposed to accelerate the process of atherogenesis by the following sequence of events (Figure 33.8). Oxidation of polyunsaturated fatty acids in LDL leads to the release of degradation products consisting of aldehydes, in the lipoprotein complex. They react with lysine residues on apoprotein B, modifying its structure and altering the charge on the molecule. As a result, apo-B is no longer recognized by the normal LDL receptors. It is, however, recognized by 'scavenger' receptors on macrophages and smooth muscle cells, and these are capable of taking up large amount of lipid. Lipid accumulation, not controlled by feedback mechanisms, leads to the formation of foam cells, which are the basic components for the development of the atheromatous plaque. Oxidized LDL itself may be cytotoxic, injuring and killing cells, resulting in the formation of necrotic extracellular lipid deposits which are characteristic of the progression of the lesion. The role of antioxidant nutrients is clearly important here, as they can block the formation of lipid oxidation products. Vitamin E and other lipid-soluble antioxidants are present in LDL, and an enhanced supply of these nutrients may have a protective effect against LDL oxidation and uptake by macrophages.

Various epidemiological studies suggest that the risk of cardiovascular disease may be lowered by increasing the intake of the antioxidant nutrients, vitamin E, carotenoids and flavonoids, although the contribution of each nutrient to the overall protective effect is unknown. Although epidemiological studies cannot establish causal relationships, but only point to them, there is a large and increasing number of such studies now, suggesting a direct preventive function of antioxidant nutrients in the development of atherosclerosis and coronary heart disease. A 1993 WHO cross-cultural study of 16 European populations, for example, showed an inverse correlation between plasma tocopherol levels and mortality from cardiovascular disease. Men who had similar blood pressure and cholesterol levels differed in tocopherol levels and heart disease death rates, with the higher tocopherol levels correlating with lower mortality.

Cancer

Cancer is a multistep disease process in man, consisting of initiation, promotion and progression through which a single cell can develop into a malignancy which can destroy the organism (see Chapter 37). Cancer is characterized by an increase in the number of abnormal cells and an invasion of adjacent tissue by these cells, followed by a spread of malignant cells to distant sites (metastasis).

It has been estimated that about 75% of human cancers are environmentally induced, with 30–40% probably related to dietary factors, about 20% associated with viral infections, and only about 2% developing as a result of inherited genetic changes. Fundamental among the events that cause a cell to become malignant is the interaction of a carcinogenic agent with DNA.

Free radical reactions are possible contributory mechanisms by which cells can become malignant, but it is unlikely that free radicals alone, or any other single mechanism, are sufficient to account for the entire process beginning with the genetic alteration, leading to mutation and resulting in disease. Free radicals may be involved in carcinogenesis through modification of DNA bases, formation of single strand breaks, activation of oncogenes or inactivation of tumour suppressor genes.

The involvement of oxidants in cell proliferation would seem to be important when considering free radicals and cancer. There is evidence that free radical mediated events can modulate growth regulation and there have been observations relating levels of lipid peroxidation to cell proliferation. It seems that DNA synthesis and tissue repair is associated with a rise in vitamin E levels and a decreased susceptibility to oxidation.

The role of free radicals in chemical carcinogenesis is perhaps easier to unravel. There is substantial evidence for the involvement of free radicals in this process, mainly based on the following findings:

(a) some chemical carcinogens are known to act through free radical intermediates
(b) some antioxidant molecules can intercept the carcinogenic action of a variety of chemical carcinogens.

Polycyclic hydrocarbons (PAH), for example, are carcinogens requiring metabolic activation before reaction with cellular macromolecules, an important step in carcinogenesis. Cytochrome P-450 dependent metabolism facilitates the binding of PAH to DNA by two pathways of activation. The major pathway is one-electron activation, to form free radical cations, whereas mono-oxygenation, to form diol epoxides, is only a minor pathway. Ninety-nine per cent of the DNA adducts formed with dimethylbenzanthracene, and 80% of those formed with benzo(a)pyrene, in mouse liver and skin cells arise via the free radical cation pathway.

The evidence that agents in the diet may help to prevent cancer has given support to the concept that oxygen-derived free radicals may be an important component of a variety of human cancers including breast, colon and prostate cancer. Epidemiological studies show that there is a low incidence of these cancers in vegetarian populations. Components of fruit and vegetables that might be responsible for this prevention are vitamin A, vitamin C, carotenoids and flavonoids.

Cataract

This is partial or complete opacity in the lens or capsule of the eye, impairing vision and finally causing blindness. Exposure to electromagnetic radiation promotes singlet oxygen generation in the lens, and this is enhanced by the action of singlet oxygen sensitizers which are known to be present in this tissue. The action of singlet oxygen and related superoxide radicals results in the cross-linking of lens proteins and disruption of ion transporting systems responsible for maintaining ion gradients between the lens and its environment (see Chapter 31).

Inflammatory diseases

Excessive activation of phagocytes and production of the superoxide radical can harm surrounding tissue. This can be seen in inflammatory diseases.

In **rheumatoid arthritis**, overproduction of the superoxide and other radicals is known to contribute to the tissue injury in the inflamed joint. In **Crohn's disease** (chronic inflammatory bowel disease), or in **ulcerative colitis**, there is increased activity of phagocytic cells in the bowel, resulting in excessive free radical production, which may contribute to the progression of large bowel cancer.

In some forms of **adult respiratory distress syndrome (ARDS)**, which are characterized by respiratory failure due to pulmonary oedema, excessive infiltration and activation of neutrophils in the lung may contribute to tissue injury. This condition often arises after tissue damage due to burns, trauma or severe infection, and is known as **'shock'**.

Diseases caused by oxygen therapy: retrolental fibroplasia and reperfusion injury

Retrolental fibroplasia

Retrolental fibroplasia (RLF) is a form of retinal damage leading to blindness. There was an abrupt increase in the incidence of this disease in the 1940s in the UK, among premature infants of low birthweight. It was subsequently attributed to damage caused by the use of hyperbaric oxygen in the incubators of these infants, and exacerbated by the fact that their antioxidant nutrient status was low. Premature infants can be deficient in vitamin E, as the placental transfer of this vitamin is very poor. Better control of oxygen tension in incubators, and administration of vitamin E to premature infants of low birthweight, has led to a decrease in the incidence of the disease.

Reperfusion injury after myocardial infarction

It is well documented that the survival of ischaemic myocardial tissue depends upon timely reperfusion, but the very treatment may lead to further tissue injury. A number of mechanisms are likely to be involved in reperfusion injury, but the major ones seem to be the generation of oxygen-derived free radicals, which are known to be cytotoxic to surrounding cells, and the inflammatory response, which involves activation of the complement system and recruitment of neutrophils.

Environmental pollution: cigarette smoking and emphysema

Cigarette smoking can cause cancer, emphysema and other respiratory diseases which manifest themselves after many years. Cigarette smoke contains carbon monoxide and free radicals which are generated because of the high temperatures (up to 900°C) at the burning tip.

The free radicals found in the tar phase are different to those in the gas phase with respect to stability and half-life. The tar phase contains long-lived and fairly stable radicals. It also contains transition metal ions which can drive the formation of the reactive hydroxyl radical. The gas phase contains short-lived reactive radicals such as nitric oxide (NO). In the presence of oxygen, this is oxidized to nitrogen dioxide radicals, which can damage lung tissue:

$$\cdot NO + \tfrac{1}{2}O_2 \rightarrow \cdot NO_2$$

Emphysema is the pathological accumulation of air in tissues or organs. Pulmonary emphysema is a condition of the lungs characterized by abnormal increase in the size of air spaces distal to the terminal bronchioles, either from dilatation of the alveoli or from destruction of their walls.

Observations on emphysema and ARDS support the suggestion that injury to lung tissue is due to unregulated host defence mechanisms. The neutrophil is a major mediator of host defence, and injury, and defects in the mechanisms which regulate neutrophil action, allow these cells to damage the lungs.

Hereditary emphysema is caused by deficiency in α_1-antitrypsin (α_1-antiproteinase, α_1-AT), a proteinase inhibitor. This results in the unopposed action of neutrophil elastase and other proteolytic enzymes, leading to destruction of the protein elastin.

In emphysema caused by cigarette smoking, the levels of α_1-AT are normal, but its ability to inhibit neutrophil elastase is decreased. This is because a methionine residue in the part of the α_1-AT that binds elastase becomes oxidized by free radicals generated in cigarette smoke. As a result, elastase is not controlled by α_1-AT and acts unopposed, elastin is degraded, and the lung tissue loses its elasticity as its destruction progresses. Emphysema resulting from cigarette smoking not only affects smokers, but also other people who are exposed to cigarette smoke, generally referred to as **'sidestream'** or **'passive'** smokers.

33.5 Antioxidants as nutrients: vitamin E

Vitamin E was discovered in 1922 as a factor needed to prevent reproductive failure in rats. Its deficiency leads to a number of well-recognized disease states in animals, but is so rare in man that it is difficult to establish the dietary requirement for this vitamin.

Structure, nomenclature and dietary sources

The structure of vitamin E was elucidated 15 years after its discovery. The term vitamin E refers to two series of compounds. The first, is a group of tocopherols, α-, β-, γ- and δ-tocopherol, which differ in the number and position of the methyl group(s) on the ring (Figure 33.9). The structural differences are reflected in differences in biological activity, with α-tocopherol being the most potent.

The second, is a group of related, but less potent, compounds, with an unsaturated side-chain, known as

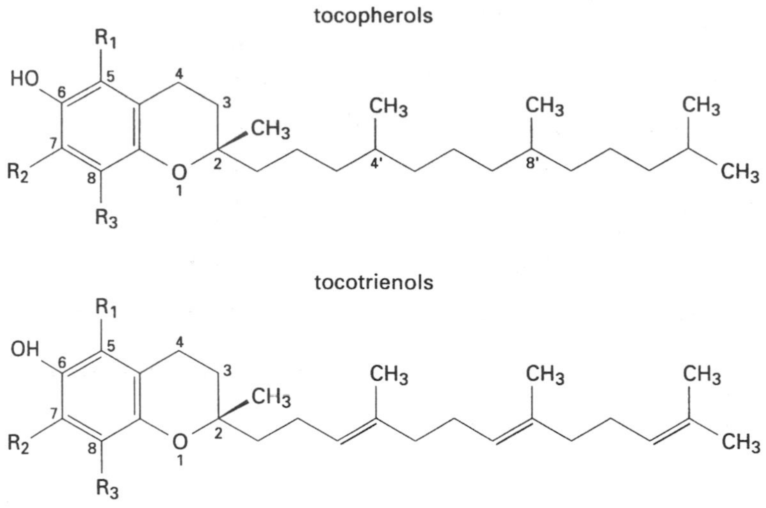

Figure 33.9 **The structures of naturally occurring tocopherols and tocotrienols:**
α-tocopherol: R_1:CH_3, R_2:CH_3, R_3:CH_3
β-tocopherol: R_1:CH_3, R_2:H, R_3:CH_3
γ-tocopherol: R_1:H, R_2:CH_3, R_3:CH_3
δ-tocopherol: R_1:H, R_2:H, R_3:CH_3

the tocotrienols. Vitamin E activity is expressed in terms of tocopherol equivalents (TE). One milligram TE corresponds to the amount of a compound with vitamin E activity equivalent to 1 mg of α-tocopherol. The activity is also often expressed in terms of international units (IU), 1 IU being equivalent to 1 mg synthetic α-tocopherol acetate.

Tocopherols are susceptible to oxidation and loss of biological activity when exposed to heat, light, alkaline conditions and divalent cations, and commercial preparations are in the form of the more stable tocopherol esters. The major dietary sources of vitamin E are fats and oils, with different tocopherols predominating in different oils. The richest sources are soya and corn oils (50–150 mg/100 g), followed by palm, and safflower oils (20–70 mg/100 g), whereas coconut and olive oils are relatively low in vitamin E content (1–10 mg/100 g).

Absorption, transport and storage

Vitamin E is absorbed mainly from the upper part of the small intestine, together with vitamin A and dietary lipid, the presence of the latter increasing the efficiency of absorption of both vitamins. It is incorporated into chylomicrons which reach the periphery, and it reaches the liver in the form of chylomicron remnants. In the liver, tocopherols are incorporated into very-low-density lipoprotein (VLDL) which is the vehicle of delivery of the vitamin to target cells. Not all tocopherols are handled in the same way. The main form reaching cells is α-tocopherol, whereas γ-tocopherol is selectively excreted into the bile. In cells, tocopherols are distributed in plasma membranes and intracellular membranes, where antioxidant activity is required.

Vitamin E requirements and deficiency

Vitamin E deficiency states in animals are well characterized and they are usually the result of cell membrane damage. The signs include not only reproductive failure as mentioned above but also myopathies, neuropathy and damage to liver cells.

As mentioned above, vitamin E deficiency in humans is so rare that it is difficult to establish a reference nutrient intake. The first documented vitamin E deficiency state was that observed in premature infants who were fed on formula feeds containing no vitamin E. They presented with haemolytic anaemia, thrombocytosis and oedema. As a result, vitamin E is now included in all commercial infant feeds.

Another presentation of vitamin E deficiency is retrolental fibroplasia (RLF), observed in some premature infants of low birthweight (less than 1.5 kg), and it is a significant cause of morbidity in such infants (see Section 33.4). Vitamin E administration by intramuscular injection protects premature infants from RLF.

Vitamin E deficiency has also been seen in patients with abetalipoproteinaemia (failure to synthesize apoB). These patients are unable to form chylomicrons and VLDL (see Chapter 10) and, as a result, vitamin E is neither absorbed via the lymphatic route, nor distributed via VLDL. Children with this defect show signs of severe progressive neuropathy and retinopathy which develop towards the end of the first decade of life. Vitamin E therapy can arrest or reverse the neuropathy. High doses of vitamin E are recommended for all patients with chronic fat malabsorption syndromes. The requirement for vitamin E depends to a large extent on the polyunsaturated fatty acid (PUFA) content of tissues, which is, in turn, determined to a large extent on the PUFA content of the diet. Setting the reference nutrient intakes, therefore, seems to be of little practical value, as the PUFA intakes of the population vary considerably from individual to individual. The UK Department of Health consider 4 mg of tocopherol equivalents to be acceptable for most people, but recognize that some individuals may require higher intakes. A level of plasma tocopherol below 11.6 μM is considered indicative of vitamin E deficiency, as erythrocytes taken from subjects whose circulating vitamin E is less than this value tend to ready haemolysis when exposed to oxidizing agents such as dilute hydrogen peroxide. It must be noted that this is a deficiency at the biochemical level, and no clinical signs are apparent at this stage. Plasma tocopherol concentrations are sometimes expressed in terms of plasma lipid concentrations, because a high level of circulating lipids causes tocopherol to leak out of cell membranes and appear in the circulation, thereby increasing plasma levels. The commonest way of expressing plasma tocopherol is as a tocopherol: cholesterol ratio.

Unlike other fat-soluble vitamins, such as A and D, vitamin E does not seem to have toxic effects. No adverse effects have been reported with daily doses ranging from 400–2000 mg after many months of intake.

A distinction has to be made between vitamin E requirements, that is, amounts which will prevent the appearance of deficiency signs, and the optimal intake of vitamin E in the diet. The reason for this is that recent epidemiological studies point to a protective role of vitamin E, when taken at levels above those required to cure deficiency symptoms, for conditions such as cardiovascular disease and some cancers (see Section 33.3). There seems to be an inverse relationship between plasma vitamin E concentration and mortality from these diseases.

33.6 Antioxidants as nutrients: vitamin C

The disease described as scurvy became prominent among seafarers setting sail from Europe on long voyages of exploration during the fifteenth century. In Vasco da Gama's voyage round the Cape of Good Hope in 1498, 100 out of his crew of 160 died of scurvy. A hundred years later, Sir Richard Hawkins reported records of 10 000 seamen dying from scurvy. Land dwellers also suffered, however, particularly in winter.

Scurvy is characterized by a tendency to bleed, especially into joints and under the skin, with the formation of petechiae, and by poor wound healing. Gums become soft and spongy and teeth become loose. Bones are weakened and anaemia and infections develop. If untreated, these infections lead to death. The reason for the widespread incidence of the disease was the poor rations carried on these voyages, and in particular the lack of fresh fruit and vegetables. This was not recognized until 1747, when the British naval surgeon James Lind, who treated patients by feeding them two oranges and a lemon a day, reported a miraculous recovery within a week. At the end of the eighteenth century, lemon or lime juice was prescribed daily for British sailors, a practice which led to the description of British sailors as 'Limeys' by the Americans.

Structure, nomenclature and dietary sources

The active principle in citrus fruits which prevents scurvy was isolated and purified in the 1920s, and was called vitamin C or ascorbic acid. Its structure is shown in Figure 33.10.

Very few animals require vitamin C in their diet, as they can synthesize it from glucose. Primates, and some other species including guinea-pigs, the Indian fruit-eating bat and some birds and fishes, lack the enzyme gulonolactone oxidase, which catalyses one of the reactions in the conversion of glucose to ascorbic acid, and consequently have a dietary requirement for ascorbic acid.

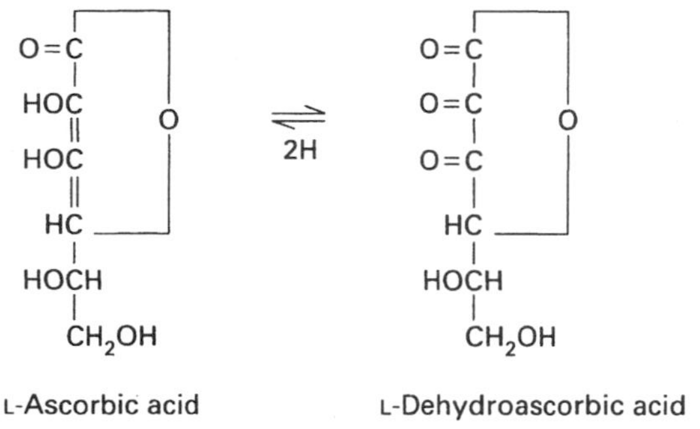

Figure 33.10 **The structure of ascorbic acid and its reversible oxidation to dehydroascorbic acid**

Ascorbic acid is a strong reducing agent, and acts as an antioxidant. It can be reversibly oxidized to L-dehydroascorbic acid (Figure 33.10), but further oxidation to 2,3-diketogulonic acid is irreversible and results in loss of function. The great sensitivity of the vitamin to oxygen, metal ions, alkaline conditions, heat or light renders vitamin C one of the most labile nutrients in the diet.

The main dietary sources of vitamin C are vegetables and fruit, especially citrus fruits, strawberries, tomatoes, spinach and potatoes. Cereals contain no vitamin C, and animal tissues and dairy produce are very poor sources.

Absorption and storage

Ascorbic acid is absorbed from the intestine and circulates in plasma, red cells and leukocytes. It is found in highest concentrations in the adrenals, the pituitary and the retina.

In general, plasma levels reflect the recent intake of vitamin C, whereas leukocyte levels are a better index of body stores. Plasma levels of 1 mg/l or less, and leukocyte levels of 70 mg/l or less, are indicative of scurvy. Malabsorption syndromes, such as tropical sprue, and excessive alcohol consumption interfere with the absorption of vitamin C and may cause scurvy.

Cigarette smoking can increase ascorbic acid turnover by up to 50%. This is thought to be due to free radical scavenging by the vitamin.

Function

Ascorbic acid functions as a reducing agent in many metabolic processes, the best understood being collagen synthesis, where it is needed for the hydroxylation of proline and lysine residues, a process essential for collagen stability (see Chapter 29).

Ascorbic acid also facilitates the absorption of non-haem iron from the intestine by reducing it to the Fe^{2+} state. In the adrenal medulla it serves as a reducing agent in hydroxylation reactions in the synthesis of adrenaline and noradrenaline (see Chapter 30).

Many other processes are influenced by ascorbic acid but its involvement is difficult to quantify. Claims first made by Linus Pauling in 1970, that, when taken in quantities of 1–4 g/day, it reduces the incidence of the common cold, have remained largely unsubstantiated, but they have influenced large numbers of the public to take doses of vitamin C far exceeding the physiological, antiscorbutic levels. Recent studies have actually shown that although the incidence of the common cold is not reduced by vitamin C, the duration of cold episodes and the severity of the symptoms can be decreased. Vitamin C may act in this respect by reacting with free radicals released by phagocytic leukocytes which become activated in an infection. This would result in a decrease in the inflammatory effects caused by these oxidants. This may have similarities with the way in which vitamin C decreases the frequency of bacterial infections in chronic granulomatous disease. It is thought to act as an antioxidant improving neutrophil migration.

Vitamin C is known to be able to reduce, *in vitro*, and probably *in vivo*, the tocopheroxyl radical produced in the process of free radical scavenging by vitamin E, and so regenerate functional tocopherol (see Section 33.2). Another potential role for vitamin C is that of an anticancer agent, suggested by the fact that *in vitro* it inhibits nitrosamine formation from naturally occurring nitrates. There is also increasing epidemiological evidence that a synergistic effect of the antioxidant vitamins E, C, and the carotenoids, may have a significant role in the prevention of cancer, cardiovascular disease and cataract formation, although the quantitative contributions of these components to the overall effect are unknown.

Dietary requirements and deficiency

The reference nutrient intake (RNI) for adults in the UK is 40 mg/day. In the USA, the recommended daily allowance is 60 mg/day for adults, with a recommendation for increase to 70 mg/day for pregnant women, 95 mg/day for lactating women, and 100 mg/day for smokers.

Controversy still exists regarding the optimal intake of vitamin C. The dietary guidelines, especially in the UK, have been established with prevention of deficiency as a goal, rather than attainment of optimal health. There seems to be no doubt that vitamin C should be taken in amounts higher than those required merely for the prevention of overt scurvy, but lack of information on the levels required to act on other systems in the body, apart from collagen metabolism, make it difficult to set higher RNIs. None of the studies which have noted a positive benefit of vitamin C supplements in the prevention or treatment of disease has defined the minimum dietary requirement needed to achieve it.

Toxicity

Vitamin C can be taken in doses of up to 2–3 g/day without obvious undesirable effects. Above these levels, however, it cannot be absorbed from the intestine, and can cause severe diarrhoea. It can also cause deposition of oxalate stones in the kidneys of subjects susceptible to stone formation, as oxalate is its major metabolite.

Another undesirable effect may be that of 'systemic conditioning'. High intakes lead to increased turnover, and sudden cessation of such intakes may cause deficiency through increased vitamin catabolism.

Chapter 34

Toxicology: general principles

34.1 Introduction

In this and the next three chapters a major focus will be on the adverse effects of chemicals on health; that is, the subject matter will come extensively from the science of **toxicology**. Historically the ideas we now associate with modern toxicology first arose from the study of poisons and their antidotes and were brought together in the fourteenth century by the Jewish philosopher Maimonides. The subject of poisons and antidotes was systematized in the early nineteenth century by the Spanish physician Orfila. More recently, many of the ideas and methods now used by toxicologists were first introduced and developed by pharmacologists. It was they who put the study of the absorption, distribution, metabolism and excretion of drugs on a firm footing. The techniques they developed to deal with the special case of drugs apply equally well to all 'foreign compounds'. Other important contributors to the science of toxicology, the occupational physicians, are sometimes overlooked. Thus in 1700 the Italian physician Ramazzini was the first to describe some of the effects of occupational exposure to lead and mercury. In his writing he showed himself to be aware that effects can sometimes be cumulative, with long periods being required before adverse reactions are seen. The contributions of physicians in recognizing occupational causes of cancer are discussed in Chapter 37.

Toxicologists may now approach their subject in three distinct ways:

1. They may seek to understand the adverse health effects produced by chemicals under differing conditions of exposure.
2. They may try to elucidate the mechanisms whereby adverse effects arise.
3. They may be called upon to give advice on possible means of avoiding adverse health effects to those individuals likely to be exposed to potentially hazardous substances, and also to the legislators who seek to control such exposure.

34.2 Environmental chemicals

Environmental chemicals may be classified into four broad categories, as shown in Table 34.1. The members of group I are described as 'desirable', i.e. beneficial, and include most natural foodstuffs. It should be borne in mind that no chemical is totally harmless: examples are cited elsewhere in the text of the toxic effects of hyperbaric oxygen and of the possible adverse health effects of the excessive intake of sucrose. Group II of the classification includes inert components of the environment and in particular of the diet. These components have no direct nutritional value and are also non-toxic; the group consists of substances not normally absorbed or absorbed only to a very limited extent, for example cellulose and sulphate ions. It should be noted that, as argued in Chapter 14, such 'inert' substance may still play a role in protecting against cancer of the colon, for example. Included in the third group of Table 34.1 are several metals, such as copper and zinc, which are essential in trace amounts but may prove toxic in excess.

Table 34.1 **Classification of chemicals of the environment**

Group	
I	Desirable, e.g. foodstuffs
II	Harmless and inert
III	Desirable in small quantities, but toxic in large quantities
IV	Toxic in moderate to small quantities

Most toxic substances are placed in the fourth group, but there can be enormous variations in the doses of specific chemicals which induce adverse acute effects in animals. One way that has been used to define acute toxicity of chemicals is by means of the **LD_{50} value**. This is defined as the dose of a chemical required to produce death in 50% of the animals exposed to it. The ethics of determining LD_{50} values forms a controversial subject, but some examples are given in Table 34.2 simply to illustrate the range of toxicity encountered in environmental molecules. A scientific, rather than ethical, criticism of acute LD_{50} determinations is that, as was appreciated by Ramazzini, in many cases of human exposure small doses are encountered over long periods of time.

In much of modern toxicology the focus tends to be on synthetic chemicals and the exposure of the population to these. It has been argued that most of the most toxic compounds are natural, not man-made, chemicals and the effects of these on human populations in terms of, for example, the promotion of tumour formation are likely to be of much greater importance. This view, however, misses the important point that while we can have relatively little control over our exposure to naturally toxic compounds we have a right to expect that everything possible is done to protect us from exposure to man-made chemicals however small is the risk they appear to pose.

34.3 Biochemical damage caused by toxic substances

A wide range of different mechanisms, with very different timecourses, may result in damage to tissues by toxic substances.

Two classes of toxic substance are generally recognized: those that are inherently toxic and those which become toxic following their metabolism in the body. The first group includes substances such as mercuric salts, while members of the second group can be represented by compounds such as cyanide glycosides which, on hydrolysis, form cyanide ions and by the polycyclic hydrocarbons that become carcinogenic following their metabolism. The latter two examples serve to illustrate the wide range of times required for effects to emerge. Thus the generation of cyanide ions can lead rapidly to death, whereas the carcinogenic effects of polycyclic hydrocarbons are seen only after years or decades.

The basis of most toxic effects is that the substances or their metabolites bind to proteins or to DNA. If the protein is an enzyme, then inactivation of a particular metabolic pathway may result. Binding to DNA may lead to mutations, the results of which become manifest only after cells have undergone many divisions. Toxicants may also act as antagonists. In these cases the substance bears sufficient structural similarity to a natural substrate that it prevents its action. For example,

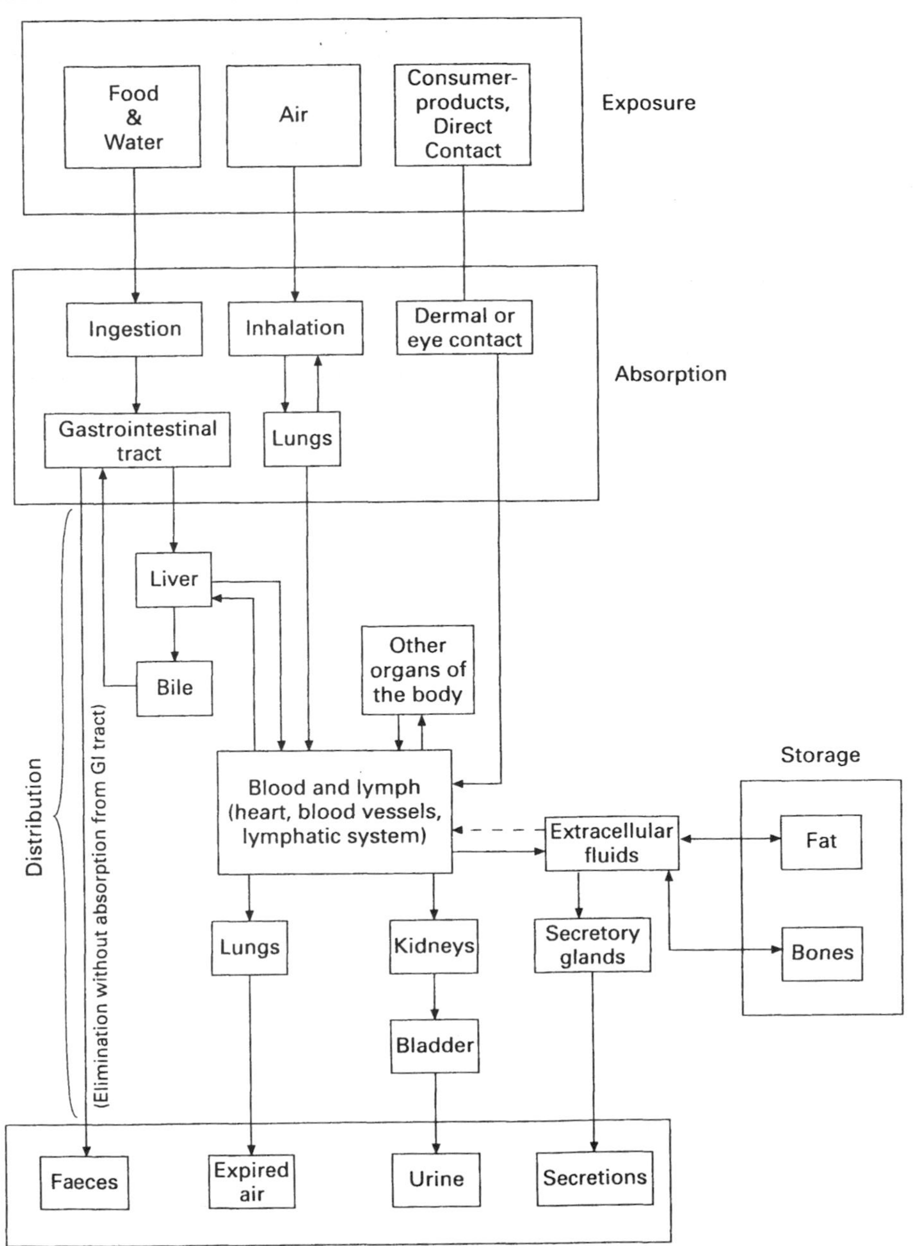

Figure 34.1 **A schematic showing how chemicals may enter, be absorbed into, distributed within and excreted from the body**

Table 34.2 **Approximate LD_{50} of a number of chemical agents**

Chemical	*Animal tested*	*Route*	*LD_{50}* (mg/kg)
Ethanol	Rat	Oral	13 700
Sodium chloride	Rat	Oral	3 750
Ferrous sulphate	Mouse	Oral	1 520
Sodium phenobarbitol	Rat	Oral	660
Morphine	Mouse	Subcutaneous	500
DDT	Rat	Oral	200
Nicotine	Rat	Oral	50
Strychnine	Rat	Oral	5
Picrotoxin	Mouse	Intravenous	4
Tetrodotoxin	Mouse	Intraperitoneal	0.01
Ricin	Mouse	Intraperitoneal	0.001
Botulinum toxin	Rat	Intravenous	0.00001

dicoumarol, which is found in sweet clover, is able to act as an anticoagulant by virtue of its ability to block the actions of vitamin K, to which it is structurally related. Similarly, goitrogens contained in many plants give rise to thiocyanate ions which compete with iodide ions for uptake into the thyroid gland.

A wide range of special proteins present in plants, snake and spider venoms and bacteria are toxic. The special term '**toxin**' is usually reserved for these proteins. The lectins produced by plants cause red cell agglutination. Snake and spider venoms contain many enzymes, important among which are phospholipases, which can cause haemolysis of red cells as their catalytic activity results in the destruction of part of the plasma membrane. Several of the bacterial toxins are neurotoxic by virtue of their ability to interfere with neurotransmission. Thus the related group of toxins produced by *Clostridium botulinum*, which are some of the most toxic substances known, act proteolytically to prevent the release of neurotransmitters from nerve terminals.

On the positive side, a knowledge of the mechanisms underlying the toxicity of certain toxins is leading to the introduction of related compounds that are targeted on particular groups of cells, e.g. on cancer cells (Chapter 37).

34.4 The ways in which the body handles foreign compounds

Figure 34.1 shows the possible fates of compounds to which we may be exposed, either in our food and water, via the air we breathe, or through dermal contacts. For many chemicals, exposure is followed by absorption and, once in the body, they undergo distribution, in many cases followed by metabolism. Most chemicals entering the gastrointestinal tract are absorbed from the intestine and are therefore presented first to the liver via the hepatic portal blood system. An exception to this are certain drugs of abuse such as cocaine which, when 'snorted' are absorbed via the nasal mucous membranes, with consequent delivery of the drug direct to the brain. Ultimately the chemicals or their metabolites will be excreted from the body, but this may be a long process if the chemical in question is lipid-soluble e.g. topically-applied anti-inflammatory steroids, or is able to be stored in bone, e.g. strontium 90.

Chapter 35

Toxic metals

35.1 Introduction

In this chapter, copper, mercury, lead, aluminium and the radionuclides are considered as examples of toxic metals. The toxic effects of iron overload are discussed in Chapter 21. Copper differs in an important respect from the other metals with which it has been grouped for the purposes of this chapter: it is an essential trace element for all forms of life. This means that there exist tight controls on its metabolism, with the consequence that failure of these regulatory mechanisms can lead to disease. In turn, it follows that the mechanisms underlying the diseases are generally concerned with the failure of specific processes. By contrast, diseases associated with the other metals dealt with in this chapter tend to arise as the result of non-specific effects, although these may involve specific types of proteins.

An understanding of the non-specific toxic effects is enhanced if there is some elementary knowledge of the binding properties of the metals (strictly speaking their ions) and how this can lead to the formation of **chelates**.

35.2 Metal ion complexes and chelates

Complexes of metal ions with ammonia and with other molecules or ions were first studied in the early years of this century by the German chemist Werner, who observed that the transition elements iron, cobalt, chromium, nickel, and elements which showed transition element behaviour, such as copper [Cu(II) but not Cu (I)], silver and zinc, readily formed complexes with ammonia. These derivatives were called 'coordination complexes', a good example being the formation of a bright blue 'cuprammonium complex' resulting from the addition of excess aqueous ammonia to the solution of a copper salt, a well-known experiment in elementary inorganic chemistry. It was Werner who pointed out that the metals which formed stable complexes possessed an unfilled electron shell immediately below the outer valence shell. He proposed that the ammonia, as a donor, provided electrons to fill these shells, so forming a stable complex in which the electron configuration around the metal closely resembles that of an inert gas.

Further studies of these metal complexes in the 1920s demonstrated another important aspect: if binding by the donor molecule could occur on two or more sites, so that a ring or rings incorporating the metal ion was formed, very stable complexes resulted. Complexes of this type were termed **'chelates'** and the donor atoms **'ligands'**. In general, the more ligand atoms that a chelating compound can assemble around the metal ion, the more stable the resulting chelate. As a corollary to this, it should be mentioned that a chelator must either possess precise built-in geometry or be sufficiently flexible so that its ligand atoms can be presented to donate their pairs of electrons towards the metal ions along geometrically-acceptable axes. The number of ligands a chelator presents is indicated by terming it 'bidentate' for two, 'tridentate' for three, etc. Comparison of the stability constants for ammonia, ethylenediamine and ethylenediaminetetraacetic acid (EDTA) shows that the avidity of binding Cu(II) rises many orders of magnitude as the number of ligand groups supplied by a single molecule rises from one to two to four. The great avidity with which EDTA and similar compounds bind metal ions means that they may be useful for the rapid sequestration of potentially toxic quantities of metals, a subject dealt with in Section 35.9.

It should be noted that biochemical (and pharmacological) usage often refers to a substance (including a metal ion) which binds to a protein as a ligand, a direct inversion of the meaning the term has in inorganic chemistry.

35.3 Protein–metal ion complexes

Many of the amino acyl side-chains of proteins provide very effective ligand or chelating atoms. Of particular importance are the thiol groups of cysteinyl residues, the carboxylate groups of glutamyl and aspartyl residues, the hydroxyl group of tyrosinyl residues and the imidazole group of histidyl residues. A list of these groups and their possible interactions with metals has been provided in Chapter 20. As was apparent from this, many different types of metal–protein interaction are possible. Whether these actually occur is frequently determined by the folding of the polypeptide chain.

If the metal ion has a high affinity for the ligands, then formation of the metal–protein complex can lead to the distortion of the protein in the course of formation of a stable chelate. Distortion of this nature may lead to loss of biological activity and ultimately to denaturation of the protein. If one of the groups bound forms part of the active centre of an enzyme, then inactivation will be rapid. On the other hand, such binding might cause the protein to assume a biologically-active conformation.

From a knowledge of metal–ion binding with simple molecules, various predictions about the interaction of metal ions with proteins can be made: thus copper, mercury, lead and zinc will bind very strongly to proteins by many different ligand groups. The nitrogen of imidazole groups and the sulphur of thiol groups are particularly important in binding these metals. On the other hand, binding to calcium, magnesium and strontium is weaker and usually by oxygen in carboxylate groups. Neither sodium nor potassium form complexes because their inner electron shells are already filled.

35.4 Copper

Introduction

The status of copper of an essential trace element puts it into a different category from that of the other metals considered in the following sections. Toxic effects of the type associated with excessive exposure to non-essential metals such as lead and mercury can occur with copper, but, in addition, deficiency states may occasionally arise. More importantly, deficient or toxic manifestations may be observed when the supply of the metal itself is neither deficient nor superabundant. Such conditions are associated with failures in metabolism of the metal. Serious conditions can arise if copper is not supplied for the enzymes that require it: these include cytochrome oxidase, superoxide dismutase, tyrosinase, dopamine β-hydroxylase and lysyl oxidase. These enzymes are all oxido-reductases.

Kinetics

The main aspects of copper kinetics are illustrated in Figure 35.1. About 2 mg of copper is ingested per day and a similar quantity is excreted in the faeces; about 0.6 mg is absorbed, about 0.1 mg being excreted in the bile, giving the true net absorption as only a very small percentage of the ingested copper. The absorbed copper is taken into the tissues to balance loss in the urine and from skin and hair. Copper is transported in the circulation complexed with albumin or amino acids, to be taken up by body tissues. The liver forms a copper-binding protein, **caeruloplasmin**. This is a glycoprotein of relative molecular mass 150 kDa which contains 0.34% copper. For a long time the major role of caeruloplasmin was believed to be that of transporting copper to the liver. It is now appreciated that caeruloplasmin also has the important property of catalysing the oxidation of Fe(II) to Fe(III) and a family in Japan has been described in which an inherited mutation to the gene for caeruloplasmin is associated with a condition of iron overload (see Chapter 21).

The low dietary requirement for copper means that true deficiency states are rarely seen in humans.

Another important aspect of copper transport should be mentioned and this concerns its movement into cells and between intracellular compartments. These processes are central to copper metabolism, because the metal ion does not readily cross plasma or other membranes by means of passive diffusion. Two similar copper-transport proteins have recently been characterized. Both are Cu^{2+}–ATPases that share common mechanistic features. One seems to transport Cu^{2+} across plasma membranes, while the other is required for the movement of the cation into organelles. The latter process is especially important, because many of the copper-dependent enzymes are in mitochondria

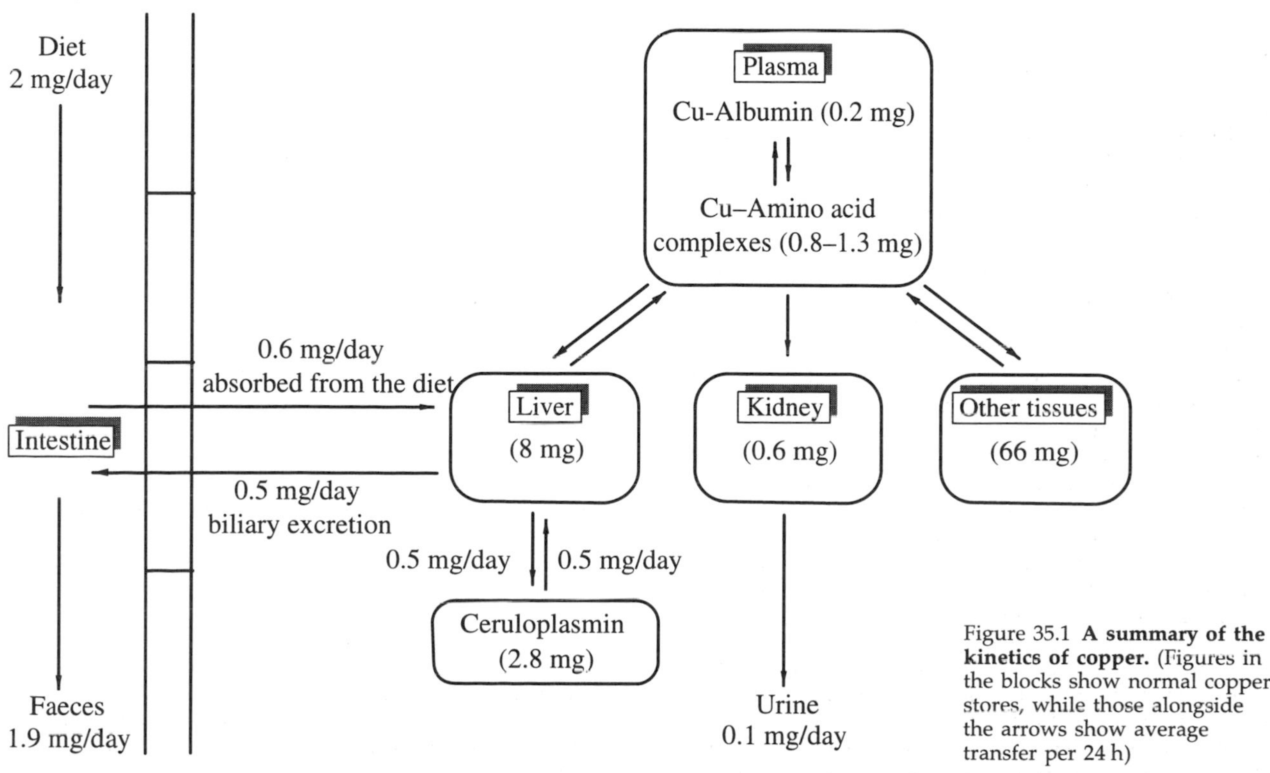

Figure 35.1 **A summary of the kinetics of copper.** (Figures in the blocks show normal copper stores, while those alongside the arrows show average transfer per 24 h)

Table 35.1 **A comparison of Menke's and Wilson's diseases**

	Menke's disease	*Wilson's disease*
Clinical chemistry findings	Decreased serum Cu, decreased serum ceruloplasmin, increased kidney/intestinal Cu, decreased hepatic Cu	Decreased serum Cu, decreased serum ceruloplasmin, increased urinary Cu, increased hepatic Cu
Defect	Intestinal Cu absorption, deficiency of Cu-dependent enzymes	Biliary Cu excretion, Cu uptake by ceruloplasmin
Treatment	None is effective	Chelating agents, e.g. penicillamine
Defective gene	Cu–ATPase (intracellular, transport into organelles)	Cu–ATPase (plasma membrane, export from cells)
Chromosomal location and inheritance	Xq13.3 Recessive	13q14.3 Recessive
Gene normally expressed in:	All tissues, except the liver	Liver, kidney and placenta

(e.g. cytochrome oxidase) or in vesicles (e.g. dopamine β-hydroxylase).

Inherited disorders of copper metabolism

The occurrence of two inherited disorders involving copper has been known for some time. **Wilson's disease** was first described in 1912. It is now known to be an autosomal recessive disease with as many as 1 in 200 of the population being heterozygous for the condition. The onset of the disease occurs in late childhood and is treatable by using metal–ion chelators (see Section 35.9). Patients present with muscular rigidity, tremor, lack of coordination, and damage to the basal ganglia and liver (which is pigmented). Characteristic Kayser–Fleischer rings are seen in the cornea. The biochemical picture is one of decreased serum copper, including that bound to caeruloplasmin, and increased urinary and hepatic copper.

The second, and more serious, inherited disease of copper metabolism is the X-linked **Menke's disease**. The onset is at birth and chelation therapy is ineffective. There is cerebral degeneration, abnormal hair growth ('kinky hair'), arterial rupture and thrombosis. Death usually ensues before the third birthday. As with Wilson's disease, serum copper is low (including that associated with caeruloplasmin). There is an increase in copper in several tissues, especially the intestine and kidney, but hepatic copper is decreased.

For the two diseases the mutations have been traced to the two genes, referred to previously, each of which encode Cu^{2+}–ATPases. The gene affected in Menke's disease is expressed in all tissues except the liver, while in Wilson's disease the major sites of expression of the aberrant gene are found to be the liver, kidneys and placenta. The Cu^{2+}–ATPase affected in Menke's disease is believed to be one responsible for intracellular transport of Cu^{2+}. Patients therefore show signs of Cu^{2+} deficiency, because the cation does not reach the enzymes that depend upon it, even when the concentration of Cu^{2+} in the cytosol, and hence in the tissue, is high. The requirement for Cu^{2+}-dependent lysyl oxidase in the process of cross-linking in elastin probably explains the occurrence of arterial rupture. The reason for the accumulation of copper in the liver is not understood.

In Wilson's disease the mutation is believed to affect a Cu^{2+}–ATPase that is responsible for the export of the cation, especially from the liver. Thus biliary excretion of Cu^{2+} and its incorporation into caeruloplasmin from extrahepatic sources are both greatly reduced.

It is now clear that Menke's disease is the more severe of the two inherited diseases of copper metabolism because, while sharing with Wilson's disease the problems of free-radical attack on the tissues caused by the accumulation of uncomplexed Cu^{2+} (see Chapter 33), there are the additional challenges resulting from hypofunctioning of copper-containing enzymes.

Table 35.1 compares the salient features of Menke's and Wilson's diseases.

35.5 Toxicity of mercury

Introduction

It should be appreciated that mercury, like several other toxic metals, can exist in three distinct forms: as metallic mercury, as inorganic mercuric salts or as part of an organic molecule. Quite often, all these forms are described as 'mercury' without qualification, but it should be noted that they possess different biochemical properties leading to very different biological effects.

Metallic mercury is of relatively low reactivity and does not normally constitute a serious health hazard. Recently, however, concern has been expressed about the danger to the health of dentists and dental technicians of preparing mercury-based amalgams for the purpose of filling teeth. In particular, the possible danger to their fetuses for pregnant dental workers has been highlighted. Nevertheless the following discussion will centre on the inorganic and organic forms of the metal.

Inorganic mercury

The toxicity of mercury salts results from the formation of stable mercury–protein complexes. Of all the metals, mercury is the one with the strongest affinity for thiol groups and will, therefore, bind extremely readily and inactivate any enzyme with such a group at its active centre. On the other hand, such binding, termed isomorphous replacement, has proved very useful in the X-ray analysis of crystalline proteins.

Organic mercury

Organo-mercury compounds are frequently used as fungicides, particularly in industries such as paper-making.

The first serious outbreak of mercury poisoning to be adequately documented occurred in the area of Minimata Bay, Kyusu, Japan where, by 1974, 800 people were poisoned and more than 100 died. The diagnosis of mercury poisoning was made and the disease subsequently controlled as a result of individual physicians seeing, and informing the authorities of, isolated cases of neurological disorders with patients presenting with paraesthesiae, ataxia, nervous stammering, and loss of vision and hearing. The work of Shoji Kitumura was especially noteworthy. The affected patients had all consumed fish or shellfish caught in the bay and containing methylmercury. The cause was traced to the release from a fertilizer factory of inorganic mercury which was converted into the organic form of methylmercury by bacteria resident in the guts of the affected seafood. The investigation of this outbreak has served as a model for subsequent cases of environmental poisoning. It is a matter of regret, therefore, that in the UK there is no equivalent for toxic chemicals of the 'green card' system used by physicians to notify adverse reactions to drugs.

A much more serious outbreak of mercury poisoning occurred in Iraq in 1972 when 6530 people were poisoned, and 459 died, as a result of eating bread made with wheat grain treated with methylmercury chloride which should only be used as fungicide at planting. When the cause of the poisoning was established, it was discovered that the treated wheat contained 3.7–14.9 μg Hg/g grain (the legal safety limits for Hg in the drinking water in the European Union is 1.0 μg/l).

It is now known that 80% of methylmercury is transported in the red cell, the remainder being found in the plasma. Little of the organic mercury taken into the tissues is converted into the inorganic form, although such a conversion appears to precede urinary excretion. Analysis of hair has proved to be a useful way of assessing the time when exposure to mercury occurred (Figure 35.2) and treatment with chelating agents such as *N*-acetylpenicillamine is useful for accelerating the clearance of the metal from the body (Figure 35.3).

The precise biochemical reason for the toxic effects of organic mercury compounds has not been established, but it has been suggested that the slow conversion to the inorganic form that precedes urinary clearance is the key. The inorganic mercury then binds to and inactivates enzymes with essential thiol groups.

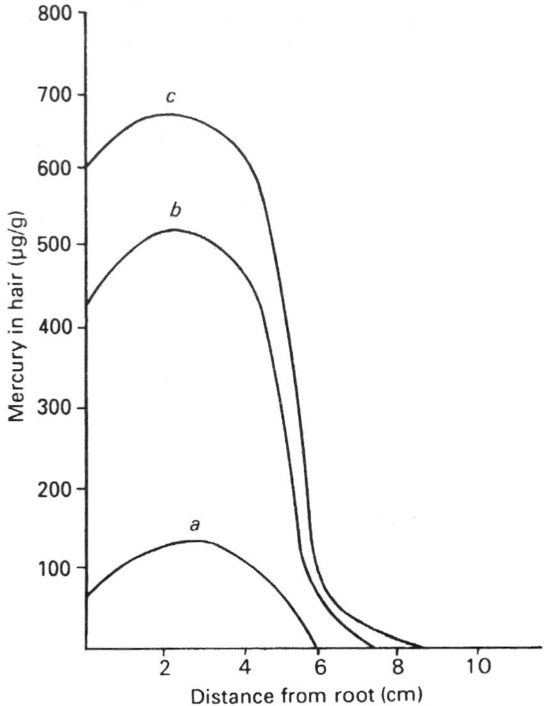

Figure 35.2 **Mercury in hair after methylmercury poisoning.** (*a*) Inorganic mercury; (*b*) methylmercury; (*c*) total mercury

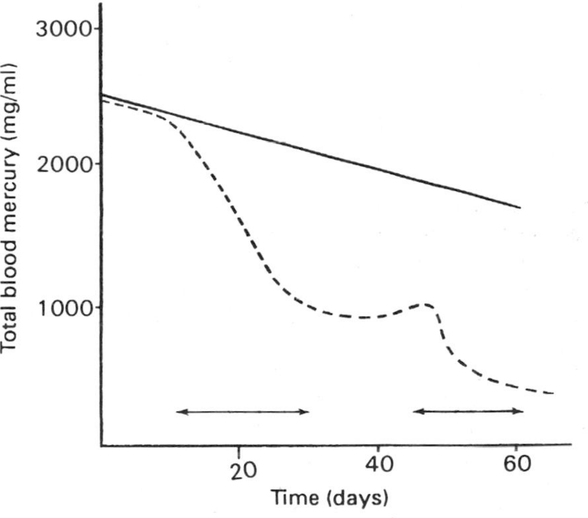

Figure 35.3 **Clearance of mercury from blood after methylmercury poisoning.** (——) Untreated patients; (– – –) patients treated with *N*-acetylpenicillamine (40 mg/kg per day) during periods indicated

35.6 Toxicity of lead

Introduction

In his writings the Roman architect Vitruvius showed that he was aware of the health danger posed by the use of lead piping for the distribution of water. This concern is still with us, because a survey in the UK carried out in 1975–76 showed that the concentration of lead in the drinking water exceeded the EC safety limit current at the time of 100 μg/l in 34% of houses in Scotland The corresponding figure in England was 7%. The large number of houses affected in Scotland reflected the

Table 35.2 **Comparison of lead content of teeth of modern and ancient man**

Age group (years)	*Lead content* (ppm)	
	Modern teeth (USA, 1975)	*Ancient teeth (AD200–600)*
0–9	11	5.3
10–19	17	3.0
20–29	19	3.7
30–39	24	1.3
40–49	27	2.0
50–59	50	
60–69	54	
70–79	55	

presence of older types of property which tend to retain much more lead piping. The dissolution of the lead from these pipes is then favoured by the generally acid water supplied in Scotland. This should now be a diminishing problem, because in 1964 the use in the UK of lead piping was banned for domestic plumbing. Nevertheless the problem of lead toxicity remains a major public issue in many industrial countries. As is the case for mercury, lead exists as the metal and also in inorganic and organic forms. Salts of lead have been widely used in paints and, in the organic form of tetraethyl lead, the metal has been added to petrol as an 'antiknock' agent. In the USA it is now illegal to use more than 600 ppm of lead in paint, but in the UK, manufacturers are required only to label paint products containing large quantities of lead.

Analyses of the lead content of both ancient and modern teeth have shown human exposure to lead to have increased dramatically between AD200–600 and the present day, as is shown in Table 35.2. This table also indicates that the quantity of lead in teeth increases throughout life.

Inorganic lead

Inorganic lead enters the body mainly through ingestion: children (who are most vulnerable to lead poisoning) can obtain toxic quantities of lead by chewing toys or furniture, but people of all ages also ingest lead that has been deposited on their food from the air or from lead-based solders on food cans. It has been suggested that lead poisoning contributed to the deaths of the members of Sir John Franklin's 1848 expedition to discover a north-west passage, between the Atlantic and Pacific Oceans, when in their ice-bound ship they were forced to rely on canned food. The lead content of plants can be relatively high in areas adjoining lead smelting works or motorways.

Lead has a high affinity for complex formation with proteins and, like mercury, binds strongly to thiol groups. Some of its toxic effects may arise in this way; however, the main symptoms of lead poisoning are manifested by changes in the central nervous system, where its ability to affect Ca^{2+}-dependent processes probably holds the key. The reason for this is that despite important differences between the inorganic chemistry of the two metals, in several biologically-important ways, Pb^{2+} seem to mimic Ca^{2+}. Thus lead is able to enter cells via hormone- or voltage-gated channels used by calcium, and inside cells lead binds to calmodulin and to other specific Ca^{2+}-binding proteins (including a Ca^{2+}-gated K^{+} channel). The ability of lead to interact with these systems, several of which are directly implicated in neurotransmission (see Chapter 30), may help to explain the neurological manifestations of lead poisoning.

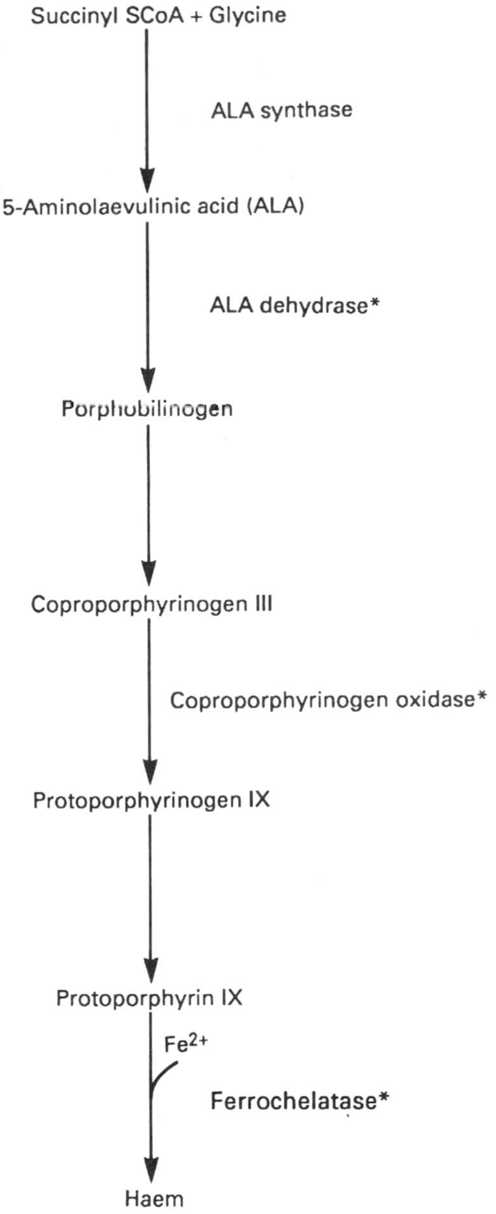

Figure 35.4 **The interference of haem synthesis by lead.** The asterisks indicate the sites of enzyme inhibition

Lead is also known to cause anaemia: it has been shown to inhibit the biosynthesis of haem by acting at several steps in the process (Figure 35.4). Particularly important are the reaction catalysed by 5-aminolaevulinate dehydrase and the step at which iron is incorporated into the porphyrin ring (a process catalysed by ferrochelatase). Lead poisoning, therefore, leads to an increased urinary excretion of 5-aminolaevulinate (ALA) and coproporphyrin with the accumulation of protoporphyrin in the red blood cells. Anaemia is a late manifestation of lead poisoning; ALA excretion is variable in lead poisoning and, therefore, the most reliable indication of exposure to lead is the accumulation of protoporphyrin in erythrocytes.

Organic lead

The form of organic lead in the environment giving most cause for concern is tetraethyl lead because of its continued (although in developed countries diminishing) use in petrol. Like the organic forms of mercury, tetraethyl lead has a very high affinity for lipids and it therefore tends to be concentrated in the lipid fraction of the central nervous system, thereby exerting its effects on this tissue. The mechanism underlying the damage to the brain remains unclear, although it has recently been suggested that a known hepatic enzyme capable of generating the free radical $Pb(C_2H_5)_3^+$ has a cerebral counterpart. This may mean that this damaging free radical is generated in the brain.

Organic derivatives of lead have no effect on haem biosynthesis.

35.7 Toxicity of aluminium

Introduction

Despite occasional nineteenth century reports to the contrary, aluminium has long been considered as a non-essential, non-hazardous element to man. In the last 30 years the view that aluminium does not constitute a toxic hazard to man has had to be re-evaluated. This re-evaluation has been prompted by several observations. Workers exposed to aluminium dust were found to be subject to lung disease and uraemic patients undergoing haemodialysis were found to suffer from aluminium-induced dementia if ordinary tap water was used to prepare their dialysis fluid. Elimination of aluminium also eliminated further cases of dialysis-induced dementia. More recently, two further findings have served to fuel concern about possible deleterious effects of the metal. First, high concentrations of aluminium have been found in the brains of people suffering from Alzheimer's disease. Secondly, in areas with acid rain, aluminium is leeched from the soil into fresh water with the consequent death of fish.

Properties and exposure

Aluminium is by far the most abundant metal in the earth's crust; indeed, after oxygen and silicon it is the third most abundant element. Its major ores are bauxite and cryolite. Aluminium is sparingly soluble in water at neutral pH values, but dissolves readily in acid or alkaline solutions. The dissolution of the metal ion from aluminium hydroxide is greatly accelerated by the formation of complexes with organic acids including amino acids, and the formation of such complexes may well facilitate the absorption of aluminium from the gastrointestinal (GI) tract. Otherwise aluminium is poorly absorbed from the GI tract. Aluminium sulphate is used in the treatment of drinking water because of its ability to cause the flocculation of complex organic material. Indeed some 100 000 tonnes of aluminium is used in this way annually in England and Wales. The presence of large quantities of aluminium sulphate at waterworks resulted in serious problems when 20 tonnes of an 8% solution of the salt was accidentally dumped in to the already treated drinking water at Camelford in Cornwall.

Another route whereby exposure to aluminium-containing products is causing concern is in the consumption of non-prescription drugs. Buffered aspirin and many antacid preparations contain aluminium salts. These may be consumed in large quantities and for long periods of time. Typically, a single dose may contain 35–200 mg of the metal. It has been calculated that at a maximum recommended dosage a person might ingest 0.8–5 g/day, thus exceeding by about 100-fold all other sources of the metal.

In the USA 5 million tonnes of aluminium are used per year in the following way: packaging (29%), building materials (22%) transportation (21%), electrical (10%), consumer durables (7%) and all other uses (11%).

Effects on health

Occupational exposure to aluminium is encountered in aluminium manufacturing workers, in those involved in the production of fireworks and explosives and the preparation of aluminium-based abrasives. Workers in these categories may develop pulmonary fibrosis, a disease in which breathlessness arises due to thickening and scarring of lung tissues around inhaled particles.

For a while it was believed that aluminium might be the cause of **Alzheimer's disease**. In this condition a deterioration of the brain is seen in the elderly. It is a progressive disease in which lapses in memory and learning ability are the first signs. Subsequently motor control is lost and death ensues. The observation of elevated concentrations of the metal ion in the brains of patients dying from Alzheimer's disease prompted the view that aluminium might cause the disease. Recently genetic analysis has identified a mutation in a gene that encodes a brain protein, named amyloid, in patients with an inherited form of the disease (see Chapter 30). This has led to the modified opinion that aluminium is an important risk factor for the disease. A recent study in Ontario, Canada has helped to substantiate this view. In this 2-year randomized study, 63 patients with probable Alzheimer's disease but living at home were divided into two groups, one of which was given the trivalent cation-binding compound desferrioxamine (a drug used for treating iron overload). The performance skill of the volunteers was assessed at home by means of video recordings. The performance scores of the subjects receiving the drug were significantly better than those of the controls.

Biochemical effects of aluminium

Much of the data on the mechanism of action of aluminium ions has been obtained in experimental animals or cells in culture and it is not certain how well this relates to man. The clinical toxicity effects of aluminium in humans usually present in the form of vitamin D-resistant osteomalacia and diminished bone remodelling activity. At the cellular level, aluminium reduces the deformability of red cells. It has been suggested that this is due to aluminium stimulation of NADPH oxidation and consequent free radical formation, but the evidence does not appear to be substantial. In addition, it seems that aluminium resembles lead in its ability to mimic and sometimes to antagonize calcium effects.

35.8 Radionuclides

An entirely different type of metal toxicity arises when any of a group of metals, described as bone-seeking isotopes, gain access to the body. These bone-seeking isotopes form divalent cations sufficiently similar to calcium as to become incorporated into the hydroxyapatite-type structure of bones. Such isotopes, some of which are listed in Table 35.3, undergo radioactive decay in the bones and the radiation released can cause mutation to bone cells ultimately leading to the development of osteosarcomas (bone cancers).

Table 35.3 **Typical bone-seeking isotopes**

Element	*Isotope*	*Type of radiation*	*Half-life*
Radium	^{224}Ra	α,β,γ	3.6 days
Radium	^{226}Ra	α,β,γ	1602 years
Thorium	^{232}Th	α,β,γ	1.4×10^{10} years
Plutonium	^{239}Pu	α,β,γ	2.4×10^{4} years
Strontium	^{90}Sr	β	28 years

Of these elements, an isotope of plutonium, ^{239}Pu, which is widely produced in nuclear power stations, is a potential hazard. A typical 1 GW nuclear power plant produces about 450 lb (19 800 Ci; 7.3×10^{13} Bq) of plutonium per annum. Under normal operating conditions, only traces of this are released into the environment, but several well-publicized disasters have led to the release of more. By comparison, the testing of nuclear weapons released some 8000 lb of the isotope (350 000 Ci; 1.3×10^{15} Bq) and, in view of its long half-life, most of this remains in the environment.

Plutonium is, in its own right, highly toxic. It has been estimated that as little as 1 μg deposited in the lung is sufficient to cause lung cancer. In addition, fission of ^{239}Pu yields many radioactive products, but only a few are sufficiently long-lived to be of potential danger to humans. One hazardous product of the decay of ^{239}Pu is strontium (^{90}Sr). Its half-life of 28 years means that it persists in the environment, where it may undergo bioconcentration. Of the many decay modes for ^{239}Pu, the one that sees 3.5% converted to krypton-90 (^{90}Kr) gives rise to ^{90}Sr by the pathway shown in Figure 35.5. Strontium possesses a very similar size and electronic configuration to calcium and is readily taken up into bone hydroxyapatite. If the quantity of ^{90}Sr becoming

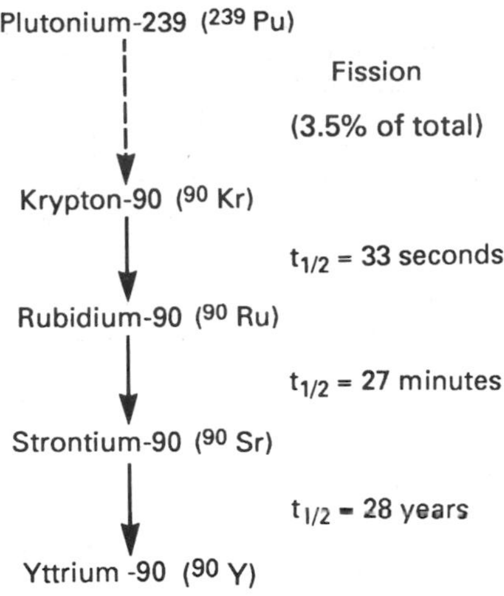

Figure 35.5 **The formation of strontium-90 (^{90}Sr) as a result of the fission of plutonium-239 (^{239}Pu). - - -> indicates that the process has several stages**

incorporated into bone is sufficiently large, serious consequences may follow. Its ionizing radiation can initiate tumorous transformation of bone cells into sarcomas and of the blood-forming cells of bone marrow into leukaemic cells. The extent to which this actually happens is the subject of much debate.

Caesium-137 (^{137}Cs) is another radioisotope that has been suggested to cause leukaemia. This too is produced by nuclear power plants and nuclear explosions. In the 1960s the concentration of ^{137}Cs in Eskimo populations was found to be 100 times higher than in people living in lower latitudes. This was shown to derive from the testing of nuclear weapons (some 500 nuclear devices were exploded world-wide in the 1950s and early 1960s prior to the test-ban treaty). Prevailing high-altitude, northerly winds had transported the isotope to the Arctic where it had been deposited on the tundra and subsequently undergone bioconcentration. Nuclear testing resulted in the release of 34 million Ci (1.37×10^{18} Bq) of ^{137}Cs, and since its half-life is 30 years, about half that quantity remains in the environment. One million Ci (3.7×10^{16} Bq) of ^{137}Cs were released over a period of 10 days in 1986 as a result of the disastrous explosion and fire at the nuclear plant at Chernobyl in the Ukraine. This estimate of ^{137}Cs released was based on the amount of material *found* on the ground; estimates based on the quantity *missing* from the reactor are about 4–5 times higher. Nevertheless, the contamination in parts of the then Soviet Union and Europe exceeded the total quantity deposited as a result of all nuclear tests. The concern is that caesium mimics potassium and is therefore distributed throughout the body, and its presence in rapidly dividing tissue, such as bone marrow, will cause cancer. Estimates of the number of cancers, in excess of those that would normally be expected, that will arise as a result of Chernobyl vary from thousands to hundreds of thousands, depending on the assumptions made about quantity of isotope released, distribution, etc.

As far as the hazards of nuclear power plants are concerned, it should be noted that the effects of the disaster at Chernobyl would have been almost entirely restricted to the site if the reactor had been contained, as is required by law in most developed countries. In the UK, the average dose of radiation to the general public from all nuclear power stations is 2 μSv/yr while the *radiation dose* from fossil fuel plants is 4 μSv/yr (27% of all electricity supplied in the UK in 1994 came from nuclear power stations).

35.9 Removal of toxic metals: chelation therapy

It would be very desirable if excess deposits of toxic metal, such as iron or copper, could be removed from the tissues and excreted. This is a difficult problem, but an understanding of the chemistry of metal chelates has resulted in the development of a number of useful compounds.

One of the first effective chelating agents to be used therapeutically was dimercaprol (2,3-*bis*mercaptopropanol) introduced during World War II as an antidote to poisoning with trivalent arsenicals. One such compound was the war gas Lewisite (chlorovinylarsine *bis*chloride). Because it was used to combat Lewisite poisoning, dimercaprol was termed 'British anti-Lewisite' (BAL). BAL has also proved to be useful in cases of mercury and gold intoxication. The presence of two thiol groups in BAL means that it is able to form a stable five-membered ring with these metals or their derivatives (Figure 35.6). By this means protection is afforded to reactive thiol groups in proteins.

As described earlier in this chapter, copper has a very high affinity for thiol groups, and this fact has led to the development of the very effective drug **penicillamine** for removing copper from the tissues in Wilson's disease, for example. The penicillamine provides two ligand atoms, sulphur and nitrogen, which allow strong binding to copper (Figure 35.7).

Such drugs are ineffective in removing bone-seeking radionuclides. These elements, however, resemble Ca^{2+} in binding to the oxygen of carboxylate groups. Thus molecules with several such groups, e.g. ethylenediaminetetraacetic acid (EDTA) and diethylenetriaminepentaacetic acid (DTPA) have proved useful for the removal of ^{90}Sr and ^{239}Pu.

Chelators of this type are, however, very toxic themselves. This is because they also bind calcium very strongly and can therefore provoke hypocalcaemia, with all of the adverse consequences that flow from this (see

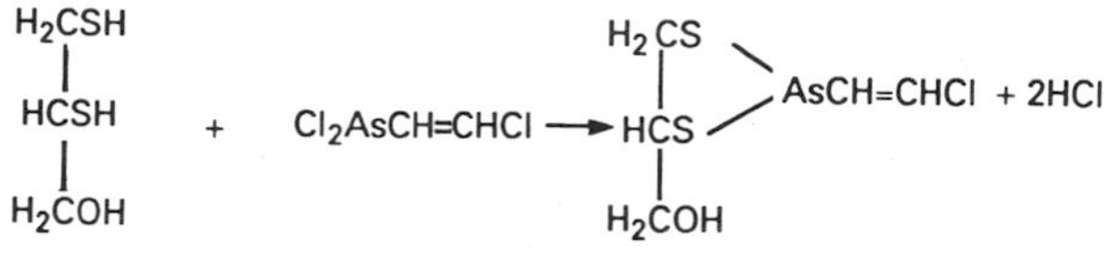

$$H_2CSH\text{–}HCSH\text{–}H_2COH + Cl_2AsCH{=}CHCl \rightarrow H_2CS/HCS{>}AsCH{=}CHCl\text{ (ring)}\text{–}H_2COH + 2HCl$$

Figure 35.6 **The binding of the trivalent arsenical war gas Lewisite (cholorovinylarsine dichloride) by dimercaprol**

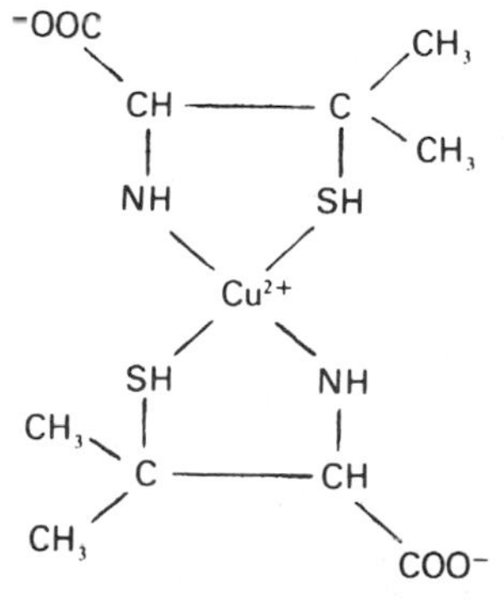

-OOC
CH — C
CH3
CH3
NH SH
Cu^{2+}
SH NH
CH3
C — CH
CH3
COO-

Figure 35.7 **The binding of copper by penicillamine.** Two molecules of penicillamine are shown

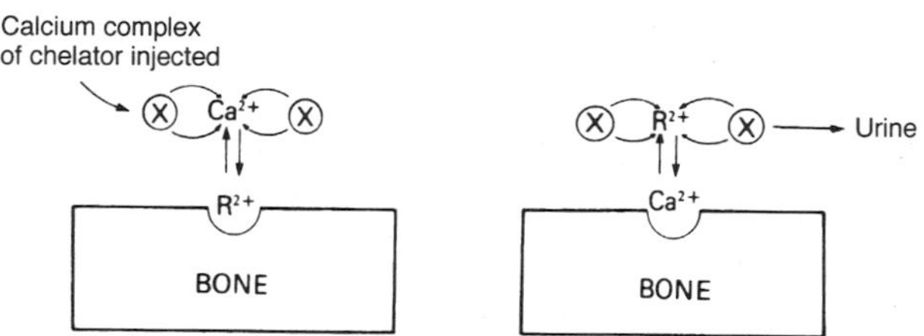

Figure 35.8 **Use of chelators for the removal of isotopes (R^{2+}) from bone.** Competition of two metals for the chelator

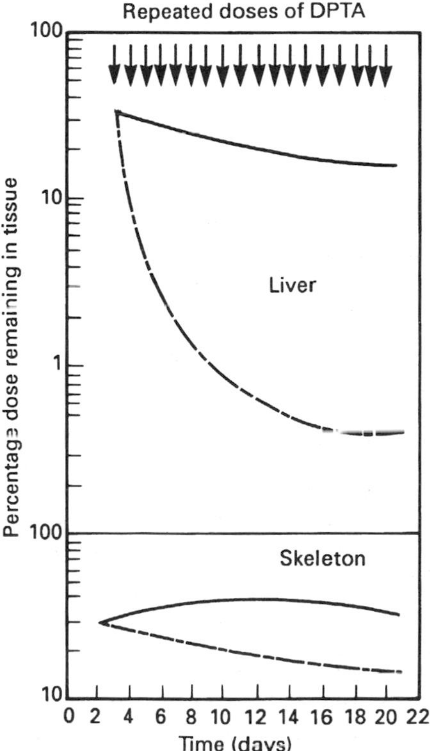

Figure 35.9 **Use of the chelator diethylenetriamine-pentaacetic acid (DTPA) for removing ^{239}Pu from the tissues.** (—) Untreated; (- - -) treated

Chapters 12 and 29). The problem is circumvented by treatment with the chelator already complexed with calcium. The uptake of another metal, for example ^{239}Pu by EDTA will depend on the relative affinity of the chelator for the two metals. If the affinity for the metal is only slightly greater than that for calcium, which is the case for most of the chelators tested, a slow and only partial excretion occurs (Figure 35.8). Much effort has, therefore, gone into the development of compounds which bind unwanted metal ions much more strongly than calcium ions. Experiments with DTPA in animals given small doses of ^{239}Pu, have been encouraging (Figure 35.9).

Chapter 36

Metabolism of foreign compounds

36.1 The nature of foreign compounds and their routes of entry into the body

Many tens of thousands of compounds that may be considered to be of 'no biological value' may gain access to the body through a variety of routes. At different times and in different contexts the terms 'foreign compound', 'anutrient' or 'xenobiotic' have been used to describe compounds to which we are all exposed, but which have no known biological value. It is usual to include drugs under this heading when the emphasis is on the way in which the body handles the drug metabolically rather than on how the drug influences bodily processes. For this reason the subject matter of the current chapter may be referred to generally as foreign compound metabolism or specifically as drug metabolism.

In the past, the metabolic processes that lead to the disposition of foreign compounds have been referred to as 'detoxication mechanisms', but now it is appreciated that there are occasions when the way the body handles a foreign compound metabolically leads to intoxication. For example, some metabolites of polycyclic hydrocarbons are carcinogenic, whereas the parent compounds have no deleterious effects. Nevertheless, as was discussed in Chapter 34, there are a wide range of compounds exposure to which carries a risk of toxic effects without the requirement for prior metabolism, and the metabolism of these compounds does indeed lead to their detoxication.

Although many of the foreign compounds to which we are exposed are 'man-made' it should not be thought that careful avoidance of such chemicals, if it were possible, would eliminate completely the problem of exposure to potentially toxic substances. There is a wide range of natural, toxic chemicals, some of which may be taken into the body, for example, when plant foods are eaten. Such substances may arise in one of two ways: the plants themselves may contain toxic chemicals, or such substances may be formed by the action of bacteria in the gut on perfectly harmless substances, for example, toxic amines may be produced by the decarboxylation of ingested amino acids.

36.2 General properties of the metabolites of foreign compounds

The first example of the metabolism of a foreign compound was described well over 100 years ago and indeed constitutes a detoxification mechanism (perhaps prompting the unfounded belief that all such metabolism would be protective). It was observed that following the ingestion by animals of phenol, phenyl sulphate could be isolated from the urine:

Phenyl sulphate is considerably less toxic than phenol. Studies of the excretory products of a wide range of foreign compounds demonstrated that, although most were metabolized to less toxic products, there were several exceptions. One such exception was the conversion of pyridine to the much more toxic *N*-methylpyridine (Table 36.1).

Two additional general principles concerning the metabolites of foreign compounds have emerged: these compounds are usually more water soluble than the parent (Table 36.2) and the majority are more acidic (Table 36.3).

36.3 Central role of the liver in the metabolism of foreign compounds

The tissues with the greatest capacity for catalysing the metabolism of foreign compounds are those at the

Table 36.1 **Toxicity of foreign compounds and their metabolites**

Toxicity or LD_{50} of foreign compound (g/kg mice)	*Drug*	*Metabolite*	*Toxicity or LD_{50} of metabolite* (g/kg mice)
2.0	Benzoic acid	→ Hippuric acid	4.15
2.85	*p*-Aminobenzoic acid	→ *p*-Aminohippuric acid	4.93
1.8	Sulphadiazine	→ Acetylsulphadiazine	0.6
1.2	Pyridine	→ N-Methylpyridine chloride	0.22

Table 36.2 **Solubility of foreign compounds and their metabolites**

Foreign compound solubility (mg/100 ml H_2O)	*Foreign compound*	*Metabolite*	*Metabolite solubility* (mg/100 ml H_2O)
184	Benzoic acid	→ Hippuric acid	463
311	*o*-Chlorophenyl acetic acid	→ *o*-Chlorophenaceturic acid	457
915	Phenylacetic acid	→ Phenaceturic acid	1145
		Phenylacetylglutamine	117
1480	Sulphanilamide	→ *N*-Acetylsulphanilamide	534

Table 36.3 **Acidity of foreign compound metabolites**

Foreign compound, pK_a			*Metabolite* (pK_a)
10.0	Phenol	→ Phenylglucuronide	3.4
4.2	Benzoic acid	→ Hippuric acid	3.7
	Benzene	→ Phenylmercapturic acid	3.7
2.9	*o*-Chlorobenzoic acid	→ *o*-Chlorohippuric acid	3.8

'portals' of entrance to or exit from the body. Predominant among these is the liver (Figure 36.1).

The metabolites produced by the liver are, generally, relatively water soluble and are passed into the blood for excretion by the kidneys. Several hepatic metabolites of foreign compounds are excreted initially in the bile, but only a limited proportion of these is excreted in the faeces, the remainder being returned to the liver via the enterohepatic circulation (see Chapter 25).

36.4 Phase I and phase II reactions

Several classes of transformation are used by the liver (and other tissues) for metabolism of foreign compounds. These include oxidation, reduction, hydrolysis and conjugation. Of these, the first and last are quantitatively the most significant. Occasionally a single transformation occurs, but more frequently two processes are involved, e.g. those of oxidation followed by conjugation. It is usual, therefore, to describe the metabolism of foreign compounds as occurring as the result of phase I and phase II reactions (Figure 36.2). **Phase I reactions** involve oxidation, reduction or hydrolysis and metabolites may be excreted without further reaction. But if the product is subsequently conjugated, a **phase II** metabolite is said to be produced. Direct conjugation may also occur without a preceding phase I reaction. Typical examples of phase I followed by phase II metabolism are shown in Figure 36.3.

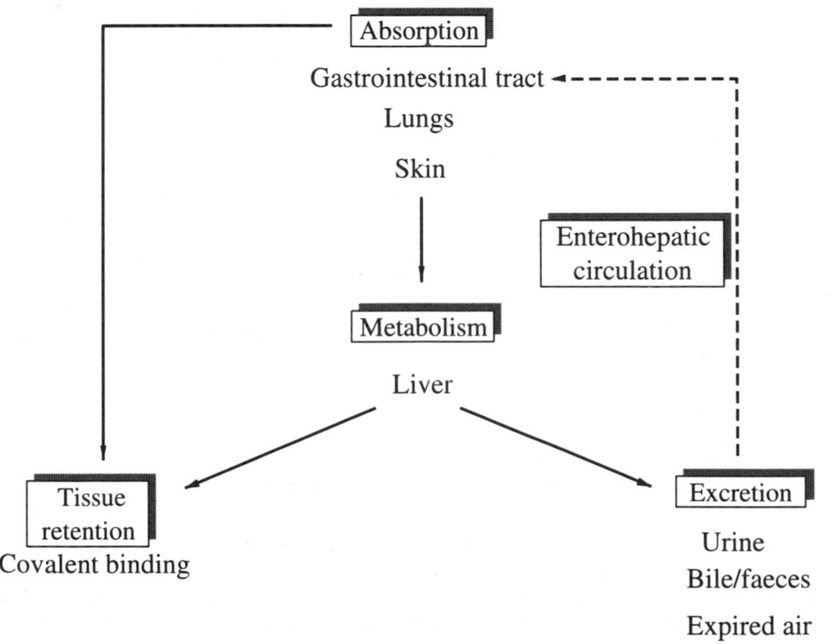

Figure 36.1 **Interrelationships in the metabolism of foreign compounds**

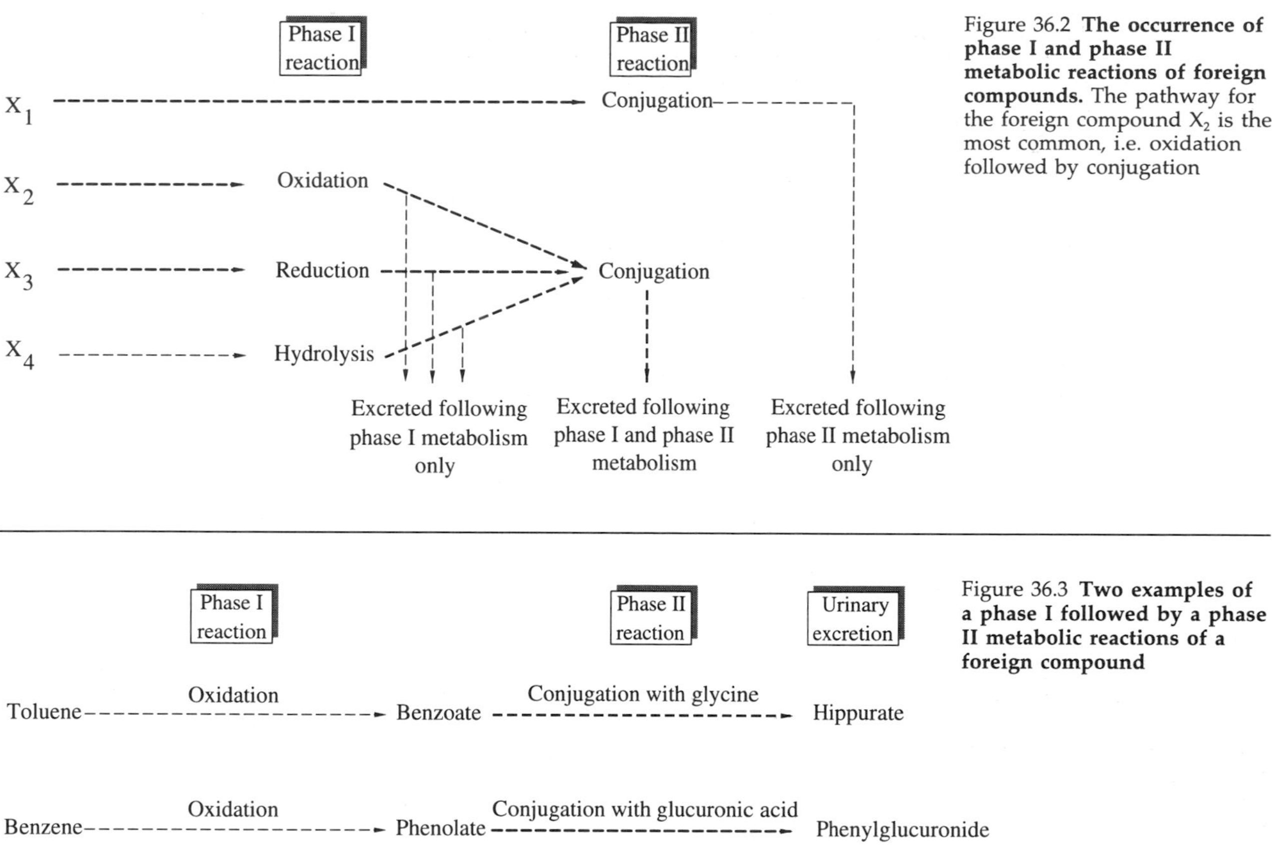

Figure 36.2 **The occurrence of phase I and phase II metabolic reactions of foreign compounds.** The pathway for the foreign compound X_2 is the most common, i.e. oxidation followed by conjugation

Figure 36.3 **Two examples of a phase I followed by a phase II metabolic reactions of a foreign compound**

Occurrence of multi-drug resistance

Recently, the observation of the phenomenon of multi-drug resistance in cancer chemotherapy has prompted some workers to refer to the process as 'phase III metabolism'. This does not appear to be appropriate, however, because no metabolic process is involved (except the hydrolysis of ATP). What is observed in multi-drug resistance is that tumour cells become resistant to a range of chemically and functionally unrelated drugs. In most cases this resistance is associated with the over-production of a membrane glycoprotein (P-glycoprotein) and this is, in turn, associated with increased efflux of the drugs from resistant cells. P-glycoprotein has been shown to bind both ATP and the range of drugs to which the cell is resistant. This has led to the proposal that P-glycoprotein functions as an ATPase that is able to harness the free energy of hydrolysis of the nucleoside triphosphate to promote the efflux from the cells of the drugs it is able to bind. Multi-drug resistance poses a major problem in the chemotherapy of cancer, while the occurrence of a similar process in infectious organisms may severely limit the use of antibiotics (see Chapter 39).

36.5 Role of reductive processes

Although oxidation is of much greater importance in the phase I metabolism of foreign compounds, a limited number of reductive reactions do occur.

Nitro reduction

Aromatic nitro compounds frequently undergo reduction to the corresponding amine. For example, the nitro group in the antibiotic chloramphenicol is reduced during the phase I metabolism of this drug. The nature of the enzyme that catalyses this reaction is not fully resolved, but may be a flavoprotein.

Azo reduction

An important example of the process of azo reduction was uncovered shortly after the introduction of the red azo dye prontosil as an antibiotic (see Chapter 39) and led to the discovery of sulphonamides. In a rather low specificity reaction, prontosil is reduced in the liver to give rise to sulphanilamide, the active bacteriostatic drug.

36.6 Hydrolysis of foreign compounds

Some foreign compounds undergo hydrolysis and this can occur before other metabolism, or as a phase II reaction subsequent to oxidation. The types of reaction include the hydrolysis of esters and amides and the hydration of epoxides.

Ester and amide hydrolysis

Foreign compounds that are esters and amides are subject to hydrolysis by a number of enzymes, including both hepatic and plasma esterases and an hepatic amidase. Thus the anaesthetic procaine is hydrolysed and inactivated rapidly in a reaction catalysed by the plasma enzyme, while the absence of an amidase in the circulation means that procainamide has to be transferred to the liver prior to inactivation. As a consequence of this the amide is longer acting with a more generalized action.

$C_6H_5\text{-}CO.O.CH_2CH_2.N(C_2H_5)_2$

Procaine

$C_6H_5\text{-}CO.NH.CH_2.CH_2.N(C_2H_5)_2$

Procainamide

Epoxide hydration

Although reactions catalysed by enzymes under this heading are not strictly hydrolytic in nature (the enzymes are classified as lyases, not hydrolases), it is convenient to include these reactions here.

Epoxides are very reactive electrophiles formed during the oxidative metabolism of aromatic and some aliphatic foreign compounds (see Section 36.7). Their hydration catalysed by epoxide hydratases results in the formation of much less reactive hydroxyl derivatives that additionally are able to undergo phase II conjugation, particularly to form glucuronides (see Section 36.8).

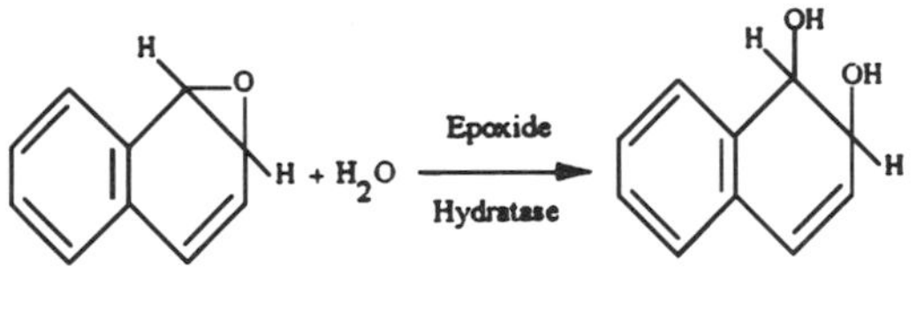

Naphthalene 1,2-oxide + H_2O —(Epoxide Hydratase)→ Naphthalene 1,2-diol

36.7 Oxidative metabolism of foreign compounds

Oxidative reactions are now understood to underlie a range of phase I metabolic reactions that were previously believed to be quite different in type, for example hydroxylation and demethylation :

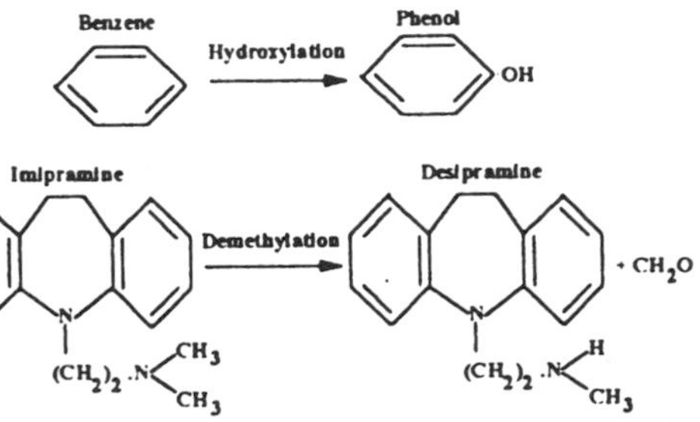

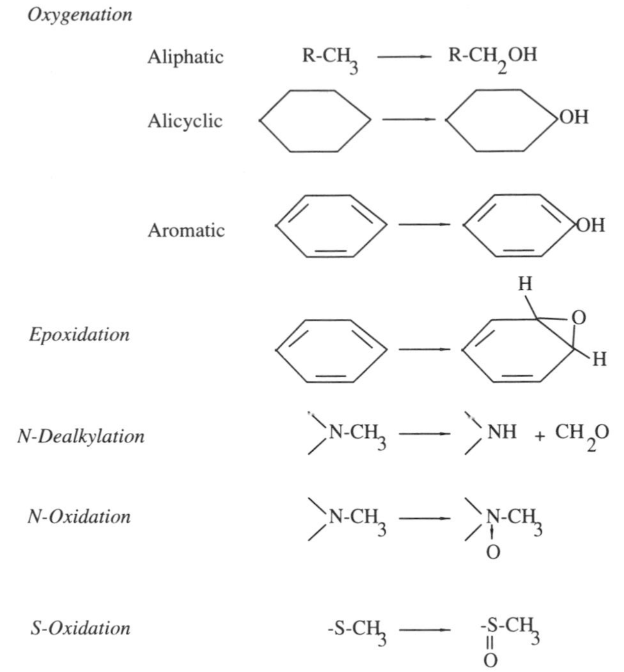

Benzene —(Hydroxylation)→ Phenol

Imipramine —(Demethylation)→ Desipramine + CH_2O

Oxygenation

- Aliphatic: $R\text{-}CH_3 \rightarrow R\text{-}CH_2OH$
- Alicyclic: cyclohexane → cyclohexanol (OH)
- Aromatic: benzene → phenol (OH)

Epoxidation: benzene → arene epoxide (H, O, H)

N-Dealkylation: $>N\text{-}CH_3 \rightarrow >NH + CH_2O$

N-Oxidation: $>N\text{-}CH_3 \rightarrow >N(\rightarrow O)\text{-}CH_3$

S-Oxidation: $\text{-}S\text{-}CH_3 \rightarrow \text{-}S(=O)\text{-}CH_3$

Figure 36.4 **Examples of a phase I oxidative metabolism of foreign compounds**

The unifying principle was found to be that for the range of oxidative reactions of the type shown in Figure 36.4 there was a common requirement for both atmospheric oxygen and NADPH to act as a reducing agent. A series of closely-related enzyme systems were all shown to use one of the two atoms of molecular oxygen to form water, while the second atom was used to form an oxidized derivative of the foreign compound undergoing reaction (in some cases, such as the demethylation of imipramine, the oxidized intermediate proved unstable and was difficult to isolate, but the formation of methanal indicated the oxidative nature of the reaction). Enzyme systems in this class are termed **'mixed function oxidases'** and the overall reaction they catalyse may be formulated:

$$NADPH + H^+ + O_2^* + XH \rightarrow X\text{–}O^*H + H_2O^* + NADP^+$$

The asterisk is used here to emphasize that one of the oxygen atoms from the dioxygen molecule is incorporated into the oxidized metabolite of the foreign compound, while the other appears in water.

Cytochrome P450

It was originally believed that the oxidation of foreign compounds implied the presence of a wide variety of specific enzymes, e.g. 'hydroxylases' or 'sulphoxidases'. However, although these names persist, it is now appreciated that these reactions are dependent on a special class of cytochromes, referred to collectively as **cytochrome P450**.

In the liver, this group of cytochromes occurs almost exclusively in the smooth endoplasmic reticulum, but in other tissues, such as the kidney or the adrenal gland, some forms of the enzyme are found in the smooth endoplasmic reticulum while others are mitochondrial.

When a suspension prepared from tissues and containing a high concentration of cytochrome P450 in the reduced form is treated with carbon monoxide, a new absorption spectrum is generated with a clearly defined maximum at 450 nm and hence the name 'cytochrome P450' is used. Cytochrome P450 proteins are able to bind a wide range of both endogenous molecules and foreign compounds.

Mode of action of cytochrome P450

After binding the foreign compound (or indeed an endogenous substrate), cytochrome P450 requires molecular oxygen and a supply of electrons from NADPH (and possibly NADH) to function as is shown in simplified form in Figure 36.5. In detail, the reaction catalysed by the enzyme complex that includes cytochrome P450 is shown in Figure 36.6. In the first stage (A) the cytochrome with the iron in an oxidation state of Fe(III) binds the foreign compound (XH). Subsequently the iron is reduced to Fe(II), the necessary electron being transferred by the flavoprotein NADPH–cytochrome P450 reductase (B). This is an unusual enzyme in that it has both FAD and FMN as its prosthetic groups. Once the iron is reduced, oxygen is able to bind to the cytochrome (C). The iron is then reoxidized by the transfer of an electron to the oxygen to form the superoxide anion ($O_2^{\cdot -}$, D). This allows a second electron to be transferred to the iron and this may also originate from NADPH or it may derive from NADH via a second flavoprotein (also with two prosthetic groups): NADH–cytochrome b_5 reductase. If this is the case the electron is donated to cytochrome P450 via cytochrome b_5 (E). Finally the hydroxylated species (the peroxyl anion, O_2^{2-}) is generated (F) and the hydroxylated foreign compound is released together with a molecule of water. In some reactions, for example the oxidation of polycyclic hydrocarbons, the attack of the active species generated by cytochrome P450 results in the formation of a cation free radial or an epoxide rather than an hydroxyl derivative.

Role of cytochrome P450

The cytochromes P450 play a central role in the metabolism of a wide range of foreign compounds, prominent among which are drugs. Some drugs, such as

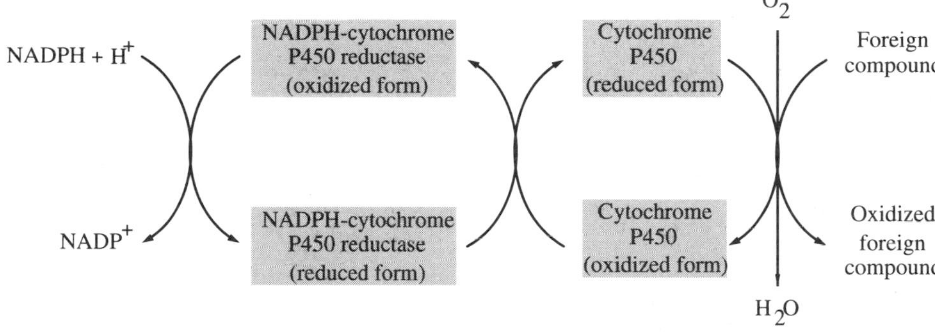

Figure 36.5 **Scheme for the oxidation of foreign compounds by cytochrome P450**

Figure 36.6 **Mode of action of cytochrome P450.** Fe^{3+} and Fe^{2+} represent the haem iron of cytochromes P450 and b_5 in the oxidized and reduced form, respectively. Proteins 1 and 2 (FAD/FMN) represent flavoproteins

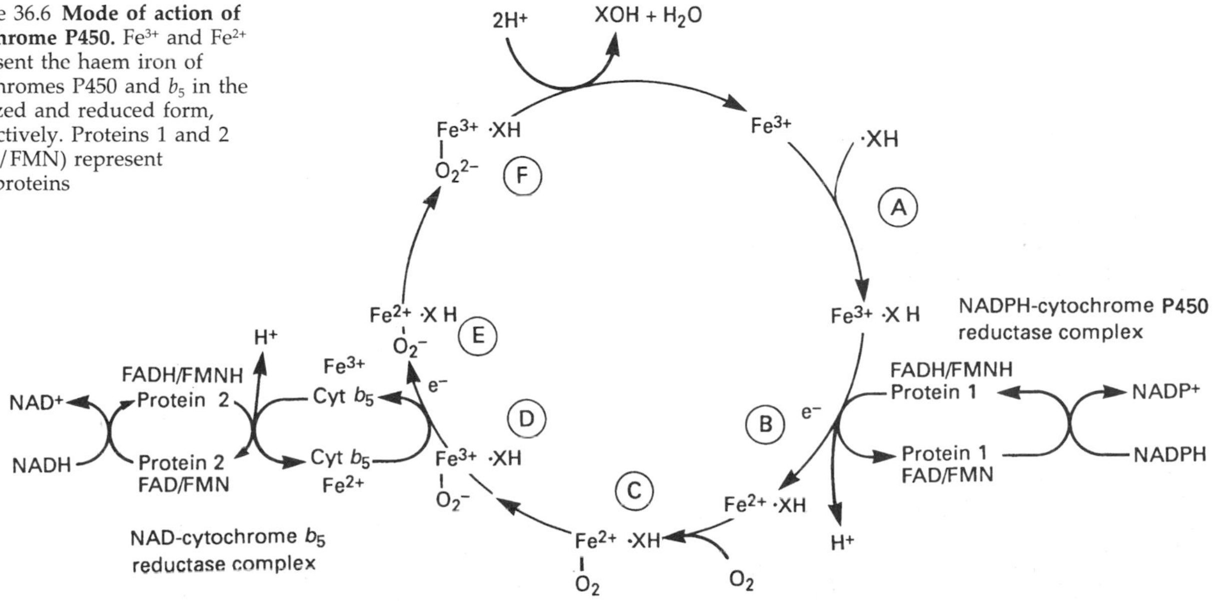

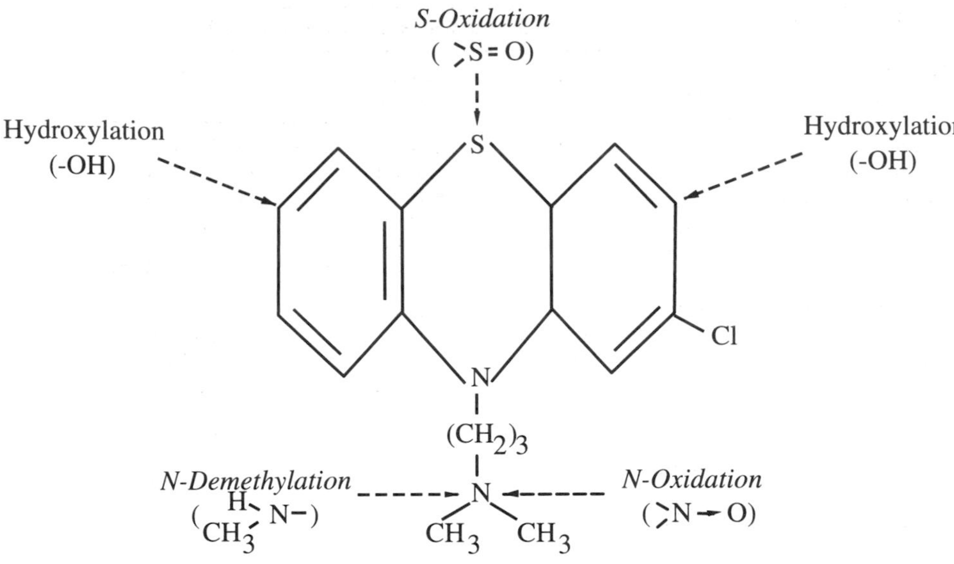

Figure 36.7 **Multiple sites for the phase I oxidation of chlorpromazine**

Table 36.4 **Roles of cytochrome P450 in intermediary metabolism (including some not further discussed in this book)**

Biosynthesis and degradation of:
1. Steroids
2. Vitamin D derivatives
3. Leukotrienes
4. Prostaglandins
5. Pheromones

chlorpromazine, undergo multiple cytochrome P450-dependent reactions (Figure 36.7).

In addition to their roles in the oxidation of foreign compounds, it should be remembered that a wide range of endogenous compounds are numbered among the substrates for the cytochrome P450 group of enzymes and some of these are listed in Table 36.4.

Specificity and multiple forms of cytochrome P450

The recent cloning of genes encoding the proteins has confirmed the view that with such a wide range of substrates there must be multiple cytochromes P450. The precise number of these has not been fully established, but the number of cytochrome P450 genes seems to vary from 30 to 200 depending on the organism. In humans, there are two cytochrome P450 families: those involved in the biosynthesis of steroids and bile acids and those that primarily handle foreign compounds. In naming the cytochrome P450 enzymes the family name is designated by CYP and a number (e.g. CYP2). Sub-families have a letter (e.g. CYP2D), with a final number reserved to indicate a specific protein (e.g. CYP2D6). Different allelic forms of the same enzyme may have an additional letter. For example, the gene for CYP2D6A has a single base-pair deletional mutation compared with CYP2D6 and this means that this protein has no catalytic activity. In humans, the major family of cytochrome P450 proteins that handle foreign compounds is that of CYP2. In particular the enzymes that have been shown to play a major role in the metabolism of drugs are CYP2A6, CYP2B6, CYP2C, CYP2D6 and CYP2E1, although CYP3A and CYP1A2 are also found. Several of these enzymes show broad range substrate specificity, for example CYP2D6 catalyses the oxidative metabolism of a range of cardiovascular and psychoactive substances and also of codeine and its derivatives. On the other hand, the catalytic activity of CYP2E1 is almost entirely restricted to the oxidation of ethanol and methanol (it forms part of the complex known as the 'microsomal ethanol-oxidizing system' – see Chapter 25).

Developmental changes in oxidative drug metabolism

There is an important physiological difference between humans and most experimental animals (rat, guinea-pig, rabbit, hamster and pig) when fetal oxidative metabolism of foreign compounds is considered: in the tissues of the animal species cited, the catalytic activity associated with cytochrome P450 molecules and their associated flavoprotein reductases is not detectable in the fetus. This observation correlates with the finding that, in these animals, the smooth endoplasmic reticulum (SER) appears only after birth. This is true not only of hepatic tissue but also for the kidneys, intestinal mucosal cells and lungs. These observations contrast strongly with those in humans for whom the SER is present after 2 months of gestation. This early emergence of the smooth membranes is reflected in the early appearance of the systems capable of catalysing the oxidative metabolism of foreign compounds. Consequently the fetal liver is capable of metabolizing a range of drugs, for example, at a rate of some 30–50% of that seen in adults. This generalized observation does not apply, however, to enzymes such as CYP1A2 that handle polycyclic hydrocarbons. Aryl hydroxylations proceed to only 2–5% of the rate observed in adults. After parturition, the rate of cytochrome P450-supported oxidation of foreign compounds increases to reach adult levels after about 3–8 weeks.

It is not clear whether the metabolism of foreign compounds by the fetus is of significance in their clearance, but *any* oxidative metabolism is potentially dangerous if the metabolite is toxic (as many are). This is especially true when the development of UDP–glucuronosyl transferases is considered, one of which would normally be expected to catalyse a phase II conjugation reaction with the phase I metabolite and thereby 'detoxify' it (see Section 36.8). These UDP–glucuronosyl transferases emerge at parturition and indeed their absence in premature infants is a cause of neonatal jaundice, because glucuronide conjugation with bilirubin (followed by biliary excretion) does not occur (see Chapter 25).

Although much is known about the perinatal development of foreign compound metabolizing enzymes, the same is not true for possible changes that might occur in senescence. In one study, the mean plasma half-life of the aminopyrine (an antipyretic drug), that undergoes *N*-demethylation, was 45% longer in geriatric patients (average age 77.6 years) than in younger patients (average 26 years). Whether the difference is due to a specific decline with age in the enzymes that catalyse phase I (and II) reactions or is due to unrelated causes such as decreased hepatic blood flow, for example, has not been established.

Microsomal FAD-containing mono-oxygenase

This is a special enzyme found in the SER of many cells that is an alternative to cytochrome P450 in that it also catalyses both *N*- and *S*-oxidations of foreign compounds. The enzyme is present in high concentration in the human liver, but there is relatively little in the livers of small experimental animals such as rodents. These, therefore, are poor models for studying this form of handling of potential drugs in humans, for this reason.

36.8 Conjugation reactions of foreign compounds

Foreign compounds may undergo conjugation reactions without prior metabolism but, as mentioned in Section 36.3, conjugation occurs more frequently as a phase II reaction subsequent to preliminary modification of the molecule by, for example, oxidation. The effect of conjugation is to make the compound more polar and, therefore, more easily excreted in the urine or bile.

The wide range of molecules used for conjugation is shown in Table 36.5, although they are not all used to the same extent. Thus conjugation with glucuronic acid is much more common than with glucose. In addition, there is a degree of species variation in the selection of the conjugating molecule: thus conjugation with ornithine is common in birds but rare in mammals, and phenylacetic acid is conjugated with glycine in mammals but with glutamine in humans. Multiple forms of conjugation occur quite frequently; for example, both sulphate and glucuronide conjugates of phenol are found in the urine of humans who have been exposed to the latter compound.

Table 36.5 **Molecules used for conjugation**

Source	*Conjugating molecule*	*Products*
Carbohydrate	Glucuronic acid	→ Glucuronide
	Glucose	→ Glucoside
	Ribose	→ Riboside
	Xylose	→ Xyloside
Amino acids	Glycine	–
	(Glutathione)	→ Mercapturic acids
	Ornithine	–
	Glutamine	–
Derived from amino acids	(Cysteine)	→ Sulphate
	(Methionine)	→ Methylated derivative
Various	AcetylCoA	→ Acetylated derivative

Where no name is specified for the product, this varies with the molecule with which conjugation occurs.

Conjugation reactions are synthetic in nature and consequently are either directly or indirectly, energy-requiring. In some cases the foreign compound is activated prior to conjugation, whereas in others the conjugating molecule itself is activated. As an example of the former, benzoic acid is converted to benzoylCoA before its conjugation with glycine to form benzoylglycine (hippuric acid). In the second category, UDP–glucuronic acid is the activated donor in the formation of glucuronides.

Although eight distinct types of conjugation have been described, the following descriptions are limited to the reactions with glucuronic acid, glutathione, sulphuric acid and ethanoic (acetic) acid, these being quantitatively the most important to occur in humans.

Formation of glucuronides

The activated intermediate for glucuronide formation, **UDP–glucuronic acid**, is formed by the oxidation of UDP–glucose (the formation of this compound as a preliminary to glycogen synthesis is described in Chapter 6). There are a group of enzymes, named **UDP–glucuronosyl transferases**, that catalyse the formation of glucuronide conjugates, with glucuronic acid being transferred (Figure 36.8) to a range of acceptor groups (Table 36.6). These enzymes also have endogenous substrates; thus the conjugation of bilirubin with two glucuronic acid residues is an important preliminary to the excretion of the former compound in the bile (see Chapter 25).

Conjugation with glutathione

The existence of sulphur-containing detoxication products in the urine was established well over 100 years ago. Sulphur occurs in several different forms in the urine, one important group of sulphur-containing compounds being described as mercapturates. These compounds were first isolated from the urine of animals that had been fed with halogenated aromatic hydrocarbons. Mercapturates were recognized as being

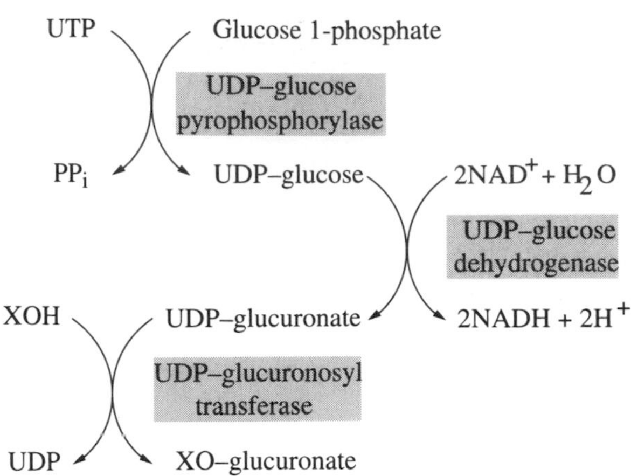

Figure 36.8 **Phase II metabolism of a foreign compound (XOH) by the formation of a glucuronate conjugate**

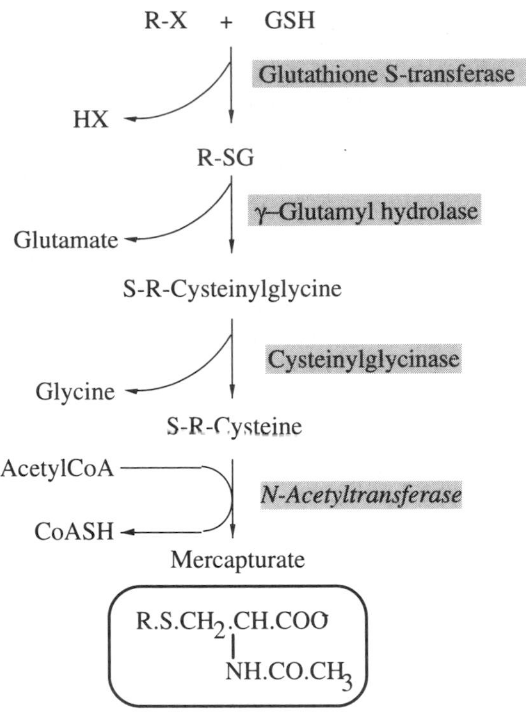

Figure 36.9 **Phase II metabolism of a foreign compound (RX) by mercapturate formation**

Table 36.6 **Examples of drugs and the groups in them that form glucuronides**

Site of attachment of glucuronate		*Example of drug*
Group		
Hydroxyl	Phenolic	Morphine
	Alcoholic	Chloramphenicol
Carboxylate		
	Aromatic	Salicylic acid
	Aliphatic	Indomethacin
Sulphydryl		Mercaptobenzothiazole
Amino		Dapsone
Imide		Sulphathiazole

unusual in that the sulphur was present as part of an organic molecule rather than as the inorganic sulphate (the major sulphur-containing species in urine).

It is now known that the sulphur atom derives from the tripeptide **glutathione**. A wide range of foreign compounds are known to be excreted following conjugation with glutathione and these include aromatic nitro and halogenated compounds, aliphatic halides and the epoxides formed in phase I oxidations of polycyclic hydrocarbons. The initial reaction leading to mercapturate formation is occasionally spontaneous (non-enzymic), with the nucleophilic glutathione reacting with an electrophilic centre in the foreign compound, but more commonly one of a group of enzymes named **glutathione *S*-transferases** is required. The importance of these enzymes is indicated by the observation that they constitute no less than 2% of the total soluble protein in the human liver. Following the initial reaction of glutathione with the foreign compound, a series of reactions gives rise to mercapturate formation. The first two of these reactions, in which sequentially the glutamyl and then the glycyl residue are lost, are those that lead to the catabolism of unconjugated glutathione. The cysteine conjugate formed then undergoes a reaction in which its amino group is acetylated to form the mercapturate and this is excreted. The formation of a mercapturate following the metabolism of a representative foreign compound is shown in Figure 36.9. Because there is only a limited quantity of glutathione in the liver, this compound can be depleted following the ingestion of large quantities of drugs such as phenacetin which undergoes a phase II reaction with glutathione following its phase I oxidative metabolism to form paracetamol. As a consequence of this depletion, the general protection against oxidative challenge afforded by the tripeptide is lost and severe hepatocellular damage can ensue. Administration of large doses of paracetamol has similar toxic effects.

In addition to dealing with electrophilic foreign compounds the glutathione *S*-transferases have some important endogenous substrates. Predominant among these is the epoxide derivative leukotriene A_4 which reacts with glutathione to give rise to the peptido-leukotrienes B_4, etc. (see Chapter 17).

Formation of sulphate esters

Phase II reactions leading to the formation of sulphate esters are catalysed by a group of enzymes named **sulphotransferases**. The sulphate donor for this reaction is the high-energy intermediate **3′-phosphoadenosine 5′-phosphosulphate** which is formed in two successive ATP-dependent reactions (Figure 36.10). Alcohols or amines, both aromatic and alipihatic, may act as acceptors for sulphotransferase reactions. Reference has already been made to the transfer of a sulphate group to phenol to form phenylsulphate as an example of a detoxication reaction.

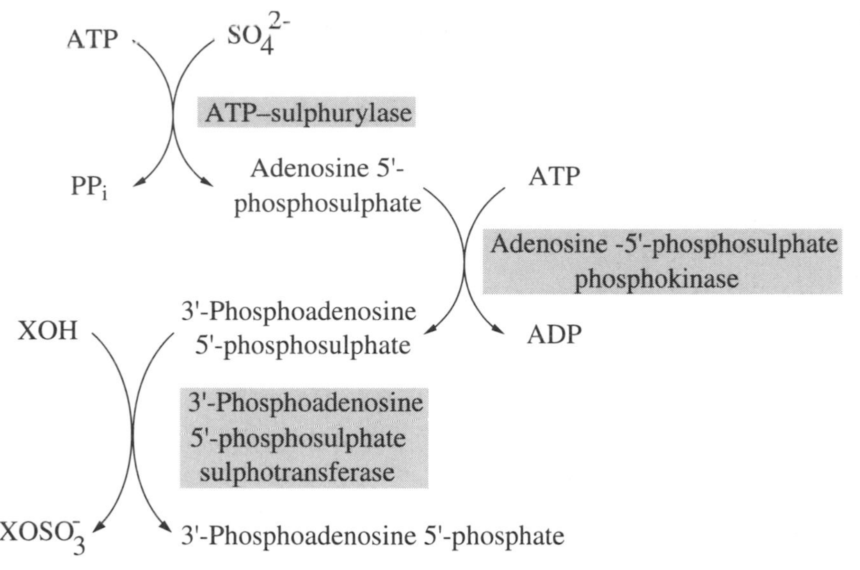

Figure 36.10 **Phase II metabolism of a foreign compound (XOH) by the formation of a sulphate conjugate**

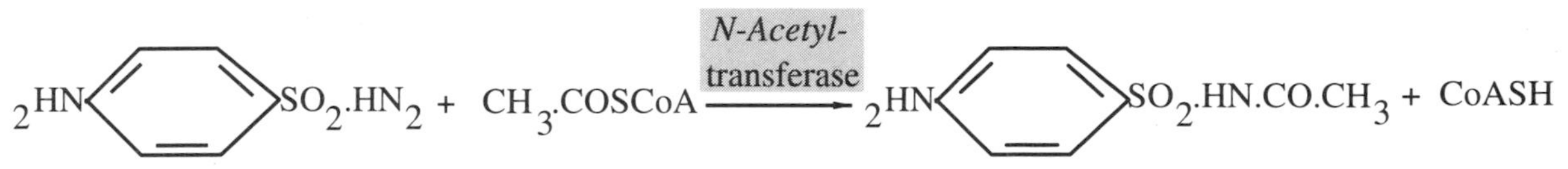

$_2HN$–C_6H_4–$SO_2.HN_2$ + $CH_3.COSCoA$ —*N-Acetyl-transferase*→ $_2HN$–C_6H_4–$SO_2.HN.CO.CH_3$ + CoASH

Sulphanilamide | Acetyl coenzyme A | *N*-Acetylsulphanilamide | Coenzyme A

Figure 36.11 **Phase II metabolism of sulphanilamide by *N*-acetylation**

As with all the phase II reactions there are also endogenous counterparts of foreign compounds that undergo sulphation reactions. An important example of such an endogenous substrate is the adrenal androgen dehydroepiandrosterone which is extensively converted into its 3-sulphate ester (Chapter 28).

Acetylation

A range of compounds with an amino group undergo *N*-acetylation in a reaction catalysed by a range of **acetyltransferases**, all of which use acetylCoA as the acetyl donor. A typical example of a phase II acetylation reaction is provided by the formation of *N*-acetylsulphanilamide (Figure 36.11). This reaction also provides a good example of an 'intoxication' reaction, since *N*-acetylsulphanilamide is less water-soluble than the drug and tends to be deposited in the kidney where it may cause renal failure in severe cases.

The anti-tuberculosis drug isoniazid also undergoes *N*-acetylation prior to its excretion.

36.9 Induction of the metabolism of foreign compounds

Clinicians have observed, over a number of years, that patients who receive prolonged drug treatment frequently develop a resistance and, consequently, require increased drug doses to elicit the same effect as the initial dose. The assumption was that the target for drug activity had become more 'resistant', but this explanation had to be discarded when 'cross- resistance' was demonstrated. In this phenomenon, patients treated with a drug A for a period showed resistance to a second drug B (with quite a different molecular target) on its introduction.

In some cases the emergence of resistance is associated with the production of the multi-drug resistance transporter P-glycoprotein (see Chapter 39). But resistance may arise in other ways and important among these are drug-mediated changes in the activity of the enzymes that catalyse phase I or II metabolic reactions.

Induction of oxidative metabolism

Several foreign compounds (most notably drugs) are able to cause a general increase in the amount of the flavoproteins that catalyse the reduction of cytochrome P450, i.e. they act as **inducers**. The result of this drug-mediated effect is that the general rate of disposition of foreign compounds is increased. This is because the flavoproteins in question are able to induce a wide variety of cytochrome P450 from different families of the enzyme. On the other hand, it is known that certain inducers affect the rate of foreign compound metabolism

Table 36.7 **Mechanisms whereby foreign compounds induce the different forms of cytochrome P450**

Cytochrome P450 form	*Example of an inducer*	*Main mechanism of induction*
CYP1A1	Dioxin	Activation of transcription
CYP1A2	3-Methylcholanthrene	Stabilization of mRNA
CYP2B1	Phenobarbitone	Activation of transcription
CYP2B2	Phenobarbitone	Activation of transcription
CYP2E1	Ethanol	Protein stabilization
CYP4A1	Clofibrate	Activation of transcription

differentially. Thus the administration of phenobarbitone causes an increase in the rate of demethylation of the drug aminopyrine (formally used as an antipyretic), while the polycyclic hydrocarbon methylcholanthrene has no effect on this process. Conversely the latter compound causes an almost three-fold increase in the rate of metabolism of benzo[a]pyrene, whereas phenobarbitone is without effect on this process. These differential effects are observed because cytochromes P450 from different families are induced by different agents and handle different ranges of foreign compounds. Thus CYP2B6 is the major form of cytochrome P450 induced by phenobarbitone, while CYP1A2 is induced by polycyclic hydrocarbons. In the case of CYP2E1 (also referred to as microsomal ethanol oxidizing enzyme) the substrate, ethanol, is able to induce this enzyme. This probably underlies, in part, the observation that chronic alcoholics are able to metabolize ethanol more rapidly than non-alcoholics.

The ability of one drug to influence the rate of metabolism of a second is of great clinical importance, as is shown in the case of patients treated simultaneously with the drugs phenylbutazone (a uricosuric agent) and warfarin (an anticoagulant – see Chapter 24). The former induces the enzyme CYP2C9 that catalyses the hydroxylation of the latter in the 7-position. This leads to the pharmacological inactivation of warfarin. If the uricosuric drug is withdrawn and the treatment with anticoagulant is continued using the same dose, patients may suffer intragastric haemorrhaging. Measurements show that the circulating concentration of warfarin is elevated outside the therapeutic range in such patients because the effect of the inducer wears off and consequently the rate of metabolism (and clearance) of warfarin decreases.

Effect of induction by foreign compounds on the oxidative metabolism of endogenous substrates

The CYP2 and CYP3 families of cytochrome P450 have endogenous (steroid) as well as exogenous substrates and some of these enzymes are inducible by foreign compounds. It follows, therefore, that such inducers may influence profoundly normal substrate/product relationships. Thus there are recorded cases of epileptic patients receiving phenytoin as their anticonvulsant therapy who have developed signs of osteomalacia. In part this relates to an increase in the hydroxylation of 25-hydroxycholecalciferol to the inactive 24,25-dihydroxycholecalciferol (see Chapters 12 and 29), although, in addition, phenytoin also acts as an inhibitor of hepatic 25-hydroxylase, thereby preventing the formation of 25-hydroxycholecalciferol and hence the biologically-active 1,25-dihydroxy-derivative.

More beneficially, the inappropriately high circulating concentrations of cortisol seen in patients with Cushing's disease have been lowered by the administration of DDE, a metabolite of the insecticide DDT. The former compound induces a 6β-hydroxylase, the activity of which eliminates the ability of the steroid to bind to the glucocorticoid receptor and produce biological effects. This hydroxylation also hastens the elimination of the steroid. Since 6β-hydroxycortisol is normally a very minor metabolite of the glucocorticoid, it has been proposed that the measurement of urinary 6β-hydroxysteroids may give an indication of the state of induction of a subject's liver.

Mechanisms that underlie the induction of oxidative metabolism

The processes involved in the induction of cytochrome P450 proteins are complex, with different foreign compounds influencing different enzymes by distinct mechanisms. These range from an increase in the rate of transcription of individual cytochrome P450 genes to stabilization of the final protein product. Table 36.7 gives some examples of the different types of inductive mechanisms.

Induction of enzymes that catalyse phase II reactions

The discussion in this section has focused on the induction of cytochrome P450 and in view of its great importance this is appropriate. It should be appreciated, however, that some of the enzymes that catalyse phase II reactions, e.g. UDP–glucuronosyl transferases and glutathione *S*-transferases, are also inducible. For example, isoenzymes from both of these classes are induced by phenobarbitone.

36.10 Inter-ethnic variation in the metabolism of foreign compounds

Many examples are known of inborn differences in the ways in which foreign compounds are handled. Much of the knowledge of the mechanisms underlying these differences and their consequences derives from the study of drugs by pharmacologists. One of the earliest examples of genetic subpopulations of drug metabolizers to be described in humans followed the introduction of **isoniazid** as an anti-tuberculosis drug (see Chapter 39). This drug is inactivated, prior to excretion, by acetylation and two classes of phenotype were recognized in the population: 'fast' or 'slow' acetylators. In the latter phenotype, the therapeutic dose of isoniazid may have toxic side-effects, including the development of peripheral neuritis. In Caucasian populations, 58% are 'slow' acetylators, while among American Indians only 19% come into this category. Genetic investigations have shown two autosomal alleles: R for 'fast' and r for 'slow' acetylation (R is dominant, while r is recessive). When whole populations are considered, it is usually not possible to distinguish between people who are homozygous (RR) from those who are heterozygous (Rr), both types being classified as 'fast' acetylators. Nevertheless, among the Canadian Inuit Eskimos, 'fast' acetylators are mostly homozygous, whereas Europeans recorded as 'fast' acetylators are mostly heterozygous. This means that when the distribution of acetylation rates in populations are plotted, the Inuit classified as 'fast' will tend to be found at the upper part of the curve and Caucasians at lower values. This may help to explain why, generally, higher doses of isoniazid are required by the Inuit than by Caucasians if therapeutic failure is to be avoided.

Another example of inter-ethnic variation in the metabolism of a drug was found when **debrisoquine** was introduced as an antihypertensive agent (it has now been withdrawn). The phase I metabolism of debrisoquine to the 4-hydroxy-derivative depends on the isoform of cytochrome P450 designated CYP2D6. Investigations of populations allowed two groups to be recognized based on the ratio of the dose of the drug excreted unchanged to that appearing as the 4-hydroxy-metabolite. In Caucasian populations, about 93% had a ratio of about 1:1 and where termed 'extensive' metabolizers (EM), while the remaining 7% had a ratio of about 20:1 and were termed 'poor' metabolizers (PM). When the population of the Peoples' Republic of China was investigated in this way, only 1% proved to be of the PM phenotype, whereas in West Africans from Nigeria and Ghana the PM phenotype was found in 13% of the population. CYP2D6 catalyses the metabolic activation of codeine to morphine and it might be expected, therefore, that therapeutic failure of codeine-induced analgesia might be seen in PM subjects. In some cases this has been true, but it was found not to be the case for the Chinese, because compensatory changes occur in the rate of the phase II conjugation of codeine with glucuronic acid (this process is in competition with the reaction leading to the formation of morphine). This means that codeine is less well converted to morphine in the Chinese PM population, but the concentration of the substrate for CYP2D6 remains higher, as less is converted in the water-soluble glucuronide for urinary excretion.

Since a range of drugs, including phenacetin and phenytoin are metabolized in phase I reactions catalysed by CYP2D6, elucidation of EM/PM phenotypes by using debrisoquine as a model compound is proving useful in predicting the likely metabolic status of individuals for these other drugs.

Chapter 37

Multiple environmental challenges: cancer

37.1 Introduction: what is cancer?

Cancer is certainly one of the most serious diseases afflicting humans. A WHO report published in 1990 estimated that each year some 6½ million people are diagnosed as having cancer and 5 million die of the disease. In these numbers there is great skewing to the developed nations in which half of these deaths are recorded, whereas the populations of these countries only account for a quarter of the world total. The situation is such that, in 1990, 25.3% of all male deaths recorded in England and Wales were due to cancer.

A **cancer** is defined as a **'malignant neoplasm'**. In 1958, Willis defined a neoplasm as 'an abnormal mass of tissue, the growth of which exceeds, and is uncoordinated with, that of normal tissues and persists in the same excessive manner after cessation of the stimulus that evoked the change'. In the past the term 'cancer' has sometimes been used only to describe neoplasms of epithelial cells, also termed **carcinomas**. These affect tissues such as the skin, and constitute some 90% of all neoplasms. However, the term cancer also describes those malignant neoplasms arising in mesodermal cells (such as those of the bone) which are termed **sarcomas**, those of haematopoietic cells termed **leukaemias** and those affecting nerve cells.

A major development of the past decade has been the proof that cancer is essentially a genetic disease. However, cancer differs in two important respects from other genetic diseases of the type described in Chapter 38. First, cancer is largely due to the occurrence of somatic mutations, whereas other genetic diseases (except mitochondrial diseases – see Chapter 5) are due solely to germline mutations. Secondly, each individual cancer requires the accumulation of mutations. The belief that there is a requirement for several mutations to accumulate in a cell for cancer to manifest itself is referred to as the **'multi-hit'** concept.

It should be noted that germline mutations predisposing to particular forms of cancer may be inherited. The disease is referred to as **familial** if there is an inherited component and **sporadic** if arising due to random somatic mutations.

37.2 Cancer as a multi-step disease

The evidence that cancer arises as the result of several mutations affecting a number of special genes is now overwhelming. This has been provided in several ways: including epidemiological studies and observations of the tumours themselves.

Epidemiology

One of the most persuasive pieces of evidence in favour of a multi-step model of cancer comes from the simple observation that the incidence rate of most common cancers increases dramatically with age. Thus a 70-year-old man is over 1000 times more likely than a 10-year-old boy to develop cancer of the colon. If a graph of the incidence rate for the most common cancers in the UK is plotted against age using a log/log scale, straight lines result the slopes of which lie between 3 and 7 (Figure 37.1). The most satisfactory explanation of this observation is that a minimum of 3–7 separate 'hits' (mutations) need to accumulate in a particular cell for malignancy to emerge full-blown, and these simply take time to occur. Because almost all of the cells in a tumour appear to be genetically identical it is commonly held that all of the 'hits' accumulate in one cell and its progeny (clonal growth). In other words tumours grow as a result of clonal selection driven by mutation (Figure 37.2).

One consequence of this age-related increase in the incidence of cancer is that, as health care improves, with people consequently living longer, more deaths due to

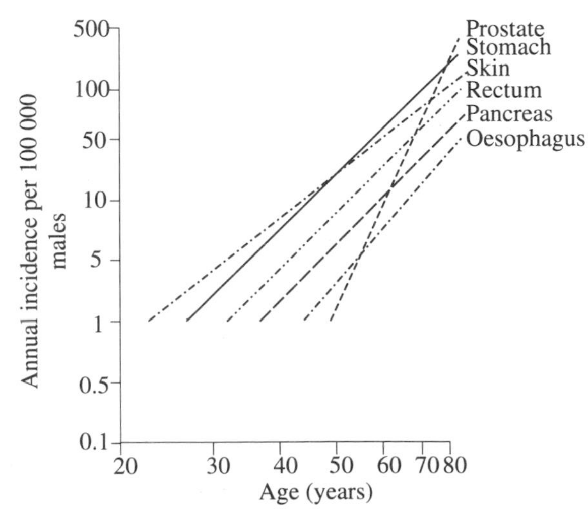

Figure 37.1 **The incidence of many cancers increases according to a power of age**

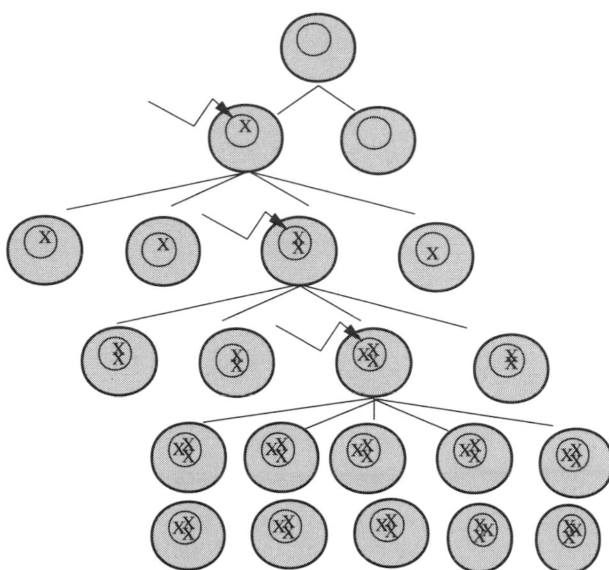

Figure 37.2 **Tumours grow by a process of clonal selection driven by somatic mutation (arrows)**

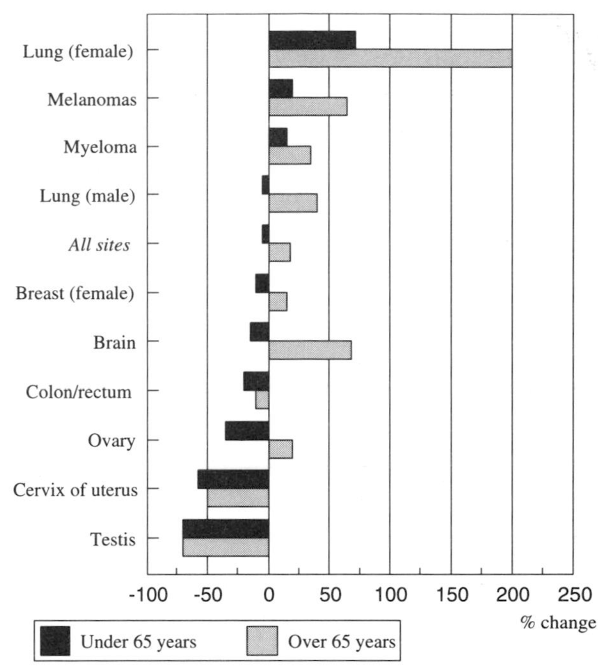

Figure 37.3 **Changes in US cancer death rates 1973–90**

cancer are expected to be recorded. In the case of lung cancer, however, even when the ageing of the population is accounted for, there has been a steady rise in deaths caused by this disease (Figure 37.3).

Further epidemiological evidence in favour of the multi-hit hypothesis comes from the study of patients suffering from ankylosing spondylitis (an autoimmune disease of the spine – 'bamboo spine') who, between 1935 and 1954, underwent X-ray therapy in an attempt to alleviate the pain associated with the disease. This group of patients showed an overall 28% increase in neoplasms compared with the normal population. The proportional increase reached a maximum of 72% between 10 and 12.4 years after treatment. The best explanation for the 'lag' in developing cancers is that the ionizing radiation caused one or a number of mutations in cells, but further mutations were required for a cancer to develop.

Studies of human cancer

Morphological observations in humans also favour the multi-hit hypothesis. In the colon, the evolutionary stages of tumours have been extensively studied. Initially small adenomas are observed in the colorectal epithelia. (**Adenomas** are benign, i.e. non-invasive, tumours of epithelial origin.) At this early stage the cells are almost normal in appearance, but with time they grow and become less organized both intra- and extracellularly. Eventually the condition evolves, so that a frank carcinoma is found. Again the interpretation of the delay is that cells in the adenomas require time for the additional mutations of the genotype to occur and drive the emergence of an invasive and metastatic

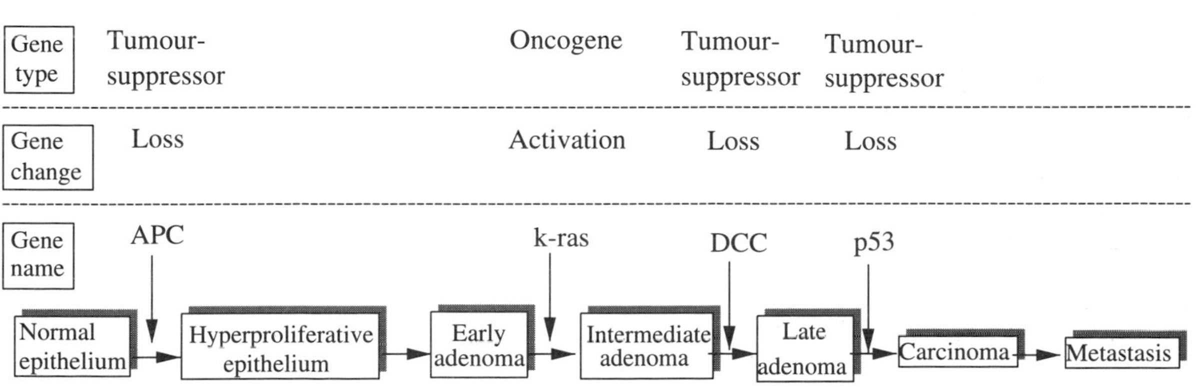

Figure 37.4 **Some of the mutations that affect colorectal epithelial cells on the way to metastasis**

phenotype. In the case of colorectal cancer, the molecular nature of the changes that underlie the **initiation** and **progression** to carcinoma have been documented. Thus, it has been shown that malignant cells within a single tumour have the same set of mutations found in the benign portion of the tumour, but with the addition of at least one further mutation.

Figure 37.4 shows in simple diagrammatic form some of the mutations that arise in colorectal epithelia in the progression to malignancy. Similar progression is observed in cervical carcinomas. The **cervical smear test** is designed to detect neoplastic cells shed from the cervix at a stage when insufficient mutations have occurred to support a full-blown malignancy. It is very important that this test is conducted properly, both by the doctor who collects the sample and by the scientist who examines it. This is because surgical intervention at this stage is nearly always life-saving, whereas once the tumour has acquired the ability to **metastasize** (i.e. to spread to other sites in the body) through further mutations, the resulting cancers are frequently incurable.

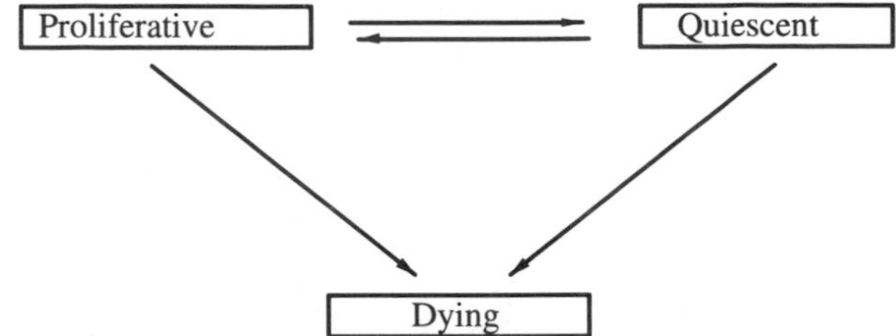

Figure 37.5 **In normal tissues, cells may be in one of three states.** In neoplasia, there is an increase in cell number (loss of control) and this may be due to: (a) an increase in proliferation, (b) a decrease in quiescence, (c) a decrease in the number of cells dying, or a combination of two or more of these events

37.3 The transformation of cells to a cancerous phenotype

A lack of balance in three processes that control the state of cells in a tissue can lead to cancer

The cells within any given tissue can be regarded as being in one of three possible states, these being **proliferative** (growing and dividing), **quiescent** (resting) or **dying**. The total number of cells in a tissue depends upon the proportion of cells in these three states (Figure 37.5). It is therefore important to appreciate that all three processes are normal physiological events, and in particular, that programmed cell death (termed **apoptosis**) is a normal event. Indeed, in certain tissues at certain times apoptosis comes to predominate. This is most clearly illustrated by the destruction of self-reacting T-lymphocytes in the thymus, B-lymphocytes in bone marrow (Chapter 32) and the large-scale cell death that occurs in the developing brain. During apoptosis, cells shrink, their chromatin is degraded and their plasma membranes lose their integrity. The resulting fragments undergo rapid phagocytosis and in this way they do not provoke an inflammatory response.

All three states of cells (proliferation, quiescence and dying) arise as the result of the operation of normal regulatory mechanisms, and a characteristic feature of these mechanisms is that there is much redundancy. This means that there are several different possible paths to a given end state. In neoplasia, control of this balance is lost and this can arise due to an increase in the proliferative rate, a reduction in the number of quiescent or dying cells or a combination of all three. How can control be lost?

Figure 37.6 reinforces the point made in Chapter 3 that any cell in a tissue is constantly receiving information

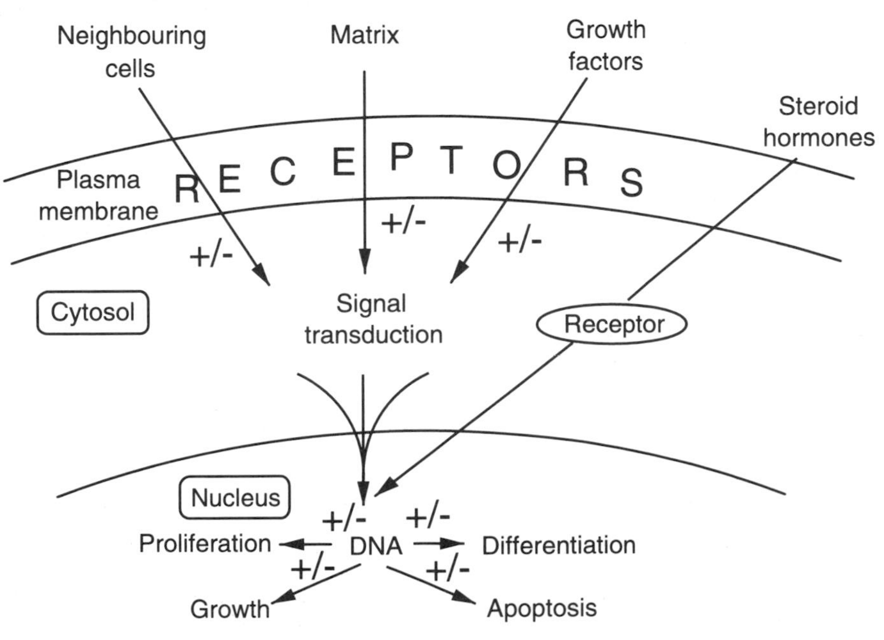

Figure 37.6 **In tissues, cells receive both positive and negative signals from many sources and the balance of these serves to control cellular growth, proliferation, differentiation and death**

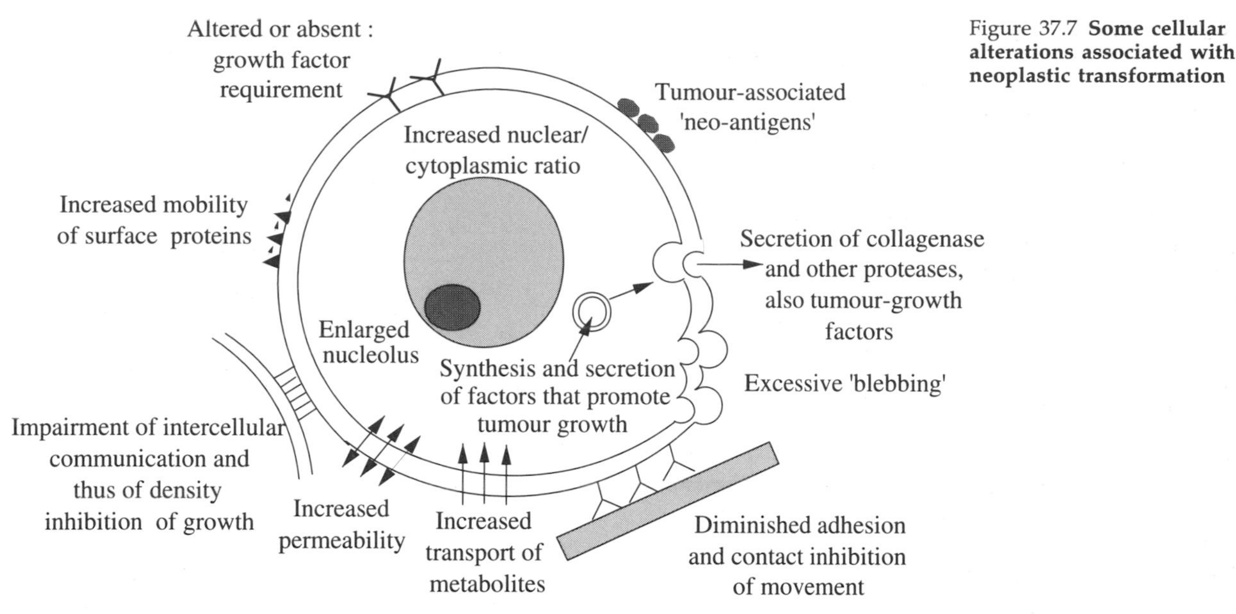

Figure 37.7 **Some cellular alterations associated with neoplastic transformation**

about its environment via receptors in its plasma membrane (or in the cytosol). Interaction of the receptors with the surfaces of other cells, with the molecules making up the matrix with soluble growth factors or with steroid hormones, all result in the increased or decreased activity of intracellular signalling pathways. Crucially, as far as cancer is concerned, changes in the activity of these pathways can affect the rate of transcription of key genes. These genes control such processes as proliferation (mitosis), differentiation and apoptosis. It is precisely these genes that are either directly or indirectly affected by the mutations which, by accumulating, give rise to malignancy. These changes in the growth properties of cells are referred to as **transformation**. Some of the cellular alterations that accompany transformation are shown in Figure 37.7.

The special name that is given to genes, whose protein products cause cellular transformation, is **oncogenes**. Oncogenes can be transforming because the gene itself has undergone mutation so that a malfunctional protein is produced. Many oncogenes, however, encode perfectly normal proteins, but the genes are expressed (transcribed and translated) too much, too little, at the wrong time (wrong developmental stage) or in the wrong place (tissue). Because these genes are part of the normal genotype of the cell and only become transforming when inappropriately activated, they are referred to as **proto-oncogenes** or **cellular oncogenes (*c-onc*)**.

Two major categories of transforming genes are now recognized: dominant transforming genes and recessive transforming genes. The former are genes which, when expressed, give rise to transformation (**gain of function**), while for the latter, transformation follows their inactivation (**loss of function**). It is the former genes that are now referred to as oncogenes. Genes in the second category are termed **tumour-suppressor genes**: their products help to prevent cellular transformation. The terms **dominant** and **recessive** are used here to emphasize that, for the former genes, change of a single allele may suffice for transformation to be initiated, but for the latter, loss of function of both alleles is required. Table 37.1 contrasts the main properties of oncogenes with those of tumour-suppressor genes.

The gradient of unrestrained growth

It is useful to consider cells in tissues as being at rest on a 'gradient of unrestrained growth', as shown in Figure

Table 37.1 **The properties of oncogenes and tumour-suppressor genes**

Property	*Oncogene*	*Tumour-suppressor gene*
1. Number of mutational events required to contribute to the cancer	One	Two
2. Function of mutant allele	Gain of function; evidence of dominance	Loss of function; recessive behaviour
3. Inheritance through the germline	No example has yet been found	Frequently has an inherited form
4. Somatic mutation contributory to cancer	Yes	Yes
5. Tissue specificity of mutational event	To a degree, but may act on a range of tissues	The inherited form frequently shows tissue specificity

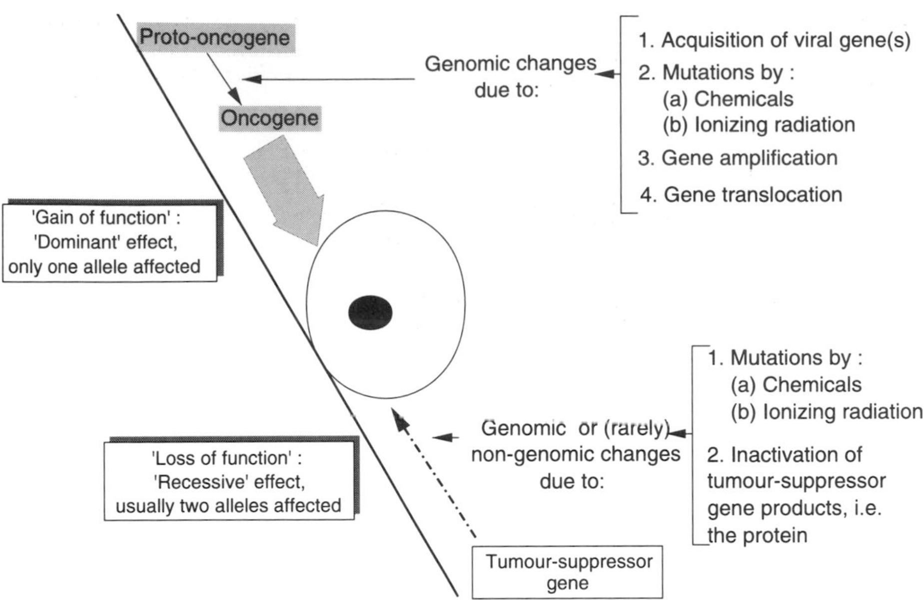

Figure 37.8 **Ways in which the normal balance between the effects of proto-oncogenes and tumour-suppressor genes may be distributed, leading to cellular transformation and cancer**

37.8 in which the cell is held in position as long as the growth-promoting effects of proto-oncogenes are exactly opposed by the restraining influence of tumour-suppressor genes. In the progression to malignancy, the cells 'roll' down the slope as a result of the combined augmentation of a 'push' in the form of oncogene expression and a removal of the constraint resulting from the inactivation of suppressor genes. Figure 37.8 shows how these changes might come about, matters that are further explored in Sections 37.4 and 37.5.

37.4 The nature of oncogenes

The nature of oncogenes was discovered some time before the recessive tumour-suppressor genes. One of the experimental approaches that led to the discovery of oncogenes was surprisingly straightforward. Cells of the mouse fibroblast cell-line 3T3 were forced to take up fragments of DNA extracted from human tumours (the process is referred to as **transfection**). The end-point for this assay was then the unrestrained ability of the host cells to proliferate in culture. Oncogenes were detected as being present in many different human cancer cells by this means. Analysis of these genes showed them to be very similar to cancer-causing genes at that time being found in **retroviruses** (see Chapter 4). The current view is that these oncogenes have been acquired from a host cell during an infection. To indicate that these oncogenes are of viral origin, they are designated *v-onc*. Although they are related to *c-onc* genes, those found in viruses have very frequently undergone mutations. One consequence of this is that the protein they encode is malfunctional. For example, the oncogene found in the Rous sarcoma virus of chickens and designated *v-src* has its cellular counterpart in *c-src*. Analysis has shown that *v-src* has 514 codons which are subject to point mutations in several positions. The corresponding *c-src* has 533 codons which are not commonly found to bear point mutations. Another difference between *v-src* and *c-src* is that the former reflects its mRNA origin and lacks introns (see Chapter 4), whereas the latter is a typical eukaryotic split gene.

The proteins encoded by oncogenes

As indicated, the proteins encoded by oncogenes may be normal cellular proteins produced in inappropriate amounts or mutated versions of such proteins. The normal proteins play central roles in the control of cell proliferation and differentiation. This implies that all of the proteins involved in the signal transducing processes shown in Figure 37.6 are potentially oncogenic.

Figure 37.9 and Table 37.2 show some known oncogenes, the nature and site of operation of their protein products and the tumours with which they have been associated.

Two examples to illustrate how mutations may give rise to cellular transformation

To show how mutations can give rise to transformation, two illustrative examples may be studied: one derives from a mutation to a growth factor receptor and the other involves a GTP-binding protein.

A major group of oncogene products have proven to be related to the catalytic receptors that are protein tyrosine kinases (Chapter 3) responsive to growth factors. An early discovered example of this was the product of the oncogene designated *v-erb* which was found to be a truncated form of the receptor for epidermal growth factor (EGF). The part of the receptor that was deleted in the mutation was the part outside the cell (the binding site for EGF). A consequence of this change was that the catalytic activity of the tyrosine kinase domain inside the cell was continually activated even in the absence of EGF. Cells expressing *v-erb* are constitutively stimulated to proliferate – one step on the path to transformation.

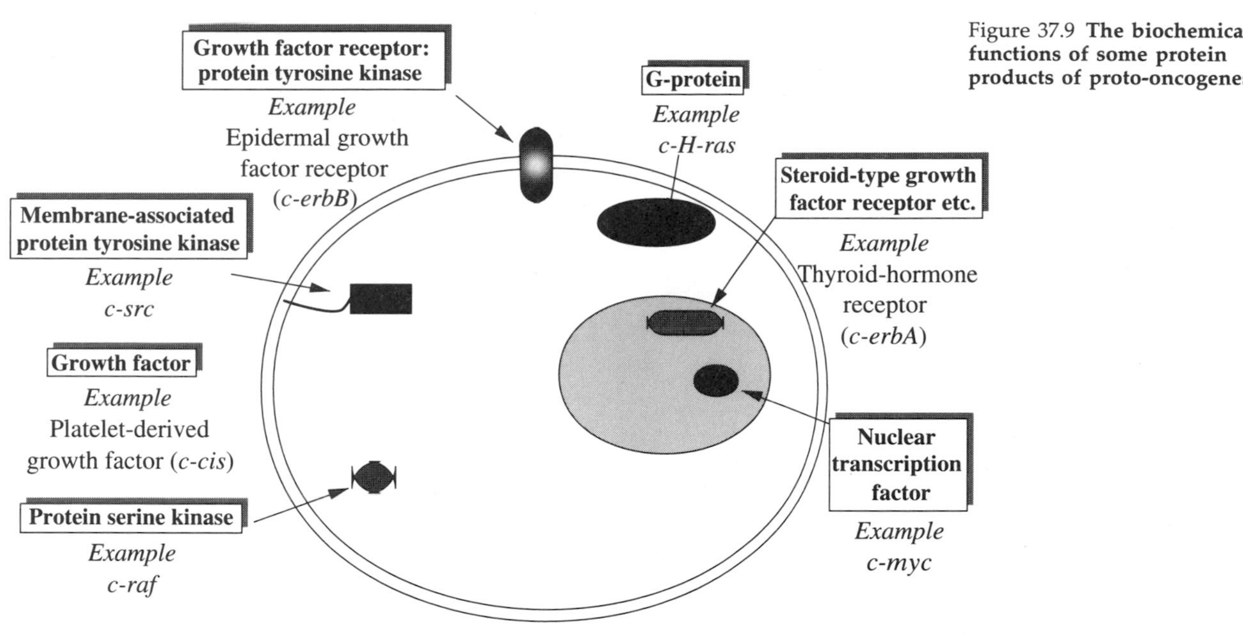

Figure 37.9 **The biochemical functions of some protein products of proto-oncogenes**

Table 37.2 **Some examples of oncogenes**

Category of molecule	*Gene affected*	*Nature of gene-product* (protein)	*Cellular location of protein*	*Tumour association*
1. Growth factor	• *cis*	Platelet-derived growth factor	Extracellular signalling molecule	Sarcomas in monkeys
2. Growth factor receptor	• *erb*-B	Epidermal growth factor receptor (protein tyrosine kinase)	Plasma membrane	Fibrosarcoma in chickens
3. Cytosolic signalling	• *ras*	GTP-binding protein	Cytosol	Sarcomas in rats
	• *src*	Protein tyrosine kinase	Moves between cytosol and (when activated) the plasma membrane	Sarcomas in chickens
4. Transcription factor	• *myc*	Binds to DNA to promote transcription	Nucleus	Sarcomas in chickens

Other oncogenes have been found to encode mutated GTP-binding proteins (G-proteins). G-proteins are intimately involved in the control of the activity of signalling systems of cells and as such are very much concerned with controlling the transduction of extracellular proliferative signals. Oncogenes that have been designated *v-ras* specify such GTP-binding proteins and the mutations have meant that they have lost their ability to catalyse the hydrolysis of GTP. This is an important mechanism for terminating the action of these proteins. Once again the expression of these oncogenes can prompt proliferation and set cells on the way to malignancy as further mutations accumulate.

Inappropriate expression of cellular oncogenes

Inappropriate expression of *c-onc* genes can arise in several ways: following point mutations, as a result of gene-amplification, following gene translocation or by retroviral insertion (Figure 37.10).

In some cases of childhood **neuroblastoma**, an expansion of the number of copies of the oncogene N-*myc* is seen, and the greater the number of copies of the gene the worse is the prognosis. A second example of a tumour-associated gene-expansion is that of the oncogene MDM2 originally found in transformed cells of the mouse *Balb/c* type, but now known to be present in humans.

In **Burkitt's lymphoma**, the cellular oncogene *c-myc* is found to have been translocated from its normal position on chromosome 8 to chromosome 14. When this happens, *c-myc* becomes divorced from its normal regulatory elements and comes under the control of the promoter for the immunoglobulin heavy chain genes. These genes, and hence *c-myc*, are expressed constitutively in affected B-lymphocytes. This leads to inappropriate growth of the cells. When this translocation is coupled with infection with the Epstein–Barr virus the development of a lymphoma becomes that much more likely (see Section 37.6)

The consequences of retroviral insertion for cell transformation are described in Section 37.6.

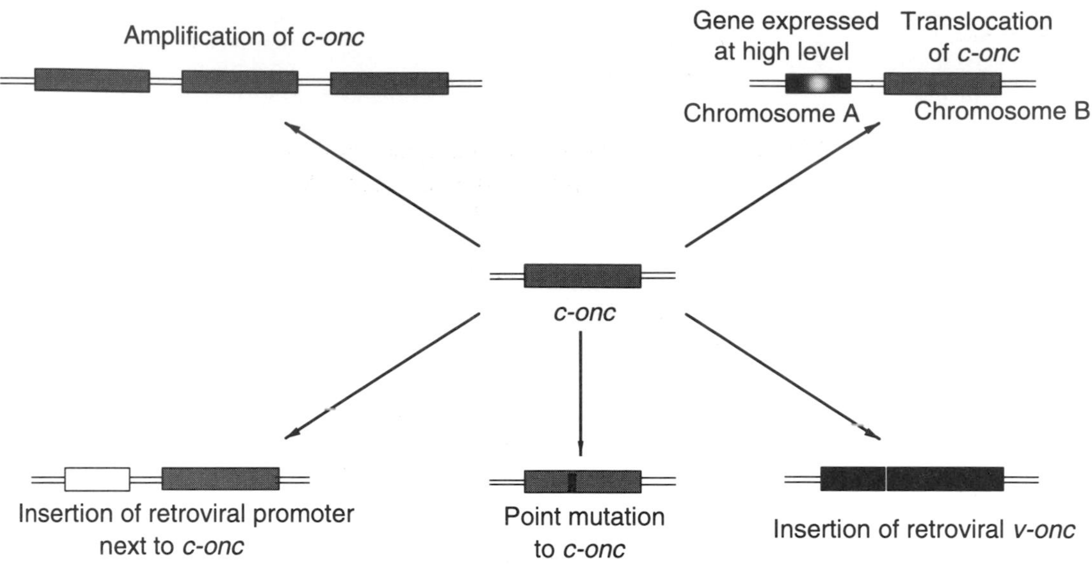

Figure 37.10 **Ways in which the proto-oncogene (*c-onc*) may be converted to an oncogene**

37.5 The nature of tumour-suppressor genes

Cell hybridization studies provided the first evidence for the existence of tumour-suppressor genes. In this work, normal cells were fused with tumour cells and the resultant hybrids were almost always non-tumorigenic. This indicated that the normal cells were bringing something to the partnership capable of suppressing the neoplastic phenotype of the tumour cell. These hybrids frequently had unstable karyotypes and shed chromosomes originating from one or other of the parent cells. If the loss was of the normal chromosomes, then the emerging cells reverted to the tumorigenic state.

Table 37.3 lists some of the established inherited tumour-suppressor genes, their gene products and the tumours in which they are found to malfunction.

The study of retinoblastoma has helped in the understanding of tumour-suppressor genes

The conclusions drawn from studies of cell hybrids *in vitro* were reinforced by observations on human tumours, in particular the rare childhood eye tumour called **retinoblastoma**. Some 20 years ago, Knudson suggested that this disease arises as the result of two lesions of the cell genome. The reason he made this suggestion was associated with the two ways in which the disease presents: in a familial (inherited) form or sporadically (randomly). A common feature of the inherited disease is that it is bilateral, involving both eyes (rarely, it is found to be 'trilateral' and then the pineal gland is also affected). By contrast, in the sporadic form only one eye is involved. Knudson argued, therefore, that for the sporadic disease both of the required lesions arise in somatic cells, affecting the retinal cell lineage and occurring long after conception. The chance of both changes occurring in the cells in both eyes is vanishingly small. Children with the familial form of the disease were viewed as having inherited a mutation to the first allele in the germ-line. For these children, functional inactivation of the second allele would lead to the disease, and bilateral affliction becomes much more likely. This is because a loss of function of the remaining allele in any retinal cell would lead to the disease. Although Knudson proposed that mutations to both alleles occur, it is now appreciated that functional inactivation of the second allele can arise due to non-mutational mechanisms.

Table 37.3 **Tumour-suppressor genes whose inheritance, in mutated form, is associated with human tumours**

Gene	*Inherited tumour*	*Likely function*
Rb	Retinoblastoma, osteosarcoma	Regulates the activity of transcription factors
p53	Osteosarcoma, adrenalcortical, breast, brain	Causes arrest of cells in the cell cycle in G_1 until damaged DNA is repaired – if repair is not possible, causes apoptosis
WT-1	Wilms' tumour of the kidney	A transcription factor
NF-1	Von-Recklinghausen's neurofibromatosis	GTPase activating protein (GAP) acts jointly with the product of the oncogene *ras*
APC	Adenomatous polyposis	Not known
DCC	No associated inherited tumour known, undergoes somatic mutation in the colon (Deleted in Colon Cancer)	A plasma membrane protein involved in cell-matrix adhesions

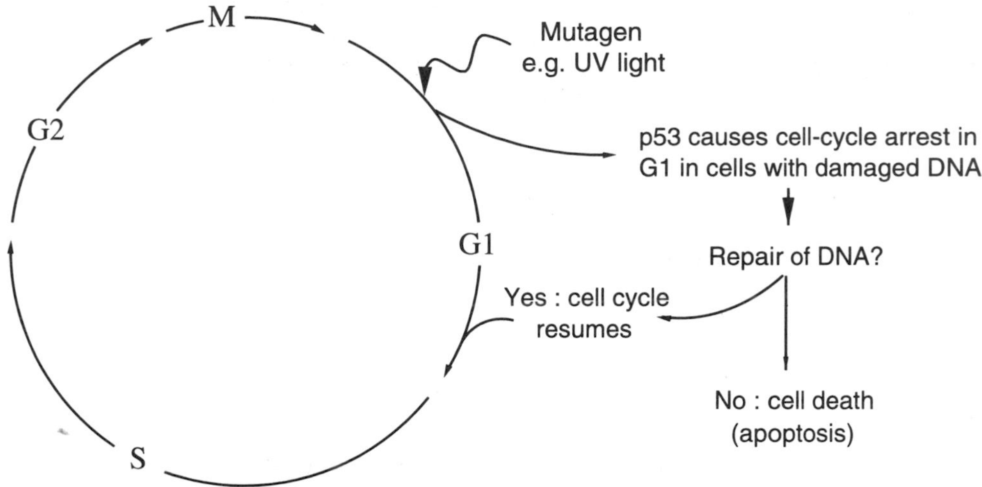

Figure 37.11 **The protein encoded by the tumour-suppressor gene p53 has two roles:** (a) as the 'guardian of the genome' it causes cell cycle arrest so that damaged DNA may be repaired; (b) as the 'guardian of tissues' it promotes apoptosis if repair proves impossible

The gene in question has now been identified and designated Rb. It is located on chromosome 13 in humans and its protein product is a DNA-binding protein.

The tumour-suppressor gene p53 is mutated, or its protein product inactivated, in many human cancers

Attention is currently focused on a particular tumour-suppressor gene known as **p53** (so named because its protein product has a RMM of 53 kD). This gene has been shown to be mutated, or its protein product inactivated, in at least half of all human malignancies.

In most cases the p53 gene is altered so that no protein is produced or the protein is malfunctional. Failure to produce the protein is frequently due to a large-scale deletion of all or part of chromosome 17. A malfunctional protein is the consequence of a point (mis-sense or non-sense) mutation.

In the remaining cases the gene product (i.e. the protein itself) is inactivated by its interaction with other proteins. An example of such an inactivating protein is the human papillomavirus (see Section 37.6) oncoprotein E6 which targets p53 for degradation by the ubiquitin-mediated proteolytic pathway (see Chapter 11).

It has been suggested, with increasingly good evidence, that the physiological role of p53 is to promote the transcription of genes that prevent the replication of damaged DNA (growth inhibitory genes) until repair can be effected; that is, they arrest the cell cycle in G_1 (see Chapter 4). If the damage is so extensive as to render repair impossible, then p53 promotes apoptosis of the affected cell (Figure 37.11). The dual ability of p53 to allow time for DNA repair to occur or to cause apoptosis has led to its being described both as the 'guardian of the genome' and as the 'guardian of the tissues' –

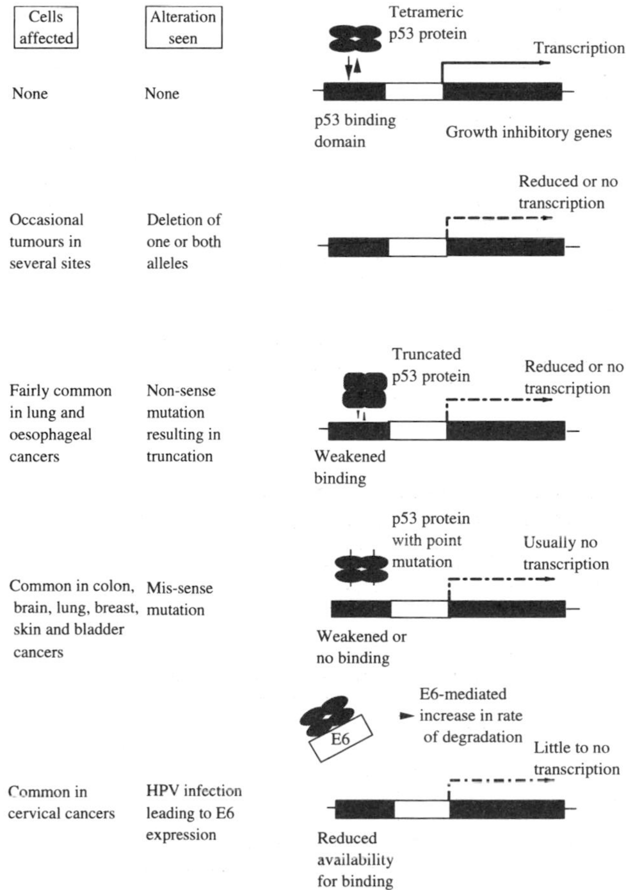

Figure 37.12 **Mechanisms that interfere with the proper functioning of the tumour-suppressor gene product p53**

protecting cells from damaged DNA and the tissues from damaged cells.

Clearly, the loss of p53 function would result in a greater likelihood of daughter cells receiving damaged DNA. Figure 37.12 shows some of the ways in which mutation of the p53 gene, or inactivation of its protein product, may lead to loss of function.

37.6 Viral oncogenesis

It is convenient to deal with environmental factors contributing to the development of cancer under the separate headings of viruses, chemical carcinogens (and also radiation), but in view of the multi-step nature of the disease, mutations brought about by a combination of these may underlie the development of a particular malignancy.

It is appropriate to start by considering the role of viruses in cancer because it has been estimated that 20% of human tumours world-wide are associated with viruses. Virus-associated cancers constitute 24% of all cases in developing countries and only 7% in developed countries. This means that viral infection is second only to tobacco consumption as a risk factor for cancer.

As has already been shown, the study of retroviruses bearing oncogenes capable of stimulating the proliferation of host cells has greatly enhanced our understanding of cancer. In addition, however, DNA viruses can be tumorigenic. Indeed, since transformation almost always requires the incorporation of viral DNA into the cell genome, only DNA viruses and retroviruses have the potential to be oncogenic.

DNA viruses and retroviruses can also act indirectly to foster the emergence of tumours. This latter is the case for the human immunodeficiency retroviruses (HIV-1,-2, etc.) which are immunosuppressive by causing the T-cell death that characterizes the disease acquired immunodeficiency syndrome (AIDS). The immunosuppressed state then favours the development of the previously rare and unrelated Kaposi's sarcoma, possibly associated with infection with a herpes virus (see Chapter 38).

RNA viruses and associated tumours

Although the important part played by the study of retroviruses in understanding the role of oncogenes and proto-oncogenes in cancer has already been referred to, it is appropriate to consider some human cancers that are positively associated with retroviral infections. Table 37.4 summarizes the important characteristics of retroviruses. Their ability to replicate depends on the gene they carry for the enzyme reverse transcriptase (RNA-dependent DNA polymerase; see Chapter 4). Another crucial property is their ability to pass from generation to generation in the germ cells.

It is now established that some, but not all, retroviruses can foster the transformation of cells. This can arise in several distinct ways. Many genomes of transforming retroviruses are delimited by the presence of special sequences termed **long-terminal repeats** (**LTRs**; Figure 37.13). LTRs have two functions: they direct the incorporation of the retroviral DNA, produced by the action of reverse transcriptase, into the host genome and once inserted they can act as promoters for transcription. A retrovirus that includes an oncogene (*v-onc*) between these LTRs will be transforming, because initiation of transcription by this repeat sequence will lead to the production of an oncoprotein. The time between infection and onset of tumour formation is short (weeks). This is the case with the Rous sarcoma virus of chickens, which carries the oncogene *v-src*.

Table 37.4 **Characteristics of retroviruses**

1. The genetic material is in the form of single-stranded (sense) RNA
2. The 5′- and 3′-ends of the genome are delimited by 'long terminal repeats' (LTRs), which act as promoters of transcription
3. They have a gene that encodes reverse transcriptase which can catalyse the formation of a double-stranded DNA (dsDNA) copy of the viral genome
4. The LTRs can direct the (random) insertion of the dsDNA copy into the host genome whereby the virus may persist in the host (endogenous virus)
5. Endogenous virus can be transmitted from generation to generation
6. They may 'swap' genes with other retroviruses
7. Endogenous retroviral insertion into the host genome may position a LTR next to a host gene, which may thereby be inappropriately expressed, as the LTR acts as a promoter. If the gene is a proto-oncogene (*c-onc*) this may lead to cellular transformation
8. Retroviruses may themselves carry viral oncogenes (*v-onc*), the expression of which may also be transforming. These oncogenes are probably acquired from host genomes, becoming subsequently mutated

A second route to retroviral transformation arises when a virion with LTRs is incorporated into the host genome next to a proto-oncogene (*c-onc*). Inappropriate transcription of the proto-oncogene promoted by the LTR can then lead to transformation. The time to tumour onset is in the intermediate range (months). The avian leukaemia virus brings a LTR next to the proto-oncogene *c-myc* in B-cells when the gene product (a DNA-binding protein) promotes a B-cell lymphoma. A third means of retroviral transformation is less direct. In this, the virus brings to the host cell a gene which, following its transcription and translation, directs the production of a protein that acts elsewhere in the host genome to promote inappropriate transcription of proto-oncogenes, thus rendering them oncogenic (this is referred to as 'action in trans'). In these cases a long time (years) is required before the appearance of a tumour. For example, the human T-cell lymphotrophic virus, type 1 (HLTV-1) carries a gene *tax* that encodes a protein which acts to increase the rate of transcription of host T-cell genes for interleukin 2 (IL-2) and its receptor. The normal role of Il-2 is to bind to its receptor and act in a controlled autocrine fashion to promote proliferation of T-cells (see Chapter 32). In 1977, the disease adult T-cell leukaemia was described in Japan and by 1980 it was shown to be associated with infection by HLTV-1. It has been estimated that as many as 20% of the population of South Japan carry the virus. The annual incidence of the disease in seropositive males is 1 in 500. It is highly probable that a cofactor is required for the disease to

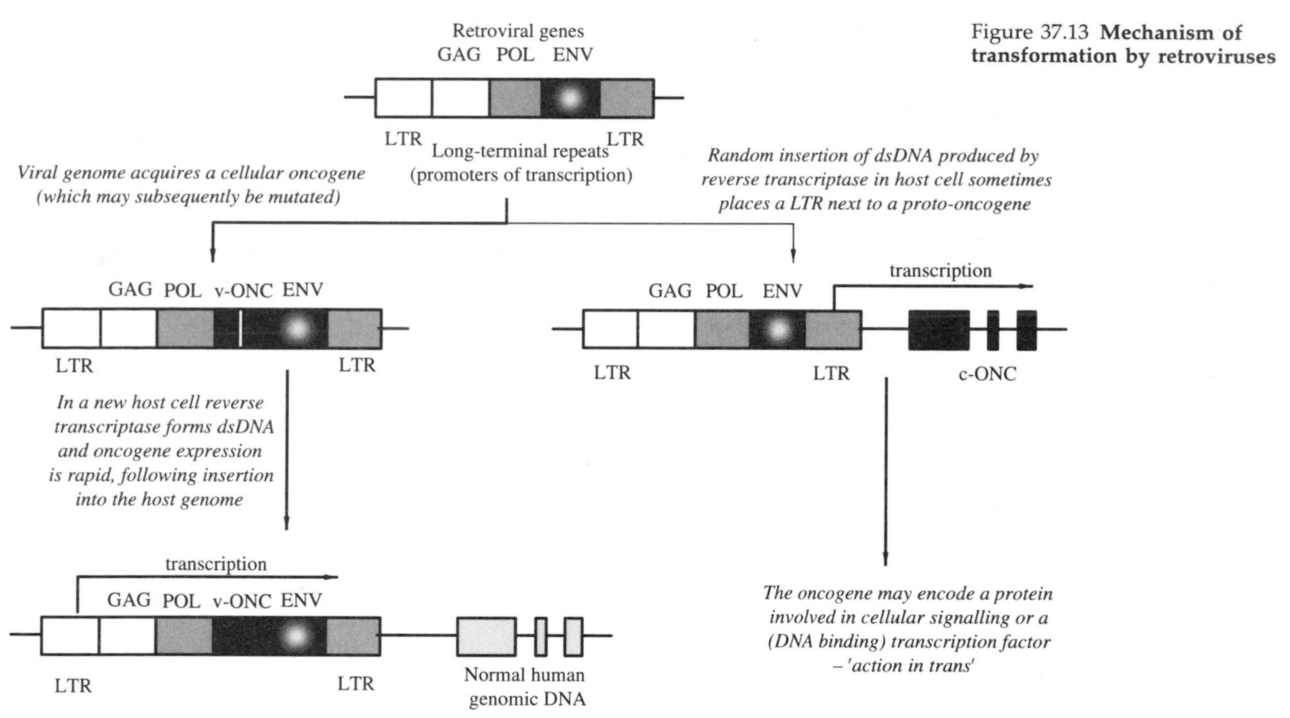

Figure 37.13 **Mechanism of transformation by retroviruses**

emerge, but this has yet to be identified. More recently, HLTV-1 has also been found to be associated with tropical spastic paraparesis (also known as HLTV-1 associated myelopathy), although this is not a cancer. This disease is prevalent in Afro-Caribbean populations including those living in the UK.

DNA viruses and associated tumours

Retroviruses are not the only type of virus that can contribute to the development of human tumours. Several groups of DNA tumour viruses have been recognized. In contrast to the retroviruses, however, DNA tumour viruses seem neither to carry viral equivalent of cellular genes (*v-onc*) nor the LTRs. Instead their genomes encode proteins that are able to interact with, and thereby inactivate, nuclear cell-cycle regulatory proteins. This may lead to transformation, but usually only in the presence of some other cofactor(s).

Human papillomaviruses (HPV) are double-stranded, DNA-containing viruses that infect and replicate in the squamous epithelium. This results in benign skin tumours such as papillomas and warts. In the presence of other cofactors, however, malignancies can arise. In people with HPV-induced epidermodysplasia verruciformis (multiple warts or scaly macules), exposure to strong sunlight (as the cofactor) can give rise to squamous cell carcinoma (SCC). HPV can also give rise to SCC in immunocompromised patients. HPV-16, which is associated with carcinoma of the cervix, has been found to carry two genes (E6 and E7) that respectively encode proteins that target the proteins p53 and RB (tumour-suppressor gene products) for catabolism (see Section 37.5).

Another DNA virus, the Epstein–Barr virus, is able to infect epithelial and lymphoid cells and in so doing can cause glandular fever. But in some parts of the world in which malaria is an endemic cofactor, for example the region west of the Nile in Africa and in Papua New Guinea, infected patients may develop Burkitt's lymphoma. The virus is not believed to cause the chromosomal translocation that is found in the disease (see Section 37.4), but the presence of the virus *and* the translocation make transformation by other cofactors more certain. The annual incidence of Burkitt's lymphoma in Uganda, for example, is about 75 per million population. In other parts of the world, predominantly in Southern China, Alaska and Singapore, the same virus interacts with chemical carcinogens believed to derive from salted food (probably nitroamines, see Section 37.7) as the cofactor, to give rise to nasopharyngeal carcinoma. The annual incidence of nasopharyngeal carcinoma in Hong Kong is 470 per million population.

A third type of DNA virus known to be associated with cancer is the hepatitis-B virus. Probably by acting together with an aflatoxin derived from contaminated food (see Section 37.7) as a cofactor, this virus helps to induce liver cancer, especially in the populations of tropical Africa and Southeast Asia.

Table 37.5 lists some of the viruses known to cause cancer in humans, the tumours they cause, the most susceptible populations and the possible cofactors involved.

37.7 Chemical carcinogenesis

Occupational carcinogens

The London physician Sir Percival Pott, working at St Bartholomew's Hospital, provided the first evidence that

Table 37.5 **Viruses and human tumours**

Virus	*Characteristics*	*Tumour*	*Areas of high incidence*	*Possible cofactors*
RNA VIRUSES				
HTLV1	Retrovirus: single-stranded (sense) RNA	Adult T-cell leukaemia	South Japan	?
Hepatitis C	Togavirus (?): single-stranded (sense) RNA	Primary liver cell cancer	Tropical Africa	Aflatoxin Alcohol
DNA VIRUSES				
Human papillomaviruses	Papovavirus: double-stranded circular DNA	1. Carcinoma of the cervix (HPV-16) 2. Juvenile laryngeal papillomas 3. Squamous cell carcinoma (e.g. HPV-5)	Global	Smoking, activation of *ras* oncogene X-rays, strong sunlight Immunocompromise
Epstein–Barr	Herpesvirus: double-stranded linear DNA	1. Burkitt's lymphoma 2. Nasopharyngeal carcinoma	Africa west of the Nile, Papua New Guinea Southern China, Alaska, Singapore, East Africa	Malaria Salted fish (nitrosamines; SE Asia), salted nuts (E. Africa), genetic
Hepatitis B	Hepadnavirus: double-stranded DNA (RNA intermediate)	Primary liver cell cancer	Taiwan, Tropical Africa	Smoking, alcohol, aflatoxin, genetic

Table 37.6 **Established occupational causes of cancer**

Chemical agent	*Site of cancer development*	*Occupation of workers*
Aromatic amines (e.g. 2-naphthylamine)	Bladder	Manufacturers of dyes, rubber products and town (coal) gas
Benzene	Marrow – erythroleukaemia	Work involving the manufacture or use of glues and varnishes
Polycyclic hydrocarbons (arise during incomplete combustion of fossil fuels)	Lungs and skin	Workers with asphalt
Vinylchloride	Liver – angiosarcoma	Plastics manufacturers
Nitrosamines	Several sites including liver	Preparers of cured (salted) food, leather workers (tanners) and manufacturers of herbicides

an environmental chemical could cause cancer when, in 1775, he drew attention to the high incidence of cancer of the scrotum among London chimney-sweeps. These young boys often had to climb up inside large chimneys to sweep all the soot down. Pott correctly associated the soot with the cancer. At the time it was not possible to visualize how such an apparently inert material as soot could cause such serious damage. Indeed, confirmation of Potts' hypothesis was not forthcoming for a further 140 years when, in 1915, the Japanese workers Katsusaburo Yamagiwa and Koichi Ichikawa found that the repeated painting of coal tar extract onto the ears of rabbits caused tumours of the skin. This work was extended by E.L. Kennaway and I. Heiger working at the Chester Beatty Research Institute in London in 1929. They isolated 1,2,5,6-dibenzanthracene from coal tar, thus obtaining the first pure chemical carcinogen.

By the end of the nineteenth century, other workers were beginning to describe occupational cancers. Ludwig Rehn in Germany noticed the occurrence of bladder cancers among workers in the aniline dye industry. This observation was placed on a firmer footing in 1937 when Wilhelm Hueper and his colleagues at the United States National Cancer Institute reported the occurrence of bladder cancers in dogs given the aromatic amine 2-naphthylamine.

Mesotheliomas of the pleural cavity among shipyard workers in the USA and hepatic angio-sarcomas among plastic factory workers in the USA and Italy were shown to be due respectively to exposure to asbestos and vinylchloride. P.N. Magee then showed that the industrially-used *N*-dimethylnitrosamine is carcinogenic.

Table 37.6 lists some of the established occupational factors contributing to cancer.

Environmental carcinogens

Chance findings in the UK and the USA led the toxicologist Gerald Wogan and the chemist George Buchi to discover the first environmental carcinogen. They followed up the observed occurrences of hepatic necrosis in turkeys and hepatoma in rainbow trout. In both cases the feed given to the animals had included spoiled

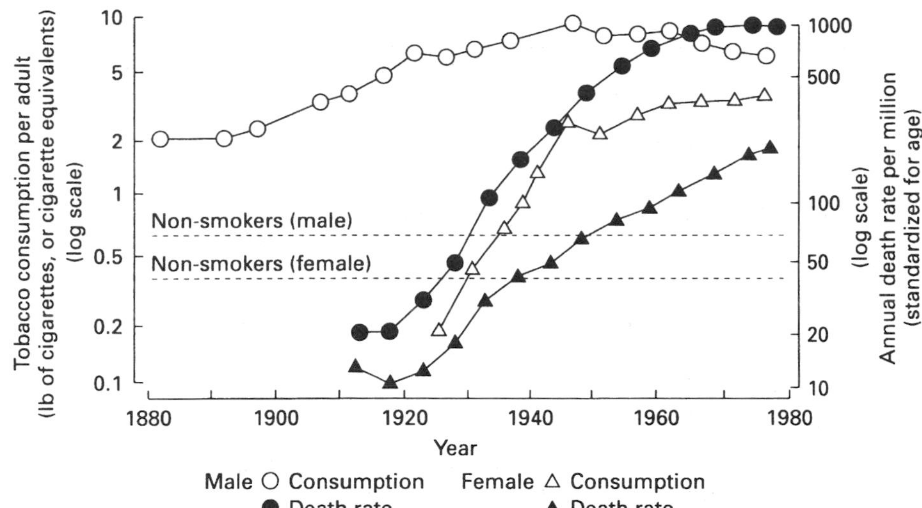

Figure 37.14 **Trends in male and female age-standardized lung cancer mortality rates in England and Wales, 1916–70 and cigarette sales per adult.** The dotted lines are lung cancer rates found in lifelong non-smokers in cohort studies conducted in the 1950s and 1960s – note that the observed national rates early this century were lower than these rates and these differences in all likelihood are a reasonable measure of the improvements in diagnostic accuracy that have been achieved over the years

peanuts that had become contaminated with the organism *Aspergillus flavus*. Wogan and Buchi were able to isolate and characterize four related molecules, they termed aflatoxins, and show these to be carcinogenic. Two of the molecules showed a blue fluorescence and were termed aflatoxins B_1 and B_2, while two with green fluorescence were designated G_1 and G_2.

The Turkish scientist A.N. Pamukcu was the first to associate the eating of bracken fern with bladder cancer in cows. The active carcinogen in this case has proved to be aquilide A.

Smoking and cancer

Since chimney sweeping had become better organized and forms of heating other than coal fires came into vogue, the carcinogenicity of polycyclic hydrocarbons uncovered by Pott, Yamagiwa and Ichikawa, Kennaway and Heiger tended to be of restricted interest until 1953 when Doll published the results of an intensive epidemiological study, which demonstrated quite clearly for the first time that cigarette smoking is a prime cause of lung cancer (Figure 37.14). Careful analysis of the smoke and tar obtained from cigarettes showed them to contain many carcinogenic hydrocarbons, the most important including benzo[a]pyrene and several aromatic amines. Thus, for the first time, a close link was established between experimentally proven cancer in animals and a very common environmental hazard. Consequently, the medical world (if not initially their patients) began to take the role of cigarette smoking in carcinogenesis very seriously. Indeed it is now appreciated that tobacco consumption is the major risk factor for cancer worldwide. Nor is lung cancer the only disease to which cigarette smokers make themselves especially vulnerable (see Chapter 33).

Cancer and diet

Many of the trace environmental carcinogens to which we are all exposed gain access to the body via the gastrointestinal tract. Exposure to some of these may be avoided; for example, the pyrolysis products formed during the barbecuing of meat have been shown to be potent carcinogens when administered to animals, but in general exposure to some trace carcinogens is inevitable. However, we can exercise some control over the composition of the food we eat. There is now abundant evidence that certain dietary constituents (such as fats) are promoters of some cancers, whilst others (β-carotene, vitamins C and E, the trace element selenium, eicosapentaenoic acid and also certain compounds found in the cruciferous vegetables broccoli, cabbage and cauliflower) are protective, at least in some forms of cancer.

Colorectal cancer

Epidemiological findings appear to support the view that the consumption of high-fat diets predisposes to the development of colorectal cancer and that the presence of dietary fibre is protective. The actual causative agent in promoting this disease may be bile salts, or their metabolites created by the microflora in the gut (see Chapter 15). It would be expected that large quantities of fat in food would result in the secretion of corresponding large amounts of bile salts for digestion and absorption.

One of the best documented cases of dietary protection is that of fibre, which the work of Denis Burkitt in Africa showed to be protective against a range of alimentary tract diseases including large-bowel cancer.

Breast cancer

For a long time during the 1980s it was held that high-fat diets consumed during adulthood greatly increase the risk of breast cancer in women. This seemed to be an important finding, suggesting that the incidence of the disease in those parts of the world in which breast cancer is a major killer (in the USA, the lifetime risk to women of developing the disease is 1 in 8 and appears to be rising) might be reduced by dietary means. However, the

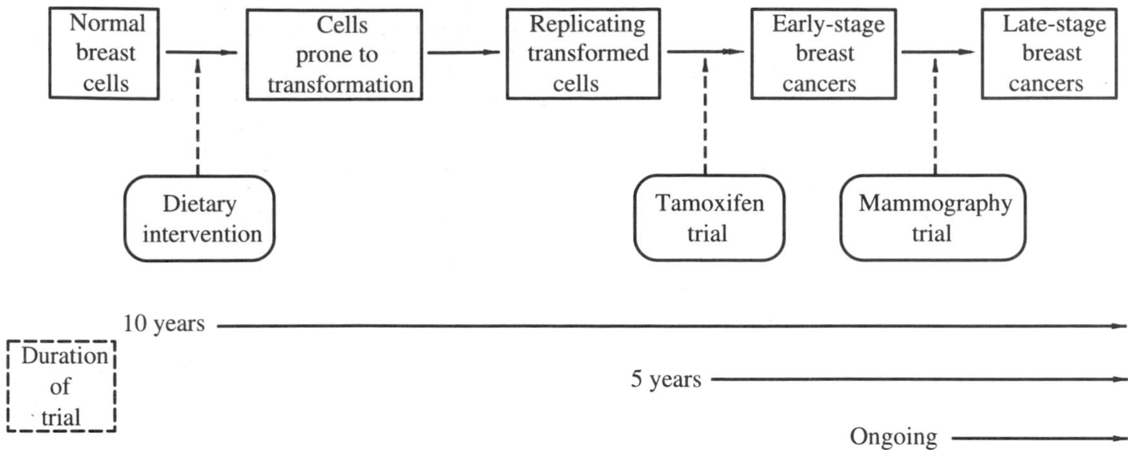

Figure 37.15 **Some current approaches to the prevention of breast cancer**

results of one of the largest and best controlled studies on this and other topics, known as the Harvard Nurses Health Study (for which 87 254 nurses have been studied from 1980), failed to produce any conclusive evidence concerning dietary fat and breast cancer for North American women. (The study is the more convincing because it *did* clearly show the relationship of dietary fat to colorectal cancer.) In this study it was found that women on extremely high-fat diets (50% energy intake as fat) are at little or no greater risk than those that adhere to an extremely lean diet (less than 29% fat). These findings have tended to refocus attention on oestrogen exposure as a risk factor. Thus in North America women reach menarche on average at age 12.8, whereas the average for women living in China is 18.0. It is well known that the incidence of breast cancer is much higher in the former than the latter women. (There were 16 times more deaths from breast cancer in US women than in Chinese women in 1989.) So perhaps the shrinking of the duration of childhood in North American women (due to improved nutrition?) has served to increase their exposure to the potentially tumour-promoting effects of oestrogens. Such ideas must be subject to much closer scrutiny before any firm conclusions are reached.

Prevention of breast cancer

Figure 37.15 outlines the strategy of three current cancer-prevention trials directed at breast cancer. They target three different stages of the progression of the disease. The first involves the institution of a diet that is very low in fat and high in fruits and vegetables (despite the discouraging results of the Harvard Nurses Health Trial, diets rich in vitamin C may well prove effective). This targets the early stage of the disease. A second approach is aimed at preventing the progression of the disease. This is believed to be oestrogen-dependent, so the women in the trial are taking tamoxifen which competes with oestrogens for their receptors. (Although this appears to be a scientifically sound approach the trial of tamoxifen was recently suspended for reasons of suspected maladministration.) The third initiative is to remove or kill recognizable cancerous lesions very early, before they cause clinical symptoms. For this purpose mammography has been carried out for some time for older women in the UK and has recently been instituted in the USA.

Metabolic activation of chemical carcinogens

The demonstration by Kennaway and Heiger that benzanthracene is carcinogenic led to the synthesis of a whole family of aromatic hydrocarbons, notably by Cook in Scotland and Fieser in the USA. The idea was that such molecules could be tested so that the structure necessary to cause cancer could be determined. In fact, these studies all proved highly inconclusive and there matters rested until the 1960s when it was shown, by Sims among others, that the compounds were not themselves carcinogens but their metabolites were. In particular, the metabolic activation of polycyclic hydrocarbons by a liver microsomal preparation in the presence of NADPH and oxygen was found to occur and the metabolites proved to bind strongly and covalently to DNA and proteins (Table 37.7). These studies showed that, although non-carcinogenic hydrocarbons such as phenanthrene were metabolised so that they reacted with proteins they reacted only poorly with DNA. Known carcinogens such as benzo[a]pyrene reacted equally well with protein and DNA.

The fact that carcinogens do react covalently with both proteins and DNA has opened up another window on carcinogen exposure with the application of **'molecular dosimetry'**. By measuring the amount of protein- (usually haemoglobin) or DNA-carcinogen adduct, an individual's exposure to the carcinogen may be assessed.

In a short time it was shown that the metabolic activation of polycyclic hydrocarbons prior to adduct formation was carried out by enzymes of the endoplasmic reticulum that are very similar to those described in Chapter 36 as being involved in the metabolism of drugs and other foreign compounds. In

Table 37.7 **Binding of metabolites of polycyclic hydrocarbons to DNA and protein**

Polycyclic hydrocarbon	*Binding to protein* (nmol/mol)		*Binding to DNA* (nmol/mol)	
	Before metabolism	*After metabolism*	*Before metabolism*	*After metabolism*
Benzo[a]pyrene	0	0.78	0	1.41
Methylcholanthrene	0	0.73	0	0.78
Phenanthrene	0	0.72	0	0.05
Dibenzanthracene	0	0.95	0	0.44

particular, an NADPH/O_2-requiring member of the cytochrome P450 family was shown to convert the hydrocarbon into a highly-reactive free-radical intermediate (see Chapter 33). This derivative is reactive enough to carry out an electrophilic attack on the bases of DNA. Binding of the electrophile to DNA subsequently causes anomalous base-pairing during replication e.g. thymine will pair with guanine alkylated on O-6 and N-7. Such base-pairing will give rise to the incorporation of incorrect nucleoside phosphates into DNA and thus cause mutations, if not repaired.

Many other carcinogens also undergo activation by oxidative metabolism. Vinylchloride, for example, forms an epoxide, and dimethylnitrosamine is oxidised to release the highly-reactive methyl radical CH_3^+. Chemicals that require prior activation before becoming carcinogens are referred to as **procarcinogens**.

Induction of the enzyme systems involved in the oxidative metabolism of carcinogens

In Chapter 36, the ability of phenobarbitone to induce the enzymes responsible for the metabolism of foreign compounds is described. Several carcinogenic hydrocarbons also have this ability, which extends to their own metabolism. This means that such molecules have the sinister property of facilitating their own conversion into carcinogens. This inducibility appears to be under genetic control, which has the implication that some individuals are more susceptible to carcinogens because their genetic make-up has resulted in enzyme systems that are more readily induced. Support for this concept has come from the study of leukocytes from two groups of cancer patients and control subjects. The inducibility of the enzyme metabolizing benzo[a]pyrene was then studied by incubating the cells with 3-methylcholanthrene (a polycyclic hydrocarbon with high ability to induce the metabolizing enzymes) and the inducibility was then rated as low, medium or high. The degree of induction in the cells from lung cancer patients was much greater than in the other two groups (Table 37.8).

Polymorphisms in enzymes that metabolize procarcinogens and carcinogens may help to determine individual susceptibility to cancer

The above findings might provide a partial answer to the question why some smokers succumb to lung cancer in middle age, whereas others survive into old age without signs of the disease. Perhaps those that contract lung cancer early possess the pattern of effective induction of the enzymes that handle the polycyclic hydrocarbons produced from tobacco.

This polymorphism in the inducibility of key enzymes is not the only way in which individual susceptibility to cancer may be determined genetically. The type of enzyme present, rather than its inducibility, is also subject to individual variation (see Chapter 35). Thus, the presence of the null allele for glutathione *S*-transferase I (no enzyme detectable) is highly significantly correlated with adenocarcinoma of the lung among smokers. This finding serves to emphasize that the presence of enzymes (such as glutathione *S*-transferase I) that catalyse the inactivation of carcinogens (phase II metabolic reactions; see Chapter 36) are just as important as those that activate procarcinogens as determinants of ability of chemicals to transform cells.

Examination of the type of mutation occurring to the tumour-suppressor gene p53 may assist in the identification of the causative carcinogen

As has been indicated previously, the tumour suppressor gene p53 has been found to have undergone mutation in about half of human tumours studied. It has been shown that carcinogen-induced cancer in different tissues is

Table 37.8 **Inducibility of leukocyte polycyclic hydrocarbon hydroxylase and incidence of lung tumours in human subjects**

No. of subjects studied	*Leukocyte donor*	*Percentage of donors in whom leukocyte hydroxylase inducibility was:*		
		Low	*Intermediate*	*High*
83	Healthy controls	44.7	45.9	9.4
46	Patients with other tumours	33.5	45.6	10.9
50	Patients with lung cancer	4.0	66.0	30.0

associated with different types of mutation of the p53 gene. Thus exposure to dietary aflatoxin B_1 is associated in hepatocellular carcinomas with a G : T transversion mutation to p53, whilst exposure to ultraviolet light is associated in skin carcinomas with the gene undergoing C : T transition mutations. One transversion found in aflatoxin-induced hepatocellular carcinoma results in the appearance of a seryl residue in a hydrophobic region of the p53 protein which disrupts its ability to bind to DNA.

Thus, in the future, the environmental causes of each cancer may be inferred by analysis of the precise nature of the mutation present.

37.8 The progression to cancer

Initiation and promotion

When carcinogenic chemicals were first tested on the skins of mice, it was often observed that they were ineffective or only weakly effective unless applied together with certain seed oils, e.g. croton oil. The reason for this was clarified by Peyton Rous, Isaac Berenblum and Philipe Shubik. They showed the oil to contain substances which, although they were not themselves carcinogenic, nevertheless acted to facilitate carcinogenesis. They termed these co-carcinogens, but now they are referred to as **promoters**. The primary carcinogen is called the **initiator** and it has been found that the sequence of treatment is important (Figure 37.16), since application of the promoter after the initiator will always induce tumours, whereas when the promoter precedes initiator no tumours develop. Some initiators can also act as promoters as well and are referred to as **complete carcinogens**, while promoter-dependent initiators are called **incomplete carcinogens**. Multiple applications of complete carcinogens or a single application of such a chemical followed by a promoter hastens the development of the cancer. This last is true only if the application of the promoter occurs sufficiently soon after the carcinogen (before DNA repair is effected?).

Both natural and synthetic chemicals may act as promoters

Although first described as acting to promote skin cancers, it is now appreciated that promotion occurs in many cancers. A wide range of promoters have been characterized. Cyclamates, saccharin and even the amino acid tryptophan can act as effective promoters for tumours of the bladder (Table 37.9). Metabolites of tyrosine, e.g. phenol and *p*-cresol and of tryptophan, e.g. indole and its acetate ester, can act as promoters in the gastrointestinal tract. It has also been suggested that the carcinogenic effect of asbestos in the lungs is due to its ability to promote rather than to initiate. This view is supported by the finding that the incidence of cancer in asbestos workers is much greater when they are smokers and therefore exposed to initiating polycyclic hydrocarbons (Table 37. 10).

Promoters may activate second-messenger signalling systems inappropriately

The mode of action of some promoters is now being unravelled. The croton oil-derived promoter has been shown to be a phorbol ester which directly activates the intracellular signalling pathway involving protein kinase C (see Chapter 3) as do the promoters teleocidin A and B (produced by blue/green algae). Diacylglycerols are physiological activators of protein kinase C and both phorbol esters and the teleocidins mimic this effect, but in an unregulated fashion. Another promoter known to activate an intracellular signalling process is thapsigargin which causes the release of intracellular stores of Ca^{2+}. It may well be that mobilization of Ca^{2+} stores is a common feature of all these promoters, as the

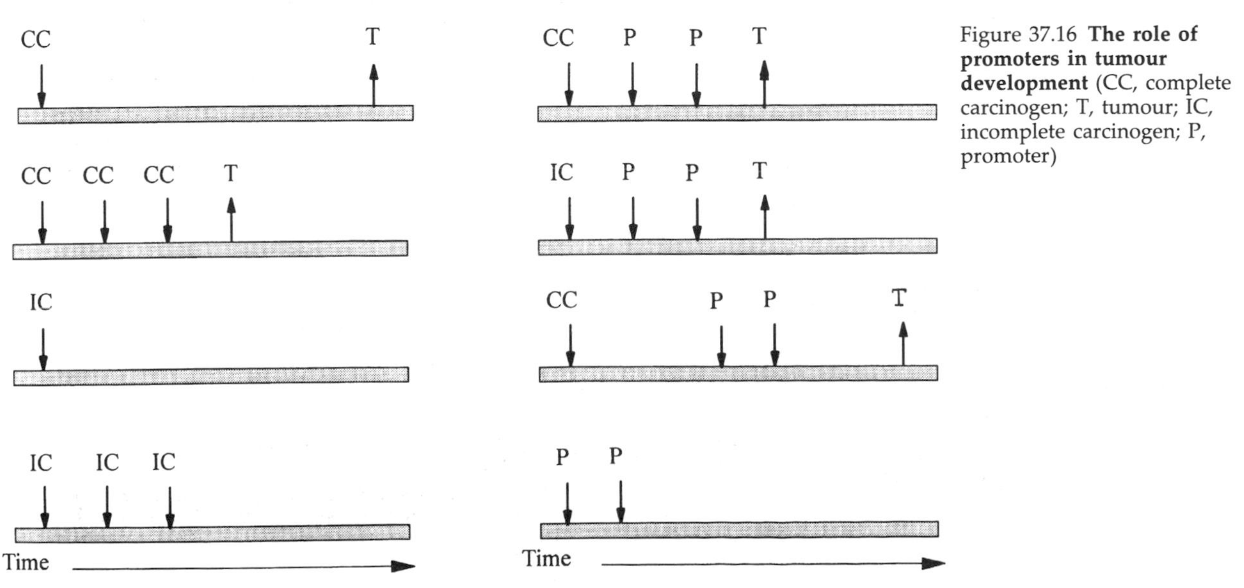

Figure 37.16 **The role of promoters in tumour development** (CC, complete carcinogen; T, tumour; IC, incomplete carcinogen; P, promoter)

Table 37.9 **Promotion of bladder tumours in rats by dietary saccharin, cyclamates or tryptophan**

Initiator	*Promoter*	*Tumour incidence (%)*
N-methyl-*N*-nitrosourea (single injection)	Dietary cyclamate (5%)	
+	–	0
–	+	2
+	+	52
N-methyl-*N*-nitrosourea (single injection)	Dietary saccharin (5%)	
+	–	0
–	+	1
+	+	58
FANFT* (2% in diet)	Dietary tryptophan (2%)	
+	–	20
–	+	0
+	+	53

*FANTF = *N*-[4-95-nitro-2-furyl)-2-thiazolyl]formamide.

Table 37.10 **Effect of exposure to asbestos and smoking on human lung cancer incidence**

Exposed to asbestos	*Smokers*	*Cancer incidence (%)*
–	–	1
+	–	5
+	+	53

ability of teleocidin to promote the carcinogenic effect of dimethylbenzanthracene is prevented by drugs that interfere with the binding of Ca^{2+} to its effector molecule calmodulin (see Chapter 3).

Fortunately a series of compounds are now being identified as acting as **antipromoters**. These molecules act to prevent the progression from initiation to malignancy. Included in the list of such compounds are the dietary components previously described as protecting against cancer: vitamins A, C and E, selenium and the cruciferous vegetables.

The prevention of progression to malignancy

Figure 37.17 shows the ways in which tumours may be initiated by carcinogens due to generation of DNA in a mutated but repressed state. Promoters are then believed to cause inappropriate expression of previously mutated genes, leading to the early stages of tumorigenesis and finally to full malignancy. The figure also shows how the progression to malignancy may be prevented at each stage. Initiation can be interfered with by detoxification of the carcinogen (or procarcinogen) or by repair of the damage to the DNA. Antipromoters then prevent the expression of mutated genes caused by promoters. Although Figure 37.17 shows initiation by carcinogen-mediated mutation, the mutated gene might equally be

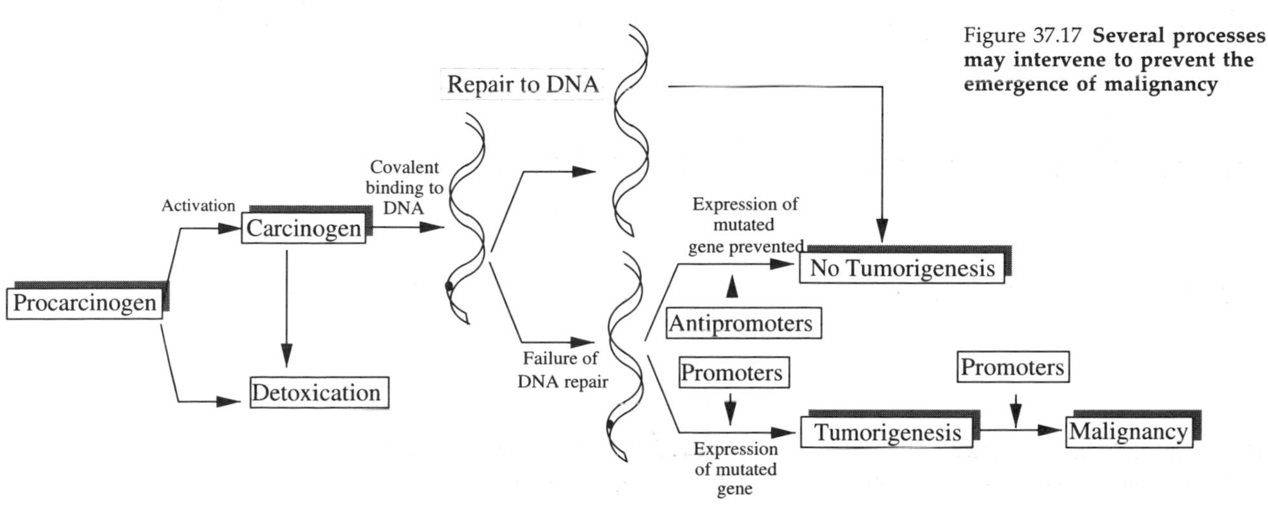

Figure 37.17 **Several processes may intervene to prevent the emergence of malignancy**

acquired from a virus. Likewise those oncogenes that encode proteins concerned with the mobilization of intracellular Ca^{2+} could activate the promotion pathways.

37.9 Testing for mutagens: the Ames test

It has been estimated that 125 000 chemicals are in daily use in agriculture, consumer products and industry within the European Union, and new chemicals are being introduced at the rate of 1000 per annum. Many times the latter number are actually synthesized each year with a view to their possible use. It would be a virtually impossible task to test all such compounds for carcinogenicity in experimental animals, even if were to be considered acceptable ethically. What is required is a simple screening procedure for mutagenicity, so that the development of frank carcinogens can be aborted at an early stage. Such a test was introduced by B.N. Ames in the early 1970s.

The Ames test is basically a **bacterial reversion** assay. For this, 'tester' strains of *Salmonella typhimurium* have been developed that are auxotrophic for the amino acid histidine (*his*⁻); that is, the cells have lost the ability to synthesize histidine as the result of a mutation to a gene on the biosynthetic pathway. Back mutation induced by mutagenic chemicals under test generates bacteria once again capable of growing and forming colonies without a histidine supplement in the culture agar (reversion). The strains currently in use show a high level of chemical mutagen-induced reversion with a low level of spontaneous reversion.

It was understood at an early stage that the assay would not detect procarcinogens requiring metabolic activation to become mutagenic. To deal with this, the Ames assay not only includes tester bacteria but also an activation system. The most common approach is to use a liver homogenate prepared from rats previously treated with the potent inducer of cytochrome P450-dependent drug-metabolizing enzymes, Aroclor 1254.

Having found the Ames test to be negative for a particular compound does not remove the necessity for testing it in experimental animals because false negatives may arise and also because it will not serve to detect promoters or molecules that act in a different way; for example polychlorinated biphenyls (PCBs) are said to mimic unregulated exposure to oestrogen in their carcinogenic actions.

37.10 The treatment of cancer

The basis of the chemotherapeutic approach to cancer is dealt with in Chapter 39, but it is useful here to look at newer methods of trying to achieve the selective death of cancer cells. There is particular interest in the development of new treatments for cancer because, whereas chemotherapeutic drugs can cure certain leukaemias and lymphomas and assist in the prolongation of the lives of patients with breast and ovarian cancers, chemotherapy is not a cure for the common forms of cancer. In developed countries, these are cancer of the lungs, breast, prostate and the colon/rectum.

Recombinant toxins

One new approach involves the use of recombinant toxins. The idea is to target a natural, usually bacterial or plant, cytotoxin to the cancer cell. To this end the toxin is attached to an antibody or a growth factor that preferentially binds to cancer cells. For example each cell in squamous carcinomas expresses 3×10^6 receptors for epidermal growth factor on its surface, and this exceeds by a factor of 10 the number of normal cells that express the receptor. Differentiation antigens (which can be recognized by antibodies from different species) are often present both on normal B-lymphocytes and on B-cell lymphomas, but not on the stem cells that produce B-cells (see Chapter 32). Thus administration of toxin-bearing antibodies targeted on the differentiation antigens will result in the death of both normal mature B-cells and lymphoma cells, but only the former will be replaced as the stem cells mature. Likewise tumour cells expressing abnormally high numbers of receptors for epidermal growth factor are found in a wide range of cancers and an exotoxin obtained from *Pseudomonas* sp. and attached to transforming growth-factor α (a ligand for the receptor) is being tested in the treatment of these cancers.

Molecules must have three properties for them to be cytotoxic: they must be able to bind to the surface of target cells, they must be capable of translocation across the plasma membrane and finally they must cause the disruption of a vital intracellular function. Bacterial and plant toxins are among the most cytotoxic molecules known. They bind to and are taken up into eukaryotic

Table 37.11 **Some recombinant toxins against human tumour cells tested in clinical trials**

Target on the surface of tumour cell	*Cancer treated*	*Recombinant toxin used*
Epidermal-growth factor receptor	Lung, head and neck, bladder, glioblastoma, breast	Transforming-growth factor α/*Pseudomonas* exotoxin (TGFα/PE)
Interleukin-2 receptor	T-cell lymphomas	IL-2/PE *Diphtheria* toxin/IL-2
Interleukin-6 receptor	Myeloma, hepatoma	IL-6/PE
CD22: B-cell surface molecule	B-cell malignancies	Anti-CD22 antibody/ricin A chain
Erb-B2 surface protein	Breast, ovarian, bladder carcinomas	Anti-Erb-B2 antibody/PE

PE: Pseudomonas extoxin

cells. Once inside they act catalytically to prevent protein synthesis. *Diphtheria* toxin causes the inactivation of protein synthesis elongation factor 2 and the plant toxin ricin causes the hydrolysis of the 28S RNA of ribosomes. Therefore if the cytotoxic part of the toxin can be isolated from those structures required for binding to and uptake into cells, it can be retargeted by covalent attachment to a growth factor or an antibody. Originally such targeted toxins were constructed by the chemical attachment of toxins to antibodies, but now the method of choice is to fuse modified toxin genes to DNA elements encoding growth factors or the combining region for antibodies. The chimerical gene is then expressed in *Escherichia coli*, and the recombinant molecule is isolated. Table 37.11 lists some toxins that have been targeted in this way, their targets and the cancers for which they have been prepared.

Nutritional approaches

Another, quite different, type of approach to the treatment of cancer has arisen as the result of work by Tisdale on the biochemical reasons for the **cachexia** (progressive loss of body weight) that accompanies about 90% of all cancers. Cachexia is an important factor in cancer mortality, accounting for 22% of cancer deaths in the UK, as well as being a contributory factor in many more deaths. It has been shown that for cancer patients with severe weight loss a ketogenic diet is associated with an increase in body weight and inhibition of tumour growth. Such a diet would be hypocaloric, with a distribution of energy-supplying fuels as follows: carbohydrate (15%), lipid (68%) and protein (17%). This effect is also produced in experimental animals bearing tumours by the administration of the ketone body 3-hydroxybutyrate.

The possibility of dietary means forming a part of cancer therapy is also pointed to by the finding that fish oils (or specifically the eicosapentaenoic acid they contain) are effective both in reducing cachexia and in inhibiting tumour growth in experimental, tumour-bearing animals. The polyunsaturated fatty acid was more effective at preventing weight loss than it was at arresting tumour growth. Thus, although dietary fat in general tends to promote some tumours, specific components of a fatty diet can be protective.

37.11 Human cancer risk assessment

For a long time, essentially the only way of assessing the risk to human populations of developing cancers was via epidemiological studies. This approach has been and continues to be a very powerful one. Now, however, those wishing to make cancer risk assessments can combine these studies with molecular information of the type deriving from laboratory studies with animals and cells in culture, and reviewed in the present chapter. This combined approach is facilitating much more penetrating analyses (Figure 37.18).

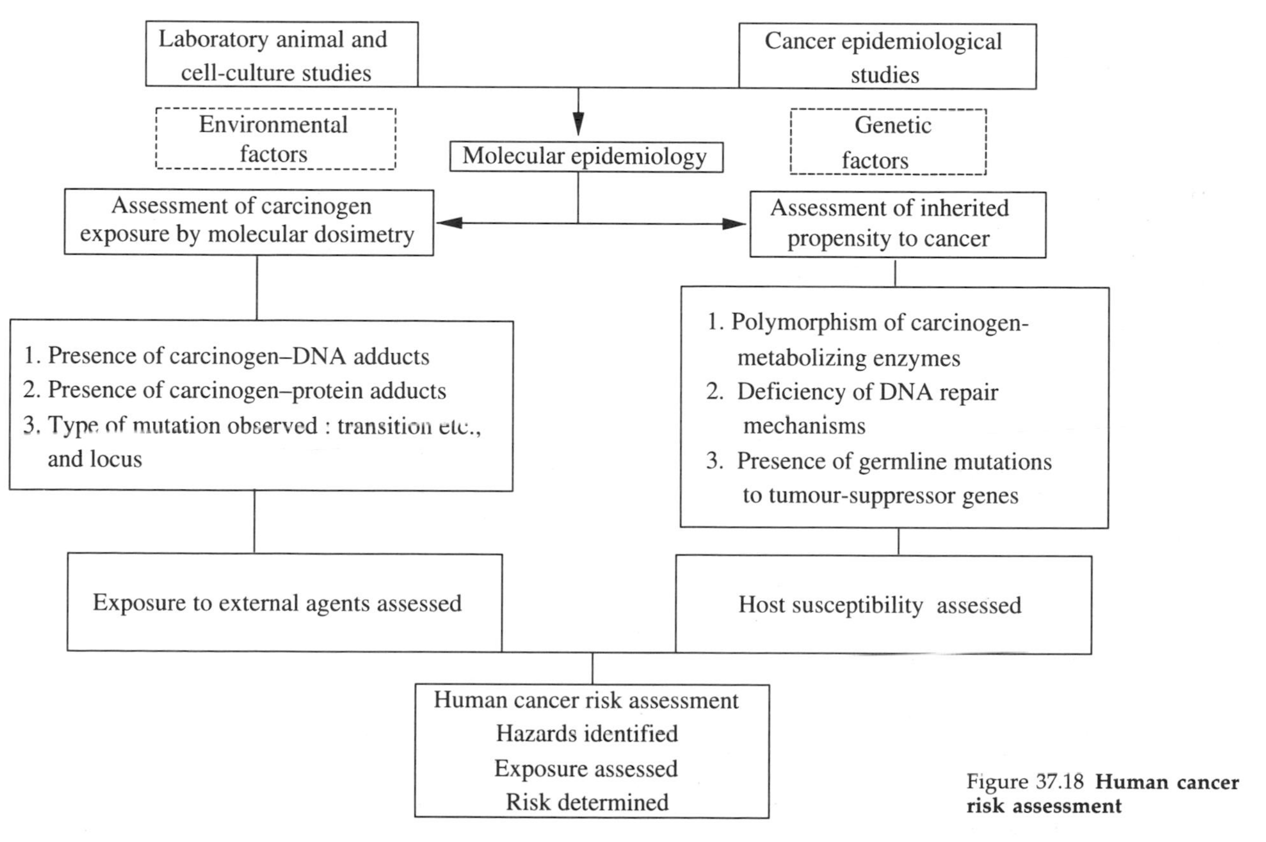

Figure 37.18 **Human cancer risk assessment**

The potential of the molecular approach was recently demonstrated when the family of former United States Vice-President Hubert Humphrey gave their permission for a sample obtained, when the diagnosis of his cancer was uncertain, to be tested for a mutation to the p53 gene. The analysis showed a mutation to be present. Had this result been available at the time the biopsy sample was collected, more vigorous treatment might have been instituted at an earlier stage. Such advances are very important, because the 1990 WHO report referred to earlier, recorded that for 16 developed countries the overall cancer rate for males has been rising for at least 18 years. Rates for females in some countries have been rising as much as they have been falling in others. The only encouraging finding was that there was an almost universal, substantial decline in the death rate for children and young adults. In general, we must conclude that the struggle to reduce cancer deaths, explicitly supported by United States President Nixon in 1971 (when he signed the US National Cancer Act) had, to 1993, resulted in the expenditure of $25 billion, but had apparently not yet begun to achieve its goal, although there are now clear indications that the new insights arising from molecular epidemiology will change this.

Part 5

A biochemical perspective on disease and its treatment

Chapter 38

Recombinant DNA and genetic engineering

38.1 Introduction

There is a sense in which the term '*re*combinant DNA' is misleading, because the generally accepted definition of recombination is 'the generation of novel genomes by the joining together of DNA *not previously joined*'. The process of recombination was first demonstrated in the course of experiments conducted by Oswald Avery and his colleagues in the 1940s to show that genetic information is stored in DNA. They showed that DNA from a dead, but previously virulent (capable of causing disease), strain of *Pneumococcus* could be taken up by an avirulent strain of the same organism. The latter strain was found to be **'transformed'** as a result of this process. That is the previously avirulent strain became virulent. It is now appreciated that what had happened is that the some cells of the avirulent strain had taken up DNA released from the cells of the virulent strain as they lysed. This DNA was incorporated into the genome of the host cell in such a way that the expression of its genetic information was able to occur. In turn, this caused the production of virulence factors (especially those associated with the formation of a capsule characteristic of virulent, but not of avirulent, strains).

These early experiments that demonstrated recombination were restricted to prokaryotic organisms, but it is now appreciated that these processes also occur in eukaryotic cells, including those of human origin. The development of an understanding of the mechanisms that underlie the processes of recombination has led to the introduction of powerful laboratory techniques for the manipulation and analysis of genetic material. These are referred to collectively as **genetic engineering**. The uses that can be made of the techniques of genetic engineering are summarized in Figure 38.1 and these form the basis for discussion in the rest of this chapter.

38.2 Techniques and enzymes used to manipulate DNA

The ability to manipulate DNA (and RNA) *in vitro* has been made possible by two key factors: the annealing properties of nucleic acids and the availability of a range of nucleic acid-specific enzymes with very special characteristics.

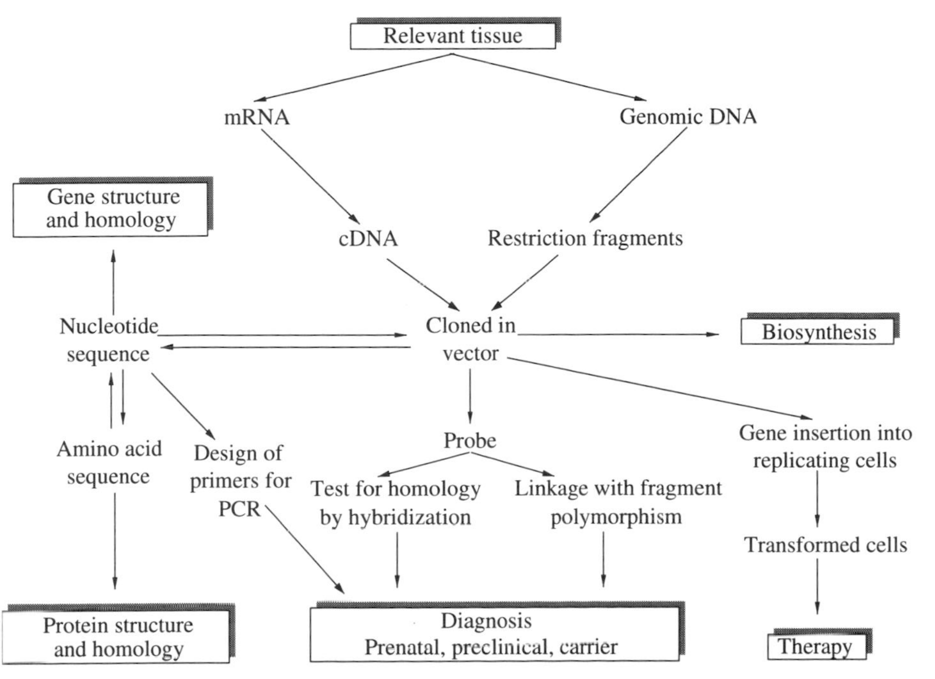

Figure 38.1 **Some examples of DNA technology and how these are approached**

Figure 38.2 **The annealing properties of nucleic acids provides a useful means of detecting genes and mRNA.** Any pieces of single-stranded DNA or RNA can bind by virtue of Watson–Crick base pairing to form duplexes. In practice, for statistical reasons, complementarity of at least 20 bases is required to identify a specific gene or mRNA from the pool of DNA or mRNA in a cell. The example shows a DNA–RNA hybrid, but RNA–RNA and DNA–DNA hybrids are possible

Separating, annealing and probing DNA molecules

Of fundamental importance in the manipulation of nucleic acids is the observation that any two pieces of single-stranded DNA or RNA will **anneal** or bind together to form **duplexes**, provided that they have continuous complementary sequences of 20 or more bases in common (in some special cases the degree of complementarity may be less). Not only can DNA bind to DNA, but also RNA to RNA and DNA–RNA hybrids may be generated (Figure 38.2).

One powerful application of the ready annealing of nucleic acids is in the process of 'probing'. In this, nucleic acid of unknown origin or composition may be examined to determine whether it contains particular sequences. These may correspond to, for example, whole genes. The **probe** is a piece of DNA that is complementary to the sequence sought and which has been labelled with a radioisotope (commonly ^{32}P is used). In using a probe to interrogate a section of double-stranded DNA for the presence of a gene of interest (Figure 38.3), the two strands of the target must first be separated and this may be achieved either by heating to about 95°C or by treatment with alkali (e.g. sodium hydroxide). The strands may be kept separate and also 'immobilized' by depositing them onto a nitrocellulose membrane followed by heating. The ability of the separated strands to bind the radioactive probe as evidenced by the retention of radioactivity by the membrane is diagnostic of the presence of the gene of interest. For example, some patients may present with a disease known as β-thalassaemia in which the gene encoding the β-chain of haemoglobin is absent from both copies of their chromosome 11 (i.e. they are homozygous for the condition). If the separated chains of genomic DNA from such a patient are deposited onto nitrocellulose and probed with a radioactive probe for the β-chain of haemoglobin, little radioactivity will be bound to the membrane.

In many cases in which DNA probing is to be carried out, the ability of the probe to hybridize with the object DNA fails to provide sufficient discrimination, e.g. for a point mutation to a gene. In such cases the object DNA is 'cut' into fragments by using a restriction enzyme (see next section) and it is the joint determination of the size of the fragment (compared with a control) and its ability to hybridize with the probe that proves diagnostic. Fragments of DNA are most readily separated according to size by electrophoresis on agarose with fragments containing from 100 to 10 000 base pairs being most

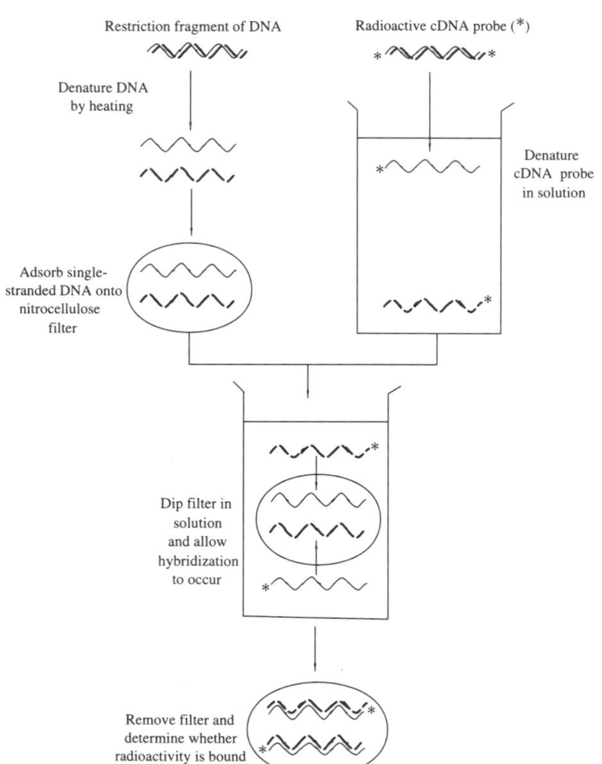

Figure 38.3 **Filter hybridization to establish whether a piece of genomic DNA and cDNA probe are complementary**

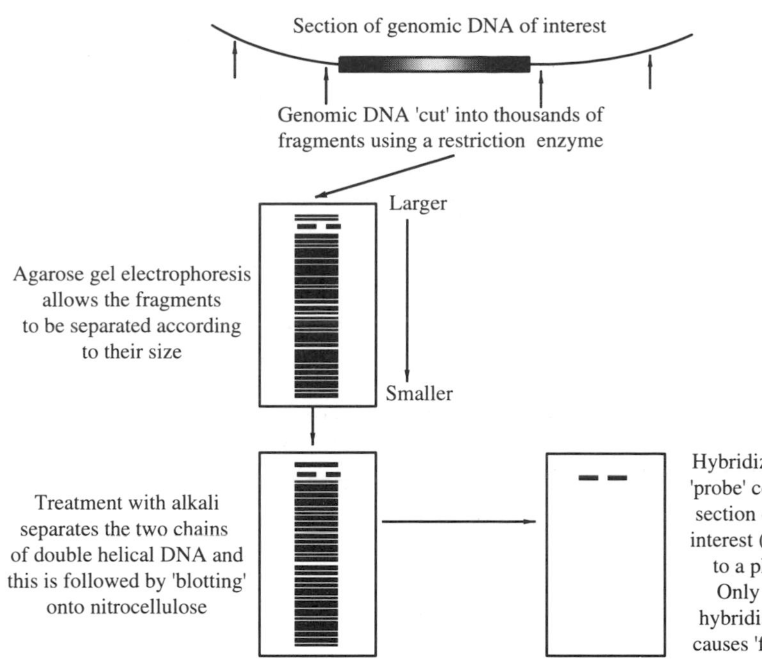

Figure 38.4 **The 'Southern blot'**: a hybridization technique for the detection of a specific sequence in a fragment from genomic DNA

easily resolved. Relative ???? may be assessed by running markers of known mass alongside the fragments to be analysed.

Southern blotting

The techniques of electrophoretic separation of DNA fragments and of radioactive probing are powerfully combined in the process known as **Southern blotting** (Figure 38.4). The name derives from Edwin Southern, who first introduced the method. In this, the double-stranded DNA to be examined is digested into fragments using a restriction endonuclease and the fragments are separated by electrophoresis. Following their separation, the fragments are 'denatured' (the complementary strands are separated) using sodium hydroxide. To render the DNA strands accessible to the probe and to prevent their re-annealing (see previous section), a nitrocellulose membrane is pressed on top of the agarose gel. This has the effect of transferring the denatured DNA to the membrane ('blotting'). The single-stranded DNA is fixed to the nitrocellulose by heating and the membrane is exposed to the radioactive probe. The probe then anneals only with fragments of complementary sequence, generating a band of radioactivity on the nitrocellulose. Finally this may be detected by placing an X-ray film on the membrane and observing the fogging of parts of the film exposed to radiation. It is important to realize that, provided the fragment of DNA being analysed in this way has a sufficiently large region of continuous complementarity with the probe that is to detect it, the whole sequence does not need to match. Thus a fragment of DNA corresponding to part or all of a gene (including any introns) may be detected by a probe that is complementary only to the exons (a cDNA probe, see later). An illustrative example (Figure 38.5) will serve to show the principle. This example also indicates that the gene encoding the α-chain of haemoglobin consists of two exons and one intron. Examination of the known coding sequence of the α-globin gene reveals that it contains just one sequence recognized by the restriction endonuclease *Hind III* (see next section), while two other recognition sites for the same enzyme flank the gene. It follows that an intronless gene, when subjected to hydrolysis by *Hind III*, should yield two fragments capable of hybridizing with a probe complementary to all of the exons of the α-globin gene. In fact, three fragments capable of hybridizing with the probe are generated by *Hind III*. This must mean that a *Hind III* recognition site not encoded in the α-globin sequence is contained in an intron.

Enzymic reactions involving nucleic acids

The reactions catalysed by the enzymes commonly used to manipulate nucleic acids are shown in Figure 38.6.

Reverse transcriptase (RNA-dependent DNA polymerase) is an enzyme encoded in the genome of RNA-containing retroviruses (see Chapter 2, 5 and 37) and catalyses the formation of DNA on an RNA template. This enzyme, in common with all DNA polymerases, requires a template (RNA) and a primer.

Terminal transferases catalyse the addition of homopolydeoxyribonucleotide sequences to either 3'-end of double-stranded DNA molecules. For these reactions there is no requirement for a template, while the free 3'-ends act as primers. There are separate enzymes for each of the four deoxyribonucleotides.

Two important types of enzyme catalyse very specific hydrolytic reactions involving nucleic acids. **Ribonuclease H** is an endonuclease that catalyses the hydrolysis of phosphodiester bonds in RNA, but only when it forms a hybrid double-stranded arrangement as

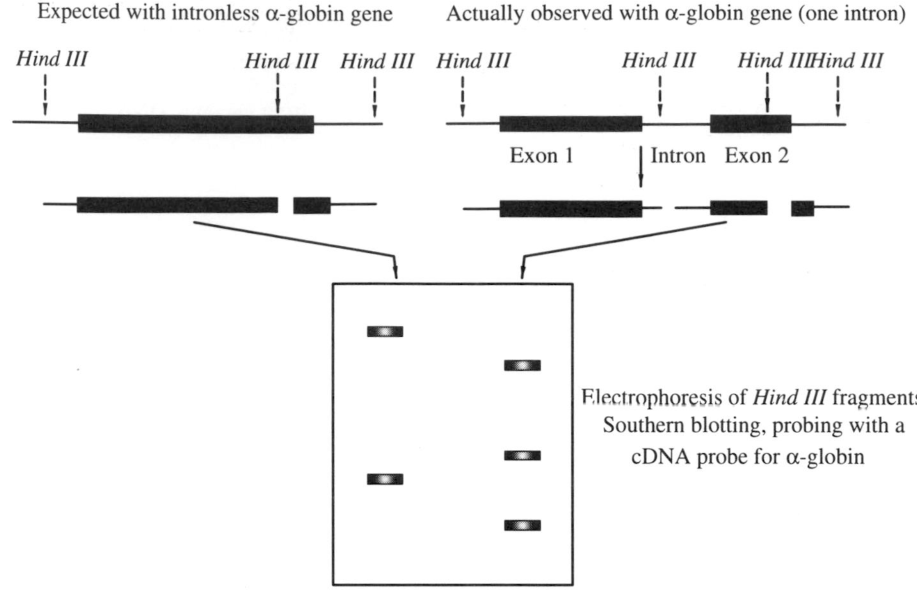

Figure 38.5 **The protein coding sequence of the gene for α-globin has only one site recognized by the restriction ezyme *Hind III*.** Nevertheless, treatment of genomic DNA with Hind III generates three, not two fragments that hybridize with a cDNA probe for α-globin

Figure 38.6 **Enzymic reactions that involve DNA**

1. Reverse transcriptase
   Catalyses the formation of DNA on an RNA template. Uses deoxyribonucleoside triphosphates. Requires a primer.

RNA template

5'-Primer

2. Terminal transferases
   Catalyse the addition of many copies of a particular nucleotide to both the 3'-ends of double-stranded DNA. Can use any dNTP as a substrate

poly(dN) Double-stranded DNA

poly(dN)

3. Ribonuclease H
   Catalyses the random hydrolysis of phosphodiester bonds in RNA, in a heteroduplex with DNA

DNA

RNA

Hydrolysis

4. Restriction endonucleases
   Catalyse the selective hydrolysis of phosphodiester bonds in DNA in 'palindromic' regions (e.g. *Hind III*)

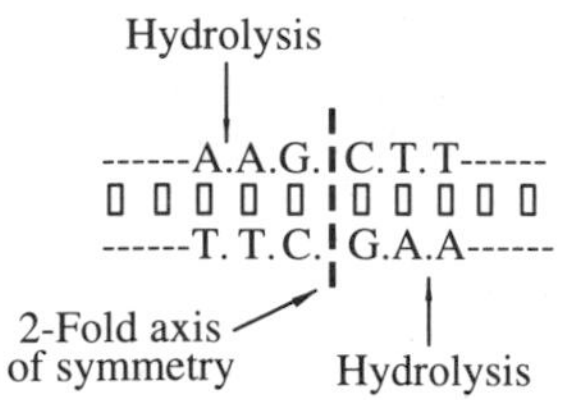

5. DNA ligases
   Catalyse the ATP (or NAD)-dependent joining of 'nicks' in double-stranded DNA

Formation of a new phosphodiester bond

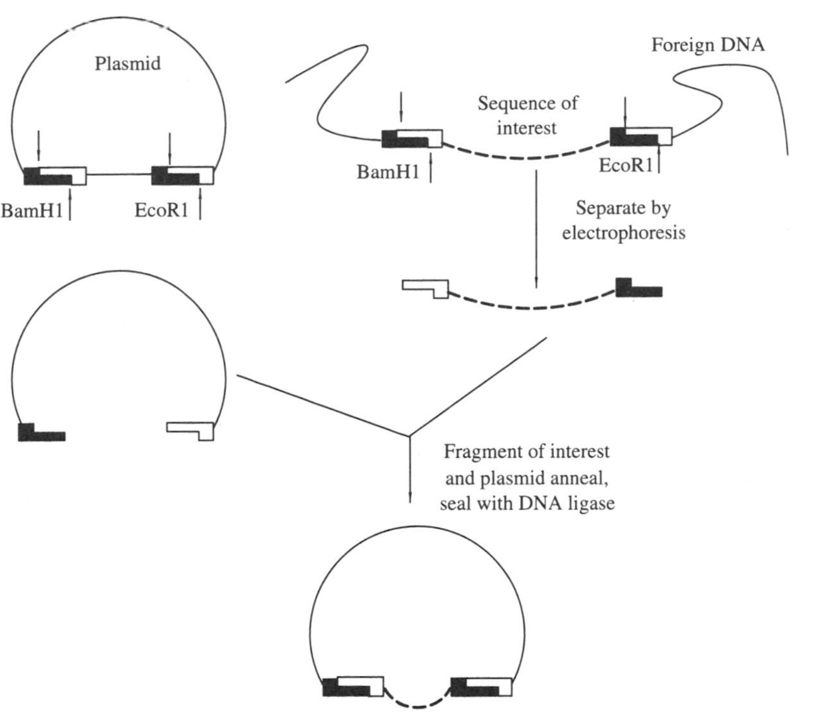

Figure 38.7 **The 'cutting' and 'pasting' of DNA**

the result of complementary base-pairing with DNA. Such structures arise naturally as a result of the activity of reverse transcriptase; indeed, when derived from some organisms, this enzyme possesses dual ribonuclease H and reverse transcriptase activities. Consequently, partial or complete hydrolysis of the template RNA accompanies the action of this type of reverse transcriptase.

Restriction endonucleases (or simply **restriction enzymes**) have been referred to previously as affording protection to bacteria (innate immunity) against infection by invading bacteriophage DNA (see Chapter 32). Enzymes in this class catalyse the simultaneous hydrolysis of double-stranded DNA molecules on both chains. The specificity of these enzymes is such that they bind to DNA at particular 'palindromic' sequences, i.e. sequences with internal complementarity. When such sequences arise, they present in the DNA a two-fold axis of symmetry. This is because on the complementary strands the sequences are identical when read in the appropriate direction (5'⇒3'). Hydrolysis catalysed by restriction enzymes of the type represented by *Hind III* cause 'staggered cuts' in the two strands. Such cuts are said to give rise to 'sticky ends'; that is, the ends of the DNA have short lengths of single-stranded DNA 'sticking out' or 'overhanging'. A practical consequence of this generation of 'sticky ends' is that fragments of DNA may be generated from larger sections of DNA in such a way that they have 'sticky ends' at both ends (Figure 38.7). These may be mixed with DNA obtained from a quite different source, for example a **plasmid** (a circular piece of extra-chromosomal double-stranded DNA found in some bacteria) that has been 'cut' with the same enzyme. Following the rules of base-pairing the two will anneal with each other and the two may be joined permanently when covalent bonds are formed by a **DNA ligase** (see Chapter 4).

It is now possible to engineer plasmids in such a way that they contain only one or two sites recognized by particular restriction enzymes in its entire sequence. The process of linearizing a plasmid and incorporating into its sequence a piece of DNA previously hydrolysed by the same restriction enzymes is referred to as 'cutting and pasting' (Figure 38.7). Plasmids designed to be used in this way are referred to as 'cloning vectors' or simply as 'vectors'.

Other restriction enzymes attack DNA at the two-fold axis of symmetry when 'blunt ends' are generated. The enzyme *Hpa I* provides an example of this type of specificity. There are many hundreds of restriction enzymes and the names given to them indicate the bacterium from which they have been isolated. *Hind III* derives from *Haemophilus influenzae* Rd, while *Hpa I* is found in *H. parainfluenzae*. Bacteria producing one or more restriction enzymes, protect their own DNA from hydrolytic attack by producing enzymes that cause covalent modification of the bases forming the palindromic recognition sequences (for example their methylation). This has the effect of preventing autodigestion of the bacterial DNA.

Complementary DNA (cDNA)

The isolation of reverse transcriptases has meant that it is now possible to use RNA molecules to act as templates for the formation of DNA molecules complementary to them (cDNA). This ability to 'copy' RNA is most useful when applied to mRNA molecules. When of eukaryotic origin these molecules (with very

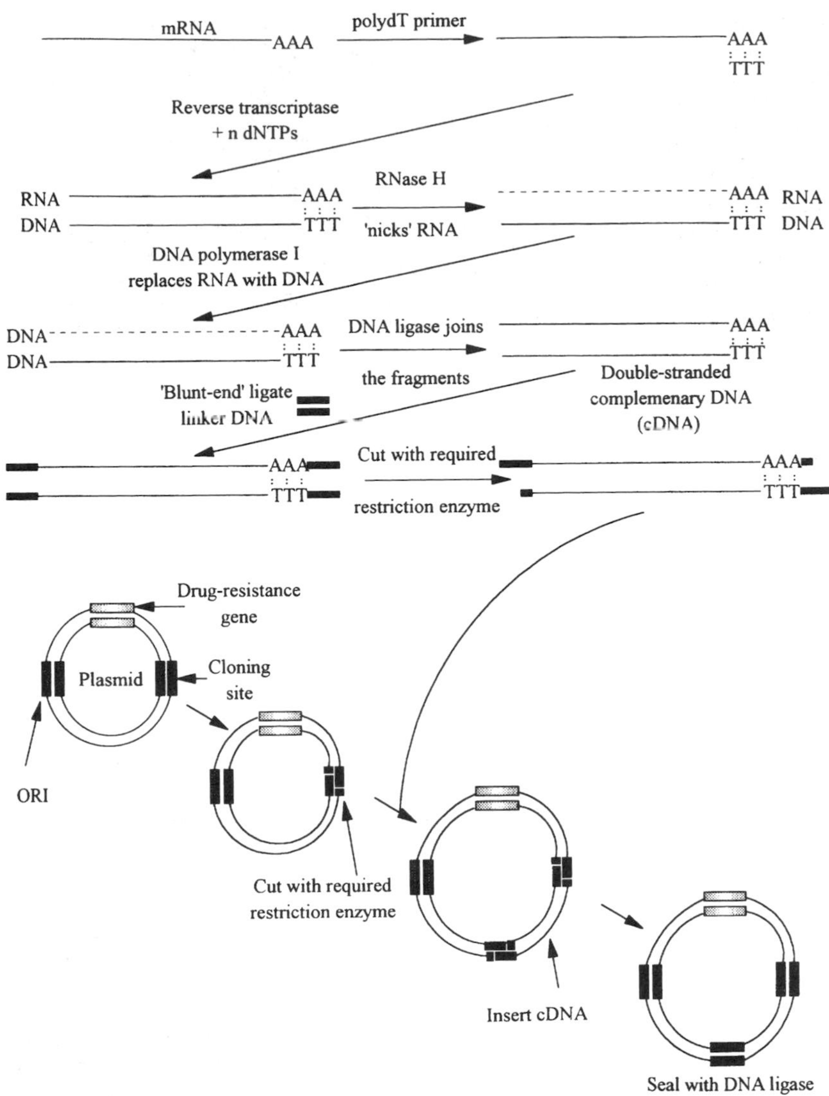

Figure 38.8 **The formation of a cDNA and its insertion into a plasmid**

few exceptions) have polyA 'tails' at their 3'-ends (see Chapter 4). This feature provides a simple way of meeting the second requirement for reverse transcriptase activity – that for a primer. This may be supplied by allowing synthetic polydeoxythymidine (polydT) to hybridize with the mRNA (Figure 38.8). Following the action of reverse transcriptase, the hetroduplex of DNA/RNA can be converted into double-stranded DNA by using others of the nucleic acid-specific set of enzymes. Ribonuclease H will 'cut' the RNA at random to yield a series of RNA oligonucleotides that are, nevertheless, still bound to the DNA. Such an arrangement, a DNA template and RNA oligonucleotide primers, meets precisely the requirements of DNA polymerases (see Chapter 4) which can therefore be used to act jointly as a polymerase to add DNA and as a 5-exonuclease to remove the RNA. Subsequently the resulting DNA oligonucleotides may be joined using DNA ligase.

The ability in this way to produce a double-stranded copy of a mRNA is further exploited in the process known as cDNA cloning.

cDNA cloning

Most mRNA molecules are present in cells in relatively low abundance and it is important, if full use is to be made of molecules produced as described above, that the quantity of cDNA obtained is amplified. This may be as a preliminary to determination of sequence, for example. This may be achieved by **cDNA cloning** and a modification of this technique can be employed so that the genetic information in the DNA may be expressed.

The basic idea behind all cDNA cloning strategies is that a desired piece of the cDNA should be 'recombined' with a vector that itself is capable of undergoing replication in a suitable host cell. The quantity of cDNA available will then expand as the host cell grows and divides and the vector is replicated. The term 'cloning' is used in this context because conditions are so arranged that a **single cell** containing the recombinant vector of interest is selected and allowed to undergo many rounds of cell division. In this way, large quantities of a unique recombinant vector (and hence its incorporated cDNA) may be obtained.

Figure 38.8 illustrates how a cDNA may be amplified according to these principles. The 'blunt-ended cDNA' is adapted by the process of blunt-ended ligation to linker DNA that includes in its sequence a site recognized by a restriction enzyme. Treatment of this extended cDNA with that restriction enzyme yields 'sticky ends'. The overhangs generated will allow the cDNA to anneal with a plasmid vector cut using the *same* restriction enzyme. Apart from a single recognition site for the restriction enzyme used, the plasmid vector used will include in its sequence two special regions. The first of these (termed Ori) allows the plasmid to be replicated by the host cell. (It is this sequence recognized for the *ori*gin of replication of the plasmid in the host.) The second region is a gene that, when expressed in the host cell, confers the ability to grow in the presence of an antibiotic (**'antibiotic-resistance gene'**). For example, if the host is a bacterium, the gene may encode an enzyme that catalyses the inactivation of ampicillin. Once the recombinant vector is taken up into the host cell, the latter is allowed to grow on an agar plate at low density such that each colony established on the plate derives from a single host cell (a clone). The process of causing host cells to take up vectors is of very low efficiency, but only cells with the vector are able to grow because the agar growth medium includes the antibiotic for which the vector provides resistance. At the end of the process the recombinant plasmid may be readily isolated and the desired cDNA excised from it using the original restriction enzyme.

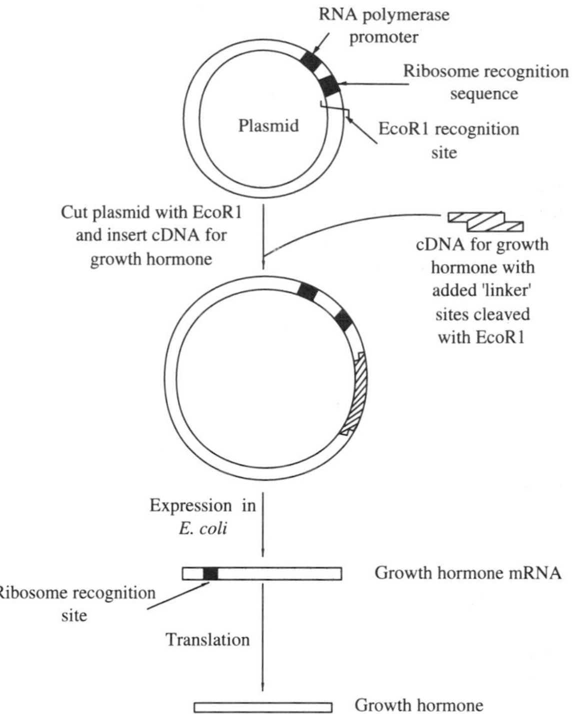

Figure 38.9 **The production of human growth hormone by bacteria**

The production of a human protein by bacteria: growth hormone

The fact that cDNA molecules were derived from mRNA molecules immediately suggested that, if suitably handled, the former ought to be able to specify the formation of the latter, i.e. cDNA 'genes' in recombinant vectors ought to be capable of transcription (and the transcripts capable of translation). This end is now routinely achieved, most beneficially when cDNA are expressed in bacteria. Several important proteins ('recombinant proteins') are produced in this way, e.g. **growth hormone**.

Genetic engineers wishing to produce growth hormone in bacterial cultures, using a cDNA for the hormone, need to pay attention to the important differences that exist between prokaryotes and eukaryotes in the ways in which transcription and translation are initiated. The bacterial enzyme apparatus used for these processes must be 'fooled' into treating the cDNA as a prokaryotic gene and the consequent transcript as a prokaryotic mRNA. In practice, this is achieved by incorporating the cDNA into the cloning vector close to a promoter for the bacterial RNA polymerase (see Chapter 4). Beyond the promoter (in the 3′ direction) there should be a sequence which, when transcribed continuously with the cDNA 'gene', will permit the binding of bacterial ribosomes to the 'mRNA' produced (a ribosome recognition site; see Chapter 6). These two plasmid DNA sequences must be located close to, but 'upstream' (in the 5′-direction) from, a site recognized by a restriction enzyme. Figure 38.9 shows the steps that need to be taken to express a linker-adapted cDNA for human growth hormone in a bacterium. (It should be noted that sequences of the cDNA that serve to target the nascent protein in eukaryotic cells, e.g. the 'signal' sequence, are removed prior to use.)

The production of growth hormone in this way was a very important goal because, unlike insulin, growth hormone must have the human sequence to be effective in treating patients of short stature. In the past the only source of the material was cadavers and the isolation procedure was very difficult.

It should be noted that the strategy of using bacteria to produce human proteins for therapeutic purposes may fail if the protein in question needs to undergo post-translational modification, such as glycosylation. This is because bacterial host cells lack the apparatus to carry this out. In such cases, a eukaryotic cell line (i.e. cells that are effectively immortal in cell culture) is used. Cells commonly used in this way have been derived from the ovaries of the Chinese hamster and for this reason are referred to as *CHO* cells. A recent success achieved by using this strategy has been in the production of **erythropoietin**. This is a protein hormone of relative molecular mass of 35 kDa, 30–50% of which is carbohydrate. This hormone is produced by the kidney and stimulates erythropoiesis (the production of erythrocytes). Patients with kidney failure, especially those undergoing renal dialysis, tend to become anaemic as their kidneys fail to sustain the requirement for the synthesis of the hormone. Until recombinant erythropoietin became available, dialysis patients required frequent blood transfusions with the attendant dangers of haemochromatosis.

The polymerase chain reaction

Many of the techniques used to obtain information about genetic diseases in humans depend on the analysis of genomic DNA, but there are restrictions inherent in all of the techniques that have been described so far for analysing DNA for diagnostic purposes. In particular, the DNA of interest invariably only forms a minute part of the total DNA obtained from cells. There are two ways of dealing with this problem of looking for a 'needle in a haystack'. The first has already been referred to and involves the use of a radioactive probe for the sequence of interest. There are two reasons why this is not entirely satisfactory: the DNA of interest may not be present in sufficient quantity, and the use of radioisotopes is potentially dangerous. In dealing with the former problem, the **polymerase chain reaction (PCR)** obviates the second. The PCR is designed vastly to amplify any selected sequence of double-stranded DNA. When this is done, the DNA may then be detected on electrophoretograms, for example, by conventional staining techniques.

The PCR process depends critically on the annealing properties of complementary DNA molecules and a remarkably thermostable form of DNA polymerase originally obtained from the thermophilic bacterium *Thermus aquaticus*, and designated *Taq* polymerase. As a preliminary to carrying out the PCR, a DNA sequence of interest is identified (this may range from 100 to 2500 base pairs) and two 'primers' are designed to be complementary to regions at the 5′ ends of the two strands of the target DNA (Figure 38.10). The target strands of DNA are separated by heating and the mixture of primers is added in molar quantities vastly in excess of the DNA to be amplified (which might come from a single cell). On cooling the mixture, the primers bind to the separate strands. *Taq* polymerase then uses the primers for DNA synthesis with the target strands as templates. The reaction is allowed to proceed for a timed period when the mixture is once more heated to 'melt' all double-stranded structures. Crucial to the whole operation is the fact that the enzyme survives this period of DNA 'melting'. A second period of cooling allows further binding of the primers and a second round of polymerization. Repetition of the cycle of events quickly leads to massive amplification of the DNA sequence of interest. It should be noted that although in the first round of the reaction the polymerase reaction would be expected to 'overrun' the desired length, a situation will quickly be arrived at in which overwhelmingly the templates will be provided by the sequence of interest.

By slight variations in the method used, it is possible to combine the PCR strategy with the use of reverse transcriptase to amplify mRNA sequences, without having recourse to cDNA cloning techniques.

The newly amplified DNA (whether generated from DNA or RNA) retains the structural features of its parents; in particular, any sequences recognized by restriction enzymes are still present. Consequently, PCR products may be cut using these enzymes and separated according to size on electrophoresis.

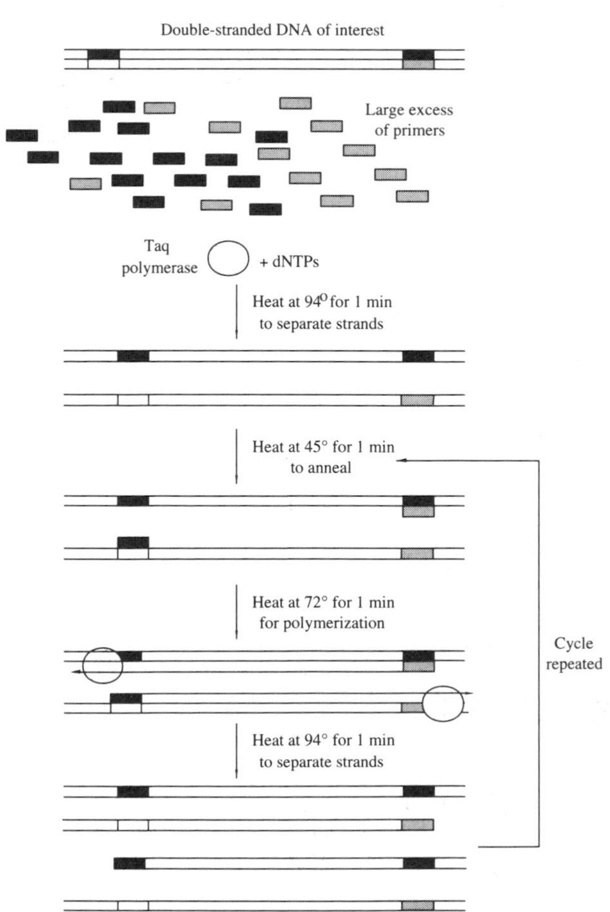

Figure 38.10 **The polymerase chain reaction for amplification of particular DNA sequences**

38.3 Molecular analysis of inherited disease

The area of medicine that has benefited most dramatically from the application of the techniques for the manipulation of DNA has been that of the diagnosis and management of genetic diseases (especially those that are inherited). (The task of making developmentally early, accurate and rapid diagnoses of infectious diseases has also recently begun to be greatly facilitated by similar techniques.)

The nature of genetic defects

It is important to remember that when we talk of genetic disorder we have in mind clinical or subclinical conditions ranging from those in which genetic defects play a dominant role, to those conditions ascribable to the environment in which there is a genetic component. Examples of the former are so-called **'single-gene defects'**, traceable through whole families. In such cases one sees the expected patterns of dominant, recessive or sex-linked inheritance. At the other end of the spectrum of genetic disorders are those conditions which are **polygenic** in nature, depending on many gene products, e.g. cancer (see Chapter 37). In addition, there are known to be genetic components in diseases such as schizophrenia and insulin dependent diabetes mellitus.

Even when one is dealing with so-called single-gene defects it should be borne in mind that a disease which may appear to be homogeneous at the clinical level may well prove to be heterogeneous at the molecular level. This is well illustrated in the occurrence of the haemoglobinopathy called β-thalassaemia. This disease is characterized by an anaemia which arises because of a failure of the synthesis of the β-chain of haemoglobin. The actual defect that gives rise to the disease may be the loss of the gene for the protein or the occurrence of frameshift or point mutations, for example. Even point mutations in the intron of the β-chain gene can result in a failure to produce mRNA from hnRNA if the affected base forms part of the sequence required for splicing (see Chapter 4).

The occurrence of restriction fragment length polymorphisms (RFLPs)

Before advances in molecular biology had made the analysis of areas of the human genome relatively straightforward, geneticists had nevertheless begun to construct what were called 'linkage maps'. Although these maps did not usually indicate the precise location of particular genes on their chromosomes, they did indicate which genes were physically close together. This was possible because of the event of 'crossing-over' that occurs between a pair of chromatids in meiosis. In this (Figure 38.11), each chromatid is broken and the fragments generated 'cross-over'. The idea is that the closer any two points are on the chromatid, the less likely is the break point to lie between them. Conversely, two genes at the extreme ends of a chromatid, will always be separated in meiosis. Thus, two closely linked genes will end up on the same chromatid and, following mitosis, will be inherited together. In the past, the only way to assess linkage was to look for the expression of the inherited genes in question. However, since a gene is only a convenient marker for a particular part of a DNA molecule, it follows that any marker on the DNA will serve equally well for assessing linkage, e.g. sites recognized by restriction enzymes. Thus, if it can be shown in a particular family that a mutated gene is linked to a mutation to a restriction enzyme recognition site, this fact may be used to used to recognize the presence of the mutated gene.

Detection of inherited diseases by linkage analysis using RFLPs

An example of the power of this approach is illustrated in Figure 38.12 which relates to the inheritance of a mutated gene for the β-chain of haemoglobin. In the particular case shown, there is a point mutation in the first exon of the β-globin gene. This has the effect of directing the formation of a globin β-chain in which the glutamic acid that should be present at position 6 of the polypeptide chain is replaced by valine. When presenting in the homozygous state the disease is known as **sickle-cell disease**. Analysis of the chromosome carrying the normal β-globin gene has revealed the presence of three sites (1, 2 and 3 in Figure 38.12) recognized by the restriction enzyme *Hpa I*. One of these sites (2) lies 'downstream' (in

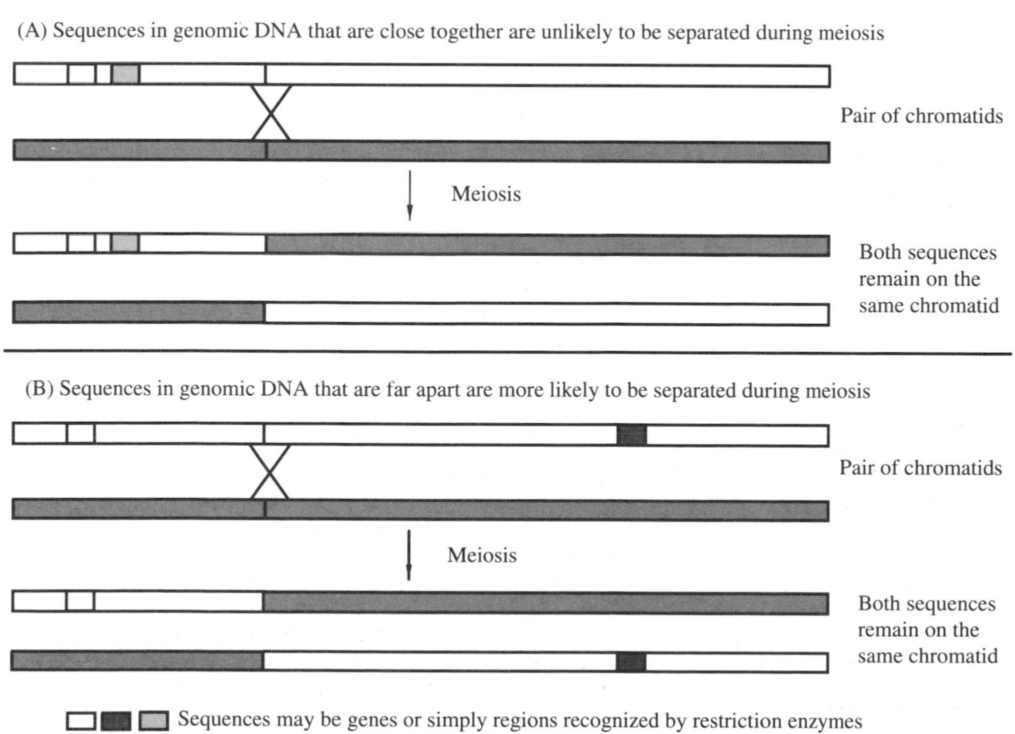

Figure 38.11 **'Linkage analysis' of marker sequences in DNA**

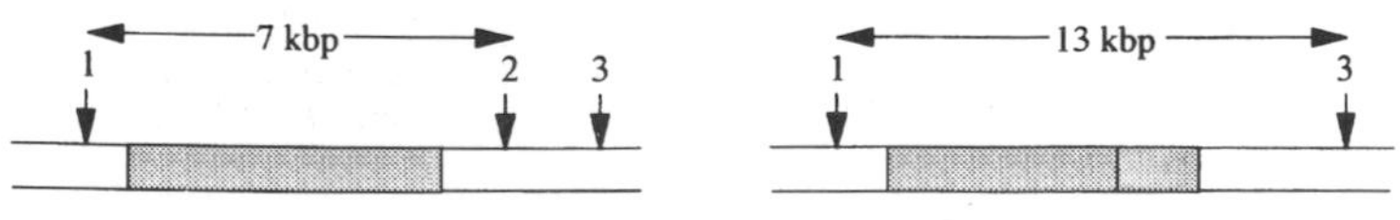

Hydrolysis by *Hpa* 1

Hpa 1 recognition site 2 is linked to the normal but not the sickle cell gene in West African populations

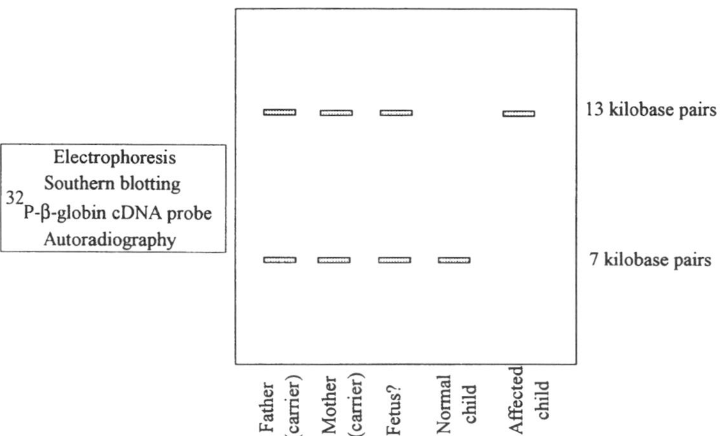

Figure 38.12 **Indirect detection of a mutation using a restriction enzyme recognition site that is linked to the gene in question**

the 3′ direction) from the gene. Investigation of certain populations in which sickle-cell disease is prevalent, e.g. those of West Africa, has shown that inheritance of the mutated gene is closely linked with the inheritance of a mutation to site B. The mutation to site B means that *Hpa I* fails to hydrolyse there. Consequently, if DNA from normal individuals is cut with *Hpa I* a fragment of 7 kilobase pairs spanning the gene of interest is generated. If the DNA comes from an individual with sickle-cell disease the fragment will be longer (13 kilobase pairs) owing to the lack of the restriction site B. If the person is heterozygous for the condition (**sickle-cell trait**), two fragments will be generated from the two chromosomes. The fragments (RFLPs) can be detected and their sizes assessed by using Southern blotting and probing with a cDNA probe for the β-chain of haemoglobin. It should be noted that the probe hybridizes equally well with both the normal and the affected gene and therefore the diagnosis depends entirely upon the affected gene being linked to the absence of restriction enzyme recognition site B. If this linkage is established for a given family, then for two sequences within DNA apart by only 6 kilobase pairs, 20 000 meioses would be required for them to be separated by crossing-over, i.e. the event is highly improbable. Hence the presence of a fragment of 13 rather than 7 kilobase pairs can be regarded as being diagnostic.

In practice, it is sometimes difficult to obtain sufficient material to carry out useful RFLP-based analyses and it is this circumstance that is provided for so well by the great amplifying power of PCR. For example, it has been established that in intron 18 of the gene for blood clotting factor VIII (see Chapter 24) there is a polymorphism associated with the disease of **haemophilia B** in which a restriction site for the enzyme *Bc II* is lost. Having established linkage of the absent restriction site with the disease, two primers are used to amplify a sequence of 142 base pairs that span the region in which the restriction site for *Bcl I* is normally found. The product of the process is then treated with *Bcl I* and if hydrolysis occurs the amplified DNA was normal, while partial digestion would indicate heterozygosity and failure to hydrolyse points to haemophilia **B**. Figure 38.13 shows the several electrophoretic patterns obtained from members of a family in which this X-linked recessive form of haemophilia **B** is known to be present. It is instructive to note that no part of the strategy devised for the diagnosis of haemophilia **B** described here is directed at the detection of the mutation itself, only the polymorphism with which it is linked.

Prenatal diagnosis

One area of diagnosis in which DNA-based techniques have made a major impact has been in the detection *in utero* of a propensity to inherited disease. Before the advent of methods of examining the whole of genomic DNA for mutations, attempts to detect pregnancies at risk from inherited disorders were limited by the type and quantity of fetal-derived material that could be obtained. Fetal cells were obtained during the first trimester of pregnancy by the process of obtaining small biopsy samples of fetal origin from the chorion (**chorionic villus biopsy**). In the second trimester, a

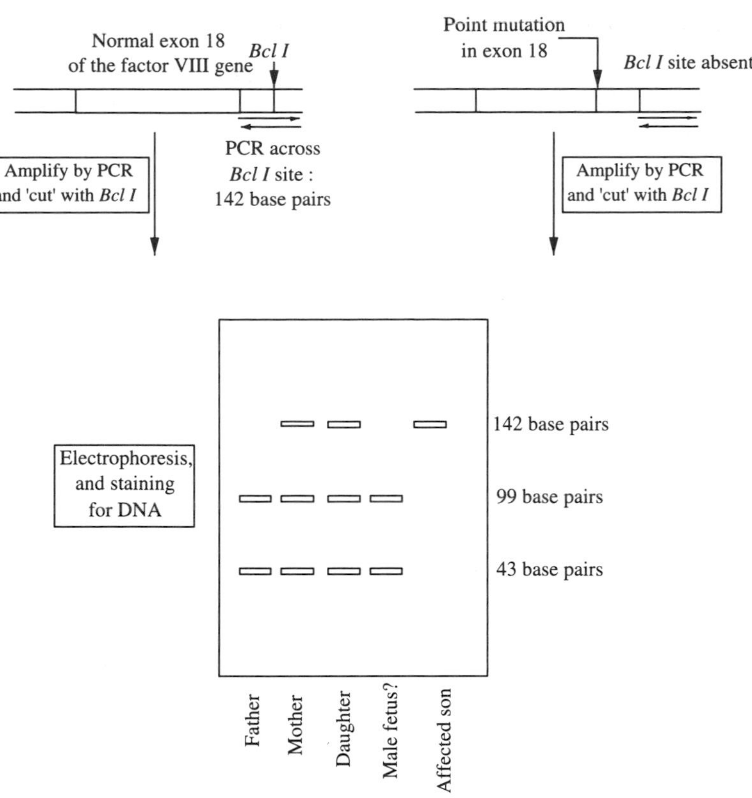

Figure 38.13 **Indirect detection of a point mutation to the gene for blood-clotting factor VIII using PCR**

needle may be introduced into the amniotic fluid and a sample of about 15–20 ml aspirated (**amniocentesis**). When such samples were obtained, however, analyses were restricted to the recognition of gross chromosomal abnormalities, e.g. the presence of three (rather than the normal two) copies of chromosome 21 in Down's syndrome or the detection of abnormal proteins. In the latter case, it was necessary that the cells isolated, actually produced the protein of interest.

The introduction of DNA-based techniques meant that, in principle, any mutation could be detected. In some cases, however, the number of cells obtained at biopsy or during amniocentesis fails to provide sufficient DNA for cDNA probe-based methods to work. For this reason the great amplifying power of PCR is increasingly forming the basis for prenatal diagnosis .

38.4 Other medical applications of the polymerase chain reaction

The power of the polymerase chain reaction as a diagnostic tool spreads well beyond the diagnosis of inherited disorders.

Monitoring the course of therapy in follicular lymphoma

This type of cancer is associated with a translocation between chromosomes 14 and 18 in such a way as to give 14–18 hybrid chromosomes (designated t[14;18]). The precise location of the break point seems to vary little between patients and consequently oligonucleotide primers may be designed to detect the translocation. One primer binds to chromosome 14 close to the break-points and the other to chromosome 18. As is shown in Figure 38.14, only in cells in which there is a translocation will there be a complete template for PCR. It is of great benefit for the oncologist to be able to assess the success of the chosen strategy for treating his/her patient and to have an early warning of treated relapse. This knowledge allows the treatment, which often involves the combined use of drugs that themselves make the patient feel very unwell, to be more precisely tailored to the need. By using the PCR reaction, it is possible to detect the DNA from one malignant cell per million normal cells.

A number of types of leukaemia are also characterized by chromosome translocations; for example, in chronic myelogenous leukaemia a translocation between chromosomes 9 and 22 is found. These translocation sites also constitute transcription points, with oncogenes (see Chapter 37) being transcribed across the join. This means that these cells will produce many copies of mRNA molecules that encode oncoproteins. By using reverse transcriptase to convert this mRNA into cDNA and then using this as a template for PCR, it is possible to detect the effect of the translocation. The sensitivity of this method exceeds that described for follicular lymphoma because in producing a mRNA the cell itself has already amplified the oncogene being sought.

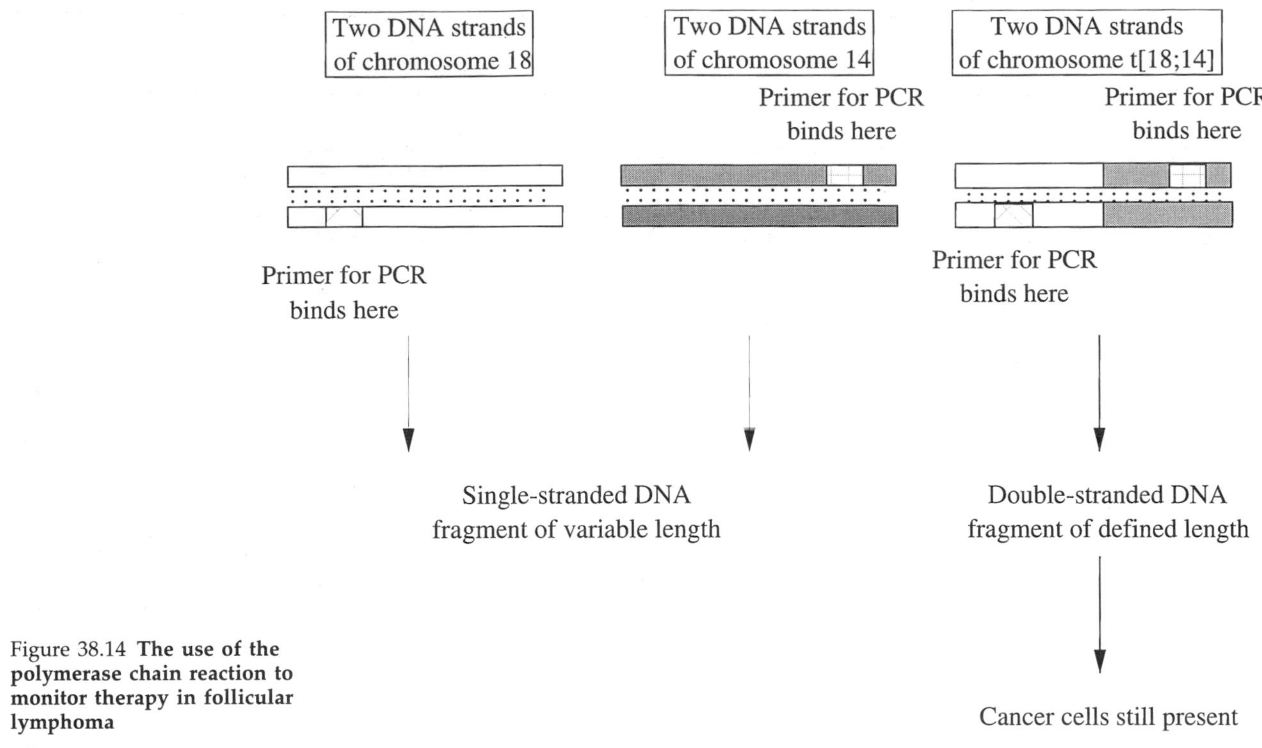

Figure 38.14 **The use of the polymerase chain reaction to monitor therapy in follicular lymphoma**

The detection of bacterial and viral infection

Because PCR reactions work equally well with DNA from any source, it is now relatively routine to design primers to recognize in the genome of infectious organisms unique sequences in the DNA, perhaps encoding organism-specific proteins or enzymes. If amplification occurs when PCR is applied to samples from suspected infected tissues, then the nature of the infection can be deduced and suitable strategies for chemotherapy instituted. Recently, the presence of a previously unknown virus of the *Herpes* family has been detected in cells taken from the tumour growing in 90% of patients with AIDS who had developed Kaposi's sarcoma (see Chapter 37).

Chapter 39

Biochemical principles underlying chemotherapy and drug resistance

39.1 Introduction

In the first two editions of this book a chapter entitled 'Principles of chemotherapy' charted the initially slow, but then rapidly accelerating discovery and synthesis of drugs with selective toxicity towards cells or organisms foreign to the mammalian body. The targets for these agents were initially largely bacteria and a few protozoa but, as biochemical knowledge has expanded, fungi, viruses and tumour cells have also been successfully attacked using cytotoxic drugs. There have been three principal sources of these chemicals: microorganisms, synthetic chemistry and plant extracts. By convention, molecules produced by microorganisms that kill other types of microorganism have been termed **antibiotics**, while synthetic drugs are referred to as **chemotherapeutic agents**. It must be said that these terms tend to be used interchangeably and either one may be used to describe plant-derived drugs.

The past decade has seen the disturbing emergence of drug resistance in many classes of organism previously controlled by drug treatment. For example, the disease tuberculosis which was once believed to have been virtually eradicated is now on the increase in the USA (especially in the large cities), but not in Europe, and this has been traced to the development of multi-drug resistance by the organism *Mycobacterium tuberculosis*. In the current chapter, therefore, it is important not only to review the mechanisms that lead to selective toxicity of drugs effective against pathogens and tumour cells, but also to pay attention to the mechanisms these cells have developed to resist the drugs. Not surprisingly these twin objectives are interrelated; for example, a gene encoding the target enzyme for a drug may be mutated in a way that means that the drug no longer acts as an inhibitor of catalysis.

At the same time, the emergence of drug resistance in pathogenic organisms has caused more attention to be paid to ways in which the development of the symptoms of diseases brought on by infection may be prevented. This strategy has its critics, but there are some distinct advantages. Because the drug does not target the organism, but only its effects, resistance is unlikely to develop. Secondly, by arresting the onset of symptoms more time is available for making a proper diagnosis and selecting an appropriate cytotoxic drug. Such a facility is likely to be particularly useful in dealing with opportunistic fungal infections (for example those affecting debilitated, hospitalized patients). The biochemical basis for the actions of drugs that interfere with the symptoms of infectious diseases is reviewed in Section 39.6.

39.2 The discovery of drugs active against infectious agents

Chemotherapeutic agents

The development of chemotherapy is of relatively recent origin. Until the twentieth century, Western physicians had depended on the same concoctions of inorganic compounds of mercury, antimony or arsenic known in the Middle Ages. These included calomel (mercuric chloride), tartar emetic (a mixture of antimony salts) and Fowler's solution (potassium arsenite). Most of these compounds appear to have been administered to Napoleon immediately prior to his death on St Helena from stomach cancer in 1826. More successful appears to have been the administration in 1865 of Fowler's solution to his leukaemia patients by the German physician Lissauer. Still working with metals, but as organic complexes to reduce their toxicity, another German, Paul Ehrlich, made important advances in the development of drugs specifically designed to combat infection. One of Ehrlich's first commissions was the production of a drug to combat the African scourge of trypanosomiasis (sleeping sickness). The disease is caused by the bite of tsetse flies carrying the protozoa *Trypanosoma gambiense* or *T. rhodesiense*. In 1905 Ehrlich synthesized the drug atoxyl (Figure 39.1) that indeed proved effective against trypanosomiasis. Despite its name, however, atoxyl proved rather toxic. A less toxic derivative, tryparsamide, was produced in 1919 (four years after Ehrlich's death). In 1910, his Japanese co-worker Hata showed another of Ehrlich's arsenical compounds, number 606 or '**Salvarsan**' (Figure 39.1), to be effective against the syphilis-causing spirochaete *Treponema pallidum* while showing acceptably low toxicity. One of the first instances of the successful use of Salvarsan was by Alexander Fleming, after he had discovered lysozyme but before his discovery of penicillin. Indeed in his army regiment, the London Scottish, Fleming was known as 'Private 606'!

Organic derivatives of metals continue to have their place in chemotherapy. For example, some US military participants in the Gulf War were bitten by sandflies, thereby becoming infected with the protozoan *Leishmania tropica*, the causative agent of leishmaniasis, and were successfully treated with **sodium stibogluconate** (**stibofen**, a derivative of antimony). In addition, derivatives of platinum, for example **cisplatin**, are important in the treatment of certain forms of cancer.

The first purely organic synthetic chemical effective against infectious agents was the red dye

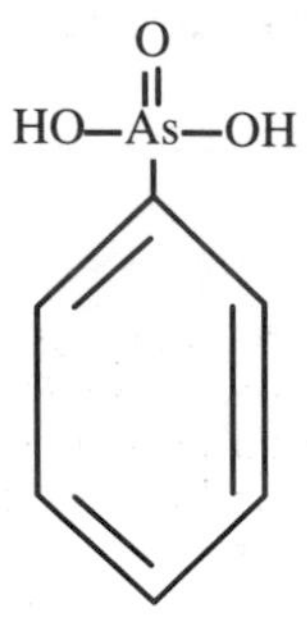

Atoxyl

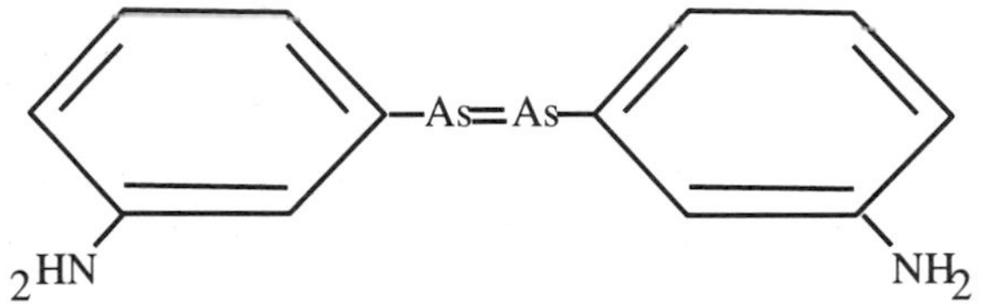

Salvarsan (Compound '606')

Figure 39.1 **Two examples of arsenical drugs**

sulphamindochrysoidine (prontosil) which in 1932 was shown by the German chemist Domagk and his co-workers to control staphylococcal septicaemia in mice but to have no effect on bacterial cultures *in vitro*. So convinced was Domagk of the efficacy of the red dye that when his daughter Hildegarde developed septicaemia he injected her with prontosil, leading to her recovery. It was not until 1935 that these results were published, and soon after this the team of Fourneau, Jaques and Therese Trefouel, Nitti and Bovet working at the Pasteur Institute in Paris synthesized 18 other derivatives of prontosil, showed most of them to share the antibacterial actions of the parent compound and crucially showed that reductive metabolism of prontosil in the host gave rise to sulphanilamide (Figure 39.2) which proved to be the effective drug. The more potent sulphonamides **sulphaguanidine** and **sulphthiazole** followed in 1939. An example of a modern sulphonamide is **sulphamethoxazine**.

Plant extracts

Chemical synthesis was not the only way that chemical compounds capable of controlling microorganisms have been obtained. There is a long history of such drugs

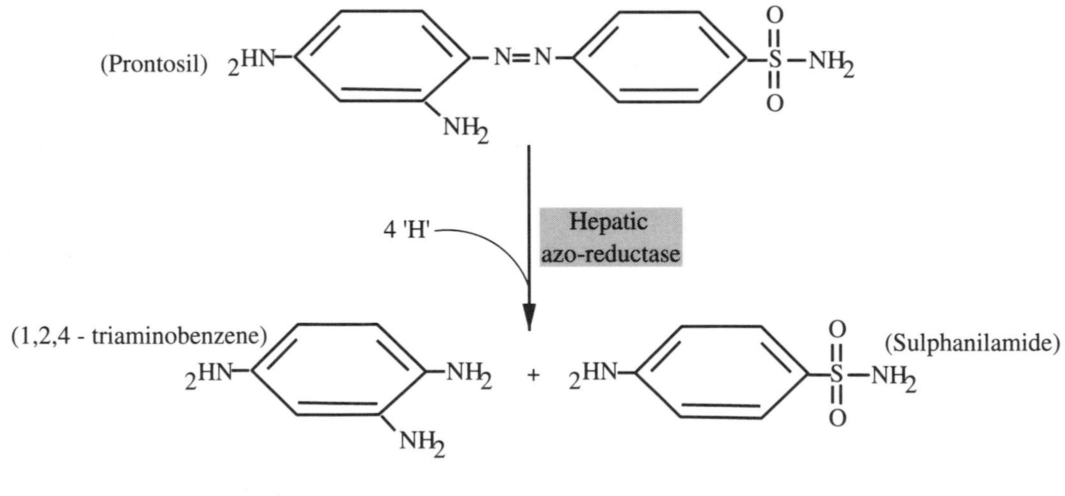

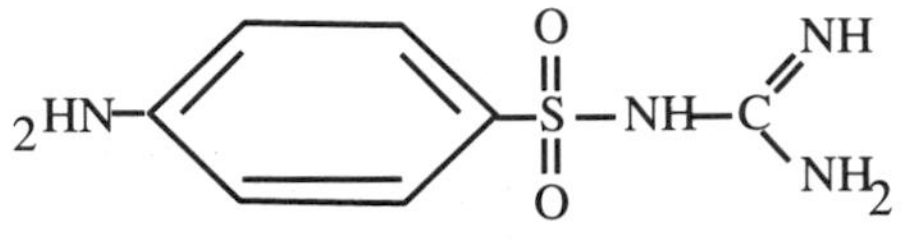

(Sulphaguanidine)

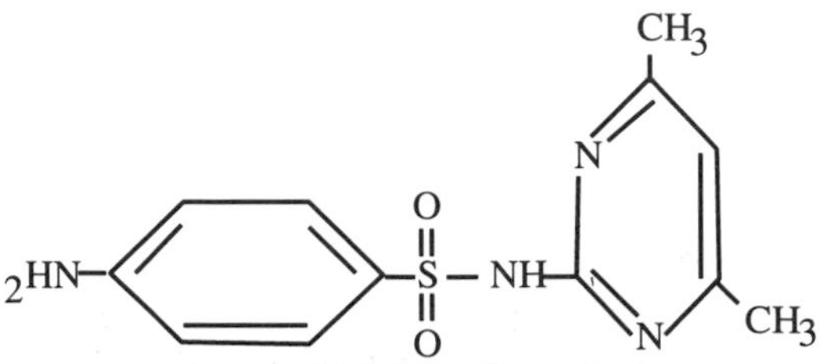

(Sulphamethoxazine)

Figure 39.2 **The metabolic conversion of prontosil into sulphanilamide and two examples of sulphonamide drugs** being extracted from plants. In the sixteenth century the Jesuits who accompanied the Spanish invaders of Peru observed the natives controlling the fever associated with the 'ague' (usually malaria) by using powdered extracts of 'Peruvian bark'. It was widely reported in Europe that 'Jesuits' powder' had been successfully used to treat the malaria of the Countess of Chinchon, the second wife of the Spanish Viceroy to Peru, but this appears to have had no basis in historical fact. Nevertheless, the report led Linnaeus to assign the tree to a new family, *Cinchona*, thereby misspelling the Countess' name but also causing the bark to be renamed 'cinchona bark'. It has been suggested that Oliver Cromwell's death in 1658 may have been prevented had he been prepared to take 'the Jesuits' powder' for his malaria (although his death appears to have resulted from septicaemia). The bark of *cinchona* contains some 30 alkaloids, the most important of which is **quinine** (the Peruvian name for the tree was *quina-quina*). This has antipyretic properties which have made it useful in treating fevers, but in addition it is effective against *Plasmodium* parasites that cause malaria. In the second half of this century the use of quinine has been superseded by more powerful synthetic derivatives such as **chloroquine** and **primaquine**. However there is evidence that *Plasmodium* parasites are becoming resistant to these two drugs. It is of interest, therefore, that the Chinese have recently treated patients with malaria by the administration of *qinghaosu*, extracted from the plant *Artemesia annua*.

A second plant extract of South American origin, used for centuries in Brazil, has proved to contain valuable chemotherapeutic agents. Extracts of the root of the shrub *Cephalis acuminata* contain the alkaloids **emetine** and **cephaline**, and these have proved very useful in the treatment of amoebiasis caused by infection with *Entamoeba histolytica*. Indeed this extract, under the name of ipecacuanha, is still to be found in the casualty departments of many hospitals where its emetic properties are used in treating cases of (usually childhood) poisoning.

Antibiotics

The discovery in 1928 of the first antibiotic, **penicillin**, is correctly attributed to Alexander Fleming. It is worth noting, however, that an extract obtained from *Penicillium glaucum* was first successfully used by Lister 44 years earlier to treat a young nurse, Ellen Jones, who had a large, non-healing abscess. How many of the 20 million deaths of the great influenza pandemic of 1918–19 might have been averted if penicillin had then been available? The victims died largely of pneumonia brought on by their infection with bacteria when in a virus-weakened state. It was not until 1941 that therapeutic quantities of penicillin were obtained from the mould *Penicillium notatum* by Florey, Chain, Abraham and Heatley. The first patient to be given this penicillin was a policeman, Albert Alexander, who developed a generalized infection with *Staphylococcus aureus* and *Streptococcus pyogenes* after suffering a facial scratch from a rose bush. Within five days his temperature was down and he was eating well. Unfortunately, despite its re-isolation from the patient's urine, the supply of penicillin was insufficient to allow a full course of treatment. Consequently, 10 days later, Alexander relapsed and died from staphylococcal septicaemia. This was an early and salutary indication of the importance of completing courses of treatment with antibiotics.

The structure of the form of penicillin produced by the mould *Penicillium chrysogenum* that was found to give the best yield of the drug, then known as penicillin G (now designated benzylpenicillin, still a commonly prescribed antibiotic), was determined by the British X-ray crystallographer Dorothy Hodgkin (Figure 39.3), and the route to its synthesis was developed by the American chemist John Sheehan.

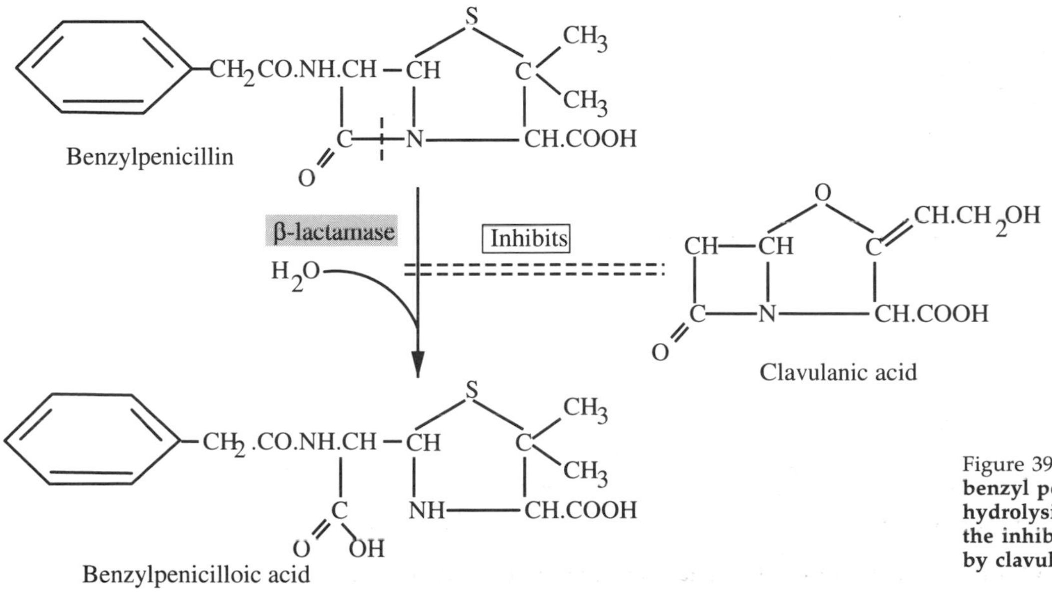

Benzylpenicillin

β-lactamase

H_2O

Inhibits

Clavulanic acid

Benzylpenicilloic acid

Figure 39.3 **The structure of benzyl penicillin, its hydrolysis by β-lactamase and the inhibition of the enzyme by clavulanic acid**

39.3 Sources of new drugs

The efficacy of chemotherapeutic agents obtained from natural sources has meant that a search for such environmental drugs continues. A recent example of such a group of drugs are the **everninomicins**. These compounds were extracted from bacteria obtained from the soil of a dried lake in Kenya. These drugs are effective against infections with enterococci and staphylococci. The first drugs isolated had problems with nephrotoxicity, but this has been avoided by the synthesis of derivatives. The great attraction is that, although the mechanism of action is not known, these drugs do not appear to interact with the three common targets for antibacterial drugs (cell wall structure, protein synthesis and nucleic acid synthesis – see Section 39.4).

As is suggested by the above example, however, increasingly the synthetic chemist and the molecular biologist have come to the fore in the search for new drugs. Initially, the work concentrated on the synthesis and testing of a wide range of analogues of drugs of known efficacy. However, as the molecular targets of many of the most effective drugs are being identified, more is being done by way of rational design. Many of the targets are enzymes and as these are crystallized and their overall architecture, including that of the active site, is determined, drugs can be produced with precisely the correct structure to allow them to bind to and inactivate the enzyme. This also applies to the enzymes that inactivate the drugs (see Section 39.6). For example, in 1986 the three-dimensional structure of the enzyme **β-lactamase** was solved by X-ray crystallography. This enzyme catalyses the hydrolysis and thus the inactivation of penicillin-related drugs (possessing the β-lactam ring). A knowledge of the arrangement of the active centre for this enzyme has led to the introduction of powerful inhibitors which are given with β-lactam-based drugs. A combination currently favoured is that of the antibiotic **amoxicillin** with the β-lactamase inhibitor **clavulanic acid** (Figure 39.3).

39.4 The mode of action of drugs effective against pathogens

Although, as will be shown later in the chapter, the types of drug that are effective in controlling infections may also be useful in the treatment of cancer, this section will concentrate on drugs used for infectious agents.

Effective chemotherapeutic agents achieve their selectivity by targeting some aspect of the biochemistry of the pathogen that has no direct counterpart in the host. In some cases the distinction is that the structure(s), and hence the enzymes that catalyse the biosynthesis of their component parts, do not exist in the host. For example, there is no mammalian equivalent of the cell wall that surrounds bacteria and fungi. Thus, classes of drug prevent the formation of cell walls or disrupt them once they are formed. For most of the remainder of clinically-effective drugs the target molecules are found in both host and pathogen, but they differ in structural detail. Thus, some antifungal drugs, for example the polyene **nystatin**, bind to ergosterol and thereby disrupt the cytoplasmic membrane of the organism. The sterol is used rather than the mammalian cholesterol as part of the membrane. Most drugs in this general category, however, depend on detailed differences in the ways hosts and pathogens carry out protein or nucleic acid synthesis.

Although some drugs are effective against different genera of pathogen, it is convenient to consider each genus of organism separately.

Drugs effective against bacterial infections

Figure 39.4 shows the various sites of action of chemotherapeutic agents effective against bacteria.

Inhibitors of cell-wall biosynthesis: penicillins, cephalosporins, vancomycin and isoniazid

The osmotically-fragile cytoplasmic membrane possessed by all bacteria is protected by additional structures

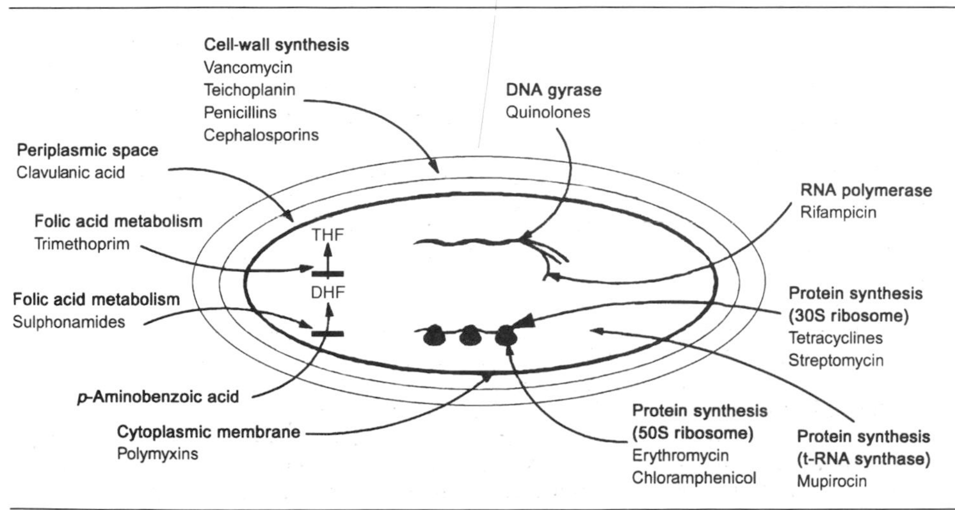

Figure 39.4 **Sites of action of some clinically-important antibiotics effective against bacterial infections**

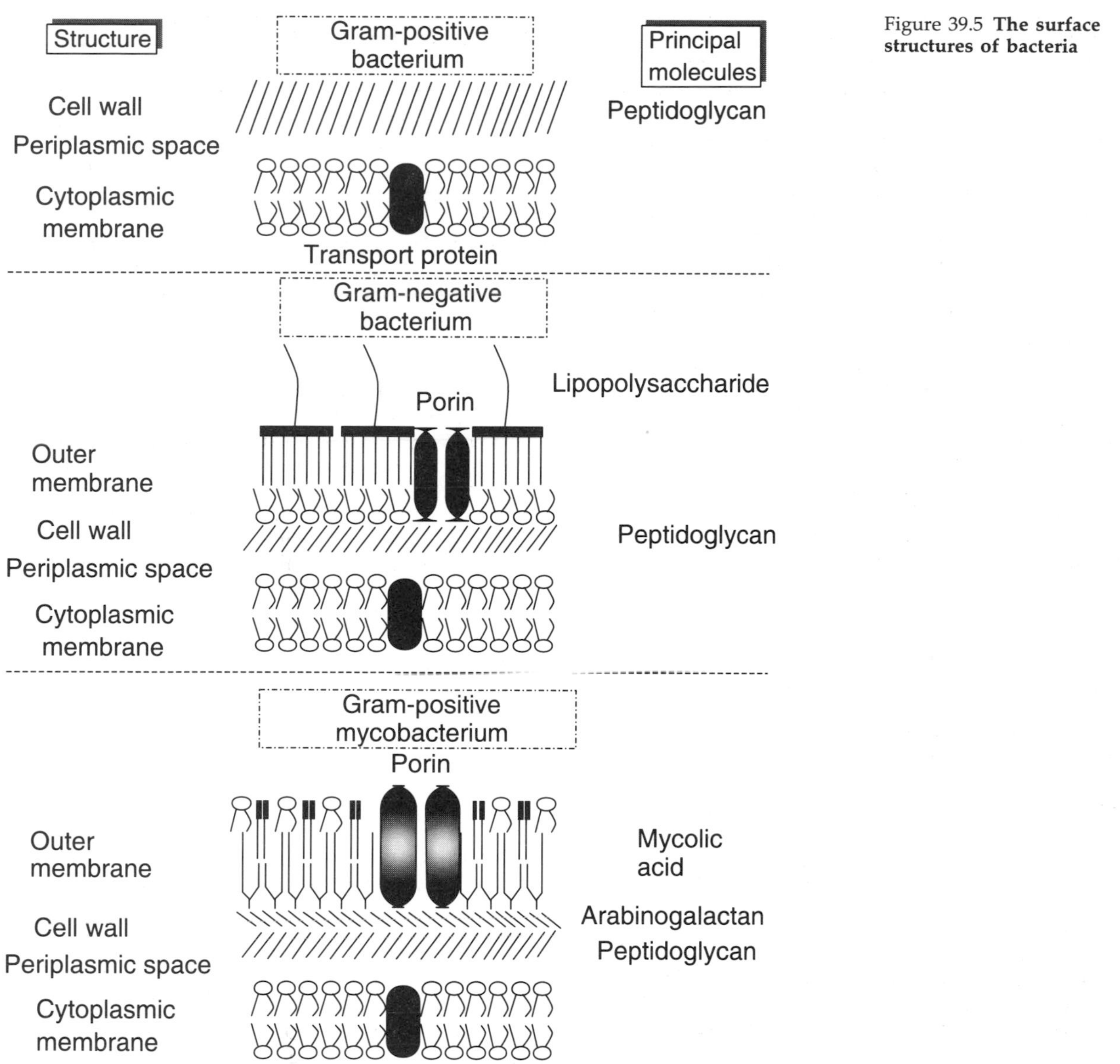

Figure 39.5 **The surface structures of bacteria**

surrounding the cell (Figure 39.5). Bacteria that take up the Gram stain (**Gram-positive**), are surrounded by a thick (up to 20 layers) **peptidoglycan cell wall**. This is mechanically very strong, but the meshwork is rather coarse and such cell walls offer very little by way of a permeability barrier to small molecules, including antibiotics. By contrast, **Gram-negative** bacteria additionally surround themselves with a second membrane (Figure 39.5) of unusual construction, outside a more limited peptidoglycan cell wall. The outer leaflet of this outer membrane is composed of lipopolysaccharide (LPS) which replaces the phosphoglycerolipids found in most other biological membranes (including in the inner leaflet of this membrane). The 6- or 7-fatty acid head groups of the LPS are uniformly saturated, with the resulting effect that this membrane is of low fluidity and therefore presents a very impermeable barrier, so much so that specialized permeation proteins, called **porins**, are found in the membrane. It is through these porins that antibiotics effective against Gram-negative bacteria must pass. Mycobacteria are Gram-positive bacteria that also have a second membrane (Figure 39.5) which is even less permeable than that formed by the LPS. In the mycobacteria, as many as several hundred of the 70 C-atom fatty acid mycolic acid are esterified to an underlying arabinogalactan (based on the pentose arabinose and the hexose galactose) which, in turn, is linked to the underlying peptidoglycan structure. Antibiotics penetrate only very slowly into these organisms.

The presence of peptidoglycan cell walls in bacteria makes their biosynthesis an ideal target for chemotherapy.

The basic repeat unit of the peptidoglycan consists of a disaccharide of *N*-acetylglucosamine in β 1–4 linkage with *N*-acetylmuramic acid (NAcMA). These polysaccharide chains are cross-linked by peptides that

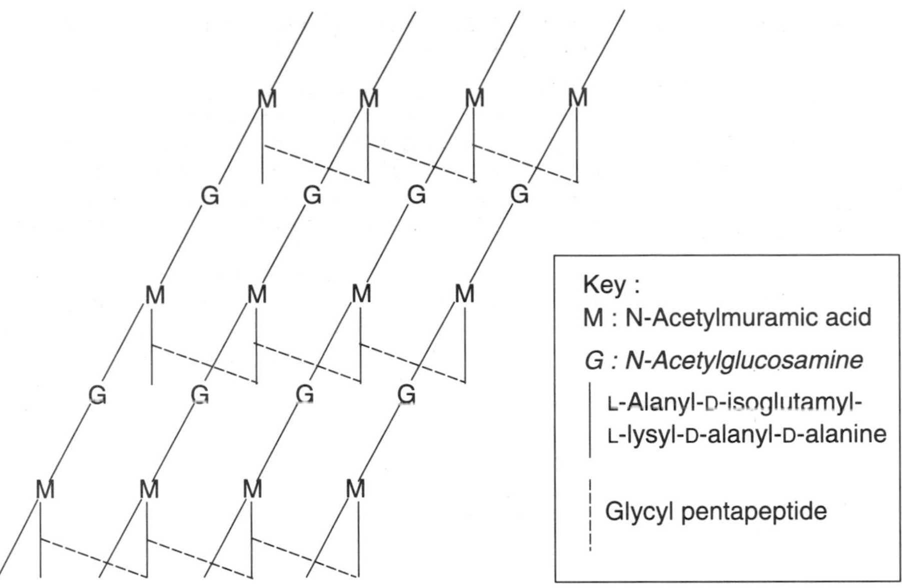

Figure 39.6 **The cross-linked structure of the cell wall of a typical Gram-positive bacterium: *Staphylococcus aureus***

include D-amino acids. The nature of the cross-linking peptides varies from bacterium to bacterium, but the peptidoglycan formed by the Gram-positive *Staphylococcus aureus* will be described for illustrative purposes (Figure 39.6). In the course of the biosynthesis of the cell wall a pentapeptide consisting of L-alanyl-D-isoglutamyl-L-lysyl-D-alanyl-D-alanine is built up attached to the lactosyl group of the NAcMA. Following the formation of the β 1–4 linkage between *N*-acetylglucosamine and the NAcMA-peptide, a second pentapeptide consisting of repeating glycyl residues is attached to the ε-amino group of the lysyl residue of the original peptide. The peptidyl disaccharide is added in further glycosidic linkage to pre-existing cell wall and the crucial enzyme **D-alanyl-D-alanine carboxypeptidase** (a **transpeptidase**) catalyses the attack of the N-terminal glycine of the glycyl pentapeptide on the peptide bond joining the two D-alanyl residues. As a consequence, parallel polysaccharide chains are cross-linked by peptide bridges and the terminal D-alanine is eliminated. (In Gram-negative bacteria the ε-amino group of the lysyl residue carries out the attack on the D-alanyl-D-alanine structure, so cross-linking is direct, with the result that the 'mesh' is less 'coarse'.)

In the cross-linking step for Gram-positive bacteria, the transpeptidase forms a covalent D-alanyl-acyl enzyme intermediate (Figure 39.7). Because of its structural resemblance to the D-alanyl-D-alanine sequence, penicillin becomes attached covalently to the transpeptidase to form a penicilloyl–enzyme complex.

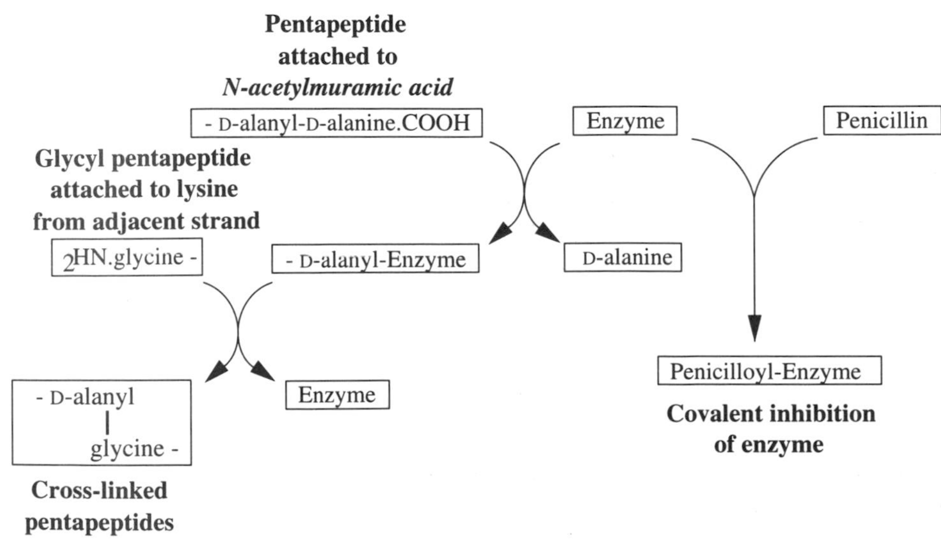

Figure 39.7 **The reaction catalysed by transpeptidase in bacterial cell wall synthesis is inhibited by the covalent binding of penicillin to the enzyme**

The presence of the strained, four-membered β-lactam ring ensures that the enzyme reacts much more readily with penicillin than with the D-alanyl-D-alanine structure, and hence the antibiotic is very efficient in inhibiting cell-wall biosynthesis. It should be noted that different bacteria produce slightly different transpeptidases and since the assay of these enzymes is very difficult their presence is recognized by their ability covalently to bind penicillin which has led to their designation as **penicillin-binding proteins (PBPs)**.

Penicillins are active against a range of Gram-positive bacteria and at higher concentrations they are also inhibitory to certain Gram-negative organisms, for example the diplococci that cause gonorrhoea and meningitis.

Cephalosporins are structurally related to the penicillins, and they appear to act in a similar way. This has led to their use as an alternative to penicillin where resistance is encountered or when patients prove allergic to penicillin. Another drug that acts to inhibit cell-wall biosynthesis and may be used when resistance or allergy to penicillin is encountered is **vancomycin** (and the related **teichoplanin**). This antibiotic is believed to inhibit the attachment of the NAcMA-peptide intermediate of cell-wall biosynthesis to a lipid carrier. Vancomycin is particularly useful because, as a glycopeptide, it in no way resembles the β-lactam antibiotics. Nevertheless it is prescribed only when severe staphylococcal infections are encountered.

Isoniazid is a synthetic drug that is one of the few useful in treating infections with *Mycobacterium* species, for example *M. tuberculosis*. Although a structural analogue of nicotinamide (Figure 39.8) it is believed that isoniazid inhibits the biosynthesis of the mycolic acid found in the outer membrane of the mycobacteria rather than by interfering with the biosynthesis of NAD, as might be expected. Because resistance to isoniazid is becoming widespread, it is administered to patients with tuberculosis in combination with **rifampicin** (an inhibitor of RNA synthesis).

Drugs that disrupt the cytoplasmic membranes of bacteria: polymixins

Several cyclic peptides such as the **polymixins** (Figure 39.8) act by increasing the permeability of the bacterial cytoplasmic membrane to small molecules; they also cause disruption of the outer membranes of Gram-negative bacteria. As can be seen from the structure of **polymixin B**, the peptide includes the unusual amino acid α,γ-diaminobutyric acid (DAB) and has a hydrophobic 'tail' consisting of 6-methyloctanoic acid esterified to a DAB residue. Polymixin B still disrupts the outer membrane when the 6-methyloctanoate is removed, but the drug then has no effect on the cytoplasmic membrane and is no longer bactericidal. It is believed that the 'tail' allows the drug to insert in the cytoplasmic membrane and the cyclic peptide structure forms a hydrophilic 'channel' that conducts small molecules into the bacteria, by passive diffusion, thereby promoting their osmotic disruption.

Inhibitors of protein synthesis: aminoglycosides, macrolides, chloramphenicol, tetracyclines and mupirocin

The fact that the ribosomes of prokaryotic organisms differ in composition and size from those of the eukaryotes (see Chapter 6) has meant that a range of antibiotics that exercise their effects by interfering with protein synthesis have been introduced.

Aminoglycosides are a group of drugs in which an array of amino groups is covalently attached to oligosaccharides. The first member of this family to be isolated was obtained by Selman Waksman and his colleagues in 1943 from the bacterium *Streptomyces griseum*, and named **streptomycin**. At the time of its discovery, the action of streptomycin against Gram-negative organisms greatly commended its use, but its relatively high toxicity has meant that it has been replaced by safer drugs. The aminoglycosides bind

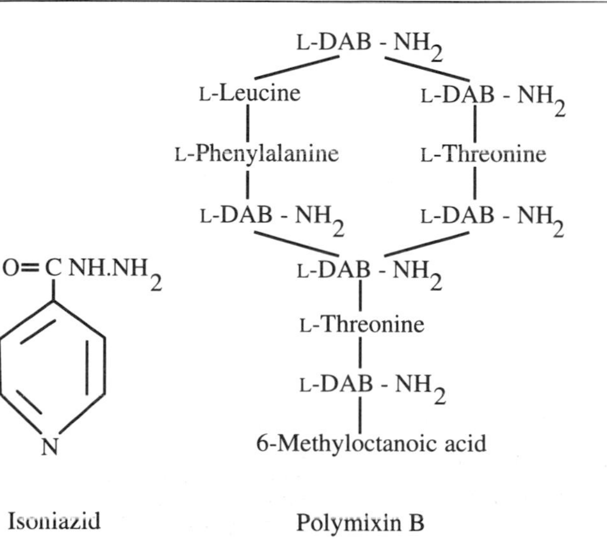

NO_2

HO C H

H C NH.CO.CH_2.Cl_2

CH_2OH

Chloramphenicol

Figure 39.8 **Some examples of antibiotics**

covalently to the 30S ribosomal subunit and prevent the movement of the ribosome relative to the mRNA that underlies the process of translation. **Gentamicin** is an aminoglycoside that is commonly used for serious urinary tract infections.

The most clinically-important drug from the group of antibiotics known as **macrolides** is **erythromycin**. This drug is effective because it interferes with the process of translocation that occurs between the A- and P-sites of the ribosome during protein synthesis, apparently by binding to the 23S ribosomal RNA found in the larger of the two ribosomal subunits. Erythromycin has proved useful in treating Legionnaires' disease caused by the Gram-negative rod *Legionella pneumophila*.

The first member of the group of antibiotics known as **tetracyclines** was obtained in 1948. Members of this family have a common fused, four-ring structure. Tetracyclines, e.g. **doxycycline**, bind to the ribosomal A-site thereby preventing the binding of aminoacyl-tRNA molecules. It has been suggested that their accumulation of tetracycline may explain why the ancient Nubians of Egypt were remarkably free from infectious diseases. The concentration of the compound found in their mummified bones is equivalent to modern therapeutic doses, and probably derived from contamination of their stored grains with *Streptomyces* species

Chloramphenicol (Figure 39.8), the first of the so-called 'broad-spectrum' antibiotics (effective against both Gram-negative and Gram-positive bacteria), was isolated from *Strep. venezuelae* in 1947 by Ehrlich, Burkholder and Gotlieb. It is effective because it inhibits the actual process of peptide bond formation in bacteria. Its toxicity in bone marrow means that the use of chloramphenicol is restricted to topical application (for eye infections, for example) or for severe systemic infections, e.g. meningitis caused by *Neisseria* and *Haemophilus* species.

Mupirocin is another useful drug that also inhibits protein synthesis, but in this case it does so by interfering with the charging of the amino acid leucine to its tRNA molecule, i.e. an **inhibitor of leucyl-tRNA synthase**. Mupirocin is applied topically, e.g. to the nose where it acts to eradicate the nasal carriage of *Staphylococcus aureus*. This is most useful because the nose-to-hand route is important for the spread of methicillin- (a semisynthetic penicillin) resistant *Staph. aureus*, as occurs in hospitals, from hospital workers to patients or from patient to patient.

Inhibitors of the biosynthesis of nucleic acids

Drugs that fall into this category may act in one of three main ways: they interfere with the processes of replication or transcription or they interfere with the biosynthesis of nucleoside triphosphates (the nucleic acid precursors; see Chapter 4). Compounds in the later category are referred to as **antimetabolites** (see Section 39.7). The interference with replication or transcription may arise because the drug binds strongly to DNA, they inhibit DNA or RNA polymerases or they interfere with the activity of accessory proteins, e.g. DNA gyrases.

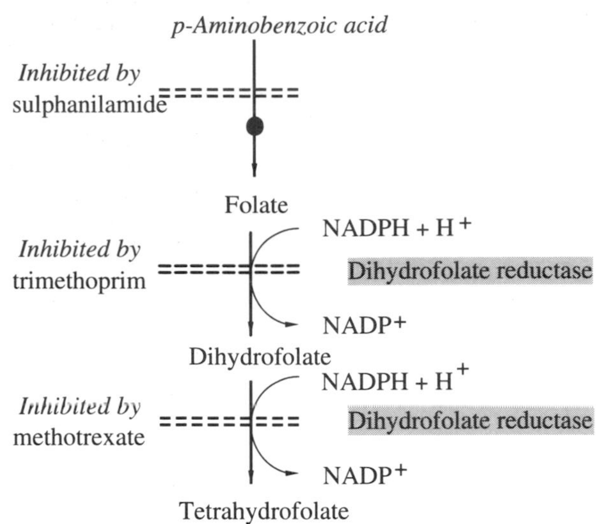

Figure 39.9 **Inhibitors of the formation of folic acid and its reduction to the tetrahydro form**

Antimetabolites

The discovery of the antibacterial activity of the **sulphonamides** has already been referred to: their mode of action was proposed in 1939 by the British scientists Woods and Fildes, thereby indicating the reason for the selective toxicity of this group of drugs. These workers proposed that certain bacteria are able to carry out the biosynthesis of the human vitamin folic acid, in which one stage requires *para*aminobenzoic acid as a substrate and **sulphanilamide** (and, in general, the sulphonamides) is a sufficiently close structural analogue for it to act as a strong, competitive inhibitor of the reaction (Figure 39.9). A major role of folic acid is in the biosynthesis of the purines, with the consequence that in the absence of the vitamin, and hence nucleoside triphosphate precursors, the biosynthesis of nucleic acids ceases. More recently the drug **trimethoprim** has been introduced. This also interferes with folic acid metabolism, but at a later stage in which folate is reduced to the dihydro- form by the enzyme **dihydrofolate reductase** (Figure 39.9). This reaction does have its counterpart in mammalian biochemistry, but the selectivity arises because trimethoprim binds approximately 5×10^4 more avidly to the bacterial or protozoal enzymes than it does to the form found in humans. Trimethoprim is combined with the sulphonamide **sulphamethoxazole** in the treatment of *Pneumocystis* pneumonia, for example. The conversion of dihydrofolate to tetrahydrofolate, which is also catalysed by dihydrofolate reductase, is inhibited by the drug **methotrexate**, which is used in cancer chemotherapy (see Section 39.7).

Direct inhibitors of nucleic acid biosynthesis

Of the many compounds that have been shown to inhibit the biosynthesis of nucleic acids, two types of drug have significant clinical use: the **quinolones** and the

Table 39.1 **Classification of some commonly-used antibiotics according to their mechanism of action**

Site of action	*Examples of antibiotics*	*Mechanism of action*
1. Inhibitors of cell wall biosynthesis	*β-Lactams* • Penicillins • Cephalosporins	Inhibit D-alanyl-D-alanine carboxypeptidases (transpeptidases or penicillin-binding proteins – PBPs)
	Vancomycin Teichoplanin	Inhibit the formation of lipid–carrier complexes
	Isoniazid	Inhibits the biosynthesis of mycocolic acid by mycobacteria
2. Disrupt cytoplasmic membranes	*Polymixins* • Polymixin B	Insertion into the cytoplasmic membrane, leading to osmotic disruption of cell
3. Inhibitors of protein biosynthesis	*Aminoglycosides* • Streptomycin • Gentamicin	Covalent binding to 30S ribosomal subunit
	Macrolides • Erythromycin	Interferes with translocation between 'A' and 'P' sites on ribosomes
	Tetracyclines • Doxycycline	Prevents the binding of aminoacyl tRNA to the ribosomal 'A' site
	Chloramphenicol Mupirocin	Inhibits peptide bond formation Inhibits leucyl tRNA synthase
4. Inhibitors of nucleic acid biosynthesis	*Antimetabolites* • Sulphamethoxazole • Trimethoprim	Interfere with the functions of folate
	Quinolones • Ciprofloxacin	Inhibits DNA gyrase
	Rifamycins • Rifampicin	Inhibits DNA-dependent RNA polymerase

rifamycins. The quinolones inhibit the enzyme DNA gyrase and thereby interfere with replication. The first quinolone to be used was **nalidixic acid**, but this has been replaced by the semi-synthetic **fluoroquinolone, ciprofloxacin. Rifampicin** is an inhibitor of the bacterial, but not the mammalian, DNA-dependent RNA polymerase, interacting with β-subunit of the enzyme.

Table 39.1 shows a classification of the commonly-used antibacterial drugs by their mechanisms of action.

Drugs effective against fungal infections

In the past 20 years the incidence of fungal infections has increased dramatically. The increase has been most marked in immunosuppressed patients (those with organ transplants, those undergoing cancer chemotherapy and those with AIDS), in patients treated with broad-spectrum antibiotics and in patients subject to invasive procedures (placements of catheters or prostheses, etc.). The situation is such that the fungus *Candida albicans* is the fourth most common blood isolate in the USA; the incidence of invasive candidiasis having increased ten-fold in the 1980s. Invasive pulmonary aspergillosis is a leading cause of mortality in bone marrow transplant patients and *Pneumocystis carinii* causes pneumonia and death in many patients with AIDS in both North America and Europe.

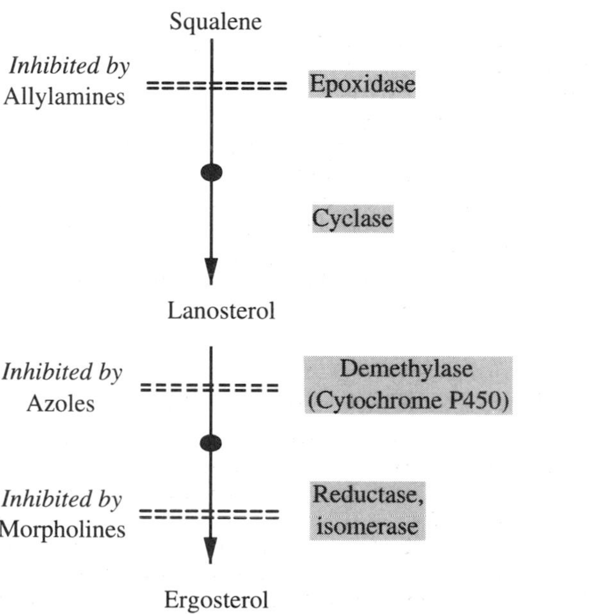

Figure 39.10 **Antifungal drugs that inhibit the biosynthesis of ergosterol**

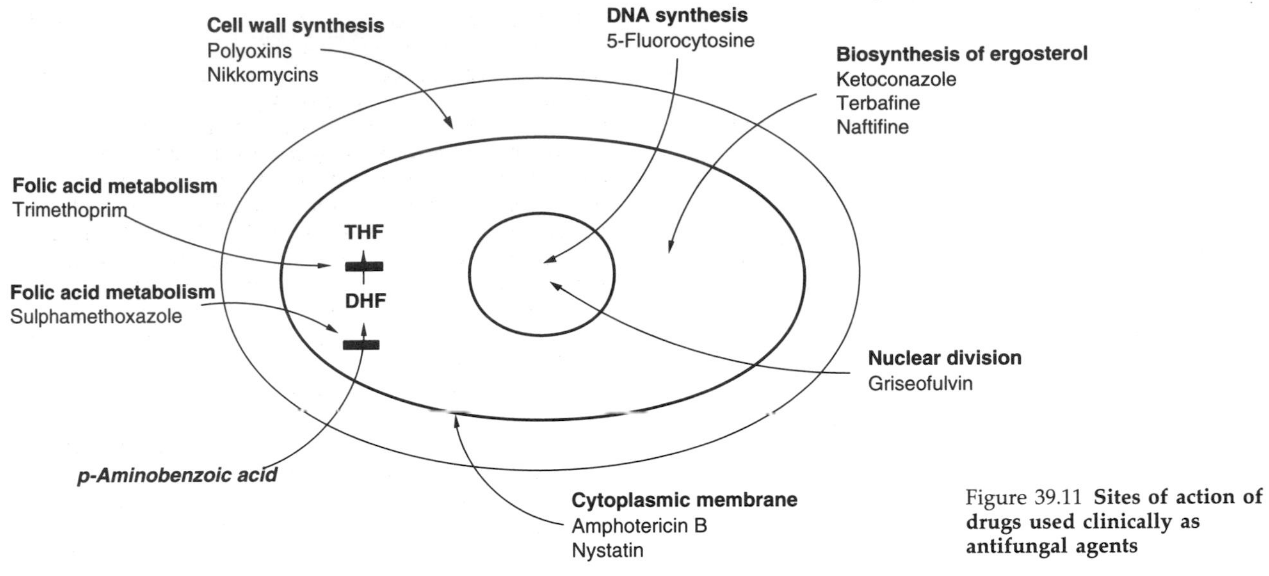

Figure 39.11 **Sites of action of drugs used clinically as antifungal agents**

Three major groups of antifungal compounds are in extensive clinical use: the **polyene** antibiotics, the **azole** derivatives and the **allylamines**. All interact with or inhibit the production of ergosterol, the major sterol of the cytoplasmic membrane of fungal cells (Figures 39.10 and 39.11). Polyenes, such as **nystati**, discovered in the 1950s by Elizabeth Hazen and Rachel Brown, and **amphotericin B** disrupt the cytoplasmic membrane of fungi by interacting with ergosterol, the fungal counterpart of the mammalian cholesterol. Both of these polyene drugs are quite toxic and are usually used topically, occasionally orally but never by injection. Recent findings suggest, however, that entrapment of the drugs in liposomes (artificially-generated lipid vesicles) is effective in reducing their toxic side-effects.

Azoles are purely synthetic drugs that inhibit ergosterol biosynthesis at the (cytochrome P450-dependent) C14 demethylation step (Figure 39.10). The overall effect of the azoles tends to be fungistatic rather than fungitoxic (i.e. growth inhibition rather than cell death). **Ketoconazole** is one such drug, but it has the drawback of inhibiting the cytochrome P450-dependent steps in steroid hormone biosynthesis in humans.

Two allylamine antifungals **naftifine** (used topically) and **terbinafine** (given orally) act to inhibit squalene epoxidase, which catalyses an early step in the biosynthesis of ergosterol (Figure 39.10).

It should be noted that none of the above drugs is effective against *Pneumocystis carinii* which incorporates cholesterol, not ergosterol, into its cytoplasmic membrane.

Other drugs useful against fungal infections include **griseofulvin** which inhibits the process of nuclear division in fungi (the agricultural fungicide **benomyl** has the same effect, but is too toxic for use in humans). Griseofulvin is an older drug, but it is still used for dermatophytosis of the scalp, body and groin. **Flucytosine**, an antimetabolite fluoro-analogue of the pyrimidine cytosine, is effective in inhibiting nucleic acid synthesis by fungi. It is often given in combination with the polyene amphotericin B for systemic fungal infections.

Some drugs, such as the **polyoxins** and **nikkomycins**, act to inhibit the biosynthesis of the fungal cell wall (and hence resemble penicillin). The cell wall in fungi consists of the polysaccharide chitin (a homopolymer of *N*-acetylglucosamine). These drugs are structural analogues for UDP-*N*-acetylglucosamine the substrate for the enzyme, **chitin synthase**, which they therefore inhibit. These drugs are of rather low efficacy because their target enzyme is located on the inner aspect of the cell cytoplasmic membrane and transport by a dipeptide permease is necessary for the enzyme and its potential inhibitor to be brought together. This happens inefficiently because natural body dipeptides effectively compete with the drug for the transport mechanism.

Figure 39.11 shows the sites of action of the principal antifungal agents.

Drugs effective against viral infections

The search for antiviral agents has proved much more difficult than that for drugs effective against bacteria. This is indicated by the fact that, in 1994, about 100 antibiotics, but only 20 antiviral drugs, were approved for use in the USA. The really effective agents number just a handful. These drugs focus in their action on DNA synthesis. Of these, a very successful example has been **acyclovir** (Figure 39.12). This drug was introduced in the 1980s for the treatment of diseases caused by three types of DNA-containing herpesviruses: herpes simplex 1 and 2 (HSV-1 and HSV-2) and herpes varicella-zoster (the causative agent in chicken pox and shingles). Acyclovir is a structural analogue of deoxyguanosine and its selectivity depends on subtle differences in activity between the two viral and mammalian enzymes that use deoxyguanosine and its triphosphate. The thymidylate kinase encoded in the HSV genome catalyses the phosphorylation of acyclovir at a rate that exceeds that

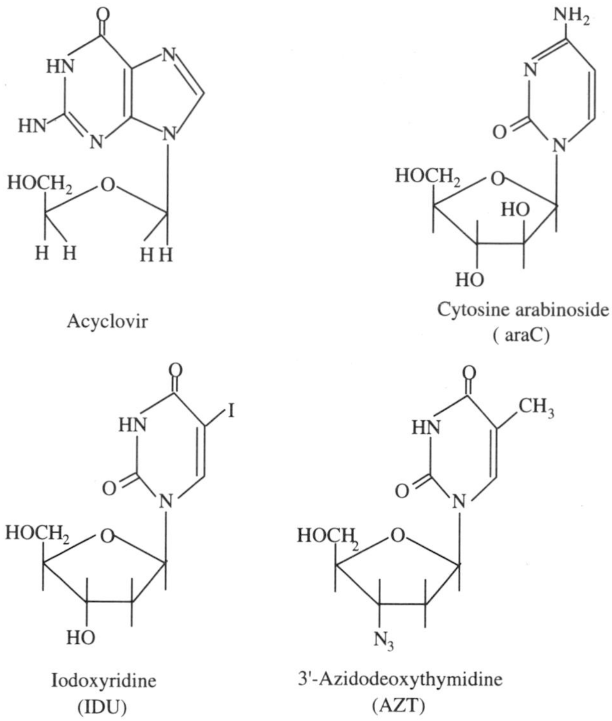

Acyclovir

Cytosine arabinoside (araC)

Iodoxyridine (IDU)

3'-Azidodeoxythymidine (AZT)

Figure 39.12 **Some drugs that inhibit viral DNA synthesis**

of the host cell enzyme by a factor of 200. For this reason, infected cells accumulate acyclovir monophosphate which is sequentially phosphorylated to the triphosphate and this is used as a substrate by the viral DNA polymerase in preference to deoxyguanosine triphosphate. Once incorporated into the growing DNA strands, the acyclovir continues to bind strongly to the viral DNA polymerase and as a consequence replication ceases in infected cells.

Other nucleoside analogues are used, usually topically, for viral infections. These include **iodoxyridine (IDU)** and **cytosine arabinoside (araC)** (Figure 39.12) but these do not have the degree of selectivity shown by acyclovir and depend for their action on virally-infected cells being very much more active than host cells in DNA synthesis.

Despite the huge efforts that have gone into the project, only one drug appears to have met with any measure of success in combating the human immunodeficiency viruses (HIV) that cause AIDS; this is the drug **AZT** also known as **zidovudine** or **3′-azidodeoxythymidine** (Figure 39.12). This drug does not cure the disease, but simply acts to retard its progress (or perhaps it reduces the incidence of opportunistic infections). As an analogue of thymidine it undergoes phosphorylation to the ′triphosphate which is incorporated by the retrovirally-encoded reverse transcriptase (RNA-dependent DNA polymerase; see Chapter 4) and this prevents further elongation of the DNA chain. AZT is about 100 times more efficient at inhibiting reverse transcriptase than it is against the host DNA polymerase. In fact, AZT inhibits HIV genome replication at a concentration of one-thousandth that required to inhibit replication of DNA in lymphocytes.

The prospects of finding a single drug for the successful treatment of AIDS seems remote, for it has been estimated that an asymptomatic patient infected with HIV harbours about 10^6 genetically-distinct virions (the number rises to 10^8 in AIDS). Under these circumstances it is almost certain that viruses with mutations allowing them to resist any drug (directed against any viral enzyme) already exist in asymptomatic subjects, and the use of a single drug will simply favour their selection.

Drugs effective against protozoa and helminths

Although estimates of the number of people in the world with diseases due to unicellular and multicellular parasites vary, a low estimate is still one of 600 million. Of these, 270 million have malaria and 200 million schistosomiasis (bilharzia or snail fever) (Table 39.2). It is estimated that parasitic diseases cause well in excess of 2 million deaths worldwide each year and of these half a million are due to childhood malaria. Most of the attempts at intervention are directed at the control of the vectors for the parasites and these fall outside the scope of the present chapter. Unfortunately the foci for the occurrence of parasitic diseases are the developing and poor countries of the world. In these there is perceived to be no substantial market for drugs and hence their development has lagged behind the need. Nevertheless, there are some drugs used widely against parasites.

Table 39.2 **Prevalence of parasitic diseases and their vectors**

Disease	*Estimated number of people infected worldwide*	*Vector*
1. Chagas' disease	18 million	Tritomid bugs
2. Filariasis	90 million	Mosquitoes, black fleas, mites
3. Guinea worm disease (dracunculiasis)	250 thousand	Water fleas
4. Leishmaniasis	12 million	Sandflies
5. Malaria	270 million	Mosquitoes
6. Schistosomiasis (bilharzia, snail fever)	200 million	Snails
7. Sleeping sickness (trypanosomiasis)	250 thousand	Tsetse fly

Protozoa

The drugs **chloroquine** and **quinacrine** are used against protozoal infections. They both act by intercalating between the bases of DNA, thus preventing replication (see Section 39.7). Chloroquine is used against the *Plasmodia* that cause malaria, while quinacrine is effective against *Giardia lamblia*, the causative agent in giardiasis. The later disease may also be treated with **metronidazole (flagyl)**. This is a widely-used antiprotozoal drug and is also useful for vaginitis caused by *Trichomonas vaginalis* or *Gardnerella vaginalis*. The mechanism of action for flagyl is uncertain, but it has been suggested that it promotes the formation of reactive oxygen species that cause damage to the DNA. **Nitrofurans** such as **furazolidine**, which act as protein synthesis inhibitors, have been used against *Trypanosoma cruzi*, the protozoan causing Chagas' disease (from which Charles Darwin is believed to have suffered, following his trips to South America).

Difluoromethylornithine (DFMO, efluornithine), a drug with a quite different action, has proved useful in treating African sleeping sickness caused by infection with *Trypanosoma gambiense*. DFMO is a 'mechanism-based' (suicide) inhibitor of the enzyme ornithine decarboxylase. This enzyme catalyses the first step in the formation of the polyamines spermidine and spermine, and in using DFMO as a substrate it generates a powerful alkylating agent that binds covalently to the active centre to inactivate it. This means that that the precursor of the polyamines, putrescine, is not formed (see Chapter 23). Polyamines are required for many cellular processes, including the packaging of DNA, but the reason for the trypanosomal toxicity of DFMO is uncertain. A degree of selectivity is achieved because the *T. gambiense* enzyme has a very long half-life compared to the mammalian enzyme: hours compared with 15 min. This means that inactivated enzyme is rapidly replaced in the host, but not in the parasite.

It should be borne in mind that protozoa may exist in different forms, e.g. as cysts, during their life-cycles and susceptibility to drugs may vary as a consequence.

Helminths

Tapeworm infections, such as might occur with *Diphyllobothrum latum*, may be controlled by using **niclosamide**. The action of this compound is quite different from that of other commonly-used drugs in that it acts to inhibit the process of oxidative phosphorylation in these helminths. The incidence of tapeworm infestations appears to be on the increase amongst *sushi* (raw fish) eaters. The schistosome parasite causing schistosomiasis is treated with antimony-based drugs such as **stibophen**. The antimony in the drug inhibits glycolysis in the helminths, probably by inactivating enzymes such as pyruvate kinase that have a thiol group in their active centre. Stibofen is frequently combined with efluornithine.

The cells of the parasites under consideration in this section are eukaryotic and this means that the drugs referred to here target processes also found in mammalian cells. Thus, these therapies are not very specific. Consequently the drugs tend to be toxic and may only be used for aggressive, short-term treatments. The known targets for some antiparasitic drugs are given in Table 39.3

Prion-related diseases

Prions cause a range of neurodegenerative diseases that in origin may be infectious, inherited or sporadic (of random occurrence).

The diseases with which they are associated include two in humans that may be inherited, namely **fatal familial insomnia** (FFI) and **Creutzfeldt–Jakob disease (CJD)**, and **kure** that has the characteristics of an infectious disease. In other animals, prions are responsible for scrapie in sheep and bovine spongiform encephalopathy (BSE, 'mad cow disease'). Recently the fear has arisen that the infectious prion, or a variant, that causes BSE may also be the cause an early-onset form of CJD in humans, but the scientific evidence is far from being convincing.

In contrast to the infectious agents discussed previously in the current chapter, prions have no nucleic acid-containing genomes. The only established constituent is the protease-resistant protein designated PrP^{Sc}. This protein is an altered version of a normal cellular protein of the central nervous system designated PrP^{C}. The intronless gene encoding PrP^{C} is found on chromosome 20 and the protein product is sensitive to proteolysis. Mutations to this gene which occur in some cases of CJD and FFI give rise to prions which consist of large (amyloid) insoluble aggregates of prion protein and glycosaminoglycans (plaques). In addition, however, the unmutated gene for PrP^{C} may be

Table 39.3 **Drugs effective in treating parasitic infections**

Disease	*Parasitic target*	*Drug used*	*Biochemical mechanism of action*
	PROTOZOA		
Malaria	*Plasmodium falciparum*	Chloroquine	Disrupts replication by intercalating in the DNA
Giardiasis	*Gardia lamblia*	Quinacrine	Disrupts replication by intercalating in the DNA
Vaginitis	*Gardnerella vaginalis*	Metronidazole (flagyl)	Promotes damage to DNA by 'reactive oxygen species'
Chagas' disease	*Trypanosoma cruzi*	Furazolidine	Protein synthesis inhibitor
Sleeping sickness	*T. gambiense*	Efluornithine	Interferes with the biosynthesis of polyamines
	HELMINTHS		
Infestation with tapeworms	*Diphyllobothrum latum*	Niclosamide	Inhibits oxidative phosphorylation
Schistosomiasis ('snail fever')	*Schistosoma haematobium*	Stibophen	Inhibits glycolysis

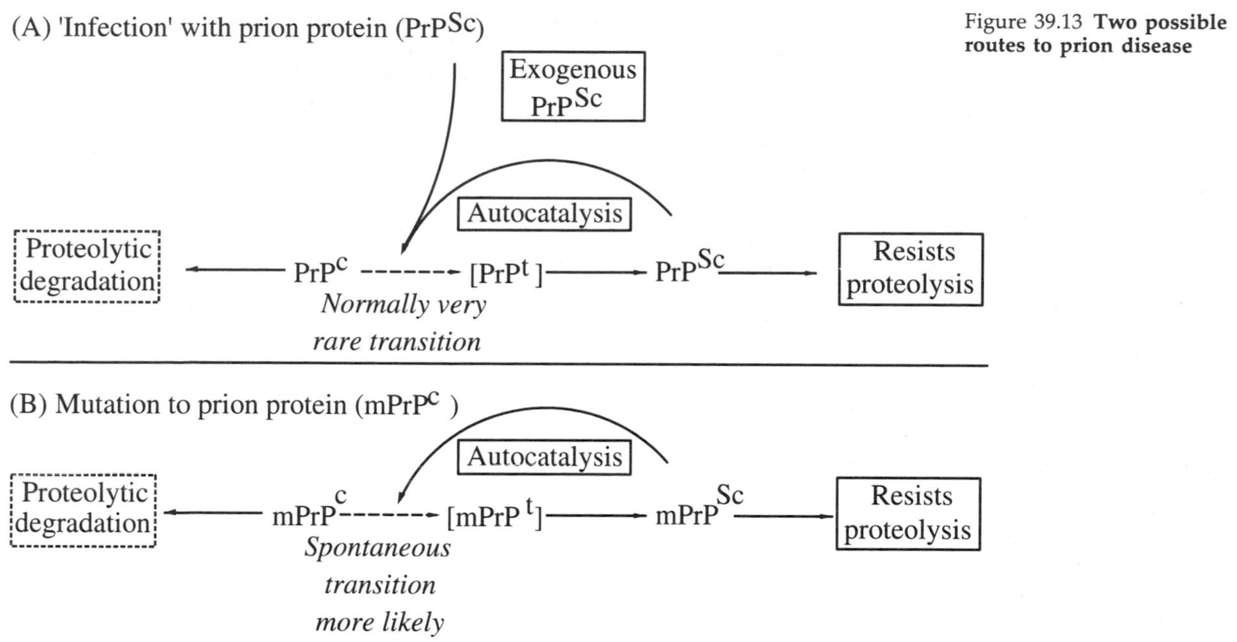

Figure 39.13 **Two possible routes to prion disease**

associated with PrP^{Sc} production and hence with prion formation. The conclusion that the transition from PrP^C to PrP^{Sc} arises due to conformational change to the former is supported by the observation that the former has very little β-sheet in its secondary organization, whereas the latter has 43%. The normal role of PrP^C in nerve cells has not been established, but it is known to be a protein of the outer aspect of the plasma membrane where it is anchored by a lipid 'tail'.

How is it possible for a protein to be infectious and how does this relate to the familial and sporadic prion diseases? Possible mechanisms are shown in Figure 39.13. This suggests that the transition from PrP^C to PrP^{Sc} can be autocatalytic, and therefore infection of cells with PrP^{Sc} coming from other cells, already containing the molecule, would cause PrP^C to be converted into PrP^{Sc}. In sporadic cases of prion disease (rare) random events leading to a PrP^C to PrP^{Sc} transition would also be autocatalytic. The occurrence of mutations to PrP^C would also favour the transition to PrP^{Sc}.

This brief description of prion disease suggests that either preventative or curative treatments are likely to be very difficult to devise. The expression of a normal gene will be difficult to target selectively, as will the uptake of the prion particle. One possible approach that is currently under investigation is via the inhibition of the folding pathway that leads to the formation of Pr^{Sc}. It has been shown that the amyloid-binding dye Congo red is able to inhibit the accumulation of Pr^{Sc}. Amyloid is rich in sulphated glycosaminoglycans, to which Congo red binds, and it is suggested that these polysaccharides assist in the folding of Pr^C to yield Pr^{Sc}. According to the amyloid hypothesis, Congo red, by binding to amyloid, competitively inhibits the binding and consequently the aberrant folding of Pr^C. This hypothesis is strengthened by the finding that certain complex sulphated glycans, of the type that bind to amyloid, are found to act prophylactically against scrapie in sheep.

39.5 The occurrence of drug resistance in bacteria

Although, as might be expected on evolutionary grounds, the development of drug-resistant organisms is encountered with all of the classes of pathogen, this section will focus on bacteria. This is both because the emergence of bacterial resistance is particularly serious and also the mechanisms underlying the resistance are well understood.

Resistant bacteria are of two kinds:

1. Unaltered strains that have inherent resistance to antibiotics. These have been selected as a result of the control of competing organisms by the often indiscriminate use of antibiotics. These bacteria, e.g. *Pseudomonas aeruginosa*, are not strongly virulent and are termed 'opportunistic pathogens', they tend to infect debilitated hospital patients. Such bacteria are said to have **intrinsic resistance.**
2. Well-known 'professional pathogens' for which drug-resistant strains have been selected by the use of antibiotics. Often they will have received their resistance to a drug from another organism via an R-(drug-resistance conferring) plasmid: the bacteria have **acquired resistance**.

Processes leading to drug resistance

Two broad categories of processes leading to drug resistance may be recognized – general and specific.

General processes leading to drug resistance

These include the occurrence of low permeability barriers (as described in Section 39.4 for Gram-negative bacteria and mycobacterium) and of active transport systems that promote the efflux of drugs from cells. Some of these active transport proteins are specific for one type of drug, but others promote the extrusion of a wide range of apparently unrelated drugs. For example, the gene designated MxA of *Pseudomonas aeruginosa* encodes a cytoplasmic membrane protein that actively pumps out tetracycline, chloramphenicol and norfloxacin (a fluoroquinolone). The drugs best transported seem to be amphipathic in character. In the case of Gram-negative bacteria, 'adapter' molecules serve to connect the transporter of the inner membrane with porins of the outer membrane.

Specific processes leading to drug resistance

These may either affect the drug itself or its target protein. In the first category, enzymes that metabolize the drug may lead to its degradation or simply to its modification, leading to inactivation. Thus, acetylation of chloramphenicol by chloramphenicol–acetyl transferase leads to inactivation of this drug. Likewise, hydrolysis of the β-lactam ring of penicillin and related drugs by β-lactamase leads to loss of activity of the antibiotic (see Figure 39.3). The drug **clavulanic acid** has been introduced specifically to protect β-lactam drugs by inhibiting β-lactamase.

In most cases, resistance to penicillin-like drugs is due to the presence of β-lactamase, but in an increasing number of instances the resistant bacteria are found to have penicillin-binding proteins (transpeptidases – see Section 39.4) with a low affinity for the drug, i.e. with an alteration to the drug target. This has been found to be the case for *Neisseria gonorrhoeae* and *Streptococcus pneumoniae*, for example, in which point mutations to PBP-encoding genes lead to the production of proteins with reduced affinity for penicillin. Resistance due to changes in PBPs tends to be poor, because most bacteria use several PBPs and each needs to be altered for resistance to be effective, for example *Strep. pneumoniae* produces four PBPs.

Although resistance to natural antibiotics usually arises due to their metabolism, another example of an altered target is seen in the case of resistance to rifampicin. In this, a number of different point mutations to the β-subunit of RNA polymerase means that the drug binds only with very low affinity to the enzyme. This type of resistance is seen in *Mycobacterium tuberculosis* and *M. leprae* (the causative agent in leprosy).

Target changes due to point mutations in the A-subunit of DNA gyrase underlie many cases of resistance to the quinolones and fluoroquinolones. For example, in *Staph. aureus* a single point mutation to the DNA gyrase A-subunit gene increases the minimum inhibitory concentration (MIC) of the drug from 16 to 128 μg/ml. Likewise, point mutations to one of the genes encoding enzymes involved in the biosynthesis of mycolic acid underlies the resistance of some strains of *M. tuberculosis* to isoniazid.

Point mutations are not the only means whereby resistance to drugs due to target changes arises: in some cases of failure of sulphonamide- or of trimethoprim-based therapies the organism is found to have acquired a new gene for dihydropteroate synthase and dihydrofolate reductase, respectively (see Section 39.4).

In other cases, target alterations occur as the result of acquisition of a gene that encodes an enzyme that catalyses the covalent modification of the target. For example, resistance to macrolide antibiotics such as erythromycin results from the activity of an enzyme that causes the methylation of the 23S ribosomal RNA molecule. As a consequence the drugs bind only weakly to ribosomes and have little effect on protein synthesis. Point mutations to the genes that encode ribosomal proteins may also lead to drug resistance.

The various mechanisms underlying bacterial drug resistance are summarized in Table 39.4. It should be

Table 39.4 **Mechanisms underlying the resistance of bacteria to some drugs**

Drug	*Mechanism of resistance*	*Location of gene associated with resistance*	*Organisms in which resistance is causing particular concern*
β-Lactams • Penicillins • Cephalosporins	Altered penicillin-binding proteins Reduced permeability β-lactamase	Bacterial main chromosome Bacterial main chromosome Bacterial main chromosome and plasmid	*Staphylococcus aureus* *Staph. epidermidis* *Pseudomonas aeruginosa* *Neisseria gonorrhoeae*
Aminoglycosides • Gentamicin	Reduced binding to ribosomes	Bacterial main chromosome	*Streptococci*
Macrolides • Erythromycin	Methylation	Bacterial main chromosome and plasmid	*Strep. pneumoniae* of rRNA
Chloramphenicol	Acetyltransferase	Bacterial main chromosome and plasmid	*Strep. pneumoniae*
Tetracyclines • Tetracycline • Doxycyclin	Efflux	Plasmid	
Rifampicin	Reduced binding to DNA polymerase	Bacterial main chromosome	*Staph. epidermidis*
Folate antimetabolites • Trimethoprim	Reduced permeability	Bacterial main chromosome	*Campylobacter jejuni*
Mupirocin	Change to leucyl-tRNA synthetase	Bacterial main chromosome	*Staph. aureus*

borne in mind that several strains of bacteria are able to resist drugs because they have several of the types of change listed. For example, it is not uncommon in hospital isolates to find resistance to β-lactam antibiotics owing jointly to the presence of β-lactamase, altered PBPs and reduced uptake. Likewise, the occurrence of multi-drug resistance is not uncommon.

39.6 Prevention of the symptoms of infectious diseases

As our understanding of the molecular biology of host responses to infection, and of how these may be perverted in disease, are being elucidated attention is being directed at the possibility of producing drugs that prevent aberrant host responses, i.e. the treatment of symptoms of infectious diseases. Almost all such drugs are directed at the responses of the immune system.

Three reasons can be recognized for trying to prevent aberrant host responses:

1. It is frequently the case that these responses themselves are life-threatening; if they can be arrested, more time is thereby made available for the selection and application of more conventional methods of chemotherapy.
2. Several quite different organisms may give rise to a particular disease and these probably will not all be susceptible to the same antibiotics. This is well illustrated by the occurrence of sepsis in which just over half of the cases are due to Gram-positive bacteria and just under half to Gram-negative bacteria (e.g. *Escherichia coli*). For these patients, who annually total half a million in the USA and of whom a third die, it is vital that there is sufficient time to make a proper diagnosis and institute the appropriate therapy.
3. As the drugs do not need to come into contact with the pathogen, drug resistance becomes much less likely.

Figure 39.14 shows the points in the cascade of immune responses evoked by bacterial infection and at which intervention is being targeted. A central plank in these strategies is the application of molecules, monoclonal antibodies or 'decoy' receptors that bind the signalling molecules of the immune system such as interleukin 1 and tumour necrosis factor α (IL-1 and TNFα; see Chapter 32) and thereby prevent them from binding to their receptors. Reservations about such approaches focus on the built-in redundancy of immune responses, and the effect of 'knocking out' responses to TNFα, for example, may simply be the amplification of the effects of IL-1.

39.7 Approaches to cancer chemotherapy

Although cancer arises as a result of the accumulation of somatic mutations in cells (see Chapter 37), it is unusual for the mutation to be such that selectively toxic drugs, of the type which are effective against bacteria, may be used. Most strategies of cancer chemotherapy therefore depend on the fact that cancer cells are undergoing many rounds of mitosis. The drugs, therefore, target rapidly-dividing cells, and consequently they must have side-effects on body cells that are themselves constantly dividing. A major focus of cancer chemotherapy is the operation of the cell cycle (see Chapter 4), with some

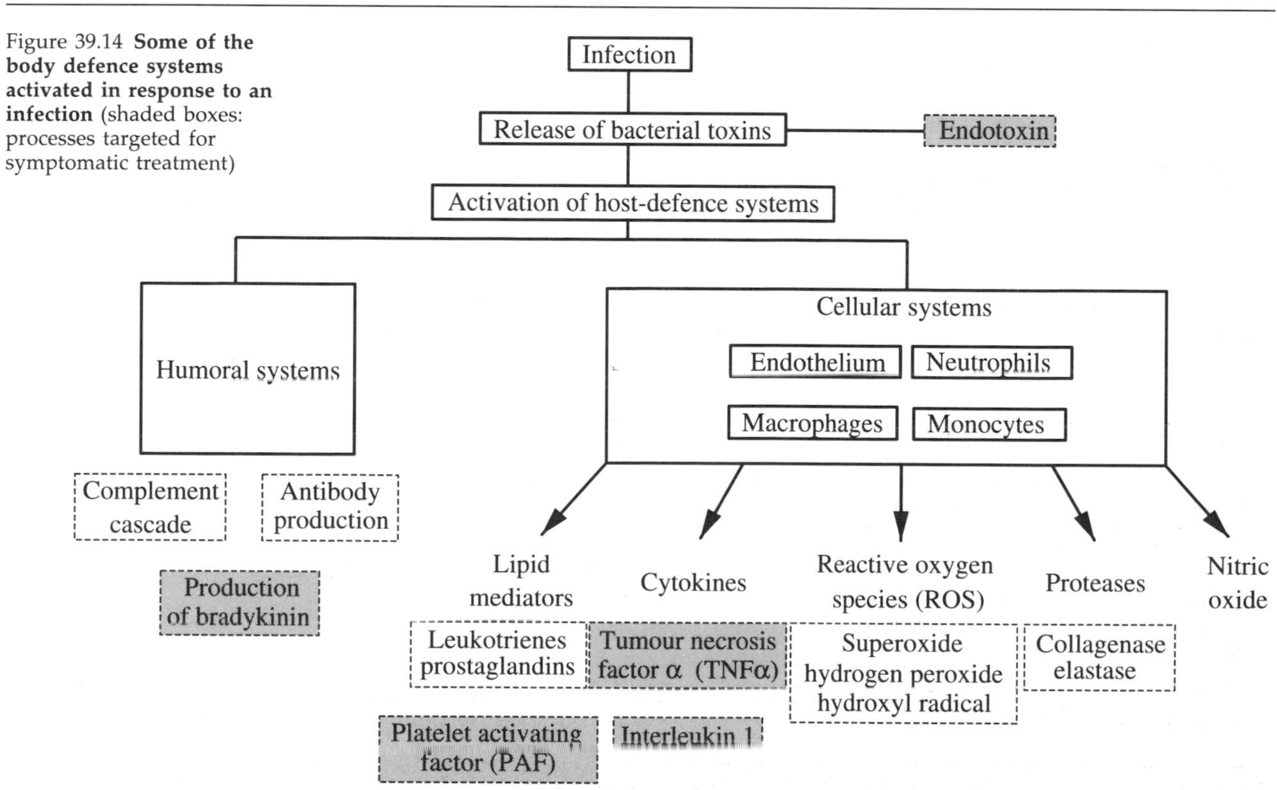

Figure 39.14 **Some of the body defence systems activated in response to an infection** (shaded boxes: processes targeted for symptomatic treatment)

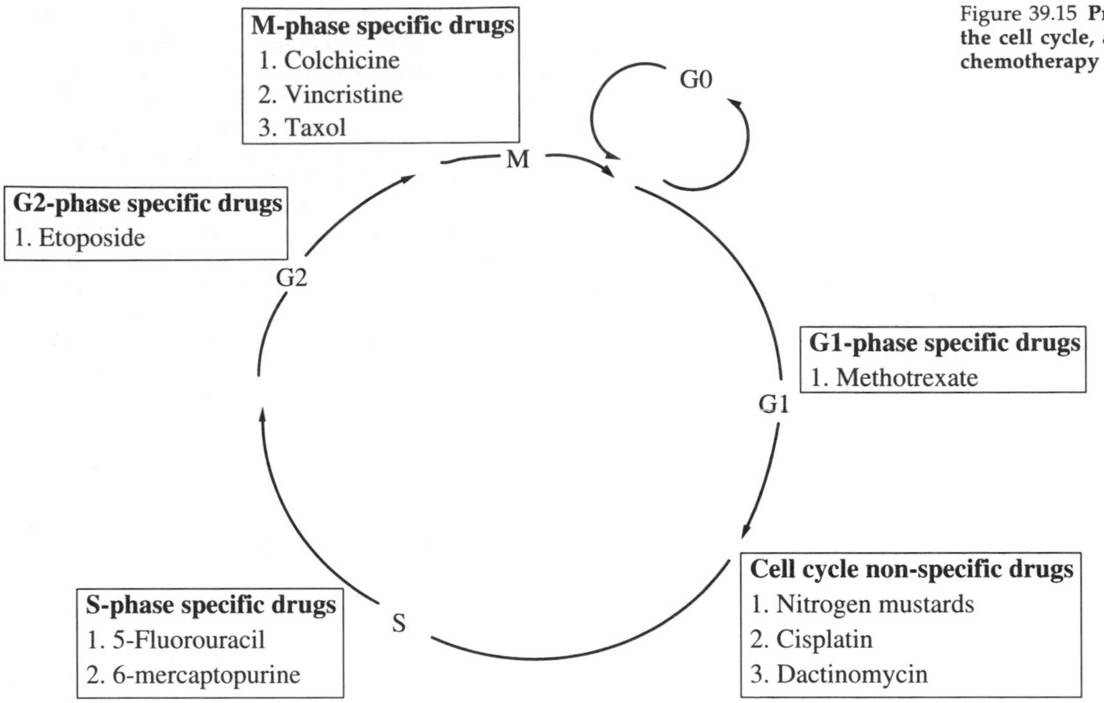

Figure 39.15 **Principal sites in the cell cycle, at which cancer chemotherapy drugs act**

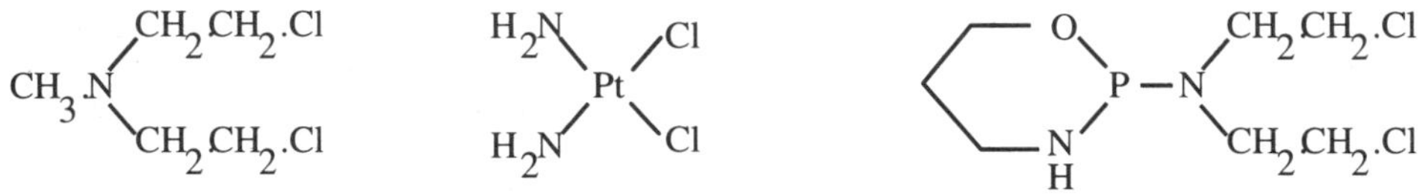

Mechlorethamine (HN2; a nitrogen mustard) | *Cis*diaminodichloroplatinum (cisplatin) | Cyclophosphamide

Figure 39.16 **Some examples of drugs used in cancer chemotherapy**

drugs being targeted on several phases of the cell cycle (**phase non-specific drugs**) whereas other drugs are **phase-specific** (Figure 39.15).

It should be borne in mind, however, that in the clinic, certain slow-growing tumours are curable by chemotherapy, whereas some rapidly-dividing cancers are not. Drugs effective against the former type of tumour may act by promoting apoptosis (programmed cell death; see Chapter 37).

Drugs that bind covalently to DNA

Mechlorethamine (**HN2**, a **nitrogen mustard**; Figure 39.16) was the first drug with proven ability to kill tumour cells to be given to cancer patients. It was tested by Gilman, Goodman, Lindskorg and Dougherty in 1942, but was introduced widely only after World War II, when its relatively high toxicity meant that it was soon replaced by derivatives that were, at the same time, less toxic and more efficacious. Of these, one of the most successful has been **cyclophosphamide** (Figure 39.16), which is an example of a drug that needs metabolic activation by the microsomal cytochrome P450 system (see Chapter 36). Either alone or more frequently in combination with other drugs it is still in use for a range of cancers. The nitrogen mustards are bis-functional alkylating agents, that is they have two groups both independently capable of interacting at the N-7 position of purine bases of DNA. If the two guanine residues form the complementary sequences 5′-G-C-3′ and 3′-C-G-5′ on the two strand, then an alkylation on the first guanine puts the second alkylating group in the correct position to interact with the second guanine and thus the DNA becomes 'cross-linked' by **inter-chain cross-linking** and (in the absence of repair) incapable of replication (Figure 39.17). Another example of an inter-chain cross-linking drug is provided by the antibiotic **mitomycin C** which requires metabolic activation, by reduction in the host (see Chapter 36), before acting rather like the nitrogen mustards. Mitomycin C has been used to improve both subjective and objective symptoms in a number of cancers, including carcinomas of the stomach, pancreas, breast and uterus. Nitrogen mustards are examples of cell-cycle phase non-specific drugs.

A different type of phase non-specific cross-linking occurs when the drug ***cis*diaminodichloroplatinum** (**cisplatin**, see Figure 39.16) is used. The two chlorines and two nitrogens and the platinum are all coplanar and the consequent geometry appears to be just right for **intra-chain cross-linking** to occur, the sequence 5′-G-G-3′ being a particular target with the drug once again

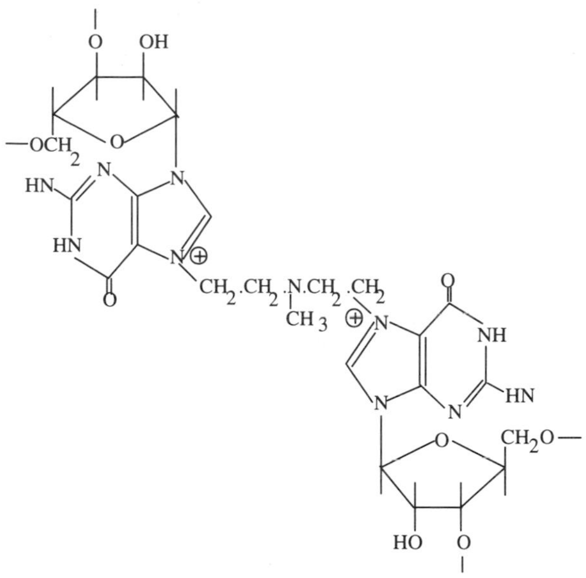

Figure 39.17 **The cross-linking of guanine residues on different chains of DNA by a nitrogen mustard**

attaching via position N-7 of the purine rings. Cisplatin is useful in patients with metastatic testicular tumours who have already been treated by surgery and/or radiotherapy. Its combination with **vinblastine** and **bleomycin** is particularly effective.

Drugs that bind non-covalently to DNA

A series of commonly used anticancer drugs are **intercalating agents**. These compounds all have planar ring regions that allow them to insert (intercalate) between the stacks of bases in DNA, usually between two deoxyguanine residues. Two intercalating drugs in routine use in the clinic are the **anthracycline** antibiotics **mitoxathrene** and **dactinomycin (actinomycin D)**. Their cytotoxicity is directly related to their ability to bind to DNA, but although their binding does lead to the inhibition of DNA and RNA polymerases and thus the synthesis of nucleic acids *in vitro*, it is unlikely that this is the mechanism of action *in vivo*. There is evidence that these drugs are capable of catalysing the generation of reactive oxygen species such as the hydroxyl radical ·OH which may damage DNA (see Chapter 33). It has been established that the cardiotoxicity of a related intercalating drug **doxorubicin** is due to free radical generation. Dactinomycin is used to treat paediatric solid tumours, e.g. Wilms' tumour of the kidney.

Antimetabolites

The term antimetabolite used in the present context refers to drugs that interfere, either directly or indirectly, with the biosynthesis, salvage or catabolism of the precursors of the nucleic acids, i.e. the nucleotides.

Methotrexate is a successful anticancer drug that acts indirectly to prevent formation of nucleotides. Its principal action is in G_1 phase of the cell cycle (although cells are arrested in mid S-phase) where it prevents the biosynthesis of the deoxyribonucleoside triphosphates necessary for replication. Methotrexate is a powerful inhibitor of dihydrofolate reductase and its action is to prevent the formation of tetrahydrofolate that is actually required for nucleotide biosynthesis (see Figure 39.9). Methotrexate is prescribed for many types of neoplasia.

Whereas methotrexate, among its actions, acts to inhibit thymidine nucleotide synthesis indirectly, other antimetabolites are direct inhibitors of the key enzyme **thymidylate synthase**. An example of a drug in this category, in current use, is **5'-fluorodeoxyuridine (FdU)**. This is converted in cells into 5'-fluorodeoxyuridine monophosphate (FdUMP; Figure 39.18). When this

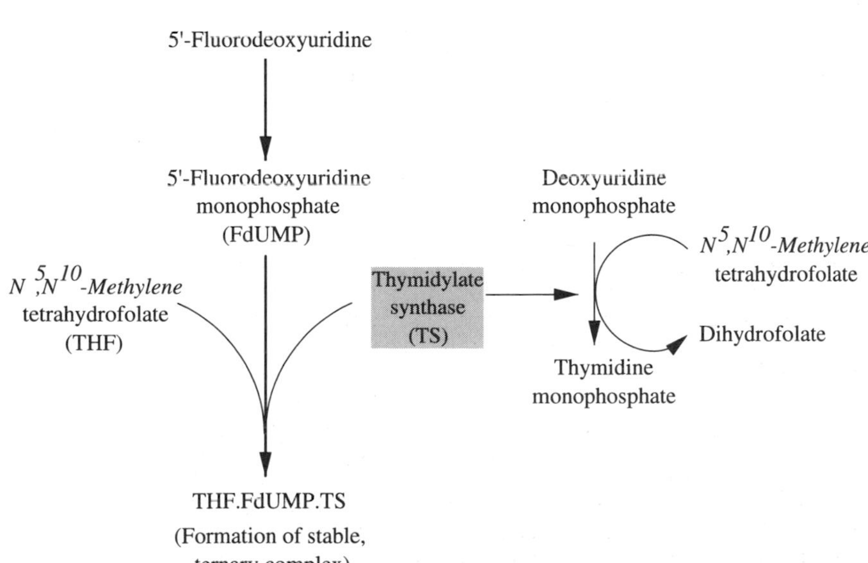

Figure 39.18 **The drug 5'-fluorodeoxyuridine inhibits the formation of thymidine monophosphate by the irreversible formation of a ternary complex with thymidylate synthase and N^5,N^{10}-tetrahydrofolate**

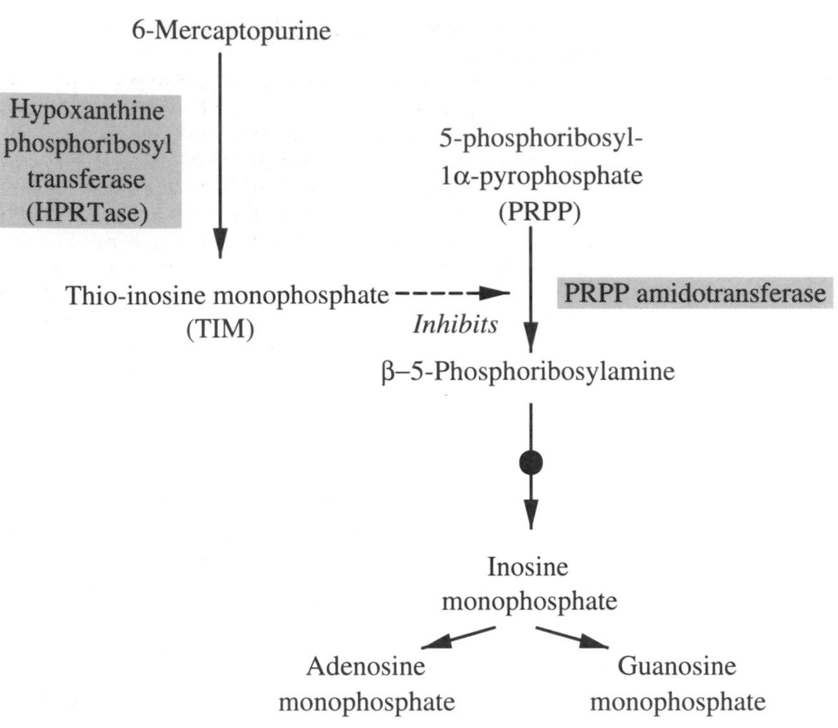

Figure 39.19 **The drug 6-mercaptopurine, following its conversion into thio-inosine monophosphate, inhibits an important step in the biosynthesis of purine nucleotides**

intermediate is used as a substrate by thymidylate synthase, a stable ternary complex is formed which traps the enzyme, making it unavailable for the methylation of its normal substrate deoxyuridine monophosphate. Whereas FdU interferes with the *de novo* pathway for pyrimidine biosynthesis, the drug **6-mercaptopurine** becomes activated following its conversion to the nucleotide thioinosine monophosphate (TIM) in a reaction of the 'salvage' pathway for purines catalysed by the enzyme hypoxanthine phosphoribosyl transferase (HPRTase; Figure 39.19). TIM acts at several stages in the *de novo* pathway for purine synthesis, but most notably at the phosphoriboside pyrophosphate amidotransferase reaction, which is rate-limiting for purine biosynthesis. 6-Mercaptopurine is a particularly useful drug because it shows a degree of selectivity towards tumour cells, although the reason for this is not clear. It is used principally to treat acute leukaemia, when its use in children is particularly successful.

A wide range of antimetabolite drugs are used for cancer chemotherapy and they all exert their principal effects in S-phase.

Drugs that interfere with chromatin function

Several drugs act in the M-phase of the cell cycle. For one group of drugs the formation of mitotic spindles is the target. **Colchicine**, from the crocus *Colchicum autumnale*, is one such drug. This binds to tubulin α,β-dimers and prevents polymerization to form microtubules (see Chapter 6). Microtubules form the basis of mitotic spindle fibres that are necessary for the correct separation of the daughter chromatids in mitosis. Colchicine was first introduced as means of alleviating the joint pains of patients with gout. The therapy was effective because the division of inflammatory cells attracted to the site of uric acid deposition was prevented. The use of colchicine for this purpose has been superseded by drugs such as allopurinol. Likewise, **vinblastine** and **vincristine**, obtained from the periwinkle *Vinca rosa*, and referred to as vinca alkaloids, have replaced colchicine in cancer chemotherapy. These two drugs also act as 'spindle poisons' in the same way as colchicine. They are used for a wide range of tumours including Hodgkin's disease.

The drug **taxol**, which has entered phase II clinical trials for the treatment of breast cancer, has the converse action to that of the vinca alkaloids. Its presence favours the polymerization of tubulin dimers and this results in the formation of abnormal, non-functional bundles of microtubules. The early results with taxol have been very encouraging, but unfortunately its source, the yew tree *Taxus brevifolia*, which yields the drug from its bark, is very slow-growing and stripping its bark kills the tree. Consequently supplies have been restricted, but recently the production of a semisynthetic analogue, the drug **taxotere**, with the same action has been achieved, starting with extracts from the (renewable) needles of a related species *T. baccata*.

Another group of drugs that interfere with chromosome dynamics are **topoisomerase inhibitors**. Topoisomerases are the biological solution to the topological problem of requiring that the DNA double helix be unwound and rewound in the course of replication (see Chapter 4). Topoisomerase II achieves this by sequentially making and then rejoining double-stranded 'cuts' in the DNA. Topoisomerase inhibitors such as **etoposide** act to inhibit, not the 'cutting', but rather the subsequent rejoining. This means that fragmentation of DNA occurs and replication in tumour cells is prevented. Etoposide has proved most useful in the treatment of testicular tumours. Cells in

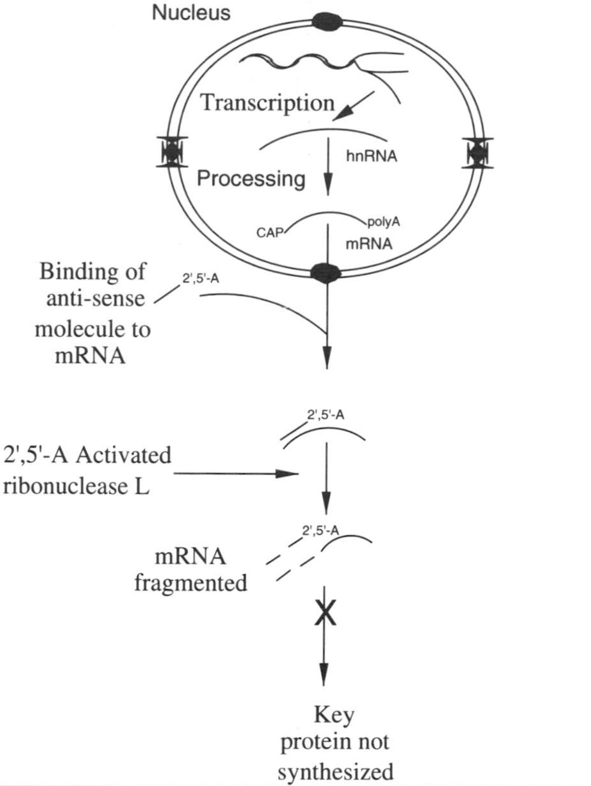

Figure 39.20 **The use of 'anti-sense' therapy to destroy tumour cells**

culture treated with the drug are arrested in the G_2 phase of the cell cycle.

Other potential targets for and means of cancer chemotherapy

Protein tyrosine kinases play an important role in the control of cell proliferation (see Chapters 3 and 37). Drugs that act as inhibitors of this group of enzymes may therefore be useful for chemotherapy, for example in chronic myelogenous leukaemia in which enhancement of protein tyrosine kinase activity is part of the disease process. One such group of drugs are the **tyrophostins**.

Molecular biologists are now developing a process known as 'anti-sense' therapy. In this, oligonucleotides are used that are complementary, in the Watson and Crick base pair sense, to specific mRNA molecules and to which the 2′,5′ cyclic nucleotide of adenosine (2′,5′-A) is attached. On application, the drugs are taken into cells, bind to a specific mRNA by base-pairing and the 2′5′-A activates ribonuclease L that destroys the mRNA. The mRNA molecules targeted would be those encoding proteins vital for cell growth and division, for example the growth factors or their receptors. Figure 39.20 suggests how anti-sense therapy might work.

39.8 Resistance to the drugs used in cancer chemotherapy

The categories of intrinsic and acquired resistance as ascribed to bacteria in Section 39.5 are also encountered in the treatment of cancer. Certain types of cancer are very responsive to chemotherapy, whereas others barely respond at all. Remission is almost invariably seen following the treatment of the acute lymphocytic leukaemia of childhood, whereas hepatocellular carcinomas rarely respond to chemotherapy. Thus, most leukaemia cells in a given case of the disease are intrinsically responsive, whereas those in the liver malignancy are not. If treatment of the leukaemia with a single drug is continued, it is found that successive rounds of treatment are progressively less effective and the disease recurs: the cells have acquired resistance. This does not mean that the cells have acquired resistance *after* institution of the therapy, although many of the drugs in use are mutagenic. It usually means that a small number of cells that had achieved resistance (by mutational or other means) *prior* to treatment are allowed to grow as the bulk of the cells that constitute the tumour are killed by the drug. The drug is said to exert **'selective pressure'**.

The actual mechanisms underlying resistance in most cases resemble those described for bacteria in Section 39.5. These include increased drug efflux, decreased drug uptake and other mechanisms summarized in Table 39.5. Some strategies have no direct bacterial counterpart. One of these is **gene amplification**. For example, the gene encoding dihydrofolate reductase may be amplified so that the genome of the tumour cells

Table 39.5 **Mechanisms underlying the development of drug resistance by tumour cells and the drugs principally affected**

Mechanism	*Drugs affected*
1. Decreased intracellular drug level due to:	
• increased drug efflux	Dactinomycin
• decreased drug uptake	Antimetabolites, nitrogen mustards
2. Increased metabolic inactivation	Antimetabolites, bleomycin
3. Decreased rate of metabolic activation	Antimetabolites
4. Increased amount of target enzyme or receptor	Methotrexate
5. Decreased affinity of target enzyme or receptor for drug	Antimetabolites
6. Increased rate of repair of drug-induced damage	Alkylating agents, *cis*platin

contains up to 180 copies. This, in turn, means that vastly more mRNA and enzyme are produced and there is not enough methotrexate to inhibit it all. The genes are often very unstable, as they reside on separate genetic elements called double minute chromosomes, and are lost if the cells undergo mitosis in the absence of the drug. A second path to treatment resistance is via failure to respond to cues prompting apoptosis.

The occurrence of drug resistance in cancer chemotherapy is a good reason for using combinations of drugs that act at different points in the cell cycle and for which there are different mechanisms for resistance. The combinations may involve either concurrent or sequential administration of the drugs. It is also important to use as large a dose of drug(s) as will be tolerated in initial treatment, so that resistance is less likely to be established. For example, highly toxic doses of methotrexate may be given followed by the 'rescue' administration of thymidine to allow the more slowly growing normal cells to survive.

Chapter 40

Metabolism in injury and trauma

40.1 Introduction: the causes and nature of injury

In England and Wales in 1986, 3.23% of all deaths recorded were attributed to external causes of injury or poisoning. This amounted to 18 484 people, of which 11 271 were males and 7213 were females. Included in these figures are the deaths of 793 children. As a result, accidents are the leading cause of childhood death (Figure 40.1). The view may be taken that these statistics represent a legal rather than a medical problem, requiring tighter laws on drinking and driving or the fitting of smoke detectors in all homes, for example. This has some validity because one-third of the 4953 people involved in motor vehicle traffic accidental deaths in 1986 were found to have a blood ethanol concentration in excess of 80 mg% (17.4 mM; the legal limit for driving in the UK). Similarly, although fire and flames claimed only 656 lives in that year, the management of burns injuries is one of the most difficult for the medical profession. This last observation also makes the point that doctors treat injured patients who will ultimately die as a result of their injuries: 20% of the injured survive for days, weeks or even months before succumbing. In this group the elderly, especially elderly females, are excessively represented.

Discussion of these fatality statistics should not obscure the fact that accidents requiring hospital admission in the UK exceed by almost 100-fold those resulting in death. There were 311 473 road accident casualties in 1987, of which about 40 000 were to children (resulting in some 10 000 with permanent disability). In fact, road traffic accidents resulted in 33% of all hospital admissions of children. As with the whole population so with the young, males are twice as likely as females to suffer an accident.

The management of trauma victims requires the application of a wide range of the principles of basic medical science. In the early stages after injury the principles of physiology and anatomy, as they apply to resuscitation, are particularly important, with their implications with regard to methods for the maintenance of airways, breathing and circulation. It is the aim of the rest of the present chapter to demonstrate that a proper appreciation of the biochemical principles underlying metabolism and its regulation can also make a vital contribution to the understanding of the support required by the injured.

Mechanical injury (whether accidental or surgical), burn injury and sepsis all markedly alter metabolism and fuel utilization. These challenges to bodily function are different in nature, but there are enough common metabolic features of the responses to them to recognize some important general principles. A consideration of these principles leads to the important conclusion that the altered pattern of fuel utilization seen in injury and

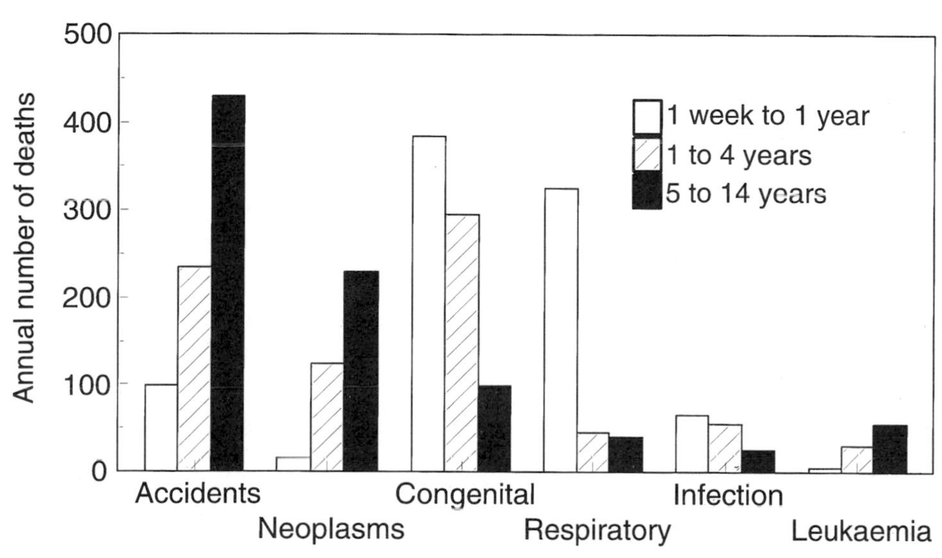

Figure 40.1 **Except for the very young, accidents are the leading cause of childhood death in the UK**

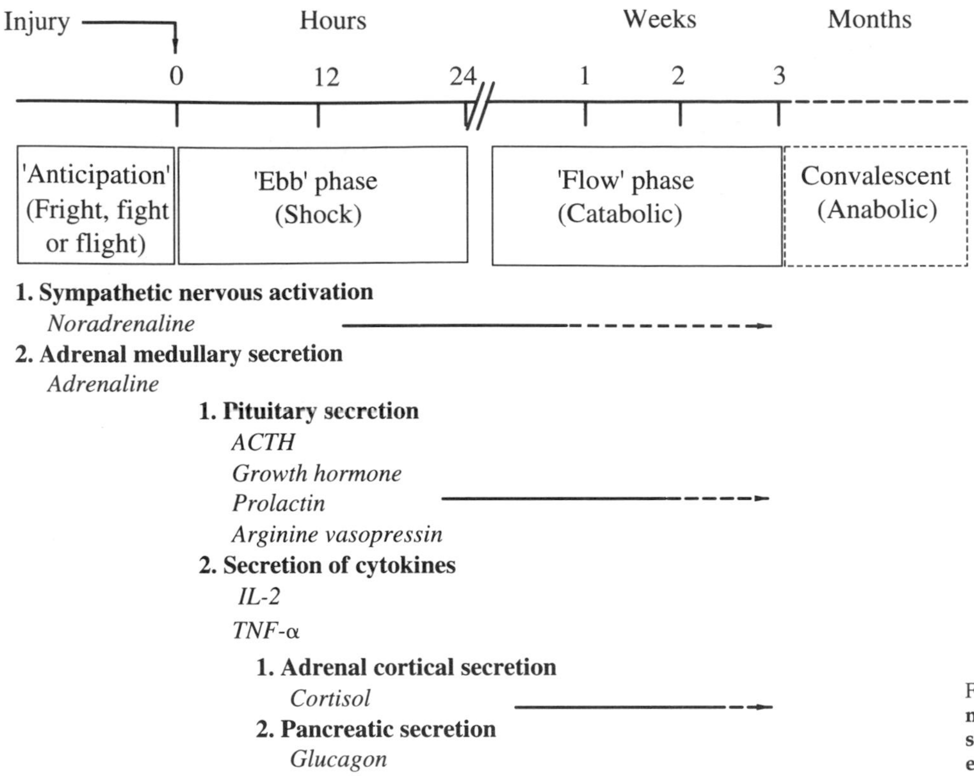

Injury

Hours | Weeks | Months

0 12 24 1 2 3

'Anticipation' (Fright, fight or flight)

'Ebb' phase (Shock)

'Flow' phase (Catabolic)

Convalescent (Anabolic)

1. Sympathetic nervous activation
Noradrenaline
2. Adrenal medullary secretion
Adrenaline

1. Pituitary secretion
ACTH
Growth hormone
Prolactin
Arginine vasopressin
2. Secretion of cytokines
IL-2
TNF-α

1. Adrenal cortical secretion
Cortisol
2. Pancreatic secretion
Glucagon

Figure 40.2 **The phases of the metabolic response to injury, showing the underlying endocrine changes**

sepsis should be regarded as physiological rather than pathological. That is, an adaptive change that should actively be supported, not a derangement to be overcome. This has important consequences as far as the nutritional management of injured patients is concerned. Nutritional therapies should be aimed at maintaining these adaptive patterns.

40.2 The phases of the metabolic response to injury

The first work to describe the metabolic responses to injury was started in 1929 by Cuthbertson, who studied patients with long bone (tibia) fractures. He described the occurrence of a hypermetabolic state following the fracture. In this changed state he noted an increase in urinary nitrogen output, in oxygen consumption, in body temperature and in pulse rate. He subsequently noted that the urinary excretion of compounds incorporating nitrogen, phosphorus and sulphur appeared to correspond with the amounts of these elements in the tissues, and was much greater than could be accounted for by the simple destruction of the tissue associated with the wound. He correctly deduced there to be a general bodily response. Further studies led Cuthbertson and others to recognize that metabolic responses can be divided into three phases: an initial **ebb** (or shock) **phase** and subsequent **flow** (or catabolic) and **convalescent** (or anabolic) **phases** (Figure 40.2). The duration (and in some cases the occurrence) of these phases depends upon the type and extent of the injury suffered.

40.3 The ebb, or shock, phase

Metabolic changes

The ebb phase of response to injury lasts from a few hours to up to 2 days and a metabolic response to injury may actually be initiated before the injury is inflicted, if it is anticipated. This is the well-known 'fright, fight or flight' response involving the activation of the sympathetic nervous system and the secretion of adrenaline from the adrenal medulla (Figure 40.3). This results in the mobilization of triacylglycerol (TAGs) and glycogen stores, from adipose tissue and the liver respectively, leading to increases in the circulating concentrations of both non-esterified fatty acids (NEFA) and glucose, together with the breakdown of glycogen in muscle. Thus the metabolic response to potential injury involves the provision of fuels for muscular activity. This is accompanied by complementary cardiovascular changes in which blood flow is diverted from the viscera to the muscles. If injury cannot be avoided, or if there is no warning, these responses continue (or are initiated) more extensively and are now supplemented by the secretion of the pituitary hormones adrenocorticotrophin (ACTH), arginine vasopressin (AVP, especially when fluid loss occurs), growth hormone (GH) and prolactin. Secretion of cytokines also occurs. ACTH acts on the adrenal cortex to promote the secretion of cortisol, and the sympathetic innervation and high circulating adrenaline act on the endocrine pancreas to cause the secretion of glucagon, and after severe injury, to inhibit the insulin response that ought to accompany the

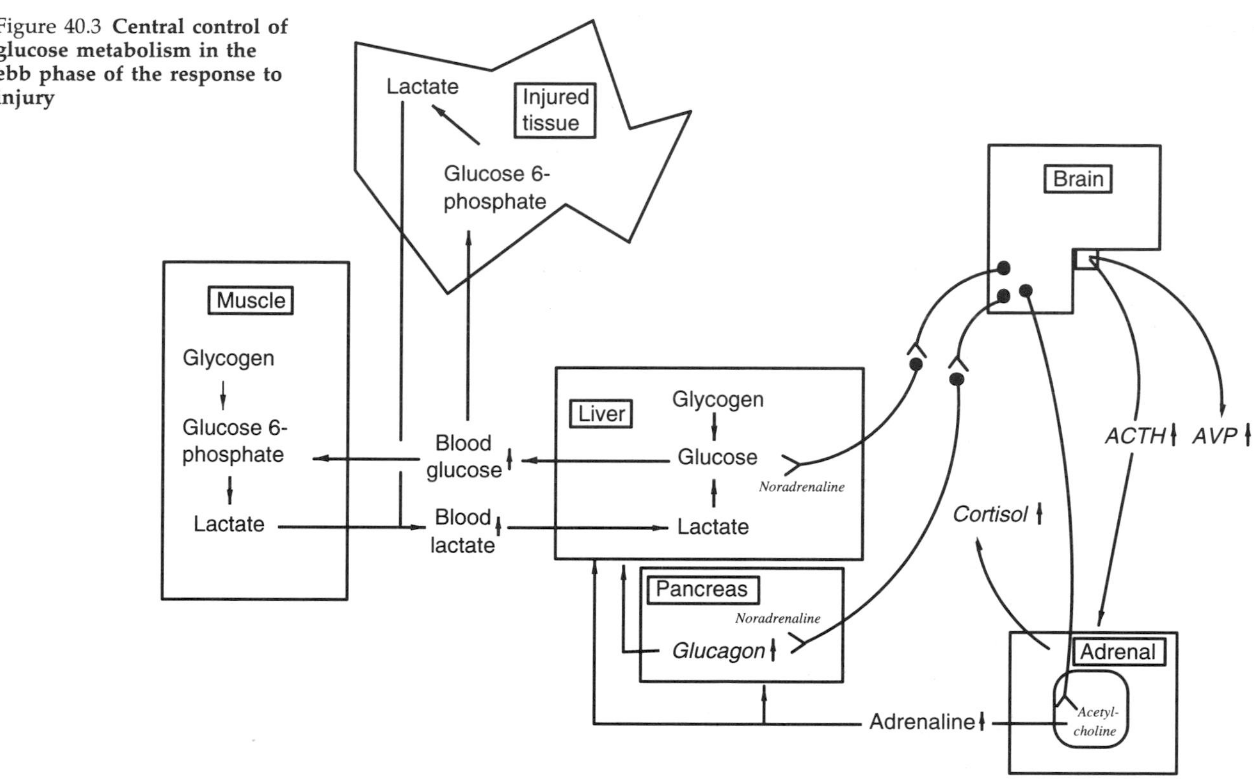

Figure 40.3 **Central control of glucose metabolism in the ebb phase of the response to injury**

evolving hyperglycaemia Prolactin is believed to activate several of the cell types of the immune system, particularly the macrophages (see Chapter 32). These changes are summarized in the Figure 40.2. The metabolic effect of these responses is one of promoting additional mobilization of NEFA from TAGs stores in adipose tissue. Acting together, the nervous and endocrine responses ensure that mobilization of TAGs and glycogen is both extensive and rapid.

As all of these changes involve fuel utilization it may appear surprising that there is no corresponding increase in the overall metabolic rate (it may actually be reduced). But a consideration of the likely consequences of the fuel and hormonal changes that occur help to explain this apparent paradox. Part of the reason is that the oxidation of carbohydrates (especially exogenous) is greatly depressed, with much of the glucose that is used by the extrahepatic tissues being converted into pyruvate via glycolysis before being reduced to lactate and released into the circulation. This lactate is taken to the liver to be reconverted to glucose via gluconeogenesis. This cycling between glucose and lactate is an inefficient way of deriving energy from glucose and accounts for a proportion of the increased energy expenditure in injury. The point is illustrated in Table 40.1, which shows the results of an investigation of groups of patients with leg burns extending over 10% (small) or greater than 50% (large) of the surface. Glucose extraction and lactate production were measured for both groups at rest. It was found that apparently more than 100% of the glucose (presumably muscle glycogen gave rise to the additional lactate) was converted into lactate across legs with small burns (i.e. very little was oxidized). The corresponding figure across the legs of patients with large burns was 88%.

The relative failure of the tissues to oxidize glucose in trauma may be ascribed to fatty acids being used in

Table 40.1 **Arterial–femoral vein differences in glucose and lactate concentrations in patients with small or large leg burns**

Extent of burn (% of surface affected)	*Blood flow* (ml/min per 100 ml leg)	*Glucose removal* (μM/min)	*Lactate release* (μM/min)	*Caclulated % of glucose appearing as lactate*
Small (10%)	4.20	2.22	6.60	150*
Large (> 50%)	8.00	18.90	33.30	88

*'Extra' lactate presumed to derive from glycogen stores in muscle.

preference to the monosaccharide as metabolic fuels in the muscles. This is a direct consequence of the elevated circulating concentrations of adrenaline and NEFA (see Chapter 10).

The net result is a prolonged hyperglycaemia, and this may be maintained even after liver glycogen depletion, when the source of the glucose must be via gluconeogenesis. Plasma glucose concentrations soon after severe injury are typically in the range 10–15 mM (the normal range being 4–5 mM) and may remain this high for as long as 24 h.

Thus in the ebb phase the injured patient will characteristically oxidize less glucose, convert less glucose to fats and oxidize more fatty acids than does the normal subject on a comparable nutritional regime.

Treatment

As indicated previously, the most important measures for intervention in this period are not directly concerned with the perturbations in metabolism but, rather, are directed at such questions as low blood pressure (either with or without reduced blood flow) which is frequently associated with hypovolaemia (reduced quantity of blood) and heart failure or sepsis.

With modern methods for the careful management of elective surgery the ebb phase is often absent and the flow phase commences immediately to reach a peak within hours and is complete in no more than a week.

40.4 The flow, or catabolic, phase

Metabolic changes

The transition from ebb phase to the flow phase is characterized by an increase in the metabolic rate and, as was noted by Cuthbertson, this is generally catabolic, involving both injured and uninjured tissue. In fact, during this period wound healing proceeds at the expense of healthy tissue. This makes good sense for the animal in the wild prevented from gathering food by its injury, but is unfortunate for the patient in hospital, and much effort is directed at minimizing this effect.

The extent of the increase in the metabolic rate in the flow phase depends on the severity of the condition (Figure 40.4), with maximal increases of around 100% following severe third-degree burns (50% or more of body surface burned). It is noteworthy that malnourished subjects have a reduced metabolic rate which means that if they sustain an injury their metabolic rate may still be less than normal. All of this means that, if the resting normal individual requires 6300 kJ/day, the range for the injured may be from 3780–12 600 kJ/day.

The duration of this catabolic phase, as assessed by the persistence of a negative nitrogen balance, is found to be related to the severity of the original injury and of the subsequent hypermetabolic state.

The main metabolic features of the flow phase are summarized in Table 40.2, and the factors that contribute to the development of this phase are listed in Table 40.3. Some of these have been noted already as characterizing the ebb phase. These include raised levels of the hormones that oppose the effects of insulin (the 'counter-regulatory' hormones: adrenaline, cortisol and glucagon). In addition, a peripheral resistance to the anabolic effects of insulin is observed. This

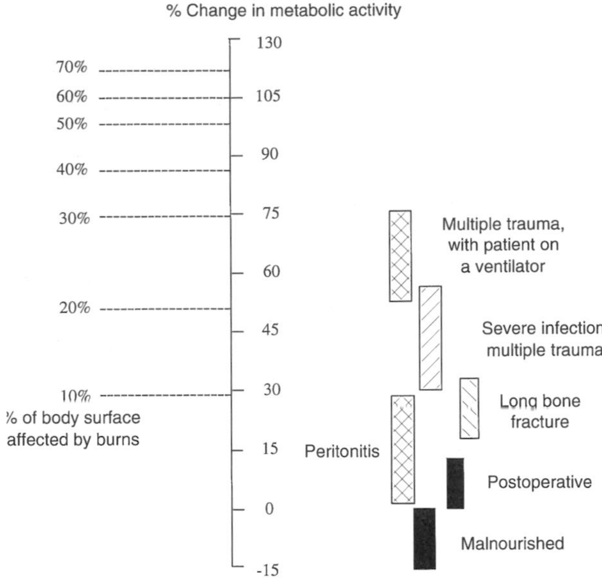

Figure 40.4 **Changes in the resting metabolic rate in the flow phase that follows injury**

Table 40.2 **The main features found in the 'flow phase' following injury or the development of sepsis**

Overt changes observed in patient
1. Elevated excretion of urinary nitrogen
2. Increase in resting metabolic rate
3. Loss of weight
4. Evidence of muscle wasting

Underlying metabolic changes occurring
1. Protein turnover increased, but catabolism dominates
2. Associated increase in gluconeogenesis
3. Ability to deal with exogenous glucose (i.e. tolerance) impaired
4. Associated oxidation of exogenous glucose reduced
5. Fat turnover increased, but lipolysis dominates
6. Associated increase in fat oxidation

Table 40.3 **The main factors contributing to the metabolic changes that arise in the 'flow phase' following injury or the development of sepsis**

1. Metabolic requirements for tissue repair
2. Metabolic requirement of lymphocytes for dealing with infection
3. Lack of appetite, leading to partial starvation
4. Increased circulating concentrations of glucagon, adrenaline, cortisol and growth hormone
5. General resistance of peripheral tissues to insulin
6. Release of mediators of immune and inflammatory responses, e.g. interleukin-2 (IL-2) and tumour necrosis factor α (TNF-α)

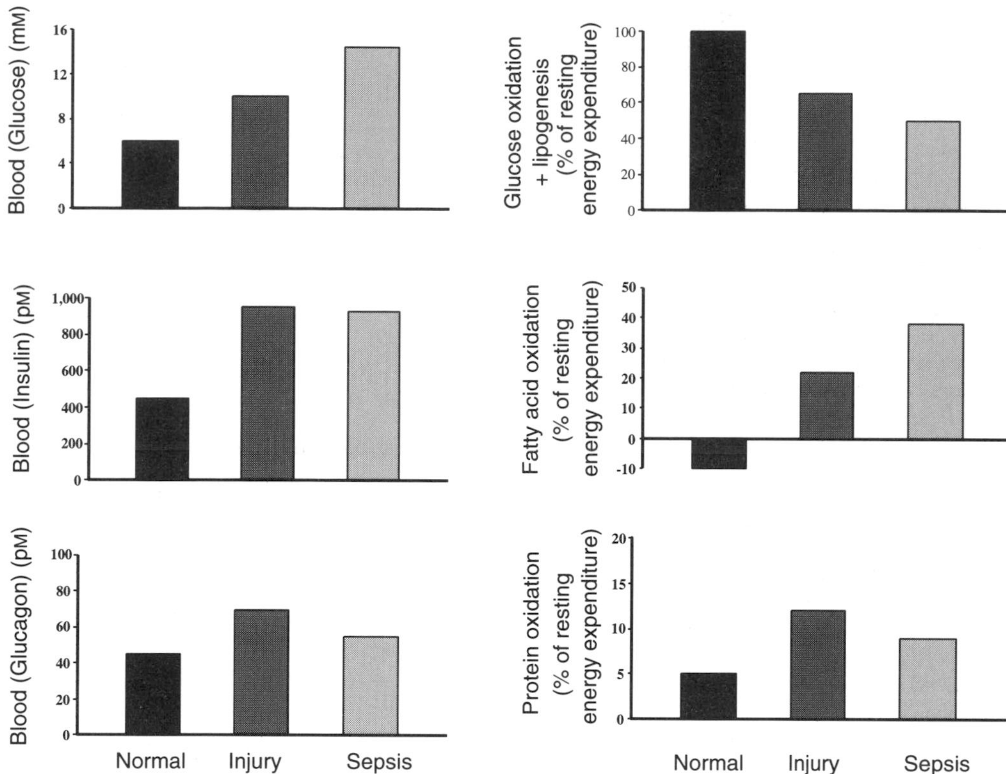

Figure 40.5 **Circulating glucose, insulin and glucagon, and fuel oxidation in the flow phase in normal subjects or following injury or sepsis.** (Subjects were infused with glucose to provide energy at the rate of 120 kJ/kg body weight per day)

resembles that seen in the maturity-onset form of diabetes mellitus (see Chapter 41), but the mechanism is not understood.

The situation is very often exacerbated in the early part of the flow phase by a condition of partial starvation that has been described as resembling anorexia in that patients lose all interest in eating. This means that patients, especially those severely ill, will continue to oxidize NEFA as well as, or even in preference to, glucose for much of the flow phase. This can best be illustrated by a study made of groups of patients and control volunteers. The first group included injured patients. This was a mixed group and comprised those with bone fractures due to falls, abdominal gunshot wounds and one patient who had a cystectomy. The second group of patients were classified as septic and had sustained either perforated duodenal ulcers or other bowel perforations. The study was started within 2 days of injury or the onset of sepsis. During the first 24 h of the study all subjects were infused with 5% glucose prior to 'baseline' metabolite and hormonal evaluation. Thereafter dextrose (glucose) was infused to supply, as the sole source of energy, 120 kJ/kg body weight per day. After a further 24 h period the metabolite and hormone determinations were repeated. The rate of infusion was chosen to exceed the total energy requirements of people in all three groups which are: control individuals, 90 kJ/kg body weight per day; for the severely injured, 105 kJ/kg body weight per day; for septic subjects, 115 kJ/kg body weight per day. The results (Figure 40.5) show that both sets of patients became hyperglycaemic, especially those with sepsis. This hyperglycaemia was sustained despite a doubling of the circulating insulin concentration in the patients. Likewise, glucagon levels, far from being suppressed by the hyperglycaemia (see Chapter 28), were actually slightly increased compared with controls. In a larger, separate study, the fasting blood glucose concentration in a group of 247 severely injured patients early in the flow phase was 9.6 ± 4.8 mM, compared with a control value of 4.5 ± 0.3 mM.

Consideration of Figure 40.5 serves to indicate some of the reasons for the differences between patients and controls. The rate of oxidation of glucose was less in the patients and, whereas glucose was used for lipogenesis in the control subjects (negative value for fatty acid oxidation), fatty acid oxidation persisted in the patients. In addition, the oxidation of protein was increased relative to the controls in the two groups of patients. From these observations it may be deduced that there is peripheral resistance to the actions of insulin and hence a reduced ability of peripheral tissues to use glucose. There is also a continued stimulation of glucagon secretion (probably by β-adrenergic stimulation of the α- (or A–) cells of the pancreas by catecholamines; see Chapter 28). The other notable feature of the energy provision data is that even when all the theoretical energy requirements are supplied as glucose, amino acids continue to be oxidized and, in the absence of exogenous amino acids, these must derive from muscle catabolism. Therefore it is not possible to prevent the muscle wasting

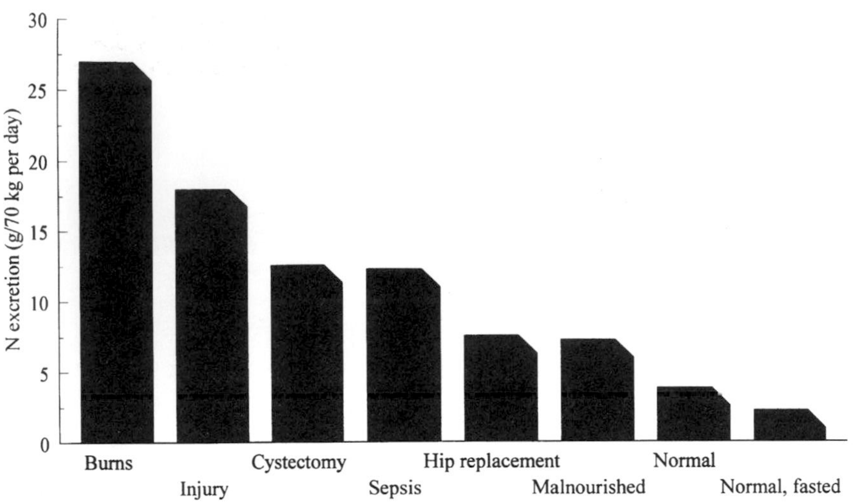

Figure 40.6 **Total urinary nitrogen excretion during 5% dextrose infusion in patients and normal volunteers**

that accompanies injury and sepsis simply by giving glucose alone, and this is borne out by Figure 40.6 which shows the output of urinary nitrogen by normal (fasted) individuals and in various other groups of patients (including injured and septic), all of whom were receiving an infusion containing 5% glucose. Indeed, it is not possible totally to prevent muscle wasting even by giving amino acids as well. This is because the anabolic influence of insulin on muscle is reduced but the catabolic influence of cortisol is fully effective.

Finally we may deduce that if glucose concentrations are maintained at a very high level by infusion of monosaccharide as the sole source of energy to the patients, then something must happen to it. There are three possible major fates: oxidation (which is occurring, but at a reduced rate), lipogenesis (which is unlikely in the face of increased fatty acid oxidation) or glycogen synthesis. It has been claimed that up to 1 kg (or more) of glycogen may accumulate in the liver under these circumstances, with the possibility of liver damage.

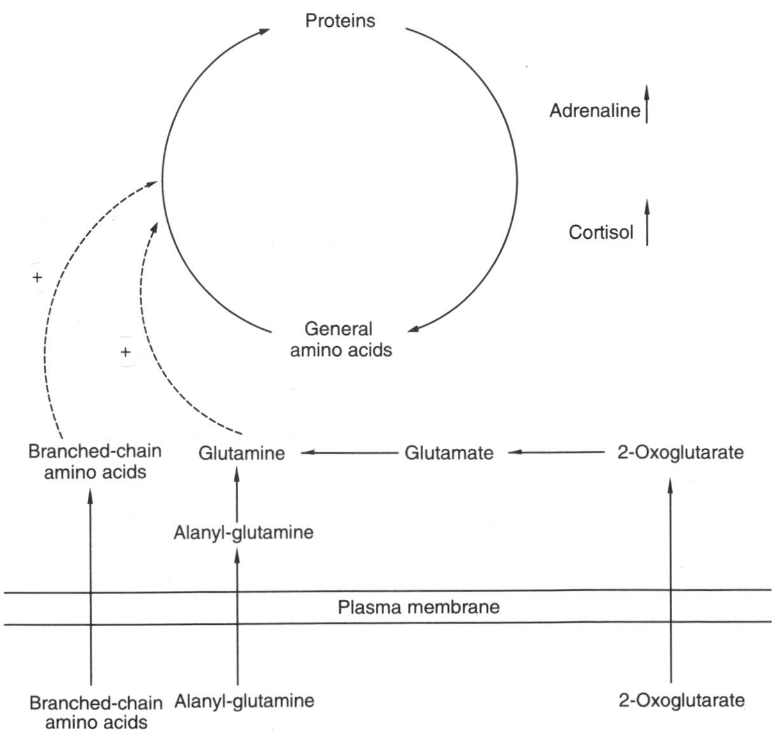

Figure 40.7 **Branched-chain amino acids, alanyl-glutamine or 2-oxoglutarate may be given to spare the muscle wasting that occurs following injury**

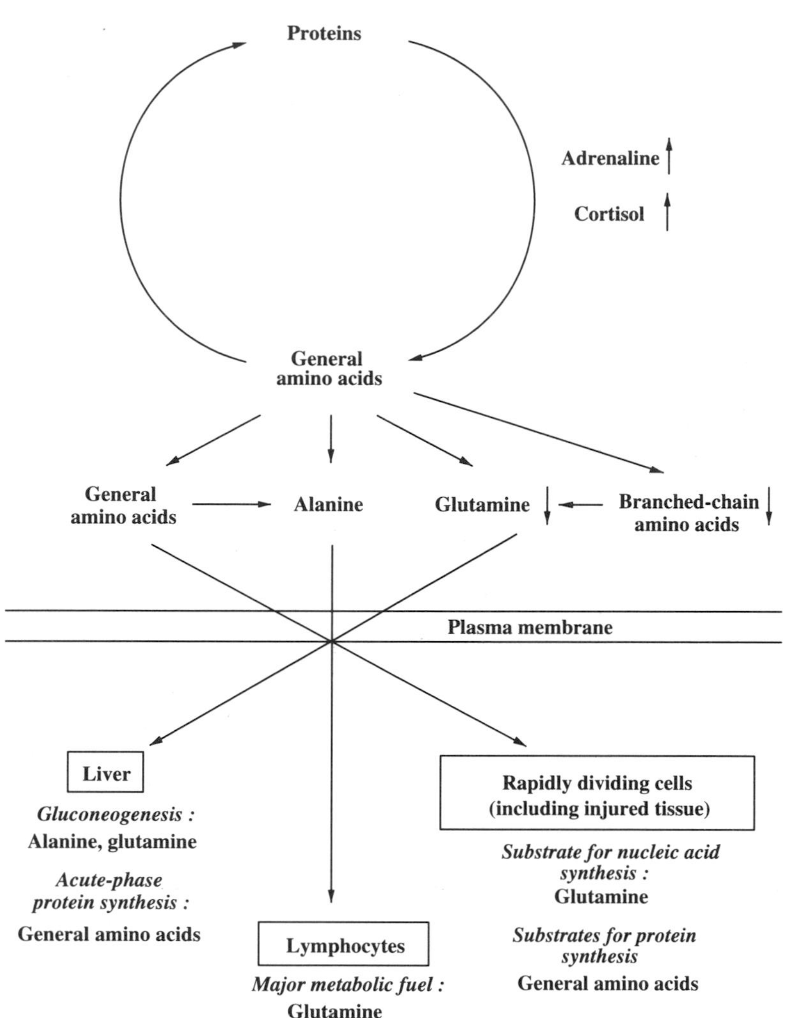

Figure 40.8 **Uses made of the amino acids mobilized in the flow phase following injury**

Returning to amino acid and protein metabolism, it has already been noted that there is a net increase in urinary nitrogen excretion with injury (Figure 40.6) and this is not totally preventable either by giving glucose or by giving amino acid mixtures together with glucose. What is the cause of this mobilization of amino acids and hence of proteins from muscle, and why does it persist even if muscle wastage becomes quite profound? As has been indicated, this is often exacerbated by the anorexia that commonly accompanies the early stages following injury. The driving forces are really the catabolic hormones adrenaline and cortisol coupled with a peripheral resistance to insulin, so that its anabolic effects are not seen to the extent the circulating concentration warrants (see Figure 40.5). This means that the rate of breakdown of muscle proteins exceeds that of resynthesis. The amino acids in the mixture produced from muscle proteins are then 'remodelled' (see Chapter 11) to give a relative increase in the amount of alanine and glutamine. The net effect is that amino acids leave the muscle. (The rate of efflux of glutamine is enhanced by the ability of cortisol to induce the production of a glutamine-transporter protein.) The efflux of amino acids further reduces the rate of protein synthesis, both because the building blocks are lost and also because glutamine (and also the branched-chain amino acids, BrCAAs) actively promote protein synthesis. These biochemical facts suggest a way that might be employed in an attempt to prevent, or at least to reduce, the mobilization of proteins from muscle that accompanies injury. Patients may usefully be given BrCAAs, the dipeptide alanyl-glutamine or 2-oxoglutarate in circumstances in which provision has to be made for total parenteral nutrition (TPN). The last two are to be preferred over glutamine itself because the amino acid is unstable in solution. Glutamine may be produced by the hydrolysis of the dipeptide or following the transamination and amidation of the oxo-acid.

TPN may be necessary, for example, following abdominal injury or surgery. An additional benefit of providing for a supply of glutamine is that patients appear to be less susceptible to sepsis of gut origin: a 'healthy gut' is maintained. The rationale for using BrCAAs and glutamine is indicated in Figure 40.7.

The amino acids released from the muscle serve a range of important functions in supplying other cells and tissues: repair of damaged tissues, energy metabolism of lymphocytes (glutamine) and in the liver gluconeogenesis and acute-phase protein synthesis (Figure 40.8; Table 40.4). This last process is particularly important in sepsis as the acute-phase protein response represents an important component of the operation of the innate immune system (see Chapter 32).

Table 40.4 **Changes occurring to the circulating concentrations of plasma proteins in the 'flow phase' following injury or the development of sepsis**

Increased	*Decreased*
'Acute phase' proteins:	• Albumin
• C-reactive protein	• Transferrin
• Mannose-binding protein	• Retinol-binding protein
• α_1-Antiproteinase	• Thyroxine-binding prealbumin
• Haptoglobin	
• Haemopexin	
• Caeruloplasmin	

Table 40.5 **Composition of a lipid emulsion (Intralipid) used to supply up to 50% of total energy requirements (about 60 kJ/kg per day) during total parenteral nutrition regimens**

- Egg yolk phospholipid : 6 g
- Glycerol : 11 g
- Soya bean oil : 100 or 50 g (triacylglycerols largely containing oleic, linoleic and linolenic acids)
- Water to 500 ml

Mixtures to be used for total parenteral nutrition

With all this biochemical information on the flow phase it is possible to formulate the composition of a mixture for TPN designed to maximize the efficiency of bodily responses that occur at that time. Energy is supplied equally by glucose (about 60 kJ/kg per day) and a lipid emulsion (Intralipid). The composition of Intralipid is given in Table 40.5. The mixture also contains required electrolytes, vitamins and other essential nutrients and may be supplemented with 2-oxoglutarate.

40.5 The convalescent, or anabolic, phase

The definition of the precise time at which a patient may be said to be convalescent has proved rather difficult. There is general agreement that by the time wound healing is complete, metabolism is characteristically anabolic, with a corresponding positive nitrogen balance. However, the signals for the switch from the catabolic to the anabolic mode have yet to be fully defined. It may be that the cells cease to produce the mediators of the catabolic state. These would include tumour necrosis factor-α, other cytokines and lipid mediators derived from arachidonic acid.

Another feature of the convalescent stage is a characteristic muscle fatigue. This is reported to be far in excess of what might be expected to result from muscle protein catabolism in the flow phase. Consequently, patients tend to remain immobile. Here again, attention must be paid to nutritional matters, in this case to ensure that obesity is not encouraged. In this context, early experiences with patients undergoing renal dialysis and therefore relatively immobile for long periods are instructive. Such patients were receiving some 180 g of glucose from their dialysis fluid per day by a route that bypassed the hepatic portal system. It was found that they suffered weight increases up to 5 kg in a few months. That this resulted from the deposition of fat in adipose tissue is indicated by the finding in one patient that the circulating triacylglycerol concentration increased from 2.6 to 3.6 g/l in 6 months.

As is true for the general population, the convalescent patient will undoubtedly benefit from regular exercise during the day and adequate rest at night (when anabolic processes tend to come to the fore).

Chapter 41

Diabetes mellitus

41.1 Introduction

In his Banting Lecture, published in 1994, C. Ronald Kahn demonstrated that medical knowledge of the disease we now call diabetes mellitus may be traced back in excess of two thousand years. He quoted from the doctrines of Ayurvedic medicine, which date from 100–200 BC. In these we find:

> The person who passes urine which is exceedingly sweet, cool, slightly viscid, turbid and resembles the juices of sugarcane ... suffers from glycosuria ... There are two types of urinary disorder – one, natural due to genetic factors, and the other due to indiscreet living or dietetic indiscretions. The patient suffering from the former is thin, pale, eats less and drinks too much ... the patient with the latter is usually obese, eats a lot, is stout, of sedentary habits and sleeps too much.

While one might now take issue with some of the details of the aetiology, this remains a remarkably accurate description of the two forms of diabetes. Throughout the history of medicine one can find similar descriptions of the disease, which initially was called 'diabetes' to indicate the frequent urination. Subsequently, 'mellitus' was added to indicate the honey-sweet nature of the urine.

A major scientific finding leading to the understanding of the disease was made in Strasbourg in 1889 by Minkowski and van Mering. They removed the pancreas from a dog in order to settle a dispute about the role of pancreatic enzymes in the digestion and absorption of fat. Minkowski noted that the house-trained dog developed polyuria and tested positively for glucose (actually reducing substances) in the urine voided on the laboratory floor. Subsequently, Hiden performed elegant experiments which showed that the secretions of what we now call the endocrine pancreas were necessary to prevent diabetes. That the molecule we now know as **insulin** is extractable from the pancreas was first established by Paulesco in Romania in 1921. That insulin could be purified to such an extent that it could be used in the treatment of diabetes mellitus was then demonstrated by Banting, Best and Collip in Canada. Thus they were the first to purify a peptide hormone for the treatment of an endocrine disorder. Thereafter, the study of insulin has coincided with several of the major advances in twentieth century science. Insulin was the first protein to have its complete amino acid sequence determined, by Sanger in the mid-1950s. This was the first substantial peptide to be sequenced. More recently, in 1979, insulin became the first hormone to be produced by recombinant DNA techniques (see Chapter 38). The study by Steiner of the processes involved in the biosynthesis of proinsulin and its processing to the active hormone in the β or B-cells of the islets of Langerhans has been a paradigm for many other peptide hormones. Finally, the **insulin receptor** has been characterized as one of a family of growth-promoting receptors, and the intracellular mechanisms involved in its actions are being unravelled.

41.2 Two types of diabetes mellitus

The key roles played by insulin and glucagon in the regulation of many aspects of fuel metabolism described in previous chapters suggest that serious consequences must ensue if there is a failure either to produce insulin in the proper amounts, or for cells properly to decode in the insulin 'message', for then the essentially catabolic roles of glucagon (and also of adrenaline and cortisol) will go unopposed. As was described in Ayurvedic doctrines, two broad manifestations of the disease diabetes mellitus are recognized. The first is called juvenile onset, type I or **insulin-dependent diabetes mellitus (IDDM)**. The second is called maturity onset, type II or **non-insulin-dependent diabetes mellitus (NIDDM)**. It should be borne in mind, however, that type II diabetes may be present in children.

In the UK, about 90% of all endocrine disorders seen by doctors turn out to be diabetes mellitus in one form or another. In Table 41.1 shows a list of the factors that distinguish the two types of the disease.

41.3 Natural history of the disease

All the evidence supports the view that IDDM diabetes mellitus is immune-mediated, but also involves the interplay of both genetic and environmental factors. Auto-antibodies to islet cells (ICABs) are found in the sera of newly diagnosed diabetics and the detection of such ICABs in an undiagnosed, prediabetic subject is strongly indicative of the development of the disease. In addition, there is a T-cell-mediated destruction of the β-cells of the islets of Langerhans. Family history of the disease is not an important factor: only about 10% of patients know of an affected relative. Nevertheless there does appear to be an inherited *predisposition* to developing the disease. An important marker of genetic influence is located in the HLA class II DR/DQ locus (see Chapter 32), but another is found in the insulin gene locus. The nature of the environmental factor is

Table 41.1 **Insulin dependent diabetes mellitus (IDDM) versus non-insulin-dependent diabetes mellitus (NIDDM)**

Main features	*IDDM*	*NIDDM*
EPIDEMIOLOGY		
Frequency in Northern Europe	0.02–0.4%	1–3%
Predominance	N. European Caucasians	Worldwide Lowest in rural areas of developing countries
CLINICAL CHARACTERISTICS		
Age of onset	<30 yr	>40 yr
Weight	Low	Normal or increased
Onset	Rapid	Slow
Ketosis	Common	Under stress
Endogenous insulin	Low / absent	Present
HLA associations	Yes	No
Islet cell antibodies	Yes	No
PATHOPHYSIOLOGY		
Aetiology	Autoimmune destruction of pancreatic islet cells	Unclear. Impaired insulin secretion and insulin resistance
Genetic associations	Polygenic	Strong
Environmental factors	Viruses and toxins implicated	Obesity, physical inactivity

From Gaw, A. *et al.* (1995) *Clinical Biochemistry*, Churchill Livingstone, Edinburgh

uncertain, but there is evidence implicating proteins of the Coxsackie B virus. These have sequence homologies with the enzyme glutamate decarboxylase (GAD; see Chapter 30) and GAD has been implicated as an autoantigen against which the cell-mediated destruction of the β-cells is directed.

The characteristics of NIDDM are quite different. The pathophysiology appears to be associated with the intersection of two processes: the insulin resistance of target cells (muscle, adipose, liver) and a failure of the β-cells of the pancreas to compensate for this by an increase in insulin secretion. Indeed, there is excellent evidence that insulin resistance both precedes and acts as a predictor of NIDDM. The reasons for the resistance are not known. Reductions in the number of insulin receptors in tissues, in their tyrosine kinase activity and in the activity of a host of 'downstream' enzymes have been described, but no reasons have been given. The frequency of mutation to these proteins does not differ from the normal. A partial exception to this is the insulin-receptor substrate (IRS; see Chapter 28) which does show more extensive sequence variation in type II diabetic patients as against normal subjects. The nature of the second intersecting factor, at the level of the β-cell, is better understood. The cells respond less well than normal to the presence of glucose. This defect, which is genetically determined, may be in the glucose transporter (Glut 2; see Chapter 2) that brings glucose into β-cells. It is the subsequent metabolism of glucose that signals the secretion of insulin (see Chapter 28).

41.4 Deranged metabolism in diabetes mellitus

The nature and causes of the metabolic disturbances seen in IDDM are summarized in Figure 41.1. The simplest explanation of the disease is that in the absence of insulin, peripheral tissues fail to take up glucose adequately and consequently hyperglycaemia ensues. This, however, pays no attention to two cardinal points. Firstly, before the introduction of insulin injections, patients were treated by heroic dietary deprivation (3.25 MJ / day, compared with a 'normal' 7.4 MJ / day) and yet they still became hyperglycaemic. Secondly there is severe ketoacidosis. These facts can only be sensibly explained by involving the liver in a model of the disease. An increase in the glucagon : insulin ratio favours lipolysis over lipogenesis in adipose tissue, and the fatty acids released reduce the dependence of muscle on glucose as an energy source. The fatty acids also pass to the liver where they become the substrates for ketogenesis, causing the ketoacidosis. The fatty acids are also used by the liver for the formation of tricylglycerols and thence VLDLs. **Diabetic ketoacidosis** is particularly severe compared with that found in starvation, for example, and patients failing to receive their injections of insulin may become comatose as the combined result of their ketoacidosis and the consequent lowering of their plasma pH. The reason for the difference between starvation and IDDM in terms of ketone body formation is that, in the former, ketone bodies cause the release of insulin, which in turn limits the release of fatty acids from adipose tissue. The inability of the pancreas to secrete insulin in IDDM means that this important regulatory mechanism cannot operate.

The other metabolic imbalance that occurs in IDDM sees the mobilization of muscle protein as amino acids. The amino acids released, together with the glycerol from adipose tissue, both support gluconeogenesis. This also fuels the hyperglycaemia.

The major differences between IDDM and NIDDM are that in the latter disease the patient does produce insulin, and a degree of tissue response to the hormone is present. Although **insulin resistance** characterizes the disease, the liver and adipose tissue seem to respond better than the muscles to the hormone. Consequently the liver is able to convert some of the excess of carbohydrate to triacylglycerols and these are transferred to the adipose tissue in the form of VLDL. The action of insulin in the adipocytes is sufficient to tip

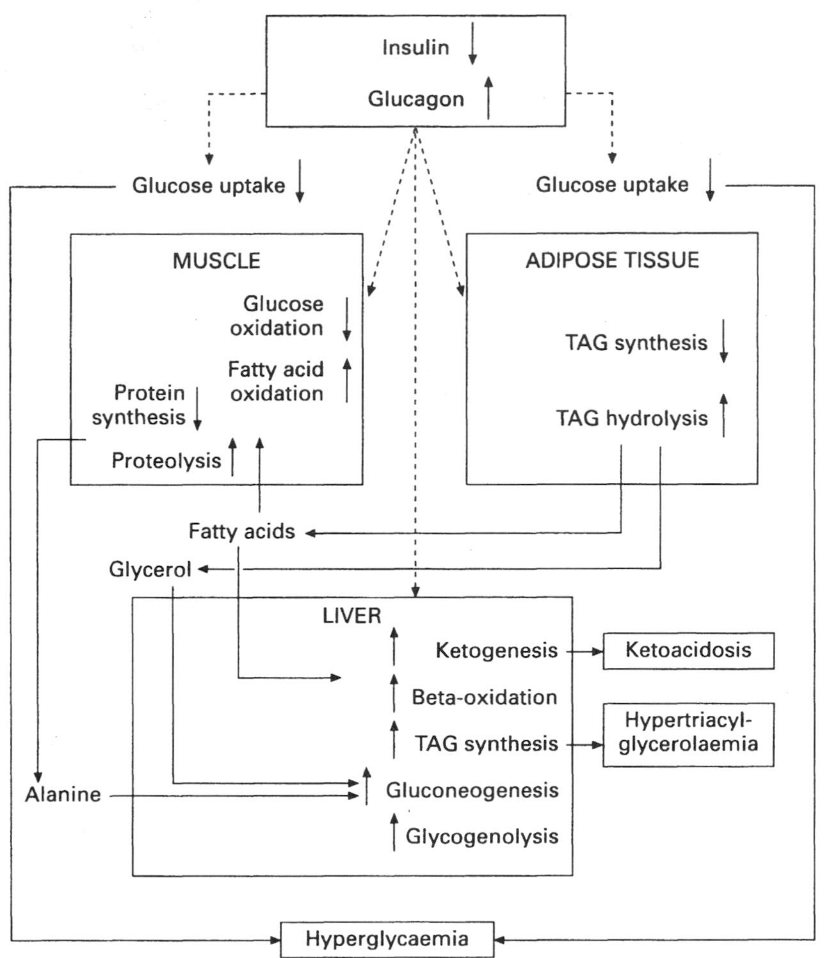

Figure 41.1 **The inappropriate glucagon : insulin ratio of insulin-dependent diabetes mellitus influences three principal tissues to produce the characteristic hyperglycaemia, ketoacidosis and hypertriacylglycerolaemia**

the balance more in favour of lipogenesis and consequently the release of fatty acids does not compare with that occurring in IDDM. In turn, this means that ketoacidosis is not usually an important component of NIDDM.

41.5 The biochemical basis for the complications of diabetes mellitus

In addition to the occurrence of hyperglycaemia, a number of late complications are associated with diabetes mellitus. These include:

1. *Microangiopathy*. This is characterized by changes in the walls of the small blood vessels, seen most clearly as a thickening of the basement membrane.
2. *Retinopathy*. Blindness is some 25 times more common in the diabetic patient and this seems to arise as a result of vitreous haemorrhages from the degenerating retinal blood vessels.
3. *Nephropathy* Chronic renal failure is 17 times more common in the diabetic patient. The damage appears to be to the glomerulus in the kidney. As a consequence, in the late stages of the disease, there is albuminuria and uraemia.
4. *Neuropathy* Evidence of nervous tissue damage in the disease includes the occurrence of diarrhoea, postural hypotension and impotence. The failure to supply adequate nutrients to peripheral nerves (resulting from the microangiopathy) means that neuropathic foot ulcers are seen.

The reason for this constellation of complications is not fully established, but there is good evidence (in the case of IDDM at least) that the onset of these can be delayed by improving the degree of control achieved over glycaemia. This suggests that long-term elevation of blood glucose itself may cause the damage. A clue to this is supplied by the observation that a minor form of haemoglobin A, designated HbA_{1c}, is present in elevated amounts in diabetic patients, even those being treated for the disease. This form of haemoglobin arises as the result of covalent attachment of glucose to the amino groups of the amino acids at the N-terminus of adult haemoglobin. This is purely a chemical reaction, and the rate will be determined by the concentration of glucose to which the haemoglobin is exposed. The normal amount of HbA_{1c} is about 4.5% of the total HbA, while in diabetic patients it may be double this amount, or more. Because the chemical modification of haemoglobin is irreversible in the red cell, measurements of glucosylated forms of haemoglobin give a good idea of the cumulative failure to maintain normoglycaemia during the life of the red cell (about 140 days), i.e. the adequacy of glycaemic control.

It seems likely that many of the membrane-dependent effects associated with diabetic complications result from the covalent attachment of glucose to membrane proteins. This may lead to functional changes in the proteins and hence to the membranes containing them. Evidence that glucosylation of proteins other than haemoglobin does occur comes from the observation that the group of proteins known as crystallins (found in the lens of the eye) are modified by glucosylation in patients with diabetes mellitus, and this might contribute to the occurrence of cataract in the disease (see Chapter 31).

41.6 Treatment

For IDDM, since its discovery and isolation, the administration of insulin has provided the means of treatment. Originally pancreatic insulin of either bovine or ovine origin were the only forms available for injection, but now human recombinant insulin may be prescribed. Much attention has focused in the means and timing of delivery, because, except in special cases, insulin cannot be given orally. These studies have been coupled, in recent years, with careful investigations of precisely what should be the goal of insulin therapy: essentially how good does **glycaemic control** need to be? It is now clear that good control, i.e. that which most closely parallels normal physiological control, has great benefits in delaying the onset and slowing the progression of the complications of retinopathy, nephropathy and neuropathy that characterize the disease. It must be said, however, that these improvements are achieved only at the cost of great dedication on the part of both patients and health care teams. Nevertheless, patients do monitor their own blood or urine glucose levels and, with the advice of the physician, adjust their doses of insulin accordingly. In addition, patients with IDDM must pay attention to diet, and exercise is known to be beneficial.

Dietary control is an important aspect of the treatment of NIDDM. Over the years there has been unanimity as to the goals of dietary control for these patients, but not about how this should be achieved. The dietary goals are the maintenance of blood glucose and lipid levels as close as possible to normal, while maintaining nitrogen balance. The majority view is that this is best achieved by aiming for a diet that is relatively high in carbohydrate (supplying 50–55% of total energy requirements) and low in fat (30% of energy requirements). The focus in this dietary strategy is on decreasing the risk of coronary artery disease, which is increased in NIDDM patients. Mono-unsaturated fats are to be favoured in the fat component of the diet. However, if adequate glycaemic control is not achieved by this type of diet, then a switch may have to be made to one in which the distribution of the contribution of carbohydrate and fats to the energy is closer to 40% : 40%.

Oral hypoglycaemic agents may be given to assist in glycaemic control. These drugs have several sites of action, but one usually dominates. Thus, **biguanides**, such as **metaformin**, appear to improve the response of target tissues to insulin and patients treated for long periods with the drug (combined with dietary control) actually have a reduction in the circulating concentration of insulin. The action of biguanides seems to be to increase the numbers of insulin-dependent glucose transporters (Glut 4; see Chapter 2) found in the surface of cells in adipose tissue and muscle. The dominant action of **sulphonylureas**, such as **glibenclamide**, is on the β-cells of the pancreas. In this tissue, it enhances the

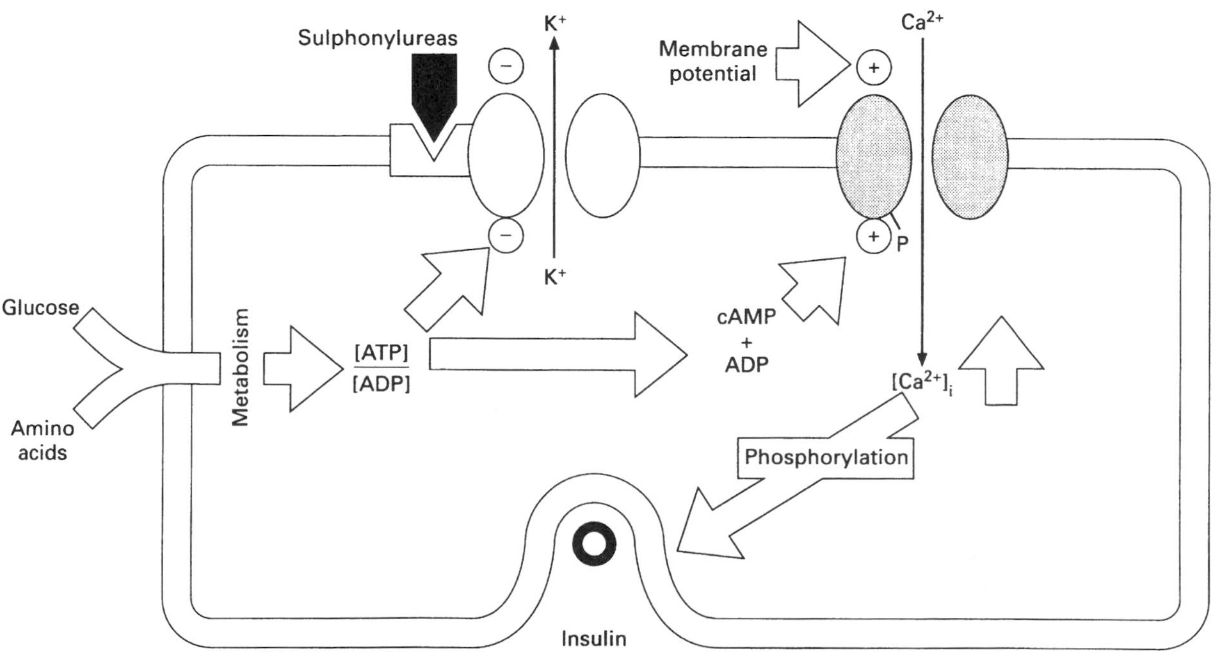

Figure 41.2 **Possible route of action of sulphonylureas in promoting the secretion of insulin**

secretion of insulin in response to glucose (and also to amino acids). Sulphonylureas appear to cause the closing of ATP-dependent K^+-channels in the plasma membrane of β-cells. In turn, this causes the membrane to depolarize, resulting in the entry of Ca^{2+} ions into the cell via voltage-gated Ca^{2+}-channels. The increase in cytosolic Ca^{2+} promotes the secretion of insulin (Figure 41.2). Since biguanides and sulphonylureas produce their effects at quite different locations, it is possible that their combined use might be beneficial in patients for whom glycaemic control is difficult to achieve by using only one drug. Such use, however, would depend on the pharmacological compatibility of the drugs.

Because of the difficulty in maintaining glycaemic control in patients with IDDM without unacceptably frequent monitoring of blood glucose concentrations, much attention is being paid to other ways of providing insulin or of preventing the disease.

An optimal answer as far as insulin delivery is concerned is the whole-organ transplant of the pancreas. Up until 1994, 5000 patients have received cadaver-derived pancreases. The operation is favoured in patients with IDDM who are showing signs of kidney failure. They are then given combined pancreatic–kidney transplants. The reason for this is that these patients will almost certainly require a kidney transplant at a later date, i.e. they must undergo surgery anyway, and thus they will also need lifelong immunosuppression. The results have been very encouraging, especially in the younger, female patient.

Several diabetes prevention trials are currently in progress throughout the world. These include a study in the USA of the effects of parenteral and oral insulin in high-risk relatives of patients with IDDM. It is suggested that the early administration of insulin helps to prevent the progression of the autoimmune destruction of the β-cells. A direct attack on the autoimmune nature of the disease was attempted in a joint Canadian–European study in which the immunosuppressive drug, cyclosporin (see Chapter 32), was given to patients within 6 weeks of commencement of insulin therapy. Although there was some evidence of short-term remission, the high dose of drug used caused signs of renal toxicity and the trial was discontinued. There is also a European Nicotinamide Diabetes Introduction Trial (ENDIT). In this, nicotinamide is given, as part of their treatment, to patients newly diagnosed as having IDDM. This B vitamin has proved successful in a mouse model of the disease, but the results in humans are not very encouraging.

Chapter 42

Asthma

42.1 Introduction

The word **asthma** derives from the Greek and literally means 'panting' or 'hard breathing'. The second century physician Galen provided an early account of the disease when he described an intermittent obstruction to breathing, which he ascribed to secretions dipping from the brain into the lungs (a view that persisted until the seventeenth century). The leading twelfth century physician Moses Maimonides made a significant observation about the disease in his 'Treatise on Asthma' when he noted that the disease frequently began in puberty. His suggested treatment of the disease by the administration of the soup prepared from fat hens was not so useful! The importance of environmental factors in asthma was foreshadowed by the Italian Girolamo Cardano who, in 1522, was able to cure an archbishop's asthma by persuading him to abandon his feather bed and pillows. This environmental view of causative agents was expanded in the last century by a Dr Salter, who noted that exercise, cold air, dust, pungent fumes and 'animal and vegetable emanations' were important causes of the disease. The contribution of the early twentieth century physician William Osler was not as fortunate because his proposal that there is a strong neurotic element in the majority of cases of bronchial asthma has tended to place an unfortunate stigma on the disease.

Asthma is found worldwide, but with variable incidence. At the two extremes the disease is very rare in American Indians and Eskimos, but over one-third of the inhabitants of the South Atlantic island of Tristan da Cunha suffer from the disease. This last finding points to the importance of inheritance in the disease because all of the inhabitants of the island are descended from fifteen ancestors, three of whom suffered from asthma. In general, asthma is a disease of developed rather than developing countries (again pointing to environmental factors). In the UK, one in seven children suffer from asthma, more boys than girls being affected. The disease is less common in adults, but between one in 10 and one in 20 adults have the disease, with both sexes being equally susceptible. All of this means that there are known to be over half a million asthma sufferers in the UK (with probably many more cases not diagnosed). For children, an acute attack of asthma is the commonest reason for their admission to hospital in the UK as it is in Australia and New Zealand. This, however, has only become true in the past 10 years. Undoubtedly this dramatic increase can in part be accounted for by improved means of diagnosis and increased awareness of the need for prompt treatment.

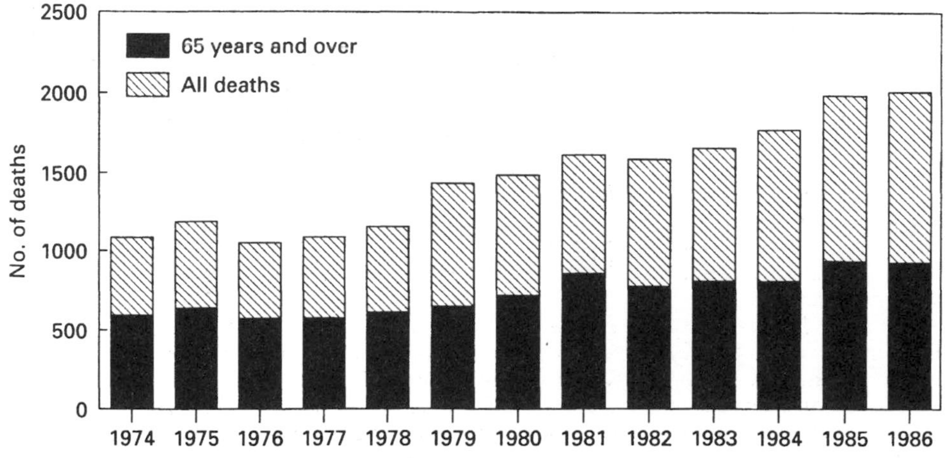

Figure 42.1 **Graph of asthma deaths in the UK**

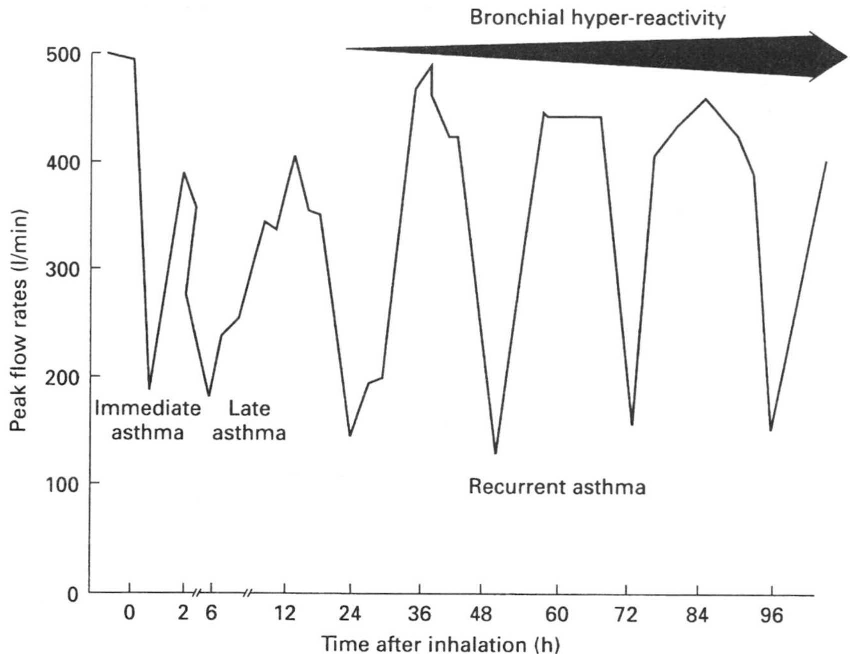

Figure 42.2 **Recurrent asthma and hyperresponsiveness**

This is not the whole story, however, for there is excellent epidemiological evidence that the disease has become both more common and more severe. The mortality figures support this conclusion, for between 1976 and 1985 the number of deaths from the disease doubled from about 1000 to 2000, with the largest increase being in those under 65 years (Figure 42.1). In New Zealand, the increase has been even more marked, with a five-fold increase in asthma-associated deaths in the last decade. The reasons for these large increases is not fully resolved, but it seems likely that environmental factors, particularly those relating to housing and motor transport emissions, are important.

42.2 The disease

The nature of the disease and of the processes underlying it can be readily understood from a description, given by Davies and Ollier in their text *Allergy*, of a patient with asthma. They described a 40-year-old arable farmer from Sussex who complained of a severe debilitating cough and shortness of breath (accompanied by wheezing) beginning within minutes of his shovelling grain and lasting for over a week. Following a week in hospital for investigation he was free of symptoms, with normal lung function both during the day and at night. He was then asked to shovel grain and within minutes he suffered a severe asthmatic attack in which his peak flow (the capacity to force-expire air from the lungs) was greatly compromised. One or two hours later, he felt better, but this was only temporary, with a second period of peak flow decline, of longer duration, being established. A second period of apparent recovery was followed by the first of four attacks on successive nights (Figure 42.2).

It is now understood that these symptoms are associated with chronic inflammation of the airway walls. This means that asthma involves many of the inflammatory cells and mediators of the types described in Chapter 32. In particular, there is an activation of the mast cells resident in the airways, increased numbers of $CD4^+$ T- lymphocytes in the lungs and an infiltration of eosinophils (which degranulate). Since these changes are often associated with inappropriate immune responses to inherently harmless agents, e.g. the faeces of the dust mite, they are classified as **allergic** and the precipitating agents are referred to as **allergens**. It should be remembered, however, that asthma is a complex disease and the symptoms, in individuals who suffer primarily from an allergy to pet dander, for example, may be brought on by a variety of 'trigger' factors, e.g. exercise, cold, emotional stress or airborne pollutants.

From the pattern of disease seen in the case of the Sussex farmer cited earlier, it is possible to describe the manifestations of an asthmatic attack as occurring in three phases: a rapid-onset spasmogenic phase (associated with smooth muscle contraction), a late sustained phase (of evolving inflammatory response) and a chronic, underlying inflammatory phase (Figure 42.3).

42.3 The causes of asthma

Figure 42.4 summarizes the complex interactions that can underlie an attack of asthma. A feature that is found in greater or lesser degree in all asthmatics is the presence of 'twitchy' airways. This means that the smooth muscles that encircle the bronchi and the alveoli of the lungs (see Figure 42.3) are functionally hyper-responsive. Consequently, in the asthmatic patient these muscles are unusually prone to contract, and it is the

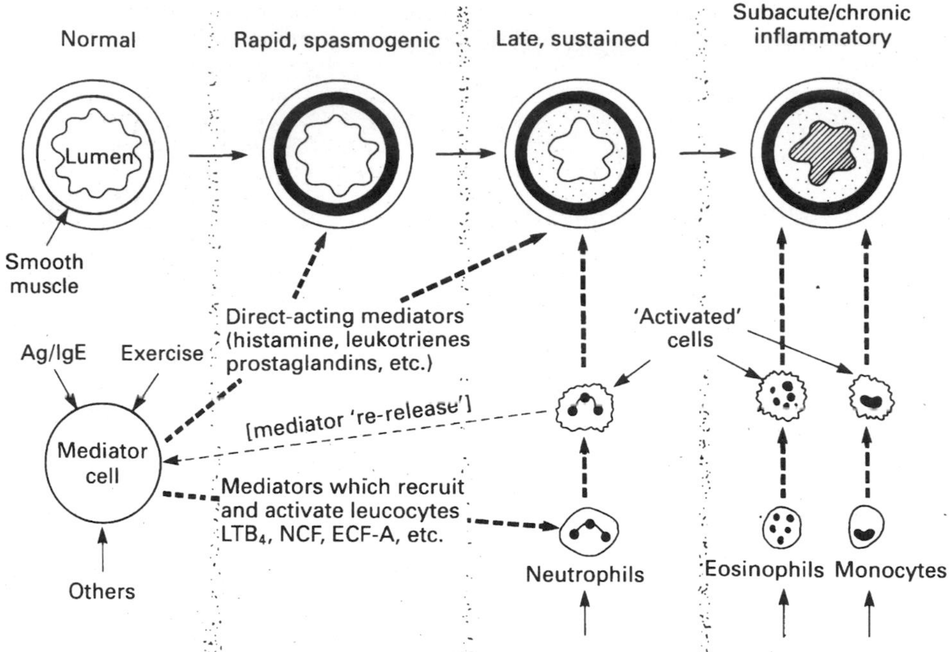

Figure 42.3 **Mediators, inflammatory cells and changes in the airway lumen during the progression of asthma.** A cross-section of a bronchus is shown, with lumen surrounded by epithelium and a deeper layer of smooth muscle (Ag/IgE, antigen/immunoglobulin reaction; LTB_4, leukotriene B_4; NCF, neutrophil chemotactic factor; ECF-A, eosinophil chemotactic factor

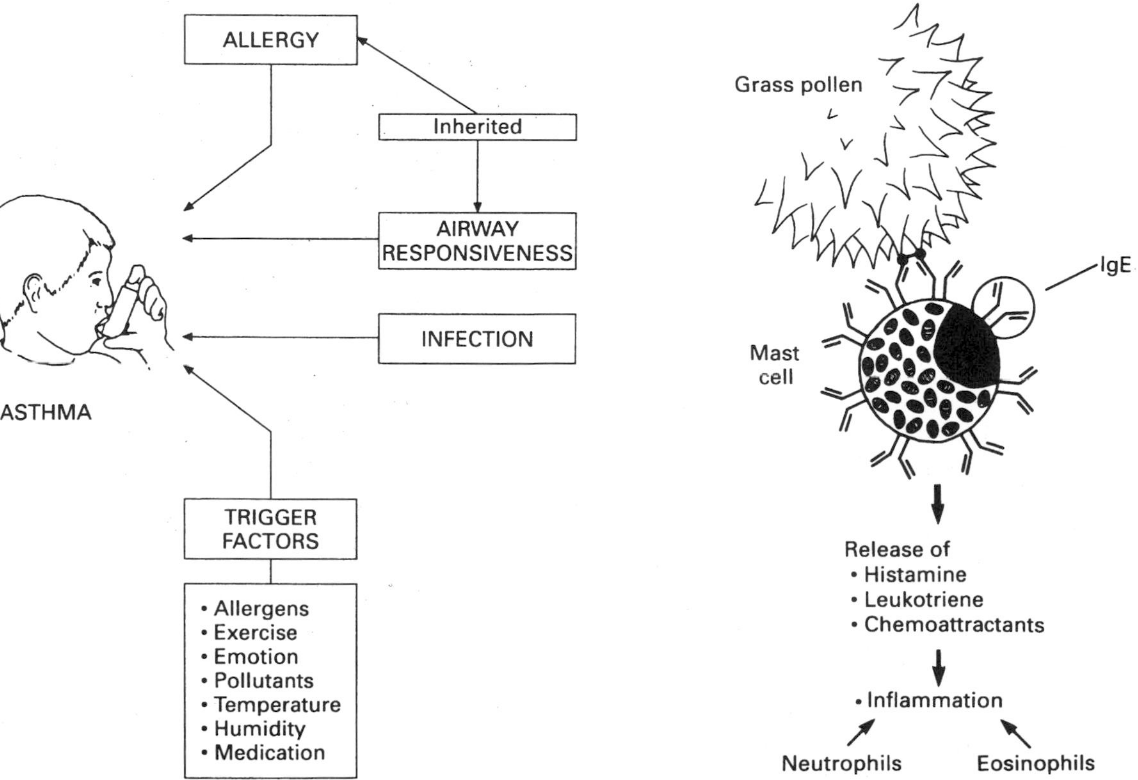

Figure 42.4 **Causes of asthma**

Figure 42.5 **The allergic reaction and inflammation**

inappropriate contraction of these muscles that is the source of the 'wheeze' that characterizes the disease. Figure 42.4 also provides a reminder that people who suffer unduly from asthma frequently have an inherited predisposition to the disease (although the nature of this is unknown).

42.4 Events triggered during the course of an asthmatic attack

Since the course of asthmatic attacks precipitated by exposure to allergens has been extensively documented, this will be used to describe the processes that may be activated in the disease. (It is not clear whether allergy is the primary cause in *all* cases of asthma.)

In the rapid spasmogenic phase of the disease the allergen binds to IgE molecules, which themselves are firmly bound to their own specific receptors on the surface of mast cells (Figure 42.5). The binding of the allergen to several IgE molecules causes rapid activation and degranulation of the mast cells, with the release of a host of vasoconstrictor and other molecules. These include histamine, prostaglandins, leukotrienes and also the mast-cell specific neutral protease tryptase. The vasoconstrictor molecules bind to their G-protein-coupled receptors on smooth muscle cells to promote the activation of phospholipase C (see Chapter 3). The consequent inositol trisphosphate-mediated release of intracellular Ca^{2+} into the cytosol leads to the Ca^{2+}/calmodulin-mediated activation of myosin light-chain kinase (see Chapter 27) which then catalyses the phosphorylation of its target protein, with consequent muscular contraction.

The spasmogenic phase may be well controlled by the use of bronchodilator aerosols, such as **Ventolin**, which contain the β-adrenoceptor agonist **salbutamide**.

The development of the later phases of an asthmatic attack depends upon the recruitment of other pro-inflammatory cells to the site of the spasmogenic response. Important among these cells are the **eosinophils** that are both attracted and activated by molecules such as platelet-activating factor. (It has been known for some time that there is a close correlation

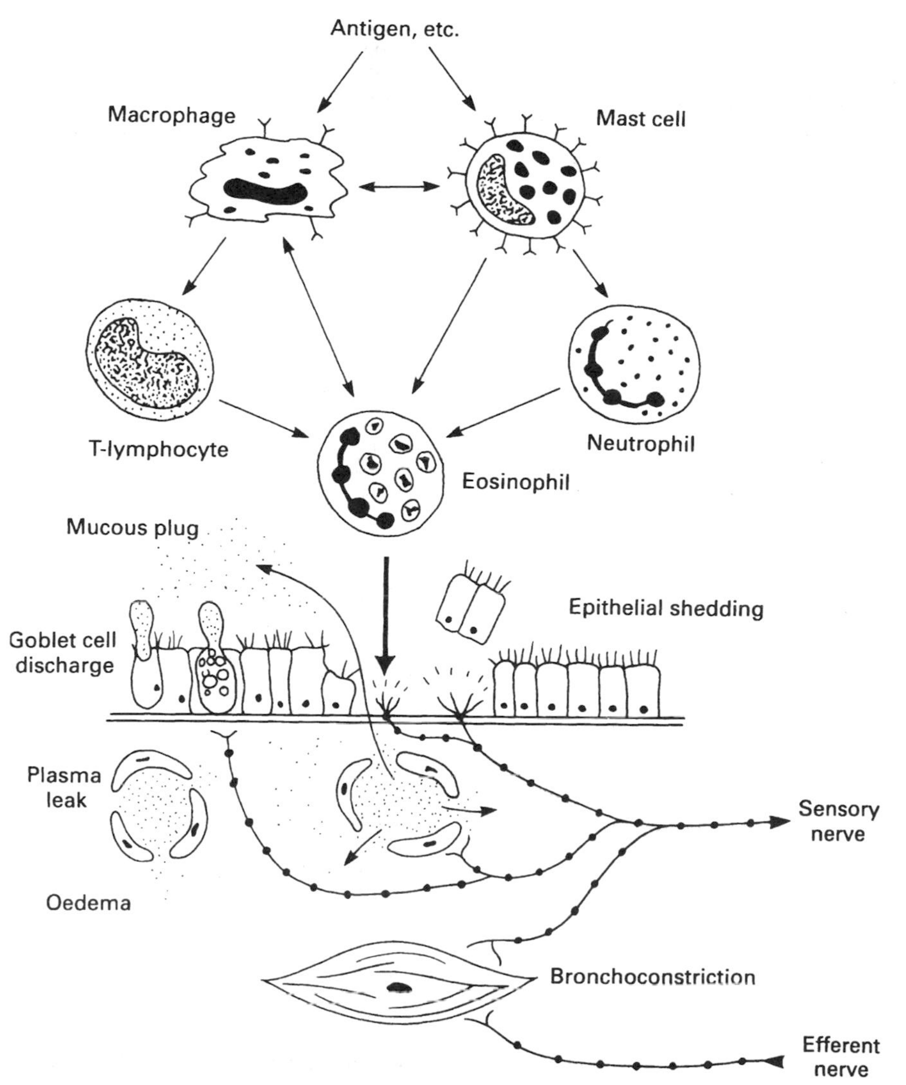

Figure 42.6 **Asthmatic inflammation involves several types of inflammatory cell which interact with each other, leading to the release of inflammatory mediators that act on target cells in the airway**

between the severity of asthma and the number of circulating eosinophils.) Once activated, eosinophils themselves release a host of inflammatory mediators (Figure 42.6) and also oxygen free radicals (see Chapter 33). Several of the mediators are highly cationic proteins which may be causative agents in the shedding of epithelial cells that is another feature of the disease. Indeed, histological examination of the lungs of patients dying in *status asthmaticus* reveals only a few areas of normal epithelium remaining. These shed cells contribute to the formation of 'mucus plugs' in the disease. The gaps arising as the epithelial cells are shed have the effect of increasing the permeability of the airway mucosae to further infiltration by allergens and bacteria, and also by inflammatory mediators released into the airway lumen. All of these factors exacerbate the inflammatory process. Another consequence of the shedding of epithelia is that underlying sensory neurons become exposed to mediators arriving from the lumen. These neurons may then release neuroactive peptides, such as neurokinin A (a potent bronchoconstrictor) and substance P, via 'axonal reflexes'. The latter peptide appears to be particularly effective in promoting plasma leakage and secretion from submucosal glands.

The multifactorial, self-perpetuating nature of the inflammatory response in asthma means that bronchodilators (which act largely on mast cells to prevent their degranulation) are ineffective at this stage. Indeed the use of bronchodilators may actually serve to disguise the evolution of the inflammatory aspect of the disease. The drugs of choice therefore are **anti-inflammatory agents**. These include **sodium cromoglycate (Intal)** and **nedocromil sodium** for use in mild asthma, and glucocorticoids that are effective in most (but not all) cases of asthma. The great virtue of glucocorticoid therapy is that the steroid (usually **prednisolone** or one of its esters, if tablets are given, or **beclomethasone** or one of its esters, if inhaled) acts to control the expression of a number of different pro- or anti-inflammatory genes and therefore addresses directly the multifactorial nature of the disease. Following its binding to its intracellular receptor, the steroid–receptor complex moves to the nucleus where it binds to **glucocorticoid response elements** (**GRE**) associated with key genes. A positive GRE enhances the expression of the gene for the anti-inflammatory protein **lipocortin**, while negative GREs reduce the expression of pro-inflammatory genes, e.g. those for the interleukins. The synthetic glucocorticoids have the disadvantage that they can act to suppress the functional activity of the hypothalamo-pituitary–adrenal axis. In this action they mimic the negative feedback effect of cortisol. This unwanted effect, caused by the administration of synthetic glucocorticoids, is avoided in the case of beclomethasone by its administration by inhalation, or by limiting the oral dose to 15 mg. The mechanism of action of sodium cromoglycate remains obscure.

There is growing evidence that the excessive use of β-agonists, such as salbutamide, to control the spasmogenic phase of asthma actually has the effect of rendering glucocorticoids progressively less effective as anti-inflammatory agents in treating the disease. In some way the cAMP produced in response to salbutamide appears to prevent the binding of the glucocorticoid–receptor complex to GREs in the DNA.

42.5 Environmental agents in asthma

The most common allergen causing asthma in the UK is the antigen, Der P1, found in the faeces of the house dust mite *Dermatophagoides pternyssinus*. The role of the **dust mite** as a cause of asthma was suggested as long ago as 1928 by the German scientist Dekker. These organisms feed on desquamated human skin, and therefore are particularly active in bedding, especially in moist conditions. Thus in London the numbers of mites (and of people allergic to them) correlates with the distance from the river Thames (or one of several underground streams). In addition, local factors, such as the number of people sleeping in a room and the size of the room, help to determine the degree of humidity (each person loses 500 ml water by evaporation each night). A relative humidity of 80%, together with a temperature of 25°C (perhaps maintained by central heating with humidifiers), are optimal for the proliferation of mites.

Allergy to the spores of the mould *Didymella exitalis* probably accounts for the occurrence of so-called 'epidemics' of asthma noted in both Britain and Spain following thunderstorms. This is because the mould, which is found on the leaves of ripening wheat and barley, is released by moisture. A more common route of release is by the formation of dew, which results in maximal sporulation at 03.00 a.m. The early hours of the morning are a common time for attacks of asthma.

Exposure to a wide range of pollutants, for example to cigarette smoke and motor vehicle exhaust fumes, is very often sufficient to provoke an asthmatic attack in the susceptible individual. The causative agents are not fully characterized, but undoubtedly include the oxides of both nitrogen and sulphur and probably ozone.

Chapter 43

Amyotrophic lateral sclerosis

43.1 Introduction

As the final study in this section, in which we have attempted to show how an appreciation of the basic principles of biochemistry covered in this book can help to illuminate an understanding of diseases and their treatment, we have chosen **amyotrophic lateral sclerosis** (ALS), also known as **motor neuron (MN) disease**. This study differs from its predecessors in this section in that, while a knowledge of biochemistry is helping in the understanding of the nature of the disease it has not yet reached the stage of providing strategies for prevention or for treatment.

ALS is a chronic progressive disorder of the nervous system characterized by the gradual disintegration of motor neurons in the cerebral cortex, the brain stem and the spinal cord, i.e. may involve upper or lower motor neurons or both. The clinical features of the disease and the underlying pathology were first described by Charcot in 1869. This represents the **'classical' or sporadic form of ALS**. It is now appreciated that a familial form of the disease exists. In addition, there is what is probably a different disease but with similar symptoms that is referred to as Western Pacific (atypical) ALS, seen in subjects living in that part of the world, most especially on the Island of Guam.

43.2 The disease

The classical disease begins asymmetrically and distally, in one limb, most often a leg. Thereafter it spreads to involve successive groups of motor neurons. However, certain groups of neurons seem to be spared, e.g. those controlling the bladder. Symptoms include progressive muscular twitching, growing weakness and wasting of body muscles. Once respiratory muscles are involved death ensues, often due to respiratory infection.

The pathology of classical ALS is distinguished by the atrophy and death of affected neurons and this involves dissolution of both the cytoplasm and the nucleus. The cytoskeleton in affected neurons is disrupted with the accumulation of neurofilaments as disordered 'spheroid' bodies in the perikaryon and in the proximal axon. Currently the most likely biochemical hypothesis concerning the disease is that is arises due to oxidative and excitotoxic injury affecting specific regions in motor neurons. The classical disease is more prevalent in temperate than in hot climates. Thus the annual number of deaths due to ALS is about 0.1 per 10 000 in Chile and Mexico, but about 0.5 per 10 000 in Scotland and the Netherlands. In contrast to multiple sclerosis, ALS is seen more frequently in men than in women, with the male : female ratio being 1.5 in Chile and 2.9 in Ireland. The maximum rate of onset occurs at about age 65 years and thereafter survival is for no more than 3–5 years.

The familial form of the disease closely resembles classical ALS, but the average age of onset is 50 years, with progression to death occurring within 3 years. About 5–10% of all cases of classical ALS are inherited. There are two forms of **familial ALS**. The predominant one is inherited as an autosomal dominant trait. In addition, there is a rare recessively inherited, childhood onset form of the disease which is associated with a very long survival. The atypical form of ALS found in the Western Pacific differs from the classical form of the disease in that other groups of neurons seem to be involved, e.g. cells in the hippocampus. The age of onset is about 20 years earlier than is seen with classical ALS, but seems to have become more delayed since it was first reported in the 1950s. Early reports suggested an incidence of 50 per 10 000, but these numbers now seem to be on the decline.

43.3 Possible causes of ALS

Any hypothesis aimed at explaining the pathogenesis of ALS must incorporate its characteristic features: the occurrence of specific neuronal cell death and the adult onset. Epidemiological evidence has suggested an environmental causative agent in ALS. The evidence for this is seen most clearly in the Western Pacific variety of the disease. There is very high incidence of ALS in the Chamarro people of Guam, the inhabitants of the Kii peninsula of Honshu Island in Japan and the aborigines living on the Australian island of Groote Eylandt in the Gulf of Carpentaria. In the latter case, the incidence of the disease seems to have increased following the enforced movement of the population into a major manganese mining area. Similarly, the manganese mines of Guam in the Pacific were worked extensively by the Japanese using local labourers in the period 1941–44. Other possible environmental agents, considered in the case of the atypical form of the disease, are plant toxins. In particular toxins such as β-*N*-methylamino-L-alanine derived from the cycads have been considered. This compound, when given to monkeys in relatively high doses, does cause degeneration of motor neurons, but the disorder induced does not resemble atypical ALS. Because the polio virus also attacks motor neurons it was thought that a member of this family of viruses may be causative, but there is no evidence for the presence of polio-virus-like RNA in tissues taken from victims of the disease. Nor do extracts from affected human tissues cause the disease when inoculated into non-human primates. This finding also tends to eliminate the possibility of a prion disease.

The involvement of autoimmunity has been considered in ALS and reports of circulating antibodies to voltage-gated calcium channels in affected people have appeared. Indeed, this has led to the use of immunosuppressive therapy, but such treatments have failed to arrest the progression of the disease.

Although all of these possible causes of ALS are under active consideration, most rapid progress in understanding the biochemical changes underlying the condition is coming from the study of the two inherited forms of ALS.

43.4 Inherited forms of ALS

In the predominant, adult-onset, familial form of ALS, the gene for the cytosolic form of **superoxide dismutase** (SOD1; see Chapter 33) has been found to be mutated in several families. It may be recalled that SOD1 is a homodimeric enzyme that has a joint requirement for copper and zinc for its catalytic activity. By 1995 no fewer than 20 different mutations had been identified, i.e. genetic heterogeneity exists. The effect of the mutations that have been studied is to reduce the catalytic activity of the enzyme by 25–50%. In view of the autosomal dominant pattern of inheritance of the disease, this finding is rather surprising. Generally, for mutations affecting enzymes the catalytic activity of the normal product of the unaffected allele is sufficient to prevent symptoms, i.e. the trait is recessive when affecting an autosome. These considerations have led to the suggestion that the mutations cause a *gain* in cytotoxic function in SOD1, and this must be considered together with a possible diminution in the ability to deal with $O_2^{\cdot -}$. For example, if the mutation reduces the affinity of SOD1 for copper, then the intracellular concentration of the free metal ion may rise above its normal value of 10^{-15} M, thereby promoting oxidative damage to the tissues. In addition, if the mutations render the protein less soluble, then it might be expected to be deposited in the cell causing damage. But since SOD1 is ubiquitous in mammalian cells, why is it only motor neurons that are affected? A clue to this comes from a study of the rare, familial, childhood-onset form of the disease in which the mutations affect the proteins of neurofilaments (these are known to be disrupted in the sporadic form of the disease). The involvement of neurofilaments would serve to target neurons for disruption. However, this, again, fails to account for the specificity in the destruction of motor neurons.

Perhaps the answer is in part anatomical, because motor neurons have some of the longest axons found. If, for some reason, axonal transport is disrupted, e.g. by deposition of neurofilament spheroids, then the supply of SOD1 to nerve terminals would be badly affected. Another aspect of the answer may be more biochemical, because not all long-axon neurons are affected in ALS, e.g. dorsal root ganglion (DRG) neurons are not. The neurotransmitter control of motor neurons and DRG neurons differ, however, in that the former are extensively controlled by glutamate. This has led to the suggestion that excitotoxicity of the type that occurs in stroke, for example, might be contributory. At the heart of the excitotoxic hypothesis of stroke is the opening of the NMDA receptor-channels which allow uncontrolled access of Ca^{2+} to the cytosol (see Chapter 30). It is

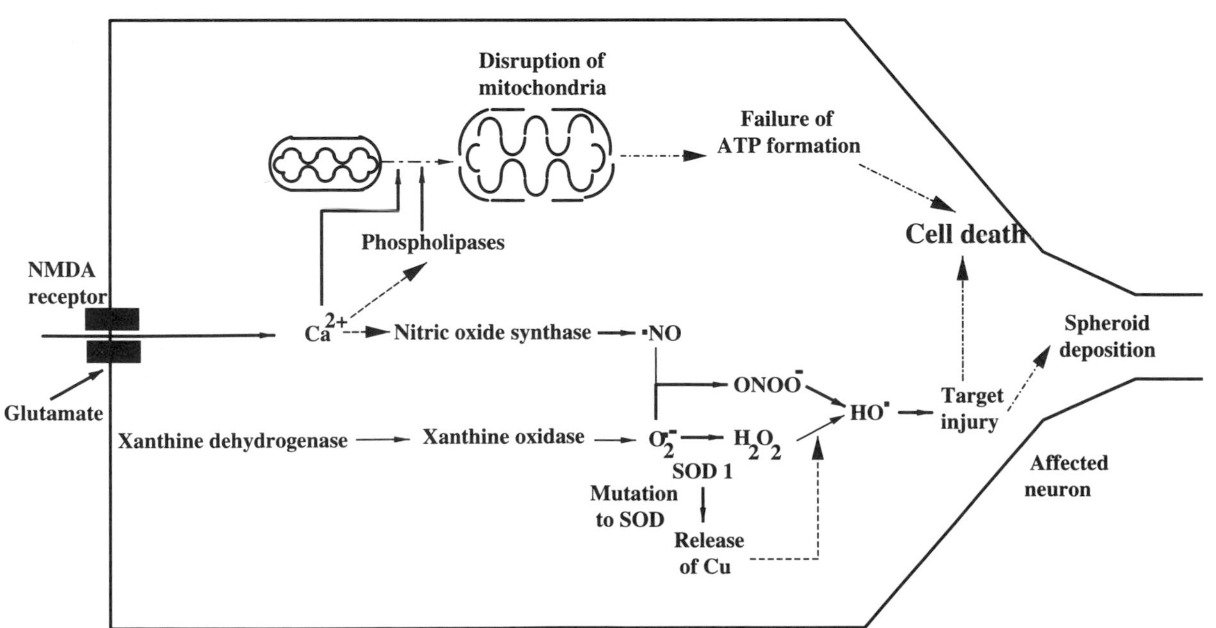

Figure 43.1 **Possible routes to the neurodegeneration seen in amyotrophic lateral sclerosis**

proposed that motor neurons may be particularly susceptible to even small perturbations of normal cellular calcium homeostasis, because the intracellular Ca^{2+}-buffering proteins calbindin-28K (see Chapter 29) and parvalbumin are either absent or present only in low concentration in these nerves.

In a current model of ALS, oxidative damage to neurofilaments and also to membranes (both plasma and intracellular) is proposed to set up a vicious cycle leading to the destruction of motor neurons. In some cases of the adult-onset form of the familial disease the oxidative challenge arises as a result of mutation to SOD1, but the nature and basis for the challenge in the sporadic form of the disease remains to be determined. Figure 43.1 proposes a model for motor neuron degeneration in familial ALS.

43.5 Prospects for treatment

Although, as has been mentioned, the possibility of arresting the progression of ALS by immunosuppression has been investigated, this has proved no more successful than any other method. Consequently there are currently no methods either to prevent or to arrest the progression of the disease. An important goal will be to determine whether the degenerative changes are necrotic or apoptotic in nature. It is believed that $O_2^{\cdot -}$ is one of the signals for apoptosis. If the latter is the case, i.e. if normal programmed cell death is activated inappropriately, there is the possibility that a new generation of anti-apoptotic drugs might be useful in arresting the progress of the disease.

Index

'A'-band
in skeletal muscle, 242
Abetalipoproteinaemia
vitamin E deficiency, 352
AcetoacetylCoA
formation in β-oxidation, 43
N-Acetylaspartate
possible role in brain development, 307
Acetylcholine
as a neurotransmitter, 293
control of adrenal medullary function, 269
dependence of the brain on glucose for, 304
Acetylcholine esterase
collagen like structure, 281
termination of action of acetylcholine, 294
Acetylcholine receptor, nicotinic
structure and function, 298
Acetylcholine receptors, muscarinic
presence in B cells in the pancreas, 263
AcetylCoA
formation in β-oxidation, 43
formation from pyruvate, 42
provision for the biosynthesis of acetylcholine, 294
N-Acetylpenicillamine
chelation therapy following mercury poisoning, 361
Acetyltransferases
in the metabolism of foreign compounds, 375
Achlorhydrasia
relationship to iron deficiency, 187
Achondroplasia
associated with short stature, 285
mutation to a receptor for fibroblast growth factor, 285
Acne vulgaris
use of retinoic acid in treatment, 174
Aconitase
reaction catalysed in mitochondria, 44
relationship to iron response factor, 191
Acquired immunity
nature, 318
Acquired immunodeficiency syndrome (AIDS)
magnitude of the problem, 317
nature of the human immunodeficiency virus (HIV), 386
use of zidovuline (AZT) in the treatment, 421
ACTH
see Adrenocorticotrophin
Actin filaments
occurrence in the cytoskeleton, 17
role in skeletal muscle, 244
Actinin
in skeletal muscle, 244
Action potential
description of the genesis of in neurons, 291
Active transport
across membranes, characteristics, 19
Acute phase proteins
C-reactive protein and mannose binding proteins, 329
Acyclovir
as an antiviral drug, 420
AcylCoA: cholesterol acyltransferase (ACAT)
biosynthesis of cholesteryl esters, 96
AcylCoA dehydrogenase
reaction catalysed, 43
AcylCoA synthase
reaction catalysed, 43
Adaptive immune responses
mediators, 318
nature, 318
Addison's disease
adrenal insufficiency, 277
Additives in food, toxicity
incidental, 130
intentional, 130
Adenosine
interaction with glucagon in the control of lipolysis, 266
Adenosine deaminase
release of ammonium ions, 102
Adenosine triphosphate (ATP)
formation during oxidative phosphorylation, 49
role in contraction of skeletal muscle, 248
Adenosine triphosphate/Adenosine diphosphate (ATP/ADP) exchange
in mitochondria, 50
Adenylate kinase
presence in skeletal muscle, 250
Adenylyl cyclase
activation by glucagon, 266
in the control of steroidogenesis, 274
mechanism of activation and inactivation, 25
role in the control of glycogenolysis in liver, 59
ADP ribosylation
action of cholera toxin, 39
modification of histones, 39
Adrenal androgens
biosynthesis, 272
as precursors of oestrogens in post menopausal women, 273
Adrenal gland
structure of, 269
Adrenal insufficiency
in Addison's disease, 277
Adrenal cortex
hormones, 270
Adrenal medulla
hormones, 269
Adrenaline
as a neurotransmitter, 296
control of glycogenolysis in muscle, 59
effect on protein synthesis in the post absorptive state, 103
metabolic and cardiovascular effects, 270
modes of action, 270
β-Adrenergic stimulation
of smooth muscle, 252
Adrenocorticotrophin (ACTH)
action in steroidogenesis, 274
in the ebb phase of the metabolic response to injury, 432
related peptides, 260
Adrenoleucodystrophy
association with the accumulation of long chain fatty acylesters of cholesterol, 97
Adult respiratory distress syndrome (ARDS)
occurrence, 351
Aflatoxin B_1
hepato-cellular carcinoma, 388–9
Agarose gels
electrophoresis of DNA, 401
Age
effect on plasma lipid levels, 97
Age-related maturation of collagen fibrils
processes, 283
Agglutination
of particulate antigens by antibodies, 324
AIDS
see acquired immunodeficiency syndrome
AIDS patients
occurrence of Karposi's sarcoma, 386
Airways
chronic inflammation in asthma, 445
Alanine
as a precursor for gluconeogenesis during short term starvation, 165
release from muscles following injury, 437

Albumin
 transport of fatty acids, 85
 transport of T_3 and T_4, 267
 transport of steroid hormones, 274
Alcohol intake
 guidelines of the USA National Research Council, 130
 UK guidelines for a healthy diet, 130
Alcoholic cirrhosis
 as a cause of haemochromatosis, 194
Alcoholism
 associated with Cushing's disease-like signs and symptoms, 277
 occurrence of Wernicke–Korsakoff syndrome, 178
Aldehyde dehydrogenase
 in the liver, 227
 variants, 227
Aldose reductase inhibitors
 possible prophylactic use in patients at high risk of cataract, 311
Aldosterone
 biological actions, 275
 biosynthesis, 272
 role in the kidney, 236
Alkoxyl radicals
 formation from lipid hydroperoxides, 347
Allergens
 exposure to, resulting in asthmatic attacks, 448
Allergic reactions
 occurrence in asthma, 445
Allergy
 role of IgE, 326
Allysine
 in collagen cross-linking, 283
Alport's syndrome
 degeneration of renal basement membranes, 232
 occurrence, 285
Alternative complement pathway
 organization, 329
Aluminium
 biochemical effects, 363–4
 effects on health, 363
 toxicity, 363
Alzheimer's disease
 nature and occurrence, 300
 presence of aluminium in the brain, 363
Ames test
 for detection of potential mutagens, 394
Amino acid and protein metabolism
 effects of insulin on, 265
 in the brain, 307
Amino acid-catabolising enzymes
 induction in the fed state, 103
Amino acid metabolism
 effect of growth hormone, 261
 in injury, 437
 organization, 60
 role of the liver, 223
Amino acid pool
 nature, 99, 142
Amino acid transporters
 in the brain, 307
Amino acids
 activation for protein synthesis, 64–5
 in urine, 233
 intestinal absorption, 122
 renal recovery, 238
AminoacyltRNA synthases
 role in the activation of amino acids prior to protein synthesis, 64
Aminoglycoside antibiotics
 gentamycin as an example, 418
 streptomycin as an example, 417
γ-Aminolaevulinate
 urinary excretion following lead poisoning, 362–3
 in haem biosynthesis, 60
Ammonium ions
 formation and excretion in the kidney, 241
Amniocentesis
 in prenatal diagnosis, 409
 in the diagnosis of inherited diseases of blood clotting, 218
Amylase
 action on amylose, 120
Amylin
 production by B cells in the pancreas, 263
Amyloid
 possible role in prion diseases, 422
Amyloid plaque
 deposition in Alzheimer's disease, 301
Amylopectin
 as a dietary polysaccharide, 155
Amylose
 as a dietary polysaccharide, 154
 hydrolysis by amylase, 120
Amylotrophic lateral sclerosis (ALS)
 world deaths from, 449
Anabolic reactions
 nature, 45
Anaemia
 associated with deficiency in red blood cell enzymes, 207
 associated with glutathione peroxidase deficiency, 208
 associated with pyruvate kinase deficiency, 208
 lead induced, 363
Anaemia, aplastic
 underproduction of erythropoietin, 241
Anaemia, haemolytic
 of the new born, 325
Anaemias
 causes, 192
Anaesthetic risks
 in obesity, 160
Anaphylatoxins
 generation in complement pathway, 328
Androstenedione
 an adrenal androgen, 273
Anergy
 'silencing' of T-cells, 336
Angiotensin-converting enzyme (ACE)
 reaction catalysed, 236
Angiotensin II
 formation and actions, 274
Angiotensin receptor
 activation of phospholipase C, 274
Angiotensinase
 reaction catalysed, 274
Anions
 in the diet, 184
Ankylosing spondylitis, treatment
 increased incidence of cancer associated with, 379
Annealing properties of DNA
 description, 400
Anomalous trichromy
 in colour blindness, 316
Anorexia
 occurrence in the injured patient, 435
Anorexia nervosa
 occurrence, 168
Anterior pituitary gland
 hormones, 260–2
Anthropometric measurements
 role in the detection of malnutrition, 132
Anti-apoptotic drugs
 possible use in the treatment of amylotrophic lateral sclerosis (ALS), 451
Antibiotic-resistance genes
 nature of in bacteria, 424
 use in DNA cloning, 405
Antibiotics
 effective against protein synthesis, 67
 nature, 411
Antibodies
 constitute the γ-globulin fraction of plasma, 321
 role in the immune response, 320
 sources of diversity, 323
 see also Immunoglobulins
Antibodies, homocytotropic
 nature, 326
Antibodies, reaginic
 IgE, 326
Anticipation
 in Huntingdon's disease, 296
Anticoagulant drugs
 actions of, 214
Antidiuretic hormone (ADH)
 renal water recovery, 235
 see also vasopressin
Antigen
 definition, 321
Antigen presentation
 roles of MHC classes I and II proteins, 334
Antigen-presenting cells
 types, 333
Antigenic variation
 examples, 320
Antigenicity
 definition, 321
Anti-metabolites
 active in bacterial infection, 418
 as anti-cancer drugs, 427
Antimycin A
 inhibition of mitochondrial complex III (ubiquinol dehydrogenase), 51
Antioxidant nutrients
 effect on the risk of cardiovascular disease, 350
 nature, 348
Antioxidants
 defence of tissue integrity, 348
Antiplasmins
 roles in blood clotting, 217
Antiport processes
 in membranes, 19
Anti-promoters
 prevention of the progression to cancer, 393
α1-Anti-proteinase (α_1-anti-trypsin)
 effect of smoking on, 351

Antisense therapy
  potential use in the treatment of cancers, 429
Antithrombin
  role in limiting the extent of blood clot formation, 216
Apolipoprotein (a)
  as the protein component of lipoprotein (a), 98
Apolipoprotein A1
  presence in HDL, 86
Apolipoprotein B-100
  modification of lysine residues, 350
  oxidative modification of structure, 347
Apolipoprotein B-100
  presence in very low density lipoproteins (VLDL), 86
Apolipoprotein B-48
  presence in chylomicrons, 86
Apolipoprotein C II
  as an activator of lipoprotein lipase, 86
Apolipoproteins
  associated with nascent high density lipoproteins (HDL), 94
Apoproteins
  presence in lipoproteins, 86
Apoptosis
  occurrence and cells affected, 380
  of bone marrow cells, 331
  of T-cells in the thymus, 333
Aquaporin, 2
  in the kidney, 262
Aquaporins
  in renal water recovery, 235
Arachidonic acid
  as a precursor of eicosinoids, 149
  status as an essential fatty acid, 147
Aromatase
  in steroid hormone biosynthesis, 271
Arsenic poisoning (trivalent)
  dimercaprol as an antidote, 365
Ascorbic acid
  see vitamin C
Aspirin
  antipyretic action, 150
  as an inhibitor of prostaglandin H_2 synthase, 211
  mechanism of action as an anti-coagulant, 211
  use following aorto-coronary by-pass surgery 211
  use to reduce the risk of heart attack and stroke, 211
Asthma
  increasing incidence in the UK, Australia and New Zealand, 444
Astrocytes
  in re-uptake of glutamate, 295
Astrocytes
  supportive role in brain, 290
Atherosclerosis
  relationship to plasma cholesterol concentrations, 97
  relationship to dietary lipid intake, 151–3
Atoxyl
  early use in the treatment of trypanosomiasis (sleeping sickness), 411
ATP
  see Adenosine triphosphate
Atracylate
  inhibition of ATP/ADP exchange across mitochondrial inner membrane, 51
Atrial natriuretic peptide (ANP)
  signalling mechanism, 24
  effect on smooth muscle, 252
  role in promoting natriuresis, 237
Atrophic gastritis
  as a cause of pernicious anaemia, 201
Autocrine effects
  nature, 21
Autoimmune disease
  myasthenia gravis, affecting the nicotinic acetylcholine receptor, 299
  insulin dependent diabetes mellitus (IDDM), 439
  nature, 318
Autolysis
  by lysosomes, 74
Axonal transport
  nature and occurrence, 307
Axons
  nature of in neurons, 289
Ayurvedic medicine
  doctrines relating to diabetes mellitus, 439
Azo-reduction
  in the phase I metabolism of foreign compounds, 369
AZT
  in the treatment of acquired immunodeficiency syndrome (AIDS), 421
  mitochondrial myopathies associated with administration, 55

B-cell
  development, 331
  interaction with T_{H2} cells, 338
  presence of prolactin receptors, 262
β-group of vitamins
  general considerations, 175
β-Oxidation of fatty acids
  reactions involved, 43
B7 receptor
  in cell mediated immunity, 336
Bacterial cell walls
  effect of penicillins, 416
  formation, 415–6
Bacterial flora
  in the intestinal tract, 124
Bacterial reversion
  as the basis of the Ames test for potential mutagens, 394
Bacteriophages
  protection of bacteria by restriction enzymes, 319
Bare lymphocyte syndrome
  occurrence, 333
Basal metabolic rate
  factors affecting, 137
  in the ebb phase of the metabolic response to injury, 433
  in the flow phase of the metabolic response to injury, 435
  nature and determination, 137
Base-pairing rules
  in DNA, 30
Basement membranes
  collagen found in, 281
  organization in the kidney, 232
Basic amino acids
  influence on absorption of dietary calcium, 107
Basophils
  roles in immune responses, 330
Beclomethasone
  use in asthma, 448
Becker muscular dystrophy
  occurrence and nature of disease, 252, 254
Benzylpenicillin (penicillin G)
  structure, 413
Beriberi
  occurrence and features, 178
Bicarbonate ion (HCO_3^-)
  handling in the kidney, 238
Bile
  composition and excretory functions, 117–8
  digestive function, 118
  formation, 117
Bile salts
  metabolites as possible carcinogens, 389
  roles in the digestion and absorption of dietary fats, 123–4
Bile salts, primary
  production in the liver, 225
Bile salts, secondary
  production in the intestine, 225
Biliary excretion
  occurrence, 226
  of cholesterol, 97
Binding protein
  for corticotrophin-releasing hormone, 259
Biochemical tests
  role in the detection of malnutrition, 133
Biopsy
  fetal muscle in the diagnosis of muscular dystrophies, 254
Biotin
  as a coenzyme for acetylCoA carboxylase, 62
  metabolic function, 181
  structure, dietary sources and causes of deficiency, 181
1,3-*Bis*phosphoglyceric acid
  formation in glycolysis, 57
*Bis*phosphonates
  in the treatment of osteoporosis, 288
Blackwater fever
  antimalarial drugs and, 207
Bleeding disorders
  occurrence and nature, 218
Blood clot
  dissolution, 217
  limitation by thrombomodulin, 217
Blood clots
  soft and hard, 216
Blood clotting
  inherited diseases of, 218
  mechanism of extrinsic pathway, 213
  mechanism of intrinsic pathway, 213
  outline of process, 211
  role of fibrin, 216
  role of high molecular weight kininogen (HMWK), 213
  role of proaccelerin, 215
  role of thrombin, 215

Blood clotting diseases
recognition following amniocentesis, 218
Blood group antigens
infection with *Helicobacter pylori*, 319
Blood transfusions
as a cause of haemochromatosis, 194
Body mass index
definition, 129
Body weight, control
relationship to energy balance, 141
Bone
action of parathyroid hormone, 108–9
endochondral ossification, 285
formation and growth, 285
intramembranous ossification, 285
matrix, nature and composition, 281
mineral composition, 279
mineral constituents, 279
repair, 287
role of vitamin D, 112
roles of lysosomes, 75
remodelling, 287
Bone calcification
theories of, 287
Bone marrow
as a primary lymphoid tissue, 331
Bone mass, peak
as a determinant of osteoporotic bone risk, 288
Bone matrix
calcium-binding proteins, 282
Bone-seeking radioisotopes
nature, 364
Bone, trabecular
new growth of, 286
Bovine spongiform encephalopathy (BSE)
as an infectious prion disease of cattle, 422
Bowman's capsule
organization, 232
Bradykinin
production from high molecular weight kininogen (HMWK), 213
Brain
cell types in, 289–90
Branched-chain amino acids
low capacity for transamination in the liver, 100–1
promotion of protein synthesis in muscles, 436
utilization in the muscles, 102
Branched-chain oxo acid dehydrogenase
role in muscle tissue, 102
Branched-chain oxo acids
oxidation in the liver, 100–1
Breast cancer
possible causes, 389–90
risks associated with diet, 128–9
strategies for prevention, 390
Brittle bone disease
see osteogenesis imperfecta
Bromocriptine
possible use as an immuno-suppressant drug, 262
Bronchodilator aerosols
use in the spasmogenic phase of an asthmatic attack, 447
Brown adipose tissue
role in thermogenesis, 139
role of thermogenin, 50,139
Burkitt's lymphoma
association with infection with Epstein–Barr virus (EBV), 387
c-myc in, 383

C-cells
in the thyroid gland, 266
c-Myc
in Burkitt's lymphoma, 383
C-peptide
production associated with the biosynthesis of insulin, 263
C-reactive protein
in innate immune responses, 329
Cachexia
occurrence in most cancers, 167–8, 395
Calcitonin
structure and biosynthesis, 269
Caesium-137 (^{137}Cs)
risk of leukaemia, 364
Calbindin
absence from certain motor neurons, 451
role in the absorption of calcium from the intestine, 112
Calcification of bone
general description, 279
Calcitonin
mode of action, 269
role in calcium homeostasis, 109
Calcium (Ca^{2+})
regulation of plasma concentration, 108
in hormone secretion, 258
plasma concentration, 106
role in blood clotting, 211
role in muscle contraction, 247
Calcium (Ca^{2+}) ATPase
role in absorption of calcium from the intestine, 112
Calcium (Ca^{2+}) channels, voltage gated
in neurotransmitter release, 293
in the β-cells of the pancreas, 442
Calcium (Ca^{2+}) in the cytosol
effect of tumour promoters, 392
Calcium (Ca^{2+}) sequestration
role of the endoplasmic reticulum, 68
Calcium (Ca^{2+}) uptake into mitochondria
control of mitochondrial enzyme activity, 49
Calcium
principal roles in the body, 106
Calcium absorption
factors influencing, 107
Calcium-binding proteins
in bone matrix, 282
Calcium homeostasis
disorders, 114
importance, 106
Calcium phosphate
in bone structure, 279
Calcium-sensing receptor
role in the parathyroid glands, 108
Caldesmon
a protein of smooth muscle, 252
Calmodulin
in smooth muscle, 252
role in Ca^{2+}-mediated signalling, 28
Calorimetry
for the assessment of energy expenditure, 136
Calpontin
a protein of smooth muscle, 252
Calsequestrin
Ca^{2+}-binding protein of skeletal muscle, 249
Cancer
dietary lipids as risk factors, 143
free radicals and, 350
risk assessment, 395
risks associated with diet, 128
CapZ
role in skeletal muscle, 245
'Capping'
of eukaryotic mRNA molecules, 38
Carbohydrate
digestion and absorption, 120–121
Carbohydrate metabolism
effects of T_3 and T_4, 268
effect of growth hormone, 261
effects of insulin, 264
in the brain, 304
role of the liver, 221
Carbohydrates, dietary
intakes in different societies, 153
protein-sparing effect, 153
Carbon monoxide
as a possible neuromodulator, 23
influence on mitochondrial respiration, 51
Carbon tetrachloride
genesis of fatty liver, 94
Carbonic anhydrase
in the kidney, 240
γ-Carboxyglutamic acid
presence in blood-clotting proteins, 214
presence in bone calcium-binding proteins, 282
Carboxypeptidase E
role in the biosynthesis of insulin, 263
Carboxypeptidases
specificities and actions, 122
Carcinogenesis
chemical, 387
and free radicals, 350
Carcinomas
as neoplasms of epithelial cells, 378
Cardiac glycosides
ouabain as an example, 237
Cardiovascular disease
effect of antioxidant nutrients on risk, 349
impact of diet, 128
relationship to dietary lipid intake, 151–3
Carnitine: acylcarnitine exchange carrier
role in β-oxidation, 43
role in mitochondria, 51
Carotenoids
as antioxidants, 174, 348
effect on the risk of cardiovascular disease, 350
Catabolic reactions
nature, 45
Catalase
presence in peroxisomes, 9
reaction catalysed, 348
Cataract
risk factors, 310
role of free radicals, 350
Catecholamine *O*-methyltransferase
inactivation of catecholamines, 297

Catecholamines
as neurotransmitters, 296
biosynthesis, 269
control of release from the adrenal medulla, 269
Cathepsins
roles in lysosomal activity, 73
cDNA
see Complementary DNA
cDNA cloning
see Complementary DNA cloning
Cell cycle
action of p53 in, 385
occurrence, 39
targets for cancer chemotherapy, 425
Cell walls
effect of penicillin, 416
formation by bacteria, 415–6
Cell-mediated immunity
general organization and components, 330
Cellulose
in the diet, 156
Centrifugation
for the separation of subcellular organelles, 7–8
Cephaline
in the treatment of amoebiasis, 413
Cephalosporins
antibiotic use in cases of penicillin resistance, 417
Ceruloplasmin
copper-binding protein, 359
role in the binding of iron to transferrin, 191
role in the oxidation of iron, 359
Ceruloplasmin gene
mutation, 359
Cervical smear test
for the early detection of cancer, 380
Chagas' disease
treatment with nitrofurans, 422
Chelation therapy
haemochromatosis, 194
mercury poisoning, 361
removal of toxic metals, 365
Chemotherapeutic agents
nature of, 411
Chemical carcinogenesis
general features, 387
Chemical carcinogens
metabolic activation, 390
Chemical score
of proteins in the diet, 144
Chemiosmotic mechanism
for oxidative phosphorylation, 47–8
Chemotatic molecules
release from mast cells during an inflammatory response, 339
Chemotherapy
for cancer, 425
Chenodeoxycholic acid
formation, 225
Chernobyl
release of caesium-137 (^{137}Cs), 364
Childhood starvation
incidence and effects, 168
Chinese hamster ovary (CHO) cells
use in genetic engineering, 405
Chitin synthase
as a target for antifungal drugs, 422
Chloramphenicol
as a 'broad spectrum' antibiotic, 418
inhibition of protein synthesis in prokaryotes, 67
metabolism by nitro reductase, 369
Chloroquine
as an antimalarial drug, 413
Cholecalciferol (1,25 dihydroxy)
mechanism of action, 111
Cholecalciferol
role in calcium homeostasis, 109–10
sources, 113
synthesis in the skin and subsequent activation, 110
Cholecystokinin
nature and actions, 118–9
role in satiety, 140
sulphation in the Golgi apparatus, 258
Cholera
toxin, mechanism of action, 39
vaccination of chickens against, 317
Cholesterol
effect of T_3 and T_4 on the circulating concentration, 268
influence on membrane fluidity, 14
relationship to coronary heart disease, 152
sources, 95
Cholesterol biosynthesis
regulation, 96
tissues making important contributions to the process, 95
Cholesterol homeostasis
maintenance, 97
summary, 97
Cholesterol-lowering drugs
use, 98
Cholesteryl ester transfer protein (apolipoprotein D)
facilitation of cholesteryl ester transfer between lipoproteins, 94
Cholesteryl esters
in the formation of cortisol, 270
Cholestyramine
as a blood cholesterol-lowering drug, 98
Cholic acid
biosynthesis from cholesterol, 225
formation, 225
Choline
Na^+-dependent uptake, 294
source for the synthesis of acetylcholine in the brain, 293
Choline acetyltransferase
biosynthesis of choline, 293
Chondrocytes
in collagen biosynthesis, 282
Chondrodystrophies
occurrence and nature, 285
Chorionic gonadotrophin
secretion from syncytiotrophoblast cells in pregnancy, 261
Chorionic villus biopsy
in prenatal diagnosis, 408
Choroidal circulation of the retina
supply of oxygen, 308
Chromaffin cells
characteristic cells of the adrenal medulla, 269
Chromatin
presence in the nucleus, 30
Chromogranins
requirement for regulated secretion of proteins from cells, 72
Chromosomes
translocations of in chronic myelogenous leukaemia, 409
translocations of in follicular lymphoma, 409
Chronic granulomatous disease
associated with a deficiency in NADPH oxidase, 345
Chronic myelogenous leukaemia
possible treatment with tyrophosphins, 429
Chronic pancreatitis
as a cause of haemochromatosis, 194
Chronic renal failure
association with weak bones, 107
Chylomicrons
disposition in the liver as remnants, 90
nature, constitution and formation, 86
Chymotrypsin
specificity and action, 122
Cigarette smoking
associated risks, 351
effect on vitamin C turnover, 354
Citrate
formation during the tricarboxylic acid cycle, 44
transport from mitochondria, 53
Citrate cleavage enzyme
formation of acetylCoA for acetylcholine biosynthesis, 294
Citrate lyase
in the cytosolic cleavage of citrate, 54
Citrate synthase
reaction catalysed in mitochondria, 44
Class switching
in immune responses, 323
Classical pathway for complement activation
organization, 328
Clathrins
role in plasma membrane retrieval of proteins, 72
Clinical examination
role in the detection of malnutrition, 132
Clofibrate
a peroxisome-proliferator drug, 9
actions as a peroxisome-proliferator drug, 98
Clonal expansion of B cells
occurrence, 323
Clonal selection
role in the development of cancer, 378
Cloning
of complementary DNA (cDNA), 404
Cloning vectors
for DNA, 403
Clostridium botulinum
toxins from, 357
Clotting factor VIII
detection of mutation associated with haemophilia B by RFLP analysis, 409
Cluster designation (CD)
system of nomenclature of lymphocyte surface proteins, 334

Cobalamin
  see vitamin B_{12}
Cocaine
  euphoric effect via blockade of the dopamine transporter, 297
Codon
  nature, 63
Coenzyme Q (ubiquinone)
  location in the inner mitochondrial membrane, 47
Congenital stationary night blindness
  nature of disease, 315
Colchicine
  as an anticancer drug, 428
Cold
  as a 'trigger factor' in asthma, 446
Cold induced thermogenesis
  nature, 139
Collagen
  biosynthesis, 282
  structures, 281
  types, 281
Collagen diseases
  Alport's syndrome, 232, 285
  chondrodystrophy, 285
  copper deficiency, 285
  Ehlers–Danlos syndrome, 285
  general, 284
  lathyrism, 285
  osteogenesis imperfecta, 284
  vitamin C deficiency (scurvy), 285
Collagen, type I
  in the cornea, 308
Collagen, type IV
  in renal basement membranes, 232
Colon cancer
  risk associated with dietary fats, 128
Colony-stimulating factor-1
  mutations associated with osteopetrosis, 288
Colorectal cancer
  diet and, 389
Colour blindness
  causes, 316
Common cold
  effect of vitamin C, 16, 354
Complement
  activation by antibodies, 328
  general description of pathways of activation, 328
  role in innate immune responses, 320
Complement protein C1q
  collagen like structure, 281
Complement proteins
  roles in inflammatory responses, 339
Complementary DNA (cDNA)
  production, 404
Complementary DNA (cDNA) cloning
  methods, 404
Complete carcinogens
  nature, 392
Complex I (NAD^+ dehydrogenase)
  deficiency in parkinsonism, 47
  reaction catalysed in mitochondria, 47
Complex II (succinate dehydrogenase)
  reaction catalysed in mitochondria, 47
Complex III (ubiquinol dehydrogenase)
  reaction catalysed in mitochondria, 47
Complex IV (cytochrome oxidase)
  reaction catalysed in mitochondria, 47
Conditionally-essential amino acids
  nature, 144
Conduction
  of impulses by neurons, 290
  velocity in neurons, 292
Cone vision
  characteristics, 316
Congenital adrenal hyperplasia
  occurrence and nature of diseases, 277
Congenital lipoid adrenal hyperplasia
  occurrence and nature, 278
Conjugation reactions (phase II)
  of foreign compounds, 373
Conn's syndrome
  primary hyperaldosteronism, 277
Contraction, muscular
  mechanism in skeletal muscle, 247
Convalescent phase
  following injury, 438
Coporphyrin
  urinary excretion following lead poisoning, 363
Copper (Cu^{2+}) -ATPases
  in the transport of copper, 359
  mutations in Wilson's and Menke's diseases, 360
Copper
  as an essential nutrient, 359
  deficiency, collagen diseases and, 285
  enzymes containing, 359
  kinetics, 359
  required by dopamine β-hydroxylase, 296
  requirement of lysine oxidase, 283
Cornea
  occurrence of Kayser Fleischer rings in Wilson's disease, 360
  occurrence of type I collagen fibres, 309
Coronary heart disease
  impact of diet, 128
Corpus luteum
  effect of hCG on in pregnancy, 261
  effect of LH, 261
Cortical bone
  remodelling, 287
Corticosterone oxidases I and II
  in the biosynthesis of aldosterone, 272
Corticotrophin-releasing hormone
  interaction with oxytocin in parturition, 259
  interaction with vasopressin in ACTH release, 259
  structure and functions, 259
Corticotrophin-releasing hormone binding protein
  role in pregnancy, 259
Cortisol
  biological actions, 275
  biosynthesis, 272
  effect on gene transcription, 23
  reductive metabolism in the liver, 226
Cortisol binding globulin (CBG)
  functional role, 274
Cortisol, urinary free
  in Cushing's disease, 232
Coxsackie B virus
  possible role in Keshan disease (selenium deficiency), 348
  possible involvement in insulin dependent diabetes mellitus (IDDM), 440
Creatine kinase
  role in skeletal muscle, 250
Cretinism
  world wide prevalence, 268
Creutzfeld–Jakob disease (CJD)
  as an inherited prion disease, 422
Cristae
  occurrence in mitochondria, 42
Crohn's disease
  as a cause of pernicious anaemia, 201
  association with free radical induced damage, 351
Cromoglycate, sodium (Intal)
  use in asthma, 448
Cushing's disease
  ectopic, 278
  hypersecretion of cortisol, 276, 278
  urinary free cortisol determination, 232
'Cutting and pasting'
  of DNA, 403
Cyanide ions
  influence on mitochondrial respiration, 50
cyclicAMP (cAMP)
  action of adrenaline, 270
  production during signalling, 25
  role in the control of glycogenolysis in liver, 59
  role in the control of steroidogenesis, 274
cyclicAMP (cAMP) responsive element binding protein
  control of POMC gene by, 302
cyclicGMP (cGMP)
  in smooth muscle, 252
  in the rod outer segment of the retina, 315
cyclicGMP (cGMP) phosphodiesterase
  role in scotopic vision, 315
Cyclins
  role in the staging of the cell cycle, 40
Cyclooxygenase
  lipid hydroperoxide formation, 344
  in eicosinoid biosynthesis, 148
  inhibition by aspirin, 150
  inhibition by ibuprofen, 150
  inhibition by indomethacin, 150
Cyclosporin A
  as an immunosuppressive drug, 338
Cystinosis
  occurrence and nature, 75
Cystinuria
  occurrence, 238
Cytochrome oxidase
  formation of the superoxide anion, 343
Cytochrome P450
  roles in the activation of cholecalciferol, 110
  as a target for antifungal drugs, 420
  developmental changes, 372
  families, 372
  formation of the superoxide anion, 343
  in polycyclic hydrocarbon carcinogenesis, 390
  inducers of, 375
  mode of action, 371
  roles in foreign compound metabolism, 371
  specificity and multiple forms, 372

Cytochrome P450$_{SCC}$
in steroidogenesis, 271, 274
Cytochromes
absence from the cornea and the lens in the eye, 308
roles in electron transport, 47
Cytokines
glucocorticoid inhibition of secretion of, 288
in bone remodelling, 287
roles in immune responses, 320
types, 337
Cytoskeleton
occurrence in eukaryotic cells, 10, 56
Cytosol
constituents, 56
general description, 10
Cytotoxic T-cells
production, 334

Dalton, John
on his colour blindness, 316
De-iodinase
role in the conversion of T_4 into T_3, 266
Deficiency
biotin, 181
essential fatty acids, 147
folate, 196
niacin, 179
pantothenic acid, 181
pyridoxine (vitamin B_6), 180
vitamin A (retinol), 175
Vitamin B_{12}, 199
vitamin C, 354
vitamin D, 113–4
vitamine E, 348
Dehydroepiandrosterone
biological actions, 275
role in the formation of oestriol in the placenta, 272
Delayed-type hypersensitivity
occurrence, 339
Deletion mutations
occurrence and nature, 64
Dementia
associated with haemodialysis, 363
Demyelination
in multiple sclerosis, 303
Dendrites
nature and roles, 289
Dense bodies
presence in smooth muscle, 251
Dense-core vesicles
functional role, 68
Dental caries
relationship to dietary sucrose, 156
Dental enamel
hydroxyapatite, 279
Deoxycholic acid
formation, 225
Depolarization
passive spread in nerve cell dendrites, 291
Deprenyl
inhibitor of monoamine oxidase B, 297
Desferrioxamine
effect in Alzheimer's disease, 363
in chelation therapy for haemochromatosis, 194
Desmin
a protein of smooth muscle, 251
Di-oxygenases
potential generators of free radicals, 344
Diabetes, steroid
in Cushing's disease, 277
Diabetes insipidus
nature of the disease, 236
nephrogenic form of the disease, 236
occurrence of, 262
Diabetes mellitus
as a risk factor in cataract, 311
maternally transmitted form, 55
renal basement membrane thickening, 232
see also Insulin dependent diabetes mellitus and Non insulin dependent diabetes mellitus
Diabetic ketoacidosis
occurrence in insulin dependent diabetes mellitus (IDDM), 440
Diacylglycerol (DAG)
as a second messenger, 25
Dichromy
pigment loss in colour blindness, 316
Dicoumarol
as an anticoagulant from sweet clover, 214
Diet
and cancer, 389
Diet-induced thermogenesis
nature, 138
Dietary carbohydrate
relationship to disease, 156
Dietary control
of non insulin dependent diabetes mellitus (NIDDM), 442
Dietary deficiencies
primary, 131
secondary, 131
Dietary fibre
and colorectal cancer, 389
Dietary iron
absorption, 187
sources, 187
Dietary lipids
relationship to cancer, 153
Dietary reference proteins
types, 145–6
Dietary reference values (DRV)
proteins, 145
UK Department of Health, 128
Dietary requirements
calcium, 106
estimation, 127
general scope, 126–7
iodide, 268
phosphate, 106
vitamin B_1 (thiamin), 178
vitamin D, 113
Diethylenetriaminepentacetic acid
chelation therapy in poisoning with strontium-90 (^{90}Sr) and plutonium-239 (^{239}Pu), 365
Difluoromethylornithine
in the treatment of sleeping sickness, 422
Digestible energy
nature, 135
Digestive organs
roles, 116
Digestive secretions
nature, 116
Dihydrofolate reductase
as a target for methotrexate, 427
inhibition by trimethoprim, 418
reaction catalysed, 418
5α-Dihydrotestosterone
biological actions, 275
formation from testosterone, 273
Dihydroxyacetone phosphate
formation in glycolysis, 56–7
Dimercaprol
antidote in trivalent arsenic poisoning, 365
2,4-Dinitrophenol
as an uncoupler of oxidative phosphorylation, 50
Dipeptidases
specificities and actions, 122
Diphtheria toxin
effect on protein synthesis, 395
effect on translation in eukaryotes, 67
Disaccharidases
roles in the brush border microvilli, 120
Disaccharides
dietary sources, 154
Discs
presence in rods in the retina, 312
Disease
associated with signalling failure, 29
associated with steroid hormone production and action, 276
possible roles for free radicals, 349
Disulfiram
in the treatment of alcoholism, 227
Diverticular disease
relationship to dietary non starch polysaccharide, 157
DNA
annealing properties, 400
presence in mitochondria, 55
recognition of palindromic sequences by restriction enzymes, 403
DNA gyrases
inhibition by quinolines, 419
DNA methylation
function in bacteria, 403
DNA polymerase (Taq)
a thermostable enzyme used in genetic engineering, 406
DNA polymerases
roles in DNA replication in eukaryotes, 36
roles in DNA replication in prokaryotes, 34
DNA replication
substrate requirement, 34
DNA tumour viruses
genomes, 387
Docosahexanoic acid
status as an essential fatty acid, 147
Dolichol
outline of role in protein *N*-glycosylation, 72
L-DOPA (dihydroxyphenylalanine)
formation in the brain, 296
in the treatment of parkinsonism, 297
L-DOPA (dihydroxyphenylalanine) decarboxylase
biosynthesis of catecholamines, 296
Dopamine
as a prolactin release inhibiting hormone, 259
as a neurotransmitter, 296

Dopamine transporter
  blockade by cocaine, 297
Down's syndrome (trisomy 21)
  a clue to the nature of Alzheimer's disease, 300
Drug-induced osteoporosis
  occurrence, 288
Drug metabolism
  by the fetal liver, 372
  general principles, 367
  in senescence, 373
Drug resistance in bacteria
  nature, 423
  occurrence, 423
Drug resistance in cancer chemotherapy
  causes, 429
Duchenne muscular dystrophy
  molecular defects associated with the disease, 253
  occurrence and nature of disease, 252–4
Dulcitol
  accumulation in the lens in cataract, 311
Dust-mite faeces
  as a precipitating agent in asthma, 448
Dynamic model
  for transfer of genetic information, 38
Dystroglycan
  in muscle cells, 253
Dystrophin
  in Duchenne muscular dystrophy, 253
  in skeletal muscle, 245

Eating disorders
  anorexia nervosa and bulimia, 168
Ebb phase
  in the metabolic response to injury, 432
EBV (Epstein–Barr virus)
  association with Burkitt's lymphoma, 387
Ectopic calcification of soft tissues
  association with hypercalcaemia, 106
Ehlers–Danlos syndrome
  mutations to lysine oxidase, 285
  mutations to procollagen proteinases, 285
Eicosinoids
  derivation from C:20 fatty acids, 149
  structures and nomenclature, 148
Electron transfer chain
  organization in mitochondria, 47
Electron transfer reactions
  occurrence in mitochondria, 47
Electrophoresis
  DNA on agarose gels, 401
Electrotonic (passive) spread of current in neurons, 291
Elongation factors
  roles in protein synthesis, 66–7
Emetine
  in the treatment of amoebiasis, 413
Emphysema
  nature and occurrence, 351
Endochondrial ossification of bones
  occurrence, 285
Endocrine axes
  hypothalamo-pituitary-adrenal cortical axis as an example, 255
  occurrence and organization, 255
Endocrine effects
  nature, 20
Endocytosis
  by macrophages, 320
  mechanism, 72
Endopeptidase of the gastrointestinal tract
  specificity and action, 122
Endoplasmic reticulum
  cellular organization, 68
  general description, 9
  protein retention, 71
Endoplasmic reticulum, rough
  in the biosynthesis of parathyroid hormone, 257
β-Endorphin
  part of the POMC molecule, 260
Endothelium
  presence of lipoprotein lipase, 89
Energy
  forms and units of measurement, 135
  supply and utilization, 135–6
  UK dietary reference values, 139
Energy balance
  control of body weight, 141
Energy content
  dietary fat, 146
  food, 135
Energy expenditure
  factors determining, 137
  methods for assessment, 136
  relationship to obesity, 160
Energy intake
  control, 140
  genetic determination, 160–1
  relationship to obesity, 161
  UK guidelines for a healthy diet, 130
Energy provision
  for muscle contraction, 247, 249
Energy requirements
  of the individual, 139
Enhancers
  in RNA transcription, 38
Environmental factors
  importance in asthma, 444
EnoylCoA dehydrogenase
  role in β-oxidation, 43
Entero-hepatic recirculation
  bile salts, 225
Enteroglucagon
  nature and actions, 119
Environmental carcinogens
  examples, 388
Environmental chemicals
  categories, 355
Enzyme polymorphisms
  in the metabolism of procarcinogens, 391
Eosinophils
  infiltration in asthma, 447
  roles in immune responses, 330
Epidermal growth factor receptor
  as a target for recombinant toxins, 394
  relationship to vErb oncoprotein, 382
Epidermolysis bullosa
  nature of the disease, 285
Epilepsy drugs
  induction of osteomalacia, 376
Epiphyses
  role in bone formation, 285
Epithelial cells
  neoplasms of, 378
Epitope
  recognition by antibodies, 321
Epoxide hydratase
  reaction catalysed, 370
Epstein–Barr virus (EBV)
  association with Burkitt's lymphoma, 387
  association with naso-pharyngeal carcinoma, 387
Ergocalciferol
  dietary sources, 109–10
Ergosterol
  drugs that interfere with, as antifungals, 419
Erythrocytes
  effects of enzyme deficiencies, 207
  functional roles, 205
Erythromycin
  inhibition of protein synthesis in prokaryotes, 67, 418
Erythropoiesis
  role of the kidney, 241
Erythropoietin
  production by recombinant DNA technology, 241, 405
  role, 241
  underproduction in aplastic anaemia, 241
  use of recombinant form in renal failure, 241
Erythropoietin receptor family
  activation by cytokines, 337
  members, 24
Essential amino acids
  conservation in the fasting state, 104
  identification, 144
Essential fatty acids
  chemical structure and occurrence, 146–7
  summary of the functional roles, 147–51
Estimated average requirements (EAR)
  definition of dietary needs, 127
Ethanol
  concentration in the blood after the consumption of alcoholic drinks, 431
  effect on redox status of hepatocytes, 228
  hepatic metabolism, 227
  tolerance, 227
Euchromatin
  nature, 32
Eukaryotic cells
  nature, 3
Eukaryotic genome
  organization, 36
Excitation
  of cells in the brain, 290
Excitatory post synaptic potentials (EPSP)
  generation, 298
Excitotoxicity
  possible contributing role on sporadic amylotrophic lateral sclerosis (ALS), 450
  possible role of NMDA receptor, 300
Exercise
  as a 'trigger factor' in asthma, 445
  effect on plasma lipid levels, 97
  effect on protein synthesis in fast twitch (type II) muscle fibres, 102

Exinuclease
role in DNA repair mechanisms, 36
Exons
occurrence in eukaryotic genes, 37
Exonuclease activity
of DNA polymerases, 34–5
Exopeptidases of the gastrointestinal tract
specificity and action, 122
Extrinsic pathway for blood clotting
general description, 213
mechanism, 213
Extrinsic sugars (dietary)
nature, 154

F actin/G actin
structural organization, 244
Facilitated transport
across membranes, characteristics, 18
Fanconi syndrome
occurrence and nature, 75
Fasting state
plasma lipid transport, 91
release of glucose, 79
Fat intake
guidelines of the USA National Research Council, 130
UK guidelines for a healthy diet, 129
Fat-soluble vitamins
basis for classification, 172
Fatal-familial insomnia (FFI)
as an inherited prion disease, 422
Fats
content in the UK diet, 146
Fatty acid biosynthesis
from fructose and lactate, 83
outline description, 60
role of citrate transport, 53
Fatty acid mobilization
in the ebb phase of the metabolic response to injury, 433
Fatty acids
biosynthesis, 60
β-oxidation reactions, 43
oxidation in skeletal muscle, 249
oxidation in the kidney, 234
Fatty acids, essential
chemical structure and occurrence, 146–7
Fatty acyl synthase
reactions catalysed, 60
Fatty liver
causes, 94
occurrence, 222
Febrile illnesses
T_3/rT_3 ratio in, 268
Fenfluramine
in the treatment of obesity, 162
Ferric hydroxide
ready formation in the earth's atmosphere, 186
Ferritin
role in iron storage, 189
Fetal carbohydrate metabolism
basic features, 82
Fetal liver
drug metabolism, 372
in phase I metabolism of foreign compounds, 370–3
lack of UDP glucosuronyl transferase, 373
Fibrin
role of in blood clotting, 215
Fibrin-stabilizing factor
role in blood clotting, 216
Fibrinogen
role of in blood clotting, 215
Fibrinolysis
occurrence, 217
Fibronectin
role in signalling, 23
FIGLU (formiminoglutamic acid)
in the diagnosis of folate deficiency, 199
FK506
as an immunosuppressive drug, 338
Flavin adenine dinucleotide (FAD)
role in the AcylCoA dehydrogenase reaction, 43
role in the succinate dehydrogenase reaction, 44
Flavin mononucleotide (FMN)
role in the reaction catalysed by NAD^+, dehydrogenase, 47
Flavonoids
effect on the risk of cardiovascular disease, 350
Flow phase
in the metabolic response to injury, 432, 434
Fluoride
in hydroxyapatite of dental enamel, 279
prevention of tooth decay, 184
intake, guidelines of the USA National Research Council, 130
Fluoride ions
as a stimulus to bone formation, 288
Fluoro (^{18}F) compounds
in PET imaging of the brain, 304
Foam cells
formation in atherosclerosis, 347, 350
Folate
absorption and distribution, 197
assessment of nutritional status, 134
reference nutrient intake (RNI), 197
principal actions, 199
Folate and vitamin B_{12}
interrelationships, 198
Folate coenzymes
roles in the biosynthesis of nucleic acids, 199
Folate deficiency
biochemical basis for the symptoms, 201
occurrence and vulnerable groups, 200
Follicle-stimulating hormone(FSH)
control of steroidogenesis, 274
functional roles, 261
Follicular lymphoma
monitoring therapy by polymerase chain reaction, (PCR), 409
Food
energy content, 135
protein content, 143–4
Food intolerance
causes, 167
Food toxicity
causative agents, 130–1
Footprints
of free radical attack on biological molecules, 347
Foreign compounds
handling in the body, 357
nature, 355
Formiminoglutamic acid (FIGLU)
in the diagnosis of folate deficiency, 199
Formate starvation
in vitamin B_{12} deficiency, 202
Free radical
definition, 343
Free radical $Pb(C_2H_5)_3^+$
as a possible toxic metabolite of tetraethyl lead, 363
Free radicals
attack by in Wilson's and Menke's disease, 360
from polycyclic hydrocarbons, role in carcinogenesis, 390
methyl, from dimethylnitrosamine in carcinogenesis, 391
possible role in stroke induced excitotoxicity, 300
possible roles in disease, 349
Fructose
dietary sources, 154
intestinal absorption, 121
metabolism, 83
Fructose-1,6-*bis*phosphatase
role in gluconeogenesis, 58
Fructose-1,6-*bis*phosphate
formation in glycolysis, 56
Fructose-2,6-*bis*phosphate
regulation of the balance between glycolysis and gluconeogenesis, 57
Fructose 6-phosphate
formation in glycolysis, 57
Fruit and vegetables
guidelines of the USA National Research Council, 130
Fuel homeostasis
the liver in, 220
Fuel utilization
patterns in injury and trauma, 431
Fumarase
reaction catalysed in mitochondria, 44
Fumarate
formation in the tricarboxylic acid cycle, 44
Fungal cell walls
as a target for antifungal therapies, 419
Fungal infections
patients especially vulnerable, 419

γ-Aminobutyrate (GABA)
as a neurotransmitter, 294
γ-Carboxyglutamic acid (GLA) residues
role of vitamin K in the formation of, 214
roles in blood clotting, 214
G-protein coupled receptors
as metabotrophic receptors, 301
hormones influencing, 255
mechanism of operation, 25
G-proteins
as proto-oncogene products, 382
GABA: 2-oxoglutarate aminotransferase (GAB_A-T)
inactivation of GABA, 296
$GABA_A$ receptor
structure and function, 298

GABAergic nerve terminals
organization of, 295
Gadolinium
an enhancer in magnetic resonance imaging (MRI), 304
Galactose
metabolism, 83
Galactosaemia
as a risk factor in cataract, 311
Galactose
intestinal absorption, 121
β-Galactosidase
see lactase
Gall stones
formation, 225
Gastric inhibitory peptide, (GIP)
nature and secretion, 119
Gastric juice
role in the digestive process, 117
Gastric ulcers
Helicobacter pylori as a causative agent, 319
Gastrin
nature and actions, 118
Gastrointestinal hormones
control of digestive secretions, 118–20
Gaucher's disease
β-glucosidase deficiency, 305
Gender
effect on plasma lipid levels, 97
Gene amplification
as a cause of drug-resistance in cancer chemotherapy, 429
Genetic code
nature and organization, 63–4
Genetic engineering
summary of the processes, 399
Genetic information
evidence for storage in DNA, 30
Glial cells
presence of glutamine synthase, 294
structures and functions, 290
Glial scars
formation following CNS injury, 290
Glibenclamide
in the treatment of non insulin dependent diabetes mellitus (NIDDM), 442
Glomerulus
structural organization, 232
Glucagon
control of glycogenolysis in liver, 58
mode of action, 266
role in plasma glucose homeostasis, 82
role in short-term starvation, 163
secretion during the ebb phase of the metabolic response to injury, 432
structure and biosynthesis, 266
Glucocorticoid-suppressible hyperaldosteronism
effect of dexamethasone, 277
Glucocorticoid therapy
risk of osteoporosis associated with, 288
Glucocorticoids
use in asthma, 448
Glucokinase (hexokinase D)
in the control of insulin secretion, 263
presence in liver, 56
role in the liver, 81
Gluconeogenesis
effects of insulin on, 264
imbalance in insulin dependent diabetes mellitus (IDDM), 440
in the ebb phase of the metabolic response to injury, 433
occurrence, 58
Gluconeogenesis in the liver
short term starvation, 164
Glucose
dependence of the brain for the formation of acetylcholine, 304
dietary sources, 154
intestinal absorption, 121
renal recovery, 238
Glucose 1-phosphate
formation in glycogenolysis, 59
Glucose-6-phosphatase
role in gluconeogenesis, 58
Glucose 6-phosphate
formation in glycolysis, 56
Glucose-6-phosphate dehydrogenase
role in the pentose phosphate pathway, 59
deficiency, incidence and effects, 208
Glucose concentration, blood
maintenance, 79
Glucose transport by Glut 4
response to insulin, 264
Glucose transporter (Glut 2)
defect in some cases of non insulin dependent diabetes mellitus (NIDDM), 440
Glucose transporter 4 (Glut 4)
insulin dependence, 82
Glucose transporters
in the kidney, 238
occurrence and nature, 19
Glucosylated haemoglobin (HbA_{1c})
in diabetes mellitus, 441
Glucuronide formation
in phase II reactions of foreign compounds, 373
Glut 1
glucose transporter, 19
Glut 2
glucose transporter, 19
Glut 4
glucose transporter, insulin dependence, 19, 264
Glutamate (AMPA) receptor
function, 299
Glutamate (NMDA) receptor
structure and function, 299
possible role in amylotrophic lateral sclerosis (ALS), 450
possible role in excitotoxicity, 299
Glutamate
as a neurotransmitter, 294
Glutamate decarboxylase (GAD)
as an antigen in the autoimmune condition insulin dependent diabetes mellitus (IDDM), 440
formation of GABA, 295
Glutaminase
formation of neurotransmitter glutamate by, 294
Glutamine
as a precursor for gluconeogenesis in short-term starvation, 165
concentration in cerebrospinal fluid, 294
oxidation in the kidney, 234
requirement of mucosal cells for nucleic acid synthesis, 100
release from muscles following injury, 437
Glutamine synthase
presence in glial cells, 294
Glutathione
in phase II conjugation reactions, 373
possible role in kidney function, 238
role in red blood cells, 207
role in the lens, 310
Glutathione peroxidase
deficiency-associated anaemia, 208
in red blood cells, 207
reaction catalysed, 348
Glutathione reductase
part of the antioxidant defences, 348
reaction catalysed in red blood cells, 206
Glutathione *S*-transferases
association with adenocarcinomas of the lung, 391
phase II conjugations with endogenous and exogenous substrates, 374
Glutathione-insulin transhydrogenase
in the catabolism of insulin, 266
Glyceraldehyde 3-phosphate
formation in glycolysis, 57
Glyceraldehyde-3-phosphate dehydrogenase
reaction catalysed in glycolysis, 57
reaction as an example of substrate level phosphorylation, 45
Glycerophospholipids
occurrence in biological membranes, 15
Glycocholic acid
formation, 225
Glycogen
dietary sources, 155
Glycogen phosphorylase
reaction catalysed, 59
Glycogen synthesis
reactions and regulation, 58
Glycogenesis
effect of insulin, 264
Glycogenolysis
reactions and regulation, 58
in the liver, in short-term starvation, 163
Glycogenosis, type II
a glycogen storage disease, 75
Glycolysis
in red blood cells, 205
reactions, 56
regulation, 57
sole energy generating pathway in the red blood cell, 308
Glycoproteins
occurrence in plasma membranes, 15
Glycosylation of proteins
in the endoplasmic reticulum/Golgi apparatus, 72
Glycyrrhizic acid
hypertensive actions, 236
Goitre
world prevalence, 268
Goitrogens
from plants, 357

Golgi apparatus
cellular organization, 68
general description, 9
in the processing of prohormones, 258
Gonadotrophin-releasing hormone
as a neuromodulator, 23
structure and functions, 259
Granulocytes
cell population, 330
free radical generation, 320
Graves' disease
as an autoimmune disease, 268
Griseofulvin
inhibition of fungal cell division, 420
Gross energy
of food, 135
Growth hormone (human)
production by bacteria, 405
Growth hormone
actions, 261
effect on carbohydrate metabolism, 261
effect on lipid metabolism, 261
effect on mineral metabolism, 262
effect on protein and amino acid metabolism, 261
hormones influencing secretion, 259
in the ebb phase of the metabolic response to injury, 432
role in skeletal bone growth, 286
structure and isolation, 261
Growth factor receptors
mechanism of action, 24
Guam disease
complex of conditions, including parkinsonism, 297, 449
Guanosine triphosphate (GTP)
formation in the tricarboxylic acid cycle, 44
roles in protein synthesis, 66
Guanylyl cyclase
activation by nitric oxide in smooth muscle, 252
activity of ANP receptor of smooth muscle, 252
in the rod outer segment of the retina, 315
Guanylyl cyclase, soluble
activation by nitric oxide, 344
Gulf War
treatment of leishmaniasis following, 411

Haem biosynthesis
effect of lead, 362
outline description, 60
Haemochromatosis
hereditary, 193
occurrence and effects, 193
treatment, 194
Haemodialysis
aluminium-induced dementia, 363
Haemoglobin
oxygen binding, 206
HaemoglobinA$_{1c}$
as an indicator of severity of diabetes mellitus, 79, 441
Haemolytic disease
riboflavin deficiency, 208
Haemolytic disease of the newborn (rhesus disease)
occurrence, 325
Haemolytic episodes
associated with glucose-6-phosphate dehydrogenase deficiency, 208
Haemophilia B
detection of mutation to clotting factor VIII by RFLP analysis, 409
Haemosiderin
possible role in iron storage, 189
Hairline fractures of bones
repair, 287
Haptens
discovery and nature, 321
Haptocorrin
see transcobalamin
Haversian canals
in bone, 279
HbA$_{1c}$ (glucosylated haemoglobin)
in diabetes mellitus, 79, 441
Health
definition, 126
Healthy diet
general guidelines, 129
UK guidelines, 129–30
Helicases
role in DNA replication, 35
Helicobacter pylori
as a causative agent in gastric ulcers, 319
Helminth infections
drugs effective against, 422
Heparan sulphate
in renal basement membranes, 232
Heparin
as a glycosaminoglycan anticoagulant, 217
prophylactic use prior to surgery, 217
Hepatocytes
oxygen consumption, 41
size, 4
Heterochromatin
nature, 32
Heteronuclear RNA (hnRNA)
as a precursor of mRNA, 38
Heteroplasmy
in mitochondrial diseases, 55
Hexokinase
reaction in glycolysis, 56
regulation of activity, 57
Hexokinase D (glucokinase)
in the control of insulin secretion, 263
Hexose monophosphate shunt
see pentose phosphate pathway
High density lipoproteins (HDL)
nature and constitution, 86
High energy compounds
nature, 45
High intakes of dietary proteins
possible harmful effects, 145
High molecular weight kininogen (HMWK)
role in blood clotting, 213
Histamine
release during an asthmatic attack, 447
release from mast cells during an inflammatory response, 339
Histochemistry
for the characterization of cells, 5
Histones
as basic DNA-associated proteins, 30
HIV (human immunodeficiency virus)
in AIDS, 386
HLTV, 1 (human T cell lymphotrophic virus, 1)
T-cell leukaemias, 386
Homocysteine methyltransferase (methionine synthase)
in the biosynthesis of methionine, 200
Homocytotrophic antibodies
nature, 326
Homoplasmy
in mitochondrial diseases, 55
Hormonal control
prostaglandin biosynthesis, 149-50
plasma amino acids, 104
Hormonal responses
in the ebb phase of the metabolic response to response to injury, 432
Hormone-responsive elements, 23
occurrence and nature
Hormone-sensitive lipase
reaction catalysed, 91
Hormone replacement therapy
prevention of osteoporosis, 288
Hormones
acting via G-protein-coupled receptors, 255
chemical groups, 255
physico-chemical properties, 20
House dust mites
faeces as an allergen, 448
Hyperlipoproteinaemias, familial
occurrence and nature, 97
HPV (human papilloma virus)
infection associated with squamous cell carcinoma, 387
Human immunodeficiency virus (HIV)
in AIDS, 386
numbers infected, 317
prospects for a vaccine, 317
Human leukocyte associated antigen (HLA)
locus and predisposition to insulin dependent diabetes mellitus (IDDM), 439
nature, 333
Humoral immune responses
nature, 320
Huntingdon's disease
occurrence and cause, 295
Hyaluronidase
release from mast cells during an inflammatory response, 339
Hybridization
of DNA and RNA molecules, 400
Hydrazaline
pro-convulsant effect of overdosage, 295
Hydrogen sulphide
influence on mitochondrial respiration, 51
Hydrolases
roles in digestion of foods, 116
Hydrolysis of esters and amides
in foreign compound metabolism, 369
Hydroperoxides, lipid
formation, 345
3-Hydroxy-3-methylglutarylCoA (HMGCoA) reductase
in cholesterol biosynthesis, 96
3-Hydroxy-3-methylglutarylCoA (HMGCoA) synthase
in cholesterol biosynthesis, 96

3-HydroxyacylCoA dehydrogenase
role in β-oxidation, 43
Hydroxyapatite
in bone structure, 279
Hydroxyl radical
formation from peroxynitrite, 344
generation in chemical reaction, 343
in lipid peroxidation, 345
11β-Hydroxylase
in the biosynthesis of aldosterone, 272
in the biosynthesis of cortisol, 272
17α-Hydroxylase
in steroid hormone biosynthesis, 271
1-Hydroxylation of 25-hydroxycholecalciferol
in the kidney, 110
25-Hydroxylation of cholecalciferol
in the liver, 110
Hydroxylysine
in collagens, 281
Hydroxyprolines
in collagens, 281
Hydroxypyridinones
possible use in chelation therapy for haemochromatosis, 195
11β-Hydroxysteroid dehydrogenase
protection of mineralocorticoid receptors in the kidney, 236
17α-Hydroxysteroid dehydrogenase
in the biosynthesis of androgens, 273
3β-Hydroxysteroid dehydrogenase
in steroid hormone biosynthesis, 271
Hyperbaric oxygen
adverse effects in infants, 351
deleterious effect, 345
Hypercalcaemia
consequences, 106
Hypercholesterolaemia, familial
occurrence, 93
Hyperglycaemia
in the ebb and flow phases of the metabolic response to injury, 432
in the flow phase of the metabolic response to injury, 435
Hyperparathyroidism
occurrence and nature of the disease, 114
Hypersensitivity
occurrence, 321
Hypersensitivity, immediate
role of IgE, 326
Hypertension
associated with steroid 11β-hydroxylase deficiency, 278
Hypocalcaemia
consequences, 106
Hypochlorous acid
generation by myeloperoxidase, 345
Hypoglycaemia
relationship to the feeling of hunger, 140
Hypolactasia, congenital
signs and symptoms, 120
Hypophyseal portal blood flow
delivery of hypothalamic hormones, 256
Hypothalamo-pituitary-adrenal cortical hormone axis
organization, 255–6
Hypothalamus
hormones, 258

'I'-band
in skeletal muscle, 242
I-cell disease
involving lysosomal targeting of enzymes, 75
Ibuprofen
inhibition cyclooxygenase, 150, 211
Immune system
cells, 330
Immunodeficiency, X linked, severe, combined
occurrence, 338
Immunocytochemistry
applications, 5
Immunogenicity
definition, 321
Immunoglobulin
folded domain, 323
Immunoglobulin
structure, 323
Immunoglobulin classes
IgA, 326
IgD, 326
IgE, 326
IgG, 325
IgM, 326
Immunological memory
laying down, 323
Immunosuppressive drugs
uses, 338
In situ hybridization
detection of mRNA in cells, 5–6
Innate immune response
nature, 318
Incomplete carcinogens
nature, 392
Indomethacin
inhibition of cyclooxygenase, 150, 211
Induction
of foreign compound metabolism, 375
of phase I metabolism of foreign compounds, 375
Induction of metabolism
of procarcinogens, 391
Infectious diseases
symptomatic treatment, 425
Inflammatory diseases
free radicals in, 351
Inflammatory response
description, 339
Inherited disease
blood clotting, 217
molecular analysis, 406
muscular dystrophies, 252
of collagen biosynthesis, 284
use of restriction fragment length polymorphism (RFLP) linkage analysis to detect, 407
Inhibitory post synaptic potentials (IPSP)
generation, 298
Initiation factors
roles in protein synthesis, 65
Initiators
of cancer, 392
Injury
energy requirements, 434
major causes in the UK, 431
Inosine
use in the storage of blood, 207
Inositol *tris*phosphate (IP_3)
as a second messenger, 28

Insertion mutations
occurrence and nature, 64
Insulin
control of secretion, 263
effects on liver, 264
effects on amino acid and protein metabolism, 265
effects on carbohydrate metabolism, 264
effects on skeletal muscle, 266
inactivation, 266
mode of action, 264
purification and first administration to patients, 439
regulation of the appearance of lipoprotein lipase in the endothelium, 88
release in the fed state, 80
structure and biosynthesis, 263
Insulin dependent diabetes mellitus (IDDM)
natural history of the disease, 439–40
Insulin receptor
structure and function, 264
Insulin receptor substrate (IRS)
identification, 264
sequence variation in some patients with non insulin dependent diabetes mellitus (NIDDM), 440
Insulin resistance
occurrence in non insulin dependent diabetes mellitus (NIDDM), 440
Insulin, human recombinant
in the treatment of insulin dependent diabetes mellitus, 442
Insulin like growth factor I
role in skeletal bone growth, 286
Insulin like growth factors
release stimulated by growth hormone, 261
Intal (sodium cromoglycate)
use in asthma, 448
Integral membrane proteins
organization, 16
Integrins
role in the absorption of dietary iron, 187
Intercalating drugs
cancer chemotherapy, 427
Interferons
production, 337
protective role, 320
receptors, 25
Interleukin-1 (IL 1)
secretion by antigen-presenting cells, 338
Interleukin-2 (IL 2)
secretion by activated T_H cells, 337
Interleukin-4 (IL 4)
role in the maturation of B-cells, 338
Interleukin-6 (IL 6)
stimulation of osteoblast formation, 288
Interleukin-7 (IL 7)
in B-cell development, 331
Interleukin genes
effects of glucorticoids on expression of, 448
Interleukins
production, 337
receptors for, 25

Intrinsic factor
  role in the absorption of vitamin B_{12} (cobalamin), 198
Intramembranous ossification of bones
  processes in, 285
Intraprandial phase
  metabolic adaptation in, 163–5
Intrinsic pathway for blood clotting
  general description, 213
  mechanism, 213
Intrinsic sugars (dietary)
  nature, 154
Intron/exon structure
  organization of, 402
Introns
  occurrence in eukaryotic genes, 37
  presence in the β-gene for haemoglobin, 402
  splicing in hnRNA processing, 38
Iodide
  dietary requirements, 268
Iodine deficiency
  relationship to cretinism, 268
Ion channel proteins
  general description, 17–18
Ion channel receptors
  types, 298
Ionising radiation
  as a risk factor in cataract, 310
  generation of free radical, 345
Ionotrophic receptors
  examples, 298–301
  definition, 24
Ipecacunha
  use an emetic in cases of childhood poisoning, 413
Iron
  assessment of nutritional status, 134
  balance, 186
  chemical properties, 186
  intracellular homeostasis, 191
  kinetics, 190
  valency changes during metabolism, 191
Iron-bearing proteins
  found in the body, 186
Iron-deficiency anaemias
  classification, 191
Iron losses from the body
  classification and sources, 186
Iron overload
  see haemochromatosis
Iron oxidation
  role of ceruloplasmin, 359
Iron-regulatory elements
  occurrence in the mRNAs for the transferrin receptor and ferritin, 191
Iron-response factor
  role in intracellular iron homeostasis, 191
Iron storage
  ferritin and haemosiderin, 189
Iron, dietary
  absorption, 187
  sources, 187
Islets of Langerhans
  pancreatic formation of somatostatin, 259
  source of pancreatic hormones, 263
Isocitrate dehydrogenase
  reaction catalysed in mitochondria, 44
Isomaltase
  role in the digestion of carbohydrates, 121
Isoniazid
  phase II acetylation reaction, 375
  use in treating mycobacterial infections, 417

JAK proteins
  role in response to growth hormone, 261
JAK protein coupled receptors
  mechanism of action, 25

K^+-channels, ATP gated
  in the B-cells of the pancreas, 442
K^+-channels (leak)
  in neurons, 291
K^+-channels (voltage gated)
  in neurons, 291
Kachin–Beck disease
  selenium deficiency, 348
Kallikrein
  role in blood clotting, 213
Karposi's sarcoma
  AIDS patients, 386
Kayser–Fleischer rings in the cornea
  in Wilson's disease, 360
Kearns–Sayre syndrome
  arising due to mutations to mitochondrial DNA, 55
Kennedy's disease
  trinucleotide repeats in the androgen receptor gene, 296
Keratan sulphate I
  role in the cornea, 309
Keshan disease
  selenium deficiency, 348
Ketoacidosis
  occurrence in insulin dependent diabetes mellitus (IDDM), 440
Ketone bodies
  formation during prolonged starvation, 166
  use in the management of cancer cachexia, 395
  role of the liver, 222
Kidney
  absorption of electrolytes, 236
  action of parathyroid hormone, 108
  basement membrane thickening in diabetes mellitus, 232
  energy provision, 233
  handling of K^+, 237
  in pH regulation, 230, 238
  reabsorption of calcium, 112
  recovery of Na^+ and Cl^-, 236
  recovery of amino acids, 238
  recovery of glucose, 238
  regulatory functions, 230
  role in erythropoiesis, 241
  roles of lysosomes, 74–5
  water reabsorption, 234
Kidney neoplasia
  polycythaemia in, 241
Kupffer cells
  roles in biliary secretion, 118
  role in phagocytosis in the liver, 227

Laber's hereditary neuropathy (LHON)
  arising due to mutations to mitochondrial DNA, 55
Lactase
  role in the digestion of lactose, 120
Lactase deficiency
  incidence and prevalence, 125
Lactate
  oxidation in the kidney, 234
  production in exercise, 250
  production in the ebb phase of the metabolic response to injury, 433
Lactate dehydrogenase
  in the reoxidation of cytosolic NADH, 57
Lactoferrin
  antibiotic properties, 188
Lactose
  dietary sources, 154
  digestion and absorption, 120
  influence on absorption of dietary calcium, 107
Lactose intolerance
  occurrence and causes, 154
Lamellar structures
  in bone, 279
Lamins
  role in the nucleus, 33
Laminin
  role in signalling, 23
  role in the glomerulus, 232
Lathyrism
  occurrence, 285
LD_{50} values
  toxic chemicals, 355
Lead
  effect on haem biosynthesis, 362
  exposure, 361
  toxicity, 362
Lecithin: cholesterol acyltransferase (LCAT)
  occurrence in high density lipoproteins (HDL), 96
Legionnaire's disease
  use of erythromycin for, 418
Lens
  structural organization, 309
Leptin
  a hormone of the adipose tissue, relationship to obesity, 161
Leukaemia
  risk associated with exposure to strontium-90 (^{90}Sr) or Caesium-137 (^{137}Cs), 364
Leukaemia, acute myelocytic (APL)
  therapy with retinoic acid, 174
Leukotriene B_4
  requirement for glutathione in formation of, 374
Leukotrienes
  in bone remodelling, 287
  release during an asthmatic attack, 447
  structures and nomenclature, 148
Levastatin
  inhibition of 3-hydroxy-3-methylglutarylCoA (HMGCoA) reductase, 98
Leydig cell
  formation of testosterone, 273
Lignin
  as a dietary non starch polysaccharide, 155
Linkage analysis of inherited disease
  use of restriction fragment length polymorphisms (RFLPs), 407

Linoleic acid
as an essential fatty acid, 146
α-Linolenic acid
status as an essential fatty acid, 146
Lipase, hormone sensitive
reaction catalysed, 91
Lipid-anchored membrane proteins
organization, 16
Lipid hydroperoxides
formation, 344, 345
Lipid levels, plasma
factors leading to their elevation, 97
Lipid metabolism
effects of T_3 and T_4, 268
effect of growth hormone, 261
in the brain, 305
role of the liver, 222
Lipid peroxidation
general mechanism, 345
Lipid
composition of plasma membranes, 12
content in the UK diet, 146
digestion and absorption, 123–4
Lipids, circulating
nature, 85
Lipids, dietary
and cancer risk, 153
relationship to cardiovascular disease, 151–3
Lipids, plasma
sources and utilization, 87
Lipoic acid
role in the pyruvate dehydrogenase reaction, 42
Lipopolysaccharide (LPS)
in Gram negative bacterial outer membrane, 415
Lipoprotein (a)
association with increased risk of atheromatous vascular disease, 98
Lipoprotein lipase
role in the handling of lipoprotein particles, 88
Lipoproteins
separation by electrophoresis or centrifugation, 85
structures, 86
Liposomes
as vehicles for the safer delivery of antifungal drugs, 420
Lipoxygenase
lipid hydroperoxide formation, 344
reaction catalysed, 148
Liver
as a storage organ, 224
effects of insulin on, 264
excretory function, 226
metabolism of ethanol, 227
reductive metabolism of cortisol, 226
reductive metabolism of steroid hormones, 226
role in amino acid metabolism, 223
role in carbohydrate metabolism, 221
role in ketone body formation, 222
role in lipid metabolism, 222
role in the metabolism of foreign compounds, 368
Liver function
fuel homeostasis, 220
Long terminal repeats (LTRs)
in retroviral genomes, 386
Low density lipoprotein (LDL)
as a risk factor in coronary heart disease, 152
metabolic fate, 93
nature and constitution, 86
possible role in cardiovascular disease, 347
Lower reference nutrient intake (LRNI)
definition, 127
Lung cancer
risk from plutonium-239 (^{239}Pu), 364
risk from smoking, 389
Luteinising hormone (LH)
functional roles, 261
in the control of steroidogenesis, 274
17,20-Lyase
a bifunctional enzyme of steroid biosynthesis, 271
see also 17α-hydroxylase
Lymphatic system
involvement in chylomicron distribution, 88
Lymphocytes
in the immune response, 320
roles in immune responses, 330
Lymphokines
nature, 25
production, 337
Lymphotrophic virus 1, human T cell (HLTV 1)
T-cell leukaemias, 387
Lysine hydroxylase
in collagen biosynthesis, 282
Lysine oxidase
in collagen fibre maturation, 283
mutations to in Ehlers–Danlos syndrome, 285
Lysosomal enzymes
nature and properties, 73
Lysosomes
functions in the tissues, 74
general description, 9
life cycle, 74
occurrence in cells, 73
possible involvement in diseases involving trinucleotide repeats, 296
storage diseases, 75
Lysozyme
discovery and reaction catalysed, 319
Lytic agents
effects on plasma membranes, 12

Macrocytic anaemia
occurrence and nature, 192
Macrolide antibiotics
erythromycin as an example, 418
Macrophages
generation of nitric oxide, 344
in T_{H1} stimulated killing, 339
presence of prolactin receptors, 262
role in immune responses, 320, 339
role in multiple sclerosis, 304
uptake of oxidised LDL, 347, 350
Magnetic resonance imaging (MRI)
in the detection of brain lesions, 304
Major histocompatibility complex (MHC)
in antigen presentation, 320
protein structures, 334
Malabsorption syndromes
occurrence and nature, 125
Malate
formation in the tricarboxylic acid cycle, 44
Malate dehydrogenase (decarboxylating)
reaction catalysed, 100
Malate dehydrogenase
involvement in gluconeogenesis, 58
reaction catalysed in mitochondria, 44
Malignant neoplasm
definition, 378
Malnutrition
causes, 131
means of detection, 132
Malondialdehyde
generation as a result of lipid peroxidation, 347
MALT (mucosal associated lymphoid tissues)
location and functions, 331
Maltase
role in the digestion of disaccharides, 120
Manganese
as a possible causative agent in amylotrophic lateral sclerosis (ALS), 449
Mannose 6-phosphate
role in targeting of proteins to the lysosomes, 72
Mannose-binding protein
in innate immune responses, 329
Maple syrup disease
association with thiamin deficiency, 178
Marfan's syndrome
mutations to the fibrillin gene, 284
Mast cells
release of heparin from, 217
role in the acute phase of an asthmatic attack, 447
role of IgE in degranulation, 327
roles in inflammatory responses, 339
Matrix, mitochondrial
location in mitochondria, 42
Matrix vesicle theory
bone calcification, 287
Measles
treatment by vitamin A supplementation, 174
Medical and dietary history
role in the detection of malnutrition, 132
Mediterranean diets
prevention of heart disease, 153
α-Melanocyte stimulating hormone
structure, 260
MELAS syndrome
arising due to mutations to mitochondrial DNA, 55
Membrane composition
in relation to essential fatty acids, 147
Membrane fluidity
determinants of, 13–14
Membrane proteins
catabolism in the endolysosomal compartment, 143
bitopic, 16
integral, 16
lipid anchored, 16
monotopic, 16
peripheral, 16
polytopic, 16

Membrane transport
facilitated, 18
Membrane-attack complex
formation from complement proteins, 329
Memory, immunological
laying down, 323
Menke's disease
occurrence and nature, 360
Mercapturates
in phase II reactions of foreign compounds, 374
Mercury
toxicity, 361
Mercury poisoning
chelation therapy, 361
Messenger RNA (mRNA)
role in protein synthesis, 62
Metabolic adaptation
to starvation, 162–6
Metabolic changes
in insulin dependent diabetes mellitus (IDDM), 440
Metabolic disorders of bone
nature, 287
Metabolic effects
adrenaline, 270
noradrenaline, 270
T_3 and T_4, 268
Metabolic pathway
definition, 11
Metabolic rate, resting
determination, 137
Metabolic responses
in the ebb phase following injury, 432
Metabolizable energy
of food, 135
Metabotrophic receptors
examples, 301–2
nature, 301
Metaformin
in the treatment of non insulin dependent diabetes mellitus (NIDDM), 442
Metal ions
complex formation, 358
roles in biological processes, 183
Metals
dietary requirements, 183
found in the human body, 182
Methaemoglobin reductase
in red blood cells, 207
MethionyltRNA
in the initiation of protein synthesis, 65
Methotrexate
as an antimetabolite anticancer drug, 427
inhibition of dihydrofolate reductase, 199
Methyl 'cap' (5' cap)
role of initiation of eukaryotic protein synthesis, 38
'Methyl trap' hypothesis
occurrence of megaloblastic anaemia in vitamin B_{12} deficiency, 202
MethylmalonylCoA mutase
role in the catabolism of branched-chain amino acids, 200
Methylmercury
poisoning, 361
Metronidazole
drug action involving reactive oxygen species (ROS), 422
MHC (major histocompatibility complex)
in antigen presentation, 320
Micelles
nature and formation, 15
Micro-organisms
contaminating food, 130–31
Microangiopathy
occurrence in diabetes mellitus, 441
Microcytic anaemia
occurrence and nature, 192
Microglia
macrophage-like role in brain, 290
β_2-Microglobulin
part of the MHC class I protein complex, 334
Microsomal ethanol metabolising system (MEOS)
reaction catalysed, 227
Microsomal FAD containing mono-oxygenase
roles in *N*- and *S*-oxidation of foreign compounds, 371
Minamata bay
organic mercury poisoning, 361
Mineral
composition of bone, 279
Mineral metabolism
effect of growth hormone, 262
Mineralocorticoid receptor
in renal function, 236
Mis-sense mutations
nature and occurrence, 64
Mitochondria
formation, 55
general description, 9
structure and functions, 42
Mitochondrial diseases
associated with drug administration, 55
Mitochondrial myopathies
occurrence in AIDS patients treated with AZT, 55
occurrence and nature, 55
Mixed function oxidases
reactions catalysed, 370
MK801
use-dependent inhibitor of the glutamate NMDA receptor, 300
Mobilferrin
possible role in the absorption of dietary iron, 187
Molecular dosimetry
uses in accessing exposure to potential carcinogens, 390
Mono-oxygenases
potential generators of free radicals, 343
Monoamine oxidase
inactivation of monoamine neurotransmitters, 297
location in the mitochondrial outer membrane, 42
Monocytes
roles in immune responses, 330
Monosaccharides
as inducers of cataract, 311
intestinal absorption, 121
Monotopic membrane proteins
organization, 16
Mortality risk
related to obesity, 159
Motor neuron disease
nature, 449
MPTP
as a cause of parkinsonism, 297
mRNA
detection in cells by *in situ* hybridization, 5–6
see also messenger RNA
Mucins
role in the absorption of dietary iron, 187
Mucosal associated lymphoid tissues (MALT)
simplified scheme for the organization of, 332
Multi-drug resistance
induction of P-glycoprotein, 375
occurrence, 369
Multi-hit
hypothesis of cancer, 378
Multiple sclerosis (MS)
occurrence and nature of the disease, 303
Mupirocin
inhibition of leucyltRNA synthase, 418
Muscarinic acetylcholine receptors
presence in B cells in the pancreas, 263
Muscle
types, 242
Muscle cells, smooth
uptake of oxidized LDL, 350
Muscle contraction
energy provision, 247
mechanism in skeletal muscle, 247–9
Muscle fibres, type I
high aerobic capacity, 250
Muscle fibres, type II
low oxidative capacity, 250
Muscle, cardiac
organization and function, 250
Muscle, smooth
structure and function, 251
Muscular dystrophies
diagnosis and possible treatment, 253
Muscular dystrophy
Becker, occurrence and nature of disease, 252, 254
Duchenne, occurrence and nature of disease, 252–4
severe autosomal childhood onset, 253
Mutations
occurrence, 36
to the huntingtin gene in Huntingdon's disease, 296
familial and sporadic, 378
nature, 64
to colony-stimulating factor 1, association with osteopetrosis, 288
to $Cu2^+$-ATPases in Wilson's and Menke's diseases, 360
to lysine oxidase in Ehlers–Danlos syndrome, 285
to the androgen receptor gene in testicular feminization syndrome, 278

Mutations (*cont.*)
to the ceruloplasmin gene, 359
to the fibrillin gene in Marfan's syndrome, 284
to mitochondrial DNA, 55
to a receptor for fibroblast growth factor in achondroplasia, 286
Myasthenia gravis
autoimmune disease of the nicotinic acetylcholine receptor, 299
Mycolic acid
inhibition of biosynthesis by isoniazid, 417
presence in mycobacteria, 415
Myelin
nature and location, 303
function, 303
Myeloperoxidase
formation of hypochlorous acid, 345
Myoclonic epilepsy
arising due to mutations to mitochondrial DNA, 55
Myofibrils
organization, 242
Myoglobin
presence in type I muscle fibres, 250
Myokinase
presence in skeletal muscle, 250
Myometrium
control of contraction during parturition, 259
Myopathies, mitochondrial
occurrence and nature, 55
Myosin
functional role in skeletal muscle, 244
isoforms in muscle tissue, 250
Myosin light chain kinase (MLCK)
activation during an asthmatic attack, 447
role in smooth muscle contraction, 251

Na^+-channels (voltage gated)
in neurons, 291
organization of polypeptide chains, 291
Na^+/K^+-ATPase
action, 19
in renal water recovery, 234
nerve cells, 291
role in rod function, 314
role in secondary active transport of choline, 294
role in secondary active transport of glucose, 19
NAD^+
see nicotinamide adenine dinucleotide
NAD^+ dehydrogenase (complex I)
deficiency in parkinsonism, 47
reaction catalysed in mitochondria, 47
NAD^+ synthesis
occurrence and cellular location, 39
$NADP^+$-dependent 5a-reductase
metabolism of testosterone by, 273
NADPH
roles in metabolism, 59
NADPH oxidase
generation of the respiratory burst of phagocytes, 344
Naso-pharyngeal carcinoma
association with infection with Epstein–Barr virus (EBV), 387

Natural killer cells
role in immune responses, 330
role in viral infections, 320
Nebulin
in skeletal muscle, 245
Nedocromil sodium
use in asthma, 448
Negative selection
of B-cells in bone marrow, 331
of T-cells in the thymus, 334
Neonatal carbohydrate metabolism
basic features, 82
Neoplasia, renal
polycythaemia in, 241
Neoplasm, malignant
definition, 378
Nephron
organization, 230
Nephropathy
occurrence in diabetes mellitus, 441
Nervous system
roles of lysosomes, 75
Neuroblastoma
N-myc expansion, 383
Neurodegenerative diseases
association with trinucleotide repeats, 296
lysosomal involvement, 75
Neurofibrillary tangles
formation in Alzheimer's disease, 300
Neurofilaments
disruption in sporadic amylotrophic lateral sclerosis (ALS), 449
Neurokinin A
in asthma, 448
Neuromodulation
by serotonin (5 HT), 302
Neurons
structure and function, 289
Neuropathy
occurrence in diabetes mellitus, 441
Neurophysins I and II
oxytocin and vasopressin binding proteins, 262
Neurotoxins
examples, 357
Neurotransmitters
biosynthesis, release and inactivation, 293
identification, 293
Neutral lipids
nature of, 12
Neutralization
of toxins by antibodies, 324
Neutrophils
roles in immune responses, 330
Niacin
clinical deficiency, 179
metabolic functions, 179
structure, dietary sources and requirements, 179
Nicotinamide
use in the treatment of insulin dependent diabetes mellitus (IDDM), 443
Nicotinamide adenine dinucleotide (NAD^+)
role in the reaction catalysed by hydroxyacylCoA dehydrogenase, 43–4
reduction during glycolysis, 57
role in the reaction catalysed by isocitrate dehydrogenase, 44
role in the reaction catalysed by 2-oxoglutarate dehydrogenase, 44
role in the reaction catalysed by malate dehydrogenase, 44
Nicotinic acetylcholine receptor
structure and function, 293
control of adrenal medullary function, 269
Nicotinic acid
therapeutic uses and toxicity, 179
Nidogen
role in the glomerulus, 232
Night blindness
occurrence, 315
vitamin A deficiency, 313
Nitric oxide
effect on guanylyl cyclase, 21
intracellular signalling, 344
production by endothelial cells associated with smooth muscle, 252
Nitric oxide synthase
reaction catalysed and isoforms, 344
Nitro reduction
in the phase I metabolism of foreign compounds, 369
Nitrogen balance
factors determining, 143
Nitrogen excretion, urinary
in the flow phase of injury, 436
Non-insulin dependent diabetes mellitus (NIDDM)
pathophysiology, 441
risk associated with obesity, 129
Non-self
immunological significance, 318
Non-starch polysaccharides (NSP), dietary
relationship to disease, 156–7
UK Department of Health recommendations, 157
sources, 155
Non-prescription drugs
high aluminium content of some, 363
Nonsense mutations
nature and occurrence, 64
Noradrenaline
as a neurotransmitter, 296
metabolic and cardiovascular effects, 270
modes of action, 270
Noradrenaline *N*-methyltransferase
role in the formation of adrenaline, 297
Normocytic anaemia
occurrence and nature, 192
Nuclear envelope
nature of, 30
Nuclear lamina
organization, 33
Nuclear pores
size and organization, 30
Nucleation theory
of bone calcification, 287
Nucleic acid biosynthesis
in bacteria, direct inhibitors, 418
see also DNA and RNA polymerases
Nucleolus
role in ribosomal RNA synthesis, 33
Nucleosomes
organization of chromatin into, 32
Nutrients
nature, 126

Nutritional approaches to the treatment of cancer
principles, 395
Nutritional status
assessment, 131–4
definition, 126

Oat cell carcinomas of the bronchus
ACTH related peptides secreted by, 278
ob Gene
obese mice, 161
Obesity
as a health risk, 159
assessment via body mass index (BMI), 158
causes, 160–1
classification and measurement, 158–9
genetic and environmental factors, 160
incidence, 158
risk associated with diet, 129
treatment, 161–2
Occupational carcinogens
examples, 387
Occupational exposure to toxic chemicals
occurrence, 355
Oedematous malnutrition
kwashiokor, 168
Oestradiol
biosynthesis, 272
Oestrogen replacement therapy
prevention of osteoporosis, 288
Oestrogens
biological actions, 274
biosynthesis, 271
breast cancer, 278
role in epiphyseal closure in bones, 286
Okazaki fragments
occurrence in DNA replication, 34
Oligodendrocytes
myelination of CNS neurons, 300
Oligomycin
inhibition of ATP formation in mitochondria, 51
Oncogenes
nature, 382
Oncogenesis
viral, 386
Opioid peptides
in satiety, 140
Oral hypoglycaemic agents
in the treatment of non insulin dependent diabetes mellitus (NIDDM), 442
Ornithine cycle
reactions, 60–1
Osteitis deformans
see Paget's disease of the bone
Osteoblasts
in bone remodelling, 287
in collagen biosynthesis, 282
roles in bone, 279
Osteocalcin, 282
Osteocytes
roles in bone, 279
Osteogenesis imperfecta
brittle bone disease, 284
Osteomalacia
induction by anti epilepsy drugs, 376
occurrence and nature of the disease, 114
Osteomalacia, vitamin D resistant
toxic effect of aluminium, 364
Osteonectin
role in bone, 282
Osteopetrosis
occurrence, 288
Osteopontin
role in bone, 282
Osteoporosis
occurrence in post menopausal women, 288
treatments for, 288
Osteosarcomas
associated with exposure to bone seeking radioisotopes, 364
Ouabain
as a cardiotonic steroid, 270
as a possible natriuretic agent, 237
inhibition of the Na^+/K^+-ATPase, 237
Ovarian steroid biosynthesis
interaction between thecal and granulosa cells, 272
Oxaloacetate
formation in the tricarboxylic acid cycle, 44
Oxidases
potential generators of free radicals, 343
β-Oxidation
of fatty acids in peroxisomes, 9
reactions, 43
Oxidative decarboxylation
in the handling of branched-chain amino acids in muscle, 103
Oxidative metabolism of foreign compounds
general description, 370
Oxidative phosphorylation
in energy conservation, 47
Oxidative stress
occurrence and nature, 349
2-Oxoglutarate
formation in the tricarboxylic acid cycle, 44
2-Oxoglutarate dehydrogenase
reaction catalysed in mitochondria, 44
Oxygen
reduction to water in mitochondria, 47
Oxygen consumption
by hepatocytes, 41
Oxygen therapy
diseases associated with, 351
Oxygen-derived free radicals
as components of the process of carcinogenesis, 350
Oxytocin
interaction with corticotrophin-releasing hormone in parturition, 259
mode of action, 263
structure, 262

P-Glycoprotein
in the induction of multi-drug resistance, 375
role in multi-drug resistance, 369
P/O ratio
definition for oxidative phosphorylation, 50
P1 and P2 proteases
formation of ACTH related peptides, 260
in prohormone processing, 258
p53
action in the cell cycle, 385
mutated in many human cancers, 385
mutations and exposure to potential carcinogens, 391
Paget's disease of the bones
occurrence, 288
Papilloma virus
oncoprotein E6 production, 385
Palindromic sequences in DNA
recognition by restriction enzymes, 402–3
Palmitate
handling in β-oxidation, 43
Pancreatic B cells
release of insulin, 80
Pancreatic juice
composition and functions, 118
Pancreatic polypeptide
production and actions, 119
production by PP cells in the pancreas, 263
Pantothenic acid
clinical deficiency, 181
metabolic functions, 181
structure and dietary sources, 181
Papilloma virus, human (HPV)
infection with associated with squamous cell carcinoma, 387
Paracetamol
metabolism-induced toxic effects, 374
Paracrine effects
nature, 21
Paramyxovirus
possible association with Paget's disease of the bone, 288
Parathyroid glands
presence of a calcium sensing receptor, 108
Parathyroid hormone (PTH)
role in calcium homeostasis, 108
biosynthesis, 257
mode of action, 269
structure, 269
Parathyroid hormone related protein (PTHrP)
role in calcium homeostasis, 109
Parkinsonism
nature of the disease, 297
Parturition
control of myometrium contraction during, 259
corticotrophin-releasing hormone interaction with oxytocin, 259
Passive diffusion
across membranes, characteristics, 18
Passive smoking
effects of, 351
PCR
see Polymerase chain reaction
Pectins
as dietary non starch polysaccharides, 156
Pellagra
occurrence, 179

Penicillin
  first uses of, 413
  inhibition of bacterial cell wall biosynthesis, 416
Penicillin-binding protein (PBPs)
  mutations leading to penicillin resistance, 424
  role in bacterial cell wall biosynthesis, 417
Pentose phosphate pathway
  in red blood cells, 205
  in the cornea, 309
  metabolic roles, 59
Pentose sugars
  dietary sources, 154
Pepsin
  formation from pepsinogen, 117
  specificity and action, 121–2
Peptide hormones
  biosynthesis, 257
Peptidoglycans
  roles in bacterial cell walls, 415
Peptidyl transferase
  role in protein synthesis, 67
Peripheral membrane proteins
  organization, 16
Pernicious anaemia
  as an autoimmune disease, 201
  first description by Addison, 196
Peroxidase
  thyroid, 266
  presence in peroxisomes, 9
Peroxidation
  of polyunsaturated fatty acids, 151
Peroxisomes
  general description, 9
Peroxynitrite
  formation and decomposition, 344
pH regulation
  role of kidney, 238
Phagocytes
  generation of superoxide anion, 344
  respiratory burst, 344
Phagocytes, professional
  roles of lysosomes, 75
Phagocytosis
  by hepatic Kupffer cells, 227
Phase I and II metabolism of foreign compounds
  occurrence, 368
Phase I metabolism of foreign compounds
  induction of, 375
Phase II reactions of foreign compounds
  energy requirements, 373
  glucuronide formation in, 373
Phenacetin
  metabolism-induced toxic effects, 372
Phenylketonuria
  effect of phenylalanine on transport of tyrosine and tryptophan into the brain, 307
Phloridzin
  inhibition of renal transport of glucose, 238
Phorbol esters
  as tumour promoters, 392
Phosphatidylcholine
  major phospholipid component of the plasma membrane, 15
Phosphatidylinositol*bis*phosphate (PIP_2)
  in intracellular signalling, 25
Phosphocreatine
  formation in skeletal muscle, 250
Phospho*enol*pyruvate carboxykinase
  role in gluconeogenesis, 58
Phosphofructokinase I
  reaction catalysed in glycolysis, 57
  regulation of activity, 57
6-Phosphogluconate
  formation in the pentose phosphate pathway, 59
Phospholipase C
  activation during an asthmatic attack, 447
  mechanism of activation, 25
Phospholipids
  role in blood clotting, 211
Phosphorylase, glycogen
  reaction catalysed, 59
Photo-oxidation of polyunsaturated fatty acids in the retina
  occurrence, 313
Physical activity, (PAR)
  relationship to energy expenditure, 137
Physical activity ratio
  definition, 138
Phytic acid (inositol hexaphosphate)
  influence on absorption of dietary calcium, 107
  metal binding effect in the diet, 154
Pituitary gland, anterior
  hormones, 262
Pituitary gland, posterior
  hormones, 262
Placental corticotrophin-releasing hormone
  possible function, 259
Placental transfer
  of IgG, 325
Plaque, atheromatous
  formation, 350
Plasma amino acids
  effect of dietary intake, 99
  hormonal regulation, 104
  nature, 99
  utilization in humans, 99
Plasma cells
  as antibody secretors, 332
Plasma lipids
  classification, 85
  factors leading to their elevation, 97
  sources and utilization, 87
Plasma membrane
  general description, 10
  nature, 12
Plasma transport
  T_3 and T_4, 267
  steroid hormones, 274
Plasmids
  as cloning vectors for DNA, 403
Plasmin inhibitors
  nature, 217
Plasmin/plasminogen
  in blood clot dissolution, 217
Platelet aggregation
  role of prostacyclin, 210
  role of thromboxane A_2, 210
Platelet-derived antiplasmin
  functional role, 217
Platelet-derived growth factor
  in bone repair, 287
Platelets
  major abnormalities, 217
  role in blood clot formation, 209
  structure, 209
Plutonium-239 (^{239}Pu)
  risk of lung cancer, 364
Pollutants, airborne
  as 'trigger factor' in asthma, 446
PolyA 'tails'
  of most eukaryotic mRNA molecules, 38
Polyamines
  outline of their biosynthesis and roles, 201
Polycistronic mRNA molecules
  occurrence in prokaryotes, 65
Polycylic hydrocarbons
  free radical intermediates in carcinogenesis, 350
Polycythaemia
  association with renal neoplasia, 241
Polymerase chain reaction (PCR)
  diagnosis of bacterial infections, 410
  diagnosis of muscular dystrophies, 254
  principles underlying, 406
  use in monitoring therapy for follicular lymphoma, 409
  diagnosis of deuteronopia in John Dalton, 316
Polymerase, RNA
  roles in transcription, 37
Polymerases, DNA
  roles in DNA replication in eukaryotes, 36
  roles in DNA replication in prokaryotes, 34
Polymixins
  effect on bacterial membranes, 417
Polymorphisms
  in enzymes that metabolise procarcinogens, 391
Polymorphonuclear (PMN) cells
  roles in immune responses, 330
Polyol
  pathway of glucose metabolism, 83
Polysaccharides, dietary
  classification, 154
Polysomes
  occurrence and nature, 67
Polytopic membrane proteins
  organization, 16
Polyunsaturated fatty acids
  influence on membrane fluidity, 14
  photo-oxidation in the retina, 313
  relationship to vitamin E requirement, 352
Porins
  transport role in bacteria, 415
Portacaval shunt operation
  as a cause of haemochromatosis, 194
Positron emission tomography (PET)
  applications, 304
Positive selection of T-cells
  in the thymus, 334
Post absorptive phase
  metabolic adaptation in, 163–5
Post menopausal women
  occurrence of osteoporosis, 288
Post-translational modification of proteins
  in the Golgi apparatus, 71

Posterior pituitary gland
hormones, 262
Potassium ions
high concentrations in soft tissues, 182
Prader–Willi syndrome
primary obesity in, 160
Precipitation
of soluble antigens by antibodies, 324
Prednisolone
use in asthma, 448
Pregnancy
a special role for calcitonin in, 269
circulating corticotrophin-releasing hormone, 259
dietary reference values for energy, 139
dietary reference values for proteins, 145
risk of mercury poisoning in dental workers, 360–4
secretion of chorionic gonadotrophin used for early detection, 261
Pregnanetriol
urinary excretion in congenital adrenal hyperplasia, 277
Pregnenolone
formation, 271
Prekallikrein
role in blood clotting, 213
Premature infants
vitamin E deficiency, 352
Pressor actions
of vasopressin, 262
Primaquine
as an antimalarial drug, 413
Primary immune response
role of IgM, 326
Primase
in DNA replication, 34
Prion-related diseases
occurrence, 422
Pro-convulsant actions of hydrazides
inhibitors of glutamate decarboxylase, 295
Proaccelerin
role in blood clotting, 215
Probing
DNA, 400
Procollagen
biosynthesis, 282
Procollagen proteinase
mutations to in Ehlers–Danlos syndrome, 285
reaction catalysed, 283
Proconvertin
role in blood clotting, 213
Proenzymes
nature and activation, 121–2
Professional phagocytes
roles of lysosomes, 75
Progesterone
biological actions, 274
biosynthesis, 271
Progression to cancer
steps, 392
Prohormone
T_4 as, 267
testosterone as, 273
Prohormone formation
general mechanisms, 258
Proinsulin
as a precursor of insulin, 263
Prokaryotic cells
nature, 3
Prolactin
hormones influencing secretion, 259
hypersecretion in autoimmune diseases, 262
in the ebb of the metabolic response to injury, 432
structure and functions, 262
Proline hydroxylases
in collagen biosynthesis, 282
Prolonged starvation
adaptive metabolic changes, 165–6
Promoter sites
role in the initiation of transcription, 36
Promoters of cancer
examples, 393
Prontosil
early use in septicaemia, 411–2
metabolic activation by azo-reductase, 369
Proof-reading
in DNA replication, 36
Pro-opiomelanocortin (POMC)
precursor of ACTH-related peptides, 260
Pro-opiomelanocortin (POMC) gene
control by cAMP response element binding protein, 302
Properdin
role in alternative complement pathway, 329
Prostacyclins
production and further metabolism, 148–9
role in platelet aggregation, 210
Prostacyclin synthase
production of anticoagulant PGI_2, 210
Prostaglandin H_2 synthase
role in platelet aggregation, 210
Prostaglandins
in bone remodelling, 287
physiological roles, 149–50
release during an asthmatic attack, 447
structures and nomenclature, 148
Proteases P1 and P2
formation of ACTH related peptides, 260
in prohormone processing, 258
Protein
dietary requirements, 142
digestion and absorption, 121–23
Protein biosynthesis
general description, 62
initiation, 65
role of the liver, 223
Protein content
of food, 143–4
Protein energy malnutrition (PEM)
associated with growth failure, 168
associated with kwashiokor, 168–70
associated with marasmus, 168–70
causes and development, 169
clinical, biochemical and pathological features, 169
incidence and prevalence, 168
treatment, 170
tests to distinguish different forms, 131
Protein intake
guidelines of the USA National Research Council, 130
UK guidelines for a healthy diet, 130
Protein kinase A
activation by cAMP, 25
Protein kinase C
activation by diacylglycerol (DAG), 28
Protein metabolism
effects of T_3 and T_4, 268
Protein orientation in membranes
classification of types, 71
Protein quality, dietary
determinants, 144
Protein requirements
dietary, 145
Protein retrieval from the plasma membrane
mechanism, 72
Protein secretion
regulated, 72
Protein-sparing
effect of dietary carbohydrate, 153
Protein targeting
control in cells, 72
Protein turnover
nature and occurrence, 99, 142
Protein tyrosine kinase receptors
mechanism of action, 24
Protein utilization, net
as a measure of the nutritional value of a protein, 144
Protein metal ion complexes
nature, 358
Proteolytic enzymes of the gastrointestinal tract characteristics, 121
Prothrombin
conversion to thrombin in blood clotting, 215
Proto-oncogenes
nature, 382
Proton transport
across the inner mitochondrial membrane, 48–9
Protoporphyrin
accumulation in red blood cells following lead poisoning, 362
Protozoal infections
drugs effective against, 421
Pseudhypoparathyroidism
nature of the disease, 114
Pseudoidiopathic hypoparathyroidism
nature of the disease, 114
Psoriasis
use of retinoic acid in, 174
Pulmonary fibrosis
occurrence in workers with aluminium, 363
Puromycin
inhibition of protein synthesis in prokaryotes and eukaryotes, 67
Pyridoxal phosphate
requirement by *L*-DOPA decarboxylase, 296
requirement of glutamate decarboxylase and GABA T in the brain, 296
Pyridoxine (vitamin B_6)
absorption, transport storage and dietary requirements, 180

Pyridoxine (vitamin B_6) (*cont.*)
assessment of dietary status, 134
clinical deficiency, 180
metabolic functions, 180
structure and dietary sources, 180
therapeutic uses and toxicity, 180
Pyruvate carboxylase
role in gluconeogenesis, 58
Pyruvate dehydrogenase
reaction catalysed, 42–3
Pyruvate kinase
deficiency associated anaemia, 208
reaction as an example of substrate-level phosphorylation, 45
reaction catalysed in glycolysis, 57

Quercetin
possible prophylactic use in patients at high risk of cataract, 311
Quinine
as an antipyretic and antimalarial drug, 413
Quinolines
inhibitors of DNA gyrases, 419

R-binders
role in the absorption of vitamin B_{12} (cobalamin), 198
Radiation-induced cataracts
occurrence, 310
Radionuclides
toxic effects, 364
Ragged-red fibre disease
arising due to mutations to mitochondrial DNA, 55
Rational design of drugs
approaches to, 414
Reactive oxygen species (ROS)
generation during the action of doxorubicin as an anticancer drug, 427
in the action of the drug metronidazole, 422
nature, 343
Reaginic antibodies
IgE, 326
Receptor-mediated endocytosis
transferrin, 190
Receptor proteins
general description, 17
Receptors
for neurotransmitters, 298
for T_3 and T_4, 267
for low density lipoprotein (LDL), 93
Receptors, steroid hormones
defects, 276
Recombinant toxins
in the treatment of cancer, 394
Recombination, genetic
definition, 399
Recommended daily intake
replacement by reference nutrient intake values, 128
Red blood cells
transport of methylmercury, 361
Redox reactions
occurrence in mitochondria, 46
Redox status of hepatocytes
effect of ethanol, 228
Reference nutrient intake (RNI)
calcium, 106
definition, 127
minerals, 184
niacin, 179
thiamin, 178
vitamin A, 175
vitamin C, 354
vitamin D, 113
Remodelling of bone
processes, 287
Renal dialysis
effect on circulating triacylglycerols, 438
Renin
in the production of aldosterone, 236
Reperfusion injury
occurrence, 351
Replication fork
in DNA synthesis, 34
Replication
of DNA, 33
Respiratory burst
phagocytes, 344
Respiratory control
regulation of oxygen mitochondrial consumption by ADP availability, 50
Respiratory distress syndrome, adult (ARDS)
occurrence, 351
Respirometers
indirect assessment of energy expenditure, 136
Responsive elements
found associated with eukaryotic genes, 38
Restriction enzymes (endonucleases)
nature, 402–3
innate immunity in bacteria, 320
Restriction fragment length polymorphisms (RFLPs)
in the diagnosis of muscular dystrophies, 254
occurrence, 407
use in linkage analysis of inherited disease, 407
Retina
functional organization, 312
oxygen requirements, 308
Retinal isomerase
reaction catalysed in the retina, 313
Retinitis pigmentosa
nature of disease, 316
Retinoblastoma
role tumour suppresser gene Rb, 384
Retinoic acid
role in cellular differentiation, 174
therapeutic uses, 174
Retinoic acid binding proteins
intracellular roles, 174
Retinol (vitamin A)
absorption, transport and storage, 173
functional roles, 174
structure, nomenclature and dietary sources, 172
role in vision, 174
Retinol dehydrogenase
in the retina, 313
Retinol equivalents
of vitamin A precursors, 172
Retinopathy
occurrence in diabetes mellitus, 441
Retrolental fibroplasia
occurrence of disease, 351, 352
Retroposons
occurrence and nature, 39
Retroviruses
cancer causing genes of, 386
nature of, 38
Reverse transcriptase (RNA-dependent DNA polymerase) inhibition by AZT, 421
of retroviruses, 386
reaction of, 38, 402
Reverse transcriptase polymerase chain reaction (RT/PCR) in monitoring therapy for chronic myelogenous leukaemia, 409
uses, 409
Reverse transcription
nature of, 38
Reversion, bacterial
as the basis of the Ames test for potential mutagens, 394
RFLP analysis
in detection of sickle cell disease, 407
RFLPs
see Restriction fragment length polymorphisms
Rhesus disease
see haemolytic disease of the new born
Rheumatoid arthritis
superoxide anion production in, 351
Rhodopsin
presence and structure in rods, 313
Riboflavin (vitamin B_2)
absorption, transport and storage, 178
assessment of dietary status, 133–4
requirements and deficiency, 178
deficiency association with haemolytic disease, 208
structure, dietary sources and function, 178
Ribonuclease H (RNAse H)
in reverse transcription, 38
in cDNA production, 402
reaction catalysed, 401
Ribonuclease L
in 'antisense' therapy for cancers, 429
Ribosomes
occurrence in mitochondria, 41
role in protein synthesis, 65
Ribulose 5-phosphate
formation in the pentose phosphate pathway, 59
Ricin
effect on protein synthesis, 394
Rickets
description of the disease, 113
occurrence and nature of the disease, 114–5
Rifampicin
inhibition of DNA dependent RNA polymerase, 417
Rigor complex
occurrence in skeletal muscle, 247
RNA dependent DNA polymerase
see Reverse transcriptase
RNA polymerase, DNA dependent
roles in transcription, 36
inhibition by rifampicin, 417
RNA synthesis
eukaryotes, 37
Rods in the retina
organization, 312

Rotenone
  inhibition of mitochondrial complex I (NAD+ dehydrogenase), 51
Rough endoplasmic reticulum
  in the biosynthesis of parathyroid hormone, 257–8
Rous sarcoma virus
  as a tumour virus, 386
RT/PCR
  see Reverse transcriptase/polymerase chain reaction
rT_3 (reverse T_3)
  formation of, 268

S-Adenosylmethionine
  in the formation of adrenaline, 297
  formation and roles, 202
Safe intake
  dietary, 127
Salbutamide
  use in the spasmogenic phase of an asthmatic attack, 447
Saliva
  role in the digestive process, 116
Salivary amylase
  role in the digestion of starch, 120
Salt (sodium) intake
  guidelines of the USA National Research Council, 130
Saltatory conduction
  in myelinated neurons, 292
Salvarsan
  early use in the treatment of syphilis, 411
Sarcomas
  as neoplasms of mesodermal cells, 378
Sarcomeres
  functional units in skeletal muscle, 242
Sarcoplasmic reticulum
  Ca^{2+} stores, 9
  storage and release of Ca^{2+}, 247
Satellite DNA
  nature and occurrence, 36
Saturated fatty acids
  examples of, 13
Saxitoxin
  binding to voltage gated Na^+-channels, 292
Scavenger receptors
  uptake of oxidized LDL, 347
Schistosomiasis
  protective role of IgE, 326
Schwann cells
  myelination of peripheral neurons, 303
Scotopic vision
  rods in, 312
Scrapie
  as an infectious prion disease of sheep, 422
Scurvy
  occurrence and nature of the disease, 285, 354
Second messenger signalling systems
  effects of some tumour promoters, 392
Secondary active transport
  of glucose from the gut, 19
Secondary lymphoid organs
  functional organization, 331
Secretase
  in the processing of amyloid precursor protein, 301
Secretin
  nature and actions, 118
Secretory vesicles
  functional roles, 68
Selectivity of drug action
  sources, 414
Selenium
  deficiency associated with human disease, 348
  presence in glutathione peroxidase, 348
Self tolerance
  development, 332
Semiconservative
  nature of DNA synthesis, 34
Senescence
  drug metabolism in, 373
Sero-mucous secretions
  presence of IgA, 326
Serotonin (5-hydroxytryptamine, 5-HT)
  detection in neurons, 5
  role in satiety, 140
  release from mast cells during an inflammatory response, 339
Sertoli cells
  in testicular spermatogenesis, 261
Sex-hormone binding globulin
  functional role, 274
Shine–Delgarno sequence
  in the binding of mRNA to prokaryotic ribosomes, 65
Short-term starvation
  adaptive metabolic changes, 163–5
Short stature
  achondroplasia associated, 286
Sickle cell disease
  detection by analysis of RFLPs, 408
Signal peptidase
  catalytic activity, 70
Signal peptide
  in the biosynthesis of parathyroid hormone, 257
  role in protein trafficking, 70
Signal recognition particle (SRP)
  in the biosynthesis of parathyroid hormone, 257
  interaction with growing polypeptide chains on ribosomes, 70
Silent mutations
  nature and occurrence, 64
Single-strand binding proteins
  role in DNA replication, 36
Singlet oxygen
  in cataract, 350
  nature and formation in the presence of ionizing radiation, 345
Skeletal muscle growth
  effects of insulin on, 266
Sleeping sickness (trypanosomiasis)
  early use of atoxyl for, 411
  early use of tryparsamide for, 411
Sliding filament model
  for muscular contraction, 247
Slimming diets
  lack of value of most, 162
Small nuclear RNA molecules
  nature of, 38
Smallpox
  global eradication, 317
Smear test, cervical
  for the early detection of cancer, 380
Smoke
  association with free radical generation, 345
Smoking cigarettes
  and lung cancer, 389
  effect on vitamin C turnover, 353
  risk associated with, 351
Smoking, passive
  possible effects, 351
Smooth endoplasmic reticulum
  functional roles, 68
  role in phase I metabolism of foreign compounds, 370
Smooth muscle
  'twitchy' (hyper-responsive) state in the alveoli of the lungs in asthma, 445
  effect of oxytocin in the breast, 263
  effect of oxytocin on in the uterus, 263
  structure and function, 251
Smooth muscle cells
  uptake of oxidized LDL, 350
Smooth muscle contraction
  myometrium, 259
Smooth muscle light chain kinase (MLCK)
  reaction catalysed, 251
Smooth muscle, vascular
  relaxation induced by nitric oxide, 344
Sodium intake
  UK guidelines for a healthy diet, 130
Solenoids
  in higher level folding of DNA, 32
Somatostatin
  formation and paracrine actions in the gastrointestinal tract, 120
  inhibition of growth hormone secretion, 259
Sorbinil
  possible prophylactic use in patients at high risk of cataract, 311
Sorbitol
  accumulation in the lens in cataract, 311
  dietary sources, 154
Southern blotting
  principles, 401
Spatial information
  exchange between cells, 20
Spermatogenesis
  role of FSH, 261
Spheroid bodies
  formation in amylotrophic lateral sclerosis (ALS) neurons, 449
Sphingolipid-activating proteins (SAPs)
  role in lysosomal activity, 73
  mutations leading the sphingolipidoses, 306
Sphingolipidoses
  affecting brain function, 305
Sphingolipids
  occurrence in biological membranes, 12
Spironolactone
  antagonist of aldosterone, 237
  use in the treatment of primary hyperaldosteronism, 237

Splanchnic nerve
  control of adrenal gland by, 269
Squalene epoxidase
  as a target for antifungal drugs, 420
Squamous cell carcinoma
  association with infection with human papilloma virus (HPV), 387
Starch
  as a major dietary polysaccharide, 154
  digestion and absorption, 120
Starch intake
  UK guidelines for a healthy diet, 130
Starvation
  clinical aspects, 167
  metabolic adaptation, 162
STAT proteins (signal transducing activators of transcription)
  description, 302
Steroid, 5a-reductase
  metabolism of testosterone, 273
Steroid diabetes
  in Cushing's disease, 276
Steroid hormone
  super family of receptors, 23
Steroid hormone production and action
  diseases associated with, 276
Steroid hormones
  biosynthesis, 270
  inactivation and elimination, 276
  plasma transport, 274
  receptor defects, 278
  reductive metabolism in the liver, 226
  sources and structures, 270–1
6β-Steroid hydroxylase
  in assessment of state of induction of foreign compound metabolizing enzymes, 376
Steroidogenesis
  control of, 274
Steroidogenic acute regulatory protein
  carrier in steroid hormone biosynthesis, 271
Steroidogenic organs
  functions, 270
Stibofen
  in the treatment of leishmaniasis, 411
'Sticky ends' of DNA
  generation by restriction enzymes, 403
Storage diseases
  lysosomal, 75
Streptokinase
  in blood clot dissolution, 217
Streptomycin
  antibiotic action, 417
  inhibition of protein synthesis in prokaryotes, 67
Stress
  as a 'trigger factor' in asthma, 446
  effect on circulating glucose concentrations, 84
  effect on plasma lipid levels, 98
Stroke
  excitotoxicity model, 300
Strontium-90 (^{90}Sr)
  risk of sarcoma and leukaemia, 364
Structural membrane proteins
  nature, 17
Sub-acute combined degeneration
  as a consequence of vitamin B_{12} deficiency, 202
Substance P
  in asthma, 448
Substitution mutations
  nature and occurrence, 64–5
Substrate-level phosphorylation
  examples, 45
Subtilisin like proteases
  formation of ACTH related peptides, 260
  in prohormone processing, 258
Succinate dehydrogenase (complex II)
  reaction catalysed in mitochondria, 44, 48
  demonstration in mitochondria, 7
SuccinylCoA thiokinase
  reaction catalysed in mitochondria, 44
Sucrase
  role in the digestion of sucrose, 120
  digestion and absorption, 120
Sugar (sucrose) intake
  UK guidelines for a healthy diet, 130
Sulphanilamide
  phase II acetylation reaction, 375
Sulphate esters
  formation in the metabolism of foreign compounds, 374
Sulphonamides
  conversion to sulphanilamide, 412
  mechanism of action, 418
Sulphonylureas
  in the treatment of non insulin dependent diabetes mellitus (NIDDM), 442
Sulphotransferases
  in the metabolism of foreign compounds, 374
Superoxide anion
  generation in chemical reactions, 343
Superoxide dismutase
  mutations in some familial cases of amylotrophic lateral sclerosis (ALS), 450
  in red blood cells, 207
  reaction catalysed, 344
  role, 348
Surfactant protein SP-A in lungs
  collagen like structure, 281
Sweet clover
  as a source of the anticoagulant dicoumorol, 214
Sweet pea seeds
  a cause of lathyrism, 285
Symport processes
  in membranes, 18–19
Symptomatic treatment
  for infectious diseases, 425
Synapse
  description of a general, 290
Synaptic cleft
  occurrence in the nervous system, 290
Syncytiotrophoblast cells
  secretion of hCG as a means of detecting pregnancy early, 261

T-cell (helper)
  activation, 338
T-cell receptor (TCR)
  structure, 333
T-cell
  development, 333
  destruction of B cells of the pancreas in insulin dependent diabetes mellitus (IDDM), 439
  interactions with other cells, 334
  presence of prolactin receptors, 262
  'silencing' of by anergy, 336
T-cells, cytotoxic
  production, 334
T-helper cell
  activation, 338
  production, 334
T_{H1} and T_{H2} cells
  production, 338
T_{H1}
  cytotoxic T-cell interactions in the immune response, 338
T_{H2} and B-cells
  interactions in the immune response, 338
T_3 and T_4
  caloric effects, 267
  inactivation, 268
  metabolic effects, 268
  mode of action, 267
  plasma transport, 267
Tamoxifen
  use in the prevention of breast cancer, 390
Taq polymerase
  use in the polymerase chain reaction, 406
Targeting
  of lysosomal enzymes, 73
Taxol
  as an anticancer drug, 428
Tay–Sachs disease
  occurrence, 305
Teeth
  detection of lead poisoning, 362
Teichoplanin
  antibiotic use in cases of penicillin resistance, 417
Teleocidins A & B
  as tumour promoters, 392
Temporal information
  exchange between cells, 20
Terminal transferase
  reaction catalysed, 401–2
Termination (release) factors
  role in protein synthesis, 67
Testes
  biosynthesis of testosterone, 273
Testicular feminization syndrome
  nature of disease, 278
Testosterone
  biological actions, 275
  biosynthesis, 273
  metabolism by steroid, 5α-reductase, 273
Tetracycline antibiotics
  action, 417–8
  inhibition of protein synthesis in prokaryotes, 67
Tetraethyl lead
  continuing use in petrol as an 'antiknock', 362
Thalassaemia
  causes, 407
  detection, 400
Thapsigargin
  as a tumour promoter, 392

Thermogenesis, cold induced
nature, 139
Thermogenesis, diet induced
nature, 138
Thermogenin
heat production by brown adipose tissue, 139
Thiamin (vitamin B_1)
absorption, transport and causes of deficiency, 177
assessment of dietary status, 133
dietary requirements, 178
metabolic functions, 177
structure and dietary sources, 177
Thiamin pyrophosphate
role in the pyruvate dehydrogenase reaction, 42
Thrombin
formation in blood clotting, 215
role in blood clotting, 215
Thrombomodulin
complex with thrombin, 217
Thromboxane A_2
production and action, 148
role in platelet aggregation, 210
Thromboxane synthase
production of pro coagulant TxA_2 in platelets, 210
Thromboxanes
structures and nomenclature, 148
Thunderstorms
occurrence of asthma epidemics following, 448
Thymine
modification in DNA by hydroxyl radical, 347
Thymus gland
as a primary lymphoid organ, 331
in T-cell development, 333
Thyroglobulin
iodination and storage, 266
Thyroid gland
calcitonin production, 269
effect of goitrogens, 357
thyroid hormone production, 266
Thyroid hormone
control of secretion by TSH, 267
quantity in store, 266
role in chondrocyte maturation, 286
Thyroid peroxidase
reaction catalysed, 266
Thyrotoxicosis
occurrence in Graves' disease, 268
Thyrotrophin
functional roles, 260
Thyrotrophin-releasing hormone
control of prolactin secretion, 259
structure and functions, 259
Thyroxine
structure and biosynthesis, 266
see also T_3 and T_4
Thyroxine binding globulin (TBG)
transport of T_3 and T_4, 267
Thyroxine binding prealbumin (TBPA)
transport of T_3 and T_4, 267
Tissue type plasminogen activator (tPA)
in blood clot dissolution, 217
Titin
in skeletal muscle, 245
Topoisomerase
roles in DNA replication, 35
Topoisomerase inhibitors
as anticancer drugs, 428
Total parental nutrition (TPN)
constituents in mixtures used for, 438
Toxic substances
damage caused, 356
Toxic substances in the diet
nature, 130–131
Toxins
neutralization by antibodies, 324
Trabecular bone
new growth of, 286
remodelling, 287
Trabecular structures
in bone, 279
Trace metals
found in the human body, 182
Trans unsaturated fatty acids
dietary status, 151
risks associated with, 128
Transamination reactions
occurrence in the liver, 101
Transcobalamins
roles in the distribution of vitamin B_{12}, 198
Transcription
general description, 36
Transcription factors
roles in RNA synthesis in eukaryotes, 37
Transcription, reverse
nature of, 38
Transcytosis
of IgA across epithelial cells, 72, 326
Transducin
as an example of a G protein, 25
G-protein of the rods, 315
Transfection of DNA
occurrence, 382
Transfer RNA (tRNA) molecules
role in protein synthesis, 65
Transferrin
determination of the degree of iron saturation for diagnosis, 188
role in determining the amount of iron absorbed from the diet, 187
role in plasma transport of iron, 188
structural organization, 189
Transformation
genes affected, 382
of bacteria, 399
to a cancerous phenotype, 380
Transition mutations
nature, 64
Translocation of chromosomes
in Burkitt's lymphoma, 383
in chronic myelogenous leukaemia, 409
in follicular lymphoma, 409
Transplantation
of the pancreas and kidneys, for diabetic patients, 443
Transport proteins
occurrence in membranes, 18
role of porins in bacteria, 415
Transporters, amino acids
in the kidney, 238
in the brain, 307
Transversion mutations
nature, 64
Treatment of obesity
surgery, 162
drug therapy, 162
Tri-iodothyronine
structure and biosynthesis, 266
see also T_3 and T_4
Triacylglycerols
biosynthesis, 88
formation from dietary lipids and distribution in chylomicrons, 88
in renal dialysis patients, 438
Tricarboxylic acid cycle
occurrence in mitochondria, 44
Trimethoprim
inhibition of dihydrofolate reductase, 418
Trinucleotide repeats
in Huntingdon's disease, 295
Tropical spastic paraparesis
association with infection with HLTV-1, 386
Tropomyosin
in skeletal muscle, 245
Troponin
in skeletal muscle, 245
Tryparsamide
early use in the treatment of trypanosomiasis (sleeping sickness), 411
Trypsin
specificity and action, 122
Tryptase
release during an asthmatic attack, 448
Tumour necrosis factor-α
role in cancer cachexia, 168
Tumour suppressor genes
nature, 381, 384
Tumours
associated with retroviral infection, 386
of the adrenal gland, in Conn's syndrome, 277
of the adrenal gland, in Cushing's disease, 276
of the anterior pituitary gland, in Cushing's disease, 276
Tyrophosphins
possible use in the treatment of chronic myelogenous leukaemia, 429
Tyrosine
as a precursor of the catecholamine neurotransmitters, 296
Tyrosine hydroxylase
rate limiting enzyme in the biosynthesis of catecholamines, 296
regulation by phosphorylation, 302

Ubiquinol dehydrogenase (Complex III)
reaction catalysed in mitochondria, 47
Ubiquinone (coenzyme Q)
location in the inner mitochondrial membrane, 49
presence and role in mitochondria, 47
Ubiquitin
role in protein turnover, 104, 143
UDP-glucuronic acid
substrate for UDP glucosuronyl transferases, 373
Ulcerative colitis
free radicals in, 351

Uncoupling
of electron transport from ATP formation in mitochondria, 50
Under-nutrition
in infectious diseases, 167
nature, 126
Unsaturated fatty acids
examples of, 13
Urea synthesis
subcellular organization, 60
Uric acid
occurrence in urine, 233
Urine
abnormal constituents, 233
amino acids, 233
electrolytes, 233
nitrogenous constituents, 233
normal composition, 232
uric acid, 233
urobilinogen, 233
volume, normal range of values, 232
Urokinase
in blood clot dissolution, 217
Utrophin
deficiency in severe childhood autosomal recessive muscular dystrophy, 253

V-D-J joining
creation of antibody diversity, 327
v Erb
relationship to epidermal growth factor receptor, 383
v Ras
relationship to GTP binding proteins, 383
Vaccines
generation using haptenic peptides, 322
Vancomycin
antibiotic use in cases of penicillin resistance, 417
Vascular smooth muscle
relaxation induced by nitric oxide, 344
Vascularization of the osteoid layer
in bone formation, 285
Vasoactive intestinal polypeptide
control of prolactin secretion, 259
formation and paracrine actions in the gastrointestinal tract, 120
Vasopressin
control of secretion, 262
interaction with corticotrophin-releasing hormone in ACTH release, 259
modes of action, 262
structure, 262
Vectors
for DNA cloning, 403
Venesection therapy
for haemochromatosis, 194
Ventolin
use in the spasmogenic phase of an asthmatic attack, 447
Very low density lipoprotein (VLDL)
metabolic fate, 92
nature and constitution, 86
Vesicular ATPase
presence in putative glutamatergic neurons, 294
role in the acidification of secretory vesicles in nerve terminals, 294
uptake of dopamine into vesicles depends on, 296
Vimentin
a protein of smooth muscle, 251
Vinblastine
as an anticancer drug, 428
Vinylchloride
a chemical carcinogen, 388
Viral infections
drugs effective against, 420–1
Vitamin A (retinol)
absorption, transport and storage, 173
deficiency, 175
functional roles, 174
role in the retina, 313
structure, nomenclature and dietary sources, 172
Vitamin B_1 (thiamin)
absorption, transport and causes of deficiency, 177
dietary requirements, 178
metabolic functions, 177
structure and dietary sources, 177
Vitamin B_{12} (cobalamin)
absorption and distribution, 198
assessment of dietary status, 134
chemical structure, 196
deficiency, 201
reference nutrient intake (RNI), 198
Vitamin B_2 (riboflavin)
absorption, transport and storage, 178
requirements and deficiency, 178
structure, dietary sources and function, 178
Vitamin B_6 (pyridoxine)
absorption, transport, storage and dietary requirements, 180
clinical deficiency, 180
metabolic functions, 180
structure and dietary sources, 180
therapeutic uses and toxicity, 180
Vitamin C
assessment of dietary status, 133
absorption and storage, 353
dietary requirements and deficiency, 354
interaction with vitamin E, 346
structure and dietary sources, 353
toxicity, 354
Vitamin D
dietary requirement in the population of the UK, 113
functions not related to calcium homeostasis 113
role in calcium homeostasis, 113
toxicity, 115
Vitamin D-resistant rickets
occurrence, 114–5
Vitamin D-resistant osteomalacia
toxic effect of aluminium, 364
Vitamin E
absorption, transport and storage, 352
deficiency in premature infants, 352
functions, 353
in the retina, 313
interaction with vitamin C, 346
presence in LDL particles, 352
requirements and deficiency, 352
structure and dietary sources, 351
Vitamin K
assessment of dietary status, 134
role in blood clotting, 214
structures of vitamins K_1 and K_2, 214
supplementary administration to infants, 214
Vitamins
classification and nomenclature, 171–2
early recognition of requirement, 171
Voltage-gated Ca^{2+}-channels
in acetylcholine release, 294
in neurons, 293
Voltage-gated K^+-channels
in neurons, 292
Voltage-gated Na^+-channels
in neurons, 291
organization of polypeptide chains, 291

Warfarin
prophylactic use in patients recovering from thrombosis, 214
use as a rat poison, 214
Water reabsorption
kidney, 234
Water-soluble vitamin
basis for classification, 171
Wernicke–Korsakoff syndrome
in thiamin deficiency, 178
Western Pacific
variety of amylotrophic lateral sclerosis (ALS), 450
Wilson's disease
nature of disease, 360

X-linked, severe, combined immuno-deficiency
occurrence of disease, 338
Xeroderma pigmentosum
occurrence and nature of the disease, 36

Yup'ik eskimoes
incidence of congenital adrenal hyperplasia, 277

'Z'-lines
in skeletal muscle, 242
Zidovudine (AZT)
mitochondrial myopathies associated with administration, 55
in the treatment of acquired immunodeficiency syndrome (AIDS), 421
Zona fasciculata
role in the adrenal biosynthesis of cortisol, 270
Zona glomerulosa
receptors for angiotensin II in, 274
role in the adrenal biosynthesis of aldosterone, 274
Zona reticularis
role in the adrenal biosynthesis of androgens, 272